DISCRETE 2012 – Third Symposium on Prospects in the Physics of Discrete Symmetries

AA008839

Journal of Physics: Conference Series Volume 447

Lisbon, Portugal
3 - 7 December 2012

ISBN: 978-1-62993-072-5
ISSN: 1742-6588

Printed from e-media with permission by:

Curran Associates, Inc.
57 Morehouse Lane
Red Hook, NY 12571

Printed by Curran Associates, Inc. (2013)

For permission requests, please contact the Institute of Physics
at the address below.

Institute of Physics
Dirac House, Temple Back
Bristol BS1 6BE UK

Phone: 44 1 17 929 7481
Fax: 44 1 17 920 0979

techtracking@iop.org

Additional copies of this publication are available from:

Curran Associates, Inc.
57 Morehouse Lane
Red Hook, NY 12571 USA
Phone: 845-758-0400
Fax: 845-758-2634
Email: curran@proceedings.com
Web: www.proceedings.com

TABLE OF CONTENTS

Peer review statement

All papers published in this volume of *Journal of Physics: Conference Series* have been peer reviewed through processes administered by the proceedings Editors. Reviews were conducted by expert referees to the professional and scientific standards expected of a proceedings journal published by IOP Publishing.

Mass hierarchy and string phenomenology in the LHC era

I. Antoniadis[1]

Department of Physics, CERN - Theory Unit, 1211 Geneva 23, Switzerland

E-mail: `ignatios.antoniadis@cern.ch`

Abstract. I will discuss the status of the mass hierarchy problem and prospects for beyond the Standard Model physics in the light of the Higgs scalar discovery at the LHC and the experimental searches for new physics. In particular, I will discuss in this context low energy supersymmetry, living with the fine tuning and strings at the TeV scale.

1. Introduction

During the last few decades, physics beyond the Standard Model (SM) was guided from the problem of mass hierarchy. This can be formulated as the question of why gravity appears to us so weak compared to the other three known fundamental interactions corresponding to the electromagnetic, weak and strong nuclear forces. Indeed, gravitational interactions are suppressed by a very high energy scale, the Planck mass $M_P \sim 10^{19}$ GeV, associated to a length $l_P \sim 10^{-35}$ m, where they are expected to become important. In a quantum theory, the hierarchy implies a severe fine tuning of the fundamental parameters in more than 30 decimal places in order to keep the masses of elementary particles at their observed values. The reason is that quantum radiative corrections to all masses generated by the Higgs vacuum expectation value (VEV) are proportional to the ultraviolet cutoff which in the presence of gravity is fixed by the Planck mass. As a result, all masses are "attracted" to about 10^{16} times heavier than their observed values.

Besides compositeness, there are two main theories that have been proposed and studied extensively during the last years, corresponding to different approaches of dealing with the mass hierarchy problem. (1) Low energy supersymmetry with all superparticle masses in the TeV region. Indeed, in the limit of exact supersymmetry, quadratically divergent corrections to the Higgs self-energy are exactly cancelled, while in the softly broken case, they are cutoff by the supersymmetry breaking mass splittings. (2) TeV scale strings, in which quadratic divergences are cutoff by the string scale and low energy supersymmetry is not needed. Both ideas are experimentally testable at high-energy particle colliders and in particular at LHC. Another option is to live with the hierarchy, addressing differently questions such as gauge coupling unification, dark matter candidate and electroweak (EW) vacuum stability. Split supersymmetry is an interesting example of such a possibility. All these ideas are experimentally testable at high-energy particle colliders and in particular at the LHC.

[1] On leave from CPHT (UMR CNRS 7644) Ecole Polytechnique, F-91128 Palaiseau.

On the other hand, the recent major discovery of the Higgs boson at the LHC with a mass around 126 GeV is so far compatible with the Standard Model within 2σ and its precision tests. It is also compatible with low energy supersymmetry, although with some degree of fine-tuning in its minimal version. Indeed, in the minimal supersymmetric Standard Model (MSSM), the lightest Higgs scalar mass m_h satisfies the following inequality:

$$m_h^2 \lesssim m_Z^2 \cos^2 2\beta + \frac{3}{(4\pi)^2}\frac{m_t^4}{v^2}\left[\ln\frac{m_{\tilde{t}}^2}{m_t^2} + \frac{A_t^2}{m_{\tilde{t}}^2}\left(1 - \frac{A_t^2}{12m_{\tilde{t}}^2}\right)\right] \lesssim (130GeV)^2\,, \tag{1}$$

where the first term in the r.h.s. corresponds to the tree-level prediction and the second term includes the one loop corrections due to the top and stop loops. Here, $m_Z, m_t, m_{\tilde{t}}$ are the Z-boson, the top and stop quark masses, respectively, $v = \sqrt{v_1^2 + v_2^2}$ with v_i the VEVs of the two higgses, $\tan\beta = v_2/v_1$, and A_t the trilinear stop scalar coupling. Thus, a Higgs mass around 126 GeV requires a heavy stop $m_{\tilde{t}} \simeq 3$ TeV for vanishing A_t, or $A_t \simeq 3m_{\tilde{t}} \simeq 1.5$ TeV in the 'best' case. These values are obviously consistent with the present LHC bounds on supersymmetry searches, but they will certainly be probed in the next run at double energy. Theoretically, they imply a fine-tuning of the EW scale at the percent to per mille level. This fine-tuning can be alleviated in supersymmetric models beyond the MSSM, as discussed in the next session.

2. MSSM Higgs sector with dimension-five and dimension-six operators

Although extremely successful, the Standard Model or its supersymmetric version (MSSM) is not a fundamental theory, and this motivated the theoretical efforts to understand the nature of new physics beyond it. This search can be done using an effective field theory approach, in which the "new physics" is parametrised by effective operators. The power of this approach resides in arranging these operators in powers of $1/M_*$ where M_* is the scale of new physics that generated them. To improve the predictive power, one considers additional organising principles, such as: **(i)** symmetry constraints that these operators should respect, often inspired by phenomenology (for example: R-parity, lepton or baryon number conservation, etc). **(ii)** a truncation of the series of operators to a given order in the power of the inverse scale $1/M_*$. The effective low-energy Lagrangian then takes the form

$$\mathcal{L} = \mathcal{L}_0 + \sum_{i,n}\frac{c_{n,i}}{M_*^n}\mathcal{O}_{n,i} \tag{2}$$

where $\mathcal{L}_0$ is the SM or the MSSM Lagrangian; $\mathcal{O}_{n,i}$ is an operator of dimension $d = n + 4$ with the index i running over the set of operators of a given dimension; $c_{n,i}$ are some coefficients of order $\mathcal{O}(1)$. This description is appropriate for scales E which satisfy $E \ll M_*$. Constraints from phenomenology can then be used to set bounds on the scale of new physics M_*.

Regarding the origin of operators $\mathcal{O}_{n,i}$, they can be generated classically or at the quantum level. At the classical level, this can happen by integration of some new massive states, via the equations of motion and one then generates an infinite series. This can happen even in 4D renormalisable theories; indeed, even though the low energy interaction looks nonrenormalisable, it may actually point to a renormalisable theory valid up to a much higher scale (a familiar example is the Fermi interaction). Such effective operators are also generated at the quantum level, for example following compactification of a higher dimensional theory, by the radiative corrections associated with momentum and winding modes of the compactification [1, 2, 3, 4, 5].

The effects of these operators on the low energy observables can be comparable to the radiative effects of light states in the SM/MSSM [6] and this shows the importance of their study. In the following we shall study these effects to the case of the MSSM Higgs sector with additional operators of dimensions $d = 5$ and $d = 6$. In particular we show that the mass of lightest SM-like Higgs can easily be increased above the LEP bound by new physics in the region of few TeV. We then discuss the nature of the "new physics" behind the effective operators.

2.1. MSSM with d=5 operators.

As an application, consider the MSSM extended by all possible $d = 5$ operators that respect R-parity, baryon and lepton number symmetry. The Lagrangian is $\mathcal{L} = \mathcal{L}_0 + \mathcal{L}^{(5)}$ where

$$\begin{aligned}
\mathcal{L}_0 &= \int d^4\theta \sum_{i=1,2} Z_i(S,S^\dagger)\, H_i^\dagger\, e^{V_i}\, H_i + \Big\{ \int d^2\theta\, \mu_0\, (1+B_0\, S)\, H_1.H_2 + h.c. \Big\} + \cdots \\
&+ \int d^2\theta \Big[Q\, \lambda_U(S)\, U^c\, H_2 - Q\, \lambda_D(S)\, D^c\, H_1 - L\, \lambda_E(S)\, E^c\, H_1 \Big] + h.c. \qquad (3)
\end{aligned}$$

The dots stand for Higgs-independent terms and $Z_i(S,S^\dagger) = 1 - c_i\, S^\dagger S$, $c_i \sim \mathcal{O}(1)$. Further

$$\begin{aligned}
\mathcal{L}^{(5)} &= \frac{1}{M_*} \int d^4\theta \Big[H_1^\dagger\, e^{V_1} Q\, Y_U U^c + H_2^\dagger\, e^{V_2} Q Y_D D^c + H_2{}^\dagger\, e^{V_2} L\, Y_E E^c \qquad (4) \\
&+ a\, D^\alpha [b\, H_2 e^{-V_1}] D_\alpha [c\, e^{V_1} H_1] + \delta(\overline{\theta}^2) [Q U^c\, T_Q\, Q D^c + Q U^c\, T_L\, L E^c + \lambda_H (H_1 H_2)^2] + h.c. \Big]
\end{aligned}$$

with a standard notation. We introduced here some spurion dependent functions: $a, b, c, Z_{1,2}, Y_F$, $F = U, D, E$, which are general functions of $(S, S^\dagger)$ while T_Q, T_L, λ_H, are holomorphic functions of S. Here $S = m_0 \theta^2$ is the spurion superfield and m_0 the supersymmetry breaking scale, with $m_0 = \langle F_{hidden} \rangle / M_P$, so supersymmetry breaking is transmitted via gravitational interaction. Any supersymmetry breaking associated with the presence of the above interactions is included using the spurion field technique. Not all operators in (4) are independent [6]. To remove the redundant operators we use field re-definitions:

$$\begin{aligned}
H_1 &\to H_1 - \frac{1}{M_*} \overline{D}^2 \Big[\Delta_1\, H_2^\dagger\, e^{V_2}\, (i\, \sigma_2) \Big]^T + \frac{1}{M_*}\, Q\, \rho_U\, U^c \\
H_2 &\to H_2 + \frac{1}{M_*} \overline{D}^2 \Big[\Delta_2\, H_1^\dagger\, e^{V_1}\, (i\sigma_2) \Big]^T + \frac{1}{M_*}\, Q\, \rho_D\, D^c + \frac{1}{M_*}\, L\, \rho_E\, E^c \qquad (5)
\end{aligned}$$

where $\rho_F = \rho_F(S)$, $F = U, D, E$, $\Delta_i = \Delta_i(S, S^\dagger)$, $i = 1, 2$, can be chosen arbitrarily. To avoid the presence of flavour changing neutral currents, the following simple ansatz can be made:

$$\begin{aligned}
T_Q(S) &= c_Q(S)\, \lambda_U(0) \otimes \lambda_D(0), \qquad T_L(S) = c_L(S)\, \lambda_U(0) \otimes \lambda_E(0), \\
\rho_F(S) &= c_F(S)\, \lambda_F(0), \qquad Y_F(S, S^\dagger) = y_F(S, S^\dagger)\, \lambda_F(0), \quad F = U, D, E. \qquad (6)
\end{aligned}$$

and, as usual $\lambda_F(S) = \lambda_F(0)\,(1 + A_F\, S)$. Using a suitable choice for the (otherwise arbitrary) coefficients of the spurion entering in $\Delta_{1,2}$, one can set $T_Q = T_L = 0$ also $a = b = c = 0$ and $Y_F \to y_F(S^\dagger)\, \lambda_F(0)$. Then one finds [6]

$$\begin{aligned}
\mathcal{L}^{(5)} &= \frac{1}{M_*} \int d^4\theta \Big[H_1^\dagger\, e^{V_1}\, Q\, Y_U'(S^\dagger)\, U^c + H_2^\dagger\, e^{V_2} Q\, Y_D'(S^\dagger)\, D^c + H_2^\dagger\, e^{V_2} L\, Y_E'(S^\dagger)\, E^c + h.c. \Big] \\
&+ \frac{1}{M_*} \int d^2\theta\, \lambda_H'(S)\, (H_1\, H_2)^2 + h.c. \qquad (7)
\end{aligned}$$

Detailed calculations show[6] that the new Yukawa couplings $Y_F^{'}(S^\dagger)$ now depend on $S^\dagger$ only, $Y_F^{'}(S^\dagger) = \lambda_F(0)\,(x_0^F + x_2^F\, S^\dagger)$. After (5), the couplings of $\mathcal{L}_0$ also acquired, at tree level, threshold corrections which depend on M_* [6]. The new form of $\mathcal{L}^{(5)}$ in (7) gives the minimal irreducible set of R-parity, B, L conserving $d{=}5$ operators that can be present beyond MSSM.

A consequence of this analysis is the generation of new couplings, beyond those in the MSSM at the tree level. For example there is a "wrong"-Higgs Yukawa coupling, that exchanges the holomorphic dependence on one Higgs by that on the hermitian conjugate of the other [8, 9].

Journal of Physics: Conference Series **447** (2013) 012001 doi:10.1088/1742-6596/447/1/012001

Such couplings do arise in the MSSM at one-loop, after integrating out massive squarks and are suppressed by $m_0^2/M_*^2 \times$(loop-factor). Here they are suppressed by m_0/M_* only, as seen below:

$$\frac{M_s}{M_*}\Big[x_2^U\,[\lambda_U(0)]_{ij}\,(h_1^\dagger\,q_{L\,i})\;u_{R\,j}^c + x_2^D\,[\lambda_D(0)]_{ij}\,(h_2^\dagger\,q_{L\,i})\;d_{R\,j}^c + x_2^E\,[\lambda_E(0)]_{ij}\,(h_2^\dagger\,l_{L\,i})\;e_{R\,j}^c + h.c.\Big] \quad (8)$$

These couplings bring a $\tan\beta$ enhancement of a prediction for a physical observable, such as the bottom quark mass, relative to bottom quark Yukawa coupling:

$$m_b = (1/\sqrt{2})\,v\cos\beta\,(\lambda_b + \delta\lambda_b + \Delta\lambda_b \tan\beta) \quad (9)$$

Here λ_b is the usual bottom quark Yukawa coupling, $\delta\lambda_b$ is its one-loop correction in MSSM and $\Delta\lambda_b$ is a "wrong"-Higgs coupling' corrections, obtained after integrating out at one-loop massive squarks in MSSM; in our case $\Delta\lambda_b$ receives an extra correction from (8), which can actually be larger than its one-loop-generated MSSM counterpart [8, 10, 11, 12]. This can bring a $\tan\beta$ enhancement of the Higgs decay rate into bottom quarks pairs (for further details see [6]).

2.2. MSSM Higgs sector with d=5 and d=6 operators

We can extend the previous discussion by including all effective operators of both dimension $d = 5$ and $d = 6$ that can exist beyond the MSSM Higgs sector. This can be motivated in various ways. The MSSM Higgs sector is a minimal construction and extension of that of the SM. It does not take into account possible non-perturbative effects [13] or additional massive states that can couple to the Higgs sector and generate, when integrated out, new contributions. The fine tuning [14, 15] needed to have the SM-like Higgs mass well above the LEP bound [16] can also be a problem and it may indicate the existence of new physics beyond the Higgs sector. Such problems may be addressed by using a model-independent approach, using the effective operators. In the leading order, new physics beyond the MSSM Higgs sector can manifest itself as operators of either $d = 5$ [6, 17, 18, 19] or $d = 6$ [7, 20] or both. If generated by the same new physics, by comparing $\mathcal{O}(1/M^*)$ and $\mathcal{O}(1/M_*^2)$ terms one can estimate when the series expansion in $1/M_*$ breaks down. The operators in the Higgs sector of dimension $d = 5$ were:

$$\begin{aligned}
\mathcal{L}_1 &= \frac{1}{M_*}\int d^2\theta\;\lambda_H'(S)\,(H_2.H_1)^2 + h.c. = 2\,\zeta_{10}\,(h_2.h_1)(h_2.F_1 + F_2.h_1) + \zeta_{11}\,m_0\,(h_2.h_1)^2 + h.c,\\
\mathcal{L}_2 &= \frac{1}{M_*}\int d^4\theta\,\Big\{a(S,S^\dagger)D^\alpha\Big[b(S,S^\dagger)\,H_2\,e^{-V_1}\Big]D_\alpha\Big[c(S,S^\dagger)\,e^{V_1}\,H_1\Big] + h.c.\Big\}
\end{aligned} \quad (10)$$

where

$$\lambda_H'(S)/M_* = \zeta_{10} + \zeta_{11}\,m_0\,\theta\theta, \qquad \zeta_{10},\,\zeta_{11} \sim 1/M_*, \quad (11)$$

$\mathcal{L}_1$ can be generated by integrating out a massive gauge singlet or $SU(2)$ triplet. Indeed, in the MSSM with a massive gauge singlet, with an F-term of type $M_*\Sigma^2 + \Sigma\,H_1.H_2$, when integrating out Σ generates $\mathcal{L}_1$. $\mathcal{L}_2$ can be generated in various ways (see Appendix A, B in [6]) but perhaps the simplest way is via an additional pair of massive Higgs doublets of mass of order M_*. As already discussed, $\mathcal{L}_2$ can be removed by general spurion-dependent field redefinitions, up to soft terms and μ term renormalisation and $\mathcal{O}(1/M_*^2)$ corrections [6].

We assume that $m_0 \ll M_*$, so that the effective approach is reliable. If this is not respected and the "new physics" is represented by "light" states (like the MSSM states), the $1/M_*$ expansion is not reliable and one should work in a setup where these are not integrated out.

The list of $d = 6$ operators is longer [21]:

Journal of Physics: Conference Series **447** (2013) 012001 doi:10.1088/1742-6596/447/1/012001

$$\begin{aligned}
\mathcal{O}_1 &= \frac{1}{M_*^2}\int d^4\theta\ \mathcal{Z}_1\ (H_1^\dagger\, e^{V_1}\, H_1)^2, & \mathcal{O}_5 &= \frac{1}{M_*^2}\int d^4\theta\ \mathcal{Z}_5\ (H_1^\dagger\, e^{V_1}\, H_1)\ H_2.\, H_1 + h.c.\\
\mathcal{O}_2 &= \frac{1}{M_*^2}\int d^4\theta\ \mathcal{Z}_2\ (H_2^\dagger\, e^{V_2}\, H_2)^2, & \mathcal{O}_6 &= \frac{1}{M_*^2}\int d^4\theta\ \mathcal{Z}_6\ (H_2^\dagger\, e^{V_2}\, H_2)\ H_2.\, H_1 + h.c.\\
\mathcal{O}_3 &= \frac{1}{M_*^2}\int d^4\theta\ \mathcal{Z}_3\ (H_1^\dagger\, e^{V_1}\, H_1)\,(H_2^\dagger\, e^{V_2}\, H_2), & \mathcal{O}_7 &= \frac{1}{M_*^2}\int d^2\theta\ \mathcal{Z}_7\, \mathrm{Tr}\, W^\alpha\, W_\alpha\, (H_2\, H_1) + h.c.\\
\mathcal{O}_4 &= \frac{1}{M_*^2}\int d^4\theta\, \mathcal{Z}_4\ (H_2.\, H_1)\,(H_2.\, H_1)^\dagger, & \mathcal{O}_8 &= \frac{1}{M_*^2}\int d^4\theta\, \mathcal{Z}_8\ (H_2\, H_1)^2 + h.c. \qquad (12)
\end{aligned}$$

where $W^\alpha = (-1/4)\,\overline{D}^2 e^{-V} D^\alpha\, e^V$ is the chiral field strength of $SU(2)_L$ or $U(1)_Y$ vector superfields V_w and V_Y respectively. Also $V_{1,2} = V_w^a(\sigma^a/2) + (\mp 1/2)\, V_Y$ with the upper (minus) sign for V_1. The remaining $d = 6$ operators involve extra space-time derivatives:

$$\begin{aligned}
\mathcal{O}_9 &= \frac{1}{M_*^2}\int d^4\theta\ \mathcal{Z}_9\, H_1^\dagger\, \overline{\nabla}^2\, e^{V_1}\, \nabla^2\, H_1 & \mathcal{O}_{12} &= \frac{1}{M_*^2}\int d^4\theta\ \mathcal{Z}_{12}\, H_2^\dagger\, e^{V_2}\, \nabla^\alpha\, W_\alpha^{(2)}\, H_2\\
\mathcal{O}_{10} &= \frac{1}{M_*^2}\int d^4\theta\ \mathcal{Z}_{10}\, H_2^\dagger\, \overline{\nabla}^2\, e^{V_2}\, \nabla^2\, H_2 & \mathcal{O}_{13} &= \frac{1}{M_*^2}\int d^4\theta\ \mathcal{Z}_{13}\, H_1^\dagger\, e^{V_1}\ W_\alpha^{(1)}\, \nabla^\alpha\, H_1\\
\mathcal{O}_{11} &= \frac{1}{M_*^2}\int d^4\theta\ \mathcal{Z}_{11}\, H_1^\dagger\, e^{V_1}\, \nabla^\alpha\, W_\alpha^{(1)}\, H_1 & \mathcal{O}_{14} &= \frac{1}{M_*^2}\int d^4\theta\ \mathcal{Z}_{14}\, H_2^\dagger\, e^{V_2}\ W_\alpha^{(2)}\, \nabla^\alpha\, H_2 \ (13)
\end{aligned}$$

Also $\nabla_\alpha\, H_i = e^{-V_i}\, D_\alpha\, e^{V_i} H_i$ and W_α^i is the field strength of V_i. To be general, in the above operators one should include spurion (S) dependence under any ∇_α, of arbitrary coefficients, to account for supersymmetry breaking effects associated to them. Finally, the wavefunction coefficients are spurion dependent and have the structure

$$(1/M_*^2)\,\mathcal{Z}_i(S, S^\dagger) = \alpha_{i0} + \alpha_{i1}\, m_0\, \theta\theta + \alpha_{i1}^*\, m_0\, \overline{\theta\theta} + \alpha_{i2}\, m_0^2\, \theta\theta\overline{\theta\theta}, \qquad \alpha_{ij} \sim 1/M_*^2. \qquad (14)$$

Regarding the origin of these operators: $\mathcal{O}_{1,2,3}$ can be generated in the MSSM by an additional, massive $U(1)'$ gauge boson or $SU(2)$ triplets, when integrated out [17]. $\mathcal{O}_4$ can be generated by a massive gauge singlet or $SU(2)$ triplet, while $\mathcal{O}_{5,6}$ can be generated by a combination of $SU(2)$ doublets and massive gauge singlet. $\mathcal{O}_7$ is essentially a threshold correction to the gauge coupling, with a moduli field replaced by the Higgs. $\mathcal{O}_8$ exists only in non-susy case, but is generated when removing the $d = 5$ derivative operator $\mathcal{L}_2$ by field redefinitions [6], so we keep it.

It can be shown that operators $\mathcal{O}_{9,\dots,14}$, can be eliminated along the lines discussed in the previous sections. For example, in the absence of gauge interactions, $\mathcal{O}_9$ is similar to the operator in eq.(7) and only brings a wavefunction renormalisation, $\mathcal{O}_9 \sim |\mu|^2/M_*^2 \int d^4\theta H_1^\dagger H_1$, and similar for $\mathcal{O}_{10}$. Regarding $\mathcal{O}_{11,12}$, in the supersymmetric case they vanish, following the definition of ∇^α and an integration by parts. Further, $\mathcal{O}_{13,14}$ are similar to $\mathcal{O}_{9,10}$, which can be seen by using the definition of $W_\alpha^{(i)}$ and the relation between ∇^2, $(\overline{\nabla}^2)$ and D^2, $(\overline{D}^2)$. In the presence of supersymmetry breaking, elimination of these operators and their supersymmetry breaking contribution is still possible, up to a renormalisation of the soft terms and μ term [6].

2.3. *Higgs mass corrections from $d = 5$ and $d = 6$ operators.*

With the remaining set of independent, effective operators $\mathcal{L}_1$, $\mathcal{O}_{1,\dots,8}$ of dimensions $d = 5$ and $d = 6$, one finds the scalar potential V and its EW minimum; this is perturbed by $\mathcal{O}(1/M_*^2)$

corrections from that of the MSSM. The expression of V is long and it is not given here (see [7] for its form). From V one computes the mass of CP-odd/even Higgs fields. One has:

$$m_A^2 = (m_A^2)_{\rm MSSM} - \frac{2\,\zeta_{10}\,\mu_0\,v^2}{\sin 2\beta} + 2\,m_0\,\zeta_{11}\,v^2 + \delta m_A^2, \quad \delta m_A^2 = \mathcal{O}(1/M_*^2) \tag{15}$$

for the pseudoscalar Higgs, with $(m_A^2)_{\rm MSSM}$ the MSSM value, with δm_A^2 due to $\mathcal{O}(1/M_*^2)$ corrections from $d=5$ and $d=6$ operators. For the CP-even Higgs one has [6, 17, 19]

$$\begin{aligned} m_{h,H}^2 &= (m_{h,H}^2)_{\rm MSSM} \\ &+ (2\,\zeta_{10}\,\mu_0)\,v^2\,\sin 2\beta\,\Big[1 \pm \frac{m_A^2+m_Z^2}{\sqrt{\tilde w}}\Big] + \frac{(-2\,\zeta_{11}\,m_0)\,v^2}{2}\,\Big[1 \mp \frac{(m_A^2-m_Z^2)\,\cos^2 2\beta}{\sqrt{\tilde w}}\Big] \\ &+ \delta m_{h,H}^2, \qquad \text{where} \qquad \delta m_{h,H}^2 = \mathcal{O}(1/M_*^2) \end{aligned} \tag{16}$$

The upper (lower) signs correspond to h (H), and $\tilde w \equiv (m_A^2+m_Z^2)^2 - 4\,m_A^2\,m_Z^2\,\cos^2 2\beta$. With this result one can show that the mass m_h can be increased above the LEP bound, also with the help of quantum corrections [6, 17, 18, 19].

Regarding the $\mathcal{O}(1/M_*^2)$ corrections of $\delta m_{h,H}^2$, δm_A^2 and $\delta m_{h,H}^2$ of (15), (16), in the general case of including all operators and their associated supersymmetry breaking, they have a complicated form. Exact expressions can be found in [7, 20]. For most purposes, an expansion of these in $1/\tan\beta$ is accurate enough. At large $\tan\beta$, $d=6$ operators bring corrections comparable to those of $d=5$ operators. The relative $\tan\beta$ enhancement of $\mathcal{O}(1/M_*^2)$ corrections compensates for the extra suppression that these have relative to $\mathcal{O}(1/M_*)$ operators (which involve both h_1, h_2 and are not enhanced in this limit). Note however that in some models only $d=6$ operators may be present, depending on the details of the "new physics" generating the effective operators.

Let us present the correction $\mathcal{O}(1/M^2)$ to $m_{h,H}^2$ for the case m_A is kept fixed to an appropriate value. The result is, assuming $m_A > m_Z$, (otherwise δm_h^2 and δm_H^2 are exchanged):

$$\begin{aligned} \delta m_h^2 &= -2\,v^2\,\Big[\alpha_{22}\,m_0^2 + (\alpha_{30}+\alpha_{40})\mu_0^2 + 2\alpha_{61}\,m_0\,\mu_0 - \alpha_{20}\,m_Z^2\Big] - (2\,\zeta_{10}\,\mu_0)^2\,v^4\,(m_A^2-m_Z^2)^{-1} \\ &+ v^2\cot\beta\Big[(m_A^2-m_Z^2)^{-1}\Big(4\,m_A^2\,(\,(2\alpha_{21}+\alpha_{31}+\alpha_{41}+2\alpha_{81})\,m_0\,\mu_0 + (2\alpha_{50}+\alpha_{60})\,\mu_0^2 + \alpha_{62}\,m_0^2) \\ &- (2\alpha_{60}-3\alpha_{70})\,m_A^2\,m_Z^2 - (2\alpha_{60}+\alpha_{70})\,m_Z^4\Big) + 8\,(m_A^2+m_Z^2)\,(\mu_0\,m_0\,\zeta_{10}\,\zeta_{11})\,v^2/(m_A^2-m_Z^2)^2\Big] \\ &+ \mathcal{O}(1/\tan^2\beta) \end{aligned} \tag{17}$$

A similar formula exists for the correction to m_H:

$$\begin{aligned} \delta m_H^2 &= \Big[-2\,(m_0\mu_0\,(\alpha_{51}+\alpha_{61}) + \alpha_{82}), m_0^2)\,v^2 + (2\,\zeta_{10}\,\mu_0)^2\,v^4(m_A^2-m_Z^2)^{-1}\Big] \\ &+ v^2\cot\beta\Big[(m_A^2-m_Z^2)^{-1}\Big(2m_A^2(2(\alpha_{11}-\alpha_{21})\,m_0\mu_0 + (\alpha_{60}-\alpha_{50})\,\mu_0^2 + (\alpha_{52}-\alpha_{62})m_0^2 - \alpha_{60}m_A^2) \\ &- \;[4\,(\alpha_{11}+\alpha_{21}+\alpha_{31}+\alpha_{41}+2\alpha_{81})\,m_0\mu_0 + 6(\alpha_{50}+\alpha_{60})\,\mu_0^2 + 2(\alpha_{52}+\alpha_{62})\,m_0^2 \\ &- (\alpha_{50}+5\alpha_{60}-2\alpha_{70})\,m_A^2]m_Z^2 - (\alpha_{50}-\alpha_{60})m_Z^4\Big) - 8\,(m_A^2+m_Z^2)(\mu_0 m_0\zeta_{10}\zeta_{11})\,v^2/(m_A^2-m_Z^2)^2\Big] \\ &+ \mathcal{O}(1/\tan^2\beta) \end{aligned} \tag{18}$$

The mass corrections in (17), (18) must be added to the rhs of eqs.(16) to obtain the full value of $m_{h,H}^2$. Together with (12), (14), these corrections identify the operators of $d=6$ with the largest contributions, which is important for model building beyond the MSSM Higgs sector. These operators are $\mathcal{O}_{2,3,4}$ in the absence of supersymmetry breaking and $\mathcal{O}_{2,6}$ when this is broken. It is preferable, however, to increase m_h^2 by supersymmetric rather than supersymmetry-breaking effects of the effective operators, because the latter are less under control in the effective approach; also, one would favour a supersymmetric solution to the fine-tuning problem associated with increasing the MSSM Higgs mass above the LEP bound. Therefore $\mathcal{O}_{2,3,4}$ are the leading operators, with the remark that $\mathcal{O}_2$ has a smaller effect, of order $(m_Z/\mu_0)^2$ relative to $\mathcal{O}_{3,4}$ (for similar α_{j0}, $j = 2, 3, 4$). At smaller $\tan\beta$, $\mathcal{O}_{5,6}$ can also give significant contributions, while $\mathcal{O}_7$ has a relative suppression factor $(m_Z/\mu_0)^2$. Note that we kept all operators $\mathcal{O}_i$ independent. By doing so, one can easily single out the individual contribution of each operator, which helps in model building, since not all operators are present in a specific model.

One limit to consider is that where the operators of $d = 6$ have coefficients such that their contributions add up to maximise δm_h^2. Since α_{ij} are not known, one can choose:

$$-\alpha_{22} = -\alpha_{61} = -\alpha_{30} = -\alpha_{40} = \alpha_{20} > 0 \tag{19}$$

In this case, at large $\tan\beta$:

$$\delta m_h^2 \approx 2\, v^2 \alpha_{20} [m_0^2 + 2\, m_0 \mu_0 + 2\, \mu_0^2 + m_Z^2] \tag{20}$$

A simple numerical example is illustrative. For $m_0 = 1$ TeV, $\mu_0 = 350$ GeV, and with $v \approx 246$ GeV, one has $\delta m_h^2 \approx 2.36\, \alpha_{20} \times 10^{11}$ (GeV)^2. Assuming $M_* = 10$ TeV and ignoring $d = 5$ operators, with $\alpha_{20} \sim 1/M_*^2$ and the MSSM value of m_h taken to be its upper classical limit m_Z (reached for large $\tan\beta$), we obtain an increase of m_h from $d = 6$ operators alone of about $\Delta m_h = 12.15$ GeV to $m_h \approx 103$ GeV. An increase of α_{20} by a factor of 2.5 to $\alpha_{20} \sim 2.5/M_*^2$ would give $\Delta m_h \approx 28$ GeV to $m_h \approx 119.2$ GeV, which is already above the LEP bound. Note that this increase is realised even for a scale M_* of "new physics" beyond the LHC reach.

The above choice of $M_* = 10$ TeV was partly motivated by the fine-tuning results [18] (for $d = 5$ operators) and on convergence grounds: the expansion parameter of our effective analysis is m_q/M_* where m_q is any scale of the theory, in particular it can be m_0. For a susy breaking scale $m_0 \sim \mathcal{O}(1)$ TeV (say $m_0 = 3$ TeV) and $c_{1,2}$ or α_{ij} of $\mathcal{Z}_i(S, S^\dagger)$ of order unity (say $c_{1,2} = 2.5$) one has for $M_* = 10$ TeV that $c_{1,2}\, m_0/M_* = 0.75$ which is already close to unity, and at the limit of validity of the effective expansion in powers of $1/M_*$. To conclude, even for a scale of "new physics" above the LHC reach, one can still classically increase m_h to the LEP bound.

2.4. Final remarks

The final step is to identify the nature of "new physics" that generated the operators with the largest correction to m_h, ideally from a renormalisable model. At the level of dimension $d = 5$ operators, this is clear from previous discussion: a massive gauge singlet can generate operator $\mathcal{L}_1$ of (10) and the needed increase of m_h, for a scale $M_* \sim 5 - 10$ TeV [18]; this can provide a solution to the little hierarchy problem, provided that one can fix dynamically the scale M_*.

For dimension-six operators, from the above discussion one finds that to increase m_h it is needed that one or more of the following conditions are satisfied:

$$\alpha_{20} > 0,\ \alpha_{30} < 0,\ \alpha_{40} < 0 \tag{21}$$

First recall that $\mathcal{O}_{1,2,3}$ can be most easily generated by integrating out a massive gauge boson $U(1)'$ or $SU(2)$ triplets [17], while $\mathcal{O}_4$ can be generated by a massive gauge singlet or $SU(2)$ triplets. Let us discuss the signs of the operators when they are generated as above:
(a): Integrating out a massive vector superfield $U(1)'$ under which Higgs fields have opposite charges (to avoid a Fayet-Iliopoulos term), one finds $\alpha_{20} < 0$ and $\alpha_{30} > 0$ (also $\alpha_{10} < 0$) [17],

which is opposite to what we need. This can be changed, if for example there are additional pairs of massive Higgs doublets also charged under new $U(1)'$; then $\mathcal{O}_3$ could be generated with $\alpha_{30} < 0$.
(b): Integrating massive $SU(2)$ triplets that couple to the MSSM Higgs sector would bring $\alpha_{20} > 0$, $\alpha_{40} < 0$, $\alpha_{30} > 0$, so the first two relations agree with what we need.
(c): Integrating a massive gauge singlet would bring $\alpha_{40} > 0$, which would instead decrease m_h.

Finally, at large $\tan\beta$, due to additional corrections that effective operators bring to the ρ parameter [22], it turns out that α_{40} and α_{30} can have the largest correction to m_h^2, while avoiding ρ-parameter constraints. The case of a massive gauge singlet or additional $U(1)'$ vector superfield (giving $\mathcal{O}_{3,4}$) have the advantage of preserving gauge couplings unification at one-loop. Following the above information, one can proceed to construct explicit models with additional states that can generate these effective operators.

Let us mention that the method provided here to reduce the fine tuning in the MSSM for m_h larger than the LEP bound, relies on introducing an additional scale in the visible sector, due to "new physics" in this sector. Other solutions to this problem may exist, which essentially rely on a low scale in the hidden sector of supersymmetry breaking [23]. In this case the quartic coupling and the mass of the SM-like Higgs are increased by a factor proportional to $(\mu^2/f)^2$, where f is the hidden sector supersymmetry breaking scale. While not without problems, the advantage of this latter method is that it does not pay the "cost" of an additional parameter (scale) in the visible sector, as models with effective operators do.

3. Strings and extra dimensions

The appropriate and most convenient framework for low energy supersymmetry and grand unification is the perturbative heterotic string. Indeed, in this theory, gravity and gauge interactions have the same origin, as massless modes of the closed heterotic string, and they are unified at the string scale M_s. As a result, the Planck mass is predicted to be proportional to M_s:

$$M_P = M_s/g\,, \tag{22}$$

where g is the gauge coupling. In the simplest constructions all gauge couplings are the same at the string scale, given by the four-dimensional (4d) string coupling, and thus no grand unified group is needed for unification. In our conventions $\alpha_{\rm GUT} = g^2 \simeq 0.04$, leading to a discrepancy between the string and grand unification scale $M_{\rm GUT}$ by almost two orders of magnitude. Explaining this gap introduces in general new parameters or a new scale, and the predictive power is essentially lost. This is the main defect of this framework, which remains though an open and interesting possibility.

The other other perturbative framework that has been studied extensively in the more recent years is type I string theory with D-branes. Unlike in the heterotic string, gauge and gravitational interactions have now different origin. The latter are described again by closed strings, while the former emerge as excitations of open strings with endpoints confined on D-branes [24]. This leads to a braneworld description of our universe, which should be localized on a hypersurface, i.e. a membrane extended in p spatial dimensions, called p-brane (see Fig. 1). Closed strings propagate in all nine dimensions of string theory: in those extended along the p-brane, called parallel, as well as in the transverse ones. On the contrary, open strings are attached on the p-brane. Obviously, our p-brane world must have at least the three known dimensions of space. But it may contain more: the extra $d_\parallel = p - 3$ parallel dimensions must have a finite size, in order to be unobservable at present energies, and can be as large as $\text{TeV}^{-1} \sim 10^{-18}$ m [25]. On the other hand, transverse dimensions interact with us only gravitationally and experimental bounds are much weaker: their size should be less than about 0.1 mm [26]. In the following, I review the main properties and experimental signatures of low string scale models [27].

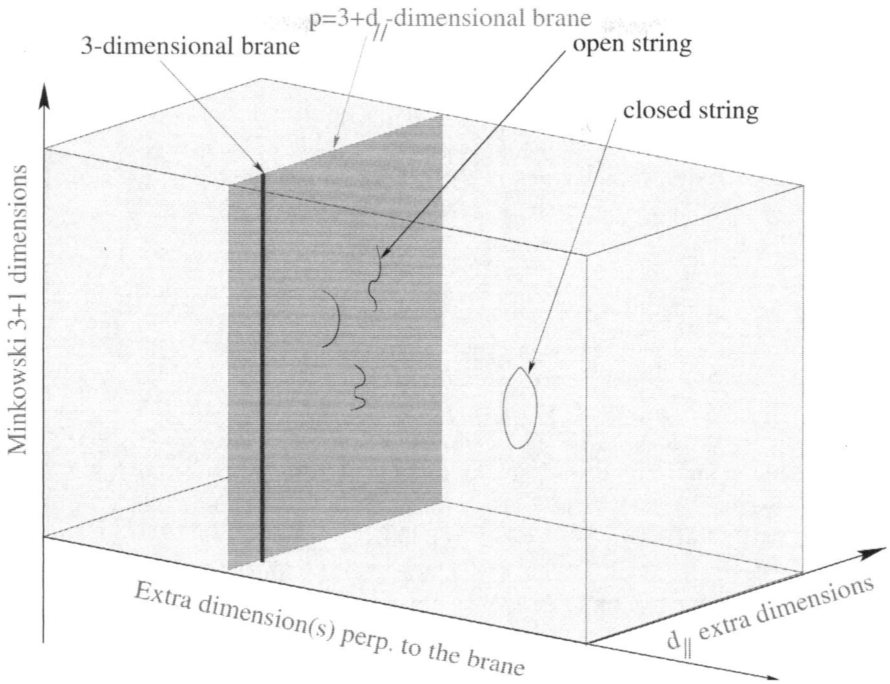

Figure 1. D-brane world universe in type I string framework.

3.1. Framework of low scale strings

In type I theory, the different origin of gauge and gravitational interactions implies that the relation between the Planck and string scales is not linear as (22) of the heterotic string. The requirement that string theory should be weakly coupled, constrain the size of all parallel dimensions to be of order of the string length, while transverse dimensions remain unrestricted. Assuming an isotropic transverse space of $n = 9 - p$ compact dimensions of common radius $R_\perp$, one finds:

$$M_P^2 = \frac{1}{g_s^2} M_s^{2+n} R_\perp^n \,, \qquad g_s \simeq g^2 \,. \tag{23}$$

where g_s is the string coupling. It follows that the type I string scale can be chosen hierarchically smaller than the Planck mass [28, 27] at the expense of introducing extra large transverse dimensions felt only by gravity, while keeping the string coupling small [27]. The weakness of 4d gravity compared to gauge interactions (ratio M_W/M_P) is then attributed to the largeness of the transverse space $R_\perp$ compared to the string length $l_s = M_s^{-1}$.

An important property of these models is that gravity becomes effectively $(4+n)$-dimensional with a strength comparable to those of gauge interactions at the string scale. The first relation of Eq. (23) can be understood as a consequence of the $(4+n)$-dimensional Gauss law for gravity, with $M_*^{(4+n)} = M_s^{2+n}/g^4$ the effective scale of gravity in $4+n$ dimensions. Taking $M_s \simeq 1$ TeV, one finds a size for the extra dimensions $R_\perp$ varying from 10^8 km, .1 mm, down to a Fermi for $n = 1, 2$, or 6 large dimensions, respectively. This shows that while $n = 1$ is excluded, $n \geq 2$ is allowed by present experimental bounds on gravitational forces [26, 29]. Thus, in these models, gravity appears to us very weak at macroscopic scales because its intensity is spread in the "hidden" extra dimensions. At distances shorter than $R_\perp$, it should deviate from Newton's law, which may be possible to explore in laboratory experiments.

3.2. Large number of species

Here, we open a parenthesis to describe that low scale gravity with large extra dimensions is actually a particular case of a more general framework, where the ultraviolate (UV) cutoff is lower than the Panck scale due to the existence of a large number of particle species coupled to gravity [30]. Indeed, it was shown that the effective UV cutoff M_* is given by

$$M_*^2 = M_P^2/N \,, \tag{24}$$

where the counting of independent species N takes into account all particles which are not broad resonances, having a width less than their mass. The derivation is based on black hole evaporation but here we present a shorter argument using quantum information storage [31]. Consider a pixel of size L containing N species storing information. The energy required to localize N wave functions is then given by N/L, associated to a Schwarzschild radius $R_s = N/LM_P^2$. The latter must be less than the pixel size in order to avoid the collapse of such a system to a black hole, $R_s \leq L$, implying a minimum size $L \geq L_{min}$ with $L_{min} = \sqrt{N}/M_P$ associated precisely to the effective UV cutoff $M_* = L_{min}$ given in eq. (24). Imposing $M_* \simeq 1$ TeV, one should then have $N \sim 10^{32}$ particle species below about the TeV scale!

In the string theory context, there are two ways of realizing such a large number a particle species by lowering the string scale at a TeV:

(i) In large volume compactifications with the SM localized on D-brane stacks, as described in the previous section. The particle species are then the Kaluza-Klein (KK) excitations of the graviton (and other possible bulk modes) associated to the large extra dimensions, given by $N = R_\perp^n l_s^n$, up to energies of order $M_* \simeq M_s$.

(ii) By introducing an infinitesimal string coupling $g_s \simeq 10^{-16}$ with the SM localized on Neveu-Schwarz NS5-branes in the framework of little strings [32]. In this case, the particle species are the effective number of string modes that contribute to the black hole bound [33]: $N = 1/g_s^2$ and gravity does not become strong at $M_s \sim \mathcal{O}(\text{TeV})$.

Note the both TeV string realizations above are compatible with the general expression (23), but in the second case there is no relation between the string and gauge couplings.

3.3. Experimental implications in accelerators

We now turn to the experimental predictions of TeV scale strings. Their main implications in particle accelerators are of three types, in correspondence with the three different sectors that are generally present:

(i) New compactified parallel dimensions; In this case $RM_s \gtrsim 1$, and the associated compactification scale $R_\parallel^{-1}$ would be the first scale of new physics that should be found increasing the beam energy [25, 34]. The main consequence is the existence of KK excitations for all SM particles that propagate along the extra parallel dimensions. These can be produced on-shell at LHC as new resonances [35].

(ii) New extra large transverse dimensions and low scale quantum gravity,. The main experimental signal is gravitational radiation in the bulk from any physical process on the world-brane [36].

(iii) Genuine string and quantum gravity effects. Direct production of string resonances in hadron colliders leads generically to a universal deviation from Standard Model in jet distribution [37]. In particular, the first Regge excitation of the gluon has spin 2 and a width an order of magnitude lower than the string scale, leading to a characteristic peak in dijet production; similarly, the first excitations of quarks have spin 3/2.

(iv) Concerning possible micro-black hole production, note that a string size black hole has a horizon radius $r_H \sim 1$ in string units, while the Newton's constant behaves as $G_N \sim g_s^2$. It follows that the mass of a d-dimensional black hole is [38]: $M_{\rm BH} \sim r_H^{d/2-1}/G_N \simeq 1/g_s^2$. Using the value of the SM gauge couplings $g_s \simeq g^2 \sim 0.1$, one finds that the energy threshold $M_{\rm BH}$ of micro-black hole production is about four orders of magnitude higher than the string scale, implying that one would produce 10^4 string states before reaching $M_{\rm BH}$.

3.4. Electroweak symmetry breaking

Non-supersymmetric TeV strings offer also a framework to realize gauge symmetry breaking radiatively. Indeed, from the effective field theory point of view, one expects quadratically divergent one-loop contributions to the masses of scalar fields. The divergences are cut off by M_s and if the corrections are negative, they can induce electroweak symmetry breaking and explain the mild hierarchy between the weak and a string scale at a few TeV, in terms of a loop factor [39]. More precisely, in the minimal case of one Higgs doublet H, the scalar potential is:

$$V = \lambda(H^\dagger H)^2 + \mu^2(H^\dagger H)\,, \tag{25}$$

where λ arises at tree-level. Moreover, in any model where the Higgs field comes from an open string with both ends fixed on the same brane stack, it is given by an appropriate truncation of a supersymmetric theory. Within the minimal spectrum of the SM, $\lambda = (g_2^2 + g'^2)/8$, with g_2 and g' the $SU(2)$ and $U(1)_Y$ gauge couplings. On the other hand, μ^2 is generated at one loop:

$$\mu^2 = -\varepsilon^2\, g^2\, M_s^2\,, \tag{26}$$

where ε is a loop factor that can be estimated from a toy model computation and varies in the region $\epsilon \sim 10^{-1} - 10^{-3}$.

Indeed, consider for illustration a simple case where the whole one-loop effective potential of a scalar field can be computed. We assume for instance one extra dimension compactified on a circle of radius $R > 1$ (in string units). An interesting situation is provided by a class of models where a non-vanishing VEV for a scalar (Higgs) field ϕ results in shifting the mass of each KK excitation by a constant $a(\phi)$:

$$M_m^2 = \left(\frac{m + a(\phi)}{R}\right)^2\,, \tag{27}$$

with m the KK integer momentum number. Such mass shifts arise for instance in the presence of a Wilson line, $a = q\oint \frac{dy}{2\pi} gA$, where A is the internal component of a gauge field with gauge coupling g, and q is the charge of a given state under the corresponding generator. A straightforward computation shows that the ϕ-dependent part of the one-loop effective potential is given by [40]:

$$V_{eff} = -Tr(-)^F \frac{R}{32\,\pi^{3/2}} \sum_n e^{2\pi i n a} \int_0^\infty dl\; l^{3/2} f_s(l)\; e^{-\pi^2 n^2 R^2 l} \tag{28}$$

where $F = 0, 1$ for bosons and fermions, respectively. We have included a regulating function $f_s(l)$ which contains for example the effects of string oscillators. To understand its role we will consider the two limits $R >> 1$ and $R << 1$. In the first case only the $l \to 0$ region contributes to the integral. This means that the effective potential receives sizable contributions only from the infrared (field theory) degrees of freedom. In this limit we would have $f_s(l) \to 1$. For example, in the string model considered in [39]:

$$f_s(l) = \left[\frac{1}{4l}\frac{\theta_2}{\eta^3}(il + \frac{1}{2})\right]^4 \to 1 \qquad \text{for} \qquad l \to 0, \tag{29}$$

and the field theory result is finite and can be explicitly computed. As a result of the Taylor expansion around $a = 0$, we are able to extract the one-loop contribution to the coefficient of the term of the potential quadratic in the Higgs field. It is given by a loop factor times the compactification scale [40]. One thus obtains $\mu^2 \sim g^2/R^2$ up to a proportionality constant which is calculable in the effective field theory. On the other hand, if we consider $R \to 0$, which by T-duality corresponds to taking the extra dimension as transverse and very large, the one-loop effective potential receives contributions from the whole tower of string oscillators as appearing in $f_s(l)$, leading to squared masses given by a loop factor times M_s^2, according to eq. (26).

More precisely, from the expression (28), one finds:

$$\varepsilon^2(R) = \frac{1}{2\pi^2}\int_0^\infty \frac{dl}{(2\,l)^{5/2}}\frac{\theta_2^4}{4\eta^{12}}\left(il+\frac{1}{2}\right)R^3\sum_n n^2 e^{-2\pi n^2 R^2 l}\,, \tag{30}$$

which is plotted in Fig. 2. For the asymptotic value $R \to 0$ (corresponding upon T-duality to

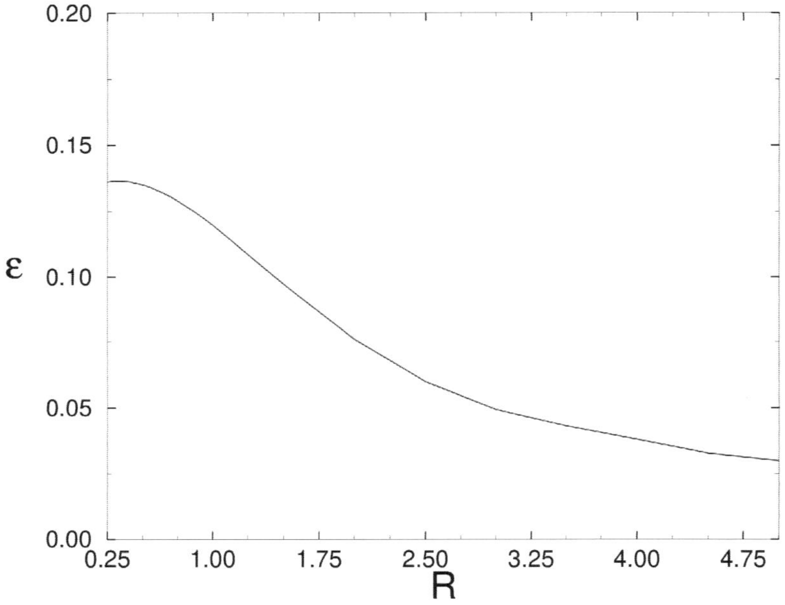

Figure 2. The coefficient ε of the one loop Higgs mass (26).

a large transverse dimension of radius $1/R$), $\varepsilon(0) \simeq 0.14$, and the effective cut-off for the mass term is M_s, as can be seen from Eq. (26). At large R, $\mu^2(R)$ falls off as $1/R^2$, which is the effective cut-off in the limit $R \to \infty$, as we argued above, in agreement with field theory results in the presence of a compactified extra dimension [41, 42]. In fact, in the limit $R \to \infty$, an analytic approximation to $\varepsilon(R)$ gives:

$$\varepsilon(R) \simeq \frac{\varepsilon_\infty}{M_s\,R}\,, \qquad \varepsilon_\infty^2 = \frac{3\,\zeta(5)}{4\,\pi^4} \simeq 0.008\,. \tag{31}$$

The potential (25) has the usual minimum, given by the VEV of the neutral component of the Higgs doublet $v = \sqrt{-\mu^2/\lambda}$. Using the relation of v with the Z gauge boson mass, $M_Z^2 = (g_2^2 + g'^2)v^2/4$, and the expression of the quartic coupling λ, one obtains for the Higgs mass a prediction which is the Minimal Supersymmetric Standard Model (MSSM) value for $\tan\beta \to \infty$ and $m_A \to \infty$: $m_H = M_Z$. The tree level Higgs mass is known to receive important radiative corrections from the top-quark sector and rises to values around 120 GeV. Furthermore, from (26), one can compute M_s in terms of the Higgs mass $m_H^2 = -2\mu^2$:

$$M_s = \frac{m_H}{\sqrt{2}\,g\varepsilon}, \tag{32}$$

yielding naturally values in the TeV range.

Acknowledgments
Work supported in part by the European Commission under the ERC Advanced Grant 226371 and the contract PITN-GA-2009-237920.

References

[1] D. M. Ghilencea and H. M. Lee, JHEP **0509** (2005) 024.
[2] D. M. Ghilencea, Phys. Rev. D **70** (2004) 045018.
[3] S. Groot Nibbelink, M. Hillenbach, Phys. Lett. B **616** (2005) 125.
[4] S. Groot Nibbelink, M. Hillenbach, Nucl.Phys. B **748** (2006) 60.
[5] J. F. Oliver, J. Papavassiliou and A. Santamaria, Phys. Rev. D **67** (2003) 125004.
[6] I. Antoniadis, E. Dudas, D. Ghilencea, P. Tziveloglou, Nucl. Phys. B **808** (2009) 155.
[7] I. Antoniadis, E. Dudas, D. Ghilencea, P. Tziveloglou, Nucl. Phys. B **831** (2010) 133.
[8] H. E. Haber, J. D. Mason, Phys. Rev. D **77** 115011 (2008).
[9] S. P. Martin, Phys. Rev. D **61** 035004 (2000).
[10] M. S. Carena, H. E. Haber, H. E. Logan, S. Mrenna, Phys. Rev. D **65** 055005 (2002).
[11] D. M. Pierce, J. A. Bagger, K. T. Matchev, R. J. Zhang, Nucl. Phys. B **491** 3-67 (1997).
[12] L. J. Hall, R. Rattazzi and U. Sarid, Phys. Rev. D **50** 7048-7065 (1994).
[13] G. F. Giudice, C. Grojean, A. Pomarol and R. Rattazzi, JHEP **0706** (2007) 045.
[14] S. Cassel, D. M. Ghilencea, G. G. Ross, Nucl. Phys. B **835** (2010) 110.
[15] S. Cassel, D. M. Ghilencea and G. G. Ross, Phys. Lett. B **687** (2010) 214.
[16] R. Barate *et al.* [LEP Working Group for Higgs boson searches and ALEPH, DELPHI, L3, OPAL Collaborations, Phys. Lett. B **565** (2003) 61; S. Schael *et al.* ALEPH, DELPHI, L3, OPAL and LEP Working Group for Higgs Boson Searches, Eur. Phys. J. C**47**(2006)547
[17] M. Dine, N. Seiberg, S. Thomas, Phys. Rev. D **76** (2007) 095004.
[18] S. Cassel, D. Ghilencea, G. G. Ross, Nucl. Phys. B **825** (2010) 203.
[19] K. Blum, C. Delaunay and Y. Hochberg, arXiv:0905.1701 [hep-ph].
[20] M. Carena, K. Kong, E. Ponton and J. Zurita, arXiv:0909.5434 [hep-ph].
M. Carena, E. Ponton and J. Zurita, arXiv:1005.4887 [hep-ph].
[21] D. Piriz and J. Wudka, Phys. Rev. D **56** (1997) 4170.
[22] A. Brignole, J. A. Casas, J. R. Espinosa and I. Navarro, Nucl. Phys. B **666** (2003) 105.
[23] I. Antoniadis, E. Dudas, D. M. Ghilencea and P. Tziveloglou, arXiv:1006.1662 [hep-ph].
[24] C. Angelantonj and A. Sagnotti, *Phys. Rept.* **371** (2002) 1 [Erratum-ibid. **376** (2003) 339] [arXiv:hep-th/0204089].
[25] I. Antoniadis, *Phys. Lett.* B **246** (1990) 377.
[26] D. J. Kapner, T. S. Cook, E. G. Adelberger, J. H. Gundlach, B. R. Heckel, C. D. Hoyle and H. E. Swanson, Phys. Rev. Lett. **98** (2007) 021101.
[27] N. Arkani-Hamed, S. Dimopoulos and G. R. Dvali, *Phys. Lett.* B **429** (1998) 263 [arXiv:hep-ph/9803315]; I. Antoniadis, N. Arkani-Hamed, S. Dimopoulos and G. R. Dvali, *Phys. Lett.* B **436** (1998) 257 [arXiv:hep-ph/9804398].
[28] J. D. Lykken, *Phys. Rev.* D **54** (1996) 3693 [arXiv:hep-th/9603133].
[29] J.C. Long and J.C. Price, *Comptes Rendus Physique* **4** (2003) 337; R.S. Decca, D. Lopez, H.B. Chan, E. Fischbach, D.E. Krause and C.R. Jamell, *Phys. Rev. Lett.* **94** (2005) 240401; R.S. Decca et al., arXiv:0706.3283 [hep-ph]; S.J. Smullin, A.A. Geraci, D.M. Weld, J. Chiaverini, S. Holmes and A. Kapitulnik, arXiv:hep-ph/0508204; H. Abele, S. Haeßler and A. Westphal, in 271th WE-Heraeus-Seminar, Bad Honnef (2002).
[30] G. Dvali, arXiv:0706.2050 [hep-th]; Int. J. Mod. Phys. A **25** (2010) 602 [arXiv:0806.3801 [hep-th]]; G. Dvali and M. Redi, Phys. Rev. D **77** (2008) 045027 [arXiv:0710.4344 [hep-th]]; R. Brustein, G. Dvali and G. Veneziano, JHEP **0910** (2009) 085 [arXiv:0907.5516 [hep-th]].

Journal of Physics: Conference Series **447** (2013) 012001 doi:10.1088/1742-6596/447/1/012001

[31] G. Dvali and C. Gomez, Phys. Lett. B **674** (2009) 303.
[32] I. Antoniadis and B. Pioline, *Nucl. Phys.* B **550** (1999) 41; I. Antoniadis, S. Dimopoulos and A. Giveon, JHEP **0105** (2001) 055 [arXiv:hep-th/0103033]; I. Antoniadis, A. Arvanitaki, S. Dimopoulos and A. Giveon, arXiv:1102.4043.
[33] G. Dvali and D. Lust, arXiv:0912.3167 [hep-th]; G. Dvali and C. Gomez, arXiv:1004.3744 [hep-th].
[34] I. Antoniadis and K. Benakli, *Phys. Lett.* B **326** (1994) 69.
[35] I. Antoniadis, K. Benakli and M. Quirós, *Phys. Lett.* B **331** (1994) 313 and *Phys. Lett.* B **460** (1999) 176; P. Nath, Y. Yamada and M. Yamaguchi, *Phys. Lett.* B **466** (1999) 100 T. G. Rizzo and J. D. Wells, *Phys. Rev.* D **61** (2000) 016007; T. G. Rizzo, *Phys. Rev.* D **61** (2000) 055005; A. De Rujula, A. Donini, M. B. Gavela and S. Rigolin, *Phys. Lett.* B **482** (2000) 195.
[36] G.F. Giudice, R. Rattazzi and J.D. Wells, *Nucl. Phys.* B **544** (1999) 3; E.A. Mirabelli, M. Perelstein and M.E. Peskin, *Phys. Rev. Lett.* **82** (1999) 2236; T. Han, J.D. Lykken and R. Zhang, *Phys. Rev.* D **59** (1999) 105006; K. Cheung, W.-Y. Keung, *Phys. Rev.* D **60** (1999) 112003; C. Balázs *et al.*, *Phys. Rev. Lett.* **83** (1999) 2112; J.L. Hewett, *Phys. Rev. Lett.* **82** (1999) 4765.
[37] L.A. Anchordoqui, H. Goldberg, D. Lust, S. Nawata, S. Stieberger and T.R. Taylor, Phys. Rev. Lett. **101** (2008) 241803 [arXiv:0808.0497 [hep-ph]].
[38] G.T. Horowitz and J. Polchinski, Phys. Rev. D **55** (1997) 6189.
[39] I. Antoniadis, K. Benakli and M. Quirós, *Nucl. Phys.* B **583** (2000) 35.
[40] I. Antoniadis, K. Benakli and M. Quiros, *New Jour. Phys.* **3** (2001) 20.
[41] I. Antoniadis, C. Muñoz and M. Quirós, *Nucl. Phys.* B **397** (1993) 515; I. Antoniadis and M. Quirós, *Phys. Lett.* B **392** (1997) 61; A. Pomarol and M. Quirós, *Phys. Lett.* B **438** (1998) 225; I. Antoniadis, S. Dimopoulos, A. Pomarol and M. Quirós, *Nucl. Phys.* B **544** (1999) 503; A. Delgado, A. Pomarol and M. Quirós, *Phys. Rev.* D **60** (1999) 095008; R. Barbieri, L. J. Hall and Y. Nomura, *Phys. Rev.* D **63** (2001) 105007.
[42] I. Antoniadis, S. Dimopoulos and G. Dvali, *Nucl. Phys.* B **516** (1998) 70; S. Ferrara, C. Kounnas and F. Zwirner, *Nucl. Phys.* B **429** (1994) 589.

Journal of Physics: Conference Series **447** (2013) 012002 doi:10.1088/1742-6596/447/1/012002

Implications of the Higgs discovery for SUSY

Abdelhak Djouadi

Laboratoire de Physique Théorique, U. Paris–Sud and CNRS, F–91405 Orsay, France.
abdelhak.djouadi@cern.ch

Abstract. The implications of the discovery of the Higgs boson at the LHC with a mass of approximately 125 GeV are discussed in the context of minimal supersymmetric extension of the Standard Model, the MSSM.

1. Introduction

The ATLAS and CMS discovery of the particle with a mass of approximately 125 GeV [1] and properties that are compatible with those of a Higgs boson [2] has far reaching consequences for new physics models, in particular, for supersymmetric theories. The latter are widely considered to be very attractive extensions as they naturally protect the Higgs mass against large radiative corrections and stabilise the hierarchy between the electroweak and Planck scales.

In the minimal supersymmetric (SUSY) extension, the Minimal Supersymmetric Standard Model (MSSM), two Higgs doublet fields are required to break the electroweak symmetry, leading to five Higgs particles: two CP–even h and H, a CP–odd A and two charged $H^{\pm}$ particles [3]. Two parameters are needed to describe the Higgs sector at the tree–level: one Higgs mass, which is generally taken to be that of the pseudoscalar boson M_A, and the ratio of vacuum expectation values of the two Higgs fields, $\tan\beta$, expected to lie in the range $1 \leq \tan\beta \leq 60$. At high M_A values, $M_A \gg M_Z$, one is in the decoupling regime in which the neutral CP–even state h is light and has almost exactly the properties of the SM Higgs boson, i.e. its couplings to fermions and gauge bosons are the same, while the other CP–even H and the charged Higgs $H^{\pm}$ states are heavy and degenerate in mass with the A state, $M_H \approx M_{H^{\pm}} \approx M_A$. In this regime, the MSSM Higgs sector thus looks almost exactly as the one of the SM with its unique Higgs boson [2].

There is, however, one major difference between the two cases: while in the SM the Higgs mass is essentially a free parameter (and should simply be smaller than about 1 TeV), the lightest CP–even Higgs particle in the MSSM is bounded from above and, depending on the SUSY parameters that enter the radiative corrections, it is restricted to values $M_h^{\rm max} \approx 110$–$135$ GeV. Hence, the requirement that the MSSM h boson coincides with the one observed at the LHC, i.e. with $M_h \approx 125$ GeV, would place very strong constraints on the MSSM parameters through their contributions to the radiative corrections to the Higgs sector.

Here, we summarise the consequences of such a value of M_h for the MSSM [4, 5, 6]. We first consider the phenomenological MSSM [7] in which the relevant soft SUSY–breaking parameters are allowed to vary freely (but with some restrictions) and constrained MSSM scenarios such as the minimal supergravity (mSUGRA) [8], gauge mediated (GMSB) [9] and anomaly mediated (AMSB) [10] supersymmetry breaking models (for a review, see again Ref. [3]). We also discuss the implications of such an M_h value for scenarios in which the supersymmetric spectrum is extremely heavy, the so–called split SUSY [11] or high–scale SUSY models [12].

2. Implications in the phenomenological MSSM

In an unconstrained MSSM, there is a large number of soft SUSY-breaking parameters but analyses can be performed in the so–called "phenomenological MSSM" (pMSSM) [7], in which CP conservation, flavour diagonal sfermion mass and coupling matrices and universality of the first and second generations are imposed. The pMSSM involves 22 free parameters in addition to those of the SM: besides $\tan\beta$ and M_A, these are the higgsino mass μ, the three gaugino masses $M_{1,2,3}$, the diagonal left– and right–handed sfermion mass parameters $m_{\tilde{f}_{L,R}}$ and the trilinear sfermion couplings A_f. Fortunately, besides $\tan\beta$ and M_A, only two of them play a major role: the SUSY breaking scale $M_S = \sqrt{m_{\tilde{t}_1} m_{\tilde{t}_2}}$ and the mixing parameter in the stop sector, $X_t = A_t - \mu\cot\beta$. The maximal value of the h mass, $M_h^{\rm max}$ is given in the leading one–loop approximation by (with some RG improvement using the running top quark mass $\bar{m}_t$)

$$M_h \overset{M_A \gg M_Z}{\to} M_Z|\cos 2\beta| + \frac{3\bar{m}_t^4}{2\pi^2 v^2 \sin^2\beta}\left[\log\frac{M_S^2}{\bar{m}_t^2} + \frac{X_t^2}{M_S^2}\left(1 - \frac{X_t^2}{12M_S^2}\right)\right] \quad (1)$$

and is obtained for the following choice of parameters: *i*) a decoupling regime with a heavy pseudoscalar Higgs boson, $M_A \sim \mathcal{O}$(TeV); *ii*) large values of the parameter $\tan\beta$, $\tan\beta \geq 10$; *iii* heavy stops, i.e. large M_S and we choose $M_S = 3$ TeV as a maximal value; *iv*) a stop trilinear coupling $X_t = \sqrt{6}M_S$, the so–called maximal mixing scenario.

An estimate of the upper bound on M_h can be obtained by adopting the scenario above. To be more precise, we use the program `Suspect` [13] which calculate the Higgs and superparticle spectrum in the MSSM including the most up–to–date information to obtain the maximal value $M_h^{\rm max}$. Following and early analysis [14], we have performed a large scan of the pMSSM 22 parameter space using `Suspect` in an uncorrelated way in the following domains:

$$\begin{gathered} 1 \leq \tan\beta \leq 60\,,\ 50\ \text{GeV} \leq M_A \leq 3\ \text{TeV}\,,\ -9\ \text{TeV} \leq A_f \leq 9\ \text{TeV}\,, \\ 50\ \text{GeV} \leq m_{\tilde{f}_L}, m_{\tilde{f}_R}, M_3 \leq 3\ \text{TeV}\,,\ 50\ \text{GeV} \leq M_1, M_2, |\mu| \leq 1.5\ \text{TeV}. \end{gathered} \quad (2)$$

The results are shown in Fig. 1 where, in the left–hand side, the obtained maximal value $M_h^{\rm max}$ is displayed as a function of the ratio of parameters X_t/M_S. The resulting values are confronted to the mass range 123 GeV $\leq M_h \leq$ 127 GeV when the parametric uncertainties from the SM inputs and the theoretical uncertainties in the determination of M_h are included.

Hence, only the scenarios with large X_t/M_S values and, in particular, those close to the maximal mixing scenario $A_t/M_S \approx \sqrt{6}$ survive. The no–mixing scenario $X_t \approx 0$ is ruled out for $M_S \leq 3$ TeV, while the typical mixing scenario, $X_t \approx M_S$, needs large M_S and moderate to large $\tan\beta$. We obtain $M_h^{\rm max}$=136, 123 and 126 GeV with maximal, zero and typical mixing.

The right–hand side of Fig. 1 shows the contours in the $[m_{\tilde{t}_1}, X_t]$ plane where we obtain the mass range 123 GeV $< M_h <$ 127 GeV from our pMSSM scan with $X_t/M_S \leq 3$; the regions in which $\tan\beta \leq 3, 5$ and 60 are highlighted. One sees again that a large part of the parameter space is excluded if the Higgs mass constraint is imposed. In particular, large $m_{\tilde{t}_1}$ values, in general corresponding to large M_S are favored. However, the possibility that $m_{\tilde{t}_1}$ is of the order of a few hundred GeV is still allowed, provided that stop mixing (which leads to a significant $m_{\tilde{t}_1}, m_{\tilde{t}_2}$ splitting) is large. Thus light stops can be still searched for at the LHC...

Squark as well as gluino masses above 1 TeV are also required by the direct searches of SUSY particles at the LHC, confirming the need of high M_S values to cope with $M_h \approx 125$ GeV. Nevertheless, relatively light stops and electroweak sparticles such as sleptons, charginos and neutralinos are still possible but the present searches are becoming more and more constraining...

Note that the $M_h^{\rm max}$ values discussed above are obtained with a heavy superparticle spectrum, for which the constraints from flavour physics and sparticle searches are evaded, and in the decoupling limit in which the h production cross sections and the decay branching ratios are those of the SM Higgs boson. In the real word, there are other (stringent) constraints on the

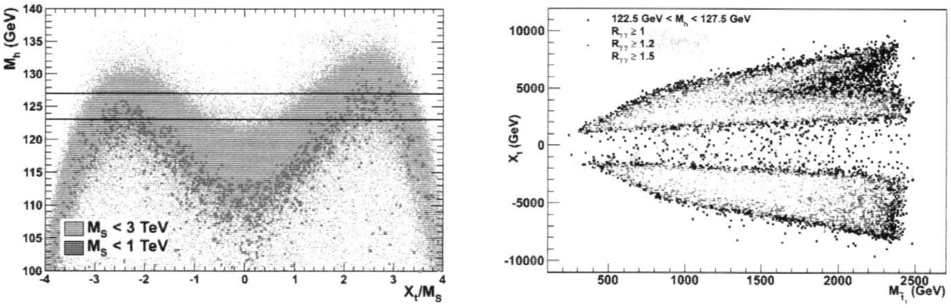

Figure 1. The maximal value of the h boson mass as a function of X_t/M_S in the pMSSM when all other soft SUSY–breaking parameters and $\tan\beta$ are scanned (left) and the contours for 123< M_h <127 GeV in the $[m_{\tilde{t}_1}, X_t]$ plane for some selected range of $\tan\beta$ values (right). From Ref. [4].

pMSSM to be included: *i)* the production/decay rates of the observed Higgs particle, *ii)* the CMS and ATLAS search limits in of the heavier MSSM bosons in the channels $pp \to A/H/(h) \to \tau\tau$ and $pp \to t\bar{t}$ with $t \to bH^+$, *iii)* the non observation of heavier Higgses in the ZZ,WW,tt channels, *iv)* the constraints from sparticle searches and eventually Dark Matter, and *v)* constraints from flavor physics such and the limits from the processes $B_s \to \mu\mu$ and $b \to s\gamma$. An example of output that can be obtained in such a case in given in the left-hand side of Fig. 2.

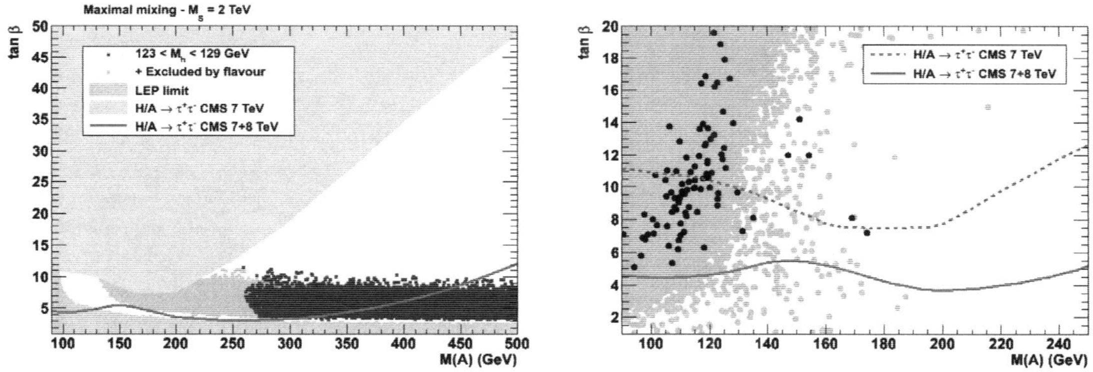

Figure 2. The maximal value of the h boson mass as a function of X_t/M_S in the pMSSM when all other soft SUSY–breaking parameters and $\tan\beta$ are scanned (left) and the contours for 123< M_h <127 GeV in the $[m_{\tilde{t}_1}, X_t]$ plane for some selected range of $\tan\beta$ values (right). From Ref. [5, 6].

In addition, we also searched for points in the parameter space in which the boson with mass $\simeq$ 125 GeV is the heavier CP–even H state which corresponds to values of M_A of order 100 GeV. Among the $\approx 10^7$ valid MSSM points of the scan, only $\approx 1.5 \times 10^{-4}$ correspond to this scenario. However, if we impose that the H cross sections times branching ratios are compatible with the SM values within a factor of 2 and include the constraints from MSSM Higgs searches in the $\tau^+\tau^-$ channel, only a few points survive. These are all excluded once flavor constraints are imposed. This is shown in the right–hand of Fig. 2

3. Implications for constrained MSSM scenarios

In constrained MSSM scenarios (cMSSM), the various soft SUSY–breaking parameters obey a number of universal boundary conditions at a high energy scale, thus reducing the number of basic input parameters to a handful. These inputs are evolved via the MSSM renormalisation group equations down to the low energy scale M_S where the conditions of proper electroweak symmetry breaking (EWSB) are imposed. Three classes of such models have been widely discussed in the literature: *i*) the minimal supergravity (mSUGRA) model in which SUSY–breaking is assumed to occur in a hidden sector which communicates with the visible sector only via flavour-blind gravitational interactions, leading to universal soft breaking terms ie a common $m_{1/2}, m_0, A_0$ values for the gaugino, sfermion masses and trilinear couplings, *ii*) the gauge mediated SUSY–breaking (GMSB) model in which SUSY–breaking is communicated to the visible sector via gauge interactions, *iii*) the anomaly mediated SUSY–breaking (AMSB) model in which SUSY–breaking is communicated to the visible sector via a super-Weyl anomaly.

These models are described by $\tan\beta$, the sign of μ and a few continuous parameters. We adopt the following ranges for the input parameters of these scenarios:

$$\begin{array}{rlll} \text{mSUGRA:} & 50\text{ GeV} \le m_0 \le 3\text{ TeV}, & 50\text{ GeV} \le m_{1/2} \le 3\text{ TeV}, & |A_0| \le 9\text{ TeV}; \\ \text{GMSB:} & 10\text{ TeV} \le \Lambda \le 1000\text{ TeV}, & 1 \le M_{\rm mess}/\Lambda \le 10^{11}, & N_{\rm mess} = 1; \\ \text{AMSB:} & 1\text{ TeV} \le m_{3/2} \le 100\text{ TeV}, & 50\text{ GeV} \le m_0 \le 3\text{ TeV}. & \end{array}$$

allow for both signs of μ, require $1 \le \tan\beta \le 60$ and, to avoid the need for excessive fine–tuning in the EWSB conditions, impose an additional bound $M_S = M_{\rm EWSB} = \sqrt{m_{\tilde{t}_1} m_{\tilde{t}_2}} < 3$ TeV.

In the case of mSUGRA, one can study special cases such as: no-scale scenario with $m_0 \approx A_0 \approx 0$, $m_0 \approx 0$ and $A_0 \approx -\frac{1}{4} m_{1/2}$ which approximately corresponds to the constrained next-to–MSSM (cNMSSM), $A_0 \approx -m_0$ which corresponds to a very constrained MSSM (VCMSSM), and a non–universal Higgs mass model (NUHM) in which the universal soft SUSY–breaking scalar mass terms are different for the sfermions and for the two Higgs doublet fields.

Using the program `Suspect`, we have performed a full scan of these scenarios. The results for $M_h^{\rm max}$ are shown in the left-hand side of Fig. 3 as a function of $\tan\beta$, the input parameter that is common to all models, and in the right-hand side of the figure as a function of M_S.

In contrast to the pMSSM, the various parameters which enter the radiative corrections to the MSSM Higgs sector are not all independent in constrained scenarios as a consequence of the relations between SUSY breaking parameters that are set at the high–energy scale and the requirement that electroweak symmetry breaking is triggered radiatively for each set of input parameters which leads to additional constraints. Hence, it is not possible to freely tune the relevant weak–scale parameters to obtain the maximal value of M_h given previously. In order to obtain a reliable determination of $M_h^{\rm max}$ in a given constrained SUSY scenario, it is necessary to scan through the allowed range of values for all relevant SUSY parameters.

In all cases, the maximal M_h value is obtained for $\tan\beta$ around 20. We observe that in the adopted parameter space of the models and with the central values of the SM inputs, the upper h mass value (rounded to the upper half GeV) is $M_h^{\rm max} = 121$ GeV in AMSB, i.e. much less that 125 GeV, while in the GMSB scenario one has $M_h^{\rm max} = 121.5$ GeV. Thus, clearly, the two scenarios are disfavoured if the lightest CP–even Higgs particle has indeed a mass in the range $123 < M_h < 127$ GeV. In the case of mSUGRA, we obtain a maximal value $M_h^{\rm max} = 128$ GeV and, thus, some parameter space of the model would still survive the M_h constraint.

The upper bound on M_h in these scenarios can be qualitatively understood by considering in each model the allowed values of the trilinear coupling A_t, which essentially determines the stop mixing parameter X_t and thus the value of M_h for a given scale M_S. In GMSB, one has $A_t \approx 0$ at relatively low scales and its magnitude does not significantly increase in the evolution down to the scale M_S; this implies that we are almost in the no–mixing scenario which gives a low value of M_h as can be seen from Fig. 1. In AMSB, one has a non-zero A_t that is fully predicted at any renormalisation scale in terms of the Yukawa and gauge couplings; however, the ratio A_t/M_S with M_S determined from the overall SUSY breaking scale $m_{3/2}$ turns out to be rather small, implying again that we are close to the no–mixing scenario. Finally, in the mSUGRA model, since we have allowed A_t to vary in a wide range as $|A_0| \leq 9$ TeV, one can get a large A_t/M_S ratio which leads to a heavier Higgs particle. However, one cannot easily reach A_t values such that $X_t/M_S \approx \sqrt{6}$ so that we are not in the maximal–mixing scenario and the higher upper bound on M_h in the pMSSM is not reached.

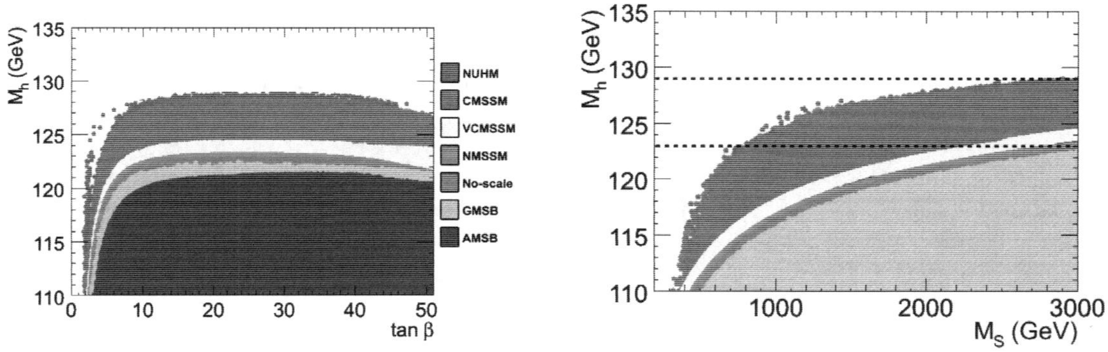

Figure 3. The maximal value of the h boson mass as a function of $\tan\beta$ (left) and M_S (right) with a scan of all other parameters in various constrained MSSM scenarios. The range $123 < M_h < 127$ GeV for the light h boson mass is highlighted. From Ref. [4].

In turn, in two particular cases of mSUGRA that we have discussed in addition, the "no–scale" and the "approximate cNMSSM" scenarios, the upper bound on M_h is much lower than in the more general mSUGRA case and, in fact, barely reaches the value $M_h \approx 123$ GeV. The main reason is that these scenarios involve small values of A_0 at the GUT scale, $A_0 \approx 0$ for no–scale and $A_0 \approx -\frac{1}{4}m_{1/2}$ for the cNMSSM. One then obtains A_t values at the weak scale that are too low to generate a significant stop mixing and, hence, one is again close to the no–mixing scenario. Thus, only a very small fraction of the parameter space of these two sub–classes of the mSUGRA model survive (in fact, those leading to the $M_h^{\max}$ value) if we impose $123 < M_h <$ 127 GeV. These models hence should have a very heavy spectrum as a value $M_S \geq 3$ TeV is required to increase $M_h^{\max}$. In the VCMSSM, $M_h \simeq 124.5$ GeV can be reached as $|A_0|$ can be large for large m_0, $A_0 \approx -m_0$, allowing at least for typical mixing.

Finally, since the NUHM is more general than mSUGRA as we have two more free parameters, the $[\tan\beta, M_h]$ area shown in Fig. 3 is larger than in the mSUGRA case. However, since we are in the decoupling regime and the value of M_A does not matter much (as long as it a larger than a few hundred GeV) and the key weak–scale parameters entering the determination of M_h, i.e. $\tan\beta, M_S$ and A_t are approximately the same in both models, one obtains a bound $M_h^{\max}$ that is only slightly higher in NUHM compared to mSUGRA. Thus, the same discussion above on the mSUGRA scenario, holds also true in the NUHM case.

In the case of these constrained scenarios and in particular in the "general" mSUGRA model, most of the points giving the correct Higgs mass correspond to the decoupling regime of the MSSM Higgs sector and, hence, to an h boson with SM cross sections and branching ratios. Furthermore, as the resulting SUSY spectrum for $M_h = 125 \pm 2$ GeV is rather heavy in constrained scenarios, one obtains very small contributions to $(g-2)_\mu$ and to B–meson physics observables such as $\mathrm{BR}(B_s \to \mu^+\mu^-)$. The correct cosmological density as required by WMAP can be easily implemented. Hence, the resulting spectrum complies with all currently available constraints.

4. Split and high–scale SUSY models

In the preceding discussion, we have always assumed that the SUSY–breaking scale is relatively low, $M_S \leq 3$ TeV, which implies that some of the supersymmetric and heavier Higgs particles could be observed at the LHC or at some other TeV collider. However, as already mentioned, this choice is mainly dictated by fine–tuning considerations which are a rather subjective matter as there is no compelling criterion to quantify the acceptable amount of tuning. One could well have a very large value of M_S which implies that, except for the lightest h boson, no other scalar particle is accessible at the LHC or at any foreseen collider.

This argument has been advocated to construct the so–called split SUSY scenario [11] in which the soft SUSY–breaking mass terms for all the scalars of the theory, except for one Higgs doublet, are extremely large, i.e. their common value M_S is such that $M_S \gg 1$ TeV (such a situation occurs e.g. in some string motivated models). Instead, the mass parameters for the spin–$\frac{1}{2}$ particles, the gauginos and the higgsinos, are left in the vicinity of the EWSB scale, allowing for a solution to the dark matter problem and a successful gauge coupling unification, the two other SUSY virtues. The split SUSY models are much more predictive than the usual pMSSM as only a handful parameters are needed to describe the low energy theory. Besides the common value M_S of the soft SUSY-breaking sfermion and one Higgs mass parameters, the basic inputs are essentially the three gaugino masses M_1, M_2, M_3 (which can be unified to a common value at $M_{\rm GUT}$ as in mSUGRA), the higgsino parameter μ and $\tan\beta$. The trilinear couplings A_f, which are expected to have values close to the EWSB scale, and thus much smaller than M_S, will in general play a negligible role.

Concerning the Higgs sector, the main feature of split SUSY is that at the high scale M_S, the boundary condition on the quartic Higgs coupling of the theory is determined by SUSY:

$$\lambda(M_S) = \frac{1}{4}\left[g^2(M_S) + g'^2(M_S)\right] \cos^2 2\beta \; . \tag{3}$$

where g and g' are the SU(2) and U(1) gauge couplings. Here, $\tan\beta$ is not a parameter of the low-energy effective theory: it enters only the boundary condition above and cannot be interpreted as the ratio of two Higgs vacuum expectation values. In this case, it should not be assumed to be larger than unity as usual and will indeed adopt the choice $1/60 \leq \tan\beta \leq 60$.

If the scalars are very heavy, they will lead to radiative corrections in the Higgs sector that are significantly enhanced by large logarithms, $\log(M_{\rm EWSB}/M_S)$, where $M_{\rm EWSB}$ is the scale set by the gaugino and higgsino masses. In order to have reliable predictions, one has to properly decouple the heavy states from the low-energy theory and resum the large logarithmic corrections; in addition, the radiative corrections due to the gauginos and the higgsinos have to be implemented. Following the early work of Ref. [11], a comprehensive study of the split SUSY spectrum has been performed in Ref. [15]. All the features of the model have been implemented in the Fortran code `SuSpect` [13] upon which the numerical analysis presented in Ref. [4] and summarised here is based.

One can adopt an even more radical attitude than in the split SUSY case and assume that the gauginos and higgsinos are also very heavy, with a mass close to the scale M_S; this is the case in the so–called high–scale SUSY model [12]. Here, one abandons not only the SUSY solution to the fine-tuning problem but also the solution to the dark matter problem by means of the LSP and the successful unification of the gauge coupling constants. However, there will still be a trace of SUSY at low energy: the matching of the SUSY and the low–energy theories is indeed encoded in the Higgs quartic coupling λ given by Eq. (3). Hence, even if broken at very high scales, SUSY would still lead to a "light" Higgs boson whose mass will contain information on M_S and $\tan\beta$.

The treatment of the Higgs sector of the high–scale SUSY scenario is similar to that of split SUSY: one simply needs to decouple the gauginos and higgsinos from the low energy spectrum (in particular remove their contributions to the renormalisation group evolution of the gauge and Yukawa couplings and to the radiative corrections to the h boson mass) and set their masses to M_S. We have adapted the version of the program `Suspect` which handles the split SUSY case to also cover the case where $M_1 \approx M_2 \approx M_3 \approx |\mu| \approx M_S$. Using this program, we have performed a scan in the $[\tan\beta, M_S]$ plane to determine the value of M_h in the split SUSY and high–scale SUSY scenarios. The values given by the PDG for the SM input parameters have been adopted and, in the case of split SUSY, we have chosen $M_{\rm EWSB} \approx \sqrt{|M_2\mu|} \approx 246$ GeV for the low scale. The results are shown in Fig. 4. In this figure M_h is displayed as a function of M_S for selected values of $\tan\beta$ in split and high–scale SUSY.

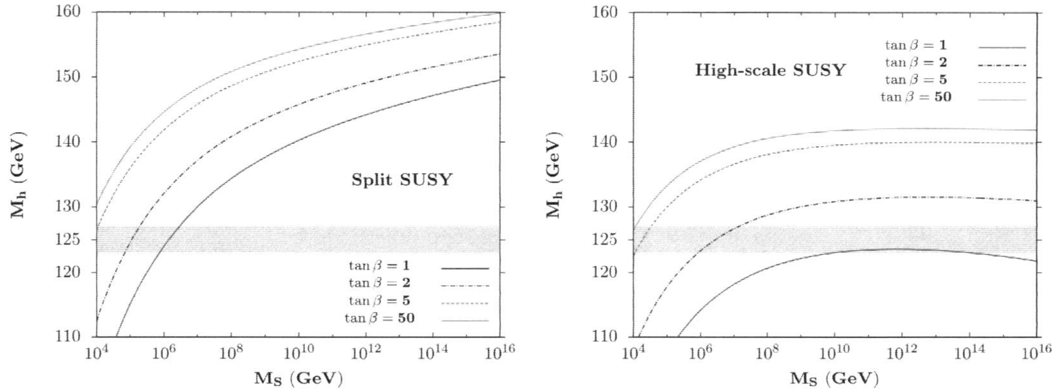

Figure 4. The value of M_h as a function of M_S for several values of $\tan\beta = 1, 2, 5, 50$ in the split SUSY (left) and high–scale SUSY (right) scenarios. From Ref. [4].

As expected, the maximal M_h values are obtained at high $\tan\beta$ and M_S values and, at the scale $M_S \approx 10^{16}$ GeV at which the couplings g and g' approximately unify in the split SUSY scenario, one obtains $M_h \approx 160$ GeV for the higher $\tan\beta = 50$ value. We do not include the error bands in the SM inputs which would lead to an uncertainty of about 2 GeV on M_h, mainly due to the 1 GeV uncertainty on the top quark mass. In addition, we have assumed the zero–mixing scenario as the parameter A_t is expected to be much smaller than M_S; this approximation might not be valid for M_S values below 10 TeV and a maximal mixing $A_t/M_S = \sqrt{6}$ would increase the Higgs mass value by up to 10 GeV at $M_S = \mathcal{O}(1$ TeV) as was discussed earlier for the pMSSM. In the high–scale SUSY scenario, we obtain a value $M_h \approx 142$ GeV (with again an uncertainty of approximately 2 GeV from the top mass) for high $\tan\beta$ values and at the unification scale $M_S \approx 10^{14}$ GeV as in Ref. [12]. Much smaller M_h values, in the 120 GeV range, can be obtained for lower scales and $\tan\beta$.

Hence, the requirement that the Higgs boson mass is in the range $123 < M_h < 127$ GeV imposes strong constraints on the parameters of these two models. For this Higgs mass range, very large scales are needed for $\tan\beta \approx 1$ in the split (high–scale) SUSY scenario, while scales not too far from $M_S \approx 10^4$ GeV are required at high $\tan\beta$. Thus, even in these extreme scenarios, SUSY should manifest itself at scales much below $M_{\rm GUT}$ if $M_h \approx 125$ GeV.

Finally, we note that in Ref. [16] we have considered the production of the heavier H, A and $H^{\pm}$ bosons of the MSSM at the LHC, focusing on the low $\tan\beta$ regime, $\tan\beta \lesssim$3–5, that is allowed if the SUSY scale is assumed to be very high, $M_S \geq 10$ TeV. We have first shown that in this case, the requirement that $M_h \approx 125$ GeV fixes the radiative corrections so that, to a good approximation, one needs only two basic input parameters in the MSSM Higgs sector even at higher orders: $\tan\beta$ and M_A. We have then shown that the searches for a heavier SM Higgs boson in the WW, ZZ final states and the search for new resonances in the $\tau\tau, t\bar{t}, hZ$ and hh channels constrain the $[\tan\beta, M_A]$ paremeter space at low $\tan\beta$ and M_A values.

5. Conclusions

We have discussed the impact of a Standard Model–like Higgs boson with a mass $M_h \approx 125$ GeV on supersymmetric theories in the context of both unconstrained and constrained MSSM scenarios. We have shown that in the phenomenological MSSM, strong restrictions can be set on the mixing in the top sector and, for instance, the no–mixing scenario is excluded unless the supersymmetry breaking scale is extremely large, $M_S \gg 1$ TeV, while the maximal mixing scenario is disfavoured for large M_S and $\tan\beta$ values.

In constrained MSSM scenarios, the impact is even stronger. Several scenarios, such as minimal AMSB and GMSB are disfavoured as they lead to a too light h particle. In the mSUGRA case, including the possibility that the Higgs mass parameters are non–universal, the allowed part of the parameter space should have large stop masses and A_0 values. In more constrained versions of this model such as the "no–scale" and approximate "cNMSSM" scenarios, only a very small portion of the parameter space is allowed by the Higgs mass bound.

Finally, significant areas of the parameter space of models with large M_S values leading to very heavy supersymmetric particles, such as split SUSY or high–scale SUSY, can also be excluded as, in turn, they tend to predict a too heavy Higgs particle with $M_h \geq 125$ GeV.

Acknowledgements

I thank the organisers of this conference for their nice invitation and their patience while waiting for the write–up of my talk. I thank the CERN theory division for the kind hospitality offered to me. This work is supported by the ERC Advanced Grant Higgs@LHC.

References

[1] ATLAS collaboration, arXiv:1207.7214 [hep-ex]; CMS collaboration, arXiv:1207.7235 [hep-ex].
[2] For a review of the SM Higgs sector, see: A. Djouadi, Phys. Rept. 457 (2008) 1.
[3] For a review of the MSSM Higgs sector, see: A. Djouadi, Phys. Rept. 459 (2008) 1.
[4] A. Arbey et al., Phys. Lett. B708 (2012) 162.
[5] A. Arbey, M. Battaglia, A. Djouadi and F. Mahmoudi, JHEP 1209 (2012) 107.
[6] A. Arbey, M. Battaglia, A. Djouadi and F. Mahmoudi, Phys. Lett. B720 (2013) 153.
[7] A. Djouadi et al. (MSSM Working Group), hep-ph/9901246.
[8] A.H. Chamseddine, R. Arnowitt and P. Nath, Phys. Rev. Lett. 49 (1982) 970; R. Barbieri, S. Ferrara and C. Savoy, Phys. Lett. B119 (1982) 343; L. Hall, J. Lykken and S. Weinberg, Phys. Rev. D27 (1983) 2359; N. Ohta, Prog. Theor. Phys. 70 (1983) 542.
[9] M. Dine and W. Fishler, Phys. Lett. B110 (1982) 227; C. Nappi and B. Ovrut, Phys. Lett. B113 (1982) 1785; L. Alvarez-Gaumé, M. Claudson and M. Wise, Nucl. Phys. B207 (1982) 96; M. Dine and A. Nelson, Phys. Rev. D48 (1993) 1277; M. Dine, A. Nelson and Y. Shirman, Phys. Rev. D51 (1995) 1362; G.F. Giudice and R. Rattazzi, Phys. Rept. 322 (1999) 419.
[10] L. Randall and R. Sundrum, Nucl. Phys. B557 (1999) 79; G. Giudice, M. Luty, H. Murayama and R. Rattazzi, JHEP 9812 (1998) 027; J. Bagger, T. Moroi and E. Poppitz, JHEP 0004 (2000) 009.
[11] N. Arkani-Hamed and S. Dimopoulos, JHEP 0506 (2005) 073; G.F. Giudice and A. Romanino, Nucl. Phys. B699 (2004) 65; J.D. Wells, Phys. Rev. D71 (2005) 015013.
[12] See L. Hall and Y. Nomura, JHEP 1003 (2010) 076; G. Giudice and A. Strumia, Nucl.Phys. B858 (2012) 63.
[13] A. Djouadi, J.L. Kneur and G. Moultaka, Comput. Phys. Commun. 176 (2007) 426.
[14] B. Allanach et al., JHEP 0409 (2004) 044.
[15] N. Bernal, A. Djouadi and P. Slavich, JHEP 0707 (2007) 016.
[16] A. Djouadi and J.Quevillon, arXiv:1304.1787 [hep-ph].

A Theory of Flavor

Rabindra N. Mohapatra

Dept. of Physics and Maryland Center for Fundamental Physics, University of Maryland, College Park, MD-20742

E-mail: rmohapat@umd.edu

Abstract. Understanding flavor is one of the fundamental puzzles of the standard model. The discovery of neutrino masses in the past decade, exhibiting mixing and mass patterns so very different from the quark sector has added an extra dimension to this problem. In this paper, I discuss a proposal that appears to provide a promising way to resolve this puzzle in a unified framework incorporating both quarks and leptons.

1. Introduction

The discovery of the Higgs boson at the LHC has solved the problem of the origin of mass for the quarks and charged leptons. This is a major accomplishment. There however remains the problem of why there are three generations of quarks and leptons and the pattern of masses and mixings between generations. In the quark sector the masses as well as mixings exhibit a hierarchical pattern i.e. for masses $m_{u,d} \ll m_{c,s} \ll m_{t,b}$; for mixing angles $V_{ub} \ll V_{cb} \ll V_{cd} \ll V_{ud,cs,tb}$. This is known as the flavor puzzle for quarks. Unravelling this puzzle has long been recognized as a challenge for physics beyond the standard model.The puzzle of flavor got more challenging after the discovery of neutrino masses and mixings. Unlike the quark sector, lepton mixings do not exhibit a hierarchical pattern. For example, neutrino mixings between generations, denoted by θ^l_{ij}, are given by $\theta^l_{23} \sim 45^o$, $\theta^l_{12} \simeq 35^o$ and $\theta^l_{13} \sim 9^0$ (as against $\theta^q_{23} \sim 2.5^o$ and $\theta^q_{12} \sim 13^o$ and $\theta^q_{13} \simeq 0.03^0$). The neutrino masses also do not exhibit as strong as strong a hierarchy as the quarks or charged leptons i.e. for a normal hierarchy for neutrinos, $m_2/m_3 \simeq V_{us} \gg m_\mu/m_\tau$. An important question now is: what kind of new physics beyond the standard model can provide a unified explanation of these observations ?

There are two possible approaches to the flavor problem: (i) treat quarks and leptons separately and understand each sector on the basis of new physics or (ii) consider an unified approach to both flavor problems in the same framework. The first approach has led to the consideration of many new discrete flavor symmetries in the lepton sector. We take the second approach in this talk and discuss the flavor problem within a grand unified framework, since such a framework is known to unify quarks and leptons. There are also other reasons for considering grand unified theories, which have many appealing features such as unification of forces, quantization of electric charges etc. More importantly for us, understanding small neutrino masses may also be pushing us in that direction,as we see below. The reason is that neutrinos are known to have very tiny masses compared to charged fermions. One popular way to understand this has been to realize that neutrinos are electrically neutral particles and could therefore be their own anti-particles, the so-called Majorana fermions and adopt a framework

Journal of Physics: Conference Series **447** (2013) 012003 doi:10.1088/1742-6596/447/1/012003

known as the seesaw mechanism[1], where this Majorana nature plays a critical role. In the seesaw mechanism, one introduces a right handed neutrino (N) to the standard model and give it a large Majorana mass, M_N. The right handed neutrino of course mixes with the familiar left-handed one (ν). The neutrino mass matrix then involves both the (ν, N). Diagonalizing the (ν, N) mass matrix then immediately gives the formula for the light neutrino mass as $m_\nu \simeq \frac{m_D^2}{M_N}$. First this predicts that the neutrinos are their own anti-particles and secondly, if we assume that $M_N \gg v_{wk}$, it explains why $m_\nu \ll m_{e,q}$.

The question then arises asto what new physics gives N a large Majorana mass ? Since the Majorana mass of N breaks B-L symmetry of the standard model, seesaw mechanism is indicative of a new symmetry of nature. Secondly, it is more appealing to consider local $B-L$ symmetry rather a global one. What we now need to know is: what is the scale of B-L symmetry breaking (or the value of the N Majorana mass) ? Secondly: what the associated new physics is. Naive analysis of the seesaw formula implies that $M_N \simeq 10^{14}$ GeV or thereabouts, although TeV scale M_N can be entertained if neutrino Yukawa couplings are chosen to be of the same order as the electron Yukawas. Since the high scale seesaw mass 10^{14} GeV is tantalizingly close to the typical grand unification scale of 10^{16} GeV, one may suspect that GUT physics may be manifesting via the neutrino mass. In fact seesaw mechanism becomes truly natural within a grand unified theory suggesting that this framework which also unifies quarks and leptons may provide a logical choice while considering a unified approach to flavor.

Clearly, the problem of quark-lepton masses and mixings becomes specially puzzling in grand unified theories due to quark-lepton unification at very high scales. However, we argue that if proper approach and a proper class of grand unified model is chosen, this problem can be resolved. Thus the hope is that cracking this code may not only move us forward in our attempt at understanding the flavor problem but it may also provide a hint of some really new exciting underlying physics. We give an example of this below.

The proposal for flavor unification described here was made in a paper by Dutta and Mimura and this author in 2010 [2] where it was found that it is realizable in the framework of SO(10) grand unified theories . We discuss the various elements of the proposal and its realization below.

As a prelude to this discussion, let us realize that one way to understand the quark mass hierarchy is to start with a rank one mass matrix for up, down quarks and charged leptons in the leading order i.e.

$$M_{u,d,l} = m_{t,b,\tau}\begin{pmatrix} 0 & 0 & 0 \\ 0 & 0 & 0 \\ 0 & 0 & 1 \end{pmatrix} \tag{1}$$

and consider corrections coming from non-leading operators. Within such a framework, the third generation masses would come from the leading order terms and the others from the nonleading terms. Second clue for our ansatz is the observation that small quark mixings could be due to the following: if we wrote the up and down quark mass matrices as a sum a "big" matrix plus a small matrix with the "big" matrix part for both sectors being proportional to each other, then in the leading order the CKM angles would vanish e.g.

$$M_{u,d} = M^0_{u,d} + \delta_{u,d} \tag{2}$$

with $M^0_d = rM^0_u$ being the "big" matrix and $\delta_{u,d}$ being the smaller part. As just mentioned, the proportionality of the large parts of the mass matrices guarantees that the mixing angles will necessarily be small since the diagonalizing matrix for the large parts are "parallel" or "aligned" and the nontrivial CKM matrix represents the "small" misalignment between the two matrices determined by the "smaller" parts of the mass matrix.

We can therefore now state our ansatz[2] which consists of two parts:

Journal of Physics: Conference Series **447** (2013) 012003 doi:10.1088/1742-6596/447/1/012003

(i) The quark and lepton mass matrices have the following general feature:

$$\begin{aligned} M_u &= M_0 + \delta_u; \\ M_d &= rM_0 + \delta_d; \\ M_l &= rM_0 + \delta_l; \\ M_\nu &= fv_L \end{aligned} \tag{3}$$

(ii) M_0 has rank one.

Note that by a choice of the lepton basis, we can make f diagonal without loss of generality. It is then clear from the first part of the ansatz (item one) that for an arbitrary form of the matrices M_0 and δ as long as $\delta_{ij} \ll M_{0,ij}$, the lepton mixing angles are large whereas the quark mixing angles are small. The second rank one property than guarantees that quark and charged lepton masses are hierarchical whereas since f matrix is arbitrary, any hierarchy in the neutrino sector is likely to be milder. Incidentally, rank one property to understand mass hierarchy has been used in the past; see for instance[3].

2. Gauge group required to implement the ansatz

The question that arises next is how to implement our ansatz with a gauge model framework. To implement the first part, it is important to notice that the up and down quark mass matrices must be proportional to each other. Such relations do not emerge from the standard model since the u_R and d_R fields are separate fields and their Yukawa couplings responsible for the up and down quark mass matrices are therefore independent of each other. The situation however changes once we expand the gauge group to the left-right symmetric group $SU(2)_L \times SU(2)_R \times U(1)_{B-L}$ group since the (u_R, d_R) form a doublet of the $SU(2)_R$ group and thus relate the up and down Yukawa matrices[4]. Since quark-lepton unifications arises naturally within grand unification theories, the obvious group to consider is the SO(10) group as we do in the next section, although the basic conditions of the ansatz could also be realized with less predictive power within the $SU(2)_L \times SU(2)_R \times SU(4)_c$ partial unification groups.

3. SO(10) realization of flavor unification ansatz

As is well known, in SO(10) models, the matter fermions belong to **16**-dim. spinor representations. To get fermion masses, we will consider SO(10) models with **10**, **126** plus possibly another **10** or **120** Higgs fields where fermion masses are generated by renormalizable Yukawa couplings [5] only and where type II seesaw[6] is responsible for neutrino masses [7]. To implement our idea, we require that one of the **10** Yukawa couplings is the dominant one contributing to up, down and charged lepton masses and has rank one with other smaller couplings providing neutrino masses as well as most of the quark lepton flavor hierarchy. We postpone the discussion of how to get rank one till later. Let us see if this model does indeed give us our ansatz. It is well known that in these models[5], we have the following form for all fermion masses:

$$\begin{aligned} Y_u &= h + r_2 f + r_3 h', \\ Y_d &= r_1(h + f + h'), \\ Y_e &= r_1(h - 3f + c_e h'), \\ Y_{\nu D} &= h - 3r_2 f + c_\nu h', \end{aligned} \tag{4}$$

where Y_a are mass matrices divided by the electro-weak vev v_{wk} and r_i and $c_{e,\nu}$ are the mixing parameters which relate the $H_{u,d}$ to the doublets in the various GUT multiplets. More precisely, the matrices h, f and h' in Y_a are multiplied by the Higgs mixing parameters when they appear in the fermion mass matrices.

Furthermore, we use the type II seesaw formula[6] for getting neutrino masses gives [7].

$$\mathcal{M}_\nu \;=\; f v_L. \tag{5}$$

Note that f is the same coupling matrix that appears in the charged fermion masses in Eq. (4), up to factors from the Higgs mixings and the Clebsch-Gordan coefficients. This helps us to connect the neutrino parameters to the quark-sector parameters. The equations (4) and (5) are the key equations in our unified approach to addressing the flavor problem and obviously satisfy our flavor unification ansatz.

4. Implementing rank one strategy

The rank one Yukawa coupling with **10** Higgs field generates the features of flavor hierarchy, and rank 1 matrices can often appear in various ways (flavor symmetry, discrete symmetry, and string models). In this section, we give an SO(10) model, where the rank one ansatz used in our discussion of flavor emerges from extra vector like spinors above the GUT scale as well as a discrete symmetry.

When the direct couplings of chiral fermions with a Higgs field are forbidden by the chosen discrete symmetry, and the effective Yukawa couplings are generated by propagating vector-like matter fields, the rank of the effective Yukawa matrix depends on the number of the vector-like fields. Actually, when there are only one pair of vector-like matter fields as a flavor singlet, the effective Yukawa matrix is rank 1.

To illustrate this in a warm-up example, we consider a model which has one extra vector-like pair of matter fields to start with with mass slightly above the GUT scale contributing to the **10** coupling (denoted by $\psi_V \equiv \mathbf{16}_V \oplus \bar{\psi}_V \equiv \overline{\mathbf{16}}_V$) and three gauge singlet fields Y_a. We add a Z_4 discrete symmetry to the model under which the fields $\psi_a \to i\psi_a$, and $Y_a \to -iY_a$. The **10**-Higgs field H is invariant under this symmetry. The gauge invariant Yukawa superpotential under this assumption is given by

$$W \;=\; \psi_V H \lambda \psi_V \;+ M_V \psi_V \bar{\psi}_V \;+\; \bar{\psi}_V \sum_a Y_a \psi_a. \tag{6}$$

When we give vevs $\langle Y_a \rangle \neq 0$, ψ_V and ψ_a are mixed. The heavy vector-like fields, $\bar{\psi}_V$ and a linear combination of ψ_V and ψ_a (i.e. $M_V\psi_V + \sum_a Y_a \psi_a$), and the effective operator below its scale and at the GUT scale is given by:

$$\mathcal{L}_{eff} \;=\; \frac{\lambda}{M_V^2 + \sum_a Y_a^2} \left[\sum_a Y_a \psi_a\right] H \left[\sum_b Y_b \psi_b\right]. \tag{7}$$

This gives rise to a rank one h coupling. We note that it does not contradict the $O(1)$ top Yukawa coupling, when $M_V^2 \sim \sum_a Y_a^2$ (or $M_V^2 < \sum_a Y_a^2$).

If we let the $\overline{\mathbf{126}}$ Higgs field transform like -1 under Z_4, it can induce the f coupling with rank three. Our final model given below builds on this but differs in details e.g. it has got two vectorlike spinor multiplets instead of one etc.

5. Making the model predictive

Simply using the above ansatz in the context of an SO(10) model with mass relations in Eq. (5), turns out to reproduce the qualitative features of the quark and lepton spectra quite well. For example in the context of a two generation model (involving the second and third generation), this simple ansatz predicts $V_{cb} \simeq (m_s/m_b + e^{i\sigma} m_c/m_t) \cot\theta$, where θ is the atmospheric mixing

Journal of Physics: Conference Series **447** (2013) 012003 doi:10.1088/1742-6596/447/1/012003

	ψ	H	$\bar{\Delta}$	H'	ϕ_1	ϕ_1	ϕ_2	ϕ_2	ϕ_3	ϕ_3	ψ_{V1}	ψ_{V1}	ψ_{V2}	ψ_{V2}
SO(10)	**16**	**10**	$\overline{\mathbf{126}}$	**10**	**1**	**1**	**1**	**1**	**1**	**1**	**16**	$\overline{\mathbf{16}}$	**16**	$\overline{\mathbf{16}}$
S_4	$\mathbf{3_2}$	$\mathbf{1_1}$	$\mathbf{1_2}$	$\mathbf{1_1}$	$\mathbf{3_2}$	$\mathbf{3_2}$	$\mathbf{3_1}$	$\mathbf{3_1}$	$\mathbf{3_2}$	$\mathbf{3_2}$	$\mathbf{1_1}$	$\mathbf{1_1}$	$\mathbf{1_2}$	$\mathbf{1_2}$
Z_n	1	ω^{-4}	ω^{-2-a}	ω^{-1}	ω^2	ω^{-2}	ω	ω^{-1}	ω^{α}	$\omega^{-\alpha}$	ω^2	ω^{-2}	ω	ω^{-1}

angle. This relation is in rough agreement with observations to leading order. In addition, we have at GUT scale $m_b \sim m_\tau$ as well as $m - \mu \sim -3m_s$ also in rough agreement with observations.

Encouraged by these results, we can be more ambitious and start using this ansatz in combination of other ideas to make as many predictions as possible. To this end, we note that in the limit vanishing Yukawa couplings, the standard model has $[U(3)]^5$ global symmetry. It is therefore quite possible that in the final understanding of flavor, a subgroup of this large symmetry does play an important role[8], specially subgroups which have three dimensional representation to fit three generations. In order to exploit this observation, one may replace all Yukawa couplings by flavon fields which transform as three dimensional representations of a subgroup of $[SU(3)]^5$ and consider the minima of the flavon theory in flavon space as determining the values of the Yukawa couplings. It turns out that there are nontrivial examples where this program is realized. In the second paper of [2], we presented an S_4 subgroup example. Below I briefly recapitulate this example.

6. The $SO(10) \times S_4 \times Z_n$ model of flavor

Recall that the S_4 group is a 24 element group describing permutations of four distinct objects and has five irreducible representations with dimensions $\mathbf{3_1} \oplus \mathbf{3_2} \oplus \mathbf{2} \oplus \mathbf{1_2} \oplus \mathbf{1_1}$. The distinction between the representations with subscripts 1 and 2 is that the later change sign under the transformation of group elements involving the odd number of permutations of S_4.

We assign the three families of **16**-dim. matter fermions ψ to $\mathbf{3_2}$-dim. representation of S_4 and the Higgs field H, $\bar{\Delta}$ and H' to $\mathbf{1_1}, \mathbf{1_2}$, and $\mathbf{1_1}$ reps, respectively. We then choose three SO(10) singlet flavons ϕ_i transforming as $\mathbf{3_2}, \mathbf{3_1}, \mathbf{3_2}$ reps of S_4 and one gauge and S_4 singlet fields s_1, s_2 transforming as $\mathbf{1_2}$ and $\mathbf{1_1}$ respectively. We further assume that at a scale slightly above the GUT scale, there are two S_4 singlet vectorlike pairs of $\mathbf{16} \oplus \overline{\mathbf{16}}$ fields denoted by ψ_V and $\bar{\psi}_V$. In order to get the desired Yukawa couplings naturally from this high scale theory, we supplement the S_4 group by an Z_n group with all the above fields belonging to representations given in the Table I. The fields and representations to generate the desired Yukawa couplings. $\omega = e^{i\frac{2\pi}{n}}$. $\alpha = 2 + a - b$. In addition to the fields below, there are two gauge and S_4 singlet $1_2, 1_1$ fields with Z_n quantum numbers ω^a and ω^b respectively.

The most general high scale Yukawa superpotential involving matter fields invariant under this symmetry is given by:

$$\begin{aligned} W = (\phi_1\psi)\bar{\psi}_{V1} \; + \; \psi_{V1}\psi_{V1}H \; + \; M_1\bar{\psi}_{V1}\psi_{V1} \\ +(\phi_2\psi)\bar{\psi}_{V2} \; + \; \frac{1}{M_P}s_1\psi_{V2}\psi_{V2}\bar{\Delta} + \; M_2\bar{\psi}_{V2}\psi_{V2} \\ + \; \frac{1}{M_P^2}s_2(\phi_3\psi\psi)\bar{\Delta} + \frac{1}{M_P}(\phi_2\psi\psi)H', \end{aligned} \tag{8}$$

where the brackets stand for the S_4 singlet contraction of flavor index. The singlet field s_i can have large vev as follows: consider its Z_n charge to be such that the only polynomial term involving the s_i in the superpotential has the form $s_i^{k_i}/M_P^{k_i-3}$ (in order to describe the essential potential, we ignore a possible $s_1^{\ell_1}s_2^{\ell_2}$ term). The dominant part of the potential in the presence of SUSY breaking has the form:

$$V(s_i) = -m_{s_i}^2|s_i|^2 + k\frac{s_i^{2k_i-2}}{M_P^{2k_i-6}} + \cdots. \tag{9}$$

Minimizing this leads to $\langle s_i\rangle \sim [m_{S_i}^2 M_P^{2k_i-6}]^{\frac{1}{2k_i-4}}$, which is above GUT scale for larger values of the integer k_i (which in turn is determined by the Z_n symmetry charge of s_i). One could also have large vevs for s_1, s_2 by using anomalous $U(1)$ charges for them using D-terms to break the $U(1)$ symmetry.

The effective theory below the scales $M_{1,2}$ and $\langle s_i\rangle$ of the vector-like pair masses and the s_i-vevs respectively is given by:

$$W = (\phi_1\psi)(\phi_1\psi)H + (\phi_2\psi)(\phi_2\psi)\bar{\Delta} + (\phi_3\psi\psi)\bar{\Delta} + (\phi_2\psi\psi)H', \tag{10}$$

where we have omitted the dimensional coupling constants to make it simple for the purpose of writing. The discrete symmetries prevent ϕ^2/M^2 corrections to these terms. So our predictions based on this effective superpotential do not receive large corrections. We note that the non-renormalizable terms in Eq.(8) can also be obtained from renormalizable couplings if we introduce further S_4-triplet vectorlike fields. Here, however we use only S_4-singlet vectorlike fields to get rank 1 contribution to h and f Yukawa couplings and that is why we need the non-renormalizable terms to be present in Eq.(8.

In order to get fermion masses, we have to find the alignment [9] of the vevs of the flavon fields $\phi_{1,2,3}$. We show below that the following choice of vevs are among the minima of the flavon superpotential provided the couplings of mixed terms between different ϕ_i's are small compared to other couplings:

$$\phi_1 = \begin{pmatrix} 0 \\ 0 \\ 1 \end{pmatrix}, \quad \phi_2 = \begin{pmatrix} 0 \\ -1 \\ 1 \end{pmatrix}, \quad \phi_3 = \begin{pmatrix} 1 \\ 1 \\ 1 \end{pmatrix}. \tag{11}$$

Clearly, there are other vacua for the flavon model that we do not choose. What is however nontrivial is that the alignments are along quantized directions. This is a consequence of supersymmetry combined with discrete symmetries in the theory. Given these vev, we find from Eq. (10) that the Yukawa coupling matrices h, f, h' have the form:

$$h \propto \begin{pmatrix} 0 & 0 & 0 \\ 0 & 0 & 0 \\ 0 & 0 & 1 \end{pmatrix}, \tag{12}$$

$$f \propto \begin{pmatrix} 0 & 0 & 0 \\ 0 & 1 & -1 \\ 0 & -1 & 1 \end{pmatrix} + \lambda \begin{pmatrix} 0 & 1 & 1 \\ 1 & 0 & 1 \\ 1 & 1 & 0 \end{pmatrix}, \tag{13}$$

$$h' \propto \begin{pmatrix} 0 & 1 & -1 \\ 1 & 0 & 0 \\ -1 & 0 & 0 \end{pmatrix}, \tag{14}$$

and the charged fermion mass matrices can then be inferred. The neutrino mass matrix in this basis has the form:

$$\mathcal{M}_\nu = \begin{pmatrix} 0 & c & c \\ c & a & c-a \\ c & c-a & a \end{pmatrix}, \tag{15}$$

Journal of Physics: Conference Series **447** (2013) 012003 doi:10.1088/1742-6596/447/1/012003

where $c/a = \lambda \ll 1$. It is diagonalized by the tri-bi-maximal matrix

$$U_{\rm TB} = \begin{pmatrix} \sqrt{\frac{2}{3}} & \sqrt{\frac{1}{3}} & 0 \\ -\sqrt{\frac{1}{6}} & \sqrt{\frac{1}{3}} & -\sqrt{\frac{1}{2}} \\ -\sqrt{\frac{1}{6}} & \sqrt{\frac{1}{3}} & \sqrt{\frac{1}{2}} \end{pmatrix}. \tag{16}$$

This is however not the full PMNS matrix which will receive small corrections from diagonalization of the charged lepton matrix, which not only make small contributions to the $\theta_{\rm atm}$ and $\theta_\odot$ but also generate a small θ_{13}.

The neutrino masses are given by $m_{\nu 3} = 2a - c$; $m_{\nu 2} = 2c$ and $m_{\nu 1} = -c$. To fit observations, we require $\lambda = c/a \simeq \sqrt{\Delta m^2_\odot/\Delta m^2_{\rm atm}} \sim 0.2$, which fixes the neutrino masses $m_{\nu 3} \simeq 0.05$ eV, $m_{\nu 2} \simeq 0.01$ eV, and $m_{\nu 1} \simeq 0.005$ eV. We will see below that λ is also the Cabibbo angle substantiating our claim that neutrino mass ratio and Cabibbo angle are related.

For the charged lepton, up and down quark mass matrices, we have:

$$\begin{aligned} M_\ell &= \frac{r_1}{\tan\beta}\begin{pmatrix} 0 & -3m_1+\delta & -3m_1-\delta \\ -3m_1+\delta & -3m_0 & 3m_0-3m_1 \\ -3m_1-\delta & 3m_0-3m_1 & -3m_0+M \end{pmatrix}, \\ M_d &= \frac{r_1}{\tan\beta}\begin{pmatrix} 0 & m_1+\delta & m_1-\delta \\ m_1+\delta & m_0 & -m_0+m_1 \\ m_1-\delta & -m_0+m_1 & m_0+M \end{pmatrix}, \\ M_u &= \begin{pmatrix} 0 & r_2m_1+r_3\delta & r_2m_1-r_3\delta \\ r_2m_1+r_3\delta & r_2m_0 & -r_2m_0+r_3m_1 \\ r_2m_1+r_3\delta & -r_2m_0+r_3m_1 & r_2m_0+M \end{pmatrix}, \end{aligned} \tag{17}$$

where $\tan\beta$ is a ratio of $H_{u,d}$ vevs. Note that $m_1/m_0 = \lambda \sim 0.2$ and of course $m_0 \ll M$. A quick examination of these mass matrices leads to several immediate conclusions:

(i) The model predicts that at GUT scale $m_b \simeq m_\tau$.

(ii) Since $(M_d)_{11} \to 0$, we get $V_{us} \simeq \sqrt{m_d/m_s}$.

(iii) The empirically satisfied relation $m_\mu m_e \simeq m_s m_d$ can be obtained by the choice of parameters $-3m_1 + \delta = (m_1+\delta)e^{i\sigma}$, where σ is a phase. Solving this equation, we find that $\delta = m_1(1 + i\cot\sigma/2)$. We obtain $V_{us} \simeq (1 - r_3/r_2)\delta/m_0$, thereby relating Cabibbo angle to the neutrino mass ratio $m_\odot/m_{\rm atm} \simeq \lambda$.

(iv) $m_\mu \sim -3m_s$.

(v) $V_{cb} \sim \dfrac{m_s}{m_b}\cot\theta_{\rm atm}$.

(vi) The masses of up and charm quarks are given by the parameters $r_{2,3}$ and are therefore not predictions of the model.

(vii) CP violation in quark sector can put in by making the parameters h' complex.

(viii) The model predicts a small amplitude for neutrino-less double beta decay from light neutrino mass: $m_{\nu_{ee}} \sim c\sin\theta^l_{12} \simeq 0.3$ meV.

The first four relations are fairly well satisfied by observations; the fifth prediction (i.e. that for U_{e3}) can be tested in upcoming reactor and long baseline experiments. Note that the deviation from tri-bi-maximal mixing pattern coming from the charged lepton mass diagonalization could be thought of as a small perturbation of the neutrino mass matrix except that we predict the form of the perturbation from symmetry considerations. The sixth prediction gives a smaller value for V_{cb} (0.02 as against observed GUT scale value of 0.03) if one uses GUT scale extrapolated value of the known b mass. However, in the MSSM there are threshold corrections to the $b-s$ quark mass mixing from gluino and wino exchange one-loop diagrams; by choosing this contribution, one could obtain the desired V_{cb}.

Note that in this model, the top quark Yukawa coupling at GUT scale arises from an effective higher dimensional operator. We have showed the effective operator in Eq.(10) by expanding ϕ/M. The more precise form for the top Yukawa coupling is $\phi^2/(M_1^2+\phi^2)h_{\psi_V\psi_V H}$, where $h_{\psi_V\psi_V H}$ is a coupling of $\psi_V\psi_V H$ term, and ϕ is the vev of ϕ_1 multiplied by $\phi_1\psi\psi_V$ coupling. This is simply because the low energy third generation field is a linear combination of the form $\cos\alpha\,\psi_3 - \sin\alpha\,\psi_V$ with the mixing angle $\sin\alpha \simeq \phi/\sqrt{M_1^2+\phi^2}$.

Therefore, in general, there is no gross contradiction to the fact that the top Yukawa coupling is order 1. However, in our case, if ϕ/M_1 becomes close to 1, the atmospheric mixing shifts from the maximal angle. Given the error in the determination of the atmospheric mixing angle, this is consistent with data and as this measurement sharpens, this is going to provide a test of this particular model. The desired smallness of the effective f and h' couplings however are more naturally obtained due to the presence of the Planck mass in the denominator. In order to make the f-coupling dominate over the h', we have to choose a small coupling for the H' Higgs field in Eq. (4). Similarly the λ term in Eq. (9) is assumed to be small compared to the coefficient of the first matrix.

Thus within these set of assumptions, this model is in good phenomenological agreement with observations. In a more complete theory, these assumptions need to be addressed. We however find it remarkable that despite these shortcomings, the model provides a very useful unification strategy of the diverse quark-lepton mixing patterns.

7. Vev alignment as minima of flavon theories

In this section, we give examples of how the minima of flavon theories can determine the Yukawa couplings of the fermions and lead to predictive flavor models. We discuss the specific case of the S_4 model at hand. This mechanism is of course applicable to any general group.

We start our discussion by giving some simple examples and discussing the flavon alignment as a prelude to the more realistic example. First thing to note is that $\mathbf{3_1}^3$ is invariant under S_4, but $\mathbf{3_2}^3$ is not. Denoting $\phi=(x,y,z)$, we see that in the first case, the singlet of $\phi^3 = xyz$. The superpotential for a $\mathbf{3_1}$ flavon field ϕ can therefore be written as

$$W = \frac{1}{2}m\phi^2 - \lambda\phi^3 = \frac{1}{2}m(x^2+y^2+z^2) - \lambda xyz. \tag{18}$$

The solution of F-flat vacua ($\phi \neq 0$) are

$$\phi = \frac{m}{\lambda}\{(1,1,1) \text{ or } (1,-1,-1) \text{ or } (-1,1,-1) \text{ or } (-1,-1,1)\}. \tag{19}$$

These vacua break S_4 down to S_3 and in the process determine the Yukawa couplings.

On the other hand, when $\mathbf{3_2}$ flavon is used (or the cubic term is forbidden by a discrete symmetry), quartic term involving the triplet is crucial for the F-flat vacua. The invariant quartic term ϕ^4 gives two linear combinations of the form $x^4+y^4+z^4$ and $x^2y^2+y^2z^2+z^2x^2$. This is because they have to be symmetric homogenous terms and invariant under the Klein's group, which is π rotation around the x,y,z axes.

Thus, the superpotential term for $\mathbf{3_2}$ field ϕ is

$$\begin{aligned} W &= \frac{1}{2}m\phi^2 - \frac{\kappa^{(1)}}{M}(\phi^4)_1 - \frac{\kappa^{(2)}}{M}(\phi^4)_2 \qquad (20)\\ &= \frac{1}{2}(x^2+y^2+z^2) - \frac{\kappa^{(1)}}{4M}(x^4+y^4+z^4) - \frac{\kappa^{(2)}}{2M}(x^2y^2+y^2z^2+z^2x^2). \end{aligned}$$

The nontrivial F-flat vacua ($\phi \neq 0$) are

$$\phi = \sqrt{\frac{mM}{\kappa^{(1)}}}\,\vec{a}, \quad \sqrt{\frac{mM}{\kappa^{(1)} + 2\kappa^{(2)}}}\,\vec{b}, \quad \sqrt{\frac{mM}{\kappa^{(1)} + \kappa^{(2)}}}\,\vec{c}, \tag{21}$$

where $\vec{a} = (0,0,\pm1)$, $(0,\pm1,0)$, $(\pm1,0,0)$, $\vec{b} = (\pm1,\pm1,\pm1)$, and $\vec{c} = (0,\pm1,\pm1)$, $(\pm1,\pm1,0)$, $(\pm1,0,\pm1)$. We note that these vectors correspond to the axes of the regular hexahedron. The vacua break S_4 down to Z_4, Z_3, and Z_2, respectively. More importantly, the vacuum states in Eq. (12) used in the analysis of fermion masses in the previous section are a subset of the above vacua.

Note that if we add a ϕ^4 term to the superpotential involving the $\mathbf{3_1}$ flavon field, $\vec{a}$ vacuum is possible, in addition to the original $\vec{b}$ vacua. However, $\vec{c}$ vacuum is absent.

A detailed numerical analysis of the model subsequent to the measurement of θ_{13} [10] and it turns out that in the strictly minimal vev alignment discussed above predicts $\theta_{13} \simeq 5^0$; however, a slight change in the vev alignment without any extension of the particle content of the model leads to the experimentally observed value[11].

8. Comments

A complete understanding of flavor is clearly a very ambitious task. Our proposal should be considered as a simple beginning towards a final theory. It should be noted that even though we have considered on SO(10) group, our general unification ansatz (without as much predictivity) in $SU(2)_L \times SU(2)_R \times SU(4)_c$ theories as well and perhaps other groups such as E_6. Similarly, one should explore other flavor models.

A second point of importance is that while we have kept only leading order terms, one should clearly consider higher order corrections to our predictions systematically. In the above mode, we have checked next order corrections and found them to be absent due to the discrete symmetries.

9. Conclusion

In summary, I have discussed a recently proposed ansatz that has the potential to provide a unified description of the diverse quark and lepton flavor patterns. This could provide the first opening into a very difficult problem of particle physics- the problem of flavor. A simple realization of this ansatz is shown to occur within a grand unified SO(10) model with type II seesaw describing the neutrino masses. The successes of that model are that it seems to provide an understanding of several observed quark-lepton mass relations such as bottom-tau mass unification, strange quark-muon mass ratio (1/3) etc. and predicts a value for θ_{13} in agreement with observations and the atmospheric mixing angle different from the maximal value. The model like most grand unified theories of neutrinos predicts a normal hierarchy and observation of inverted hierarchy will therefore rule out this model (as well as most simple grand unified theories). Under certain reasonable approximations, this also seems to explain why $m_{solar}/m_{atm} \sim \theta_C$. It also predicts a value of 0.3 meV for the effective neutrino mass in neutrinoless double beta decay experiments.

Acknowledgement

This work is supported by the US National Science Foundation under grant No. PHY-0968854. The author is grateful to Y. Mimura and B. Dutta in whose collaboration this work was done and to Prof. G. Branco for hospitality at IST, Lisbon.

References

[1] P. Minkowski, Phys. Lett. B **67** (1977) 421; T. Yanagida in *Workshop on Unified Theories, KEK Report 79-18*, p. 95, 1979; M. Gell-Mann, P. Ramond and R. Slansky, *Supergravity*, p. 315. Amsterdam: North Holland,

1979; S. L. Glashow, *1979 Cargese Summer Institute on Quarks and Leptons*, p. 687. New York: Plenum, 1980; R. N. Mohapatra and G. Senjanovic, Phys. Rev. Lett, **44** (1980) 912.

[2] B. Dutta, Y. Mimura and R. N. Mohapatra, Phys. Rev. D **80**, 095021 (2009); JHEP **1005**, 034 (2010).

[3] B. S. Balakrishna, A. L. Kagan and R. N. Mohapatra, Phys. Lett. B **205**, 345 (1988); K. S. Babu and R. N. Mohapatra, Phys. Rev. Lett. **64**, 2747 (1990); B. A. Dobrescu and P. J. Fox, JHEP **0808**, 100 (2008); L. Ferretti, S. F. King and A. Romanino, JHEP **0611**, 078 (2006).

[4] K. S. Babu, B. Dutta and R. N. Mohapatra, Phys. Rev. D **60**, 095004 (1999).

[5] K. S. Babu and R. N. Mohapatra, Phys. Rev. Lett. **70**, 2845 (1993).

[6] G. Lazarides, Q. Shafi and C. Wetterich, Nucl. Phys. B **181**, 287 (1981); J. Schechter and J. W. F. Valle, Phys. Rev. D **22**, 2227 (1980); R. N. Mohapatra and G. Senjanovic, Phys. Rev. D **23**, 165 (1981).

[7] B. Bajc, G. Senjanovic and F. Vissani, Phys. Rev. Lett. **90**, 051802 (2003).

[8] G. Altarelli and F. Feruglio, Nucl. Phys. B **741**, 215 (2006) [hep-ph/0512103]; G. Altarelli, F. Feruglio and C. Hagedorn, JHEP **0803**, 052 (2008) [arXiv:0802.0090 [hep-ph]]; F. Feruglio, C. Hagedorn and L. Merlo, JHEP **1003**, 084 (2010) [arXiv:0910.4058 [hep-ph]]; S. Morisi and E. Peinado, Phys. Rev. D **80**, 113011 (2009) [arXiv:0910.4389 [hep-ph]]; M. C. Chen and S. F. King, JHEP **0906**, 072 (2009) [arXiv:0903.0125 [hep-ph]]; Y. Cai and H. B. Yu, Phys. Rev. D **74**, 115005 (2006) [hep-ph/0608022]; F. Bazzocchi and S. Morisi, Phys. Rev. D **80**, 096005 (2009) [arXiv:0811.0345 [hep-ph]]; H. Ishimori, Y. Shimizu and M. Tanimoto, Prog. Theor. Phys. **121**, 769 (2009) [arXiv:0812.5031 [hep-ph]]; F. Bazzocchi, L. Merlo and S. Morisi, Phys. Rev. D **80**, 053003 (2009) [arXiv:0902.2849 [hep-ph]]; G. Altarelli, F. Feruglio and L. Merlo, JHEP **0905**, 020 (2009) [arXiv:0903.1940 [hep-ph]]; W. Grimus, L. Lavoura and P. O. Ludl, J. Phys. G **36**, 115007 (2009) [arXiv:0906.2689 [hep-ph]]; G. J. Ding, Nucl. Phys. B **827**, 82 (2010) [arXiv:0909.2210 [hep-ph]]; Y. Daikoku and H. Okada, arXiv:0910.3370 [hep-ph]; M. K. Parida, Phys. Rev. D **78**, 053004 (2008) [arXiv:0804.4571 [hep-ph]]; C. Luhn, S. Nasri and P. Ramond, Phys. Lett. B **652**, 27 (2007) [arXiv:0706.2341 [hep-ph]]; C. Hagedorn, M. A. Schmidt and A. Y. Smirnov, Phys. Rev. D **79**, 036002 (2009) [arXiv:0811.2955 [hep-ph]]; F. Bazzocchi and I. de Medeiros Varzielas, Phys. Rev. D **79**, 093001 (2009) [arXiv:0902.3250 [hep-ph]]; A. Adulpravitchai, A. Blum and C. Hagedorn, JHEP **0903**, 046 (2009) [arXiv:0812.3799 [hep-ph]]; S. F. King and C. Luhn, Nucl. Phys. B **820**, 269 (2009) [arXiv:0905.1686 [hep-ph]]; M. C. Chen and K. T. Mahanthappa, Phys. Lett. B **681**, 444 (2009) [arXiv:0904.1721 [hep-ph]]; arXiv:0910.5467 [hep-ph]; For some earlier models, see K. S. Babu, E. Ma and J. W. F. Valle, Phys. Lett. B **552**, 207 (2003) [hep-ph/0206292]; K. S. Babu and X. G. He, hep-ph/0507217; R. N. Mohapatra, S. Nasri and H. B. Yu, Phys. Lett. B **639**, 318 (2006) [hep-ph/0605020].
For recent reviews, see G. Altarelli and F. Feruglio, arXiv:1002.0211 [hep-ph]; H. Ishimori, T. Kobayashi, H. Ohki, H. Okada, Y. Shimizu and M. Tanimoto, arXiv:1003.3552 [hep-th]; S. F. King and C. Luhn, arXiv:1301.1340 [hep-ph].

[9] S. F. King and G. G. Ross, Phys. Lett. B **520**, 243 (2001) [hep-ph/0108112]; S. F. King, JHEP **0508**, 105 (2005) [hep-ph/0506297]; I. de Medeiros Varzielas and G. G. Ross, Nucl. Phys. B **733**, 31 (2006) [hep-ph/0507176]; I. de Medeiros Varzielas, S. F. King and G. G. Ross, Phys. Lett. B **644**, 153 (2007) [hep-ph/0512313]; Phys. Lett. B **648**, 201 (2007) [hep-ph/0607045]. S. F. King and C. Luhn, Nucl. Phys. B **832**, 414 (2010); C. Hagedorn, S. F. King and C. Luhn, JHEP **1006**, 048 (2010).

[10] F. P. An *et al.* [DAYA-BAY Collaboration], Phys. Rev. Lett. **108**, 171803 (2012) [arXiv:1203.1669 [hep-ex]]; arXiv:1210.6327 [hep-ex]; J. K. Ahn *et al.* [RENO Collaboration], Phys. Rev. Lett. **108**, 191802 (2012) [arXiv:1204.0626 [hep-ex]]. K. Abe *et al.* [T2K Collaboration], Phys. Rev. Lett. **107**, 041801 (2011) [arXiv:1106.2822 [hep-ex]]; Y. Abe et al. (Double Chooz Collaboration), Phys. Rev. Lett. **108**, 131801 (2012); P. Adamson *et al.* [MINOS Collaboration], [arXiv:1301.4581 [hep-ex]].

[11] P. S. Bhupal Dev, B. Dutta, R. N. Mohapatra and M. Severson, Phys. Rev. D **86**, 035002 (2012) [arXiv:1202.4012 [hep-ph]].

Journal of Physics: Conference Series **447** (2013) 012004 doi:10.1088/1742-6596/447/1/012004

Neutrino mass, mixing and discrete symmetries

Alexei Y. Smirnov [1]

International Centre for Theoretical Physics, Strada Costiera 11, 34014 Trieste, Italy

E-mail: smirnov@ictp.it

Abstract. Status of the discrete symmetry approach to explanation of the lepton masses and mixing is summarized in view of recent experimental results, in particular, establishing relatively large 1-3 mixing. The lepton mixing can originate from breaking of discrete flavor symmetry G_f to different residual symmetries G_ℓ and G_ν in the charged lepton and neutrino sectors. In this framework the *symmetry group condition* has been derived which allows to get relations between the lepton mixing elements immediately without explicit model building. The condition has been applied to different residual neutrino symmetries G_ν. For generic (mass independent) $G_\nu = \mathbf{Z}_2$ the condition leads to two relations between the mixing parameters and fixes one column of the mixing matrix. In the case of $G_\nu = \mathbf{Z}_2 \times \mathbf{Z}_2$ the condition fixes the mixing matrix completely. The non-generic (mass spectrum dependent) G_ν lead to relations which include mixing angles, neutrino masses and Majorana phases. The symmetries G_ℓ, G_ν, G_f are identified which lead to the experimentally observed values of the mixing angles and allow to predict the CP phase.

1. Introduction

A possibility to use the discrete flavor symmetries for understanding fermion masses and mixing had been proposed long time ago [1]. To a large extend the recent developments of this approach [2] was motivated by the tri-bimaximal (TBM) mixing [3]. The special values of elements of the TBM mixing matrix,

$$|U_{e3}|^2 = \sin^2\theta_{13} = 0, \quad |U_{\mu 3}|^2 = |U_{\tau 3}|^2 = \sin^2\theta_{23} = \frac{1}{2}, \quad |U_{e2}|^2 = \sin^2\theta_{12} = \frac{1}{3}, \tag{1}$$

indicated possible geometric and group-theoretical origins of the mixing. The values in (1) also implied that there is no relation between mixing and masses (mass ratios). In this connection the most appealing framework is the one in which mixing originates from different ways of the flavor symmetry, G_f, breaking in the charged lepton and neutrino Yukawa sectors [4]:

$$G_f \to \text{breaking} \to \begin{cases} G_\nu & \text{neutrinos} \\ G_l & \text{charged leptons} \end{cases}. \tag{2}$$

The *residual* symmetries G_ν and G_l of the neutrino and charged lepton mass matrices are different. G_ν and G_l should be generic symmetries which exist for arbitrary values of masses. In this case appearance of TBM is, indeed, maximally controlled by symmetry.

[1] Talk given at the Symposium Discrete 2012, IST, Lisboa, Portugal, December 3 - 7, 2012.

Realizations of this framework in specific models are, however, rather complicated and not convincing. One should construct the Lagrangian invariant under symmetry G_f with certain field content and assignment of charges. Additional auxiliary symmetries are needed to forbid some interactions. Further model building is needed to achieve required vacuum alignment. Essentially, two different scalar sector should be introduced for spontaneous symmetry breaking in the neutrino and charged lepton sectors. Finally, after spontaneous symmetry breaking the mass matrices with certain residual symmetries emerge. Accidental symmetries and relations between mixing elements may show up in models which are not related to the original flavor symmetry. Complicated structure of models, many *ad hoc* parameters and new fields, additional symmetries, difficulty to include quarks, *etc.* cast doubt upon whole the approach [5].

Recent measurements of the neutrino oscillation parameters, and especially discovery of rather large 1-3 mixing [6], [7], [8], [9], [10], contrary to (1), have further disfavored TBM and the discrete symmetries behind. The TBM can be accidental and whole discrete flavor symmetry approach – phenomenologically irrelevant.

Further developments along this line were in two directions: (i) introduction of large corrections to TBM to reproduce results of measurements [11], (ii) modification of symmetries in such a way that they are consistent with nonzero 1-3 mixing [12]. In some situations these two lines coincide giving the same results. With this the symmetry effects become rather hidden.

In this connection in the framework of residual symmetries (2) a formalism has been developed [13], [14], [15] which allows to obtain consequences of flavor symmetries for mass and mixing without model building. The formalism allows to explore mixing patterns for wide class of different residual and covering symmetries. Inversely, it can be used to perform the "symmetry building" for a required mixing pattern. It helps to understand various features of mixing that are related to symmetries.

Symmetries and their consequences are the nicest part of the program, model building is the ugly one. In this review we will focus on the former: on the model independent part. The formalism [13, 14] will be explained in details and various related issues will be clarified. The applications to different residual neutrino symmetries will be described. Still one should remember that the problems with model building remain.

The paper is organized as follows. In Sec. 2 we summarize the relevant information on neutrino masses and mixing and its possible implications for discrete flavor symmetries. Next possible step in developments of the field will be outlined. In Sec. 3 the symmetry group conditions are derived which lead to relations between the lepton mixing elements without model building. Applications of these symmetry group conditions to different residual neutrino symmetries are described in Sec. 4. Discussion and conclusions are presented in Sec. 5.

2. Masses and mixing: status and implications

2.1. Data and Observations.

Results and observations from global fits [16], [17], [18] relevant for the discrete flavor symmetries can be summarized in the following way.

1). The 1-3 mixing deviates substantially from zero:

$$\sin^2\theta_{13} = 0.022 - 0.024. \tag{3}$$

2). There are indications of significant deviation of the 2-3 mixing from maximal: $\sin^2\theta_{23} = 0.38 - 0.42$, which can be characterized by

$$d_{23} \equiv 0.5\cos 2\theta_{23} = 0.08 - 0.12. \tag{4}$$

The result follows from (i) the atmospheric neutrino data (included in the global fit); (ii) from MINOS direct measurements [19]: $\sin^2 2\theta_{23} = 0.950^{+0.035}_{-0.036}$ (1σ), and (iii) from comparison of the T2K result [6] on $\nu_\mu \to \nu_e$ oscillations ($\propto \sin^2\theta_{23}\sin^2 2\theta_{13}$) and the reactor disappearance results ($\propto \sin^2 2\theta_{13}$). The atmospheric neutrino and T2K data being sensitive to $\sin^2\theta_{23}$ favor the first quadrant, $\theta_{23} < \pi/4$, at least for the normal mass hierarchy (NH). In the case inverted hierarchy (IH) different global fits give different results.

3). With increase of accuracy the deviation of the 1-2 mixing from the TBM value 1/3 becomes stronger and more significant statistically:

$$\sin^2\theta_{12} = 0.30 - 0.31. \tag{5}$$

4). First glimpses on the CP-violating phase show up in different global fits: $\delta = 180°$ from [16], $\delta \sim 300°$ for both hierarchies from [17], and $\delta = 145°$ (NH) $\delta = 0$ (IH) from [18]. Being at 1σ level or even lower all these are statistically insignificant. In some analyses one can trace the physical effects which favor certain value δ, in other cases that can be artifact of the global fit without any clear reason behind or just fluctuation.

The deviation of mixing from TBM can be characterized by the values of elements of the third column of the PMNS matrix, $U_{\alpha 3} \equiv \{U_{e3}, U_{\mu 3}, U_{\tau 3}\}$:

$$U_{\alpha 3}^{TBM} = \{0,\ 0.71,\ 0.71\}, \quad \text{whereas} \quad U_{\alpha 3}^{exp} = \{0.15,\ 0.62,\ 0.77\}, \tag{6}$$

and for the "numerology bookkeeping": $U_{\tau 3} - U_{\mu 3} \approx U_{e3}$.

2.2. The 1-3 mixing: relations and implications.

The same value of 1-3 mixing can be connected to other observables in various ways which have different theoretical implications.

(i) "Naturalness":

$$\sin^2\theta_{13} = A\frac{\Delta m_{21}^2}{\Delta m_{32}^2}, \tag{7}$$

where $A = 0.78$ ($\sim 1 - \sin\theta_C$) for the best fit value of θ_{13}. This relation follows from "naturalness" of mass matrix [20]: the fact that there are two large mixings connecting neighboring generations, and from the following two assumptions: (i) normal mass hierarchy, (ii) absence of fine tuning between different elements of the mass matrix (e.g., $|m_{e\mu} - m_{e\tau}| \sim |m_{e\mu}|$).

(ii) "QLC":

$$\sin\theta_{13} \approx \frac{1}{\sqrt{2}}\sin\theta_C(1 - V_{cb}\cos\delta) - V_{ub} \approx \frac{\theta_C}{\sqrt{2}}, \tag{8}$$

where θ_C is the Cabibbo angle [21]. Varying δ we have from (8) $\sin^2\theta_{13} = 0.025 \pm 0.003$ in agreement with observations (3, 4). Approximately, $\sin^2\theta_{13} \approx \frac{1}{2}\sin^2\theta_C$. This prediction is essentially result of permutation of the matrices of the maximal mixing 2-3 rotation and the 1-2 rotation on the Cabibbo angle:

$$U_{12}(\theta_C)U_{23}\left(\frac{\pi}{4}\right).$$

Here $U_{12}(\theta_C)$ can follow from the charged leptons, whereas $U_{23}\left(\frac{\pi}{4}\right)$ – from neutrinos. The former implies a kind of quark-lepton symmetry, or unification, or common horizontal symmetry which lead to equality of the 1-2 mixings in the quark and lepton sectors. The second ($\pi/4$) rotation comes from the neutrino sector and can be part of the bi-maximal [21] or tri-bimaximal [22] mixings. The latter case is referred as the tribimaximal-Cabibbo mixing. The permutation is needed to reduce the mixing matrix to the standard form. The relation was first realized at the purely phenomenological level [23] and then in the Quark-Lepton Complementarity (QLC) framework [21]. Appearance of the bi-maximal or tri-bimaximal mixings can be a consequence of discrete flavor symmetries.

(iii) The $\nu_\mu - \nu_\tau$ symmetry breaking:

$$\sin^2\theta_{13} = \frac{1}{2}\cos^2 2\theta_{23} = 2d_{23}^2 \approx 0.022 \tag{9}$$

in perfect agreement with measurements (3,4). The relation (9) may follow from "universal" violation of the $\nu_\mu - \nu_\tau$ symmetry. (In the case of exact symmetry $\sin^2\theta_{13} = d_{23} = 0$.)

(iv) The self-complementarity [24]:

$$\theta_{12} + \theta_{13} = \theta_{23}. \tag{10}$$

It is purely leptonic relation which is also reproduced by QLC.

(v). "Anarchy":

$$\sin^2\theta_{13} > 0.025. \tag{11}$$

The inequality is realized at 1σ level in the anarchy approach, in which values of mixing angles appear as random numbers [25].

(vi) "Quark-lepton universality":

$$\theta_{13} \approx \frac{1}{2}\theta_{12}\theta_{23}.$$

This equality is similar to the one in the quark sector: $V_{ub} = \frac{1}{2}V_{us}V_{cb}$. It may indicate that the mixing patterns of quarks and leptons are organized in the same way but with certain rescaling. That would testify for (i) a kind of Fritzsch ansatz for mass matrices; (ii) normal mass hierarchy; (iii) relation between masses and mixing; (iv) flavor ordering or alignment in the mass matrix. The later means that values of elements of the neutrino mass matrix in the flavor basis gradually decrease from $m_{\tau\tau}$ to m_{ee}. In fact, due to deviation of the 2-3 mixing from maximal sharp difference of the elements of the dominant $\mu\tau$-block and the sub-dominant $e-$line is washed out. It seems that neutrino mass matrix is rescaling of the charged fermion mass matrices with weaker hierarchy of the elements. This can originate from power dependence of elements on large expansion parameter $\lambda \sim 0.7 - 0.8$. It looks like another complementarity: $\lambda_\nu = 1 - \lambda_q \approx \theta_C$. The power dependence could testify for a kind of Froggatt-Nielsen mechanism.

No one of these results explicitly testifies for discrete flavor symmetries, although, as we will see later, consequences of symmetry can be hidden in rather complicated relations between mixing parameters. It seems that relations (i), (iv) and (vi) have no connection to discrete symmetries at all, the relation (ii) may require the symmetry to explain the maximal 2-3 rotation from neutrino sector, finally (iii) may or may not be a consequence of symmetry.

Existence of sterile neutrinos with the eV-scale mass can strongly affect our considerations. Their mixing with active neutrinos is not a small perturbation of the $3\nu-$ pattern: all theoretical constructions should be reconsidered unless further complications and fine tunings are introduced [26].

2.3. Race for the mass hierarchy.

Future developments in the field will be related (apart from better measurements of known parameters) to

- clarification of the situation with sterile neutrinos;
- checks of the claim of the neutrinoless double beta decay observation in the Heidelberg-Moscow experiment; and
- establishing neutrino mass hierarchy.

All this will have serious impact on our understanding of mixings and masses and relevance of the discrete flavor symmetries.

As far as the mass hierarchy is concerned, in the 2ν approximation changing the hierarchy means flip of the sign: $\Delta m_{31}^2 \to -\Delta m_{31}^2$. The discrete symmetry transformation can be introduced which makes this flip. The case of inverted hierarchy is the special since it implies strong degeneracy of the two mass states ν_1 and ν_2: $\Delta m/m \sim \Delta m_{21}^2/2\Delta m_{31}^2 \sim 1.6 \cdot 10^{-2}$. This strong degeneracy can be a consequence of certain discrete symmetry. However it is not accompanied by maximal mixing, which leads to certain tension.

A number of proposals are participating in the "race" for the mass hierarchy. Many are related to influence of matter effects on the 1-3 mixing: In uniform or nearly uniform medium the resonance enhancement of oscillations driven by the 1-3 mixing is in the neutrino channel for NH and in the antineutrino channel in the case of IH. To some extend this can be explored using the beam experiments, e.g., in NOvA [27]. The measured relatively large 1-3 mixing ensures that in supernova the 1-3 level crossing (H resonance) is highly adiabatic. Consequently, conversion effects in the cases of NH and IH are strongly different [28]. That determines also the oscillation effects of SN neutrinos in the matter of the Earth. Observation of the oscillation effect in the antineutrino channel would be the proof of NH [29].

Huge atmospheric neutrino detectors with low ($\sim$ 1 GeV) energy threshold such as PINGU [30] and ORCA have good chance to be the first and the cheapest ones [31]. These experiments have also good potential to measure precisely the 2-3 mixing and mass splitting.

3. Discrete symmetries and patterns of lepton mixing.

We assume that neutrinos are Majorana particles, and that the discrete flavor symmetry can be embedded into $SU(3)$. The latter requires that determinants of the transformation matrices should be 1.

3.1. Residual symmetries and transformations.

In what follows we will consider the framework of the residual symmetries [4] in details. Let S_ν and T be the transformations of the neutrino, M_ν, and charged leptons, M_ℓ, mass matrices which correspond to the residual symmetries G_ν and G_l. The invariance means that

$$S_\nu^T M_\nu S_\nu = M_\nu \,, \tag{12}$$

$$T M_\ell M_\ell^\dagger T^\dagger = M_\ell M_\ell^\dagger . \tag{13}$$

The diagonalization of M_ν and M_ℓ gives mixing matrix which reflects symmetries G_ν and G_l. In fact, S_ν and T can be used as the generating elements of G_ν and G_l.

The invariance conditions (12) and (13) should be satisfied in the basis of neutrino and charged lepton states in which the charged current (CC) interactions are diagonal. Only in this basis all the information about mixing is contained in the mass matrices. This basis is not unique: it is determined upto equal rotations of the left handed components of neutrinos and charged leptons. (The flavor basis is one possibility.) Only in this basis one can obtain information about mixing and masses exploring the mass matrices only [2].

Most of the symmetry bounds on lepton mixing in the framework (2, 12, 13) have been obtained via *explicit model building*. That is, via construction of the Lagrangian invariant under G_f, spontaneous breaking of this symmetry, and diagonalization of the generated mass matrices. However, consequences of symmetry for mixing (as well as masses) can be obtained *without explicit model building* immediately from (2, 12, 13), *i.e.*, from the conditions that

[2] Obviously, in the neutrino and charged lepton mass bases all the information on mixing is in the charged current interactions, and the diagonal mass matrices do not contain information about mixing.

Journal of Physics: Conference Series **447** (2013) 012004 doi:10.1088/1742-6596/447/1/012004

- the mass matrices of neutrinos and charged leptons (in the basis where the CC interactions are diagonal) have certain "residual" symmetries, and
- that these symmetries originate from breaking of the original flavor symmetry (or inversely, that they can be embedded into this flavor symmetry).

In other words, the relations between the mixing matrix elements can be obtained from the fact that the neutrino and charged lepton mass matrices are invariant under transformations which generate the flavor symmetry group. Relations obtained in specific models must coincide with relations obtained here, once G_ν and G_l and the covering group G_f are the same.

3.2. Symmetry group condition.
The transformations S_ν and T in the CC diagonal basis encode information about mixing (and masses) that originate from the flavor symmetry. The relations between mixing matrix elements follow from the conditions that S_ν and T belong to the same *finite discrete* group G_f [13, 14]. This means that the product of S_ν and T

$$W \equiv S_\nu \cdot T \tag{14}$$

also belongs to G_f. Furthermore, since G_f is a *finite* group, there must exist integers n, m and p such that $S_\nu^n = \mathbf{I}$, $T^m = \mathbf{I}$ and

$$W^p = (S_\nu \cdot T)^p = \mathbf{I}. \tag{15}$$

We will call this the *symmetry group condition.* The relations

$$S_\nu^n = T^m = W^p = \mathbf{I} \tag{16}$$

form a presentation of G_f which corresponds to the von Dyck group $D(n, m, p)$. The condition for the group $D(n, m, p)$ to be finite is

$$\frac{1}{n} + \frac{1}{m} + \frac{1}{p} > 1\,, \tag{17}$$

and complete list of the finite von Dyck groups which have irreducible representation **3** includes $\mathbf{D}_n = D(2, 2, n)$, $\mathbf{A}_4 = D(2, 3, 3)$, $\mathbf{S}_4 = D(2, 3, 4)$ and $\mathbf{A}_5 = D(2, 3, 5)$.

For definiteness, let us consider the problem in the *flavor basis*, which means that the charged lepton mass matrix is diagonal and we will use the same notation for its symmetry transformations T as before. We denote by $S_{\nu U}$ the transformation matrix that lefts invariant the neutrino mass matrix in the flavor basis:

$$S_{\nu U}^T M_{\nu U} S_{\nu U} = M_{\nu U}. \tag{18}$$

(The same result can be obtained in any basis where the CC interactions are diagonal).

It is the condition (15), that connects symmetry transformations of the neutrinos and charged leptons, that leads to relations between the mixing matrix elements. To see this we introduce S_m - the symmetry transformation of the neutrino mass matrix in the neutrino mass basis:

$$S_m^T m_\nu S_m = m_\nu\,, \qquad m_\nu \equiv \mathrm{diag}\{m_1, m_2, m_3\}. \tag{19}$$

Let us express $S_{\nu U}$ in terms of S_m. In the flavor basis the neutrino mass matrix can be represented as

$$M_{\nu U} = U_{PMNS}^* m_\nu U_{PMNS}^\dagger\,, \tag{20}$$

where U_{PMNS} is the PMNS lepton mixing matrix. Using Eqs. (18), (19) and (20) it is easy to show that the transformation matrix $S_{\nu U}$ equals

$$S_{\nu U} = U_{PMNS} S_m U^{\dagger}_{PMNS} \,. \tag{21}$$

Since S_m does not depend on mixing, whole information on the mixing is explicit in (21). Inserting $S_\nu = S_{\nu U}$ into (15) we obtain

$$[U_{PMNS} S_m U^{\dagger}_{PMNS} T]^p = \mathbf{I}. \tag{22}$$

This is the main relation which connects the mixing matrix and generating elements of the group in the mass basis. It should give relations between the mixing matrix elements in terms of the group parameters. Let us stress that restrictions on mixing follow from the symmetry group condition which includes both the neutrino and charged lepton transformations.

One can obtain the relations for the matrix elements immediately from the Eq. (22), however simpler way to solve it and to analyze solutions is the following. It can be shown [13] that solution of (22) is equivalent to the solution of equation

$$\mathrm{Tr}\left[U_{PMNS} S_m U^{\dagger}_{PMNS} T\right] = a \,, \tag{23}$$

where a is the sum of three p-th roots of unity, λ_α:

$$a = \sum_{\alpha=1}^{3} \lambda_\alpha, \qquad (\lambda_\alpha)^p = 1. \tag{24}$$

The p-th roots of unity are a finite set of complex numbers, and in general (especially for large p) several complex values for a can be found [3]. In fact, λ_α are the eigenvalues of W they can be parameterized [32] as

$$\lambda_\alpha = e^{i\xi_\alpha}, \qquad \xi_\alpha = 2\pi \frac{s_\alpha}{p}, \qquad (\alpha = e, \mu, \tau), \tag{25}$$

where integers s_α satisfy inequality $s_\alpha < p$. Since $\mathrm{Det}[W] = \lambda_e \lambda_\mu \lambda_\tau = 1$, we have $s_\tau = -s_e - s_\mu$, and consequently, $a = a(p, s_e, s_\mu)$. For known a and given S_m and T, Eq. (23) provides a complex condition that the entries of U_{PMNS} must satisfy.

Notice that instead of (14) one can consider some other products of S_ν and T or even two products simultaneously. The latter should not lead though to new relations between the mixing elements but will impose constrains on the group parameters (see below).

The symmetries G_ν and G_l are broken in whole the theory, and therefore one expects corrections to the results obtained in the symmetry limit. Corrections can not be taken into account in this model independent formalism. If however, the residual symmetry of mass matrix exists (or imposed) after the corrections are taken into account, then the relations exist for mixing parameters with corrections.

3.3. Generic invariance of mass matrices.

There are certain symmetries of M_ν and M_ℓ which are always present. They play important role in our consideration. For arbitrary values of masses the neutrino mass matrix $m_\nu \equiv \mathrm{diag}\{m_1, m_2, m_3\}$ is invariant under the transformations

$$S_1 = \mathrm{diag}\{1, -1, -1\}\,, \quad S_2 = \mathrm{diag}\{-1, 1, -1\}\,, \tag{26}$$

[3] Here we use the same definition of a as in the paper [14] which has an opposite sign with respect to definition given in [13].

and $S_3 = S_1 S_2$ with $\mathrm{Det}[S_i] = 1$:

$$S_i^T m_\nu S_i = m_\nu \,. \tag{27}$$

The transformations S_1 and S_2 satisfy conditions $S_i^2 = \mathbf{I}$ and generate the Klein group $\mathbf{Z}_2 \otimes \mathbf{Z}_2$. Then in the flavor basis the mass matrix $M_{\nu U}$ is invariant under the transformation $S_{iU} = U_{PMNS} S_i U_{PMNS}^\dagger$.

The charged lepton mass matrix in the flavor basis, $m_\ell = \mathrm{diag}(m_e,\ m_\mu,\ m_\tau)$, has a full $[U(1)]^3$ symmetry. We assume that the residual discrete symmetry is a $\mathbf{Z}_m$ subgroup of $[U(1)]^3$. So, the corresponding transformation which satisfies the condition $T^m = \mathbf{I}$, can be written as

$$T \equiv \mathrm{diag}\{e^{i\phi_e},\ e^{i\phi_\mu},\ e^{i\phi_\tau}\}, \tag{28}$$

where

$$\phi_\alpha \equiv 2\pi \frac{\kappa_\alpha}{m}\,, \qquad \alpha = e,\ \mu,\ \tau\ , \tag{29}$$

with $\kappa_\alpha \le m$. Since the flavor group is a subgroup of $SU(3)$, we have $\mathrm{Det}[T] = 1$, which gives

$$\phi_e + \phi_\mu + \phi_\tau = 0, \tag{30}$$

or equivalently, $\kappa_\tau = -\kappa_e + \kappa_\mu$.

Notice that T and W enter the formalism in similar way and can be interchanged with substitution $m \leftrightarrow p$, $\kappa_\alpha \leftrightarrow s_\alpha$.

Recall that symmetries generated by S_{iU} and T are always present and they can not be broken. The invariance does not depend on specific values of masses. Therefore the generic symmetries lead to relations between mixing parameters without connection to masses. In particular, it is for this reason the generic symmetry based on $\mathbf{Z}_2 \otimes \mathbf{Z}_2$ was used in models of the TBM mixing. The symmetry transformations are present, but may or may not form the flavor group G_f, that is, they may or may not satisfy the symmetry group condition (15).

4. Mixing patterns for different G_ν

In what follows we will use T as in (28), and consider different G_ν.

4.1. $G_\nu = \mathbf{Z}_2$

Let us take one $\mathbf{Z}_2$ as G_ν, *i.e.* only one generic ("mass independent") symmetry of the neutrino mass matrix. This means that only one transformation, e.g., S_{1U} satisfies the group condition $(S_{1U} \cdot T)^p = 1$, whereas two others - don't: $(S_{iU} \cdot T)^q \neq 1$, $i = 2, 3$ for any q. They are outside of the group and do not lead to bounds on the mixing matrix. Obviously, $S_i^2 = \mathbf{I}$ and therefore $n = 2$ in the presentation (16).

With T from (28) the condition (23) can be written explicitly as

$$\sum_\alpha e^{i\phi_\alpha}(2|U_{\alpha i}|^2 - 1) = a, \tag{31}$$

or introducing $a \equiv a_R + i a_I$, as

$$\sum_\alpha \cos\phi_\alpha (2|U_{\alpha i}|^2 - 1) = a_R, \qquad \sum_\alpha \sin\phi_\alpha (2|U_{\alpha i}|^2 - 1) = a_I. \tag{32}$$

These relations

- depend on the *moduli* of matrix elements, which is related to the diagonal form of the generating elements S_j and T;

- relate elements of a *single column* j, and this index j is determined by the index j of the neutrino transformation matrix S_j;
- determine the column j *completely*, since two relations (for real and imaginary parts of a) plus unitarity are imposed.

Explicitly the relations read:

$$
\begin{aligned}
|U_{ej}|^2 &= \frac{a_R \cos\frac{\phi_e}{2} + \cos\frac{3\phi_e}{2} - a_I \sin\frac{\phi_e}{2}}{4\sin\frac{\phi_{e\mu}}{2}\sin\frac{\phi_{\tau e}}{2}}, \\
|U_{\mu j}|^2 &= \frac{a_R \cos\frac{\phi_\mu}{2} + \cos\frac{3\phi_\mu}{2} - a_I \sin\frac{\phi_\mu}{2}}{4\sin\frac{\phi_{e\mu}}{2}\sin\frac{\phi_{\mu\tau}}{2}}, \\
|U_{\tau j}|^2 &= \frac{a_R \cos\frac{\phi_\tau}{2} + \cos\frac{3\phi_\tau}{2} - a_I \sin\frac{\phi_\tau}{2}}{4\sin\frac{\phi_{\tau e}}{2}\sin\frac{\phi_{\mu\tau}}{2}},
\end{aligned}
\tag{33}
$$

where

$$\phi_{\alpha\beta} \equiv \phi_\alpha - \phi_\beta\,, \quad \alpha,\, \beta = e\,, \mu,\, \tau\,. \tag{34}$$

Due to the unitarity, $\sum_\alpha |U_{\alpha j}|^2 = 1$, there are two independent relation. For specific S_j, the equations give the absolute values of the j-th column of U_{PMNS}, if the values of m, p, κ_e, κ_μ and a are given. In turn, $a = a(p, s_e, s_\mu)$, so that mixing elements are determined by 6 integer parameters (here $n = 2$) which fix eigenvalues of T and W: $(m,\ \kappa_e,\ \kappa_\mu,\ p,\ s_e,\ s_\mu)$. A choice of these parameters is, however, restricted by the fact that T and W form a finite group, see Sec. 4.2 and [32]. Substituting the standard parametrization for U_{PMNS} in Eqs. (33) one obtains the two conditions that the mixing angles and the CP phase must satisfy.

If one of the charged leptons has the T-charge zero, $\kappa_\alpha = 0$ (and so two others have opposite signs), the expressions (33) simplify:

$$|U_{\alpha j}|^2 = \eta, \tag{35}$$

$$|U_{\beta j}|^2 = |U_{\gamma j}|^2 = \frac{1-\eta}{2}, \qquad \beta,\, \gamma \neq \alpha, \tag{36}$$

where

$$\eta \equiv \frac{1+a}{4\sin^2\left(\frac{\pi k}{m}\right)}. \tag{37}$$

Recall that here α corresponds to the charged lepton which is invariant under T transformation, j corresponds to S_j. All the mixing parameters are determined by single quantity η which, in turn, is the function of the group parameters $a(p)$, m, κ_e, and κ_μ. In other words η is determined by the group assignment: $\{p, m, \kappa_e, \kappa_\mu, a\}$. For small p, the parameter a is determined by p uniquely.

For finite groups we have found from (33) and (35 - 37) the following phenomenologically interesting possibilities

$$\frac{1}{6}\begin{pmatrix}4\\1\\1\end{pmatrix}, \quad \frac{1}{3}\begin{pmatrix}1\\1\\1\end{pmatrix}, \quad \frac{1}{4}\begin{pmatrix}1\\2\\1\end{pmatrix}. \tag{38}$$

The first possibility corresponds to $\{p, m, \kappa_e, \kappa_\mu, a\} = \{4, 3, 0, 1, 1\}$ and is called the trimaximal mixing-1. It can be used as the first column ($j = 1$) of the mixing matrix. The second one called the trimaximal mixing-2 can be obtained for $\{p, m, \kappa_e, \kappa_\mu, a) = (3, 3, 0, 1, 0\}$. It is viable solution for $j = 2$. Also the last possibility realized for $\{p, m, \kappa_e, \kappa_\mu, a\} = \{3, 4, 3, 0, 1\}$ with certain corrections can be used for $j = 2$. Another class of possibilities can be obtained by $(m \leftrightarrow p)$ permutation which corresponds to the exchange $T \leftrightarrow W$ in a group presentation [14].

Notice that the first as well as the second columns in (38) coincide with columns of the TBM mixing. This means that TBM is special point in the parameter space of the solutions determined by one of the columns.

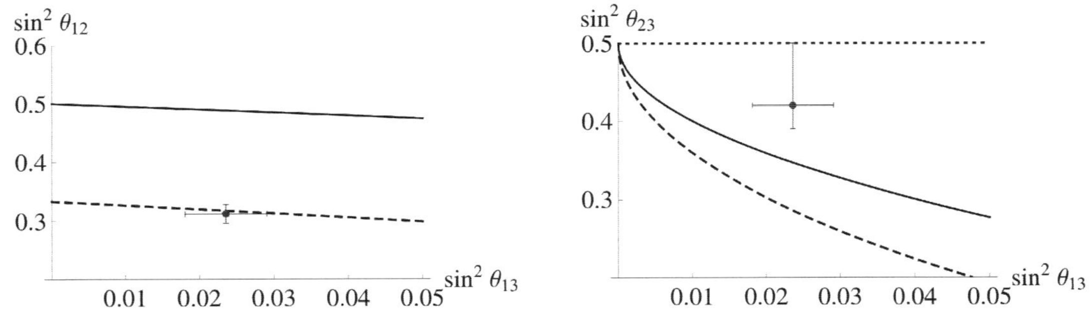

Figure 1. Symmetry relations between the mixing angles in the case of $G_f = \mathbf{S}_4$ and $G_\nu = \mathbf{Z_2}$. Shown are dependences of $\sin^2\theta_{12}$ on $\sin^2\theta_{13}$ (left panel), and $\sin^2\theta_{23}$ on $\sin^2\theta_{13}$ for $\delta = \pi$ (right pannel) for two different symmetry assignment: $\{p, m, \kappa_e, \kappa_\mu, a\} = \{4, 3, 0, 1, 1\}$ and $j = 1$ (dashed lines) and $\{p, m, \kappa_e, \kappa_\mu, a\} = \{3, 4, 0, 3, 0\}$ (solid lines). The dotted line in the right panel corresponds to $\delta = \pi/2$ for both symmetry assignments. Crosses show the experimental results.

Using the standard parameterization for the PMNS matrix one can find from (35 - 37) relations between the mixing angles and the CP - phase. Eqs. (33) and (35) impose two relations on 4 parameters. Therefore using two known mixing angles one can predicts the CP-phase and the angle which is not yet well known (e.g., θ_{23}).

As an example, we show in Fig. 1 the relations for two different symmetry assignments. The assignment $\{p, m, \kappa_e, \kappa_\mu, a\} = \{4, 3, 0, 1, 1\}$ with $j = 1$ (it corresponds to $G_f = \mathbf{S}_4$ and trimaximal-1) gives dependences which agree very well with the experimental data. Furthermore, according to the right panel of Fig. 1, one expects $\delta = 90^\circ - 120^\circ$.

4.2. Finite subgroups of the infinite von Dyck groups.

Number of possibilities which lead to finite groups is very restricted. It can be extended if we consider big (> 5) numbers m, n, p. This, however, makes group infinite, and therefore in general condition (16) can not be imposed. In order to render the group finite, an additional relation between the generating elements S_{iU} and T should be added to Eq. (16). One possibility is the relation of the form

$$X^q = 1\,, \quad X \equiv S_{jU}T^{-1}S_{jU}T \tag{39}$$

or explicitly

$$(S_{iU}T^{-1}S_{iU}T)^q = 1. \tag{40}$$

Together with (16), that would correspond to certain modular groups: $PSL(2, \mathbf{Z}_7)$ if $\{p, m\} = \{7, 3\}$ or $\{7, 4\}$, $\Delta(96)$ if $\{p, m\} = \{8, 3\}$ and $\Delta(384)$ for $\{p, m\} = \{16, 3\}$ [33].

Using the same procedure as before for the matrix W, we obtain that Eq. (39) leads to the condition

$$\mathrm{Tr}[X] = x, \tag{41}$$

where

$$x = \sum_\beta \lambda'_\beta, \qquad \lambda'^q_\beta = 1. \tag{42}$$

The equation for X gives relations between the mixing elements of the same column j (since the same S_j is involved) which have already been completely fixed by the condition for W. The relations should be consistent. Therefore the condition Eq. (39) does not add new constraints on mixing angles, but imposes constrains on the group parameters to be consistent with conditions from W. Indeed, $\mathrm{Tr}[S_{iU}T^{-1}S_{iU}T]$ (and therefore x) can be expressed in terms of $\mathrm{Tr}[W] = a$ and other group parameters as [14]

$$x = |a|^2 + a\mathrm{Tr}[T^\dagger] + a^\dagger \mathrm{Tr}[T] + A, \tag{43}$$

where

$$A = 2 + e^{i\phi_{e\mu}} + e^{-i\phi_{e\mu}} + e^{i\phi_{\mu\tau}} + e^{-i\phi_{\mu\tau}} + e^{i\phi_{\tau e}} + e^{-i\phi_{\tau e}}\,. \tag{44}$$

The equation (43) should be considered as condition for x, a and phases $\phi_\alpha = \phi_\alpha(m, k_\alpha)$. So, it is essentially a condition for the matrix T that fixes the values of k_e and k_μ and makes the group finite. Instead of k_α one can impose bounds on a or s_α. (See discussion in [32] where the bounds on these parameters have been obtained by scanning of the groups.) Imposing the second condition gives systematic way to get bounds on the group parameters

Let us consider two examples [14].

1). The group $PSL(2, \mathbf{Z}_7)$ is a subgroup of the infinite von Dyck group with the presentation (see Eq. (16))

$$S_{iU}^2 = T^7 = (S_{iU}T)^3 = \mathbf{I} \tag{45}$$

(*i.e.*, $m = 7$) and with an additional condition

$$X^4 = \mathbf{I} \tag{46}$$

(Eq. (39) with $q = 4$). For symmetry assignment $\{p, m, \kappa_e, \kappa_\mu, a\} = \{3, 7, 5, 3, 0\}$, we obtain $x = 0$ and values of the mixing elements

$$|U_{\mu j}|^2 = \frac{1}{4\left[1 + \sin\frac{\pi}{14}\right]}, \qquad |U_{\tau j}|^2 = \frac{1}{4\left[1 + \cos\frac{\pi}{7}\right]}, \tag{47}$$

and $|U_{ej}|^2 = 1 - |U_{\mu j}|^2 - |U_{\tau j}|^2$. For $j = 2$ this gives good description of the experimental data. The corresponding relations for mixing angles are shown in Fig. 2. According to the right panel within 1σ allowed region of mixing angles the CP phase equals $\delta = (80^\circ - 90^\circ)$.

2). The group $\Delta(384)$ is a subgroup of the infinite von Dyck group with presentation

$$S_{iU}^2 = T^{16} = (S_{iU}T)^3 = \mathbf{I}\,, \tag{48}$$

and additional relation

$$X^3 = \mathbf{I}. \tag{49}$$

Again, we have $x = 0$. Viable mixing pattern can be obtained in the case of $W \leftrightarrow T$ permuted version, when W is the symmetry of the charged leptons and $T = S_{iU}W$ gives the symmetry group condition. Now $p' = m$, $m' = p$, $a' = \mathrm{Tr}[T]$ and $\kappa_\alpha \to \xi_\alpha$. For the set of the group parameters

$$\{p', m', s_e, s_\mu, a'\} = \left\{16, 3, 1, 1, \frac{-1+i}{\sqrt{2}}\right\} \tag{50}$$

we obtain

$$\{|U_{ej}|^2,\, |U_{\mu j}|^2,\, |U_{\tau j}|^2\} = \left\{\frac{4+\sqrt{2}+\sqrt{6}}{12},\, \frac{4+\sqrt{2}-\sqrt{6}}{12},\, \frac{2-\sqrt{2}}{6}\right\}. \tag{51}$$

If $j = 3$, this model is able to fit all mixing angles for $\delta \sim 50^\circ - 60^\circ$.

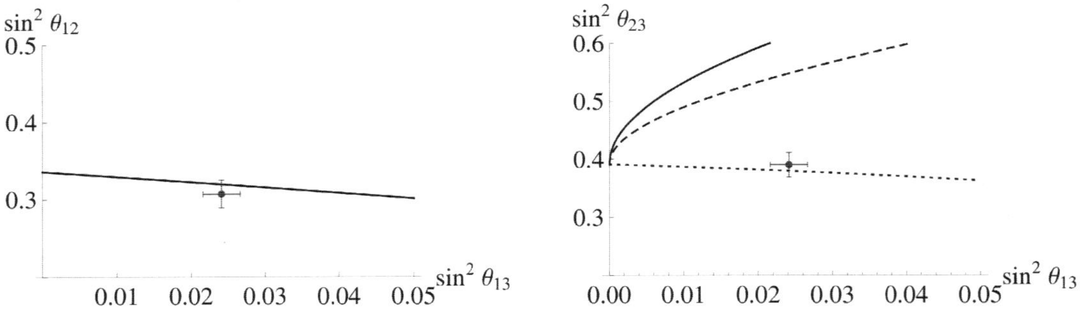

Figure 2. Symmetry relations between mixing angles in the case of $G_f = PSL(2, \mathbf{Z}_7)$ and $G_\nu = \mathbf{Z_2}$. *Left panel:* dependence of $\sin^2\theta_{12}$ on $\sin^2\theta_{13}$, *right panel:* dependence of $\sin^2\theta_{23}$ on $\sin^2\theta_{13}$ for $\delta = 0$ (solid line), $\delta = \pi/4$ (dashed line) and $\delta = \pi/2$ (dotted lines). The symmetry assignment is $\{p, m, \kappa_e, \kappa_\mu, a\} = \{3, 7, 5, 3, 0\}$ and $j = 2$. Crosses show the experimental results.

4.3. $G_\nu = \mathbf{Z}_2 \otimes \mathbf{Z}_2$.

Let us consider the complete generic symmetry given by the Klein group $G_\nu = \mathbf{Z}_2 \otimes \mathbf{Z}_2$. A presentation of the group is given (for definiteness we take S_1, S_2 as generating elements) by

$$S_{1U}^2 = S_{2U}^2 = T^m = W_{1U}^{p_1} = W_{2U}^{p_2} = \mathbf{I}, \quad [S_{1U}, S_{2U}] = 0. \tag{52}$$

They impose two sets of conditions on U_{PMNS} which can be written as

$$\mathrm{Tr}[W_{jU}] = \sum_\alpha e^{i\phi_\alpha}(2|U_{\alpha j}|^2 - 1) = a_j\,, \qquad j = 1,\, 2, \tag{53}$$

where $a_j \equiv \sum_k \lambda_k^{(j)}$ are sums of three p_j-th roots of unity. Now there are four relations between the mixing matrix elements, corresponding to two columns of the mixing matrix, which determine U_{PMNS} completely [14]. An example of this case is given by two columns in Eq. (38) with $\{p_1, p_2, m\} = \{4, 3, 3\}$. The only mixing matrix compatible with first two columns in (38) is the TBM. In a sense the TBM is special being related to maximal generic neutrino symmetry group.

Complete scan of the discrete finite groups with order less than 1536 and $G_\nu = \mathbf{Z}_2 \otimes \mathbf{Z}_2$ has been performed in [34] and for these groups the mixing pattens has been computed [4].

The formalism allows to explain an observation [33] that in some cases, the mixing matrix derived from Eqs. (53) has the property that the absolute values of entries of two column were equal up to a permutation:

$$|U_{\alpha i}|^2 = |U_{f(\alpha) j}|^2, \quad i \neq j. \tag{54}$$

Here $f(\alpha)$ is the permutation operation of the flavor indices. An example, which is realized in the case of $PSL(2,\, \mathbf{Z}_7)$ group is

$$|U_{\alpha i}|^2 = \begin{pmatrix} c_1 & c_2 & c_3 \\ c_3 & c_1 & c_2 \\ c_2 & c_3 & c_1 \end{pmatrix} \tag{55}$$

[4] The approach in [34] is, however, different from the one presented here: the starting point in [34] is the explicit form of the generators of the residual symmetries for leptons and neutrinos in the basis of diagonal CC interaction. The generators satisfy presentation of the full selected flavor group. The generators are diagonalized by rotations Ω_ℓ and Ω_ν. Then, as can be shown, the mixing matrix is given by $\Omega_\ell^\dagger \Omega_\nu$. See also discussion in [32].

with

$$c_2 = \frac{1}{4(1+\sin\frac{\pi}{14})}, \qquad c_3 = \frac{1}{4(1+\cos\frac{\pi}{7})} \qquad c_1 = 1 - c_2 - c_3. \tag{56}$$

4.4. Non-generic neutrino symmetry.

We can introduce a non-generic neutrino symmetry G_ν which includes transformation, Q_m, under which neutrino mass matrix in the mass basis is invariant only for some specific values of masses. Inversely, introduction of such a symmetry in the flavor symmetry imposes conditions on neutrino masses. The symmetry group condition for Q_m reads $(Q_{mU} \cdot T)^r = 1$, where r is integer and $Q_{mU} = U_{PMNS} Q_m U_{PMNS}^\dagger$ is the symmetry transformation in the flavor basis. Also some of generic symmetries $\mathbf{Z}_2$ could be a part of G_f. In the latter case one should impose also the relation $(Q_{mU} \cdot S_U)^q = (Q_m \cdot S)^q = 1$. The mixing matrix disappears in the last equality and therefore it should be considered as a consistency condition.

5. Conclusion

1. The appealing discrete flavor symmetry framework (inspired by TBM) is based on idea that the lepton mixing originates from different ways of the discrete flavor symmetry breaking in the neutrino and charged lepton Yukawa sectors. These different ways lead to different residual symmetries of the neutrino and charged lepton mass matrices which ensure special forms of these matrices, and consequently, special form of the mixing matrix. Here the symmetry transformation are generic transformations valid for arbitrary masses.

2. Recent measurements of the neutrino oscillation parameters show substantial deviations from the TBM mixing. This may further indicate that TBM is accidental and whole discrete flavor symmetry approach is phenomenologically irrelevant. Experimental value of the 1-3 mixing can be connected with other observables in various ways which have different implications to theory. It is not clear however if these relations imply discrete symmetries: some – certainly not, although symmetry effects can be rather hidden.

Further developments in the field can be related to establishing the neutrino mass hierarchy. "Race" for the mass hierarchy and CP has started: studies of the atmospheric neutrinos with multi-megaton mass detectors having low energy thresholds can provide fast, inexpensive and reliable answer.

3. Discrete symmetries still may play important role in formation of the lepton flavor structures. TBM can be treated as the lowest order structure which requires corrections. Discrete symmetries can be consistent with the non-zero 1-3 mixing and deviation of the 2-3 mixing from maximal.

4. Consequences of symmetries for mixing, and in general, for masses in this framework can be obtained immediately from the *symmetry group condition(s)*, without explicit model building. For this the knowledge of (assumption about) the residual symmetries and the covering group is enough. The symmetry group condition includes the PMNS matrix and the generating elements of the residual symmetries in the mass basis.

From the point of view of specific models, here it is assumed that the relevant model-building has already been done and the mass matrices of neutrinos and charged leptons with certain symmetries obtained.

5. The symmetry group conditions have been applied to different residual neutrino symmetries and full flavor groups. (i) The generic $G_\nu = \mathbf{Z}_2$ with finite von Dyck group as well as finite subgroup of the infinite von Dyck group as G_f lead to two relations between the mixing parameters. This fixes one of the columns of mixing matrix. (ii) The Klein group $\mathbf{Z}_2 \times \mathbf{Z}_2$ as G_ν fixes the mixing matrix completely. (iii) The formalism can be generalized to include non-generic neutrino symmetries G_ν which lead to relations between mixings and neutrino masses.

6. The relations between mixing parameters obtained here include the CP-phase and therefore the phase can be predicted using measured values of mixing angles. In the examples we discussed δ can be around 60° or 90°. So, future measurements of the phase can provide important test of the framework.

7. The formalism presented here can be used in various ways. In specific model, once the residual symmetries are identified the consequences on mixing and masses can be obtained immediately. The formalism allows to analyze systematically possible mixing patterns that can be extracted from the finite groups or subgroups of the von Dyck type. Inversely, for a given pattern of mixing the formalism allows to perform the "symmetry building" - identify the residual and covering symmetry groups.

The formalism allows also to explain various features of mixing matrix which follow from symmetries. In particular, in the case of Klein group in the neutrino sector it can explain appearance of mixing patterns in which two or three columns of U_{PMNS} have equal but permuted elements.

8. Further generalizations of the formalism are possible. On the other hand, discrete symmetries can be realized in some other way in frameworks which differs from (2). Mixing (or its zero order structure) can originate from different nature of the mass terms of the charged leptons (Dirac) and neutrinos (Majorana), or from neutrino mixing with new degrees of freedom (e.g., singlets of SM).

5.1. Acknowledgments

Large part of this talk is based on papers written with in collaboration with D. Hernandez. I am grateful to C. Hagedorn, A. Pilaftsis and R. Mohapatra for useful discussions during the Symposium.

References

[1] S. Pakvasa and H. Sugawara, Phys. Lett. B **73** (1978) 61. G. C. Branco, Phys. Lett. B **76** (1978) 70. Y. Yamanaka, H. Sugawara and S. Pakvasa, Phys. Rev. D **25** (1982) 1895 [Erratum-ibid. D **29** (1984) 2135]; T. Brown, S. Pakvasa, H. Sugawara and Y. Yamanaka, Phys. Rev. D **30** (1984) 255.

[2] G. Altarelli, F. Feruglio, Rev. Mod. Phys. **82** (2010) 2701; S. F. King and C. Luhn, Rept. Prog. Phys. **76** (2013) 056201 [arXiv:1301.1340 [hep-ph]]; H. Ishimori, T. Kobayashi, H. Ohki, Y. Shimizu, H. Okada, M. Tanimoto, Prog. Theor. Phys. Suppl. **183** (2010) 1. [arXiv:1003.3552 [hep-th]]; P. O. Ludl, arXiv:0907.5587 [hep-ph].

[3] P. F. Harrison, D. H. Perkins and W. G. Scott, *Phys. Lett.* B **458**, 79 (1999), *Phys. Lett.* B **530**, 167 (2002); L. Wolfenstein, *Phys. Rev.* D **18**, 958 (1978).

[4] C. S. Lam, *Phys. Rev.* D **78**, 073015 (2008); C. S. Lam, Phys. Lett. **B656** (2007) 193; C. S. Lam, Phys. Rev. Lett. **101** (2008) 121602; C. S. Lam, [arXiv:1003.0498 [hep-ph]]; W. Grimus, L. Lavoura and P. O. Ludl, *J. Phys.* G **36**, 115007 (2009); W. Grimus and L. Lavoura, *JHEP* **0904**, 013 (2009).

[5] A. Y. Smirnov, J. Phys. Conf. Ser. **335** (2011) 012006 [arXiv:1103.3461 [hep-ph]].

[6] K. Abe *et al.* [T2K Collaboration], Phys. Rev. Lett. **107** (2011) 041801 [arXiv:1106.2822 [hep-ex]]. K. Abe *et al.* [T2K Collaboration], arXiv:1304.0841 [hep-ex].

[7] F. P. An *et al.* [DAYA-BAY Collaboration], arXiv:1203.1669 [hep-ex], F. P. An *et al.* [Daya Bay Collaboration], Chin. Phys. C **37** (2013) 011001 [arXiv:1210.6327 [hep-ex]].

[8] J. K. Ahn *et al.* [RENO Collaboration], Phys. Rev. Lett. **108** (2012) 191802 [arXiv:1204.0626 [hep-ex]].

[9] Y. Abe *et al.* [DOUBLE-CHOOZ Collaboration], Phys. Rev. Lett. **108** (2012) 131801 [arXiv:1112.6353 [hep-ex]]. Y. Abe *et al.* [Double Chooz Collaboration], arXiv:1301.2948 [hep-ex].

[10] P. Adamson *et al.* [MINOS Collaboration], Phys. Rev. Lett. [arXiv:1301.4581 [hep-ex]].

[11] G. Altarelli, F. Feruglio, L. Merlo and E. Stamou, JHEP **1208** (2012) 021 [arXiv:1205.4670 [hep-ph]]. M. -C. Chen, J. Huang, J. -M. O'Bryan, A. M. Wijangco and F. Yu, JHEP **1302** (2013) 021 [arXiv:1210.6982 [hep-ph]].

[12] S. F. Ge, D. A. Dicus and W. W. Repko, Phys. Rev. Lett. **108** (2012) 041801 [arXiv:1108.0964 [hep-ph]]. S. F. Ge, D. A. Dicus and W. W. Repko, Phys. Lett. B **702** (2011) 220 [arXiv:1104.0602 [hep-ph]].

[13] D. Hernandez and A. Y. Smirnov, Phys. Rev. D **86** (2012) 053014 [arXiv:1204.0445 [hep-ph]].

[14] D. Hernandez and A. Y. Smirnov, arXiv:1212.2149 [hep-ph].

[15] D. Hernandez and A. Y. Smirnov, arXiv:1304.7738 [hep-ph].

[16] G. L. Fogli, E. Lisi, A. Marrone, D. Montanino, A. Palazzo and Phys. Rev. D **86** (2012) 013012 [arXiv:1205.5254 [hep-ph]].
[17] M. C. Gonzalez-Garcia, M. Maltoni, J. Salvado and T. Schwetz, JHEP **1212** (2012) 123 [arXiv:1209.3023 [hep-ph]].
[18] D. V. Forero, M. Tortola and J. W. F. Valle, Phys. Rev. D **86** (2012) 073012 [arXiv:1205.4018 [hep-ph]].
[19] P. Adamson *et al.* [MINOS Collaboration], arXiv:1304.6335 [hep-ex].
[20] E. K. Akhmedov, G. C. Branco and M. N. Rebelo, Phys. Rev. Lett. **84** (2000) 3535 [hep-ph/9912205]. W. Rodejohann, M. Tanimoto and A. Watanabe, Phys. Lett. B **710** (2012) 636 [arXiv:1201.4936 [hep-ph]].
[21] A. Y. Smirnov, arXiv:hep-ph/0402264; M. Raidal, *Phys. Rev. Lett.* **93** 161801 (2004); H. Minakata and A. Y. Smirnov, *Phys. Rev.* D **70**, 073009 (2004).
[22] S. F. King, Phys. Lett. B **718** (2012) 136 [arXiv:1205.0506 [hep-ph]].
[23] C. Giunti and M. Tanimoto, Phys. Rev. D **66** (2002) 113006 [hep-ph/0209169].
[24] X. Zhang and B. -Q. Ma, Phys. Lett. B **710** (2012) 630 [arXiv:1202.4258 [hep-ph]], Y. Zhang, X. Zhang and B. -Q. Ma, Phys. Rev. D **86** (2012) 093019 [arXiv:1211.3198 [hep-ph]].
[25] A. de Gouvea and H. Murayama, arXiv:1204.1249 [hep-ph], G. Altarelli, F. Feruglio, I. Masina and L. Merlo, JHEP **1211** (2012) 139 [arXiv:1207.0587 [hep-ph]].
[26] J. Barry, W. Rodejohann and H. Zhang, JHEP **1107** (2011) 091 [arXiv:1105.3911 [hep-ph]].
[27] R. B. Patterson [NOvA Collaboration], arXiv:1209.0716 [hep-ex].
[28] A. S. Dighe and A. Y. Smirnov, Phys. Rev. D **62** (2000) 033007 [hep-ph/9907423]. C. Lunardini and A. Y. Smirnov, Nucl. Phys. B **616** (2001) 307 [hep-ph/0106149].
[29] C. Lunardini and A. Y. Smirnov, JCAP **0306** (2003) 009 [hep-ph/0302033].
[30] D.J. Koskinen, Mod.Phys.Lett. A26 (2011) 2899.
[31] E. K. Akhmedov, S. Razzaque and A. Y. Smirnov, JHEP **02** (2013) 082 [arXiv:1205.7071 [hep-ph]].
[32] C. S. Lam, arXiv:1301.3121 [hep-ph].
[33] R. de Adelhart Toorop, F. Feruglio and C. Hagedorn, Nucl. Phys. B **858** (2012) 437 [arXiv:1112.1340 [hep-ph]].
[34] M. Holthausen, K. S. Lim and M. Lindner, Phys. Lett. B **721** (2013) 61 [arXiv:1212.2411 [hep-ph]].

Journal of Physics: Conference Series **447** (2013) 012005 doi:10.1088/1742-6596/447/1/012005

Time reversal violation for entangled neutral mesons

J Bernabéu

Department of Theoretical Physics, University of Valencia, and IFIC, Joint Centre Univ. Valencia-CSIC

Jose.Bernabeu@uv.es

Abstract. A direct evidence for Time Reversal Violation (TRV) means an experiment that, considered by itself, clearly shows TRV independent of, and unconnected to, the results for CP Violation. No existing result before the recent BABAR experiment with entangled neutral B mesons had demonstrated TRV in this sense. There is a unique opportunity for a search of TRV with unstable particles thanks to the Einstein-Podolsky-Rosen (EPR) Entanglement between the two neutral mesons in B, and PHI, Factories. The two quantum effects of the first decay as a filtering measurement and the transfer of information to the still living partner allow performing a genuine TRV asymmetry with the exchange of "in" and "out" states. With four independent TRV asymmetries, BABAR observes a large deviation of T-invariance with a statistical significance of 14 standard deviations, far more than needed to declare the result as a discovery. This is the first direct observation of TRV in the time evolution of any system.

1. Introduction

I was asked by the organizers to report about the recent discovery of Time Reversal Violation (TRV) in the Entangled Neutral B-system and the conceptual basis for such observation. This subject has had an important impact in the scientific community and media. Under the title "The arrow of time", The Economist devoted its central page on Science and Technology in the number of September 1st 2012 [1] to TRV, as shown in figure 1. In particle physics one expects that not all processes run in the same

Figure 1. Illustration of "The Economist" for discussing the observed asymmetry for motion reversal of a neutral B-meson in a transition between two definite states: going forwards is thus not the same as going backwards.

way forwards in time as they do backwards. The direct observation of this phenomenon by the BABAR Collaboration was reported [2] in November 2012 with a high significance result, and the journals Nature [3] and Physics Today [4] presented it stating, as shown in figures 2 and 3, that

Figure 2. Nature 491,640(2012) reports that a cornerstone of theoretical particle physics -the idea that not all processes run in the same way forwards in time as they do backwards- has been observed directly for the first time.

Figure 3. This report in Physics Today emphasizes that, in transitions between neutral B states not connected by CP, BABAR finds transition rates that depend on temporal direction in a way that can only be attributed to T violation.

TRV has finally been clearly seen. Additionally, Physics World revealed its top breakthroughs for physics in 2012 and the first three results appear to be [5] the Higgs-like boson discovery at CERN, a Majorana fermion excitation in solid state physics and Time-Reversal Violation, as shown in figure 4.

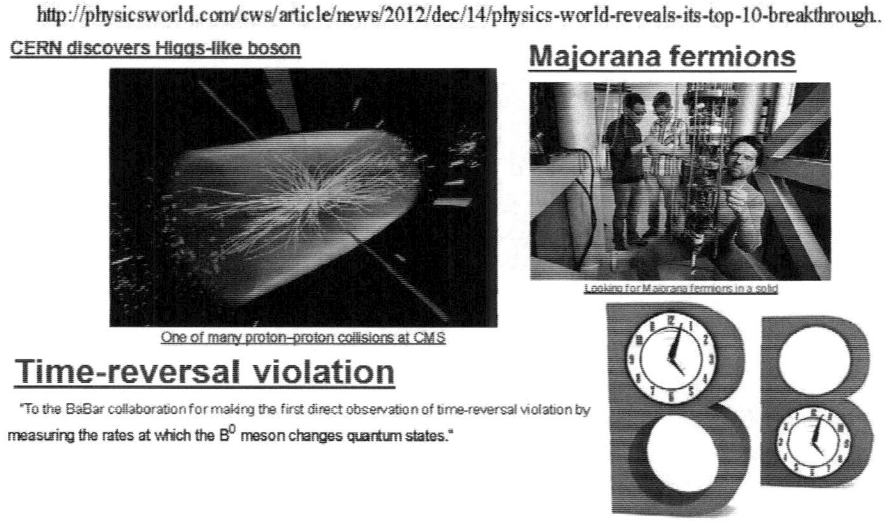

Figure 4. Together with the Physics World award for the 2012 Breakthrough of the Year to the discovery of a Higgs-like particle at the LHC, the journal includes the first evidence of the elusive Majorana fermion in a solid and the first direct observation of Time Reversal Violation.

The main point associated to a genuine test of T-invariance is that one needs an interchange between in-states and out-states for a given process, a request particularly difficult to accomplish for particles that decay. The interest in the unstable neutral meson $K^0 \to \overline{K}^0$ and $B^0 - \overline{B}^0$ systems stems from the fact that Violation of CP invariance (CPV) has been observed for them [6]. By virtue of the CPT theorem [7], as imposed by any local quantum field theory with Lorentz invariance and Hermiticity, one then expects that T violation should also appear in those systems. A direct evidence for TRV would mean an experiment that, considered by itself, clearly shows TRV independent of, and unconnected to, the results for CPV. No existing result before the BABAR experiment [2] had clearly demonstrated TRV in this sense. For particles in a decaying state, T-transformation is not defined, because the image under T is not a physical state. It looks like the decay prevents a true test of T-symmetry. What is the conceptual basis that is able to bypass this argument? Everything started with the papers [8] that used the quantum Einstein-Podolsky-Rosen (EPR) [9] Entanglement to transfer the information, lost in the irreversible decay of one particle, to its still living orthogonal partner. This correlation allows the preparation, in the quantum mechanical sense, of the state of the second neutral meson as a filtering measurement does. These ideas were scrutinized by several authors, including Wolfenstein [10], Quinn [11] and many others, with the conclusion that "it appears to be a true TRV effect".

The original ideas and calculations [8] were more recently transformed into a definite experimental proposal [12] and its feasibility was demonstrated by a full simulation using the realistic statistics available in the B Factories. In this presentation I discuss the conceptual basis, the methodology and the experimental results leading to the direct evidence of TRV in the time evolution of the neutral B meson in the interval between the two decays to definite flavour and definite CP of the decay products, for the entangled B meson system produced by the Υ(4S) decay. In Section 2 the fundamental distinction between the "arrow of time" for complex systems and the TRV in the fundamental laws of Physics is discussed. Section 3 is devoted to the problematics of unstable systems to be used for a test of T-symmetry, and its solution by means of the EPR-Entanglement for the transfer of information to the still living meson at the time of the first decay. The foundations of the experimental analysis are presented in Section 4, whereas the results and the T-violating asymmetries are given in Section 5. Section 6 summarizes the main conclusions.

2. T symmetry in the laws of physics

We are interested in Microscopic T-symmetry Violation. Effects in particle physics odd under the change of sign of time $t \leftrightarrow -t$ are not necessarily T-violating. Theses observables can occur in theories with exact T-symmetry and are called T-odd effects, like those induced by absorptive components of the transition amplitude. On the other hand, for complex physical systems, well known time asymmetries are the Universe t-asymmetry and the macroscopic t-asymmetry called the "arrow of time". But none of these t-asymmetries is a test of TRV in the fundamental Laws of Physics.

T-Violation exists in the Standard Model or any field theoretic extension of it. All local quantum field theories with Lorentz invariance and Hermiticity satisfy the CPT theorem [7], so that they establish an automatic connection between CP Violation (CPV) and a related TRV. CPV has been observed in the $K^0 \to \overline{K}^0$and $B^0 - \overline{B}^0$ systems [6]. All experimental results are in agreement with the Standard Cabibbo-Kobayashi-Maskawa (CKM) mechanism in the ElectroWeak Theory. As a consequence, one also expects to have TRV in these neutral meson systems. How to observe it? T and CPT symmetries are implemented by Antiunitary Operators in Quantum Mechanics, with the algebraic commutation rules left invariant. **A genuine TRV Observable means an Asymmetry under the interchange of in $\leftrightarrow$ out states**. The antiunitary character, rather than unitary, introduces many intriguing subtleties.

Figure 5. This Map tracing the oldest light in the sky has been produced by Europe's Planck Surveyor satellite [13].

There is no doubt that the Universe is expanding, even accelerating at present cosmological age. This natural t-asymmetry t $\leftrightarrow$ -t is perfectly compatible with fundamental laws of physics that are Time Reversal-symmetric. It is due to the initial condition for our Universe, like Inflation. This asymmetry is similar to the fact that in our Universe we have a privileged reference frame, the one associated with the Cosmic Microwave Background (CMB) radiation at a definite temperature with fluctuations. In figure 5 we show the result that allows fundamental measurements of cosmological parameters, compatible with the ΛCDM Model. The CMB radiation has a thermal black body spectrum at a temperature of 2.725 K, as shown in Figure 6. The spectrum peaks in the microwave range frequency of 160.2 BHz, corresponding to a 1.9 mm wavelength, in the intensity per unit frequency. This privileged reference frame in our Universe does not mean a violation of Lorentz invariance in the fundamental laws of physics. Similarly for the Universe time-asymmetry: it does not mean a violation of Time Reversal invariance in the fundamental laws of physics.

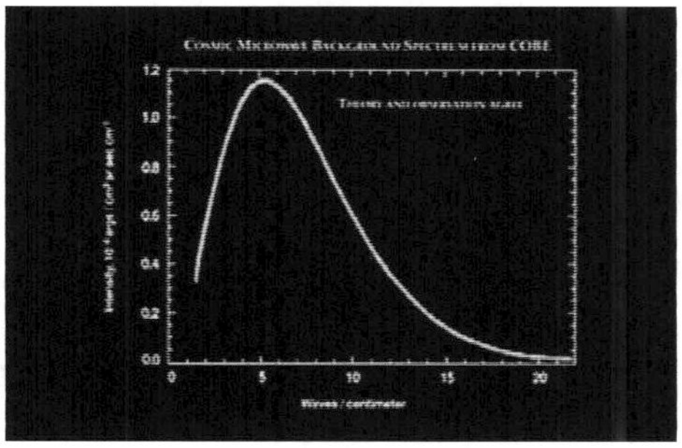

Figure 6. Graph of cosmic microwave background spectrum, the most precisely measured black body spectrum in nature.

For complex systems, their t-asymmetric behaviour is in the nature of Thermodynamics. As discussed by Eddington [14], figure 7, the "arrow of time" is a property of entropy alone, as a measurement of the disorder. It says that time is asymmetric with respect to the amount of order in isolated systems. **The arrow indicates the direction of progressive increase of the random element**. There is probably a connection between the cosmological t-asymmetry and the "arrow of time" for complex systems: it would say that the initial condition in the evolution of the Universe was more ordered and thus highly improbable.

Figure 7. The "arrow of time" is a concept developed in 1927 by Arthur Eddington involving the one-way direction of asymmetry of time. At the macroscopic level there is an obvious flow of time generated by entropy alone.

None of these t-asymmetries is related to a fundamental TRV of the physical laws.

3. Direct evidence for TRV in unstable particles?

A direct evidence for TRV means an experiment that, considered by itself, clearly shows TRV independent of, and unconnected to, the results for CPV. Two types of experiments can do it:

1) A non-zero expectation value of a T-odd operator for a non-degenerate stationary state. This is the case for an electric dipole moment, which is a P-odd, C-even, T-odd quantity. It can be generated by either strong T-violation, with the non-perturbative θ-term $\varepsilon_{\mu\nu\tau\sigma} F^{\mu\nu} F^{\tau\sigma}$ of the tensor gluon field with its dual, unless it is rotated away by a Peccei-Quinn symmetry [15] leaving the axion as remnant, or by T-violation in weak interactions. In the standard model, with the CKM mechanism, a non-vanishing electric dipole moment of the neutron only appears to three loop amplitudes. The present experimental status is summarized in [16].
2) For transitions, as discussed before, the antiunitary character of the T-operator demands an asymmetry under the in ↔ out exchange, i.e., the comparison between $<f|S|i>$ and $<-i|S|-f>$, where $|-i>$, $|-f>$ indicate the T-transformed of $|i>$, $|f>$ respectively. Transitions in the $K^0 \to \overline{K}^0$ and $B^0 - \overline{B}^0$ systems have demonstrated the existence of CPV. It is then natural to search for TRV in those systems.

The Kabir asymmetry $K^0 \to \overline{K}^0$ vs $\overline{K}^0 \to K^0$ was measured in 1998 by the CPLEAR experiment [17] at CERN with a non-vanishing value and a statistical significance near 4 standard deviations. But the interpretation of this observable as a direct evidence of TRV has generated some controversy. It is based on several facts associated with this flavour-flavour transition:

- Taking $K^0 \to \overline{K}^0$ as Reference, and calling (X,Y) the observed decays at times t_1 and t_2, with $\Delta t = t_2 - t_1 > 0$, which tag the flavour for initial and final states, the CP, T and CPT transformed transitions, as well as the Δt-reversal of the observed decays, are given in Table 1:

Table 1. Symmetry transformations applied to $K^0 \to \overline{K}^0$

Transition	$K^0 \to \overline{K}^0$	$\overline{K}^0 \to K^0$	$\overline{K}^0 \to K^0$	$K^0 \to \overline{K}^0$	$K^0 \to \overline{K}^0$
(X,Y)	(l^-, l^-)	(l^+, l^+)	(l^+, l^+)	(l^-, l^-)	(l^-, l^-)
Transformation	Reference	CP	T	CPT	Δt

As a consequence, assuming that T can be defined for these transitions, there is no way to distinguish it from CP, they are experimentally identical. Even if CPT is violated, there is no way to observe it in this transition.

- The Kabir asymmetry, as a consequence of being associated to flavour mixing transitions, needs the interference of CP mixing $im_{\not{CP}}$ with the "initial state interaction" to generate the effect which is proportional to ΔΓ, as shown in the diagrams of figure 8. **The decay plays then an essential role**. For decaying states, the T-operator is not defined: its time reverse is not a physical state [18].

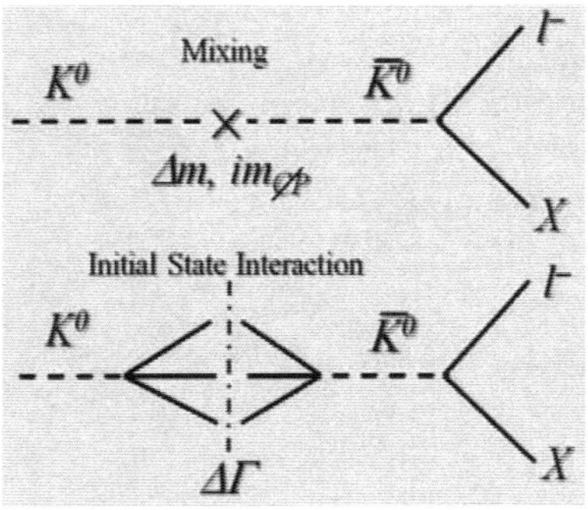

Figure 8. The CP mixing asymmetry for $K^0 \to \overline{K}^0$ is generated by the interference of CP-violation in the mass matrix with the "initial state interaction" given by ΔΓ.

- The time evolutions of $K^0 \to \overline{K}^0$ and $\overline{K}^0 \to K^0$ are in fact equal and the asymmetry becomes time independent. In the Weisskopf-Wigner approach [19], the entire effect comes from the overlap of the non-orthogonal "stationary" K_L, K_S states. If the stationary states were orthogonal, the asymmetry vanishes.

All these arguments hint to the claim [18] by Wolfenstein that "it is not as direct a test of TRV as one might like".

The previous reasoning is not ruling out all tests of T-symmetry for particles that decay, as long as $\Delta\Gamma$ is not needed for the observable asymmetry. This is the situation when the interference which leads to an asymmetry is taking place for amplitudes with and without mixing. The corresponding transition is associated with flavour and CP decay products. But still we have the problem of the in↔out exchange required for a genuine test of T-symmetry. The fact that particles decay looks like it prevents performing such an exchange.

The solution arises [8] from the Quantum Mechanical Entanglement imposed by the EPR correlation between the neutral mesons produced in the B, or Φ, Factories. This idea will give us the opportunity to have separate tests of CP, T and CPT-symmetries, depending on the selection of the decay channel.

The $B^0 - \overline{B}^0$ EPR-Entanglement is due to Particle Identity, because $B^0\ \overline{B}^0$ are two states of a unique (complex) field. These two states are connected by charge conjugation C, so that the symmetry requirement for bosons implies $C\mathcal{P} = +$, where $\mathcal{P}$ is the permutation operation. In B-Factories, $B^0\ \overline{B}^0$ are produced by the Υ(4S)-decay, with J = 1, S = 0, L = 1 quantum numbers. For this particle-antiparticle system, C = - is well defined, so we conclude that $\mathcal{P} = -$: the wave function has to be antisymmetric.

At the moment of the $\Upsilon \to B^0\ \overline{B}^0$ decay, the neutral meson system is in the state

$$|\,i\rangle = \frac{1}{\sqrt{2}}\left[B^0(t_1)\overline{B}^0(t_2) - \overline{B}^0(t_1)B^0(t_2)\right] \tag{1}$$

where the states 1, 2 are defined by the time ordering of the decays, with $t_1 < t_2$. The times t_1 and t_2 in equation (1) are not time dependences, but labels to characterize the states. The antisymmetric character of equation (1) is crucial for the time evolution of $|i>$ when the mixing $B^0 - \overline{B}^0$ is included. In spite of the mixing, only terms $B^0\ \overline{B}^0$ appear at any time. This behaviour is perfect for a Flavour-Tag mechanism. As it is well known, the observation of the decay of one meson to a positive lepton l^+, for example, at time t_1, signals that the (still living) partner meson state at this time is the B-state not decaying to l^+, i.e., $\overline{B}^0$. The decay has filtered the B^0-state, and the orthogonal state $\overline{B}^0$ is then tagged at time t_1. For $t > t_1$, we have a single state time evolution from $\overline{B}^0$.

But, for the entangled state of the two mesons, the individual state of each neutral meson is not defined before its collapse as a filter imposed by the observation of the decay. This quantum mechanical entanglement was in fact argued by EPR [9] as being against local realism and, as such, that the quantum theory was not "complete". This epistemological EPR "paradox" was later converted to physics by John Bell and it is now the foundation of quantum information and quantum computing.

One can rewrite the same $|i>$ state in equation (1) in terms of any other pair of orthogonal states of the individual neutral B-mesons: a linear combination of B^0 and $\overline{B}^0$, and its orthogonal. One may considerer the states B_+ and B_- of the neutral mesons, where B_+ is the state not decaying to the decay product $J/\psi\ K_+$ and K_+ is the neutral K-meson filtered by its decay to $\pi\pi$. The orthogonal state B_- is thus the neutral B-meson filtered by the decay to the CP = - final state. The observation of the decay to this CP-eigenstate at time t_1 generates an automatic transfer of information to the (still living) partner meson. We may call the quantum preparation of the initial state at t_1, using the filter imposed by a first observation of this decay, a "CP-tag" [20]. The same entangled state is then better to write it as

$$|\,i\rangle = \frac{1}{\sqrt{2}}\left[B_+(t_1)B_-(t_2) - B_-(t_1)B_+(t_2)\right] \tag{2}$$

The main question is now: If B_- is the B-state filtered by the CP = - decay J/ψ K_+, what is the orthogonal state B_+ experimentally? For these CP-eigenstate decay products, the condition to filter a definite state is [12] that the decay amplitude has a single weak phase. The state B_+ is that filtered by the CP = + decay in the same system J/ψ K_-, where K_- is the neutral K-meson filtered by its decay to $\pi^0\pi^0\pi^0$. If, for the B-system, one neglects the small CP-violation of the K-system, one can associate B_- with J/ψ K_S decay and B_+ with J/ψ K_L decay.

This last association is the foundation of the experimental performance of the Time-Reversal Transformation for a transition of the neutral B-meson between a first flavour decay and a second decay to a CP-eigenstate. This T-transformation is illustrated in figure 9 for the comparison between $\overline{B}^0 \underrightarrow{\Delta\tau} B_-$ and $B_- \underrightarrow{\Delta\tau} \overline{B}^0$ transitions. As seen, for the first transition one has to observe the decays (l^+, J/ψ K_S) in this time ordering, whereas the T-reversed transition corresponds to (J/ψ K_L, l^-) for the two decays. This nontrivial T-reversal is thus not given by the t-reversal exchanging the two decays. Entanglement has been essential for the quantum preparation of the initial state of the neutral B-meson in one transition and its T-reverse transition. The problem of particle unstability for a T-symmetry test has been bypassed. Experimentally, we need a very good time resolution for disentangling the ordering of the two decays.

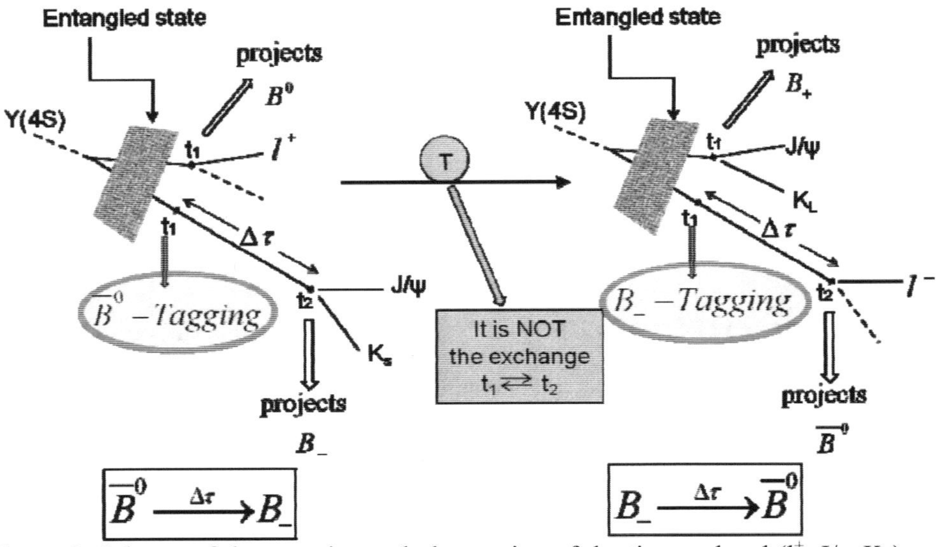

Figure 9. Scheme of the experimental observation of the time ordered (l^+, J/ψ K_S) and (J/ψ K_L, l^-) decays for performing the study of the time evolution of the transition $\overline{B}^0 \underrightarrow{\Delta\tau} B_-$ and its T-reverse $B_- \underrightarrow{\Delta\tau} \overline{B}^0$.

4. Genuine Observables not needing ΔΓ

We may now proceed to a partition of the complete set of events into four categories, defined by the tag in the first decay at time $t_1 : B_+, B_-, B^0$ or $\overline{B}^0$, so we have eight different Decay-Intensities at our disposal, as function of $\Delta t = t_2 - t_1 > 0$. Each one of these eight Flavour-CP transitions has an Intensity given by

$$\mathrm{I}_i(\Delta\tau) \sim e^{-\Gamma\Delta\tau}\{ C_i \cos(\Delta m \Delta\tau) + S_i \sin(\Delta m \Delta\tau) + \mathrm{C'}_i \cosh(\Delta\Gamma\Delta\tau) + S'_i \sinh(\Delta\Gamma\Delta\tau) \} \quad (3)$$

where Γ is the average width.

For a genuine test of a symmetry, one has to compare the I_i (Δt) of a transition with its transformed by the symmetry operation. For the case of T, one builds the in ↔ out exchange asymmetry.

We notice that the Intensities (3) contain terms independent of $\Delta\Gamma$, in such a way that these asymmetries, contrary to Kabir symmetry, do not need a non-vanishing $\Delta\Gamma$ associated to the decay properties. In fact, the interference in the time evolution between the two decays is built between mixing and no mixing.

Up to now, for CPV analyses in B-factories, BABAR & BELLE had assumed CPT-invariance and $\Delta\Gamma = 0$. In this case, there is a theorem which is operating [21]: Then $\Delta t \leftrightarrow -\Delta t$ exchange, i.e., the exchange of the two decay products at t_1 and t_2, which is not a T-symmetry operation, becomes equivalent to T in the sense that one invariance implies the other. In this case, as CP ~T ~ Δt are theoretically connected, only two independent Intensities remain to be compared. Alternatively, one may establish $S_i \neq 0$ for a single transition, proving that it is Δt-asymmetric. The coefficients C_i and S_i are then related for the eight transitions. In the Standard Model (SM), the CKM Mixing Matrix [22] conveys the fact that the quarks with definite properties under charged current weak interactions are linear combinations of the quark mass eigenstates. For three families, the unitarity conditions are represented by triangles in the complex plane. For the $B^0 - \overline{B}^0$ system, the unitarity triangle is given by

$$V_{ud}\,V_{ub}^{*} + V_{cd}\,V_{cb}^{*} + V_{td}\,V_{tb}^{*} = 0 \tag{4}$$

and its consistency with all existing experimental results on Flavour Mixing and CP Violation is shown in Figure 10

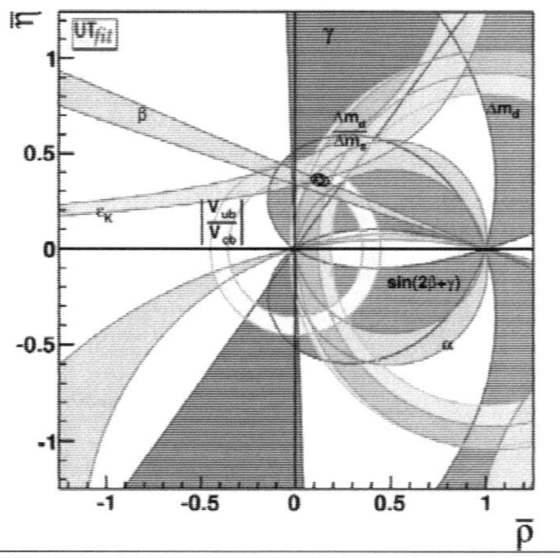

Figure 10. The intersection of different experimental inputs from Flavour Mixing and CP-violation, analyzed within the SM unitarity triangle, provides a consistent fit.

For our selected transitions with the associated interference of Mixing x Decay, with the two decays to Flavour and CP eigenstates for $b \to c\bar{c}s$, all the 8 S_i coefficients in equation (3) are related to a single value $\sin(2\beta) = 0.67 \pm 0.02$, where β is the CP-phase between the t- and c-sides of the unitarity triangle (4).

The SM connection is not followed here for the study of observables and, in a general context using quantum mechanics only, the eight Intensities (3) are experimentally independent. We proceed now to study the effects of the three symmetry operations CP, T and CPT separately:

1) Take $B^0 \to B_+$ as the Reference transition and call (X,Y) the observed decay products at times t_1 and t_2, respectively. The CP, T and CPT transformed transitions are given in Table 2, together with the so-called Δt-operation (not a symmetry!) exchanging X↔Y

Table 2. Symmetry Transformations applied to $B^0 \to B_+$

Transition	$B^0 \to B_+$	$\overline{B}^0 \to B_+$	$B_+ \to B^0$	$B_+ \to \overline{B}^0$	$B_- \to \overline{B}^0$
(X,Y)	(l^-, J/ΨK_L)	(l^+ ,J/ΨK_L)	(J/ΨK_S, l^+)	(J/ΨK_S, l^-)	(J/ΨK_L, l^-)
Transformation	Reference	CP	T	CPT	Δt

As you may check, all transitions are experimentally independent. It is important to point out that the two sets of events, called sometimes "Δt > 0" and "Δt < 0", for the same two decay products l^- and J/ψK_L, experimentally included in the same sample of events, are not connected by any symmetry.

2) Take $B^0 \to B_-$ as the Reference transition. The CP, T and CPT transformed transitions are given in Table 3

Table 3. Symmetry Transformation applied to $B^0 \to B_-$

Transition	$B^0 \to B_-$	$\overline{B}^0 \to B_-$	$B_- \to B^0$	$B_- \to \overline{B}^0$	$B_+ \to \overline{B}^0$
(X,Y)	(l^-, J/ΨK_S)	(l^+ ,J/ΨK_S)	(J/ΨK_L, l^+)	(J/ΨK_L, l^-)	(J/ΨK_S, l^-)
Transformation	Reference	CP	T	CPT	Δt

Therefore a second Asymmetry for each of the 3 symmetry transformations can be built. Again the result of the Δt-operation is a different transition from the other four transitions connected by the symmetries.

3) Select now as Reference the $B_- \to \overline{B}^0$ transition, obtained by the choice (Y, X) of decay products in the Reference 1) and proceed with the symmetry transformations.
4) Select as Reference the $B_+ \to \overline{B}^0$ transition, obtained by the choice (Y, X) of decay products in the Reference 2), and proceed with the symmetry transformations.

We thus conclude that there are 4 Model-Independent Asymmetries for each of the 3 symmetry transformations CP, T and CPT. They are now explicitly written as

-Genuine CPV-Asymmetries

$$
\begin{aligned}
A_{CP,1} &= \frac{\Gamma(l^-, J/\Psi K_L) - \Gamma(l^+, J/\Psi K_L)}{+} \\
A_{CP,2} &= \frac{\Gamma(l^-, J/\Psi K_S) - \Gamma(l^+, J/\Psi K_S)}{+} \\
A_{CP,3} &= \frac{\Gamma(J/\Psi K_L, l^-) - \Gamma(J/\Psi K_L, l^+)}{+} \\
A_{CP,4} &= \frac{\Gamma(J/\Psi K_S, l^-) - \Gamma(J/\Psi K_S, l^+)}{+}
\end{aligned}
\qquad (5)
$$

(Δt connects $A_{CP,1}$ with $A_{CP,3}$, and $A_{CP,2}$ with $A_{CP,4}$.)

-Genuine TRV-Asymmetries

$$\begin{aligned}
A_{T,1} &= \frac{\Gamma(l^-, J/\Psi K_L) - \Gamma(J/\Psi K_S, l^+)}{+} \\
A_{T,2} &= \frac{\Gamma(l^-, J/\Psi K_S) - \Gamma(J/\Psi K_L, l^+)}{+} \\
A_{T,3} &= \frac{\Gamma(J/\Psi K_L, l^-) - \Gamma(l^+, J/\Psi K_S)}{+} \\
A_{T,4} &= \frac{\Gamma(J/\Psi K_S, l^-) - \Gamma(l^+, J/\Psi K_L)}{+}
\end{aligned} \tag{6}$$

-Genuine CPTV-Asymmetries

$$\begin{aligned}
A_{CPT,1} &= \frac{\Gamma(l^-, J/\Psi K_L) - \Gamma(J/\Psi K_S, l^-)}{+} \\
A_{CPT,2} &= \frac{\Gamma(l^-, J/\Psi K_S) - \Gamma(J/\Psi K_L, l^-)}{+} \\
A_{CPT,3} &= \frac{\Gamma(l^+, J/\Psi K_L) - \Gamma(J/\Psi K_S, l^+)}{+} \\
A_{CPT,4} &= \frac{\Gamma(l^+, J/\Psi K_S) - \Gamma(J/\Psi K_L, l^+)}{+}
\end{aligned} \tag{7}$$

In equations (5), (6) and (7) the asymmetries connected by Δt-exchange are pointed out.

These Asymmetries in the time dependent decay rates for any pair of symmetry-conjugated transitions would be apparent through differences between the corresponding coefficients S_i or C_i in equation (3). In the analysis for the $B^0 - \overline{B}^0$system, we will take $\Delta\Gamma = 0$. In our notation for $S^{\pm}_{\alpha,\beta}(C^{\pm}_{\alpha,\beta})$, α will indicate the flavour decay l^+ or l^- and β the CP-eigenstate decay J/ψ K_L or J/ψ K_S. The superindex "+" is for time ordering (Flavour, CP), whereas "-" is for the opposite time ordering of the two decays. For T-symmetry, a measure of TRV in the time evolution between the two decays is given by the asymmetry parameters

$$\begin{aligned}
\Delta S_T^+ &= S^-_{l^-,K_L} - S^+_{l^+,K_S} \qquad \Delta C_T^+ = C^-_{l^-,K_L} - C^+_{l^+,K_S} \\
\Delta S_T^- &= S^+_{l^-,K_L} - S^-_{l^+,K_S} \qquad \Delta C_T^- = C^+_{l^-,K_L} - C^-_{l^+,K_S}
\end{aligned} \tag{8}$$

Similarly for the asymmetry parameters $\Delta S^{\pm}_{CP}\,(\Delta C^{\pm}_{CP})$ and $\Delta S^{\pm}_{CPT}\,(\Delta C^{\pm}_{CPT})$ which measure CPV and CPTV, respectively. The $\Delta S^{\pm}, \Delta C^{\pm}$ parameters for the three symmetries CP, T and CPT are represented in figure 11 on top of the Intensities for the 8 independent transitions we are considering. These transitions are characterized by the flavour $l^{\pm}$, the CP eigenstate J/ψ K_S (K_L) and the time ordering Δt>0 (Δt<0).

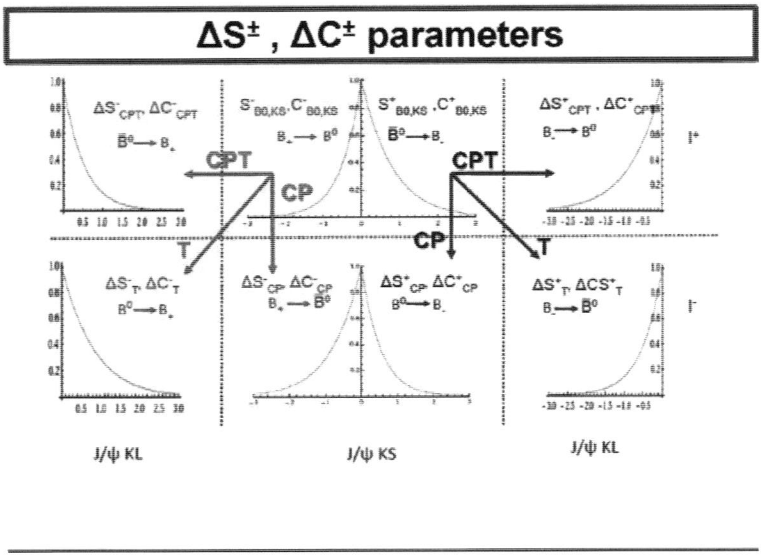

Figure 11. The eight Intensities, obtained for different flavours $l^{\pm}$, different CP eigenstates J/ψ K_S (K_L) and different sign of Δt, are connected by the symmetry transformations CP, T and CPT. There are two sets of asymmetry parameters.

One should notice that a genuine test of T implies the comparison of: 1) "Opposite Δt sign", i.e., in↔out; 2) Different CP eigenstates, J/ψ K_L vs. J/ψ K_S; and 3) Opposite flavor states, $B^0 vs. \bar{B}^0$.

In the SM, all 8 coefficients are related as a consequence of CPT invariance and $\Delta\Gamma = 0$, and given by the value of sin(2β)

$$S = S^+_{l^+,K_S} = -S^+_{l^-,K_S} = -S^-_{l^+,K_S} = S^-_{l^-,K_S} = -S^+_{l^+,K_L} = S^+_{l^-,K_L} = S^-_{l^+,K_L} = -S^-_{l^-,K_L} \approx 0.67 \tag{9}$$

$$C = C^+_{l^+,K_S} = -C^+_{l^-,K_S} = C^-_{l^+,K_S} = -C^-_{l^-,K_S} = C^+_{l^+,K_L} = -C^+_{l^-,K_L} = C^-_{l^+,K_L} = -C^-_{l^-,K_L} \approx 0$$

5. Experimental Results

The details of the experimental analysis by the BABAR Collaboration may be consulted in reference [2]. For the $\bar{B}^0 \to B_-$ transition, neglecting reconstruction effects, we have for the TRV asymmetry

$$A_T(\Delta t) \approx \frac{\Delta S_T^+}{2}\sin(\Delta m \Delta t) + \frac{\Delta C_T^+}{2}\cos(\Delta m \Delta t) \tag{10}$$

The other three T-violating asymmetries can be written similarly. Figure 12 shows the four observed asymmetries, overlaid with the projection of the best fit results to the Δt distribution with and without the eight T-invariance restrictions: $\Delta S_T^{\pm} = \Delta C_T^{\pm} = 0$, $\Delta S_{CP}^{\pm} = \Delta S_{CPT}^{\pm}$ and $\Delta C_{CP}^{\pm} = \Delta C_{CPT}^{\pm}$.

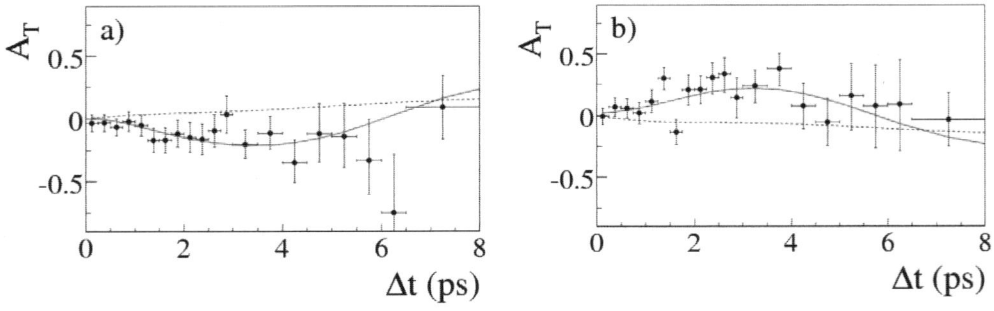

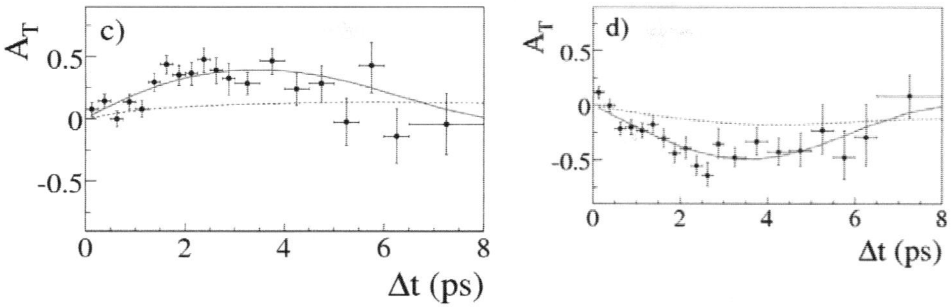

Figure 12. Experimental results for the four raw time dependent asymmetries associated with TRV.

The measured values of the T, CP and CPT-asymmetry parameters are given in Table 4, together with the values of reference coefficients of the time dependent intensities (at the bottom).

Table 4. Measured values of the asymmetry parameters{ $\Delta S^{\pm}, \Delta C^{\pm}$ } for each of the three symmetries T, CP and CPT.

Parameter	Final result	SM expected val.
ΔS_T^+	$-1.37 \pm 0.14 \pm 0.06$	-1.34
ΔS_T^-	$1.17 \pm 0.18 \pm 0.11$	1.34
ΔC_T^+	$0.10 \pm 0.14 \pm 0.08$	0.
ΔC_T^-	$0.04 \pm 0.14 \pm 0.08$	0.
ΔS_{CP}^+	$-1.30 \pm 0.11 \pm 0.07$	-1.34
ΔS_{CP}^-	$1.33 \pm 0.12 \pm 0.06$	1.34
ΔC_{CP}^+	$0.07 \pm 0.09 \pm 0.03$	0.
ΔC_T^-	$0.08 \pm 0.10 \pm 0.04$	0.
ΔS_{CPT}^+	$0.16 \pm 0.21 \pm 0.09$	0.
ΔS_{CPT}^-	$-0.03 \pm 0.13 \pm 0.06$	0.
ΔC_{CPT}^+	$0.14 \pm 0.15 \pm 0.07$	0.
ΔC_{CPT}^-	$0.03 \pm 0.12 \pm 0.08$	0.
$S_{\ell+,K_S^0}^+$	$0.55 \pm 0.09 \pm 0.06$	0.67
$S_{\ell+,K_S^0}^-$	$-0.66 \pm 0.06 \pm 0.04$	-0.67
$C_{\ell+,K_S^0}^+$	$0.01 \pm 0.07 \pm 0.05$	0.
$C_{\ell+,K_S^0}^-$	$-0.05 \pm 0.06 \pm 0.03$	0.

The first uncertainty is statistical and the second systematic. In the last column, the SM expected value is also given, with such a good precision that its error is well below the present experimental uncertainty.

The significance of the T-violation signal is obtained from the CL contours shown in figure 13

Journal of Physics: Conference Series **447** (2013) 012005 doi:10.1088/1742-6596/447/1/012005

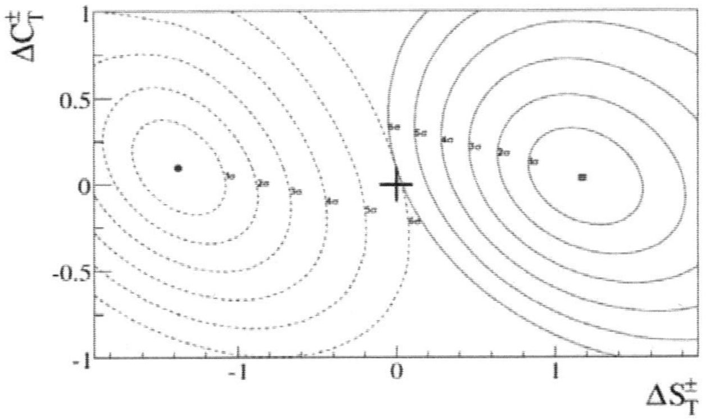

$$\Delta S_T^+ = -1.37 \pm 0.14 \pm 0.06$$
$$\Delta S_T^- = 1.17 \pm 0.18 \pm 0.11$$
$$\Delta C_T^+ = 0.10 \pm 0.16 \pm 0.08$$
$$\Delta C_T^- = 0.04 \pm 0.16 \pm 0.08$$

Figure 13. Confident Level (CL) contours for the T-asymmetry parameters { $\Delta S_T^{\pm}, \Delta C_T^{\pm}$ }

These CL contours are given in two dimensions for the T-asymmetry parameters $(\Delta S_T^+, \Delta C_T^+)$ and $(\Delta S_T^-, \Delta C_T^-)$. Assuming Gaussian errors, the result corresponds to a significance equivalent to 14 standard deviations, and thus constitutes a direct observation of T violation.

The significance of CP and CPT violation is determined analogously from the CL contours shown in figure 14,

$$\Delta S_{CP}^+ = -1.30 \pm 0.10 \pm 0.07$$
$$\Delta S_{CP}^- = 1.33 \pm 0.12 \pm 0.06$$
$$\Delta C_{CP}^+ = 0.07 \pm 0.10 \pm 0.03$$
$$\Delta C_{CP}^- = 0.08 \pm 0.09 \pm 0.04$$

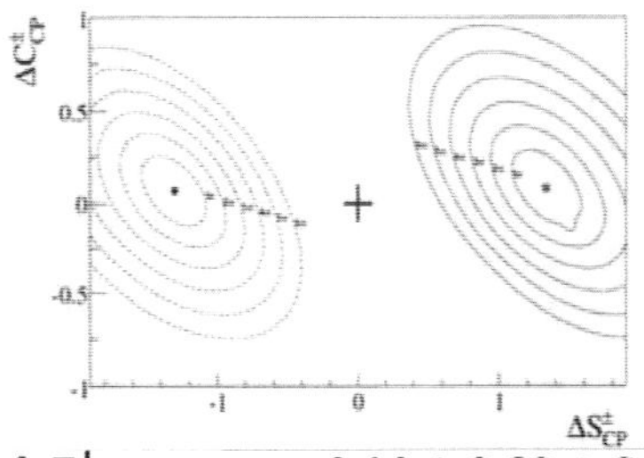

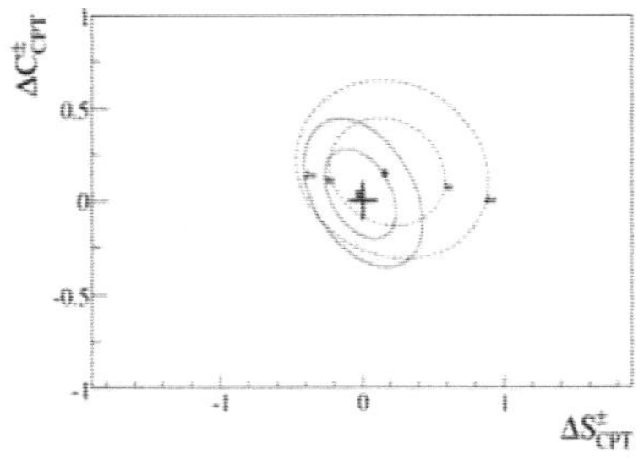

$$\Delta S_{CPT}^+ = 0.16 \pm 0.20 \pm 0.09$$
$$\Delta S_{CPT}^- = -0.03 \pm 0.13 \pm 0.06$$
$$\Delta C_{CPT}^+ = 0.15 \pm 0.17 \pm 0.07$$
$$\Delta C_{CPT}^- = 0.03 \pm 0.14 \pm 0.08$$

Figure 14. Confident Level (CL) contours for the CP and CPT asymmetry parameters { $\Delta S^{\pm}, \Delta C^{\pm}$ }

obtaining a result equivalent to 17 and 0.3 standard deviations, respectively, consistent with CP violation and CPT invariance.

6. Conclusion

We have discussed the conceptual basis, the methodology and the experimental result for a direct evidence of Time-Reversal Violation in the fundamental laws of physics. The observed time-asymmetries in complex systems, like de Arrow of Time, are not T-violating, but a property of Entropy alone.

A genuine TRV means an asymmetry under the exchange in $\leftrightarrow$ out of states in a transition, a requirement which looks impossible to be satisfied for particles that decay. The interest in the unstable $B^0 - \overline{B}^0$ (and $K^0 - \overline{K}^0$) system originates in the important CP violating effects observed in the Mixing x Decay interference, so that the CPT theorem predicts a related T-violation in those systems.

A unique opportunity for bypassing the problem of T-symmetry tests in unstable particles is provided by the Einstein-Podolsky-Rosen Entanglement between the two neutral mesons in B, and Φ, factories. The information transfer from the first decay, used as a filtering measurement, to the still living partner allows a quantum mechanical preparation, the flavour or CP tag, of the appropriate state for performing the T-symmetry study.

Using the channels for the two decays to Flavour and CP eigenstates, we find 8 different Intensities for the time evolution of the neutral B-meson between the two decays. In appropriate combinations, each CP, T, CPT symmetry can be tested separately using 4 genuine independent Asymmetries between the time dependent Decay Rates. These results have been expressed in terms of independent Asymmetry Parameters $\Delta S^{\pm}, \Delta C^{\pm}$ for each symmetry transformation.

BABAR has measured the time dependent Asymmetries and has extracted the Asymmetry parameters. The experimental result shows a large deviation of T-invariance with a significance of 14 standard deviations, far more than needed to declare a Discovery. In turn, the results are consistent with CPT invariance in the time evolution of the neutral B-meson between the two decays, connecting CPV and TRV in different transitions.

This experimental result constitutes a direct evidence of Time Reversal Violation in the time evolution of the neutral B-meson.

This discovery was made possible thanks to the spectacular quantum properties of EPR entangled states: the reality of two entangled B's is much more than the sum of two separate B local realities.

Acknowledgments
DISCRETE 2012 was a great scientific event. I would like to thank the Organizers for their careful preparation of this succesful Symposium and for the opportunity to discuss the subject of TRV with all participants. This work was supported by the funded Projects MINECO FPA 2011/23596 and Generalitat Valenciana PROMETEO 2008/004 and GVISIC 2012/020.

References

[1] "The arrow of time. Backward ran sentences...", *The Economist*, Sept. 1st 2012, http://www.economist.com/node/21561111
[2] Lees JP et al. BABAR Collaboration *Phys. Rev. Lett.* **109**(2012)211801.
[3] Nature Research Highlights: "Time's arrow in B mesons", *Nature* **491**(2012)640, Nov 2012, http://www.nature.com/nature/journal/v491/n7426/full/491640a.html
[4] "Time Reversal asymmetry in particle physics has finally been clearly seen" *Physics Today* Nov 2012 pag 16 http://www.physicstoday.org/resource/1/phtoad/v65/i11/p16_s1
[5] "Physics World reveals its top 10 breakthroughs for 2012" Dec 14 2012 http://physicsworld.com/cws/article/news/2012/dec/14/physics-world-reveals-its-top-10-breakthroughs-for-2012
[6] Christenson JH et al. *Phys. Rev. Lett.* **13**(1964)138
Aubert B et al. BABAR Collaboration *Phys. Rev. Lett.* **87**(2001)091801
Abe K et al. (Belle Collaboration) *Phys. Rev. Lett.* **87**(2001)091802
[7] Luders G *Annals Phys* **2**(1957)1
Pauli W in "Niels Bohr and the Development of Physics" Pergamon Press New York 1995 p30
[8] Bañuls MC and Bernabéu J *Phys. Lett. B* **464**(1999) 117
Nucl. Phys. B **590**(2000)19
[9] Einstein A, Podolsky B and Rose N *Phys. Rev.* **47**(1935)777
[10] Wolfenstein L *Int. J. Mod. Phys.* **8**(1999)501
[11] Quinn HR *Proc. DISCRETE'08 J. Phys. Conf. Serv.* **171**:011001(2009)

Bernabéu J *Proc. DISCRETE2010 J. Phys. Conf. Ser.* **335**:012011(2011)
[12] Bernabéu J, Martínez-Vidal F, Villanueva-Pérez P *JHEP* **1208**(2012)064
[13] Ade PAR et al. Planck Collaboration *Astronomy & Astrophysics* manuscript no. PlanckMission2013, March 22, 2013
[14] See Callender C "Thermodynamic Asymmetry in Time" The Stanford Encyclopedia of Philosophy E.N. Zalta ed (2008)
[15] Peccei RD and Quinn HR *Phys. Rev. Lett.* **38**(1977)1440
Phys. Rev. D **16**(1977)1791
[16] See Rathmann F article in this DISCRETE2012 Proc.
[17] Angelopoulos A et al. CPLEAR Collaboration *Phys. Lett. B* **444**(1998)43
Kabir PK *Phys. Rev. D* **2**(1970)540
[18] Wolfenstein L *Phys. Rev. Lett.* **83**(1999)911
[19] Weisskoph VF and Wigner EP *Z. Phys.* **63**(1930)54
ibid **65**(1930)18
[20] Bañuls MC and Bernabéu J, *JHEP* **9906**(1999)032
[21] Bernabéu J and Bañuls MC TAUP 99 Paris *Nucl. Phys. Proc. Suppl.* **87**(2000)315
Bernabéu J and Espinoza C Venice 2009 Neutrino Telescopes p.159
[22] Cabibbo N *Phys. Rev. Lett.* **10**(1963)531
Kobayashi M and Maskavo T Prog. Theor. Phys. **49**(1973)652

Journal of Physics: Conference Series **447** (2013) 012006 doi:10.1088/1742-6596/447/1/012006

Dark Matter Indirect searches: phenomenological and theoretical aspects

Marco Cirelli

Institut de Physique Théorique, CNRS, URA 2306 & CEA/Saclay, F-91191 Gif-sur-Yvette, France

E-mail: marco.cirelli@cea.fr

Abstract. I present a phenomenological, data-centered and data-driven assessment of the status of indirect searches for particle Dark Matter, with the main focus on some recent "anomalies" and of the main implications for DM model building. Tantalizing hints and stringent constraints seem to continuosly chase each other, as a number of well performing experiments keep delivering their data. This year may be the one in which we finally see the resolution of these dark puzzles, or it may not.

[Report number: SACLAY-T13/021]

1. Introduction

Cosmology and astrophysics provide several convincing evidences of the existence of Dark Matter (DM). The observation that some mass is missing to explain the internal dynamics of galaxy clusters and the rotations of galaxies dates back respectively to the '30s and the '70s. The observations from weak lensing, for instance in the spectacular case of the so-called 'bullet cluster', provide evidence that there is mass where nothing is optically seen. More generally, global fits to a number of cosmological datasets (Cosmic Microwave Background, Large Scale Structure and also Type Ia Supernovae) allow to determine very precisely the amount of DM in the global energy-matter content of the Universe at $\Omega_{\rm DM}h^2 = 0.1123 \pm 0.0035$ [1][1].

All these signals pertain to the gravitational effects of Dark Matter at the cosmological and extragalactical scale. Searches for explicit manifestation of the DM particles that are supposed to constitute the halo of our own galaxy (and the large scale structures beyond it) have instead so far been giving negative results, but this might be on the point of changing.

Indirect searches for Dark Matter aim at detecting the signatures of the annihilations or decays of DM particles in the fluxes of Cosmic Rays (CRs), intended in a broad sense: charged particles (electrons and positrons, protons and antiprotons, deuterium and antideuterium), photons (gamma rays, X-rays, synchrotron radiation), neutrinos. In general, a key point of all these searches is to look for channels and ranges of energy where it is possible to beat the background from ordinary astrophysical processes. This is for instance the basic reason why searches for charged particles focus on fluxes of antiparticles (positrons, antiprotons,

[1] Here $\Omega_{\rm DM} = \rho_{\rm DM}/\rho_c$ is defined as usual as the energy density in Dark Matter with respect to the critical energy density of the Universe $\rho_c = 3H_0^2/8\pi G_N$, where H_0 is the present Hubble parameter. h is its reduced value $h = H_0/100$ km s^{-1}Mpc^{-1}.

antideuterons), much less abundant in the Universe than the corresponding particles, and searches for photons or neutrinos have to look at areas where the DM-signal to astro-noise ratio can be maximized.

Pioneering works have explored indirect detection (ID) as a promising avenue of discovery since the late-70's. Since then, innumerable papers have explored the predicted signatures of countless particle physics DM models. In the past 3 years or so, however, the field has experienced a significant burst of activity, mainly due to the results presented by a few very well performing experiments, above all the PAMELA satellite, the FERMI satellite and the HESS telescope. It is fair to say that the field has passed, for better or for worse, from a theory-driven state to a data-driven phase.

2. Charged cosmic rays and the PAMELA, FERMI, HESS anomaly

2.1. Data

There has been a flurry of positive results from a few indirect detection experiments looking at the fluxes of charged cosmic rays. In particular, the signals pointed to an excess of electrons and positrons at the TeV and sub-TeV scale:

- Data from the PAMELA satellite [2] showed a steep increase in the energy spectrum of the positron fraction $e^+/(e^+ + e^-)$ above 10 GeV up to 100 GeV, compatibly with previous hints from HEAT [3] and AMS-01 [4].
- Recently, these findings have been confirmed with an independent measurement by the FERMI satellite [5], and extended to about 200 GeV.
- Data from PAMELA [6] also showed no excess in the $\bar{p}/p$ energy spectrum compared with the predicted background.
- The balloon experiments ATIC-2 [7] and PPB-BETS [8] were reporting the presence of a peak in the $e^+ + e^-$ energy spectrum at around 500-800 GeV.
- This sharp feature has been later questioned and superseded by the results of the FERMI satellite [9]: while an excess with respect to the expected background is confirmed, the $e^+ + e^-$ spectrum is found to be instead reproduced by a simple power law.
- The HESS telescope also reports the measurement of the $e^+ + e^-$ energy spectrum above energies of 600 GeV [10], showing a power law spectrum in agreement with the one from FERMI and eventually a steepening at energies of a few TeV.

The data are displayed in fig. 1, together with the expected astrophysical 'backgrounds'. The latter ones are uncertain and are an interesting subject of study by themselves in CR physics, but I will not discuss them further here. The signals presented above are striking because they imply the existence of a source of 'primary' e^+ (and e^-) other than the ordinary astrophysical ones. This unknown new source can well be itself of astrophysical nature, e.g. one or more pulsar(s) / pulsar wind nebula(æ), supernova remnants etc [11]. It is however very tempting to try and read in these 'excesses' the signature of DM.

2.2. Dark Matter interpretation

As already mentioned above, the DM particles that constitute the DM halo of the Milky Way are expected to annihilate into pairs of primary SM particles (such as $b\bar{b}$, $\mu^+\mu^-$, $\tau^+\tau^-$, W^+W^- and so on) which, after decaying and through the processes of showering and hadronizing, give origin to fluxes of energetic cosmic rays: $e^-, e^+, \bar{p}$ (and also γ-rays, ν...), denoted dN_f/dE. Depending on which one has been the primary SM particle, the resulting spectra differ substantially in the details. Generically, however, they feature a 'bump'-like shape, characterized by a high-energy cutoff at the DM particle mass and, for $e^\pm$ in particular, a softly decreasing tail at lower energies (see e.g. the examples in fig. 2). It is thus clear that it is very natural to expect a DM source to

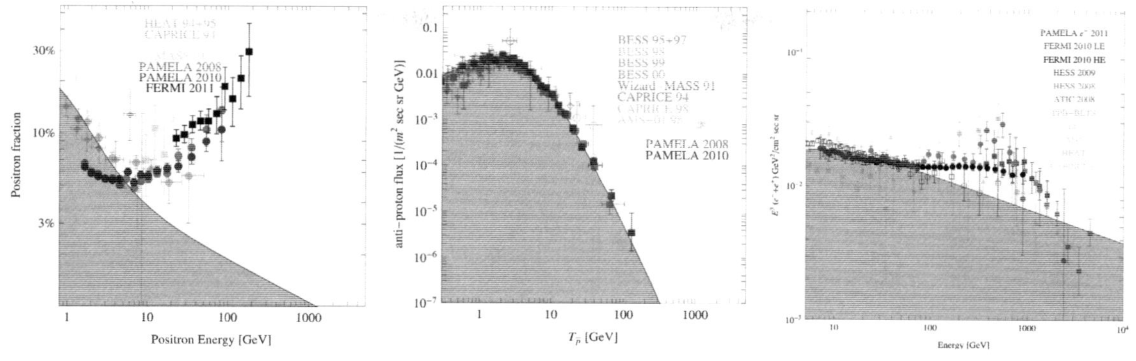

Figure 1. A compilation of recent and less recent data in charged cosmic rays, superimposed on plausible but uncertain astrophysical backgrounds from secondary production. Left: positron fraction. Center: antiproton flux. Right: sum of electrons and positrons.

'kick in' on top of the secondary background and explain the $e^\pm$ excesses. The energy range, in particular, is tantalizingly right: the theoretically preferred TeV-ish DM would naturally give origin to TeV and sub-TeV bumps and rises.

The e^-, e^+ and $\bar{p}$ produced in any given point of the halo propagate immersed in the turbulent galactic magnetic field. The field consists of random inhomogeneities that act as scattering centers for charged particles, so that their journey can effectively be described as a diffusion process from an extended source (the DM halo) to some final given point (the location of the Earth, in the case of interest). While diffusing, charged CRs experience several other processes, and in particular energy losses due to synchrotron radiation, Inverse Compton Scattering (ICS) on the low energy photons of the CMB and starlight, Coulomb losses, bremsstrahlung, nuclear spallations.... Quantitatively, the steady-state number density $n_f(\vec{x}, E)$ per unit energy E of the cosmic ray species f $(= e^+, e^-, \bar{p})$ in any given point $\vec{x}$ obeys to a diffusion-loss equation [12]

$$\begin{aligned} - \mathcal{K}(E) \cdot \nabla^2 n_f - \frac{\partial}{\partial E} \left(b(E, \vec{x})\, n_f \right) &+ \frac{\partial}{\partial z} \left(\mathrm{sign}(z)\, V_{\mathrm{conv}}\, n_f \right) \\ &= Q(E, \vec{x}) - 2h\, \delta(z)\, \Gamma\, n_f\, . \end{aligned} \quad (1)$$

The first term accounts for diffusion, with a coefficient conventionally parameterized as $\mathcal{K}(E) = \mathcal{K}_0 (E/\mathrm{GeV})^\delta$. The second term describes energy losses: the coefficient b is position-dependent since the intensity of the magnetic field (which determines losses due to synchrotron radiation) and the distribution of the photon field (which determines losses due to ICS) vary across the galactic halo. The third term deals with convection while the last term accounts for nuclear spallations, that occur with rate Γ in the disk of thickness $h \simeq 100$ pc. The different processes described above have a different importance depending on the particle species: the journey of electrons and positrons is primarily affected by synchrotron radiation and inverse Compton energy losses, while for antiprotons these losses are negligible and convection and spallation dominate.

The source, DM annihilations, is given by $Q = 1/2\, (\rho(\vec{x})/m_{\rm DM})^2 \sum_i \mathrm{BR}_i \langle \sigma v \rangle\, (dN^i_f/dE)$, where $m_{\rm DM}$ is the DM mass, σv is the total annihilation cross section and the sum runs over all primary channels i in which the cosmic ray species f is produced. $\rho(\vec{x})$ is the DM density distribution in the galactic halo. What to adopt for the latter is another one of the main open problems in the field. Based on the results of increasingly more refined numerical simulations or on direct observations, profiles that differ even by several orders of magnitude at the Galactic Center are

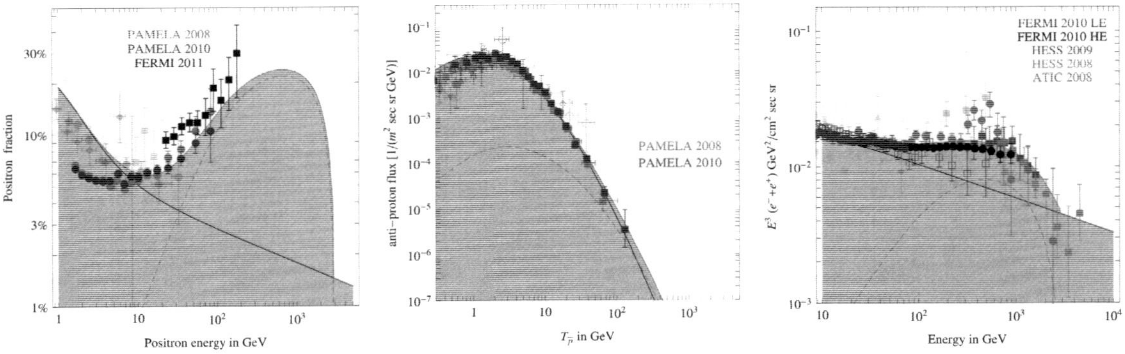

Figure 2. Charged cosmic ray data interpreted in terms of Dark Matter annihilations: the flux from the best fit DM candidate (a 3 TeV DM particle annihilating into $\tau^+\tau^-$ with a cross section of $2 \cdot 10^{-22}\,\mathrm{cm}^3/\mathrm{sec}$) is the lower dashed line and is summed to the supposed background, giving the pink flux which fits the data. Left, center and right like in fig. 1.

routinely adopted: e.g. the classical Navarro-Frenk-White (NFW) or the Einasto one, which exhibit a cusp at the galactic center, or the truncated isothermal or the Burkert one, which feature a central core. All profiles, on the other hand, are roughly normalized at the same value at the location of the Earth (≈ 0.3 GeV/cm^3). These features generically imply that observables which depend mostly on the local DM density (for instance, the flux of high energy positrons, which cannot come from far away due to energy losses) will not be very affected by the choice of profile, while those that are sensitive to the density at the GC will be affected the most (e.g. gamma rays observations of regions close to the GC).
Eq. (1) is usually solved numerically in a diffusive region with the shape of a solid flat cylinder that sandwiches the galactic plane, with height $2L$ in the z direction and radius $R = 20\,\mathrm{kpc}$ in the r direction. The location of the solar system corresponds to $\vec{x}_\odot = (r_\odot, z_\odot) = (8.33\,\mathrm{kpc}, 0)$. Boundary conditions are imposed such that the number density n_f vanishes on the surface of the cylinder, outside of which the charged cosmic rays freely propagate and escape. The values of the propagation parameters δ, K_0, V_{conv} and L are deduced from a variety of (ordinary) cosmic ray data and modelizations.

The datasets listed in Sec. 2 pin-point the properties of the DM particle needed to interpret them in terms of annihilations quite precisely. The DM has to be:

- ▷ With a mass of a *few TeV*, in order to reproduce the feature in the $e^+ + e^-$ spectrum.
- ▷ *Leptophilic*, i.e. annihilating almost exclusively into leptonic channels, otherwise the antiproton measurements would be exceeded.
- ▷ With a *very large annihilation cross section*, of the order of $10^{-23}\,\mathrm{cm}^3/\mathrm{sec}$ or more (for the masses under consideration), much larger than the thermal one, in order to produce a large enough flux that can fit the positron rise and the $e^+ + e^-$ bump.

Fig. 3 illustrates these points in a systematic way. On the left, it shows how the DM DM $\rightarrow \tau^+\tau^-$ has the best χ^2; other leptonic channels (e.g. $\mu^+\mu^-$) can give acceptable fits, but all other annihilations into quarks, vector and Higgs bosons are significantly disfavored. The value of the required annihilation cross section as a function of the DM mass is illustrated in fig. 3 (right). The actual best fit case is illustrated in fig. 2: it consists of a candidate with a mass of 3 TeV and annihilating into $\tau^+\tau^-$, a channel which produces smooth leptonic spectra, with a cross section of $2 \cdot 10^{-22}\,\mathrm{cm}^3/\mathrm{sec}$.

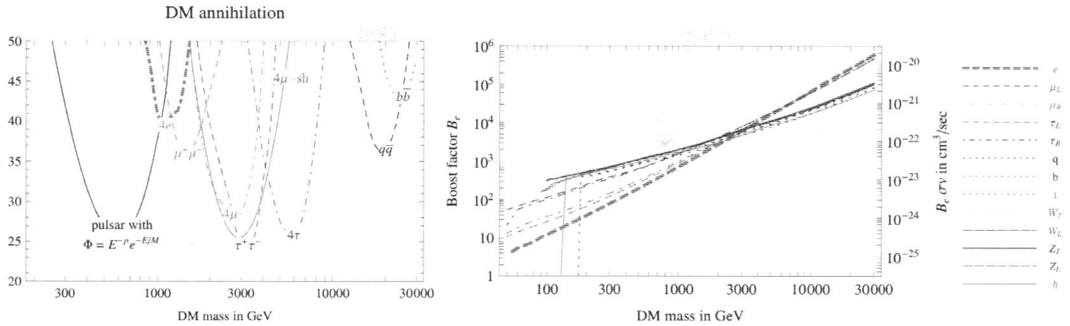

Figure 3. Left: Global fit of different DM annihilation channels to the PAMELA, FERMI and HESS data. The labels on each curve indicate the primary channel (figure from [13] (2009); the fit results remain essentially valid even through the subsequent data updates; the four-lepton lines refer to exotic channels discussed later). Right: Values of $B_e \cdot \sigma v$ (right axis) and of the boost factor B_e (left axis, for $\sigma v = 3\ 10^{-26}\text{cm}^3/\text{sec}$) needed to fit the data (figure from [14]).

The appearence of a small but visible flux of antiprotons from a DM DM $\to \tau^+\tau^-$ annihilation mode in fig. 2 (center) may be at first sight surprising. It is due to the fact that these fluxes are computed including electroweak corrections, i.e. the radiation from the initial $\tau^\pm$ of EW gauge bosons ($W^\pm$, Z) which then decay into many other SM particles, including quarks that hadronize into antiprotons. More generally, the importance of such corrections has been appreciated only relatively recently, in a string of papers with varying scopes and levels of accuracy [15]. Without entering in the details, it is enough for my purposes to remind that (i) the corrections are particularly relevant for large DM masses (above a TeV); (ii) they can alter significantly the ID fluxes, both in their spectral shape and in their amplitude, affecting especially the low energies portion [16], and (iii) in some cases they can also largely modify the annihilation cross section itself, since they can lift the helicity suppression into light fermions [17].

This concludes my overview of the phenomenological interpretation of charged CR data. I now move to a discussion of how natural or preposterous the properties in page 4 are and of what it takes to realize them.

2.3. DM model building

Inspired by the flurry of data discussed above, the field of DM model building has experienced a huge surge of activity in the latest few years. This is essentially because the DM properties individuated by the charged CR signals, and the associated constraints, pose a tough challenge for the traditional DM candidates. Let us take the supersymmetric neutralino as a strawman and see how it fares as a candidate for the DM phenomenology outlined in page 4: (i) The data require a multi-TeV DM mass, but this fits rather uncomfortably in a theory, such as SuSy, whose typical scale should be close to the mass of the higgs boson, if naturalness is a valid criterion. (ii) The data require leptophilicness, but typical neutralino annihilation channels do not distinguish between leptons and quarks, and often couple to gauge bosons, so a dangerous component of hadrons is generically expected. (iii) The data require a huge annihilation cross section into relatively light final states such as leptons, but the neutralino is a Majorana particle and, as such, its (s-wave) annihilation cross section is helicity suppressed by a large factor of $(m_f/m_{\rm DM})^2$, where m_f is the fermion mass. The scorecard for the neutralino does not look great. Of course these generic arguments can be circumvented in specific situations. Indeed several works have argued that it *is* possible to explain the CR excesses insisting on an MSSM DM candidate [18, 19, 20, 21, 22]. This, however, happens at the price of finely tuning at least some aspect: the properties of the $\bar{p}$ background, the position of a nearby DM clump, assuming (uplifted) resonances, finding just few configurations in a scatter plot, explaining only the positron rise and leaving the rest to ad-hoc astrophysics... . These examples are therefore perceived as somewhat anecdotical. Similar arguments would apply for other 'traditional' DM frameworks, e.g. Kaluza-Klein (extradimensional) DM.

This is why the community has preferred to explore new model building possibilities and a lot of works have been published since 2008. Before sketching the main directions of this activity, let us first address an aspect which finds application in several models: how to obtain the large flux.

2.3.1. Tools for enhancements So the question is: is it possible to reconcile the very large value of the annihilation rate (corresponding to $\langle\sigma v\rangle \gtrsim 10^{-23}\,\mathrm{cm}^3/\mathrm{sec}$) required 'today' to fit the CR excesses and the smaller value (corresponding to $\langle\sigma v\rangle \simeq 3\,10^{-26}\,\mathrm{cm}^3/\mathrm{sec}$) individuated by the paradigm of DM production as a thermal relic in the Early Universe? More generically, and thinking beyond the current CR excesses, a large flux would greatly increase our hopes for detection: are there ways to naturally obtain large rates?

The ingenuity of theorists has found at least three positive possible answers to these questions. Yes, it is possible:...

A) Via an astrophysical boost factor: the presence, in today's galactic halo, of DM overdensities predicted by numerical simulations boosts the annihilation rate (proportional to the squared density of DM particles). If the boost could reach a $\mathcal{O}(10^3)$ value, this would explain why the rate is much larger today, without modifying the cross section itself. The typical realistic values, however, have been proven to be $\mathcal{O}(10)$ at most [23].

B) Annihilating via a resonance [14, 24, 25, 26]: if the resonance mass is just below twice the DM mass, the annihilation cross section becomes sensitive to the details of the velocity distribution of the DM particles; since (on average) DM particles are slower today than in the Early Universe, many more of them meet the conditions of resonant annihilation and therefore the rate is enhanced, provided that the relevant parameters are appropriately fine tuned. Typically one needs the mass of the resonance to differ by less than 1% from twice the DM mass and the width of the resonance to be of the order of 10^{-5} of its mass, in order to obtain an enhancement of $\mathcal{O}(10^3)$.

C) Thanks to the Sommerfeld enhancement (see [27, 28, 14], and then [34, 29, 30]), a non-perturbative effect which modifies the annihilation cross section in the regime of *small relative velocity* of the annihilating particles and in presence of an effectively *long-range force* between them. Indeed this well known quantum mechanical effect, first discussed by Sommerfeld in the context of positronium e^+e^- annihilations under the effect of resummed γ exchanges, can occur in DM annihilations if the two annihilating particles exchange an interaction mediated by a force carrier of mass m_V and with a coupling constant $g \approx \sqrt{4\pi\,\alpha}$ such that $\alpha M_{\rm DM}/m_V \gtrsim 1$. For very heavy DM particles ($\gtrsim$ 10 TeV), the exchange of SM weak bosons can mediate the effect, in which case α is just the one of weak interactions and $m_V \approx m_{W^\pm,Z} \approx 100$ GeV. If a new force exist, however, mediated by a particle with mass $m_V \approx 1$ GeV and gauge-like coupling strength with DM particles only, even DM particles of mass $\lesssim 1$ TeV would enjoy the Sommerfeld enhancement. The details of the enhancement of the cross-section are model dependent (see [31, 32, 33] for some examples), but some general features can be identified: in particular, the enhancement shows an inverse proportionality with the relative velocity of the two particles, and it typically saturates to a maximum value when that is $\beta \lesssim 10^{-3}$.

While in some models the enhancement factors discussed above are natural or even unavoidable [2],

[2] See e.g. the Sommerfeld enhancement in [28, 35], published before the CR data made an enhancement needed,

it is fair to say that in many cases they have been used quite liberally in model building, along the lines of 'If I miss a factor of N to fit the data, I throw in A, B or C, or a combination thereof, until I get N' [3]. Perhaps precisely for this reason, they have been studied in detail and they have become kind of a tool in the model builder toolbox. So they are likely to stay with us in the near future.

2.3.2. Recent theory directions A possible categorization of the intense model building work of the past few years would identify the following classes:

★ Minimalistic Dark Matter models. These are models loosely identified by the fact that they aim at providing a viable DM candidate insisting on introducing the minimal set of new particles beyond the Standard Model. They arised originally in opposition to the mainstream direction of obtaining DM as a byproduct of a more ambitious and comprehensive theory, such as SuSy or Extra Dimensions. The namesake Minimal Dark Matter (MDM) [37] falls in this class, as well as less fundamentalist theories such as the model in [38], the hidden vector [43], the Inert Doublet Model (IDM) [39, 40] and others.
Thanks to their relative simplicity, the models in this class are often free of many free parameter and therefore are quite predictive: their ID signatures can be computed and compared univocally with data. The MDM model, featuring, in its minimal realization, a 9.6 TeV DM particle annihilating almost exclusively into W^+W^-, initially excited its authors since it had predicted the size and shape of the positron rise (and the $\bar{p}$ null result) in PAMELA, provided that an astrophysical boost factor of ~ 50 was adopted. Later it has been disfavored by the FERMI+HESS $e^\pm$ data, which prefer a lower mass, and γ-ray data, see e.g. [41] (it remains a viable DM candidate –albeit obviously not an explanation for the CR excesses– for a smaller –and, by the way, more realistic– astrophysical boost factor). The phenomenology of the IDM shares a similar history [42].

★ Models with new dark forces or, more generically, a rich Dark Sector. In this class fall most of the models whose construction has been directly stimulated by the CR excesses.
The model which undoubtedly has most attracted attention and has best spelled out the ingredients is presented in [34], although similar ideas have been proposed before or around the same time [44, 25, 45, 46, 47, 48, 49]. The model in [34] features a TeV-ish DM particle which is sterile under the SM gauge group but which interacts with itself via a new force-carrying boson ϕ (with the strength of typical gauge couplings). The DM annihilation therefore proceeds through DM DM $\to \phi\phi$. A small mixing between ϕ and the electromagnetic current assures that ϕ eventually decays. Therefore the process of DM annihilation occurs in 2 steps: first two DM's go into two ϕ's and then each ϕ's, thanks to its mixing with a photon, goes into a couple of SM particles. The crucial ingredient is that the mass of ϕ is chosen to be light, of the order of $\lesssim 1$ GeV. This simple assumption, remarkably, kills two birds with a stone. On one side, the exchange of ϕ realizes the Sommerfeld enhancement discussed in Sec. 2.3.1, thus providing a very large annihilation cross section today but preserving the thermal production of DM in the Early Universe. On the other side, ϕ can only decay into SM particles lighter than a GeV, i.e. electrons, muons and possibly pions, but not protons: this assures that the annihilation is leptophilic, for a simple kinematical reason. The model therefore fulfils all the requirements listed in page 4. The construction can then be complicated *ad libitum*, e.g. assuming that the dark gauge group is non-abelian and the DM sits in a multiplet of such group, with small splitting

just to mention an example that I know well.
[3] See e.g. the boost factor of ~ 50 in [35, 36], introduced after the CR data showed that a little bit more of an enhancement was needed, just to mention an example that I know well.

Journal of Physics: Conference Series **447** (2013) 012006 doi:10.1088/1742-6596/447/1/012006

between the components. This allows to accommodate other experimental anomalies, not discussed here.
The kinematical argument is not the only one available to justify a leptophilic nature for DM. In the literature, variations have been proposed in which DM is coupled preferentially to leptons because it carries a lepton number [57], because it shares a quantum number with a lepton [14, 58], because quarks live on another brane [59] or... 'because I say so' [60].[4]
The Indirect Detection phenomenology of these theories has of course been worked out in detail [50, 13, 51]. In short, they can easily provide a fit to the charged CR excesses as good as, or better than, the one from ordinary annihilating DM, also thanks to the fact that the 2-step annihilation softens the spectrum of final $e^{\pm}$ and allows a better agreement with the data. [5] They are also subject to the same constraints from neutrinos and especially gamma rays: the bottom line of a series of analyses [52, 53, 54, 55] is that, while the bounds are somewhat alleviated (essentially thanks to the smaller yield of γ-rays and the softening of the spectra mentioned above), there still remain a tension. The conclusion of, e.g., [56], which also considers cosmological constraints, is that it is possible for these models to constitute an explanation of the CR excesses while remaining consistent with the constraints, albeit barely and at the condition of assuming a specific admixture of $4e$ and 4μ final states and quite a specific mass for ϕ.

★ Decaying Dark Matter. The possibility that Dark Matter consists of a particle that actually decays on a very long time scale has been considered since a long time, e.g. in the context of gravitino DM with R-parity violation [61].
More recently, this option has gained steam (see e.g. [62, 63, 64, 65, 66, 67, 68]), precisely because of the cosmic ray excesses and the difficulty of explaining them with annihilating DM. Indeed, if the decay half-life is tuned to $\approx 10^{26}$ seconds (a figure possibly motivated by some high enegy physics scale suppressed operators) [6] and if the production of hadrons is adequately suppressed by some a priori unrelated mechanism, the features needed to fit the data are obtained. A crucial advantage is also that gamma ray (and neutrino) constraints are in general less severe in these scenarios, as they are proportional to the first power of the DM density (and not the second, like for annihilations).
The ID phenomenology of decaying DM has been studied in detail [13, 65, 70]. Generally speaking, it is not much different from the phenomenology of annihilating DM, modulo a few points: i) decay channels other than the ordinary particle-antiparticle pair become possible for fermionic DM (e.g. DM $\to W^{\pm}\ell^{\mp}$, DM $\to \ell^+\ell^-\nu$); ii) the local spectrum of charged CR is often somewhat harder for decaying DM than for annihilating DM, since in the latter case a sizable fraction of low energy particles manage to diffuse to the Earth from the high production regions in the inner Galaxy, softening the spectrum. All in all, it is possible to find fits to the data as good as, or even better than, those of ordinary annihilating DM. On the other hand, as anticipated, the gamma ray constraints from the Galaxy are not as stringent as in the case of annihilating DM. The most stringent bounds come from the isotropic γ flux [71, 54, 74] and, recently, from galaxy clusters in FERMI data [72, 73]: in both cases decay half-lives shorter than $\sim$ few $\cdot 10^{26}$ sec or even up to 10^{27} sec (with the precise value depending on the $m_{\rm DM}$ and the decay channel) are excluded [7]. This represents a serious blow for the decaying DM interpretation of the CR excesses: [74] shows that they

[4] Strictly speaking, ref.s [59] and [60] should not fall in the class of models with new dark forces, since the former works in an extradimensional setup and the latter works in the framework of an effective field theory.
[5] See however [11] for some points of criticism.
[6] Of course, this value of $\tau_{\rm dec}$ is so much longer than the age of the Universe that the slow decay does not make a dent in the overall cosmological DM abundance and does not spoil the agreement with a number of astrophysical and cosmological observations, including the Cosmic Microwave Background [69].
[7] The more stringent constraints from clusters derived in [75] are irrealistic.

Journal of Physics: Conference Series **447** (2013) 012006 doi:10.1088/1742-6596/447/1/012006

are ruled out unless very conservative choices are adopted. The CMB constraints turn out to be not competitive for decaying DM [76, 77].

3. Gamma rays and the Fermi 135 GeV line anomaly

3.1. Hints of DM in gamma rays

Actually, sporadically, claims of evidence of DM in γ-rays data from the Fermi satellite have been made.

Evidences for DM have been reported at the GC in [78, 79, 80],[81], in the isotropic flux [82], in the Virgo cluster [83], possibly in the MW halo [84]... Most notably and most recently an evidence for one (or two) γ-ray line(s) around 130 GeV has been reported, as discussed below.

Setting aside for a moment this latter claim, it is fair to say that none of the other ones, so far, has gathered enough consensus to be considered more than a tentative hint. The crucial points of criticism often have to do with the assumptions made for the astrophysical background (which can, at this stage of significance, mimic a DM signal) or the lack of independent confirmations of the same signal in other channels or in other locations.

In this respect there are two strings of papers that illustrate well the difficulty of the analyses and of the debate. (i) The claim in [79] in favor of light ($\mathcal{O}(10)$ GeV) DM annihilations at the GC has been questioned in [85, 86], which essentially affirm that known astrophysical sources (e.g. millisecond pulsars, MSP) can explain it away. But it has then been re-claimed in [80, 87], saying that those sources are too few, too point-like, too dim or too soft to produce the emission. The latest installment consists of [81], which reaffirms the MSP hypothesis while entertaining the DM one as well. (ii) The claim in [83] of DM annihilations in the Virgo cluster has been refuted first in [88] and then by an extended set of authors of the original paper in [93]. In particular, the latter found that, removing previously-unaccounted-for astrophysical point sources, the evidence in [83] evaporates and actually one can impose stringent bounds (possibly the most stringent ones) using the very same Virgo cluster observation.

Let's now come to '130 GeV line' claim (for a more thorough review see [89]). Originally spotted by [101] and, above all, by [90] in the publicly available Fermi data from an extended region including the GC, it has later found support in other analyses [91, 94, 95, 96], with varying degrees of accuracy and claimed significance. [91, 96] have seen it in what could possibly be DM subhaloes of the MW, and there might be two lines, at 111 GeV and 129 GeV [92, 94]. [95] has seen it in galaxy clusters too. For a response, [97, 98, 99] challenged the analyses in a number of ways, suggesting that the line(s) could be due to unidentified instrumental, statistical or astrophysical origin. [8] If however the line(s) is (are) from DM, it is plausible to expect an associated γ-ray continuum (and possibly a flux of other CRs, e.g. anti-protons) at lower energies, originating from the annihilations of DM into other SM particles; if the cross section into $\gamma\gamma$ is normalized to the one required by Fermi data ($\mathcal{O}(10^{-27})\,\mathrm{cm}^3/\mathrm{s}$), the inferred flux in the continuum poses problems in a variety of cases [102, 103, 104, 105, 106]. However, the details are model dependent: there is no airtight exclusion and working exceptions can be built [107].

In any case, it is clear that this is one of the most interesting areas of development and no claim should be dismissed without deep scrutiny. It might well be that one of these claims will prove to be the harbinger of a full-fledged discovery!

In general, some of our best hopes for clarifying the situation lie in the Fermi satellite itself, of which the prelaunch predicted sensitivities [108] let us believe that a DM with thermal cross section will be probed at 3σ up to a mass of several tens or even hundreds of GeV, depending

[8] By the way, can astrophysics produce a γ-ray line (something which has been advertised for decades as a 'smoking gun' evidence for DM)? Yes, of course, it sufficed to ask and pulsars (always pulsars...) took on themselves the task of producing lines at 100 GeV or somewhat above, via cold electron winds impinging on environmental gamma radiation [100]. It is challenging, though, to see how this process could give rise to a spatially extended emission.

on the annihilation channel and the chosen target. Another set of upcoming experiments will further improve the sensitivity on γ-ray lines, e.g. HESS-II which will soon be operational, CTA which should start in a few years and GAMMA-400 which is being advocated for. In the merit of the 130 GeV line signal, they will provide precious information [109].

3.2. DM interpretation and model building

The findings briefly introduced in the previous Subsection have started inspiring some model building activity in the latest few months (for a more thorough review see [89]). The exercise is: produce a $\gamma\gamma$ line (or something narrow enough to look like a line) with relative cross section $\langle\sigma v\rangle_{\gamma\gamma} \sim 10^{-27}\,\text{cm}^3/\text{sec}$. The starting point is well known: since DM is electrically neutral, it does not directly couple to photons and therefore one needs some mediating mechanism in order to produce the $\gamma\gamma$ pair: (i) via a loop of charged particles, which can be SM ones [110], SuperSymmetric ones [111, 112, 113] or from a new sector [114, 115, 116]; (ib) via Chern-Simons terms [117], the remnants of loops that have been integrated-away; (ii) via a normally subdominant DM-photon coupling, e.g. a magnetic dipole [107, 118, 119]; (iii) via axions [120]... As another option, one may produce a 'top-hat' gamma ray spectrum by having DM annihilate into two metastable states ϕ that then decay into $\gamma\gamma$ (see e.g. [121, 122]): the spectrum is flat between the edges $E_\pm = m_{\rm DM}/2\left(1 \pm \sqrt{1 - m_\phi^2/m_{\rm DM}^2}\right)$, so for m_ϕ close enough to $m_{\rm DM}$ the top-hat fakes a line. As yet another option one may interpret the feature at 130 GeV not as a proper line but rather as the peak of internal bremsstrahlung radiation [123, 124].

Next, one is confronted to two possible (possibly related) problems: the added 'mediator' physics usually implies that the $\gamma\gamma$ process is suppressed with respect to other processes (for instance, the tree level annihilation into the particles running in the loops of (i)), but the needed $\langle\sigma v\rangle_{\gamma\gamma}$ required by FERMI data is relatively large and therefore the 'other processes' are expected to have an even larger $\langle\sigma v\rangle$, with the result that: (1) they might be constrained by other observations (e.g. by FERMI data themselves); (2) they might tend to give a too large annihilation in the Early Universe and therefore an insufficient yield of the thermal DM abundance. This latter point is similar to the one already encountered for the charged CR anomalies (see Sec. 2.3.1), so that some of the tools already explored might apply. Additional proposed solutions include: (a) assuming that the tree level process is not accessible kinematically (e.g. the particles in the loop are heavier than the c.o.m. energy, i.e. the DM mass [110]) [9], (b) engineering other mechanisms which obtain a correct relic abundance but which turn off today (e.g. coannihilations), (c) decoupling the relic DM from the one that produces signal, in model with multi-component DM, (d) giving up altogether the thermal relic mechanism... See [107] for a lucid summary.

Another quite generic 'theory' consequence is that, wherever you can attach a γ emission you can usually also attach a Z or a h emission, with the result that another line is expected at an energy $E = m_{\rm DM}(1 - m_{Z,h}^2/4m_{\rm DM}^2)$ (see e.g. [92]).

4. Conclusions

Like possibly other fields in particle physics in these days, the one of Dark Matter Indirect Detection is currently characterized by an exciting mix of tantalizing hints, increasingly stringent constraints and ever rising hopes. According to my taste, and to the material presented above, I would classify:

△ **Hints**: the e^+ and $e^\pm$ excesses in PAMELA, FERMI and HESS.

[9] Actually, in a model like [110], one postulates that (or selects a scenario in which) DM annihilates (only) into top quarks via a Z'. Demanding the correct relic abundance leads to requiring (a). Which naturally solves (1) and (2) and produces line-like features via (i).

Journal of Physics: Conference Series **447** (2013) 012006 doi:10.1088/1742-6596/447/1/012006

If interpreted in terms of DM annihilations, these point to a rather preposterous particle: multi-TeV, leptophilic and with a huge annihilation cross section.

the '130 GeV line' in FERMI data.

If interpreted in terms of DM annihilations, this seems to point to particle which reassures at first and then puzzles: its mass-scale and its spatial distribution are consistent with the phenomenologists' best dreams, but the large cross section is unsettling.

▽ **Constraints**: the γ-ray measurements from FERMI and HESS (and, to a smaller extent, VERITAS and the other Čerenkov telescopes), the neutrino measurements from SUPERKAMIOKANDE and ICECUBE and the CMB bounds based on WMAP.
These rule out most or all of the parameter space for the preposterous DM above. They also put in a difficult spot some of the more exotic models discussed in Sec. 2.3.2.
On the other hand, they show how far-reaching these kinds of searches can be. They are starting to explore the parameter space of ordinary (thermally produced) DM and already rule out the lower mass portion of the tens-of-GeV range.

◇ **Hopes**: from the point of view of data, the γ-ray (and, probably to a smaller extent, the neutrino) telescopes will continue to play a major role: they will keep up their march across the parameter space and at some point they may see something totally convincing. The AMS-02 detector, onboard of the International Space Station, will probably deliver data in 2013: if the $e^{\pm}$ excesses are of astrophysical origin (as it is probably judicious to believe) then they constitute a formidable background to any future signal from DM in this channel; we have to place our hopes in another channel: the $\bar{p}$ one looks promising, and the $\bar{d}$ one (also explored by the GAPS experiment) may reserve surprises.
From the point of view of theory, the single most striking result of the recent activity is that the community has diversified its interests and explored many new directions, both in terms of phenomena that might enhance the DM signal (see Sec. 2.3.1) and in terms of model-building (see Sec. 2.3.2 and 3.2). The hope is that one or more of these interesting ramifications will bear fruit soon.

Acknowledgments

I would like to thank the organizers of the DISCRETE 2012 conference. This work is supported by the European Research Council (ERC) under the EU Seventh Framework Programme (FP7/2007-2013) / ERC Starting Grant (agreement n. 278234 - 'NewDark' project). This work is also supported in part by the French national research agency ANR under contract ANR 2010 BLANC 041301 and by the EU ITN network UNILHC.

References

[1] E. Komatsu *et al.* [WMAP Coll.], Astrophys. J. Suppl. **192** (2011) 18 [arXiv:1001.4538 [astro-ph.CO]].
[2] O. Adriani *et al.* [PAMELA Collaboration], Nature 458, 607-609, 2009, arXiv:0810.4995.
See also: O. Adriani, G. C. Barbarino, G. A. Bazilevskaya, R. Bellotti, M. Boezio, E. A. Bogomolov, L. Bonechi, M. Bongi *et al.*, Astropart. Phys. **34** (2010) 1, arXiv:1001.3522.
[3] S. W. Barwick *et al.* [HEAT Coll.], Astrophys. J. **482** (1997) L191 [astro-ph/9703192].
[4] M. Aguilar *et al.* [AMS-01 Collaboration], Phys. Lett. B **646** (2007) 145 [astro-ph/0703154].
[5] M. Ackermann *et al.* [The Fermi LAT Collaboration], arXiv:1109.0521 [astro-ph.HE].
[6] O. Adriani, G. Barbarino, G. Bazilevskaya, R. Bellotti, M. Boezio, E. Bogomolov, L. Bonechi and M. Bongi *et al.*, Phys. Rev. Lett. **102** (2009) 051101 [arXiv:0810.4994 [astro-ph]].
O. Adriani *et al.* [PAMELA Collaboration], Phys. Rev. Lett. **105** (2010) 121101 [arXiv:1007.0821].
[7] J. Chang, J. H. Adams, H. S. Ahn, G. L. Bashindzhagyan, M. Christl, O. Ganel, T. G. Guzik and J. Isbert *et al.*, Nature **456** (2008) 362.
A. D. Panov, V. I. Zatsepin, N. V. Sokolskaya, J. H. Adams, Jr., H. S. Ahn, G. L. Bashindzhagyan and J. Chang *et al.*, Astrophys. Space Sci. Trans. **7** (2011) 119 [arXiv:1104.3452 [astro-ph.HE]].
See also the talk by J. Isbert at the 'TANGO in PARIS" workshop, 4-6 may 2009, IAP, Paris; website: irfu.cea.fr/Meetings/tangoinparis.
[8] S. Torii *et al.* [PPB-BETS Collaboration], arXiv:0809.0760 [astro-ph].

[9] A. A. Abdo *et al.* [The Fermi LAT Collaboration], Phys. Rev. Lett. **102** (2009) 181101, arXiv:0905.0025.
M. Ackermann *et al.* [Fermi LAT Collaboration], Phys. Rev. D **82** (2010) 092004, arXiv:1008.3999.
[10] F. Aharonian *et al.* [H.E.S.S. Collaboration], Phys. Rev. Lett. **101** (2008) 261104 [arXiv:0811.3894 [astro-ph]].
F. Aharonian *et al.* [H.E.S.S. Collaboration], Astron. Astrophys. **508** (2009) 561 [arXiv:0905.0105].
[11] P. D. Serpico, arXiv:1108.4827 [astro-ph.HE], in press on Astroparticle Physics.
[12] P. Salati, *Indirect and direct dark matter detection*, Proceedings of the 2007 Cargèse Summer School: Cosmology and Particle Physics Beyond the Standard Models, PoS(cargese)009, pos.sissa.it/cgi-bin/reader/conf.cgi?confid=49
[13] P.Meade, M.Papucci, A.Strumia, T.Volansky, Nucl. Phys. B **831** (2010) 178, arXiv:0905.0480.
[14] M.Cirelli, M.Kadastik, M.Raidal, A.Strumia, Nucl. Phys. B **813** (2009) 1, arXiv:0809.2409.
[15] In M. Kachelriess and P. D. Serpico, Phys. Rev. D **76** (2007) 063516 [arXiv:0707.0209 [hep-ph]] the authors first point out the potential relevance of the effect.
See also: N. F. Bell, J. B. Dent, T. D. Jacques and T. J. Weiler, Phys. Rev. D **78** (2008) 083540 [arXiv:0805.3423 [hep-ph]].
More recent analyses and a more complete appraisal of the phenomenological importance of the effect have been performed in: M. Kachelriess, P. D. Serpico and M. A. Solberg, Phys. Rev. D **80** (2009) 123533 [arXiv:0911.0001 [hep-ph]];
P. Ciafaloni and A. Urbano, Phys. Rev. D **82** (2010) 043512 [arXiv:1001.3950 [hep-ph]].
The antiproton constraints (especially for leptophilic channels) have been considered in M. Garny, A. Ibarra and S. Vogl, JCAP **1107** (2011) 028 [arXiv:1105.5367 [hep-ph]] and M. Garny, A. Ibarra and S. Vogl, arXiv:1112.5155 [hep-ph].
The related phenomenon of gluon emissions is also addressed in M. Asano, T. Bringmann and C. Weniger, arXiv:1112.5158 [hep-ph].
[16] P.Ciafaloni, D.Comelli, A.Riotto, F.Sala, A.Strumia, A.Urbano, JCAP **1103** (2011) 019, arXiv:1009.0224.
[17] P. Ciafaloni, M. Cirelli, D. Comelli, A. De Simone, A. Riotto and A. Urbano, JCAP **1106** (2011) 018, arXiv:1104.2996 [hep-ph].
The latter, among other things, corrects some mistakes of previous publications: N. F. Bell, J. B. Dent, T. D. Jacques and T. J. Weiler, Phys. Rev. D **83** (2011) 013001, arXiv:1009.2584 [hep-ph] and N. F. Bell, J. B. Dent, T. D. Jacques and T. J. Weiler, Phys. Rev. D **84** (2011) 103517, arXiv:1101.3357 [hep-ph].
Similarly, the correction (of the mistakes) is addressed in N.F.Bell, J.B.Dent, A.J.Galea, T.D.Jacques, L.M.Krauss and T.J.Weiler, Phys. Lett. B **706** (2011) 6, arXiv:1104.3823.
The peculiar case of *initial* state EW radiation is discussed in P. Ciafaloni, M. Cirelli, D. Comelli, A. De Simone, A. Riotto and A. Urbano, JCAP **1110** (2011) 034, arXiv:1107.4453 [hep-ph] and P. Ciafaloni, D. Comelli, A. De Simone, A. Riotto and A. Urbano, arXiv:1202.0692 [hep-ph].
[18] P. Grajek, G. Kane, D. Phalen, A. Pierce, S. Watson, Phys. Rev. D **79** (2009) 043506, arXiv:0812.4555.
[19] G. Kane, R. Lu and S. Watson, Phys. Lett. B **681** (2009) 151, arXiv:0906.4765 [astro-ph.HE].
[20] D.Hooper, A.Stebbins, K.Zurek, Phys. Rev. D**79** (2009) 103513, arXiv:0812.3202 [hep-ph].
[21] K. Kadota, K.Freese, P.Gondolo, Phys. Rev. D**81** (2010) 115006, arXiv:1003.4442.
[22] R. Cotta, J. Conley, J. Gainer, J. Hewett and T. Rizzo, JHEP **1101** (2011) 064, arXiv:1007.5520.
[23] J.Lavalle, J.Pochon, P.Salati, R.Taillet, Astron. Astrophys. **462**(2007) 827, astro-ph/0603796.
J. Lavalle, Q. Yuan, D. Maurin, X. J. Bi, Astron. Astrophys. **479** (2008) 427, arXiv:0709.3634.
L. Pieri, J. Lavalle, G. Bertone, E. Branchini, Phys. Rev. D **83** (2011) 023518, 0908.0195.
[24] M.Ibe, H.Murayama, T.Yanagida, Phys.Rev.D**79** (2009) 095009, arXiv:0812.0072 [hep-ph].
See also: W. -L. Guo, Y. -L. Wu, Phys. Rev. D **79** (2009) 055012, arXiv:0901.1450 [hep-ph].
[25] D. Feldman, Z. Liu and P. Nath, Phys. Rev. D **79** (2009) 063509, arXiv:0810.5762 [hep-ph].
[26] H. An and M. Pospelov, arXiv:1206.2371 [hep-ph].
[27] A.Sommerfeld, "Über die Beugung und Bremsung der Elektronen", Ann. Phys. 403, 257 (1931).
J.Hisano, S.Matsumoto, M.Nojiri, Phys. Rev. Lett. 92 (2004) 031303, arXiv:hep-ph/0307216.
J.Hisano, S.Matsumoto, M.Nojiri, O.Saito, Phys. Rev. D 71 (2005) 063528, hep-ph/0412403.
See also previous work in K. Belotsky, D. Fargion, M. Khlopov and R. V. Konoplich, Phys. Atom. Nucl. 71 (2008) 147 [arXiv:hep-ph/0411093] and references therein.
[28] M. Cirelli, A. Strumia, M. Tamburini, Nucl. Phys. B **787** (2007) 152, arXiv:0706.4071.
[29] M. Lattanzi and J. I. Silk, Phys. Rev. D **79** (2009) 083523, arXiv:0812.0360 [astro-ph].
J. D. March-Russell, S. M. West, Phys. Lett. B **676** (2009) 133, arXiv:0812.0559 [astro-ph].
L. Pieri, M. Lattanzi and J. Silk, arXiv:0902.4330 [astro-ph.HE].
R. Iengo, JHEP **0905** (2009) 024, arXiv:0902.0688 [hep-ph].
[30] J. Bovy, Phys. Rev. D **79** (2009) 083539 arXiv:0903.0413 [astro-ph.HE].
[31] T. R. Slatyer, JCAP **1002** (2010) 028, arXiv:0910.5713 [hep-ph].
[32] J. Feng, M. Kaplinghat, H.-B. Yu, Phys. Rev. Lett. **104** (2010) 151301, arXiv:0911.0422.

J. Feng, M. Kaplinghat, H.-B. Yu, Phys. Rev. D **82** (2010) 083525, arXiv:1005.4678 [hep-ph].
[33] A. Hryczuk, Phys. Lett. B **699** (2011) 271, arXiv:1102.4295 [hep-ph].
A. Hryczuk, R. Iengo and P. Ullio, JHEP **1103** (2011) 069, arXiv:1010.2172 [hep-ph].
A. Hryczuk and R. Iengo, arXiv:1111.2916 [hep-ph].
[34] N. Arkani-Hamed, D. P. Finkbeiner, T. R. Slatyer and N. Weiner, Phys. Rev. D **79** (2009) 015014 [arXiv:0810.0713 [hep-ph]].
[35] M. Cirelli, R. Franceschini, A. Strumia, Nucl. Phys. B **800** (2008) 204, arXiv:0802.3378.
[36] M. Cirelli and A. Strumia, PoS IDM **2008** (2008) 089, arXiv:0808.3867 [astro-ph].
[37] M. Cirelli, N. Fornengo and A. Strumia, Nucl. Phys. B **753** (2006) 178, hep-ph/0512090.
M. Cirelli and A. Strumia, New J. Phys. **11** (2009) 105005, arXiv:0903.3381 [hep-ph].
See also: T. Hambye, F.-S. Ling, L. Lopez Honorez and J. Rocher, JHEP **0907** (2009) 090 [Erratum-ibid. **1005** (2010) 066], arXiv:0903.4010 [hep-ph].
[38] R. Mahbubani and L. Senatore, Phys. Rev. D **73** (2006) 043510, hep-ph/0510064.
[39] R. Barbieri, L. J. Hall and V. S. Rychkov, Phys. Rev. D **74** (2006) 015007, hep-ph/0603188.
See also N. G. Deshpande and E. Ma, Phys. Rev. D **18** (1978) 2574. where the model is introduced, but without explicitly addressing the DM problem.
[40] T. Hambye and M. H. G. Tytgat, Phys. Lett. B **659** (2008) 651, arXiv:0707.0633 [hep-ph].
[41] M. Pato, L. Pieri, G. Bertone, Phys. Rev. D **80** (2009) 103510, arXiv:0905.0372 [astro-ph.HE].
[42] E. Nezri, M. H. G. Tytgat, G. Vertongen, JCAP **0904** (2009) 014, arXiv:0901.2556 [hep-ph].
[43] T. Hambye, JHEP **0901** (2009) 028, arXiv:0811.0172 [hep-ph].
[44] M.Pospelov, A.Ritz, M.B.Voloshin, Phys. Lett. B **662** (2008) 53, arXiv:0711.4866 [hep-ph].
M. Pospelov, A. Ritz, Phys. Lett. B **671** (2009) 391 [arXiv:0810.1502 [hep-ph]].
[45] I.Cholis, L.Goodenough, N.Weiner, Phys. Rev. D **79** (2009) 123505, arXiv:0802.2922.
[46] A. E. Nelson and C. Spitzer, JHEP **1010** (2010) 066, arXiv:0810.5167 [hep-ph].
[47] I.Cholis, D.P.Finkbeiner, L.Goodenough, N.Weiner, JCAP **0912** (2009) 007, 0810.5344.
[48] Y. Nomura and J. Thaler, Phys. Rev. D **79** (2009) 075008, arXiv:hhref0810.5397 [hep-ph].
[49] Y. Bai and Z. Han, Phys. Rev. D **79** (2009) 095023, arXiv:0811.0387 [hep-ph].
[50] I. Cholis, G. Dobler, D. P. Finkbeiner, L. Goodenough and N. Weiner, Phys. Rev. D **80** (2009) 123518, arXiv:0811.3641 [astro-ph].
[51] J. Mardon, Y. Nomura, D. Stolarski, J. Thaler, JCAP **0905** (2009) 016, arXiv:0901.2926.
[52] L. Bergstrom, G. Bertone, T. Bringmann, J. Edsjo and M. Taoso, Phys. Rev. D **79** (2009) 081303, arXiv:0812.3895 [astro-ph].
[53] P. Meade, M. Papucci and T. Volansky, JHEP **0912** (2009) 052, arXiv:0901.2925 [hep-ph].
[54] M. Papucci and A. Strumia, JCAP **1003** (2010) 014, arXiv:0912.0742 [hep-ph].
[55] M. Cirelli and J. M. Cline, Phys. Rev. D **82** (2010) 023503, arXiv:1005.1779 [hep-ph].
[56] D. P. Finkbeiner, L. Goodenough, T. R. Slatyer, M. Vogelsberger and N. Weiner, JCAP **1105** (2011) 002, arXiv:1011.3082 [hep-ph].
[57] D.J.Phalen, A.Pierce, N.Weiner, Phys. Rev. D **80** (2009) 063513, arXiv:0901.3165 [hep-ph].
[58] P. J. Fox and E. Poppitz, Phys. Rev. D **79** (2009) 083528, arXiv:0811.0399 [hep-ph].
[59] S. C. Park and J. Shu, Phys. Rev. D **79** (2009) 091702, arXiv:0901.0720 [hep-ph].
[60] R. Harnik and G. D. Kribs, Phys. Rev. D **79** (2009) 095007, arXiv:0810.5557 [hep-ph].
[61] W.Buchmuller, L.Covi, K.Hamaguchi, A.Ibarra, T.Yanagida, JHEP **0703** (2007) 037, hep-ph/0702184.
[62] C. -R. Chen and F. Takahashi, JCAP **0902** (2009) 004, arXiv:0810.4110 [hep-ph].
[63] P.-f.Yin, Q.Yuan, J.Liu, J.Zhang, X.-j.Bi, S.-h.Zhu, Phys. Rev. D **79** (2009) 023512, arXiv:0811.0176.
[64] K. Ishiwata, S. Matsumoto, T. Moroi, Phys. Lett. B **675** (2009) 446, arXiv:0811.0250.
[65] A. Ibarra and D. Tran, JCAP **0902** (2009) 021, arXiv:0811.1555 [hep-ph].
[66] C. -R. Chen, M. M. Nojiri, F. Takahashi and T. T. Yanagida, Prog. Theor. Phys. **122** (2009) 553, arXiv:0811.3357 [astro-ph].
[67] E. Nardi, F. Sannino and A. Strumia, JCAP **0901** (2009) 043 [arXiv:0811.4153 [hep-ph]].
[68] A. Arvanitaki, S. Dimopoulos, S. Dubovsky, P. W. Graham, R. Harnik and S. Rajendran, Phys. Rev. D **79** (2009) 105022, arXiv:0812.2075 [hep-ph].
[69] R. Cen, Astrophys.J. 546 (2001) L77, [astro-ph/0005206].
M.Oguri, K.Takahashi, H.Ohno, K.Kotake, Astrophys. J. **597** (2003) 645 astro-ph/0306020.
K. Ichiki, P. M. Garnavich, T. Kajino, G. J. Mathews, M. Yahiro, Phys. Rev. **D68** (2003) 083518. [astro-ph/0210052].
K.Takahashi, M.Oguri, K.Ichiki, Mon.Not.Roy.Astron.Soc. **352**(2004)311, astro-ph/0312358.
K. Ichiki, M. Oguri, K. Takahashi, Phys. Rev. Lett. **93** (2004) 071302. [astro-ph/0403164].
[70] A. Ibarra and D. Tran, JCAP **0906** (2009) 004, arXiv:0904.1410 [hep-ph].
[71] M.Cirelli, P.Panci, P.D.Serpico, Nucl. Phys. B **840** (2010) 284, arXiv:0912.0663.

[72] X. Huang, G. Vertongen and C. Weniger, arXiv:1110.1529 [hep-ph].
[73] L. Zhang, C. Weniger, L. Maccione, J. Redondo and G. Sigl, JCAP **1006** (2010) 027, arXiv:0912.4504 [astro-ph.HE].
[74] M. Cirelli, E. Moulin, P. Panci, P. D. Serpico and A. Viana, arXiv:1205.5283 [astro-ph.CO].
[75] L.Dugger, T.E.Jeltema, S.Profumo, JCAP **1012** (2010) 015, arXiv:1009.5988 [astro-ph.HE].
[76] X. -L. Chen and M. Kamionkowski, Phys. Rev. D **70** (2004) 043502, astro-ph/0310473.
L. Zhang, X. Chen, M. Kamionkowski, Z. -g. Si and Z. Zheng, Phys. Rev. D **76** (2007) 061301, arXiv:0704.2444 [astro-ph].
[77] M. Cirelli, F. Iocco and P. Panci, unpublished.
[78] L. Goodenough and D. Hooper, arXiv:0910.2998 [hep-ph].
[79] D. Hooper and L. Goodenough, Phys. Lett. B **697** (2011) 412 [arXiv:1010.2752 [hep-ph]].
[80] D. Hooper and T. Linden, arXiv:1110.0006 [astro-ph.HE].
[81] K. N. Abazajian and M. Kaplinghat, arXiv:1207.6047 [astro-ph.HE].
[82] G. Hutsi, A. Hektor and M. Raidal, JCAP **1007** (2010) 008 [arXiv:1004.2036 [astro-ph.HE]].
[83] J. Han, C. S. Frenk, V. R. Eke, L. Gao and S. D. M. White, arXiv:1201.1003 [astro-ph.HE].
[84] T. Lin, D. P. Finkbeiner, G. Dobler, Phys. Rev. D **82** (2010) 023518, arXiv:1004.0989.
[85] K. N. Abazajian, JCAP **1103** (2011) 010, arXiv:1011.4275 [astro-ph.HE].
[86] A.Boyarsky, D.Malyshev, O.Ruchayskiy, Phys. Lett. B **705** (2011) 165, arXiv:1012.5839.
[87] D. Hooper, arXiv:1201.1303 [astro-ph.CO].
[88] O. Macias-Ramirez, C. Gordon, A. M. Brown and J. Adams, arXiv:1207.6257 [astro-ph.HE].
[89] T. Bringmann and C. Weniger, arXiv:1208.5481 [hep-ph].
[90] C. Weniger, JCAP **1208** (2012) 007 [arXiv:1204.2797 [hep-ph]].
[91] E. Tempel, A. Hektor and M. Raidal, arXiv:1205.1045 [hep-ph].
[92] A. Rajaraman, T. M. P. Tait and D. Whiteson, arXiv:1205.4723 [hep-ph].
[93] J. Han, C. S. Frenk, V. R. Eke, L. Gao, S. D. M. White, A. Boyarsky, D. Malyshev and O. Ruchayskiy, arXiv:1207.6749 [astro-ph.CO].
[94] M. Su and D. P. Finkbeiner, arXiv:1206.1616 [astro-ph.HE].
[95] A. Hektor, M. Raidal and E. Tempel, arXiv:1207.4466 [astro-ph.HE].
[96] M. Su and D. P. Finkbeiner, arXiv:1207.7060 [astro-ph.HE].
[97] A. Boyarsky, D. Malyshev and O. Ruchayskiy, arXiv:1205.4700 [astro-ph.HE].
[98] N. Mirabal, arXiv:1208.1693 [astro-ph.HE].
[99] A. Hektor, M. Raidal and E. Tempel, arXiv:1208.1996 [astro-ph.HE].
[100] F. Aharonian, D. Khangulyan and D. Malyshev, arXiv:1207.0458 [astro-ph.HE].
[101] T. Bringmann, X. Huang, A. Ibarra, S. Vogl and C. Weniger, arXiv:1203.1312 [hep-ph].
[102] W. Buchmuller and M. Garny, arXiv:1206.7056 [hep-ph].
[103] T. Cohen, M. Lisanti, T. R. Slatyer and J. G. Wacker, arXiv:1207.0800 [hep-ph].
[104] I. Cholis, M. Tavakoli and P. Ullio, arXiv:1207.1468 [hep-ph].
[105] X. -Y. Huang, Q. Yuan, P. -F. Yin, X. -J. Bi and X. -L. Chen, arXiv:1208.0267 [astro-ph.HE].
[106] D. Hooper and T. Linden, arXiv:1208.0828 [astro-ph.HE].
[107] S. Tulin, H. -B. Yu and K. M. Zurek, arXiv:1208.0009 [hep-ph].
[108] E. A. Baltz, B. Berenji, G. Bertone, L. Bergstrom, E. Bloom, T. Bringmann, J. Chiang and J. Cohen-Tanugi *et al.*, JCAP **0807** (2008) 013, arXiv:0806.2911 [astro-ph].
[109] L. Bergstrom, G. Bertone, J. Conrad, C. Farnier and C. Weniger, arXiv:1207.6773 [hep-ph].
[110] C. B. Jackson, G. Servant, G. Shaughnessy, T. M. P. Tait and M. Taoso, JCAP **1004**, 004 (2010) [arXiv:0912.0004 [hep-ph]].
[111] B. Kyae and J. -C. Park, arXiv:1205.4151 [hep-ph].
[112] B. S. Acharya, G. Kane, P. Kumar, R. Lu and B. Zheng, arXiv:1205.5789 [hep-ph].
[113] D. Das, U. Ellwanger, P. Mitropoulos, JCAP **1208** (2012) 003 [arXiv:1206.2639 [hep-ph]].
[114] J. M. Cline, arXiv:1205.2688 [hep-ph].
[115] J. -C. Park and S. C. Park, arXiv:1207.4981 [hep-ph].
[116] N. Weiner and I. Yavin, arXiv:1209.1093 [hep-ph].
[117] E. Dudas, Y. Mambrini, S. Pokorski and A. Romagnoni, arXiv:1205.1520 [hep-ph].
[118] J. H. Heo and C. S. Kim, arXiv:1207.1341 [astro-ph.HE].
[119] J. M. Cline, A. R. Frey and G. D. Moore, arXiv:1208.2685 [hep-ph].
[120] H. M. Lee, M. Park and W. -I. Park, arXiv:1205.4675 [hep-ph].
[121] A. Ibarra, S. Lopez Gehler and M. Pato, JCAP **1207** (2012) 043 [arXiv:1205.0007 [hep-ph]].
[122] J. Fan and M. Reece, arXiv:1209.1097 [hep-ph].
[123] L. Bergstrom, arXiv:1208.6082 [hep-ph].
[124] B. Shakya, arXiv:1209.2427 [hep-ph].

Journal of Physics: Conference Series **447** (2013) 012007 doi:10.1088/1742-6596/447/1/012007

Advances in Leptogenesis

Apostolos Pilaftsis

Consortium for Fundamental Physics, School of Physics and Astronomy, University of Manchester, Manchester M13 9PL, United Kingdom

E-mail: apostolos.pilaftsis@manchester.ac.uk

Abstract. We briefly review the scenario of baryogenesis through leptogenesis. Leptogenesis is an appealing scenario that may relate the observed baryon asymmetry in the Universe to the low-energy neutrino data. Particular emphasis will be laid on recent developments on the field, such as the flavourdynamics of leptogenesis, resonant leptogenesis near the electroweak phase transition and thermal effects. It is illustrated how these recent developments enable the modelling of phenomenologically predictive scenarios that can directly be tested at the LHC and indirectly in low-energy experiments of lepton-number and lepton-flavour violation.

1. Introduction

An elegant framework to consistently address the observed Baryon Asymmetry in the Universe (BAU) [1, 2] is leptogenesis [3]. According to the standard paradigm of leptogenesis [4], minimal extensions of the Standard Model (SM) realize heavy Majorana neutrinos of masses related to the Grand Unified Theory (GUT) scale $M_{\rm GUT} \sim 10^{16}$ that decay out of equilibrium and create a net excess of lepton number (L). This excess in L gets then reprocessed into the observed baryon number (B), through the $(B+L)$-violating sphaleron interactions [5]. The attractive feature of such a scenario is that the GUT-scale heavy Majorana neutrinos could also explain the observed smallness in mass of the SM light neutrinos by means of the so-called seesaw mechanism [6].

The original GUT-scale leptogenesis scenario, however, runs into certain difficulties, when one attempts to explain the flatness of the Universe and other cosmological data [2] within supergravity models of inflation. To avoid overproduction of gravitinos $\widetilde{G}$ whose late decays may ruin the successful predictions of Big Bang Nucleosynthesis (BBN), the reheat temperature $T_{\rm reh}$ of the Universe should be lower than 10^9–10^6 GeV, for $m_{\widetilde{G}} = 8.0$–0.2 TeV [7]. This implies that the heavy Majorana neutrinos should accordingly have masses as low as $T_{\rm reh} \lesssim 10^9$ GeV, thereby rendering the relation of these particles with GUT-scale physics less natural. On the other hand, it proves very difficult to directly probe the heavy-neutrino sector of such a model at high-energy colliders, e.g. at the LHC or ILC, or in any other foreseeable experiment.

A potentially interesting solution to the above problems may be obtained within the framework of resonant leptogenesis (RL) [8]. The key aspect of RL is that self-energy effects dominate the leptonic asymmetries [9], when two heavy Majorana neutrinos happen to have a small mass difference with respect to their actual masses. If this mass difference becomes comparable to the heavy neutrino widths, a resonant enhancement of the leptonic asymmetries takes place that may reach values $\mathcal{O}(1)$ [8, 10]. An indispensable feature of RL models is that flavour effects due to the light-to-heavy neutrino Yukawa couplings [11, 12, 14] play a dramatic role and can modify the predictions for the BAU by many orders of magnitude [11, 12]. Most importantly, these flavour effects enable the modelling [12, 13] of minimal RL scenarios with electroweak-scale heavy Majorana neutrinos that could be tested at the LHC [15, 16, 17] and in other non-accelerator experiments, while maintaining agreement with the low-energy neutrino data. Many variants of RL have been proposed in the literature [18, 19], including soft leptogenesis [20] and radiative leptogenesis [21].

In this review we give a brief exposition of selected topics related to recent advancements of the field. In detail, in Section 2, we remind ourselves about Sakharov's basic conditions for baryogenesis and list a few classical scenarios that have been suggested in the literature, including leptogenesis. In Section 3 we discuss the importance of the various sources of flavour effects on leptogenesis. Section 4 presents the field theory of RL, including a discussion of its flavourdynamics and thermal effects. Section 5 discusses the potential particle-physics implications of RL. Finally, Section 6 summarizes the main points of this review.

2. Matter–Antimatter Asymmetry and Leptogenesis

There are two pieces of observational evidence that establish the domination of the observable matter over antimatter in our Universe. The first stems from the analysis of the power spectrum of the cosmic microwave background [2]. In particular, the baryon-to-photon ratio η_B of number densities is found to be

$$\eta_B^{\rm CMB} \;=\; \frac{n_B}{n_\gamma} \;=\; 6.1^{+0.3}_{-0.2} \times 10^{-10}\,. \qquad (1)$$

The second piece of evidence originates from the concordance of the light elements and big bang nucleosynthesis, which gives [22]

$$\eta_B^{\rm BBN} \;=\; 3.4\text{–}6.9 \times 10^{-10}\,. \qquad (2)$$

In spite of their different origin, the two deduced values for η_B are in good agreement with one another.

More than fifty years ago [1], Sakharov has studied the conditions for which a baryon asymmetry may successfully be generated. These basic conditions are:

- the theory should have B-violating interactions.
- the interactions should violate both C and CP.
- the processes of a net baryon-number generation should have a degree of irreversibility due to an out-of-thermal equilibrium dynamics.

There are many scenarios that have been proposed in the existing literature. A few classical examples are:

- *Baryogenesis through the decay of a heavy particle.* This scenario is based on the out-of-equilibrium B-violating decay of a heavy GUT particle, e.g. in SO(10) theory [23]. Serious constraints on these scenarios come from the stringent lower limit on proton's long lifetime [22] and the sphaleron erasure, as the dominant decay modes usually violate $(B+L)$-number as well (see also discussion below). The latter problem might be circumvented by introducing a certain set of non-renormalizable operators into the GUT theory [24].
- *Baryogenesis at the electroweak phase transition.* In this scenario, BAU is generated by $(B + L)$-violating sphaleron interactions at $T \sim T_c \approx 140$ GeV, through a first order

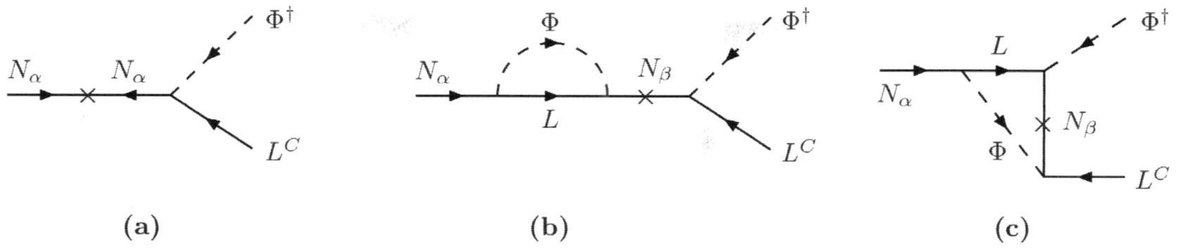

Figure 1. Feynman diagrams contributing to the L-violating decays of heavy Majorana neutrinos, $N_\alpha \to L^C\Phi^\dagger$, where L and Φ represent lepton and Higgs-boson iso-doublets, respectively: (a) tree-level graph, and (b) self-energy and (c) vertex graph.

phase transition [5]. However, such a scenario cannot be realized within the Standard Model (SM) [25], since the phase transition is not sufficiently strong enough given the current observation of the Higgs-boson H with mass $M_H \simeq 126$ GeV. This last feature changes drastically in the MSSM [26]. However, the present limits on the Thalium, neutron and Mercury Electric Dipole Moments [27], as well as constraints on squark and slepton masses, significantly squeeze the allowed parameter space for electroweak baryogenesis in the MSSM [28].

- *Baryogenesis through leptogenesis.* This scenario combines the two key ideas of the above two scenarios. Leptogenesis is based on the out-of-equilibrium L-violating decays of heavy Majorana neutrinos. These decays produce a net lepton asymmetry, which is converted into the observed BAU through $(B+L)$-violating sphaleron interactions [3].

Several variants of leptogenesis have been discussed so far in the literature [29]. These include scenarios of hierarchical leptogenesis [30], Dirac leptogenesis [31], non-thermal leptogenesis [32], Affleck-Dine and spontaneous leptogenesis [33, 34], and RL which will be discussed more extensively in Section 4.

3. The Flavourdynamics of Leptogenesis

It is important to appreciate the special role that flavour effects can play in leptogenesis. This special role of flavour effects on scenarios of baryogenesis is not new [35] and has already been recognized in the past [36, 37]. The key observation here is that sphalerons do not only conserve $B-L$, but also the individual quantum numbers:

$$\frac{1}{3}B - L_{e,\mu,\tau}\ .$$

Thus, a generated baryon asymmetry can be protected from erasure if a particular lepton flavour combination or an individual lepton number $L_{e,\mu,\tau}$ is out of thermal equilibrium. For instance, this idea has been used in [36] to show how specific R-parity-violating supersymmetric scenarios do not wash out any baryon asymmetry that could eventually account for the observed matter–antimatter asymmetry in the Universe.

The new aspect of flavourdynamics in leptogenesis is that the BAU can now be both *generated from* and *protected in* a *single* lepton flavour [11], e.g. in the τ-lepton number L_τ. In order to better understand this, we need first to clarify the two sources of flavour effects:

- *flavour effects due to charged-lepton Yukawa couplings* $h_{e,\mu,\tau}$. These effects are related to the interactions mediated by charged-lepton Yukawa couplings [38]. If these interactions

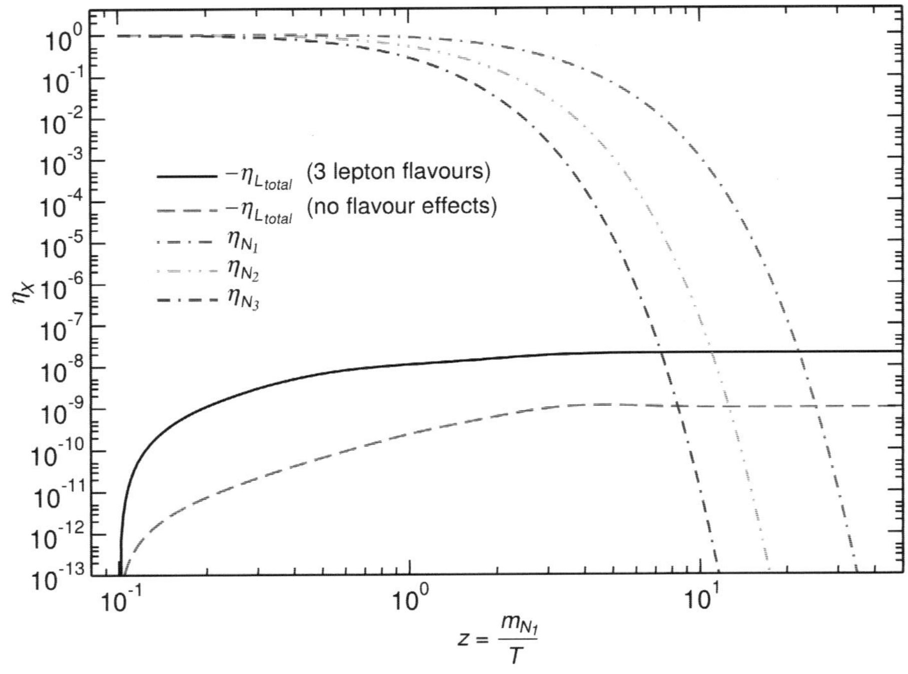

Figure 2. Numerical estimates for a scenario with $m_{N_1} = 10^{10}$ GeV, $m_{N_2} = 2m_{N_1}$, $m_{N_3} = 3m_{N_1}$.

are in or out of thermal equilibrium at the scale, at which leptogenesis takes place, then the predicted value for the BAU gets modified significantly, typically up to one order of magnitude [39]. Thus, if the leptogenesis scale is higher than 10^9 TeV, then all charged-lepton Yukawa interactions are out of thermal equilibrium, whereas if the scale is below 10 TeV, all charged-lepton Yukawa couplings are in thermal equilibrium [37] and their full impact on the BAU needs be taken into account [12].

- *flavour effects due to heavy-neutrino Yukawa couplings* $h^{\nu}_{l\alpha}$. These effects are governed by heavy-neutrino Yukawa-coupling interactions [11, 14]. In this case, one may consider scenarios that would generate an excess in an individual lepton number, e.g. L_τ, and protect the resulting asymmetry from potentially large wash-out effects due to sphalerons. For example, one could arrange the neutrino Yukawa couplings $h^{\nu}_{l\alpha}$, such that all heavy Majorana neutrinos $N_{1,2,3}$ decay out of equilibrium in L_τ, N_1 decays out of equilibrium in $L_{e,\mu,\tau}$, but $N_{2,3}$ could decay very rapidly into $L_{e,\mu}$. In Fig. 2, we give predictions for an hierarchical scenario with and without neutrino Yukawa coupling effects included [12, 14]. The two predictions can differ by up to one order of magnitude. As we will discuss in Section 4, however, flavour effects due to $h^{\nu}_{l\alpha}$ may modify the BAU predictions by many orders of magnitude, e.g. bigger than 10^6, in RL models, thereby rendering their inclusion indispensable [11, 12].

We close this section by mentioning that flavour effects do also play an important role in collision terms of the Boltzmann equations that describe scatterings [10, 12]. These are $\Delta L = 1$ Yukawa and gauge-mediated scatterings, as well as $\Delta L = 0, 2$ scatterings mediated by heavy Majorana neutrinos.

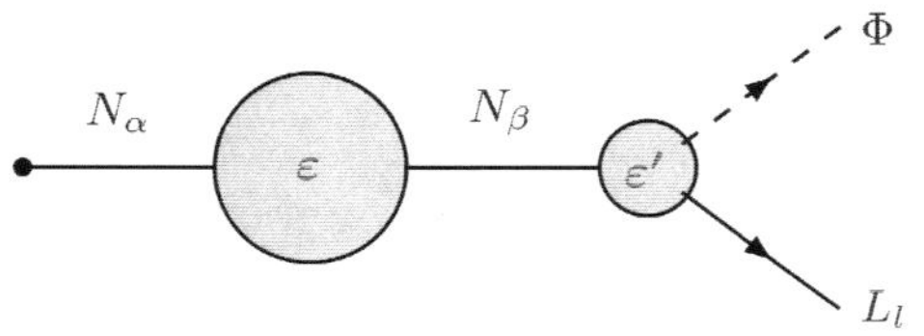

$$\times\,[S_{\alpha\alpha}(\not p)]^{-1}\,u_{N_\alpha}(p) \;=\; \mathcal{T}_{N_\alpha}$$

Figure 3. LSZ-type approach to mixing and decay of heavy Majorana neutrinos

4. Resonant Leptogenesis

A scenario inspired by the dynamics of the $K^0\bar{K}^0$ system [40] is resonant leptogenesis [8, 10]. Resonant leptogenesis is based on the observation that self-energies dominate the lepton asymmetries if $|m_{N_1} - m_{N_2}| \ll m_{N_{1,2}}$ [9]. However, as we will see below, the proper inclusion of heavy-neutrino width effects is necessary in order to obtain a non-singular well-behaved analytic expression for the leptonic asymmetries.

The are several variants of RL:

- *Soft RL [20].* In this scenario, leptogenesis results from sneutrino decays in the MSSM. CP-violating soft SUSY-breaking parameters may split the degeneracy within a single generation of right-handed sneutrino states, which leads to a resonant enhancement of the leptonic asymmetries.
- *Radiative RL [21].* In this case, the heavy Majorana neutrinos are exactly degenerate at the GUT scale. Then, small mass differences among the heavy Majorana neutrino states are generated via renormalization group (RG) running from the GUT scale down to the leptogenesis scale. These small mass differences give rise to RL.
- *Coherent RL (via sterile neutrino oscillations) [41, 42].* This scenario relies on the possibility that sterile neutrinos with masses well below the critical temperature of the electroweak phase transition, e.g. of order 10 GeV, maintain the coherence of their CP asymmetric oscillations and can thus produce an enhanced leptonic asymmetry of the amount needed to create the observed BAU.

We will now briefly review the field theory developed in [8, 10] for thermal RL. This is based on an effective LSZ-type formalism [43] with the aim to incorporate both the mixing and the decay of the heavy Majorana neutrinos. As shown in 3, one needs to resum self-energy graphs, whose absorptive parts regularize the tree-level mass singularity when $m_{N_\alpha} \to m_{N_\beta}$. For a two-generation heavy-neutrino model, the resummed propagator matrix reads:

$$S_{\alpha\beta}(\not p) \;=\; \begin{pmatrix} \not p - m_{N_1} + \Sigma_{11}(\not p) & \Sigma_{12}(\not p) \\ \Sigma_{21}(\not p) & \not p - m_{N_2} + \Sigma_{22}(\not p) \end{pmatrix}^{-1} . \tag{3}$$

Detailed discussion of this formalism may be found in [8, 10].

Making use of the corresponding $K^0\bar{K}^0$ terminology, there are two types of CP violation in the leptonic decays of the heavy Majorana neutrinos:

- *ε'-type CP violation.* In this case, the required CP violation arises from the interference of the tree-level graph with the absorptive of the self-energy effects.

$$\varepsilon'_{N_\alpha} \;=\; \frac{\mathrm{Im}\,(h^{\nu\dagger}\,h^{\nu})^2_{\alpha\beta}}{(h^{\nu\dagger}\,h^{\nu})_{\alpha\alpha}\,(h^{\nu\dagger}\,h^{\nu})_{\beta\beta}}\left(\frac{\Gamma_{N_\beta}}{m_{N_\beta}}\right)\, f\left(\frac{m^2_{N_\beta}}{m^2_{N_\alpha}}\right) , \tag{4}$$

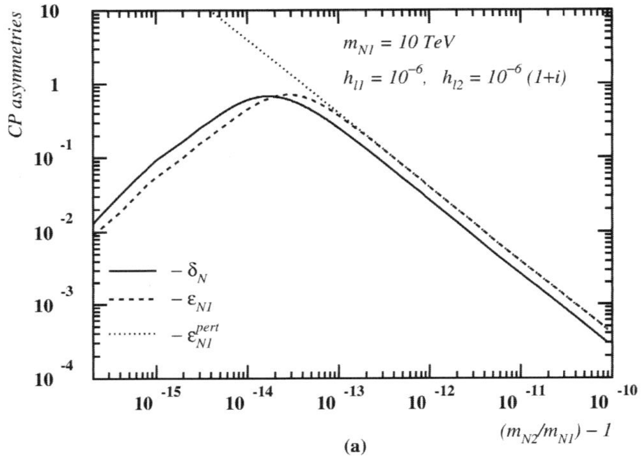

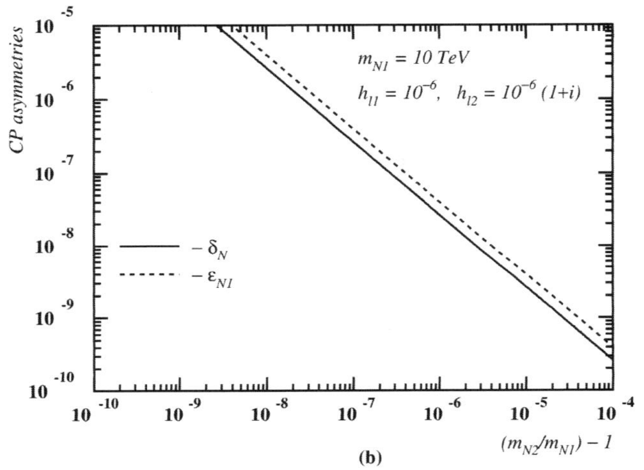

Figure 4. ε- and ε'-type contributions to the leptonic asymmetries in a model with two heavy Majorana neutrinos, with $\delta_{N_{1,2}} \approx \varepsilon_{N_{1,2}} + \varepsilon'_{N_{1,2}}$ and $\delta_N \approx \delta_{N_1} + \delta_{N_2}$.

where

$$\Gamma_{N_\beta} = \frac{(h^{\nu\dagger} h^\nu)_{\beta\beta}}{8\pi} m_{N_\beta}$$

is the tree-level decay width of N_β, and $f(x) = \sqrt{x}[1 - (1+x)\ln(1+1/x)]$ is the Fukugita–Yanagida loop function [3].

- *ε-type CP violation.*

$$\varepsilon_{N_\alpha} = \frac{\mathrm{Im}\,(h^{\nu\dagger} h^\nu)^2_{\alpha\beta}}{(h^{\nu\dagger} h^\nu)_{\alpha\alpha}\,(h^{\nu\dagger} h^\nu)_{\beta\beta}} \left(\frac{\Gamma_{N_\beta}}{m_{N_\beta}}\right) \frac{(m^2_{N_\alpha} - m^2_{N_\beta})\, m_{N_\alpha} m_{N_\beta}}{(m^2_{N_\alpha} - m^2_{N_\beta})^2 + m^2_{N_\alpha} \Gamma^2_{N_\beta}} \tag{5}$$

Note that ε_{N_α} dominate over the ε'_{N_α} when $m_{N_\alpha} - m_{N_\beta} \ll m_{N_{\alpha,\beta}}$. In the limit $m_{N_\alpha} \to m_{N_\beta}$, the would-be singularity is regularized by the decay width of the heavy neutrino N_β.

From (5), the following *two* conditions for $\mathcal{O}(1)$ leptonic asymmetries may be obtained [8]:

$$\text{(i)}\quad m_{N_\alpha} - m_{N_\beta} \;\sim\; \frac{1}{2}\,\Gamma_{N_{\alpha,\beta}}\,, \qquad \text{(ii)}\quad \frac{\mathrm{Im}\,(h^{\nu\dagger}\,h^\nu)^2_{\alpha\beta}}{(h^{\nu\dagger}\,h^\nu)_{\alpha\alpha}\,(h^{\nu\dagger}\,h^\nu)_{\beta\beta}} \;\sim\; 1\,. \tag{6}$$

Figure 4 shows the impact of the above two conditions on the leptonic asymmetries for a low-scale seesaw model. The generic feature of the resonant conditions remains valid even in the presence of flavour effects [12].

As was already mentioned in Section 3, flavour effects due to neutrino Yukawa couplings play an important role in models of RL. As an illuminating example, we will consider a scenario of RL, which uses the τ-lepton number to generate the BAU and has the potential to be directly tested as we will see in the next section. In such a model, the smallness of the light neutrinos is not due to the usual seesaw mechanism [6], but thanks to the approximate breaking of lepton flavour symmetries [44, 45, 15, 11, 13]. In detail, the heavy neutrino sector is initially assumed to be SO(3), and gets broken down to the identity $\mathbf{I}$ via the neutrino Yukawa couplings in two steps [13]:

$$\mathrm{SO}(3) \;\stackrel{\sim h_\tau}{\longrightarrow}\; \mathrm{U}(1)_{L_e+L_\mu} \times \mathrm{U}(1)_{L_\tau} \;\stackrel{\sim h_e}{\longrightarrow}\; \mathbf{I}\,.$$

As a consequence, the neutrino Yukawa-coupling matrix h^ν takes on the general form

$$h^\nu \;=\; \begin{pmatrix} \varepsilon_e & a\,e^{-i\pi/4} & a\,e^{i\pi/4} \\ \varepsilon_\mu & b\,e^{-i\pi/4} & b\,e^{i\pi/4} \\ \varepsilon_\tau & \kappa_1\,e^{-i(\pi/4-\gamma_1)} & \kappa_2\,e^{i(\pi/4-\gamma_2)} \end{pmatrix}, \tag{7}$$

with $a \sim b \sim 10^{-2} \sim h_\tau$ and $|\varepsilon_{e\,\mu,\tau}| \sim 10^{-7} \sim h_e$ and $\kappa_{1,2} \ll a,\, b$. RG effects violate the SO(3)-invariant form of the Majorana mass matrix $m_M(M_X) = m_N \mathbf{1}_3$ by the 3-by-3 matrix Δm_M. In the leading-log approximation, the SO(3)-breaking matrix Δm_M receives the following radiative corrections:

$$\Delta m_M \;=\; -\frac{m_N}{8\pi^2}\,\ln\left(\frac{M_X}{m_N}\right)\,\mathrm{Re}\left[h^{\nu\dagger}(M_X)h^\nu(M_X)\right]\,. \tag{8}$$

In Fig. 5, we give numerical estimates of the baryon asymmetry η_B for different initial conditions in the resonant τ-leptogenesis (RτL) model described above. The heavy Majorana mass scale is $m_N = 250$ GeV. Our results show that the baryon asymmetry η_B is almost independent on the initial conditions. This property is a consequence of the fixed-point dynamics which exhibits a system that is in quasi-thermal equilibrium. As can be seen from the lower panel in Fig. 5, this almost independence of η_B on the initial conditions persists, even if the heavy Majorana mass m_N is as low as 100 GeV, even though this feature becomes less pronounced and the dynamics of leptogenesis at the electroweak phase transition needs to be carefully considered. In this case, one should take into consideration contributions to the leptonic asymmetries that originate from the transverse polarizations of the W- and Z-boson gauge fields [47].

There have been a number of studies in the recent literature that go beyond the classical approach to Boltzmann equations, introducing thermal effects [48] and aspiring to describe transport phenomena from first principles [49, 50, 51]. The aim of the recent studies is to derive thermally-corrected Boltzmann equations for leptogenesis from the Kadanoff–Baym equations in the Schwinger–Keldysh Closed Time Path (CTP) formalism. Particular approximations and certain assumptions were made to define particle number densities and solve the quantum transport equations, such as the quasi-particle approximation and gradient expansion. In this context, it is worth mentioning a new formalism for a perturbative non-equilibrium thermal field theory [52], which makes use of physically meaningful particle number densities that are directly derivable from the Noether charge. The formalism also enables to resum all orders of gradients, thus capturing more accurately the early non-Markovian regime of non-equilibrium dynamics.

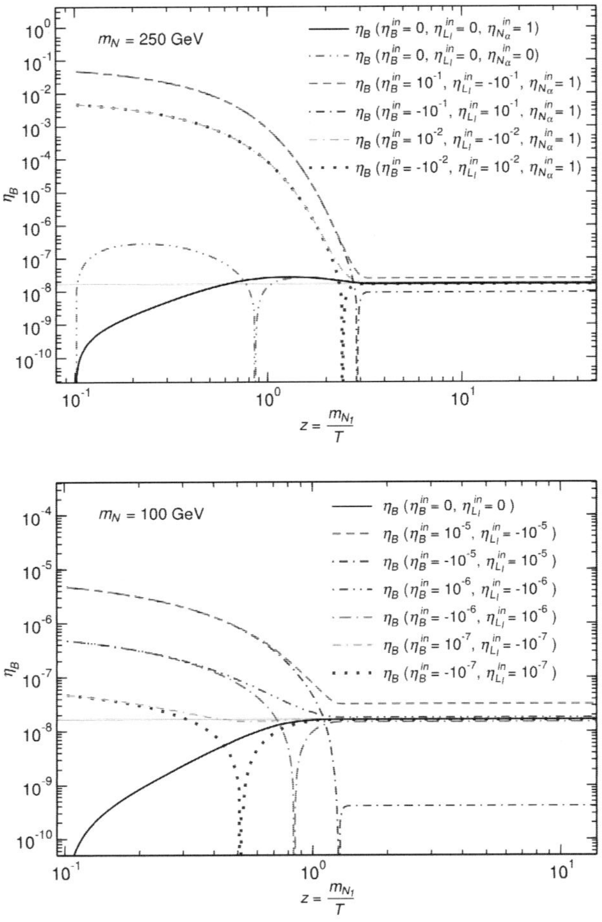

Figure 5. η_B versus $z = m_{N_1}/T$ in the RτL model with $m_N = 250$ GeV (upper panel) and $m_N = 100$ GeV (lower panel).

5. Particle-Physics Phenomenology of Resonant Leptogenesis

RL models can give rise to a number of phenomenologically testable signatures. Here, we will present the generic predictions of RτL models for the $0\nu\beta\beta$ decay, and for processes of lepton flavour violation (LFV), such as $\mu \to e\gamma$, $\mu \to eee$ and $\mu \to e$ conversion in nuclei. Finally, we give numerical estimates of production cross sections of heavy Majorana neutrinos at the LHC.

5.1. $0\nu\beta\beta$ *Decay*

Neutrinoless double beta decay ($0\nu\beta\beta$) corresponds to a process in which two single β decays [53] occur simultaneously in one nucleus. As a consequence of this, a nucleus (Z, A) gets converted into a nucleus $(Z + 2, A)$, i.e.

$$^A_Z X \ \rightarrow \ ^{A}_{Z+2} X \ + \ 2e^- \ .$$

Evidently, this process violates L-number by two units and can naturally take place in minimal RL models, in which the observed light neutrinos are Majorana particles. The observation of such a process would provide further information on the structure of the light neutrino mass matrix $\mathbf{m}^\nu$. In particular, the transition rate of $0\nu\beta\beta$ decay is governed by the effective Majorana neutrino mass $\langle m \rangle$. For the RτL model with inverted hierarchy, the effective neutrino mass is given by

$$|\langle m \rangle| \;=\; |(\mathbf{m}^\nu)_{ee}| \;=\; \frac{v^2}{2m_N}\left|\frac{\Delta m_N}{m_N}\,a^2 \;-\; \varepsilon_e^2\right| \;\approx\; 0.013\ \mathrm{eV}\,, \tag{9}$$

where $\Delta m_N = 2(\Delta m_M)_{23} + i[(\Delta m_M)_{33} - (\Delta m_M)_{22}]$. Such a prediction lies within the reach of future experiments which will be sensitive to values of $|\langle m \rangle|$ of order 10^{-2}, such as SuperNEMO [54].

5.2. $\mu \to e\gamma$, $\mu \to eee$ and $\mu \to e$ conversion

Quantum effects due to heavy Majorana neutrinos may induce LFV couplings to the photon (γ) and the Z boson. These couplings give rise to LFV decays, such as $\mu \to e\gamma$ [55], $\mu \to eee$ [56] and $\mu \to e$ conversion [57]. More details on the actual calculation of these processes may be found in [56, 58]. Also, related phenomenological analyses of LFV effects in the SM with singlet neutrinos may be found in [59, 60, 61, 62].

Let us first consider the decay $\mu \to e\gamma$. In the RτL model, the branching fraction for $\mu \to e\gamma$ is given by

$$B(\mu \to e\gamma) \;=\; 9 \cdot 10^{-4} \,\times\, \frac{|a|^2\,|b|^2\,v^4}{m_N^4}\,. \tag{10}$$

The theoretical prediction should be contrasted with the experimental upper limit [22]:

$$B_{\rm exp}(\mu \to e\gamma) \;<\; 2.4 \times 10^{-12}\,. \tag{11}$$

For the particular scenario considered in Section 4, we find $B(\mu \to e\gamma) \sim 10^{-12}$, which falls well within reach of the sensitivity reported by the MEG collaboration at the 10^{-14} level [63]. In Fig. 6, we exhibit the allowed parameter space of a Minimal RτL model studied in [13], where the required mass splitting among the three heavy neutrinos was induced by RG effects [21] from the GUT scale down to the heavy Majorana mass scale $m_N = 120$ GeV.

We then consider the decay $\mu \to eee$. Assuming an electroweak-scale Majorana mass $m_N = 250$ GeV, its branching ratio $B(\mu \to eee)$ may be related to $B(\mu \to e\gamma)$ through [56]:

$$B(\mu \to eee) \;\simeq\; 1.4 \cdot 10^{-2} \times B(\mu \to e\gamma) \;\simeq\; 1.4 \cdot 10^{-14} \tag{12}$$

This value is a factor ~ 70 below the present experimental bound [22]: $B_{\rm exp}(\mu \to eee) < 1.0 \times 10^{-12}$. It would be therefore encouraging, if higher sensitivity experiments could be designed to further probe this observable.

Finally, one of the most sensitive experiments to probe LFV is the coherent conversion of $\mu \to e$ in nuclei. Specifically, for the ${}^{48}_{22}$Ti case, $B_{\mu e}(26, 22)$ can be also directly linked to $B(\mu \to e\gamma)$ through the relation [64, 65]:

$$B_{\mu e}(26, 22) \;\simeq\; 5 \times B(\mu \to e\gamma)\,, \tag{13}$$

for $m_N = 250$ GeV.

On the experimental side, the upper bound on the $\mu \to e$ conversion rate in ${}^{48}_{22}$Ti [66] is

$$B_{\mu e}^{\rm exp}(26, 22) \;<\; 4.3 \times 10^{-12}\,, \tag{14}$$

at the 90% CL. However, the proposed experiment PRISM/PRIME [67] will be sensitive to conversion rates of order 10^{-18}. Hence, the prediction in (13) falls well within reach of the sensitivity proposed by the PRISM/PRIME collaborations.

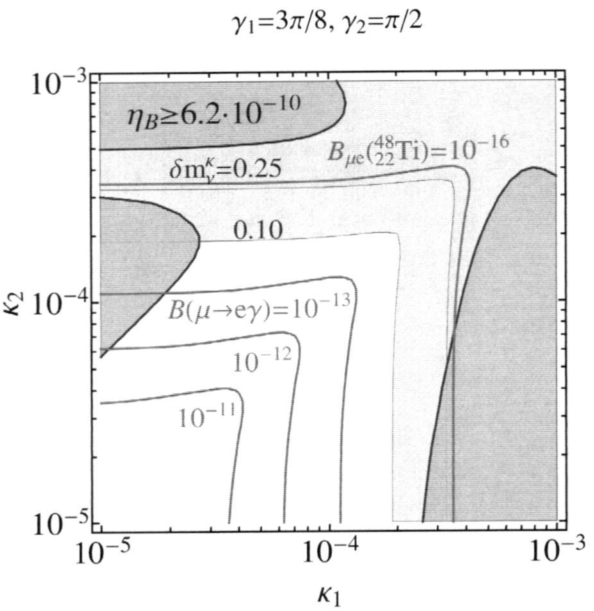

Figure 6. Contour plots of κ_1 versus κ_2 in a Minimal RτL model, with $m_N = 120$ GeV [13].

As opposed to other scenarios of baryogenesis, leptogenesis models in general do not suffer from predicting a too large contribution to the electron electric dipole moment (EDM). This is because EDM effects first arise at two loops and are suppressed either by higher powers of small Yukawa couplings and/or by small factors, such as $(m_{N_1} - m_{N_{2,3}})/m_N$ in RL models [8, 68]. Finally, other manifestations of heavy Majorana neutrino effects are LFV and universality-breaking effects in the Z-boson decays [60, 61] and non-unitarity effects in neutrino oscillations [69].

5.3. Heavy Majorana Neutrino Production at the LHC

Electroweak-scale RL models, such as the RτL model, predict heavy Majorana neutrinos with masses and couplings within the range that could be produced at the LHC. The dominant production channel is through the $W^\pm$-boson exchange graph shown in Fig. 7 [15, 16, 70, 71, 72]. The characteristic signature is the observation of like-sign dileptons with no missing transverse momentum [73].

According to the recent studies [70, 71], the LHC may be rather sensitive to the light-to-heavy neutrino mixing expression $b^2v^2/(2m_N^2) \sim 10^{-3}$–$10^{-4}$, for $m_N = 100$–150 GeV. A recent experimental analysis [74] confirms this expectation. Hence, a possible discovery of heavy Majorana neutrinos at the LHC will unravel the scale of lepton-flavour violation, thus putting the predictions of RτL models into direct test. Finally, low-scale seesaw models of the type considered here could also predict sizeable signatures of simultaneous lepton flavour and lepton number violation, such as $pp \to e^+\mu^+,\ e^-\mu^-,\ e^-\tau^- \ldots$, as well as resonantly enhanced $\mathcal{O}(1)$ CP asymmetries [75]. Thus, a complete exploration of the phenomenological consequences of RL models would be interesting.

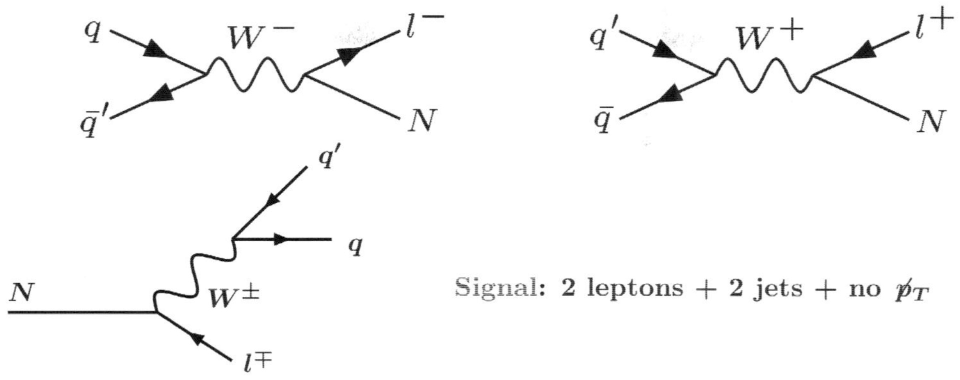

Figure 7. Production and decay of heavy Majorana neutrinos at the LHC.

6. Conclusions

Leptogenesis provides an interesting mechanism for explaining our observable matter–antimatter asymmetric Universe, which may relate to the origin of the neutrino mass. In this brief review, we have paid particular attention to the recent developments on the field, such as the flavourdynamics of leptogenesis and resonant leptogenesis near the electroweak phase transition. In particular, flavourdynamics plays an important role in the predictions for the BAU and in the model-building of phenomenologically testable scenarios. It has been explicitly demonstrated how the observed matter–antimatter asymmetry in the Universe may be successfully explained by thermal resonant leptogenesis at the electroweak scale, independently of any non-zero primordial baryon asymmetry. In fact, any primordial baryon asymmetry will be erased and the actual value of the BAU will be set by the actual mechanism itself, as a consequence of the quasi-in-thermal equilibrium conditions that govern resonant leptogenesis.

Models of resonant leptogenesis may have rich collider phenomenology. Electroweak-mass heavy Majorana neutrinos may give rise to observable signatures of lepton number and lepton flavour violation at the LHC. There may also give rise to correlated predictions for low-energy observables, such as $B(\mu \to e\gamma)$, coherent $\mu \to e$ conversion in nuclei, $B(\mu \to eee)$ and the effective neutrino mass in the $0\nu\beta\beta$ decay. Thus, an exciting era of new discoveries at the LHC and in other low-energy experiments may lie just ahead of us, revealing a greater picture that unifies particle physics and cosmology.

Acknowledgements

I am grateful to the organisers of the conference DISCRETE'12 for their hospitality. The work of AP has been supported in part by the Lancaster-Manchester-Sheffield Consortium for Fundamental Physics under STFC grant ST/J000418/1, and by an IPPP associateship from Durham University.

References

[1] A.D. Sakharov, JETP Lett. 5 (1967) 24.
[2] J. Dunkley *et al.* [WMAP Collaboration], Astrophys. J. Suppl. **180** (2009) 306.
[3] M. Fukugita and T. Yanagida, Phys. Lett. B **174** (1986) 45.

[4] For a comprehensive review, see,
W. Buchmuller, R. D. Peccei and T. Yanagida, Ann. Rev. Nucl. Part. Sci. **55** (2005) 311.
[5] V. A. Kuzmin, V. A. Rubakov and M. E. Shaposhnikov, Phys. Lett. B **155** (1985) 36.
[6] P. Minkowski, Phys. Lett. B **67** (1977) 421;
M. Gell-Mann, P. Ramond and R. Slansky, in *Supergravity*, eds. D.Z. Freedman and P. van Nieuwenhuizen (North-Holland, Amsterdam, 1979);
T. Yanagida, in Proc. of the *Workshop on the Unified Theory and the Baryon Number in the Universe*, Tsukuba, Japan, 1979, eds. O. Sawada and A. Sugamoto;
S.L. Glashow, in *Quarks and Leptons, Cargèse 1979*, eds. M. Lévy, et al., (Plenum 1980 New York), p. 707;
R. N. Mohapatra and G. Senjanović, Phys. Rev. Lett. **44** (1980) 912;
J. Schechter and J. W. F. Valle, Phys. Rev. D **22** (1980) 2227.
[7] M. Y. Khlopov and A. D. Linde, Phys. Lett. B **138** (1984) 265;
J. R. Ellis, J. E. Kim and D. V. Nanopoulos, Phys. Lett. B **145** (1984) 181;
J. R. Ellis, D. V. Nanopoulos and S. Sarkar, Nucl. Phys. B **259** (1985) 175;
J. R. Ellis, G. B. Gelmini, J. L. Lopez, D. V. Nanopoulos and S. Sarkar, Nucl. Phys. B **373** (1992) 399;
R. H. Cyburt, J. R. Ellis, B. D. Fields and K. A. Olive, Phys. Rev. D **67** (2003) 103521;
J. R. Ellis, K. A. Olive and E. Vangioni, Phys. Lett. B **619** (2005) 30;
M. Kawasaki, K. Kohri and T. Moroi, Phys. Lett. B **625** (2005) 7; Phys. Rev. D **71** (2005) 083502.
[8] A. Pilaftsis, Phys. Rev. D **56** (1997) 5431; Int. J. Mod. Phys. A **14** (1999) 1811.
[9] J. Liu and G. Segré, Phys. Rev. D **48** (1993) 4609;
M. Flanz, E.A. Paschos and U. Sarkar, Phys. Lett. B **345** (1995) 248;
L. Covi, E. Roulet and F. Vissani, Phys. Lett. B **384** (1996) 169.
[10] A. Pilaftsis and T. E. J. Underwood, Nucl. Phys. B **692** (2004) 303.
[11] A. Pilaftsis, Phys. Rev. Lett. **95** (2005) 081602.
[12] A. Pilaftsis and T. E. J. Underwood, Phys. Rev. D **72** (2005) 113001.
[13] F. Deppisch and A. Pilaftsis, Phys. Rev. D **83** (2011) 076007.
[14] T. Morozumi and Z. h. Xiong, Prog. Theor. Phys. **111** (2004) 123;
P. Di Bari, Nucl. Phys. B **727** (2005) 318;
O. Vives, Phys. Rev. D **73** (2006) 073006.
[15] A. Pilaftsis, Z. Phys. C **55** (1992) 275.
[16] A. Datta, M. Guchait and A. Pilaftsis, Phys. Rev. D **50** (1994) 3195.
[17] J. Kersten and A. Y. Smirnov, Phys. Rev. D **76** (2007) 073005.
[18] T. Hambye, J. March-Russell and S. M. West, JHEP **0407** (2004) 070;
C. H. Albright and S. M. Barr, Phys. Rev. D **70** (2004) 033013;
E. K. Akhmedov, M. Frigerio and A. Y. Smirnov, JHEP **0309** (2003) 021;
E. J. Chun, Phys. Rev. D **72** (2005) 095010;
S. M. West, Mod. Phys. Lett. A **21** (2006) 1629;
Z. z. Xing and S. Zhou, Phys. Lett. B **653** (2007) 278;
K. S. Babu, A. G. Bachri and Z. Tavartkiladze, Int. J. Mod. Phys. A **23** (2008) 1679;
S. Uhlig, JHEP **0711** (2007) 066;
V. Cirigliano, A. De Simone, G. Isidori, I. Masina and A. Riotto, JCAP **0801** (2008) 004;
T. Asaka and S. Blanchet, Phys. Rev. D **78** (2008) 123527.
[19] For studies of RL in higher-dimensional theories, see
A. Pilaftsis, Phys. Rev. D **60** (1999) 105023;
A. D. Medina and C. E. M. Wagner, JHEP **0612** (2006) 037;
T. Gherghetta, K. Kadota and M. Yamaguchi, Phys. Rev. D **76** (2007) 023516;
M. T. Eisele, Phys. Rev. D **77** (2008) 043510.
[20] Y. Grossman, T. Kashti, Y. Nir and E. Roulet, Phys. Rev. Lett. **91** (2003) 251801;
G. D'Ambrosio, G. F. Giudice and M. Raidal, Phys. Lett. B **575** (2003) 75;
C. S. Fong and M. C. Gonzalez-Garcia, JHEP **0806** (2008) 076 [hep-ph].
[21] R. Gonzalez Felipe, F. R. Joaquim and B. M. Nobre, Phys. Rev. D **70** (2004) 085009;
K. Turzynski, Phys. Lett. B **589** (2004) 135;
G. C. Branco, R. Gonzalez Felipe, F. R. Joaquim and B. M. Nobre, Phys. Lett. B **633** (2006) 336;
G. C. Branco, A. J. Buras, S. Jager, S. Uhlig and A. Weiler, JHEP **0709** (2007) 004;
G. C. Branco, R. Gonzalez Felipe, M. N. Rebelo and H. Serodio, Phys. Rev. D **79** (2009) 093008.
[22] J. Beringer et al. (Particle Data Group), Phys. Rev. D **86** (2012) 010001.
[23] M. Yoshimura, Phys. Rev. Lett. **41** (1978) 281;
S. Dimopoulos and L. Susskind, Phys. Rev. D **18** (1978) 4500.

[24] K. S. Babu and R. N. Mohapatra, Phys. Rev. Lett **109** (2012) 091803.
[25] M. B. Gavela, P. Hernandez, J. Orloff, O. Pene and C. Quimbay, Nucl. Phys. B **430** (1994) 382;
V. A. Rubakov and M. E. Shaposhnikov, Usp. Fiz. Nauk **166** (1996) 493 [Phys. Usp. **39** (1996) 461].
[26] M. S. Carena, J. M. Moreno, M. Quiros, M. Seco and C. E. M. Wagner, Nucl. Phys. B **599** (2001) 158;
M. S. Carena, M. Quiros, M. Seco and C. E. M. Wagner, Nucl. Phys. B **650** (2003) 24;
T. Konstandin, T. Prokopec, M. G. Schmidt and M. Seco, Nucl. Phys. B **738** (2006) 1;
D. J. H. Chung, B. Garbrecht, M. J. Ramsey-Musolf and S. Tulin, Phys. Rev. Lett. **102** (2009) 061301.
[27] J. R. Ellis, J. S. Lee and A. Pilaftsis, JHEP **0810** (2008) 049.
[28] T. Cohen, D. E. Morrissey and A. Pierce, Phys. Rev. D **86** (2012) 013009;
M. Carena, G. Nardini, M. Quiros and C. E. M. Wagner, JHEP **1302** (2013) 001.
[29] For a review, see,
R.N. Mohapatra *et al*, Rev. Prog. Phys. **70** (2007) 1757.
[30] S. Davidson, A. Ibarra, Phys. Lett. B **535** (2002) 25;
W. Buchmuller, P. Di Bari, M. Plumacher, Annals Phys. **315** (2005) 305;
J. Racker, M. Pena and N. Rius, JCAP **1207** (2012) 030.
[31] K. Dick, M. Lindner, M. Ratz, D. Wright, Phys. Rev. Lett. **84** (2000) 4039.
[32] G. Lazarides and Q. Shafi, Phys. Lett. B **258** (1991) 305;
H. Murayama, H. Suzuki, T. Yanagida and J. Yokoyama, Phys. Rev. Lett. **70** (1993) 1912; Phys. Rev. D **50** (1994) 2356;
T. Asaka, K. Hamaguchi, M. Kawasaki and T. Yanagida, Phys. Lett. B **464** (1999) 12.
[33] For a review, see,
M. Dine and A. Kusenko, Rev. Mod. Phys. **76** (2004) 1.
[34] N. Mavromatos and S. Sarkar, arXiv:1211.0968.
[35] S. Y. Khlebnikov and M. E. Shaposhnikov, Nucl. Phys. B **308** (1988) 885;
J. A. Harvey and M. S. Turner, Phys. Rev. D **42** (1990) 3344;
M. Laine and M. E. Shaposhnikov, Phys. Rev. D **61** (2000) 117302.
[36] H. Dreiner, G.G. Ross, Nucl. Phys. B **410** (1993) 188.
[37] J.M. Cline, K. Kainulainen, K.A. Olive, Phys. Rev. D **49** (1994) 6394.
[38] R. Barbieri, P. Creminelli, A. Strumia and N. Tetradis, Nucl. Phys. B **575** (2000) 61.
[39] Y. Nir, J. Racker and E. Roulet, JHEP **0601** (2006) 068;
A. Abada, S. Davidson, F. X. Josse-Michaux, M. Losada and A. Riotto, JCAP **0604** (2006) 004.
[40] T.D. Lee, R. Oehme and C.N. Yang, Phys. Rev. **106** (1957) 340.
[41] E. K. Akhmedov, V. A. Rubakov and A. Y. Smirnov, Phys. Rev. Lett. **81** (1998) 1359.
[42] T. Asaka and M. Shaposhnikov, Phys. Lett. B **620** (2005) 17;
M. Shaposhnikov, JHEP **0808** (2008) 008.
[43] A. Pilaftsis, Nucl. Phys. B **504** (1997) 61.
[44] R.N. Mohapatra and J.W.F. Valle, Phys. Rev. D **34** (1986) 1642;
S. Nandi and U. Sarkar, Phys. Rev. Lett. **56** (1986) 564.
[45] G. C. Branco, W. Grimus and L. Lavoura, Nucl. Phys. B **312** (1989) 492.
[46] For an updated analysis, see,
M. Maltoni, T. Schwetz, M. A. Tortola and J. W. Valle, New J. Phys. **6** (2004) 122.
[47] A. Pilaftsis, Phys. Rev. D **78** (2008) 013008.
[48] G. F. Giudice, A. Notari, M. Raidal, A. Riotto and A. Strumia, Nucl. Phys. B **685** (2004) 89.
[49] M. Garny, A. Hohenegger, A. Kartavtsev and M. Lindner, Phys. Rev. D **81** (2010) 085027.
[50] M. Beneke, B. Garbrecht, C. Fidler, M. Herranen and P. Schwaller, Nucl. Phys. B **843** (2011) 177.
[51] A. Anisimov, W. Buchmüller, M. Drewes and S. Mendizabal, Annals Phys. **326** (2011) 1998.
[52] P. Millington and A. Pilaftsis, arXiv:1211.3152.
[53] M. Doi, T. Kotani and E. Takasugi, Prog. Theor. Phys. Suppl. **83** (1985) 1; H.V. Klapdor-Kleingrothaus and U. Sarkar, Mod. Phys. Lett. A **16** (2001) 2469;
H.V. Klapdor-Kleingrothaus, H. Päs and A. Yu. Smirnov, Phys. Rev. D **63** (2001) 073005;
W. Rodejohann, Nucl. Phys. B **597** (2001) 110; J. Phys. G **28** (2002) 1477;
H. Minakata and H. Sugiyama, Phys. Lett. B **532** (2002) 275;
S.. Pascoli, S. T. Petcov and W. Rodejohann, Phys. Lett. B **549** (2002) 177;
S. Pascoli, S.T. Petcov and T. Schwetz, Nucl. Phys. B **734** (2006) 24;
F. Deppisch and H. Pas, Phys. Rev. Lett. **98** (2007) 232501.
[54] H. Ohsumi [NEMO and SuperNEMO Collaborations], J. Phys. Conf. Ser. **120** (2008) 052054.
[55] T. P. Cheng and L. F. Li, Phys. Rev. Lett. **45** (1980) 1908.
[56] A. Ilakovac and A. Pilaftsis, Nucl. Phys. B **437** (1995) 491.
[57] J. Bernabéu, E. Nardi and D. Tommasini, Nucl. Phys. B **409** (1993) 69;

T.S. Kosmas, G.K. Leontaris and J.D. Vergados, Prog. Part. Nucl. Phys. **33** (1994) 397;
H.C. Chiang, E. Oset, T.S. Kosmas, A. Faessler and J.D. Vergados, Nucl. Phys. A **559** (1993) 526.

[58] A. Ioannisian and A. Pilaftsis, Phys. Rev. D **62** (2000) 066001.

[59] A. Pilaftsis, Mod. Phys. Lett. A **9** (1994) 3595;
M. C. Gonzalez-Garcia and J. W. F. Valle, Mod. Phys. Lett. A **7** (1992) 477; Erratum **9** (1994) 2569;
L. N. Chang, D. Ng and J. N. Ng, Phys. Rev. D **50** (1994) 4589;
G. Bhattacharya, P. Kalyniak and I. Mello, Phys. Rev. D **51** (1995) 3569;
D. Tommasini, G. Barenboim, J. Bernabeu and C. Jarlskog, Nucl. Phys. B **444** (1995) 451;
A. Ilakovac, B. A. Kniehl, and A. Pilaftsis, Phys. Rev. D **52** (1995) 3993;
A. Ilakovac, Phys. Rev. D **54** (1996) 5653;
S. Fajfer and A. Ilakovac, Phys. Rev. D **57** (1998) 4219;
M. Raidal and A. Santamaria, Phys. Lett. B **421** (1998) 250;
M. Czakon, M. Zralek and J. Gluza, Nucl. Phys. B **573** (2000) 57;
J. I. Illana and T. Riemann, Phys. Rev. D **63** (2001) 053004;
G. Cvetic, C. Dib, C. S. Kim and J. D. Kim, Phys. Rev. D **66** (2002) 034008; Phys. Rev. D **71** (2005) 113013;
A. Masiero, S. K. Vempati and O. Vives, New J. Phys. **6** (2004) 202;
F. Deppisch and J. W. F. Valle, Phys. Rev. D **72** (2005) 036001; Nucl. Phys. B **752** (2006) 80.

[60] J. Bernabeu, A. Santamaria, J. Vidal, A. Mendez and J. W. F. Valle, Phys. Lett. B **187** (1987) 303;
J. G. Körner, A. Pilaftsis and K. Schilcher, Phys. Lett. B **300** (1993) 381.

[61] J. Bernabéu, J. G. Körner, A. Pilaftsis and K. Schilcher, Phys. Rev. Lett. **71** (1993) 2695.

[62] C. P. Burgess, S. Godfrey, H. Konig, D. London and I. Maksymyk, Phys. Rev. D **49** (1994) 6115;
E. Nardi, E. Roulet and D. Tommasini, Phys. Lett. B **327** (1994) 319;
S. Bergmann and A. Kagan, Nucl. Phys. B **538** (1999) 368;
F. del Aguila, J. de Blas and M. Perez-Victoria, Phys. Rev. D **78** (2008) 013010.

[63] S. Ritt [MEG Collaboration], Nucl. Phys. Proc. Suppl. **162** (2006) 279.

[64] A. Ilakovac and A. Pilaftsis, Phys. Rev. D **80** (2009) 091902.

[65] R. Alonso, M. Dhen, M. B. Gavela and T. Hambye, JHEP **1301** (2013) 118.

[66] C. Dohmen et al. (SINDRUM II Collaboration), Phys. Lett. **B317** (1993) 631.

[67] Y. Kuno, Nucl. Phys. Proc. Suppl. **149** (2005) 376.

[68] D. Ng and J.N. Ng, Mod. Phys. Lett. A **11** (1996) 211;
J.P. Archaumbault, A. Czarnecki and M. Pospelov, Phys. Rev. D **70** (2004) 073006;
W.-F. Chang and J.N. Ng, New J. Phys. **7** (2005) 65.

[69] For a recent studies on related models, see,
S. Goswami and T. Ota, Phys. Rev. D **78** (2008) 033012;
M. Malinsky, T. Ohlsson and H. Zhang, Phys. Rev. D **79** (2009) 073009.

[70] T. Han and B. Zhang, Phys. Rev. Lett. **97** (2006) 171804;
A. Atre, T. Han, S. Pascoli and B. Zhang, JHEP **0905** (2009) 030.

[71] F. del Aguila, J. A. Aguilar-Saavedra and R. Pittau, JHEP **0710** (2007) 047.

[72] S. Bar-Shalom, G. Eilam, T. Han and A. Soni, Phys. Rev. D **77** (2008) 115019.

[73] W. Y. Keung and G. Senjanovic, Phys. Rev. Lett. **50** (1983) 1427.

[74] U.-K. Yang and J. Almond (on behalf of the Manchester–ATLAS Group), ATLAS-CONF-2012-139.

[75] S. Bray, J. S. Lee and A. Pilaftsis, Nucl. Phys. B **786** (2007) 95.

Testing discrete symmetries with kaons: status and perspectives

Antonio Di Domenico
Dipartimento di Fisica, Sapienza Università di Roma,
and I.N.F.N. Sezione di Roma, P.le A. Moro, 2, I-00185 Rome, Italy
E-mail: antonio.didomenico@roma1.infn.it

Abstract. The status of present experiments and future projects with kaons is reviewed, focusing on prospects for discrete symmetries tests.

1. Introduction

Since their discovery in 1947, K mesons, or kaons, turned out to be one of the most interesting and promising system to study discrete symmetries in particle physics. The discovery of the $\tau - \theta$ anomaly [1] and of CP violation in K^0 decays [2], and the subsequent theory development are two bright examples of the outcomes of these studies.

The present report focuses on the status of present experiments and future projects on this topic; for general and historical reviews the reader is referred to several excellent papers and books in the literature [3].

2. Kaon experiments at hadron machines

2.1. The discovery of direct CP violation

Tens of years after CP violation was established in the $K^0 - \bar{K}^0$ mixing, the ϵ'/ϵ parameter was definitely measured to be different from zero, representing the first indication of direct CP violation in the decay amplitudes, and verifying the Cabibbo-Kobayashi-Maskawa (CKM) theory. This important result was accomplished by two experiments: KTeV at Fermilab and NA48 at CERN. They were third-generation experiments, both taking data at the end of the 90s, using high-energy proton beams to produce long- and short-lived neutral kaons on a fixed target.

2.2. KTeV experiment at FNAL

In the KTeV experiment [4], two neutral kaon beams were formed from the secondary particles produced by 800 GeV/c protons colliding on a beryllium oxide target using a system of collimators, absorbers and sweeping magnets. The neutral kaon decays were detected in 110 - 158 m range from the production target. One of the beams passed through an active regenerator, made of scintillator, which produced a coherent mixture of K_S and K_L states. The regenerator alternated between the two neutral beams in order to reduce systematic differences between K_L and K_S decays.

The charged decay products were detected in a drift chamber spectrometer. The spectrometer was equipped with two chambers before and two after an analyzing magnet. Each chamber measured charged particle tracks in horizontal and vertical views. The neutral decay products were measured in a calorimeter composed by an array of 3100 Cesium Iodine (CsI) crystals, located after the spectrometer. A nearly hermetic photon veto system (up to 100 mrad) rejected background events for the $\pi^0\pi^0$ mode coming from interactions in the regenerator, semileptonic and $K_L \to \pi^0\pi^0\pi^0$ decays.

2.3. NA48 experiment at CERN

The main feature of the NA48 experiment [5] was the two simultaneous, almost collinear beams of neutral kaons derived from proton beams from the CERN SPS, and delivered to two fixed targets. The kaon beams had a common decay region and decays from both beams were recorded with the same detector. Both targets were made from beryllium and had a length of 400 mm and a diameter of 2 mm. The far target was located 126 m before the beginning of the decay region while the near target was only 6 m up-stream of the fiducial region and displaced by 7.2 cm in the vertical direction from the axis of the far-target beam. The two beam axes had an angle of 0.6 mrad with respect to each other and crossed at the longitudinal position of the electromagnetic calorimeter, 120 m down-stream of the near target. In both beams, charged particles were deflected by sweeping magnets. Particle decays from the far target were almost exclusively K_L decays, while decays originating from the near target were mainly K_S (and neutral hyperon) decays with, however, a small component of K_L decays. The NA48 detector was designed to measure with high precision momenta of both charged and neutral particles. The charged particle reconstruction was provided by a magnetic spectrometer, with 4 drift chambers and a magnet. The reconstruction of photon energy, direction, time, and position was given by a LKr calorimeter.

2.4. The legacy of NA48 and KTeV experiments

The two collaborations published the final measurement of $\Re(\epsilon'/\epsilon)$ as $(19.2 \pm 2.1) \times 10^{-4}$ (KTeV) [6] and $(14.7 \pm 2.2) \times 10^{-4}$ (NA48) [7] where $\Re(\epsilon'/\epsilon)$ is obtained from the double ratio of decay rates:

$$\frac{\Gamma(K_L \to \pi^0\pi^0)/\Gamma(K_S \to \pi^0\pi^0)}{\Gamma(K_L \to \pi^+\pi^-)/\Gamma(K_S \to \pi^+\pi^-)} = \frac{|\eta_{00}|^2}{|\eta_{+-}|^2} \simeq 1 - 6\Re(\epsilon'/\epsilon) , \tag{1}$$

with

$$\begin{aligned} \eta_{00} &\equiv |\eta_{00}|e^{i\phi_{00}} = \frac{\langle\pi^0\pi^0|T|K_L\rangle}{\langle\pi^0\pi^0|T|K_S\rangle} = \epsilon - 2\epsilon' \\ \eta_{+-} &\equiv |\eta_{+-}|e^{i\phi_{+-}} = \frac{\langle\pi^+\pi^-|T|K_L\rangle}{\langle\pi^+\pi^-|T|K_S\rangle} = \epsilon + \epsilon' . \end{aligned} \tag{2}$$

Combining all the recent measurements, the world average is $\Re(\epsilon'/\epsilon) = (16.8 \pm 2.0) \times 10^{-4}$ [8], clearly demonstrating the existence of the CP violation in decay. However, due to theoretical uncertainties in the hadronic matrix elements, to get information from $\Re(\epsilon'/\epsilon)$ on the Standard Model (SM) and New Physics (NP) beyond it, is difficult and remains a challenge to theoretical calculations. Recent progresses in lattice QCD calculations promise to solve this issue [9].

The KTeV collaboration measured also the $\Im(\epsilon'/\epsilon)$ parameter, exploiting the coherent regeneration phenomenon to modify the pure K_L beam into a coherent superposition of K_L and K_S. The fit of the measured $\pi^+\pi^-$ (and $\pi^0\pi^0$) decay intensity downstream the regenerator yielded the best CPT tests in two pion decays [6]: $\phi_{00} - \phi_{+-} \simeq -3\Im(\epsilon'/\epsilon) = (0.30 \pm 0.35)^\circ$ and

$\phi_\epsilon - \phi_{SW} = (0.40 \pm 0.56)^\circ$, with $\phi_\epsilon \simeq (\phi_{00} + 2\phi_{+-})/3$, $\phi_{SW} = \arctan(2\Delta m/\Delta\Gamma)$ the so called *superweak* phase, and consistent with no CPT violation.

The unitarity relation, originally derived by Bell and Steinberger [10]:

$$\left(\frac{\Gamma_S + \Gamma_L}{\Gamma_S - \Gamma_L} + i \tan\phi_{SW}\right)\left[\frac{\Re\bar{\epsilon}}{1+|\bar{\epsilon}|^2} - i\Im\delta\right] = \\ = \frac{1}{\Gamma_S - \Gamma_L}\sum_f \mathcal{A}^*(\mathrm{K_S} \to f)\mathcal{A}(\mathrm{K_L} \to f) \equiv \sum_f \alpha_f \ , \qquad (3)$$

where the sum runs over all accessible final states f appearing in the decay amplitudes $\mathcal{A}(\mathrm{K_{S,L}} \to f)$, can be used to bound the CPT violating parameter[1] $\Im\delta$ (and other parameters), after having provided all the α_i parameters, Γ_S, Γ_L, and ϕ_{SW} as inputs.

The analysis including the latest results by the KTeV collaboration [6] yields significant limits on $\Im\delta$ and the neutral kaon mass difference (tests of the CPT symmetry): $\Im\delta = (-1.5 \pm 1.6) \times 10^{-5}$ and $|m_{\mathrm{K}^0} - m_{\bar{\mathrm{K}}^0}| < 4.8 \times 10^{-19}$ GeV at 95 % CL, slightly improving the results obtained in a previous analysis [12].

2.5. NA48/2 experiment at CERN

The NA48/2 collaboration at CERN operated with simultaneous K^+ and K^- beams of 60 GeV/c in 2003-2004. The charge asymmetry of the Dalitz plot linear slopes $A_g = (g_+ - g_-)/(g_+ + g_-)$ was measured for both $\mathrm{K}^\pm \to \pi^\pm\pi^+\pi^-$ and $\mathrm{K}^\pm \to \pi^\pm\pi^0\pi^0$ decays, being $A_g \neq 0$ a manifestation of CP violation in decay. The charge asymmetries were measured to be $A_g^c = (-1.5 \pm 2.2) \times 10^{-4}$ with 3.11×10^9 $\mathrm{K}^\pm \to \pi^\pm\pi^+\pi^-$ decays, and $A_g^n = (1.8 \pm 1.8) \times 10^{-4}$ with 9.13×10^7 $\mathrm{K}^\pm \to \pi^\pm\pi^0\pi^0$ decays [13], still consistent with no CP violation.

2.6. Searches for rare kaon decays at LHCb

It's worth mentioning that searches for rare kaon decays are also possible at LHC, at least for the cleanest decay channels with a closed kinematics and a controllable background. This holds for the $\mathrm{K_S} \to \mu^+\mu^-$ decay, which is a Flavour Changing Neutral Current (FCNC) transition that has not yet been observed. This decay is suppressed in the SM, with an expected branching fraction [14] $\mathrm{BR}(\mathrm{K_S} \to \mu^+\mu^-) = (5.0 \pm 1.5) \times 10^{-12}$, while the current experimental upper limit is 3.2×10^{-7} at 90% confidence level (C.L.) [15]. Although the dimuon decay of the $\mathrm{K_L}$ meson is known to be $\mathrm{BR}(\mathrm{K_L} \to \mu^+\mu^-) = (6.84 \pm 0.11) \times 10^{-9}$ [8], in agreement with the SM, effects of new particles can still be observed in $\mathrm{K_S} \to \mu^+\mu^-$ decays.

The LHCb detector, described in detail in Ref.[16], is a single-arm forward spectrometer covering the pseudorapidity range $2 < \eta < 5$. A search for the decay $\mathrm{K_L} \to \mu^+\mu^-$ is performed, based on a data sample of 1.0 fb^{-1} of pp collisions at $\sqrt{s} = 7$ TeV collected by the LHCb experiment at the Large Hadron Collider. The observed number of candidates is consistent with the background only hypothesis, yielding an upper limit of $\mathrm{BR}(\mathrm{K_S} \to \mu^+\mu^-) < 11(9) \times 10^{-9}$ at 95 (90)% confidence level. This limit is a factor of thirty below the previous measurement [17, 18].

3. Future kaon experiments at hadron machines

3.1. The intensity frontier for NP searches

The present collider searches at high transverse momenta exclude wide regions of the parameter space for NP thus increasing the importance of the study of rare processes sensitive to very high

[1] The two physical states can be expressed as $|\mathrm{K_S}\rangle \propto [|\mathrm{K_1}\rangle + \epsilon_S|\mathrm{K_2}\rangle]$ and $|\mathrm{K_L}\rangle \propto [|\mathrm{K_2}\rangle + \epsilon_L|\mathrm{K_1}\rangle]$, with ϵ_S and ϵ_L two small complex parameters describing the CP impurity in the physical states. One can equivalently define $\bar{\epsilon} \equiv (\epsilon_S + \epsilon_L)/2$, and $\delta \equiv (\epsilon_S - \epsilon_L)/2$; adopting a suitable phase convention (e.g. the Wu-Yang phase convention [11]) $\bar{\epsilon} \neq 0$ implies T violation, $\delta \neq 0$ implies CPT violation, while $\delta \neq 0$ or $\bar{\epsilon} \neq 0$ implies CP violation.

energy. This is true, in particular, for the $K \to \pi\nu\bar{\nu}$ decays, whose SM predictions are known with very high precision [19]. The branching ratio expectation of the CP-violating $K_L \to \pi^0\nu\bar{\nu}$ channel is: $BR(K_L \to \pi^0\nu\bar{\nu}) = (0.243 \pm 0.039 \pm 0.006) \times 10^{-10}$, while for the charged mode is: $BR(K^+ \to \pi^+\nu\bar{\nu}) = (0.781 \pm 0.075 \pm 0.029) \times 10^{-10}$. In both cases, the first error summarizes the parametric uncertainties, while the second one the remaining theoretical error. The extraction of the hadronic part of the amplitude is obtained from the very well measured $K_{\ell 3}$ decay. The main uncertainties are due to the knowledge of the CKM matrix elements.

Processes mediated by FCNC are suppressed in the SM by the GIM mechanisms [20]. Further suppression can originate from hierarchies in CKM matrix or helicity. In some models of NP large enhancement with respect to SM rates are possible. In this sense the $K \to \pi\nu\bar{\nu}$ modes play a key role in seeking NP beyond the SM. This is clear from Fig.1, which reproduces the BR values predicted by the SM and by some models of NP [21]. Apart from establishing a direct signal of NP, the correlation of the BR of the two modes can be exploited to probe the flavour structure of NP theories and, therefore, to distinguish among different classes of NP scenarios. The exclusion regions given by the Grossman-Nir consistency conditions [22], limiting the BR ratio of the two decay modes, and by the only measurement of the charged mode [23] are also reported in Fig.1.

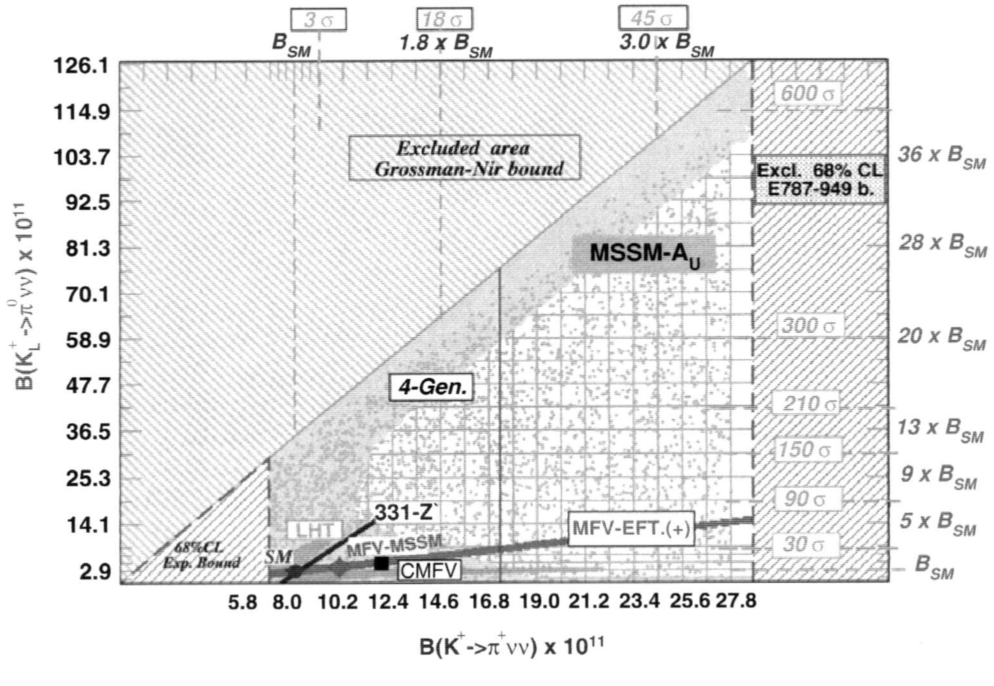

Figure 1. Correlation between the branching ratios of the charged and neutral $K \to \pi\nu\bar{\nu}$ decays showing the predictions of SM and some NP models [21]

The measurement $BR(K^+ \to \pi^+\nu\bar{\nu}) = (1.73 \pm 1.15) \times 10^{-10}$ is based on 7 events observed by E959 and E787 experiments at BNL [23], and is compatible with the SM prediction. The CP-violating neutral $K_L \to \pi^0\nu\bar{\nu}$ decay has not been observed yet. An upper limit to the BR was set by the E391a collaboration at KEK [24]: $BR(K_L \to \pi^0\nu\bar{\nu}) < 260 \times 10^{-10}$ (90% C.L.).

3.2. NA62 experiment at CERN

The main goal of the NA62 experiment at CERN [25] is to collect about 100 $K^+ \to \pi^+\nu\bar{\nu}$ events with 10% of background in two years of data taking. This implies to collect more than 10^{13} K decays with a background rejection factor of at least 10^{12}, assuming a signal acceptance of 10%.

The experimental method relies on exploiting a decay-in-flight technique with an intense charged kaon beam at high energy. This requires a beam of un-separated charged hadrons (with 6% K^+), a long decay region with an extended detector and the event-by-event measurement of kaon momenta. Positive charged kaons will be used in order to get, at the same proton energy and flux, higher kaon fluxes and lower pion background.

The incoming kaon is measured by the Gigatracker system in the beam. The charged decay particle is measured by the straw-chamber spectrometer and is identified by the Ring Imaging Cerenkov (RICH) detector and the muon-veto sampling calorimeter. The LKr calorimeter, originally built for NA48, is used as a veto for forward photons. Photons at large angles are intercepted by a series of 12 ring-shaped veto counters constructed using lead-glass blocks from the OPAL electromagnetic barrel calorimeter.

The NA62 analysis strategy is based on the accurate kinematic reconstruction of all the particles detected in the event to disentangle the signal from the huge amount of background processes, a precise timing to associate correctly the π^+ with the parent K^+, a system of efficient vetoes to reject events with γ and μ in the final state, effective particle identification systems to identify K^+ among other particles of the intense hadron beam and to distinguish π^+ from μ^+ and e^+ in the final states.

At present, the construction of the experimental apparatus is very advanced (see Ref. [26]). A commissioning run has taken place in the second half of 2012, with a subset of the detector and of the data acquisition and read-out systems. Other technical runs are foreseen during 2013, while completing the installation of detectors, which will continue in 2013 and 2014. The first physics run is expected at the end of 2014 (after CERN machines shutdown).

It's worth mentioning that at an early stage of the NA62 project, in 2007-08, physics data have been collected using an optimized kaon beam line and the former NA48/2 detector to measure the ratio $R_K = \Gamma(K^\pm \to e\nu_e)/\Gamma(K^\pm \to \mu\nu_\mu)$. The result $R_K = (2.488 \pm 0.010) \times 10^{-5}$ is the most precise measurement to date and agrees with the SM predictions [27].

3.3. KOTO experiment at J-PARC

The KOTO experiment [28, 29] at the new high-intensity proton accelerator facility J-PARC (Japan Proton Accelerator Research Complex) [30] aims to observe for the first time the rare decay $K_L \to \pi^0\nu\bar{\nu}$. The challenge consists in being a so-called *nothing in - nothing out* experiment, with both initial state and final decay products completely neutral.

A decay volume for K_L is surrounded by particle detectors. The signature of a $K_L \to \pi^0\nu\bar{\nu}$ decay is that there are two photons from a π^0 decay and no other visible particles in the final state. An electromagnetic calorimeter is placed downstream of the decay volume to detect the two photons. All the K_L decay modes except $K_L \to \pi^0\nu\bar{\nu}$ and $K_L \to \gamma\gamma$ have at least two charged particles, or two or more extra photons in the final state. These decays can be rejected by detecting additional particles with the surrounding detectors. The $K_L \to \gamma\gamma$ decays can be rejected by requiring a finite transverse momentum for the two photon system. In the case of the $K_L \to \pi^0\nu\bar{\nu}$ decay, the two photon system has a finite transverse momentum, because the undetected two neutrinos take some momentum away. The decay vertex is calculated and the $K_L \to \pi^0\nu\bar{\nu}$ decay reconstructed from two photons in the calorimeter, with the assumption that the two photons come from a π^0 decay on the line of flight.

The new neutral beam-line has been built at the Hadron Experimental Hall of J-PARC. The CsI crystal calorimeter (made of the crystals used in the past by the KTeV experiment), the charged-particle veto counters (CV) in front of it, and the main-barrel photon veto counters (MB) which surround the decay volume in the vacuum vessel, have been built, while the construction of the front-end photon counters is being completed.

After the major Earthquake in Japan on 11 March 2011 (no damages to beam line and KOTO detector) J-PARC started re-commissioning on December 2011 and resumed operations on January 2012. Commissioning and engineering runs have been scheduled during beam times at the beginning of 2013. In the first physics run, expected for May - June 2013, the KOTO experimental sensitivity will cross the Grossman-Nir bound allowing new physics search to start. After the long shutdown of the J-PARC accelerators from August 2013 to January 2014 - with an upgrade of the linac - the KOTO experiment is expected to start a series of long physics runs, reaching the final sensitivity at the level of the SM prediction of BR($K_L \to \pi^0 \nu\bar{\nu}$), therefore extensively exploring the parameter space of NP models.

3.4. ORKA experiment at FNAL

ORKA is a proposed experiment [31] aiming to detect about 1000 $K^+ \to \pi^+ \nu\bar{\nu}$ events and to measure the corresponding branching ratio with 5% precision using the Fermilab Main Injector high intensity proton source. The detector design is based on the former BNL E787/E949 experiments, which detected seven candidate events [23]. Two orders of magnitude improvement in sensitivity relative to the BNL experiments comes from enhancements to the beam line and the detector acceptance.

The ORKA collaboration has the support of the FNAL PAC and the FNAL directorate. The collaboration expects to begin construction of ORKA in 2014 and begin collecting $K^+ \to \pi^+ \nu\bar{\nu}$ decays in 2017.

3.5. TREK experiment at J-PARC

In the $K^+ \to \pi^0 \mu^+ \nu$ decay, the transverse muon polarization p_t (the perpendicular component of the muon spin vector relative to the decay plane determined by the momentum vectors of the muon and the pion in the K^+ rest frame) is a T -odd quantity. The SM prediction of p_t is almost vanishing ($\sim 10^{-7}$) and a non-zero value for such quantity at a level above 10^{-4} would be an indication of time-reversal violation in NP, still allowed by the current limit ($p_t < 5 \times 10^{-3}$) set by the KEK E246 experiment [32].

A successor to E246 is another new kaon experiment at J-PARC, E06 TREK [33], aiming at a p_t sensitivity of 10^{-4}. This experiment is conducted in conjunction with the E36 one aiming to measure the R_K ratio with a relative uncertainty of $\sim 2.5 \times 10^{-3}$, which is about half of the current world record. The physics run of E36 is expected in 2014-15.

3.6. Other projects

Project X is a proposed proton accelerator complex at Fermilab that would provide particle beams to multiple experiments searching for rare and hard-to-detect phenomena [34]. Construction for Project X could begin in 2016 and be completed in 2021. The plans include ultra-rare $K^\pm$ and K_L decay measurements, and neutral kaon interferometry and CPT tests at a very high level of sensitivity.

The KLOD R&D program for a K_L measurement at Protvino should also be mentioned, as well as the OKA project, in the same laboratory, using a separated charged kaon beam.

Journal of Physics: Conference Series **447** (2013) 012008 doi:10.1088/1742-6596/447/1/012008

4. Kaon experiments at e^+e^- collider

4.1. Neutral kaon interferometry at ϕ-factory

At a ϕ-factory neutral kaons are produced in entangled pairs with the ϕ-meson quantum numbers $J^{PC} = 1^{--}$:

$$
\begin{aligned}
|i\rangle &= \frac{1}{\sqrt{2}}\{|\mathrm{K}^0\rangle|\bar{\mathrm{K}}^0\rangle - |\bar{\mathrm{K}}^0\rangle|\mathrm{K}^0\rangle\} \\
&= \frac{\mathcal{N}}{\sqrt{2}}\{|\mathrm{K_S}\rangle|\mathrm{K_L}\rangle - |\mathrm{K_L}\rangle|\mathrm{K_S}\rangle\}\ , \qquad (4)
\end{aligned}
$$

where $\mathcal{N} = \sqrt{(1+|\epsilon_S|^2)(1+|\epsilon_L|^2)/(1-\epsilon_S\epsilon_L)} \simeq 1$ is a normalization factor.

The observable quantity is the double differential decay rate of the state $|i\rangle$ into decay products f_1 and f_2 at proper times t_1 and t_2, respectively. After integration on (t_1+t_2) at fixed time difference $\Delta t = t_1 - t_2$, the decay intensity can be written as follows [35]:

$$
\begin{aligned}
I(f_1, f_2; \Delta t \geq 0) = C_{12}\{&|\eta_1|^2 e^{-\Gamma_L \Delta t} + |\eta_2|^2 e^{-\Gamma_S \Delta t} \\
&-2|\eta_1||\eta_2| e^{-\frac{(\Gamma_S+\Gamma_L)}{2}\Delta t}\cos[\Delta m \Delta t + \phi_2 - \phi_1]\}\ . \qquad (5)
\end{aligned}
$$

This expression is valid for $\Delta t \geq 0$, while for $\Delta t < 0$ the substitutions $\Delta t \to |\Delta t|$ and $1 \leftrightarrow 2$ have to be applied, and with $\Delta m = m_L - m_S$,

$$
\eta_i \equiv |\eta_i| e^{i\phi_i} = \frac{\langle f_i|T|\mathrm{K_L}\rangle}{\langle f_i|T|\mathrm{K_S}\rangle}\ , \qquad (6)
$$

$$
C_{12} = \frac{|\mathcal{N}|^2}{2(\Gamma_S+\Gamma_L)}|\langle f_1|T|\mathrm{K_S}\rangle\langle f_2|T|\mathrm{K_S}\rangle|^2\ . \qquad (7)
$$

Due to the huge difference in the lifetimes of the physical states ($\tau_L \gg \tau_S$), for $t_1 \gg t_2, \tau_S$ (or $t_2 \gg t_1, \tau_S$) the decay intensity in eq.(5) behaves like the initial state were an incoherent mixture of states $|\mathrm{K_S}\rangle|\mathrm{K_L}\rangle$ and $|\mathrm{K_L}\rangle|\mathrm{K_S}\rangle$. Hence the detection of a kaon at large times *tags* a $\mathrm{K_S}$ in the opposite direction. This is a unique feature at a ϕ-factory, not possible at fixed target experiments, that can be exploited to select very pure $\mathrm{K_S}$ beams.

4.2. KLOE experiment at DAΦNE

DAΦNE, the Frascati ϕ-factory is an e^+e^- collider working at a center of mass energy of $\sqrt{s} \sim 1020$ MeV, corresponding to the peak of the ϕ resonance. The ϕ-meson production cross section is $\sim 3\mu$b, and its decay into $\mathrm{K}^0\bar{\mathrm{K}}^0$ has a branching fraction of 34%, yielding $\sim 10^6$ $\mathrm{K}^0\bar{\mathrm{K}}^0$ pairs per pb^{-1} of integrated luminosity.

The KLOE detector is a 4π detector setup, which is able to measure both charged and neutral particles. It consists of a large volume drift chamber [36], which provides excellent momentum and vertex reconstruction for charged particles, and a barrel shaped electromagnetic calorimeter with two end-caps [37], made from lead and scintillating fibers, which surrounds the drift chamber. The energy deposits of charged and neutral particles in the calorimeter are measured with very good time resolution, which allows for the identification of charged particles based on their time of flight. Drift chamber and calorimeter are enclosed in a superconducting solenoid, providing an axial 0.52 T magnetic field.

The KLOE experiment at DAΦNE completed its first data taking campaign in March 2006 with a total integrated luminosity of ~ 2.5 fb^{-1}, corresponding to a production of $\sim 7.5 \times 10^9$ ϕ-mesons and $\sim 2.5 \times 10^9$ $\mathrm{K}^0\bar{\mathrm{K}}^0$ pairs.

The quantum interference between the two kaon decays in the CP violating channel $\phi \to \mathrm{K_S K_L} \to \pi^+\pi^-\pi^+\pi^-$ has been observed for the first time by KLOE [38]. This decay mode is very rich in physics; in fact several kinds of possible decoherence and CPT violation mechanisms, which in some cases might be justified in a quantum gravity framework, could show up as a deviation from the quantum mechanical prediction in eq.(5), especially in the region at $\Delta t \approx 0$, where $I(\pi^+\pi^-, \pi^+\pi^-; \Delta t = 0) = 0$. No deviation from the expectations of quantum mechanics and CPT symmetry is observed, while the precision on the measurements of the corresponding parameters is very high, and - in some cases - reaches the interesting Planck scale region. These results, together to other CPT symmetry tests in $\mathrm{K_S} \to \pi e \nu$ decays, have been thoroughly reviewed in Refs. [35, 39, 40]. In the following very recent KLOE results on CP and CPT tests are reported, while a detailed discussion is referred elsewhere [41, 42, 43].

4.3. Search for CP violation in $\mathrm{K_S}$ *decay*

The decay $\mathrm{K_S} \to 3\pi^0$ violates CP invariance. The parameter η_{000}, defined as the ratio of $\mathrm{K_S}$ to $\mathrm{K_L}$ decay amplitudes, can be written as:

$$\eta_{000} = \frac{\langle 3\pi^0|T|\mathrm{K_S}\rangle}{\langle 3\pi^0|T|\mathrm{K_L}\rangle} = \epsilon_S + \epsilon'_{000} \; , \tag{8}$$

where ϵ'_{000} is due to a direct CP-violating term. Since we expect $\epsilon'_{000} \ll \epsilon_S$ (at lowest order in Chiral Perturbation Theory one has [44, 45]: $\epsilon'_{000} = -2\epsilon'$, with ϵ' the direct CP violation parameter in $\pi\pi$ decays), it follows that $\eta_{000} \sim \epsilon_S$, and therefore in the SM one has (assuming CPT invariance, i.e. $\epsilon_S = \epsilon$) $\mathrm{BR}(\mathrm{K_S} \to 3\pi^0) \sim 1.9 \times 10^{-9}$ to an accuracy of a few %, making the direct observation of this decay quite a challenge.

The best upper limit on $\mathrm{BR}(\mathrm{K_S} \to 3\pi^0)$ comes from the analysis of 450 pb^{-1} data collected by the KLOE experiment in years 2001-2002 [46]: $\mathrm{BR}(\mathrm{K_S} \to 3\pi^0) < 1.2 \times 10^{-7}$ at 90% C.L.

In a new improved analysis, using 1.7 fb^{-1} of data collected in years 2004-2005, $n_s = 0$ candidate events for the signal are found, with $n_{bkg} = 0$ background events expected from Monte Carlo with an effective statistics of two times that of the data. Hence the final KLOE upper limit is [41, 42]:

$$\mathrm{BR}(\mathrm{K_S} \to 3\pi^0) < 2.6 \times 10^{-8} \quad \text{at 90\% C.L.,} \tag{9}$$

which is almost five times lower than the latest published result [46]. This limit can be directly translated into a limit on $|\eta_{000}|$:

$$\begin{aligned} |\eta_{000}| &= \sqrt{\frac{\tau_L}{\tau_S} \frac{\mathrm{BR}(\mathrm{K_S} \to 3\pi^0)}{\mathrm{BR}(\mathrm{K_L} \to 3\pi^0)}} \\ &< 0.0088 \quad \text{at 90\% C.L.} \; . \end{aligned} \tag{10}$$

4.4. CPT and Lorentz symmetry test using neutral kaon interferometry

CPT invariance holds for any realistic Lorentz-invariant quantum field theory. However a very general theoretical possibility for CPT violation is based on spontaneous breaking of Lorentz symmetry, as developed by Kostelecký [47, 48, 49], which appears to be compatible with the basic tenets of quantum field theory and retains the property of gauge invariance and renormalizability (Standard Model Extensions - SME). In SME for neutral kaons, CPT violation manifests to lowest order only in the mixing parameter δ, (e.g. vanish at first order in the decay amplitudes), and exhibits a dependence on the 4-momentum of the kaon:

$$\delta \approx i \sin\phi_{SW} e^{i\phi_{SW}} \gamma_K (\Delta a_0 - \vec{\beta_K} \cdot \Delta\vec{a})/\Delta m \tag{11}$$

where γ_K and $\vec{\beta_K}$ are the kaon boost factor and velocity in the observer frame, and Δa_μ are four CPT- and Lorentz-violating coefficients for the two valence quarks in the kaon.

Following Ref. [48], the time dependence arising from the rotation of the Earth can be explicitly displayed in eq.(11) by choosing a three-dimensional basis $(\hat{X}, \hat{Y}, \hat{Z})$ in a non-rotating frame, with the $\hat{Z}$ axis along the Earth's rotation axis, and a basis $(\hat{x}, \hat{y}, \hat{z})$ for the rotating (laboratory) frame (see Fig.2). The CPT violating parameter δ may then be expressed as:

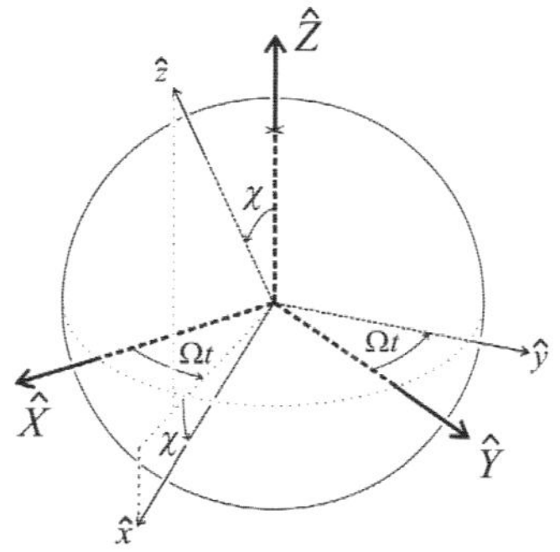

Figure 2. Basis $(\hat{x}, \hat{y}, \hat{z})$ for the rotating frame, and basis $(\hat{X}, \hat{Y}, \hat{Z})$ for the fixed non-rotating frame. The laboratory frame precesses around the Earth's rotation axis $\hat{Z}$ at the sidereal frequency Ω. The angle between $\hat{Z}$ and the positron beam direction $\hat{z}$ defined in the laboratory frame of KLOE is $\chi \simeq 113°$.

$$\begin{aligned}
\delta(\vec{p}, t_{sid}) \quad = \quad & \frac{i \sin\phi_{SW} e^{i\phi_{SW}}}{\Delta m} \gamma_K \{\Delta a_0 \\
& +\beta_K \Delta a_Z \cos\theta \cos\chi \\
& -\beta_K \Delta a_Z \sin\theta \cos\phi \sin\chi \\
& -\beta_K \Delta a_X \sin\theta \sin\phi \sin\Omega t_{sid} \\
& +\beta_K \Delta a_X \cos\theta \sin\chi \cos\Omega t_{sid} \\
& +\beta_K \Delta a_X \sin\theta \cos\phi \cos\chi \cos\Omega t_{sid} \\
& +\beta_K \Delta a_Y \cos\theta \sin\chi \sin\Omega t_{sid} \\
& +\beta_K \Delta a_Y \sin\theta \cos\phi \cos\chi \sin\Omega t_{sid} \\
& +\beta_K \Delta a_Y \sin\theta \sin\phi \cos\Omega t_{sid}\} \qquad (12)
\end{aligned}$$

where $\vec{p}$ is the kaon momentum, t_{sid} is the sidereal time, Ω is the Earth's sidereal frequency, $\cos\chi = \hat{z} \cdot \hat{Z}$; θ and ϕ are the conventional polar and azimuthal angles of the kaon momentum defined in the laboratory frame about the $\hat{z}$ axis. The sensitivity to the four Δa_μ parameters can be very different for fixed target and collider experiments, showing complementary features [48]. At a fixed target experiment usually the kaon momentum direction is fixed, while $|\vec{p}|$ might vary within a certain interval. On the contrary, at a ϕ-factory kaons are emitted in all directions with the characteristic *p-wave* angular distribution $dN/d\Omega \propto \sin^2\theta$, while $|\vec{p}|$ is almost fixed[2].

At KLOE the analysis strategy to measure the four Δa_μ parameters is based on exploiting neutral kaon interferometry. In particular when $f_1 = f_2 = \pi^+\pi^-$ the corresponding η_i parameters can be slightly different for the two kaons due to the momentum dependence of the CPT violation effects as come from eq.(12):

$$\begin{aligned}
\eta_1 &= \epsilon_L + \epsilon' = \bar{\epsilon} - \delta(\vec{p_1}, t_{sid}) + \epsilon' \\
\eta_2 &= \epsilon_L + \epsilon' = \bar{\epsilon} - \delta(\vec{p_2}, t_{sid}) + \epsilon' \ , \qquad (13)
\end{aligned}$$

[2] At DAΦNE $|\vec{p}|$ is not fixed because of a small ϕ meson momentum $\vec{p_\phi}$ in the laboratory frame ($|\vec{p_\phi}| \simeq 13$ MeV/c).

with $\vec{p_2} = \vec{p_\phi} - \vec{p_1}$. The distribution $I(f_1, f_2; \Delta t)$ is sensitive to any deviation from unity of the ratio η_1/η_2 in the interference region (i.e. $\Delta t \approx 0$). Therefore a suitable analysis of the decays $\phi \to \mathrm{K_S K_L} \to \pi^+\pi^-,\ \pi^+\pi^-$ as a function of sidereal time and kaon momenta can provide a measurement of the four parameters Δa_μ.

The two kaons are distinguished by their emission in the forward ($\cos\theta > 0$) or backward ($\cos\theta < 0$) hemispheres, as sketched in Fig.3. The data sample is divided in two subsets in which the kaons going in the forward direction ($\cos\theta > 0$) are emitted in a quadrant along ($\cos\phi > 0$) or opposite ($\cos\phi < 0$) to the ϕ momentum $\vec{p_\phi}$, thus having a higher (or lower) value of γ_K than the companion kaons emitted in the backward direction (see Fig.3). Moreover the data are divided into four bins of sidereal time. In this way fitting simultaneously the corresponding eight $I(\pi^+\pi^-, \pi^+\pi^-; \Delta t)$ distributions one is able to observe possible modulation effects induced by the CPT-violating parameter δ in eq.(12) as a function of sidereal time and kaon momentum.

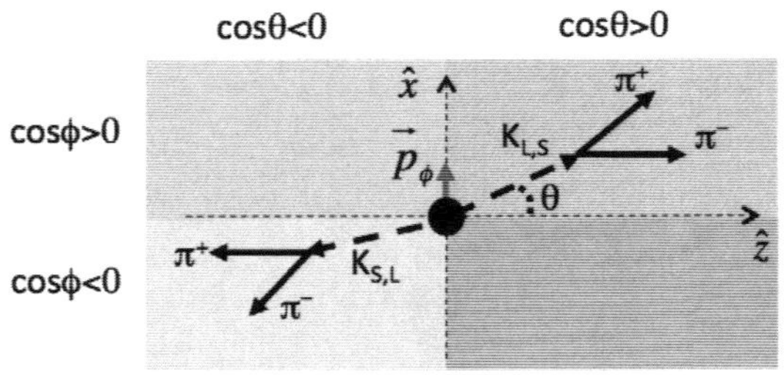

Figure 3. Sketch of the quadrant subdivisions as seen in a top view of the KLOE detector; $\vec{p_\phi}$ is directed along the $\hat{x}$ axis.

It is worth noting that the presence of the small momentum $\vec{p_\phi}$ makes $\gamma_{K,1} \neq \gamma_{K,2}$ on an event-by-event basis, which is a necessary condition in order to have the $I(\pi^+\pi^-, \pi^+\pi^-; \Delta t)$ distribution sensitive to the CPT violation effects induced by the Δa_0 parameter.

Using 1.7 fb^{-1} of data collected in years 2004-2005, the preliminary results are [43]:

$$\begin{aligned}
\Delta a_0 &= (-6.2 \pm 8.2_{\text{stat}} \pm 3.3_{syst}) \times 10^{-18}\ \text{GeV} \\
\Delta a_X &= (+3.3 \pm 1.6_{\text{stat}} \pm 1.5_{syst}) \times 10^{-18}\ \text{GeV} \\
\Delta a_Y &= (-0.7 \pm 1.3_{\text{stat}} \pm 1.5_{syst}) \times 10^{-18}\ \text{GeV} \\
\Delta a_Z &= (-0.7 \pm 1.0_{\text{stat}} \pm 0.3_{syst}) \times 10^{-18}\ \text{GeV}
\end{aligned} \tag{14}$$

They can be compared to similar results obtained in the B meson system [50], where an accuracy on the Δa_μ^B parameters of $\mathcal{O}(10^{-13}\text{GeV})$ has been reached.

5. Future kaon experiments at e^+e^- collider

5.1. The KLOE-2 experiment at DAΦNE improved in luminosity

After a successful experimental test [51], DAΦNE has been upgraded implementing an innovative collision scheme based on a *crab-waist* configuration, providing an improvement in the peak luminosity of a factor ~ 3.

The KLOE-2 experiment aims to continue and extend the physics program of its predecessor by collecting $\mathcal{O}(10\ \text{fb}^{-1})$ of data at the upgraded DAΦNE with an improved KLOE detector. The KLOE-2 physics program has been described in detail in Ref. [52] and among the main issues includes neutral kaon interferometry and tests of discrete symmetries and quantum mechanics. Improvements of about one order of magnitude in almost all present limits on CPT violation and decoherence parameters are expected [35, 39, 40, 52].

The upgrade of the KLOE detector would consist of the addition of (i) an inner tracker based on cylindrical GEM technology for the improvement of tracking and decay vertex resolution close to the interaction point (IP), (ii) a $e^{\pm}$ tagging system for the $\gamma\gamma$ physics, and (iii) two calorimeters in the final focusing region to improve acceptance and efficiency for photons coming from the IP and neutral kaon decays inside the detector volume.

The KLOE solenoid constitutes a strong perturbation on the machine optics, modifying the DAΦNE working conditions with respect to the test experiment described in Ref.[51]. The commissioning phase of the upgraded machine is in progress and is expected to be concluded by the year 2013. In the meanwhile the construction of the KLOE detector upgrades have been completed, their installation started, and is foreseen to be completed for mid 2013.

5.2. Direct test of Time reversal symmetry at a ϕ-factory

In the context of a local quantum field theory with Lorentz invariance and Hermiticity, as in the Standard Model, the validity of the CPT theorem ensures an automatic theoretical connection between any source of CP violation and a corresponding T (time reversal) violation. Even though CPT invariance has been confirmed by all present experimental tests, particularly in the neutral kaon system with stringent limits to possible CPT violation effects [8, 38, 39, 40], the theoretical connection between CP and T symmetries does not imply an experimental identity between them. In fact a direct evidence of T violation should result from an experiment clearly showing the violation, independently from the results of CP violation and without any connection with them. For unstable systems, the associated irreversibility looks like it prevents a true test of T symmetry [53].

In case of a transition process, due to the antiunitarity of the operator implementing the symmetry transformation, T invariance requires that the rate for the reaction $i \to f$ equals that of the reaction $f_T \to i_T$, with *in* and *out* states exchanged and T *inverted* (for spinless particles this corresponds to the reaction $f \to i$).

In the past the measurement of a non-zero value of the Kabir asymmetry [54], comparing the rates of the process $\mathrm{K}^0 \to \bar{\mathrm{K}}^0$ and its T conjugated one $\bar{\mathrm{K}}^0 \to \mathrm{K}^0$, has been presented as a proof for T violation [55, 56]. However, this process has the feature that $\mathrm{K}^0 \to \bar{\mathrm{K}}^0$ is a CPT even transition, so that T and CP transformations are identical in this case, and the corresponding observables are not independent. Therefore it is impossible to separate T violation from CP violation in the Kabir asymmetry.

In order to overcome this difficulty it has been suggested to exploit the Einstein-Podolsky-Rosen (EPR) [57] entanglement of neutral mesons produced at a ϕ-factory (or B-factory) [58, 59]. In fact in this case the neutral kaon pair can be rewritten in terms of any pair of orthogonal states K_+ and K_-, e.g. the CP eigenstates:

$$|i\rangle = \frac{1}{\sqrt{2}}\{|\mathrm{K}^0\rangle|\bar{\mathrm{K}}^0\rangle - |\bar{\mathrm{K}}^0\rangle|\mathrm{K}^0\rangle\} = \frac{1}{\sqrt{2}}\{|\mathrm{K}_+\rangle|\mathrm{K}_-\rangle - |\mathrm{K}_-\rangle|\mathrm{K}_+\rangle\} \,. \tag{15}$$

Thus, exploiting the perfect anticorrelation of the states implied by Eq. (15), it is possible to have a "flavor-tag"or a "CP-tag", i.e. to infer the flavor (K^0 or $\bar{\mathrm{K}}^0$) or the CP (K_+ or K_-) state of the still alive kaon by observing a specific flavor decay ($\pi^+\ell^-\nu$ or $\pi^-\ell^+\bar{\nu}$) or CP decay ($\pi^+\pi^-$ or $\pi^0\pi^0\pi^0$) of the other (and first decaying) kaon in the pair, as explained in detail in Ref. [60]. In this way one can experimentally access other transitions than $\mathrm{K}^0 \to \bar{\mathrm{K}}^0$ or $\bar{\mathrm{K}}^0 \to \mathrm{K}^0$. These new accessible processes can be divided into four categories of events, as summarized in Table 1.

For instance, one can consider $\mathrm{K}^0 \to \mathrm{K}_+$ as the reference process, where the initial state K^0 is identified at time t_1 with the flavor tag, and the final state K_+ is identified at a subsequent time $t_2 \geq t_1$ with a CP decay.

Journal of Physics: Conference Series **447** (2013) 012008 doi:10.1088/1742-6596/447/1/012008

Reference	T-conjug.	CP-conjug.	CPT-conjug.
$K^0 \to K_+$	$K_+ \to K^0$	$\bar{K}^0 \to K_+$	$K_+ \to \bar{K}^0$
$K^0 \to K_-$	$K_- \to K^0$	$\bar{K}^0 \to K_-$	$K_- \to \bar{K}^0$
$\bar{K}^0 \to K_+$	$K_+ \to \bar{K}^0$	$K^0 \to K_+$	$K_+ \to K^0$
$\bar{K}^0 \to K_-$	$K_- \to \bar{K}^0$	$K^0 \to K_-$	$K_- \to K^0$

Table 1. Scheme of possible reference transitions and their associated T, CP or CPT conjugated processes accessible at a ϕ-factory.

I) The T transformed process is $K_+ \to K^0$; any asymmetry in the rate between $K^0 \to K_+$ and $K_+ \to K^0$ would be a genuine T violating effect.

II) The CP transformed process is $\bar{K}^0 \to K_+$; any asymmetry in the rate between $K^0 \to K_+$ and $\bar{K}^0 \to K_+$ would be a genuine CP violating effect.

III) The CPT transformed process is $K_+ \to \bar{K}^0$; any asymmetry in the rate between $K^0 \to K_+$and $K_+ \to \bar{K}^0$ would be a genuine CPT violating effect.

One may check that the events used for the asymmetries in I), II), and III) are completely independent from each other.

Based on a similar strategy, the first direct observation of T violation, in the sense discussed above, has been accomplished in the neutral B meson system by the Babar collaboration [61, 62]. The same methodology, with some differences, can be implemented at a ϕ-factory, and the KLOE-2 experiment at DAΦNE could make in the next future the first significant T symmetry test in the neutral K system, as discussed in detail in Ref. [60].

5.3. Bells inequality and discrete symmetries

Entanglement and its consequences - in particular the violation of Bell inequalities, which defies our concepts of realism and locality [63] - have been proven to play key roles in Nature by many experiments for various quantum systems.

A recent formulation of the Bell's inequality for entangled neutral kaons produced at a ϕ-factory [64] overcomes drawbacks in the past formulations, making the inequality experimentally testable also for the $K^0 - \bar{K}^0$ system. It's worth mentioning that - surprisingly - CP violation in this case turns out to be a key ingredient to make the inequality violated by Quantum Mechanics. Feasibility studies to evaluate the experimental sensitivity of the test at KLOE-2 are in progress.

6. Conclusions

After direct CP violation has been firmly established in neutral kaon decays, in a couple of years, the next generation of kaon experiments at hadron machines (NA62 and KOTO) will be operational, definitely entering a new precision era in probing the SM, and in searching for New Physics in a way complementary to the LHC experimentation.

Other kaon experiments planned in the future will continue this program or address other specific issues, e.g. the transverse muon polarization in K^+ decays.

The neutral kaon system still constitutes an excellent laboratory for the study of discrete symmetries, and a ϕ-factory represents a unique opportunity to push forward these studies. CPT symmetry and quantum mechanics tests will be one of the key issues at KLOE-2, and their precision will be further improved, while new ideas - e.g. the direct T-symmetry test - might also be implemented.

Acknowledgements
The author would like to thank G. Branco, M. N. Rebelo, and all the organizing committee for the invitation to the DISCRETE 2012 symposium, and the very pleasant stay in Lisbon.

References
[1] M. Baldo-Ceolin et al., Nuovo Cim. **6** (1957) 84.
[2] J. H. Christenson et al., Phys. Rev. Lett. **13** (1964) 138.
[3] see e.g. G. C. Branco, L. Lavoura, and J. P. Silva, *CP violation*, Oxford Univ. Press, 1999;
M. Sozzi, *Discrete symmetries and CP violation*, Oxford Univ. Press, 2008;
T. K. Komatsubara, Progr. Part. Nucl. Phys. **67** (2012) 995;
M. Sozzi, J. Phys. Conf. Ser. 335 (2011) 012002.
[4] A. Alavi-Harati et al., KTeV collaboration, Phys. Rev. D **67** (2003) 012005.
[5] V. Fanti et al., NA48 Collaboration, Nucl. Instrum. Meth. A **574** (2007) 433.
[6] E. Abouzaid et al., KTeV collaboration, Phys. Rev. D **83** (2011) 092001.
[7] J. R. Batley, et al., Phys. Lett. B **544** (2002) 97.
[8] J. Beringer et al. (Particle Data Group), Phys. Rev. D **86** (2012) 010001.
[9] T. Blum et al., Phys. Rev. Lett. **108** (2012) 141601.
[10] J.S. Bell and J. Steinberger, *Proc. Oxford Int. Conf. on Elementary Particles* (1965).
[11] T. T. Wu, C. N. Yang, Phys. Rev. Lett. **13**, 380 (1964).
[12] F. Ambrosino et al., KLOE collaboration, JHEP **12** 011 (2006).
[13] J.R. Batley et al., NA48/2 Collaboration, Eur. Phys. J. C **52** (2007) 875.
[14] G. Ecker and A. Pich, Nucl. Phys. B **366** (1991) 189;
G. Isidori and R. Unterdorfer, JHEP 01 (2004) 009.
[15] S. Gjesdal et al., Phys. Lett. B **44** (1973) 217.
[16] A. Augusto Alves, Jr. et al., LHCb collaboration, JINST 3 (2008)S08005
[17] R. Aaij et al., LHCb Collaboration, JHEP 01 (2013) 090.
[18] S. Oggero, these proceedings.
[19] J. Brod, M. Gorbahn and E. Stamou, Phys. Rev. D **83** (2011) 034030.
[20] S.L. Glashow, J. Iliopoulos, L. Maiani, Phys. Rev. D **2** (1970) 1285.
[21] C. Smith, arXiv:1012.3698v1[hep-ph] (2010).
[22] Y. Grossman and Y. Nir, Phys. Lett. B **398** (1997) 163
[23] A.V. Artamonov et al., Phys. Rev. D **79** (2009) 092004.
[24] J. K. Ahn et al., Phys. Rev. D **81** (2010) 072004.
[25] G. Anelli et al., CERN-SPSC-2005-013, CERN-SPSC-P-326 (2005);
NA62 Collaboration, CERN-SPSC-2007-035, SPSC-M- 760 (2007).
[26] S. Balev, these proceedings.
[27] C Lazzeroni et al, NA62 Collaboration, Phys. Lett. B **719** (2013) 326.
[28] H. Watanabe, for the J-PARC E14 KOTO collaboration, PoS ICHEP 2010 (2010) 274;
see also http://koto.kek.jp/.
[29] M. Yamauchi, these proceedings.
[30] http://j-parc.jp/.
[31] E.T. Worcester for the ORKA Collaboration, Nucl. Phys. B (Proc. Suppl.) **233** (2012) 285.
[32] M. Abe et al, E246 Collaboration, Phys. Rev. Lett. **93** (2004) 131601.
[33] J. Imazato, PoS KAON09 (2009) 007;
see also http://trek.kek.jp/.
[34] http://projectx.fnal.gov/.
[35] A. Di Domenico (Editor) , Handbook on Neutral Kaon Interferometry at a ϕ-factory, Frascati Phys. Ser. **43** (2007).
[36] M. Adinolfi et al., KLOE collaboration, Nucl. Instr. and Meth. A **488** (2002) 51.
[37] M. Adinolfi et al., KLOE collaboration, Nucl. Instr. and Meth. A **482** (2002) 364.
[38] F. Ambrosino et al., KLOE collaboration, Phys. Lett. B **642**, 315 (2006).
[39] A. Di Domenico and the KLOE collaboration, J. Phys. Conf. Ser. **171**, 012008 (2009).
[40] A. Di Domenico and the KLOE collaboration, Found. Phys. **40**, 852 (2010).
[41] D. Babusci et al., arXiv:1301.7623 [hep-ex], submitted to Phys. Lett. B.
[42] M. Silarski, these proceedings.
[43] A. De Santis, these proceedings.
[44] L. F. Li and L. Wolfenstein, Phys. Rev. D **21**, 178 (1980).
[45] L. Maiani, N. Paver *CP violation in* $K \to 3\pi$ *decays*, and

G. D'Ambrosio, G. Isidori, A. Pugliese *CP and CPT measurements at DAΦNE*, in The second DAPHNE physics handbook, ed. L. Maiani, G. Pancheri, N. Paver, Vol. 1, p.51-62 and 63-95, INFN-LNF, Frascati, 1995.

[46] F. Ambrosino et al., Phys. Lett. **619**, 61 (2005).
[47] V. A. Kostelecký, Phys. Rev. Lett. **80**, 1818 (1998).
[48] V. A. Kostelecký, Phys. Rev. D **61**, 016002 (1999).
[49] V. A. Kostelecký, Phys. Rev. D **64**, 076001 (2001).
[50] B. Aubert, et al., BABAR collaboration, Phys. Rev. Lett. **100**, 131802 (2008).
[51] M. Zobov et al., Phys. Rev. Lett. **104**, 174801 (2010).
[52] G. Amelino-Camelia et al., Eur. Phys. J. C **68**, 619 (2010).
[53] L. Wolfenstein, Int. J. Mod. Phys. E **8**, 501 (1999).
[54] P. K. Kabir, Phys. Rev. D **2**, 540 (1970).
[55] A. Angelopoulos et al., CPLEAR coll., Phys. Lett. B **444**, 43 (1999).
[56] A. Angelopoulos et al., CPLEAR coll., Eur. Phys. C **22**, 55 (2001).
[57] A. Einstein, B. Podolski and N. Rosen, Phys. Rev. **47**, 777 (1935).
[58] M. C. Banuls and J. Bernabeu, Phys. Lett. B **464**, 117 (1999); Nucl. Phys. B **590**, 19 (2000).
[59] J. Bernabeu, J. Phys. Conf. Ser. **335** 012011 (2011).
[60] J. Bernabeu, A. Di Domenico, P. Villanueva-Perez, Nucl. Phys. B. **868** (2013) 102.
[61] J. P. Lees *et al.*, Babar Collaboration, Phys.Rev.Lett. **109** (2012) 211801.
[62] J. Bernabeu, F. Martínez-Vidal, P. Villanueva-Pérez, JHEP **08**, 064(2012).
[63] J. Bell, *Speakable and unspeakable in quantum mechanics*, Cambridge Univ. Press, 1987.
[64] B. C. Hiesmayr, A. Di Domenico, et al., Eur. Phys. J. C **72** (2012) 1856.

DISCRETE 2012 – Third Symposium on Prospects in the Physics of Discrete Symmetries IOP Publishing
Journal of Physics: Conference Series **447** (2013) 012009 doi:10.1088/1742-6596/447/1/012009

Lorentz and CPT violation

Robertus Potting

CENTRA and Department of Physics, FCT, University of the Algarve, Faro, Portugal

E-mail: rpotting@ualg.pt

Abstract. During the past two decades there has been a growing interest in the possibility that Lorentz and/or CPT might not be exact symmetries of Nature. In this short review, we present the current state of affairs, addressing both theoretical and experimental/observational issues. We pay particular attention to the role that has been played by the so-called Standard Model Extension.

1. CPT and Lorentz invariance violation

Invariance under Lorentz and CPT symmetry is a fundamental ingredient of both quantum field theory and General Relativity. This has been the major reason physicist have been, and to a large degree still are, reluctant to consider any violation of these symmetries. Nevertheless, in the last two decades, there has been growing interest in the possibility that Lorentz symmetry may not be an exact symmetry of Nature, or, at least, in testing whether or not this is the case.

Roughly, there are two major reasons as to why this has happened.

On the theoretical side, it turns out that many candidate theories of quantum gravity involve Lorentz invariance violation (LIV) as a possible effect. For example, the possibility of four-dimensional LIV has been investigated in string theory [1], non-commutative geometry [2], loop quantum gravity [3] and warped brane worlds [4]. Also other ideas, including emergent gauge bosons [5, 6] and emergent gravity [7] include LIV. We will discuss some of them in the next section.

On the experimental and phenomenological side, a very important development has been the formulation of low-energy effective field theories with LIV. In particular, the so-called Standard Model Extension (SME) has prompted much interest in the experimental testing of Lorentz and CPT symmetry. It includes all possible terms in which the Standard Model fields are coupled in a Lorentz-covariant way to constant tensor coefficients, while maintaining consistency requirements like gauge invariance, causality, etc. The value of these coefficients can in principle be measured (or bound) in experiments. This has allowed a systematic search for a large range of possible Lorentz-violating effects. In section 4 we will outline the main features of the SME.

CPT invariance is an issue that is closely related to Lorentz invariance. Indeed, the CPT theorem [8] states that any Lorentz-invariant, local quantum field theory with a hermitian Hamiltonian must have CPT symmetry. Conversely it can be shown [9] that any unitary interacting theory that violates CPT necessarily violates Lorentz invariance.[1] Thus violation of

[1] See the talk by Nick Mavromatos for a scenario with CPT violation that bypasses the assumptions of this theorem.

CPT is always accompanied by Lorentz violation. However, the reverse is not true: it is possible to have Lorentz violation while maintaining CPT invariance!

2. Models with Lorentz invariance violation

Let us consider some types of explicit models that exhibit LIV. Roughly we can consider two types. First of all, there are fundamental models that are basically Lorentz invariant by construction, but where LIV arises at low energy, usually as a small effect. There are also models that are not Lorentz invariant by construction, but where Lorentz invariance arises as an approximate symmetry at low energy. In the next two subsections we will consider examples of both categories.

2.1. Fundamental models with Lorentz invariance violation

2.1.1. Spontaneous symmetry breaking with Lorentz invariance violation. Consider a model with a Dirac spinor ψ and a vector field B^μ which includes an axial spinor-vector coupling

$$\mathcal{L} \supset \lambda B^\mu \bar{\psi} \gamma_5 \gamma_\mu \psi. \tag{1}$$

Suppose moreover that the dynamics of the vector field is such that the latter acquires a vacuum expectation value $\langle B^\mu \rangle = b^\mu \neq 0$. This could happen in the presence of an effective nonderivative potential for the vector field

$$V_{eff}(B^\mu B_\mu) \tag{2}$$

in which V_{eff} has a minimum for nonzero argument. When expanding the fields around this vacuum the kinetic term of the fermion will acquire the additional term

$$\mathcal{L} \supset \lambda b^\mu \bar{\psi} \gamma_5 \gamma_\mu \psi \tag{3}$$

which evidently violates Lorentz invariance.

The example just described is an example of a so-called bumblebee model [5]. Note that the Lagrangian itself is Lorentz invariant, while the LIV arises by spontaneous symmetry breaking. Usually it is assumed that the vacuum expectation value b^μ is a very small effect. More general models containing Lorentz tensors and fermions assume the presence of couplings of the form:

$$\mathcal{L} \supset \lambda m_{\rm pl}^{-k} T \cdot \bar{\psi} \Gamma (i\partial)^k \psi \tag{4}$$

with $m_{\rm pl}$ some fundamental mass scale (presumably the Planck mass) and T a higher rank tensor contracted with a fermion bilinear containing a combination of spacetime derivatives and Dirac gamma matrices. If T acquires some vacuum expectation value, (4) generates the following Lorentz-violating contribution to the fermion inverse propagator:

$$\Delta K(p) = \lambda m_{\rm pl}^{-k} \langle T \rangle \cdot \Gamma p^k. \tag{5}$$

One could wonder whether models that include non-derivative potentials like (2) are realistic, as they seem to be incompatible with gauge invariance. However, there are indications that such potentials might arise in the context of string field theory [1]. Various authors have also shown effective potentials that could allow for spontaneous Lorentz violation can arise in the context of fermion models, where fermion bilinears carrying Lorentz indices acquire nonzero vacuum expectation values [6].

2.1.2. Cosmologically varying scalars. Rather than having a Lorentz-violating expectation value arise through the expectation value of a vector field, the same could be due to a vacuum expectation value of a scalar that has a (slow) variation as a function of spacetime [10]. In other words, a preferred direction is selected through a non-zero value for the gradient of a scalar. For instance, one can take an axion coupling to the electromagnetic field:

$$\mathcal{L} \supset a(x) F\tilde{F}. \tag{6}$$

Integrating (6) by parts yields the contribution

$$\mathcal{L}' \supset -k^{\mu} A^{\nu} \tilde{F}_{\mu\nu} \tag{7}$$

which violates Lorentz invariance as well as CPT. As we will see below, the Lagrangian term (7) is incorporated in the SME.

2.1.3. Noncommutative geometry. Consider spacetime where the coordinates are taken to be noncommuting quantities [2]:

$$[x_\alpha, x_\beta] = i\frac{1}{\Lambda_{NC}}\theta_{\alpha\beta}. \tag{8}$$

Here $\theta_{\alpha\beta}$ is a tensor-valued set of coefficients of $\mathcal{O}(1)$, while Λ_{NC} denotes the noncommutative energy scale.

Noncommutative quantum field theories can be constructed by taking an ordinary quantum field theory and replacing the ordinary multiplication of fields with Moyal products:

$$f \star g(x) = \exp\left(\tfrac{1}{2} i\theta^{\mu\nu}\partial_{x^\mu}\partial_{y^\nu}\right) f(x)g(y)\,|_{x=y}\,. \tag{9}$$

The definition of gauge transformations must be adapted analogously.

It is possible to re-express resulting noncommutative field theory in terms of a conventional one, by use of the so-called Seiberg-Witten map [13]. It expresses the non-commutative fields in terms of ordinary gauge fields. For instance, by applying this map to non-commutative Quantum Electrodynamics one obtains the following Lorentz-violating expression, at lowest nontrivial order in $1/\Lambda_{NC}$:

$$\begin{aligned} S = {} & \frac{i}{2}\bar{\psi}\gamma^{\mu}\overleftrightarrow{D}_{\mu}\psi - m\bar{\psi}\psi - \frac{1}{4}F_{\mu\nu}F^{\mu\nu} + \frac{q}{8}\frac{\theta^{\alpha\beta}}{\Lambda_{\mathrm{NC}}}\Big[-iF_{\alpha\beta}\bar{\psi}\gamma^{\mu}\overleftrightarrow{D}_{\mu}\psi \\ & + 2iF_{\alpha\mu}\bar{\psi}\gamma^{\mu}\overleftrightarrow{D}_{\beta}\psi + 2mF_{\alpha\beta}\bar{\psi}\psi - 4F_{\alpha\mu}F_{\beta\nu}F^{\mu\nu} + F_{\alpha\beta}F_{\mu\nu}F^{\mu\nu}\Big]. \end{aligned} \tag{10}$$

The effect of the Lorentz-violating contributions is amenable to experimental detection, or can be bounded. They are an example of (non-minimal) Standard Model Extension terms.

A study of the perturbative dynamics of noncommuting field theories has revealed that high energies of virtual particles in loops produce non-analyticity at low momentum, questioning the consistency of the low energy effective action [11]. Supersymmetric gauge theories exhibit better behaviour in the infrared [12]. This issue has been the object of extensive recent study.

2.1.4. LIV from topology. Another interesting idea that has been explored [14] is the possibility that LIV arise from nontrivial spacetime topology. Consider a spacetime in which one of the dimensions is compact with a large radius R. Then the vacuum fluctuations along this dimension have periodic boundary conditions. This defines a preferred direction in the vacuum. Applying this to electrodynamics once again yields the lagrangian term given by eq. (7), where k^{μ} points in the direction of the compact dimension and has a size of order R^{-1} [14].

2.2. *Models with approximate Lorentz invariance at low energy*

A second class of models are those that have no Lorentz invariance built in at the fundamental level at all, but nevertheless exhibits approximate Lorentz invariance at low energy.

An example is given by the recently proposed *Horava-Lifshitz gravity* [15], which has received much attention in the literature. Here the fundamental role of local Lorentz invariance is abandoned and instead it is assumed that this appears only at low energies as an approximate symmetry. Spacetime is endowed with a preferred foliation by 3-dimensional spacelike surfaces, which defines the splitting of the coordinates into space and time, which are thus assumed to be on a fundamentally different footing. The Hilbert-Einstein action of general relativity (GR) is completed with higher spatial derivatives of the metric which improve the UV behavior of the graviton propagator, allowing for a power-counting renormalizable theory. However, the time derivatives of the action are maintained at second order, thus avoiding any problems with ghosts. A required key property is thus that the theory flow to GR in the infrared limit.

It has been pointed out that the original proposal by Horava contains an extra propagating scalar degree of freedom with pathological behavior. An extended model has been proposed where this problem is claimed to be cured [16]. In the infrared limit this the model reduces to a Lorentz-violating scalar-tensor gravity theory. Other "projectable" versions of Horava-Lifshitz gravity have also been considered. It should be noted that this remains very much an open field.

3. Kinematic frameworks

Historically most relativity tests have only considered kinematic aspects, that is, they investigated deviations of single-particle motion from the expected Lorentz-invariant dispersion relation. In this section we will consider various tests of relativity of this kind that have been proposed and used in the past.

3.1. *Modified dispersion relations*

One of the most simple approaches is to assume that Lorentz-violating effects modify the usual relativistic dispersion relation $E^2 = p^2 + m^2$ to the more general relation $E^2 = F(p, m)$. It is natural to expand $F(p, m)$ in a Taylor series:

$$E^2 = m^2 + p^2 + m_{\rm pl} f_i^{(1)} p^i + f_{ij}^{(2)} p^i p^j + \frac{f^{(3)}}{m_{\rm pl}} p^i p^j p^k + \ldots \tag{11}$$

with dimensionless coefficients $f^{(n)}$, that depend on the particle species. The order n of the first nonzero coefficient depends on the underlying fundamental theory. The coefficients $f^{(n)}$, while arbitrary, are presumably such that Lorentz violation is a small effect. Note that this will be automatically the case for $n \geq 3$ whenever $p \ll m_{\rm pl}$ if the $f^{(n)}$ are of order 1, due to the inverse powers of the Planck mass accompanying those. Thus no special suppression mechanism is necessary for those terms. This will be discussed in more detail below. It might be noted that terms with odd powers of p tend to have problems with coordinate invariance, causality and/or positivity [17].

Much of the relevant literature assumes rotational invariance, turning (11) into

$$E^2 = m^2 + p^2 + m_{\rm pl} f^{(1)} |p| + f^{(2)} p^2 + \frac{f^{(3)}}{m_{\rm pl}} |p|^3 + \ldots \tag{12}$$

For an example of an application to the photon sector, see [18].

It has been suggested that stochastic or foamy spacetime structure can lead to modifications of spacetime structure that modify over time. In such frameworks the particle dispersion is taken to fluctuate according to a model-dependent probability distribution [19].

Journal of Physics: Conference Series **447** (2013) 012009 doi:10.1088/1742-6596/447/1/012009

3.2. The Robertson-Mansouri-Sexl framework
This method assumes a preferred frame where the speed of light is isotropic. The Lorentz transformation to other frames is generalized with respect to the conventional boosts [20]:

$$t' = a^{-1}(t - \vec{\epsilon}\cdot\vec{x}) \tag{13}$$

$$\vec{x}' = d^{-1}\vec{x} - (d^{-1} - b^{-1})\frac{\vec{v}(\vec{v}\cdot\vec{x})}{v^2} - a^{-1}\vec{v}t \tag{14}$$

with a, b, c, d, e functions of the relative speed v. Without Lorentz violation and Einstein clock synchronization we have $a = b^{-1} = \sqrt{1-v^2}$, $d = 1$, and $\vec{\epsilon} = \vec{v}$. Modifying the values of the parameters results in a variable speed of light, assuming experiments that use a fixed set of rods and clocks.

The Robertson-Mansouri-Sexl framework is incorporated in the SME.

3.3. The c^2 model
The c^2 model is a test model developed for application to studies of Lorentz invariance, as a limiting case of the TH$\epsilon\mu$ formalism. It is described by a lagrangian that considers the motion of test particles in an electromagnetic field [21]. It assumes a preferred frame in which the limiting speed of particles is considered to be 1, but the speed of light $c \neq 1$.

Also this framework can be incorporated in the SME [42].

3.4. Doubly Special Relativity
Doubly Special Relativity (DSR) is a recently developed idea [22]. It is assumed that the Lorentz transformations act in a modified way on the physical four-momentum such that both c as well as a special energy scale E_{DSR} are invariant. The physical energy/momentum are taken to be given by

$$E = \frac{\epsilon}{1+\lambda_{\mathrm{DSR}}\epsilon}, \qquad p = \frac{\pi}{1+\lambda_{\mathrm{DSR}}\pi}, \tag{15}$$

where $\lambda_{\mathrm{DSR}} = E_{\mathrm{DSR}}^{-1}$, in terms of the pseudo energy/momentum ϵ and π, which transform normally under Lorentz boosts. The dispersion relation becomes

$$E^2 - p^2 = \frac{m^2(1-\lambda_{\mathrm{DSR}}E)^2}{(1-\lambda_{\mathrm{DSR}}m)^2}. \tag{16}$$

Also this framework can be incorporated in the SME [23].

It should be noted that the physical meaning of the quantities E and p, and of DSR itself, has been questioned.

4. Effective field Theory
What should be a suitable dynamical framework for describing LIV? We can list a few general criteria that it should satisfy:

(i) *Observer coordinate independence:* The physics it describes should be independent of "observer" coordinate transformation, that is, changes of coordinates used be the observer to describe the same physical situation;

(ii) *Realism:* the framework must incorporate known physics, while allowing for a suitable parametrization of LIV effects;

(iii) *Generality:* the framework should ideally be as general as possible, to maximize reach.

A framework that satisfies all of these criteria is the *Standard Model Extension* (SME) [24]. It is an effective field theory which incorporates:

(i) The Standard Model of particle physics as well as General Relativity;
(ii) Any scalar term formed by contracting operators for Lorentz violation with tensor-valued coefficients controlling size of the LIV effects;
(iii) Possibly additional requirements like gauge invariance, locality, stability, and renormalizability can be included as well.

The SME includes, in principle, terms of any mass dimension (starting at dimension three). Imposing power-counting renormalizability limits one to terms of dimension four. This is usually referred to as the minimal SME (mSME). The mSME has a finite number of LIV parameters, while the number of LIV parameters in the full SME is in principle unlimited. The SME leads not only to breaking of Lorentz symmetry, but also to that of CPT, for about half of its terms.

Example: the free fermion sector of the SME:

$$\mathcal{L} = \bar{\psi}(i\Gamma^\mu \partial_\mu - M)\psi \tag{17}$$

$$\Gamma^\mu = \gamma^\mu + c^{\mu\nu}\gamma_\nu - d^{\mu\nu}\gamma^\nu\gamma_5, \qquad M = m + \not{a} - \not{b}\gamma_5 + \tfrac{1}{2}H^{\mu\nu}\sigma_{\mu\nu} \tag{18}$$

A separate set of coefficients exists for every elementary particle.

As the SME is to be considered an effective field theory, one can relax the requirement of renormalizability. This means, that the coefficients of the mSME become generalized to higher mass dimensions. For instance, the $c^{\mu\nu}$ coefficients in (18) can be generalized to include arbitrary higher dimensions as follows:

$$c^{\mu\nu} \rightarrow \hat{c}^{\mu\nu} \equiv i^{d-4} \sum_{d=4}^{\infty} c^{(d)\mu\nu\alpha_1\ldots\alpha_{(d-4)}} \partial_{\alpha_1} \ldots \partial_{\alpha_{(d-4)}} \tag{19}$$

The coefficients $c^{(d)}$ have mass dimension $4-d$. Thus they can naturally be expected to be of order $m_{\rm pl}^{4-d}$, so that their contribution to $\hat{c}^{\mu\nu}$ is naturally suppressed at low energies.

Table 1. The $SU(3) \times SU(2) \times U(1)$ fields and coupling constants.

Leptons	$L_A = \binom{\nu_A}{l_A}_L$,	$R_A = (l_A)_R$		$(A = e, \mu, \tau)$
Quarks	$Q_A = \binom{u_A}{d_B}_L$,	$U_A = (u_A)_R$, $D_B = (d_B)_R$		$(A = u, c, t)$ $(B = d, s, b)$
Gauge fields	G_μ,	W_μ,	B_μ	
Higgs doublet	ϕ			
Gauge couplings	g_3,	g,	g'	
Yukawa couplings	G_L,	G_U,	G_D	

Let's consider more in detail at the construction of the mSME. In table 1 we define the $SU(3) \times SU(2) \times U(1)$ Standard Model fields and coupling constants. The Standard Model consists of:

$$\mathcal{L}_{\rm lepton} = \tfrac{1}{2}i\bar{L}_A\gamma^\mu \overleftrightarrow{D_\mu} L_A + \tfrac{1}{2}i\bar{R}_A\gamma^\mu \overleftrightarrow{D_\mu} R_A \tag{20}$$

$$\mathcal{L}_{\rm quark} = \tfrac{1}{2}i\bar{Q}_A\gamma^\mu \overleftrightarrow{D_\mu} Q_A + \tfrac{1}{2}i\bar{U}_A\gamma^\mu \overleftrightarrow{D_\mu} U_A + \tfrac{1}{2}i\bar{D}_A\gamma^\mu \overleftrightarrow{D_\mu} U_A \tag{21}$$

$$\mathcal{L}_{\rm Yukawa} = -(G_L)_{AB}\bar{L}_A\phi R_B - (G_U)_{AB}\bar{Q}_A\phi^c U_B - (G_D)_{AB}\bar{Q}_A\phi D_B \tag{22}$$

$$\mathcal{L}_{\rm Higgs} = (D_\mu\phi)^\dagger D^\mu\phi + \mu^2\phi^\dagger\phi - \tfrac{\lambda}{6}(\phi^\dagger\phi)^2 \tag{23}$$

$$\mathcal{L}_{\rm gauge} = -\tfrac{1}{2}{\rm Tr}(G_{\mu\nu}G^{\mu\nu}) - \tfrac{1}{2}{\rm Tr}(W_{\mu\nu}W^{\mu\nu}) - \tfrac{1}{4}(B_{\mu\nu}B^{\mu\nu}) \tag{24}$$

In the minimal Standard Model Extension, to this are added:

- *Fermion sector*

$$\mathcal{L}^{\mathsf{CPT-even}}_{\mathsf{lepton}} = \tfrac{1}{2}i(c_L)_{\mu\nu\,AB}\bar{L}_A\gamma^\mu \overleftrightarrow{D^\nu} L_B + \tfrac{1}{2}i(c_R)_{\mu\nu\,AB}\bar{R}_A\gamma^\mu \overleftrightarrow{D^\nu} R_B \tag{25}$$

$$\mathcal{L}^{\mathsf{CPT-odd}}_{\mathsf{lepton}} = -(a_L)_{\mu\,AB}\bar{L}_A\gamma^\mu L_B - (a_R)_{\mu\,AB}\bar{R}_A\gamma^\mu R_B \tag{26}$$

$$\mathcal{L}^{\mathsf{CPT-even}}_{\mathsf{quark}} = \tfrac{1}{2}i(c_Q)_{\mu\nu\,AB}\bar{Q}_A\gamma^\mu \overleftrightarrow{D^\nu} Q_B + \tfrac{1}{2}i(c_U)_{\mu\nu\,AB}\bar{U}_A\gamma^\mu \overleftrightarrow{D^\nu} U_B + \tfrac{1}{2}i(c_D)_{\mu\nu\,AB}\bar{D}_A\gamma^\mu \overleftrightarrow{D^\nu} D_B \tag{27}$$

$$\mathcal{L}^{\mathsf{CPT-odd}}_{\mathsf{quark}} = -(a_Q)_{\mu\,AB}\bar{L}_A\gamma^\mu Q_B - (a_R)_{\mu\,AB}\bar{U}_A\gamma^\mu U_B - (a_D)_{\mu\,AB}\bar{D}_A\gamma^\mu D_B \tag{28}$$

- *Higgs sector*

$$\mathcal{L}^{\mathsf{CPT-even}}_{\mathsf{Higgs}} = \tfrac{1}{2}(k_{\phi\phi})^{\mu\nu}(D_\mu\phi)^\dagger D_\nu\phi - \tfrac{1}{2}(k_{\phi B})^{\mu\nu}\phi^\dagger\phi B_{\mu\nu} - \tfrac{1}{2}(k_{\phi W})^{\mu\nu}\phi^\dagger W_{\mu\nu}\phi \tag{29}$$

$$\mathcal{L}^{\mathsf{CPT-even}}_{\mathsf{Yukawa}} = -\tfrac{1}{2}(H_L)_{\mu\nu\,AB}\bar{L}_A\phi\sigma^{\mu\nu}R_B - \tfrac{1}{2}(H_U)_{\mu\nu\,AB}\bar{Q}_A\phi\sigma^{\mu\nu}U_B - \tfrac{1}{2}(H_D)_{\mu\nu\,AB}\bar{Q}_A\phi\sigma^{\mu\nu}D_B \tag{30}$$

- *Gauge sector*

$$\mathcal{L}^{\mathsf{CPT-even}}_{\mathsf{gauge}} = -\tfrac{1}{2}(k_G)^{\kappa\lambda\mu\nu}\mathsf{Tr}(G_{\kappa\lambda}G_{\mu\nu}) - \tfrac{1}{2}(k_W)^{\kappa\lambda\mu\nu}\mathsf{Tr}(W_{\kappa\lambda}W_{\mu\nu}) - \tfrac{1}{2}(k_B)^{\kappa\lambda\mu\nu}B_{\kappa\lambda}B_{\mu\nu} \tag{31}$$

$$\mathcal{L}^{\mathsf{CPT-odd}}_{\mathsf{gauge}} = (k_3)_\kappa\epsilon^{\kappa\lambda\mu\nu}\mathsf{Tr}(G_\lambda G_{\mu\nu} + \tfrac{2}{3}ig_3G_\lambda G_\mu G_\nu) + (k_1)_\kappa\epsilon^{\kappa\lambda\mu\nu}B_\lambda B_{\mu\nu} + (k_0)_\kappa B^\kappa + (k_2)_\kappa\epsilon^{\kappa\lambda\mu\nu}\mathsf{Tr}(W_\lambda W_{\mu\nu} + \tfrac{2}{3}ig_3W_\lambda W_\mu W_\nu \tag{32}$$

In much the same way conventional quantum electrodynamics (QED) can be obtained from the usual standard model, an extended LIV version of QED can be obtained from the SME. After electroweak symmetry breaking, one sets to zero the gluon fields, the weak bosons and the Higgs field. The (uncharged) neutrinos can be disregarded, leaving only the photon, the charged leptons and the quarks. Considering here the simplified case of only photons and a single fermion one obtains

$$\begin{aligned}\mathcal{L}_{\mathsf{fermion}} = {} & \tfrac{1}{2}i\bar{\psi}\gamma^\mu \overleftrightarrow{D_\mu} \psi - m\bar{\psi}\psi - \tfrac{1}{4}F_{\mu\nu}F^{\mu\nu} \\ & - a_\mu\bar{\psi}\gamma^\mu\psi - b_\mu\bar{\psi}\gamma_5\gamma^\mu\psi + \tfrac{1}{2}ic_{\mu\nu}\bar{\psi}\gamma^\mu \overleftrightarrow{D^\nu} \psi + \tfrac{1}{2}id_{\mu\nu}\bar{\psi}\gamma_5\gamma^\mu \overleftrightarrow{D^\nu} \psi \\ & - \tfrac{1}{2}H_{\mu\nu}\bar{\psi}\sigma^{\mu\nu}\psi - \tfrac{1}{4}(k_F)_{\kappa\lambda\mu\nu}F^{\kappa\lambda}F^{\mu\nu} + \tfrac{1}{2}(k_{AF})^\kappa\epsilon_{\kappa\lambda\mu\nu}A^\lambda F^{\mu\nu}.\end{aligned} \tag{33}$$

For every fermion there is a separate set of coefficients a_μ, b_μ, $c_{\mu\nu}$, $d_{\mu\nu}$ and $H_{\mu\nu}$. Sometimes one also considers the additional terms

$$\mathcal{L}^{\mathsf{extra}}_{\mathsf{fermion}} = \tfrac{1}{2}ie_\mu\bar{\psi} \overleftrightarrow{D^\mu} \psi - \tfrac{1}{2}if_\mu\bar{\psi}\gamma_5 \overleftrightarrow{D^\mu} \psi + \tfrac{1}{4}ig_{\lambda\mu\nu}\bar{\psi}\sigma^{\lambda\mu} \overleftrightarrow{D^\nu} \psi \tag{34}$$

that are compatible with renormalizability but that cannot be obtained as a reduction from the SME. We note here that for phenomenological purposes protons and neutrons are treated as fundamental constituents with their own set of

The SME fields can also be coupled to gravity [25]. As an example, consider the lepton sector. In the Standard Model it couples through the Lagrangian density

$$\mathcal{L}_{\mathsf{lepton}} = \tfrac{1}{2}ie\,e^\mu{}_a\bar{L}_A\gamma^a \overleftrightarrow{D_\mu} L_A + \tfrac{1}{2}ie\,e^\mu{}_a\bar{R}_A\gamma^a \overleftrightarrow{D_\mu} R_A. \tag{35}$$

Here e_a^μ is the vierbein, which is used to convert local Lorentz indices to curved spacetime indices: $b_\mu = e_\mu{}^a b_a$. The flat-space LIV lepton sectors sectors can similarly be coupled to gravity, for example:

$$\mathcal{L}_{\text{lepton}}^{\text{CPT-even}} = -\tfrac{1}{2}i(c_L)_{\mu\nu AB} e e^\mu{}_a \bar{L}_A \gamma^a \overset{\leftrightarrow}{D^\nu} L_B - \tfrac{1}{2}i(c_R)_{\mu\nu AB} e e^\mu{}_a \bar{R}_A \gamma^a \overset{\leftrightarrow}{D^\nu} R_B. \tag{36}$$

Also for the pure gravity sector itself a LIV extension can be defined. The minimal sector is described by the Lagrangian density

$$\mathcal{L}_{e,\omega}^{\text{LV}} = e(k_T)^{\lambda\mu\nu} T_{\lambda\mu\nu} + e(k_R)^{\kappa\lambda\mu\nu} R_{\kappa\lambda\mu\nu} + e(k_{TT})^{\alpha\beta\gamma\lambda\mu\nu} T_{\alpha\beta\gamma} T_{\lambda\mu\nu} + e(k_{DT})^{\kappa\lambda\mu\nu} D_\kappa T_{\lambda\mu\nu}. \tag{37}$$

Here $R_{\kappa\lambda\mu\nu}$ is the Riemann curvature tensor, while $T_{\lambda\mu\nu}$ is the torsion tensor. In the Riemannian limit (no torsion) the action becomes a generalization of the Einstein-Hilbert action:

$$S_{e,\omega,\Lambda} = \frac{1}{2\kappa} \int d^4x\, e \left[(1-u)R - 2\Lambda + s^{\mu\nu} R_{\mu\nu} + t^{\kappa\lambda\mu\nu} R_{\kappa\lambda\mu\nu} \right]. \tag{38}$$

As has been alluded to above, different types of *scaling behaviour* can be expected for the LIV coefficients of the SME.

- For the minimal SME, with terms of mass dimension less than or equal to four, one expects the model to be renormalizable with coefficients running logarithmically as a function of energy. Indeed, a renormalization group study that has been done to one loop and to first order in LIV coefficients for the QED sector of the mSME bears this out [26]. Thus no suppression with power of energy scale occurs for these coefficients, and so any suppression that occurs should be present already at very high energy scales, creating a naturalness problem. A toy model study with a scalar field with Planck scale Lorentz-violating cutoff at high energy yields percent-level LIV at low energy [27].
- Higher (≥ 5) dimension operators can be expected to scale with inverse powers of the Planck mass. However, there are higher dimension operators that mix with dimension three or four operators, eliminating the suppression mechanism. Studies have been done identifying dimension five operators that do not exhibit such mixing [28]. It has been claimed that in a Lorentz-violating minimal supersymmetric extension of the Standard Model can only contain higher-dimension operators, but this depends on whether or not the superalgebra sector involving Q and P_μ operators is perturbed [29].

5. Phenomenology

5.1. Free particles: modified dispersion relations

The SME implies in general modified dispersion relations of the form (11), where the coefficients $f^{(n)}$ are determined by the relevant parameters of the SME effective lagrangian. One important possible consequence of modified dispersion relations is the occurrence of shifted reaction thresholds. In particular, normally allowed processes may become forbidden, while normally forbidden processes may become allowed in certain regions of phase space. For example, it is possible within the SME that charged massive particles exhibit Čerenkov radiation [30, 31]. Normally this can only happen in a refractive optical medium, but if the dispersion relation of the massive particle allows a maximum speed above the (possibly modified) speed of light, *vacuum* Čerenkov radiation can take place. Nonobservance of vacuum Čerenkov radiation for LEP electrons has lead to bounds on SME parameters in QED sector [31]. Another example is photon decay into, for instance, a positron-electron pair which can become a kinematically allowed process. Nonobservance of photon decay in Tevatron photons has lead to bounds on SME parameters [32].

5.2. Mesons

Meson systems have long provided tests for CP and CPT invariance [33]. In the contact of the SME they provide a test for the a_μ coefficients. To see this, consider the Schrödinger equation for a neutral meson system:

$$i\partial_t \Psi = \Lambda\Psi \tag{39}$$

where Ψ is a linear combination of the neutral meson and the anti-meson states (K, D, B_d or B_s). $\Lambda = M - i\Gamma/2$ is the effective 2×2 Hamiltonian, with eigenvalues $\lambda_S \equiv m_S - \frac{i}{2}\gamma_S$ and $\lambda_L \equiv m_L - \frac{i}{2}\gamma_L$. It is possible to show the following simple relation with the SME coefficients a_μ:

$$\Delta\Lambda \approx \beta^\mu \Delta a_\mu, \qquad \Delta\Lambda \equiv \Lambda_{11} - \Lambda_{22}, \quad \beta^\mu \equiv (\gamma, \gamma\vec{\beta}). \tag{40}$$

One commonly introduces the dimensionless parameter $\xi = \Delta\Lambda/\Delta\lambda \approx 2\delta$ that parametrizes CPT violation. Note that ξ depends explicitly on the meson four-velocity.

Experimental sensitivities have been obtained for Δa_μ in the K system of order 10^{-17} to 10^{-20} GeV [34],[2] in the D system of order 10^{-15} GeV [35], and in the B_d system of order 10^{-15} GeV [36].

5.3. Neutrinos

The SME suggests many possible observable consequences in the neutrino sector [37]. For example, there are LIV terms in the SME that provoke neutrino oscillations. This in turns yields very precise tests of LIV.

At leading order, LIV in the neutrino sector is described by the following effective two-component Hamiltonian acting on neutrino- antineutrino state vector:

$$\begin{aligned} h_{\text{eff}} = |\vec{p}| \begin{pmatrix} 1 & 0 \\ 0 & 1 \end{pmatrix} + \frac{1}{2|\vec{p}|} \begin{pmatrix} \tilde{m}^2 & 0 \\ 0 & (\tilde{m}^2)^* \end{pmatrix} \\ + \frac{1}{|\vec{p}|} \begin{pmatrix} (a_L)^\mu p_\mu - (c_L)^{\mu\nu} p_\mu p_\nu & -i\sqrt{2} p_\mu (\epsilon_+)_\nu (g^{\mu\nu\sigma} p_\sigma - H^{\mu\nu})\mathcal{C} \\ -i\sqrt{2} p_\mu (\epsilon_+)^*_\nu \left[(g^{\mu\nu\sigma} p_\sigma - H^{\mu\nu})\mathcal{C}\right]^* & \left[-(a_L)^\mu p_\mu - (c_L)^{\mu\nu} p_\mu p_\nu\right]^* \end{pmatrix}. \end{aligned} \tag{41}$$

Hamiltonian (41) leads to various potential signals:

- Oscillations with unusual energy dependences can be expected (oscillation length may grow rather than shrink with energy);
- Anisotropies can arise from breakdown of rotational invariance, implying sidereal variations in observed fluxes.

Many bounds on SME parameters in the neutrino sector have been deduced by analysis of LSND, MiniBooNe and MINOS (and other) data [38]. It is also interesting to note that SME-inspired models have been proposed that reproduce current observations and may help resolve the LSND anomaly [39].

5.4. QED sector

Not surprisingly, the sharpest laboratory tests have been carried out in systems where the predominant interactions are described by QED.

Starting with the QED sector of the SME, lagrangian (33), an effective Hamiltonian can be constructed using perturbation theory for small LIV, such that $i\partial_t\chi = \hat{H}\chi$. In the

[2] See also the talk by Antonio De Santis at this conference.

non-relativistic approximation, by using the Foldy-Wouthuysen approach and making field redefinitions one finds for a massive fermion [40]

$$\hat{H}_{\mathsf{pert}} = a_\mu \gamma^0 \gamma^\mu - b_\mu \gamma_5 \gamma^0 \gamma^\mu - c_{00} m \gamma^0 - i(c_{0j} + c_{j0}) D^j + i(c_{00} D_j - c_{jk} D^k) \gamma^0 \gamma^j \\ - d_{j0} m \gamma_5 \gamma^j + i((d_{0j} + d_{j0}) D^j \gamma_5 + i(d_{00} D_j - d_{jk} D^k) \gamma^0 \gamma_5 \gamma^j + \tfrac{1}{2} H_{\mu\nu} \gamma^0 \sigma^{\mu\nu} \tag{42}$$

This expression assumes fixed non-rotating axes. In the usual convention one employs a sun-centered frame with celestial equatorial coordinates, denoted by uppercase X, Y, Z, T. Passing to a frame of rotating, earth-fixed laboratory axes implies using an appropriate mapping. For instance, for the combination

$$\tilde{b}^e_j \equiv b^e_j - m d^e_{j0} - \tfrac{1}{2} \epsilon_{jkl} H^e_{kl} \tag{43}$$

one finds

$$\begin{aligned} \tilde{b}^e_1 &= \tilde{b}^e_X \cos\chi \cos\Omega t + \tilde{b}^e_Y \cos\chi \sin\Omega t - \tilde{b}^e_Z \sin\chi \\ \tilde{b}^e_2 &= -\tilde{b}^e_X \sin\Omega t + \tilde{b}^e_Y \cos\Omega t \\ \tilde{b}^e_3 &= \tilde{b}^e_X \sin\chi \cos\Omega t + \tilde{b}^e_Y \sin\chi \sin\Omega t + \tilde{b}^e_Z \cos\chi \end{aligned} \tag{44}$$

Here the earth's rotation axis is chosen along Z, while the angle χ is between the $j = 3$ lab axis and the Z axis. Ω is the angular frequency corresponding to a sidereal day ($\Omega \approx 2\pi/(23\mathsf{h}\ 56\mathsf{m})$).

5.5. Photons

The most general photon kinetic term in the mSME follows from lagrangian (33):

$$\mathcal{L}_{\mathsf{photon}} = -\tfrac{1}{4} F_{\mu\nu} F^{\mu\nu} - \tfrac{1}{4} (k_F)_{\kappa\lambda\mu\nu} F^{\kappa\lambda} F^{\mu\nu} + \tfrac{1}{2} (k_{AF})^\kappa \epsilon_{\kappa\lambda\mu\nu} A^\lambda F^{\mu\nu} \tag{45}$$

which exhibits two LIV terms.

The k_{AF} *term* is CPT violating and leads to birefringence. It can give rise to vacuum Čerenkov radiation [30].

The k_F *term* is CPT conserving. It has 19 LIV degrees of freedom, some being parity-even, others parity-odd. One degree of freedom, denoted κ_{tr}, is rotationally invariant. Ten degrees of freedom lead to birefringence, the remaining nine (including κ_{tr}) don't. Also the k_F term can give rise to vacuum Čerenkov radiation [31].

5.6. Bounds on higher dimensional LIV operators

Much less work has been done on bounding higher dimensional operators. Laboratory experiments are concerned with low energies, thus best suited for mSME. Higher-dimension operators scale with energy, giving an a-priori advantage to astrophysical tests.

Higher-dimensional operators in photon sector have been obtained by considering the most general SME photon Lagrangian [58]

$$\mathcal{L} = -\tfrac{1}{4} F_{\mu\nu} F^{\mu\nu} - \tfrac{1}{4} F^{\kappa\lambda} (\hat{k}_F)_{\kappa\lambda\mu\nu} F^{\mu\nu} + \tfrac{1}{2} \epsilon_{\kappa\lambda\mu\nu} A^\lambda (\hat{k}_{AF})^\kappa F^{\mu\nu} \tag{46}$$

where

$$(\hat{k}_F)^{\kappa\lambda\mu\nu} = \sum_{d=2,4,6\ldots} (k_F^{(d)})^{\kappa\lambda\mu\nu\alpha_1\ldots\alpha_{(d-4)}} \partial_{\alpha_1} \ldots \partial_{\alpha_{(d-4)}}\,, \tag{47}$$

$$(\hat{k}_{AF})_\kappa = \sum_{d=1,3,5\ldots} (k_{AF}^{(d)})_\kappa{}^{\alpha_1\ldots\alpha_{(d-3)}} \partial_{\alpha_1} \ldots \partial_{\alpha_{(d-3)}}\,. \tag{48}$$

Journal of Physics: Conference Series **447** (2013) 012009 doi:10.1088/1742-6596/447/1/012009

Here $(k_F^{(d)})^{\kappa\lambda\mu\nu\alpha_1\ldots\alpha_{(d-4)}}$ and $(k_{AF}^{(d)})_\kappa{}^{\alpha_1\ldots\alpha_{(d-3)}}$ are constant coefficients with mass dimension $4-d$.

By analyzing polarization changes due to birefringence in CMB radiation, various $k_{AF}^{(5)}$ coefficients have been bound to $\mathcal{O}(10^{-19}\,\text{GeV}^{-1})$ and various $k_F^{(6)}$ coefficients have been bound to $\mathcal{O}(10^{-9}\,\text{GeV}^{-2})$ [58]. from analyzing dispersion relations (time of flight differences) in gamma ray bursts various $k_F^{(6)}$ coefficients have been bound to $\mathcal{O}(10^{-22}\,\text{GeV}^{-2})$ [58, 59].

6. Experimental tests of LIV

The sensitivity of experimental tests of Lorentz/CPT violation stems from their ability to detect anomalous energy shifts in various systems. Experiments are most effective when all energy levels are scrutinized for possible anomalous shifts. In past decade a number of new Lorentz/CPT signatures have been identified in addition to more classical tests that were known before, such as the Michelson-Morley experiment. Roughly one can distinguish two types of laboratory tests:

- Lorentz tests, which generally scrutinize sidereal time variations in energy levels;
- CPT tests, which measure differences in particle/antiparticle energy levels.

In the next subsections we discuss some examples of experiments that have been carried out recently or are under way.

6.1. Penning traps

Penning traps have been used recently in experiments with electrons and positrons. High precision measurements are done of the anomaly frequency ω_a and the cyclotron frequency ω_c of trapped particles. One can show to lowest order in the mSME parameters that [49]

$$\omega_c^{e^-} = (1 - c_{00}^e - c_{XX}^e - c_{YY}^e)\omega_c^{e,0} >, \tag{49}$$

$$\omega_a^{e^\pm} = \omega_a^{e,0} \pm 2b_Z^e + 2d_{Z0}^e m_e + 2H_{XY}^e. \tag{50}$$

Comparing the anomaly frequencies for electrons and positrons yields the bound $|\vec{b}^e| \lesssim 3 \times 10^{-25}\,\text{GeV}$ [50].

6.2. Clock comparison experiments

The classic Hughes-Drever experiments amount to spectroscopic tests of isotropy of mass and space [51]. They typically examine hyperfine or Zeeman transitions, and provide many of the sharpest LIV bounds for the neutron and proton. For example, a recent Hughes-Drever-type test of Lorentz/CPT for the neutron which used a ^{3}He/^{129}Xe co-magnometer has yielded the bound $|\tilde{b}_J^n| \lesssim 10^{-33}\,\text{GeV}$ for $J = X, Y$ [52]. A bound of $|\tilde{b}_Z^p| \lesssim 2 \times 10^{-27}\,\text{GeV}$ for $J = X, Y$ on Lorentz/CPT violation in the proton sector has been obtained by using a H maser [53]. Clock-comparison experiments conducted in space can provide access to many unmeasured coefficients for Lorentz and CPT violation. The orbital configuration of a satellite platform and the relatively large velocities attainable in a deep-space mission would permit a broad range of tests with Planck-scale sensitivity.

6.3. Hydrogen and antihydrogen

(Anti)hydrogen is the simplest (anti)atom, and it provides for some of the cleanest possible Lorentz and CPT tests involving protons or electrons [54]. The ALPHA and ATRAP experiments underway at CERN intend to make high precision spectroscopic measurements of 1S-2S transitions in H and anti-H and attain a frequency comparison at level of 10^{-18}. Inclusion of magnetic field provides leading order sensitivity to Lorentz/CPT. The ASACUSA experiment intends to analyze ground state Zeeman hyperfine transitions which has leading order LIV corrections in the SME.[3]

6.4. Muon experiments
Several different types of experiments with muons have been conducted such as muonium experiments and $g - 2$ experiments. In the former, frequencies of ground-state Zeeman hyperfine transitions are measured in strong magnetic fields [55] which have yielded a bound $|\tilde{b}^{\mu}_J| \leq 2 \times 10^{-23}$ GeV. Analysis of relativistic $g-2$ experiments using positive muons with large boost parameter have yielded bounds at a level of 10^{-24} GeV [56].

6.5. Spin polarized torsion pendulum
Experiments with a spin-polarized torsion pendulum at the University of Washington provide currently the sharpest bounds on Lorentz/CPT violation in electron sector. It is built out of a stack of toroidal magnets with a huge number of electron spins (8×10^{22}). The aparatus is suspended on a rotating turntable and time variations of the twisting pendulum are measured and analysed for sidereal time variations that would indicate LIV effects. Bounds have been obtained at the levels of $|\tilde{b}^{e}_J| \lesssim 10^{-31}$ GeV for $J = X, Y$ and $|\tilde{b}^{e}_Z| \lesssim 10^{-30}$ GeV [57].

6.6. The photon sector
An extensive body of work exists trying to bound the degrees of freedom of the LIV coefficients $k_F^{\alpha\beta\gamma\delta}$ and k^{μ}_{AF} (see (45)).

Cosmological sources with known polarization permit searching for energy-dependent polarization changes either from distant sources or from CMB. This has yielded the extremely tight bound $|(k_{AF})^{\mu}| \leq 10^{-42}$ GeV [41].

Regarding the components of k_F, bounds have been established on the degrees of freedom that lead to birefringence by cosmological observations. This implies that the polarization plane of photons rotates between emission and observation. Thus cosmological sources with known polarization can be used to verify model-dependent polarization changes. This way the bound $|k_F^{\alpha\beta\gamma\delta}| \leq 2 \times 10^{-32}$ has been obtained [47].

The rotationally-invariant $\kappa_{\rm tr}$ has been bound by a variety of lab experiments. Best laboratory bounds from LEP data are $\mathcal{O}(10^{-15})$ [32, 43], while the best astrophysical bound (absence of vacuum Čerenkov radiation in cosmic rays) is of $\mathcal{O}(10^{-19})$ [46].

The remaining eight (non-birefringent) coefficients have been bounded up to $\mathcal{O}(10^{-17})$ by studying sidereal effects in optical or microwave cavities [44] and up to $\mathcal{O}(10^{-12})$ by an experiment studying sidereal effects in Compton edge photons [45].

A complete updated list of currently existing bounds on the SME parameters is kept on the archives [48].

7. Conclusions
Fundamental theories may allow for Lorentz-invariance violation (LIV), typically assumed to occur at the Planck scale as a result of quantum-gravity effects. This makes LIV an attractive testing ground for search for new physics.

We described a series of testing schemes, both kinematical ones as well as effective field theories. We considered in particular the Standard Model Extension, which offers a comprehensive parametrization of Lorentz and CPT violation at low energy, allowing for systematic experimental testing. We discussed a series of current and experiments of Lorentz and CPT violating parameters.

[3] See also the talk by Chloe Malbrunot in this conference.

Financial support by the Portuguese Fundação para a Ciência e a Tecnologia is gratefully acknowledged.

References

[1] Kostelecky V A and Samuel C 1989 *Phys. Rev.* D **39** 683; Kostelecky V A and Potting R 1991 *Nucl. Phys.* B **359** 545; Kostelecky V A and Potting R 1996 *Phys. Lett.* B **381** 89; Hashimoto K and Murata M 2012 (*Preprint:* arXiv:1211.5949 [hep-th])
[2] Connes A and Kreimer D 1998 *Commun. Math. Phys.* **199** 203
[3] Gambini R and Pullin J 1999 *Phys. Rev.* D **59** 124021
[4] Burgess C P, Cline J M, Filotas E, Matias J and Moore G D 2002 *JHEP* **0203** 043
[5] Bluhm R and Kostelecky V A 2005 *Phys. Rev.* D **71** 065008;
[6] Bjorken J 2001 (*Preprint*: hep-th/0111196); Kraus P and Tomboulis E T 2002 *Phys. Rev.* D **66** 045015; Chkareuli J L, Froggatt C D, Jejelava J G and Nielsen H B 2008 *Nucl. Phys.* B **796** 211; Tomboulis E T 2012 *Phys. Rev.* D **85** 024042
[7] Kostelecky V A and Potting R 2005 *Gen. Rel. Grav.* **37** 1675; Kostelecky V A and Potting R 2009 *Phys. Rev.* D **79** 065018
[8] Bell 54, Pauli 55 Schwinger J S 1951 *Proc. Nat. Acad. Sci.* **37** 452; *Proc. Nat. Acad. Sci.* **37** 455; Lüders G 1954 *Kong. Dan. Vid. Sel. Mat. Fys. Med.* **28N5** 1; Jost R 1957 *Helv. Phys. Acta* **30** 409
[9] Greenberg O W 2002 *Phys. Rev. Lett.* **89** 231602
[10] Kostelecky V A, Lehnert R and Perry M J 2003 *Phys. Rev.* D **68** 123511; Arkani-Hamed N, Cheng H-C, Luty M A and Mukohyama S 2004 *JHEP* **0405** 074
[11] Minwalla S, Van Raamsdonk M and Seiberg N 2000 *JHEP* **0002** 020
[12] Nishimura M 2003 *Phys. Lett.* B **570** 105
[13] Seiberg N and Witten E 1999 *JHEP* **9909** 032; Carroll S N, Harvey J A, Kostelecky V A, Lane C D and Okamoto T 2001 *Phys. Rev. Lett.* **87** 141601
[14] Klinkhamer F R 2000 *Nucl. Phys.* B **578** 277
[15] Horava P 2009 *Phys. Rev.* D **79** 084008
[16] Blas D, Pujolas O and Sibiryakov S 2010 *Phys. Rev. Lett.* **104** 181302
[17] Lehnert R 2003 *Phys. Rev.* D **68** 085003
[18] Amelino-Camelia G, Ellis J R, Mavromatos N E, Nanopoulos D V and Sarkar S 1998 *Nature* **393**; Ellis J R, Mavromatos N E, Nanopoulos D V, Sakharov A S and Sarkisyan E K G 2006 *Astropart. Phys.* **25** 402 [*Astropart. Phys.* **29** 158]
[19] Ng Y J and Van Dam H 1994 *Mod. Phys. Lett.* A **9** 335; Ng Y J and Van Dam H 2000 *Phys. Lett.* B **477** 429; Shiokawa K 2000 *Phys. Rev.* D **62** 024002; Dowker F, Henson J and Sorkin R D 2004 *Mod. Phys. Lett.* A **19** 1829; Alexandre J, Farakos K, Mavromatos N E and Pasipoularides P 2008 *Phys. Rev.* D **77** 105001
[20] Robertson H P 1949 *Rev. Mod. Phys.* **21** 378; Mansouri R and Sexl R U 1977 *Gen. Rel. Grav.* **8** 497
[21] Eardley D M, Lee D L, Lightman A P, Wagoner R V and Will C M 1973 *Phys. Rev. Lett.* **30** 884; Will C M 2001 *Living Rev. Rel.* **4** 4
[22] Amelino-Camelia G 2001 *Phys. Lett.* B **510** 255; Magueijo J and Smolin L 2003 *Phys. Rev.* D **67** 044017
[23] Kostelecky V A and Mewes M 2009 *Phys. Rev.* D **80** 015020
[24] Colladay D and Kostelecky V A 1997 *Phys. Rev.* D **55** 6760; Colladay D and Kostelecky V A 1998 *Phys. Rev.* D **58** 116002
[25] Kostelecky V A 2004 *Phys. Rev.* D **69** 105009
[26] Kostelecky V A, Lane C D and Pickering A G M 2002 *Phys. Rev.* D **65** 056006
[27] Collins J, Perez A, Sudarsky D, Urrutia L and Vucetich H 2004 *Phys. Rev. Lett.* **93** 191301
[28] Myers R C and Pospelov M 2003 *Phys. Rev. Lett.* **90** 211601; Bolokhov P A and Pospelov M 2008 *Phys. Rev.* D **77** 025022
[29] Berger M S and Kostelecky V A 2002 *Phys. Rev.* D **65** 091701; Groot Nibbelink S and Pospelov M 2005 *Phys. Rev. Lett.* **94** 081601; Bolokhov P A, Groot Nibbelink S and Pospelov M 2005 *Phys. Rev.* D **72** 015013; Colladay D and McDonald P 2011 *Phys. Rev.* D **83** 025021; Pujolas O and Sibiryakov S 2012 *JHEP* **1201** 062
[30] Lehnert R and Potting R 2004 *Phys. Rev. Lett.* **93** 110402; Lehnert R and Potting R 2004 *Phys. Rev.* D **70** 125010 [*Erratum-ibid.* D **70** 129906]
[31] Altschul B 2010 *Phys. Rev.* D **82** 016002
[32] Hohensee M A, Lehnert R, Phillips D F and Walsworth R L 2009 *Phys. Rev. Lett.* **102** 170402; Hohensee M A, Lehnert R, Phillips D F and Walsworth R L 2009 *Phys. Rev.* D **80** 036010;
[33] Kostelecky V A and Potting R 1995 *Phys. Rev.* D **51** 3923
[34] Di Domenico A, KLOE Collaboration 2010 *Found. Phys.* **40** 852; Di Domenico A, KLOE Collaboration

2009 *J. Phys. Conf. Ser.* **171** 012008; Bossi F et al, KLOE Collaboration 2008 *Riv. Nuov. Cim.* **031** 531; Nguyen H, KTeV Collaboration 2002 *CPT and Lorentz Symmetry II* ed V A Kostelecky (World Scientific, Singapore) (*Preprint* hep-ex/0112046)

[35] Link J et al, FOCUS Collaboration 2003 *Phys. Lett.* B **556** 7 (*Preprint* hep-ex/0208034)

[36] Aubert B et al, BaBar Collaboration 2008 *Phys. Rev. Lett.* **100** 131802; Aubert B et al, BaBar Collaboration 2006 (*Preprint* hep-ex/0607103); Kostelecky V A and Van Kooten R 2010 *Phys. Rev.* D **82** 101702

[37] Kostelecky V A and Mewes M 2004 *Phys. Rev.* D **70** 031902; Kostelecky V A and Mewes M 2004 *Phys. Rev.* D **70** 076002; Diaz J S, Kostelecky V A and Mewes M 2009 *Phys. Rev.* D **80** 076007; Kostelecky V A and Mewes M 2012 *Phys. Rev.* D **85** 096005

[38] Kostelecky V A and Mewes M 2004 *Phys. Rev.* D **69** 016005;

[39] Diaz J S and Kostelecky V A 2012 *Phys. Rev.* D **85** 016013

[40] Kostelecky V A and Lane C D 1999 *J. Math. Phys.* **40** 6245

[41] Carroll S M, Field G B and Jackiw R 1990 *Phys. Rev.* D **41** 1231

[42] Kostelecky V A and Mewes M 2001 *Phys. Rev. Lett.* **87** 251304; Kostelecky V A and Mewes M 2002 *Phys. Rev.* D **66** 056005

[43] Altschul B 2009 *Phys. Rev.* D **80** 091901

[44] Muller H, Chiow S-W, Herrmann S, Chu S and Chung K-W 2008 *Phys. Rev. Lett.* **100** 031101

[45] Bocquet J-P, Moricciani D, Bellini V, Beretta M, Casano L, D'Angelo A, Di Salvo R and Fantini A et al 2010 *Phys. Rev. Lett.* **104** 241601

[46] Klinkhamer F R and Risse M 2008 *Phys. Rev.* D **77** 016002 (*Addendum* 117901)

[47] Kostelecky V A and Mewes M 2001 *Phys. Rev. Lett.* **87** 251304; Kostelecky V A and Mewes M 2006 *Phys. Rev. Lett.* **97** 140401

[48] Kostelecky V A and Russell N 2011 *Rev. Mod. Phys.* **83** 11 (*Preprint* arXiv:0801.0287 [hep-ph])

[49] Bluhm R, Kostelecky V A and Russell N 1998 *Phys. Rev.* D **57** 3932

[50] Mittleman R K, Ioannou I I, Dehmelt H G and Russell R 1999 *Phys. Rev. Lett.* **83** 2116

[51] Hughes V W, Robinson H G and Beltran-Lopez V 1960 *Phys. Rev. Lett.* **4** 342; Drever R W P 1961 *Phil. Mag.* **6** 683; Kostelecky V A and Lane C D 1999 *Phys. Rev.* D **60** 116010

[52] Gemmel C et al 2010 *Phys. Rev.* D **82** 111901

[53] Phillips D F et al 2001 *Phys. Rev.* D **63** 111101; Humphrey M A et al 2003 *Phys. Rev.* A **68** 063807

[54] Bluhm R and Kostelecky V A 2000 *Phys. Rev. Lett.* **84** 1381; Altschul B 20120 *Phys. Rev.* D **81** 041701; Fujiwara M C et al, ALPHA collaboration 2011 *CPT and Lorentz Symmetry V* ed V A Kostelecky (World Scientific, Singapore)

[55] Hughes V W, Grosse Perdekamp M, Kawall D, Liu W, Jungmann K and zu Putlitz G 2001 *Phys. Rev. Lett.* **87** 111804

[56] Bennett G W et al, Muon (g-2) Collaboration 2009 *Phys. Rev.* D **80** 052008

[57] Heckel B R, Adelberger E G, Cramer C E, Cook T S, Schlamminger S and Schmidt U 2008 *Phys. Rev.* D **78** 092006

[58] Kostelecky V A and Mewes M 2009 *Phys. Rev.* D **80** 015020

[59] Vasileiou V, Fermi GBT and LAT Collaborations, in *CPT and Lorentz Symmetry V* ed V A Kostelecky (World Scientific, Singapore) (*Preprint* [arxiv:1008.2913]); Abdo A A et al, Fermi LAT and Fermi GBM Collaborations 2009 *Science* **323** 1688; Aharonian F et al, H. E. S. S. Collaboration, *Phys. Rev. Lett.* **101** 170402; Albert J et al, MAGIC Collaboration 2008 *Phys. Lett.* B **668** 253; Boggs S E, Wunderer C B, Hurley K and Coburn W 2004 *Astrophys. J.* **611** L77

Journal of Physics: Conference Series **447** (2013) 012010 doi:10.1088/1742-6596/447/1/012010

Measurement of the properties of the 125 GeV Higgs boson with the CMS detector

João Varela[1]
Laboratório de Instrumentação e Partículas
Av. Elias Garcia, 1000-149 Lisbon, Portugal
Deputy Spokesperson, CMS Experiment
E-mail: joao.varela@cern.ch

Abstract. The measurement of the properties of the recently discovered boson is central to the LHC physics program. In this contribution we review preliminary measurements of the properties of the new 125 GeV boson performed by the CMS experiment using the full proton-proton dataset collected in 2011-12 (~25 fb^{-1}). In the $H \rightarrow ZZ$ ($4l$) channel, a signal significance of 6.7σ is now observed. In the other high-resolution mode, $H \rightarrow \gamma\gamma$, updated results were obtained on the signal strength, $\mu = \sigma/\sigma_{SM}$, which is now measured to be ~0.8 ± 0.3. The two high-resolution modes allowed independent determinations of the Higgs mass: 125.8±0.6 GeV, in $H \rightarrow ZZ$ (4l); and 125.4±0.8 GeV, in $H \rightarrow \gamma\gamma$. The four-lepton channel permitted tests of the spin-parity of the new boson. From these studies, the pure pseudoscalar hypothesis is excluded at 99.8% C.L. and, for the first time, simple spin 2 models are excluded with greater than 98.5% C.L. Significantly, strong evidence is seen in a fermionic decay mode of the Higgs for the first time, namely in the $H \rightarrow \tau\tau$ channel, which is reported with a significance of nearly 3σ. The new measurements of the spin-parity (J^P) assignments for this particle, coupled with the measured strength of the interaction of this particle with other particles, strongly indicates that the new particle is a Higgs boson, responsible for the Electroweak Symmetry Breaking. While all of these measurements are consistent with values predicted for a SM Higgs boson, they still fall far short of the requisite precision to rule out all BSM scenarios. Additional data from Run 2 of the LHC and HL-LHC will be needed to firmly establish this conclusion.

1. Introduction

The Higgs boson is the quantum of a scalar field that is postulated to give mass to W^+, W^- and Z^0 bosons and to explain the masses of the fundamental quarks and leptons. It has been the target of searches at accelerators around the world for more than a quarter of a century. In the context of a Higgs boson search in proton-proton collisions at the Large Hadron Collider at CERN [1], the ATLAS and CMS experiments have observed a new particle with a statistical significance of five standard deviations using the first 10 fb^{-1} of data at 7 and 8 TeV [2,3].

The LHC has accelerated and collided beams of protons at energies of $\sqrt{s}$ = 7 (8) TeV in 2011 (2012). The LHC peak luminosity in 2012 was approximately 7.7×10^{33} $cm^{-2}s^{-1}$, achieved with 1368

[1] On behalf of the CMS Collaboration

proton bunches per beam at 50 ns separation with up to ~1.5×10^{11} protons per bunch squeezed to a transverse size of about 20 μm at the interaction point. At this luminosity each bunch-crossing yields on average 39 proton-proton (pp) collisions known as pileup.

The Standard Model Higgs production cross section increases with $\sqrt{s}$ and decreases with m_H. At $\sqrt{s} = 8$ TeV and for $m_H = 125$ GeV, the probability of SM Higgs boson production per pp collision is ~10^{-10}. For every fb^{-1} of integrated luminosity about 20,000 SM Higgs bosons with mass 125 GeV are expected to be produced. The majority of these decay to final states that have large background, making identification difficult or impossible. CMS developed dedicated methods to exploit channels with lower decay fractions by selecting kinematical regions of the decay products where the signal-to-background ratio is sufficiently large to make the observation of a SM Higgs boson possible.

Following the observation of the new boson, the measurement of the properties of this new particle became central to the CMS physics program. In this contribution we review preliminary measurements of the properties of the new 125 GeV boson performed by the CMS experiment using the full pp dataset collected in 2011-12 (~25 fb^{-1}). Five main decay modes are studied involving bosons ($\gamma\gamma$, ZZ, W^+W^-) and the heaviest accessible fermions (b^-b^+, $\tau^+\tau^-$).[2]

2. CMS detector

The CMS detector measures particles produced in high-energy proton-proton and heavy-ion collisions [4]. The central feature of the detector is a superconducting solenoid 13 m long, with an internal diameter of 6 m. Within its volume it generates a uniform 3.8 T magnetic field along the axis of the LHC beams. Within the field volume are a silicon pixel and strip tracker, a lead tungstate ($PbWO_4$) scintillating crystal electromagnetic calorimeter and a brass/scintillator hadron calorimeter. Muons are identified and measured in gas-ionization detectors embedded in the outer steel magnetic-flux-return yoke. The detector is subdivided into a cylindrical barrel part, and endcap disks on each side of the interaction point. Forward calorimeters complement the coverage provided by the barrel and endcap detectors. The CMS detector has a large angular acceptance, detecting particles over the full azimuthal range and with θ larger than 0.8 degrees, where θ is the polar angle relative to the beam axis.

The silicon tracker has 66M pixels and 9.3M strips in 13 (11) layers in the central (forward) regions for > 99% tracking efficiency and 1.5-3% p_T resolution at 100 GeV. Decay vertices are reconstructed with high efficiency and precision. Electrons and photons are detected in ECAL's 75,848 $PbWO_4$ crystals and measured with a precision of ~1% for the photons and electrons considered for the results presented here. Similarly, the muon system has transverse momentum (p_T) resolution better than 1% for the muons used in these analysis.

CMS uses a two-level online trigger system to reduce the event rate from the bunch crossing frequency to ~500 Hz of events recorded in disk by selecting events with leptons, photons, jets, high energy (ΣE_T), missing transverse energy (E_T^{miss}) or combinations of these features. Level 1 uses custom electronics to analyze calorimeter and muon information and has an output rate ≤ 95 kHz. The High Level Trigger uses 13,000 processors to study data from all sub-detectors. Billions of events are recorded each year and sent to computing centers around the world for reconstruction and analysis.

3. Properties of the 125 GeV boson

In 2011 CMS collected 5.1 fb^{-1} of data at 7 TeV. The LHC collision energy was raised to 8 TeV in 2012, leading to ~25% increase in SM Higgs boson production in the low mass region. By the end of 2012, CMS collected 19.6 fb^{-1} data at $\sqrt{s} = 8$ TeV.

At the LHC the SM Higgs boson is produced mainly through gluon-gluon fusion. Two additional important production processes are vector boson fusion (VBF) and associated production (VH), where

[2] We do not show particle charges in the remainder of the text.

V = W or Z. The latter two modes have better signal-to-background (S/B), but have much lower rates [5–9]. VBF events involve high-energy forward jets that can help identify this production mode.

Large samples of simulated events are used to refine the analysis techniques, estimate backgrounds and predict the expected significance for the observation of new particles [10–13]. Whenever possible backgrounds are derived from data control samples.

3.1 Decay mode $H \rightarrow \gamma\gamma$

The $H \rightarrow \gamma\gamma$ decay channel provides a clean final state topology which allows the mass to be reconstructed with high precision. At mass m_H =125 GeV, $H \rightarrow \gamma\gamma$ is one of the most sensitive channels for the Higgs measurement despite its low branching fraction of the order of ~ 0.2% [14].

The search results are presented for an analysis that uses Multi-Variate-Analysis (MVA) techniques [16, 17] both for photon identification and event classification, and extracts the signal from the background using a fit to the diphoton mass spectrum [15]. To achieve the best sensitivity to a Standard Model Higgs boson decaying to two photons, the events are separated into classes. The classes have successively looser quality criteria as determined by mass resolution, photon kinematics, absence/presence of conversions, isolation etc. Unconverted photons tend to have higher quality and higher S/B for SM Higgs, particularly for large $p_T(\gamma\gamma)$. Additional event classes are defined to identify the events from specific production mechanisms, selecting events based on the presence of additional objects in the final state. The presence of two forward jets selects events produced by the vector boson fusion (VBF) production mechanism while events with a muon, an electron, or missing transverse energy (E_T^{miss}) target production in association with a vector boson (VH).

The analysis searches for a localized excess of diphoton events over a smoothly falling background due to prompt diphoton production and to events with at least one jet misidentified as photon. Energy resolution and scale are thus the critical parameters. An *in-situ* ECAL calibration uses $\pi^0,\eta^0 \rightarrow \gamma\gamma$ decays and electrons from W and Z decays. Radiation causes a small loss of transparency in ECAL crystals. This is monitored in real-time by a laser system to obtain corrections enabling a response uniformity of ±0.2% as verified by the E/p ratio for electrons from $W \rightarrow e\nu$ decays. $Z \rightarrow ee$ data are used for the final energy resolutions and scales.

The diphoton mass resolution is not strongly affected by angular resolution of the photon trajectories provided that the interaction point is known to better than ~1 cm. This can be a challenge in a high pileup environment, particularly since photons cannot be tracked unless they convert to an e^+e^- pair. To find the correct vertex, a variety of vertex-related information such as Σp_T^2 and overall p_T balance of tracks relative to the momentum of the $\gamma\gamma$ system is used in a multivariate analysis based on boosted decision trees. Information from conversions is also used when it is available. Boosted decision trees are also used to identify photons and to extract their energies and uncertainties. The full set of selection criteria are provided in ref. [15] for the data considered here.

The description of the Higgs boson signal is obtained from MC simulation using the next-to-leading order (NLO) matrix-element generator POWHEG [18, 19] interfaced with PYTHIA [20]. For the dominant gluon fusion process events have been reweighted to reproduce the NNLL+NNLO distributions computed by the HQT program [21, 22]. The SM Higgs boson cross sections and branching fractions used are taken from ref. [23]. The background estimation is entirely data driven. The background models are obtained by fitting the observed diphoton mass distributions in each of the event classes over the range 100 -180 GeV. Polynomials of orders from 2 to 5 are used to model the background distributions in the various categories. By design, the choice of the background model induces a bias on the fitted signal strength less than five times the statistical uncertainty on the background [24]. Therefore the systematic associated to the background shape can be safely neglected.

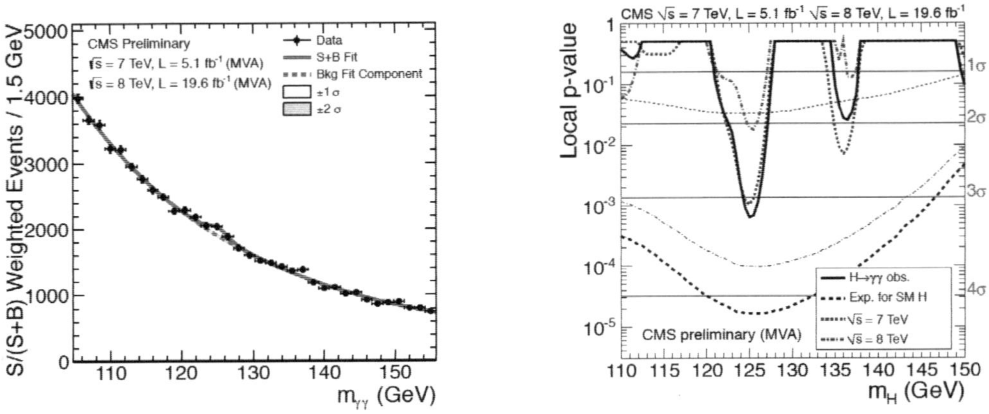

Figure 1. Left: Combination of mγγ event class distributions weighted by S/(S+B). Solid red lines show fits to S+B assuming a 125 GeV SM Higgs. Dashed red lines show background-only with 1 and 2σ bands in yellow and green, respectively. Right: The observed probability (local p-value) as a function of m_H for the background to fluctuate to the excess observed.

Figure 1 shows the mass spectrum where the event classes are weighted by S/(S+B) and combined. An excess of events is seen at 125 GeV that is consistent with expectations for a SM Higgs. The background mostly involves SM processes producing two real photons. A smaller fraction involves a real photon and a fake photon from a misidentified jet. The magnitude of the excess above the background for all categories combined has a significance of 3.2 σ relative to the background-only hypothesis, where a local significance of 4.2 σ is expected from a Standard Model Higgs boson. Consistent excesses are present in 2011 and 2012 data. The best fit signal strength corresponding to the largest signal like fluctuation at 125 GeV is $\sigma/\sigma_{SM} = 0.78^{+0.28}_{-0.26}$.

The calibration of the energy scale is established with reference to the known Z boson mass. The two main sources of systematic uncertainty are the imperfect simulation of the detector response to electrons/photons and the mis-modeling of the extrapolation from the Z to the Higgs energy scale. Figure 2 shows the 2D 68% and 95% CL contours for the signal strength and the mass of the observed particle (left) and the scan of the negative-log-likelihood as a function of the mass (right). From the later, the mass of the observed boson is measured to be 125.4 ± 0.5(stat.) ± 0.6(syst.) GeV.

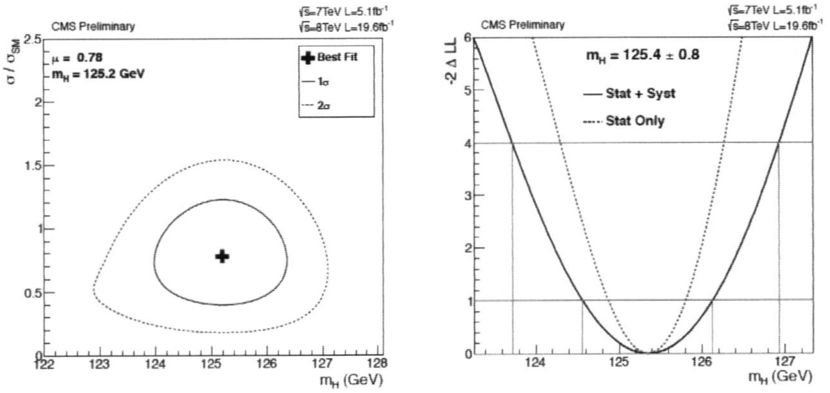

Figure 2. Left: The 2D 68% confidence level region for the signal strength and the mass of the observed particle. Right: The scan of the negative-log-likelihood as a function of the mass.

The four main Higgs boson production mechanisms can be associated with either top-quark couplings (gluon fusion and ttH) or vector boson couplings (VBF and VH). Figure 3 shows the 68% and 95% CL contours for the signal strength modifiers associated with the gluon fusion plus ttH and for VBF plus VH production mechanisms. The observed values are consistent with the expectations from the production of a SM Higgs boson.

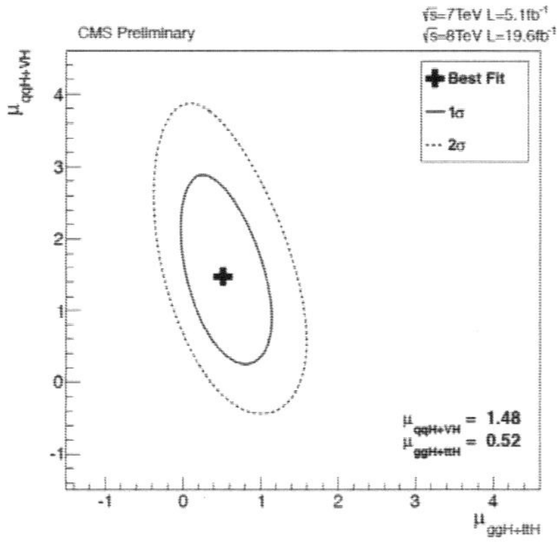

Figure 3. The 68% and 95% CL intervals for the signal strength modifiers associated with the gluon fusion plus ttH and for VBF plus VH production mechanisms. The cross indicates the best-fit value.

3.2 Decay mode $H \rightarrow ZZ^* \rightarrow 4l$

The branching ratio B($H \rightarrow ZZ^*$) is ~2.6% for a 125 GeV Higgs where there is at least one off-shell Z. The decays $Z \rightarrow ll$ where l = e, μ have smallest backgrounds, producing 2e2μ, 4e and 4μ final states. In this contribution, we present an update of the properties of the new boson measured in this channel using the whole 2011-12 dataset. The properties reported here are the mass, the signal strength relative to the expectation for the SM Higgs boson and the spin-parity quantum numbers. A more complete account of the results can be found in [25, 31].

Collision events are selected by the trigger system that requires the presence of a pair of electrons or a pair of muons, or a triplet of electrons. The minimal momenta of the first and second lepton are 17 and 8 GeV, respectively, for the double lepton triggers, while they are 15, 8 and 5 GeV for the triple electron trigger. The trigger efficiency within the acceptance of this analysis is greater than 98% for a Higgs boson signal with $m_H > 120$ GeV.

Offline selection cuts go down to 5 (7) GeV for μ (e) in $|\eta| < 2.4$ (2.5). We require the presence of four isolated leptons that originate from the same interaction vertex. Both Z boson candidates are required to decay to two same-flavor leptons of opposite charge and the invariant mass of the dileptons produced in the Z boson decays must be in the range 40–120 GeV for the heavier of the two, and 12–120 GeV for the lighter one.

The invariant mass of the ZZ system can be reconstructed and measured with good accuracy in CMS from the four-lepton momenta. Hence, the presence of a Higgs boson in the data should manifest itself as a peak in the ZZ invariant mass spectrum in the presence of a small continuum background. SM background processes include irreducible, non-resonant ZZ* production which is estimated via Monte Carlo (MC). Other backgrounds such as individual Z bosons produced in association with heavy-flavor jets and tt production are estimated from data.

MC samples for the SM Higgs boson signal and for background processes are used to optimize the event selection and to evaluate the acceptance and systematic uncertainties. The Higgs boson signals are generated with POWHEG [26] at next-to-leading order (NLO). Signal samples with alternative spin-parity scenarios are generated with JHUGen [27] at Leading Order (LO). At low mass, the analysis is carried out in the framework of the narrow-width approximation, describing the Higgs lineshape with a Breit-Wigner distribution. At high mass the lineshape is corrected to match the results presented in [28–30]. Events at generator level are reweighted according to the total cross section σ(pp→H) containing contributions up to next-to-next-to-leading order (NNLO). The total cross section is scaled by the branching fraction B(H→4*l*) calculated with PROPHECY4F. These higher order corrections are described in the references quoted in [31]. The SM background contribution from ZZ production via qq is generated at NLO with POWHEG, while other diboson processes (WW, WZ) are generated with MADGRAPH [32] with cross sections rescaled to NLO predictions. The gg → ZZ contribution is generated with GG2ZZ [33].

The mass spectrum for selected events is shown in Figure 4. A lepton can radiate a photon at an early stage, potentially leading to an underestimate of its energy if this is not taken into account. Photons consistent with final state radiation are thus included in the lepton energy calculation where appropriate. Figure 4 displays a statistically significant peak near 125 GeV confirming the results reported in [2]. There is also a clear Z peak at 91 GeV from Z→ 4*l* where one lepton pair comes from the conversion of an energetic virtual photon.

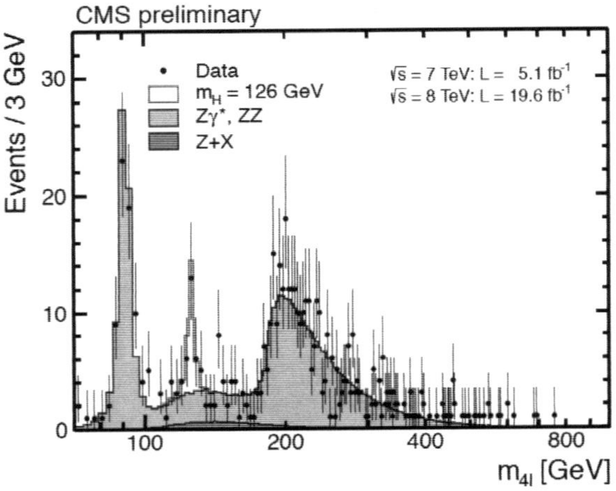

Figure 4. Distribution of invariant mass for 4e, 4μ and 2e2μ with expectations for background and 126 GeV SM Higgs indicated.

Five angles and the invariant masses of the lepton pairs fully describe the kinematics of the decay process at a given mass of the four-lepton system. These observables provide significant discriminating power between signal and background. We use a matrix element likelihood approach [2] to construct a kinematic discriminant K_D based on the probability ratio of the signal and background hypotheses, where the leading-order matrix elements define the probabilities for each value of m_{4l}. Several choices of matrix elements have been studied for signal and background, including analytical parameterization [27,35,35], JHUGEN [27, 34], MCFM [36–38] implemented within the MELA framework [2] and MADGRAPH [32] implemented within the MEKD framework [39]. Different matrix elements were found to provide nearly identical performance. MVA techniques such as boosted decision trees or Bayesian neural networks were also investigated and provided similar results as the matrix element approaches. The distributions of the kinematic discriminant K_D versus the four-lepton reconstructed mass m_{4l} are shown for the selected events and compared to expectations in Fig. 5. The distribution of events is seen to agree well with the SM expectations in the high mass range, while the accumulation of events seen at high K_D is consistent with the expected signal for m_H = 126 GeV.

The local p-values, representing the significance of local excesses relative to the background expectation, are shown as a function of m_H in Fig. 6. The minimum is reached around m_{4l} = 125 GeV, and corresponds to a local significance of 6.7σ (for an expectation of 7.1 σ). This constitutes an observation of the new boson in the four-leptons channel alone.

The signal strength μ, relative to the expectation for the SM Higgs boson, is measured to be $\mu=0.91^{+0.30}_{-0.24}$ at 125.8 GeV. Using simulation it is found that the distribution of the kinematic discriminant K_D for the signal at a mass around m_H = 126 GeV is similar for a scalar, pseudo-scalar, vector, pseudo-vector or a spin-two resonance with the minimal couplings [27]. Therefore the results presented are nearly model-independent in the low-mass region.

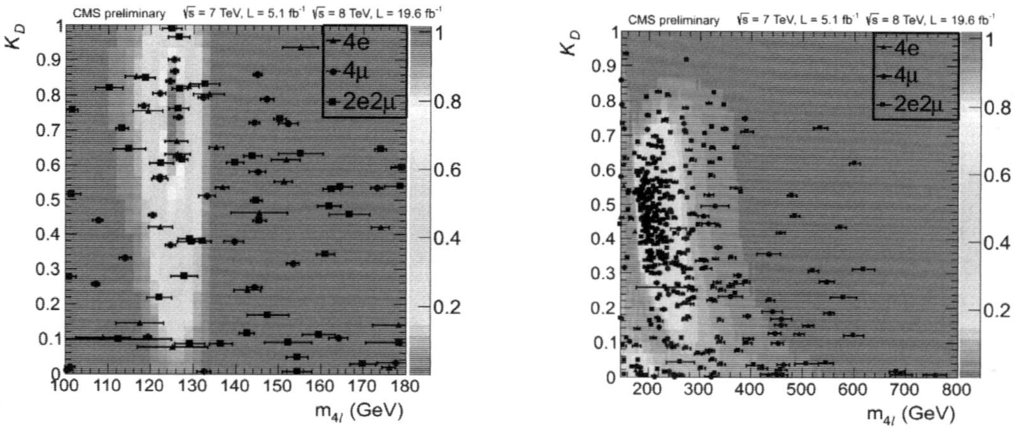

Figure 5. Distribution of the kinematic discriminant K_D versus the four-lepton reconstructed mass m_{4l} in the low-mass range [100-180] GeV (left) and in the high-mass range [180-800] GeV (right). The points representing the individual events are shown together with their reconstructed mass uncertainties. The contours represent the signal expectation for m_H = 126 GeV (left) and the SM background expectation (right).

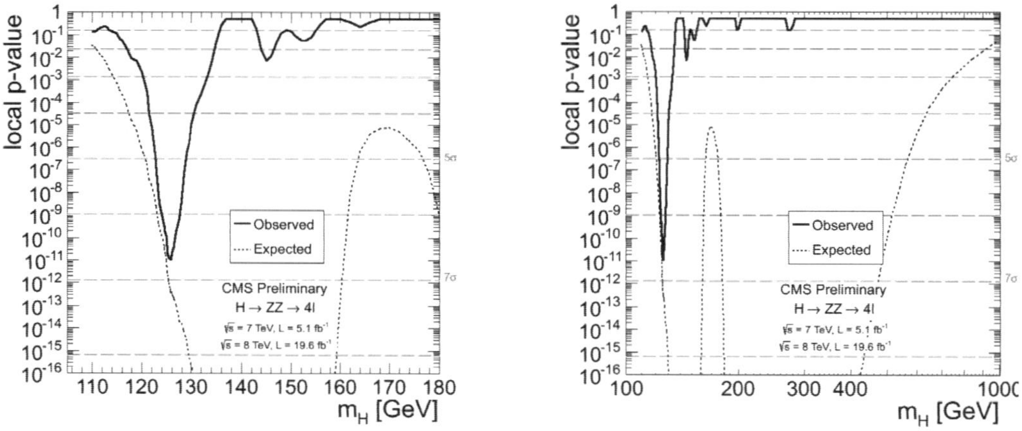

Figure 6. Significance of the local excess with respect to the standard model background expectation as a function of the Higgs boson mass for inclusive analysis. The results are shown for the full data sample in the range [110-180] GeV (left) and [110-1000] GeV (right) for the sum of the 4e, 4μ, and 2e2μ channels.

The mass measurement of the new resonance is performed with a three-dimensional fit using for each event the four-lepton invariant mass, the associated per-event mass error, and the kinematic discriminant. Figure 7 (left) shows the one-dimensional likelihood scan versus the Higgs boson mass.

The resulting fit gives $m_H = 125.8 \pm 0.5(\text{stat.}) \pm 0.2(\text{syst.})$ GeV. The systematic uncertainty accounts for the effect of the lepton momentum scale and resolution.

Categorization of events according to the number of jets and the utilization of variables sensitive to vector boson fusion are used to disentangle the production mechanisms of the observed new state. The production mechanisms are split into two categories depending on whether the production is induced by vector bosons (VBF, ZH, WH) or fermions (gluon fusion loop with quarks, ttH). Two respective signal strength modifiers (μ_F, μ_V) are introduced as scale factors to the SM expected cross section. The result of a two dimensional fit for the two signal strength modifiers assuming a mass hypothesis of m_H=125.8 GeV is shown in figure 7 (right). The measured values are consistent with the expectations from the production of a SM Higgs boson.

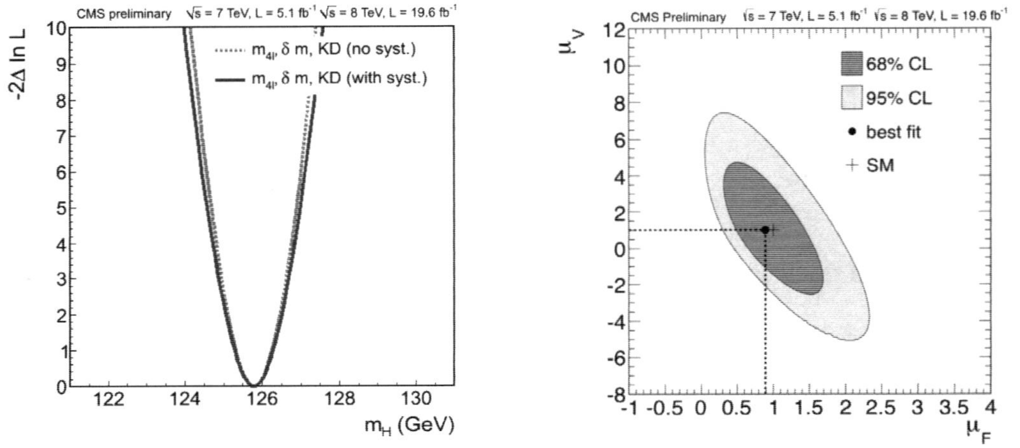

Figure 7. Left: One-dimensional likelihood scan versus the Higgs boson mass in the channel $H \rightarrow ZZ^* \rightarrow 4l$. Right: Likelihood contours on the signal strength modifiers associated with fermions (μ_F) and vector bosons (μ_V) shown at 68% and 95% CL.

The kinematics of the new boson decay to the ZZ*(4l) final state is sensitive to its spin and properties [27, 34]. We use this fact to determine its spin-parity following a methodology with a kinematic discriminant similar to the one used before but instead of the signal-to-background probability ratio we construct the probability ratio for two signal hypotheses. The discriminant for signal hypothesis testing is constructed as $D_{JP}=P_{SM}/(P_{SM}+P_{JP})$, where P_{SM} is the probability distribution for the SM Higgs boson hypothesis, and P_{JP} is the probability for an alternative model. The second observable is $D_{bkg} = P_{sig}/(P_{sig} + P_{bkg})$, where the probabilities P_{bkg} and P_{sig} also include the m_{4l} parameterizations of background and signal with m_H = 126 GeV. Simulation studies showed that the D_{bkg} distributions are very similar between the SM and alternative hypotheses but differ significantly from background.

The separation of the SM Higgs boson model from the pseudoscalar (0^-) or minimal coupling spin-2 resonance produced in gluon fusion (2^+_{mgg}) has been presented by CMS [25], with data strongly disfavoring the pure pseudoscalar hypothesis. We present here additional tests of spin-parity hypotheses and consider the models $J^P = 0^+$, 0^+_h, 0^-, 2^+_{mgg} , 2^+_{mqq}, 1^-, 1^+, respectively the SM scalar, scalar with higher dimension operators, pseudoscalar, minimal coupling spin-2 resonance produced in gluon fusion and produced in quark-antiquark annihilation, exotic vector, and exotic pseudovector. In Figure 8 the D_{JP} distributions are shown in the mass range 106 to 141 GeV used to perform this measurement, for data and simulation.

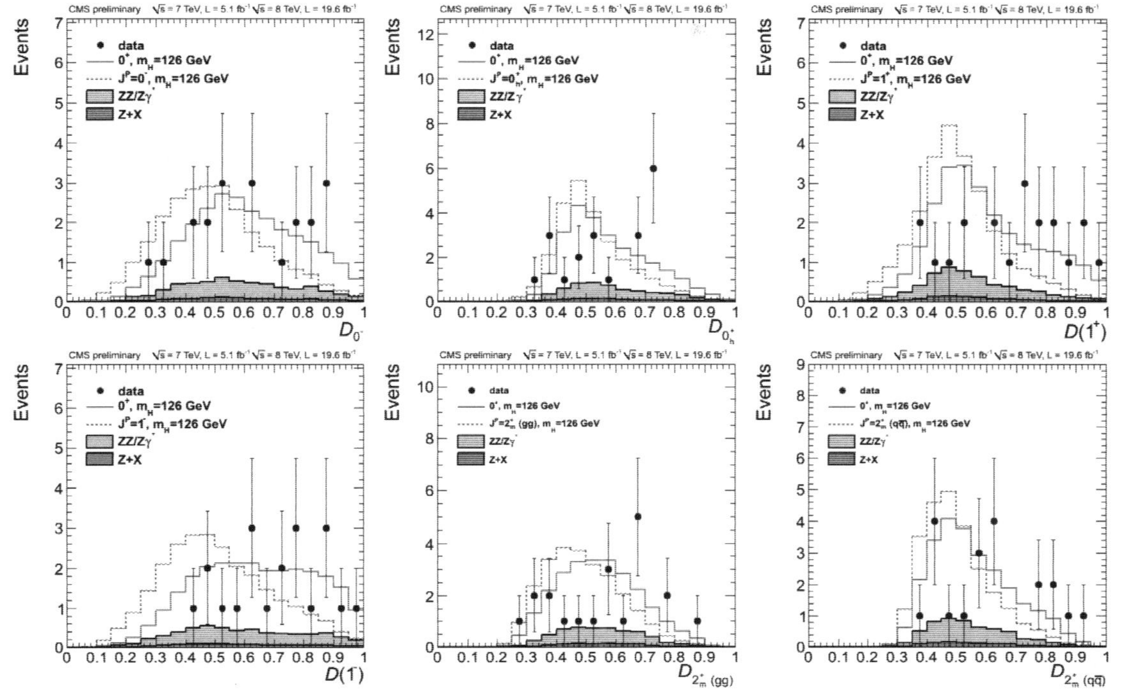

Figure 8. Distributions of D_{JP} with a requirement $D_{bkg} > 0.5$, in data (points with error bars) and expectations for background and signal. The six alternative signal hypotheses described in the text are tested.

A likelihood fit of the ensemble of selected events to the expected 2D distributions (D_{JP}, D_{bkg}) is performed with data and simulated samples, allowing the signal rates to float independently for each signal type. The distribution of the likelihood ratio $q = -2\ln(\mathcal{L}_{JP}/\mathcal{L}_{SM})$ is obtained with generated samples of background and signal of seven types (SM 0^+ and six J^P) for m_H=126 GeV. The expected and observed values of q are shown in Figure 9 and the results are summarized in Table 1. We define a CL_S criterion as the ratio of the probabilities to observe, under the J^P and the 0^+ hypotheses, a value of the test statistics q equal or larger than the one in the data. The data disfavors the alternative hypotheses J^P with a CL_S value in the range 0.1–10%.

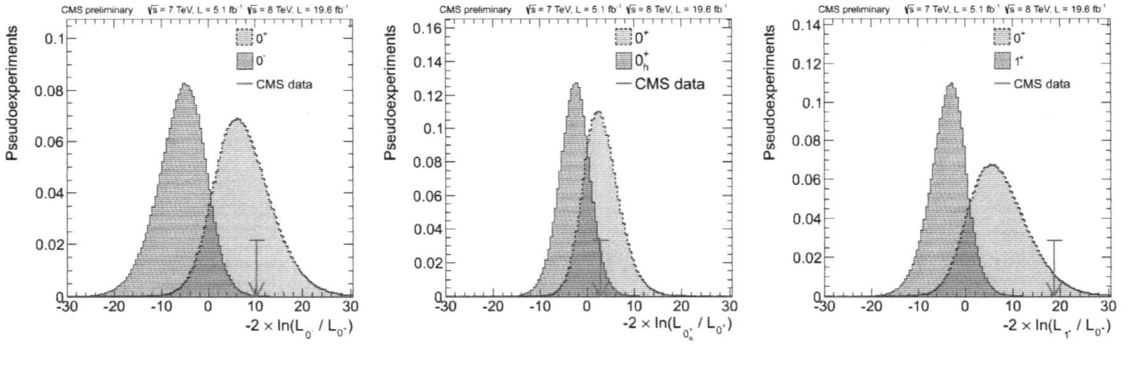

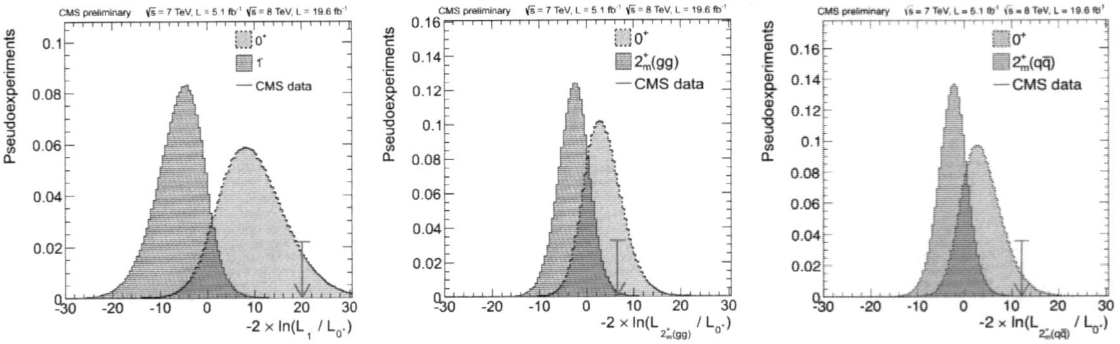

Figure 9. Distributions of $q = -2\ln(\mathcal{L}JP/\mathcal{L}SM)$ for two signal types (0^+ represented by the yellow histogram and alternative J^P hypothesis by the blue histogram) for m_H =126 GeV obtained with a large number of generated pseudo experiments. The arrow indicates the value observed in data. Six alternative hypotheses, described in the main text, are evaluated.

Table 1. List of models used in spin-parity analysis. The expected separation is quoted for two scenarios, when the signal strength is pre-determined from the fit to data and when events are generated with SM expectation for the signal yield (μ = 1). The observed separation quotes the difference between the observation and the expected average of the 0^+ model or the J^P model expressed in standard deviations, and corresponds to the scenario where the signal strength is pre-determined from the fit to data. The last column quotes CLs criterion for the J^P model.

J^P	production	comment	expect (μ=1)	obs. 0^+	obs. J^P	CLs
0^-	$gg \to X$	pseudoscalar	2.6σ (2.8σ)	0.5σ	3.3σ	0.16%
0^+_h	$gg \to X$	higher dim operators	1.7σ (1.8σ)	0.0σ	1.7σ	8.1%
2^+_{mgg}	$gg \to X$	minimal couplings	1.8σ (1.9σ)	0.8σ	2.7σ	1.5%
2^+_{mqq}	$qq \to X$	minimal couplings	1.7σ (1.9σ)	1.8σ	4.0σ	<0.1%
1^-	$qq \to X$	exotic vector	2.8σ (3.1σ)	1.4σ	$>4.0\sigma$	<0.1%
1^+	$qq \to X$	exotic pseudovector	2.3σ (2.6σ)	1.7σ	$>4.0\sigma$	<0.1%

3.3 Decay mode $H \to WW$

For $H \to WW$, the most sensitive final states contain two opposite-sign leptons (ee, eμ or $\mu\mu$) and significant E_T^{miss} due to the undetected neutrinos. In contrast to the $\gamma\gamma$ and ZZ modes the mass cannot be reconstructed precisely. The signal is therefore expected to appear as an excess over background that extends over a broad mass range. Multivariate analysis techniques are used to optimize the sensitivity to signal. Events are categorized by lepton flavor content and jet multiplicity with different backgrounds and sensitivities. Jet identification and E_T^{miss} require some care in a high pileup environment where energy deposits from multiple pp collisions can become intertwined. Pileup effects are largely neutralized by associating charged particles to their correct interaction vertices and by means of MVA techniques that use jet shape variables to separate real jets from clusters of pileup energy deposits. The final impact of pileup on the correct accounting of events with zero additional jets is negligible, as seen in Figure 10.

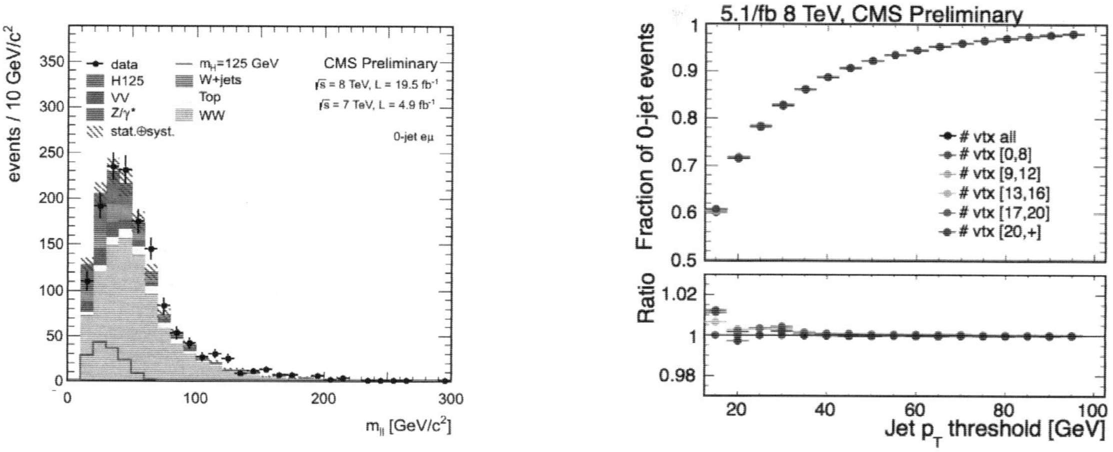

Figure 10. Left: Invariant mass of lepton pairs for the zero-jet eμ category in the search for $H \rightarrow WW$. The 125 GeV SM Higgs boson signal is shown added to the background. Right: Fraction of events in the 0-jet category as a function of jet p_T threshold for 5 exclusive ranges of the number of primary vertices.

We select events in which the most (2nd most) energetic lepton has $p_T > 20$ (10) GeV and typically $E_T^{miss} > 20$ GeV. Estimation of SM backgrounds makes use of data control samples complemented by MC samples to extrapolate into the signal region. Upper limits are derived for the product of the Higgs boson production cross section and the $H \rightarrow WW$ branching fraction, $\sigma_H \times BR(H \rightarrow WW)$, relative to the SM Higgs expectation, σ/σ_{SM}. Combining all classes from 2011 and 2012 data, we obtain the 95% observed and median expected C.L. upper limits shown in Figure 11 (left). An excess of events is observed for low Higgs boson masses. Due to the poor mass resolution of this channel the excess extends over a large mass range. The significance of the excess at mass 125 GeV with respect to the background only hypothesis is 4.0 standard deviations for the combination of the 7 and 8 TeV data. The observed σ/σ_{SM} value for $m_H = 125$ GeV is 0.76 ± 0.13 (stat.) ± 0.16 (syst.). A more complete account of the results in this channel can be found in [40].

3.4 Decay mode $H \rightarrow \tau\tau$

Under the assumption that its coupling is proportional to mass, we search for decays of this new particle to the heaviest accessible fermions – namely pairs of τ leptons and b quarks. The detection of τ leptons is challenging because they have a multitude of decay modes to leptons ($\tau \rightarrow l\nu\nu$, $l = e,\mu$), or to one or three $\pi^{\pm}$, possibly accompanied by π^0's, and a ν_τ. CMS reconstructs the leptonic decays and several classes of hadronic decays with good efficiency and low fake rates as verified with $Z \rightarrow \tau\tau$ data [41].

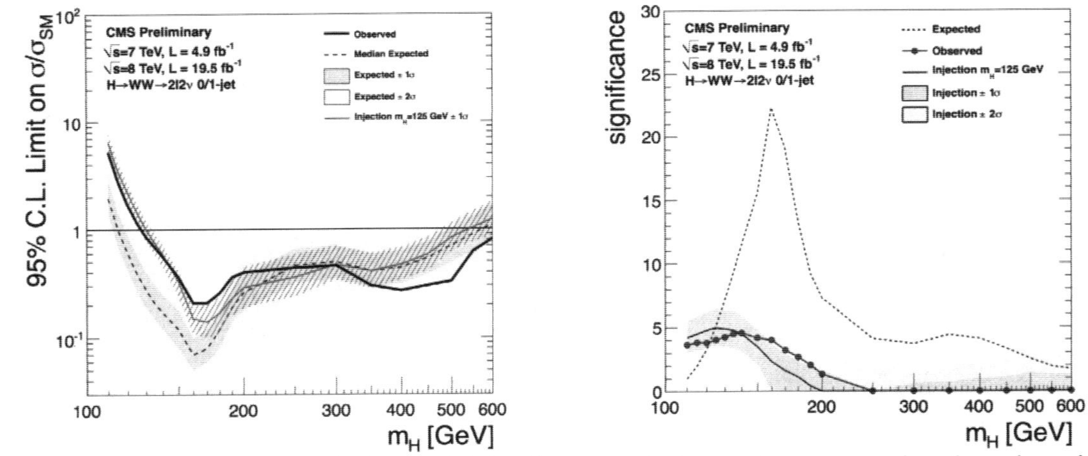

Figure 11. Left: Expected and observed 95% CL upper limits on the cross section times branching fraction, relative to the SM Higgs expectation, using the 7 and 8 TeV data. The expected limits in the presence of the Higgs with $m_H = 125$ GeV and its associated uncertainty are also shown (red). Right: The dashed line shows the significance expected for the search for a Higgs boson with a mass m_H if a Higgs boson of this mass exists. The solid line and colored bands show the expected significance at a mass m_H if the true Higgs boson mass is 125 GeV. The dots show the observed significance of the excess for the search for a Higgs boson with a mass m_H.

The challenge here is the understanding of the backgrounds. They are mostly extracted using control samples in data. Events are classified by five pairings ($\tau_e\tau_\mu$, $\tau_e\tau_h$, $\tau_\mu\tau_h$, $\tau_\mu\tau_\mu$, $\tau_h\tau_h$) of tau decays and by the presence of jets. The classification by the number of additional jets in the final state is used to enhance the contribution of different Higgs boson production mechanisms.

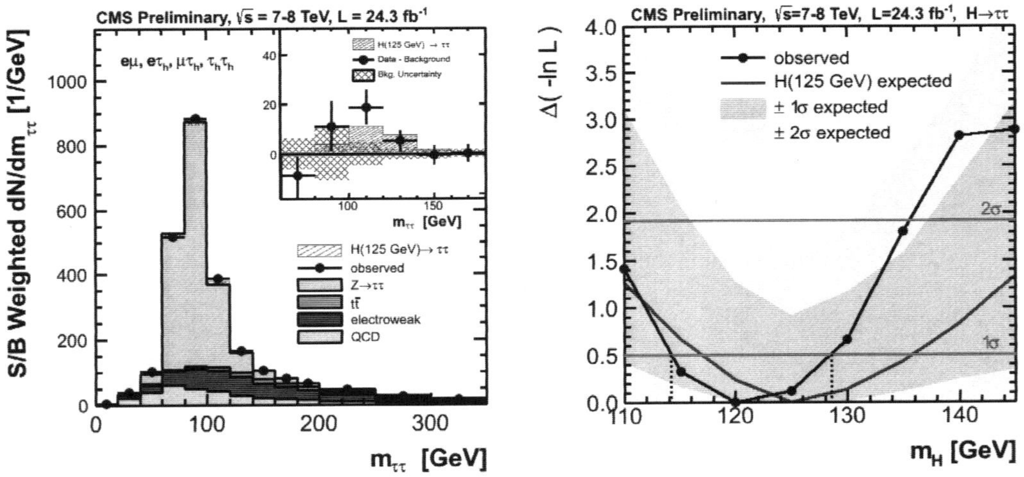

Figure 12. Left: Combined observed and expected $m_{\tau\tau}$ distributions for the $\mu\tau_h$, $e\tau_h$, $e\mu$ and $\tau_h\tau_h$ channels. The distributions obtained in each category of each channel are weighted by the ratio between the expected signal and background yields in the category. The insert shows the corresponding difference between the observed data and expected background distributions, together with the expected signal distribution for a standard-model Higgs signal at $m_H = 125$ GeV, with a focus on the signal region. Right: Log likelihood versus SM Higgs boson mass, combining all search channels. The black curve shows the result from the observed spectrum of tau pair invariant mass, and

the blue curve shows the result expected if the observed spectrum was equal to that expected from the background plus a SM Higgs boson signal at a mass of 125 GeV.

The one-jet category selects primarily signal events with a Higgs boson produced by gluon fusion, or in association with a W or Z boson decaying hadronically. Events in the vector-boson fusion (VBF) category are required to have two jets separated by a large rapidity gap, which mainly selects signal events with a Higgs boson produced by vector-boson fusion and strongly enhances the signal contribution. Different categories of events have different backgrounds and signal sensitivities. In each channel, the signal is separated from the background, and in particular from the irreducible $Z \rightarrow \tau\tau$, using the τ-pair mass reconstructed from the four-momenta of the visible decay products of the two τ leptons and from the missing transverse energy. The presence of neutrinos implies that the signal would appear as a broad enhancement over background.

Figure 12 (left) shows the observed and expected $m_{\tau\tau}$ distributions combined for different channels. An excess of events is seen in the mass region where a standard-model Higgs signal at m_H = 125 GeV is expected. The best-fit value for the signal strength combining all channels is $\mu = 1.1 \pm 0.4$ at m_H =125 GeV. The minimum p-value of the excess is observed at m_H = 120 GeV, corresponding to a significance of 2.9 standard deviations. This is an indication that the new boson discovered in 2012 couples to τ leptons, with a strength compatible to the one predicted by the standard model. Figure 12 (right) shows the fit likelihood as a function of the Higgs mass. The best fit is obtained for a SM Higgs boson of $m_H = 120^{+9}_{-7}$ (stat+syst) GeV. A more complete account of the results in this channel can be found in [42].

3.5 Decay mode $H \rightarrow bb$

A search for the standard model Higgs boson H decaying to bb when produced in association with a weak vector boson (V) was performed for the following modes: W(μν)H, W(eν)H, Z(μμ)H, Z(ee)H and Z(νν)H [43]. The search was performed in data samples corresponding to integrated luminosities of 5.0 fb^{-1} at $\sqrt{s}$ =7 TeV and 12.1 fb^{-1} at $\sqrt{s}$ =8 TeV, recorded by the CMS experiment at the LHC.

Upper limits, at 95% confidence level, on the VH production cross section times the $H \rightarrow bb$ branching ratio, with respect to the expectations for a standard model Higgs boson, were derived for a Higgs boson in the mass range 110-135 GeV. In this range, the observed upper limits vary from 1.0 to 4.2 times the standard model prediction; the corresponding expected limits vary from 0.9 to 1.9. An excess of events is observed above the expected background with a local significance of 2.2 standard deviations, which is consistent with the expectation from the production of the standard model Higgs boson.

4. Conclusions

The CMS experiment is traversing an historic period in particle physics following the discovery of a new boson in 2012. The investigation of the nature of this new particle is being intensively pursued. Improved measurements of the strength of the interaction of this particle with other particles together with comparisons of data with expectations of a variety of spin-parity models, reported here, strongly indicate that indeed we have observed a Higgs boson that is responsible for Electroweak Symmetry Breaking.

References

1. L. Evans, P. Bryant (editors), *LHC Machine*, JINST 03 (2008) S08001, doi:10.1088/1748-0221/3/08/S08001
2. CMS Collaboration, *Observation of a new boson at a mass of 125 GeV with the CMS experiment at the LHC*, Phys. Lett. B 716 (2012) 30–61, doi:10.1016/j.physletb.2012.08.021, arXiv:1207.7235.

3. ATLAS Collaboration, *Observation of a new particle in the search for the Standard Model Higgs boson with the ATLAS detector at the LHC,* Phys.Lett.B 716 (2012) 1-29.
4. CMS Collaboration, *The CMS experiment at the CERN LHC,* JINST 03 (2008) S08004, doi:10.1088/1748-0221/3/08/S08004.
5. J.R. Ellis, M.K. Gaillard, D.V. Nanopoulos, *A phenomenological profile of the Higgs boson*, Nucl. Phys. B 106 (1976) 292, doi:10.1016/0550-3213(76)90382-5.
6. H.M. Georgi, S.L. Glashow, M.E. Machacek, D.V. Nanopoulos, *Higgs Bosons from Two Gluon Annihilation in Proton Proton Collisions*, Phys. Rev. Lett. 40 (1978) 692, doi:10.1103 /PhysRevLett.40.692.
7. S.L. Glashow, D.V. Nanopoulos, A. Yildiz, *Associated Production of Higgs Bosons and Z Particles*, Phys. Rev. D 18 (1978) 1724, doi:10.1103/PhysRevD.18.1724.
8. S. Alioli, P. Nason, C. Oleari, E. Re, *NLO Higgs boson production via gluon fusion matched with shower in POWHEG*, JHEP 04 (2009) 002, doi:10.1088/1126-6708/2009/04/002.
9. P. Nason, C. Oleari, *NLO Higgs boson production via vector boson fusion matched with shower in POWHEG*, JHEP 02 (2010) 037, doi:10.1007/JHEP02(2010)037.
10. T. Sjöstrand, S. Mrenna, P.Z. Skands, *PYTHIA 6.4 physics and manual*, JHEP 05 (2006) 026, doi:10.1088/1126-6708/2006/05/026.
11. S. Gieseke, D. Grellscheid, K. Hamilton, A. Ribon, P. Richardson, et al., *Herwig++ 2.0 Release Note (2006)*, arXiv:hep-ph/0609306.
12. J. Alwall, P. Demin, S. de Visscher, R. Frederix, M. Herquet, et al., *MadGraph/MadEvent v4: the new web generation*, JHEP 09 (2007) 028, doi:10.1088/1126-6708/2007/09/028, arXiv:0706.2334.
13. S. Agostinelli, et al., *GEANT4—a simulation toolkit*, Nucl. Instrum. Meth. A 506 (2003) 250, doi:10.1016/S0168-9002(03)01368-8.
14. S. Actis et al., Nucl. Phys. B 811 (2009) 182, doi:10.1016/j.nuclphysb.2008.11.024, arXiv:0809.3667.
15. CMS Collaboration, *Updated measurements of the Higgs boson at 125 GeV in the two photon decay channel,* CMS PAS HIG-13-001, CDS Record 1530524 (2013).
16. H. B. Prosper, *Multivariate methods in particle physics: Today and tomorrow*, 2008, PoS ACAT08 (2008) 010.
17. P. C. Bhat, *Multivariate Analysis Methods in Particle Physics*, Ann. Rev. Nucl. Part. Sci. 61 (2011) 281-309.
18. S. Alioli et al., *NLO Higgs boson production via gluon fusion matched with shower in POWHEG*, JHEP 04 (2009) 002, doi:10.1088/1126-6708/2009/04/002.
19. P. Nason and C. Oleari, *NLO Higgs boson production via vector-boson fusion matched with shower in POWHEG*, JHEP 02 (2010) 037, doi:10.1007/JHEP02(2010)037.
20. T. Sjostrand, S. Mrenna, and P. Z. Skands, *PYTHIA 6.4 Physics and Manual*, JHEP 0605 (2006) 026, doi:10.1088/1126-6708/2006/05/026.
21. G. Bozzi et al., *The q(T) spectrum of the Higgs boson at the LHC in QCD perturbation theory*, Phys. Lett. B 564 (2003) 65, doi:10.1016/S0370-2693(03)00656-7.
22. G. Bozzi et al., *Transverse-momentum resummation and the spectrum of the Higgs boson at the LHC*, Nucl. Phys. B 737 (2006) 73, doi:10.1016/j.nuclphysb.2005.12.022.
23. LHC Higgs Cross Section Working Group Collaboration, *Handbook of LHC Higgs Cross Sections: 1. Inclusive Observables*, CERN Report CERN-2011-002, (2011).
24. CMS Collaboration, *Evidence for a new state decaying into two photons in the search for the standard model Higgs boson in pp collisions*, CDS Record 1460419 (2012).
25. CMS Collaboration, *Study of the Mass and Spin-Parity of the Higgs Boson Candidate via Its Decays to Z Boson Pairs*, Phys. Rev. Lett. 110 (2013) 081803, doi:10.1103/PhysRevLett.110.081803, arXiv:1212.6639.
26. S. Frixione, P. Nason, and C. Oleari, *Matching NLO QCD computations with Parton Shower simulations: the POWHEG method*, JHEP 11 (2007) 070, doi:10.1088/1126-6708/2007/11/070, arXiv:0709.2092.
27. Y. Gao et al., *Spin determination of single-produced resonances at hadron colliders*, Phys. Rev. D 81 (2010) 075022, doi:10.1103/PhysRevD.81.075022, arXiv:1001.3396.

28. G. Passarino, C. Sturm, and S. Uccirati, *Higgs Pseudo-Observables, Second Riemann Sheet and All That*, Nucl. Phys. B 834 (2010) 77, doi:10.1016/j.nuclphysb.2010.03.013, arXiv:1001.3360.
29. S. Goria, G. Passarino, and D. Rosco, *The Higgs Boson Lineshape*, Nucl. Phys. B 864 (2012) 530–579, doi:10.1016/j.nuclphysb.2012.07.006, arXiv:1112.5517.
30. N. Kauer and G. Passarino, *Inadequacy of zero-width approximation for a light Higgs boson signal*, JHEP 1208 (2012) 116, doi:10.1007/JHEP08(2012)116, arXiv:1206.4803.
31. CMS Collaboration, *Properties of the Higgs-like boson in the decay H to ZZ to 4l in pp collisions at sqrt s =7 and 8 TeV,* CMS PAS HIG-13-002, CDS Record 1523767 (2013).
32. J. Alwall et al., *MadGraph/MadEvent v4: The NewWeb Generation*, JHEP 09 (2007) 028, doi:10.1088/1126-6708/2007/09/028, arXiv:0706.2334.
33. T. Binoth, N. Kauer, and P. Mertsch, *Gluon-induced QCD corrections to* $pp \rightarrow ZZ \rightarrow llll$, in Proceedings of the XVI Int. Workshop on Deep-Inelastic Scattering and Related Topics (DIS'07). 2008. arXiv:0807.0024. doi:10.3360/dis.2008.142.
34. S. Bolognesi et al., *On the spin and parity of a single-produced resonance at the LHC*, Phys. Rev. D 86 (2012) 095031, doi:10.1103/PhysRevD.86.095031, arXiv:1208.4018.
35. Y. Chen, N. Tran, and R. Vega-Morales, *Scrutinizing the Higgs Signal and Background in the* $2e2\mu$ *Golden Channel*, arXiv:1211.1959.
36. J. M. Campbell and R. K. Ellis, *MCFM for the Tevatron and the LHC*, Nucl. Phys. Proc. Suppl. 205 (2010) 10, doi:10.1016/j.nuclphysbps.2010.08.011, arXiv:1007.3492.
37. J. M. Campbell and R. K. Ellis, *An update on vector boson pair production at hadron colliders*, Phys. Rev. D 60 (1999) 113006, doi:10.1103/PhysRevD.60.113006, arXiv:hep-ph/9905386.
38. J. M. Campbell, R. Ellis, and C. Williams, *Vector boson pair production at the LHC*, JHEP 07 (2011) 018, doi:10.1007/JHEP07(2011)018, arXiv:1105.0020.
39. P. Avery et al., *Precision studies of the Higgs boson decay channel* $H \rightarrow ZZ* \rightarrow 4l$ *with MEKD*, accepted by Phys.Rev.D. (2012) arXiv:1210.0896.
40. CMS Collaboration, *Evidence for a particle decaying to W+W- in the fully leptonic final state in a standard model Higgs boson search in pp collisions at the LHC,* CMS PAS HIG-13-003, CDS Record 1523673 (2013).
41. CMS Collaboration, *Z inclusive cross-section via decays to tau pairs at 7 TeV*, JHEP 08 (2011) 117, doi:10.1007/JHEP08(2011)117., arXiv:1104.1617.
42. CMS Collaboration, *Search for the Standard-Model Higgs boson decaying to tau pairs in proton-proton collisions at sqrt(s) = 7 and 8 TeV,* CMS PAS HIG-13-004, CDS Record 1528271 (2013).
43. CMS Collaboration, *Search for the standard model Higgs boson produced in association with W or Z bosons, and decaying to bottom quarks for HCP 2012,* CMS PAS HIG-12-044, CDS Record 1493618 (2012).

The search for electric dipole moments of light ions in storage rings

Frank Rathmann
Institut für Kernphysik, Forschungszentrum Jülich GmbH, 52425 Jülich, Germany

Artem Saleev and N.N. Nikolaev
L.D. Landau Institute for Theoretical Physics
1423432 Chernogolovka, Moscow disctrict, Russia

on behalf of the JEDI and srEDM collaborations

E-mail: `f.rathmann@fz-juelich.de`

Abstract. The Standard Model (SM) of Particle Physics is not capable of accounting for the apparent matter-antimatter asymmetry of our universe. Physics beyond the SM is required and is searched for by *(i)* employing highest energies (e.g., at LHC), and *(ii)* striving for ultimate precision and sensitivity (e.g., in the search for electric dipole moments (EDMs)). Permanent EDMs of particles violate both time reversal (T) and parity (P) invariance, and are via the CPT-theorem also CP-violating. Finding an EDM would be a strong indication for physics beyond the SM, and reducing upper limits further provides crucial tests for any corresponding theoretical model, e.g., SUSY. Direct searches for proton and deuteron EDMs bear the potential to reach sensitivities beyond 10^{-29} e·cm. For an all-electric proton storage ring, this goal is pursued by the US-based **srEDM** collaboration [1], while the newly founded Jülich-based **JEDI** collaboration [2] is pursuing an approach using a combined electric-magnetic lattice, which shall provide access to the EDMs of protons, deuterons, and ^{3}He ions in the same machine. In addition, JEDI has recently proposed making a direct measurement of the proton and/or deuteron EDM at COSY using resonant techniques involving Wien filters.

1. Introduction

Electric dipole moments (EDMs) are one of the keys to understand the origin of our universe. The universe as we know it has a microscopic net baryon number of about 0.2 baryons per cubic meter, or $\sim 10^{-10}$ of the density of relic photons. In 1967 Andrei Sakharov formulated three conditions for baryogenesis [3]:

(i) Early in the evolution of the universe, the baryon number conservation must have been violated sufficiently strongly.

(ii) The C and CP invariances, and T invariance thereof, must have been violated.

(iii) At the point in time when the baryon number is generated, the evolution of the universe must have been far from thermal equilibrium.

CP violation in kaon decays is known since 1964, it has been observed in B-decays and in charmed meson decays, and based on the existing data can be described by the CP-violating phase in the Cabibbo-Kobayashi-Maskawa matrix [4] (for a review of recent developments in the charmed sector, see Ref. [5]). CP and P violation entail non-vanishing P and T violating EDMs of elementary particles. Although extremely successful in many aspects, the SM has at least two weaknesses: neutrino oscillations call for extra parameters, and, most importantly, the SM mechanisms fail miserably in the expected baryogenesis rate: the CKM matrix of the SM amounts to the relevant effective coupling $\leq 10^{-20}$ [6], much too feeble to account for the observed baryon density $n_{\rm B}/n_\gamma = (6.19 \pm 0.15) \times 10^{-10}$ [7]. Simultaneously, the SM predicts an exceedingly small electric dipole moment of nucleons $10^{-33}\,\mathrm{e}\cdot\mathrm{cm} < d_n < 10^{-31}\,\mathrm{e}\cdot\mathrm{cm}$, way below the current upper bound for the neutron EDM, $d_n < 2.9 \times 10^{-26}$ e·cm [8], and also beyond the reach of future EDM searches [9]. In the quest for physics beyond the SM one may either follow the high-energy trail or look into new methods, which offer very high precision and sensitivity. Supersymmetry is one of the most attractive extensions of the SM as S. Weinberg emphasized in 1992 [10]: "Endemic in supersymmetric (SUSY) theories are CP violations that go beyond the SM. For this reason it may be that the next exciting thing to come along will be the discovery of a neutron electric dipole moment." The SUSY predictions span typically a range of $10^{-29}\,\mathrm{e}\cdot\mathrm{cm} < d_n < 10^{-24}\,\mathrm{e}\cdot\mathrm{cm}$ and precisely this range is targeted in the new generation of EDM searches [9] discussed here.

There is consensus among theorists that measuring the EDM of the proton, deuteron and helion is as important as that of the neutron. The EDMs could have a non-trivial isospin dependence and $d_d \neq d_p + d_n$, even if the CP-violation comes from the isoscalar QCD θ-term [11]. Furthermore, it was argued some 25 years ago that T-violating nuclear forces could substantially enhance nuclear EDMs [12, 13]. At the moment, there are no significant *directly* determined upper bounds available on d_p or d_d.

Non-vanishing EDMs give rise to a precession of the spin of a particle in an electric field. While ultra-cold electrically neutral atoms and neutrons can conveniently by stored in traps, direct measurements of EDMs of charged particles have hitherto been impossible because the required new class of dedicated, primarily electrostatic storage rings does not yet exist.

An ambitious quest for a measurement of the EDM of the proton with envisioned sensitivity down to $d_p \sim 10^{-29}$ e·cm per year has been proposed at BNL [14]. The principal idea is to store protons with longitudinal polarization in a purely electrostatic ring. The EDM would cause a precession around the radial electric field and thus lead to a build-up of transverse polarization, which could be measured by employing standard polarimetry [15]. The new idea proposed by the srEDM collaboration is to freeze the horizontal spin motion, *i.e.*, the particles' spins point always along the direction of motion, which cancels the $(g-2)$ precession. During the spin coherence time, the observed build-up of a vertical polarization component in the beam would be interpreted in terms of an EDM limit of the orbiting particles [14].

The proposed new method employs radial electric (and magnetic) fields to steer the particle beam in the ring, magnetic or electric quadrupole magnets to form a strong focusing lattice (e.g., FODO), and internal polarimeters to probe the beam polarization as a function of storage time. An RF-cavity and sextupole magnets will be used to prolong the spin coherence time of the beam. For protons, it requires building a storage ring with a highly uniform radial electric field with strength of $\approx 17\,\mathrm{MV/m}$ between field plates about $\approx 2\,\mathrm{cm}$ apart. The bending radius will be $\approx 30\,\mathrm{m}$ (see Table 1). At the so-called magic momentum of $700.740\,\mathrm{MeV}/c$ ($232.792\,\mathrm{MeV}$) the $(g-2)$ precession frequency vanishes.

Related ideas on dedicated storage rings for the proton, deuteron, and helion EDM are pursued at IKP of Forschungszentrum Jülich by the newly founded JEDI collaboration [2], aiming at the determination of direct EDM limits for the proton and other charged particles in a storage ring [16].

Particle	p (GeV/c)	$G = \frac{g-2}{2}$	E (MV/m)	B (T)
Proton	0.701	1.792847	13.977	0.000
Deuteron	1.000	-0.142987	-3.361	0.135
^{3}He	1.211	-4.183963	13.943	-0.042

Table 1. Parameters for the radial electric and the vertical magnetic fields required to freeze the spin in an EDM storage ring of radius $r = 30\,\mathrm{m}$.

2. Status of EDM searches

The question whether particles possess permanent EDMs has a long-standing history, starting from the first measurement by Smith, Purcell, and Ramsey for the neutron [17] as a signature for parity (P) and time-reversal (T or CP) violation, which, over the last 50 years or so, resulted in ever decreasing upper limits [18]. In Table 2, current and anticipated EDM bounds and sensitivities for nucleons, atoms, and the deuteron are given. It should be noted that the upper bound for the proton EDM as part of a nucleus in an electrically neutral atom, $|d_p| < 7.9 \times 10^{-25}\,e\,\mathrm{cm}$, derives from the theoretical reinterpretation of the upper bound for the EDM of ^{199}Hg [19]. The second-to-last column of Table 2 provides a rough measure of the probing power relative to that of the neutron (d_n). At this level, storage ring EDM measurements bear the potential of an order of magnitude higher sensitivity than the currently planned neutron EDM experiments at SNS (Oak Ridge), ILL (Grenoble-France), and PSI (Villigen, Switzerland) [20].

Particle	Current limit	Goal	d_n equivalent	Ref.
Electron	$< 1.85 \times 10^{-27}$	$\approx 10^{-29}$		[21]
Neutron	$< 2.9 \times 10^{-26}$	$\approx 10^{-28}$	10^{-28}	[8]
^{199}Hg	$< 3.1 \times 10^{-29}$	10^{-29}	10^{-26}	[19]
^{129}Xe	$< 6.0 \times 10^{-27}$	$\approx 10^{-30}$ to 10^{-33}	$\approx 10^{-26}$ to 10^{-29}	[22]
Proton	$< 7.9 \times 10^{-25}$	$\approx 10^{-29}$	10^{-29}	[19]
Deuteron		$\approx 10^{-29}$	$\approx 3 \times 10^{-29}$ to 5×10^{-31}	

Table 2. Current EDM limits in units of [e·cm], long-term goals for the neutron, ^{199}Hg, ^{129}Xe, proton, and deuteron are given. Neutron equivalent values indicate the value for the neutron to provide the same physics reach as the indicated system.

3. Direct search for EDMs at COSY

Before jumping into the design and construction of dedicated storage rings, it is imperative to get to first base by performing a direct measurement of proton and/or deuteron EDM at an existing machine such as the cooler synchrotron COSY at Forschungszentrum Jülich. COSY is a world-wide unique facility, which provides essentially everything necessary for EDM searches: polarized beams, spin manipulators and polarimeters. Here we review two ideas for so-called precursor experiments which could be performed at COSY, and which are subject to very modest additions or modifications to the existing machine.

Wien filters, either static or radio-frequency driven, provide large horizontal (radial) electric fields. They exert no Lorentz force on the beam particles, i.e, are EDM *transparent*, because the net electric field in a Wien filter is zero. The interaction of the magnetic dipole moment

(MDM) with the combined direct and motional magnetic fields, however, causes a kick in the spin precession phase. In conjunction with the additional interaction of the EDM in the motional electric field of COSY, this MDM kick rotates the spin around the radial axis and generates a CP-violating horizontal in-plane polarization, which increases linearly with time. Without the Wien filter, the same EDM interaction in the COSY ring causes only a small tilt which does not grow with time. Therefore, such a device would allow one to accumulate during the spin coherence time the coherent build-up of the EDM-generated P and T non-invariant in-plane polarization which can be determined from the up-down asymmetry of the scattering of stored particles on suitable polarimeters (see e.g., Ref. [15]). Unless show stoppers pop up, one could theoretically aim for an upper bound for the proton or deuteron of $d < 10^{-24}$ e·cm, where it should be noted that in particular for the deuteron, such a limit would be as valuable as the existing upper bounds on d_n [12, 13].

The accumulation time of a storage ring EDM experiment is limited by the spin coherence time (SCT). Therefore, it is of prime importance to find ways to make the SCT as long as possible. To this end, a series of investigations is presently carried out at COSY [16, 23, 24].

3.1. Precursor experiments at COSY

In Ref. [25] we have reviewed a number of possible first *direct* measurements of an upper limit for the proton and/or deuteron EDM using a normal magnetic storage ring like COSY. In the subsequent two sections we present numerical estimates for the two most promising ideas. Both ideas are based on the exploitation of spin resonances, i.e., the motion of the particle on the orbit is in sync with its spin motion.

3.1.1. RF Wien filter The RF Wien filter provides a radial electric (E_x) and a vertical magnetic (B_y) field. It is operated on some harmonic K of the spin motion,

$$f_{\rm HV} = (K + G\gamma) f_{\rm rev} \,, \tag{1}$$

where $f_{\rm HV}$ is the frequency of the harmonically excited RF Wien filter and $f_{\rm rev}$ is the orbit frequency. The Wien filter shall avoid the coherent excitation of betatron oscillations in the machine. Imperfections in the alignment and field quality of the Wien filter, and also in the alignment of the magnetic elements in COSY play an important role for the systematics involved. We have started to study their effects, but here we will sketch the basic idea. The calculation is carried out for deuterons injected with vertical polarization, $\vec{P} = (P_x, P_y, P_z) = (0, 1, 0)$. The main parameters are summarized in Table 3.

The accumulation of the polarization components P_x, P_z, and the total in-plane polarization $P_{xz} = \sqrt{P_x^2 + P_z^2}$ as function of time is illustrated in Fig. 1, top panel. Assuming a deuteron EDM of $d = 10^{-22}$ e·cm, the extrapolation of the initial slope of the in-plane polarization to a total accumulation time of $t = 1000$ s under the conditions listed in Table 3 yields a value of $P_{xz} = 0.10$. It should be noted that technically, one can measure P_{xz} down to 10^{-5} to 10^{-6} per year of operation [15].

3.1.2. Static Wien filter on the $G\gamma = 2$ imperfection resonance Instead of using an RF Wien filter to excite the spin resonance, for protons it is also possible to operate COSY on a so-called imperfection resonance, for which the resonance condition is given by $G\gamma = n$, where $n = 2, 3, \ldots$ Imperfection resonances for deuterons are outside the energy range of COSY. In that case and only for protons, the Wien filter can be operated in a static mode and the pattern of accumulation of the polarization components is similar to the one of the RF Wien filter (see Fig. 1, bottom panel). Operating in static mode, larger electric fields are possible compared to the RF Wien filter, for an electric field of 10 MV/m, the extrapolation of the accumulated

	RF Wien filter	Static Wien filter
particle	deuteron	proton
length of element [m]	3	3
assumed EDM limit [e·cm]	10^{-22}	10^{-23}
beam kinetic energy [MeV]	250	108.412
$\gamma \cdot G$	-0.162	2.000
electric field E_x [MV/m]	1	10
magnetic field B_y [T]	7.089×10^{-3}	7.526×10^{-2}
Extrapolated in-plane polarization P_{xz} to $t = 1000$ s	0.10	0.06

Table 3. Parameters used to evaluate the magnitude of the CP-violating in-plane polarization components P_{xz} in the two Wien filter scenarios with the device operated in RF and static mode.

in-plane polarization component after $t = 1000$ s yields $P_{xz} = 0.06$. The operation of a machine on an imperfection resonance requires to reduce the strength of this resonance by making the closed orbit close to perfect, and by avoiding alignment errors of the magnetic elements.

4. Conclusion

Direct mesurements of proton and deuteron EDMs do not yet exist. When COSY is supplemented with an RF Wien filter, the CP violating in-plane polarization component of the deuteron, accumulated for a time interval of 1000 s, could be as large as $P_{xz} = 0.10$, corresponding to an upper bound of $d_d = 10^{-22}$ e·cm. For protons, an option to use a static Wien filter on the imperfection $G\gamma = 2$ resonance, where larger electric fields are possible, a proton EDM limit of $d_p = 10^{-23}$ e·cm corresponds to an in-plane polarization of $P_{xz} = 0.06$.

Regarding the systematics of such experiments, we have presently touched only the tip of the iceberg in a very crude analytic approach and much more scrutiny is required. Specifically, one badly needs spin tracking tools capable of handling with controlled precision up to $\sim 10^{11}$ turns in a realistically modeled machine. With all reservations, the Wien filter experiments at COSY look promising. We especially emphasize again here the importance of in-situ studies at COSY using very slow RF magnetic Wien filters to study systematic effects.

References

[1] Storage Ring Electric Dipole Moment Collaboration http://www.bnl.gov/edm/
[2] Jülich Electric Dipole Moment Investigations Collaboration http://collaborations.fz-juelich.de/ikp/jedi/
[3] Sakharov A 1967 *Pisma Zh.Eksp.Teor.Fiz.* **5** 32–35
[4] Aaij R *et al.* (LHCb Collaboration) 2012 (*Preprint* 1211.1230)
[5] Polycarpo E 2012 (*Preprint* 1212.6341)
[6] Riotto A and Trodden M 1999 *Ann.Rev.Nucl.Part.Sci.* **49** 35–75 (*Preprint* hep-ph/9901362)
[7] Komatsu E *et al.* (WMAP Collaboration) 2011 *Astrophys.J.Suppl.* **192** 18 (*Preprint* 1001.4538)
[8] Baker C *et al.* 2007 *Phys.Rev.Lett.* **98** 149102 (*Preprint* 0704.1354)
[9] Roberts B L and Marciano W J (eds) 2010 *Lepton Dipole Moments* (*Advanced Series on Directions in High Energy Physics* vol 20) (World Scientific)
[10] Weinberg S 1993 *AIP Conf.Proc.* **272** 346–366 (*Preprint* hep-ph/9211298)
[11] Guo F K and Meissner U G 2012 *JHEP* **1212** 097 (*Preprint* 1210.5887)
[12] Flambaum V, Khriplovich I and Sushkov O 1986 *Nucl.Phys.* **A449** 750
[13] Flambaum V, Khriplovich I and Sushkov O 1985 *Phys.Lett.* **B162** 213–216
[14] Semertzidis Y K (Storage Ring EDM Collaboration) 2009 *AIP Conf.Proc.* **1182** 730–736
[15] Brantjes N, Dzhordzhadze V, Gebel R, Gonnella F, Gray F *et al.* 2012 *Nucl.Instrum.Meth.* **A664** 49–64

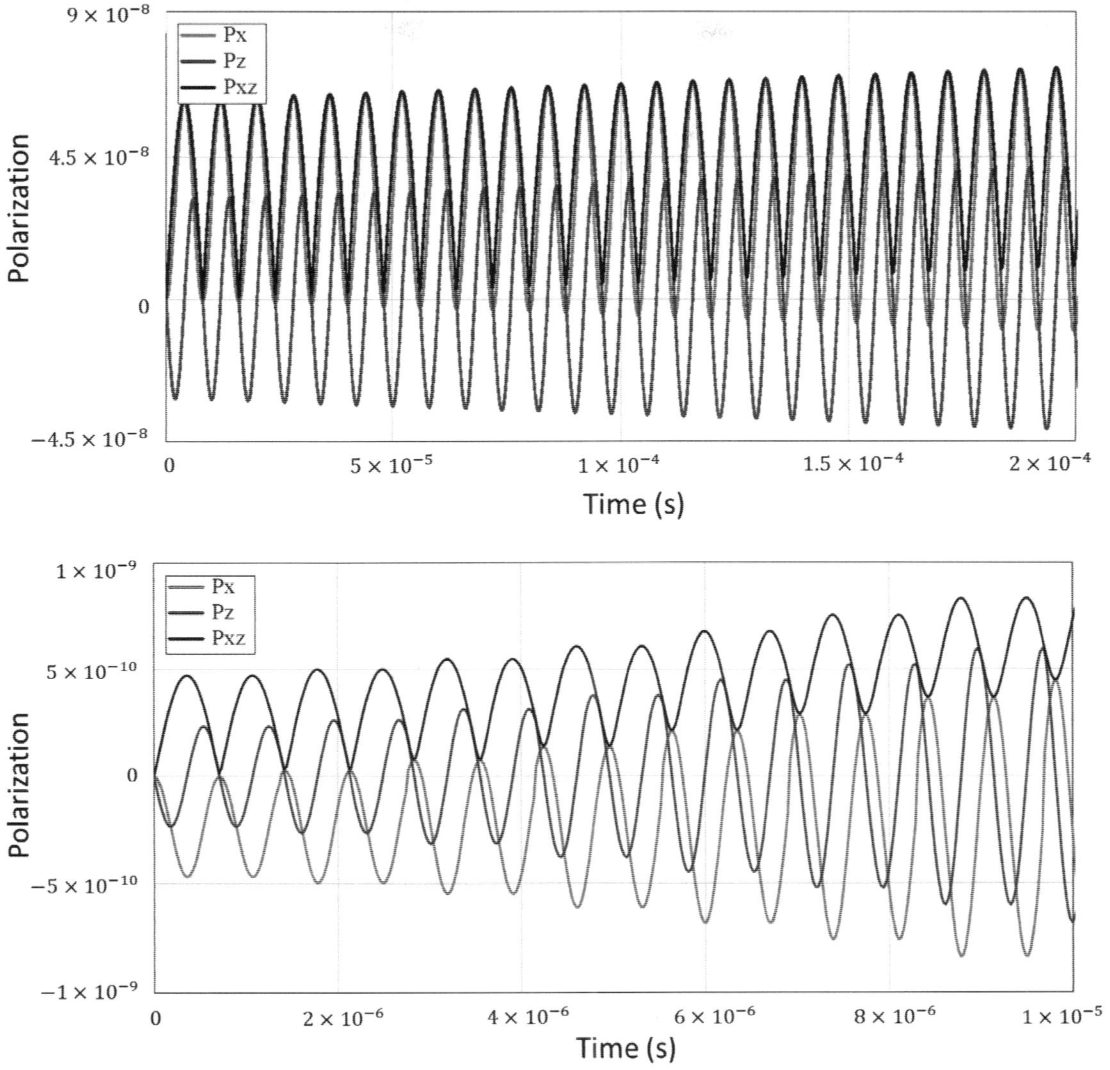

Figure 1. Top panel: Accumulation of the in-plane polarization component P_{xz} (black line) when an *RF Wien* filter is operated for deuterons at a beam energy of $T = 250$ MeV. **Bottom panel:** Accumulation of the in-plane polarization component for protons P_{xz} (black line) when a *static Wien filter* is operated on the $G\gamma = 2$ imperfection resonance ($T = 108.412$ MeV). Other conditions for both cases are listed in Table 3.

[16] Spokespersons, Lehrach A, Pretz J and Rathmann F 2012 Search for Permanent Electric Dipole Moments at COSY, Step 1: Spin coherence and systematic error studies `http://donald.cc.kfa-juelich.de/wochenplan/List_of_all_COSY-Proposals.shtml`
[17] Smith J, Purcell E and Ramsey N 1957 *Phys.Rev.* **108** 120–122
[18] Reich E 2012 *Nature* **482** 17
[19] Griffith W *et al.* 2009 *Phys.Rev.Lett.* **102** 101601
[20] Lamoreux S K and Golub D 2010 *Lepton Dipole Moments* (*Advanced Series on Directions in High Energy Physics* vol 20) (World Scientific) chap 15, p 583

[21] Kara D, Smallman I, Hudson J, Sauer B, Tarbutt M *et al.* 2012 *New J.Phys.* **14** 103051 (*Preprint* 1208.4507)
[22] Rosenberry M A and Chupp T E 2001 *Phys. Rev. Lett.* **86** 22–25
[23] Benati P, Chiladze D, Dietrich J, Gaisser M, Gebel R *et al.* 2012 *Phys.Rev.ST Accel.Beams* **15** 124202
[24] Spokespersons, Stephenson E and Lenisa P 2012 The horizontal spin coherence time is investigated in the COSY proposals 176.4 to 176.6 http://donald.cc.kfa-juelich.de/wochenplan/List_of_all_COSY-Proposals.shtml
[25] Rathmann F and Nikolaev N 2011 *PoS* **STORI11** 029

Journal of Physics: Conference Series **447** (2013) 012012 doi:10.1088/1742-6596/447/1/012012

Top quark physics: A tool for discoveries

Michele Gallinaro

Laboratório de Instrumentação e Física Experimental de Partículas
LIP Lisbon
Av. Elias Garcia, 14 - 1000-149 Lisboa - Portugal

E-mail: michgall@cern.ch

Abstract. The top quark, the heaviest known elementary particle discovered at the Fermilab Tevatron almost twenty years ago, has taken a central role in the study of fundamental interactions. The top quark behaves differently from all other quarks due to its large mass and its correspondingly short lifetime. Its large mass suggests that it may play a special role in nature. The top quark decays before it hadronizes, passing its spin information on to its decay products. Therefore, it is possible to measure observables that depend on the top quark spin, providing a unique environment for tests of the standard model and for new physics searches. With approximately 10 fb^{-1} of luminosity delivered to each experiment at the Tevatron, and about 20 fb^{-1} collected by the ATLAS and CMS experiments at the Large Hadron Collider in the first three years of operation, top quark physics is at a turning point from first studies to precision measurements with sensitivity to new physics processes. This report summarizes the latest experimental measurements and studies of top quark properties and rare decays.

1. Introduction

The long search for the top quark engaged researchers at laboratories around the world for many years, and came to a successful conclusion in February 1995 with the announcement that the top quark had been observed in two experiments at the Tevatron proton-antiproton collider at Fermilab [1]. The top quark was clearly present in the data and the CDF and D0 collaborations published papers presenting overwhelming proof the top quark had been finally found. Its mass was measured to be $m_t = 176 \pm 13$ GeV. It was expected that the top quark must exist since 1977, when its partner, the bottom quark, was discovered. The weak isospin partner of the bottom quark had been missing since then. The top quark completed the three generation structure of the standard model (SM).

Many years after its discovery, the top quark still plays a fundamental role in the program of particle physics. The study of its properties has been extensively carried out in high energy hadron collisions. However, a few important questions still remain unanswered. Why is it so heavy? Is its mass generated by the Higgs mechanism? What is the role the top quark plays in the electroweak symmetry breaking (EWSB) mechanism? Does the top quark play a role in non-SM physics? Are the couplings affected? Since the shutdown of the Tevatron in September 2011, the Large Hadron Collider (LHC) is now the only place where it is possible to study the top quark production mechanism. Before the start of the LHC, most of the measurements were limited by the small number of top quarks available. In comparison, the LHC is a top factory which produced one million of top quark events per experiment in 2011 alone. The LHC has performed extremely well in the first three years of operation, and the large samples of top quarks

Journal of Physics: Conference Series **447** (2013) 012012 doi:10.1088/1742-6596/447/1/012012

collected offer the opportunity to improve the results and enter the era of precision measurements and study rare processes. It is worth noticing that studies performed at the Tevatron and at the LHC are sometime complementary due to the different energies and production mechanisms.

The top quark is the heaviest of all known elementary particles. With a mass close to that of an atom of gold, it has a mass about 40 times heavier than the b-quark and it is heavier than the W boson, the top quark decays to $t \to Wb$ with a branching fraction close to 100%. The top quark is a fundamental particle with a mass close to the EWSB scale, and it may play an important role in the understanding of the EWSB mechanism. Furthermore, the large top quark mass implies a large coupling to the Higgs boson, thus establishing a privileged link to the Higgs sector. Due to its short lifetime, the top quark decays before hadronization and it offers the unique opportunity to study the properties of a bare quark which are preserved in the decay chain and transferred to its decay products. The top quark has charge $Q = +2/3e$ and weak isospin $T_3 = +1/2$, as the partner of the bottom quark in the third generation quark doublet.

At hadron colliders, top quarks are mostly produced in pairs through strong interactions, or individually (single-top production) via electroweak interaction. Depending on the production mode, the top quark therefore allows different tests of the participating forces. The top quark is present in the higher-order diagrams and it provides - within the electroweak theory of particle interactions - indirect constraints on the Higgs boson mass, together with the W boson mass. Top quark production also plays an important role in many scenarios in the search for new physics beyond the SM. Several models predict the existence of new particles decaying to (or with large couplings to) the top quark. Therefore, the study of top quarks may provide hints to the presence of new physics processes. Furthermore, top quark production constitutes a large background to many of the searches for new physics processes, and it is therefore important to understand the properties and the characteristics of top production and decay mechanisms, and the level of precision will have an impact on the constraints on new physics processes.

For nice and complete reviews on this topic, please see Refs. [2].

2. Top quark production

Several properties of the top quark have already been studied both at the Tevatron and at the LHC. These include studies of the kinematical properties of top production, measurement of the production cross section, reconstruction of $t\bar{t}$ pairs in the fully hadronic final states, study of τ decays of the top quark, and reconstruction of hadronic decays of the W boson from top decays.

2.1. Top pair production

At the Tevatron top quark pairs are predominantly produced in quark-antiquark annihilation (90%), whereas at the LHC the top quark pair production mechanism is dominated by gluon fusion process ($\simeq$ 80% at $\sqrt{s}$ = 7 TeV). This is due to the large gluon density in the proton at small-x. The production cross section has been measured in many different final states. Deviation of the cross section from the predicted SM value may indicate new physics processes.

In each top quark pair event, there are two W bosons and two bottom quarks. From the experimental point of view, top quark pair events are classified according to the decay mode of the two W bosons: the all-hadronic final state, in which both W bosons decay into quarks, the lepton+jet final state, in which one W decays leptonically and the other to quarks, and the dilepton final state, in which both W bosons decay leptonically. The word "lepton" here refers to electrons and muons, whereas τs are somehow classified differently, and they are generally treated separately. In the dilepton channel, the final state consists of two charged leptons, missing transverse energy, and at least two bottom jets. The branching ratio is small (5%) but the background (mostly Z+jets) is also small, which makes the dilepton the best final state to select a clean sample of top quark events. The all-hadronic final state has an experimental signature with at least 6 jets, of which two are from bottom quarks, with a large background, mostly from QCD multi-jet events. The measurement of the cross section in this channel is rather difficult, despite the large branching fraction (44%). The lepton+jet final state offers a compromise with a reasonably large branching fraction (36%) and a moderate background, mostly from W+jet events. The signature consists of one charged lepton, missing transverse energy, and at least four jets (two of them b-jets).

Cross section measurements have been performed both at the Tevatron and at the LHC and the accuracy of the experimental results rivals that of theory expectations [3]. At the Tevatron ($\sqrt{s}$ = 1.96 TeV), the top quark pair production cross section was measured very precisely as $\sigma = 7.50 \pm 0.48(\text{stat.} + \text{syst.})$pb [4] (6% precision). The first top quark pair candidates at the LHC were already reported in the summer of 2010, after a few months of data-taking at $\sqrt{s} = 7$ TeV. After two years, thousands of top quark events have already been selected. Measurement of the inclusive top quark pair production cross section have been performed at the LHC in the dilepton and lepton+jet channels using electrons and muons and provide the most precise results. Most of the cross section measurements are already limited by the systematic uncertainties. As an example, the cross section in the lepton+jet final state is determined with a simultaneous maximum likelihood fit to the number of jets, the number of b-tagged jets and the invariant mass of the tracks associated with the secondary vertex (to allow discriminating light and heavy quark contributions). The simultaneous fit in the jet and b-jet multiplicities allow constraining the top quark pair signal and the W+light (and heavy) flavour composition of the background.

Measurement are also performed in the tau+lepton, tau+jets, and all-hadronic channels. The interest of determining the cross section in all channels is mainly to check the consistency of the measurements, and check for deviations. For example, the measurement of the cross section in the tau+lepton (as well as the tau+jets) final state is important because a deviation of the measured cross section from the expected value may provide a hint for new physics. The tau+lepton channel, i.e. $t\bar{t} \to (\ell\nu_\ell)(\tau\nu_\tau)b\bar{b}$ (with $\ell = e, \mu$) is of particular interest because the existence of a charged Higgs with a mass smaller than the top quark mass $m_H < m_t$ could give rise to anomalous tau lepton production directly observable in this decay channel, via $t \to H^+b$. As in the other channels, the tau+lepton cross section results [5] are consistent with the cross sections measured in the other final states, and the measurement can be used to set stringent limits on charged Higgs production [6].

2.2. Differential distributions

The large number of top quark events collected makes also possible the measurement of differential cross sections, $d\sigma/dX$, for the relevant variable X. For instance, variables of relevance may be related to the kinematics of the top or $t\bar{t}$ systems, such as p_T, $M_{t\bar{t}}$, $t\bar{t} + N$ jets. These distributions may be used to validate given MC models as well as to check specific higher order QCD calculations. Deviations could signal contribution from new physics. Differential measurements are performed in the dilepton and lepton+jet channels, after reconstruction of the event kinematics. The measurement of the $t\bar{t} + N$ jet distribution assesses the theoretical predictions and the simulation in the recoil of the $t\bar{t}$ system and the modelling of additional quark and gluon radiation in $t\bar{t}$ production. Experimental data are needed to validate the simulated samples and to reduce the uncertainties. The overall agreement between data and simulation is remarkable, although the uncertainty is not yet at the level to be able to distinguish between different signal models. The electroweak couplings of the top quark can also be studied in the associated production to a gauge boson, such as $t\bar{t}\gamma$, $t\bar{t}W$, and $t\bar{t}Z$ events. Among those, the $t\bar{t}\gamma$ production is large enough to be measured already with the available data samples. The leading-order (LO) cross section for this process is about 1 pb at $\sqrt{s} = 7$ TeV, for a photon transverse momentum $p_T >$ 8GeV. In this process, the photon is radiated from off-shell top quarks or incoming partons, or from on-shell top quark or one of its decay products (such as the

W). The measurement in the lepton+jet channel is performed with a template fit to signal and background, where the templates are derived from data. Results agree well with SM predictions.

3. Top quark mass

The top quark mass m_t is a fundamental parameter of the SM, and it is linked to the W and Higgs boson masses. Through its measurement it is possible to constrain indirectly, together with the W mass, the Higgs boson mass value.

It is measured precisely at the Tevatron with an accuracy of about 0.5%. The combined value from the CDF and D0 experiments yields $m_t = 173.18 \pm 0.94$ GeV [7]. Direct measurements of m_t are also performed at the LHC. Thanks to the large samples of top quarks available, stringent selections and improved analysis techniques, and to an excellent performance and good understanding of the detectors, the precision of the LHC measurements are close to the precision reached at the Tevatron already after the first few years of data-taking [8]. The mass is measured in the dilepton, lepton+jet, and in the all-hadronic channels. The first measurement at the LHC was performed in the dilepton channel [9] from the kinematic characteristics of the events with a full kinematic analysis, and with an analytical matrix weighting technique using distributions derived from simulated samples. The reconstruction of m_t from dilepton events leads to an under-constrained system, since the dilepton channel contains at least two neutrinos in the final state. The lepton+jet channel provides a fully constrained system, and it is (so far) the "golden" channel as it yields the best accuracy in the mass measurement among all final states. Many techniques have been used, and the most accurate single measurement at the LHC is performed with the "Ideogram" method, in which a constrained kinematic fit is performed for all jet-parton assignment combinations. For each event, a likelihood is calculated as function of the top mass (with two terms, one for signal and one for background) corresponding to the probability for the event to be either signal or background. The signal and background probabilities are parametrized using analytic functions, derived from simulation. An overall likelihood is constructed by multiplying all event likelihoods. In general, the mass measurements are limited by systematic uncertainties, and the dominant source is the jet energy scale uncertainty, i.e. the absolute scale, ISR/FSR, fragmentation, and single particle response in the calorimeter. In the lepton+jet channel, reduced uncertainty can be achieved with an *in-situ* calibration of the W mass from the untagged jets, using the $W \to qq'$ decays.

Direct measurements of m_t rely on the reconstruction of kinematic observables sensitive to m_t. These direct measurements depend on the details of the kinematics, reconstruction, and calibration. Furthermore, the measurement is performed in a particular definition of m_t which does not correspond to a specific renormalization scheme. Alternatively, m_t can be derived indirectly from the cross section measurement. Therefore, measurements of the cross section are used to extract m_t in a well-defined renormalization scheme, such as the pole mass (m_{pole}) or $\overline{MS}$ definitions. The measured inclusive $t\bar{t}$ production cross section is compared with fully inclusive higher-order perturbative QCD computations where the top quark mass parameter is unambiguously defined. For instance, the extraction of m_{pole} from the measured $t\bar{t}$ cross section provides complementary information compared to direct methods that rely explicitly on the details of the kinematic mass reconstruction. This extraction also tests the internal consistency of perturbative QCD calculations in a well-defined renormalization scheme, and provides an important cross check of the direct measurements.

Direct measurement of a mass difference between particle and anti-particle would indicate a violation of the CPT symmetry. Quarks carry color charge and cannot be observed directly as they hadronize to colorless particles before decaying. One exception is the top quark, as it decays before hadronization due to its short lifetime. In the measurement, carried out in the lepton+jet channel, most of the systematic uncertainties cancel out. The mass difference between top and anti-top quarks Δm_t is measured and no significant deviation from zero is found.

Journal of Physics: Conference Series **447** (2013) 012012 doi:10.1088/1742-6596/447/1/012012

4. Single top production

Single top is produced through electroweak interactions in the s- and t-channels, and in association with a W boson (tW-channel). With a cross section which is a fraction ($\approx 50\%$) of the top quark pair production, single top has a much larger relative background, and it is therefore more complicated to identify. Due to the relatively large background, it was first observed at the Tevatron only in 2009 with the help of multi-variate analyses [10].

The dominant production mode is through the t-channel both at the Tevatron and at the LHC; it has the cleanest signature with a light quark jet recoiling against the top quark. It is characterized by one central isolated lepton and missing transverse energy, a b-jet and a forward high-p_T recoiling jet. The main background is due to QCD multi-jet and W+jet production. Single top production in the tW associated production cross section is smaller and it can be studied in the dilepton channel where both W bosons decay into a charged lepton and a neutrino, or in the lepton+jet channel. The tW cross section is relatively larger at the LHC than at the Tevatron because of the larger gluon fraction in the initial partons. Among the distinctive features of the tW channel are the presence of a single b-jet, and the balance between the top and the W. Even though the observation of the single top production at the Tevatron was based on a combination of t- and s-channels, the s-channel has not yet been observed individually. With a very small cross section, the s-channel is rather interesting as it is sensitive to various new physics processes. In particular, it is directly proportional to the Cabibbo-Kobayashi-Maskawa (CKM) matrix element $|V_{tb}|$ and it is sensitive to the presence of a W' boson or flavor changing neutral current processes. At leading order, the final state consists of a top and a bottom quark. With the exception of the s-channel, which so far has escaped direct detection, the agreement between experimental results and theoretical predictions is remarkably good.

5. Top quark properties

Properties of the top quark are not only interesting to better characterize this fundamental particle, but could also give indication of new physics. Measurement of the top quark charge, the ratio of branching fractions R=BR($t \to Wb$)/BR($t \to Wq$), top quark pair spin correlation, W polarization in top decays, asymmetry in $t\bar{t}$ events are some interesting properties.

5.1. Measurement of R

In the SM, the top quark is expected to decay as $t \to Wb$ with a branching fraction close to 100%, as top quark decays to a W boson and a quark of different isospin are strongly suppressed. The magnitude of the CKM matrix element $|V_{tb}|$ is expected to be close to unity as a consequence of unitarity and a deviation from this prediction could arise from a fourth quark generation, or simply due to different decay modes. Both CDF and D0 experiments have measured R = BR($t \to Wb$)/BR($t \to Wq$), where BR($t \to Wq$) is the branching fraction of the top quark to a W boson and a q quark (q=b, s, d). A recent result from the D0 experiment indicates some tension between the SM prediction and the data, in particular for the dilepton channel where both W bosons decay leptonically. At the LHC, the measurement is performed by analyzing the b-tagging jet multiplicity in $t\bar{t}$ dilepton events which are expected to be pure in $t\bar{t}$ signal. The residual background and the model for the measured b-tag multiplicity are derived from data and account not only for b-tag and mistag probabilities, but also include the probability to fully reconstruct the decay products from top quark decays. Results are in good agreement with the SM prediction, R=$1.023^{+0.036}_{-0.034}$(stat.+syst.) [11], with a lower limit R>0.945 at 95%CL using the Feldman-Cousins approach. Assuming a unitary, three-generation CKM matrix, $|V_{tb}| = 1.011^{+0.018}_{-0.017}$ (stat.+syst.) is measured and $|V_{tb}|$ >0.972 is obtained at 95% CL.

5.2. $t\bar{t}$ spin correlation

In $t\bar{t}$ production, top quarks are unpolarized but their spins are correlated. Due to the short lifetime of the top quark, which is smaller than the hadronization scale, the information of the spin correlation is preserved in the decay products. It is possible to measure the spin correlation of the top quark pair from the angular correlation of the decay products. Many models of new physics predict different angular distributions from the SM predictions. Because of differences in the production mechanisms ($p\bar{p}$ at the Tevatron vs. pp at the LHC) and energy, spin correlation measurements are complementary at the Tevatron and at the LHC. Charged leptons and down-type quarks are most sensitive to the spin correlation measurements. However, due to the difficult experimental separation between up- and down-type quarks, spin correlations are measured using dilepton events. In particular, the correlation coefficient is defined as the fractional difference in the number of events with the spins of the top quarks correlated and those with the spin anti-correlated, $A = \frac{N(\uparrow\uparrow)+N(\downarrow\downarrow)-N(\uparrow\downarrow)-N(\downarrow\uparrow)}{N(\uparrow\uparrow)+N(\downarrow\downarrow)+N(\uparrow\downarrow)+N(\downarrow\uparrow)}$. A recent Tevatron measurement indicates a 3.1σ evidence for spin correlation [12]. The LHC results, obtained for a given reference frame (either "maximal" or "helicity") are in good agreement with the SM predictions and exclude the hypothesis of zero spin correlation [13].

5.3. W polarization

The measurement of the polarization of the W boson from top quark decays is interesting as new physics may lead to an anomalous Wtb coupling. Since the spin information is preserved in top decay products and the bottom quark mass is small compared to the top and W masses, the SM predicts the W boson to be mostly longitudinally polarized ($F_0 \simeq 69\%$) or left-handed ($F_L \simeq 31\%$) through the V-A coupling. These fractions may significantly change in the presence of anomalous couplings, and may be inferred experimentally from the angular distribution between lepton from the W decay and the b-jet from the same top decay. The observed distribution has to be corrected for detector effects, such as acceptance and resolution, and theoretical predictions have to account for ISR/FSR, among others. Measurements show good agreement with the SM predictions, and the results are used to set limits [14].

5.4. Charge asymmetry in $t\bar{t}$ production

In $t\bar{t}$ events, the difference in rapidity (or other) distributions of top and anti-top quarks is usually known as charge asymmetry, which is sensitive to new physics models. For example, it probes perturbative QCD predictions and provides tests of new physics models where top quark pairs are produced through the exchange of new heavy particles, such as axigluons with anomalous axial-vector coupling of gluons to quarks, Z' bosons, or Kaluza-Klein excitations of gluons.

At the Tevatron, top quarks are emitted in the direction of the incoming quark, anti-top quarks in the direction of the incoming anti-quark. Therefore, due to the asymmetric initial state, the asymmetry manifests as a forward-backward asymmetry in the rapidity difference Δy between top and anti-top quarks as $A = \frac{N^+-N^-}{N^++N^-}$, where $N^+(N^-)$ is the number of events with positive (negative) values of Δy. Recent Tevatron results [15] yield values larger (a 3σ discrepancy) than the SM predictions. At the LHC, in pp collisions, there is no forward-backward asymmetry as the initial state is symmetric. The quantity of interest is the charge asymmetry and it shows as a preferential production of top quark quarks in the forward direction due to the fact that the anti-quarks (from the proton's sea) carry a lower momentum fraction. Differential measurements have been obtained as a function of p_T, y, and invariant mass $M_{t\bar{t}}$ of the top quark pair. Measurements are compatible with the SM predictions [16].

Journal of Physics: Conference Series **447** (2013) 012012 doi:10.1088/1742-6596/447/1/012012

6. Search for new physics using top quarks

Top quarks are present in many models of new physics beyond the SM. Some examples include new particles decaying into top quark pairs, flavor changing neutral currents, anomalous missing transverse energy, same-sign top pair production, charged Higgs production.

6.1. Top quark pair invariant mass distribution

Many extensions to the SM predict interactions with enhanced coupling to the top quarks, resulting as resonances in the $t\bar{t}$ pairs. Absence of a resonance in the first two generations is not significant as the coupling could be small in that case. New particles could be spin-0 scalars or pseudo-scalars, or spin-1 vector or axial-vector particles, such as a Z' boson, a Kaluza-Klein gluon or axigluon, or also spin-2 particles. Specific analysis tools are developed for the searches in the high-mass regions where the top quarks are highly boosted and the decay products tend to be collimated. Searches spanning from narrow (with a width of 1-3%) to wide ($\simeq$ 10%) resonances, both in the low- (up to 1 TeV) and in the high-mass (up to several TeV) regions, result in exclusion limits for new particle production rate.

6.2. Flavor Changing Neutral Currents

Searches for rare decays of top quarks are possible thanks to the large number of top events collected. Top quarks decay to W boson and a bottom quark with a branching fraction of about 100%. However, some extensions of the SM predict that the top quark may also decay to a Z boson and a quark, $t \to Zq$, where q is a u or a c quark. The latter is a decay predicted with a small branching fraction of the order of 10^{-14}, which is beyond the current experimental reach. Therefore, detection of a signal could indicate deviations from the SM predictions. Search for FCNC processes is sought in the tri-lepton final state, where one top $t \to Zq \to \ell\ell q$ is produced. FCNC can also be sought in the single top production; however, this is experimentally very challenging as the final signature $qg \to t \to W(\to \ell\nu)b$ has one isolated lepton, one b-jet, and missing transverse energy, with a large background from W+jet events. Measurements indicate that branching fractions BR$(t \to Zq) > 0.24\%$ are excluded at 95%CL.

6.3. Same-sign top quark production

New models put forth to explain the larger-than-expected forward-backward asymmetry measurement at the Tevatron require FCNC in the top sector mediated by the t-channel exchange of a new massive Z' boson. These mechanisms would generate same-sign top quark pair production. However, the LHC results disfavor the region of parameter space consistent with the Tevatron A_{FB} measurement.

7. Summary and outlook

After almost twenty years since its discovery, top quark still remains an interesting probe of the SM. Large samples of top quarks have allowed detailed tests of the SM up to levels of precision which are challenging the theory. However, many fundamental questions still remain, and some of the difficult questions are yet unanswered. Some measurements are rapidly approaching (or already did) the grey zone of being limited by systematic uncertainties.

With even larger data samples, additional studies may shed light on important open questions. Associated production of a Higgs boson with the top quark pair $t\bar{t}H$ is important as it would provide direct determination of the top-Higgs couplings, separately for the different Higgs decay modes. Supersymmetry could also affect top quark production. Due to the large top quark mass, the lightest scalar top quark $\tilde{t}_1$ can be the lightest scalar quark and even lighter than the top quark itself. Presence of light top and bottom squarks, charginos and neutralinos could alter the predicted rates through direct stop pair production, through the processes $\tilde{t}_1 \to \tilde{\chi}_1^+ b$

or $\tilde{t}_1 \to t\tilde{\chi}_1^0$. For instance, for a scalar top quark lighter than the top quark, the decay channel $\tilde{t}_1 \to \tilde{\chi}_1^+(\to \tilde{\chi}_1^0 W)b$ has a similar signature to $t\bar{t}$ events apart from the presence of the neutralinos, whose experimental signature mimics that of a neutrino.

Acknowledgments

To my friends and colleagues both at the Tevatron and at the LHC who contributed to producing an incredible wealth of results in the past almost twenty years, to those who were present at the time of the discovery and before. A special thanks to the organizers of the Discrete workshop for a nice environment and convivial atmosphere which allowed stimulating discussions and exchange of ideas.

References

[1] F. Abe *et al.* [CDF Collaboration], Phys. Rev. D **50**, 2966 (1994), Phys. Rev. Lett. **73**, 225 (1994); F. Abe *et al.* [CDF Collaboration], Phys. Rev. Lett. **74** (1995) 2626; S. Abachi *et al.* [D0 Collaboration], Phys. Rev. Lett. **74** (1995) 2632.

[2] J. R. Incandela, A. Quadt, W. Wagner and D. Wicke, Prog. Part. Nucl. Phys. **63** (2009) 239; F. Deliot and D. A. Glenzinski, Rev. Mod. Phys. **84** (2012) 211; F. P. Schilling, Int. J. Mod. Phys. A **27** (2012) 1230016.

[3] N. Kidonakis, Phys Rev. D82, 114030 (2010); M. Cacciari, M. Czakon, M.L. Mangano, A. Mitov, and P. Nason, Phys. Lett. B **710**, 612 (2012); V. Ahrens, M. Neubert, B.D. Pecjak, A. Ferroglia, and L.L. Yang, Phys. Lett. B **703**, 135 (2011); M. Aliev, H. Lacker, U. Langenfeld, S. Moch, P. Uwer *et al.*, Comput. Phys. Commun. **182**, 1034 (2011).

[4] CDF Collaboration, Conf. Note 9913 (2009); V. M. Abazov *et al.* [D0 Collaboration], Phys. Lett. B **704** (2011) 403.

[5] S. Chatrchyan *et al.* [CMS Collaboration], Phys. Rev. D **85** (2012) 112007; G. Aad *et al.* [ATLAS Collaboration], Phys. Lett. B **717**, 89 (2012).

[6] S. Chatrchyan *et al.* [CMS Collaboration], JHEP **1207** (2012) 143, CMS-PAS-HIG-12-052 (2012); G. Aad *et al.* [ATLAS Collaboration], JHEP **1206** (2012) 039.

[7] T. Aaltonen *et al.* [CDF and D0 Collaborations], Phys. Rev. D **86**, 092003 (2012).

[8] S. Chatrchyan *et al.* [CMS Collaboration], CMS-PAS-TOP-11-018 (2012).

[9] S. Chatrchyan *et al.* [CMS Collaboration], JHEP **1107** (2011) 049; S. Chatrchyan *et al.* [CMS Collaboration], Eur. Phys. J. C **72**, 2202 (2012).

[10] T. Aaltonen *et al.* [CDF Collaboration], Phys. Rev. Lett. **103** (2009) 092002; V. M. Abazov *et al.* [D0 Collaboration], Phys. Rev. Lett. **103** (2009) 092001.

[11] S. Chatrchyan *et al.* [CMS Collaboration], CMS-PAS-TOP-12-035 (2013).

[12] V. M. Abazov *et al.* [D0 Collaboration], Phys. Rev. Lett. **108** (2012) 032004; T. Aaltonen *et al.* [CDF Collaboration], Phys. Rev. D **83**, 031104 (2011).

[13] G. Aad *et al.* [ATLAS Collaboration], Phys. Rev. Lett. **108** (2012) 212001; S. Chatrchyan *et al.* [CMS Collaboration], CMS-PAS-TOP-12-004 (2012).

[14] T. Aaltonen *et al.* [CDF Collaboration], Phys. Rev. Lett. **105** (2010) 042002; G. Aad *et al.* [ATLAS Collaboration], JHEP **1206**, 088 (2012); S. Chatrchyan *et al.* [CMS Collaboration], CMS-PAS-TOP-11-020 (2012).

[15] T. Aaltonen *et al.* [CDF Collaboration], Phys. Rev. D **83** (2011) 112003; CDF Note 10975 (2013).

[16] S. Chatrchyan *et al.* [CMS Collaboration], Phys. Lett. B **709** (2012) 28; S. Chatrchyan *et al.* [CMS Collaboration], Phys. Lett. B **717**, 129 (2012); G. Aad *et al.* [ATLAS Collaboration], Eur. Phys. J. C **72** (2012) 2039.

Journal of Physics: Conference Series **447** (2013) 012013 doi:10.1088/1742-6596/447/1/012013

CP violation at the B factories

John Walsh, on behalf of the Babar and Belle Collaborations
INFN, Sezione di Pisa, Largo Pontecorvo 3, 56127 Pisa, Italy
E-mail: john.walsh@pi.infn.it

Abstract. One of the primary goals of the B factory experiments was to measure and study CP violation in B meson decays, a goal which has been met very successfully. This paper reports on recent measurements of CP violation at the Babar and Belle experiments. Particular focus is placed on the measurements of the three angles of the Unitarity Triangle, β, α and γ.

1. Introduction

The B factories, asymmetric e^+e^- colliders operating mainly at center-of-mass energy of 10.58 GeV, corresponding to the mass of the $\Upsilon(4S)$ resonance, were constructed in the late 1990's at SLAC in California and at KEK in Japan. The principal objective of the B factory experiments, Babar and Belle, was the study CP violation in B meson decays. Up to that time, CP violation had only been seen in kaon decays [1] and for many years the study of CP violation was confined to the kaon system, with its inherent difficulties owing to hadronic uncertainties. These difficulties were greatly reduced in the B meson system, due to the much larger mass of the B meson. Furthermore, CP violation was expected to be much larger in the B system, allowing a large number of precise measurements.

The numerous measurements of CP violation made by the Babar and Belle experiments have largely confirmed the mechanism by which CP violation occurs in the Standard Model (SM). The SM has been tightly overconstrained by the CP violation measurements and it has withstood this exacting test: no indications of new physics (NP) beyond the SM have been found.

In these proceedings, I will review CP violation measurements at the B factories. After a brief introduction to CP violation within the Standard Model, followed by some general comments on experimental techniques, I will discuss selected key measurements at Babar and Belle. The focus will be on the determination of the three angles of the Unitarity Triangle, β, γ and α. In Belle papers, these angles are referred to as ϕ_1, ϕ_3 and ϕ_2, respectively. I will use the Babar convention in these proceedings.

2. The CKM matrix and the Unitarity Triangle

The Cabibbo-Kobayashi-Maskawa (CKM) matrix [2, 3] describes the relations among the quark mass eigenstates and the weak interaction eigenstates. Thus, it describes the coupling strength of quarks to the charged weak interaction. In the SM, the CKM matrix is a 3×3 unitary matrix. Its complex values V_{ij} are not specified by the SM, but must be determined by experiment. The property of unitarity leads to constraints which reduce the number of independent parameters to four. A convenient parametrization of the CKM matrix from Wolfenstein [4] shows explicitly

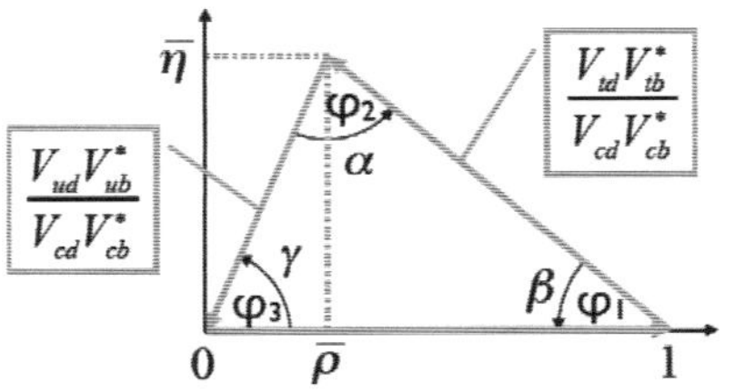

Figure 1. Unitarity triangle.

the dependence on only four independent quantities:

$$V_{CKM} = \begin{pmatrix} 1-\lambda^2/2 & \lambda & A\lambda^3(\rho - i\eta) \\ -\lambda & 1-\lambda^2/2 & A\lambda^2 \\ A\lambda^3(1-\rho-i\eta) & -A\lambda^2 & 1 \end{pmatrix} + O(\lambda^4) \tag{1}$$

At the time of B factory construction, the parameters λ and A were well measured, while ρ and η were essentially unknown.

One of the six unitarity constraints on the CKM matrix may be written as:

$$V_{ud}V_{ub}^* + V_{cd}V_{cb}^* + V_{td}V_{tb}^* = 0. \tag{2}$$

We may divide the above relation by $V_{cd}V_{cb}^*$ and express the result in the complex plane as a triangle, known as the Unitarity Triangle (UT), as shown in Figure 1. The apex of the triangle is positioned at $\bar{\rho}+i\bar{\eta}$ (the barred versions of ρ and η take into account higher orders of λ). The UT angles β, γ and α give rise to CP-violating asymmetries, which can be measured at the B factories. Measurements of these three angles are discussed in the following sections. The two (non-trivial) sides can also be measured at the B factories, using semileptonic and rare decays, as well as $B^0\bar{B}^0$ mixing.

3. Experimental considerations

The Babar and Belle detectors have been described in detail elsewhere [5, 6] and will not be discussed here. Each of the two experiments does an excellent job of reconstructing charged tracks and decay vertices, detecting photons even down to low energy ($\sim$ 30 MeV) and performing particle identification to reconstruct electrons, muons, pions, kaons and protons.

The measurement of time-dependent CP asymmetries is the main tool by which the angles β and α are measured at the B factories. When the $\Upsilon(4S)$ decays to a $B^0\bar{B}^0$ pair, one of the B's is typically fully reconstructed in a CP-eigenstate, for example $B^0 \to J/\psi K_S^0$. The flavor of the other B in the event is "tagged" by identifying some of its decay products, typically a lepton or kaon. The flavor at the time of decay can be deduced from the charge of the lepton or kaon. The third ingredient in the measurement of time-dependent CP-aymmetries is a measurment of the time elapsed between the two B decays. This is determined by precisely reconstructing the decay vertices of both B's in the event and using the known boost of the $\Upsilon(4S)$ to determine the proper time. The CP-asymmetry is then computed at a function of the time:

$$A_{CP}(t) = \frac{N(\bar{B}^0 \to f_{CP}) - N(B^0 \to f_{CP})}{N(\bar{B}^0 \to f_{CP}) + N(B^0 \to f_{CP})} \tag{3}$$

$$= S_f \sin\Delta mt - C_f \cos\Delta mt \tag{4}$$

Where the coefficients S_f and C_f are related to UT angles in a mode-dependent way. This asymmetry, as we shall see, can be related to the UT angles β and α. For the third angle γ, the most powerful methods use time-integrated asymmetries of the tree-level decays $B \to D^{(*)}K^{(*)}$.

Journal of Physics: Conference Series **447** (2013) 012013 doi:10.1088/1742-6596/447/1/012013

Table 1. Signal yields in the different modes used in Belle's $\sin 2\beta$ measurement.

Final state	Signal yield
$J/\psi K_S^0$	12649 ± 114
$\psi(2S)(\ell^+\ell^-)K_S^0$	904 ± 31
$\psi(2S)(J/\psi\pi^+\pi^-)K_S^0$	1067 ± 33
$\chi_{c1}K_S^0$	940 ± 33
$J/\psi K_L^0$	10040 ± 154

4. Measurement of β

The angle β is the most precisely measured of the three UT angles, due to the very favourable decay mode which is sensitive to it: $B^0 \to J/\psi K_S^0$. Additionally, $\sin 2\beta$ can be measured using $b \to s(q\bar{q})$ decays ($q = u, d, s$), also known as "gluonic penguins". I will discuss measurements of both types in this section.

4.1. $B^0 \to (c\bar{c})K^0$ modes

The decay mode $B^0 \to J/\psi K_S^0$, which has a relatively large branching fraction, is fully reconstructed with very low background due to the distinctive decay mode $J/\psi \to \ell^+\ell^-$ where ℓ is either a muon or an electron. Furthermore, this channel has a clean theoretical interpretation as well. Since only one decay amplitude contributes to this mode, the coefficients of Eq. 4 can be simply expressed as

$$S_f = \sin 2\beta \quad (5)$$

$$C_f = \frac{1 - |\lambda_f|^2}{1 + |\lambda_f|^2} \quad (6)$$

where $\lambda_f = (q/p)(\bar{A}_f/A_f)$, p and q are complex constants that relate the B flavour eigenstates to the mass eigenstates, and $\bar{A}_f/A_f$ is the ratio of the decay amplitudes of $\bar{B}^0$ and B^0 to the final state f. For this channel, C_f is expected to be zero, but it is typically allowed free in the fit as a consistency check.

Belle's final measurement of $\sin 2\beta$ using the $B^0 \to (c\bar{c})K^0$ modes is based on a dataset of 772 million $B^0\bar{B}^0$ pairs [7]. Table 1 shows the final states analyzed along with the event yield for each channel, while Fig. 2 shows the beam-constrained mass (or reconstructed B momentum for the $J/\psi K_L^0$ mode) for selected candidates. The selected samples are divided into two sub-samples, depending on whether the tagged B meson was a B^0 or $\bar{B}^0$. Figure 3 shows the distribution of Δt for the B^0- and $\bar{B}^0$-tagged sub-samples superimposed. The sinusoidal shape of the time-dependent CP-asymmetry (Eq. 4) is clearly evident as shown in the lower half of the figure. The CP-violating parameters, $S_f = \sin 2\beta$ and C_f, obtained from a fit to the Δt distributions, are shown in Table 2. For comparison, we also show the results of a similar analysis from Babar [8], which was based on a somewhat smaller data sample (465×10^6 $B\bar{B}$ events).

4.2. $b \to s(q\bar{q})$ modes

In the Standard Model the $b \to s(q\bar{q})$ modes are dominated by the so-called penguin amplitude, which has the same weak phase as the $b \to s(c\bar{c})$ amplitude. Figure 4 shows the Feynman diagram for the SM decay. These modes are particularly interesting since they are in principle sensitive to New Physics, with new heavy particles participating in the penguin loop. However, the situation with these decays is more complicated because suppressed SM diagrams cannot be neglected, contributing from 1–10% to the CP-violating parameter S_f, which gets shifted relative the same quantity measured in $b \to s(c\bar{c})$ decays:

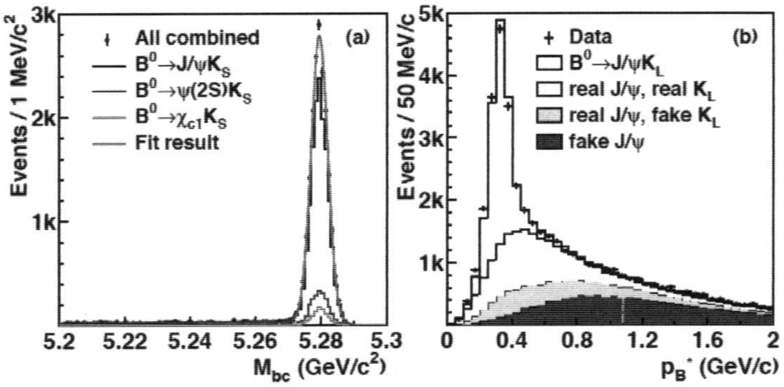

Figure 2. The beam-constrained mass for the $(c\bar{c})K_S^0$ modes (left) and the reconstructed B momentum (center-of-mass frame) for the $J/\psi K_L^0$ mode (right).

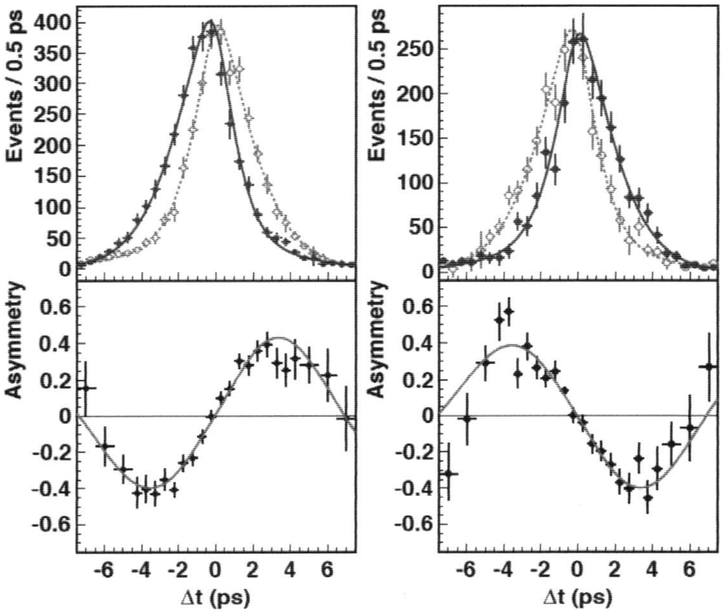

Figure 3. The distribution of Δt (top) for $\bar{B}^0$ decays (blue) and B^0 decays (red). The bottom shows the time-dependent CP asymmetry. CP-odd modes are shown on the left, while the CP-even mode is on the right.

Table 2. Fit results of CP-violating parameters from $B^0 \to (c\bar{c})K^0$ modes from Belle and Babar. The first uncertainty quoted is statistical, the second systematic. In Belle notation the angle β is called ϕ_1 and $\mathcal{A}_{CP} = C_f$ (defined in Eq. 6).

Experiment	$\sin 2\beta$	C_f
Belle	$0.667 \pm 0.023 \pm 0.012$	$-0.006 \pm 0.016 \pm 0.012$
Babar	$0.678 \pm 0.028 \pm 0.012$	$0.024 \pm 0.020 \pm 0.016$

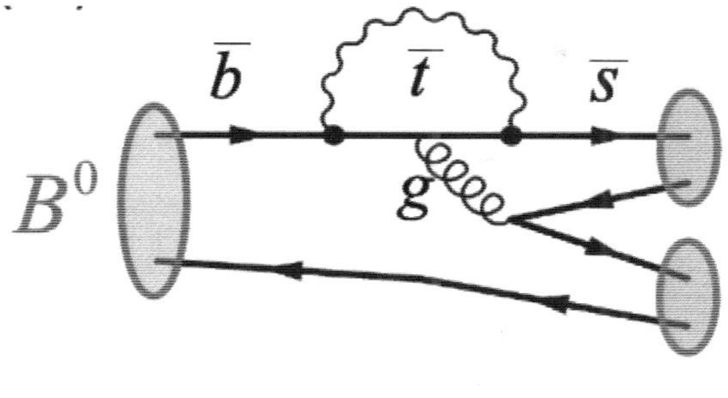

Figure 4. Feynman diagram of $b \to s(q\bar{q})$ transition.

Table 3. Results on $\sin 2\beta_{\text{eff}}$ from the time-dependent Dalitz plot analysis of $B^0 \to K^+K^-K_S$.

Renonance	$\sin 2\beta_{\text{eff}}$
ϕK_S	$(21 \pm 6 \pm 2)^\circ$
$f_0 K_S$	$(18 \pm 6 \pm 4)^\circ$
non–ϕ, non–f_0	$(20.3 \pm 4.3 \pm 1.2)^\circ$

$$S_{sc\bar{c}} = S_{sq\bar{q}} + \Delta S_{SM} = -\eta_{CP} \sin 2\beta \tag{7}$$

Any New Physics effects would give an additional contribution to $S_{sq\bar{q}}$. The quantity $-\eta_{CP}S_{sq\bar{q}}$ is often referred to as $\sin 2\beta_{\text{eff}}$.

Early measurements of $\sin 2\beta_{\text{eff}}$ using several channels seemed to indicate a discrepancy between $\sin 2\beta$ and $\sin 2\beta_{\text{eff}}$, at the level of approximately 2σ [9]. Since then analysis techniques for these modes have progressed significantly. One such advanced analysis is Babar's measurement of $\sin 2\beta_{\text{eff}}$ using a Dalitz plot analysis of the decay $B^0 \to K^+K^-K_S$ [10]. This final state is in fact a superposition of several resonant modes, for example $B^0 \to \phi K_S$ and $B^0 \to f_0 K_S$ and the Dalitz structure is used to separate the different contributions to CP violation. The Dalitz plot analysis is also able to measure $\cos 2\beta_{\text{eff}}$, which eliminates an ambiguity between β_{eff} and $90^\circ - \beta_{\text{eff}}$. Details on the Dalitz plot model can be found in [10].

Figure 5 shows the Dalitz plot distribution of selected $B^0 \to K^+K^-K_S$ events. The complex amplitude coefficients for the different resonances included in the model are derived from this fit. Flavour tagging of the other B meson in the event is applied in the usual way, allowing the separation of B^0 and $\bar{B}^0$ decays. Figure 6 shows the resulting time-dependent asymmetry, both for the $\phi(1020)$ region (left) and the region where the $\phi(1020)$ is excluded (right). Results for $\sin 2\beta_{\text{eff}}$ were extracted for the events consistent with $B^0 \to \phi K_S$, $B^0 \to f_0 K_S$ and the region excluding these two decay mode. The results are presented in Table 3.

5. Measurement of γ

The angle γ of the Unitarity Triangle is generally measured using the tree-level decays $B \to D^{(*)}K^{(*)}$. There are three primary methods used to measure γ: GGSZ, GLW and ADS. The methods generally expoit the interference between two different amplitudes that can give rise to a particular final state. I will describe analyses using the GGSZ and GLW methods in the following.

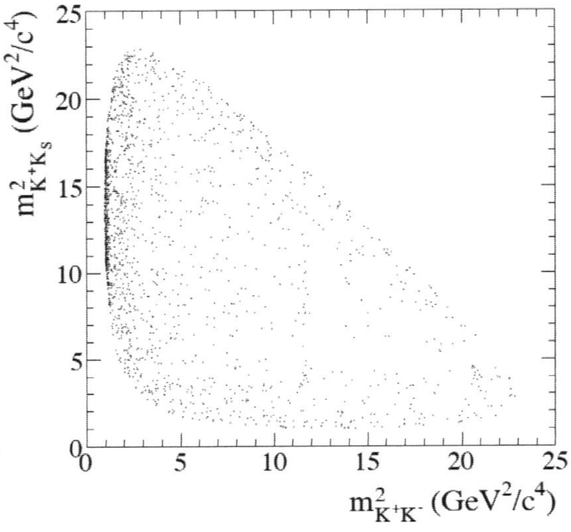

Figure 5. Dalitz plot for $B^0 \to K^+K^-K^0_S$ decays.

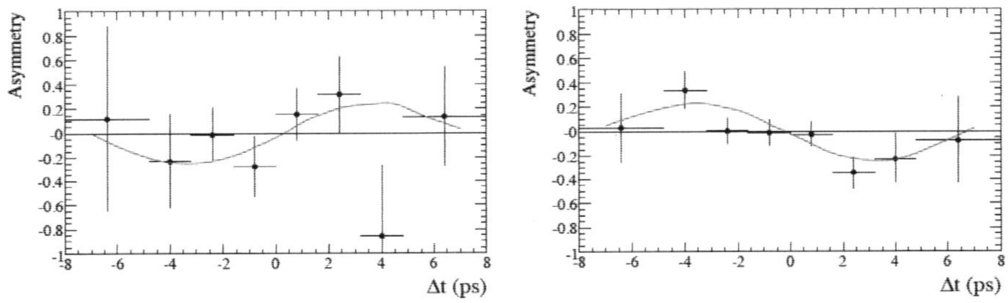

Figure 6. Time-dependent CP-asymmetry for $B^0 \to K^+K^-K^0_S$ events in the $\phi(1020)$ region (left) and for events outside the $\phi(1020)$ region (right).

5.1. GGSZ (Dalitz) method

The GGSZ method [11] exploits the interference between color-favoured and color-suppressed amplitudes in $B \to D^{(*)}K^{(*)}$ decays. In the case of $B^- \to DK^-$ channel, the color-favoured $B^- \to D^0K^-$ ($b \to c\bar{u}s$) and the color-suppressed $B^- \to \bar{D}^0K^-$ ($b \to u\bar{c}s$) amplitudes interfere when the D^0 and $\bar{D}^0$ decay to the same final state. The amplitudes for the B^- and B^+ decays may be written:

$$A_-(m^2_-, m^2_+) = |A(B^- \to D^0K^-)|[A_D(m^2_-, m^2_+) + r_B e^{i\delta_B} e^{-i\gamma} A_D(m^2_+, m^2_-)] \quad (8)$$

$$A_+(m^2_-, m^2_+) = |A(B^+ \to \bar{D}^0K^-)|[A_D(m^2_+, m^2_-) + r_B e^{i\delta_B} e^{+i\gamma} A_D(m^2_-, m^2_+)] \quad (9)$$

The position in the Dalitz plot is specified by $m_\pm = m(K_S, h^\pm)$ where $h = \pi, K$. The two contributions to each amplitude differ by a factor $r_B e^{i(\delta_B \pm \gamma)}$, where r_B is the magnitude of the ratio of the amplitudes $A(B^- \to \bar{D}^0K^-)$ and $A(B^- \to \bar{D}^0K^-)$, and δ_B is the relative strong phase. The analysis presented in [12] considers the self-conjugate D^0 decays to $K_S\pi^+\pi^-$ and $K_SK^+K^-$.

A Dalitz plot analysis of D^0 and $\bar{D}^0$ decays to these final states allows access to the complex amplitude ratios, and thus to the weak and strong phases and r_B. It is useful to express

Journal of Physics: Conference Series **447** (2013) 012013 doi:10.1088/1742-6596/447/1/012013

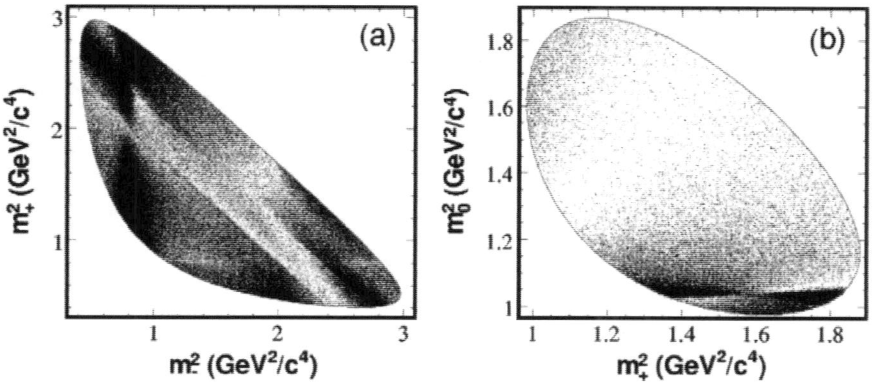

Figure 7. Dalitz plot distributions for a) $D^0 \to K_S\pi^+\pi^-$ and b) $D^0 \to K_SK^+K^-$.

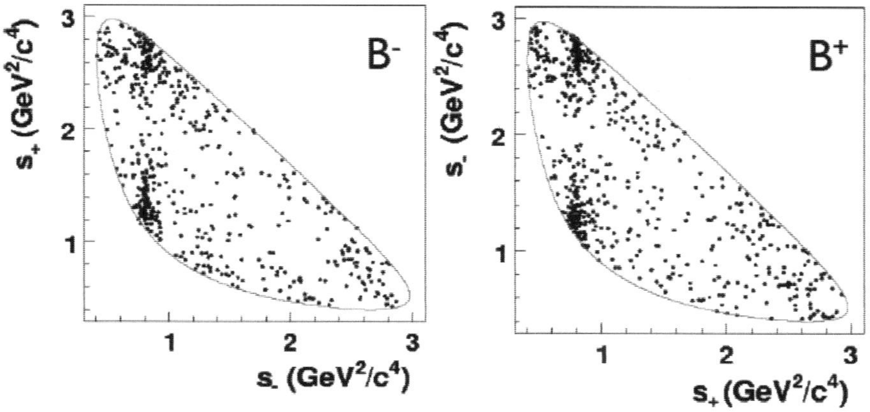

Figure 8. Dalitz plot distributions for (left) $B^- \to D^0K^-$ and (right) $B^+ \to D^0K^+$, with $D^0 \to K_S\pi^+\pi^-$.

the unknowns γ, r_B and δ_B in terms of the so-called cartesian variables, which are Gaussian-distributed and hence better behaved statistically:

$$x_\pm = r_B \cos(\delta_B \pm \gamma) \tag{10}$$
$$y_\pm = r_B \sin(\delta_B \pm \gamma) \tag{11}$$
$$z_\pm = x_\pm + iy_\pm \tag{12}$$

The D^0 amplitudes, specified as A_D in Equations 8 and 9 above, are determined from a Dalitz plot analysis performed on a high statistics sample of D^0 mesons produced in the decay $D^{*+} \to D^0\pi^-$ produced in $e^+e^- \to c\bar{c}$ events. The charge of the soft pion in the D^* decay identifies the flavour of the D meson. Figure 7 shows the Dalitz plot distributions for the two final states. The Dalitz analysis performed on these plots provides the values of $A_D(m_-^2, m_+^2)$. These are then used when analyzing the Dalitz plots corresponding to the $B \to DK$ decays, shown in Figure 8. A similar analysis was performed on the related decays $B \to D^*K$ and $B \to DK^*$. The results are summarized in Figure 9, which depicts the measurements

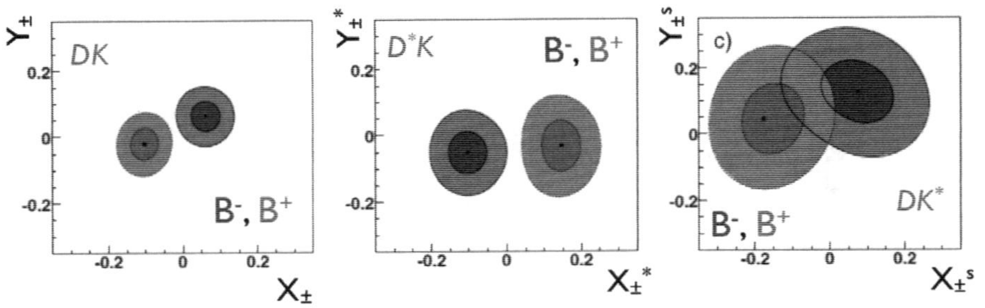

Figure 9. Results, in the $z_{\pm}$ plane of Babar's GGSZ analysis. The three plots show, from left to right, the results for the DK, D^*K and DK^* final states, respectively. The B^- results are shown in blue, the B^+ results in red.

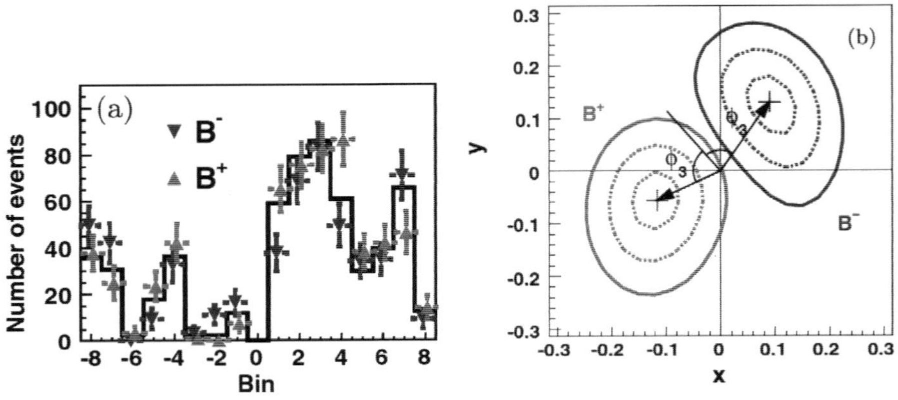

Figure 10. The yields of B^- and B^+ decays into different bins in the Dalitz plot (left), and the resulting values of the cartesian coordinates x and y (right).

of the cartesian variables for the different decay modes. Separation between the B^- and B^+ measurements in the $z_{\pm}$ plane is an indication of CP violation. For the numerical results for the cartesian variables the reader is referred to reference [12].

The Belle experiment has performed an interesting variation [13] on the GGSZ analysis using the same final states as the Babar analysis. Belle, however, employs a model-independent Dalitz analysis by symmetrically binning the Dalitz plane and forgoing the use of a complicated Dalitz plot model. This method allows the reduction of the primary systematic uncertainty, which comes from the Dalitz model employed. With this method the expected number of events in each Dalitz bin, i is given by:

$$N_i^{\pm} = hK_i + r_B^2 K_{-i} + 2\sqrt{K_i K_{-i}}(x_{\pm}c_i + y_{\pm}s_i) \qquad (13)$$

where $-i$ indicates the symmetrically opposite bin to i. K_i is the number of events in bin i from a flavour-tagged sample of $D^{*\pm} \to D\pi^{\pm}$. The coefficients c_i and s_i, which are related to the strong phase in bin i, are obtained from data taken at the $D^0\bar{D}^0$ threshold by the CLEO experiment.

The results are shown in Figure 10. On the left is plotted the number of B^- and B^+ events found in each of the Dalitz plot bins, with the black histogram representing the expection for both B^- and B^+ in the absence of CP violation. The right plot shows the resulting cartesian

Table 4. Signal yields in different decay modes for Babar's GLW $B \to DK$ analysis. The flavour eigenstate $D^0 \to K^-\pi^+$ is used as the normalization for $R_{CP\pm}$.

D^0 mode	$N(B^\pm \to DK^\pm)$
K^+K^-	367 ± 27
$\pi^+\pi^-$	110 ± 9
$K^0_S\pi^0$	338 ± 24
$K^0_S\omega$	116 ± 9
$K^0_S\phi$	52 ± 4
$K^-\pi^-$	3361 ± 82

variables. These results can also be expressed as:

$$\gamma = (77.3^{+15.1}_{-14.9} \pm 4.1 \pm 4.3)^\circ \quad (14)$$

$$r_B = 0.145 \pm 0.030 \pm 0.010 \pm 0.011 \quad (15)$$

$$\delta_B = (129.9 \pm 15.0 \pm 3.8 \pm 4.7)^\circ \quad (16)$$

where the quoted uncertainties are statistical, systematic, and the error associated with c_i and s_i, respectively.

5.2. GLW method

The GLW method for extracting γ [14] uses decay modes of the D^0 into CP-eigenstates: K^+K^- and $\pi^+\pi^-$, which are CP-even, and $K_S\pi^0$, $K_S\omega$ and $K_S\phi$, which are CP-odd. There are four observables involving the rates of B^- and B^+ decays to CP-odd and CP-even final states:

$$R_{CP\pm} = \frac{\Gamma(B^- \to D^0_{CP\pm}K^-) + \Gamma(B^+ \to D^0_{CP\pm}K^+)}{2\Gamma(B^- \to D^0K^-)} = 1 \pm 2r_b \cos\gamma \cos\delta_b + r_b^2 \quad (17)$$

$$A_{CP\pm} = \frac{\Gamma(B^- \to D^0_{CP\pm}K^-) - \Gamma(B^+ \to D^0_{CP\pm}K^+)}{\Gamma(B^- \to D^0_{CP\pm}K^-) + \Gamma(B^+ \to D^0_{CP\pm}K^+)} = \pm 2r_b \sin\gamma \sin\delta_b / R_{CP\pm} \quad (18)$$

The GLW observables may also be expressed in cartesian variables:

$$A_{CP\pm} = \pm\frac{x_- - x_+}{1 + |z|^2 \pm (x_- + x_+)} \quad (19)$$

$$R_{CP\pm} = 1 + |z|^2 \pm (x_- + x_+) \quad (20)$$

Figure 11 shows the distribution of the discriminating variable ΔE for the Babar GLW analysis of $B \to DK$ [15]. The event yields for the individual channels are shown in Table 4.

Babar finds the following results for the observables:

$$A_{CP+} = 0.25 \pm 0.06 \pm 0.02 \quad (21)$$

$$A_{CP-} = -0.09 \pm 0.07 \pm 0.02 \quad (22)$$

$$R_{CP+} = 1.18 \pm 0.09 \pm 0.05 \quad (23)$$

$$R_{CP0} = 1.07 \pm 0.08 \pm 0.04 \quad (24)$$

where the first uncertainty is statistical and the second systematic.

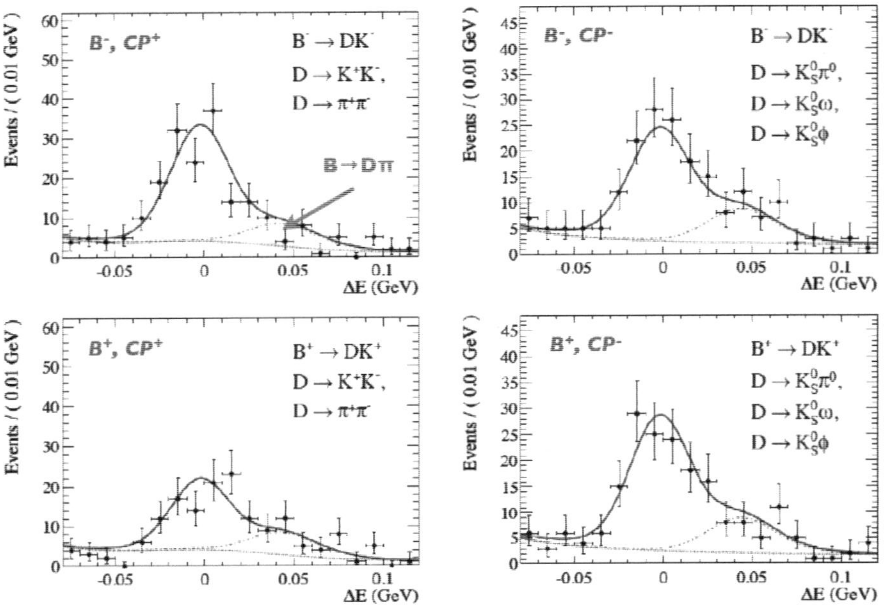

Figure 11. The ΔE distributions of B^- (top row) and B^+ (bottom row) with D decays to CP-even eigenstates (left column) and CP-odd eigenstates (right column).

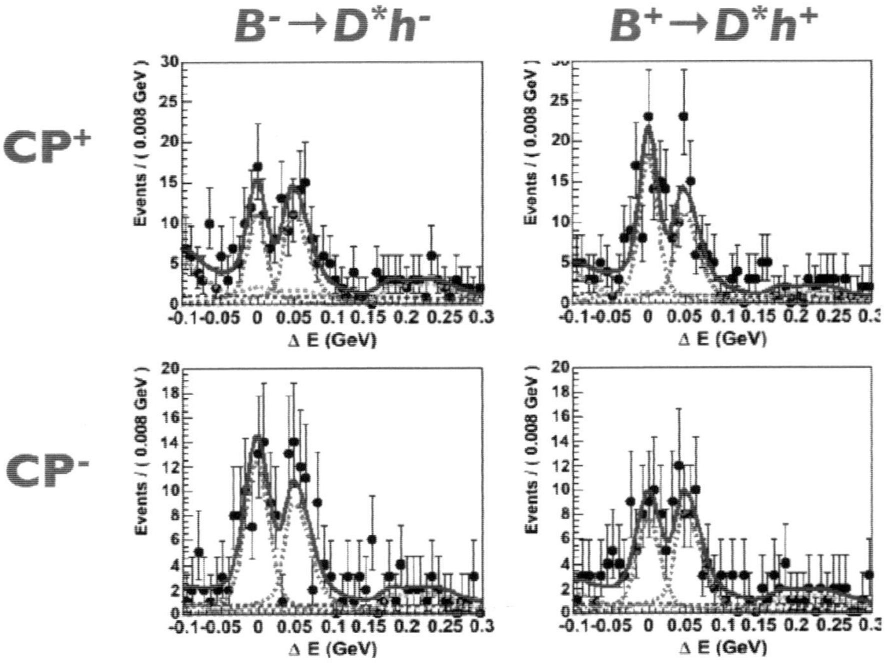

Figure 12. The ΔE distributions of B^- (left column) and B^+ (right column) with D^* decays to CP-even eigenstates (top row) and CP-odd eigenstates (bottom row).

Table 5. Determination of γ achieved by combining measurements.

Experiment	Combined γ (degrees)
Babar	69^{+17}_{-16}
Belle	68^{+15}_{-14}
LHCb	$71.1^{+16.6}_{-15.7}$

Belle has recently performed a GLW analysis using the decay mode $B \to D^*K$ [16]. Figure 12 shows the selected events for B^- and B^+ decays via CP-even and CP-odd D^* decays. Belle finds the following results for the GLW observables:

$$R_{CP+} = 1.19 \pm 0.13 \pm 0.03 \quad (25)$$
$$R_{CP-} = 1.03 \pm 0.13 \pm 0.03 \quad (26)$$
$$A_{CP+} = -0.14 \pm 0.10 \pm 0.01 \quad (27)$$
$$A_{CP-} = 0.22 \pm 0.11 \pm 0.01 \quad (28)$$

5.3. Combination of γ results

The Babar and Belle collaborations have performed dedicated studies to combine all the relevant decay modes and methods to extract the most precise values of γ. The results are first translated into the cartesian variables $x_\pm$ and $y_\pm$. Each channel DK, D^*K and DK^* has its own set of cartesian variables, each of which depends on the common parameter γ. These variables are averaged across the three analysis methods (GGSZ, GLW and ADS), by combining the likelihood functions. As a last step, γ is extracted from the best fit values of the cartesian variables. The results, as tabulated by HFAG [17] are shown in Table 5. Though not covered in this talk, we quote the current LHCb result [18] for comparison.

6. Measurement of α

The UT angle α is measured, in analogy with the β measurements, using time-dependent CP asymmetries in decays to CP eigenstates, for example $B^0 \to \pi^+\pi^-$. However, the situation with α is not so straightforward as with β due to the sizable contribution from more than a single amplitude. In the case of $B^0 \to \pi^+\pi^-$, in addition to the tree diagram, there is a significant contribution from the penguin diagram, with a different CKM structure. The result is that the time-dependent asymmetry measurement yields $\sin 2\alpha_{\text{eff}}$, with $\alpha_{\text{eff}} = \alpha + \delta$. It turns out that δ can be extracted from an isospin analysis by measuring the branching fractions and time-integrated CP asymmetries of all the channels in the $B \to \pi\pi$ system.

Belle has recently peformed an analysis of the $B^0 \to \pi^+\pi^-$ channel, including the associated isospin analysis [19], based on 772 million $B\bar{B}$ pairs. Figure 13 shows the Δt distributions for B^0- and $\bar{B}^0$-tagged events, along with the time-dependent CP asymmetry. The CP-violating parameters obtained from the fit are:

$$\mathcal{A}_{CP}(B^0 \to \pi^+\pi^-) = +0.33 \pm 0.06 \pm 0.03 \quad (29)$$
$$\mathcal{S}_{CP}(B^0 \to \pi^+\pi^-) = -0.64 \pm 0.08 \pm 0.03 \quad (30)$$

where the quoted uncertainties are statistical and systematic. With the current level of statistics, the isospin analysis of $B \to \pi\pi$ is not very effective in constraining the value of α. Belle excludes the region $23.8^\circ < \alpha < 66.8^\circ$ at the 1σ level.

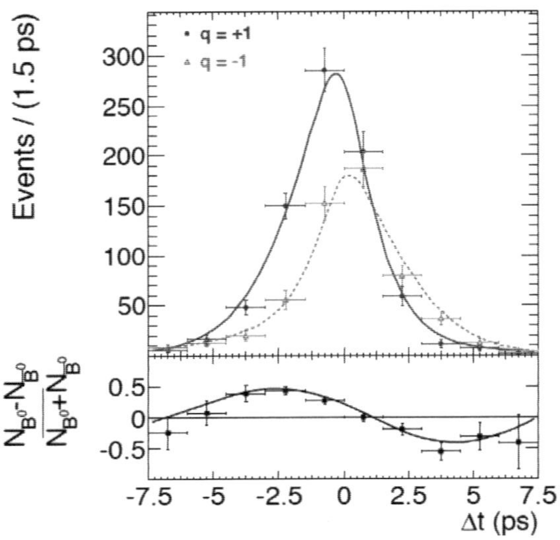

Figure 13. Time-dependent fit results for Belle's $B^0 \to \pi^+\pi^-$ analysis. The top portion shows the background-subtracted δt distribution separately for B^0 (blue) and $\bar{B}^0$ (red) tags. The lower portion of the plot shows the CP asymmetry.

To obtain more stringent constraints on α one must combine results from different modes and different experiments. In particular, the modes $B \to \rho\rho$ and $B \to \rho\pi$ contribute significantly to our knowledge of α. The CKMFitter collaboration has extracted a world average value for α using these measurements from these three channels from the two B factory experiments. They find [20]:

$$\alpha = (88.5^{+4.7}_{-4.4})^\circ \tag{31}$$

7. Global CKM fits

Perhaps the primary goal of the B factories was to make enough measurements in the B physics sector to overconstrain the CKM matrix, with the hope of finding inconsistencies, which would be an indication of physics beyond the Standard Model. Two collaborations, CKMfitter [21] and UTfit [22], have undertaken to combine the many relevant measurents from the B factories, and from other flavour experiments as well, to put stringent tests on the Standard Model. The two collaborations adopt somewhat different statistical methods, while generally using the same inputs. The results obtained are generally in good agreement.

Both collaborations provide averages for the experimental quantities, by averaging over different measurement techniques, decay modes and experiments. Table 6 shows the world averages using direct measurements of the three Unitarity Triangle angles. Additional solutions are present for α (CKMfitter) and γ (UTfit), see the individual collaboration websites for details.

More precise deteriminations of the angles can be made by including the full set of flavour measurements in a global CKM fit. Non-CP violating measurements of the sides of the UT come from semileptonic B decays and mixing in the B^0_d and B^0_s systems. The parameter $|\epsilon_K|$, a measurement of CP violation in kaon decays, is also used in the fit. Figure 14 show the results of such global fits performed by the two collaborations. Table 7 shows the results for the three UT angles and for $(\bar{\rho}, \bar{\eta})$, the coordinates of the apex of the triangle. This fit assumes the validity of the Standard Model, so, for example, the sum $\alpha + \beta + \gamma$ is constrained to 180°.

The fits can be performed without assuming the validity of the Standard Model, indeed,

Table 6. World averages of the UT angles using direct measurments. Measurements are given in degrees.

Collaboraton	β	α	γ
CKMfitter	$21.38^{+0.79}_{-0.77}$	$88.7^{+9.4}_{-8.6}$	66 ± 12
UTfit	$21.42^{+0.91}_{-0.89}$	90.9 ± 8.0	70.8 ± 7.8

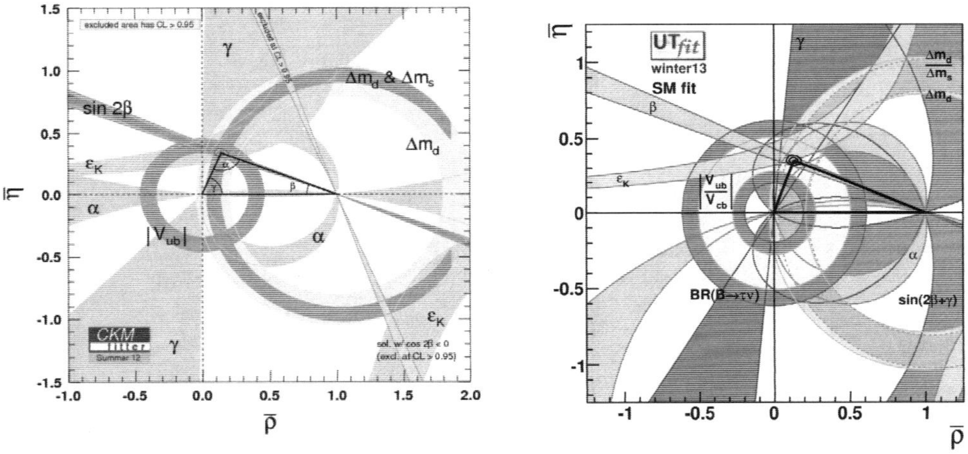

Figure 14. Global CKM fit results.

Table 7. Best fit values of the angles and apex of the Unitarity Triangle. Angle measurements are in degrees.

	CKMfitter	UTfit
β	$21.73^{+0.78}_{-0.74}$	21.95 ± 0.87
α	$90.5^{+4.3}_{-4.1}$	88.7 ± 3.1
γ	$67.7^{+4.1}_{-4.3}$	69.2 ± 3.2
$\bar{\rho}$	$0.140^{+0.027}_{-0.026}$	0.132 ± 0.021
$\bar{\eta}$	$0.343^{+0.015}_{-0.014}$	0.350 ± 0.014

specific fits to allow for generic New Physics effects have been performed to check the consistency of the full set of flavour measurements. Although there have been some hints of inconsistencies over the years, it appears that set of measurments made at the B factories and other flavour experiments are statistically consistent with Standard Model predictions.

8. Conclusions

Measurements of CP violation in B meson decays have been at the core of the B factories physics programs since their inception. A large number of independent measurements of CP-violating processes have been made, leading to precise determinations of the angles of the Unitarity Triangle. Together with non-CP-violating measurements, which tend to constrain the sides of the UT, these measurements have essentially verified the CKM mechanism of CP violation in the Standard Model. CP violation in B decays will continue to play an important role in flavour physics in the coming years. Many important measurements in the flavour sector are already being made at LHCb, while we can expect a significant improvement in the precision of many other measurements when the Belle 2 experiment turns on in a few years.

References

[1] J.H. Christenson, et al., Phys. Rev. Lett. **13**, 138 (1964).
[2] N. Cabibbo, Phys. Rev. Lett. **10**, 531 (1963).
[3] M. Kobayashi and T. Maskawa, Prog. Theor. Phys. **49**, 652 (1973).
[4] L. Wolfenstein, Phys. Rev. Lett. **51**, 1945 (1983).
[5] B. Aubert, et al. (Babar Collaboration), Nucl. Instr. Methods A **479**, 1 (2002).
[6] A. Abashian, et al. (Belle Collaboration), Nucl. Instr. Methods A **479** 117 (2002).
[7] I. Adachi, et al. (Belle Collaboration), Phys. Rev. Lett. **108**, 171802 (2012).
[8] B. Aubert, et al. (Babar Collaboration), Phys. Rev. D **79**, 072009 (2009).
[9] `http://www.slac.stanford.edu/xorg/hfag/triangle/ichep2004/index.shtml#sin2b`
[10] J.P. Lees, et al. (Babar Collaboration), Phys. Rev. D **85**, 112010 (2012).
[11] D. Atwood, et al., Phys. Rev. Lett. **78**, 3275 (1997),
A. Giri, et al., Phys. Rev. D **68**, 054018 (2003).
[12] P. del Ama Sanchez, et al. (Babar Collaboration), Phys. Rev. Lett. **105**, 121801 (2010).
[13] H. Aihara, et al. (Belle Collaboration), Phys. Rev. D **85**, 112014 (2012).
[14] M. Gronau, D. London and D. Wyler, Phys. Lett. B **253**, 483 (1991).
[15] P. del Amo Sanchez, et al. (Babar Collaboration), Phys. Rev. D **82**, 072004 (2010).
[16] Talk presented by K. Trabelsi at the CKM2012 Workshop, `http://www.physics.uc.edu/~schwartz/CKM2012/program/`
[17] `http://www.slac.stanford.edu/xorg/hfag/triangle/summer2012/index.shtml#gamma_comb`
[18] `LHCb-CONF-2012-032, http://cds.cern.ch/record/1481337?ln=en`
[19] I. Adachi, et al. (Belle Collaboration), BELLE-CONF-1211, arXiv:1302.0551.
[20] `http://ckmfitter.in2p3.fr/www/studies/plots_ckm12/ckm_plots_ckm12.html`
[21] `http://ckmfitter.in2p3.fr/www/html/ckm_main.html`
[22] `http://utfit.org/UTfit/`

Overview of Recent ATLAS Physics Results[1]

Patricia Conde Muíño on behalf of the ATLAS Collaboration
LIP, Elias Garcia 14, 1000-149 Lisbon, PT

E-mail: pconde@lip.pt

Abstract. The ATLAS experiment is situated at the Large Hadron Collider at CERN (Geneva). It was designed to cover a wide range of physics topics, such as precision measurements of Standard Model physics, the determination of the Electroweak Symmetry Breaking mechanism through the search for the Higgs boson, and searches for new physics beyond the Standard Model, like super-symmetric particles. ATLAS started operation at the LHC at the end of 2009, collecting poton-proton collisions at 7 and 8 TeV centre of mass energies. In this paper a summary of some of the most interesting recent physics ATLAS results will be presented.

1. Introduction

ATLAS [1] is one of the four Large Hadron Collider (LHC) experiments at CERN. It is also the result of more than 20 years of continuous work executed by more than 3000 physicists from about 40 different countries around the world. ATLAS aims at exploring the unprecedented high energy proton-proton (*pp*) collisions delivered by the LHC, with focus on precise measurements of Standard Model predictions, search for the Higgs boson, search for super-symmetry (SUSY) and any other expected or unexpected new physics process that may appear.

ATLAS started operation in 2009, when the LHC delivered the first *pp* collisions. Since then it has collected a total integrated luminosity of about 5 fb^{-1} of 7 TeV *pp* collisions and about 21 fb^{-1} of 8 TeV *pp* collisions. In addition, over 150 μb^{-1} of lead-lead collisions were collected at 2.76 TeV centre of mass energy per nucleon, for a dedicated heavy ion physics program[2]. The ATLAS detector operated with very high efficiency, with above 99% of the detector channels working. From the delivered luminosity, 93.7% was collected and had good quality for physics analysis. These data have been used to study many different physics processes. In this proceedings some of the most relevant ATLAS recent results will be reviewed.

2. Jet production cross section

Jet production is the most abundant high transverse momentum (p_T) process at the LHC, constituting an important tool to understand the strong interaction, to determine the proton structure and to search for physics beyond the SM. ATLAS has measured the jet production cross section in 7 TeV *pp* collisions using 37 pb^{-1} of integrated luminosity [2], covering a wide range in rapidity[3], up to $|y| < 4.4$.

[2] Not covered in this paper.

[3] The ATLAS coordinate system is a right-handed system with the x axis pointing towards the centre of the LHC ring, the z axis following the direction of the beam and the y axis pointing upwards. The azimuthal angle,

Journal of Physics: Conference Series **447** (2013) 012014 doi:10.1088/1742-6596/447/1/012014

Jets are defined using the anti-k_t [3] algorithm, as implemented in the FASTJET [4] package, with two different radius parameters R=0.4 and R=0.6. The resulting jets are corrected by all detector and reconstruction effects. Truth jets in the Monte Carlo (MC) simulation are identified using the anti-k_t algorithm and are built from stable particles, which are defined as those with a proper lifetime longer than 10 ps. This definition includes muons and neutrinos from decaying hadrons.

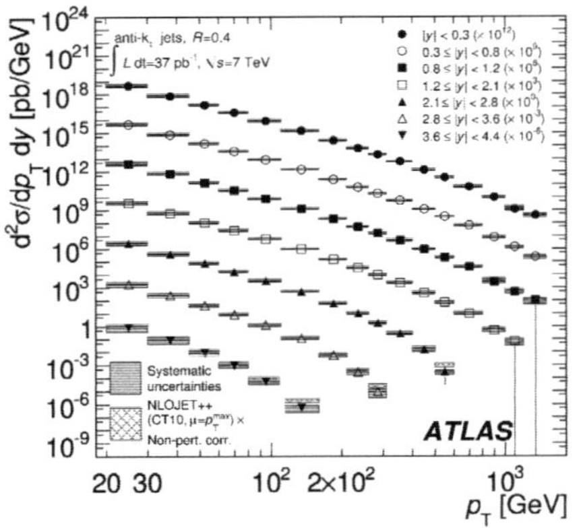

Figure 1. Inclusive jet double differential cross section as a function of the jet p_T and in bins of rapidity for jets identified with the anti-k_t algorithm with radius R=0.4 [2].

The inclusive jet production cross section as a function of p_T and in bins of rapidity, for jet $p_T > 20$ GeV and for a jet radius of $R = 0.4$ is shown in figure 1. The term inclusive means that all the jets in the event have been considered. This measurement extends up to a jet p_T of 1.5 TeV, the highest ever measured, and spans over ten orders of magnitude in cross section. The error bars, which are usually not visible, represent the statistical uncertainty. The total systematic uncertainty, dominated by the jet energy scale systematic uncertainty, is shown with shaded area. There is an additional 3.4% systematic uncertainty due to the luminosity that is not shown. Data is compared to the NLO perturbative QCD calculations using NLOJET++ [5] to which non-perturbative corrections have been applied. The CT10 NLO parton density function (PDF) set [6] has been used. The data to MC ratios are shown in figure 2. The results are also compared with predictions obtained using the MSTW 2008 [7], NNPDF 2.1 [8,9] and HERAPDF 1.5 [10] PDF sets. There is an overall agreement between the MC predictions and the data. However, the data shows slightly smaller cross section values that become worse for the most forward bins in rapidity. This difference is of the same order of the total systematic uncertainty on the measurement, including both experimental and theoretical uncertainties, providing some sensitivity to test the theoretical predictions.

These results constitute a comprehensive test of perturbative QCD in a large kinematic regime and may be used to constrain the parton distribution functions at large x and high momentum transfer.

3. Search for di-jet mass resonances at the LHC

Several extensions beyond the SM predict new heavy particles, accessible at LHC energies, that can decay into two energetic partons, such as excited quarks q^* that would be the result of quark compositeness [11–13], axigluons predicted by the chiral colour models [14, 15] or technicolor particles. ATLAS has performed a model independent search for di-jet mass resonances, using

ϕ is measured from the positive x axis and the polar angle θ is measured from the positive z axis. When dealing with massive jets and particles, the rapidity $y = \frac{1}{2}ln(\frac{E+p_z}{E-p_z})$ is used, where E is the jet energy and p_z is the z-component of the jet momentum.

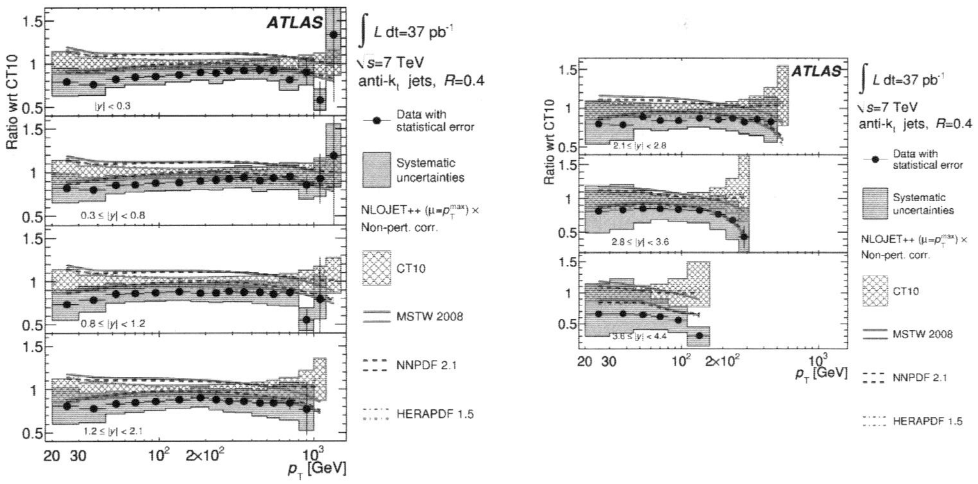

Figure 2. Ratios of inclusive jet double-differential cross section to the theoretical prediction obtained using NLOJET++ with the CT10 PDF set [2]. The ratios are shown as a function of jet p_T in different regions of rapidity.

13 fb^{-1} of pp collisions at 8 TeV centre of mass energy [16]. This study used as a benchmark the excited quark model, that predicts accessible cross sections at the LHC, imposing limits on the acceptance times cross section ($\sigma \times A$) for resonant states with intrinsic widths narrower than the experimental resolution. The analysis technique consists of searching for a peak in a continuous falling background. New di-jet resonances would be visible as a significant excess over the fitted background.

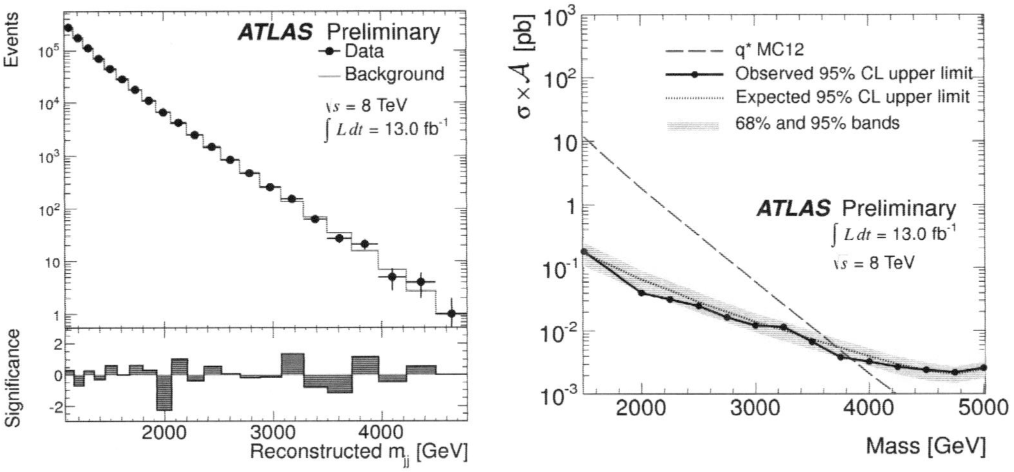

Figure 3. Left: di-jet invariant mass distribution for 13 fb^{-1} of 8 TeV pp collisions. The background fitting function is shown with a solid line [16]. Right: The 95% CL upper limit on the production cross section times acceptance of a new di-jet mass resonance as a function of di-jet invariant mass (black filled circles). The blue dashed curve represents the excited-quark cross section times acceptance prediction [16].

The di-jet mass spectrum is shown in figure 3 (left) for 13 fb^{-1} of pp collisions at 8 TeV. The background fitted function is shown with a solid line. The BumpHunter algorithm [17, 18] is used to test the presence of resonances in the dijet mass spectrum, assuming Poisson statistics and taking proper account of the *look-elsewhere* effect [19]. No significant excess was observed and, in the absence of a signal, a Bayesian method is applied to calculate the 95% Confidence Level (CL) limits on the cross section times acceptance of new resonances [16]. The resulting limits are shown in figure 3 (right). The observed (expected) lower limit on the mass of excited quarks is 3.84 TeV (3.70 TeV).

4. Top quark physics

With a top quark pair production cross section twenty times larger than the Tevatron, the LHC is a top quark factory, providing ideal conditions to measure precisely the top quark properties.

The top quark is the heaviest of all the quarks, with a mass similar to an atom of gold. Its Yukawa coupling to the Higgs is close to unity, which frequently leads to speculation that it may play a special role in the Electroweak Symmetry Breaking mechanism. Due to its very short lifetime, it decays before hadronization, preserving the properties of the bare quark, like the spin information that can be visible through the angle correlation of the decay products.

The top quark plays also a special role in Beyond the Standard Model scenarios, where new particles may decay to top quark pairs, be produced in its decay or modify its properties. Therefore, precise measurements of the properties of the top quark are very important to verify the SM prediction and search for new physics beyond it.

The top quark decays almost always to a W boson and a b quark in the SM. Depending on the decay of the W boson the final states will contain leptons and missing transverse energy due to the escaping neutrino, b-jets and/or additional jets.

4.1. $t\bar{t}$ production cross section

ATLAS has measured the $t\bar{t}$ production cross section in 7 TeV pp collisions with a precision of about 7% [20] using the semileptonic channel (only one of the W bosons decays to a lepton[4] and a neutrino), the dilepton channel (both W decay semileptonically) and the all hadronic channel (both W decay hadronically). The result of this combination is a value $\sigma_{t\bar{t}} = 177 \pm 3$ (stat) $^{+8}_{-7}$ (syst) ± 7 (lumi) pb that exceeds the theoretical prediction of $\sigma_{t\bar{t}} = 167^{+17}_{-18}$ pb [21] by about 6%, being still compatible with it within one sigma.

A new measurement has recently been performed [21] in the one-lepton plus jets channel in which one of the leptons contains a soft muon from the b-quark decay. This measurement can be sensitive to processes of cascade decays of non-SM bottom-flavoured objects, such as $b \to \tilde{b}X \to \mu c X'$ in SUSY, that would enhance the cross section value and might modify the kinematic and charge distribution of the soft muons. The measured cross section value is $\sigma_{t\bar{t}} = 165 \pm 2$ (stat) ± 17 (syst) ± 3 (lumi) pb, in good agreement with the theoretical expectation.

The $t\bar{t}$ production cross section has also been measured with 5.8 fb^{-1} [22] of pp collisions at 8 TeV centre of mass energy, using the semileptonic channel. The event selection requires a good quality isolated lepton (electron or muon), at least 3 jets and at least one of the jets b-tagged. A multivariate likelihood technique is used to enhance the signal to background ratio. The number of $t\bar{t}$ events is determined from a fit to the likelihood discriminant variable. The resulting cross section is $\sigma_{t\bar{t}} = 241 \pm 2$ (stat) ± 31 (syst.) ± 9 (lumi) pb, in good agreement with the theoretical expectation of 238^{+22}_{-24} pb [22].

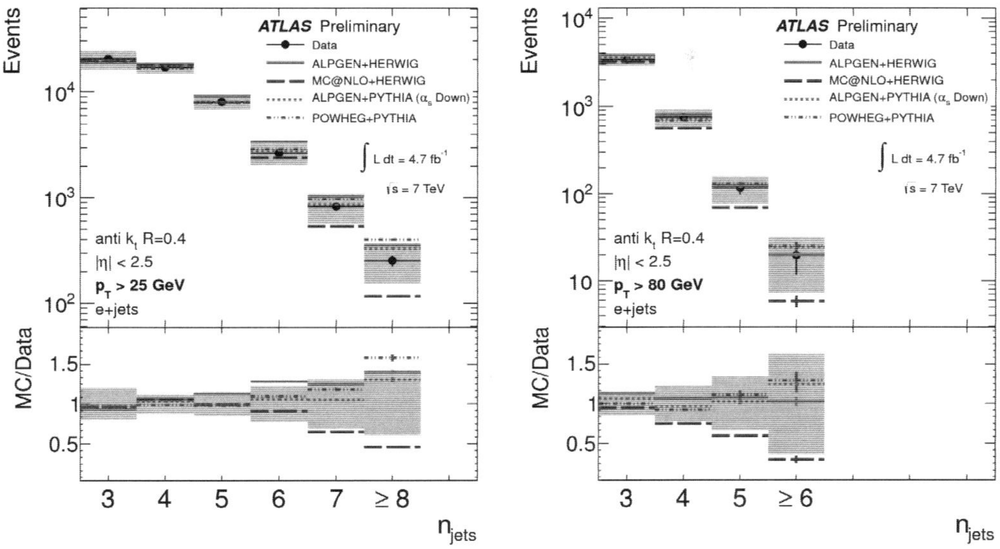

Figure 4. Jet multiplicities in $t\bar{t}$ events for two different jet p_{T} thresholds: 25 GeV (left) and 80 GeV (right) [23].

4.2. Jet multiplicities in $t\bar{t}$ events

The measurement of the $t\bar{t}$ production with additional jets as a function of the jet transverse momentum can constrain initial and final state radiation (ISR/FSR) models and tests perturbative QCD at the LHC energies. The jet multiplicity has been measured in $t\bar{t}$ events for different jet p_{T} thresholds, in the single-lepton channel, within the kinematic range accessible at ATLAS [23]. The multiplicities, fully corrected for the effects of the detector efficiencies, resolution and biases, are shown in figure 4 for two different jet p_{T} thresholds (25 and 80 GeV). Data is compared to the prediction obtained by different MC simulations (ALPGEN [24] with HERWIG [25] or PYTHIA to simulate the showering and fragmentation , MC@NLO [26] with HERWIG and POWHEG [27] with PYTHIA). The data points are shown with error bars that correspond to the statistical uncertainty, while the total uncertainty (systematic plus statistics) is shown as a shaded area. In the MC, the jets are defined using the same algorithm as in data (anti-k_t with radius parameter $R = 0.4$) running on all the stable particles except muons and neutrinos. Particles from the underlying event were included in this definition, whereas particles from overlaid inelastic events were not. For the lower p_{T} thresholds, all the MC generators describe the data multiplicities within the uncertainties, while MC@NLO underestimates the data for multiplicities higher than 5 jets. For the higher p_{T} threshold, MC@NLO underestimates the data almost for all the bins, suggesting that the jet p_{T} spectrum it generates is too soft. All the other predictions are in agreement with the data within the uncertainties, for both jet p_{T} thresholds. The present measurement is limited by systematic uncertainties, dominated by the background modelling at lower jet multiplicities and by the jet energy scale at higher jet multiplicities.

4.3. Top quark polarization in $t\bar{t}$ events

According to the SM, top quarks are expected to be produced unpolarized at the LHC, due to QCD parity conservation and to the unpolarized initial state. New physics models might result

[4] Only muons or electrons are considered.

in a sizeable polarization, measurable at the LHC. The top quark polarization is preserved in its decay and can be extracted from the W decay products by studying the polar angle of the lepton in the rest frame of its parent top quark. In order to do so, the full reconstruction of the $t\bar{t}$ system kinematics is necessary. A likelihood fit is used to determine the neutrino momentum and the assignment of final state jets to the top and antitop quark decays [28]. The polar angle distribution, $\cos\theta_l$, is fitted using templates obtained from the MC simulation of signal events with positive and negative polarization. The fraction of positively polarized top quarks

$$f = \frac{1}{2} + \frac{N(\cos\theta_l > 0) - N(\cos\theta_l < 0)}{N(\cos\theta_l > 0) + N(\cos\theta_l < 0)} \tag{1}$$

is obtained from the fit.

The fraction f has been measured using 4.66 fb^{-1} of 7 TeV pp collisions recorded in 2011, using $t\bar{t}$ events in the semileptonic decay channel. The result of the template fit is a fraction of positively polarized top quarks of $f = 0.470 \pm 0.009$ (stat) $^{+0.023}_{-0.032}$ (syst) in agreement with the SM prediction of $f_{SM} = 0.5$. This result is dominated by the systematic uncertainty, which mainly comes from the determination of the jet energy scale and the effect of the top quark mass on the signal modelling.

4.4. Top quark spin correlations

At the LHC, $t\bar{t}$ pairs are produced mainly through gluon-gluon fusion, which is dominated at low invariant mass of the $t\bar{t}$ pair by the fusion of like-helicity gluon pairs. This produces top quarks in the left-left or right-right helicity configurations [29]. New physics models may change the $t\bar{t}$ spin correlations while keeping the total cross section unchanged. When the t and $\bar{t}$ decay, if both W bosons decay semileptonically, the spin correlations can be observed in the correlation of the azimuthal angle of both leptons in the laboratory frame, $\Delta\phi$. In this case, it is not necessary to reconstruct the full kinematics of the $t\bar{t}$ pair.

The degree of correlation, A, is defined as the fractional difference between the number of events where the top and antitop quark spin orientations are aligned and those where they have opposite alignment

$$A = \frac{N(\uparrow\uparrow) + N(\downarrow\downarrow) - N(\uparrow\downarrow) - N(\downarrow\uparrow)}{N(\uparrow\uparrow) + N(\downarrow\downarrow) + N(\uparrow\downarrow) + N(\downarrow\uparrow)} \tag{2}$$

where the arrows denote the spin of the top/antitop quarks with respect to the quantization axis.

The $\Delta\phi$ distribution is fitted to templates obtained with the MC simulation to determine the fraction A. Two different bases are used: the helicity basis, using the direction of flight of the top quark in the $t\bar{t}$ rest frame and the maximal basis, which is optimized for $t\bar{t}$ production from gg fusion [30]. The SM predictions for A, obtained with MC@NLO MC at matrix element level, are the following: $A^{SM}_{\text{helicity}} = 0.31$ and $A^{SM}_{\text{maximal}} = 0.44$.

The measured asymmetry values, obtained with 2 fb^{-1} of pp collisions at 7 TeV in the centre of mass, are: $A_{\text{helicity}} = 0.40 \pm 0.04$ (stat) $^{+0.08}_{-0.07}$ (syst) and $A_{\text{maximal}} = 0.57 \pm 0.06$ (stat) $^{+0.12}_{-0.20}$ (syst). MC pseudo-experiments are used to calculate the probability that the observed degree of correlation or larger is measured, assuming no spin correlation. The spin zero hypothesis is excluded with a significance of 5.1 standard deviations (expected value 4.2).

5. Diboson production

The SM predicts Triple Gauge boson Couplings (TGC) fully constrained by the Electroweak symmetry, with some vertices that are allowed (like $WW\gamma$ and WWZ) and others that are forbidden ($ZZ\gamma$, ZZZ). The inclusive and differential di-boson production rates are sensitive

Journal of Physics: Conference Series **447** (2013) 012014 doi:10.1088/1742-6596/447/1/012014

to the TGC, that could be modified by new physics models introducing anomalous TGC, new particle decays or a composite structure of the Z and W bosons.

5.1. *WW production cross section*

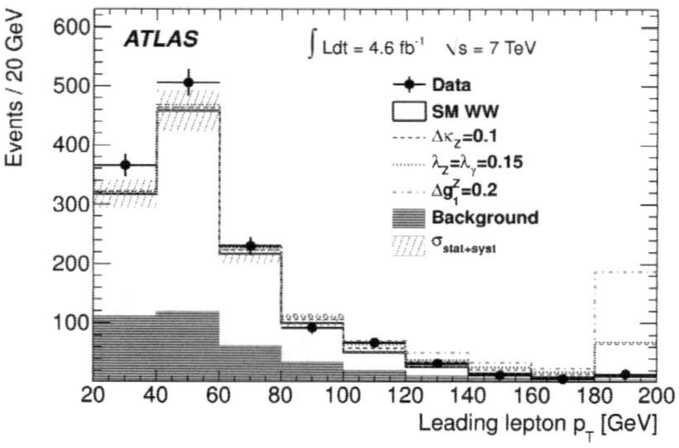

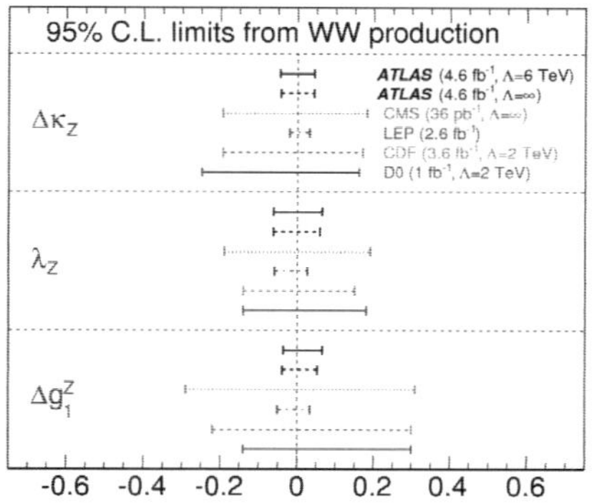

Figure 5. Left: normalized differential WW fiducial cross section as a function of the leading lepton p_T [31]. Right: limits on the anomalous WWZ and $WW\gamma$ anomalous triple gauge boson couplings Δk_Z, λ_Z and Δg_1^Z [31].

The WW Standard Model production cross section in pp collisions at 7 TeV centre of mass energy is predicted to be $44.7^{+2.1}_{-1.9}$ pb [32]. A 126 GeV Higgs boson decaying into WW^* would increase this value by approximately 3%. The measured value of the WW production cross section for 4.6 fb^{-1} of data is $\sigma_{WW} = 51.9 \pm 2.0$ (stat)± 3.9 (syst) ± 2.0 (lumi) pb [31]. It exceeds the theoretical prediction by more than one standard deviation.

The leading lepton p_T distribution is sensitive to anomalous TGC that would manifest as an increase in the high p_T tail of the distribution. The Lorentz invariant Lagrangian describing the WWZ and $WW\gamma$ interactions has fourteen independent coupling parameters [33]. Assuming electromagnetic gauge invariance and C and P conservations, the number of independent parameters reduces to five: g_1 , k_Z, k_γ , λ_Z and λ_γ. In the SM, these parameters have the values: $g_1 = k_Z = k_\gamma = 1$ and $\lambda_Z = \lambda_\gamma = 0$. Form factors that vanish at high parton centre of mass energy are introduced to ensure unitarity at high energies. These form factors depend on a cutoff parameter Λ.

The measured leading lepton p_T distribution is shown in figure 5 (left). The data distribution is compared to the SM prediction and to the distributions that would be obtained in the case of anomalous triple gauge boson couplings. The leading lepton p_T distribution is corrected by the experimental effects due to the detector acceptance, resolution and reconstruction efficiency. A test likelihood function that considers the observed number of candidate events, the expected signal as a function of the anomalous TGC and the estimated number of background events in each p_T bin is defined and used to estimate the anomalous TGC. With the assumption that the WWZ and $WW\gamma$ couplings are equal, the 95% CL limits on the deviations of the parameters from their SM value, Δk_Z and λ_Z, are $[-0.061, 0.093]$ and $[-0.062, 0.065]$ respectively, for a cutoff scale $\Lambda = 6$ TeV, which is the largest value preserving unitarity. These limits are compared to the CMS [34], CDF [35], D0 [36] and LEP [37–40] results in figure 5 (right). They are better than the limits obtained at Tevatron due to the higher production cross section at the LHC and they are approaching the precision obtained by the combined LEP results.

Journal of Physics: Conference Series **447** (2013) 012014 doi:10.1088/1742-6596/447/1/012014

5.2. $W\gamma$ and $Z\gamma$ production cross section

The $W\gamma$ and $Z\gamma$ (called collectively $V\gamma$) production cross sections have been measured for 1.02 fb^{-1} of pp collisions at 7 TeV centre of mass energy, as a function of the photon transverse energy [41]. Two different cross sections have been studied: inclusive, where there is no requirement on additional jets, and exclusive, where there is no hard jet in the event. In the inclusive $V\gamma$ cross section there is a significant contribution from final state radiated photons from the fragmentation process, while in the exclusive production the photons come primarily from the initial state radiation or from triple gauge boson vertices of the type $WW\gamma$ in the $W\gamma$ events.

Figure 6 shows the inclusive and exclusive production cross sections for $W\gamma$ (left) and $Z\gamma$ (right) as a function of the photon transverse energy. The data values are compared to the SM predictions obtained with the MCFM [42] MC generator. The data to MC ratios are also shown. There is a good agreement of the measured exclusive cross sections with the MCFM predictions. The inclusive production cross sections differ from the MCFM prediction at high photon transverse energy. This is probably due to the fact that MCFM only includes real parton emission up to one radiated quark or gluon and therefore it misses higher order QCD corrections, resulting in an underestimation of the inclusive $V\gamma$ cross sections.

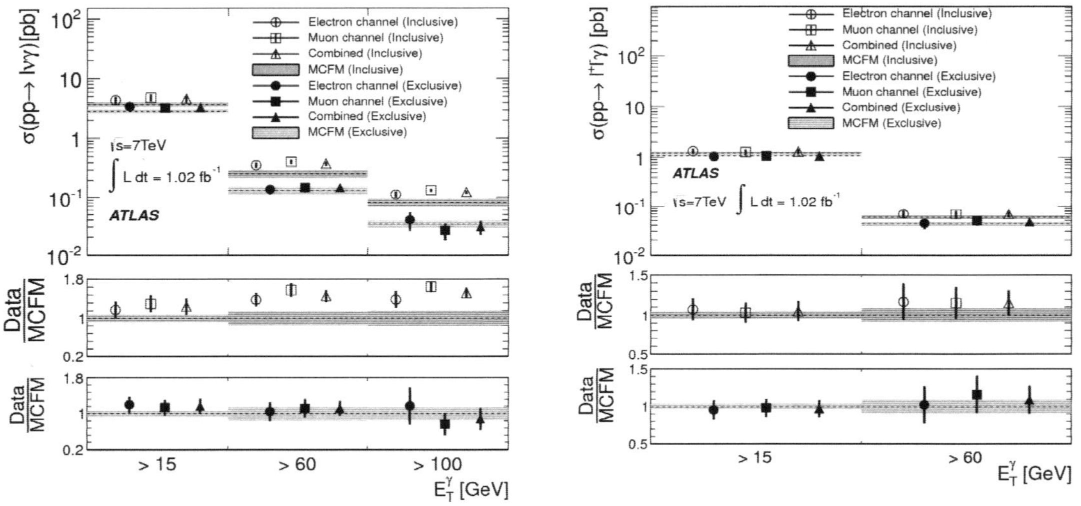

Figure 6. Inclusive and exclusive production cross section of $W\gamma$ (left) and $Z\gamma$ pairs as a function of the photon transverse energy, in 7 TeV pp collisions [41].

Assuming C and P conservation separately, the anomalous TGC are normally parameterized as $\Delta k_\gamma = k_\gamma - 1$ and λ_γ for the $WW\gamma$ vertex and h_3^V and h_4^V for the $ZV\gamma$ vertices (see [41] and references therein). The measurement of exclusive $V\gamma$ production for high photon transverse energy has been used to search for anomalous TGC. No evidence of new physics beyond the SM has been observed, and therefore limits on anomalous triple gauge boson couplings were derived. These limits, shown in figure 7, are compatible with the ones obtained at LEP and Tevatron.

5.3. ZZ production cross section

The ZZ production cross section has been measured by ATLAS for 4.6 fb^{-1} of 7 TeV pp collision [43] and for 5.8 fb^{-1} of 8 TeV pp collisions [44] using the channels $ZZ \rightarrow llll$ and $ZZ \rightarrow ll\nu\nu$ for the 7 TeV analysis and the four lepton channel for the 8 TeV analysis.

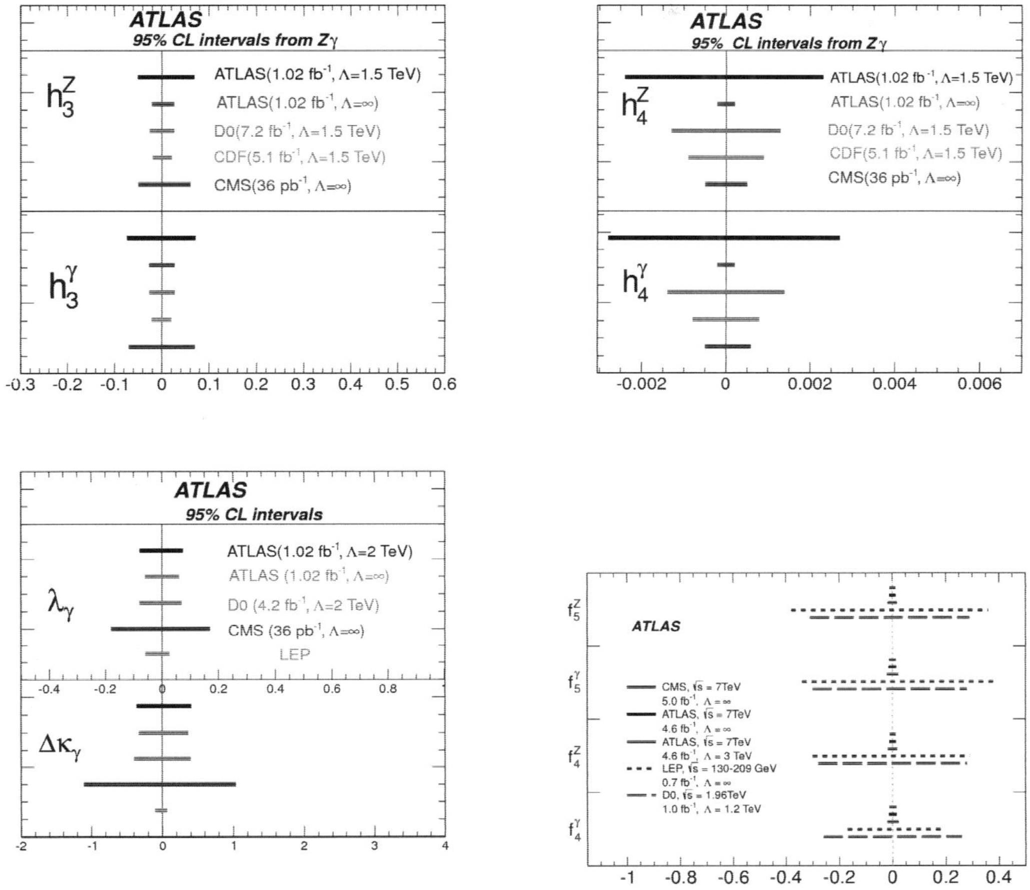

Figure 7. Limits on the anomalous TGC Δk_γ, λ_γ, h_3^W, h_3^Z, h_4^W, h_4^Z, f_4^γ, f_4^Z, f_5^γ and f_5^Z in comparison to the values obtained by D0, CDF and the LEP experiments (see [41, 43] and references therein).

The non resonant ZZ pairs are produced in the SM mainly through the t-channel and s-channel quark-antiquark annihilation process, with a small contribution from gluon-gluon fusion. No TGC $ZZ\gamma$ and ZZZ are allowed in the SM. The predicted SM cross sections for 7 and 8 TeV pp collisions are $5.89^{+0.22}_{-0.18}$ pb and 7.4 ± 0.4 pb, respectively, when the two Z bosons are on shell. The contribution of a 126 GeV Higgs boson would be of the order of 3% (not included before).

The measured cross sections are: $\sigma_{ZZ} = 6.7 \pm 0.7$ (stat) $^{+0.4}_{-0.3}$ (stat) ± 0.3 (lumi) pb and $\sigma_{ZZ} = 9.3^{+1.1}_{-1.0}$ (stat) $^{+0-4}_{-0.3}$ (syst) ± 0.3 (lumi) pb for 7 and 8 TeV pp collisions, respectively. Both results are consistent within errors with the SM expectation.

Neutral TGC can be parameterized in terms of two CP violating (f_4^γ and f_4^Z) and two CP conserving (f_5^γ and f_5^Z) complex parameters (see [43] and references therein). A form factor that depends on a cutoff parameter Λ is introduced to ensure unitarity at high energies, in the same way as before.

The leading Z boson p_{T} distribution is used to impose limits on anomalous neutral triple gauge boson couplings using a maximum profile likelihood ratio [43]. The resulting 95% confidence limits are shown in figure 7 (bottom right), in comparison with previous results. The ATLAS limits improve the limits obtained at LEP and at Tevatron by about a factor five.

Journal of Physics: Conference Series **447** (2013) 012014 doi:10.1088/1742-6596/447/1/012014

6. Higgs searches

In July 2012, the ATLAS and CMS collaborations announced the discovery of a new boson observed in the Higgs searches at the LHC [45]. The ATLAS result was based on three search channels: $H \to WW \to l\nu l\nu$, $H \to ZZ \to llll$ and $H \to \gamma\gamma$ that used 5.8-5.9 fb^{-1} of 8 TeV pp collisions and 4.6-4.8 fb^{-1} of 7 TeV pp collisions. In the same analysis, any other SM Higgs-boson was excluded with 95% CL in a large mass range, from 111-122 and 131-559 GeV. Since this discovery, the focus of the ATLAS collaboration has been the measurement of this new boson's properties, in order to understand if it is indeed the SM Higgs boson, and in particular its detection in all the possible decay and production modes.

In what follows, an overview of new ATLAS results obtained from July till December 2012, using 13 fb^{-1} of 8 TeV pp collisions, will be presented. They refer to three different searches: an update of the $H \to WW$ results, the search for the Higgs decay to b-quark pairs and to τ lepton pairs. In all the cases a blind analysis was performed, i.e., the analysis strategy was fixed based on the MC simulation and data control regions before looking at the signal region in data.

In the SM the Higgs boson is produced mainly through gluon-gluon fusion, with a cross section of the order of a few pb for the low mass range. The second most important production mechanism, with a cross section about an order of magnitude smaller, is the vector boson fusion process (VBF), that can be identified using the signature of the two forward/backward jets. The associated production of a vector boson (W or Z) and a Higgs, has an even smaller cross section, but the clear signatures provided by the leptonic decays of the W or Z boson can be used to reject background in complicated decay channels such as the $H \to bb$. Finally, the associated production mode of a Higgs boson with a $t\bar{t}$ pair has the smallest cross section, with reduced sensitivity for the current integrated luminosities [46].

In the low mass range where the new boson was observed, the dominant decay branching ratio is to b-quark pairs, followed by WW pairs. The branching ratios to ZZ or $\gamma\gamma$ are one or two orders of magnitude smaller but the clean experimental signatures, together with the very good mass resolution, mean that these channels could still provide the greatest sensitivity. The $H \to bb$ and $H \to \tau\tau$ are both affected by very large backgrounds, providing the smallest sensitivity.

6.1. $H \to WW$

Events with two isolated, high p_{T} leptons of different flavour ($e\mu$ or μe) and large missing transverse energy were selected for this analysis [47]. The event selection was further optimized to reduce the main backgrounds: W+jets, top quark production, di-boson production and Z/Drell-Yan. Where possible, data driven techniques were considered to estimate the background yield (top, WW and W+jets). Different analysis categories that are sensitive to different production mechanisms and affected by different background compositions were considered. The categories were selected requiring in addition zero or one jet with p_{T} above 25 GeV.

In figure 8 (left) the transverse mass, m_T, distribution of the selected events with a leading electron is shown. A broad excess of events, compatible with being produced by a 125 GeV Higgs boson, is visible.

The m_T distributions for 0- and 1-jet categories are combined in a profile likelihood fit to estimate the signal strength[5] μ, to measure the probability that a background fluctuation is equal or larger than the data observation (i.e., the p_0 value) and the 95% confidence exclusion limits. The minimum p_0 value, 3×10^{-3}, is obtained for a mass of 111 GeV and corresponds to 2.8 standard deviations. Due to the limited mass resolution of the $H \to WW$ analysis, the minimum of the p_0 distribution is very broad, so its value at 125 GeV is still 4×10^{-3}, corresponding to 2.6 standard deviations.

[5] The ratio of the observed cross section times branching ratio with respect to the predicted SM values.

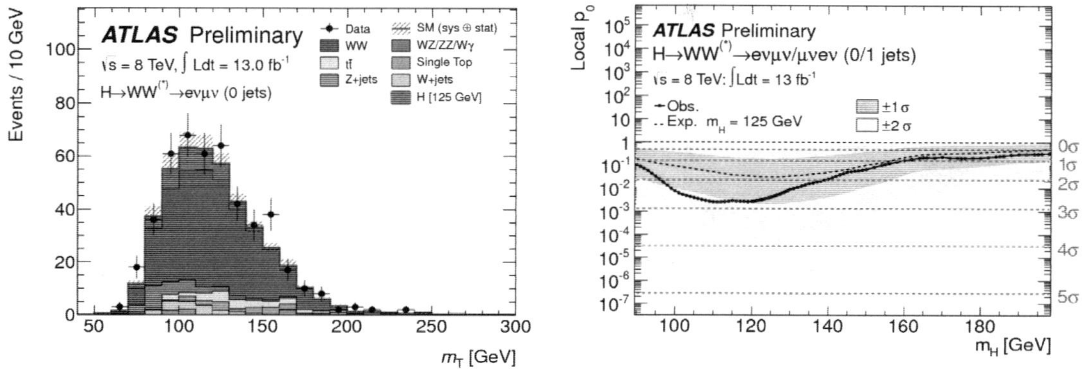

Figure 8. Left: invariant mass distribution of events with a leading electron [47]. Right: p_0 value obtained in the $H \to WW$ analysis [47].

The fitted signal strength, μ, at 125 GeV is $\mu = 1.48^{+0.35}_{-0.33}$ (stat) $^{+0.41}_{-0.36}$ (syst theor) $^{+0.28}_{-0.27}$ (syst exp) ± 0.05 (lumi). It is compatible with the SM hypothesis within the uncertainty, which is dominated by the theoretical uncertainty on the normalization of the WW background.

6.2. $H \to b\bar{b}$

Due to the overwhelming QCD background of b-quark pairs, the search for the $H \to b\bar{b}$ is performed in the associated production channel of a vector boson and a Higgs [48]. Three different channels are considered: $WH \to l\nu b\bar{b}$, $ZH \to llb\bar{b}$ and $ZH \to \nu\nu b\bar{b}$. The event topologies are characterized by high $p_{\rm T}$ isolated leptons, large missing transverse energy, $E_{\rm T}^{miss}$ (in the case of the WH and $ZH \to llb\bar{b}$) and exactly two b-tagged jets. Different categories, based on the $p_{\rm T}$ of the vector boson or $E_{\rm T}^{miss}$, are considered. The invariant mass distribution for the most sensitive category is shown in figure 9.

The different analysis categories are combined in a profile likelihood ratio to calculate the exclusion limits and the p_0 value. The limits are exercised by searching for the di-boson signal WZ and ZZ when one of the Z bosons decays to a $b\bar{b}$ pair and the other vector boson decays leptonically. This di-boson signal has a cross section that is about 5 times larger than the Higgs, with softer $p_{\rm T}^{Z}$ spectrum and peaking at a lower di-b-jet invariant mass. The result of the fit in the search for the di-boson signal yielded a strength parameter $\mu_{\rm di-boson} = 1.09 \pm 0.20$ (stat) ± 0.22 (syst), in good agreement with the SM expectations.

The 95% CL obtained in the fit to search for the SM Higgs boson in the associated production with a vector boson and decaying to a $b\bar{b}$ pair is shown in figure 10 (left). The expected sensitivity at 125 GeV is around 1.9 standard deviations, while the observed limit is 1.8σ, corresponding to a p_0 value of 0.64.

6.3. $H \to \tau\tau$

The search for the Higgs boson decay to τ-lepton pairs is done dividing the analysis in different channels, depending if the $\tau\tau$ pair is accompanied by zero, one or two jets with $p_{\rm T}$ above 25 GeV [49]. The zero-jet category is more sensitive to the gluon-gluon production mechanism, while the 2-jet final state is more sensitive to the VBF mechanism.

τ-leptons decay about 55% of the time to a narrow jet made of one or three charged pions plus several neutral pions. A multivariate analysis technique is used to select the hadronically decaying τ-leptons. Approximately 36% of the time, taus decay to another lepton and a neutrino. In that case, the invariant mass of the τ is reconstructed using the missing mass calculator, that

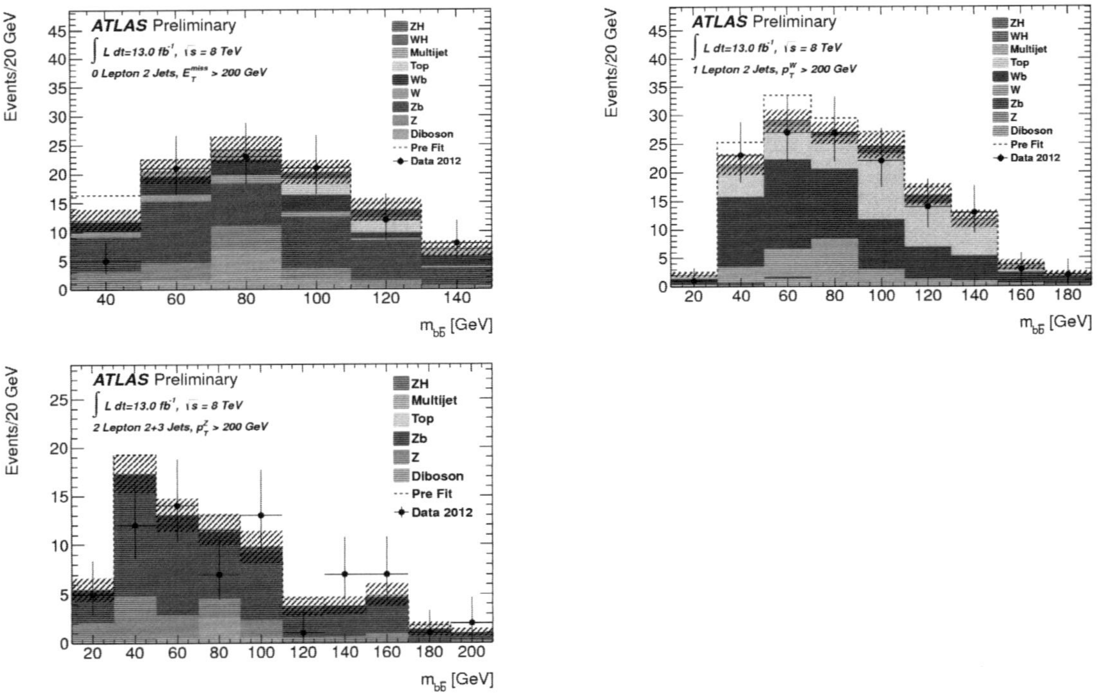

Figure 9. Invariant mass distribution of the di-b-jet pair in the search for $H \rightarrow b\bar{b}$ in the associated production mode with a vector boson [48]. The distributions show the most sensitive categories in each of the channels considered (zero, one and two leptons in the final state).

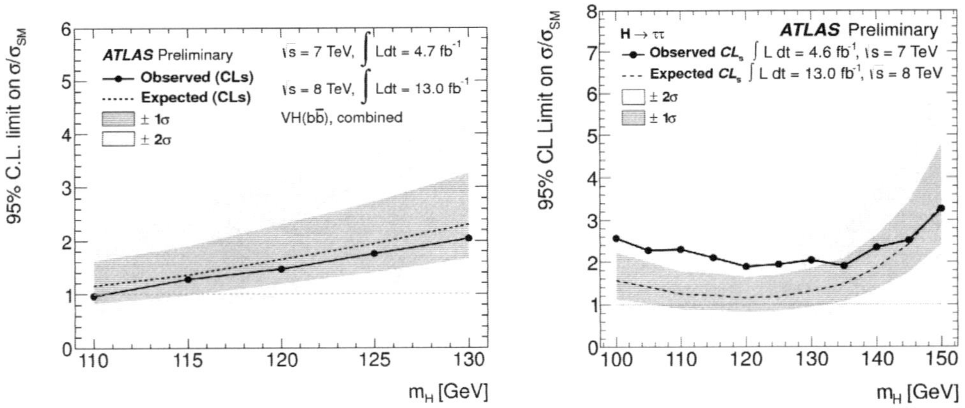

Figure 10. 95% combined CL exclusion limits as a function of the mass of the Higgs, obtained in the Higgs search in the associated production with a vector boson when the Higgs decays to $b\bar{b}$ pairs [48](left) and in the $H \rightarrow \tau\tau$ search [49](right).

constrains the neutrino momentum using the tau decay kinematics [50]. In this analysis, all the possible final state τ-lepton topologies are considered [49].

Due to the softer p_{T} spectrum of the lepton produced in the semileptonic τ decays, the large

Journal of Physics: Conference Series **447** (2013) 012014 doi:10.1088/1742-6596/447/1/012014

backgrounds affecting the hadronic signatures and the limited mass resolution, this channel has limited sensitivity, similar to the $H \rightarrow b\bar{b}$ channel.

The combined 95% confidence level exclusion limits in the $H \rightarrow \tau\tau$ channel are shown in figure 10 (right). The combined expected limits vary between 1.2 and 3.4 times the predicted SM cross-section times branching ratio for the mass range 100-150 GeV, while the observed values vary between 1.3 and 3.3 times the predicted value. The expected (observed) limit at 125 GeV is 1.2 (1.9) times the SM cross section times branching ratio. For this mass, the local significance is 1.7 standard deviations and the best fitted signal strength is $\mu = 0.7 \pm 0.7$.

6.4. Combination

A combined fit on all the available ATLAS results has been performed in [51]. The best fit signal strength parameter obtained was: $\mu = 1.3 \pm 0.3$ for 126 GeV Higgs boson signal. The compatibility of the different results is at the 36% level, while the compatibility with the SM hypothesis is 23%. The excess with respect to the SM expectation is driven by the $H \rightarrow \gamma\gamma$ results [45] (not presented here).

7. Summary and conclusions

A wide range of physics processes has been measured by the ATLAS Collaboration with the data collected so far, comprising precision measurements of Standard Model processes and searches for physics beyond it. Overall, the results are consistent with the SM, with no clear evidence of new physics.

The search for the Standard Model Higgs boson, responsible for the Electroweak Symmetry breaking mechanism, lead to the discovery in July 2012 of a new boson with a mass of 126 GeV. The focus of the ATLAS collaboration goes now into understanding the nature of this particle and measuring its properties. Three decay channels were updated in between July and December 2012 and have been presented in this paper. The results show that, within the limited statistics available for the moment, the new boson is still compatible with the SM hypothesis.

Acknowledgments

The author acknowledges the support by FCT co-financed by QREN/COMPETE of European Union ERDF fund, Portugal.

References

[1] ATLAS Collaboration 2008 *JINST* **3** S08003
[2] ATLAS Collaboration 2012 *Phys. Rev. D* **86** 014022
[3] Cacciar M, Salam G P and Soyez G 2008 *JHEP 0804:063*
[4] Cacciar M, Salam G P and Soyez G 2011 *CERN* CERN-PH-TH/2011-297
[5] Nagy Z 2003 *Phys. Rev.* **D68** 094002
[6] Lai H-L *et al.* 2010 *Phys. Rev.* **D82** 074024
[7] Martin A, Stirling W, Thorne R and Watt G 2009 *Eur. Phys. J.* **C63** 189
[8] Ball R D *et al.* 2010 *Nucl. Phys.* **B838** 136
[9] Forte S, Laenen E, Nason P and Rojo J 2010 *Nucl. Phys.* **B834** 116
[10] H1 and ZEUS Collaborations Herapdf 1.5 h1prelim-10-142, ZEUS-prel-10-018
[11] Eichten E, Lane K D and Peskin M E 1983 *Phys. Rev. Lett.* **50** 811
[12] Baur U, Hinchliffe I and Zeppenfeld D 1987 *Int. J. Mod. Phys.* **A2** 1285
[13] Baur U, Spira M and Zerwas P M 1990 *Phys. Rev.* **D42** 815
[14] Frampton P H and Glashow S L 1987 *Phys. Lett.* **B190** 157
[15] Bagger J, Schmidt C and King S 1988 *Phys. Rev. D* **37** 1188
[16] ATLAS Collaboration 2012 Search for new phenomena in the dijet mass distribution updated using 13/fb of *pp* collisions at $\sqrt{s}$ = 8 TeV collected by the ATLAS detector ATLAS-CONF-2012-148 (https://cds.cern.ch/record/1493487)
[17] CDF Collaboration 2009 *Phys. Rev.* **D79** 011101
[18] Choudalakis G 2011 On hypothesis testing, trials factor, hypertests and the BumpHunter arXiv:1101.0390

[19] Lyons L 2008 *Ann. Appl. Stat.* **2** 887
[20] ATLAS Collaboration 2012 Statistical combination of top quark pair production cross-section measurements using dilepton, single-lepton, and all-hadronic final states at $\sqrt{s} = 7$ TeV with the ATLAS detector ATLAS-CONF-2012-024 (https://cds.cern.ch/record/1430733?ln=es)
[21] ATLAS Collaboration 2012 Measurement of the top quark pair production cross section with ATLAS in *pp* collisions at $\sqrt{s} = 7$ TeV in the single-lepton channel using semileptonic b decays ATLAS-CONF-2012-131 (https://cds.cern.ch/record/1478370)
[22] ATLAS Collaboration 2012 Measurement of the top quark pair production cross section in the single-lepton channel with ATLAS in proton-proton collisions at 8 TeV using kinematic fits with b-tagging ATLAS-CONF-2012-149 (http://cdsweb.cern.ch/record/1493488)
[23] ATLAS Collaboration 2012 Measurement of the jet multiplicity in top-anti-top final states produced in 7 TeV protonproton collisions with the ATLAS detector ATLAS-CONF-2012-155 (http://cdsweb.cern.ch/record/1493494)
[24] Mangano M L *et al.* 2003 *JHEP* **0307** 001
[25] Corcella G *et al.* 2001 *JHEP* **0101** 010
[26] Frixione S and Webber B R 2002 *JHEP* **0206** 029
[27] Alioli S, Nason P, Oliari C and Re E 2010 *JHEP* **1006** 043
[28] ATLAS Collaboration 2012 Measurement of top quark polarisation in tt events with the ATLAS detector in proton-proton collisions at $\sqrt{s} = 7$ TeV ATLAS-CONF-2012-133 (http://cdsweb.cern.ch/record/1478373)
[29] Mahlon G and Parke S 2010 *Phys. Rev.* **D81** 074024
[30] ATLAS Collaboration 2012 *Phys. Rev. Lett.* **108** 212001
[31] ATLAS Collaboration 2012 Measurement of WW production in *pp* collisions at $\sqrt{(s)} = 7$ TeV with the ATLAS detector and limits on anomalous WWZ and $WW\gamma$ couplings arXiv:1203.6232, submitted to Phys. Rev. D (http://arxiv.org/abs/1210.2979)
[32] Campbell J M *et al.* 2009 *Phys. Rev.* **D80** 054023
[33] Hagiwara K, Peccei R D, Zeppenfeld D and Hikasa K 1987 *Nucl. Phys.* **B282** 253
[34] CMS Collaboration 2011 *Phys. Lett.* **B699** 25
[35] CDF Collaboration 2010 *Phys. Rev. Lett.* **104** 201801
[36] D0 Collaboration 2009 *Phys. Rev. Lett.* **103** 191801
[37] DELPHI Collaboration 2008 *Eur. Phys. J.* **C54** 345
[38] L3 Collaboration 2004 *Phys. Lett.* **B586** 151
[39] OPAL Collaboration 2004 *Eur. Phys. J.* **C33** 463
[40] ALEPH Collaboration 2005 *Phys. Lett.* **B614** 7
[41] ATLAS Collaboration 2012 *Phys. Lett.* **B717** 49
[42] Campbell J M and Ellis R 2010 *Nuclear Physics B - Proceedings Supplements* **205** 10
[43] ATLAS Collaboration 2012 Measurement of ZZ production in *pp* collisions at $\sqrt{(s)} = 7$ TeV with the ATLAS detector and limits on anomalous ZZZ and $ZZ\gamma$ couplings arXiv:1211.6096, submitted to JHEP (http://arxiv.org/abs/1211.6096)
[44] ATLAS Collaboration 2012 Measurement of the ZZ production cross section in the four-lepton channel using 5.8 fb$^-$1 of ATLAS data at $\sqrt{(s)} = 8$ TeV ATLAS-CONF-2012-090 (http://cdsweb.cern.ch/record/1460409)
[45] ATLAS Collaboration 2012 *Phys. Lett.* **B716** 1
[46] ATLAS Collaboration 2012 Search for the standard model Higgs boson produced in association with top quarks in proton-proton collisions at $\sqrt{s} = 7$ TeV using the ATLAS detector ATLAS-CONF-2012-135 (http://cdsweb.cern.ch/record/1478423)
[47] ATLAS Collaboration 2012 Update of the $H \to WW^{(*)} \to e\nu\mu\nu$ analysis with 13.0 fb^1 of $\sqrt{s} = 8$ TeV data collected with the ATLAS detector ATLAS-CONF-2012-158 (http://cdsweb.cern.ch/record/1493601)
[48] ATLAS Collaboration 2012 Search for the standard model Higgs boson produced in association with a vector boson and decaying to bottom quarks with the ATLAS detector ATLAS-CONF-2012-161 (http://cdsweb.cern.ch/record/1493625)
[49] ATLAS Collaboration 2012 Search for the standard model Higgs boson in $H \to \tau^+\tau^-$ decays in proton-proton collisions with the ATLAS detector ATLAS-CONF-2012-160 (http://cdsweb.cern.ch/record/1493624)
[50] Elagin A, Murat P, Pranko A and Safonov A 2011 *Nucl. Instrum. Meth.* **A654** 481
[51] ATLAS Collaboration 2012 Updated ATLAS results on the signal strength of the Higgs-like boson for decays into WW and heavy fermion final states ATLAS-CONF-2012-162 (http://cdsweb.cern.ch/record/1494183)

Journal of Physics: Conference Series **447** (2013) 012015 doi:10.1088/1742-6596/447/1/012015

$t\bar{t}$ charge asymmetry, family and friends

J A Aguilar-Saavedra, M Pérez-Victoria

Departamento de Física Teórica y del Cosmos, Universidad de Granada, 18071 Granada

E-mail: jaas@ugr.es,mpv@ugr.es

Abstract. We present the current status of the Tevatron charge asymmetry and its sister asymmetry at the LHC. The relation between both is elucidated, using as framework the collider-independent asymmetries they originate from. Other related observables, such as the $t\bar{t}$ differential distribution and top polarisation, are also discussed.

1. Introduction

The measurement of an anomalous forward-backward (FB) asymmetry by the CDF and D0 experiments at the Tevatron [1, 2] has fueled a plethora of proposals addressing the apparent deviation from the predictions of the Standard Model (SM). In most cases, these proposals involve new physics beyond the SM. But the simplest explanations for these departures, including those invoking higher-order SM corrections, also predict an enhancement of the charge asymmetry at the Large Hadron Collider (LHC). Such an enhancement has not been observed, though it is not excluded either, given the present experimental uncertainties. To make the situation even more puzzling, new physics models addressing the Tevatron excess often, but not always, predict related effects in $t\bar{t}$ production, such as enhancements in the high $t\bar{t}$ invariant mass tail, or a net polarisation of the top (anti)quark. None of these effects have been observed. In the following, we critically review these issues and discuss the intriguing status of the subject.

2. The charge asymmetry at the Tevatron

It is well known that the differential cross section for $q\bar{q} \to t\bar{t}$, with $q = u, d$, is not invariant under the interchange of the t and $\bar{t}$ momenta. At the Tevatron, the most commonly used observable to measure this difference is the asymmetry

$$A_{FB} = \frac{N(\Delta y > 0) - N(\Delta y < 0)}{N(\Delta y > 0) + N(\Delta y < 0)}, \tag{1}$$

where $\Delta y = y_t - y_{\bar{t}}$ is the difference between the rapidities of the top quark and antiquark, taking the z axis in the proton direction. This rapidity difference is invariant under boosts in the beam direction. The definition above exploits the fact that, to a good extent, in $p\bar{p}$ collisions the directions of the initial quark and antiquark are known: they are, respectively, the directions of the proton and the antiproton. This asymmetry is equivalent to an asymmetry in the polar angle θ between the top momentum and the quark direction in the centre of mass (CM) frame,

$$A_{FB} = \frac{N(\cos\theta > 0) - N(\cos\theta < 0)}{N(\cos\theta > 0) + N(\cos\theta < 0)}. \tag{2}$$

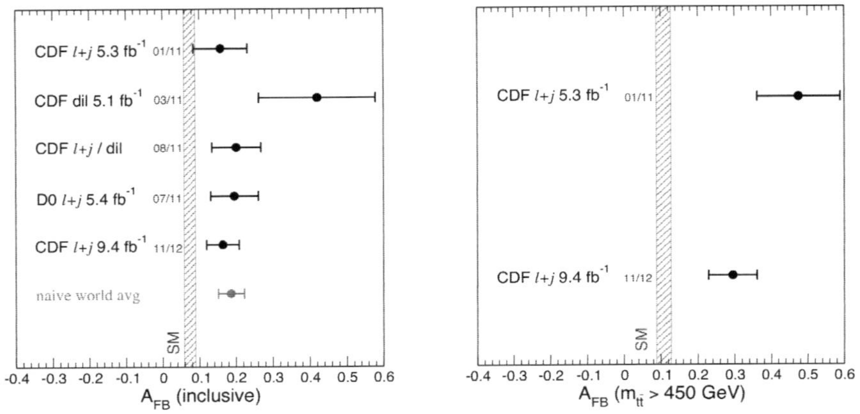

Figure 1. Measurements of A_{FB} at the Tevatron: inclusive (left) and at high $m_{t\bar{t}}$ (right).

In the SM this asymmetry is small and arises, at the lowest order in perturbation theory, from the interference of tree-level and one-loop diagrams for $q\bar{q} \to t\bar{t}$, plus some contributions from extra jet radiation. For some time, the two Tevatron experiments have consistently measured an asymmetry above the SM expectation, in the semileptonic and dilepton $t\bar{t}$ decay channel. A summary of the most recent inclusive unfolded measurements is shown in Fig. 1 (left), to be compared with the SM predictions [3–7], which range from 0.058 to 0.089. The naive world average of the latest measurements in each experiment and $t\bar{t}$ decay channel gives an asymmetry of 0.187 ± 0.036, which is 2.7σ above the closest of those SM predictions. In contrast with other recent anomalies, the trend of the measurements with time and increased luminosity does not approach the SM prediction but the positive excess persists and it does not look like a statistical fluctuation. At high $t\bar{t}$ invariant mass $m_{t\bar{t}} \geq 450$ GeV (Fig. 1 right) the departure from the SM has been reduced with respect to the first measurement [8], but the deviation is still significant.

These consistent discrepancies have motivated a number of papers proposing new physics explanations. As the excess in A_{FB} is of the same size as the one-loop QCD asymmetry, if this excess results from new physics it is likely to enter at the tree level in $q\bar{q} \to t\bar{t}$. The types of renormalisable new physics that can enter $q\bar{q} \to t\bar{t}$ can be classified by using group theory, requiring that the Lagrangian is invariant under $\mathrm{SU}(3)_C \times \mathrm{SU}(2)_L \times \mathrm{U}(1)_Y$. This requirement gives a total of 18 possibilities, 10 for spin-1 vector bosons and 8 for scalars [9]. Among them, the most popular ones are a colour octet G_μ (also called "gluon" hereafter) exchanged in the s channel [10], a Z' [11], W' [12] or scalar doublet ϕ [13] in the t channel, and a colour sextet Ω^4 or colour triplet scalar ω^4 [14] in the u channel. These "simple" models are phenomenological in the sense that their goal is to explain the anomaly simply by adding to the SM a new particle.[1] Still, they provide a good basis to test if:

(i) A_{FB} can be enhanced without spoiling the good agreement of the total $t\bar{t}$ cross section with the SM prediction.
(ii) The inclusive and high-mass measurements of A_{FB} can be reproduced.
(iii) The explanation of the Tevatron anomalies is compatible with other data, in particular from the LHC.

Once we have a model that fulfills these conditions, one can go further and build a more complete

[1] In addition to these simple models, proposals have been made with new physics entering at one loop level [15] that give similar predictions.

new physics model that explains the Tevatron measurements of the asymmetry and other related ones. We now concentrate on the first two tests, which involve Tevatron collider.

The first test that these models have to pass concerns their ability to correctly fit the observed total cross section and asymmetry. This is nontrivial since the SM prediction of the cross section, $\sigma_{\rm SM} = 7.5 \pm 0.5$ pb [16] is very close to the measured one, $\sigma_{\rm exp} = 7.68 \pm 0.4$ [17]. If one expands the cross section in the presence of new physics as $\sigma = \sigma_{\rm SM} + \delta\sigma_{\rm int} + \delta\sigma_{\rm quad}$, this implies that the interference between SM and new physics, $\delta\sigma_{\rm int}$, and the new physics quadratic term, $\delta\sigma_{\rm quad}$, have to fulfill $\delta\sigma_{\rm int} + \delta\sigma_{\rm quad} \simeq 0$. There are two possibilities for this:

(i) The quadratic term is large, which implies that the interference is also large and both terms nearly cancel each other. For this to happen, a fine-tuning of the new physics coupling is needed, since the interference and quadratic terms depend linearly and quadratically, respectively, on this coupling. This is the case, for example, of t-channel models (Z', W', ϕ). If this cancellation is arranged to happen at the Tevatron energy, then it is not expected to take place anymore at the LHC.

(ii) The quadratic term is small. The total interference $\delta\sigma_{\rm int}$ must also be small but with sizeable contributions in the forward and backward hemispheres, $\delta\sigma_{\rm int}^F \simeq -\delta\sigma_{\rm int}^B$. This is the case, for example, of an s-channel heavy gluon with axial coupling to either light quarks or to the top quark, in which case the interference identically vanishes. If one drops the condition $\delta\sigma_{\rm int}^F + \delta\sigma_{\rm int}^B \simeq 0$, as for example in u-channel colour sextet models, the generated asymmetries have to be smaller due to the total cross section constraint.

In both cases, it is evident that one needs that the interference with the SM is non-zero [18] which, in principle, can be achieved with all the types of new particles that can contribute to $q\bar{q} \to t\bar{t}$ [9].

A second Tevatron test for the new physics proposals concerns whether they can accommodate the inclusive and high-mass measurements. This is accomplished for most models, as it is shown in Fig. 2. The coloured regions show the model predictions obtained by a parameter space scan, subject to some loose constraints from the total $t\bar{t}$ cross section at the Tevatron and the high-mass tail at the LHC [19]. We only consider positive contributions to the asymmetry, which are summed to the SM one [7] in all cases. For the colour octet we consider a very heavy axigluon represented by four-fermion operators [9, 20–22]. Most of these simple models can reproduce very well the inclusive and high-mass asymmetries, except a new Z' which overpredicts the asymmetry, especially at high $t\bar{t}$ invariant mass mass. (For this and other reasons —see section 5— this model will not be considered further here as a viable candidate.) But a more revealing outcome of this comparison is the observation that the inclusive and high-mass measurements of the asymmetry are "naturally consistent" or, in other words, it does not take a contrived model to reproduce both.[2] A recent analysis of the polar angle dependence of the cross section [24] also supports the internal consistence of the deviation. The differential cross section can be expanded in terms of Legendre polynomials,

$$\frac{d\sigma}{d\cos\theta} = \sum_l a_l P_l(\cos\theta) \,. \tag{3}$$

The CDF Collaboration finds good agreement of all a_l with the SM prediction, except for the term with $l = 1$ —corresponding to a term linear in $\cos\theta$— that deviates more than 2σ from the SM. This pattern can be nicely fitted with an s-channel colour octet, for example, which enhances a_1 while keeping higher l coefficients small.

[2] More complicated models, for example with a number of s-channel coloured resonances, can reproduce complicated profiles of A_{FB} versus $m_{t\bar{t}}$ [23].

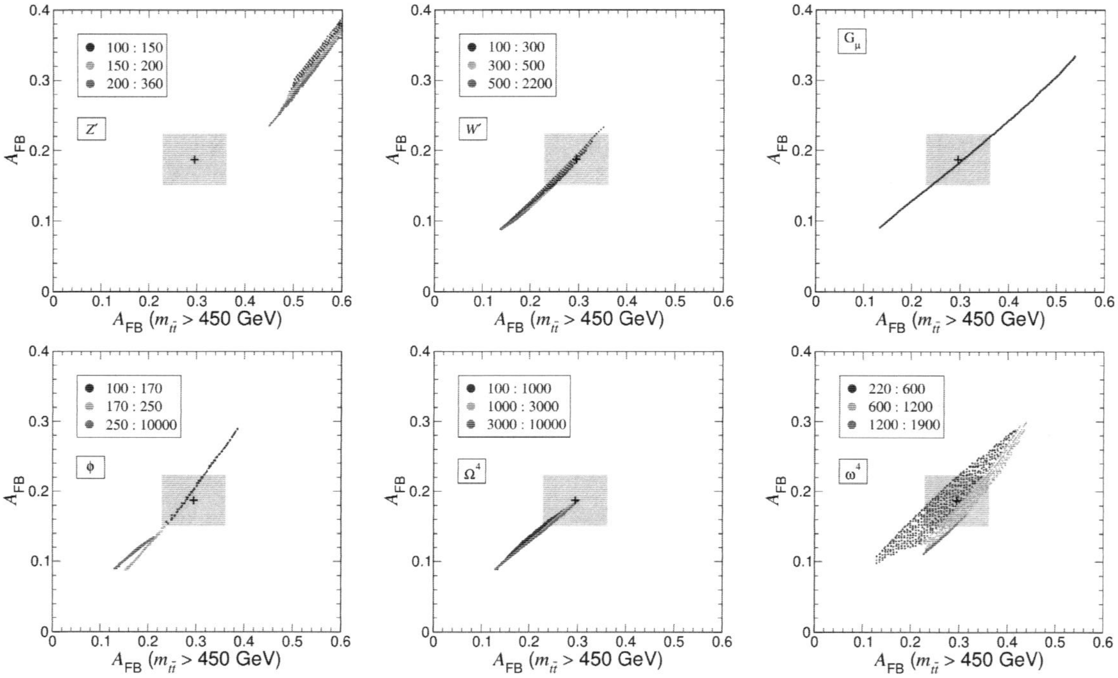

Figure 2. Inclusive versus high-mass asymmetries at the Tevatron, for several new physics models. The numbers in the legends indicate the mass range for the new particle, in GeV. The crosses correspond to the experimental measurements in Fig. 1, with the shaded boxes corresponding to the 1σ uncertainty.

Finally, at the Tevatron there are additional constraints from the cross section at the high-$m_{t\bar{t}}$ tail, but the measurements in that region are not very precise and the new physics contributions may have a much lower detection efficiency than SM top pair production, due to the limited detector acceptance [25]. This is the case, for example, of light t-channel mediators. For this reason we do not use the measurements in that region as a further constraint.

3. The younger sister: the LHC charge asymmetry

At a pp collider as the LHC, the symmetry of the initial state implies that, for a fixed choice of z axis to measure rapidities, the FB asymmetry in Eq. (1) vanishes. Then, one has to consider different observables to test an asymmetry in $q\bar{q} \to t\bar{t}$. This can be done, for example, by exploiting the fact that valence quarks have larger average momentum fraction than antiquarks, leading to a non-vanishing asymmetry [26]

$$A_{FB} = \frac{N(\Delta|y| > 0) - N(\Delta|y| < 0)}{N(\Delta|y| > 0) + N(\Delta|y| < 0)}, \tag{4}$$

with $\Delta|y| = |y_t| - |y_{\bar{t}}|$. This asymmetry has been measured by the ATLAS and CMS Collaborations, with the results shown in Fig. 3. The naive average of the latest measurements is $A_C = 0.013 \pm 0.011$, consistent with the SM predictions [6, 7, 27].

It is important to stress here a fact that should be obvious: the Tevatron A_{FB} and the LHC A_C are *not* the same observable. Therefore, a measurement of A_C consistent with the SM is not in conflict with a Tevatron excess [28]. On the other hand, comparing predictions for A_{FB}

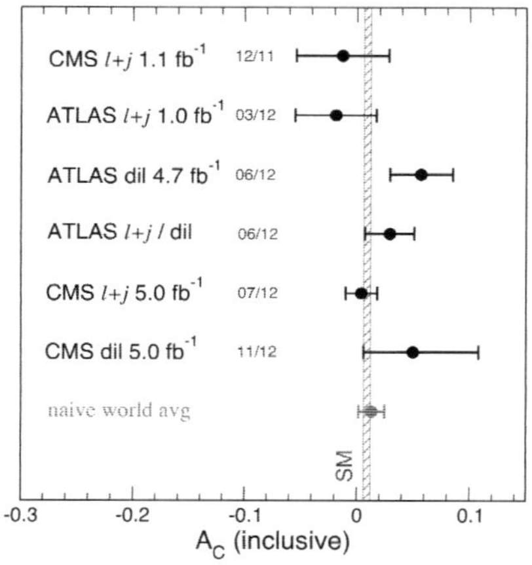

Figure 3. Measurements of A_C (inclusive) at the LHC.

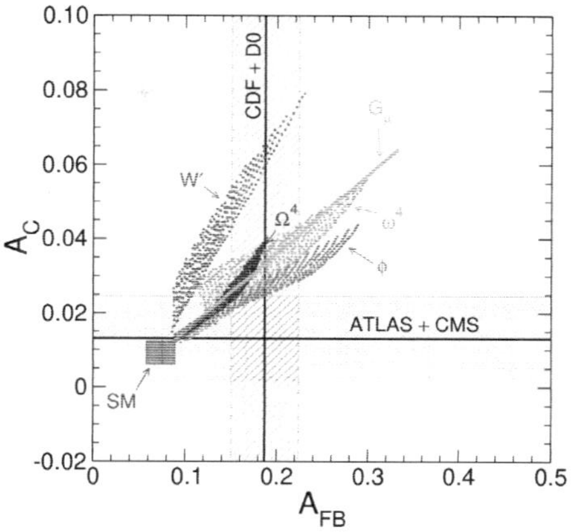

Figure 4. Comparison of predictions for the inclusive asymmetries A_{FB} and A_C for several simple models.

and A_C *within a given model* brings important consequences for the model [29], as it is clearly depicted in Fig. 4. In particular, the W' models are clearly disfavoured, as they predict values of A_C more than 3σ above present data when accommodating the Tevatron asymmetry. For the rest of simple models the fate is uncertain as they are consistent with data at the 2σ level. The difficulty to simultaneously reproduce the central values of A_{FB} and A_C has motivated the appearance of several less simple models [30, 31] that can accommodate the central values of the Tevatron and LHC asymmetries, by introducing some type of cancellation among contributions. (See also Refs. [32, 33].)

4. The parents: the collider-independent asymmetries.

The Tevatron asymmetry A_{FB} and the LHC asymmetry A_C originate from the "intrinsic" partonic asymmetries in $u\bar{u} \to t\bar{t}$, $d\bar{d} \to t\bar{t}$, which will be denoted hereafter as A_u, A_d, respectively [28]. At leading order (LO), these asymmetries only depend on the partonic CM energy $\hat{s}$. As a consequence of this, for a suitably narrow interval of $m_{t\bar{t}}$ the asymmetries A_u, A_d are nearly the same at the Tevatron and the LHC. The "daughter" asymmetries A_{FB}, A_C can be regarded as different combinations of A_u and A_d, the differences arising because

- at these two colliders the importance of $u\bar{u} \to t\bar{t}$ and $d\bar{d} \to t\bar{t}$, relative to the total $t\bar{t}$ production rate, changes due to parton density functions (PDFs);
- at the LHC the asymmetry A_C suffers from a "dilution" because not always the initial valence quark has larger momentum fraction than the sea antiquark. In case that the antiquark has larger momentum fraction, a "forward" event, that is, with the top quark in the direction of the incoming quark ($\cos\theta > 0$), has $\Delta|y| < 0$ and contributes negatively to A_C.

Then, a possible experimental test of the consistency of A_{FB} (higher than the SM prediction) with A_C (consistent with the SM) would be to measure the "collider-independent" asymmetries A_u and A_d.[3] In experiments, the intrinsic asymmetries A_u, A_d can be extracted by exploiting

[3] Although we use the name "collider-independent" for A_u and A_d, to be precise it must be noted that at next-to-leading order (NLO) some differences are introduced, of little relevance from a practical point of view. These are mainly originated from the need to replace a fixed partonic CM energy $\hat{s}$ by a narrow $m_{t\bar{t}}$ interval, which introduces some deviations due to a residual dependence on PDFs. Besides, the asymmetries in $gq \to t\bar{t}j$ are

the dependence of A_{FB} and A_C on the velocity of the $t\bar{t}$ pair [35]

$$\beta = \frac{|p_t^z + p_{\bar{t}}^z|}{E_t + E_{\bar{t}}}\,, \tag{5}$$

because A_u and A_d are independent of this variable.[4] Thus, A_u and A_d can be extracted from a fit to

$$\begin{aligned} A_{FB}(\beta) &= A_u F_u(\beta) + A_d F_d(\beta)\,, \\ A_C(\beta) &= A_u F_u(\beta) D_u(\beta) + A_d F_d(\beta) D_d(\beta)\,, \end{aligned} \tag{6}$$

where $F_{u,d}$ are the fractions of $u\bar{u}$ and $d\bar{d}$ events, respectively, and $D_{u,d}$ are factors to reflect the dilution of the asymmetries in pp collisions we have mentioned above. Both $F_{u,d}$ and $D_{u,d}$ can be computed with a Monte Carlo within the SM and used as input to extract A_u and A_d from data. The SM predictions for these asymmetries are given in Fig. 5. The left panel shows the asymmetries in 50 GeV bins, without any restriction on the transverse momentum of the $t\bar{t}$ pair $p_T^{t\bar{t}}$. It is important to remark here that, compared with their expected experimental uncertainty, the differences between the Tevatron and LHC asymmetries A_u, A_d are irrelevent and justifiy labelling these asymmetries as "collider-independent". The right panel shows the asymmetries with a cut $p_T^{t\bar{t}} < 30$ GeV that practically eliminates the deviations between the Tevatron and LHC asymmetries.

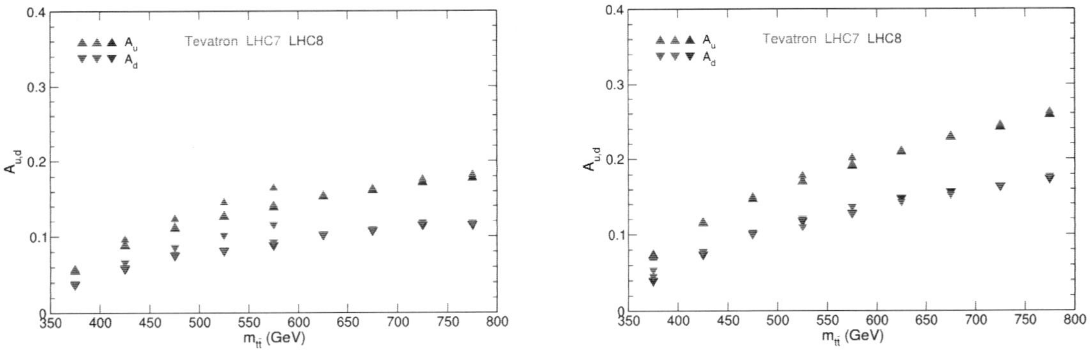

Figure 5. Asymmetries A_u, A_d in the SM, for the Tevatron and the LHC. Left: without a cut on $p_T^{t\bar{t}}$. Right: for $p_T^{t\bar{t}} < 30$ GeV.

The experimental measurement of A_u and A_d is very challenging, as it may require a three-dimensional data unfolding in β, $m_{t\bar{t}}$ and Δy ($\Delta|y|$). But the interest of this measurement may well be worth the effort. In the first place, if these asymmetries are measured, it can be tested whether the Tevatron and LHC results are consistent. We present in Fig. 6 a potential result in the bin $m_{t\bar{t}} \leq 400$ GeV, with the crosses representing the NLO SM result and the ellipses corresponding to the 1σ statistical uncertainty. In this example the ellipses intersect by construction, since the SM NLO values of A_{FB} and A_C have been used as input. But, if there is an unknown systematic effect in either collider, this may not be the case. If, on the other hand, the Tevatron and LHC determinations of the asymmetries are consistent, one can combine

irrelevant. For an extended discussion see Ref. [34].

[4] Strictly speaking, they are independent at LO and with fixed $\hat{s}$. At NLO, or in finite $m_{t\bar{t}}$ bins, one can see that the dependence is rather mild [34].

Journal of Physics: Conference Series **447** (2013) 012015 doi:10.1088/1742-6596/447/1/012015

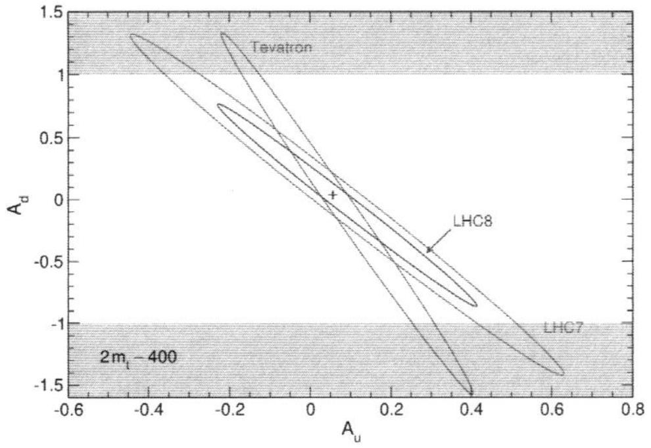

Figure 6. Example of a potential outcome of the measurement of A_u and A_d at the Tevatron and the LHC.

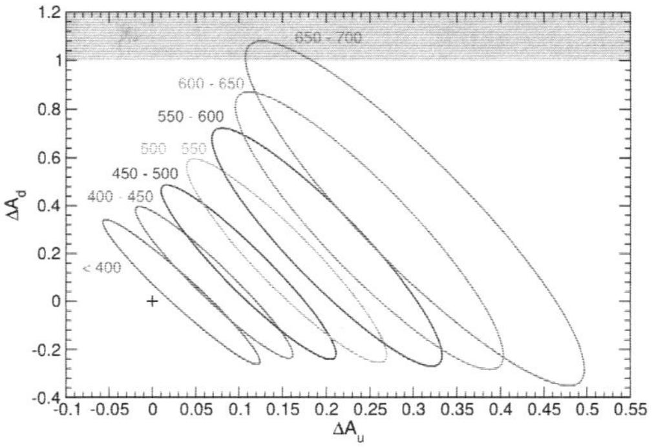

Figure 7. Example of a potential combination of the measurements of A_u and A_d at the Tevatron and the LHC.

both and test whether the combined measurement is compatible with the SM prediction. We show in Fig. 7 a potential result of a Tevatron-LHC combination, taking a heavy axigluon as a new physics benchmark. The ellipses correspond to the 1σ combined limits with the SM asymmetries subtracted, so that the SM point in this plot corresponds to zero. It is apparent that the combination of Tevatron and LHC results on A_u, A_d has a much higher significance than the individual measurements.

5. An old friend: the high-mass tail at the LHC

One of the first and most universal predictions for models explaining the Tevatron excess is an enhancement of the $t\bar{t}$ differential distribution at high $m_{t\bar{t}}$ [9, 36]. The LHC experiments have not found any sign of tail enhancement [37, 38]. The most stringent limits result from the ATLAS analysis, which measures a cross section slightly below the SM prediction, $d\log\sigma/dm_{t\bar{t}} \simeq 7 \pm 2$ PeV^{-1} for $m_{t\bar{t}} \geq 950$ GeV, compared to a SM prediction of 9 PeV^{-1}. This leads to an upper limit $\sigma/\sigma_{\rm SM} \leq 1.3$ in this mass bin, with a 99% confidence level. With all caveats that apply to a naive interpretation of an experimental result, we observe that this limit is similar to the one projected in Ref. [19], $\sigma/\sigma_{\rm SM} \leq 1.5$ for $m_{t\bar{t}} \geq 1$ TeV. Imposing the latter constraint as well as a minimum of $\sigma/\sigma_{\rm SM} \geq 0.5$, the allowed areas for the Tevatron asymmetries in Fig. 2 shrink to the corresponding ones in Fig. 8. The Z' model, already discarded because it overpredicts the Tevatron excess, does not give a positive contribution to the Tevetron asymmetry after imposing the LHC tail constraint. The W' model, which predicted too large values for A_C, cannot reproduce the Tevatron asymmetries either. The rest of models are, in principle, compatible with $t\bar{t}$ cross section measurements at the LHC and the Tevatron. Among them, the "least disturbing" one for the differential distribution is an s-channel colour octet, as long as it is heavy (or light) enough. However, if the resonance is at kinematical reach the differential cross section enhancement is much larger, and the new particle will appear as a peak (or bump, if it is very wide) in the $t\bar{t}$ invariant mass spectrum. Current LHC searches put heavy gluon models into trouble as the mass scales probed go higher, because the couplings required get too large and nonperturbative. However, these constraints can be avoided by "light" gluons with a mass of few hundreds of GeV [23, 39–41].

6. A new friend: the top polarisation

Another common signal of new physics contributions to $t\bar{t}$ production, which in general may couple differently to t_L and t_R, is a change in the polarisation of the produced top (anti)quarks.

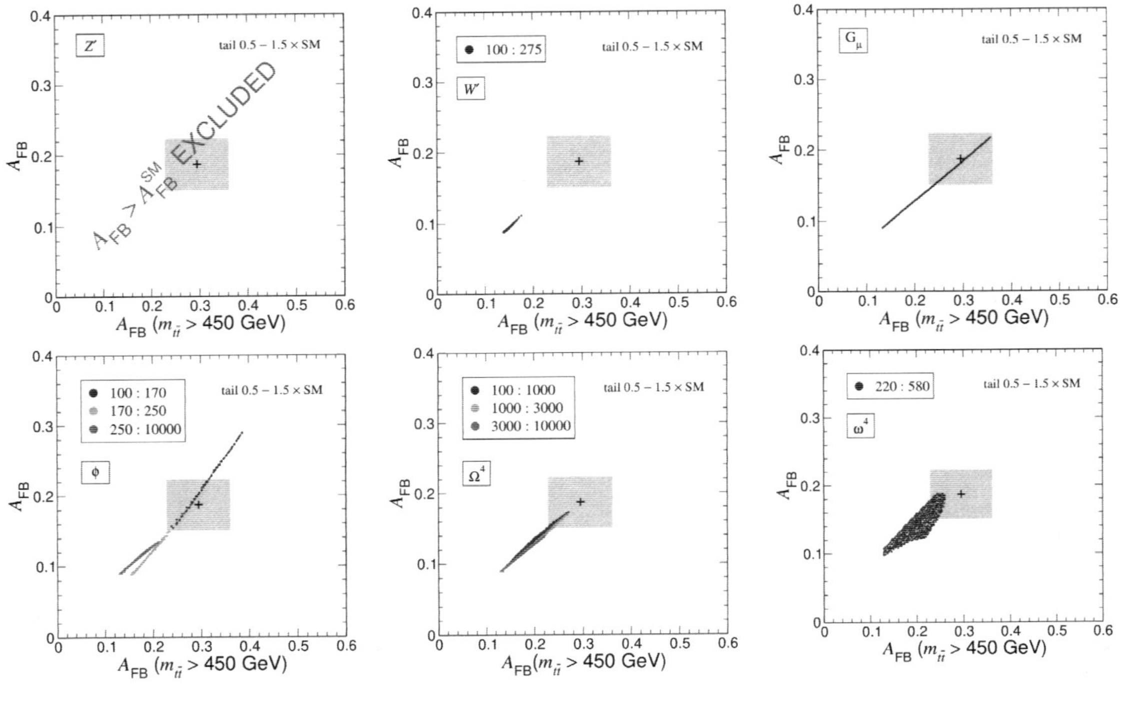

Figure 8. Inclusive versus high-mass asymmetries at the Tevatron, for several new physics models, imposing a tight constraint on the LHC high-mass tail (see the text).The numbers in the legends indicate the mass range for the new particle, in GeV. The crosses correspond to the experimental measurements in in Fig. 1, with the shaded boxes corresponding to the 1σ uncertainty.

The double angular distribution for the production of a $t\bar{t}$ pair is

$$\frac{1}{\sigma}\frac{d\sigma}{d\cos\theta_t\, d\cos\theta_{\bar{t}}} = \frac{1}{4}\left[1 + B_t\cos\theta_t + B_{\bar{t}}\cos\theta_{\bar{t}} + C\cos\theta_t\cos\theta_{\bar{t}}\right] . \tag{7}$$

with θ_t, $\theta_{\bar{t}}$ the angles between the top (antitop) momenta in the zero momentum frame (ZMF) with respect to some chosen spin axes [42, 43]. The constants B_t, $B_{\bar{t}}$ correspond to the polarisation of the top and antitop, respectively.[5] In the SM they vanish at the tree level —that is, top (anti-)quarks are produced unpolarised— due to the vector structure of the QCD coupling, and they are small at higher orders. The C constant measures the spin correlation between the top and antitop, and is nonzero for a suitable choice of spin axes.

The spin correlation coefficient C has been measured at the Tevatron, in the "beamline" basis [44–46] (Fig. 9, upper left panel), giving a naive average of $C = 0.68 \pm 0.26$, in good agreement with the SM prediction $C_{\rm SM} = 0.79$ [43]. Defining $\Delta C = C - C_{\rm SM}$, the experimental measurement $\Delta C = -0.11 \pm 0.26$ already sets some constraints on the possible contributions to A_{FB} from new physics [47] (Fig. 9, upper right panel).

At the LHC, the spin correlation has been determined in the helicity basis [48, 49], with an average measurement $C = 0.32 \pm 0.06$ (Fig. 9, middle left panel). Defining again $\Delta C = C - C_{\rm SM}$, with $C_{\rm SM} = 0.31$ [43], this average corresponds to $\Delta C = 0.01 \pm 0.06$. The measurement of

[5] Provided CP is conserved, $B_{\bar{t}} = -B_t$ if the same axis is chosen to measure the top and antitop spins. Choosing the helicity basis for each quark, $B_t = B_{\bar{t}}$.

Journal of Physics: Conference Series **447** (2013) 012015 doi:10.1088/1742-6596/447/1/012015

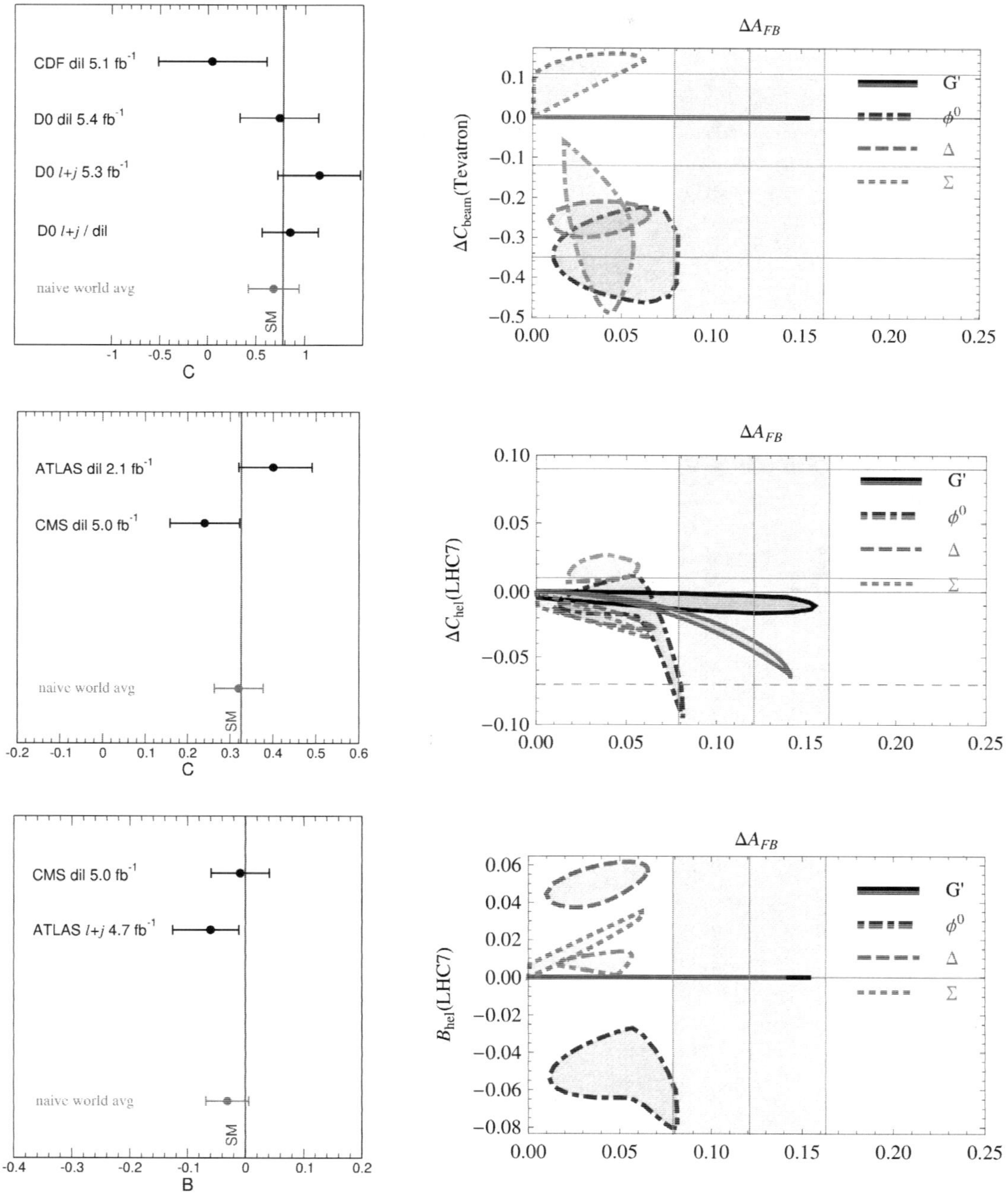

Figure 9. Left: experimental measurements of spin observables. Right (taken from Ref. [47]): theoretical predictions within several new physics models. The horizontal axis represents the increase in A_{FB} from new physics sources and the vertical axis the increase in the corresponding (B or C) coefficient, with respect to the SM value. The legends Δ, Σ correspond to ω^4, Ω^4 in our notation.

ΔC at the LHC has little effect on the allowed parameter space for models reproducing the Tevatron A_{FB} (Fig. 9, middle right panel). More restrictive is the measurement of the top polarisation, *i.e.* the B_t parameter [50, 51] (Fig. 9, lower left panel). The average of the ATLAS and CMS measurements, $B = -0.03 \pm 0.04$, disfavours at the $\sim 2\sigma$ level the explanation of the asymmetry by colour sextet and triplet scalars, which couple to t_R and predict a positive top quark polarisation (Fig. 9, lower right panel).

For completeness, it is worth mentioning that lepton-based FB asymmetries have also been measured at the Tevatron [2, 52],

$$\begin{aligned} A_{FB}^{\ell} &= \frac{N(Q\cdot\eta>0)-N(Q\cdot\eta<0)}{N(Q\cdot\eta>0)+N(Q\cdot\eta<0)}\,, \\ A^{\ell\ell} &= \frac{N(\Delta\eta>0)-N(\Delta\eta<0)}{N(\Delta\eta>0)+N(\Delta\eta<0)}\,, \end{aligned} \tag{8}$$

with η the rapidity of the charged leptons and Q their charge. These asymmetries include information from the $t\bar{t}$ FB asymmetry and the top polarisation [53, 54] —the latter is found in agreement with the SM— and provide a complementary experimental handle to probe new contributions to $t\bar{t}$ production. The measurements of A_{FB}^{ℓ} (see Fig. 10) are more precise and exhibit a positive excess with respect to SM predictions, whereas the only measurement of the dilepton asymmetry, $A^{\ell\ell} = 0.053 \pm 0.084$, is compatible with the SM prediction $A_{\text{SM}}^{\ell\ell} = 0.062$ [7].

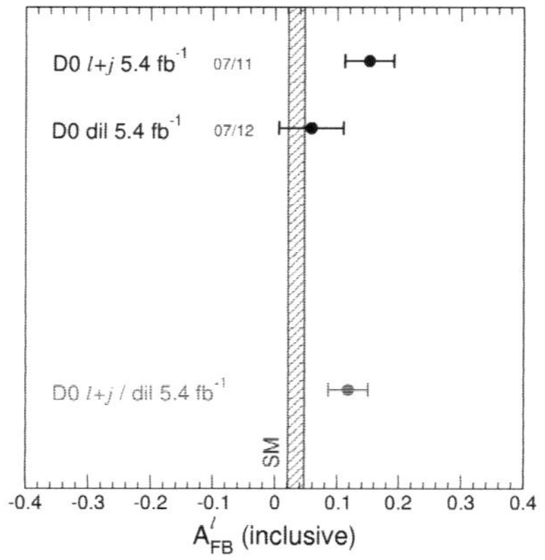

Figure 10. Measurements of the lepton asymmetry A_{FB}^{ℓ} at the Tevatron.

7. Discussion

More than two years after the measurement [8] that trigged the wide interest in the FB asymmetry at the Tevatron, there have been various new measurements, updated SM predictions and plenty of proposals for new physics explanations. But the question still remains whether the Tevatron anomaly corresponds to new physics, an unknown higher-order SM correction, or some kind of systematic effect.

The first possibility, of new physics in $t\bar{t}$ production, is certainly exciting, but it is not clear which form this new physics could have. Among the six simple models that we considered as candidates to explain the excess in the Tevatron asymmetry, only two of them (a colour octet and a scalar doublet) have survived after imposing just constraints from other observables in $t\bar{t}$ production at the Tevatron and LHC. But, in addition, there are some other measurements that put the minimal implementation of these models in trouble.

Journal of Physics: Conference Series **447** (2013) 012015 doi:10.1088/1742-6596/447/1/012015

- Light gluons mediate dijet pair [55] and four-top production [56], neither observed. The former constitute a serious constraint for lighter masses, which can be softened by assuming a large gluon width —due for example to decays into some new particle. The latter may become relevant for a wide mass range with increased luminosity. In addition, there are constraints from low-energy B physics [57, 58] and electroweak precision data [59] that are very model-dependent.
- For scalar doublets there are also stringent constraints from B physics [60], atomic parity violation [61] and possibly from associated production with a top quark.

These difficulties, in any case, may just reflect our current inability to propose a compelling model that explains the asymmetry measurements at the Tevatron and the LHC, fulfilling the constraints from other $t\bar{t}$ observables and collider data without ad-hoc assumptions and model fine-tuning. In this regard, we have stressed that a Tevatron A_{FB} excess and a small A_C at the LHC are compatible in general. By using Eqs. (6) and considering A_u, A_d as free independent parameters ranging between -1 and 1, one can obtain predictions for the correlated asymmetries A_{FB}, A_C within each $m_{t\bar{t}}$ bin. This is shown in Fig. 11. A positive excess at the Tevatron is compatible with a zero, or even negative, asymmetry at the LHC, provided there is some cancellation between the contributions from A_u and A_d.

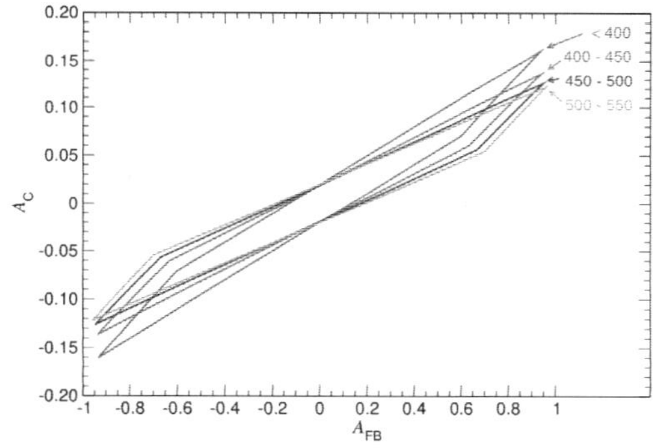

Figure 11. Model-independent prediction for the correlation between A_C and A_{FB} in several $m_{t\bar{t}}$ bins, based on the collider-independent asymmetries.

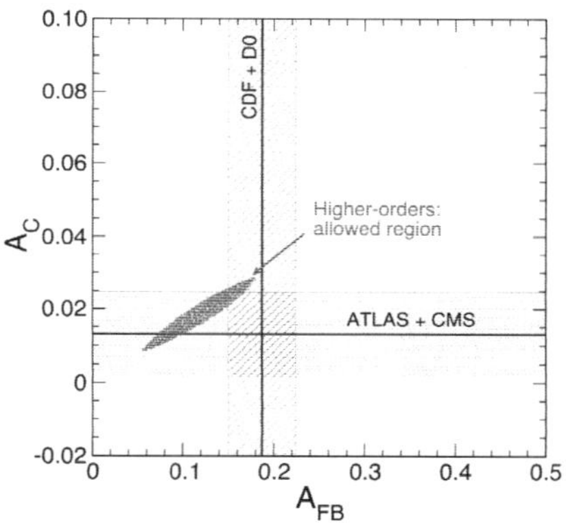

Figure 12. Estimated correlation between A_C and A_{FB} from higher-order QCD effects, based on the collider-independent asymmetries.

Explanations of the Tevatron excess by higher-order QCD effects are unlikely to fit data well, if the trend of current measurements persists. A simple reckoning indicates that, if there were large missing QCD corrections that shifted A_{FB} to its Tevatron average, they would also shift A_C away from the LHC average. In the absence of a proper next-to-next-to-leading order calculation, one can give a back-of-the-envelope estimate for this correlation by using again Eqs. (6) and varying A_u, A_d around the NLO SM value.[6] The resulting prediction for A_C versus A_{FB} in Fig. 12 shows that one cannot simultaneously fit the central values of Tevatron and LHC asymmetries with this kind of corrections to the SM. A similar type of argument is expected to hold for other QCD-based explanations of the anomalies [62, 63].

The third option, of unknown systematic errors in Tevatron or LHC experiments, is hard to understand since the two experiments at each collider provide similar results. Still, unknown systematics are unknown by definition, and little can be said in this direction at the moment.

[6] We thank W. Bernreuther and Z.-G. Si for providing us with these data.

In summary, the A_{FB} puzzle is far from being solved. There are still good hopes that there is some type of new physics in $t\bar{t}$ production, which might (or might not) be visible by precision measurements of the LHC charge asymmetry, the $t\bar{t}$ differential distributions and the top polarisation. Fortunately, the upcoming LHC and D0 measurements will provide important information to help approach a solution.[7]

Acknowledgements
This work has been supported by MICINN by projects FPA2006-05294 and FPA2010-17915, Junta de Andalucía (FQM 101, FQM 03048 and FQM 6552) and Fundação para a Ciência e Tecnologia (FCT) project CERN/FP/123619/2011.

References

[1] Aaltonen T *et al.* (CDF Collaboration) 2012 *Preprint* arXiv:1211.1003 [hep-ex].
[2] Abazov V M *et al.* (D0 Collaboration) 2011 *Phys. Rev.* D **84** 112005.
[3] Campbell J M and Ellis R K 2012 *Preprint* arXiv:1204.1513 [hep-ph].
[4] Ahrens V, Ferroglia A, Neubert M, Pecjak B D and Yang L L 2011 *Phys. Rev.* D **84** 074004.
[5] Hollik W and Pagani D 2011 *Phys. Rev.* D **84** 093003.
[6] Kuhn J H and Rodrigo G 2012 *JHEP* **1201** 063.
[7] Bernreuther W and Si Z G 2012 *Phys. Rev.* D **86** 034026.
[8] Aaltonen T *et al.* (CDF Collaboration) 2011 *Phys. Rev.* D **83** 112003.
[9] Aguilar-Saavedra J A and Pérez-Victoria M 2011 *JHEP* **1105** 034.
[10] Djouadi A, Moreau G, Richard G and Singh R K 2010 *Phys. Rev.* D **82**, 071702.
[11] Jung S, Murayama H, Pierce A and Wells J D 2010 *Phys. Rev.* D **81**, 015004.
[12] Cheung K, Keung W Y and Yuan T C 2009 *Phys. Lett.* B **682** 287.
[13] Nelson A E, Okui T and Roy T S 2011 *Phys. Rev.* D **84**, 094007.
[14] Shu J, Tait T M P and Wang K 2010 *Phys. Rev.* D **81**, 034012.
[15] Gabrielli E and Raidal M 2011 *Phys. Rev.* D **84** 054017.
[16] Aliev M, Lacker H, Langenfeld U, Moch S, Uwer P and Wiedermann M 2011 *Comput. Phys. Commun.* **182** 1034.
[17] Aaltonen T *et al.* (CDF Collaboration) 2012 *CDF note 10926*; Abazov V M *et al.* (D0 Collaboration) 2012 *D0 Note 6363-CONF.*
[18] Grinstein B, Kagan A L, Trott M and Zupan J 2011 *Phys. Rev. Lett.* **107** 012002.
[19] Aguilar-Saavedra J A and Pérez-Victoria M 2011 *JHEP* **1109** 097.
[20] Aguilar-Saavedra J A 2011 *Nucl. Phys.* B **843** 638
[21] Degrande C, Gerard J M, Grojean C, Maltoni F and Servant G 2011 *JHEP* **1103** 125.
[22] Aguilar-Saavedra J A and Pérez-Victoria M 2011 2011 *Phys. Lett.* B **701** 93.
[23] Aguilar-Saavedra J A and Pérez-Victoria M 2011 *Phys. Lett.* B **705** 228.
[24] Aaltonen T *et al.* (CDF Collaboration), *CDF note 10947.*
[25] Gresham M I, Kim I W and Zurek K M 2011 *Phys. Rev.* D **83** 114027.
[26] Diener R, Godfrey A and Martin T A W 2009 *Phys. Rev.* D **80** 075014.
[27] Frixione S and Webber B R 2002 *JHEP* **0206** 029.
[28] Aguilar-Saavedra J A and Juste A 2012 *Phys. Rev. Lett.* **109** 211804.
[29] Aguilar-Saavedra J A and Pérez-Victoria M 2011 *Phys. Rev.* D **84** 115013.
[30] Ko P, Omura Y and Yu C 2012 *Preprint* arXiv:1205.0407 [hep-ph].
[31] Drobnak J, Kamenik J F and Zupan J 2012 *Phys. Rev.* D **86** 054022.
[32] Drobnak J, Kagan A L, Kamenik J F, Perez G and Zupan J 2012 *Phys. Rev.* D **86** 094040.
[33] Alvarez E and Leskow E C 2012 *Phys. Rev.* D **86** 114034.
[34] Aguilar-Saavedra J A, Bernreuther W and Si Z G 2012 *Phys. Rev.* D **86** 115020.
[35] Aguilar-Saavedra J A, Juste A and Rubbo F 2012 *Phys. Lett.* B **707** 92.
[36] Delaunay C, Gedalia O, Hochberg Y, Perez G and Soreq Y 2011 *JHEP* **1108** 031.
[37] The ATLAS Collaboration 2012 *Preprint* arXiv:1207.5644 [hep-ex].
[38] The CMS Collaboration 2012 *Preprint* arXiv:1211.2220 [hep-ex].
[39] Barcelo R, Carmona A, Masip M and Santiago J 2012 *Phys. Lett.* B **707** 88.
[40] Marques Tavares G and Schmaltz M 2011 *Phys. Rev.* D **84** 054008.

[7] Definitely, *"When you have eliminated the impossible, whatever remains, however improbable, must be the truth."* [64].

[41] Krnjaic G Z 2012 *Phys. Rev.* D **85** 014030.
[42] Bernreuther W, Brandenburg A, Si Z G and Uwer P 2004 *Nucl. Phys.* B **690** 81.
[43] Bernreuther W and Si Z G 2010 *Nucl. Phys.* B **837** 90.
[44] Aaltonen T *et al.* (CDF Collaboration), *CDF note 10719.*
[45] Abazov V M *et al.* (D0 Collaboration) 2011 *Phys. Rev. Lett.* **107** 032001,
[46] Abazov V M *et al.* (D0 Collaboration) 2012 *Phys. Rev. Lett.* **108** 032004.
[47] Fajfer S, Kamenik J F and Melic B 2012 *JHEP* **1208** (2012) 114.
[48] The ATLAS Collaboration 2012 *Phys. Rev. Lett.* **108** 212001.
[49] The CMS Collaboration 2012 *CMS-PAS-TOP-12-004.*
[50] The CMS Collaboration 2012 *CMS-PAS-TOP-12-016.*
[51] The ATLAS Collaboration 2012 *ATLAS-CONF-2012-133.*
[52] Abazov V M *et al.* (D0 Collaboration) 2012 *Preprint* arXiv:1207.0364 [hep-ex].
[53] Berger E L, Cao Q H, Chen C R, Yu J H and Zhang H 2012 *Phys. Rev. Lett.* **108** 072002.
[54] Falkowski A, Mangano M L, Martin A, Perez G and Winter J 2012 *Preprint* arXiv:1212.4003 [hep-ph].
[55] Gross C, Marques Tavares G, Schmaltz M and Spethmann C 2013 *Phys. Rev.* D **87** 014004.
[56] Aguilar-Saavedra J A and Santiago J 2012 *Phys. Rev.* D **85** 034021.
[57] Bai Y, Hewett J L, Kaplan J and Rizzo T G 2011 *JHEP* **1103** 003.
[58] Ipek S 2013 *Preprint* arXiv:1301.3990 [hep-ph].
[59] Gresham M I, Shelton J and Zurek K M 2012 *Preprint* arXiv:1212.1718 [hep-ph].
[60] Blum J Hochberg Y and Nir Y 2011 *JHEP* **1110** 124.
[61] Gresham M I, Kim I W, Tulin S and Zurek K M 2012 *Phys. Rev.* D **86** 034029.
[62] Skands P, Webber B and Winter J 2012 *JHEP* **1207** 151.
[63] Brodsky S J and Wu X G 2012 *Phys. Rev.* D **86** 014021.
[64] Doyle A C 1890 *The Sign of Four* (London: Spencer Blackett)

Journal of Physics: Conference Series **447** (2013) 012016 doi:10.1088/1742-6596/447/1/012016

Violation of CPT Invariance in the Early Universe and Leptogenesis/Baryogenesis

Nick E. Mavromatos

Theoretical Particle Physics and Cosmology Group, Department of Physics, King's College London, Strand, London WC2R 2LS, UK.
Also currently at: CERN, Theory Division, Physics Department, Geneva 23 CH 1211, Switzerland.

E-mail: Nikolaos.Mavromatos@kcl.ac.uk

Abstract. In this talk, I review some plausible scenarios entailing violation of CPT symmetry in the early Universe, due to space-time backgrounds which do not respect some of the assumptions for the validity of the CPT theorem (here considered will be Lorentz invariance and/or Unitarity). The key point in all these models is that the background induces different populations of fermions as compared to antifermions, and hence CPT Violation (CPTV), already in thermal equilibrium. Such populations may freeze out at various conditions depending on the details of the underlying microscopic model, thereby leading to leptogenesis and baryogenesis. Among the considered scenarios is a stringy one, in which the CPTV is associated with a cosmological background with torsion provided by the Kalb-Ramond antisymmetric tensor field (axion) of the string gravitational multiplet. We also discuss briefly (Lorentz Violating) CPTV models that go beyond the *local* effective lagrangian framework, such as a stochastic Finsler metric and D-particle foam. In the latter, the CPTV is induced by the interactions of stringy matter with populations of stochastically fluctuating point-like space-time defects (D-particles) that can be encountered in string/brane Cosmologies. The preferential rôle of neutrinos is singled out in this latter scenario as a consequence of gauge symmetry (charge) conservation.

1. Introduction and Summary

The universe is overwhelmingly made up of matter rather than anti-matter. According to the standard Big Bang theory, matter and antimatter have been created in equal amounts in the early universe. However, the observed charge-parity (CP) violation in particle physics [1], prompted A. Sakharov [2] to conjecture that for Baryon Asymmetry in the universe (BAU) we need:

- Baryon number violation to allow for states with $\Delta B \neq 0$ starting from states with $\Delta B = 0$ where ΔB is the change in baryon number.
- If C or CP conjugate processes to a scattering process were allowed with the same amplitude then baryon asymmetry would disappear. Hence C and CP need to be broken.
- Chemical equilibrium does not permit asymmetries. Hence Sakharov required that chemical equilibrium does not hold during an epoch in the early universe.

Hence non-equilibrium physics in the early universe together with baryon number (B), charge (C) and charge-parity (CP) violating interactions/decays of anti-particles in the early universe, may result in the observed BAU. In fact there are two types of non-equilibrium

Journal of Physics: Conference Series **447** (2013) 012016 doi:10.1088/1742-6596/447/1/012016

processes in the early universe that can produce this asymmetry: the first type concerns processes generating asymmetries between leptons and antileptons (*leptogenesis*), while the second produces asymmetries between baryons and antibaryons (*baryogenesis*). The observed asymmetry today, between baryon (n_B) and anti-baryon ($n_{\overline{B}}$) number densities is estimated in the Big-Bang theory [3] to imply:

$$\Delta n(T \sim 1 \text{ GeV}) = \frac{n_B - n_{\overline{B}}}{n_B + n_{\overline{B}}} \sim \frac{n_B - n_{\overline{B}}}{s} = (8.4 - 8.9) \times 10^{-11} \quad (1)$$

at the early stages of the expansion, e.g. for times $t < 10^{-6}$ s and temperatures $T > 1$ GeV. In the above formula s denotes the entropy density. Unfortunately, the observed CP violation within the Standard Model (SM) of particle physics (found to be of order $\epsilon = O(10^{-3})$ in the neutral Kaon experiments [1]) induces an asymmetry much less than that in (1) [4]. There are several ideas that go beyond the SM (*e.g.* grand unified theories, supersymmetry, extra dimensional models *etc.*) which involve the decays of right handed sterile neutrinos. For relevant important works on this see [5, 6, 7, 8, 9, 10, 11]. These ideas lead to extra sources for CP violation that could generate the observed BAU. Some degree of fine tuning and somewhat *ad hoc* assumptions are involved in such scenarios and the quest for an understanding of the observed BAU still needs further investigation. An example of fine tuning is provided by the choice of the hierarchy of the right-handed Majorana neutrino masses. For instance, enhanced CP violation, necessary for BAU, can be achieved in models with three Majorana neutrinos, by assuming two of these neutrinos are nearly degenerate in mass.

The requirement of non-equilibrium is on less firm ground [12] than the other two requirements of Sakharov, *e.g.* if the non-equilibrium epoch occurred prior to inflation then its effects would be hugely diluted by inflation. A basic assumption in the scenario of Sakharov is that *CPT symmetry* [13] (where T denotes time reversal operation) holds in the very early universe which leads to the production of matter and antimatter in equal amounts. Such *CPT invariance* is a cornerstone of all known *local* effective *relativistic* field theories without gravity, which current particle-physics phenomenology is based upon. It should be noted that the necessity of non-equilibrium processes in CPT invariant theories can be dropped if the requirement of CPT is relaxed [14]. This violation of CPT (denoted by CPTV) may be the result of a breakdown of Lorentz symmetry (which might happen at ultrahigh energies [15]) or of Unitarity, in the sense of quantum-gravity-induced decoherence in the matter sector, which might be strong in the Early Universe [16, 17][1]. For many models with CPTV, in the time line of the expanding universe, CPTV generates first lepton asymmetries (*leptogenesis*); subsequently through sphaleron processes [21] or Baryon-Lepton (B-L) number conserving processes in Grand Unified Theories (GUT), the lepton asymmetry can be communicated to the baryon sector to produce the observed BAU. Thus, CPTV in the early universe may also obviate the need for including extra sources of CP violation, such as sterile neutrinos and/or supersymmetry, in order to obtain the observed BAU.

The structure of the talk, which is based on work appeared in [22], will be as follows: in the next section 2 we shall briefly review some relevant existing models for fermionic asymmetry, which entail CPTV-induced differences in the dispersion relations between particles and antiparticles propagating in curved gravitational backgrounds in the early universe. In section 3, we review a new model for gravitational leptogenesis, proposed in [22], which follows broadly an earlier framework [23, 24], but differs crucially in that the full gravitational multiplet [25] that arises in string theory is used. The leptogenesis in this model is due to CPTV dispersion

[1] Locality violations, through non-local mass terms [18], have also been considered in models of CPTV in the neutrino sector but will not be discussed here, due to limitations in space. It must be stressed, though, that such models of CPTV are independent from those due to Lorentz and Unitarity violation [19, 20].

relations between fermions/antifermions, induced by the (constant) *torsion* associated with the antisymmetric Kalb-Ramond tensor field in the gravitational multiplet of the string. In section 4, we consider another scenario for gravitational leptogenesis that involves a non-Riemannian Finsler metric, with stochastically fluctuating parameters, a variant of which appears in our string/brane ("*D-foam*") model in the following section 5. In section 5 we present our D-foam model for the stringy universe and the induced CPTV and leptogenesis/baryogenesis. The model is characterised by CPTV and matter dominance over antimatter *naturally*, without the need for an adjustment of the sign of the lepton/antilepton asymmetry. This is an interesting feature of our model which differentiates it from earlier proposals on gravitational leptogenesis/baryogenesis [26, 27, 28, 29, 24, 23, 30, 31, 32], where the sign of the asymmetry is implicitly chosen. In common with earlier discussions of CPTV in baryogenesis [33], this class of models involves violation of Lorentz symmetry, but only because of a non-zero *variance* of the stochastic parameter. Moreover, neutrinos are singled out as interacting predominantly with the D-foam, thereby feeling the CPTV effects of the background most strongly, as a result of charge conservation. This is an exclusively stringy feature that goes beyond local effective field theories.

2. Models of Background-induced CPT Violation (CPTV) in the Early Universe

In this section we shall review some existing models of CPTV induced asymmetry between matter and antimatter in the early universe, which can be contrasted with our approach in sections 3-5 later in the article. We shall be brief in our exposition, referring the interested reader to the relevant literature for more details.

2.1. CPTV Models with Particle-Antiparticle Mass Difference

The simplest possibility [34] for inducing CPTV in the early universe is through particle-antiparticle mass differences $m \neq \overline{m}$. These would affect the particle phase-space distribution function $f(E,\mu)$, $f(E,\mu) = [\exp(E-\mu)/T) \pm 1]^{-1}$, $E^2 = \vec{p}^2 + m^2$, and antiparticle phase-space distribution function $f(\overline{E},\bar{\mu}) = [\exp(\bar{E}-\bar{\mu})/T) \pm 1]^{-1}$, $\bar{E}^2 = \vec{p}^2 + \bar{m}^2$, with $\vec{p}$ being the 3−momentum. (Our convention will be that an overline over a quantity will refer to an antiparticle, + will denote a fermionic (anti-)particle and − will denote a bosonic (anti-)particle). (Anti)particles will have, by definition, (negative) positive energies. Mass differences between particles and antiparticles, $\overline{m} - m \neq 0$, generate a matter-antimatter asymmetry in the relevant number densities n and $\overline{n}$ $n - \overline{n} = g_{d.o.f.} \int \frac{d^3p}{(2\pi)^3}[f(E,\mu) - f(\overline{E},\bar{\mu})]$, where $g_{d.o.f.}$ denotes the number of degrees of freedom of the particle species under study. In the case of spontaneous Lorentz violation [33] there is a vector field A_μ with a non-zero time-like expectation value which couples to a global current J^μ such as baryon number through an interaction lagrangian density

$$\mathcal{L} = \lambda A_\mu J^\mu. \tag{2}$$

This leads to $m \neq \bar{m}$ and $\mu \neq \bar{\mu}$. Alternatively, following [34] we can make the assumption that the dominant contributions to baryon asymmetry come from quark-antiquark mass differences, and that their masses "run" with the temperature i.e. $m \sim gT$ (with g the QCD coupling constant). One can provide estimates for the induced baryon asymmetry on noting that the maximum quark-antiquark mass difference is bounded by the current experimental bound on the proton-antiproton mass difference, $\delta m_p (= |m_p - \overline{m}_p|)$, known to be less than $2 \cdot 10^{-9}$ GeV. Taking $n_\gamma \sim 0.24\,T^3$ (the photon equilibrium density at temperature T) we have [34]:

$$\beta_T = \frac{n_B}{n_\gamma} = 8.4 \times 10^{-3}\,\frac{m_u\,\delta m_u + 15 m_d\,\delta m_d}{T^2}\ , \quad \delta m_q = |m_q - \overline{m}_q|\ . \tag{3}$$

β_T is too small compared to the observed one. To reproduce the observed $\beta_{T=0} \sim 6 \cdot 10^{-10}$ one would need $\delta m_q(T = 100\text{ GeV}) \sim 10^{-5} - 10^{-6}$ GeV $\gg \delta m_p$, which is somewhat unnatural.

Journal of Physics: Conference Series **447** (2013) 012016 doi:10.1088/1742-6596/447/1/012016

However, active (*light*) neutrino-antineutrino mass differences alone may reproduce BAU; some phenomenological models in this direction have been discussed in [35], considering, for instance, particle-antiparticle mass differences for active neutrinos compatible with current oscillation data. This leads to the result

$$n_B = n_\nu - n_{\overline{\nu}} \simeq \frac{\mu_\nu T^2}{6} \tag{4}$$

yielding $n_B/s \sim \frac{\mu_\nu}{T} \sim 10^{-11}$ at $T \sim 100$ GeV, in agreement with the observed BAU. (Here s, n_ν, and μ_ν are the entropy density, neutrino density and chemical potential respectively.)

2.2. CPTV Decoherence Models

Particle-antiparticle mass differences however may not be the only way by which CPT is violated. As discussed in [16, 17], quantum gravity fluctuations in the structure of space-time, may be strong in the early universe; the fluctuations may act as an *environment* inducing decoherence for the (anti-)neutrinos. However the couplings between the particles and the environment are different for the neutrino and antineutrino sectors. Once there is decoherence for an observer with a low energy (compared to the Planck scale $M_P \sim 10^{19}$ GeV), the effective CPT symmetry generator may be *ill-defined* as a quantum mechanical operator, according to a theorem by R. Wald [36], leading to an intrinsic violation of CPT symmetry. This type of violation may characterise models of quantum gravity with stochastic space-time fluctuations due, for instance, to gravitational space-time defects, as is the case of certain brane models [37, 38, 39]. In such a case, a slight mismatch in the strength of the stochastic space-time fluctuations between particle and antiparticle sectors, can lead to different decoherence parameters to describe the interaction of the gravitational environment with matter.

In [16, 17], simple models of Lindblad decoherence [40], conjectured to characterise quantum-gravity-induced CPTV decoherent situations [41, 42], have been considered for neutrinos [43]. It was assumed on phenomenological grounds, that non-trivial decoherence parameters were *only* present in the antiparticle sector: this is consistent with the lack of any experimental evidence to date [44, 45, 46] for vacuum decoherence in the particle sector. The antineutrino decoherence parameters (with dimension of energy) had a mixed energy dependence. The model of [16, 17] assumes a diagonal $\mathcal{L}_{\mu\nu}$. A diagonal Lindblad decoherence matrix for three-generation neutrinos requires eight coefficients $\overline{\gamma}_i$. Some of the eight coefficients were assumed for simplicity in [16, 17] to be proportional to the antineutrino energy $\overline{\gamma}_i = \frac{T}{M_P} E \;, \quad i = 1, 2, 4, 5$, while the remaining (subdominant) ones were inversely proportional to it $\overline{\gamma}_j = \frac{10^{-24}\,(\mathrm{GeV})^2}{E} \;, \; j = 3, 6, 7, 8$. The model was proposed without any microscopic justification; its choice was originally motivated by fitting the LSND "anomalous data" in the antineutrino sector [47] with the rest of the neutrino data. and this required T to be $T/M_P \sim 10^{-18}$,i.e. in the temperature range of electroweak symmetry breaking. One can derive [16, 17] an active (light) $\nu - \overline{\nu}$ asymmetry of order

$$\mathcal{A} = \frac{|n_\nu - n_{\overline{\nu}}|}{n_\nu + n_{\overline{\nu}}} = \frac{\overline{\gamma}_1}{\sqrt{\Delta m^2}} = \frac{T}{M_P} \cdot \frac{E}{\sqrt{\Delta m^2}} \;, \tag{5}$$

where Δm^2 denotes the (atmospheric) neutrino mass squared difference, which plays the role of a characteristic low mass scale in the problem. This lepton number violation is communicated to the baryon sector by means of baryon number (B) plus lepton number (L) conserving sphaleron processes (in fact, in this case one needs an antilepton excess in order to produce a baryon excess). These processes lead to an estimate [16] for the current value of B to be

$$B = \frac{n_\nu - n_{\overline{\nu}}}{s} \sim \mathcal{A}\,\frac{n_\nu}{g^\star\, n_\gamma} \tag{6}$$

with n_γ the photon number density, $g^\star$ the effective number of degrees of freedom (at the temperature where the asymmetry developed, i.e. the electroweak symmetry breaking temperature in the model of [16]). $g^\star$ depends on the matter content of the model (with a typical range $g^\star \in [10^2 - 10^3]$). For such parameter values $\mathcal{A} \sim 10^{-6}$ and so the observed BAU may be reproduced in this case without the need for extra sources of CP violation e.g. sterile neutrinos. Such models, however, do not provide an underlying microscopic understanding. In particular there is missing an understanding of the preferential role of the neutrino compared to other particles of the Standard Model in the CPT violating decoherence process. Within some microscopic models of space-time foam, involving populations of point-like brane defects (D-particles) puncturing three(or higher)-spatial dimension brane worlds [37, 38, 39], such a preferred role may be justified as we shall discuss in section 5.

2.3. CPTV-induced by Curvature effects in Background Geometry

Although the role of gravity was alluded to in the last subsection, associated features of space-time were not discussed. In the literature the role of gravity has been explicitly considered within a local effective action framework which is essentially that of (2) A coupling to scalar curvature R [27, 23, 30, 28] through a CP violating interaction Lagrangian $\mathcal{L}$:

$$\mathcal{L} = \frac{1}{M_*^2} \int d^4x \sqrt{-g}\, (\partial_\mu R)\, J^\mu \tag{7}$$

where M_* is a cut-off in the effective field theory and J^μ could be the current associated with baryon (B) number. There is an implicit choice of sign in front of the interaction (7), which has been fixed so as to ensure matter dominance.

It has been shown that [27] $\frac{n_{B-L}}{s} = \frac{\dot{R}}{M_*^2 T_d}$, T_d being the freeze-out temperature for $B - L$ interactions. The idea then is that this asymmetry can be converted to baryon number asymmetry provided the $B + L$ electroweak sphaleron interaction has not frozen out. To leading order in M_*^{-2} we have $R = 8\pi G(1 - 3w)\rho$ where ρ is the energy density of matter and the equation of state is $p = w\rho$ where p is pressure. For radiation $w = 1/3$ and so in the radiation dominated era of the Friedmann-Robertson-Walker cosmology $R = 0$. However w is precisely $1/3$ when $T^\mu_\mu = 0$. In general $T^\mu_\mu \propto \beta(g) F^{\mu\nu} F_{\mu\nu}$ where $\beta(g)$ is the beta function of the running gauge coupling g in a $SU(N_c$ gauge theory with N_c colours. This allows $w \neq 1/3$. Further issues in this approach can be found in [27, 23, 30, 28].

Another approach involves an axial vector current [24, 29, 31, 32] instead of J_μ. The scenario is based on the well known fact that fermions in curved space-times exhibit a coupling of their spin to the curvature of the background space-time.The Dirac Lagrangian density of a fermion can be re-written as:

$$\mathcal{L} = \sqrt{-g}\, \overline{\psi} \left(i\gamma^a \partial_a - m + \gamma^a \gamma^5 B_a\right) \psi \,, \quad B^d = \epsilon^{abcd} e_{b\lambda} \left(\partial_a e^\lambda_{\ c} + \Gamma^\lambda_{\nu\mu} e^\nu_{\ c} e^\mu_{\ a}\right) \,, \tag{8}$$

in a standard notation, where $e^\mu_{\ a}$ are the vielbeins, $\Gamma^\mu_{\ \alpha\beta}$ is the Christoffel connection and Latin (Greek) letters denote tangent space (curved space-time) indices. The space-time curvature background has, therefore, the effect of inducing an "axial" background field B_a which can be non-trivial in certain anisotropic space-time geometries, such as Bianchi-type cosmologies or in regions of space-time near rotating (Kerr) primordial black holes [24, 29, 31, 32]. For an application to particle-antiparticle asymmetry it is necessary for this axial field B_a to be a constant in some local frame. The existence of such a frame has not been demonstrated. As before if it can be arranged that $B_a \neq 0$ for $a = 0$ then for constant B_0 CPT is broken: the dispersion relation of neutrinos in such backgrounds differs from that of antineutrinos. Explicitly we have

$$E = \sqrt{(\vec{p} - \vec{B})^2 + m^2} + B_0 \,, \quad \overline{E} = \sqrt{(\vec{p} + \vec{B})^2 + m^2} - B_0 \,. \tag{9}$$

Journal of Physics: Conference Series **447** (2013) 012016 doi:10.1088/1742-6596/447/1/012016

The relevant neutrino asymmetry emerges on following the same steps used when there was an explicit particle-antiparticle mass difference, As a consequence the following neutrino-antineutrino density difference is found in Bianchi II Cosmologies [24, 29, 31, 32]:

$$\Delta n_\nu \equiv n_\nu - n_{\overline{\nu}} \sim g^\star T^3 \left(\frac{B_0}{T}\right) \tag{10}$$

with $g^\star$ the number of degrees of freedom for the (relativistic) neutrino. An excess of particles over antiparticles is predicted only when $B_0 > 0$, which had to be assumed in the analysis of [24, 29, 31, 32]; we should note, however, that the sign of B_0 and its constancy have not been justified in this phenomenological approach[2].

At temperatures $T < T_d$, with T_d the decoupling temperature of the lepton-number violating processes, the ratio of the net Lepton number ΔL (neutrino asymmetry) to entropy density (which scales as T^3) remains constant,

$$\Delta L(T < T_d) = \frac{\Delta n_\nu}{s} \sim \frac{B_0}{T_d} \tag{11}$$

which, for $T_d \sim 10^{15}$ GeV and $B_0 \sim 10^5$ GeV, implies a lepton asymmetry (leptogenesis) of order $\Delta L \sim 10^{-10}$, in agreement with observations. The latter can then be communicated to the baryon sector to produce the observed BAU (baryogenesis) by a B-L conserving symmetry in the context of either Grand Unified Theories (GUT) [24], or sphaleron processes in the standard model.

Before closing this section we also mention that the splitting (9) of the energy levels between neutrinos and antineutrinos may also lead to neutrino-antineutrino mixing and thus oscillations (in a manner envisaged originally by Pontecorvo) or affect the magnitude of the neutrinoless double-beta decay amplitudes [31, 32]. These effects can lead to additional constraints on B^0, since, *e.g.* the neutrino-antineutrino oscillations may affect the relative abundance of different neutrino flavours in the Universe. Indeed, the increased oscillation length, predicted [32] due to the gravitational CPTV effects in early epochs of the Universe, may lead to higher production of muon neutrinos than that estimated without the gravitational effect, with profound consequences for early Universe cosmology, unless B^0 is appropriately constrained. Moreover, the neutrinoless double-beta decay amplitudes are modified in the presence of B^0 as: $\mathcal{A} \propto \sqrt{(B^0)^2 + m_{\nu_e}^2}$, where m_{ν_e} is the electron-neutrino Majorana mass.

3. CPTV-induced in (String-Inspired) Background Geometry with Torsion

In this section we will discuss the case of a constant B^0 "axial" field that appears due to the interaction of the fermion spin with a string-theory background geometry with *torsion.* This is a novel observation, which (as far as we are aware) was discussed for first time in [22]. In the presence of torsion the Christoffel symbol contains a part that is antisymmetric in its lower indices: ${\Gamma^\lambda}_{\mu\nu} \neq {\Gamma^\lambda}_{\nu\mu}$. Hence the last term of the right-hand side of the Eqn.(8) is *not* zero. Since the torsion term is of gravitational origin it couples universally to all fermion species. The effect of the coupling to neutrinos will be clarified below.

The massless gravitational multiplet in string theory contains the dilaton (spin 0, scalar, the "trace" of graviton), Φ, the graviton (spin 2, symmetric tensor), $g_{\mu\nu}$, and the spin 1 antisymmetric tensor $B_{\mu\nu}$. The (Kalb-Ramond) field B appears in the string effective action only through its totally antisymmetric field strength, $H_{\mu\nu\rho} = \partial_{[\mu} B_{\nu\rho]}$, where $[\ldots]$ denotes antisymmetrization of

[2] The above considerations concern the dispersion relations for any fermion, not only neutrinos. However, when one considers matter excitations from the vacuum, as relevant for leptogenesis, we need chiral fermions to get non trivial CPTV asymmetries in *populations* of particle and antiparticles, because $< \psi^\dagger \gamma^5 \psi > = - < \psi_L^\dagger \gamma^5 \psi_L > + < \psi_R^\dagger \gamma^5 \psi_R >$.

the indices within the brackets. The calculation of string amplitudes [48] shows that $H_{\mu\nu\rho}$ plays the role of *torsion* in a generalised connection

$$\overline{\Gamma}^{\lambda}_{\ \ \mu\nu} = \Gamma^{\lambda}_{\ \ \mu\nu} + e^{-2\Phi} H^{\lambda}_{\mu\nu} \equiv \Gamma^{\lambda}_{\ \ \mu\nu} + T^{\lambda}_{\ \ \mu\nu} \, . \tag{12}$$

In ref. [49] exact solutions to the conformal invariance conditions (to all orders in α') of the low energy effective action of strings have been presented. In four "large" (uncompactified) dimensions of the string, the antisymmetric tensor field strength can be written uniquely as

$$H_{\mu\nu\rho} = e^{2\Phi} \epsilon_{\mu\nu\rho\sigma} \partial^{\sigma} b(x) \tag{13}$$

with $\epsilon_{0123} = \sqrt{g}$ and $\epsilon^{\mu\nu\rho\sigma} = |g|^{-1}\epsilon_{\mu\nu\rho\sigma}$, with g the metric determinant. The field $b(x)$ is a "pseudoscalar " *axion*-like field. The dilaton Φ and axion b fields are fields that appear as Goldstone bosons of spontaneously broken scale symmetries of the string vacua, and so are exactly massless classically. In the effective string action such fields appear only through their derivatives. In the "physical" *Einstein frame* , relevant for cosmological observations, the temporal component of the metric is normalised to $g_{00} = +1$ by an appropriate change of the time coordinate. In this setting, the solution of [49] leads to a Friedmann-Robertson-Walker (FRW) metric, with scale factor $a(t) \sim t$, where t is the FRW cosmic time. Moreover, the dilaton field Φ behaves as $-\ln t + \phi_0$, with ϕ_0 a constant, and the axion field $b(x)$ is linear in t. There is an underlying world-sheet conformal field theory with central charge $c = 4 - 12Q^2 - \frac{6}{n+2} + c_I$ where $Q^2 (> 0)$ is the central-charge deficit and c_I is the central charge associated with the world-sheet conformal field theory of the compact "internal" dimensions of the string model [49]. The condition of cancellation of the world-sheet ghosts that appear because of the fixing of reparametrisation invariance of world-sheet co-ordinates requires that $c = 26$. The solution for the axion field is

$$b(x) = \sqrt{2} e^{-\phi_0} \sqrt{Q^2} \frac{M_s}{\sqrt{n}} t \, , \tag{14}$$

where M_s is the string mass scale and n is a positive integer, associated with the level of the Kac-Moody algebra of the underlying world-sheet conformal field theory. For non-zero Q^2 there is an additional dark energy term in the effective target-space time action of the string [49] of the form $\int d^4x \sqrt{-g} e^{2\Phi}(-4Q^2)/\alpha'$. The linear axion field (14) *remains* a non-trivial solution *even* in the *static* space-time limit with a constant dilaton field [49]. In such a case space time is an Einstein universe with positive cosmological constant and constant positive curvature proportional to $6/(n+2)$. For the solutions of [49], the covariant torsion tensor $e^{-2\Phi} H_{\mu\nu\rho}$ is *constant.* (This follows from (12) and (13) since the exponential dilaton factors cancel out in the relevant expressions.) Only the spatial components of the torsion are nonzero in this case, $T_{ijk} \sim \epsilon_{ijk} \dot{b} = \epsilon_{ijk} \sqrt{2Q^2} e^{-\phi_0} \frac{M_s}{\sqrt{n}}$, where the overdot denotes derivative with respect to t. As discussed in [22], in the framework of the target-space effective theory, the relevant Lagrangian terms for fermions (to lowest order in α') will be of the form (8), with only the temporal component B^0 of the B^d vector being nonzero:

$$B^0 \sim \sqrt{2Q^2} e^{-\phi_0} \frac{M_s}{\sqrt{n}} \, \text{GeV} > 0. \tag{15}$$

We follow the conventions of string theory for the sign of B^0 . From phenomenological considerations M_s and $g_s^2/4\pi$ are taken to be larger than $\mathrm{O}(10^4)$ GeV and about $1/20$ respectively. A stringent constraint on the Kac-Moody level is not imposed from the requirement of CPTV-induced leptogenesis. We observe from (15) that the central charge deficit Q^2 of the underlying conformal field theory determines the order of particle-antiparticle asymmetries in this model. The particle-antiparticle asymmetry occurs already in thermal equilibrium, due to the background-induced [24] difference in the dispersion relations between particles and antiparticles. For CPT-Violating leptogenesis at the GUT scales involving right-handed neutrinos, for instance, decaying to Higgs and active neutrinos and the corresponding antiparticles[3], one would require [24], $B^0 \sim 10^5$ GeV. However, since the coupling of fermions to torsion is *universal*, the axion background would also couple to quarks and charged leptons. Hence a constant torsion-induced B^0 could lead directly to *baryogenesis* at the quark decoupling temperature $T_q \sim 100$ MeV, provided B^0 assumes an appropriate value much lower than 10^5 GeV. In such a scenario, the observed BAU could be realised directly through the quark-H-torsion interactions; unfortunately, the corresponding neutrino asymmetry would be too high, given that the neutrino decoupling temperatures are at 1 MeV scale. Note that standard model B-L conserving sphaleron processes freeze-out much earlier (at temperatures of order 100 GeV), but such interactions cannot produce sufficient baryon asymmetry in the universe without extra sources for CP violation [4].

A phenomenologically viable alternative scenario is one in which B^0 is not constant with time. In such a scenario the universe undergoes a phase transition at GUT scales ($\sim 10^{15}$ GeV). Above the GUT temperature heavy right-handed Majorana neutrinos could decouple or lepton number violating (but B-L conserving) processes could occur. At the phase transition the value of the H-torsion background could change from a large to a smaller value. This approach is based on the possibility that conformal field theories with different central charges characterise different epochs of the early universe [49]. Transitions between such conformal (fixed) points in moduli space correspond to non-conformal (Liouville) time evolution [50] and represent phase transitions in the early universe. Following the idea of Ref.[24], this leptogenesis can be communicated to the baryon sector by means of B-L conserving sphaleron processes. The difference between the fermion and antifermion populations is due to the interaction with a CPTV background torsion and occurs already in thermal equilibrium. Provided the background torsion is constant in time over the appropriate epoch of the evolution of the universe (until the freeze out of the pertinent processes characterising the epoch), such differences cannot be eroded during the expansion of the universe. No enhanced CP violation in the lepton sector is essential for leptogenesis in this approach, in contrast to the case of conventional leptogenesis [51, 52, 53, 5, 54]. it should be stressed, though, that this observation does not completely remove the need for right-handed sterile neutrinos.The latter may be essential, for instance, for explaining the smallness of the active neutrino masses, through seesaw mechanisms, or may play the role of dark matter [5, 54].

4. CPTV in Stochastic Finsler Geometries

Although all the models displaying CPTV that we have considered so far are based on local effective field theories, there is no compelling reason for a restriction to such a framework. In fact a microscopic model involving space-time defects based on string-brane theory suggests the use of a non-Riemannian metric background similar to that which occurs in Finsler geometry [55, 56].(This model will be discussed in the following section.) Independently there has been much interest in Finsler geometry [57, 58, 59, 60, 61, 62, 63, 64, 65, 66] for characterising the Early universe [67, 68, 69, 70, 71] and for descriptions of modified dispersion relations for particle probes [72, 68, 56, 73]. Finsler geometry has a metric which, in addition to space-time coordinates, depends also on "*velocities*". Lorentz symmetry is broken through some fixed vectors in the metric. We explore the consequences of making such vectors having components which are stochastic with possibly zero mean. This is a feature that arises in the defect model of D-foam [72, 38, 37] that we will consider in the next section. The defects stochastically fluctuate, due to both statistical and quantum stringy effects in large populations of such D-particles that can populate eras of the early universe. As we shall discuss below, the result of the

[3] Such processes are allowed due to the Yukawa coupling terms of right-handed Majorana Neutrinos to the lepton sector of the Standard Model, which also yield Majorana mass terms to the active neutrinos via seesaw [5].

interaction of neutrinos with these defects, leads to stochastically fluctuating Finsler-like metrics. However we wish to consider the consequences for CPTV and matter-antimatter asymmetry of this stochasticity in a general context. This underlying model provides our main motivation to study this class of space-times in this section and to contrast our findings on the induced CPTV for such cases with the corresponding ones for the D-foam model.

Below we shall be brief in our description of the stochastic CPTV backgrounds, concentrating only on the main features leading to modified dispersion relations that are different between particles and antiparticles. For the interested reader details can be found in [22].

We focus on on-shell neutrinos (which are now known to have small masses). Again, this is motivated by our desire to discuss leptogenesis in such geometries. Moreover, for reasons that will become clear in section 5, it is neutrinos that play a preferential rôle in interacting non-trivially with the D-particle foam background, which induces stochastically fluctuating Finsler-like space times. As our main motivation is to compare the generic Finsler-like case with the D-foam model, as far as CPTV is concerned, we restrict our attention here on the effects of stochastically fluctuating Finsler geometries on dispersion relations of neutrinos and antineutrinos. We shall consider a particular type of Finsler metric on a manifold M which is known as the Randers metric [55] [4]. It was noted in [76] that the geodesics of this metric coincided with the minimum time trajectories of a particle moving on a Riemannian manifold in the presence of a time independent drift given by a vector field. This is similar to Fermat's principle for propagation in refractive media. We mention in passing that similarities of D-particle foam to a refracting medium have been discussed in the literature [72, 77, 78]. If we were to assume that the result on minimum time trajectories was true for a pseudo-Riemannian situation and the drift was given by collisions due to D-particle scattering, then at a heuristic level a stochastic drift could be a reasonable generic phenomenological model of the back-reaction of low dimensional recoiling branes on matter.

The metric in this models is given by [22] (below, x^μ denotes space-time coordinates, while the variables y^ν play the rôle of "velocities" in the Finsler framework)

$$g_{\mu\nu}(x,y) = r_{\mu\nu}(x) + \phi^2(x)\, l_\mu l_\nu + \left(\frac{r_{\mu\nu}(x)}{\alpha(x,y)} - \frac{r_{\mu\varrho}(x)\, r_{\nu\sigma}(x)\, y^\varrho y^\sigma}{\alpha(x,y)^3}\right)\phi(x)\, l_c y^c$$
$$+ \frac{1}{\alpha(x,y)}\left(r_{\mu\varrho}(x)\, y^\varrho \phi(x)\, l_\nu + r_{\nu\varrho}(x)\, y^\varrho \phi(x)\, l_\mu\right)\ ,\ \alpha(x,y) = \sqrt{r_{\mu\nu}(x)\, y^\mu\, y^\nu}\ ,$$

where l_μ is a constant vector. In our model $\phi(x)$ will be a gaussian stochastic variable. On average the metric will be like a Riemannian metric if the mean of $\langle\phi\rangle$ vanishes.

We shall consider a situation with $r_{\mu\nu}(x) = \eta_{\mu\nu}$ where $\eta_{\mu\nu}$ is the diagonal Minkowski matrix with entries $(1, -1, -1, -1)$. (The summation convention of repeated indices will be always understood unless explicitly stated otherwise.) Within the framework of a Robertson-Walker metric we shall ignore effects on the time-scale of the inverse expansion rate. We have assumed a homogeneous ϕ with ϕ being x independent. The mass shell conditions of generalised plane-wave solutions for particles in this background $h^{\mu\nu}\omega_\mu\omega_\nu = m^2$, where $\omega_\mu = g_{\mu\nu}(x,y)y^\nu$ is a "phase-space" variable, dual to y^μ, and $h^{\mu\nu}(x,\omega)$ is the inverse of the Finsler metric in phase-space [22], lead to the generalised dispersion relations. In the model it is possible to choose l_μ. Not all choices will lead to asymmetric population distributions between particles and anti-particles. The

[4] We mention, for completeness, that the other popular class of Finsler geometries, that appears in the General Relativistic version [59, 62, 60, 61] of the so-called Very Special Relativety Model [74], and cosmological extensions thereof [67, 69] are *not* characterised by CPTV in the dispersion relations, nor of the spin-curvature type discussed in section 2.3. In fact such VSR-related models have been prop[osed in the past as candidates for the generation of Lepton-number conserving neutrino masses [75], and hence our Lepton-number violating considerations in this work do not apply.

space-like choice $l_0 = 0$ gives a degenerate spectrum for particle and anti-particle and hence no CPTV in dispersion relations. Therefore this case cannot be used for Leptogenesis in our framework. More generally the dispersion relation is [22]

$$\omega_0 = \pm\frac{\phi}{m} l_0 \left(2\vec{\omega}^2 + m^2\right) + \mathfrak{K}\left(\phi, \omega, m\right) \tag{16}$$

where $\mathfrak{K}\left(\phi, \omega, m\right) = \left(\vec{\omega}^2 + m^2 + \frac{2\phi}{m}\left(2\vec{\omega}^2 + m^2\right)\left(l_1\omega_1 + l_2\omega_2 + l_3\omega_3\right)\right)^{1/2}$. The +sign is for the particle and the −sign is for the antiparticle. For the "time-like" case $l_0 = 1$, $l_1 = l_2 = l_3 = 0$ the dispersion relation reduces to e $\omega_0 = \sqrt{\vec{\omega}^2 + m^2} \pm \frac{\phi}{m}\left(2\vec{\omega}^2 + m^2\right)$. The sign of l_0 can be reabsorbed in ϕ. For the "null" case $l_0 = l_1 = 1$ and $l_2 = l_3 = 0$ the dispersion relation reduces to $\omega_0 = \sqrt{\vec{\omega}^2 + m^2}\left(1 + \frac{\phi}{m}\left(2 - \frac{m^2}{\vec{\omega}^2 + m^2}\right)\omega_1\right) \pm \frac{\phi}{m}\left(2\vec{\omega}^2 + m^2\right)$. The parameters $\vec{\omega}$ play the rôle of momenta $\vec{p}$ in our case of neutrinos of mass $m = m_\nu$ propagating in these space-times, and the Finsler metric may be seen as sort of back reaction on the space-time of such a propagation (to better appreciate this, the reader is invited to the discussion in the next section 5, where a particular model of D(efect)-foam is considered as a medium for neutrino propagation in the early universe, leading to Finsler-like metric distortions as a consequence of medum/particle interactions). Corresponding to such models involving D-foam, the parameter ϕ is modelled as a stochastic gaussian process with a mean $\mathfrak{a}$ and standard deviation σ. The fermion number distribution n from equilibrium statistical mechanics is given by $n = g_{d.o.f.}\int \frac{d^3p}{(2\pi)^3}\frac{1}{\exp(\beta(\omega_0 - \mu)) + 1}$, where we have ignored degeneracy factors. First of all, it immediately follows from the corresponding dispersion relation that for the "time-like" case, when $\mathfrak{a} = 0$, there is no particle/antiparticle asymmetry. This is to be expected, given that $\mathfrak{a} \neq 0$ corresponds in a sense to an averaged Lorentz violation in this stochastic geometry, and hence one of the basic assumptions for CPT Invariance of the effective theory of neutrinos in this "medium" is relaxed. For the "time-like" case, when $\mathfrak{a} \neq 0$ and β small, we obtain to leading order in $T/m \gg 1$ [22]: $\ll \Delta n \gg \sim -\mathfrak{a}\, g_{d.o.f.} T^3 \frac{7\pi^2}{15}\left(\frac{T}{m}\right)$. We require $\mathfrak{a} < 0$ in order to have a particle-antiparticle asymmetry where the particle distribution dominates the antiparticle distribution. This yields the following Lepton (neutrino) asymmetry, assumed to freeze at the neutrino decoupling temperature T_d

$$\Delta L(T \sim T_d) = \frac{\Delta n_\nu}{s} \sim -10\,\mathfrak{a}\,\frac{T_d}{m_\nu} \tag{17}$$

where, as usual, s denotes the entropy density, which for relativistic species is assumed to be $s \sim g_{d.o.f.}\frac{2\pi^2}{45}T^3$. Thus, from (17) we see that there is no CPTV asymmetry in the Lorentz invariant (on average) case $\mathfrak{a} = 0$.

To obtain the phenomenologically correct value of $\Delta L(T \sim T_d) \sim 10^{-10}$, which is then communicated to the baryon sector via B-L conserving sphaleron processes, or B-L conserving grand unified models (assumed appropriately embedded in such space-time geometries), one needs to fix appropriately the value of $\mathfrak{a}$, consider appropriate conditions for freeze-out, that depend on the underlying microscopic model, and take into account that, according to current data, the masses of the active neutrinos that are assumed to participate in (17) must be smaller than $m_\nu < 0.2$ eV. For example, in GUT-scale lepton-number violation models, as we discussed before, $T_d \sim 10^{15}$ GeV, which implies that one needs only an extremely small in magnitude violation of Lorentz symmetry on average in this stochastic Finsler space time, $\mathfrak{a} \sim -10^{-36}$, in order to reproduce the observed Baryon Asymmetry in the universe. The assumption of fixing the sign of $\mathfrak{a}$ is considered as fine tuning, and is a feature that is common in the models of gravitational leptogenesis/baryogenesis that exist in the current literature, as discussed briefly above [26, 27, 24, 29, 31, 32, 23, 30, 28]. One can calculate the asymmetry for the null case with results similar to the time-like case considered above [22].

5. Stringy-Defect(D-)foam-induced CPTV and Leptogenesis

In this section we shall consider a population of D0-branes (or lower dimensional compactified D-branes which are effectively point-like from from the point of view of a brane world observer [5]) interacting with neutral fermions such as the neutrino and anti-neutrino. This interaction leads to different dispersion relations for neutrinos and anti-neutrinos which in turn leads to an excess of the population of neutrinos over anti-neutrinos. The freeze-out of neutrinos at the decoupling temperature of neutrinos leads to leptogenesis given by the standard cosmological considerations. The latter results, through standard Baryon (B) and Lepton (L) number violating sphaleron processes or B-L conserving interactions in grand unified models describing matter excitations on the brane, to the observed Baryon Asymmetry in the Unverse, with complete dominance of matter over antimatter, in a rather natural way, as we shall discuss below. Moreover, as we shall explain below, in this model of D(efect)-foam, the prevalence of matter over antimatter, *i.e.* the *positive sign* of the asymmetry $\Delta n > 0$, follows naturally, as a consequence of loss of energy of neutrinos during their interactions with the space-time defects, due to recoil of the latter. Thus, the sign of the induced asymmetry need not be fixed by hand, unlike the cases of gravitational leptogenesis discussed in previous sections. For instructive purposes, we first discuss the properties of the foam model, in the next subsection, before moving onto issues of CPTV and leptogenesis. D-foam models [37, 38, 39] are stringy models of space-time foamy geometries, which involve brane universes, propagating in higher-dimensional bulk geometries. The bulk contains point-like D-brane defects ("D-particles" or D0 branes) whose population density is constrained by the amount of CPTV that we observe. In many string theories (such as bosonic and type IIA string theories) they are stable zero-dimensional defects. However for our purposes we will consider them to be present in string theories of phenomenological interest [77] since, even when elementary D-particles cannot exist consistently, as is the case of type IIB string models, there can be effective D-particles formed by the compactification of higher dimensional D-branes [78] (*e.g.* three-branes wrapped around three-cycles, with relatively small radii).

The preferential rôle of neutrinos in feeling the full effects of D-foam, and hence the CPTV, is attributed to electric charge conservation: the representation of SM particles as open strings, with their ends attached to the brane worlds, prevents capture and splitting of open strings carrying electric fluxes by the D-particles. (We should recall that in string theory the electric charge is at the end point of an open string.) D-particles are electrically neutral and thus electric charge would not have been conserved if such processes had taken place. This is also consistent with the effective D-particles which may have formed as a result of nucleation [26]. Hence, the D-particle foam is transparent to charged excitations of the SM, leaving neutral particles, in particular neutrinos, susceptible to the foam effects. This different behaviour of neutrinos from charged leptons implies a background-induced breaking of the SU(2) gauge symmetry of the standard model. In type IIB string theories, effective D-particles [78] can interact, but in a suppressed manner, with the entire SU(2) lepton doublet, so the SU(2) symmetry is not broken. However, in such scenarios these interactions are suppressed compared to photons, as discussed in [78]. The D-foam interactions with sterile right-handed Majorana neutrinos, though, remain unsuppressed, since the sterile neutrinos do not have any standard model charges. Such heavy states can participate in lepton-number violating processes freezing out at GUT scales. For our purposes in this work we may consider that statistically significant populations of D-particles existed in the early eras of the brane universe [79]. As the time elapses, the brane universe, which propagates in the higher-dimensional bulk, enters regions characterised by D-particle depletion, in such as way that the late eras cosmology of the universe is not affected. Nevertheless, as we shall discuss below, the early D-particle populations may still have important effects in generating

[5] It should be remarked that for the effective compactified D-"particles" the interactions with the charged matter excitations are suppressed relative to the neutral ones [78]. Hence, even in this case, it is the electrically neutral excitations which interact primarily with the D-foam.

neutrino-antineutrino populations differences (asymmetries), which are then communicated to the baryon sector via the standard sphaleron processes [21] or B-L conserving GUT symmetries in unified particle physics models.

To this end, we need to consider the *effective dispersion relation* of a (anti)neutrino field in a brane space-time punctured with statistically significant populations of D-particles. The latter is a dynamical population, consisting of defects crossing the brane all the time, thereby appearing to a brane observer as flashing "on" and "off" space-time "foamy" structures. The (anti)neutrino excitations are represented as matter open strings with their ends attached on the brane.The number density of (anti) neutrinos on the brane world is limited by the requirement that they do not overclose the universe. In [22] we estimated the *modification of the dispersion relations* of neutrinos in such a "media" of D-particles in the early universe. The interaction of a string with a D-particle implies that at least one of the ends of the string is attached to the D-particle defect. Furthermore, the simultaneous creation of virtual strings stretched between the defect and the brane, describes the recoil of the D-particle. During the interaction time, the D-particle undergoes motion characterized by non-trivial "recoil" velocities, $u_\parallel = \frac{g_s}{M_s}\Delta p_i = \frac{g_s}{M_s} r_i\, p_i$ along the brane longitudinal dimensions, where r_i denotes the proportion of the incident neutrino momentum that corresponds to the momentum transfer Δp_i during the scattering, and $v_\perp$ in directions transverse to the brane world [80]. As discussed in detail in [80, 39], one may assumed that r_i are stochastic, that is

$$u_i = \frac{g_s}{M_s}\Delta p_i \to g_s\, r_i \frac{p_i}{M_s}\ \text{, no sum over i }, \quad \ll r_i \gg = 0\ , \quad \ll\ r_i r_j \gg = \Delta^2 \delta_{ij}\ .$$

The dispersion relations then of neutrino and antineutrinos of mass m propagating on such a deformed isotropic stochastically fluctuating space-time background can be found to be [22]:

$$\ll E_\nu \gg = \sqrt{p^2 + m_\nu^2}\left(1 + \frac{1}{2}\sigma^2\right) - \frac{1}{2}\frac{M_s}{g_s}\sigma^2\ , \ll E_{\overline{\nu}} \gg = \sqrt{p^2 + m_\nu^2}\left(1 + \frac{1}{2}\sigma^2\right) + \frac{1}{2}\frac{M_s}{g_s}\sigma^2 \quad (18)$$

with $E_{\overline{\nu}} > 0$ representing the positive energy of a physical antiparticle and $\sigma^2 = \frac{g_s^2}{M_s^2}\Delta^2 p^2$. In our analysis above we have made the symmetric assumption that the recoil-velocities fluctuation strengths are the same between particle and antiparticle sectors. (Scenarios for which this symmetry was not asssumed have also been considered in an early work [39].) There can thus be *local* CPTV in the sense that the effective dispersion relation between neutrinos and antineutrinos are different. This is a consequence of the local violation of Lorentz symmetry (LV), as a result of the non-trivial recoil velocities of the D-particle [22]. The discussion of CPTV in such foamy universes now follows the line of argument adopted by others and mentioned previously: the difference in the dispersion relations between particles and antiparticles will imply differences in the relevant populations of neutrinos (n) and antineutrinos ($\overline{n}$), (*cf.* the dispersion (18)). This difference between neutrino and antineutrino phase-space distribution functions in D-foam backgrounds generates a matter-antimatter lepton asymmetry in the relevant densities, which decreases with decreasing temperature up to a freeze-out point, which occurs at temperatures T_d at which the Lepton-number violating processes decouple. The lepton asymmetry in the D-foam scenario has the form [22]

$$\Delta L(T < T_d) = \frac{\Delta n_\nu}{s} = \frac{2\Delta^2 g_s T_d}{M_s}. \quad (19)$$

The phenomenologically realistic scenario in string theory frameworks, like the present one, seems to be the one in which one has GUT scale lepton-number violating processes, which freeze out the lepton-antilepton number at temperatures $T_d \sim 10^{15}$ GeV, which characterises GUT scale

Lepton-Violating processes, (*e.g* those involving heavy Majorana neutrinos $N_I \to H\,\nu,\, \overline{H}\,\overline{\nu}$). In the context of D-particle brane Universe one may assume that at such early epochs one has a sufficiently dense D-particle gas in the bulk and on the brane, so that phenomenologically realistic Leptogenesis takes place. The latter then can be communicated via B-L conserving GUT processes to the baryon sector. After the GUT scale freeze out the bulk population of D-particles can be depleted, in the sense that the brane world representing our Universe passes through an area with scarce D-particle populations, so that late eras of the Universe are characterised by a more-or-less conventional Cosmology. From (19), we observe that for a GUT-like freeze-out temperature $T_d \sim 10^{15}$ GeV, the phenomenological value $\Delta L \sim 10^{-10}$ is attained for $\frac{M_s}{g_s} \sim 10^{25}\,\Delta^2\,\text{GeV}$. For $\Delta^2 \sim 10^{-6}$ a Planck size D-particle mass $M_s/g_s \sim 10^{19}$ GeV is required so that the D-foam provides the physically observed Lepton and, thus, Baryon Asymmetry. For the unnaturally small $\Delta^2 < 10^{-21}$ one arrives at $M_s/g_s \sim 10$ TeV. Unfortunately, for $\Delta^2 \sim \mathcal{O}(1)$ transplanckian D-particle masses are required. We stress once more that the D-foam approach to leptogenesis is distinguished from others in that a local effective field theoretical description is not adopted.

Acknowledgments

N.E.M. thanks the organisers of DISCRETE 2012 (Lisbon, Portugal) for the invitation to give a plenary talk. This work was supported in part by the London Centre for Terauniverse Studies (LCTS), using funding from the European Research Council via the Advanced Investigator Grant 267352, and by STFC UK under the research grant ST/J002798/1.

References

[1] Christenson J, Cronin J, Fitch V and Turlay R 1964 *Phys.Rev.Lett.* **13** 138–140
[2] Sakharov A 1967 *Pisma Zh.Eksp.Teor.Fiz.* **5** 32–35 reprinted in *Kolb, E.W. (ed.), Turner, M.S. (ed.): The early universe* 371-373, and in *Lindley, D. (ed.) et al.: Cosmology and particle physics* 106-109, and in Sov. Phys. Usp. 34 (1991) 392-393 [Usp. Fiz. Nauk 161 (1991) No. 5 61-64]
[3] Gamow G 1946 *Phys.Rev.* **70** 572–573
[4] Kuzmin V, Rubakov V and Shaposhnikov M 1985 *Phys.Lett.* **B155** 36
[5] Shaposhnikov M 2009 *Prog.Theor.Phys.* **122** 185–203
[6] Shaposhnikov M and Tkachev I 2006 *Phys.Lett.* **B639** 414–417 (*Preprint* hep-ph/0604236)
[7] Lindner M, Merle A and Niro V 2011 *JCAP* **1101** 034 (*Preprint* 1011.4950)
[8] Kusenko A, Takahashi F and Yanagida T T 2010 *Phys.Lett.* **B693** 144–148 (*Preprint* 1006.1731)
[9] Randall L and Sundrum R 1999 *Phys.Rev.Lett.* **83** 3370–3373 (*Preprint* Randall:1999ee)
[10] Merle A and Niro V 2011 *JCAP* **1107** 023 (*Preprint* 1105.5136)
[11] Barry J, Rodejohann W and Zhang H 2011 *JHEP* **1107** 091 (*Preprint* 1105.3911)
[12] Carmona J M, Cortes J L, Das A K, Gamboa J and Mendez F 2006 *Mod.Phys.Lett.* **A21** 883–892 (*Preprint* hep-th/0410143)
[13] Streater R and Wightman A 1989 *PCT, spin and statistics, and all that* (Princeton University Press)
[14] Bertolami O, Colladay D, Kostelecky V A and Potting R 1997 *Phys.Lett.* **B395** 178–183 (*Preprint* hep-ph/9612437)
[15] Mavromatos N E 2010 and references therein *Int.J.Mod.Phys.* **A25** 5409–5485 (*Preprint* 1010.5354)
[16] Barenboim G and Mavromatos N E 2004 *Phys.Rev.* **D70** 093015 (*Preprint* hep-ph/0406035)
[17] Barenboim G and Mavromatos N E 2005 *JHEP* **0501** 034 (*Preprint* hep-ph/0404014)
[18] Barenboim G and Lykken J D 2003 *Phys.Lett.* **B554** 73–80 (*Preprint* hep-ph/0210411)
[19] Chaichian M, Dolgov A D, Novikov V A and Tureanu A 2011 *Phys.Lett.* **B699** 177–180 (*Preprint* 1103.0168)
[20] Chaichian M, Fujikawa K and Tureanu A 2012 *Phys.Lett.* **B712** 115–118 (*Preprint* 1203.0267)
[21] Arnold P and McLerran L 1987 *Phys. Rev. D* **36**(2) 581–595
[22] Mavromatos N E and Sarkar S 2012 (*Preprint* 1211.0968)
[23] Lambiase G and Mohanty S 2007 *JCAP* **0712** 008 (*Preprint* astro-ph/0611905)
[24] Debnath U, Mukhopadhyay B and Dadhich N 2006 *Mod.Phys.Lett.* **A21** 399–408 (*Preprint* hep-ph/0510351)
[25] Zwiebach B 2004 *A First Course in String Theory* (Cambridge: Cambridge Univ. Press)
[26] Dvali G and Gabadadze G 1999 *Phys.Lett.* **B460** 47–57 (*Preprint* hep-ph/9904221)
[27] Davoudiasl H, Kitano R, Kribs G D, Murayama H and Steinhardt P J 2004 *Phys.Rev.Lett.* **93** 201301 (*Preprint* hep-ph/0403019)

[28] Li H, Li M z and Zhang X m 2004 *Phys.Rev.* **D70** 047302 (*Preprint* hep-ph/0403281)
[29] Mukhopadhyay B 2005 *Mod.Phys.Lett.* **A20** 2145–2156 (*Preprint* astro-ph/0505460)
[30] Lambiase G and Mohanty S 2011 *Phys.Rev.* **D84** 023509 (*Preprint* 1107.1213)
[31] Mukhopadhyay B 2007 *Class.Quant.Grav.* **24** 1433–1442 (*Preprint* gr-qc/0702062)
[32] Sinha M and Mukhopadhyay B 2008 *Phys.Rev.* **D77** 025003 (*Preprint* 0704.2593)
[33] Carroll S M and Shu J 2006 *Phys.Rev.* **D73** 103515 (*Preprint* hep-ph/0510081)
[34] Dolgov A 2010 *Phys.Atom.Nucl.* **73** 588–592 (*Preprint* 0903.4318)
[35] Barenboim G, Borissov L, Lykken J D and Smirnov A 2002 *JHEP* **0210** 001 (*Preprint* hep-ph/0108199)
[36] Wald R M 1980 *Phys.Rev.* **D21** 2742–2755
[37] Ellis J R, Mavromatos N E and Westmuckett M 2005 *Phys.Rev.* **D71** 106006 (*Preprint* gr-qc/0501060)
[38] Ellis J R, Mavromatos N E and Westmuckett M 2004 *Phys.Rev.* **D70** 044036 (*Preprint* gr-qc/0405066)
[39] Bernabeu J, Mavromatos N E and Sarkar S 2006 *Phys.Rev.* **D74** 045014 (*Preprint* hep-th/0606137)
[40] Lindblad G 1976 *Commun.Math.Phys.* **48** 119
[41] Ellis J R, Hagelin J, Nanopoulos D V and Srednicki M 1984 *Nucl.Phys.* **B241** 381
[42] Ellis J R, Lopez J L, Mavromatos N and Nanopoulos D V 1996 *Phys.Rev.* **D53** 3846–3870 (*Preprint* hep-ph/9505340)
[43] Benatti F and Floreanini R 2001 *Phys.Rev.* **D64** 085015 (*Preprint* hep-ph/0105303)
[44] Fogli G, Lisi E, Marrone A, Montanino D and Palazzo A 2007 *Phys.Rev.* **D76** 033006 (*Preprint* 0704.2568)
[45] Lisi E, Marrone A and Montanino D 2000 *Phys.Rev.Lett.* **85** 1166–1169 (*Preprint* hep-ph/0002053)
[46] Barenboim G, Mavromatos N E, Sarkar S and Waldron-Lauda A 2006 *Nucl.Phys.* **B758** 90–111 (*Preprint* hep-ph/0603028)
[47] Aguilar A *et al.* (LSND Collaboration) 2001 *Phys.Rev.* **D64** 112007 (*Preprint* hep-ex/0104049)
[48] Gross D J and Sloan J H 1987 *Nucl.Phys.* **B291** 41
[49] Antoniadis I, Bachas C, Ellis J R and Nanopoulos D V 1989 *Nucl.Phys.* **B328** 117–139
[50] Ellis J R, Mavromatos N, Nanopoulos D V and Westmuckett M 2006 *Int.J.Mod.Phys.* **A21** 1379–1444 (*Preprint* gr-qc/0508105)
[51] Pilaftsis A and Underwood T E 2004 *Nucl.Phys.* **B692** 303–345 (*Preprint* hep-ph/0309342)
[52] Pilaftsis A 2005 *Phys.Rev.Lett.* **95** 081602 (*Preprint* hep-ph/0408103)
[53] Pilaftsis A and Underwood T E 2005 *Phys.Rev.* **D72** 113001 (*Preprint* hep-ph/0506107)
[54] Boyarsky A, Ruchayskiy O and Shaposhnikov M 2009 *Ann.Rev.Nucl.Part.Sci.* **59** 191–214 (*Preprint* 0901.0011)
[55] Bao D, Chern S and Shen Z 2000 *An Introduction to Riemann-Finsler Geometry* Graduate Texts in Mathematics (Springer) ISBN 9780387989488 URL http://books.google.fr/books?id=m11kEYr_V_AC
[56] Girelli F, Liberati S and Sindoni L 2007 *Phys.Rev.* **D75** 064015 (*Preprint* gr-qc/0611024)
[57] Vacaru S I 2002 (*Preprint* hep-th/0211068)
[58] Vacaru S I 2003 (*Preprint* hep-th/0310132)
[59] Bogoslovsky G Y and Goenner H 1998 *Phys.Lett.* **A244** 222–228 (*Preprint* gr-qc/9804082)
[60] Bogoslovsky G Y 2007 (*Preprint* 0706.2621)
[61] Bogoslovsky G 2007 (*Preprint* 0712.1718)
[62] Gibbons G, Gomis J and Pope C 2007 *Phys.Rev.* **D76** 081701 (*Preprint* 0707.2174)
[63] Sindoni L 2008 *Phys.Rev.* **D77** 124009 (*Preprint* 0712.3518)
[64] Anastasiei M and Vacaru S I 2009 *J.Math.Phys.* **50** 013510 (*Preprint* 0710.3079)
[65] Vacaru S I 2012 (*Preprint* 1206.4012)
[66] Vacaru S I 2011 *Class.Quant.Grav.* **28** 215001 (*Preprint* 1008.4912)
[67] Kouretsis A, Stathakopoulos M and Stavrinos P 2009 *Phys.Rev.* **D79** 104011 (*Preprint* 0810.3267)
[68] Magueijo J and Smolin L 2004 *Class.Quant.Grav.* **21** 1725–1736 (*Preprint* gr-qc/0305055)
[69] Kouretsis A, Stathakopoulos M and Stavrinos P 2010 *Phys.Rev.* **D82** 064035 (*Preprint* 1003.5640)
[70] Stavrinos P C and Vacaru S I 2012 (*Preprint* 1206.3998)
[71] Vacaru S I 2012 *Int.J.Mod.Phys.* **D21** 1250072 (*Preprint* 1004.3007)
[72] Ellis J R, Mavromatos N and Nanopoulos D V 2000 *Gen.Rel.Grav.* **32** 127–144 (*Preprint* gr-qc/9904068)
[73] Skakala J and Visser M 2009 *J.Phys.Conf.Ser.* **189** 012037 (*Preprint* 0810.4376)
[74] Cohen A G and Glashow S L 2006 *Phys.Rev.Lett.* **97** 021601 (*Preprint* hep-ph/0601236)
[75] Cohen A G and Glashow S L 2006 (*Preprint* hep-ph/0605036)
[76] Gibbons G, Herdeiro C, Warnick C and Werner M 2009 *Phys.Rev.* **D79** 044022 (*Preprint* 0811.2877)
[77] Ellis J R, Mavromatos N and Nanopoulos D 2008 *Phys.Lett.* **B665** 412–417 (*Preprint* 0804.3566)
[78] Li T, Mavromatos N E, Nanopoulos D V and Xie D 2009 *Phys.Lett.* **B679** 407–413 (*Preprint* 0903.1303)
[79] Mavromatos N E, Mitsou V A, Sarkar S and Vergou A 2012 *Eur.Phys.J.* **C72** 1956 (*Preprint* 1012.4094)
[80] Mavromatos N and Sarkar S 2005 *Phys.Rev.* **D72** 065016 (*Preprint* hep-th/0506242)

Journal of Physics: Conference Series **447** (2013) 012017 doi:10.1088/1742-6596/447/1/012017

Flavour and CP Violation

Ulrich Nierste
Institut für Theoretische Teilchenophysik
Karlsruhe Institute of Technology
Engesserstrae 7
76131 Karlsruhe, Germany

E-mail: ulrich.nierste@kit.edu

Abstract. In this talk I review the status of $\mathrm{B}-\overline{\mathrm{B}}$ mixing in the Standard Model and the room for new physics in $\mathrm{B_s}-\overline{\mathrm{B}}_\mathrm{s}$ and $\mathrm{B_d}-\overline{\mathrm{B}}_\mathrm{d}$ mixing in the light of recent LHCb data.

1. Introduction

The only source of flavour-changing transition in the Standard Model (SM) is the Yukawa interaction of the Higgs doublet H with the fermion fields. The Yukawa lagrangian of the quark fields reads

$$-L_Y = Y^d_{jk}\,\overline{Q}^j_L\,H\,d^k_R \;+\; Y^u_{jk}\,\overline{Q}^j_L\,\widetilde{H}\,u^k_R \;+\; \text{h.c.} \tag{1}$$

Here Q^j_L denotes the SU(2) doublet of the left-handed quark fields of the j-th generation and d^j_R, u^j_R are the corresponding right-handed singlet fields. The 3×3 Yukawa matrices $Y^{d,u}$ and the Higgs field vacuum expectation value $v = 174$ GeV combine to the quark mass matrices $M^{d,u} = Y^{d,u}v$. The gauge interactions of the quark fields do not change under unitary rotations of any of Q^j_L, d^j_R, and u^j_R in flavour space. Using these unphysical transformations one can bring Y^u and Y^d to the form

$$Y^u = \widehat{Y}^u = \begin{pmatrix} y_u & 0 & 0\\ 0 & y_c & 0\\ 0 & 0 & y_t\end{pmatrix} \qquad \text{and} \qquad Y^d = V^\dagger \widehat{Y}^d \qquad \text{with} \quad \widehat{Y}^d = \begin{pmatrix} y_d & 0 & 0\\ 0 & y_s & 0\\ 0 & 0 & y_b\end{pmatrix} \tag{2}$$

and $y_i \geq 0$. With the choice adopted in Eq. (2) the mass matrix M^u of up, charm, and top quark is diagonal. The diagonalisation of M^d requires the additional rotation $d^j_L = V_{jk} d^{k\prime}_L$, which puts the unitary Cabbibbo-Kobayashi-Maskawa (CKM) matrix V into the W boson vertices: $W_\mu \overline{u}^j_L \gamma^\mu d^j_L = W_\mu V_{jk} \overline{u}^j_L \gamma^\mu d^{k\prime}_L$. Eq. (2) defines a basis of weak eigenstates which is particularly suited to display the smallness of flavour violation in the SM. All flavour-changing transitions originate from Y^d which reads

$$Y^d = V^\dagger \widehat{Y}^d \;=\; \begin{pmatrix} 10^{-5} & -7\cdot 10^{-5} & (12+6i)\cdot 10^{-5}\\ 4\cdot 10^{-6} & 3\cdot 10^{-4} & -6\cdot 10^{-4}\\ (2+6i)\cdot 10^{-8} & 10^{-5} & 2\cdot 10^{-2}\end{pmatrix}, \tag{3}$$

Journal of Physics: Conference Series **447** (2013) 012017 doi:10.1088/1742-6596/447/1/012017

Figure 1. Left: SM box diagram describing $B_s - \overline{B}_s$ mixing, a $|\Delta B| = 2$ process. Right: $b \to s$ penguin amplitude, a $|\Delta B| = 1$ transition.

if all Yukawa couplings are evaluated at the scale $\mu = m_t$. The off-diagonal element largest in magnitude is $V_{ts}^* y_b \equiv V_{32}^* y_b = -6 \cdot 10^{-4}$. Moreover, in the SM flavour-changing neutral current (FCNC) transitions involve an additional loop suppression, making FCNC processes an extremely sensitive probe of physics beyond the SM. This feature puts flavour physics into a win-win position: If CMS and ATLAS find new particles, FCNC observables will be used to probe the flavour pattern of the BSM theory to which these particles belong. If instead CMS and ATLAS do not find any BSM particles, flavour physics will indirectly probe new physics to scales exceeding 100 TeV, well beyond the center-of-mass energy of the LHC.

The weak interaction makes the neutral $K \sim \bar{s}d$, $D \sim c\bar{u}$, $B_d \sim \bar{b}d$ and $B_s \sim \bar{b}s$ mesons mix with their antiparticles, $\overline{K}$, $\overline{D}$, $\overline{B}_d$, and $\overline{B}_s$, respectively. This means that, say, a B_d meson evolves in time into a quantum-mechanical superposition of a B_d and a $\overline{B}_d$. This feature is a gold-mine for new-physics searches, as it permits to probe CP phases in any decay to a final state which is accessible from both the B_d and the $\overline{B}_d$ component of the decaying state. In a tree-level decay like $\overset{(-)}{B}_d \to J/\psi K_S$ the time-dependent CP asymmetry probes new physics in the mixing amplitude itself. If instead a rare decay (with FCNC decay amplitude) is studied, the time-dependent CP asymmetry may reveal a new physics contribution to the decay amplitude. Meson-antimeson mixing is a $|\Delta F| = 2$ process, meaning that the relevant flavour quantum numbers change by two units. In the case of $B_s - \overline{B}_s$ mixing depicted in Fig. 1 these flavour quantum numbers are beauty B and strangeness S. Weak decays are $|\Delta F| = 1$ processes, Fig. 1 shows a $|\Delta B| = |\Delta S| = 1$ FCNC decay amplitude. It is instructive to compare the reach of $|\Delta F| = 2$ and $|\Delta F| = 1$ amplitudes to new physics: The flavour violation in the SM the amplitudes $A_{\rm SM}^{|\Delta B|=2}$ and $A_{\rm SM}^{|\Delta B|=1}$ in Fig. 1 is governed by the small CKM element V_{ts} and both diagrams scale as $1/M_W^2$. A new-physics (NP) contribution may instead involve some new flavour-violating parameter $\delta_{\rm FCNC}$. If the scale of NP is Λ and the NP contribution enters at the one-loop level, one finds

$$\frac{|A_{\rm NP}^{|\Delta B|=2}|}{|A_{\rm SM}^{|\Delta B|=2}|} = \frac{|\delta_{\rm FCNC}|^2}{|V_{ts}|^2} \frac{M_W^2}{\Lambda^2} \qquad \text{and} \qquad \frac{|A_{\rm NP}^{|\Delta B|=1}|}{|A_{\rm SM}^{|\Delta B|=1}|} = \frac{|\delta_{\rm FCNC}|}{|V_{ts}|} \frac{M_W^2}{\Lambda^2}. \tag{4}$$

One sees that $B_s - \overline{B}_s$ mixing is more sensitive to generic NP than an FCNC $b \to s$ decay amplitude, if $|\delta_{\rm FCNC}| > |V_{ts}| \approx 0.04$. The estimate in Eq. (4) applies, for example, to the (N)MSSM with generic flavour structure in the bilinear SUSY-breaking terms, which induce new $b_R \to s_R$ or $b_L \to s_L$ transitions through squark-gluino loops [1, 2]. But $|\Delta B| = 1$ transitions can be more sensitive probes of NP in models in which $b_R \to s_L$ or $b_L \to s_R$ transitions are parametrically enhanced over their SM counterparts. This situation occurs in the (N)MSSM with a large value of the parameter $\tan\beta$ or large trilinear SUSY-breaking terms. Theories with such chirally enhanced FCNC transitions are efficiently probed with the radiative decays $b \to s\gamma, d\gamma$ [3–10], leptonic decays like the recently observed [11] decay $B_s \to \mu^+\mu^-$ [7, 12–21] and the chromomagnetic contribution to non-leptonic decays like $B_d \to \phi K_S$ [18].

Journal of Physics: Conference Series **447** (2013) 012017 doi:10.1088/1742-6596/447/1/012017

2. $B-\overline{B}$ mixing: formalism and Standard-Model prediction

$B_q-\overline{B}_q$ mixing with $q=d$ or $q=s$ is governed by the 2×2 matrix $M^q - i\Gamma^q/2$, with the hermitian mass and decay matrices M^q and Γ^q. Due to non-vanishing off-diagonal elements M_{12}^q and Γ_{12}^q a meson tagged as a B_q at time $t=0$ will for $t>0$ evolve into a superposition of B_q and $\overline{B}_q$. In the SM M_{12}^s is calculated from the dispersive part of the box diagram in Fig. 1, which amounts to discarding the imaginary part of the loop integral. Γ_{12}^s is instead obtained from the absorptive part, meaning that the real part of the loop integral is dropped. M_{12}^d and Γ_{12}^d are calculated in the same way from the box diagram with the external s quarks replaced by d quarks. M_{12}^q is dominated by the top contribution, while Γ_{12}^q stems from the diagrams with internal up and charm. Γ_{12}^q is made up of all final states into which both B_q and $\overline{B}_q$ can decay. There are three physical quantities in $B_q-\overline{B}_q$ mixing:

$$|M_{12}^q|, \quad |\Gamma_{12}^q|, \quad \phi_q \equiv \arg\left(-\frac{M_{12}^q}{\Gamma_{12}^q}\right).$$

By diagonalising $M - i\,\Gamma/2$ one finds the two mass eigenstates B_q^H and B_q^L, where the superscripts stand for "heavy" and "light". These eigenstates differ in their mass and width with

$$\Delta m_q \;=\; M_H^q - M_L^q \;\simeq\; 2|M_{12}^q|, \qquad\qquad \Delta\Gamma_q \;=\; \Gamma_L^q - \Gamma_H^q \;\simeq\; 2|\Gamma_{12}^q|\cos\phi_q\,.$$

The mass difference Δm_q simply equals the frequency at which B_q and $\overline{B}_q$ oscillate into each other. The width difference $\Delta\Gamma_s$ is sizable, so that in general untagged B_s decays are governed by the sum of two exponentials. There is no useful data on the tiny width difference $\Delta\Gamma_d$ in the B_d system yet, which is predicted to be around 0.5% of $\Gamma_{L,H}^d$ [22–25]. The CP-violating phase ϕ_q can be determined by measuring the CP asymmetry in flavour-specific decays:

$$a_{\rm fs}^q \;=\; \frac{|\Gamma_{12}^q|}{|M_{12}^q|}\sin\phi_q\,.$$

A decay $B_q \to f$ is called flavour-specific, if the decay $\overline{B}_q \to f$ is forbidden. The standard method to determine $a_{\rm fs}^q$ uses semileptonic decays, so that $a_{\rm fs}^q$ is often called semileptonic CP asymmetry. While this measurement simply requires to count the numbers of positive and negative leptons in B_q decays, it is still very difficult, because $|\Gamma_{12}^q| \ll |M_{12}^q|$ renders $|a_{\rm fs}^q|$ small, even if NP contributions to M_{12}^q enhance ϕ_q over its small SM value.

2.1. *Mass difference Δm_s*

The theoretical prediction of Δm_q requires the separation of short-distance and long-distance QCD effects. To this end one employs an operator product expansion, which factorises M_{12}^q as

$$M_{12}^q = \left(V_{tq}^* V_{tb}\right)^2\, C\langle B_q|Q|\overline{B}_q\rangle. \tag{5}$$

The Wilson coefficient C comprises the short-distance physics, with all dependence on the heavy particle masses. QCD corrections to C have been calculated reliably in perturbation theory [26]. Since the CKM elements are factored out in Eq. (5), $C = C(m_t, M_W, \alpha_s)$ is real in the SM.

$$Q \;=\; \overline{q}_L\gamma_\nu b_L\, \overline{q}_L\gamma^\nu b_L \tag{6}$$

is a local four-quark operator describing a point-like $|\Delta B| = 2$ transition, see Fig. 2. The formalism is the same for the B_d and B_s complex; for definiteness I only discuss the case $q=s$ in the following. The average of the CDF and LHCb measurements of $\Delta m_s \simeq 2|M_{12}^s|$ is [27]

$$\Delta m_s^{\rm exp} = (17.719 \pm 0.043)\,{\rm ps}^{-1}, \tag{7}$$

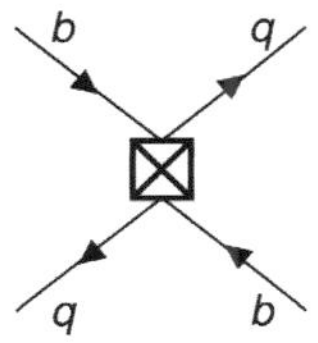

Figure 2. The $|\Delta B| = 2$ operator Q of Eq. (6). It is pictorially obtained by shrinking the $\mathrm{B_q}-\overline{\mathrm{B}}_\mathrm{q}$ mixing box diagram to a point.

meaning that $|M_{12}^s|$ is known very precisely. However, we need the hadronic matrix element $\langle B_s|Q|\overline{B}_s\rangle$ to confront Eq. (7) with the SM. The latter is usually parametrised as

$$\langle B_s|Q|\overline{B}_s\rangle \quad = \quad \frac{2}{3} M_{B_s}^2 \, f_{B_s}^2 \, B_{B_s}.$$

Here M_{B_s} and f_{B_s} are mass and decay constant of the B_s meson, respectively. The hadronic parameter B_{B_s} depends on the renormalisation scheme and scale used to define Q, in this talk $B_{B_s} \approx 0.85$ is understood to be evaluated at the scale $\mu = m_b$ in the $\overline{\mathrm{MS}}$ scheme. Noting that $|V_{ts}|$ is fixed by CKM unitarity from the well-measured element $|V_{cb}|$, we can write

$$\Delta m_s = \left(\, 18.8 \pm 0.6_{\,V_{cb}} \pm 0.3_{\,m_t} \pm 0.1_{\,\alpha_s} \right) \, \mathrm{ps}^{-1} \, \frac{f_{B_s}^2 \, B_{B_s}}{(220\,\mathrm{MeV})^2}, \tag{8}$$

where the uncertainties from the experimental errors of the input parameters indicated. Averaging various calculations from lattice gauge theory one finds [28]

$$f_{B_s}^2 B_{B_s} = [(211 \pm 9)\,\mathrm{MeV}]^2 \, . \tag{9}$$

Inserting this result into Eq. (8) implies

$$\Delta m_s = (17.3 \pm 1.5)\,\mathrm{ps}^{-1} \tag{10}$$

complying excellently with the experimental result in Eq. (7). However, the average in Eq. (9) mainly involves different calculations of f_{B_s}, which are combined with two fairly old results for B_{B_s}. With the recent preliminary Fermilab/MILC result [29]

$$f_{B_s}^2 \, B_{B_s} = 0.0559(68)\,\mathrm{GeV}^2 \simeq [(237 \pm 14)\,\mathrm{MeV}]^2$$

one finds instead

$$\Delta m_s = (21.7 \pm 2.6)\,\mathrm{ps}^{-1}. \tag{11}$$

Therefore more effort on lattice-QCD calculations of $f_{B_s}^2 \, B_{B_s}$ is highly desirable. As long as modern calculations of this quantity are unavailable, I recommend to inflate the error in Eq. (10) to $2.5\,\mathrm{ps}^{-1}$. One concludes that the precise measurement in Eq. (7) still permits a NP contribution of 15% in Δm_s.

2.2. CP phase from $B_s \to J/\psi\phi$

While Δm_s fixes the magnitude of M_{12}^s, the phase of M_{12}^s can be probed through the mixing-induced CP asymmetry in $B_s \to J/\psi\phi$. The final state contains two vector mesons, so that the orbital angular momentum L can assume the values 0,1, and 2. The final states $(J/\psi\,\phi)_{L=0,2}$

Journal of Physics: Conference Series **447** (2013) 012017 doi:10.1088/1742-6596/447/1/012017

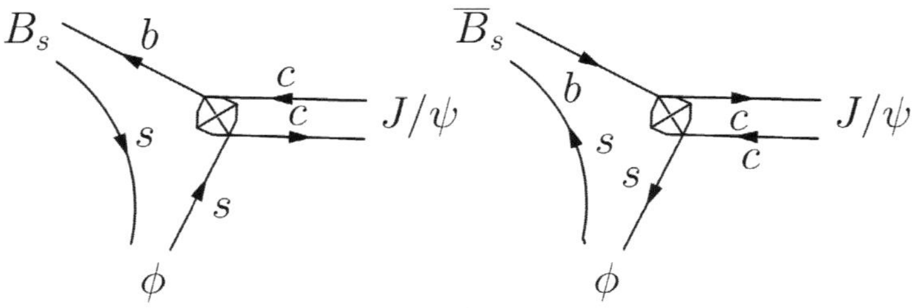

Figure 3. Amplitudes of $B_s \to J/\psi\,\phi$ and $\overline{B}_s \to J/\psi\,\phi$. The cross denotes the W-mediated $b \to c\bar{c}s$ decay.

have CP quantum number $\eta_{CP} = 1$, while $(J/\psi\,\phi)_{L=1}$ is CP-odd. Denoting a meson which was a B_s at time $t = 0$ by $B_s(t)$ (with an analogous definition of $\overline{B}_s(t)$), one can define the time-dependent CP asymmetry

$$a_{CP}(t) = \frac{\Gamma(B_s(t) \to (J/\psi\,\phi)_L) - \Gamma(\overline{B}_s(t) \to (J/\psi\,\phi)_L)}{\Gamma(B_s(t) \to (J/\psi\,\phi)_L) + \Gamma(\overline{B}_s(t) \to (J/\psi\,\phi)_L)}. \tag{12}$$

The two interfering amplitudes giving rise to $a_{CP}(t)$ are shown in Fig. 3. The analytical result reads

$$a_{CP}(t) = \eta_{CP}\frac{\sin(2\beta_s)\sin(\Delta m_s t)}{\cosh(\Delta\Gamma_s t/2) - \cos(2\beta_s)\sinh(\Delta\Gamma_s t/2)}. \tag{13}$$

The CP phase entering $a_{CP}(t)$ is $\beta_s = \arg(-V_{tb}^* V_{ts}/(V_{cb}^* V_{cs}))$, in the standard phase convention of the CKM matrix β_s is essentially just the phase of $-V_{ts}$. It should be mentioned that Eq. (12) is a theoretical definition, in practice the experimental separation of the $L = 0, 1, 2$ amplitudes involves a complicated angular analysis of the J/ψ and ϕ decays.

The CKM matrix can be parametrised in terms of the four Wolfenstein parameters λ,A,$\overline{\rho}$, and $\overline{\eta}$ [30,31]. Expanding to leading order in $\lambda = 0.225$ one finds $\sin(2\beta_s) \simeq 2\lambda^2\overline{\eta}$. The parameter $\overline{\eta}$ is the height of the CKM unitarity triangle defined by $\overline{\rho}+i\overline{\eta} = -V_{ub}^* V_{ud}/(V_{cb}^* V_{cd})$. The suppression by λ^2 renders $\sin(2\beta_s)$ small; determining the CKM elements from a global fit to the data gives $\overline{\eta} = 0.343 \pm 0.015$ and leads to the SM prediction [28]

$$2\beta_s = 2.1^\circ \pm 0.1^\circ. \tag{14}$$

The combination of CDF, D0, ATLAS, and LHCb data on $B_s \to J/\psi\phi$ and of LHCb data on $B_s \to J/\psi\pi^+\pi^-$ gives [27]

$$2\beta_s^{\rm exp} = 0.7^\circ{}^{+5.2^\circ}_{-4.8^\circ}. \tag{15}$$

Of course, this average is dominated by the LHCb data. At this conference Jeroen van Leerdam has reported $2\beta_s = 0.1^\circ \pm 4.8^\circ_{\rm stat} \pm 1.5^\circ_{\rm syst}$ from the combined LHCb analysis of $B_s \to J/\psi K^+K^-$ and $B_s \to J/\psi\pi^+\pi^-$ [32]. In Eq. (13) an additional small contribution, the "penguin pollution" has been neglected. This –presently uncalculable– contribution inflicts an additional error of order 1° on $2\beta_s^{\rm exp}$ in Eq. (15).

2.3. Decay matrix Γ_{12}^q

The calculation of Γ_{12}^q, $q = d, s$, is needed for the width difference $\Delta\Gamma_q \simeq 2|\Gamma_{12}^q|\cos\phi_q$ and the semileptonic CP asymmetry $a_{\rm fs}^q = \frac{|\Gamma_{12}^q|}{|M_{12}^q|}\sin\phi_q$. The theoretical results of Refs. [22, 23, 33] have recently been updated to [25]

$$\begin{aligned} \phi_s &= 0.22^\circ \pm 0.06^\circ, & & \phi_d = -4.3^\circ \pm 1.4^\circ, \\ a_{\rm fs}^s &= (1.9 \pm 0.3)\cdot 10^{-5}, & \text{and} \quad & a_{\rm fs}^d = -(4.1 \pm 0.6)\cdot 10^{-4}. \end{aligned} \tag{16}$$

The prediction of Γ_{12}^q involves new operators in addition to Q in Eq. (6). However, the matrix element of Q comes with the largest coefficient, so that the ratio $\Delta\Gamma_q/\Delta m_q = |\Gamma_{12}^q|/|M_{12}^q|$ suffers from smaller hadronic uncertainties than $\Delta\Gamma_q$. Predicting $\Delta\Gamma_q$ with the help of the experimental values $\Delta m_d^{\rm exp} = 0.507\,{\rm ps}^{-1}$ and $\Delta m_s^{\rm exp}$ in Eq. (7) one infers from Ref. [25]:

$$\begin{aligned} \Delta\Gamma_d &= \frac{\Delta\Gamma_d}{\Delta m_d}\Delta m_d^{\rm exp} = (27 \pm 5)\cdot 10^{-4}\,{\rm ps}^{-1} \\ \Delta\Gamma_s &= \frac{\Delta\Gamma_s}{\Delta m_s}\Delta m_s^{\rm exp} = (0.090 \pm 0.018)\,{\rm ps}^{-1} \end{aligned} \tag{17}$$

The more recent update in Ref. [34] differs from Ref. [25] in two respects: The quoted results are obtained in a different renormalisation scheme and use the preliminary lattice results of [29] instead of the averages from [28]. Ref. [34] finds for $\Delta\Gamma_s$:

$$\Delta\Gamma_s = \frac{\Delta\Gamma_s}{\Delta m_s}\Delta m_s^{\rm exp} = (0.078 \pm 0.018)\,{\rm ps}^{-1}, \tag{18}$$

which is consistent with Eq. (17). In Eqs. (17) and (18) all uncertainties are added in quadrature, which may not be a conservative estimate of the overall error. The ranges comply with the LHCb measurement [32]

$$\Delta\Gamma_s^{\rm LHCb} = [0.116 \pm 0.018_{\rm stat} \pm 0.006_{\rm syst}]\,{\rm ps}^{-1} \tag{19}$$

and the world average [27]:

$$\Delta\Gamma_s^{\rm exp} = [0.089 \pm 0.012]\,{\rm ps}^{-1}. \tag{20}$$

3. New physics in $\mathrm{B}-\overline{\mathrm{B}}$ mixing

3.1. A model-independent analysis

If some NP amplitude adds to the $\mathrm{B_q}-\overline{\mathrm{B}}_\mathrm{q}$ mixing box diagram, both $|M_{12}^q|$ and ϕ_q may deviate from their SM predictions. The DØ experiment has measured [35–37]

$$\begin{aligned} A_{\rm SL}^{\rm D0} &= (0.532 \pm 0.039)a_{\rm fs}^d + (0.468 \pm 0.039)a_{\rm fs}^s \\ &= (-7.87 \pm 1.72 \pm 0.93)\cdot 10^{-3}, \end{aligned} \tag{21}$$

which is 3.9σ off the SM prediction inferred from Eq. (16),

$$A_{\rm SL} = (-0.20 \pm 0.03)\cdot 10^{-3}. \tag{22}$$

Adding a NP contribution $\phi_{d,s}^\Delta$ to either ϕ_d or ϕ_s in Eq. (16) may enhance $|A_{\rm SL}^{\rm D0}|$, but the very same contribution will also affect the value of β_s in Eq. (15) and the value of $\beta = \arg = \arg(-V_{tb}V_{td}^*/(V_{cb}V_{cd}^*))$ found from $a_{CP}(t)(B_d \to J/\psi K_S)$. One can parametrise NP in $\mathrm{B_d}-\overline{\mathrm{B}}_\mathrm{d}$ mixing and $\mathrm{B_s}-\overline{\mathrm{B}}_\mathrm{s}$ mixing by two complex parameters Δ_d and Δ_s:

$$M_{12}^q \equiv M_{12}^{\rm SM,q}\cdot\Delta_q, \qquad \Delta_q \equiv |\Delta_q|e^{i\phi_q^\Delta}.$$

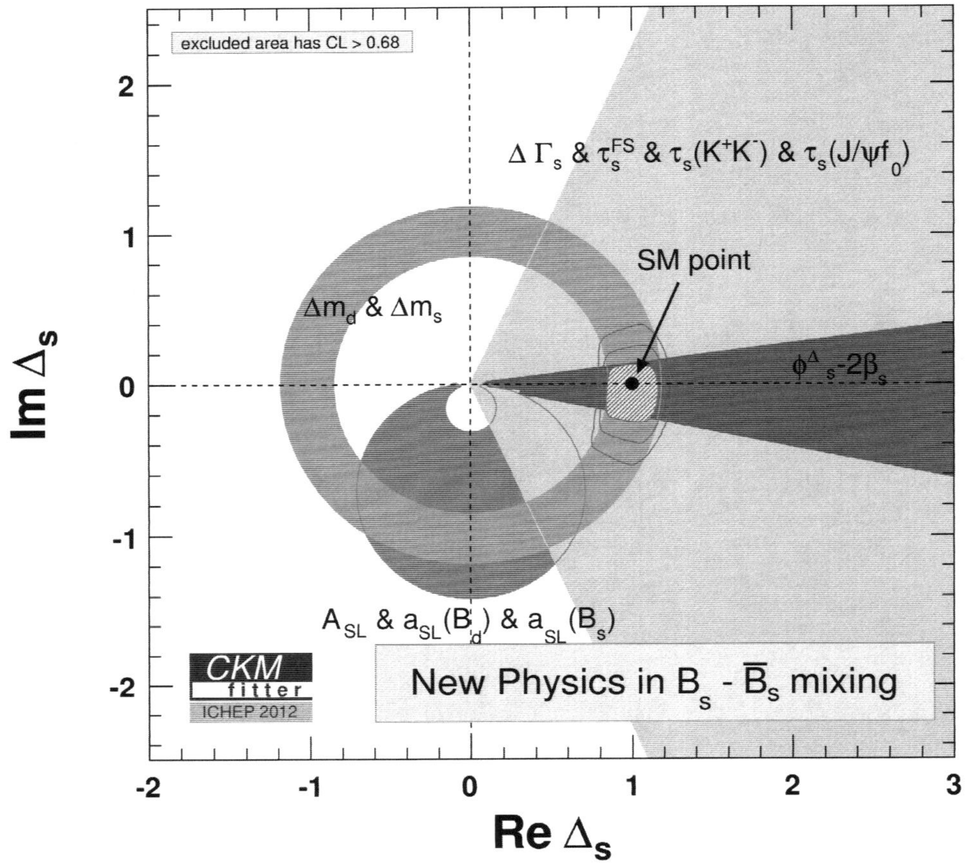

Figure 4. Allowed range for Δ_s, see [28, 39] for details.

In summer 2010, before the advent of precision data from LHCb, a global fit of all relevant flavour data to the CKM elements and Δ_d and Δ_s has resulted in a 3.6σ evidence of NP, with a large negative NP phase ϕ_s^Δ [38]. In spring 2012 this fit has been repeated [39]; Figs. 4 and 5 show an update of the results in Ref. [39] with the data of 2012 summer conferences [28]. The fit tries to accomodate $A_{\rm SL}$ in Eq. (21) and a slightly high value of the world average for $B(B \to \tau\nu)$ through $\phi_d^\Delta < 0$. Since $a_{CP}(t)(B_d \to J/\psi K_S)$ precisely fixes $2\beta + \phi_d^\Delta = 42.8^\circ \pm 1.6^\circ$, $\phi_d^\Delta < 0$ comes with a higher value of β than in the SM. ($\beta > \beta^{\rm SM}$ entails a larger value of $|V_{ub}|$ which governs $B(B \to \tau\nu)$.) Contrary to the situation in 2010, the Standard Model point $\Delta_s = \Delta_d = 1$ is merely disfavoured by 1 standard deviation, consistent with natural statistical fluctuations. At the best-fit point the problem with $A_{\rm SL}$ is only marginally alleviated.

One could relax the problem with $A_{\rm SL}$ without affecting $a_{CP}(t)(B_d \to J/\psi K_S)$ and $a_{CP}(t)(B_s \to J/\psi\phi)$ by postulating new physics in Γ_{12}^d or Γ_{12}^s [39–41]. Since Γ_{12}^s originates from Cabibbo-favoured tree-level decays, it can hardly be changed in a significant way without spoiling the value of the average decay width $\Gamma_s = (\Gamma_H^s + \Gamma_L^s)/2$ [38,39]. The LHCb measurement

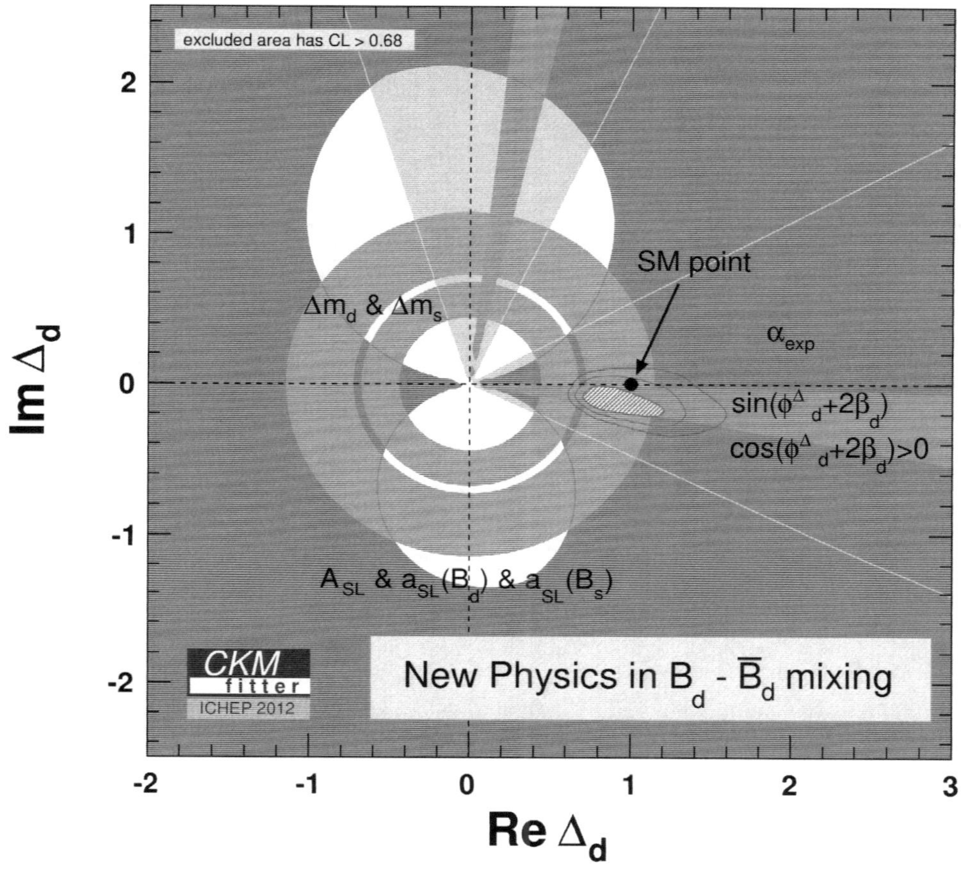

Figure 5. Allowed range for Δ_d, see [28, 39] for details.

$\Gamma_s^{\rm LHCb} = (0.6580 \pm 0.0054 \pm 0.0066)\,{\rm ps}^{-1}$ [32] implies

$$\frac{\Gamma_d}{\Gamma_s} = \frac{\tau_{B_s}}{\tau_{B_d}} = 0.997 \pm 0.013 \qquad (23)$$

in excellent agreement with the SM prediction $\tau_{B_s}/\tau_{B_d} = 0.998 \pm 0.003$ [25, 42]. NP in the doubly Cabibbo-suppressed quantity Γ_{12}^d is phenomenologically only poorly constrained, but requires a somewhat contrived model of NP.

3.2. A supersymmetric SO(10) model

The Minimal Supersymmetric Standard Model (MSSM) has many new sources of flavour violation, which all reside in the supersymmetry-breaking sector. It is no problem to get a big effect in a chosen FCNC process, but rather to suppress big effects elsewhere. This *supersymmetric flavour problem* is substantially alleviated with the lower bounds on the squark massed placed by ATLAS and CMS. Grand Unified Theories (GUT) offer the possibility to have "controlled" deviations from the CKM pattern of flavour violation in the quark sector: In GUTs quarks and leptons are unified in common symmetry multiplets, opening the possibility to observe the large lepton-flavour mixing encoded in the Pontecorvo-Maki-Nakagawa-Sakata (PMNS) matrix $U_{\rm PMNS}$ in quark flavour physics [43, 44]. Consider SU(5) multiplets:

$$\overline{\mathbf{5}}_{\mathbf{1}} = \begin{pmatrix} d_R^c \\ d_R^c \\ d_R^c \\ e_L \\ -\nu_e \end{pmatrix}, \qquad \overline{\mathbf{5}}_{\mathbf{2}} = \begin{pmatrix} s_R^c \\ s_R^c \\ s_R^c \\ \mu_L \\ -\nu_\mu \end{pmatrix}, \qquad \overline{\mathbf{5}}_{\mathbf{3}} = \begin{pmatrix} b_R^c \\ b_R^c \\ b_R^c \\ \tau_L \\ -\nu_\tau \end{pmatrix}.$$

If the observed large atmospheric neutrino mixing angle stems from a rotation of $\overline{\mathbf{5}}_{\mathbf{2}}$ and $\overline{\mathbf{5}}_{\mathbf{3}}$ in flavour space, it will induce a large $\tilde{b}_R - \tilde{s}_R$-mixing. Contrary to the situation with right-handed quark fields, rotations of right-handed squark fields in flavour space are physical because of the soft SUSY-breaking terms. The Chang–Masiero–Murayama (CMM) model has implemented this idea in a GUT based on the symmetry breaking chain SO(10)$\rightarrow$ SU(5) $\rightarrow$ SU(3)$\times$SU(2)$_L\times$U(1)$_Y$ [45–47]. In some weak basis the Yukawa matrix of down (s)quarks is diagonalised as

$$\mathsf{Y}_d = V_{\rm CKM}^* \begin{pmatrix} y_d & 0 & 0 \\ 0 & y_s & 0 \\ 0 & 0 & y_b \end{pmatrix} U_{\rm PMNS}$$

and the right-handed down squark mass matrix has the following diagonal form:

$$\mathsf{m}_{\tilde{d}}^2(M_Z) = \mathrm{diag}\left(m_{\tilde{d}}^2, m_{\tilde{d}}^2, m_{\tilde{d}}^2 - \Delta_{\tilde{d}}\right).$$

The real parameter $\Delta_{\tilde{d}}$ is calculated from renormalisation-group effects driven by the top-Yukawa coupling.

Rotating Y_d to diagonal form puts the large atmospheric neutrino mixing angle into $\mathsf{m}_{\tilde{d}}^2$:

$$U_{\rm PMNS}^\dagger\, \mathsf{m}_{\tilde{d}}^2 U_{\rm PMNS} = \begin{pmatrix} m_{\tilde{d}}^2 & 0 & 0 \\ 0 & m_{\tilde{d}}^2 - \frac{1}{2}\Delta_{\tilde{d}} & -\frac{1}{2}\Delta_{\tilde{d}} e^{i\xi} \\ 0 & -\frac{1}{2}\Delta_{\tilde{d}} e^{-i\xi} & m_{\tilde{d}}^2 - \frac{1}{2}\Delta_{\tilde{d}} \end{pmatrix} \tag{24}$$

The CP phase ξ affects CP violation in $\mathrm{B_s}-\overline{\mathrm{B}}_\mathrm{s}$ mixing!

In Ref. [47] we have confronted the CMM model with flavour data. The analysis involves seven parameters, which we have fixed by choosing values for the squark masses $M_{\tilde{u}}$, $M_{\tilde{d}}$ of right-handed up and down squarks, the trilinear term a_1^d of the first generation, the gluino mass $m_{\tilde{g}_3}$, $\tan\beta$, and the sign of the higgsino mass parameter μ. We have considered $\mathrm{B_s}-\overline{\mathrm{B}}_\mathrm{s}$ mixing, $b \to s\gamma$, $\tau \to \mu\gamma$, vacuum stability bounds, lower bounds on sparticle masses and the lower bound on the lightest Higgs boson. From these inputs first universal SUSY-breaking terms defined at a fundamental scale near the Planck scale have been determined through the renormalisation group equations (RGE). Subsequently we have used the RGE to determine all low-energy parameters, with the MSSM as the low-energy theory.

Two experimental results of the year 2012 put the CMM model under pressure: First, the sizable neutrino mixing angle θ_{13} leads to an unduly large effect in $B(\mu \to e\gamma)$. In Eq. (24) I have tacitly assumed a $U_{\rm PMNS}$ with tri-bimaximal mixing, corresponding to $\theta_{13} = 0$. For the actual value $\theta_{13} \approx 8^\circ$ the $(1,2)$ element of the charged slepton mass matrix gets large, and one has to resort to much larger sfermion masses than those considered in Ref. [47]. Second, the Higgs mass of 126 GeV challenges the model and we are unable to find parameters which simultaneously satisfy the Higgs mass constraint and the experimental upper bound on $B(\mu \to e\gamma)$ [48]. The problem with the Higgs mass could be circumvented by considering the NMSSM as the low-energy theory.

4. Conclusions

LHCb has provided us with a significantly better insight into the $B_s - \overline{B}_s$ mixing complex. Δm_s and $\Delta\Gamma_s$ comply with the SM, but we need better lattice data for the hadronic matrix elements involved. Theoretical uncertainties still permit an $\mathcal{O}(20\%)$ NP contribution to the $B_s - \overline{B}_s$ amplitude M_{12}^s. While in 2010 the DØ result for A_{SL} could be explained in scenarios with NP only in $M_{12}^{d,s}$, the LHCb data on $B_s \to J/\psi\phi$ now prohibit this solution. An alternative explanation invoking new physics in Γ_{12}^s is not viable, because this will spoil the ratio Γ_s/Γ_d which agrees well with the SM prediction. Maybe it is worthwile to look at NP in Γ_{12}^d, although this possibility leads to somewhat contrived models. Models of GUT flavour physics with $\widetilde{b} \to \widetilde{s}$ transition driven by the atmospheric neutrino mixing angle are under pressure from the large value of θ_{13}, which induces a too large $B(\mu \to e\gamma)$. Moreover, in the studied CMM model it seems impossible to accomodate the measured value of the lightest Higgs mass, if one insists on the MSSM as the low-energy theory.

Acknowledgments

I am grateful for the invitation to this stimulating and enjoyable conference! The work presented in this talk is supported by BMBF grant 05H12VKF.

5. References

[1] Gabbiani F, Gabrielli E, Masiero A and Silvestrini L 1996 *Nucl.Phys.* **B477** 321–352 (*Preprint* hep-ph/9604387)
[2] Virto J 2009 *JHEP* **0911** 055 (*Preprint* 0907.5376)
[3] Bertolini S, Borzumati F, Masiero A and Ridolfi G 1991 *Nucl.Phys.* **B353** 591–649
[4] Oshimo N 1993 *Nucl.Phys.* **B404** 20–41
[5] Carena M S, Garcia D, Nierste U and Wagner C E 2001 *Phys.Lett.* **B499** 141–146 (*Preprint* hep-ph/0010003)
[6] Degrassi G, Gambino P and Giudice G 2000 *JHEP* **0012** 009 (*Preprint* hep-ph/0009337)
[7] Buras A J, Chankowski P H, Rosiek J and Slawianowska L 2003 *Nucl.Phys.* **B659** 3 (*Preprint* hep-ph/0210145)
[8] Baer H, Brhlik M, Castano D and Tata X 1998 *Phys.Rev.* **D58** 015007 (*Preprint* hep-ph/9712305)
[9] Crivellin A and Nierste U 2010 *Phys.Rev.* **D81** 095007 (*Preprint* 0908.4404)
[10] Crivellin A and Mercolli L 2011 *Phys.Rev.* **D84** 114005 (*Preprint* 1106.5499)
[11] Aaij R *et al.* (LHCb Collaboration) 2013 *Phys.Rev.Lett.* **110** 021801 (*Preprint* 1211.2674)
[12] Babu K and Kolda C F 2000 *Phys.Rev.Lett.* **84** 228–231 (*Preprint* hep-ph/9909476)
[13] Huang C S, Liao W, Yan Q S and Zhu S H 2001 *Phys.Rev.* **D63** 114021 (*Preprint* hep-ph/0006250)
[14] Dedes A, Dreiner H K and Nierste U 2001 *Phys.Rev.Lett.* **87** 251804 (*Preprint* hep-ph/0108037)
[15] Isidori G and Retico A 2001 *JHEP* **0111** 001 (*Preprint* hep-ph/0110121)
[16] Buras A J, Chankowski P H, Rosiek J and Slawianowska L 2002 *Phys.Lett.* **B546** 96–107 (*Preprint* hep-ph/0207241)
[17] Gorbahn M, Jager S, Nierste U and Trine S 2011 *Phys.Rev.* **D84** 034030 (*Preprint* 0901.2065)
[18] Hofer L, Nierste U and Scherer D 2009 *JHEP* **0910** 081 (*Preprint* 0907.5408)
[19] Altmannshofer W, Buras A J, Gori S, Paradisi P and Straub D M 2010 *Nucl.Phys.* **B830** 17–94 (*Preprint* 0909.1333)
[20] Crivellin A 2011 *Phys.Rev.* **D83** 056001 (*Preprint* 1012.4840)
[21] Altmannshofer W, Carena M, Shah N R and Yu F 2013 *JHEP* **1301** 160 (*Preprint* 1211.1976)
[22] Beneke M, Buchalla G, Lenz A and Nierste U 2003 *Phys.Lett.* **B576** 173–183 (*Preprint* hep-ph/0307344)
[23] Ciuchini M, Franco E, Lubicz V, Mescia F and Tarantino C 2003 *JHEP* **0308** 031 (*Preprint* hep-ph/0308029)
[24] Lenz A and Nierste U 2007 *JHEP* **0706** 072 (*Preprint* hep-ph/0612167)
[25] Lenz A and Nierste U 2011 *Proceedings, 6th International Workshop, CKM 2010, Warwick, UK, September 6-10, 2010,* ed Gershon T (eConf C100906) (*Preprint* 1102.4274)
[26] Buras A J, Jamin M and Weisz P H 1990 *Nucl.Phys.* **B347** 491–536
[27] The Heavy Flavor Averaging Group (HFAG) 2012 http://www.slac.stanford.edu/xorg/hfag
[28] CKMfitter Group (J Charles et al) 2012 Eur. Phys. J. C41, 1-131 (2005) [hep-ph/0406184], updated results from http://ckmfitter.in2p3.fr
[29] Bouchard C, Freeland E, Bernard C, El-Khadra A, Gamiz E *et al.* 2011 *PoS* **LATTICE2011** 274 (*Preprint* 1112.5642)

[30] Wolfenstein L 1983 *Phys.Rev.Lett.* **51** 1945
[31] Buras A J, Lautenbacher M E and Ostermaier G 1994 *Phys.Rev.* **D50** 3433–3446 (*Preprint* hep-ph/9403384)
[32] van Leerdam J (LHCb Collaboration) 2013
[33] Beneke M, Buchalla G, Greub C, Lenz A and Nierste U 1999 *Phys.Lett.* **B459** 631–640 (*Preprint* hep-ph/9808385)
[34] Nierste U 2013 *to appear in: Proceedings, 7th International Workshop, CKM 2012, Cincinnati, Ohio, USA, 28 Sep - 2 Oct 2012* ed Schwartz A J (*Preprint* 1212.5805)
[35] Abazov V M *et al.* (D0 Collaboration) 2010 *Phys.Rev.* **D82** 032001 (*Preprint* 1005.2757)
[36] Abazov V M *et al.* (D0 Collaboration) 2010 *Phys.Rev.Lett.* **105** 081801 (*Preprint* 1007.0395)
[37] Abazov V M *et al.* (D0 Collaboration) 2011 *Phys.Rev.* **D84** 052007 (*Preprint* 1106.6308)
[38] Lenz A, Nierste U, Charles J, Descotes-Genon S, Jantsch A *et al.* 2011 *Phys.Rev.* **D83** 036004 (*Preprint* 1008.1593)
[39] Lenz A, Nierste U, Charles J, Descotes-Genon S, Lacker H *et al.* 2012 *Phys.Rev.* **D86** 033008 (*Preprint* 1203.0238)
[40] Dighe A, Kundu A and Nandi S 2010 *Phys.Rev.* **D82** 031502 (*Preprint* 1005.4051)
[41] Bobeth C and Haisch U 2013 *APP* **B44** 127 (*Preprint* 1109.1826)
[42] Keum Y Y and Nierste U 1998 *Phys.Rev.* **D57** 4282–4289 (*Preprint* hep-ph/9710512)
[43] Moroi T 2000 *JHEP* **0003** 019 (*Preprint* hep-ph/0002208)
[44] Moroi T 2000 *Phys.Lett.* **B493** 366–374 (*Preprint* hep-ph/0007328)
[45] Chang D, Masiero A and Murayama H 2003 *Phys.Rev.* **D67** 075013 (*Preprint* hep-ph/0205111)
[46] Harnik R, Larson D T, Murayama H and Pierce A 2004 *Phys.Rev.* **D69** 094024 (*Preprint* hep-ph/0212180)
[47] Girrbach J, Jager S, Knopf M, Martens W, Nierste U *et al.* 2011 *JHEP* **1106** 044 (*Preprint* 1101.6047)
[48] Nierste U and Stockel J 2013 work in progress

Journal of Physics: Conference Series **447** (2013) 012018 doi:10.1088/1742-6596/447/1/012018

Combining Flavour and CP Symmetries

Ferruccio Feruglio

Dipartimento di Fisica e Astronomia 'G. Galilei', Università di Padova,
and INFN, Sezione di Padova,
Via Marzolo 8, I-35131 Padua, Italy

E-mail: `feruglio@pd.infn.it`

Abstract. I shortly review the impact of the most recent neutrino oscillation data on our attempts to construct a realistic model for neutrino masses and mixing angles. Models based on anarchy and its variants remain an open possibility, reinforced by the latest experimental findings. Many models based on discrete symmetries no longer work in their simplest realizations. I illustrate several proposals that can rescue discrete symmetries. In particular I discuss the possibility of combining discrete flavour symmetries and CP, and I describe a recently proposed symmetry breaking pattern that allows to predict all mixing parameters, angles and phases, in terms of a single real unknown. I analyze several explicit examples of this construction, providing new realistic mixing patterns.

1. Introduction

The discovery of neutrino oscillations brought lot of excitement in particle physics. It still represents the first and, in some respect, unique evidence of new physics. The smallness of neutrino masses evokes the breaking of the total lepton number, thus making neutrinos a privileged observatory to investigate energy scales that would otherwise be unaccessible. Moreover lepton mixing properties together with those of the quark sector can shed a new light on the flavour mystery. Perhaps form this information we can identify some principle allowing us to describe in a more economic and rational way the multitude of Yukawa interactions needed to account for fermion masses and mixing angles.

The early data on neutrino oscillations were compatible with a maximal atmospheric mixing angle and a vanishing reactor angle. The whole mixing pattern was consistent with the very simple ansatz [1]:

$$\sin^2\theta_{23} = \frac{1}{2} \quad , \qquad \sin^2\theta_{12} = \frac{1}{3} \quad , \qquad \sin^2\theta_{13} = 0 \quad . \tag{1}$$

Despite the large experimental errors affecting, until recently, both the atmospheric and the reactor mixing angles, this circumstance was taken by many of us as evidence for a symmetry principle beyond the data. Discrete symmetries based on small groups such as S_3, A_4 [2] and S_4 [3] were soon recognized at the basis of efficient mechanisms able to reproduce the tribimaximal (TB) pattern in eq. (1). There was also a certain confidence that deviations from the TB pattern had to be small, since the solar mixing angle was soon measured to a very good precision, about a couple of degrees, in impressive agreement with eq. (1). On this basis we could hope that the TB ansatz were correct to few degrees. We have now evidence at the 10σ level that θ_{13} is non-vanishing [4]. Its size is comparable to that of the Cabibbo angle. We also have a first hint for a non-maximal atmospheric mixing angle. Results from a recent global fit to neutrino oscillations are reported in table 1.

$\sin^2\theta_{12}$	0.30 ± 0.013
$\sin^2\theta_{23}$	$0.41^{+0.037}_{-0.025} \oplus 0.59^{+0.021}_{-0.022}$
$\sin^2\theta_{13}$	0.023 ± 0.0023
Δm^2_{21}	$(7.50 \pm 0.185) \times 10^{-5}\ eV^2$
$\Delta m^2_{31}(N)$	$(2.47^{+0.069}_{-0.067}) \times 10^{-3}\ eV^2$
$\Delta m^2_{31}(I)$	$-(2.43^{+0.042}_{-0.065}) \times 10^{-3}\ eV^2$

Table 1. Results of a global fit to neutrino oscillations, from ref. [5] There are two best fit values for the atmospheric mixing angle θ_{23} due to the presence of two minima in the χ^2 function. The labels N and I refer to normal and inverted ordering, respectively.

There are still several open questions concerning neutrino properties. We do not know whether the total lepton number L is violated or not, whether the neutrino mass ordering is normal or inverted. The Dirac phase δ_{CP} is essentially unconstrained and the absolute scale of neutrino masses is still undetermined. Despite the remaining unknowns, our knowledge of neutrinos has greatly improved thanks to the most recent experiments and many models have been ruled out. We might expect that some coherent theoretical description had already become apparent from the data. Unfortunately this is not the case. Present data can still be described by widely different approaches.

2. Anarchy

One possible interpretation of the current results is in term of anarchy [6], which does not recognise any special pattern in the neutrino data. Lepton mixing angles and neutrino mass ratios are generic order one parameters, the smallness of θ_{13} and $\Delta m^2_{sol}/\Delta m^2_{atm}$ being accidental features with no special meaning. The actual size of θ_{13} and the indications in favour of a non maximal θ_{23} have strengthened this point of view.

It is worth saying that the idea of anarchy has several good aspects. Anarchy can be easily incorporated in valuable theoretical frameworks. For instance it can be realized in SU(5) Grand Unified Theories (GUTs), by the inclusion of a Froggatt-Nielsen $U(1)_{FN}$ group [7] under which the three generations of pentaplets $\bar{5}$ are assumed to have the same charge. It can also be realized in models with extra dimensions (ED). For example, standard model fermions in the bulk of one extra dimension, spanning an interval of finite length, develop zero modes whose profiles are controlled by bulk mass parameters. If the Higgs doublet is localized at one of the two edges of the five dimensional interval, the Yukawa couplings mimic those of a Froggatt-Nielsen setup, with the role of the Froggatt-Nielsen charges played by the fermion bulk masses. Anarchy is also compatible with the known solutions to the hierarchy problem, such as supersymmetry (SUSY) and warped ED. Finally, models based on a Froggatt-Nielsen $U(1)_{FN}$ symmetry are flexible enough to allow for the implementation of several variants of anarchy, where neutrino $U(1)_{FN}$ charges are not necessarily equal. Such variants proved successful in reproducing, at the level of order of magnitudes, all fermion masses and mixing angles [8].

The main drawback of anarchy and its variants is the difficulty to identify a quantitative test of the idea. Models based on a Froggatt-Nielsen $U(1)_{FN}$ symmetry contain a large number of order-one independent parameters, thus preventing predictions beyond the order-of-magnitude accuracy. By accepting the principle of anarchy, the best we can do is to estimate probability distributions for the physical observables and evaluate the likelihood of our universe. We loose all the potentiality offered by the excellent experimental precision with which fermion masses and mixing angles are known today. Moreover, if new degrees of freedom carrying flavour charges are present at the TeV scale, as expected in the known solutions to the hierarchy problem, new sources of flavour change and/or CP violation appear and additional mechanisms should be invoked to avoid conflict with the present data.

3. Discrete Symmetries

An alternative description of the data is based on flavour symmetries. There are many types of flavour symmetries, global or local, continuous or discrete. There is no evidence for exact flavour symmetries and one of the most important aspects in model building is represented by the symmetry breaking. There is a large freedom related to the choice of the symmetry breaking sector and to the characteristic symmetry breaking scale. It is impossible to give here even a short account of all types of models. A special class of models is the one based on discrete flavour symmetries [9], adopted to reproduce some simple pattern U^0_{PMNS}, which provides a first approximation to the observed lepton mixing matrix U_{PMNS}. We have

$$U_{PMNS} = U^0_{PMNS} + O(u) \quad , \tag{2}$$

where $O(u)$ denotes a set of small corrections, proportional to some adimensional parameter u. Such an approach was well motivated before 2012. Several examples of leading order (LO) patterns U^0_{PMNS} have been suggested. A well-known example is the TB one, given in eq. (1) and described by a mixing matrix U_{TB} of the type

$$U_{TB} = \begin{pmatrix} \frac{2}{\sqrt{6}} & \frac{1}{\sqrt{3}} & 0 \\ -\frac{1}{\sqrt{6}} & \frac{1}{\sqrt{3}} & -\frac{1}{\sqrt{2}} \\ -\frac{1}{\sqrt{6}} & \frac{1}{\sqrt{3}} & \frac{1}{\sqrt{2}} \end{pmatrix} \approx \begin{pmatrix} 0.82 & 0.58 & 0 \\ -0.41 & 0.58 & -0.71 \\ -0.41 & 0.58 & 0.71 \end{pmatrix} \quad . \tag{3}$$

This pattern can be easily derived from small discrete groups such as A_4 or S_4.

The general mechanism allowing to constrain U^0_{PMNS} by a discrete symmetry is very simple. The underlying model is assumed to be invariant under a discrete flavour symmetry G_f, broken down in such a way that neutrino and charged lepton sectors have different residual symmetries, at least in a LO approximation where small effects are neglected. The combination $m_l^\dagger m_l$ of the charged lepton mass matrix m_l is invariant under the subgroup G_e of G_f, while the neutrino mass matrix m_ν is invariant under the subgroup G_ν. It is interesting to note that, if neutrinos are of Majorana type, as assumed here, the most general group leaving m_ν invariant (and the individual masses m_i unconstrained) is $Z_2 \times Z_2$, a finite group. The subgroup G_e can be continuous, but G_e discrete remains the simplest option. We require a sufficiently large G_e to distinguish the three charged leptons. For instance we can choose $G_e = Z_n$ $(n \geq 3)$ or $G_e = Z_2 \times Z_2$. Once G_e and G_ν have been chosen inside G_f, the embedding automatically fixes the relative alignment of $m_l^\dagger m_l$ and m_ν in flavour space. Lepton masses are unconstrained and U^0_{PMNS} is determined up to Majorana phases and up to permutations of rows and columns. This freedom apart, we can predict the three mixing angles θ^0_{ij} and the Dirac phase δ^0_{CP}. In most concrete models, where symmetry breaking is achieved via vacuum expectation values (VEVs) of a set of flavons, the LO results are modified by small corrections, as in eq. (2). In the specific case $U^0_{PMNS} = U_{TB}$ these corrections were expected to be very small, of the order of few percent, not to spoil the good agreement in the predicted value of the solar mixing angle. On this basis many models reproducing U_{TB} at the leading order predicted θ_{13} not larger than few degrees, later proved wrong by experiments. Discrete flavour symmetries can also be extended to quarks and even incorporated in GUTs, but in the existing constructions the symmetry has to be badly broken in the quark sector, leaving very few hints at the level of physical quantities.

Due to the sizable deviation of the latest data from the TB ansatz, many people have contemplated several modifications of the simplest models based on discrete symmetries. If we keep adopting $U^0_{PMNS} = U_{TB}$ as LO approximation, perhaps the most economic way to reproduce the actual value of θ_{13} is to introduce large correction terms, $O(u) \approx 0.2$. This is also viable in some scheme where U^0_{PMNS} differs substantially from U_{TB}, such as the so-called bimaximal (BM) mixing. Introducing large corrections has the disadvantage that beyond the LO the number of independent contributions is generally quite large. If their typical size is about 0.2, all mixing angles tend to be affected by generic corrections of this type and predictability is lost [10]. Moreover large correction terms are dangerous if new sources of flavour changing and/or CP violation are present at the TeV scale.

Another possibility is to look for alternative LO approximations where θ_{13} is closer to the measured value. Remarkably, several groups G_f giving rise to more realistic LO approximations have been found. Of particular interest are the groups leading to special form of trimaximal (TM) mixing:

$$U^0_{PMNS} = U_{TB}\, U_{13}(\alpha) \tag{4}$$

where $U_{13}(\alpha)$ describes a rotation in the 13 plane by an angle α. Early examples are the groups of the series $\Delta(6n^2)$ [11]. The angle α is fixed by n. For $n = 4(8)$ we have $\alpha = \pm 1/12(\pm 1/24)$ and $\sin^2\theta^0_{13} = 0.045(0.011)$. The Dirac phase is zero (modulo π).

A further possibility is to relax the symmetry requirements. It is worth mentioning that the smallest group reproducing TB mixing through the breaking down to $G_e = Z_3$ and $G_\nu = Z_2 \times Z_2$ is S_4. In the basis where charged leptons are diagonal, we can identify one of the two parities in G_ν with that generated by the so-called $\mu\tau$ exchange symmetry, directly responsible for the vanishing of θ_{13} and for θ_{23} being maximal. If the residual symmetry G_ν is reduced from $Z_2 \times Z_2$ down to Z_2 by eliminating the $\mu\tau$ exchange symmetry, the transformations belonging to G_e and G_ν only generate A_4, not the whole S_4. Assuming such a breaking pattern we find that the predicted mixing is again of TM type, as in eq. (4), but $U_{13}(\alpha)$ generalises to a unitary matrix, parametrized by a rotation angle α and a phase, both unconstrained [12]. We obtain a testable sum rule, which for small θ_{13} reads

$$\sin^2\theta_{23} = \frac{1}{2} + \frac{1}{\sqrt{2}}\sin\theta_{13}\cos\delta_{CP} + O(\sin^2\theta_{13}) \quad . \tag{5}$$

Explicit models based on A_4 realizing such a breaking pattern were indeed proposed before the measurement of θ_{13} [13]. The possibility of reducing the residual symmetry G_ν to Z_2 can be systematically investigated [14]. As in the previous example, this framework only predicts two combinations of the three mixing angles.

4. Combining Discrete Symmetries and CP

A more recent possibility, which I would like to illustrate more in detail in this talk, is to combine discrete and CP symmetries and explore the symmetry breaking patterns such a combination can give rise to. This idea is not new and relies on many existing examples and suggestions. A well-known example is that of the so-called $\mu\tau$ reflection symmetry [15, 16] (not to be confused with the $\mu\tau$ exchange symmetry), which exchanges a muon (tau) neutrino with a tau (muon) antineutrino in the charged lepton mass basis. If such a symmetry is imposed, the atmospheric mixing angle is predicted to be maximal, while θ_{13} is in general non-vanishing for a maximal Dirac phase δ. Models combining S_4 and CP can be found in the recent literature [17, 18]. Other approaches dealing with discrete symmetries and CP are illustrated in [19–21].

In ref. [22] we have proposed a general formalism which combines CP with a discrete flavour symmetry. Consider a flavour symmetry group G_f and a set of fields ϕ transforming in some representation ρ of G_f:

$$\phi'(x) = \rho(g)\,\phi(x) \qquad\qquad g \in G_f \quad , \tag{6}$$

where $\rho(g)$ is a unitary matrix. We can define a CP transformation on ϕ as follows [23]

$$\phi'(x) = X\ \phi^*(x_{CP}) \qquad x_{CP} \equiv (x^0, -\vec{x}) \quad , \tag{7}$$

where X is a matrix in flavour space. If the fields ϕ are fermions, the well-known action of CP on spinor indices is understood. The choice of X is not arbitrary, but has to fulfill certain consistency conditions. First, we require the matrix X to be unitary and symmetric,

$$XX^\dagger = XX^* = \mathbb{1}\ . \tag{8}$$

In the presence of a flavour symmetry, requiring X symmetric is not the most general option, but if we do so then $CP^2 = 1$ automatically holds. A second condition arises by considering the action of CP combined with some transformation of G_f [22, 24]. For any element g of the group G_f, an element g' belonging to G_f should exist such that

$$\left(X^{-1}\rho(g)\ X\right)^* = \rho(g') \quad . \tag{9}$$

Notice that in general g and g' are distinct. As has been shown in [22], the mathematical structure of the group G_{CP} comprising G_f and CP is of the form $G_{CP} = G_f \rtimes H_{CP}$ with H_{CP} being the parity group generated by CP.

In our proposal we consider a theory invariant under G_{CP} with the three generations of lepton doublets l in some representation ρ of G_f

$$l'(x) = \rho(g)\ l(x) \quad . \tag{10}$$

Under CP we have

$$l'(x) = X\ l^*(x_{CP}) \quad , \tag{11}$$

where X satisfies both eq. (8) and eq. (9). We assume that in some limit of the theory G_{CP} is broken to the subgroups G_e and G_ν in the charged lepton and neutrino sectors, respectively. G_e is a subgroup of G_f generated by a set of elements Q_i, while $G_\nu = Z_2 \times CP$ is the direct product of a parity contained in G_f, generated by the element Z, and CP, generated by X. Notice that while Q_i, Z and X are defined in the representation ρ, for simplicity we make no distinction between (Q_i, Z, X) and the corresponding abstract elements of G_{CP} they represent. The subgroup G_ν involves the direct product between Z_2 and CP, and thus the transformations described by Z and X should commute. This gives rise to the condition

$$\left(X^{-1}Z\ X\right)^* = Z \quad , \tag{12}$$

a special version of eq. (9) with $g = g'$. Given a discrete group G_f and a parity subgroup generated by Z, eqs. (8), (9) and (12) can be read as a set of constraints on X, i.e. on the possible CP definitions we can adopt to realize the desired symmetry breaking pattern. The residual symmetries G_e and G_ν imply the following conditions on $m_l^\dagger m_l$ and m_ν:

$$Q_i^\dagger(m_l^\dagger m_l)Q_i = (m_l^\dagger m_l) \quad , \qquad Z^T m_\nu Z = m_\nu \quad , \qquad X m_\nu X = m_\nu^* \quad . \tag{13}$$

From these conditions we can derive the mixing matrix U^0_{PMNS}, and we find that in general it can be parametrized by one real parameter θ, ranging from 0 to π:

$$U^0_{PMNS} = U^0_{PMNS}(Q_i, Z, X, \theta) \qquad 0 \le \theta \le \pi \quad . \tag{14}$$

Mixing angles and phases, both Dirac and Majorana, are then predicted as a function of θ, modulo the ambiguity related to the freedom of permuting rows and columns and to the intrinsic parity of neutrinos. The formalism is completely invariant under any change of basis in field space. The physical results only depend on G_{CP} and the residual symmetries specified by (Q_i, Z, X).

5. The case $G_f = S_4$

To exemplify our results, in ref. [22] we have perform an exhaustive analysis of the case $G_f = S_4$. The group S_4 can be defined in terms of three generators S, T and U [25] which fulfill the following relations

$$\begin{aligned} &S^2 = E\ , \ \ T^3 = E\ , \ \ U^2 = E\ , \\ &(ST)^3 = E\ , \ \ (SU)^2 = E\ , \ \ (TU)^2 = E\ , \ \ (STU)^4 = E \end{aligned} \tag{15}$$

with E being the neutral element of S_4. The generators S and T alone give rise to the group A_4. The group S_4 has five irreducible representations: $\mathbf{1}$, $\mathbf{1'}$, $\mathbf{2}$, $\mathbf{3}$ and $\mathbf{3'}$. We assign the three generations of left-handed leptons to the faithful representation $\mathbf{3'}$ (equivalent results are obtained by choosing the representation $\mathbf{3}$) and for this representation we adopt a basis where the elements S, T and U are represented by the real matrices [25]

$$S = \begin{pmatrix} -1 & 0 & 0 \\ 0 & 1 & 0 \\ 0 & 0 & -1 \end{pmatrix}\ , \quad T = \frac{1}{2}\begin{pmatrix} 1 & \sqrt{2} & 1 \\ \sqrt{2} & 0 & -\sqrt{2} \\ -1 & \sqrt{2} & -1 \end{pmatrix}\ , \quad U = \begin{pmatrix} 1 & 0 & 0 \\ 0 & 1 & 0 \\ 0 & 0 & -1 \end{pmatrix}\ . \tag{16}$$

We have considered all possible choices of G_e and G_ν matching our symmetry breaking pattern. We have found that all the independent physical results are exhausted by considering the representative cases listed in table 2.

G_e	Q_i
Z_3	T
Z_4	STU
$Z_2 \times Z_2$	(TST^2S, UT^2)

G_ν	Z	X
$Z_2 \times CP$	S	$X_i\ (i = 1, ..., 6)$
$Z_2 \times CP$	SU	$X_i\ (i = 1, ..., 4)$
$Z_2 \times CP$	U	$X_i\ (i = 1, ..., 4)$

Table 2. Representative choice of generators Q_i for the subgroup G_e and (Z, X) for the subgroup G_ν. For any Z chosen in S_4 only a finite number of CP transformations X_i satisfying eqs. (8,9,12) are allowed. Their expressions in the representation $3'$ are given in eq. (17).

For any Z chosen in S_4 only a finite number of CP transformations X_i satisfying eqs. (8,9,12) are allowed. For Z chosen as in table 1, the admissible CP transformations X_i $(i = 1, ..., 6)$ are given by:

$$\begin{aligned} &X_1 = \begin{pmatrix} 1 & 0 & 0 \\ 0 & 1 & 0 \\ 0 & 0 & 1 \end{pmatrix}\ , \quad X_2 = \begin{pmatrix} -1 & 0 & 0 \\ 0 & 1 & 0 \\ 0 & 0 & -1 \end{pmatrix}\ , \quad X_3 = \begin{pmatrix} 1 & 0 & 0 \\ 0 & 1 & 0 \\ 0 & 0 & -1 \end{pmatrix}\ , \\ &X_4 = \begin{pmatrix} -1 & 0 & 0 \\ 0 & 1 & 0 \\ 0 & 0 & 1 \end{pmatrix}\ , \quad X_5 = \begin{pmatrix} 0 & 0 & -1 \\ 0 & -1 & 0 \\ -1 & 0 & 0 \end{pmatrix}\ , \quad X_6 = \begin{pmatrix} 0 & 0 & 1 \\ 0 & -1 & 0 \\ 1 & 0 & 0 \end{pmatrix}\ . \end{aligned} \tag{17}$$

These transformations are defined up to an irrelevant overall phase. X_1 is the canonical CP transformation.

Several interesting cases arises when $G_e = Z_3$. Two of them are given in table 3, in terms of the generators (Q_i, Z, X).

case	Q_i	Z	X
I	T	S	X_1
IV	T	SU	X_1

Table 3. Choice of (Q_i, Z, X) defining cases I and IV. The subgroup G_e generated by T is Z_3.

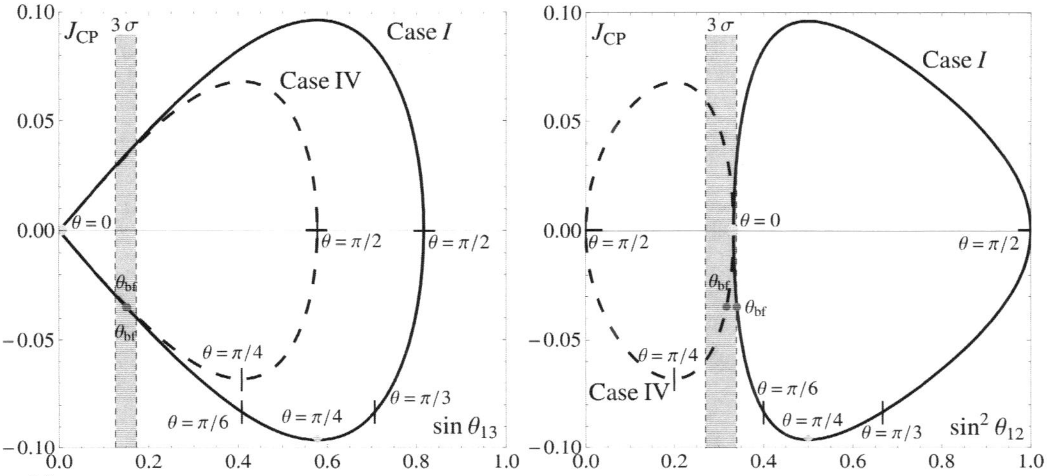

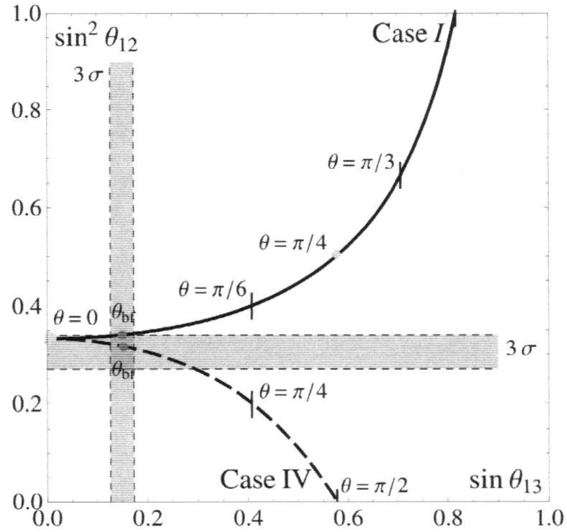

Figure 1: Results for the mixing parameters $\sin\theta_{13}$, $\sin^2\theta_{12}$ and J_{CP} for Case I (straight line) and Case IV (dashed line). We mark the value $\theta_{\rm bf}$ of the parameter θ for which the χ^2 functions have a global minimum with a red dot. For better guidance of the eye we also mark $\theta = \pi/n$, $n = 6, 4, 3, 2$ on the curves. The shown $3\,\sigma$ ranges for the mixing angles are taken from [5].

These two cases correspond to two particular realizations of the $\mu\tau$ reflection symmetry. They predict a maximal atmospheric mixing angle, a maximal Dirac phase and vanishing Majorana phases. The solar and reactor angles are given by

$$\sin^2\theta_{12} = \begin{cases} \dfrac{1}{2+\cos 2\theta} & \text{(case I)} \\ \dfrac{\cos^2\theta}{2+\cos^2\theta} & \text{(case IV)} \end{cases} \qquad \sin^2\theta_{13} = \begin{cases} \dfrac{2}{3}\sin^2\theta & \text{(case I)} \\ \dfrac{1}{3}\sin^2\theta & \text{(case IV)} \end{cases} \quad . \tag{18}$$

The correlations between $\sin^2\theta_{12}$, $\sin\theta_{13}$ and the CP invariant J_{CP} are displayed in fig. 1. By optimizing the choice of θ through the minimization of the χ^2 function, we find a reasonable

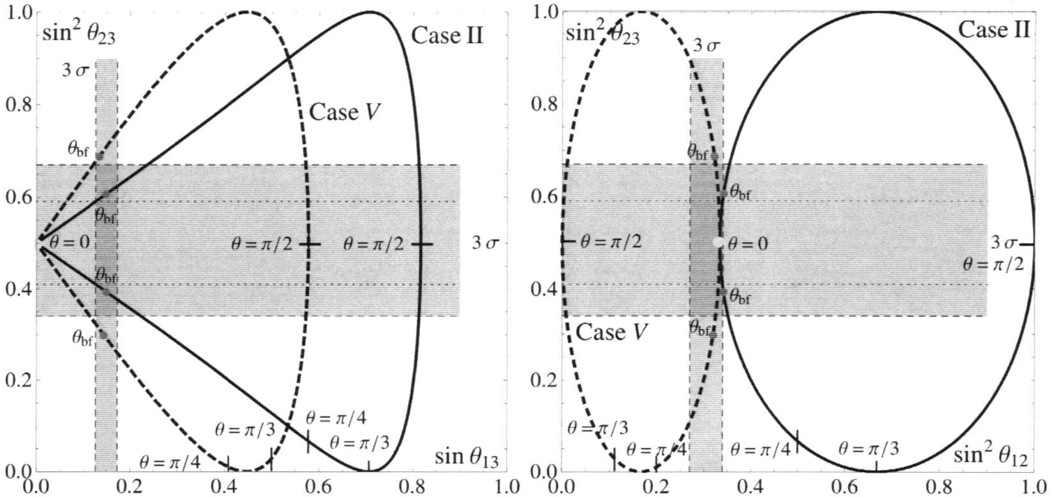

Figure 2: Results for the atmospheric, solar and reactor mixing angles for Case II (straight line) and Case V (dashed line). We mark the value $\theta_{\rm bf}$ of the parameter θ for which the χ^2 functions have a global minimum with a red dot. For better guidance of the eye we also mark $\theta = \pi/n$, $n = 4, 3, 2$ on the curves. The shown 3σ ranges for the mixing angles and the best fit values of the atmospheric mixing angle are taken from [5]. The plot for $\sin^2\theta_{12}$ and $\sin\theta_{13}$ is the same as for Case I and Case IV and can be found in figure 1.

agreement with the data. Notice that the $\mu\tau$ reflection symmetry is not explicit in the chosen basis, where the CP transformation X_1 in the $3'$ representation is canonical. To make a direct contact with the $\mu\tau$ reflection symmetry we have to move to the basis where the T generator and the combination $m_l^\dagger m_l$ are diagonal. In a concrete model realizing either case I or case IV the predicted value of the Dirac phase would only depend on the symmetry breaking pattern, and not on the specific values of Lagrangian parameters realizing it. In this sense cases I and IV could provide examples of "geometrical" CP violation.

Not all the cases considered in our analysis necessarily lead to non-trivial CP phases. For instance in cases II and V, defined in table 4, we find that all phases are trivial.

case	Q_i	Z	X
II	T	S	X_3
V	T	SU	X_2

Table 4. Choice of (Q_i, Z, X) defining cases II and V. The subgroup G_e generated by T is Z_3.

Such a result is counterintuitive, since CP is not assumed to be part of the residual symmetry of the charged lepton sector and thus we might expect that CP is always broken in our construction. Trivial CP phases can however arise from a CP symmetry of accidental type. In cases II and IV it happens that the charged lepton sector, invariant under G_e, is also accidentally invariant under the CP transformation associated to X_3. In case II this suffices to conclude that CP is conserved, at least at the level of the lepton masses and mixing angles. In case IV we should further notice that X_3 is also accidentally a symmetry of the neutrino sector, even though we originally asked invariance under the CP transformations generated by X_2. Cases II and V make the same prediction of the solar and reactor angles as cases I and IV, respectively. They are given

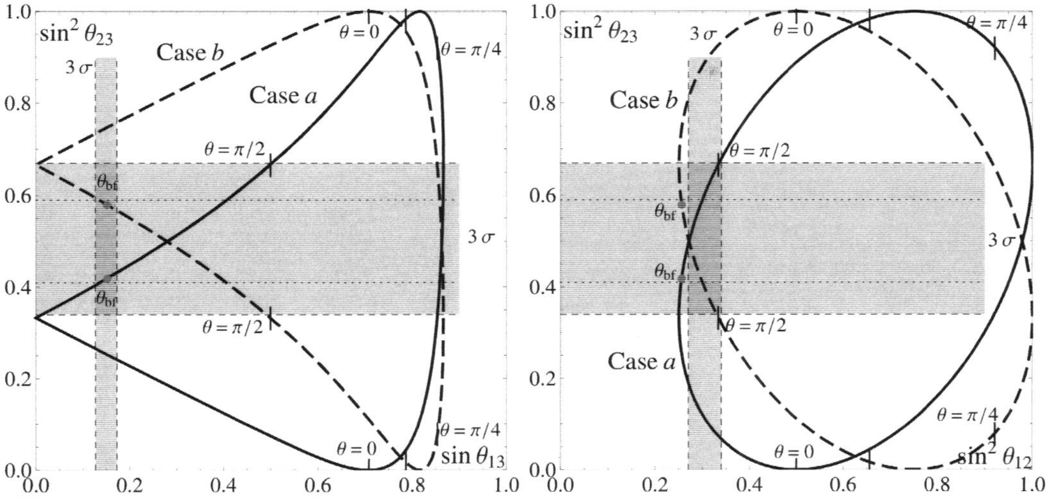

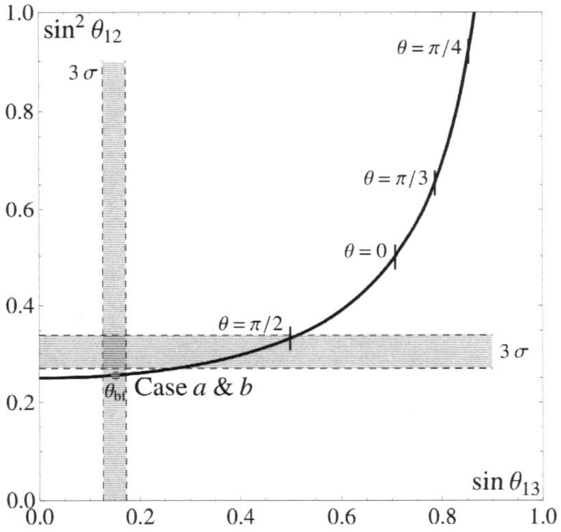

Figure 3: Results for the mixing angles $\sin\theta_{13}$, $\sin^2\theta_{12}$ and $\sin^2\theta_{23}$ for Case a (straight line) and Case b (dashed line) for $G_e = Z_4$ or $G_e = Z_2 \times Z_2$. We mark the value $\theta_{\rm bf}$ of the parameter θ for which the χ^2 functions have a global minimum with a red dot. For better guidance of the eye we also mark $\theta = 0, \pi/4, \pi/3, \pi/2$ on the curves. The shown $3\,\sigma$ ranges for the mixing angles and the best fit values of the atmospheric mixing angle are taken from [5].

in eq. (18). The atmospheric angle is related to the other angles by

$$\sin^2\theta_{23} = \begin{cases} \frac{1}{2}\left(1 - \frac{\sqrt{3}\sin 2\theta}{2+\cos 2\theta}\right) & \text{(case II)} \\ \frac{1}{2}\left(1 - \frac{2\sqrt{6}\sin 2\theta}{5+\cos 2\theta}\right) & \text{(case V)} \end{cases} . \tag{19}$$

Despite the absence of CP violating effects form non-trivial phases, cases II and V provide a nice fit to the mixing angles, once the free parameter θ is optimized by minimizing the χ^2 function. Such a fit is illustrated in fig. 2.

Other two cases very similar to cases II and V are those called VI and VII and defined in table 5, through the choice of (Q_i, Z, X). At variance with the previous cases, the residual symmetry in the charged lepton sector is $Z_4(Z_2 \times Z_2)$ for case VI(VII). Also in these cases trivial CP phases are predicted as a consequence of accidental CP symmetries. Case VI and case VII predict the same mixing angles:

$$\sin^2\theta_{13} = \frac{1}{4}\left(\sqrt{2}\cos\theta + \sin\theta\right)^2 \tag{20}$$

Journal of Physics: Conference Series **447** (2013) 012018 doi:10.1088/1742-6596/447/1/012018

$$\sin^2\theta_{12} = \frac{2}{5-\cos 2\theta - 2\sqrt{2}\sin 2\theta} \tag{21}$$

$$\sin^2\theta_{23} = \begin{cases} \dfrac{4\sin^2\theta}{5-\cos 2\theta - 2\sqrt{2}\sin 2\theta} & \text{(case a)} \\ 1-\dfrac{4\sin^2\theta}{5-\cos 2\theta - 2\sqrt{2}\sin 2\theta} & \text{(case b)} \end{cases} . \tag{22}$$

The two possibilities a and b arise from exchanging the second and third rows of U^0_{PMNS}. The correlations among the three mixing angles are shown in fig. 3. We see that also in these cases the experimental mixing angles can be well approximated by a suitable choice of the parameter θ.

case	Q_i	Z	X
VI	STU	U	X_2
VII	(TST^2S, UT^2)	U	X_1

Table 5. Choice of (Q_i, Z, X) defining cases VI and VII. The subgroup G_e generated by STU is Z_4, while that generated by (TST^2S, UT^2) is $Z_2 \times Z_2$.

We have not mentioned other cases leading to unrealistic lepton mixing angles. Among them we have also found possibilities where both the angles and the CP phases carry a non-trivial dependence from the parameter θ.

6. Conclusion

In neutrino physics we have recently witnessed a decisive progress on the experimental side. The reactor angle is now precisely measured and it is away from zero by many standard deviations. We also have a first indication favoring a non-maximal atmospheric mixing angle. While these steps have been effective in ruling out many models of fermion masses and mixing angles, it is fair to say that no compelling and unique theoretical picture has emerged so far. Perhaps the most disturbing aspect is that present data can still be described within widely different frameworks. The new data have strengthened the case of anarchy. We cannot exclude that neutrino mass ratios and mixing angles are just random $O(1)$ quantities, reflecting no special pattern. Flavour symmetries remain important tools to address the flavour problem, especially when quarks and leptons are analyzed in a common framework. Models based on discrete symmetries, tailored to reproduce the features of the early data from neutrino oscillations, are less supported by data now. Modifications of the simplest realizations are required. A lot of theoretical effort has been made in this direction. Several possibilities to accommodate a large θ_{13} have been devised. The simplest one consists in adding large corrections to the schemes predicting a vanishing θ_{13} at the LO. In alternative, several discrete groups leading to realistic LO mixing matrices have been identified. It is also possible to relax symmetry requirements to avoid $\theta_{13} = 0$ as LO prediction. In this talk we have reviewed a promising approach where CP is included in the symmetry breaking pattern. From a theory viewpoint, combining CP and a discrete symmetry G_f leads to non-trivial consistency conditions that the admissible definitions of CP should satisfy. In our construction we have assumed residual symmetries G_e and $G_\nu = Z_2 \times CP$, G_e and Z_2 being subgroups of a discrete group G_f, and shown that such a requirement determines all mixing angles and phases in terms of a single real parameter. By making a comprehensive analysis of the case $G_f = S_4$ we have identified several new realistic mixing patterns, thus proving the viability of our approach.

Acknowledgments
Many thanks to the organizers of Discrete 2012 for providing an extremely nice environment in Lisbon. I would like to thank warmly my collaborators Guido Altarelli, Claudia Hagedorn, Isabella Masina, Luca Merlo, Robert Ziegler for the pleasant collaboration on which the material of this talk is based. I have been partly supported by the European Programme "Unification in the LHC Era", contract PITN-GA-2009-237920 (UNILHC).

References

[1] P. F. Harrison, D. H. Perkins and W. G. Scott, Phys. Lett. B **530** (2002) 167 [hep-ph/0202074]; Z. -z. Xing, Phys. Lett. B **533** (2002) 85 [hep-ph/0204049].
[2] G. Altarelli and F. Feruglio, Nucl. Phys. B **741** (2006) 215 [hep-ph/0512103]; I. de Medeiros Varzielas, S. F. King and G. G. Ross, Phys. Lett. B **644** (2007) 153 [hep-ph/0512313]; X. -G. He, Y. -Y. Keum and R. R. Volkas, JHEP **0604** (2006) 039 [hep-ph/0601001]; E. Ma, Phys. Rev. D **70** (2004) 031901 [hep-ph/0404199].
[3] C. S. Lam, Phys. Lett. B **656** (2007) 193 [arXiv:0708.3665 [hep-ph]]; C. S. Lam, Phys. Rev. Lett. **101** (2008) 121602 [arXiv:0804.2622 [hep-ph]]; C. S. Lam, Phys. Rev. D **78** (2008) 073015 [arXiv:0809.1185 [hep-ph]]; E. Ma, Phys. Lett. B **632** (2006) 352 [hep-ph/0508231].
[4] K. Abe *et al.* [T2K Collaboration], Phys. Rev. Lett. **107** (2011) 041801 [arXiv:1106.2822 [hep-ex]]; P. Adamson *et al.* [MINOS Collaboration], Phys. Rev. Lett. **107** (2011) 181802 [arXiv:1108.0015 [hep-ex]]; Y. Abe *et al.* [Double-Chooz Collaboration], Phys. Rev. Lett. **108** (2012) 131801 [arXiv:1112.6353 [hep-ex]]; F. P. An *et al.* [Daya-Bay Collaboration], Phys. Rev. Lett. **108** (2012) 171803 [arXiv:1203.1669 [hep-ex]]; J. K. Ahn *et al.* [RENO Collaboration], Phys. Rev. Lett. **108** (2012) 191802 [arXiv:1204.0626 [hep-ex]]; F. P. An *et al.* [Daya-Bay Collaboration], arXiv:1210.6327 [hep-ex].
[5] M. C. Gonzalez-Garcia, M. Maltoni, J. Salvado and T. Schwetz, JHEP **1212** (2012) 123 [arXiv:1209.3023 [hep-ph]].
[6] L. J. Hall, H. Murayama, and N. Weiner, Phys. Rev. Lett. **84** (2000) 2572–2575 [arXiv:9911341 [hep-ph]]; N. Haba and H. Murayama, Phys. Rev. **D63** (2001) 053010, [arXiv:0009174 [hep-ph]]; A. de Gouvea and H. Murayama, Phys. Lett. **B573** (2003) 94–100; [arXiv:0301050 [hep-ph]]; A. de Gouvea and H. Murayama, [arXiv:1204.124 [hep-ph]].
[7] C. D. Froggatt and H. B. Nielsen, *Hierarchy of Quark Masses, Cabibbo Angles and CP Violation*, Nucl. Phys. **B147** (1979) 277.
[8] G. Altarelli, F. Feruglio, I. Masina and L. Merlo, JHEP **1211** (2012) 139 [arXiv:1207.0587 [hep-ph]].
[9] G. Altarelli, F. Feruglio, Rev. Mod. Phys. **82** (2010) 2701-2729 [arXiv:1002.0211 [hep-ph]]; H. Ishimori, T. Kobayashi, H. Ohki, Y. Shimizu, H. Okada, M. Tanimoto, Prog. Theor. Phys. Suppl. **183** (2010) 1-163 [arXiv:1003.3552 [hep-th]].
[10] G. Altarelli, F. Feruglio and L. Merlo, Fortschr. Phys. 61, No. 4 5, 507 534 (2013).
[11] R. de Adelhart Toorop, F. Feruglio and C. Hagedorn, Phys. Lett. B **703** (2011) 447 [arXiv:1107.3486 [hep-ph]]; R. de Adelhart Toorop, F. Feruglio and C. Hagedorn, Nucl. Phys. B **858** (2012) 437 [arXiv:1112.1340 [hep-ph]].
[12] X. G. He and A. Zee, Phys. Lett. B **645** (2007) 427 [arXiv:hep-ph/0607163]; W. Grimus and L. Lavoura, JHEP **0809**, 106 (2008) [arXiv:0809.0226 [hep-ph]]; C. H. Albright, W. Rodejohann, Eur. Phys. J. **C62** (2009) 599-608 [arXiv:0812.0436 [hep-ph]]; Y. Lin, Nucl. Phys. B **824** (2010) 95 [arXiv:0905.3534 [hep-ph]]; W. Grimus, L. Lavoura and A. Singraber, Phys. Lett. B **686**, 141 (2010) [arXiv:0911.5120 [hep-ph]].
[13] Y. Lin, Nucl. Phys. B **824** (2010) 95 [arXiv:0905.3534 [hep-ph]].
[14] S. -F. Ge, D. A. Dicus and W. W. Repko, Phys. Lett. B **702** (2011) 220 [arXiv:1104.0602 [hep-ph]]; S. -F. Ge, D. A. Dicus and W. W. Repko, Phys. Rev. Lett. **108** (2012) 041801 [arXiv:1108.0964 [hep-ph]]; D. Hernandez and A. Y. Smirnov, Phys. Rev. D **86** (2012) 053014 [arXiv:1204.0445 [hep-ph]]; D. Hernandez and A. Y. Smirnov, arXiv:1212.2149 [hep-ph].
[15] P. F. Harrison and W. G. Scott, Phys. Lett. B **535** (2002) 163 [hep-ph/0203209]; P. F. Harrison and W. G. Scott, Phys. Lett. B **547** (2002) 219 [hep-ph/0210197]; P. F. Harrison and W. G. Scott, Phys. Lett. B **594** (2004) 324 [hep-ph/0403278].
[16] W. Grimus and L. Lavoura, Phys. Lett. B **579** (2004) 113 [hep-ph/0305309].
[17] R. Krishnan, P. F. Harrison and W. G. Scott, arXiv:1211.2000 [hep-ph].
[18] R. N. Mohapatra and C. C. Nishi, Phys. Rev. D **86** (2012) 073007 [arXiv:1208.2875 [hep-ph]].
[19] G. C. Branco, J. M. Gerard and W. Grimus, Phys. Lett. B **136** (1984) 383; I. de Medeiros Varzielas and D. Emmanuel-Costa, Phys. Rev. D **84** (2011) 117901 [arXiv:1106.5477 [hep-ph]]; I. de Medeiros Varzielas, D. Emmanuel-Costa and P. Leser, Phys. Lett. B **716** (2012) 193 [arXiv:1204.3633 [hep-ph]]; I. de Medeiros

Varzielas, JHEP **1208** (2012) 055 [arXiv:1205.3780 [hep-ph]]; G. Bhattacharyya, I. de Medeiros Varzielas and P. Leser, Phys. Rev. Lett. **109** (2012) 241603 [arXiv:1210.0545 [hep-ph]].
[20] K. S. Babu and J. Kubo, Phys. Rev. D **71** (2005) 056006 [hep-ph/0411226]; K. S. Babu, K. Kawashima and J. Kubo, Phys. Rev. D **83** (2011) 095008 [arXiv:1103.1664 [hep-ph]].
[21] M. -C. Chen and K. T. Mahanthappa, Phys. Lett. B **681** (2009) 444 [arXiv:0904.1721 [hep-ph]]; A. Meroni, S. T. Petcov and M. Spinrath, Phys. Rev. D **86** (2012) 113003 [arXiv:1205.5241 [hep-ph]].
[22] F. Feruglio, C. Hagedorn and R. Ziegler, arXiv:1211.5560 [hep-ph].
[23] G. Ecker, W. Grimus and H. Neufeld, J. Phys. A **20** (1987) L807; H. Neufeld, W. Grimus and G. Ecker, Int. J. Mod. Phys. A **3** (1988) 603.
[24] W. Grimus and M. N. Rebelo, Phys. Rept. **281** (1997) 239 [hep-ph/9506272].
[25] C. Hagedorn, S. F. King and C. Luhn, JHEP **1006** (2010) 048 [arXiv:1003.4249 [hep-ph]].

Experimental status of particle and astroparticle searches for supersymmetry

Vasiliki A Mitsou

Instituto de Física Corpuscular (IFIC), CSIC – Universitat de València,
Parc Científic de la U.V., C/ Catedrático José Beltrán 2, E-46980 Paterna (Valencia), Spain

E-mail: `vasiliki.mitsou@ific.uv.es`

Abstract. An overview of supersymmetry searches is presented, covering collider experiments, direct and indirect searches for supersymmetric dark matter. Recent LHC experimental results are reviewed, and the constraints from B-meson decays are reported. Implications for supersymmetry of the latest direct and indirect searches are thoroughly discussed. The focus is on the complementarity of the various probes — particle and astrophysical — for constraining Supersymmetry.

1. Introduction

The Standard Model provides the current most accurate description of elementary particle physics. It has been experimentally tested up to the TeV scale with remarkably successful results. Nevertheless, there are pieces of evidence pointing to physics beyond the SM such as the existence of dark matter (DM), the matter-antimatter asymmetry, the neutrino masses, and the hierarchy problem. Several theories have been proposed to address these issues and Supersymmetry (SUSY) [1] is one of the most favoured.

SUSY postulates a discrete symmetry between fermions and bosons, thus assigning a new particle, a *superpartner*, to each SM field. Since no SUSY particles have been found yet, SUSY must be a broken symmetry, leading to numerous proposed models and various symmetry-breaking mechanisms. SUSY is characterised by a large number of free parameters, e.g. more than 100 parameters are introduced in the Minimal Supersymmetric Standard Model (MSSM). Nonetheless, a variety of experimental and observational handles are at our disposal to search for SUSY and constrain it. The different facets of SUSY manifestation and the current status of the hunt for SUSY is discussed hereupon.

This paper is structured as follows. Section 2 provides a brief introduction to dark matter and its implications for SUSY, also discussing alternative SUSY models yielding a non-trivial DM density. It reviews both direct and indirect DM detection methods and recent results. In Section 3 the features of collider experiments that play a central role in exploring SUSY are highlighted. There we concisely discuss the SUSY constraints originating from B-physics experiments, whilst strong emphasis is given on recent attempts to discovering supersymmetry at the Large Hadron Collider (LHC) [2] and the ensuing exclusion limits, covering both R-parity conserving and R-parity breaking scenarios — the latter receiving special attention. The implications for SUSY arising from the observation of a Higgs-like particle are also addressed. The paper concludes with a summary in section 4.

2. SUSY as dark matter

Unveiling the nature of dark matter [3, 4, 5] is a quest in both Astrophysics and Particle Physics. According to observations over the past two decades — obtained by combining a variety of astrophysical data, such as type-Ia supernovae [6], cosmic microwave background (CMB) [7], baryon oscillations [8] and weak lensing data [9] —, most of our Universe energy budget consists of unknown entities: $\sim 23\%$ is dark matter and $\sim 72\%$ is dark energy, a form of ground-state energy. Dark matter existence is inferred from gravitational effects on visible matter, but is undetectable by emitted or scattered electromagnetic radiation. The most precise measurement comes from anisotropies of the cosmic microwave background [7, 10]; the third peak in the temperature power spectrum, shown in Fig. 1, is used to extract information about dark matter.

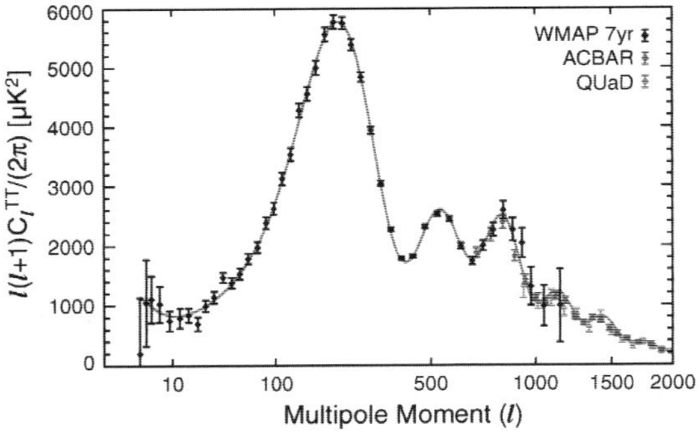

Figure 1. The WMAP 7-year temperature power spectrum, along with the temperature power spectra from the ACBAR and QUaD experiments. The solid line shows the best-fitting 6-parameter flat CDM model to the WMAP data alone. From [7].

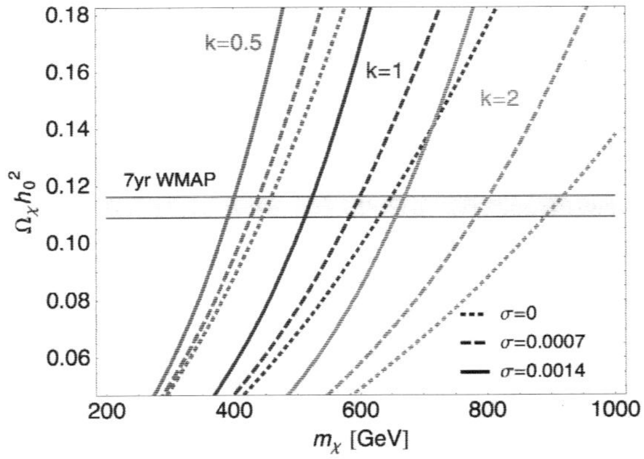

Figure 2. Predicted values of $\Omega_\chi h^2$ as a function of the DM particle mass m_χ for various deviations k from the weak coupling and for different values of foam fluctuations σ. The yellow horizontal band represents the WMAP7 value for $\Omega_\chi h^2$ [7]. From [11].

Weakly-interacting massive particles (WIMPs), where the typical (weak-scale) annihilation cross sections are of the same order of magnitude as the thermally-averaged DM annihilation cross section, constitute the most popular class of dark matter candidates. Supersymmetry is a theoretical scenario that inherently proposes such dark matter candidates. Within SUSY with R-parity conservation, the particle that plays the role of dark matter is the lightest supersymmetric particle (LSP); LSPs that have the right properties are the axino, the gravitino (as superWIMPs) and the lightest neutralino (as WIMP). Such scenarios — should they indeed explain dark matter — are severely constrained from the Big-Bang nucleosynthesis, the measured DM abundance and the results of direct DM detection experiments and indirect detection telescopes.

2.1. DM relic density and SUSY

WIMPs were created thermally in the early universe, hence their relic abundance, Ω_χ, and their properties, such as mass and couplings, are closely related through the averaged annihilation cross section. For a specific SUSY model, the predicted neutralino relic density, calculated by taking into account all possible LSP annihilation processes, is compared to the measured Ω_χ, imposing constraints on the model parameters.

The DM relic density estimation is based on the Standard Cosmological Model (ΛCDM) [12], involving cold DM as the dominant DM species, and a positive cosmological constant $\Lambda > 0$.

Nevertheless, the possibility for different theoretical scenarios is open, that modify the estimated DM relic abundance for given cosmological observations. For instance, the presence of space-time (D-particle) foam in string/brane-theory leads to Finsler-type metric distortions, induced by interactions of the DM particle(s) with the defects in the foam [11]. These metrics lead to modifications in the pertinent Boltzmann equation and consequently to enhancement of the relic abundances, as displayed in Fig. 2. However stringent constraints on the defect density in D-foam models can also be imposed by astrophysical telescopes on the arrival times of high energy cosmic photons [13].

The presence of the time-dependent dilaton may also affect the dark matter relic density calculation, since it modifies the Boltzmann equations. In this case, a *dilution* in the neutralino density of $\mathcal{O}(10)$ is predicted [14], widening the allowed parameter space of supersymmetry at collider searches. Furthermore, other astrophysical observations such as type-Ia supernovae and galaxy ages set tight constraints on the parameters of these models through their prediction for the dark energy contribution [15].

2.2. Direct WIMP detection

In direct detection low background underground experiments, one attempts to observe the nuclear recoil produced by WIMP scattering off nucleons [16]. The expected signal features a recoil spectrum which falls exponentially with energy and extends to a few tens of keV only. Various techniques applied to observe the recoil with uncorrelated systematic uncertainties: scintillator NaI, cryogenic, noble liquids (Ar, Xe), bubble chamber (superheated liquid).

The current experimental landscape is shown in Fig. 3 in terms of WIMP-nucleus elastic cross section versus WIMP mass [17]. The positive results shown represent the annual modulation observed by DAMA/LIBRA [18] and CoGeNT [19], as well as unidentified excesses of events seen by CoGeNT [19] and CRESST-II [20]. CDMS does not see any hint of annual modulation [21] although their energy threshold is higher than CoGeNT. The XENON100 data [17] lead to the strongest exclusion limit so far for constraining the DM-nucleon cross section to below 10^{-44} cm^2 for $m_\chi \sim 100$ GeV. For comparison, other experimental results are also shown [22, 19, 20], together with the $1\sigma/2\sigma$ regions preferred by supersymmetric (Constrained MSSM) models [23], still untouched by the XENON limits.

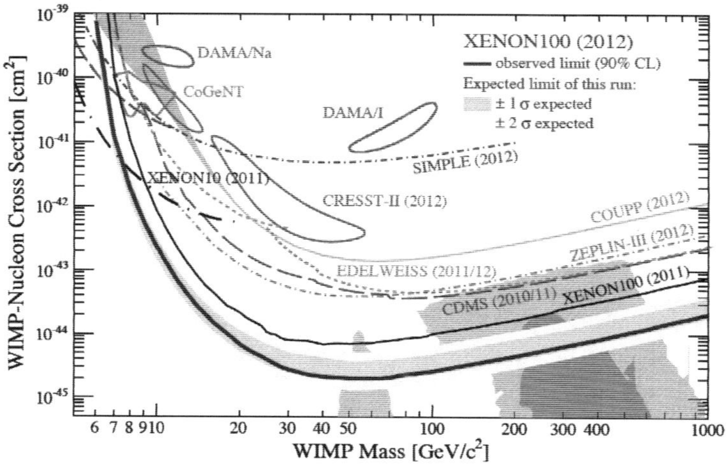

Figure 3. Spin-independent WIMP-nucleon scattering from XENON100: The expected sensitivity is shown by the green/yellow band ($1\sigma/2\sigma$) and the resulting exclusion limit (90% CL) in blue. From [17].

It is worth noting that these results strongly depend on astrophysical uncertainties, such as the local DM density and DM velocity distribution and on assumptions made regarding the WIMP interaction with the detector active material. Finally, systematic uncertainties in the nuclear modeling can affect the comparison between different detection techniques.

2.3. Indirect searches for DM

WIMPs can annihilate and their annihilation products can be detected; these include neutrinos, γ-rays, positrons, antiprotons, and antinuclei [24]. "Smoking gun" signals for indirect detection are neutrinos coming from the center of the Sun or Earth, and mono-energetic photons from WIMP annihilation in space. These techniques could distinguish between different coupling scenarios and the nature of WIMPs: neutralinos, Kaluza-Klein states, etc.

The observation of 14 dwarf spheroidal galaxies with the Fermi Gamma-Ray Space Telescope showed no significant γ-ray emission above 100 MeV from the candidate dwarf galaxies [25]. Therefore upper limits to the γ-ray flux assuming representative spectra from WIMP annihilation were set, as demonstrated in Fig. 4, also compared with predictions from mSUGRA, MSSM with a reduced set of parameters [26], Kaluza-Klein dark matter in universal extra dimensions (UED) and wino-like dark matter in AMSB. The upper limits on the γ-ray flux are already competitive for MSSM models, provided that they correspond to low thermal relic density. Furthermore, these flux upper limits already disfavour AMSB models with masses < 300 GeV.

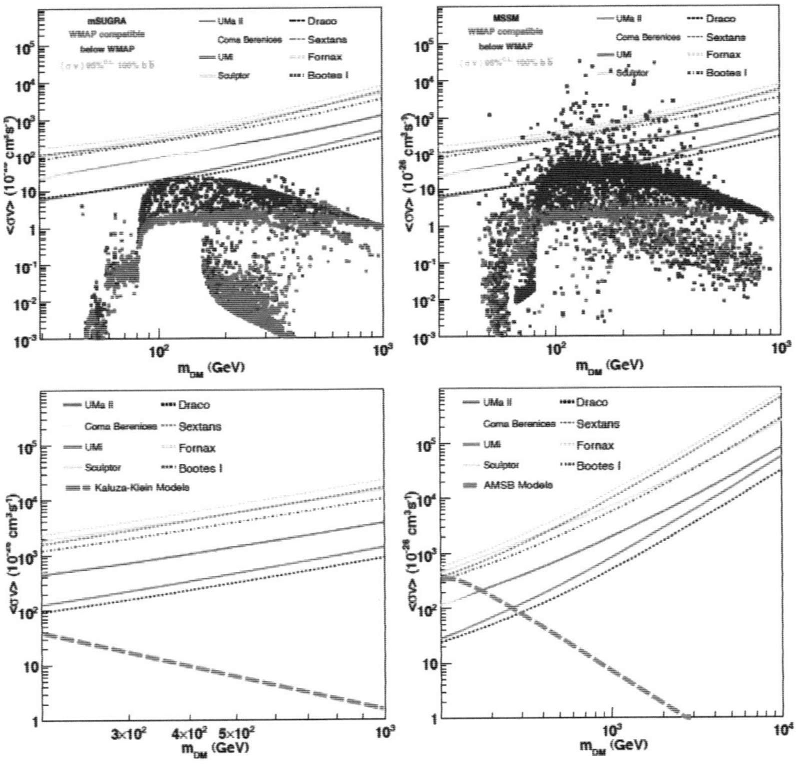

Figure 4. mSUGRA (upper left), MSSM (upper right), Kaluza-Klein UED (lower left) and anomaly mediated (lower right) models in the $(m_{\mathrm{WIMP}}, \langle\sigma v\rangle)$ plane. Red points have a Ω_χ compatible with the inferred DM density (blue points have a lower DM density). The lines indicate the Fermi 95% upper limits from likelihood analysis on the selected dwarfs. From [25].

Controversial hints of WIMP annihilation have been seen so far in data by PAMELA [27], EGRET, Fermi-LAT [28], which may be attributed to the large astrophysical uncertainties that enter in the data analysis. At present, much activity and excitement surrounds a tentative line signal at $E_\gamma \simeq 135$ GeV [29]. The required annihilation cross section to explain this signal is very large, however supersymmetric models with such large signals do exist. Further progress to determine if the line signal is robust and to improve sensitivities for both continuum and line searches is sure to come from continued running of existing and upcoming experiments.

To recapitulate, direct and indirect DM detection techniques set severe constraints on supersymmetric models, yet accelerator experiments are required to provide a measurement of different model parameters exploiting uncorrelated systematics. This is concisely depicted in Fig. 5, where the coverage of the direct and indirect detection with photons are complemented by the LHC and neutrino telescopes [30]. In the following years we expect a continuous interplay between particle physics experiments and astrophysical/cosmological observations.

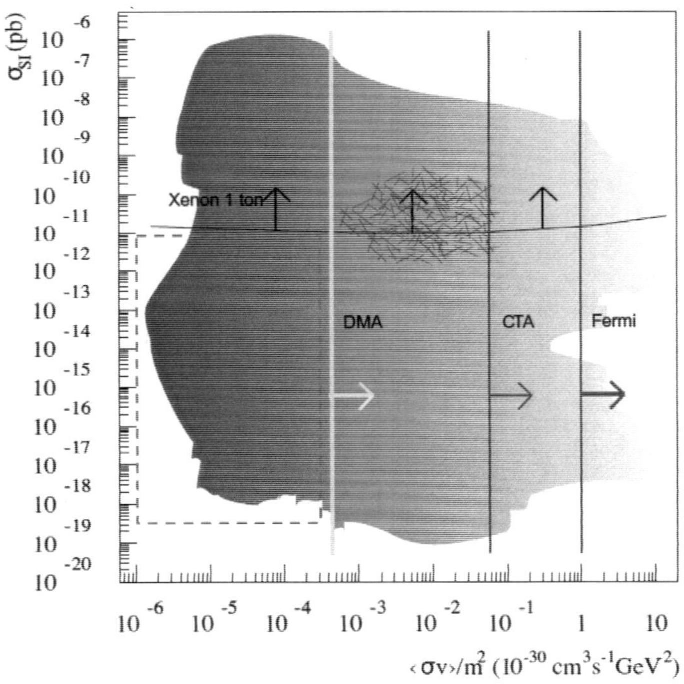

Figure 5. Illustration of the reach of direct and indirect dark matter detection experiments for γ-ray detection. The shaded region is the approximate range of WMAP-compatible MSSM model space. The smaller hatched region is a rough estimate of where Kaluza-Klein models reside [31]. The dashed rectangular indicates the region that may be covered by LHC and neutrino telescopes. The MSSM results are based on Ref. [32], to which the reader is directed for information on the mass range, composition, etc. From [30].

3. Searches in colliders

The stringent exclusion limits set on supersymmetric models and parameters come from collider experiments, such as the LEP, Tevatron and LHC detectors. The LHC currently in operation at CERN in Geneva, Switzerland, is an ideal machine for discovering supersymmetry. Although here we focus on direct searches for SUSY signals, discussed in Section 3.2, SUSY is also constrained indirectly through the (non-)observation of rare decays of B mesons.

3.1. B-physics constraints

Precise experimental measurements and theoretical predictions have been achieved for the B-meson systems in the past decade and stringent constraints due to considerable beyond-SM contributions to many observables have been acquired [33]. The rare decay $BR(B_s \to \mu^+\mu^-)$ deserves special attention as new results have been recently announced by the LHCb collaboration [34] using pp collisions at $\sqrt{s} = 8$ TeV. An excess of $B_s \to \mu^+\mu^-$ candidates with respect to the background expectation is observed with a 3.5σ significance [35]. This result is in agreement with the Standard Model expectation, therefore limits on new-physics models are set.

Here, we review constraints obtained with the LHCb results using 1 fb^{-1}of data at $\sqrt{s} =$ 7 TeV, where a stringent 95% CL limit on the branching ratio $BR(B_s \to \mu^+\mu^-) < 4.5 \cdot 10^{-9}$ has been obtained [36]. The decay $B \to K^+\mu^+\mu^-$, on the other hand, provides a variety of complementary observables as it gives access to angular distributions in addition to the differential branching fraction. In addition to the above observables, $B \to X_s\gamma$, $B \to \tau\nu$, $B \to D\tau\nu_\tau$, $B \to X_s\mu^+\mu^-$ and $D_s \to \tau\nu_\tau$ are also very sensitive to SUSY.

A comparison between different flavour observables in the plane $(m_{1/2}, m_0)$ is given in Fig. 6, where limits from $B \to X_s\gamma$, $B \to \tau\nu$, $R_{l23}(K \to \mu\nu_\mu)$, $B \to D\tau\nu_\tau$, $B \to X_s\mu^+\mu^-$ and $D_s \to \tau\nu_\tau$ are also shown [37]. The LHCb limit strongly constrains the CMSSM with large $\tan\beta = 50$, however for $\tan\beta = 30$, as can be seen from the figure, the flavour constraints and in particular $B_s \to \mu^+\mu^-$, are inferior to those from direct searches. On the other hand, $B \to X_s\gamma$ and $B \to K^+\mu^+\mu^-$ related observables and in particular the forward-backward asymmetry could play a complementary role in the intermediate $\tan\beta$ regime.

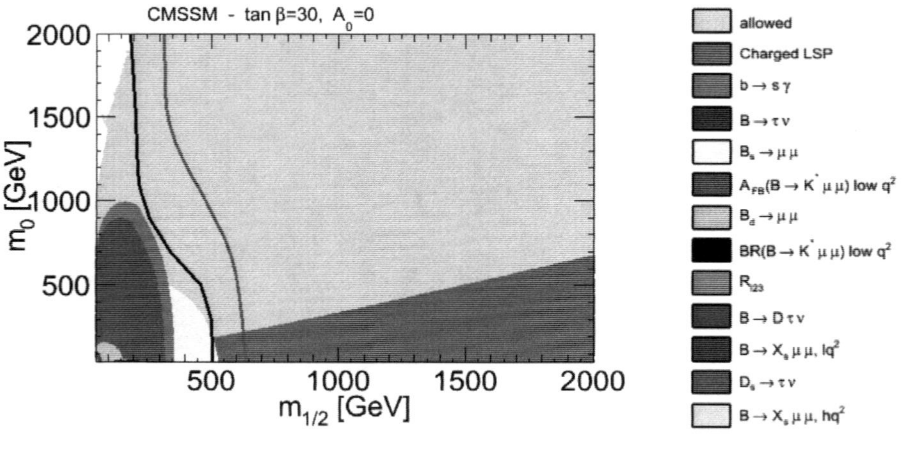

Figure 6. Constraints from flavour observables in CMSSM in the plane $(m_{1/2}, m_0)$ for $\tan\beta = 30$ and $A_0 = 0$ with the 2011 results. The black line corresponds to the CMS exclusion limit with 1.1 fb^{-1} of data [38] and the red line to the CMS exclusion limit with 4.4 fb^{-1} of data [39]. From [37].

3.2. SUSY direct searches at LHC

SUSY searches in ATLAS [40] and CMS [41] experiments typically focus on events with high transverse missing energy ($E_{\mathrm{T}}^{\mathrm{miss}}$) which can arise from (weakly interacting) LSPs, in the case of R-parity conserving SUSY, or from neutrinos produced in LSP decays, when R-parity is broken. Hence, the event selection criteria of inclusive channels are based on large $E_{\mathrm{T}}^{\mathrm{miss}}$, no or few leptons ($e$, μ), many jets and/or b-jets, τ-leptons and photons. In addition, kinematical variables such as the transverse mass, M_T, and the effective mass, $M_{\rm eff}$, assist in distinguishing further SUSY from SM events, whilst the *effective transverse energy* [42] can be useful to cross-check results, allowing a better and more robust identification of the SUSY mass scale, should a positive signal is found. Although the majority of the analysis simply look for an excess of events over the SM background, there is an increasing application of distribution shape fitting techniques [43].

Typical SM backgrounds are top-quark production — including single-top —, W/Z in association with jets, dibosons and QCD multi-jet events. These are estimated using semi- or fully data-driven techniques. Although the various analyses are motivated and optimised for a specific SUSY scenario, the interpretation of the results are extended to various SUSY models or topologies.

Analyses exploring R-parity conserving (RPC) SUSY models are currently divided into inclusive searches for: (a) squarks and gluinos, (b) third-generation fermions, and (c) electroweak production ($\tilde{\chi}^0$, $\tilde{\chi}^{\pm}$, $\tilde{\ell}$). Recent results from each category of CMS searches are presented in Ref. [44]. It is stressed that, although these searches are designed to look for RPC SUSY, interpretation in terms of R-parity violating models is also possible (cf. Section 3.4).

Strong SUSY production is searched in events with large jet multiplicities and large missing transverse momentum, with and without leptons. Various channels fall into this class of searches; here the 0-lepton plus three jets plus $E_T^{\rm miss}$ analysis from CMS is highlighted. The results are interpreted in terms of the CMSSM (Fig. 7) excluding squark masses up to 1400 GeV, gluino mass up to 900 GeV, or $m_{\tilde{q}} \sim m_{\tilde{g}} \sim 1400$ GeV [45]. A different approach is to present the limits in terms of a simplified topology, namely $\tilde{q}\tilde{q} \to q\tilde{\chi}_1^0 q\tilde{\chi}_1^0$, assuming that $m_{\tilde{g}} \gg m_{\tilde{q}}$ [46]. The results in the ($m_{\tilde{q}}$, $m_{\rm LSP}$) plane are shown in Fig. 8.

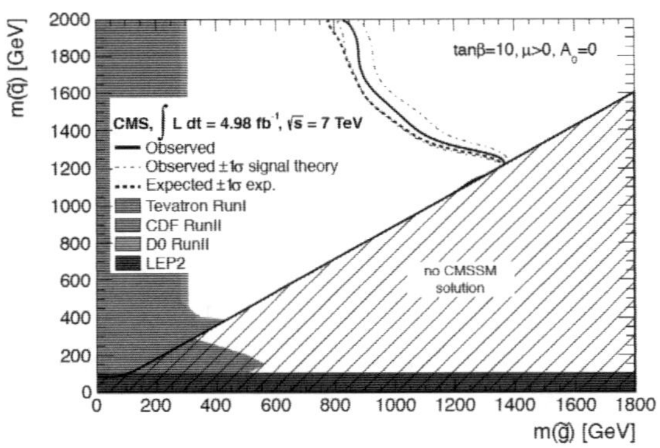

Figure 7. Observed and expected 95% CL lower limits in the CMSSM ($m_{\tilde{g}}, m_{\tilde{q}}$) plane, for $\tan\beta = 10$, $\mu > 0$, and $A_0 = 0$. The yellow-shaded region shows the 1σ variation in the expected limit, while the dot-dashed curves show the variation in the observed limit when the signal cross section is varied by its theoretical uncertainties. Limits from earlier searches by other experiments derived with different models or parameter choices are also shown. From [45].

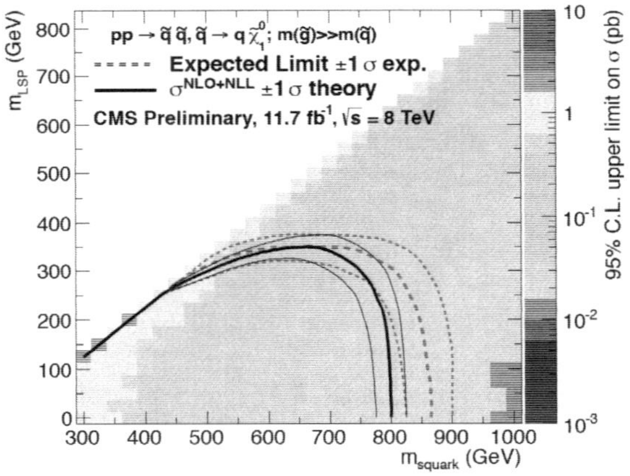

Figure 8. Upper limit on cross section at 95%CL as a function of $m_{\tilde{q}}$ and $m_{\rm LSP}$ for 1-step $\tilde{q}\tilde{q} \to q\tilde{\chi}_1^0 q\tilde{\chi}_1^0$ simplified model. The solid thick black line indicates the observed exclusion region assuming NLO+NLL SUSY production cross section. The thin black lines represent the observed excluded region when varying the cross section by its theoretical uncertainty. The dashed purple lines indicate the median (thick line) $\pm 1\sigma$ (thin lines) expected exclusion regions. From [46].

Apart from inclusive analyses targeting phenomenological models, a wide spectrum of searches is dedicated on specific sparticle decay chains predicted in well-motivated theoretical supersymmetric models. For instance, in gauge mediated supersymmetry breaking (GMSB) models, the SUSY breaking is mediated from a hidden sector by gauge interactions. Their phenomenology is characterised by a light gravitino LSP with the NLSP mass, nature and lifetime determining the final states and therefore the analysis strategy with which to probe this model [47]. A higgsino-like neutralino NLSP, for instance, preferably decays to a gravitino and a Z boson or a photon. Such signatures can be detected by looking for events with a leptonically-decaying Z, high $E_{\mathrm{T}}^{\mathrm{miss}}$ and jets. The ATLAS analysis [48] yielded the exclusion limits depicted in Fig. 9 in the context of the general gauge mediation (GGM) model.

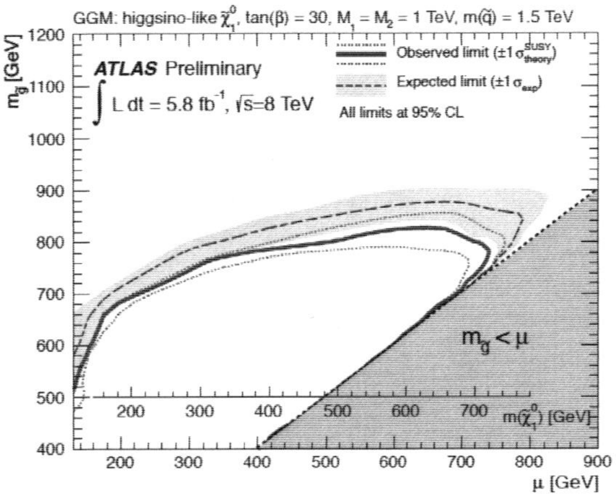

Figure 9. Expected and observed 95% CL exclusion limits for SR1 on the $m_{\tilde{g}}$ and μ parameters for GGM models with $\tan\beta = 30$, $M_1 = M_2 = 1$ TeV, $c\tau_{\mathrm{NLSP}} < 0.1$ mm, and $m_{\tilde{q}} = 1.5$ TeV. The light grey area indicates the region where the NLSP is the gluino, which is not considered in this analysis. From [48].

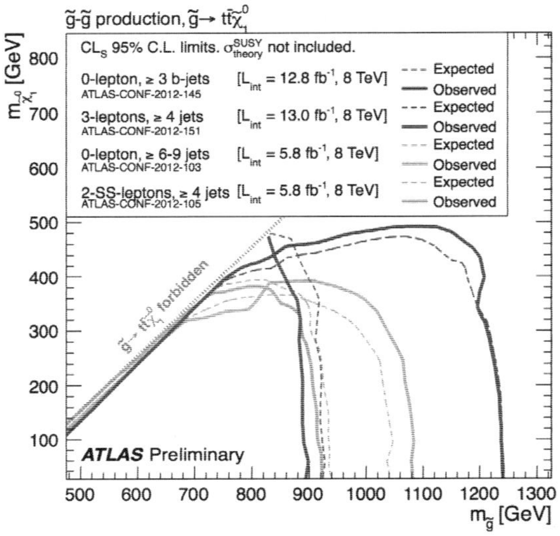

Figure 10. Exclusion limits at 95% CL for 8 TeV analyses in the $(m_{\tilde{g}}, m_{\tilde{\chi}_1^0})$ plane for the Gtt simplified model where a pair of gluinos decays via off-shell stop to four top quarks and two neutralinos (LSP). From [49].

The mixing of left- and right-handed gauge states which provides the mass eigenstates of the scalar quarks and leptons can lead to relatively light 3rd generation particles. Stop ($\tilde{t}_1$) and sbottom ($\tilde{b}_1$) with a sub-TeV mass are favoured by the naturalness argument, while the stau ($\tilde{\tau}_1$) is the lightest slepton in many models. Therefore these could be abundantly produced either directly or through gluino production and decay. Such events are characterised by several energetic jets (some of them b-jets), possibly accompanied by light leptons, as well as high $E_{\mathrm{T}}^{\mathrm{miss}}$.

Both main LHC experiments, ATLAS and CMS, have embarked in analyses dedicated to unveil third-generation sparticle production covering a wide spectrum of signatures. The results extracted by ATLAS with up to 13 fb^{-1} of data at $\sqrt{s} = 8$ TeV, as far as the gluino-mediated production is concerned, are summarised in Fig. 10. Gluino masses of up to 1240 GeV and lightest neutralino masses up to 500 GeV in the Gtt simplified model have been excluded. More updated results on the third-generation are available at Refs. [50, 49], for CMS and ATLAS, respectively.

The electroweak SUSY production at LHC proceeds through weak gaugino (charginos and neutralino) or slepton pair production and is typified by low cross section compared to the strong processes. These events are characterised by multiple charged leptons and moderately high $E_{\mathrm{T}}^{\mathrm{miss}}$. Their kinematics is quite different from the long cascade decays typically found in SUSY processes through strong interactions. Therefore, kinematic variables other than the effective mass may be deployed such as the *visible transverse energy/momentum*, to improve the power of the analysis and also give way to the determination of a combination of sparticle masses [51].

In certain SUSY breaking scenarios, characteristic signatures are expected involving heavy sparticles with a long lifetime. A non-exhaustive list of examples of such (meta-)stable particles includes long-lived sleptons in GMSB models and R-hadrons with long-lived gluinos or squarks in split SUSY. These particles traverse the entire detector and, due to their high mass, they move slowly ($\beta < 1$), yielding high specific ionisation energy loss, $\mathrm{d}E/\mathrm{d}x$, and a long time of flight (TOF). The result of a search for heavy long-lived charged particles produced in pp collisions at $\sqrt{8}$ TeV with 5 fb^{-1} of CMS data [52] is shown in Fig. 11. The inner tracking detectors were used to define a sample of events containing high-p_{T} tracks and high $\mathrm{d}E/\mathrm{d}x$. A second sample of events, which have high-momentum tracks satisfying muon identification requirements in addition to meeting high-ionisation and long time-of-flight requirements, was analysed independently. In both cases, the results are consistent with the expected background estimated from data, thus establishing cross-section limits as a function of mass within the context of models with long-lived gluinos, scalar top quarks and scalar taus. Lower limits at 95% confidence level on the mass of gluinos (stops) are found to be 1098 (737) GeV. A limit of 928 (626) GeV is set for a gluino (stop) that hadronizes into a neutral bound state before reaching the muon detectors. The lower mass limit for a pair produced $\tilde{\tau}$ is found to be 223 GeV. It is worthy to mention that the MoEDAL [53] experiment at LHC is specifically designed to explore such high-ionisation signatures.

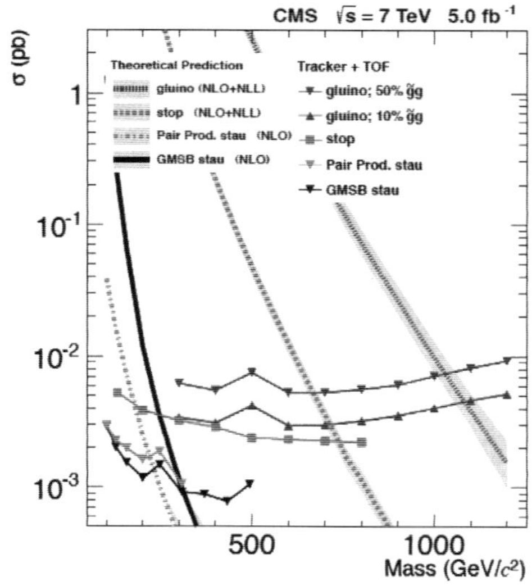

Figure 11. Predicted theoretical cross sections with associated uncertainties and observed 95% CL upper limits on the cross section for the different signal models considered: production $\tilde{t}_1$, $\tilde{g}$, and $\tilde{\tau}_1$; different fractions of R-gluonball states, using the tracker and TOF information. From [52].

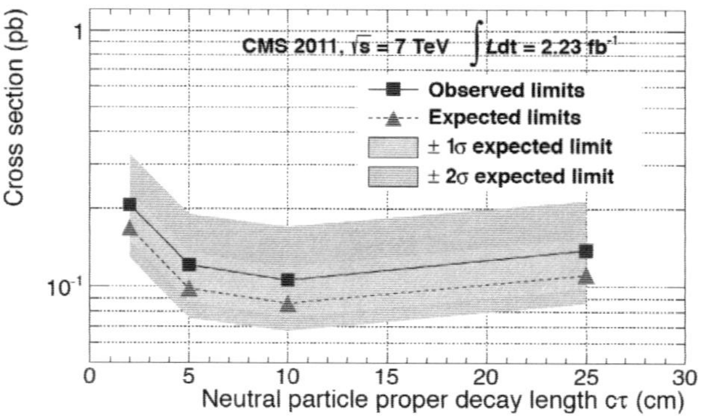

Figure 12. 95% CL upper limits on the pair production cross section for neutral particles, each of which decays into a photon and invisible particles, as a function of the neutral particle proper decay length. The observed values as a function of mass are shown by the solid line. The dashed line indicates the expected median of results for the background-only hypothesis, while the green (dark) and yellow (light) bands indicate the ranges that are expected to contain 68% and 95% of all observed excursions from the median, respectively. From [54].

GMSB scenarios, on the other hand, predict signatures involving high-p_{T} photons which may or may not point back to the primary collision vertex. These final states arise in the decay of a long-lived $\tilde{\chi}_1^0$ decaying to a photon and a gravitino. CMS carried out a search using events containing photons, $E_{\mathrm{T}}^{\mathrm{miss}}$, and jets with 2.23 fb^{-1} of data at $\sqrt{s}$ = 7 TeV [54]. The impact parameter of the photon relative to the beam-beam collision point can be reconstructed using converted photons. The method is sensitive to lifetimes of the order of 0.1 to 1 ns. Cross-section limits on pair production for such particles, each of which decays into a photon and invisible particles, are presented in Fig. 12. The observed 95% CL limits vary between 0.11 and 0.21 pb, depending on the neutral particle lifetime.

The spectrum of SUSY searches at the LHC is huge both in terms of the variety of tested models as well as the explored signatures and it was merely highlighted here. A summary of the SUSY-related analysis results from CMS [50] is shown in Fig. 13. A similar synopsis of results from the ATLAS experiment can be found in Ref. [49]. In conclusion, LHC probes supersymmetry in scales between 100 GeV and 1 TeV, without having seen yet some evidence of its existence. Experiments in upcoming e^+e^- colliders, such as the ILC [55] and CLIC [56], are expected to discover and constrain SUSY in the future.

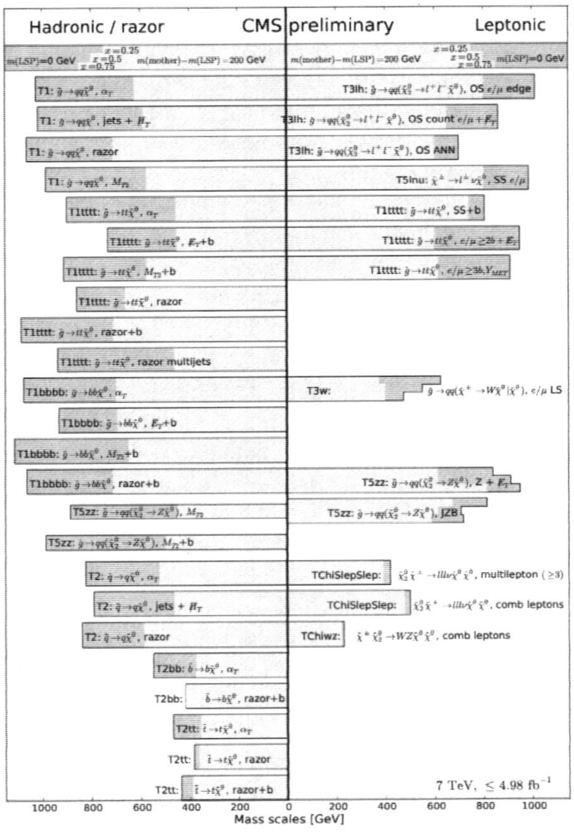

Figure 13. Exclusion limits for the masses of the mother particles, for $m_{\tilde{\chi}_1^0}$ = 0 GeV (dark blue) and $m_{\mathrm{mother}} - m_{\tilde{\chi}_1^0}$ = 200 GeV (light blue), for each analysis, for the hadronic (left) and leptonic (right) results. In this plot, the lowest mass range is m_{mother} = 0, but results are available starting from a certain mass depending on the analyses and topologies. From [50].

3.3. Implications from the Higgs boson

The recent discovery by ATLAS [57, 58] and CMS [59] of a new boson with properties consistent with those of a SM-like Higgs boson shed new light in the supersymmetric landscape [60]. The newly observed boson having a mass of 125 − 126 GeV is rather heavy for being the lightest supersymmetric Higgs, h, however it still lies in the allowed range. If h is SM-like, i.e. in the decoupling limit, the heaviest neutral SUSY Higgs bosons, H and A are pushed to the 1 TeV scale. In this case, high values of $\tan\beta$ are favoured, together with high stop mixing and heavy squarks.

Some (minimal) models are disfavoured, while other remain viable even with other constraints present, e.g. the NMSSM [61]. The singlet that is introduced in this model to evade the μ problem, leads to a richer Higgs sector than the MSSM that can accommodate a relatively heavy Higgs mass [62]. This is demonstrated in Fig. 14, where the blue points denote the models with Higgs boson masses compatible with the measured Higgs mass of 125 GeV within ±3 GeV. Moreover the black points also reproduce the observed $\gamma\gamma$ event rate.

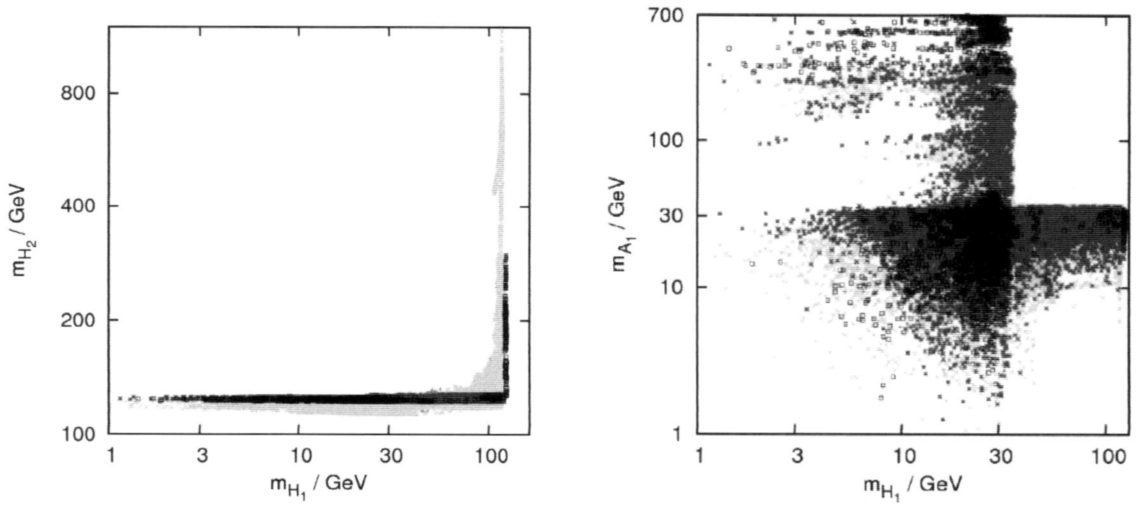

Figure 14. Masses of the Higgs (pseudo-)scalars H_1, H_2 (left) and H_1, A_1 (right). Red points are ruled out either by `HiggsBounds` constraints or the ATLAS 1 fb^{-1} jets and $E_T^{\rm miss}$ SUSY search. Green points have no Higgs with a mass in the 122 − 128 GeV interval, blue points have a Higgs (H_1 and/or H_2) within this mass range, and black points have a Higgs such that $R_{gg\gamma\gamma} > 0.4$. From [62].

3.4. R*-parity violation*

Throughout the previous discussion, we assumed that R-parity — a multiplicative quantum number defined as $R = (-1)^{3(B-L)+2s}$, where B, L and s are the baryon number, the lepton number and the spin, respectively — was conserved. This supposition is hinted but not required by proton stability; rapid proton decay is avoided if one of B or L is conserved. Allowing R-parity violating (RPV) terms [63] in the superpotential has decisive implications for SUSY phenomenology in colliders. First of all, single sparticle production is possible. The LSP may be charged and/or carry colour and, most importantly, decays leaving open the possibility for new discovery signatures and for the LSP-mass reconstruction through the invariant mass of its decay products. Furthermore, the $E_T^{\rm miss}$ may or may not be high depending on the LSP-decay products and the underlying RPV model.

Among the various ways to break R-parity, the violation through bilinear terms is especially interesting for its connection with neutrino physics. In this model the spontaneous breaking of R-parity gives vacuum expectation values (vevs) to sneutrinos providing a *vev-seesaw* mechanism that leads to neutrino masses [64]. Below the scale of these vevs, R-parity breaking is explicit through bilinear L-violating terms and their corresponding soft breaking terms. The same parameters that induce neutrino masses and mixings are also responsible for the decay of the LSP which induces a relation between some decay modes of the LSP and neutrino mixing angles. The latest exclusion limits set on this model, when the bilinear terms are embedded in mSUGRA, were obtained with 5 fb^{-1} of ATLAS data at $\sqrt{s} = 7$ TeV by searching for events with high jet multiplicity, $E_{\mathrm{T}}^{\mathrm{miss}}$ and one lepton [65] and are shown in Fig. 15.

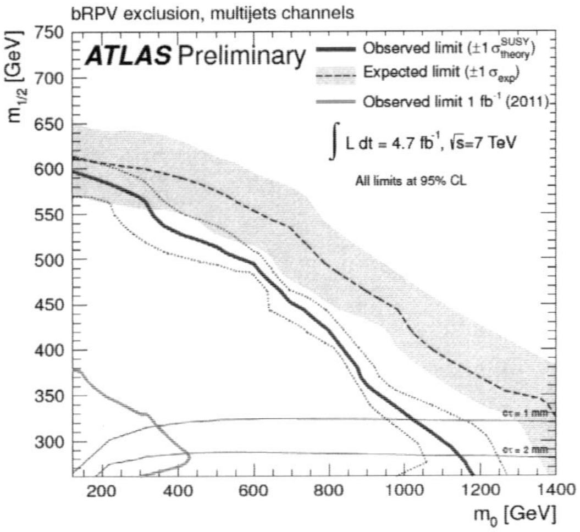

Figure 15. Expected and observed 95% CL exclusion limits in the bilinear R-parity violating model obtained by combining the e and μ channels. The band around the median expected limit shows the $\pm 1\sigma$ variations on the median expected limit, including all uncertainties except theoretical uncertainties on the signal. The dotted lines around the observed limit indicate the sensitivity to $\pm 1\sigma$ variations on these theoretical uncertainties. The thin solid black contours show the LSP lifetime. The result from the previous ATLAS search [66] is also shown. From [65].

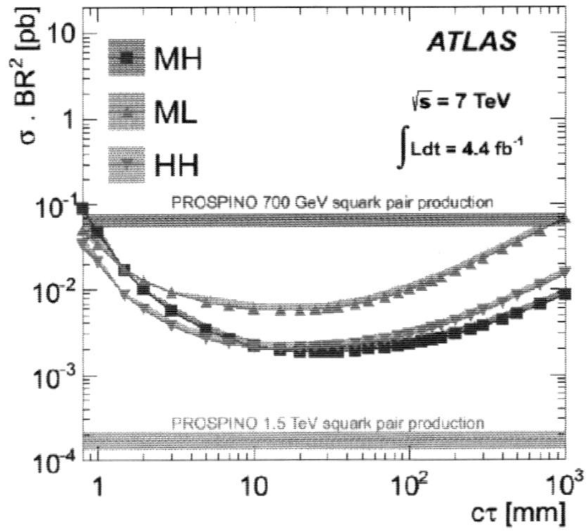

Figure 16. Upper limits at 95% CL on $\sigma \cdot BR^2$ vs. the neutralino lifetime for different combinations of squark and neutralino masses, based on the observation of zero events satisfying all criteria in a 4.4 fb^{-1} data sample. The shaded areas around these curves represent the $\pm 1\sigma$ uncertainty bands on the expected limits. The horizontal lines show the cross sections calculated from PROSPINO for squark masses of 700 GeV and 1500 GeV. The shaded regions around these lines represent the uncertainties on the cross sections. From [67].

Moreover, the potentially long LSP lifetime may give rise to displaced vertices (DV) in the detector. For instance, an RPV SUSY scenario, where the non-zero RPV coupling λ'_{2ij} allows the decay $\tilde{\chi}_1^0 \to \mu q \bar{q}'$, would give rise to a multi-track DV that contains a high-p_{T} muon at a distance between millimeters and tens of centimeters from the pp interaction point. The results of such a search [67] performed by ATLAS are shown in Fig. 16. Fewer than 0.03 background events are expected in the data sample of 33 pb^{-1}, and no events are observed. Based on this null observation, upper limits are set on the supersymmetry production cross-section $\sigma \times B$ of the simulated signal decay chain for different combinations of squark and neutralino masses and for different values of the neutralino lifetime, $c\tau$.

The occurrence of DVs is inherent in the "μ from ν" supersymmetric standard model ($\mu\nu$SSM) [68], which solves the naturalness problem of the MSSM, the so-called μ-problem, and explains the origin of neutrino masses by simply using right-handed neutrino superfields. The model is characterised by three singlets (as opposed to one in NMSSM) and bilinear RPV couplings, giving rise to a rich phenomenology with many Higgs bosons and neutralinos. It has been demonstrated [69] that a 125-GeV Higgs boson while decaying to a pair of unstable long-lived neutralinos, can lead to a distinct signal with non-prompt multileptons. This signal provides an unmistakable signature of the model, pronounced with light neutralinos. Evidence of this signal is well envisaged with sophisticated displaced vertex analysis, as shown in Fig. 17.

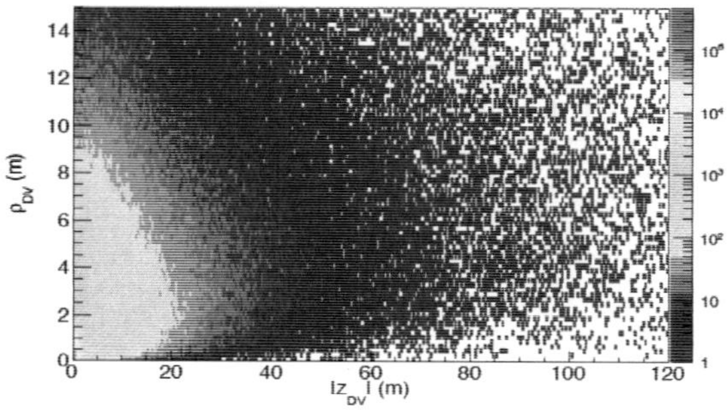

Figure 17. Cylindrical coordinates $\rho_{\rm DV}$ versus $|z_{\rm DV}|$ for the $\tilde{\chi}^0_4 \to \tau\tau\nu$ decay, when a $\mu\nu$SSM Higgs is produced at pp collisions at $\sqrt{s} = 8$ TeV and an integrated luminosity of 20 fb^{-1}. From [69].

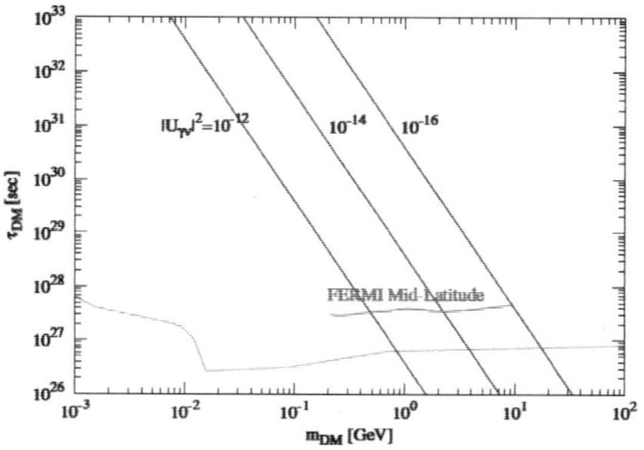

Figure 18. Constraints on $\tilde{G}$ DM lifetime vs. mass in $\mu\nu$SSM. The region below the magenta (red) line is excluded by γ-ray observations (Fermi). Black lines denote predictions for several values of $|U_{\tilde{\gamma}\nu}|^2$. From [70].

As a last remark, we address the issue of (not necessarily cold) dark matter in RPV SUSY models. These seemingly incompatible concepts *can* be reconciled in models with a gravitino [71] or an axino [72] LSP with a lifetime exceeding the age of the Universe. In both cases, RPV is induced by bilinear terms in the superpotential that can also explain current data on neutrino masses and mixings without invoking any GUT-scale physics. Decays of the next-to-lightest superparticle occur rapidly via RPV interaction, and thus they do not upset the Big-Bang nucleosynthesis, unlike the R-parity conserving case. Such gravitino DM is proposed in the context of $\mu\nu$SSM with profound prospects for detecting γ rays from their decay [70]. The constraints on $\tilde{G}$ lifetime versus mass set by Fermi and γ-ray observations are summarised in Fig. 18. Values of the gravitino mass larger than 10 GeV are disfavoured, as well as lifetimes smaller than about 3 to 5×10^{27} s.

4. Summary

The experimental and observational landscape of SUSY searches is multifaceted. In the dark matter front, the cosmological evidence helps orientate the SUSY model building, while the exploration through direct and indirect detection continues with intriguing results. Concerning the high-energy colliders, the LHC is in the forefront: powerful direct searches probe various manifestations of SUSY, whilst indirect information is collected. No signal has been observed so far, thus strong limits have been set up to a SUSY mass scale of ~ 1 TeV. The search strategies keep evolving to explore the most theoretically-attractive models and new ideas. In both Particle and Astroparticle fronts, the quest for Supersymmetry continues; many possibilities are still open.

Acknowledgments

The author is grateful to the DISCRETE2012 organisers for the kind invitation and support. Thanks to them, a warm, friendly and intellectually stimulating atmosphere was enjoyed by the speakers and participants throughout the Symposium. This work was supported in part by the Spanish Ministry of Economy and Competitiveness (MINECO) under the project FPA2009-13234-C04-01 and by the Spanish Agency of International Cooperation for Development under the PCI project A1/035250/11.

References

[1] Aitchison I J R 2009 Supersymmetry in Particle Physics. An Elementary Introduction, Cambridge, UK: Univ. Pr. (2007) 222 p
[2] Evans L and Bryant P 2008 LHC Machine *J. Instrum.* **3** S08001
[3] For a pedagogical introduction, see e.g.: Hooper D 2009 TASI 2008 Lectures on Dark Matter *Preprint* arXiv:0901.4090 [hep-ph]
[4] Mitsou V A 2011 Dark matter searches at LHC *J. Phys.: Conf. Ser.* **335** 012003 (*Preprint* arXiv:1104.1149 [hep-ph])
[5] Cirelli M 2013 Dark Matter searches: phenomenological and theoretical aspects *in these proceedings*
[6] Amanullah R *et al* 2010 Spectra and Light Curves of Six Type Ia Supernovae at $0.511 < z < 1.12$ and the Union2 Compilation *Astrophys. J.* **716** 712 (*Preprint* arXiv:1004.1711 [astro-ph.CO])
[7] Komatsu E *et al* 2011 Seven-Year Wilkinson Microwave Anisotropy Probe (WMAP) Observations: Cosmological Interpretation *Astrophys. J. Suppl.* **192** 18 (*Preprint* arXiv:1001.4538 [astro-ph.CO])
[8] Eisenstein D J *et al* [SDSS Collaboration] 2005 Detection of the Baryon Acoustic Peak in the Large-Scale Correlation Function of SDSS Luminous Red Galaxies *Astrophys. J.* **633** 560 (*Preprint* astro-ph/0501171) and references therein
[9] Broadhurst T, Umetsu K, Medezinski E, Oguri M and Rephaeli Y 2008 Comparison of Cluster Lensing Profiles with Lambda CDM Predictions *Astrophys. J.* **685** L9 (*Preprint* arXiv:0805.2617 [astro-ph]) and references therein
[10] Larson D *et al* 2011 Seven-Year Wilkinson Microwave Anisotropy Probe (WMAP) Observations: Power Spectra and WMAP-Derived Parameters *Astrophys. J. Suppl.* **192** 16 (*Preprint* arXiv:1001.4635 [astro-ph.CO])
[11] Mavromatos N E, Mitsou V A, Sarkar S and Vergou A 2012 Implications of a Stochastic Microscopic Finsler Cosmology *Eur. Phys. J.* C **72** 1956 (*Preprint* arXiv:1012.4094 [hep-ph])
[12] Carroll S M, Press W H and Turner E L 1992 The Cosmological constant *Ann. Rev. Astron. Astrophys.* **30** 499
[13] Amelino-Camelia G, Ellis J R, Mavromatos N E, Nanopoulos D V and Sarkar S 1998 Potential Sensitivity of Gamma-Ray Burster Observations to Wave Dispersion in Vacuo *Nature* **393** 763
Ellis J R, Farakos K, Mavromatos N E, Mitsou V A and Nanopoulos D V 2000 Astrophysical probes of the constancy of the velocity of light *Astrophys. J.* **535** 139
Ellis J R, Mavromatos N E, Nanopoulos D V, Sakharov A S and Sarkisyan E K G 2006 Robust Limits on Lorentz Violation from Gamma-Ray Bursts *Astropart. Phys.* **25** 402 [*Erratum-ibid* 2008 **29** 158]
[14] Dutta B, Gurrola A, Kamon T, Krislock A, Lahanas A B, Mavromatos N E and Nanopoulos D V 2009 Supersymmetry Signals of Supercritical String Cosmology at the Large Hadron Collider *Phys. Rev.* D **79** 055002 (*Preprint* arXiv:0808.1372 [hep-ph])
[15] Ellis J R, Mavromatos N E, Mitsou V A and Nanopoulos D V 2007 Confronting dark energy models with astrophysical data *Astropart. Phys.* **27** 185 (*Preprint* arXiv:astro-ph/0604272)
Mavromatos N E and Mitsou V A 2008 Observational Evidence for Negative-Energy Dust in Late-Times Cosmology *Astropart. Phys.* **29** 442 (*Preprint* arXiv:0707.4671 [astro-ph])
Basilakos S, Mavromatos N E, Mitsou V A and Plionis M 2012 Dynamics and constraints of the Dissipative Liouville Cosmology *Astropart. Phys.* **36** 7 (*Preprint* arXiv:1107.3532 [astro-ph.CO])
[16] Akimov D 2011 Techniques and Results for the Direct Detection of Dark Matter (Review) *Nucl. Instrum. Meth.* A **628** 50 and references therein
[17] Aprile E *et al* [XENON100 Collaboration] 2012 Dark Matter Results from 225 Live Days of XENON100 Data *Phys. Rev. Lett.* **109** 181301 (*Preprint* arXiv:1207.5988 [astro-ph.CO])

[18] Bernabei R *et al* [DAMA Collaboration] 2008 First results from DAMA/LIBRA and the combined results with DAMA/NaI *Eur. Phys. J.* C **56** 333 (*Preprint* arXiv:0804.2741 [astro-ph])

[19] Aalseth C E *et al* [CoGeNT Collaboration] 2011 Results from a Search for Light-Mass Dark Matter with a P-type Point Contact Germanium Detector *Phys. Rev. Lett.* **106** 131301
Aalseth C E *et al* [CoGeNT Collaboration] 2012 CoGeNT: A Search for Low-Mass Dark Matter using p-type Point Contact Germanium Detectors *Preprint* arXiv:1208.5737 [astro-ph.CO]

[20] Angloher G *et al* 2012 Results from 730 kg days of the CRESST-II Dark Matter Search *Eur. Phys. J.* C **72** 1971 (*Preprint* arXiv:1109.0702 [astro-ph.CO])

[21] Ahmed Z *et al* [CDMS Collaboration] 2012 Search for annual modulation in low-energy CDMS-II data *Preprint* arXiv:1203.1309 [astro-ph.CO]

[22] Savage C, Gelmini G, Gondolo P and Freese K 2009 Compatibility of DAMA/LIBRA dark matter detection with other searches *J. Cosmol. Astropart. Phys.* JCAP04(2009)010
Ahmed Z *et al* [CDMS-II Collaboration] 2010 Dark Matter Search Results from the CDMS II Experiment *Science* **327** 1619
Ahmed Z *et al* [CDMS-II Collaboration] 2011 Results from a Low-Energy Analysis of the CDMS II Germanium Data *Phys. Rev. Lett.* **106** 131302
Armengaud E *et al* [EDELWEISS Collaboration] 2011 Final results of the EDELWEISS-II WIMP search using a 4-kg array of cryogenic germanium detectors with interleaved electrodes *Phys. Lett.* B **702** 329
Angle J *et al* [XENON10 Collaboration] 2011 A search for light dark matter in XENON10 data *Phys. Rev. Lett.* **107** 051301
Felizardo M *et al* 2012 Final Analysis and Results of the Phase II SIMPLE Dark Matter Search *Phys. Rev. Lett.* **108** 201302
Behnke E *et al* [COUPP Collaboration] 2012 First Dark Matter Search Results from a 4-kg CF_3I Bubble Chamber Operated in a Deep Underground Site *Phys. Rev.* D **86** 052001
Akimov D Y *et al* 2012 WIMP-nucleon cross-section results from the second science run of ZEPLIN-III *Phys. Lett.* B **709** 14
Armengaud E *et al* [EDELWEISS Collaboration] 2012 A search for low-mass WIMPs with EDELWEISS-II heat-and-ionization detectors *Phys. Rev.* D **86** 051701

[23] Strege C, Bertone G, Cerdeno D G, Fornasa M, Ruiz de Austri R and Trotta R 2012 Updated global fits of the cMSSM including the latest LHC SUSY and Higgs searches and XENON100 data *J. Cosmol. Astropart. Phys.* JCAP03(2012)030
Fowlie A, Kazana M, Kowalska K, Munir S, Roszkowski L, Sessolo E M, Trojanowski S and Tsai Y-L S 2012 The CMSSM Favoring New Territories: The Impact of New LHC Limits and a 125 GeV Higgs *Phys. Rev.* D **86** 075010
Buchmueller O *et al* 2012 Higgs and Supersymmetry *Eur. Phys. J.* C **72** 2020

[24] Morselli A 2011 Indirect detection of dark matter, current status and recent results *Prog. Part. Nucl. Phys.* **66** 208 and references therein

[25] Abdo A A *et al* [Fermi-LAT Collaboration] 2010 Observations of Milky Way Dwarf Spheroidal galaxies with the Fermi-LAT detector and constraints on Dark Matter models *Astrophys. J.* **712** 147

[26] Gondolo P, Edsjo J, Ullio P, Bergstrom L, Schelke M and Baltz E A 2004 DarkSUSY: Computing supersymmetric dark matter properties numerically *J. Cosmol. Astropart. Phys.* JCAP07(2004)008

[27] Adriani O *et al* [PAMELA Collaboration] 209 An anomalous positron abundance in cosmic rays with energies 1.5-100 GeV *Nature* **458** 607 (*Preprint* arXiv:0810.4995 [astro-ph])

[28] Abdo A A *et al* [Fermi LAT Collaboration] 2009 Measurement of the Cosmic Ray e+ plus e- spectrum from 20 GeV to 1 TeV with the Fermi Large Area Telescope *Phys. Rev. Lett.* **102** 181101

[29] Weniger C 2012 A Tentative Gamma-Ray Line from Dark Matter Annihilation at the Fermi Large Area Telescope *J. Cosmol. Astropart. Phys.* JCAP08(2012)007 (*Preprint* arXiv:1204.2797 [hep-ph])

[30] Bergström L 2013 Dark matter and imaging air Cherenkov arrays *Astropart. Phys.* **43** 44

[31] For a review, see e.g. Hooper D and Profumo S 2007 Dark matter and collider phenomenology of universal extra dimensions *Phys. Rept.* **453** 29 (*Preprint* arXiv:hep-ph/0701197)

[32] Bergstrom L, Bringmann T and Edsjo J 2011 Complementarity of direct dark matter detection and indirect detection through gamma-rays *Phys. Rev.* D **83** 045024 (*Preprint* arXiv:1011.4514 [hep-ph])

[33] Harnew N 2013 Selected highlights from LHCb *in these proceedings*
Oggero S 2013 Very rare decays at LHCb *in these proceedings*

[34] Alves A A Jr *et al* [LHCb Collaboration] 2008 The LHCb Detector at the LHC *J. Instrum.* **3** S08005

[35] Aaij R *et al* [LHCb Collaboration] 2013 First evidence for the decay $B_s \to \mu^+\mu^-$ *Phys. Rev. Lett.* **110** 021801 (*Preprint* arXiv:1211.2674)

[36] Aaij R *et al* [LHCb Collaboration] 2012 Strong constraints on the rare decays $B_s \to \mu^+\mu^-$ and $B^0 \to \mu^+\mu^-$ *Phys. Rev. Lett.* **108** 231801 (*Preprint* arXiv:1203.4493 [hep-ex])

[37] Mahmoudi F 2012 Direct and indirect searches for New Physics *Preprint* arXiv:1205.3099 [hep-ph]
[38] Chatrchyan S *et al* [CMS Collaboration] 2011 Search for Supersymmetry at the LHC in Events with Jets and Missing Transverse Energy *Phys. Rev. Lett.* **107** 221804 (*Preprint* arXiv:1109.2352 [hep-ex])
[39] CMS Collaboration 2012 Search for supersymmetry with the razor variables at CMS *Physics Analysis Summary* CMS-PAS-SUS-12-005
[40] Aad G *et al* [ATLAS Collaboration] 2008 The ATLAS Experiment at the CERN Large Hadron Collider *J. Instrum.* **3** S08003
[41] Adolphi R *et al* [CMS Collaboration] 2008 The CMS experiment at the CERN LHC *J. Instrum.* **3** S08004
[42] Cabrera M E and Casas J A 2012 Understanding and improving the Effective Mass for LHC searches *Preprint* arXiv:1207.0435 [hep-ph]
[43] Cabrera M E, Casas J A, Mitsou V A, Ruiz de Austri R and Terron J 2012 Histogram comparison as a powerful tool for the search of new physics at LHC. Application to CMSSM *J. High Energy Phys.* JHEP04 (2012)133 (*Preprint* arXiv:1109.3759 [hep-ph])
[44] Barela J 2013 Overview of CMS Physics Results *in these proceedings*
[45] Chatrchyan S *et al* [CMS Collaboration] 2012 Search for new physics in the multijet and missing transverse momentum final state in proton-proton collisions at $\sqrt{s} = 7$ TeV *Phys. Rev. Lett.* **109** 171803
[46] CMS Collaboration 2012 Search for supersymmetry in final states with missing transverse energy and 0, 1, 2, 3, or at least 4 b-quark jets in 8 TeV *pp* collisions using the variable AlphaT *Physics Analysis Summary* CMS-PAS-SUS-12-028
[47] Giangiobbe V 2013 Review of Photon + X analysis in ATLAS *in these proceedings*
Abdallah J 2013 Search for new physics using jets at ATLAS *in these proceedings*
[48] ATLAS Collaboration 2012 Search for supersymmetry in final states with jets, missing transverse momentum and a Z boson at $\sqrt{s} = 8$ TeV with the ATLAS detector *CONF Note* ATLAS-CONF-2012-152
[49] ATLAS SUSY twiki: `http://twiki.cern.ch/twiki/bin/view/AtlasPublic/SupersymmetryPublicResults`
[50] CMS SUSY results: `http://twiki.cern.ch/twiki/bin/view/CMSPublic/PhysicsResultsSUS`
[51] Cabrera M E, Casas J A and Zaldivar B 2012 New techniques for chargino-neutralino detection at LHC *Preprint* arXiv:1212.5247 [hep-ph]
[52] Chatrchyan S *et al* [CMS Collaboration] 2012 Search for heavy long-lived charged particles in *pp* collisions at $\sqrt{s} = 7$ TeV *Phys. Lett.* B **713** 408 (*Preprint* arXiv:1205.0272 [hep-ex])
[53] MoEDAL Collaboration 2009 Technical Design Report of the MoEDAL Experiment *CERN Report* CERN-LHC-2009-006 *MoEDAL TDR* MoEDAL-TDR-1.1
[54] Chatrchyan S *et al* [CMS Collaboration] 2012 Search for new physics with long-lived particles decaying to photons and missing energy in *pp* collisions at $\sqrt{s} = 7$ TeV *J. High Energy Phys.* JHEP11(2012)172
[55] Battaglia M 2009 The role of an e^+e^- linear collider in the study of cosmic dark matter *New J. Phys.* **11** 105025
Aarons G *et al* [ILC Collaboration] 2007 International Linear Collider Reference Design Report Volume 2: Physics at the ILC *Preprint* arXiv:0709.1893 [hep-ph]
[56] Accomando E *et al* [CLIC Physics Working Group] Physics at the CLIC multi-TeV linear collider *Preprint* arXiv:hep-ph/0412251
[57] Aad G *et al* [ATLAS Collaboration] 2012 Observation of a new particle in the search for the Standard Model Higgs boson with the ATLAS detector at the LHC *Phys. Lett.* B **716** 1 (*Preprint* arXiv:1207.7214 [hep-ex])
[58] Gonçalo R 2013 Higgs Boson Searches with ATLAS *in these proceedings*
[59] Chatrchyan S *et al* [CMS Collaboration] 2012 Observation of a new boson at a mass of 125 GeV with the CMS experiment at the LHC *Phys. Lett.* B **716** 30 (*Preprint* arXiv:1207.7235 [hep-ex])
[60] Djouadi A 2013 Some implications of the LHC Higgs results *in these proceedings*
[61] For a review, see e.g.: Ellwanger U, Hugonie C and Teixeira A M 2010 The Next-to-Minimal Supersymmetric Standard Model *Phys. Rept.* **496** 1 (*Preprint* arXiv:0910.1785 [hep-ph])
[62] Vasquez D A, Belanger G, Boehm C, Da Silva J, Richardson P and Wymant C 2012 The 125 GeV Higgs in the NMSSM in light of LHC results and astrophysics constraints *Phys. Rev.* D **86** 035023
[63] Barbier R *et al* 2005 *R*-parity violating supersymmetry *Phys. Rept.* **420** 1 (*Preprint* arXiv:hep-ph/0406039)
[64] Hirsch M and Valle J W F 2004 Supersymmetric origin of neutrino mass *New J. Phys.* **6** 76
Romao J C, Diaz M A, Hirsch M, Porod W and Valle J W F 2000 A Supersymmetric solution to the solar and atmospheric neutrino problems *Phys. Rev.* D **61** 071703
Hirsch M, Diaz M A, Porod W, Romao J C and Valle J W F 2000 Neutrino masses and mixings from supersymmetry with bilinear *R* parity violation: A Theory for solar and atmospheric neutrino oscillations *Phys. Rev.* D **62** 113008 [*Erratum-ibid* 2002 D **65** 119901]
Diaz M A, Hirsch M, Porod W, Romao J C and Valle J W F 2003 Solar neutrino masses and mixing from bilinear *R* parity broken supersymmetry: Analytical versus numerical results (*Preprint* Phys. Rev. D **68** 013009 [*Erratum-ibid* 2005 D **71** 059904]

Porod W, Hirsch M, Romao J and Valle J W F 2001 Testing neutrino mixing at future collider experiments *Phys. Rev.* D **63** 115004

[65] ATLAS Collaboration 2012 Search for supersymmetry at $\sqrt{s} = 7$ TeV in final states with large jet multiplicity, missing transverse momentum and one isolated lepton with the ATLAS detector *CONF Note* ATLAS-CONF-2012-140

[66] Aad G *et al* [ATLAS Collaboration] 2012 Search for supersymmetry in final states with jets, missing transverse momentum and one isolated lepton in $\sqrt{s} = 7$ TeV pp collisions using 1 $\text{fb}^{-1}$of ATLAS data *Phys. Rev.* D **85** 012006 (*Preprint* arXiv:1109.6606 [hep-ex])

[67] Aad G *et al* [ATLAS Collaboration] 2013 Search for long-lived, heavy particles in final states with a muon and multi-track displaced vertex in proton-proton collisions at $\sqrt{s} = 7$ TeV with the ATLAS detector *Phys. Lett.* B **719** 280 (*Preprint* arXiv:1210.7451 [hep-ex])

[68] López-Fogliani D E and Muñoz C 2006 Proposal for a new minimal supersymmetric standard model *Phys. Rev. Lett.* **97** 041801 (*Preprint* arXiv:hep-ph/0508297)
Escudero N, López-Fogliani D E, Muñoz C and de Austri R R 2008 Analysis of the parameter space and spectrum of the $\mu\nu$SSM *J. High Energy Phys.* JHEP12(2008)099 (*Preprint* arXiv:0810.1507 [hep-ph])

[69] Ghosh P, Lopez-Fogliani D E, Mitsou V A, Munoz C and de Austri R R 2012 Displaced multileptons at the LHC – probing a 125 GeV new boson in $\mu\nu$SSM *Preprint* arXiv:1211.3177 [hep-ph]

[70] Choi K-Y, Lopez-Fogliani D E, Munoz C and de Austri R R 2010 Gamma-ray detection from gravitino dark matter decay in the $\mu\nu$SSM *J. Cosmol. Astropart. Phys.* JCAP03(2010)028

[71] Takayama F and Yamaguchi M 2000 Gravitino dark matter without R-parity *Phys. Lett.* B **485** 388
Hirsch M, Porod W and Restrepo D 2005 Collider signals of gravitino dark matter in bilinearly broken R-parity *J. High Energy Phys.* JHEP03(2005)062
Buchmuller W, Covi L, Hamaguchi K, Ibarra A and Yanagida T 2007 Gravitino dark matter in R-parity breaking vacua *J. High Energy Phys.* JHEP03(2007)037

[72] Chun E J and Kim H B 2006 Axino Light Dark Matter and Neutrino Masses with R-parity Violation *J. High Energy Phys.* JHEP10(2006)082 (*Preprint* arXiv:hep-ph/0607076)

Journal of Physics: Conference Series **447** (2013) 012020 doi:10.1088/1742-6596/447/1/012020

Charm mixing and CP violation at LHCb

Artur Ukleja
National Centre for Nuclear Studies, Warsaw, PL
for the LHCb Collaboration
E-mail: artur.ukleja@fuw.edu.pl

Abstract. In pp collisions at the LHC the LHCb experiment has collected the world's largest sample of charmed hadrons. Using data of an integrated luminosity of 1.0 fb^{-1} recorded in 2011 a measurement of direct and indirect CP violation in the charm sector and of D^0 mixing parameters was performed. Results from several decay modes are presented with complementary time-dependent and time-integrated analyses.

1. Introduction

The charm sector of particle physics is a promising place to probe for the presence of Physics beyond the Standard Model (NP) because CP violation in the charm sector is expected to be small in the Standard Model (SM). Since evidence of $D^0 - \bar{D}^0$ oscillations was first reported [1, 2, 3, 4] there was growing interest in this subject. No single measurement reached a precision above 5σ significance in combinations of all measurements however, a 10σ significance of $D^0 - \bar{D}^0$ oscillations could be reached [5]. Discovery in a single measurement was presented for the first time by the LHCb experiment [6]. Details of this measurement are discussed in Section 3. The time integrated CP asymmetry measurements in multi-body charm decays with pions and/or kaons in the final states are discussed for two, three and four body decays in Sections 4, 5 and 6, respectively.

2. LHCb detector

The LHCb detector [7] is a single-arm forward spectrometer covering the pseudorapidity range $2 < y < 5$, designed for the study of particles containing b or c quarks. The detector has a very good particle identification for π and K, where misidentification is smaller than 5%. Displaced vertices of c-hadron decays can be measured with 20 μm resolution. A decay time resolution 0.1 of the D meson lifetime using a silicon vertex locator is achieved. The tracking system measures the charged particles with momentum resolution $\Delta p/p$ that varies from 0.4% at 5 GeV to 0.6% at 100 GeV corresponding to a typical mass resolution of approximately 8 MeV for a two body charm meson decay. The $c\bar{c}$ cross-section in pp collisions of ~6 mb [8] measured with the LHCb detector at $\sqrt{s} = 7$ TeV is about 20 times larger than the $b\bar{b}$ cross-section. Therefore, the LHCb experiment has a great potential in charm physics studies.

3. First observation of $D^0 - \bar{D}^0$ oscillations

In order to measure $D^0 - \bar{D}^0$ oscillations the D flavour has to be identified at both the production and decay. The flavour of the D^0 meson of production time is identified in the process of

Journal of Physics: Conference Series **447** (2013) 012020 doi:10.1088/1742-6596/447/1/012020

$D^{*+} \to D^0(\to hh)\pi_s^+$ decays, with $h = \pi$ or K, in which the sign of the slow pion π_s tags the initial D^0 or $\bar{D}^0$. The flavour at decay is tagged by the charge of the K meson. The $D^0 \to K^+\pi^-$ decays called wrong-sign (WS) and the $D^0 \to K^-\pi^+$ decays called right-sign (RS) are considered. The RS decays are dominated by the Cabibbo-favoured (CF) decay amplitude, whereas the WS amplitude includes contributions from both the doubly-Cabibbo-suppressed (DCS) $D^0 \to K^+\pi^-$ decay and $D^0 - \bar{D}^0$ mixing followed by the favoured $\bar{D}^0 \to K^+\pi^-$ decay. In the limit of small mixing and assuming negligible CP violation, the time-dependent ratio, R, of WS to RS decay rates is approximated by [9]

$$R(t) = R_D + \sqrt{R_D}y'\frac{t}{\tau} + \frac{x'^2 + y'^2}{4}(\frac{t}{\tau})^2, \qquad (1)$$

where t/τ is the decay time expressed in units of the average D^0 lifetime, τ, R_D is the ratio of DCS to CF decay rates, $x' = x\cos\delta + y\sin\delta$, $y' = y\cos\delta - x\sin\delta$, where δ is the strong phase difference between the DCS and the CF amplitudes and $x = \Delta m/\Gamma$, $y = \Delta\Gamma/2\Gamma$, where Δm and $\Delta\Gamma$ are the mass and decay width differences between the two mass eigenstates and Γ is the average D^0 decay width.

3.1. Data set and selection

The measurement is performed with 1.0 fb^{-1} of data recorded by the LHCb detector during 2011. Events are selected by the signature of charm decays based on two oppositely charged tracks to form a D^0 candidate with a decay vertex well separated from the associated pp collision vertex. Further criteria on the quality of the reconstructed tracks are applied and described in [6].

The time-integrated $D^0\pi_s^+$ mass distributions, $M(D^0\pi_s^+)$, for the selected RS and WS candidates are shown in Fig. 1. A binned χ^2 fit is used to separate the D^{*+} signal and the background component. From the fits about 3.6×10^4 WS and 8.4×10^6 RS decays are found.

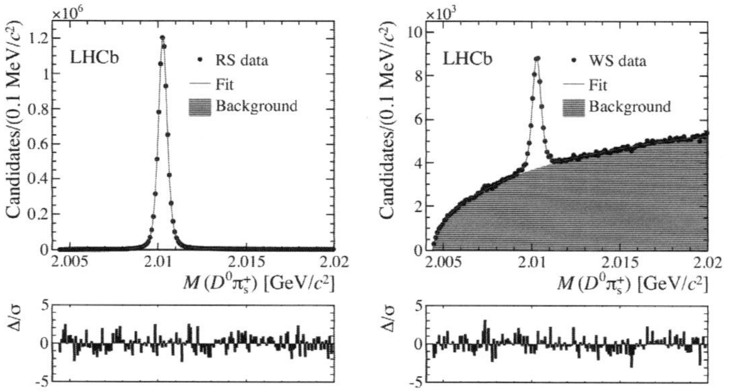

Figure 1. Time-integrated $D^0\pi_s^+$ mass distributions for the selected RS $D^0 \to K^-\pi^+$ (left) and WS $D^0 \to K^+\pi^-$ (right) candidates with fit projections overlaid. The bottom plots show the residuals between the data points and the fits.

3.2. Results

The measurement of the time-dependent WS/RS ratio is performed in bins of D^0 decay time. The data is divided into thirteen bins, chosen to have a similar number of candidates in each. The number of RS and WS decays are determined using fits to the $M(D^0\pi_s^+)$ distributions in the thirteen bins to calculate the WS/RS ratios. The measured WS/RS ratios are shown in

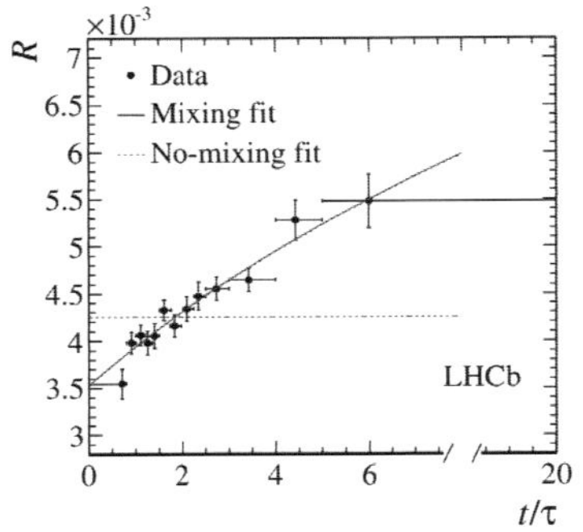

Figure 2. Decay time dependence of the ratio, R, of $D^0 \to K^+\pi^-$ to $D^0 \to K^-\pi^+$ decays with the projection of the mixing allowed (solid line) and no-mixing (dashed line) fits overlaid.

Table 1. Results of the time-dependent fit to the ratio of WS and RS decays. The uncertainties include statistical and systematic sources.

Fit type (χ^2/ndf)	Parameter	Fit results (10^{-3})	Correlation coefficient R_D	y'	x'^2
Mixing (9.5/10)	R_D	3.52 ± 0.15	1	-0.954	$+0.882$
	y'	7.2 ± 2.4		1	-0.973
	x'^2	-0.09 ± 0.13			1
No mixing (98.1/12)	R_D	4.25 ± 0.04			

Fig. 2 [6]. From the binned χ^2 fit to the time-dependence given in Eq. 1 (solid line in Fig. 2) the mixing parameters are obtained and are listed in Table 1.

Most of the systematic uncertainties cancel in the ratio of WS and RS since the WS and RS events are expected to have the same decay time acceptance. Two main sources of systematic uncertainties have been identified. One of them is related to the secondary D decays. It was found that when the secondary component is not subtracted the measured WS/RS ratio could be biased less than 3%.

A second source of the systematic uncertainty is related to background from charm decays where both daughters are reconstructed with the wrong particle type which gives a peak in $M(D^0\pi_s^+)$. This is not accounted for in the mass fit. The background is suppressed by the use of tight particle identification and two body mass requirements. It was found that the dominant peaking background is from RS events that survive the requirements of the WS selection. They are estimated to constitute $(0.4 \pm 0.2)\%$ of the WS signal. The measurements are dominated by the statistical uncertainties.

The fit to the WS/RS ratio with the no-mixing hypothesis does not describe the data (dashed line in Fig. 2). The χ^2 difference between mixing and no-mixing assumption excludes the no-mixing hypothesis with a probability corresponding to 9.1 standard deviations. This is the first observation of $D^0 - \bar{D}^0$ oscillations in a single measurement. In Fig. 3 the 1σ, 3σ and 5σ confidence regions of the measured mixing parameters x'^2 and y' are shown. They are compatible with and have substantially better precision than those from previous measurements [1, 3, 10].

4. Evidence for CPV in $D^0 \to h^-h^+$ decays

The time-integrated CP asymmetry has two contributions: an indirect and a direct component. The indirect component is universal for CP eigenstates in the SM and is expected to be

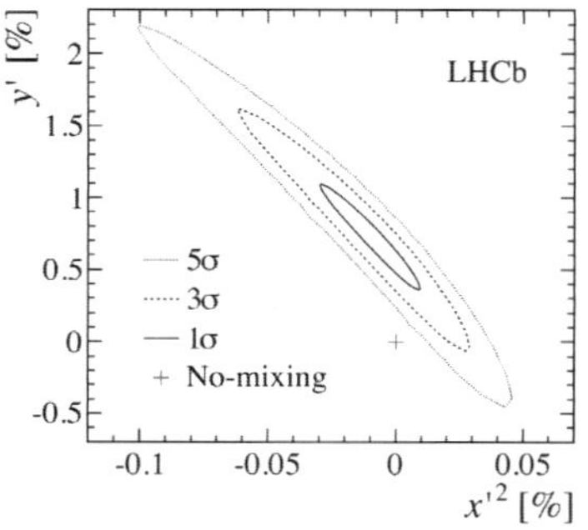

Figure 3. Estimated confidence-level (CL) regions in the (x'^2, y') plane for $1 - \mathrm{CL} = 0.317(1\sigma)$, $2.7 \times 10^{-3}(3\sigma)$ and $5.73 \times 10^{-7}(5\sigma)$. The cross indicates the no-mixing point.

small. The direct component depends in general on the final state. In the SM direct CPV in $D^0 \to K^-K^+$ and the $D^0 \to \pi^-\pi^+$ decays is expected to be 10^{-3} or less [11]. In the presence of the NP, however, the rate of CPV could be enhanced. Similar to the $D^0 - \bar{D}^0$ oscillation measurements tagged D^0 decays are analysed.

The measured raw time-integrated asymmetry may be written as a sum of various components both from physics and detector effects

$$A_{RAW}(f) = A_{CP}(f) + A_D(f) + A_D(\pi_s^+) + A_P(D^{*+}), \tag{2}$$

where $A_{CP}(f)$ is an intrinsic physics CP asymmetry, $A_D(f)$ is the asymmetry for selecting the D^0 decay into the final state f, $A_D(\pi_s^+)$ is the asymmetry for selecting a slow pion from the D^{*+} decay chain. The $A_P(D^{*+})$ is the production asymmetry of the D^{*+} mesons.

Since the decays $D^0 \to K^-K^+$ and $D^0 \to \pi^-\pi^+$ are symmetric the D^0 detection asymmetries cancel. The asymmetries $A_D(\pi_s^+)$ and $A_P(D^{*+})$ are independent of the final state and therefore the terms cancel in the subtraction. For this reason, the difference $\Delta A_{CP} = A_{RAW}(K^-K^+) - A_{RAW}(\pi^-\pi^+)$ is measured and the quantity ΔA_{CP} is a difference of CP asymmetries, $\Delta A_{CP} = A_{CP}(K^-K^+) - A_{CP}(\pi^-\pi^+)$.

4.1. Data set and selection

The measurement is performed with 0.62 fb^{-1} of LHCb data. The selection criteria applied are related to the decay time to isolate the decays of interest, the track quality, the vertex quality, the transverse momentum, the angle between the D^0 momentum in the lab frame and its daughters' momenta in the rest frame. The full selection description can be found in [12].

The invariant mass spectra of the selected $D \to K^-K^+$ and $D \to \pi^-\pi^+$ candidates are shown in Fig. 4. The half width at half maximum of the signal line shape is 8.6 MeV/c^2 for the $D \to K^-K^+$ and 11.2 MeV/c^2 for the $D \to \pi^-\pi^+$. The mass difference spectra, $\delta m \equiv m(h^-h^+\pi_s^+) - m(h^-h^+) - m(\pi^+)$, $h = K, \pi$, of selected candidates are shown in Fig. 5. The candidates are required to lie inside a wide δm window of $0 - 15$ MeV and a D^0 mass signal window of $1844 - 1884$ MeV/c^2. Approximately 1.44×10^6 candidates are found in the K^-K^+ sample and 0.38×10^6 in the $\pi^-\pi^+$ sample.

4.2. Results

The production and the detection asymmetries vary with p_T, pseudorapidity, η, and the detection efficiency of the two different D^0 decays. Therefore, the analysis is performed in bins of p_T, the η of the D^{*+} candidates and the momentum of the slow pion. In total, there are 216 statistically independent measurements for each decay mode. A value of ΔA_{CP} is determined in each bin.

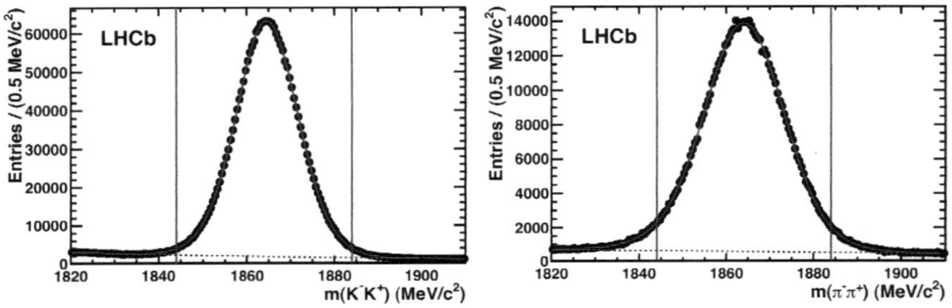

Figure 4. Fits to the $m(K^-K^+)$ (left) and the $m(\pi^-\pi^+)$ (right) spectra of D^{*+} candidates. The dashed line corresponds to the background component in the fit and the vertical lines indicate the signal window of $1844 - 1884$ MeV/c^2.

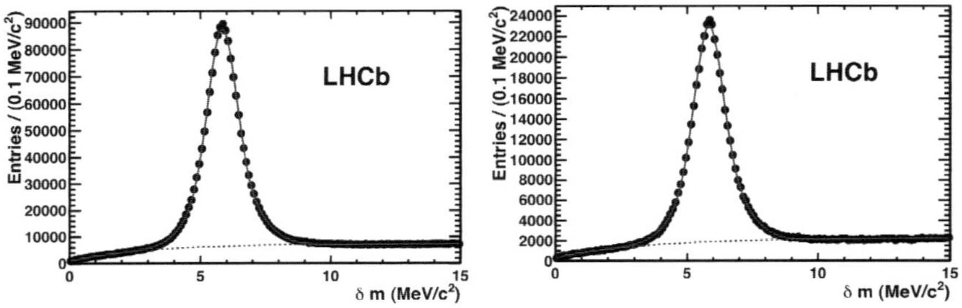

Figure 5. Fits to the mass difference spectra, $\delta m \equiv m(h^-h^+\pi_s^+) - m(h^-h^+) - m(\pi^+)$ for $h = K, \pi$, where the D^0 is reconstructed in the final states K^-K^+ (left) and $\pi^-\pi^+$ (right), with a D^0 mass lying in the window $1844 - 1884$ MeV/c^2.

No systematic dependence of ΔA_{CP} is observed with respect to the kinematic variables. The weighted average yields

$$\Delta A_{CP} = [-0.82 \pm 0.21(stat) \pm 0.11(syst)]\%.$$

Systematic uncertainties are assigned by changing the selection criteria, repeating the analysis with the asymmetry extracted through the sideband subtraction in δm instead of a fit, removing all candidates but one (chosen at random) in events with multiple candidates and comparing with the result obtained without kinematic binning.

Combining statistical and systematic uncertainties in quadrature, the result is consistent at the 1σ level with the current HFAG world average [5]. Dividing the central value by the sum in quadrature of the statistical and systematic uncertainties, the significance of the measured deviation from zero is 3.5σ. This is the first evidence for CPV in the charm sector.

4.3. Result interpretation

The CP asymmetry of each final state, $A_{CP}(f)$, can be expressed in terms of two contributions, a direct component associated with CPV in the decay amplitudes and an indirect component associated with CPV in the mixing or in the interference between mixing and decay. It can be written to the first order as [13]

$$A_{CP}(f) = a_{CP}^{dir}(f) + \frac{\langle t \rangle}{\tau} a_{CP}^{ind}, \qquad (3)$$

where $a_{CP}^{dir}(f)$ is a direct CPV for the decay, $\langle t \rangle$ is the average proper time of the reconstructed sample, τ is a D^0 lifetime, and a_{CP}^{ind} is an indirect CPV originating from the mixing and/or the interference between mixing and decay.

The measured quantity ΔA_{CP} can be related to the direct and indirect CPV

$$\Delta A_{CP} \equiv A_{CP}(K^-K^+) - A_{CP}(\pi^-\pi^+) = [a_{CP}^{dir}(K^-K^+) - a_{CP}^{dir}(\pi^-\pi^+)] + \frac{\Delta\langle t \rangle}{\tau} a_{CP}^{ind}, \quad (4)$$

where $\Delta\langle t \rangle$ is the difference in average proper time of the D^0 mesons in the K^-K^+ and the $\pi^-\pi^+$ samples. In the limit that $\Delta\langle t \rangle$ vanishes, ΔA_{CP} probes the difference in direct CPV between the two decays. Since indirect CPV is universal (up to 10^{-2}) [14], its contribution would cancel in the difference. The fractional difference $\Delta\langle t \rangle/\tau$ is obtained as $\Delta\langle t \rangle/\tau = [9.83 \pm 0.22(stat) \pm 0.19(syst)]\%$, the measured ΔA_{CP} is primarily sensitive to direct CPV.

5. Search for CPV in $D^+ \to K^-K^+\pi^+$ decays

Multi body charm meson decays are dominated by intermediate resonant states. The interference between resonances in the two dimensional phase space can lead to observable CP asymmetries which vary across the Dalitz plot for the Cabibbo suppressed decay $D^+ \to K^-K^+\pi^+$. Previously, CPV asymmetries in the Dalitz plot in three body charm decays were investigated by BABAR [15], CLEO-c [16] and BELLE [17]. No CPV has been observed in D decays.

A direct model independent comparison between the D^+ and the D^- Dalitz plots can be done on a bin by bin basis. For each bin, a local CP asymmetry is defined as [18]

$$S_{CP}^i = \frac{N^i(D^+) - \alpha N^i(D^-)}{\sqrt{N^i(D^+) + \alpha^2 N^i(D^-)}}, \quad (5)$$

where $N^i(D^+)$ and $N^i(D^-)$ are the numbers of the D^+ and the D^- candidates in the ith bin and the parameter α accounts for a global asymmetry which is the ratio between the total reconstructed D^+ and D^- candidates, $\alpha = N_{tot}(D^+)/N_{tot}(D^-)$.

In the absence of the Dalitz plot dependent asymmetries, the S_{CP}^i values are distributed according to a Gaussian distribution with zero mean and unit width. CPV manifests itself by deviations from this behaviour. The numerical comparison is performed with a χ^2 test, $\chi^2 = \Sigma(S_{CP}^i)^2$, and a $p-$value calculation.

5.1. Data set and selection

These measurements are performed with 35 pb^{-1} of LHCb data collected in 2010. The event selection is based on many criteria, the relative log-likelihood for pion and kaon hypothesis, the fit quality for each particle reconstructed in the final state, the primary and secondary vertex reconstruction and the flight distance variable [19].

The reconstructed invariant mass distribution of the $K^-K^+\pi^+$ is shown in Fig. 6. From a simultaneous fit to the invariant mass distributions of the D^+ and the D^-, a total number of 403894 candidates (signal and background) is obtained. Within a D^+ mass window of $1856.7 - 1882.1$ MeV/c^2, about 8.6% of events are background and 200336 D^+ and 203558 D^- candidates.

5.2. Results

The measured S_{CP}^i distribution in $D^+ \to K^-K^+\pi^+$ decays is shown in Fig. 7 for two binning schemes: uniform and adaptive. For the adaptive, two different numbers of different size bins were used, 25 bins (Adaptive I) and 106 bins (Adaptive II). For the uniform, two different numbers of equal size bins were used, 200 bins (Uniform I) and 530 bins (Uniform II). Results

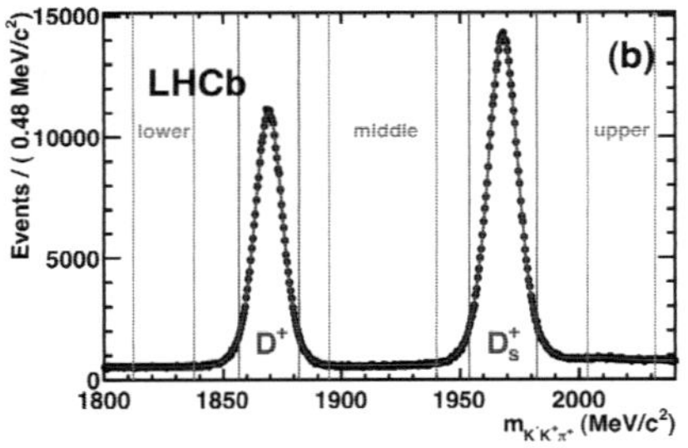

Figure 6. The fitted mass distribution of $D^+ \to K^- K^+ \pi^+$ candidates.

are listed in Table 2. The calculated $p-$values are above 10% and the fits are consistent with Gaussian distributions with zero mean and unit width. The $p-$values indicate that there is no evidence for CPV.

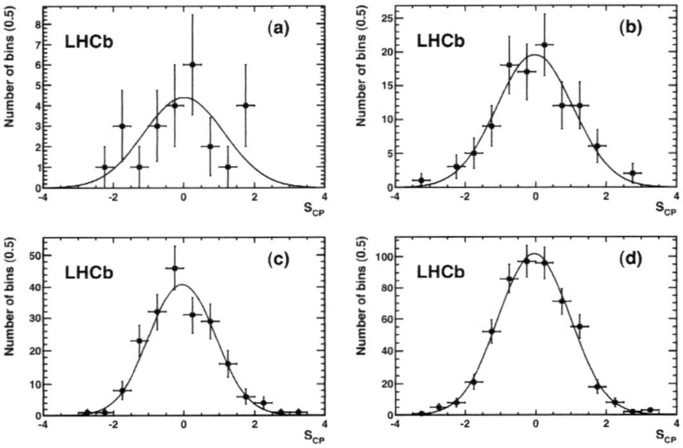

Figure 7. Distribution of S^i_{CP} fitted to Gaussian functions for (a) Adaptive I, (b) Adaptive II, (c) Uniform I and (d) Uniform II.

Table 2. Fitted means and widths, χ^2/ndf and p-values for $D^+ \to K^- K^+ \pi^+$ decays with four different binning schemes.

Binning	Fitted mean	Fitted width	χ^2/ndf	p-value (%)
Adaptive I	0.01 ± 0.23	1.13 ± 0.16	32.0/24	12.7
Adaptive II	-0.024 ± 0.010	1.078 ± 0.074	123.4/105	10.6
Uniform I	-0.043 ± 0.073	0.929 ± 0.051	191.3/198	82.1
Uniform II	-0.039 ± 0.045	1.011 ± 0.034	519.5/529	60.5

6. Search for CPV in D$^+ \to \pi^- \pi^+ \pi^- \pi^+$ decays

While the three body decay kinematics can be described fully by a two dimensional Dalitz plot, a four body decay needs a five dimensional phase space description. In this study, the same model independent method is used as for three body charm decays (Section 5). The analysis is performed with 1 fb^{-1} of LHCb data collected in 2011. About 180k $D^+ \to \pi^-\pi^+\pi^-\pi^+$ candidates with a signal purity of 95.8% have been found. The analysis has been performed on three different adaptive binnigs 15, 29 and 66. The S_{CP} distribution for 66 adaptive bins is shown in Fig. 8. The p-values are 97.1%, 95.6% and 99.8%, respectively [20] and the measurements are consistent with the hypothesis of no CPV.

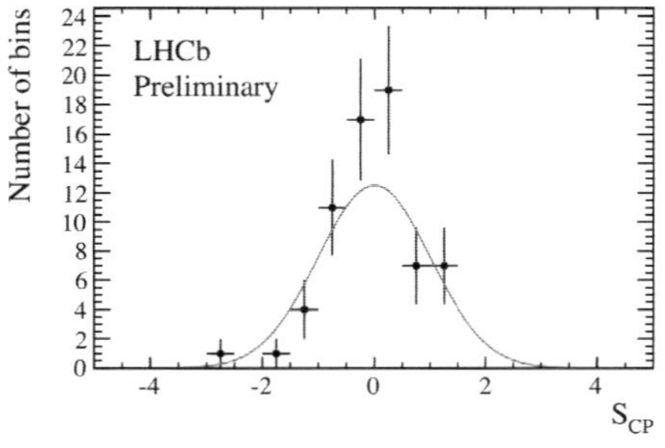

Figure 8. The S_{CP} distribution of $D^+ \to \pi^-\pi^+\pi^-\pi^+$ decays for 66 adaptive bins. The full line shows Gaussian distribution with the assumption of no CPV.

7. Conclusions

Using data of 1 fb^{-1} collected with the LHCb experiment in 2011 $D^0 - \bar{D}^0$ oscillations are observed for the first time in a single measurement evaluating the time-dependence of the ratio of $D^0 \to K^+\pi^-$ and $D^0 \to K^-\pi^+$ decays. The no-mixing hypothesis is excluded at 9.1 standard deviations. The time-integrated difference in CP asymmetry of $D^0 \to K^-K^+$ and the $D^0 \to \pi^-\pi^+$ decays has been measured using 0.62 fb^{-1} of data

$$\Delta A_{CP} = [-0.82 \pm 0.21(stat) \pm 0.11(syst)]\%.$$

The measured ΔA_{CP} is mainly sensitive to the presence of direct CP violation.

No evidence for CP violation is found with a model independent search for direct CP violation in the Cabibbo suppressed decays, $D^+ \to K^-K^+\pi^+$ with 35 pb^{-1} (2010 data) and $D^+ \to \pi^-\pi^+\pi^-\pi^+$ with 1 fb^{-1} (2011 data).

References

[1] Aubert B *et al.* (BABAR Collaboration) 2007 *Phys.Rev.Lett.* **98** 211802 (*Preprint* hep-ex/0703020)
[2] Staric M *et al.* (Belle Collaboration) 2007 *Phys.Rev.Lett.* **98** 211803 (*Preprint* hep-ex/0703036)
[3] Aaltonen T *et al.* (CDF Collaboration) 2008 *Phys.Rev.Lett.* **100** 121802 (*Preprint* 0712.1567)
[4] Aubert B *et al.* (BABAR Collaboration) 2009 *Phys.Rev.* **D80** 071103 (*Preprint* 0908.0761)
[5] Asner D *et al.* (Heavy Flavor Averaging Group) 2010 (*Preprint* 1010.1589)
[6] Aaij R *et al.* (LHCb Collaboration) 2012 (*Preprint* 1211.1230)
[7] Alves A Augusto J *et al.* (LHCb Collaboration) 2008 *JINST* **3** S08005
[8] Aaij R *et al.* (LHCb Collaboration) 2010 *LHCb-CONF-2010-013*
[9] Bianco S, Fabbri F, Benson D and Bigi I 2003 *Riv.Nuovo Cim.* **26N7** 1–200 (*Preprint* hep-ex/0309021)
[10] Zhang L *et al.* (BELLE Collaboration) 2006 *Phys.Rev.Lett.* **96** 151801 (*Preprint* hep-ex/0601029)
[11] Grossman Y, Kagan A L and Nir Y 2007 *Phys.Rev.* **D75** 036008 (*Preprint* hep-ph/0609178)
[12] Aaij R *et al.* (LHCb Collaboration) 2012 *Phys.Rev.Lett.* **108** 111602 (*Preprint* 1112.0938)
[13] Bigi I I, Paul A and Recksiegel S 2011 *JHEP* **1106** 089 (*Preprint* 1103.5785)
[14] Gersabeck M 2012 *Mod.Phys.Lett.* **A27** 1230026 (*Preprint* 1207.2195)
[15] Aubert B *et al.* (BABAR Collaboration) 2005 *Phys.Rev.* **D71** 091101 (*Preprint* hep-ex/0501075)
[16] Rubin P *et al.* (CLEO Collaboration) 2008 *Phys.Rev.* **D78** 072003 (*Preprint* 0807.4545)
[17] Staric M *et al.* (Belle Collaboration) 2011 (*Preprint* 1110.0694)
[18] Bediaga I *et al.* 2009 *Phys. Rev. D* **80** 096006
[19] Aaij R *et al.* (LHCb Collaboration) 2011 *Phys.Rev.* **D84** 112008 (*Preprint* 1110.3970)
[20] Aaij R *et al.* (LHCb Collaboration) 2012 *LHCb-CONF-2012-019*

$\mathcal{CP}$ violation in semi-leptonic B decays at LHCb

Thomas Bird *(On behalf of the LHCb collaboration)*

The School of Physics and Astronomy, The University of Manchester, Oxford Road, Manchester, M13 9PL, UK

E-mail: thomas.bird@cern.ch

Abstract. Three methods of measuring the $\mathcal{CP}$ violating quantities a^s_{fs} and a^d_{fs} are discussed. The LHCb experiment is used to measure semi-leptonic B_d and B_s decays from pp collisions created by the Large Hadron Collider. The flavour specific decays of $B_s \to D_s(\phi\pi)\mu\nu$ are selected and the final state charge asymmetry is measured, which to first order is proportional to a^s_{fs}. This is done by fitting the $KK\pi$ invariant mass distributions to determine the yield of both $D^+_s\mu^-$ and $D^-_s\mu^+$ events. The charge-dependant efficiency for selecting this decay is measured using data-driven techniques. Using this information, the yields are corrected and the final state charge asymmetry is obtained, resulting in $a^s_{fs} = (-0.24 \pm 0.54 \pm 0.33)\%$. Finally two proposed LHCb measurements are described: a time-dependent difference and sum measurement $a^s_{fs} \pm a^d_{fs}$ and a time-dependent measurement of a^d_{fs}

1. Introduction

There are two neutral B mesons, the B_s meson and the B_d meson. In the following these will generically be denoted as B_q. The B_q meson has two flavour eigenstates $|B^0\rangle$ and $|\overline{B^0}\rangle$, the time evolution of which is governed by the following equation:

$$i\frac{d}{dt}\begin{pmatrix} |B^0\rangle \\ |\overline{B^0}\rangle \end{pmatrix} = \left(\hat{M} - \frac{i}{2}\hat{\Gamma}\right)\begin{pmatrix} |B^0\rangle \\ |\overline{B^0}\rangle \end{pmatrix}.$$

In the absence of mixing one expects $\hat{M}$ and $\hat{\Gamma}$ to be diagonal matrices; however off-diagonal elements M_{12} and Γ_{12} arise from the Feynman diagrams shown in figure 1. Diagonalisation of $\hat{M}$ and $\hat{\Gamma}$ gives two mass eigenstates: a light one, $|B_L\rangle$, and a heavy one, $|B_H\rangle$. These are related to the flavour eigenstates by

$$|B_L\rangle = p|B^0\rangle + q|\overline{B^0}\rangle \quad \text{and} \quad |B_H\rangle = p|B^0\rangle - q|\overline{B^0}\rangle.$$

Three observables describe $\mathcal{CP}$ violation and mixing in B_q mesons. First is the mass difference between the heavy and the light eigenstates, $\Delta m_q = M_H - M_L = 2|M_{12}|$, then the difference in decay widths of the two eigenstates, $\Delta\Gamma_q = \Gamma_L - \Gamma_H \approx 2|\Gamma_{12}|\cos\phi$, where ϕ is the phase between M_{12} and Γ_{12}, and finally the flavour specific asymmetry, $a^q_{fs} = |\Gamma_{12}|/|M_{12}|\sin\phi \approx -2\left(|q/p| - 1\right)$, which will be the focus of the following analyses. The Standard Model predicts [1] these two observables to be close to zero, specifically

$$a^s_{fs} = (1.9 \pm 0.3) \times 10^{-5} \quad \text{and} \quad a^d_{fs} = -(4.1 \pm 0.6) \times 10^{-4}.$$

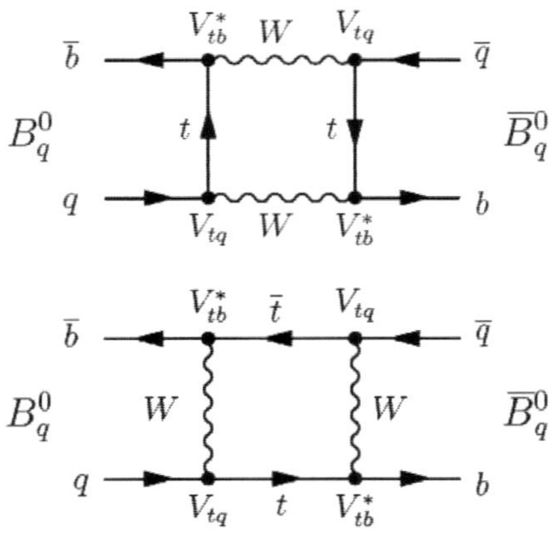

Figure 1: The two box diagrams which contribute most to B-mixing.

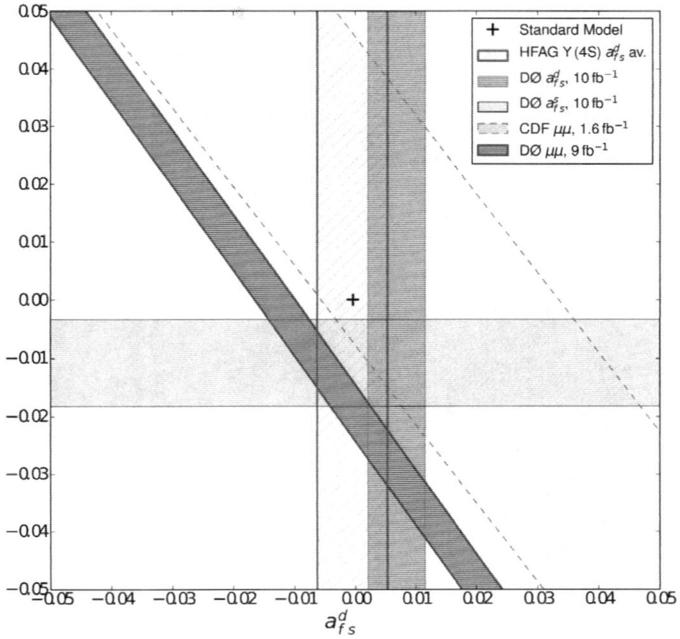

Figure 2: Plot in the $a^s_{fs} - a^d_{fs}$ plane showing CDF and DØ measurements along with an average of the B-factory measurements from HFAG and the Standard Model prediction. The horizontal (vertical) lines are direct measurements of a^s_{fs} (a^d_{fs}) and the diagonal lines show a measurement of the linear combination of a^s_{fs} and a^d_{fs}.

This asymmetry is also called a^q_{sl} when measured using semi-leptonic decays such as:

$$a^q_{fs} \to a^q_{sl} \propto \frac{N(\overline{B^0_q}(t) \to D^+_q \mu^- \overline{\nu}) - N(B^0_q(t) \to D^-_q \mu^+ \nu)}{N(\overline{B^0_q}(t) \to D^+_q \mu^- \overline{\nu}) + N(B^0_q(t) \to D^-_q \mu^+ \nu)}.$$

2. Previous measurements

There have been many different measurements of a^s_{fs} and a^d_{fs}. Figure 2 shows a selection of them. The horizontal (vertical) lines are measurements of a^s_{fs} (a^d_{fs}) and the diagonal lines are measurements of a linear combination of a^s_{fs} and a^d_{fs}. Of these measurements the most interesting is the DØ di-muon asymmetry [2], which is a linear combination of a^s_{fs} and a^d_{fs}. DØ measured $A^b_{sl} = (-0.787 \pm 0.172 \pm 0.093)\% = (0.594 \pm 0.022)a^d_{fs} + (0.406 \pm 0.022)a^s_{fs}$, which differs from the Standard Model prediction by 3.9 standard deviations.

3. LHCb detector

LHCb is a single-arm spectrometer [3] at the Large Hadron Collider (LHC), a proton-proton collider which in 2011 operated at a centre-of-mass energy of $\sqrt{s} = 7\,\text{TeV}$. A schematic of the LHCb detector is shown in figure 3, where it can be seen that the acceptance covers high pseudorapidities, compared to the general purpose detectors, specifically $2 < \eta < 5$. The Vertex Locator (VELO) is able to achieve an average resolution of $\sim 10\,\mu m$ ($\sim 60\,\mu m$) on the primary vertex position [4] in the direction perpendicular (parallel) to the beam pipe. This is used to measure the flight distance of a particle, which is proportional to the decay time. LHCb has a magnet, the polarity of which can be changed. This is used to help correct for charge asymmetries caused by different detection efficiencies for positive or negative particles. LHCb is also able to discriminate between pions and kaons using particle identification provided mainly by the Ring Imaging Cherenkov (RICH) detectors. Finally the muon stations are used to trigger events containing semi-leptonic decays and to identify muons.

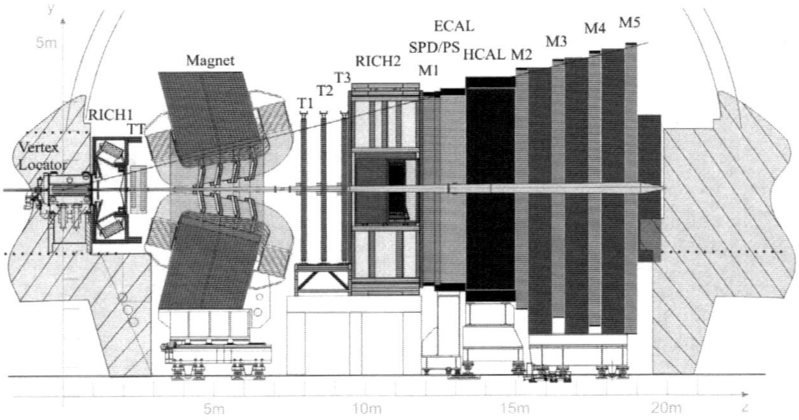

Figure 3: A cross-section of the LHCb detector. The following can be seen: the Vertex Locator (VELO), the Ring Imaging Cherenkov (RICH) detectors, the tracking stations (TT,T1-3), the magnet, the Scintillating Pad Detector (SPD), the Preshower Detector (PS), the hadronic and electromagnetic calorimeters (HCAL and ECAL) and the muon stations (M1-5).

4. Time-integrated a^s_{fs}

4.1. Analysis strategy

In this analysis decays of $B^0_s \to D_s\mu\nu$ are selected, where $D_s \to \phi(K^+K^-)\pi^{\pm}$. Using these events the un-tagged final state charge asymmetry A^s_{meas} is measured as in equation 1:

$$A^q_{meas} = \frac{N\left(D^-_q\mu^+\right) - N\left(D^+_q\mu^-\right) \times \frac{\varepsilon\left(D^-_q\mu^+\right)}{\varepsilon\left(D^+_q\mu^-\right)}}{N\left(D^-_q\mu^+\right) + N\left(D^+_q\mu^-\right) \times \frac{\varepsilon\left(D^-_q\mu^+\right)}{\varepsilon\left(D^+_q\mu^-\right)}}. \tag{1}$$

Here $N\left(D^{\mp}_q\mu^{\pm}\right)$ is the number of $D_q\mu$ combinations with the given charges. The ratio of efficiencies, $\varepsilon\left(D^-_q\mu^+\right)/\varepsilon\left(D^+_q\mu^-\right)$, corrects for any differences in efficiency for the two channels. The measured asymmetry is related to a^s_{fs} by

$$A^q_{meas} = \frac{a^q_{fs}}{2} + \left(A_p + \frac{a^q_{fs}}{2}\right)\frac{\int_0^\infty e^{-\Gamma_q t}\cos\left(\Delta m_q t\right)\varepsilon(t)dt}{\int_0^\infty e^{-\Gamma_q t}\cosh\left(\frac{\Delta\Gamma_q t}{2}\right)\varepsilon(t)dt}, \tag{2}$$

where $\varepsilon(t)$ is the efficiency of selecting an event with decay time, t. For semi-leptonic B_s decays in LHCb, the ratio of integrals in equation 2 is approximately 0.002, as determined from simulation [5]. The B_s production asymmetry, A_p, is $\sim 1\%$ [6] and so the the second term in equation 2 is of order 10^{-5}. Thus this term is assumed to be negligible, which leads to the approximation

$$A^s_{meas} = \frac{N\left(D_s^-\mu^+\right) - N\left(D_s^+\mu^-\right) \times \frac{\varepsilon\left(D_s^-\mu^+\right)}{\varepsilon\left(D_s^+\mu^-\right)}}{N\left(D_s^-\mu^+\right) + N\left(D_s^+\mu^-\right) \times \frac{\varepsilon\left(D_s^-\mu^+\right)}{\varepsilon\left(D_s^+\mu^-\right)}} \approx \frac{a^s_{fs}}{2}. \tag{3}$$

4.2. Signal Yield

In order to count the number of selected signal events in the sample, fits are performed to the $KK\pi$ mass spectra. The sample is split by charge of the muon and by the magnet polarity to correct for any charge asymmetry caused by the detector. The two mass peaks are described by triple Gaussians while the backgrounds are described by second-order polynomials. The selection requires the invariant mass of the KK combinations to be within 20 MeV of the ϕ mass.

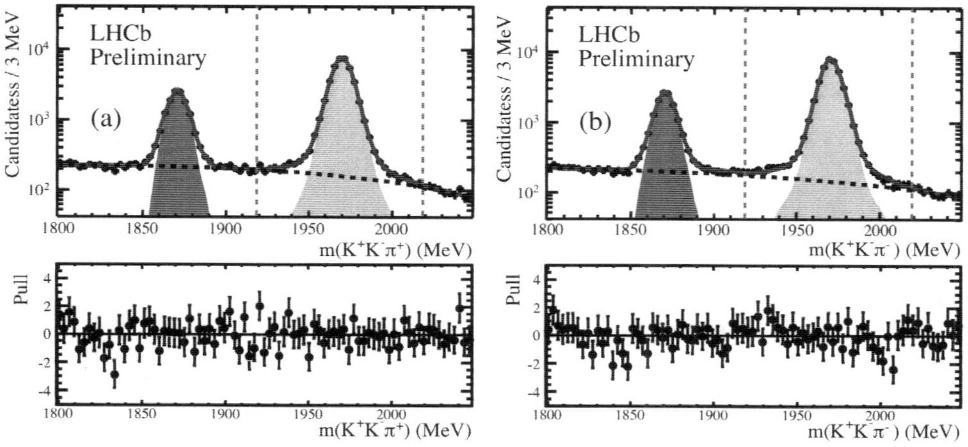

Figure 4: Invariant mass of (a) the $K^+K^-\pi^+$ combinations and (b) the $K^+K^-\pi^-$ combinations, recorded with magnet up polarity. The red dashed lines show the signal regions. The yellow shaded area is the D_s signal PDF, the red shaded region is the $D^\pm$ mass peak PDF, the black dashed line shows the background PDF and the blue line is the sum of these three.

4.3. Detection Efficiencies

The results of the fits allow $N\left(D_s^\mp\mu^\pm\right)$ to be measured, which needs to be corrected by the efficiency ratio $\varepsilon\left(D_s^-\mu^+\right)/\varepsilon\left(D_s^+\mu^-\right)$. To do this, the efficiency is broken down into several parts:

$$\varepsilon\left(D_s^\mp\mu^\pm\right) = \varepsilon_{\mathsf{id}}\left(K^\pm K^\mp\right) \times \varepsilon_{\mathsf{tr}}\left(K^\pm K^\mp\right) \times \varepsilon_{\mathsf{id}}\left(\pi^\mp\right) \times \varepsilon_{\mathsf{tr}}\left(\pi^\mp\right) \times \varepsilon_{\mathsf{id}}\left(\mu^\pm\right) \times \varepsilon_{\mathsf{tr}}\left(\mu^\pm\right) \times \varepsilon_{\mathsf{tg}}\left(D_s^\mp\mu^\pm\right). \tag{4}$$

Due to the similarity between the K^+ and K^- momentum spectra, from the decay of the ϕ resonance, the kaon identification efficiency, $\varepsilon_{\mathsf{id}}\left(K^\pm K^\mp\right)$, is the same for both $D_s^+\mu^-$ and $D_s^-\mu^+$. Thus these terms cancel in the ratio. The kaon tracking efficiencies, $\varepsilon_{\mathsf{tr}}\left(K^\pm K^\mp\right)$, also cancel. The pion identification efficiency, $\varepsilon_{\mathsf{id}}\left(\pi^\mp\right)$, is unity since no particle identification requirements are made on the pion. The other efficiencies do not cancel between charge states leaving the pion tracking efficiency, $\varepsilon_{\mathsf{tr}}\left(\pi^\mp\right)$, the muon identification efficiency, $\varepsilon_{\mathsf{id}}\left(\mu^\pm\right)$, the muon tracking efficiency, $\varepsilon_{\mathsf{tr}}\left(\mu^\pm\right)$, and the trigger efficiency, $\varepsilon_{\mathsf{tg}}\left(D_s^\mp\mu^\pm\right)$, to be studied in more detail.

To measure the pion tracking efficiency one can select partially reconstructed decays of $D^{*\pm} \rightarrow \pi^+ D^0(K^-\pi^+\pi^+\pi^-)$ [7], where one of the pions from the D^0 is missing or ignored, and then fully reconstruct the decay using only kinematics. The pion tracking efficiency is then the fraction of such events in which the additional pion is actually detected. Figure 5 shows that

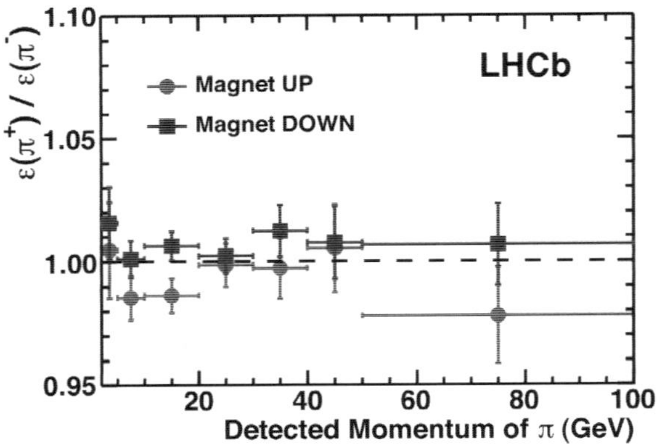

Figure 5: The ratio of π^+ and π^- tracking efficiencies as a function of pion momentum, split by magnet polarity. Data with red circles were taken with the magnet polarity in the up direction, while data with blue squares were taken with the magnet polarity in the down direction.

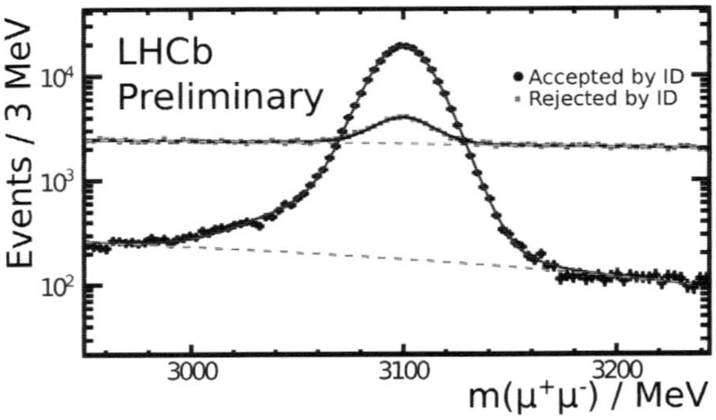

Figure 6: Invariant mass of the di-muon combination, split by those that pass or fail the muon identification cuts. The dashed green line shows the background PDF used and the solid lines are the sum of background and signal PDFs. The black points and blue line represent events which pass the muon identification cuts, while the red squares and black line represent those events which failed the same cuts.

the pion tracking efficiency ratio does not depend on the pion momentum and thus any tracking efficiency difference between the π^+ and π^- will cancel with the tracking efficiency difference between the μ^- and μ^+. Thus $\varepsilon\left(D_s^{\mp}\mu^{\pm}\right)$ is reduced to $\varepsilon_{\mathsf{id}}\left(\mu\right) \times \varepsilon_{\mathsf{tg}}\left(D_s^{\mp}\mu^{\pm}\right)$, since all other terms cancel.

The remaining two efficiencies are measured using a tag-and-probe technique. Two samples of $J/\psi \rightarrow \mu^+\mu^-$ candidates are selected using two different selection algorithms. These samples and the signal sample have different momentum spectra, so the data are divided into 50 bins of p, p_x and p_y, in which $\varepsilon_{\mathsf{id}}\left(\mu\right)$ and $\varepsilon_{\mathsf{tg}}\left(D_s^{\mp}\mu^{\pm}\right)$ are measured. In the J/ψ candidate selection, one of the muons is identified as a muon; this is the tag muon. The trigger is required to be independent of the second muon, known as the probe muon. This means the event would have been triggered regardless of whether the probe muon was present. The probe muon is then required to be identified as a muon. The number of J/ψ signal events which pass and fail this last requirement are counted, from which $\varepsilon_{\mathsf{id}}\left(\mu\right)$ is measured. The J/ψ events are counted by fitting the invariant mass of the di-muon combinations as shown in figure 6. A similar process is performed for each level of the trigger. Figure 7 shows the relative efficiency of μ^+ compared to μ^- for the hardware trigger as a function of the muon momentum.

Journal of Physics: Conference Series **447** (2013) 012021 doi:10.1088/1742-6596/447/1/012021

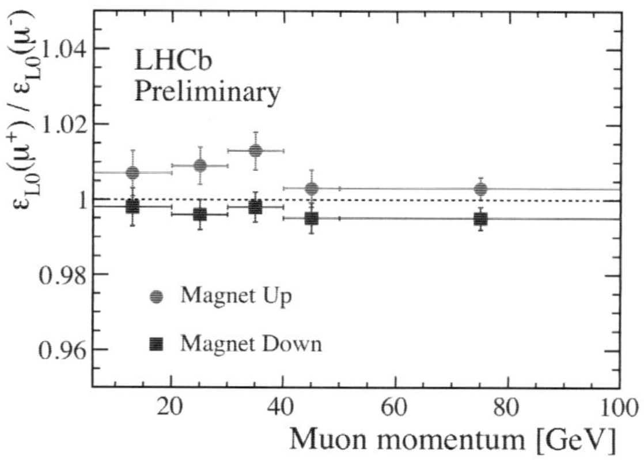

Figure 7: The ratio of μ^+ and μ^- hardware trigger efficiencies as a function of muon momentum, split by magnet polarity. Data with red circles were taken with the magnet polarity in the up direction, while data with blue squares were taken with the magnet polarity in the down direction.

Table 1: Sources of systematic uncertainties on the measured value of A^s_{meas}.

Systematic uncertainty source	$\sigma\,(A^s_{meas})$, %
Statistical uncertainty on efficiency ratios	0.10
Signal modelling in D_s^+ mass fit	0.06
Momentum difference between π and μ	0.06
Background from other b-hadrons	0.05
Muon Corrections	0.05
Bias in muon trigger	0.05
Momentum difference between K^+ and K^-	0.02
Varying run conditions	0.01
Muon mis-ID	0.01
Total	0.16

4.4. Systematic Uncertainties

Table 1 lists all effects which contribute to the total systematic uncertainty. The largest contribution of 0.10% comes from the statistical uncertainty on the efficiency correction ratios. The next largest uncertainty comes from varying the $KK\pi$ fit model. This causes variations in the number of signal events, resulting in a change of the central value by up to 0.06%. Momentum spectra of the $\mu^\pm$ and $\pi^\mp$ are different, so the tracking efficiencies may not cancel exactly, resulting in a 0.06% uncertainty. Small amounts of background are selected along with the signal; most important are decays of $b \to c\bar{c}s$ where a D_s originates from a virtual W decay and the muon comes from a semi-leptonic decay of a charmed hadron, for which a 0.05% uncertainty is assigned. The uncertainty on the muon correction is estimated by comparing the differences between the two J/ψ samples, resulting in a 0.05% uncertainty. Studies of a muon software trigger has shown a bias which was corrected with a 0.09% uncertainty. In the signal sample 60% of the events pass this trigger, so an uncertainty of 0.05% is assigned.

4.5. Results

LHCb collected 447 pb^{-1} of magnet up data and 595 pb^{-1} of magnet down data in 2011. The corrected asymmetry is measured in each kinematic bin and a weighted average is taken for each magnet polarity. The arithmetic average is taken between the two magnet polarities, as this has the effect of cancelling any possible residual biases after correction. The resulting asymmetry is measured to be:

$$A^s_{meas} = (-0.12 \pm 0.27 \pm 0.17)\%,$$

giving

$$a^s_{fs} = (-0.24 \pm 0.54 \pm 0.33)\%.$$

This result is compatible with the Standard Model prediction. Figure 8 compares this result to the Standard Model prediction and previous measurements.

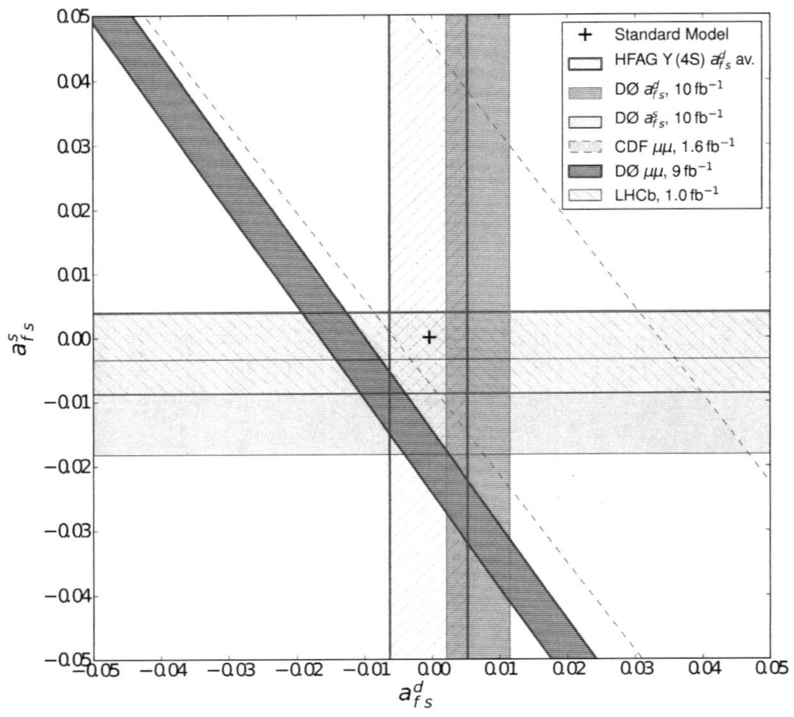

Figure 8: Plot in the $a^s_{fs} - a^d_{fs}$ plane showing CDF and DØ measurements along with an average of the B-factory measurements from HFAG and the Standard Model prediction. On top of these the LHCb time-integrated a^s_{fs} result is shown in green.

5. Time-dependent $a^s_{fs} \pm a^d_{fs}$

In a future analysis the difference and sum of a^s_{fs} and a^d_{fs} will be measured using $B^0_d \to D\mu\nu$ and $B^0_s \to D_s\mu\nu$ decays with $D_q \to KK\pi$. To measure the sum and difference, the data are fitted in three dimensions: decay time of the B_q, $KK\pi$ mass and muon charge. However there are a neutrinos in these semi-leptonic B_q decays, which cannot be detected by LHCb, so the momentum of the B cannot be fully reconstructed. To correct for this and measure the decay time of the B_q, a statistical Monte Carlo based correction is used, called k-factor [8]. The decay time distributions of the B_q are fitted with,

$$\Gamma^d_{\pm}(t) = e^{-\Gamma_d t}\left[(2 \pm x^d_1)\cosh\left(\frac{\Delta\Gamma_d t}{2}\right) \pm x^d_3 \cos\left(\Delta m_d t\right)\right] \quad \text{and} \quad \Gamma^s_{\pm}(t) = e^{-\Gamma_s t}\left[(2 \pm x^s_1)\cosh\left(\frac{\Delta\Gamma_s t}{2}\right)\right],$$

where the $\pm$ indicates the muon charge and three asymmetries are measured: a time-averaged B_s asymmetry, x^s_1, a time-averaged B_d asymmetry, x^d_1, and an asymmetry in the B_d oscillations, x^d_3. These are related to a^q_{fs} by $x^s_1 = 2A_c + a^s_{fs}$, $x^d_1 = 2A_c + a^d_{fs}$ and $x^d_3 = 2A_p - a^d_{fs}$, where A_c is the detection asymmetry and A_p is the production asymmetry. These measured asymmetries are combined in different ways to extract the physics quantities. The difference, $\Delta = (x^s_1 - x^d_1)/2 = (a^s_{fs} - a^d_{fs})/2$, eliminates the detector asymmetry. The sum, $\Sigma = (x^s_1 + x^d_1)/2 = 2A_c + (a^s_{fs} + a^d_{fs})/2$, however requires a measurement of the detector asymmetry to extract the physics parameter. Finally the asymmetry in the B_d oscillations, x_3^d, is measured directly, but will have the largest uncertainty due to the the relatively large uncertainty of the production asymmetry.

These three measurements can be combined to give a strong constraint on the $a_{fs}^s - a_{fs}^d$ plane. Furthermore, by including B flavour tagging [9], a B_q mixing measurement can also be made with these events to determine Δm_q.

6. Time-dependent a_{fs}^d

Another future analysis will select $B_d^0 \to D\mu\nu$, with $D \to K\pi\pi$. This D decay is Cabibbo allowed, unlike the $D \to KK\pi$ decay in the time-dependent $a_{fs}^s \pm a_{fs}^d$ analysis, and so more B_d events are available to this analysis. A similar method is used to the time-integrated a_{fs}^s measurement, where the final state charge asymmetry is measured, which is shown in equations 1 and 2. However due to the slow oscillations of the B_d, the integral ratio in the second term of equation 2 is no longer small. This means the B decay time distribution must be fitted, as with the $a_{fs}^s \pm a_{fs}^d$ analysis. The ratio $\varepsilon\left(D^-\mu^+\right)/\varepsilon\left(D^+\mu^-\right)$ needs to be measured as before, but this time $\varepsilon(K^\pm\pi^\mp)$ also needs to be measured since the D no longer decays via $\phi \to KK$. To do this the relative efficiency of $K^\pm\pi^\mp$ is measured using the decays shown in equation 5:

$$\frac{\varepsilon\left(K^+\pi^-\right)}{\varepsilon\left(K^-\pi^+\right)} = \frac{N\left(D^- \to K^+\pi^-\pi^-\right)}{N\left(D^+ \to K^-\pi^+\pi^+\right)} \times \frac{N\left(D^+ \to K_s^0\pi^+\right)}{N\left(D^- \to K_s^0\pi^-\right)}. \tag{5}$$

This efficiency correction is applied along with the pion and muon corrections, which are determined as in the time-integrated analysis. These three efficiencies are combined to give $\varepsilon\left(D^-\mu^+\right)/\varepsilon\left(D^+\mu^-\right)$, which is input to the B_d decay time fitter to determine a_{fs}^d.

7. Conclusion

The $\mathcal{CP}$ violating parameter $a_{fs}^s = (-0.24 \pm 0.54 \pm 0.33)\%$ has been determined using a time-integrated method. This measurement is compatible with previous measurements as well as the Standard Model prediction. LHCb is pursuing two further measurements on the $a_{fs}^s - a_{fs}^d$ plane, which will cross-check the current and previous measurements.

References

[1] A. Lenz 2011 A simple relation for B_s-mixing *Phys. Rev.* D **84** 031501
[2] V. M. Abazov *et al.* DØ Collaboration 2011 Measurement of the anomalous like-sign dimuon charge asymmetry with 9 fb^{-1} of $p\bar{p}$ collisions *Phys. Rev.* D **84** 052007
[3] R. Antunes-Nobrega, *et al.* LHCb collaboration 2003 *LHCb Reoptimised Detector Design and Performance, Technical Design Report 9* CERN LHCC 2003-030
[4] P. M. Bjørnstad 2011 Performance of the LHCb Vertex Locator *JINST* **6** C12024
[5] R. Aaij *et al.* LHCb collaboration 2012 Measurement of the flavour-specific CP violating asymmetry a_{sl}^s in B_s decays *PoS(ICHEP2012)*
[6] R. Aaij *et al.* LHCb collaboration 2012 First evidence of direct CP violation in charmless two-body decays of B_s^0 mesons *Phys. Rev. Lett.* **108** 201601
[7] R. Aaij *et al.* LHCb collaboration 2012 Measurement of the D_s^+ - D_s^- production asymmetry in 7 TeV pp collisions *Phys. Lett.* B **713** 186-95
[8] A. Abulencia *et al.* CDF collaboration 2006 Measurement of the B_s^0-$\bar{B}_s^0$ oscillation frequency *Phys. Rev. Lett.* **97** 062003
[9] R. Aaij *et al.* LHCb collaboration 2012 Opposite-side flavour tagging of B mesons at the LHCb experiment *Eur. Phys. J.* C. **72** 2022

Journal of Physics: Conference Series **447** (2013) 012022 doi:10.1088/1742-6596/447/1/012022

Measurement of the CP-violating phase in B_s^0 mixing

Jeroen van Leerdam (on behalf of the LHCb collaboration)
Nikhef, Science Park 105, 1098 XG Amsterdam, The Netherlands

E-mail: `jeroen.van.leerdam@cern.ch`

Abstract. Determination of the CP-violating phase ϕ_s in $B_s^0 \to J/\psi K^+K^-$ decays is one of the key goals of the LHCb experiment. Its value is predicted to be very small in the Standard Model but can be significantly enhanced in many other models. We present the most precise measurement of ϕ_s to date and the first observation of a non-zero $\Delta\Gamma_s$ based upon 1 fb^{-1} of data collected at LHCb during 2011. Also the measurement of ϕ_s with the decay mode $B_s^0 \to J/\psi \pi^+\pi^-$ and a combination of the $J/\psi K^+K^-$ and $J/\psi \pi^+\pi^-$ results are reported. Additionally we present first results from the measurement of the decay $B_s^0 \to J/\psi \overline{K^{*0}}$, which is potentially interesting for controlling currently neglected higher order corrections in $B_s^0 \to J/\psi K^+K^-$.

1. Introduction

Mixing of B_s^0 and anti-B_s^0 ($\overline{B}_s^0$) mesons is governed by the processes shown in figure 1 in the Standard Model of particle physics. Violation of symmetry under a combined charge conjugation and parity reversal (CP) transformation induced by these processes is expected to be small. However, significant violation of CP symmetry may be introduced by corrections from presently unknown contributions.

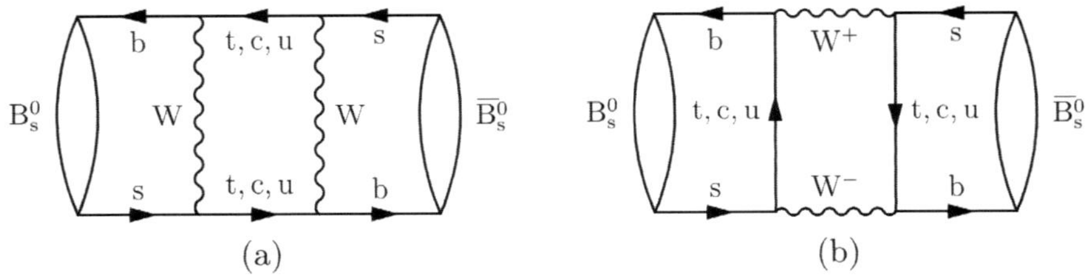

Figure 1. Feynman diagrams for B_s^0–$\overline{B}_s^0$ mixing in the Standard Model.

CP violation in B_s^0–$\overline{B}_s^0$ mixing may appear in two ways [1]. Firstly, it can cause a difference between the probabilities for $\overline{B}_s^0 \to B_s^0$ and $B_s^0 \to \overline{B}_s^0$ transitions (CP violation in mixing). Secondly, it can contribute to the phase difference between the interfering B_s^0 ($\overline{B}_s^0$) decay processes with and without a transition to $\overline{B}_s^0$ (B_s^0). In general, this phase is also affected by CP violation in the decay amplitudes.

This work focuses on measurement of the CP-violating phase between mixed and unmixed decays, which is denoted by ϕ_s. Decays of the type $B_s^0 \to J/\psi(\to \mu^+\mu^-)\,h^+h^-$ are studied, where h^+h^- can be K^+K^- or $\pi^+\pi^-$ (see figure 2). The resonant components of these processes are CP eigenstates, which are particularly suitable for a measurement of ϕ_s. While it is assumed that the J/ψ is the only component of the $\mu^+\mu^-$ system, the h^+h^- system is a sum of multiple resonant and non-resonant states.

$B_s^0 \to J/\psi\, h^+h^-$ decays are $b \to c\bar{c}s$ transitions and hence the measured CP-violating phase is labelled $\phi_s^{c\bar{c}s}$. The decay process is dominated by the diagram shown in figure 2a. In the approximation that this tree-level amplitude is the only contribution, there is no CP violation directly in the decay and $\phi_s^{c\bar{c}s}$ is a clean measurement of CP violation induced by B_s^0–$\overline{B}_s^0$ mixing [2].

In the Standard Model, assuming the tree-level approximation, $\phi_s^{c\bar{c}s}$ is approximately equal to $-2\beta_s \equiv -2\arg(-V_{ts}\,V_{tb}^*/V_{cs}\,V_{cb}^*)$, where the V_{ij} are elements of the CKM matrix [3]. An indirect determination with a global fit to present experimental data yields $-2\beta_s = (-0.0363^{+0.0015}_{-0.0016})$ rad [4]. Physics beyond the Standard Model may significantly affect the value of $\phi_s^{c\bar{c}s}$ determined in a direct measurement [5, 6].

Beyond the tree-level approximation, contributions from Standard Model penguin diagrams need to be considered (see for example the diagram in figure 2b). These amplitudes are doubly Cabibbo suppressed relative to the tree-level amplitude. Resulting corrections to $\phi_s^{c\bar{c}s}$ depend on the considered decay mode and may be controlled with measurements of decays in which similar penguin contributions are not Cabibbo suppressed, like $B_s^0 \to J/\psi\, K^-\pi^+$ [7].

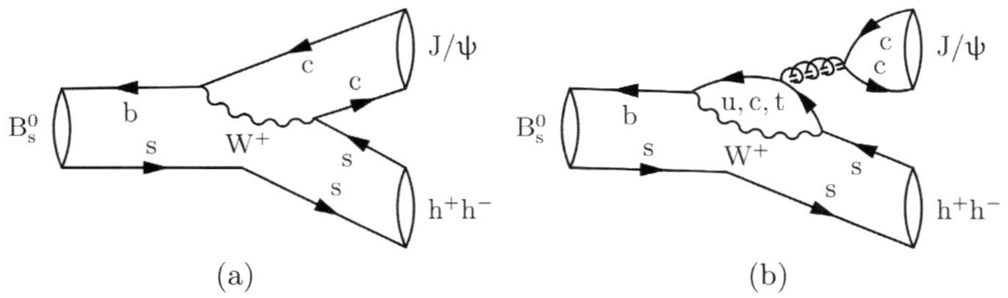

Figure 2. Feynman diagrams contributing to the decay $B_s^0 \to J/\psi\, h^+h^-$ in the Standard Model ($h^+h^- \in \{K^+K^-,\ \pi^+\pi^-\}$). (a): tree-level diagram; (b): penguin diagram.

In addition to $\phi_s^{c\bar{c}s}$, also other mixing and decay parameters of the B_s^0–$\overline{B}_s^0$ system can be measured, such as the mean decay width (Γ_s), the differences between the masses and decay widths of the mass eigenstates (Δm_s and $\Delta\Gamma_s$, respectively) and the relative amplitudes of different contributions to the decay. While Δm_s is constrained to the estimate from a dedicated LHCb measurement (17.63 ± 0.11 ps^{-1} [8]), values for the other parameters are determined with the measurements described here.

The quantities $\phi_s^{c\bar{c}s}$, Γ_s and $\Delta\Gamma_s$ are measured as parameters in the distribution of the time between production and decay of B_s^0 or $\overline{B}_s^0$ mesons [1, 2]. A model of this decay time distribution is fitted to the experimental data. The main sensitivity to $\phi_s^{c\bar{c}s}$ comes from an oscillation that modulates an (approximately) exponential shape. The frequency of this oscillation is the mass difference Δm_s and its amplitude is proportional to $\sin\phi_s^{c\bar{c}s}$.

The preliminary results in the $B_s^0 \to J/\psi\, K^+K^-$ decay channel presented here are an update of previous LHCb results [9]. The parameters $\phi_s^{c\bar{c}s}$ and $\Delta\Gamma_s$ were also measured by the D0 [10], CDF [11] and ATLAS [12] collaborations and $\Delta\Gamma_s$ by the CMS collaboration [13]. All measurements are compatible with the Standard Model expectations $\phi_s^{c\bar{c}s} \approx -2\beta_s$ and $\Delta\Gamma_s = (0.087 \pm 0.021)$ ps^{-1} [14].

2. Experimental aspects

The LHCb detector is described in reference [15]. It is designed for studies of particles containing b or c quarks produced in the proton–proton collisions of the Large Hadron Collider (LHC [16]). Excellent charged particle tracking and decay vertex finding enable a precise determination of the decay time, which is measured as the distance between the B meson production and decay vertices. Different decay modes are distinguished by the use of particle identification techniques.

Selection of decay events is based on finding $J/\psi \to \mu^+\mu^-$ decays, which are subsequently combined with oppositely charged hadron pairs. Kinematic restrictions are applied to candidate events in order to reject background from four particles that do not originate from a $B_s^0 \to J/\psi\, h^+h^-$ decay. Background that is remaining after selection is separated from the signal $B_s^0 \to J/\psi\, h^+h^-$ decay on a statistical basis, by exploiting the difference in invariant $J/\psi\, h^+h^-$ mass distributions.

The decay time measurement has a finite resolution, which effectively gives a dilution of the oscillation amplitude in its distribution and hence reduces the sensitivity to $\phi_s^{c\bar{c}s}$. This effect is modelled by smearing the theoretical distribution [17, 18]. The resolution is estimated for each decay event individually.

The resolution estimate is calibrated with a measurement of the decay time distribution of background events in which a real J/ψ is combined with two unrelated hadrons. All particles in these events are assumed to come directly from the proton–proton interaction point (prompt background), which makes their "true decay time" equal to zero. The resolution is determined from the width of the peak in the measured decay time distribution around zero. The effective measured resolution is 45 fs. This is precise enough to resolve the oscillation, which has a period of $2\pi/\Delta m_s \approx 350$ fs.

Since the oscillation in the decay time distribution is exactly opposite for B_s^0 decays and $\overline{B}_s^0$ decays, determination of the initial meson flavour is required. This flavour tag has a probability to be wrong, which leads to a partial cancellation of the oscillations in the measured distributions and gives another dilution. Also the wrong-tag probability is estimated on an event-by-event basis.

The estimated wrong-tag probability is calibrated with the decay mode $B^\pm \to J/\psi\, K^\pm$, for which the initial meson flavour is known from the final state [19]. The effective tagging efficiency is determined to be $\varepsilon_{tag}\langle\mathcal{D}^2\rangle \approx 2.3\%$ [17, 18]. This number can be interpreted as the effective fraction of events with a perfect flavour tag.

3. Analysis of $B_s^0 \to J/\psi\, K^+K^-$ decays

The most promising decay mode for the measurement of $\phi_s^{c\bar{c}s}$ is $B_s^0 \to J/\psi\, K^+K^-$ [2, 5, 6]. The preliminary LHCb measurement with the 2011 data set of 1 fb^{-1} of integrated luminosity is described in reference [17]. Approximately 21,200 signal events are extracted.

Several intermediate angular momentum states of the $J/\psi\, K^+K^-$ system contribute to the decay, each with a different distribution of final state particles decay angles [2, 20]. If expressed in terms of CP eigenstates, these contributions are either even or odd under a CP transformation. CP-even states, CP-odd states and their interferences have different decay time distributions, which are separated by an angular analysis to enhance the sensitivity of the measurement.

Analysis of the data is restricted to events with invariant K^+K^- masses in the range 1008–1032 MeV/c^2, for which the system is dominated by the $\phi(1020)$ resonance. While the K^+K^- system is in a $L = 1$ orbital angular momentum state (P-wave) for the $\phi(1020)$, there is also a small $L = 0$ (S-wave) contribution of approximately 1% [21].

The sums of the measured B_s^0 and $\overline{B}_s^0$ time and angular distributions are shown in figure 3. Only events with a measured decay time greater than 0.3 ps are selected to reject the large number of prompt background events that were used earlier to determine the decay time resolution.

Journal of Physics: Conference Series **447** (2013) 012022 doi:10.1088/1742-6596/447/1/012022

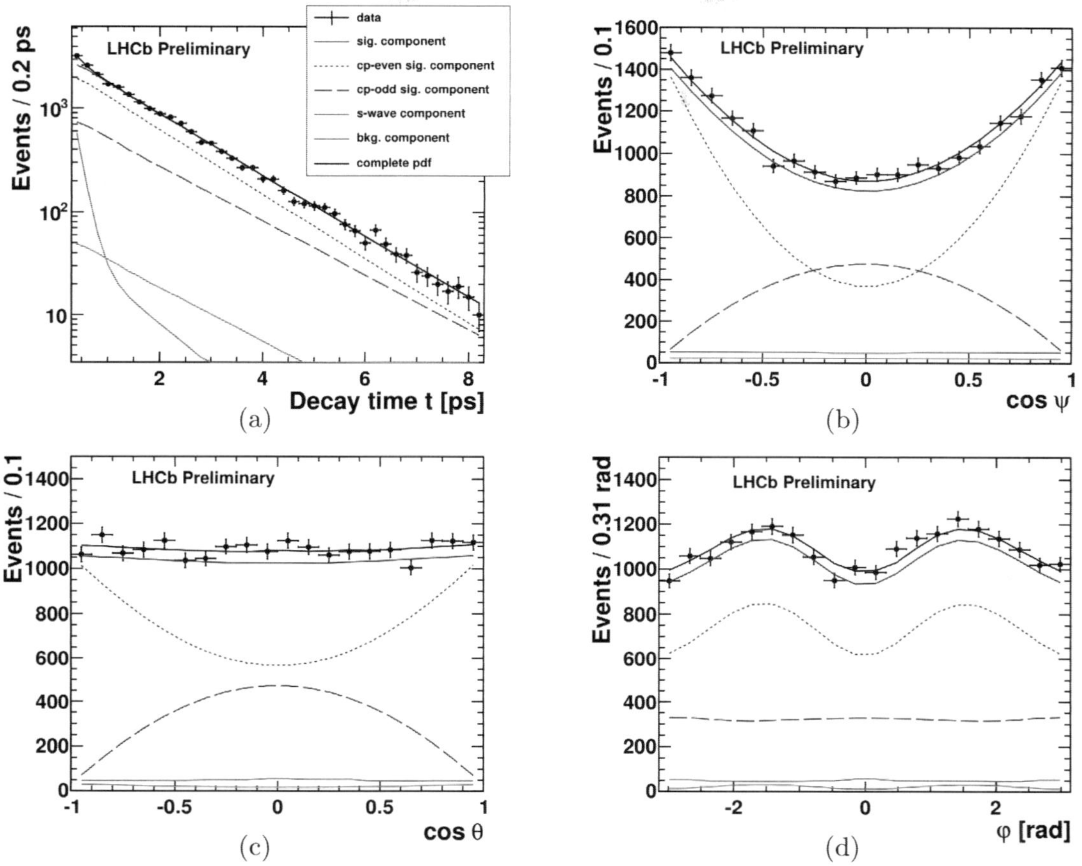

Figure 3. Measured distributions and the fitted model for the decay time (a) and angles (b, c, d) in $B_s^0 \to J/\psi\, K^+K^-$. See reference [2] for definition of the decay angles. The data are shown as black points with error bars. The sum of the signal contributions is shown as the blue, solid line, background as the red, solid line and the total sum as the black, solid line. The dotted and dashed blue contributions are the CP-even and CP-odd $B_s^0 \to J/\psi\, \phi$ angular momentum components, respectively. The K^+K^- S-wave component is shown as the magenta, solid line.

Preliminary results for the principal physics parameters in the 1 fb^{-1} $B_s^0 \to J/\psi\, K^+K^-$ measurement are

$$\begin{aligned} \phi_s^{c\bar{c}s} &= (-0.001 \pm 0.101 \pm 0.027)\ \text{rad} \\ \Delta\Gamma_s &= (+0.116 \pm 0.018 \pm 0.006)\ \text{ps}^{-1} \\ \Gamma_s &= (\ 0.6580 \pm 0.0054 \pm 0.0066)\ \text{ps}^{-1}\ , \end{aligned}$$

where the first uncertainty is statistical and the second systematic. This is the most sensitive measurement of ϕ_s to date and the first direct observation of a non-zero decay width difference in the B_s^0–$\overline{B}_s^0$ system. The results are compatible with Standard Model expectations.

A parameter transformation exists that takes the values ($\phi_s^{c\bar{c}s}$, $\Delta\Gamma_s$) to ($\pi - \phi_s^{c\bar{c}s}$, $-\Delta\Gamma_s$), but leaves the model for the decay time and angular distributions invariant. This gives an alternative set of parameter values, which describes the measured data equally well as the set quoted above. The transformation also affects the phases of the decay amplitudes for the $B_s^0 \to J/\psi\, \phi$ and K^+K^- S-wave contributions, which can be used to resolve this ambiguity in the results [20].

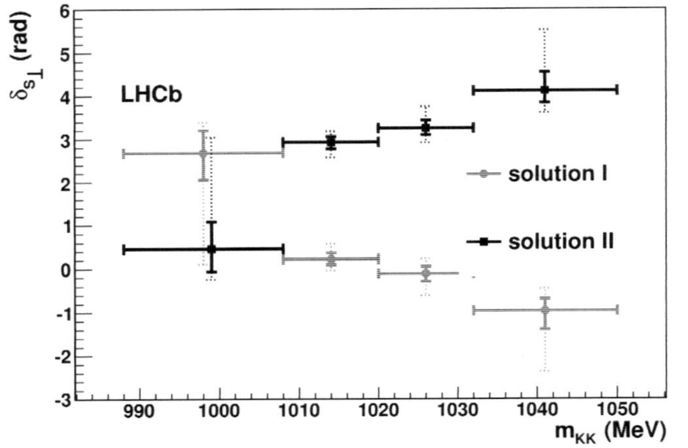

Figure 4. Measured difference between the phases of the K^+K^- S-wave and one of the $B_s^0 \to J/\psi\,\phi$ components of the $B_s^0 \to J/\psi\,K^+K^-$ decay ($\delta_{S\perp}$) in bins of the K^+K^- invariant mass. The expected trend of a decreasing phase difference with increasing mass is shown by the blue circles. The increasing trend (black squares) represents an alternative set of parameter values resulting from an ambiguity in the decay model (see text).

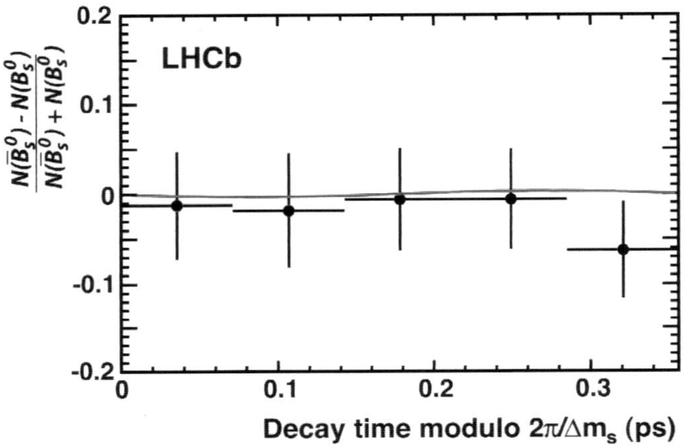

Figure 5. Measured asymmetry between the decay time distributions of decay events tagged as $\overline{B}_s^0$ and decay events tagged as B_s^0 for the $J/\psi\,\pi^+\pi^-$ final state. Data are shown in bins (black points), the fitted decay model is overlaid (red line). The asymmetry is shown as a function of decay time modulo $2\pi/\Delta m_s$ to enhance the significance of a possible oscillation.

The phase of the $J/\psi\,\phi$ contribution is expected to increase rapidly with increasing K^+K^- mass across the $\phi(1020)$ resonance, while the phase of the K^+K^- S-wave is expected to vary slowly. This leads to a decreasing trend in the phase difference between the S-wave and $J/\psi\,\phi$, which can be observed. One of the sets of parameter values gives this expected decreasing trend, while the alternative set instead gives an increasing trend. An LHCb study in a wider K^+K^- mass interval of 988–1050 MeV/c^2 has demonstrated that the set with $\phi_s \approx 0$ and $\Delta\Gamma_s > 0$ gives the expected behaviour [22] (see figure 4).

4. Analysis of $B_s^0 \to J/\psi\,\pi^+\pi^-$ decays

For invariant $\pi^+\pi^-$ masses in the range 775–1550 MeV/c^2, the $B_s^0 \to J/\psi\,\pi^+\pi^-$ decay almost exclusively proceeds via intermediate states that are odd under a CP transformation. The fraction of these contributions is found to be greater than 0.977 at 95% confidence level by an LHCb analysis of the resonant components of the decay [23]. Since CP-odd states and their interferences have identical decay time distributions, the measurement of $\phi_s^{c\bar{c}s}$ in this decay mode does not require an analysis of the decay angles to separate the different contributions.

Because of the CP-odd nature of $B_s^0 \to J/\psi\,\pi^+\pi^-$ and small CP violation, the decay is dominated by the heavy mass eigenstate of the B_s^0–$\overline{B}_s^0$ system. As a consequence, the average and the difference of the decay widths of the light and heavy eigenstates are almost fully anti-correlated and cannot be determined both in this measurement. Therefore, the values of Γ_s and

$\Delta\Gamma_s$ are constrained to the $J/\psi\, K^+K^-$ estimates from the measurement described in reference [9].

The $B_s^0 \to J/\psi\, \pi^+\pi^-$ measurement is described in reference [18]. It uses the 1 fb^{-1} data set collected in 2011, from which 7421 ± 105 signal events are extracted in a region of ± 20 MeV/c^2 around the nominal B_s^0 mass. The resulting estimate for the value of $\phi_s^{c\bar{c}s}$ is

$$\phi_s^{c\bar{c}s} = (\,-0.019^{+0.173+0.004}_{-0.174-0.003}\,)\ \text{rad}\ ,$$

where the first uncertainty is statistical and the second systematic. Also this result is compatible with Standard Model expectations.

Figure 5 shows the measured asymmetry between the decay time distributions of decay events tagged as $\overline{B}_s^0$ and decay events tagged as B_s^0. The amplitude of the oscillation in the time development of this asymmetry provides a measurement of $\sin\phi_s^{c\bar{c}s}$. The absence of a significant oscillation is consistent with the estimated value of $\phi_s^{c\bar{c}s}$, which is compatible with zero.

5. Combination of results for $\phi_s^{c\bar{c}s}$ and $\Delta\Gamma_s$

Results from the $B_s^0 \to J/\psi\, K^+K^-$ and $B_s^0 \to J/\psi\, \pi^+\pi^-$ measurements can be combined by extracting an estimate for $\phi_s^{c\bar{c}s}$ from a simultaneous fit of the decay models to both data samples. The preliminary combined result for the 1 fb^{-1} data set is [17]

$$\phi_s^{c\bar{c}s} = (\,-0.002 \pm 0.083 \pm 0.027\,)\ \text{rad}\ ,$$

where the first uncertainty is statistical and the second systematic.

A combination of the ATLAS, CDF, D0 and LHCb measurements of $\phi_s^{c\bar{c}s}$ and $\Delta\Gamma_s$ by the Heavy Flavor Averaging Group yields [24]

$$\phi_s^{c\bar{c}s} = (\,-0.013^{+0.083}_{-0.090}\,)\ \text{rad} \qquad \Delta\Gamma_s = (\,+0.089^{+0.011}_{-0.013}\,)\ \text{ps}^{-1}\ .$$

6. Measurement of the $B_s^0 \to J/\psi\, \overline{K}^{*0}$ branching and polarization fractions

Contributions from higher order diagrams to $B_s^0 \to J/\psi\, h^+h^-$ decays are neglected in the decay models used for the presented $\phi_s^{c\bar{c}s}$ measurements. This is assumed to be a good approximation with the current experimental precision, but for future measurements the effect of penguin contributions (figure 2b) will need to be considered.

Combination of the tree-level and penguin amplitudes causes CP violation in the $B_s^0 \to J/\psi\, h^+h^-$ decay, which gives a small contribution to the phase $\phi_s^{c\bar{c}s}$. To determine the contribution of CP violation induced by B_s^0–$\overline{B}_s^0$ mixing, both measurements of $\phi_s^{c\bar{c}s}$ and estimates of the size of penguin amplitudes relative to the tree-level amplitude are required. For the $B_s^0 \to J/\psi\, \phi$ component of $B_s^0 \to J/\psi\, K^+K^-$, the size of penguin contributions can be estimated with the decay $B_s^0 \to J/\psi(\to \mu^+\mu^-)\, \overline{K}^{*0}(\to K^-\pi^+)$, where K^{*0} is the $K^*(892)^0$ meson [7].

$B_s^0 \to J/\psi\, \phi$ and $B_s^0 \to J/\psi\, \overline{K}^{*0}$ are both decays into a J/ψ and a second vector meson and they are related by SU(3) flavour symmetry. However, penguin contributions in $B_s^0 \to J/\psi\, \overline{K}^{*0}$ are not Cabibbo suppressed relative to the tree-level contribution, which enables a measurement of their size for each of the three intermediate angular momentum states of the decays. SU(3)-breaking effects need to be taken into account when the resulting estimates are used to interpret a measurement in $B_s^0 \to J/\psi\, \phi$.

The experimental precision with only the LHCb data collected in 2011 is not sufficient to obtain a meaningful estimate of penguin contributions. As a first step, the B_s^0 branching fraction into $J/\psi\, \overline{K}^{*0}$ and the fractions of the three angular momentum states of the decay (f_L, $f_\parallel$ and $f_\perp$) are measured. The first part of the 2011 data set is used, which corresponds to 0.37 fb^{-1} of integrated luminosity. The results are $\mathcal{B}(B_s^0 \to J/\psi\, \overline{K}^{*0}) = (4.4^{+0.5}_{-0.4} \pm 0.8) \times 10^{-5}$, $f_L = 0.50 \pm 0.08 \pm 0.02$ and $f_\parallel = 0.19^{+0.10}_{-0.08} \pm 0.02$ [25]. The third polarization fraction is given by $f_\perp = 1 - f_L - f_\parallel$.

7. Summary and outlook

The LHCb collaboration have measured B_s^0–$\overline{B}_s^0$ mixing and decay parameters in the decay channels $B_s^0 \to J/\psi K^+K^-$ and $B_s^0 \to J/\psi \pi^+\pi^-$ with the 2011 data set of 1 fb^{-1} of integrated luminosity. A preliminary combination of the two measurements yields $\phi_s^{c\bar{c}s} = (-0.002 \pm 0.083 \pm 0.027)$ rad for the CP-violating phase in the interference between mixed and unmixed decays. The preliminary estimate $\Delta\Gamma_s = (+0.116 \pm 0.018 \pm 0.006)$ ps^{-1} is the first direct observation of a non-zero decay width in the B_s^0–$\overline{B}_s^0$ system. A first step towards controlling penguin contributions is made by measuring the branching and polarization fractions of the decay $B_s^0 \to J/\psi \overline{K}^{*0}$.

There is room for improvement of the preliminary 1 fb^{-1} results for the $B_s^0 \to J/\psi K^+K^-$ decay and updates of both this measurement and the combination with $B_s^0 \to J/\psi \pi^+\pi^-$ are planned. The next step will be a combination of the 1 fb^{-1} 2011 data set with the 2 fb^{-1} of integrated luminosity collected in 2012 to further improve the precision of the CP violation measurement.

References

[1] Carter A B, Sanda A I 1981 *Phys. Rev. D* **23** 1567
Bigi I I, Sanda A I, 1981 *Nucl. Phys. B* **193** 85

[2] Dighe A S, Dunietz I, Lipkin H J, Rosner J L 1996 *Phys. Lett. B* **369** 144
Dighe A S, Dunietz I, Fleischer R 1999 *Eur. Phys. J. C* **6** 647
Dunietz I, Fleischer R, Nierste U 2001 *Phys. Rev. D* **63** 114015

[3] Cabibbo N 1963 *Phys. Rev. Lett.* **10** 531
Kobayashi M, Maskawa T, 1973 *Prog. Theor. Phys.* **49** 652

[4] Charles J *et al.* (The CKMfitter Group) 2011 *Phys. Rev. D* **84** 033005

[5] Buras A J 2009 *PoS* **EPS-HEP2009** 024

[6] Chiang C-W *et al.* 2010 *JHEP* **1004** 031

[7] Faller S, Fleischer R, Mannel T 2009 *Phys. Rev. D* **79** 014005

[8] Aaij R *et al.* (LHCb collaboration) 2012 *Phys. Lett. B* **709** 177

[9] Aaij R *et al.* (LHCb collaboration) 2012 *Phys. Rev. Lett.* **108** 101803

[10] Abazov V M *et al.* (D0 collaboration) 2012 *Phys. Rev. D* **85** 032006

[11] Aaltonen T *et al.* (CDF collaboration) 2012 *Phys. Rev. Lett.* **109** 171802

[12] The ATLAS collaboration 2012 *Preprint* arXiv:1208.0572 [hep-ex]

[13] The CMS collaboration 2012 *Conference report* CMS-PAS-BPH-11-006

[14] Lenz A, Nierste U 2007 *JHEP* **06** 072
Badin A, Gabbiani F, Petrov A A 2007 *Phys. Lett. B* **653** 230
Lenz A, Nierste U 2011 *Preprint* arXiv:1102.4274 [hep-ph]

[15] Alves Jr A A *et al.* (LHCb collaboration) 2008 *JINST* **3** S08005

[16] Evans L, Bryant P (editors) 2008 *JINST* **3** S08001

[17] The LHCb collaboration 2012 *Conference report* LHCb-CONF-2012-002

[18] Aaij R *et al.* (LHCb collaboration) 2012 *Phys. Lett. B* **713** 378

[19] The LHCb collaboration 2012 *Conference report* LHCb-CONF-2012-026
Aaij R *et al.* (LHCb collaboration) 2012 *Eur. Phys. J. C* **72** 2022

[20] Xie Y, Clarke P, Cowan G, Muheim F 2009 *JHEP* **09** 074

[21] Aaij R *et al.* (LHCb collaboration) 2013 *Preprint* arXiv:1302.1213 [hep-ex]

[22] Aaij R *et al.* (LHCb collaboration) 2012 *Phys. Rev. Lett.* **108** 241801

[23] Aaij R *et al.* (LHCb collaboration) 2012 *Phys. Rev. D* **86** 052006

[24] Amhis Y *et al.* (Heavy Flavor Averaging Group) 2012 *Preprint* arXiv:1207.1158 [hep-ex] (and online update at `http://www.slac.stanford.edu/xorg/hfag`)

[25] Aaij R *et al.* (LHCb collaboration) 2012 *Phys. Rev. D* **86** 071102(R)

Journal of Physics: Conference Series **447** (2013) 012023 doi:10.1088/1742-6596/447/1/012023

Test of Time Reversal Invariance and COSY

Dieter Eversheim[1], Bernd Lorentz[2] and Yury Valdau[1,2]
for the PAX collaboration

1 Helmholz Institute für Strahlen and Kernphysik, University Bonn, Germany
2 Institute für Kernphysik, Forschungszentrum Jl̈ich, Germany

Abstract. Test of Time Reversal Invariance in *pd* scattering is planned at the COoler-SYnchrotron COSY-Jülich using the PAX internal target station. Feasibility test, performed in September 2012, have shown possibility to perform such an experiment using COSY and PAX. In parallel to the PAX upgrade a new high precision beam current measurement system will be constructed by the end of 2014. After these modifications, it will be possible to improve the current upper limit on a strength of T-odd P-even NN potential by an order of magnitude.

E-mail: `y.valdau@fz-juelich.de`

Introduction

The Standard Model of fundamental particles is not able to explain the Baryon Asymmetry of the Universe. A pleasurable explanation for the Baryon Asymmetry problem has been suggested in Ref. [1] and implies presence of additional sources of the CP-violation in nature, which are not yet observed and implemented into the Standard Model. All sources of the CP-violation, known up to now, are incorporated in to the Standard Model of fundamental interaction using the Kobayashi–Maskawa matrix.

An electric dipole moment (EDM) of an elementary particle [2] is one of the possible new sources of the CP-violation which provides very high sensitivity to the physics beyond the Standard Model. Only upper limits for the EDM of elementary particles or atoms have been reported in literature up to now. However, the electric dipole moment of elementary particle violates P and T-symmetries simultaneously, and hence additional theoretical model must be used in order to link it to the CP-violation [3].

Invariance of T-symmetry has been tested in different experiments using different approaches. Violations of T-odd P-even symmetry so far have been only observed in decays of K^0 [4] and, recently, B [5] mesons. There are plans to perform a similar kind of tests in the decay of the entangled ϕ mesons [6]. However, for the NN systems only upper limits on the strength of the T-odd P-even potential are reported. At presenet, upper limit on the strength of the potential is obtained from the measurements of neutron scattering off a tensor polarised ^{165}Ho and hence contains some model dependence [7].

An experiment to test T-invariance using *pd* scattering, the simplest NN system suitable for such a test, is in preparation at a COoler-SYnchrotron COSY-Jülich. In this experiment, COSY will serve as an: accelerator, detector and ideal zero degree spectrometer. The Time Reversal Invariance experiment at COSY (TRIC) aims for an improvement of the upper limit on the

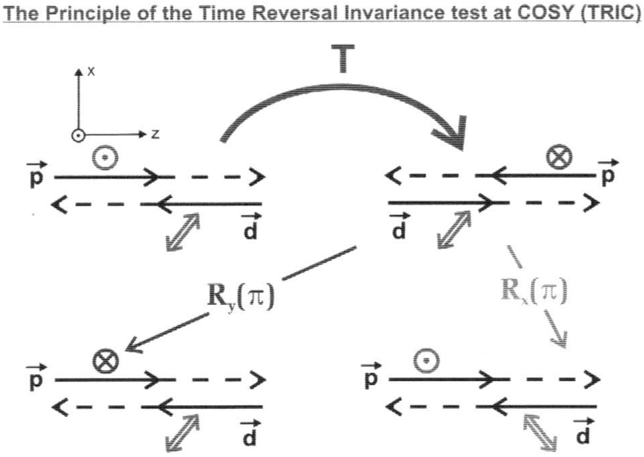

Figure 1. Principle of the TRIC experiment. A vector polarized beam of spin-half particles interacts with a tensor polarized spin-one target.

strength of the NN T-odd potential by one order of magnitude compared to present value by measuring the $A_{y,xz}$ scattering amplitude [8].

Principe of the experiment

In Ref. [9] theorem has been proven which say that exist no null test of time reversal invariance with two particles in and two particles out which is independent on dynamical assumptions. However, this theorem is not applicable for the total cross section, as it was pointed out in Ref. [10]. Based on this exception, a concept of a transmission experiment in pd scattering has been suggested which aims to measure the true T-odd null observable $A_{y,xz}$. In case of scattering of a vector polarised proton beam on a tensor polarised deuterium target the only observable from the scattering amplitude which remains is the $A_{y,xz}$. Thus, in this case the measurement of the total cross section is equivalent to the measurements of $A_{y,xz}$ and serves as a true T-odd null observable.

The principle of the proposed experiment in the center of mass system is illustrated in Fig. 1. A vector polarised proton beam interacts with a tensor polarised deuterium target. If one applies the T-symmetry operator to the initial system it is transformed to a system where a tensor polarised deuteron beam interacts with a vector polarised proton target. However, since the system is invariant under rotation around the X and Y-axes, these two situations are equivalent to a simple flip of the direction of beam and target polarisation vectors.

Experiment at COSY

At the COoler-SYncrotron COSY Jülich experiment to perform such an transmission experiment has been proposed [8]. In this measurement vector polarised proton beam is circulating in the storage ring and interact with a tensor polarised deuterium target. A decrease of the beam current, due to the beam-target interaction, which is directly proportional to the total cross section in pd scattering, will be measured using a high precision beam current measurement system implemented in the COSY ring.

At COSY there is a lot of experience in preparation and handling of unpolarised and polarised beams of protons and deuterons up to the maximum beam momentum of 3.7 GeV/c. In the storage ring electron- and stochastic-cooling systems are available and allow not only to compensate for the beam energy loss due to the interaction with the target but also to control the transversal dimensions of the beam. This allows to use a storage cell systems of relatively small diameter (up to 10 mm) at the internal target station and hence provide a high areal-thickness of a polarised atomic gas target. The EDDA internal polarimeter, located in one of the straight sections of the ring, allows to measure the beam polarisation online using a thin solid target. Thus, the necessary spin gymnastics in order to overcome depolarising resonances of the machine can be performed, and hence high degree of polarisation in the beam up to the maximum COSY beam energy can be preserved.

The experiment will take place at the PAX internal target station, which is very well suited for the proposed measurement [11]. The target chamber is placed in the low-beta section of the COSY accelerator, and hence a storage cell of only 10 mm diameter can be used for the experiment, providing target densities up to 6×10^{13} 1/s×cm^2. The PAX facility is equipped with a Breit-Rabi polarimeter which allows to monitor the target polarisation on-line. A specially designed system of coils, provides with a holding field in the storage cell region, and allows to preserve the target polarisation along the cell. After upgrade in 2014, PAX will be equipped with: openable storage cell, multipurpose detector, and openable flow limiters. These modifications should allow to perform a TRIC experiment and reach the desired goals after four weeks of beam time [8].

In Ref. [8] different options for the hight precision beam current measurement systems have been considered. It was shown that in order to reach the desired statistical precision in $A_{y,xz}$ after one month of measurement at 135 MeV it is necessary to perform the experiment with a bunched beam. Normally, COSY operates with a DC beam and the RF cavity is not used after the acceleration. The electron cooler at COSY, although commissioned and constructed to operate at 100 kV, was never in operation at voltages higher then 65 kV. Therefore, it was necessary to perform a commissioning run at T_p=135 MeV with electron cooler and bunched beam in order to find out if it is possible to perform the TRIC experiment or substantial modifications are needed. Furthermore, the initial motivation for building the PAX facility was a spin filtering experiment [11], which has to be performed at the lowest possible energy of the proton beam. Therefore, it was not planned to accelerate with the PAX magnets and a closed storage cell in the target chamber. Thus, the goal of the first beam time was to shown the feasibility of the proposed experiment at COSY using the PAX facility.

Feasibility test

During the first commissioning beam time in September 2012 following milestones have been achieved:

- It was shown that it is possible to accelerate the COSY proton beam up to the 135 MeV with the PAX magnets and a closed storage cell (length 400 mm, diameter 10 mm) in the PAX target chamber.
- The electron cooler was operated at 74 kV during the beam time without major problems. Tests at 100kV have been done successfully.
- After careful adjustment of COSY orbit and improvements of the vacuum conditions the beam life time of 10000 s has been regularly observed with electron cooled proton beam at the energy of the experiment.
- Bunching with the COSY RF system at the flat top reduces the beam life time by 10-20 % depending on the amplitude in the bunching cavity.

Journal of Physics: Conference Series **447** (2013) 012023 doi:10.1088/1742-6596/447/1/012023

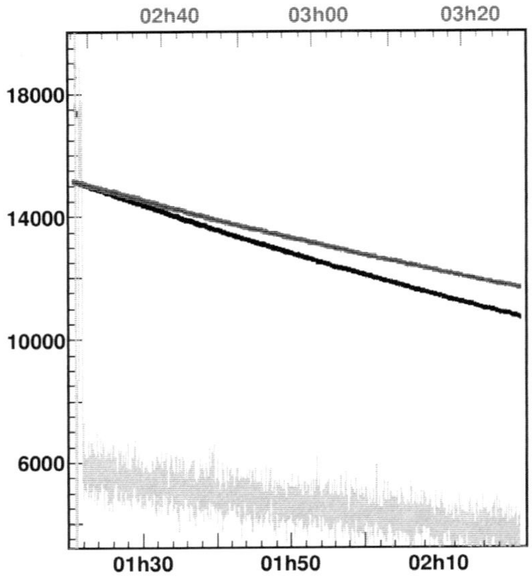

Figure 2. Beam current as a function of time measured at 135 MeV during a cycle with (black line) and without (red line) PAX target. Rate of H^0's measured in H^0-monitor is presented by the green line. The shape of this distribution represents different settings of the electron cooler used during the experiment.

- Bunching with the Barrier Backet system of COSY allows to produce a bunched beam of sufficient quality and almost do not influence beam life-time.
- A beam size of 2.2 mm after cooling at the energy of experiment has been measured using a beam profile monitor.
- Using a frame system installed in the PAX target chamber the COSY ring acceptance of 6.45 mrad has been measured at 135 MeV.
- A COSY cycle with 4×10^9 unpolarised protons in a bunch, suitable for the TRIC experiment, was prepared and regularly observed during all the beam time
- The PAX target and COSY rest gas thicknesses were measured using the Schottky system. According to the first estimates the target thickness was 5.5×10^{13} 1/s×cm^2.
- A beam life time of the order of 10000 s has been observed without target in the storage cell. Switching the PAX target on change the beam life time by 20%.

In Fig. 2 COSY beam current as a function of time measured at 135 MeV with bunched beam is presented for the two cycles with (black line) and without (red line) PAX target. The green line at the bottom shows the H^0 rate detected in H^0-monitor used for the adjustment of electron cooler. Beam life times for the cycles with and without target are 8510 and 11050 s, respectively.

Conclusions

A test beam time, performed in September 2012, have shown that COSY can provide a bunched beam of sufficient quality with the PAX low-beta section. The PAX target and the storage cell have shown designed performance and can be used for the real TRIC beam time after the PAX deuterium source commissioning in 2014. A new high precision beam current measurement system should be developed in order to achieve final goals of the planned TRIC experiment.

Journal of Physics: Conference Series **447** (2013) 012023 doi:10.1088/1742-6596/447/1/012023

References

[1] A. Sakharov, Pisma Zh.Eksp.Teor.Fiz., **5** (1967) 3235.
[2] F. Rathmann *et all*, Proceedings of DISCRETE 2013.
[3] A. Kurylov, G. C. McLaughlin and M. J. Ramsey-Musolf, Phys. Rev. D **63** 076007.
[4] A. Angelopoulus *et all* (CPLEAR Collaboration), Phys. Lett. B **444** (1998) 43.
[5] J. P. Lees *et all*, Phys. Rev. Lett. **109** (2012) 211801.
[6] J. Bernabeu, A. Di Domenico and P. Villanueva-Perez, NPB **868** (2013) 102119.
[7] P. R. Huffman *et all*, Phys. Rev. C **55** (1997) 2684.
[8] P.D. Eversheim, B. Lorentz, and Yu. Valdau, *Test of Time-Reversal Invariance in Proton-Deuteron Scattering at COSY*, COSY proposal #**215** (2012).
[9] F. Arash, M.J. Moravcsik, and G.R. Goldstein, Phys. Rev. Lett. **54** (1985) 2649.
[10] M. Beyer, Nucl. Phys. A **560** (1993) 895.
[11] W. Augustyniak *et all*, Phys. Lett. B **718** (2012) 64.

Journal of Physics: Conference Series **447** (2013) 012024 doi:10.1088/1742-6596/447/1/012024

Direct observation of time.reversal violation in B mesons at BABAR

Pablo Villanueva-Pérez

IFIC, Universitat de València-CSIC, E-46071, Paterna (Valencia), Spain and Department of Theoretical Physics, Universitat de València, E-46100 Burjassot (Valencia)

E-mail: pablo.villanueva.perez@ific.uv.es

Abstract. Although CP violation in the B meson system has been well established by the B factories, there has been no direct observation of time-reversal violation. The decays of entangled neutral B mesons into definite flavour states (B^0 or $\overline{B}^0$), and $J/\psi K^0_L$ or $c\bar{c}K^0_S$ final states (referred to as B_+ or B_-), allow comparisons between the probabilities of four pairs of T-conjugated transitions, for example, $\overline{B}^0 \to B_-$ and $B_- \to \overline{B}^0$, as a function of the time difference between the two B decays. Using 468 million $B\overline{B}$ pairs produced in $\Upsilon(4S)$ decays collected by the *BABAR* detector at SLAC, T-violating parameters in the time evolution of neutral B mesons have been measured, yielding $\Delta S^+_T = -1.37 \pm 0.14$ (stat.) $\pm\, 0.06$ (syst.) and $\Delta S^-_T = 1.17 \pm 0.18$ (stat.) $\pm\, 0.11$ (syst.) [1]. These non-zero results represent the first direct observation of T violation through the exchange of initial and final states in transitions that can only be connected by a T-symmetry transformation.

1. Time-reversal violation searches

The observations of CP-symmetry breaking, first in neutral K decays [2] and more recently in B mesons [3, 4], are consistent with the standard model (SM) mechanism of the three-family Cabibbo-Kobayashi-Maskawa (CKM) quark-mixing matrix being the dominant source of CP violation [5]. Local Lorentz invariant quantum field theories imply CPT invariance [6], in accordance with all experimental evidence [7, 8]. In spite of experimental evidence of CPT invariance, the theoretical connection between CP and T does not imply an identity, except for processes which are identical under CPT transformation.

We aim experiments that could provide direct evidence of T non-invariance, without using an observation which also violates CP, i.e., without any theoretical neither experimental connection with CP. Experiments in stable systems involve either non-vanishing expectation values of T-odd observables, or the exchange of initial and final states, which are not CP conjugates to each other, in the time evolution for transition processes. Among the former, there exist upper limits for electric dipole moments of the neutron and the electron [10]. The latter, requiring neutrinos or unstable particles, are particularly difficult to implement. In the case of unstable systems, since the SM is CPT invariant, it predicts T-violating effects in parallel to each CP-violation effect. These may appear in three different ways: T violation in decay; T violation in the mixing of neutral states; and T violation that arises from the interference between decay with and without mixing.

T violation matched to *CP* violation in decay has not been observed, because the difficulties of the preparation of the time-reversed decay process. Let us see the example [11] of the rare weak decay of neutral-B mesons to $K^+\pi^-$, for which direct *CP* violation is well established [4] (different decay rates R_1 and R_2 for $B^0 \to K^+\pi^-$ and $\overline{B}^0 \to K^-\pi^+$, respectively). T violation as implied by this result, combined with *CPT* invariance, tells us that the rates for the inverse processes $K^+\pi^- \to B^0$ and $K^-\pi^+ \to \overline{B}^0$ should be R_2 and R_1, respectively. However, there is little chance to measure these inverse rates and check directly this prediction since the weak interaction production mechanism is highly suppressed ($B^0 \to K^+\pi^-$ branching ratio of order 10^{-5}) and the strong interaction would completely swamp the feeble weak process.

T violation associated to *CP* violation in the mixing has been experimentally analysed in K [12] and B mesons [13]. Here one looks whether the rate for a neutral-K (B) meson tagged at its production as K^0 (B^0) and identified afterwards as $\overline{K}^0$ ($\overline{B}^0$) is equal to the rate for the neutral particle tagged at its production as $\overline{K}^0$ ($\overline{B}^0$) and identified later as K^0 (B^0). Any difference in this case is both *CP* and T violating, because *CP* and T are experimentally identical for this process. The experimental results for kaons yielded a T-violating difference in these rates. Such a difference is proportional not only to the T-violating term of the $K^0\overline{K}^0$ matrix that defines the mass eigenstates in terms of the flavour eigenstates, but also to the width difference $\Delta\Gamma$ between the two mass eigenstates. Therefore, this asymmetry shows T violation proportional to $\Delta\Gamma$, time independent, experimentally identical to *CP* violation, thus it is not an independent T non-invariance test as one might like [14].

Finally, we could also consider searches for T violation arising from the interference between mixing and decay in neutral B mesons. This is the place where the largest *CP*-violating asymmetry in Nature has been found between the rate for $B^0 \to J/\psi K^0_S$ and the *CP*-conjugate rate for $\overline{B}^0$ to decay to the same *CP*-eigenstate [3, 15]. Therefore, we expect the largest T violation effect. However, these *CP* violation results cannot be directly interpreted as T violation since those results are obtained invoking *CPT* invariance and $\Delta\Gamma = 0$, and not the reversal of time and the exchange of initial and final states, as required for a direct probe of T non-invariance.

2. Method description and results

Up to now as we have delineated there is not a positive direct observation of T unconnected to *CP* violation. We report the direct observation of T violation in the B meson system, through the exchange of initial and final states in transitions that can only be connected by a T-symmetry transformation [1]. The method is described in Ref. [16], based on the concepts proposed in Ref. [17] and further discussed in Refs. [11, 18, 19].

The B factories give us a unique opportunity to perform a test of T as described above due to the EPR entanglement produced by the decay of the $\Upsilon(4S)$, yielding to an entangled, antisymmetric system of orthogonal states. This two-body state is usually written in terms of flavour eigenstates, such as B^0 and $\overline{B}^0$, $|i\rangle = 1/\sqrt{2}[B^0(t_1)\overline{B}^0(t_2) - \overline{B}^0(t_1)B^0(t_2)]$, but can be expressed in terms of any linear combinations of B^0 and $\overline{B}^0$, such as the B_+ and B_- states introduced in Ref. [16], $|i\rangle = 1/\sqrt{2}[B_+(t_1)B_-(t_2) - B_-(t_1)B_+(t_2)]$. The B_+ and B_- states are defined as the neutral B states filtered by the decay to *CP*-eigenstates $J/\psi K^0_L$ (*CP*-even) and $J/\psi K^0_S$, with $K^0_S \to \pi\pi$ (*CP*-odd), respectively. They are orthogonal to each other when there is only one weak phase involved in the B decay amplitude, as it occurs in B decays to $J/\psi K^0$ final states [20], and *CP* violation in neutral kaons is neglected.

For this analysis we use events in which one B candidate is reconstructed in a B_+ or B_- state, and the flavour of the other B is identified, referred to as flavour identification (ID). We generically denote reconstructed final states that identify the flavour of the B as ℓ^-X for $\overline{B}^0$ and ℓ^+X for B^0. The notation (f_1, f_2) is used to indicate the flavour or *CP* final states that are reconstructed at corresponding times t_1 and t_2, where $t_2 > t_1$, i.e., $B_1 \to f_1$ is the first decay in the event and $B_2 \to f_2$ is the second decay. For later use in Eq. (1), we define $\Delta\tau = t_2 - t_1 > 0$. Once the B_1 state is filtered at time t_1, the living partner B_2 is prepared ("tagged") by entanglement as its orthogonal state. The notation $B_2(t_1) \to B_2(t_2)$ describes the transition of the B which decays at t_2, having tagged its state at t_1. For example, an event reconstructed in the time-ordered final states $(\ell^+X, J/\psi K^0_S)$ identifies the transition $\overline{B}^0 \to B_-$ for the second B to decay. We compare the rate for this transition to its T-reversed $B_- \to \overline{B}^0$ (exchange of initial and final states) by reconstructing the final states $(J/\psi K^0_L, \ell^- X)$. Any difference in these two rates is evidence for T-symmetry violation. There are three other independent comparisons that can be made between $B_+ \to B^0$ $(J/\psi K^0_S, \ell^+X)$, $\overline{B}^0 \to B_+$ $(\ell^+X, J/\psi K^0_L)$, and $B_- \to B^0$ $(J/\psi K^0_L, \ell^+X)$ transitions and their T-conjugates, $B^0 \to B_+$ $(\ell^-X, J/\psi K^0_L)$, $B_+ \to \overline{B}^0$ $(J/\psi K^0_S, \ell^-X)$, and $B^0 \to B_-$ $(\ell^-X, J/\psi K^0_S)$, respectively. Similarly, four different CP (CPT) comparisons can be made, e.g., between the $\overline{B}^0 \to B_-$ transition and its CP (CPT)-transformed $B^0 \to B_-$ ($B_- \to B^0$) [16].

Assuming $\Delta\Gamma_d = 0$, each of the eight transitions has a general time-dependent decay rate given by

$$g^{\pm}_{\alpha,\beta}(\Delta\tau) = e^{-\Gamma_d \Delta\tau}\{1 + S^{\pm}_{\alpha,\beta}\sin(\Delta m_d \Delta\tau) + C^{\pm}_{\alpha,\beta}\cos(\Delta m_d \Delta\tau)\}, \tag{1}$$

where indices $\alpha = \ell^+, \ell^-$ and $\beta = K^0_S, K^0_L$ stand for ℓ^+X, ℓ^-X and $c\bar{c}K^0_S, J/\psi K^0_L$ final states, respectively, and the symbol $+$ or $-$ indicates whether the decay to the flavour final state α occurs before or after the decay to the CP final state β. Here, Γ_d is the average decay width, Δm_d is the mass difference between the neutral B mass eigenstates, and $C^{\pm}_{\alpha,\beta}$ and $S^{\pm}_{\alpha,\beta}$ are model-independent coefficients. The sine term, expected to be large in the SM, results from the interference between direct decay of the neutral B to the $J/\psi K^0$ final state and decay after B^0-$\overline{B}^0$ oscillation, while the cosine term arises from the interference between decay amplitudes with different weak and strong phases, and is expected to be negligible [20]. T violation would manifest itself through differences between the $S^{\pm}_{\alpha,\beta}$ or $C^{\pm}_{\alpha,\beta}$ values for T-conjugated processes, for example between $S^{+}_{\ell^+,K^0_S}$ and $S^{-}_{\ell^-,K^0_L}$.

B_- states are reconstructed into the $J/\psi K^0_S$, $\psi(2S)K^0_S$, $\chi_{c1}K^0_S$ final states (denoted generically as $c\bar{c}K^0_S$), while B_+ are into $J/\psi K^0_L$. We also reconstruct a large sample of self-flavour tagging neutral B decays into open charm and charmonium final states, $B^0 \to D^{(*)-}[\pi^+, \rho(770)^+, a_1(1260)^+]$ and $B^0 \to J/\psi K^{*0}(\to K^+\pi^-)$, which are used for calibration of the time resolution function and the performance of the inclusive B flavour (B^0 or $\overline{B}^0$) identification. Finally we reconstruct a large sample of charged B decays into charmonium, $B^{\pm} \to J/\psi K^{\pm}$, $\psi(2S)K^{\pm}$, $J/\psi K^{*\pm}$, which are used as control sample. We use the standard kinematic constraints available at B factories from the beam energies to reconstruct the mass and the energy difference of the B mesons, $m_{ES} = \sqrt{(E^*_{beam}) - (p^*_B)^2}$ and $\Delta E = E^*_B - E^*_{beam}$, where E^*_B, p^*_B are the energy and momentum of the B in c.m., respectively. We also exploit the different topology of signal and $q\bar{q}$ events to reject continuum background. The final sample contains 7796 B_- signal events with purities ranging from 87 to 96%, and 5813 B_+ signal events with purity about 56%.

We perform a simultaneous, unbinned maximum likelihood fit to the reconstructed, signed difference of proper times between the two B decays, $\Delta t \equiv t_\beta - t_\alpha$, to flavor idenfied, $c\bar{c}K^0_S$ and $J/\psi K^0_L$ events. The signal probability density function (PDF) is [16]

$$\begin{aligned}\mathcal{H}_{\alpha,\beta}(\Delta t) \;\propto\; & g^{+}_{\alpha,\beta}(\Delta t_{\rm true})H(\Delta t_{\rm true}) \otimes \mathcal{R}(\delta t; \sigma_{\Delta t}) + \\ & g^{-}_{\alpha,\beta}(-\Delta t_{\rm true})H(-\Delta t_{\rm true}) \otimes \mathcal{R}(\delta t; \sigma_{\Delta t}),\end{aligned} \tag{2}$$

where $\Delta t_{\rm true}$ is the *signed* difference of proper times in the limit of perfect Δt resolution, H is the Heaviside step function, $\mathcal{R}(\delta t; \sigma_{\Delta t})$ with $\delta t = \Delta t - \Delta t_{\rm true}$ is the resolution function, and

$g^{\pm}_{\alpha,\beta}$ are given by Eq. (1). Note that $\Delta t_{\rm true}$ is equivalent to $\Delta\tau$ when the flavour of the B is identified at t_1, determining the flavour of the still living partner (flavour tag), or equivalent to $-\Delta\tau$ when at t_1 we reconstruct a *CP* final state, filtering the still living B meson to the orthogonal state projected by the reconstructed decay (*CP* tag) [17]. Because of the convolution with the resolution function, the distribution for $\Delta t > 0$ contains predominantly true flavour-tagged events, with a contribution from true *CP*-tagged events at low Δt, and conversely for $\Delta t < 0$. Backgrounds are accounted for by adding terms to Eq. (2) [21]. Events are assigned signal and background probabilities based on the $m_{\rm ES}$ or ΔE distributions, for $c\bar{c}K^0_S$ or $J/\psi K^0_L$ events, respectively.

We obtain in total 16 signal coefficients [16] which allow us to construct six pairs of independent asymmetry parameters $(\Delta S^{\pm}_T, \Delta C^{\pm}_T)$, $(\Delta S^{\pm}_{CP}, \Delta C^{\pm}_{CP})$, and $(\Delta S^{\pm}_{CPT}, \Delta C^{\pm}_{CPT})$, as shown in Table 1. The T-asymmetry parameters have the advantage that T-symmetry breaking would directly manifest itself through any nonzero value of $\Delta S^{\pm}_T$ or $\Delta C^{\pm}_T$, or any difference between $\Delta S^{\pm}_{CP}$ and $\Delta S^{\pm}_{CPT}$, or between $\Delta C^{\pm}_{CP}$ and $\Delta C^{\pm}_{CPT}$ (analogously for *CP*- or *CPT*-symmetry breaking). The measured values for the asymmetry parameters [1] are reported in Table 1. There is another two times three pairs of T-, *CP*-, and *CPT*-asymmetry parameters, but they are not independent and can be derived from Table 1 or Ref. [16].

Table 1. Measured values of the T-, *CP*-, and *CPT*-asymmetry parameters, defined as the differences in $S^{\pm}_{\alpha,\beta}$ and $C^{\pm}_{\alpha,\beta}$ between symmetry-transformed transitions. The values of reference coefficients are also given at the bottom. The first uncertainty is statistical and the second systematic. The indices ℓ^-, ℓ^+, K^0_S, and K^0_L stand for reconstructed final states that identify the B meson as $\overline{B}^0$, B^0, B_-, and B_+, respectively.

Parameter	Result
$\Delta S^+_T = S^-_{\ell^-,K^0_L} - S^+_{\ell^+,K^0_S}$	$-1.37 \pm 0.14 \pm 0.06$
$\Delta S^-_T = S^+_{\ell^-,K^0_L} - S^-_{\ell^+,K^0_S}$	$1.17 \pm 0.18 \pm 0.11$
$\Delta C^+_T = C^-_{\ell^-,K^0_L} - C^+_{\ell^+,K^0_S}$	$0.10 \pm 0.14 \pm 0.08$
$\Delta C^-_T = C^+_{\ell^-,K^0_L} - C^-_{\ell^+,K^0_S}$	$0.04 \pm 0.14 \pm 0.08$
$\Delta S^+_{CP} = S^+_{\ell^-,K^0_S} - S^+_{\ell^+,K^0_S}$	$-1.30 \pm 0.11 \pm 0.07$
$\Delta S^-_{CP} = S^-_{\ell^-,K^0_S} - S^-_{\ell^+,K^0_S}$	$1.33 \pm 0.12 \pm 0.06$
$\Delta C^+_{CP} = C^+_{\ell^-,K^0_S} - C^+_{\ell^+,K^0_S}$	$0.07 \pm 0.09 \pm 0.03$
$\Delta C^-_{CP} = C^-_{\ell^-,K^0_S} - C^-_{\ell^+,K^0_S}$	$0.08 \pm 0.10 \pm 0.04$
$\Delta S^+_{CPT} = S^-_{\ell^+,K^0_L} - S^+_{\ell^+,K^0_S}$	$0.16 \pm 0.21 \pm 0.09$
$\Delta S^-_{CPT} = S^+_{\ell^+,K^0_L} - S^-_{\ell^+,K^0_S}$	$-0.03 \pm 0.13 \pm 0.06$
$\Delta C^+_{CPT} = C^-_{\ell^+,K^0_L} - C^+_{\ell^+,K^0_S}$	$0.14 \pm 0.15 \pm 0.07$
$\Delta C^-_{CPT} = C^+_{\ell^+,K^0_L} - C^-_{\ell^+,K^0_S}$	$0.03 \pm 0.12 \pm 0.08$
$S^+_{\ell^+,K^0_S}$	$0.55 \pm 0.09 \pm 0.06$
$S^-_{\ell^+,K^0_S}$	$-0.66 \pm 0.06 \pm 0.04$
$C^+_{\ell^+,K^0_S}$	$0.01 \pm 0.07 \pm 0.05$
$C^-_{\ell^+,K^0_S}$	$-0.05 \pm 0.06 \pm 0.03$

We build time-dependent asymmetries $A_T(\Delta t)$ to visually demonstrate the T-violating effect. For transition $\overline{B}^0 \to B_-$,

$$A_T(\Delta t) \equiv \frac{\mathcal{H}^-_{\ell^-,K^0_L}(\Delta t) - \mathcal{H}^+_{\ell^+,K^0_S}(\Delta t)}{\mathcal{H}^-_{\ell^-,K^0_L}(\Delta t) + \mathcal{H}^+_{\ell^+,K^0_S}(\Delta t)}\,, \tag{3}$$

where $\mathcal{H}^{\pm}_{\alpha,\beta}(\Delta t) = \mathcal{H}_{\alpha,\beta}(\pm\Delta t)H(\Delta t)$. With this construction, $A_T(\Delta t)$ is defined only for positive Δt values [1]. Neglecting reconstruction effects, $A_T(\Delta t) \approx \frac{\Delta S^+_T}{2}\sin(\Delta m_d \Delta t) + \frac{\Delta C^+_T}{2}\cos(\Delta m_d \Delta t)$. We introduce the other three T-violating asymmetries similarly. Figure 1 shows the four observed asymmetries, overlaid with the projection of the best fit results to the Δt distributions with and without the eight T-invariance restrictions: $\Delta S^{\pm}_T = \Delta C^{\pm}_T = 0$, $\Delta S^{\pm}_{CP} = \Delta S^{\pm}_{CPT}$, and $\Delta C^{\pm}_{CP} = \Delta C^{\pm}_{CPT}$ [16].

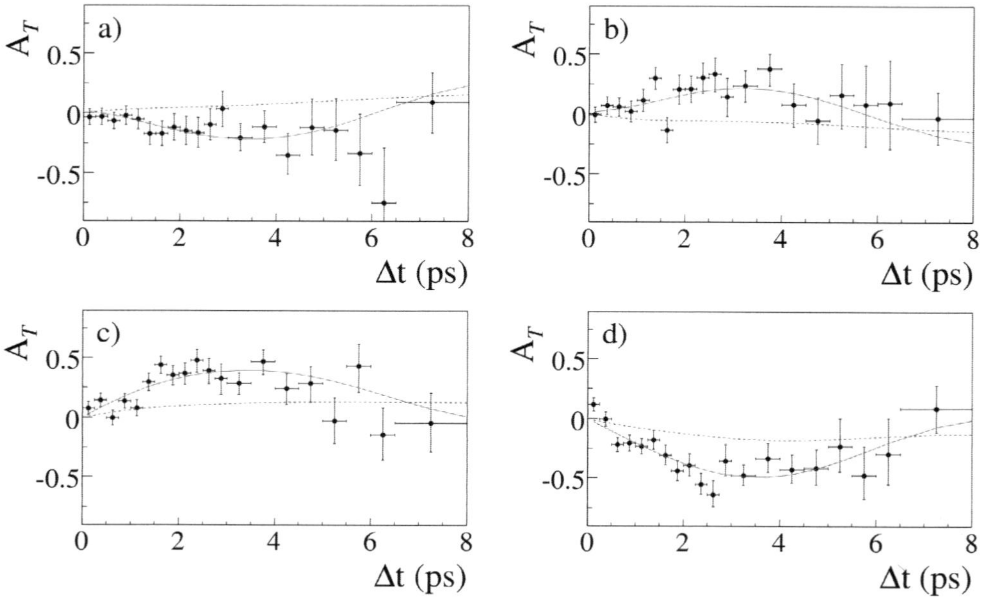

Figure 1. The four independent T-violating asymmetries for transition a) $\overline{B}^0 \to B_-$ $(\ell^+X, c\overline{c}K^0_S)$, b) $B_+ \to B^0$ $(c\overline{c}K^0_S, \ell^+X)$, c) $\overline{B}^0 \to B_+$ $(\ell^+X, J/\psi K^0_L)$, d) $B_- \to B^0$ $(J/\psi K^0_L, \ell^+X)$, for combined flavour categories with low misID (leptons and kaons), in the signal region ($5.27 < m_{\rm ES} < 5.29$ GeV/c^2 for $c\overline{c}K^0_S$ modes and $|\Delta E| < 10$ MeV for $J/\psi K^0_L$). The points with error bars represent the data, the red solid and dashed blue curves represent the projections of the best fit results with and without T violation, respectively.

Using large samples of MC simulated data, we determine that the asymmetry parameters are unbiased and have Gaussian errors. Fitting a single pair of (S, C) coefficients, reversing the sign of S under $\Delta t \leftrightarrow -\Delta t$, or $B_+ \leftrightarrow B_-$ or $B^0 \leftrightarrow \overline{B}^0$ exchanges, and the sign of C under $B^0 \leftrightarrow \overline{B}^0$ exchange, we obtain identical results to those obtained in Ref. [21]. Performing the analysis with B decays to $c\overline{c}K^{\pm}$ and $J/\psi K^{*\pm}$ final states instead of the signal $c\overline{c}K^0_S$ and $J/\psi K^0_L$, respectively, we find that all the asymmetry parameters are consistent with zero.

In evaluating systematic uncertainties in the asymmetry parameters, we follow the same procedure as in Ref. [21], with small changes [16]. The total systematic uncertainties are shown in the extra material of Ref. [1].

The significance of the T-violation signal is evaluated based on the change in negative log-likelihood with respect to the maximum ($-2\Delta\ln\mathcal{L}$). We reduce $-2\Delta\ln\mathcal{L}$ by a factor $1+\max\{m_i^2\} = 1.61$ to account for systematic errors in the evaluation of the significance. Here, $m_i^2 = -2(\ln\mathcal{L}_i - \ln\mathcal{L})/s^2$, where $\ln\mathcal{L}$ is the maximum log-likelihood, $\ln\mathcal{L}_i$ is the log-likelihood with asymmetry parameter i fixed to its total systematic variation and maximized over all other parameters, and $s^2 \approx 1$ is the change in $2\ln\mathcal{L}$ at 68% confidence level (CL) for one degree of freedom (d.o.f). Figure 2 shows CL contours calculated from the change $-2\Delta\ln\mathcal{L}$ in two dimensions for the T-asymmetry parameters $(\Delta S_T^+, \Delta C_T^+)$ and $(\Delta S_T^-, \Delta C_T^-)$. The difference in the value of $2\ln\mathcal{L}$ at the best fit solution with and without T violation is 226 with eight d.o.f., including systematic uncertainties. Assuming Gaussian errors, this corresponds to a significance equivalent to 14.0 standard deviations (σ), and thus constitutes direct observation of T violation. The significance of CP and CPT violation is determined analogously, obtaining 307 and 5, respectively, equivalent to 17σ and 0.3σ, consistent with CP violation and CPT invariance.

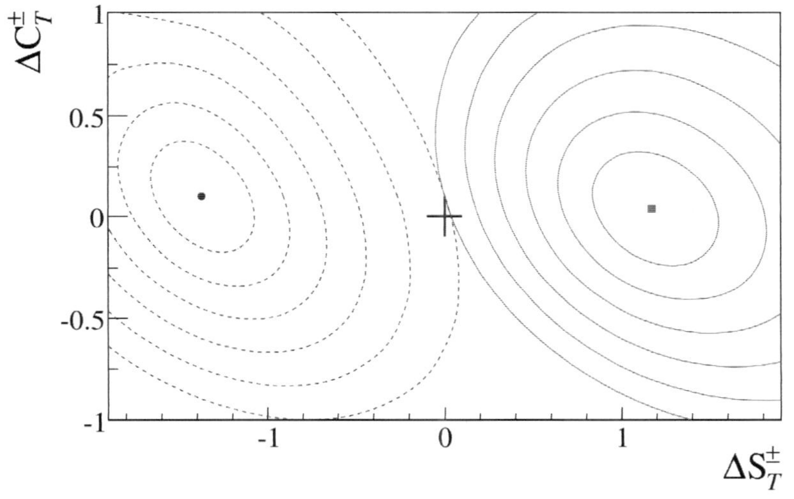

Figure 2. The central values (blue point and red square) and two-dimensional CL contours for $1-\text{CL} = 0.317$, 4.55×10^{-2}, 2.70×10^{-3}, 6.33×10^{-5}, 5.73×10^{-7}, and 1.97×10^{-9}, calculated from the change in the value of $-2\Delta\ln\mathcal{L}$ compared with its value at maximum ($-2\Delta\ln\mathcal{L} = 2.3, 6.2, 11.8, 19.3, 28.7, 40.1$), for the pairs of T-asymmetry parameters $(\Delta S_T^+, \Delta C_T^+)$ (blue dashed curves) and $(\Delta S_T^-, \Delta C_T^-)$ (red solid curves). Systematic uncertainties are included. The T-invariance point is shown as a + sign.

3. Conclusions

We have measured T-violating parameters in the time evolution of neutral B mesons, by comparing the probabilities of $\overline{B}^0 \to B_-$, $B_+ \to B^0$, $\overline{B}^0 \to B_+$, and $B_- \to B^0$ transitions, to their T conjugate. We determine for the main T-violating parameters $\Delta S_T^+ = -1.37 \pm 0.14$ (stat.) $\pm\ 0.06$ (syst.) and $\Delta S_T^- = 1.17 \pm 0.18$ (stat.) $\pm\ 0.11$ (syst.), and observe directly for the first time a departure from T invariance in the B meson system, with a significance equivalent to 14σ. Our results are consistent with current CP-violating measurements obtained invoking CPT invariance. They constitute the first direct observation of T violation in any system through the exchange of initial and final states in transitions that can only be connected by a T-symmetry transformation.

References

[1] J.P. Lees *et al.* (*BABAR* Collaboration), Phys. Rev. Lett. **109**, 211801 (2012).
[2] J.H. Christenson *et al.*, Phys. Rev. Lett. **13**, 138 (1964).
[3] B. Aubert *et al.* (*BABAR* Collaboration), Phys. Rev. Lett. **87**, 091801 (2001); K. Abe *et al.* (Belle Collaboration), Phys. Rev. Lett. **87**, 091802 (2001).
[4] B. Aubert *et al.* (*BABAR* Collaboration), Phys. Rev. Lett. **93**, 131801 (2004); Y. Chao *et al.* (Belle Collaboration), Phys. Rev. Lett. **93**, 191802 (2004).
[5] N. Cabibbo, Phys. Rev. Lett. **10**, 531 (1963); M. Kobayashi and T. Maskawa, Prog. Theor. Phys. **49**, 652 (1973).
[6] G. Lüders, Math. Fysik. Medd. Kgl. Danske Akad. Ved. Volume 28, 1954, p. 5; J.S. Bell, Birmingham University thesis (1954); W. Pauli, in W. Pauli, ed., Niels Bohr and the Development of Physics (McGraw-Hill, NY, 1955).
[7] R. Carosi *et al.*, Phys. Lett. B **237**, 303 (1990); A. Alavi-Harati *et al.*, Phys. Rev. D **67**, 012005 (2003); B. Schwingenheuer *et al.*, Phys. Rev. Lett. **74**, 4376 (1995).
[8] See "Tests of conservation laws" review in [9].
[9] K. Nakamura *et al.* (Particle Data Group), *J. Phys. G: Nucl. Part. Phys.* 37, 075021 (2010).
[10] J.J. Hudson *et al.*, Nature **473**, 493-496 (2011); C.A. Baker *et al.*, Phys. Rev. Lett. **97**, 131801 (2006).
[11] H. R. Quinn, J. Phys. Conf. Ser. **171**, 011001 (2009).
[12] A. Angelopoulus *et al.* (CPLEAR Collaboration), Phys. Lett. B **444**, 43 (1998).
[13] B. Aubert *et al.* (*BABAR* Collaboration), Phys. Rev. Lett. **96**, 251802 (2006), Phys. Rev. Lett. **92**, 181801 (2004); E. Nakano *et al.* (Belle Collaboration), Phys. Rev. D **73**, 112002 (2006); V.M. Abazov *et al.* (D0 Collaboration), Phys. Rev. Lett. **105**, 081801 (2010), Phys. Rev. Lett. **98**, 151801 (2007); F. Abe *et al.* (CDF Collaboration), Phys. Rev. D **55**, 2546 (2006).
[14] L. Wolfenstein, Phys. Rev. Lett. **83**, 911 (1999).
[15] M. Antonelli *et al.*, Phys. Rept. **494**, 197 (2010).
[16] J. Bernabeu, F. Martinez-Vidal, P. Villanueva-Perez, JHEP **1208**, 064 (2012).
[17] M. C. Bañuls and J. Bernabeu, Phys. Lett. B **464**, 117 (1999); Nucl. Phys. B **590**, 19 (2000).
[18] L. Wolfenstein, Int. Jour. Mod. Phys. **8**, 501 (1999).
[19] J. Bernabeu, J. Phys. Conf. Ser. **335**, 012011 (2011).
[20] See "*CP* violation in meson decays" review in [9].
[21] B. Aubert *et al.* (*BABAR* Collaboration), Phys. Rev. D **79**, 072009 (2009).

Journal of Physics: Conference Series **447** (2013) 012025 doi:10.1088/1742-6596/447/1/012025

CP violation at ATLAS

Adam Barton, On behalf of the ATLAS Collaboration.
Physics Department, Lancaster University, LANCASTER, LA1 4YB, UK
E-mail: abarton@cern.ch

Abstract. A measurement of several properties of the B_s meson, including the CP-violating weak phase ϕ_s and the mixing-induced width difference $\Delta\Gamma_s$, is performed using the decay $B_s \to J/\psi(\mu^+\mu^-)\phi(K^+K^-)$, from a dataset of 4.9 fb^{-1} of integrated luminosity collected in 2011 by the ATLAS detector at the LHC. The measured parameters are consistent with the world average values and theoretical expectations; in particular ϕ_s is within 1 σ of the expected value in the Standard Model.

1. Introduction

New phenomena beyond the Standard Model (SM) may alter CP violation in B-decays. The CP violation present in the SM cannot account for the dominance of matter in the observable universe, providing a strong motivation for investigating sources of CP violation as they become amenable to experimental investigation. In the $B_s \to J/\psi(\mu^+\mu^-)\phi(K^+K^-)$ decay the CP-violating phase ϕ_s is the weak phase difference between the $B_s^0 - \bar{B}_s^0$ mixing amplitude and the $b \to c\bar{c}s$ decay amplitude. The SM predicts a small value of $\phi_s \simeq -2\beta_s = -0.0368 \pm 0.0018$ rad [2]. Another quantity involved in $B_s^0 - \bar{B}_s^0$ mixing is the width difference $\Delta\Gamma_s = \Gamma_L - \Gamma_H$ between heavy (B_H) and light (B_L) eigenstates. Physics beyond the SM is not expected to affect $\Delta\Gamma_s$ significantly [3]. Extracting $\Delta\Gamma_s$ from data is nevertheless useful as it allows theoretical predictions to be confirmed. The presented analysis uses data collected by the ATLAS experiment in pp collisions at $\sqrt{s} = 7$ TeV in 2011, corresponding to an integrated luminosity of 4.9 fb^{-1}. The measurement of ϕ_s, the average decay width $\Gamma_s = (\Gamma_L + \Gamma_H)/2$ and the value of $\Delta\Gamma_s$, were performed without any attempt to distinguish the between the B_s and $\bar{B}_s$ (flavour tagging). The CP states were separated statistically through the time-dependence of the decay and angular correlations amongst the final-state particles.

2. ATLAS potential in B_s mixing and CP violation studies

ATLAS has measured the cross sections of the inclusive $b\bar{b} \to J/\psi$ and $pp \to J/\psi$ and found that given the conditions in ATLAS the $b\bar{b} \to J/\psi$ cross section is approximately $0.38 \pm 0.01(\text{stat.}) \pm 0.47(\text{syst.}) \pm 0.03(\text{spin}) \pm 0.01(\text{lumi.})\mu b$, while the $pp \to J/\psi$ cross section is $0.98 \pm 0.02(\text{stat.}) \pm 0.13(\text{stat.}) \pm 0.15(\text{spin}) \pm 0.03(\text{lumi.})\mu b$ [9]. The high ratio of $R = (bb \to J/\psi)/(pp \to J/\psi) = 0.39$ makes it viable to collect all the $(pp \to J/\psi)$ background and fit it in the maximum likelihood, since the alternative method of applying a proper decay time cut to remove the background has the side effect of complicating the fitting proceeding and introducing another type of systematic error.

3. Candidate Selection

ATLAS is a multipurpose cylindrical particle detector at the Large Hadron Collider [4]. The tracking and muon systems are of particular importance in the reconstruction of B mesons . The inner tracking detector consists of a silicon pixel detector, a silicon microstrip detector and a transition radiation tracker; all surrounded by a superconducting solenoid providing a 2 T axial field. The muon spectrometer consists of three superconducting toroids, a system of tracking chambers, and detectors for triggering. The dimuon triggers used for the analysis are based on identification of a $J/\psi \to \mu^+\mu^-$ decay, with either a 4 GeV transverse momentum (p_T) threshold for each muon or an asymmetric configuration that applies a higher p_T threshold (4-10 GeV) to one of the muons. The pairs of muon tracks are refitted to a common vertex and are accepted if the fit χ^2/d.o.f. < 10. The invariant mass of the J/ψ is required to fall in the range (2.959 - 3.229) GeV when both muons entered the barrel ($\mid \eta \mid < 1.05$) or (2.852-3.332) GeV when both muon entered the endcap ($1.05 <\mid \eta \mid < 2.5$) or (2.913-3.273) GeV otherwise. Candidates for $B_s \to J/\psi(\mu^+\mu^-)\phi(K^+K^-)$ are sought by fitting the four tracks to a common vertex resulting in χ^2/d.o.f.< 3, while the invariant mass of the two muons are fixed to the J/ψ mass [5]. In total 131k B_s^0 candidates are collected within a mass range of $\phi \to K^+K^-$ (1.0085-1.0305) GeV and $5.15 < m(B_s^0) < 5.65$ GeV, figure 1. For each B_s^0 candidate the proper decay t is determined by the expression: $t = L_{xy}M_B/(cp_{T_B})$, where p_{T_B} is the transverse momentum of the B_s^0 candidate and M_B is the mass of the B_s^0 meson (5.3663 GeV)[5]. L_{xy} is the displacement in transverse plane of the B_s^0 decay vertex with respect to the primary vertex (PV) projected onto the direction of p_{T_B}. The position of the PV is refitted following the removal of the tracks used to reconstruct the B_s^0 candidate. For the selected events the average number of pileup interactions is 5.6. The PV with the smallest d_0 is chosen for the proper decay time calculation, this is a three-dimensional impact parameter d_0, calculated as the distance between the line extrapolated from the B_s^0 vertex in the direction of the B_s^0 momentum and each PV candidate.

4. Analysis and results

The $B_s \to J/\psi(\mu^+\mu^-)\phi(K^+K^-)$ decay involves three CP states of the $J/\psi\phi$ system, combined into three polarisation amplitudes, A_0, $A_\parallel$ and $A_\perp$. The first two states are CP-even, while the last state is CP-odd. Other CP-odd states can be produced by a non-resonant K^+K^- pair or by the decay of the spin-0 f_0 meson, resulting in another independent amplitude, the S-wave A_s. The time evolution of the four decay amplitudes along with six inference terms is fitted simultaneously with the angular distributions and the reconstructed proper decay time of the cascade decay $B_s \to J/\psi(\mu^+\mu^-)\phi(K^+K^-)$. Three independent angles $\Omega = (\theta_T, \psi_T, \phi_T)$, are defined in the transversity basis [6]. The interesting parameters of the $B_s \to J/\psi(\mu^+\mu^-)\phi(K^+K^-)$ decay are $\delta_\parallel = \arg(A_\parallel, A_0)$, $\delta_\perp = \arg(A_\perp, A_0)$ and $\delta_s = \arg(A_s, A_0)$. For untagged analysis all terms involving the mixing frequency Δm_s in the time-dependant amplitudes cancel out [1]. Also the time-dependent amplitudes depending on $\delta_\perp$ are multiplied by a small value $\sin\phi_s$, hence the untagged analysis is not sensitive to $\delta_\perp$. A Gaussian constraint to the external data, $\delta_\perp = (2.95 \pm 0.39)$ rad is therefore applied [8]. An unbinned maximum likelihood fit uses information of the reconstructed B_s^0 candidate mass m, proper decay time t, their uncertainties σ_m and σ_t, and the angles $\Omega = (\theta_T, \psi_T, \phi_T)$. In total 26 parameters are determined, including the eight physics parameters of interest mentioned above, while the other quantities describe the $J/\psi\phi$ mass distribution, the proper decay time and the angular distributions of the background. The event likelihood has the form:

$$\mathcal{L} \propto w_i \cdot f_s \cdot \mathcal{F}_s i + f_s \cdot f_{B_i^0} + (1 - f_s \cdot (1 + f_{B^0})) \cdot \mathcal{F}_{bkgi} \qquad (1)$$

where the index i is used for the variables specific for each event, f_s is fraction of signal candidates, $\mathcal{F}_{si}$ and $\mathcal{F}_{bkgi}$ are probability density functions (PDF) modelling the signal and background. The backgrounds $B^0 \to J/\psi K^*$ and $B^0 \to J/\psi K^+\pi^-$ are parametrised separately by $\mathcal{F}_{B_i^0}$ with f_{B^0} being the fraction of this background events. The weighting factor w_i accounts for a small decay-time dependency of the acceptance, related to a limited resolution in the on-line track reconstruction [1]. The PDF describing the signal, $\mathcal{F}_s$, has the form:

$$\mathcal{F}_{si} = P_s(m_i \mid \sigma_{m_i}) \cdot P_s(\sigma_{m_i}) \cdot P_s(\Omega_i, t_i \mid \sigma_{t_i}) \cdot A(\Omega_i, p_{T_i}) \cdot P_s(p_{T_i}) \tag{2}$$

The signal mass density $P_s(m_i \mid \sigma_{m_i})$ is modelled as a Delta function smeared by a Gaussian with a per-candidate mass resolution σ_{m_i}. Similarly, each of the ten terms of the signal time and angular dependence, $P_s(\Omega_i, t_i \mid \sigma_{t_i})$, is convoluted with a Gaussian with a per-candidate resolution σ_t. The sculpting of the detector and kinematic cuts on the angular distributions is included in the likelihood function through $A(\Omega_i, p_{Ti})$. This is calculated using a four-dimensional binned acceptance method, applying an event-by-event acceptance according to the transversity angles $(\theta_T, \psi_T \phi_T)$ and the p_{T_B}. The acceptance is calculated from the $B_s \to J/\psi(\mu^+\mu^-)\phi(K^+K^-)$ MC events. The background PDF has the following composition:

$$\mathcal{F}_{bkgi} = P_b(m_i) \cdot P_b(\sigma_{m_i} \cdot P_b(\sigma_{m_i}) \cdot P_b(t_i \mid \sigma_{t_i}) \cdot P_b(\theta_T) \cdot P_b(\psi_T) \cdot P_b(\phi_T) \cdot P_b(\sigma_{t_i}) \cdot P_b(p_{T_i}) \tag{3}$$

The proper decay time function $P_b(t_i \mid \sigma_{t_i})$ is parameterised as a Delta function, two positive exponentials and a negative exponential. These functions are smeared with the same resolution function as the signal proper decay time-dependence. The prompt peak describes the combinatorial background events, which are expected to have reconstructed lifetimes distributed around zero. The two positive exponentials represent the fraction of longer-lived backgrounds with non-prompt J/ψ, combined with hadrons from the primary vertex or from a B/D hadron in the same event. The negative exponential takes into account events with a particularly badly measured vertex. The shapes of the background angular distributions, $P_b(\theta_T)$, $P_b(\psi_T)$, and $P_b(\phi_T)$ arise primarily from detector and kinematic sculpting are described by the empirical functions (equations 4 to 6) with nuisance parameters determined in the fit. The correlations between the background angular shapes are neglected, but a systematic error arising from this simplification was evaluated. The background mass model, $P_b(m)$ is a linear function. Mis-reconstructed $B^0 \to J/\psi K^*$ and $B^0 \to J/\psi K^+\pi^-$ (non-resonant) decays, are parametrised separately. The fractions of these components are fixed in the likelihood fit to values $(6.5 \pm 2.4)\%$ and $(4.5 \pm 2.8)\%$, calculated using MC events. Mass and angles have fixed shapes determined from the MC studies. The proper decay time is described by an exponential, convoluted by per-candidate Gaussian errors. Finally, the terms $P_{s,b}(\sigma_{m_i})$, $P_{s,b}(\sigma_{t_i})$ and $P_{s,b}(p_{T_i})$ are introduced to account for differences between signal and background per-candidate mass and proper decay time uncertainties and values of transverse momenta [1].

$$f(\cos\theta_T) = \frac{a_0 - a_1 cos^2(\theta_T)}{2a_0 - 2a_1/3} \tag{4}$$

$$f(\varphi_T) = \frac{1 + b_1 cos(2\varphi_T + b_0)}{2\pi} \tag{5}$$

$$f(\cos\psi_T) = \frac{c_0 + c_1 cos^2(\psi_T)}{2c_0 + 2c_1/3} \tag{6}$$

Systematic uncertainties are assigned by considering effects not accounted for in the likelihood fit. The impact of inner detector residual misalignments were estimated using events simulated with perfect and distorted geometries. Systematics due to limitations of the fit model were determined by 1000 pseudo-experiments generated with variations in the signal and background mass model, resolution model, background lifetime and background angles models. Systematics

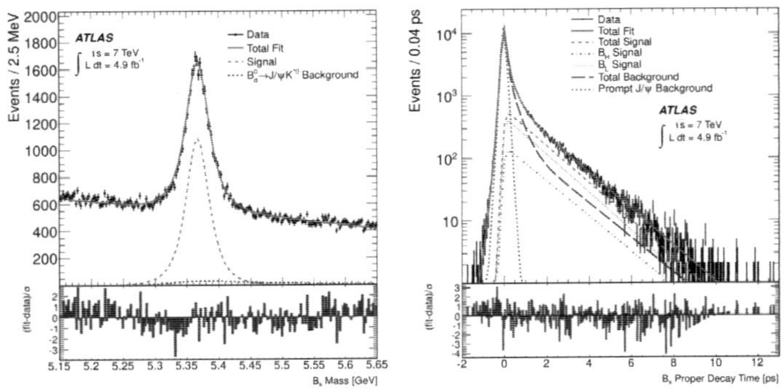

Figure 1. Mass and proper decay time fit projections for the B_s^0 candidates. The pull distribution at the bottom shows the difference between data and fit value normalised to the data uncertainty.[1]

due to $B^0 \rightarrow J/\psi K^{*0}$ and $B^0 \rightarrow J/\psi K\pi$ arise from the uncertainties of the PDG decay probabilities [5].

In the absence of initial state flavour tagging the PDF is invariant under the simultaneous transformations: $\{\phi_s, \Delta\Gamma_s, \delta_\perp, \delta_\|, \delta_s\} \rightarrow \{-\phi_s, \Delta\Gamma_s, \pi - \delta_\perp, -\delta_\|, -\delta_s\}$ leading to a fourfold ambiguity. As the constraint on $\delta_\perp$ is taken from the LHCb measurement [7], that quotes only two solutions with a positive ϕ_s and two $\Delta\Gamma_s$ values symmetric around zero, two of the four minima fitted in the present non-flavour tagged analysis are excluded from the results presented here. Additionally a solution with negative $\Delta\Gamma_s$ is excluded following the LHCb measurement [8] which determines the $\Delta\Gamma_s$ to be positive. The measured values, for the single minimum resulting from these constraints, are given in Table 1.

Par.	Value	Stat.	Syst.
ϕ_s(rad)	0.22	0.41	0.10
$\Delta\Gamma_s(ps^{-1})$	0.053	0.21	0.010
$\Gamma_s(ps^{-1})$	0.0677	0.007	0.004
$\mid A_0(0) \mid^2$	0.528	0.006	0.009
$\mid A\|(0) \mid^2$	0.220	0.008	0.007
$\mid A_s(0) \mid^2$	0.02	0.02	0.02

Table 1. Fitted values for the physics parameters along with their statistical and systematic uncertainties.[1]

The strong phase of the S-wave component is fitted relative to $\delta_\perp$, as $\delta_\perp - \delta_s = (0.03 \pm 0.13)$ rad. The fraction of S-wave KK or f_0 contamination is measured to be consistent with zero at $\mid A_s(0) \mid^2 = 0.02 \pm 0.02$. The two-dimensional likelihood contours in the $\phi_s - \Delta\Gamma_s$ plane for the 68%, 90% and 95% confidence intervals are produced using a profile likelihood method as shown in figure 3. The systematic errors are not included in figure 3 but as seen from table 1 they are small compared to the statistical errors. The ATLAS measured parameters are consistent with the world average values and with theoretical expectation in particular ϕ_s is within 1σ of the expected value in the SM.

Journal of Physics: Conference Series **447** (2013) 012025 doi:10.1088/1742-6596/447/1/012025

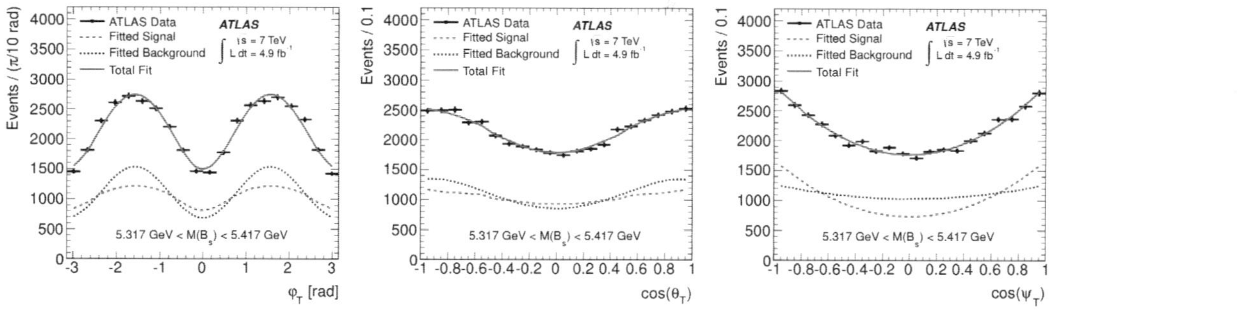

Figure 2. Fit projections for the 3 transversity angles measured in the analysis.[1]

5. Comparison with other results

The heavy flavour averaging group (HFAG) has included the ATLAS result in their world average, bringing the world average into close agreement with the SM prediction (see figure 3). The comparison shown in figure 3 of Γ_s and $\Delta\Gamma_s$ measurements display agreement with other measurements, theoretical prediction and show a competitive accuracy is obtained.

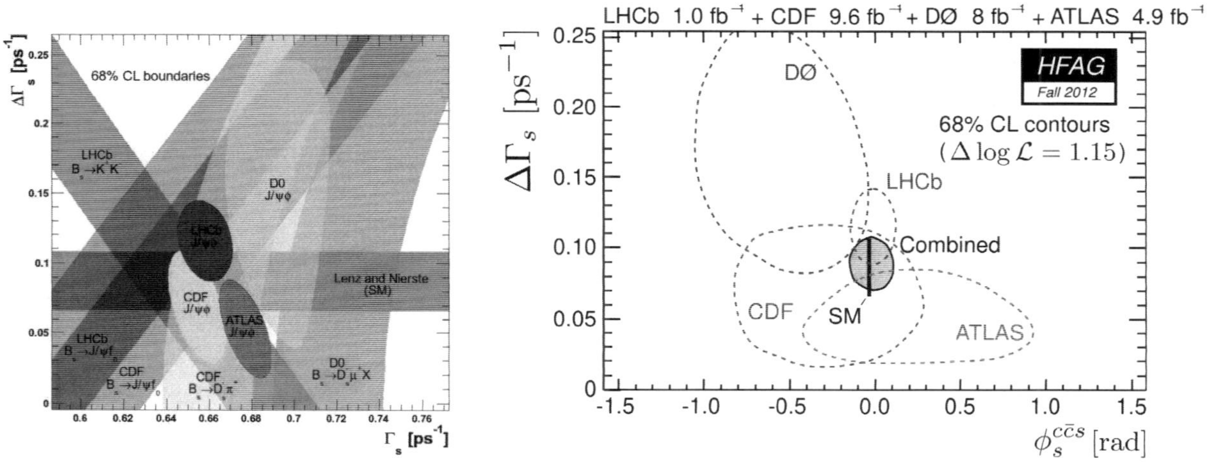

Figure 3. Contour Plots comparing the parameters with those measured at other experiments. Note the other experiments include tagging while the ATLAS contour does not.[1]

Future publications released by ATLAS will increase accuracy by including tagging information, measuring the angles defined in the helicity basis (which produces simpler acceptance sculpting) and incorporating subsequent datasets to increase statistical information.

References

[1] ATLAS Collaboration, Aad, Georges et, 2012, *Time-dependent angular analysis of the decay $B_s \to J/\psi\phi$ and extraction of $\Delta\Gamma_s$ and the CP-violating weak phase ϕ_s by ATLAS*, CERN-PH-EP-2012-182, arXiv:1208.0572

[2] UTfit Collaboration, M. Bona et al., Constraints on new physics from the quark mixing unitarity triangle, Phys.Rev.Lett. 97 (2006) 151803, [hep-ph/0605213].

[3] A. Lenz and U. Nierste, Theoretical update of $B_s\bar{B}_s$ mixing, JHEP 0706 (2007) 072, [hep-ph/0612167].

[4] ATLAS Collaboration, The ATLAS Experiment at the CERN Large Hadron Collider, JINST 3 (2008) S08003.

[5] Particle Data Group Collaboration, K. Nakamura et al., Review of particle physics, J.Phys.G G37 (2010) 075021. (And 2011 partial update for the 2012 edition).

[6] A. S. Dighe, I. Dunietz, and R. Fleischer, Extracting CKM phases and $B_s\bar{B}_s$ mixing parameters from angular distributions of nonleptonic B decays, Eur.Phys.J. C6 (1999) 647662, [hep-ph/9804253].

[7] LHCb Collaboration, R. Aaij et al., Measurement of the CP-violating phase s in the decay $B_s \to J/\psi\phi$, Phys.Rev.Lett. **108** (2012) 101803, [arXiv:1112.3183].

[8] LHCb Collaboration, R. Aaij et al., Determination of the sign of the decay width difference in the Bs system, Phys.Rev.Lett. **108** (2012) 241801, [arXiv:1202.4717].

[9] ATLAS Collaboration, Measurement of the differential cross-sections of inclusive, prompt and non-prompt J/ψ production in protonproton collisions at $\sqrt{s} = 7$ TeV. Nucl. Phys. B 850 (2011) 387-344

Journal of Physics: Conference Series **447** (2013) 012026 doi:10.1088/1742-6596/447/1/012026

Measurements of the CKM angle γ in tree-dominated B decays at LHCb

L Carson, on behalf of the LHCb Collaboration

Physics Department, Imperial College London, Blackett Laboratory, Prince Consort Rd, LONDON, SW7 2BW, United Kingdom

E-mail: l.carson@imperial.ac.uk

Abstract. The LHCb experiment is a general purpose forward spectrometer operating at the Large Hadron Collider, optimized for the study of B and D hadrons. LHCb collected $\sim$1 fb^{-1} of integrated luminosity during 2011 data taking, which provides an unprecedentedly large sample of B hadron decays to final states involving charmed hadrons. These decays offer many complementary ways to measure the angle γ of the CKM triangle. We present a selection of new world-leading measurements of CP-violating observables in these decays, using both time-independent and, for the first time, time-dependent analyses. A first LHCb combination of the time-independent measurements to provide a value of γ is also presented.

1. Introduction

The CKM parameter γ is currently the least well-constrained angle of the CKM triangle. Combinations by the CKMFitter [1] and UTFit [2] collaborations, using all existing measurements, give $\gamma = (66 \pm 12)^\circ$ and $\gamma = (76 \pm 10)^\circ$ respectively. Measurements of γ can be made using B decays that are mediated only by tree-level transitions, for example B decays to open charm mesons. Since these transitions are unlikely to be affected by New Physics, they provide a "standard candle" to measure the Standard Model (SM) value of γ. The value of γ measured in this way can then be compared to values measured using B decays mediated by loop-level transitions, for example B decays to light charmless hadrons. A significant difference between these values would indicate a New Physics contribution to the loop processes.

This document presents the latest results from the LHCb experiment [3] measuring CP-violating observables in hadronic B decays to open charm. Results from time-independent analyses of $B^+ \to \overline{D}^0 K^+$ and $B^+ \to \overline{D}^0 \pi^+$ decays, with various final states of the $\overline{D}^0$ meson, are reported. These results are combined to give a first LHCb value for γ. A further measurement, not included in the γ combination, is also reported: the first measurement of the CP-violating observables in $B^0_s \to D_s^\mp K^\pm$, using a time-dependent analysis. All results are based on data corresponding to $\sim$1 fb^{-1} of integrated luminosity, collected in 2011 at a centre-of-mass energy of 7 TeV.

2. Time-independent analyses of $B^+ \to \overline{D}^0 K^+$ and $B^+ \to \overline{D}^0 \pi^+$

In general, time-independent methods of measuring γ from hadronic B decays to open charm take advantage of the interference between $b \to c$ and $b \to u$ transitions at tree-level, where the D final state is accessible to both D^0 and $\overline{D}^0$. Aside from the weak phase γ, these analyses also measure the hadronic parameters $r_{B(D)}$ and $\delta_{B(D)}$, where the ratio of the favoured to suppressed $B(D)$ decay amplitudes is $r_{B(D)}e^{i(\delta_{B(D)})}$. The method to extract these hadronic parameters and γ depends on the D final state. The three methods discussed in this document are usually referred to by the initials of the authors of the paper(s) that first proposed the method: the "GLW" method [4], the "ADS" method [5] and the "GGSZ" method [6].

2.1. GLW Analysis

The GLW method uses D decays to CP-eigenstates. The most experimentally accessible such decays are $D^0 \to K^+K^-$ and $D^0 \to \pi^+\pi^-$. Since these states have a positive CP-eigenvalue, the decays are denoted by D_{CP+}. The two main observables in the GLW method are the average partial rate R_{CP+}, and the asymmetry A_{CP+}, defined as

$$R_{CP+} = 2\frac{\Gamma(B^- \to D_{CP+}K^-) + \Gamma(B^+ \to D_{CP+}K^+)}{\Gamma(B^- \to D^0 K^-) + \Gamma(B^+ \to \overline{D}^0 K^+)},$$

$$A_{CP+} = \frac{\Gamma(B^- \to D_{CP+}K^-) - \Gamma(B^+ \to D_{CP+}K^+)}{\Gamma(B^- \to D_{CP+}K^-) + \Gamma(B^+ \to D_{CP+}K^+)},$$

where the D^0 or $\overline{D}^0$ refers to a quasi-flavour-specific D final state, which in practice is usually $D^0 \to K^-\pi^+$ or $\overline{D}^0 \to K^+\pi^-$. These observables are related to the physics parameters via

$$R_{CP+} = 1 + r_B^2 + 2C_f r_B \cos\delta_B \cos\gamma, \quad A_{CP+} = \frac{2C_f r_B \sin\delta_B \sin\gamma}{R_{CP+}},$$

where C_f is a coherence factor that is equal to 1 for two-body D final states (such as K^+K^- or $\pi^+\pi^-$), and is between 0 and 1 for multi-body D final states (such as $K^+K^-\pi^+\pi^-$). For a multi-body final state, $C_f = 1$ corresponds to the case where the final state is always reached via intermediate resonances (for example $\phi\rho^0$), such that it can be considered as a two-body decay. The value $C_f = 0$ corresponds to the case where the final state has no contributions from intermediate resonances. For multi-body D decays, the C_f value typically lies somewhere between these two extremes. In this case the various paths to the final state can interfere, which is why the decay is not "coherent". Equivalent expressions for R_{CP+} and A_{CP+} exist for the $B^+ \to D_{CP+}\pi^+$ decay, but the asymmetry there is expected to be negligible, due to the very small value of r_B. LHCb has measured the GLW observables using the K^+K^- and $\pi^+\pi^-$ final states of the D [7]. The mass spectra of selected events in the GLW D final states (split by bachelor particle type and charge) are shown in Fig. 1. Stringent particle identification (PID) requirements are used to suppress background from the Cabibbo-favoured $D\pi$ final state to the DK final state. In Figs. 1-3, the green line represents the $D\pi$ events (including the $D\pi$ background to DK remaining after the PID requirements), the red line represents the DK events, and the shaded contribution represents the partially reconstructed background. In Figs. 2 and 3, the pink dashed line represents the background from $B_s^0 \to DK^\mp\pi^\pm$ events where the pion is not reconstructed. A clear asymmetry between B^+ and B^- can be seen in the DK final state, for both D final states. After correcting for the (small) detector and production asymmetries, and averaging over the D final states, the average partial rate and asymmetry for DK are measured to be

$$R_{CP+} = 1.007 \pm 0.038 \pm 0.012, \quad A_{CP+} = 0.145 \pm 0.032 \pm 0.010,$$

where the first error is statistical, and the second systematic. The A_{CP+} value is nonzero with a significance of 4.5 standard deviations.

Journal of Physics: Conference Series **447** (2013) 012026 doi:10.1088/1742-6596/447/1/012026

2.2. ADS Analysis

The ADS method uses D decays to quasi-flavour-specific final states, where the decay to one charge combination is Cabibbo-favoured, and the decay to the opposite charge combination is doubly-Cabibbo-suppressed. The most experimentally accessible such pair are the two-body decays $D^0 \to K^-\pi^+$ (Cabibbo-favoured) and $D^0 \to K^+\pi^-$ (doubly-Cabibbo-suppressed). The two main observables in the ADS method are defined similarly to those in the GLW method:

$$
\begin{aligned}
R_{ADS} &= \frac{\Gamma(B^- \to (K^+\pi^-)_D K^-) + \Gamma(B^+ \to (K^-\pi^+)_D K^+)}{\Gamma(B^- \to (K^-\pi^+)_D K^-) + \Gamma(B^+ \to (K^+\pi^-)_D K^+)}, \\
A_{ADS} &= \frac{\Gamma(B^- \to (K^+\pi^-)_D K^-) - \Gamma(B^+ \to (K^-\pi^+)_D K^+)}{\Gamma(B^- \to (K^+\pi^-)_D K^-) + \Gamma(B^+ \to (K^-\pi^+)_D K^+)},
\end{aligned}
$$

are are related to the physics parameters via

$$
\begin{aligned}
R_{ADS} &= r_B^2 + r_D^2 + 2C_f r_B r_D \cos(\delta_B + \delta_D)\cos\gamma, \\
A_{ADS} &= \frac{2C_f r_B r_D \sin(\delta_B + \delta_D)\sin\gamma}{R_{ADS}},
\end{aligned}
$$

where C_f is again a coherence factor analogous to the GLW case. Equivalent expressions exist for the decay with the bachelor pion. LHCb has measured the ADS observables using the $K^-\pi^+$ and $K^+\pi^-$ final states of the D [7]. The mass spectra of selected events in the ADS D final states (split by bachelor particle type and charge) are shown in Fig. 2. A clear asymmetry between B^+ and B^- can be seen for the doubly-Cabibbo-suppressed D final state, in both the

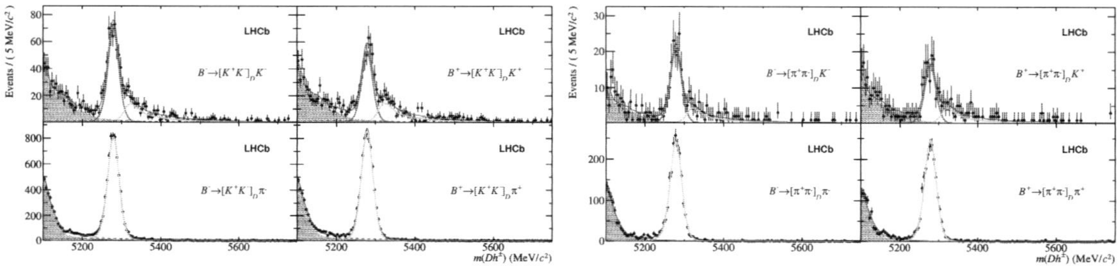

Figure 1. Left: Invariant mass distributions for $B^\pm \to [K^+K^-]_D h^\pm$ candidates, where h represents a charged kaon or pion. The top (bottom) plots show the $h = K (h = \pi)$ final state, split by charge. Right: the equivalent plots for $B^\pm \to [\pi^+\pi^-]_D h^\pm$ candidates.

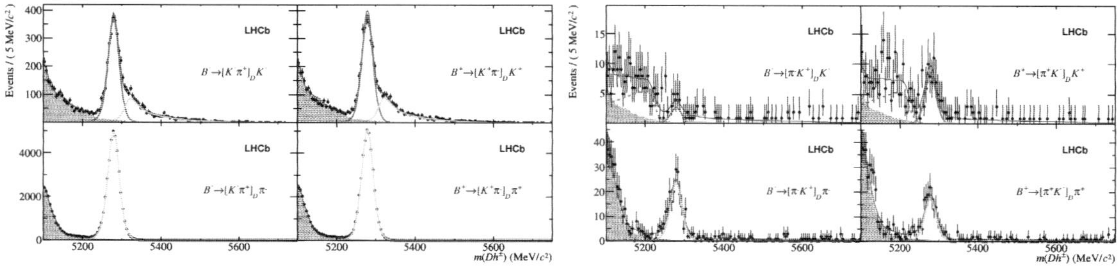

Figure 2. Left: Invariant mass distributions for $B^\pm \to [K^\pm\pi^\mp]_D h^\pm$ candidates, where h represents a charged kaon or pion. The top (bottom) plots show the $h = K (h = \pi)$ final state, split by charge. Right: the equivalent plots for $B^\pm \to [K^\mp\pi^\pm]_D h^\pm$ candidates.

DK and $D\pi$ final states. After correcting for the (small) detector and production asymmetries, the average partial rates and asymmetries for DK and $D\pi$ are measured to be

$$R_{ADS(\pi)} = 0.00410 \pm 0.00025 \pm 0.00005, \quad A_{ADS(\pi)} = 0.143 \pm 0.062 \pm 0.011,$$
$$R_{ADS(K)} = 0.0152 \pm 0.0020 \pm 0.0004, \quad A_{ADS(K)} = -0.52 \pm 0.15 \pm 0.02,$$

where the first error is statistical, and the second systematic. The DK decay with the doubly-Cabibbo-suppressed D final state is observed for the first time, with a significance of 10 standard deviations. The $A_{ADS(\pi)}(A_{ADS(K)})$ value is nonzero with a significance of 2.4(4.0) standard deviations. Considering the $A_{ADS(K)}$ measurement together with the A_{CP+} measurement from the GLW analysis, CP-violation is observed in $B^{\pm} \to DK^{\pm}$ decays for the first time, with a significance of 5.8 standard deviations.

The ADS method can also be applied to the multi-body decays $D^0 \to K^-\pi^+\pi^-\pi^+$ (Cabibbo-favoured) and $D^0 \to K^+\pi^-\pi^-\pi^+$ (doubly-Cabibbo-suppressed). Since the D decay parameters are different to the two-body case, complementary information is gained by performing this analysis in addition to the two-body analysis. LHCb has measured the ADS observables using the $K^-\pi^+\pi^-\pi^+$ and $K^+\pi^-\pi^-\pi^+$ final states of the D [8]. The mass spectra of selected events in the doubly-Cabibbo-suppressed D final state (split by bachelor particle type and charge) are shown in Fig. 3. There is some indication of an asymmetry between B^+ and B^-, in both the DK and $D\pi$ final states. After correcting for the (small) detector and production asymmetries, the average partial rates and asymmetries for DK and $D\pi$ are measured to be

$$R^{K3\pi}_{ADS(\pi)} = 0.00639 \pm 0.00036, \quad A^{K3\pi}_{ADS(\pi)} = 0.13 \pm 0.10,$$
$$R^{K3\pi}_{ADS(K)} = 0.0124 \pm 0.0027, \quad A^{K3\pi}_{ADS(K)} = -0.42 \pm 0.22,$$

where the error includes both statistical and systematic uncertainties. The $D\pi$ and DK decays with the four-body doubly-Cabibbo-suppressed D final state are both observed for the first time, with a significance of >10 standard deviations for $D\pi$, and 5.1 standard deviations for DK.

2.3. GGSZ Analysis

The GGSZ method uses D decays to self-conjugate three-body final states, with the simplest cases being $D \to K_S\pi^+\pi^-$ and $D \to K_S K^+K^-$. The analysis compares the Dalitz plots of the D decay for $B^+ \to DK^+$ and $B^- \to DK^-$. This can be thought of as combining the GLW and ADS analyses into one simultaneous analysis, since the final state includes contributions

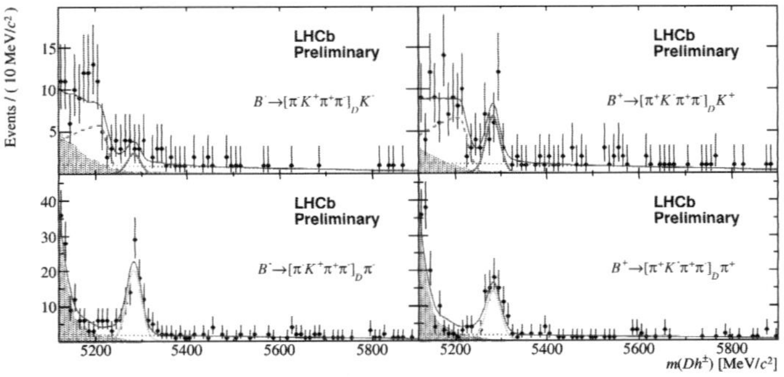

Figure 3. Invariant mass distributions for $B^{\pm} \to [K^{\mp}\pi^{\pm}\pi^{\mp}\pi^{\pm}]_D h^{\pm}$ candidates, where h represents a charged kaon or pion. The top (bottom) plots show the $h = K (h = \pi)$ final state, split by charge.

from decays with different levels of Cabibbo-suppression (for example the Cabibbo-favoured $D^0 \to K^* - \pi^+$ and the doubly-Cabibbo-suppressed $D^0 \to K^* + \pi^-$). This gives the analysis access to enough observables to allow an unambiguous determination of γ and r_B (which is not the case for the GLW or ADS analyses alone).

To measure physics parameters (such as γ) using this method, information is needed on how the D decay amplitude varies over the Dalitz plot. The LHCb analysis described here uses CLEO-c measurements [9] of the strong phase variation as input. The Dalitz plot is binned in regions of similar strong phase. The bins are numbered from $-N$ to N, where $N = 8$ for $D \to K_S\pi^+\pi^-$ and $N = 2$ for $D \to K_S K^+K^-$. The number of events in the ith bin of the Dalitz plot is given by

$$\begin{aligned} N^+_{\pm i} &= h_{B^+}[K_{\mp i} + (x_+^2 + y_+^2)K_{\pm i} + 2\sqrt{K_i K_{-i}}(x_+ c_{\pm i} \mp y_+ s_{\pm i})] \quad \text{for B}^+, \\ N^-_{\pm i} &= h_{B^-}[K_{\pm i} + (x_-^2 + y_-^2)K_{\mp i} + 2\sqrt{K_i K_{-i}}(x_- c_{\pm i} \mp y_- s_{\pm i})] \quad \text{for B}^-, \end{aligned}$$

where $h_{B^\pm}$ is a normalisation factor, the K_i are the number of events in bin i of the Dalitz plot of a flavour-tagged $D^0 \to K_S\pi^+\pi^-$ decay, and the $c_i(s_i)$ are the cos(sin) of the strong phase difference in bin i, taken from the CLEO-c measurement. The analysis thus measures the observables $x_\pm$ and $y_\pm$, which are related to the physics parameters via

$$x_\pm = r_B \cos(\delta_B \pm \gamma), \quad y_\pm = r_B \sin(\delta_B \pm \gamma).$$

LHCb has performed the GGSZ Dalitz analysis using the $K_S\pi^+\pi^-$ and $K_S K^+K^-$ final states of the D [10]. The mass spectra of selected events are shown in Fig. 4.

The Dalitz plots of the selected events in the signal region for the $B^\pm \to [K_S\pi^+\pi^-]_D K^\pm$ final state are shown for B^+ and B^- in Fig. 5. Combining the $K_S\pi^+\pi^-$ and $K_S K^+K^-$ final states,

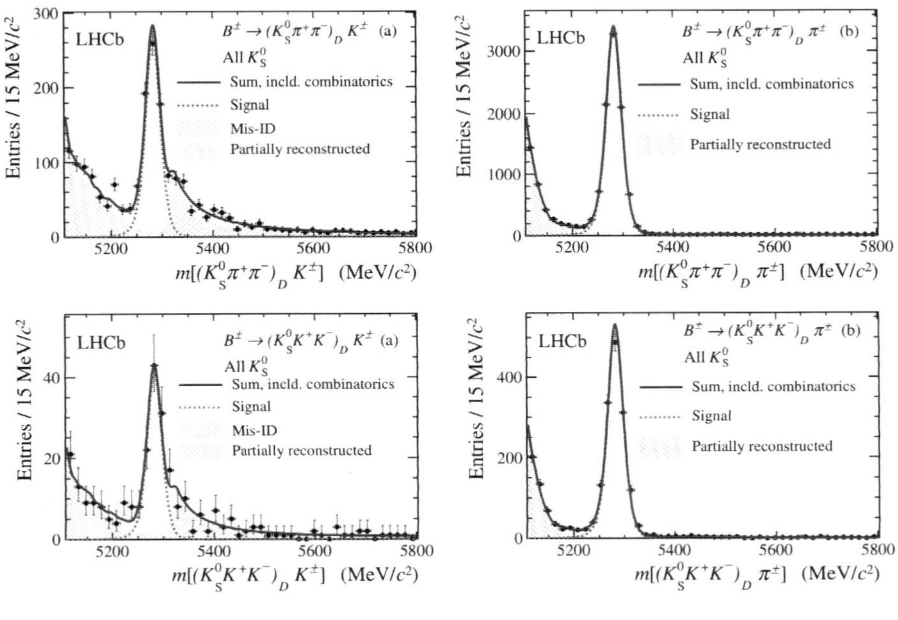

Figure 4. Invariant mass distributions for $B^\pm \to [K_S\pi^+\pi^-]_D h^\pm$ (top) and $B^\pm \to [K_S K^+K^-]_D h^\pm$ (bottom) candidates, where h represents a charged kaon or pion. The left (right) plots show the $h = K (h = \pi)$ final state. The fit components are indicated in the legend.

the measured values of $x_\pm$ and $y_\pm$ are

$$\begin{aligned} x_- &= (0.0 \pm 4.3 \pm 1.5 \pm 0.6) \times 10^{-2}, \quad y_- = (2.7 \pm 5.2 \pm 0.8 \pm 2.3) \times 10^{-2}, \\ x_+ &= (-10.3 \pm 4.5 \pm 1.8 \pm 1.4) \times 10^{-2}, \quad y_+ = (-0.9 \pm 3.7 \pm 0.8 \pm 3.0) \times 10^{-2}, \end{aligned}$$

where the first error is statistical, the second is systematic and the third is due to the uncertainties on the c_i and s_i parameters. The total precision on $x_\pm$ and $y_\pm$ from this analysis is similar to that acheived from the equivalent analyses at the B-Factories. From these $x_\pm$ and $y_\pm$ values, the physics parameters can be extracted:

$$r_B = 0.07 \pm 0.04, \quad \gamma = (44^{+43}_{-38})^\circ,$$

which have less precision than the equivalent B-Factory results, due to the lower central value of r_B measured in this sample compared to the B-Factory results.

3. Combined Value of γ

The LHCb measurements from the two-body ADS and GLW analyses, the four-body ADS analysis, and the GGSZ analysis have been combined, using a frequentist approach, to obtain a combined LHCb value for γ [11]. In addition to the LHCb measurements, a CLEO-c measurement of the strong phase in $D \to K\pi\pi\pi$ is also used [12]. The experimental likelihoods are combined as

$$\mathcal{L}(\vec{\alpha}) = f_i(\vec{A}_i^{\text{obs}} | \vec{A}_i(\vec{\alpha}_i)),$$

where $\vec{A}$ are the experimental observables (such as, R_{CP+}, x_+, etc), and $\vec{\alpha}$ are the physics parameters (such as γ, r_B, etc). Confidence intervals are then obtained from this likelihood in a frequentist way. The confidence intervals for γ are shown in Fig. 6, both from a combination of only the DK measurements, and from a combination of the DK and $D\pi$ measurements. Using only the DK measurements, the 68% confidence interval for γ is $(71.1^{+16.6}_{-15.7})^\circ$, which has similar precision to the existing constraints from B-Factory measurements. Using the DK and $D\pi$ measurements, the 68% confidence interval for γ is $[61.8, 67.8]^\circ \cup [77.9, 92.4]^\circ$.

4. Time-dependent analysis of $B_s^0 \to D_s^\mp K^\pm$

Since tree diagrams of similar magnitude exist for both B_s^0 and $\overline{B}_s^0$ decays to $D_s^- K^+$, large interference can occur between them. This allows the channel $B_s^0 \to D_s^\mp K^\pm$ to be used to

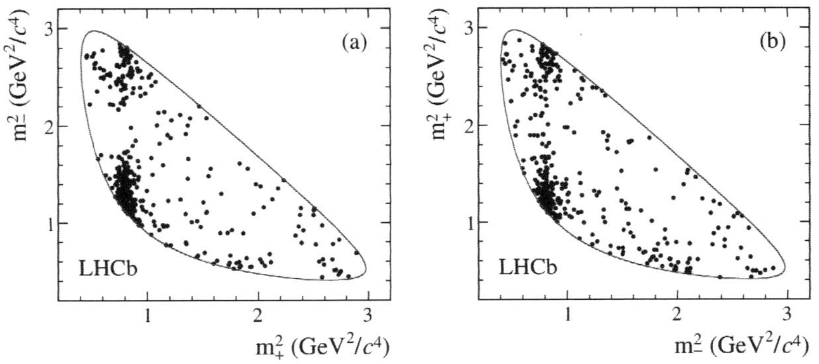

Figure 5. Dalitz plot distributions for $B^\pm \to [K_S\pi^+\pi^-]_D K^\pm$ candidates. The left (right) plot shows $B^+(B^-)$ candidates.

measure γ, via a time-dependent and flavour-tagged analysis [13]. For the general case of an initially-produced B_s^0 or $\overline{B}_s^0$ meson decaying to a final state f, the decay rates are given by

$$\frac{d\Gamma_{B_s^0\to f}(t)}{dt} = \tfrac{1}{2}|A_f|^2(1+|\lambda_f|^2)e^{-\Gamma_s t} \Big[\cosh\left(\frac{\Delta\Gamma_s t}{2}\right) - D_f \sinh\left(\frac{\Delta\Gamma_s t}{2}\right) + C_f\cos(\Delta m_s t) - S_f \sin(\Delta m_s t)\Big],$$

$$\frac{d\Gamma_{\bar{B}_s^0\to f}(t)}{dt} = \tfrac{1}{2}|A_f|^2\left|\tfrac{p}{q}\right|^2(1+|\lambda_f|^2)e^{-\Gamma_s t} \Big[\cosh\left(\frac{\Delta\Gamma_s t}{2}\right) - D_f \sinh\left(\frac{\Delta\Gamma_s t}{2}\right) - C_f\cos(\Delta m_s t) + S_f \sin(\Delta m_s t)\Big],$$

where A_f is the decay amplitude of a B_s^0 decaying to a final state f, and $\lambda_f \equiv (q/p)(\bar{A}_f/A_f)$. The decays rate for decays to the CP-conjugate state $\overline{f}$ can be obtained by replacing A_f by $\bar{A}_{\bar{f}} = \langle\bar{f}|T|\bar{B}_s^0\rangle$, λ_f by $\bar{\lambda}_{\bar{f}} \equiv (p/q)(A_{\bar{f}}/\bar{A}_{\bar{f}})$ and $|p/q|^2$ by $|q/p|^2$. For a B_s^0 decay, the D_f and $D_{\overline{f}}$ terms are experimentally accessible, due to the relatively large value of $\Delta\Gamma_s$. In the particular case of $B_s^0 \to D_s^{\mp}K^{\pm}$, the observables C ($\equiv C_f = C_{\overline{f}}$), S_f, D_f, $S_{\overline{f}}$ and $D_{\overline{f}}$ are related to the physics parameters via

$$C = \frac{1-r_{D_sK}^2}{1+r_{D_sK}^2},$$

$$D_{f(\bar{f})} = \frac{2r_{D_sK}\cos(\Delta-(+)(\gamma-2\beta_s))}{1+r_{D_sK}^2},$$

$$S_{f(\bar{f})} = \frac{2r_{D_sK}\sin(\Delta-(+)(\gamma-2\beta_s))}{1+r_{D_sK}^2},$$

where $r_{D_sK} \equiv |A(\overline{B}_s^0 \to D_s^- K^+)/A(B_s^0 \to D_s^- K^+)|$, is the ratio of the magnitudes of the decay amplitudes of the interfering tree diagrams, Δ is the strong phase difference between these amplitudes, and β_s is the B_s^0 mixing phase. Since there are five observables, the parameters r_{D_sK}, Δ and $\gamma - 2\beta_s$ can be unambiguously extracted from this analysis. The value of γ can then be determined using external measurements of β_s, for example from the channel $B_s^0 \to J/\psi\phi$ [14].

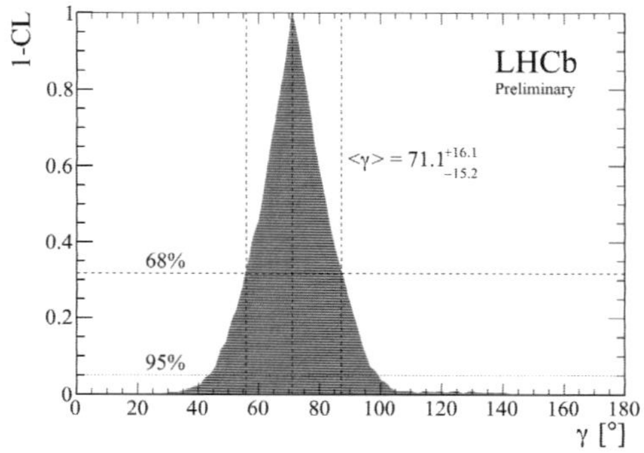

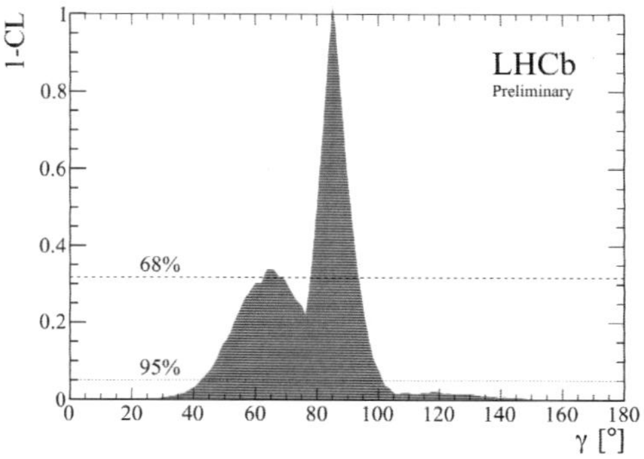

Figure 6. Confidence intervals for γ, resulting from the combination of measurements with DK only (left), and with both DK and $D\pi$ (right).

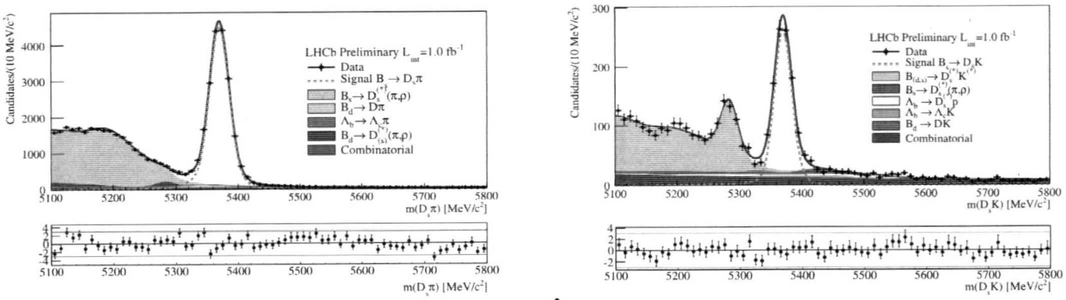

Figure 7. Invariant mass distributions for $B_s^0 \to D_s^{\mp}\pi^{\pm}$ (left) and $B_s^0 \to D_s^{\mp}K^{\pm}$ (right) candidates. The fit components are indicated in the legend. The pull distributions are shown below each mass plot.

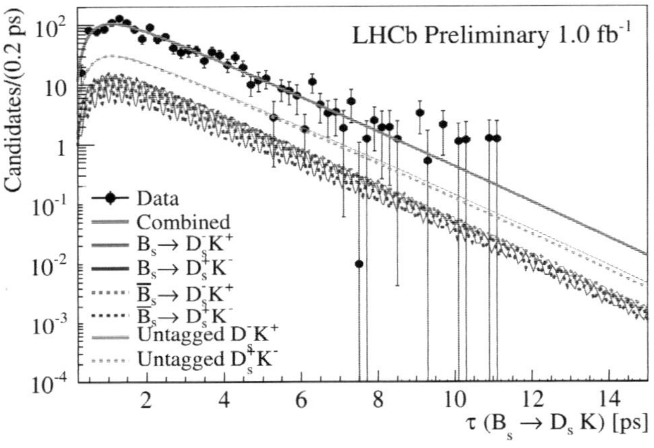

Figure 8. Decay time distribution for $B_s^0 \to D_s^{\mp}K^{\pm}$ candidates. The fit components are indicated in the legend.

LHCb has performed this time-dependent and flavour-tagged analysis of $B_s^0 \to D_s^{\mp}K^{\pm}$ decays [15]. The Cabibbo-favoured $B_s^0 \to D_s^{\mp}\pi^{\pm}$ decay is used as a control channel. Three separate decay modes of the D_s^+ are considered — $K^+K^-\pi^+$, $K^+\pi^-\pi^+$ and $\pi^+\pi^-\pi^+$. The mass spectra of selected events in the $D_s^{\mp}\pi^{\pm}$ and $D_s^{\mp}K^{\pm}$ final states, including all three final states of the D_s, are shown in Fig. 7. The decay time distribution for selected $B_s^0 \to D_s^{\mp}K^{\pm}$ events is shown in Fig. 8. The distribution for all events is shown, along with the flavour-tagged events. Only opposite-side flavour tagging is used, resulting in an effective tagging power of $(1.9 \pm 0.3)\%$. From the fit to this decay time distribution, the CP-asymmetry observables are measured to be

$$
\begin{aligned}
C &= 1.01 \pm 0.50 \pm 0.23, \\
D_f &= -1.33 \pm 0.60 \pm 0.26, \quad D_{\bar{f}} = -0.81 \pm 0.56 \pm 0.26, \\
S_f &= -1.25 \pm 0.56 \pm 0.24, \quad S_{\bar{f}} = 0.08 \pm 0.68 \pm 0.28,
\end{aligned}
$$

where the first error is statistical, and the second systematic. This is the first measurement of these observables in the $B_s^0 \to D_s^{\mp}K^{\pm}$ decay. A value for γ has not been extracted from the current analysis, due to the need to understand the correlation matrix for the systematic uncertainties, which is found to substantially affect the determination of γ.

References

[1] Charles J *et al.* 2005 *Eur. Phys. J.* C **41** 1
[2] Bona M *et al.* 2006 *JHEP* **10** 081
[3] Alves Jr. A A *et al.* 2008 *JINST* **3** S08005
[4] Gronau M and London D 1991 *Phys. Lett.* B **253** 483
Gronau M and Wyler D 1991 *Phys. Lett.* B **265** 172
[5] Atwood D, Dunietz I and Soni A 1997 *Phys. Rev. Lett.* **78** 3257
Atwood D, Dunietz I and Soni A 2001 *Phys. Rev.* D **63** 036005
[6] Giri A, Grossman Y, Soffer A and Zupan J 2003 *Phys. Rev.* D **68** 054018
Poluektov A *et al.* 2004 *Phys. Rev.* D **70** 072003
[7] Aaij R *et al.* 2012 *Phys. Lett.* B **712** 203
[8] Aaij R *et al.* 2012 LHCb-CONF-2012-030
[9] Libby J *et al.* 2010 *Phys. Rev.* D **82** 112006
[10] Aaij R *et al.* 2012 *Phys. Lett.* B **718** 43
[11] Aaij R *et al.* 2012 LHCb-CONF-2012-032
[12] Lowrey N *et al.* 2009 *Phys. Rev.* D **80** 031105
[13] Fleischer R 2003 *Nucl. Phys.* B **671** 459
[14] Aaij R *et al.* 2012 LHCb-CONF-2012-002
[15] Aaij R *et al.* 2012 LHCb-CONF-2012-029

Journal of Physics: Conference Series **447** (2013) 012027 doi:10.1088/1742-6596/447/1/012027

Recent results on *CP* violation in charm and τ decays from *BABAR*

Nicola Neri

INFN - Sezione di Milano, Milano, Italy

E-mail: nicola.neri@mi.infn.it

Abstract. In the Standard Model *CP* violation in charm decays is expected to be very small, at the level of 0.1% or below, while *CP* violation in τ decays is predicted to be negligible. A significant excess of *CP* violation with respect to the Standard Model predictions would be a signature of new physics. We report on recent searches for *CP* violation in charm meson decays and in τ lepton decays at *BABAR*, using a data sample corresponding to an integrated luminosity of about 470 fb^{-1}. In particular, we report on searches for *CPV* in the decay modes with a K_S^0 in the final state, such as $D^+ \to K_S^0 K^+$, $D_s^+ \to K_S^0 K^+$, $D_s^+ \to K_S^0 \pi^+$ and $\tau^+ \to \pi^+ K_S^0 \overline{\nu}_\tau$, and in the 3-body $D^+ \to K^+ K^- \pi^+$ decay. A lifetime ratio analysis of $D^0 \to K^+K^-, \pi^+\pi^-$ with respect to $D^0 \to K^-\pi^+$ decays, which is sensitive to D^0-$\overline{D}^0$ mixing and *CP* violation, is also presented here.

1. Introduction

In the Standard Model (SM) *CP* violation (*CPV*) is accommodated by the CKM [1, 2] mechanism which regulates the mixing of the three families of quarks. This mechanism has been proved to work well according to the experimental results that have been provided mostly by the *B*-Factory experiments, *BABAR* and Belle, during the last decade. However, the CKM mechanism is not sufficient to describe the absence of antimatter in the universe, and so this represents an open question for both experimental and theoretical physicists. Other sources of *CP* violation are currently under investigation, and the charm and τ sectors offer interesting probes for this purpose.

In the charm meson decays *CPV* is expected to be at the level of 0.1% or less [3, 4], however the predictions are affected by large theoretical uncertainties due to long distance interactions. The recent evidence of *CPV* in the difference of time-integrated asymmetries between $D^0 \to K^+K^-$ and $D^0 \to \pi^+\pi^-$ singly Cabibbo-suppressed (SCS) decays, from the LHCb [5] and the CDF [6, 7] experiments, has renewed the interest for searching new physics (NP) in the charm sector. While the observed asymmetries need to be further assessed [1], the present results are marginally compatible with the SM but are not conclusive for establishing NP [10, 11, 12]. The study of other SCS decays with identical quark transitions should help in understanding the source of any observed *CPV* in charm decays, while evidence of indirect *CP* violation in D^0-$\overline{D}^0$ mixing would be a clear sign of NP. In τ decays *CPV* is expected to be negligible in the SM, and any

[1] At the time of the writing of this document, new results were presented by the LHCb Collaboration [8, 9] that do not confirm evidence for *CPV* in the difference of time-integrated asymmetries between $D^0 \to K^+K^-$ and $D^0 \to \pi^+\pi^-$ decays.

evidence for it would be an indication of NP. Throughout the following discussion the use of charge conjugate reactions is implied, unless otherwise indicated.

2. Search for *CP* violation in $D^+ \to K^0_S K^+$ and $D^+_s \to K^0_S K^+, K^0_S \pi^+$ decays

In D meson decays with a K^0_S in the final state, *CP*-violating asymmetries defined as

$$A_{CP} = \frac{\Gamma(D^+_{(s)} \to K^0_S h^+) - \Gamma(D^-_{(s)} \to K^0_S h^-)}{\Gamma(D^+_{(s)} \to K^0_S h^+) + \Gamma(D^-_{(s)} \to K^0_S h^-)} = A^{\Delta C}_{CP} + A^{K^0_S}_{CP}, \quad (1)$$

can receive contributions from *CPV* in $\Delta C = 1$ quark transitions ($A^{\Delta C}_{CP}$), and from *CPV* in K^0-$\overline{K}^0$ mixing ($A^{K^0_S}_{CP}$). In the SM, $A^{\Delta C}_{CP}$ is difficult to calculate precisely, however it is expected to be very small, at the level of 10^{-3} or less [3, 4]. The value for the contribution from K^0-$\overline{K}^0$ mixing is precisely determined to be $A^{K^0_S}_{CP} = [\pm 0.332 \pm 0.006]\%$ [13], where the $\pm$ sign depends on whether a K^0 (+) or a $\overline{K}^0$ (-) is produced in the decay. The SM prediction has to be corrected for the detector acceptance as a function of the decay time [14], and the correction is at the level of few percent at the B factories. A sizeble deviation of the measured A_{CP} value from the $A^{K^0_S}_{CP}$ predicted value would indicate *CPV* in the $\Delta C = 1$ quark transition, and might indicate new physics effects.

The *BABAR* experiment has recently measured *CP* asymmetries in the $D^+_{(s)} \to K^0_S K^+$ and $D^+_s \to K^0_S \pi^+$ decay modes [15] using a data sample corresponding to an integrated luminosity of 469 fb^{-1}. For this analysis a technique similar to that used in the search for *CPV* in $D^+ \to K^0_S \pi^+$ [16] was employed.

The reconstructed asymmetry A_{rec} is defined as

$$A_{rec} = \frac{N_{D^+_{(s)}} - N_{D^-_{(s)}}}{N_{D^+_{(s)}} + N_{D^-_{(s)}}} = A_{CP} + A_{FB} + A_\epsilon \quad (2)$$

where $N_{D^+_{(s)}}$ ($N_{D^-_{(s)}}$) is the number of $D^+_{(s)}$ ($D^-_{(s)}$) decays determined from the fit to the reconstructed invariant mass distribution of the $D^\pm_{(s)}$ mesons, A_{FB} is the forward-backward (FB) asymmetry, and A_ϵ is the detector-induced charge reconstruction asymmetry; A_{FB} originates from the FB asymmetry in $e^+e^- \to c\bar{c}$ production, coupled with the asymmetric acceptance of the detector, and is measured on data together with A_{CP}. The fits to the $m(K^0_S h)$ distributions yield $(159.4 \pm 0.8) \times 10^3$ signal events for $D^+ \to K^0_S K^+$, $(288.2 \pm 1.1) \times 10^3$ for $D^+_s \to K^0_S K^+$, and $(14.33 \pm 0.31) \times 10^3$ for $D^+_s \to K^0_S \pi^+$. Corrections for the detector-induced charge reconstruction asymmetry are estimated using a data control sample. The method exploits the fact that $\Upsilon(4S) \to B\overline{B}$ events provide a sample of evenly populated positive and negative tracks, free of any physics-induced asymmetries, allowing the determination of detector-related asymmetries in the reconstruction of charged-particle tracks.

The *CP*-violating asymmetries A_{CP} for the $D^+ \to K^0_S K^+$, $D^+_s \to K^0_S K^+$, and $D^+_s \to K^0_S \pi^+$ decays are determined to be $[0.13 \pm 0.36(\text{stat}) \pm 0.25(\text{syst})]\%$, $[-0.05 \pm 0.23(\text{stat}) \pm 0.24(\text{syst})]\%$, and $[0.6 \pm 2.0(\text{stat}) \pm 0.3(\text{syst})]\%$, respectively. The primary sources of systematic uncertainty are due to the statistical uncertainties in the determination of the charge asymmetry in track reconstruction efficiency.

The contribution to the *CP* asymmetries due to the $\Delta C = 1$ transition is measured to be $[0.46 \pm 0.36(\text{stat}) \pm 0.25(\text{syst})]\%$, $[0.28 \pm 0.23(\text{stat}) \pm 0.24(\text{syst})]\%$, and $[0.3 \pm 2.0(\text{stat}) \pm 0.3(\text{syst})]\%$ for the respective decay processes. The results are consistent with zero, and with the SM predictions within one standard deviation.

Journal of Physics: Conference Series **447** (2013) 012027 doi:10.1088/1742-6596/447/1/012027

3. Search for direct *CP* violation in $D^+ \to K^+K^-\pi^+$ decay

The *BABAR* experiment has recently searched for *CPV* in the singly Cabibbo-suppressed $D^+ \to K^+K^-\pi^+$ decay using a data sample corresponding to an integrated luminosity of 476 fb^{-1} [17]. The 3-body decay proceeds mainly through quasi-two-body decays with resonant intermediate states, which allows the investigation of the Dalitz plot substructure for asymmetry in both magnitude and phase for each intermediate state. In the search for *CPV*, 5 different approaches were adopted involving: a measurement of the integrated *CP* asymmetry, a measurement of the *CP* asymmetry in four regions of the Dalitz plot, a comparison of the binned D^+ and D^- Dalitz plots, a comparison of the Legendre-polynomial-moment weighted distributions in the K^+K^- and $K^-\pi^+$ systems, and a comparison of the results of a parameterized fit to the D^+ and D^- Dalitz plots. Only the last is model-dependent, while the previous four approaches are model-independent.

The center-of-mass (CM) momentum of the D^+ meson is required to be between 2.4 and 5.0 GeV/c. Particle identification is applied to two of the three tracks, requiring the presence of two oppositely charged kaon candidates. At low momentum, track reconstruction efficiency is reduced, and differences in nuclear cross sections can lead to asymmetries in the reconstruction of charged-particle tracks. Hence, the pion candidate is required to have $p_t > 300\,\mathrm{MeV}/c$, where p_t is the magnitude of the momentum projection onto the plane perpendicular to the e^+e^- collision axis. The signal yield is about 223,700 events, with signal purity of about 92%. In this case, the detector-induced charged-particle reconstruction asymmetry is estimated using a high purity control sample of $e^+e^- \to \tau^+\tau^-$ events for which $\tau^\pm \to \mu^\pm\nu_\mu\nu_\tau$ or $\tau^\pm \to h^\pm h^\mp h^\pm \nu_\tau$, where two of the three hadrons (h) are required to be consistent with pions from ρ^0 decays.

The *CP*-violating decay rate asymmetry, A_{CP}, was determined to be $(0.37 \pm 0.30(\mathrm{stat}) \pm 0.15(\mathrm{syst}))\%$. The *CP* asymmetries in different regions of the Dalitz plot, defined by the reconstructed invariant mass squared values $m^2(K^-K^+)$ and $m^2(K^-\pi^+)$, are reported in Table 1.

Dalitz plot region	A_{CP} (%)
Below $\bar{K}^*(892)^0$ (A)	$-0.7 \pm 1.6(\mathrm{stat}) \pm 1.7(\mathrm{syst})$
$\bar{K}^*(892)^0$ (B)	$-0.3 \pm 0.4(\mathrm{stat}) \pm 0.2(\mathrm{syst})$
$\phi(1020)$ (C)	$-0.3 \pm 0.3(\mathrm{stat}) \pm 0.5(\mathrm{syst})$
Above $\bar{K}^*(892)^0$ and $\phi(1020)$ (D)	$1.1 \pm 0.5(\mathrm{stat}) \pm 0.3(\mathrm{syst})$

Table 1. *CP* asymmetry in regions of the Dalitz plot shown in Fig. 1. The first error is statistical and the second is systematic.

Model-independent techniques were used to search for *CPV* in the Dalitz plot. These were based on a comparison of the binned D^+ and D^- Dalitz plots, and on a comparison of the Legendre-polynomial-moment weighted distributions in the K^+K^- and $K^-\pi^+$ systems. The distributions of normalized residuals in equally populated bins ($\sim$ 1000 events per bin) of the D^+ and D^- Dalitz plots were fitted with a Gaussian function. The fit yielded a mean of 0.08 ± 0.15 and an r.m.s. deviation of 1.11 ± 0.15, which corresponds to a probability of 72% that the two Dalitz plots are consistent with no *CPV*. The comparison of Legendre-polynomial-moments for the K^+K^- and $K^-\pi^+$ systems separately was found to be consistent with no *CPV* with a probability of 11% and 13%, respectively.

A model-dependent technique based on a comparison of parameterized fits to the two Dalitz plots was also used to search for *CPV*. The D^+ decay amplitude was parameterized as a coherent sum of amplitudes describing the relevant two-body intermediate states (16 resonances) plus a constant amplitude over the Dalitz plot for the non-resonant (NR) contribution. The fit fractions

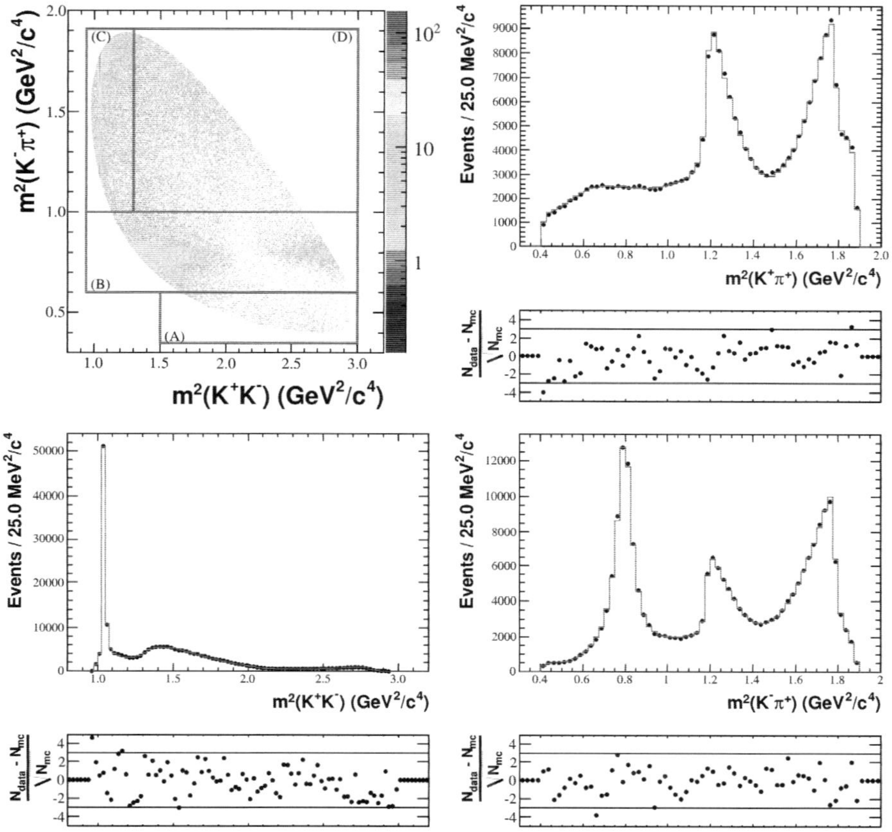

Figure 1. $D^+ \to K^+K^-\pi^+$ Dalitz plot fit projections assuming no *CPV*, with the regions used for model-independent comparisons also indicated as boxes. The data are represented by points with errors, the fit results by the histogram. The normalized residuals below each histogram, defined as $(N_{Data} - N_{MC})/\sqrt{N_{MC}}$, lie between $\pm 5\sigma$.

of the resonant and NR amplitudes are reported in Table 2. The results of the fit to the D^+ and D^- Dalitz plots do not show evidence of *CPV* for the following amplitudes: $\bar{K}^*(892)^0$, $\bar{K}^*(1430)^0$, $\phi(1020)$, NR, $\kappa(800)$, $a_0(1450)^0$, $f_0(980)$, $f_0(1370)$.

4. Measurement of D^0-$\overline{D}^0$ mixing, and search for indirect *CP* violation in $D^0 \to K^+K^-$ and $D^0 \to \pi^+\pi^-$ decays

The *BABAR* experiment has recently updated the measurement of the mixing parameter y_{CP} and the *CP*-violation parameter ΔY [18]. [2] The definitions of $\Delta Y$and A_Γ are the following:

$$\Delta Y = \frac{\Gamma^+ - \bar{\Gamma}^+}{2\Gamma} = (1 + y_{CP})A_\Gamma, \tag{3}$$

[2] Note that this definition for ΔYuses a different sign convention than that used in previous *BABAR* publications [19, 20].

Journal of Physics: Conference Series **447** (2013) 012027 doi:10.1088/1742-6596/447/1/012027

Resonance	Fraction (%)
$K^*(892)^0$	21.15 ± 0.20
$\phi(1020)$	28.42 ± 0.13
$\bar{K}_0^*(1430)^0$	25.32 ± 2.24
NR	6.38 ± 1.82
$\kappa(800)$	7.08 ± 0.63
$a_0(1450)^0$	3.84 ± 0.69
$f_0(980)$	2.47 ± 0.30
$f_0(1370)$	1.17 ± 0.21
$\phi(1680)$	0.82 ± 0.12
$\bar{K}_1^*(1410)$	0.47 ± 0.37
$f_0(1500)$	0.36 ± 0.08
$a_2(1320)$	0.16 ± 0.03
$f_2(1270)$	0.13 ± 0.03
$\bar{K}_2^*(1430)$	0.06 ± 0.02
$\bar{K}^*(1680)$	0.05 ± 0.16
$f_0(1710)$	0.04 ± 0.03
$f_2'(1525)$	0.02 ± 0.01
Sum	97.92 ± 3.09

Table 2. Fit fractions of the resonant and NR amplitudes in the Isobar model fit of the data. The uncertainties are statistical only.

$$A_\Gamma = \frac{\bar{\tau}^+ - \tau^+}{\bar{\tau}^+ + \tau^+}, \tag{4}$$

where $\tau^+ = 1/\Gamma^+$ ($\bar{\tau}^+ = 1/\bar{\Gamma}^+$) are the effective lifetimes for D^0 ($\overline{D}^0$) decaying to the *CP*-even final states K^+K^- and $\pi^+\pi^-$. In this analysis *CP* conservation in the decay is assumed, and results are averaged over the K^+K^- and $\pi^+\pi^-$ modes.

The measurements are based on the ratio of lifetimes extracted simultaneously from a sample of D^0 mesons produced through the flavor-tagged process $D^{*+} \to D^0\pi^+$, where the D^0 decays to $K^-\pi^+$, K^-K^+, or $\pi^-\pi^+$; additional samples of untagged decays $D^0 \to K^-\pi^+$ and $D^0 \to K^-K^+$ are used for the measurement of y_{CP}. These have about 4 times the statistics of the corresponding flavor-tagged samples, but have lower purity.

The main selection criteria require that the center-of-mass momentum of the D^0 (p^*) be greater than 2.5 GeV/c, that D^0 daughter tracks be identified as kaons and pions, and require for the flavor-tagged sample that the reconstructed $\Delta m = m(D^{*+}) - m(D^0)$ be close to the value of 0.1455 GeV. The selection criterion on p^* is used to reject D mesons from B decays, and to improve the signal significance.

The proper time t and proper-time error σ_t are obtained from the reconstruction of the 3-dimensional flight length ($\vec{L}$) and the momentum of the D^0 ($\vec{p}$) according to the relation $t = m/|\vec{p}|^2\vec{L} \cdot \vec{p}$, where m is the nominal mass of the D^0. The flight length is reconstructed by means of a simultaneous kinematic fit to the decay vertex and production vertex of the D^0, the latter being constrained to originate within the e^+e^- collision region. The most probable σ_t value is about 40% of the nominal D^0 lifetime, and candidates with $\sigma_t < 0.5$ ps are retained for the fit.

The *BABAR* experiment measures $y_{CP} = [0.72 \pm 0.18(\text{stat}) \pm 0.12(\text{syst})]\%$ and $\Delta Y = [0.09 \pm 0.26(\text{stat}) \pm 0.06(\text{syst})]\%$ using a data sample corresponding to an integrated luminosity of 468 fb^{-1} [18]. The systematic uncertainties on y_{CP} and ΔY are reported in Table 3. The

Category	Δy_{CP} (%)	$\Delta(\Delta Y)$ (%)
Fit region	0.057	0.022
Signal model	0.022	0.000
Charm bkg	0.045	0.001
Combinatorial bkg	0.079	0.002
Selection	0.059	0.054
Total	0.124	0.058

Table 3. Systematic uncertainties for y_{CP} and ΔY at *BABAR*. The total is the sum-in-quadrature of the entries in each column.

y_{CP} measurement provides evidence for D^0-$\overline{D}^0$ mixing at the 3.3σ level, and is the most precise measurement to date. The Belle experiment has recently presented an update of the corresponding analysis, and reports $y_{CP} = [1.11 \pm 0.22(\text{stat}) \pm 0.11(\text{syst})]\%$ and $A_\Gamma = [-0.03 \pm 0.20(\text{stat}) \pm 0.08(\text{syst})]\%$ using a data sample of 976 fb^{-1} [21]. The Belle measurement of y_{CP} corresponds to evidence for D^0-$\overline{D}^0$ mixing at the 4.5σ level. The measurements of ΔY from *BABAR*, and A_Γ from Belle are compatible within error and consistent with no *CPV*. The new measurements supersede the previous *BABAR* [19, 20] and Belle results [22]. The updated HFAG averages, including the new results, are $y_{CP} = (0.866 \pm 0.155)\%$ and $A_\Gamma = (-0.022 \pm 0.161)\%$. The comparison with the previous HFAG average values $y_{CP} = (1.064 \pm 0.209)\%$ and $A_\Gamma = (0.026 \pm 0.231)\%$ indicates significant improvement in precision and a lower central value for y_{CP}.

5. Search for *CP* violation in $\tau^- \to \pi^- K^0_S (\geq 0\ \pi^0)\nu_\tau$ decays

Bigi and Sanda [23] predicted that, in the SM, the decay of the τ lepton to final states containing a K^0_S will exibit a nonzero decay rate asymmetry due to *CP* violation in K^0-$\overline{K}^0$ mixing. The decay rate asymmetry defined as

$$A_Q = \frac{\Gamma(\tau^+ \to \pi^+ K^0_S \overline{\nu}_\tau) - \Gamma(\tau^- \to \pi^- K^0_S \nu_\tau)}{\Gamma(\tau^+ \to \pi^+ K^0_S \overline{\nu}_\tau) + \Gamma(\tau^- \to \pi^- K^0_S \nu_\tau)} \tag{5}$$

was predicted to be $(0.33 \pm 0.01)\%$, and a significant deviation from this value would be evidence of NP. As pointed out by Grossman and Nir [14], the SM prediction for A_Q has to be corrected for a factor due to the K^0_S-K^0_L interference, depending on the reconstruction efficiency of the K^0_S as a function of the K^0_S decay time. The correction factor was experimentally determined to be (1.08 ± 0.01) so that the corrected value for A_Q is $(0.36 \pm 0.01)\%$. The signal events include different τ decay modes with a K^0_S in the final state, *i.e.* $\tau^- \to \pi^- K^0_S(\geq 0\ \pi^0)\nu_\tau$, where the number of reconstructed π^0's is less than 3, and the value of the decay rate asymmetry is expected to be the same in the SM for the different τ decay modes. This analysis is based on a data sample corresponding to an integrated luminosity of 476 fb^{-1} [24]. The events were reconstructed from $e^+e^- \to \tau^+\tau^-$ reactions, where the τ pairs are produced back-to-back in the e^+e^- CM system. Each event is divided into two hemispheres with respect to the event thrust axis [25, 26]: the "signal" hemisphere, where one charged prompt track and a $K^0_S \to \pi^-\pi^-$ candidate are reconstructed, and the "tag" hemisphere, with an identified prompt lepton (e, μ). After all selection criteria have been applied, 199,064 events in the e-tag sample, and 141,602 events in the μ-tag sample are obtained. The sample composition is described in Table 4. It contains events from two other classes of decay mode: $\tau^- \to K^- K^0_S(\geq 0\ \pi^0)\nu_\tau$ and $\tau^- \to \pi^- K^0 \overline{K}^0 \nu_\tau$, that also have K^0_S mesons in the final states. The $\tau^- \to \pi^- K^0 \overline{K}^0 \nu_\tau$ events satisfy the selection criteria if one of the neutral kaons decays into $\pi^+\pi^-$ and the other decays into $2\pi^0$ or appears as a K^0_L meson. A control sample of $\tau^- \to h^- h^- h^+(\geq 0\ \pi^0)\nu_\tau$ ($h = K, \pi$),

Journal of Physics: Conference Series **447** (2013) 012027 doi:10.1088/1742-6596/447/1/012027

Source	Fractions (%)	
	e-tag	μ-tag
$\tau^- \to \pi^- K^0_S(\geq 0\ \pi^0)\nu_\tau$	78.8 ± 4.0	18.4 ± 4.0
$\tau^- \to K^- K^0_S(\geq 0\ \pi^0)\nu_\tau$	4.2 ± 0.3	4.1 ± 0.3
$\tau^- \to \pi^- K^0 \overline{K}^0 \nu_\tau$	15.7 ± 3.7	15.9 ± 3.7
Other background	1.40 ± 0.06	1.55 ± 0.07

Table 4. Composition of the τ sample after all selection criteria have been applied.

with no K^0_S, is used to verify that no significant asymmetry is induced by the detector or the selection criteria. The decay rate asymmetries measured in the control sample agree with zero within the errors, 0.12% for the e tag and 0.08% for the μ tag, which are taken as systematic uncertainties on the signal asymmetries.

The measured asymmetry $\mathcal{A}$ is related to the signal asymmetry A_1, and the background asymmetries A_2 and A_3, by

$$\mathcal{A} = \frac{f_1 A_1 + f_2 A_2 + f_3 A_3}{f_1 + f_2 + f_3} = \frac{f_1 - f_2}{f_1 + f_2 + f_3} A_Q \qquad (6)$$

where f_1, f_2 and f_3 are, respectively, the fractions of $\tau^- \to \pi^- K^0_S(\geq 0\ \pi^0)\nu_\tau$, $\tau^- \to K^- K^0_S(\geq 0\ \pi^0)\nu_\tau$ and $\tau^- \to \pi^- K^0 \overline{K}^0 \nu_\tau$ in the total selected sample, composed as shown in Table 4. In the SM, $A_1 = -A_2$ since in the $\tau^- \to \pi^- K^0_S(\geq 0\ \pi^0)\nu_\tau$ decays a $\overline{K}^0$ is produced, while in the $\tau^- \to K^- K^0_S(\geq 0\ \pi^0)\nu_\tau$ decays is produced a K^0. For the $\tau^- \to \pi^- K^0 \overline{K}^0 \nu_\tau$ decay, $A_3 = 0$ since the asymmetries of K^0 and $\overline{K}^0$ cancel each other. The measured asymmetry is $\mathcal{A} = (-0.27 \pm 0.18 \pm 0.08)\%$, and dividing it by $(f_1 - f_2)/(f_1 + f_2 + f_3) = (0.75 \pm 0.04)\%$, we obtain the decay rate asymmetry $A_Q = (-0.36 \pm 0.23 \pm 0.11)\%$, where the errors are statistical and systematic, respectively. The most relevant systematic errors are reported in Table 5. The

Source	e-tag (%)	μ-tag (%)
Detector and selection bias	0.12	0.08
Background subtraction	0.05	0.06
$K^0/\overline{K}^0$ interaction	0.01	0.01
Total	0.13	0.10

Table 5. Summary of the systematic uncertainties in the decay-rate asymmetry A_Q, separately for the e-tag and μ-tag samples.

decay-rate asymmetry $A_Q = [-0.36 \pm 0.23(\text{stat}) \pm 0.11(\text{syst})]\%$, has been measured for the first time, and is 2.8 standard deviations from the SM prediction of $(0.36 \pm 0.01)\%$ [24].

6. Conclusions

We have presented recent *BABAR* results relevant to searches for *CPV* in charm meson and tau decays, and also a measurement of the D^0-$\overline{D}^0$ mixing parameter y_{CP} and the *CPV* violation parameter ΔY. The *CPV* asymmetries for the $D^+ \to K^0_S K^+$, $D^+_s \to K^0_S K^+$, and $D^+_s \to K^0_S \pi^+$ decays are determined to be $[0.13 \pm 0.36(\text{stat}) \pm 0.25(\text{syst})]\%$, $[-0.05 \pm 0.23(\text{stat}) \pm 0.24(\text{syst})]\%$, and $[0.6 \pm 2.0(\text{stat}) \pm 0.3(\text{syst})]\%$, respectively. These results are consistent with no *CPV* and with the SM predictions within one standard deviation. The *CPV* asymmetry in the SCS $D^+ \to K^+ K^- \pi^+$ decay was measured to be $[0.37 \pm 0.30(\text{stat}) \pm 0.15(\text{syst})]\%$, and no evidence for *CPV* was found in the Dalitz plot using model-independent and model-dependent techniques. In a lifetime ratio analysis of the decays to *CP*-even eigenstates $D^0 \to K^-K^+, \pi^-\pi^+$ relative to $D^0 \to K^+\pi^+$ decays, the mixing parameter was measured to be $y_{CP} = [0.72 \pm 0.18(\text{stat}) \pm 0.12(\text{syst})]\%$ and the *CPV* parameter value $\Delta Y = [0.09 \pm 0.26(\text{stat}) \pm 0.06(\text{syst})]\%$ was obtained. The former represents the most precise measurement of the mixing parameter y_{CP} to date. Finally, we presented the first measurement of the decay-rate asymmetry in $\tau^- \to \pi^- K_S^0(\geq 0\ \pi^0)\nu_\tau$ decays, obtaining the value $A_Q = [-0.36 \pm 0.23(\text{stat}) \pm 0.11(\text{syst})]\%$. This measurement is 2.8 standard deviations from the SM prediction of $(0.36 \pm 0.01)\%$.

Acknowledgments
I would like to thank the conference organizers for the invitation to such a wonderful place, and my *BABAR* colleagues for providing excellent results and very useful feedback in preparing my contribution to the conference.

References
[1] N. Cabibbo, Phys. Rev. Lett. **10**, 531 (1963).
[2] M. Kobayashi and T. Maskawa, Prog. Theor. Phys. **49**, 652 (1973).
[3] S. Bianco, F. L. Fabbri, D. Benson and I. Bigi, Riv. Nuovo Cim. **26N7**, 1 (2003).
[4] Y. Grossman, A. L. Kagan and Y. Nir, Phys. Rev. D **75**, 036008 (2007).
[5] R. Aaij *et al.* (LHCb Collaboration), Phys. Rev. Lett. **108**, 111602 (2012).
[6] T. Aaltonen *et al.* (CDF Collaboration), Phys. Rev. D **85**, 012009 (2012).
[7] T. Aaltonen *et al.* (CDF Collaboration), Phys. Rev. Lett. **109**, 111801 (2012).
[8] R. Aaij *et al.* (LHCb Collaboration), arXiv:1303.2614.
[9] R. Aaij *et al.* (LHCb Collaboration), submitted to "48th Rencontres de Moriond on QCD and High Energy Interactions", La Thuile, Italy, 9-16 March, 2013, CERN-LHCb-CONF-2013-003.
[10] Y. Grossman, A. L. Kagan and J. Zupan, Phys. Rev. D **85**, 114036 (2012).
[11] E. Franco, S. Mishima and L. Silvestrini, JHEP **1205**, 140 (2012).
[12] G. Isidori, J. F. Kamenik, Z. Ligeti and G. Perez, Phys. Lett. B **711**, 46 (2012).
[13] J. Beringer *et al.* (Particle Data Group), Phys. Rev. D **86**, 010001 (2012).
[14] Y. Grossman and Y. Nir, JHEP **1204**, 002 (2012).
[15] J. P. Lees *et al.* (*BABAR* Collaboration), arXiv:1212.3003.
[16] P. del Amo Sanchez *et al.* (*BABAR* Collaboration), Phys. Rev. D **83**, 071103 (2011).
[17] J. P. Lees *et al.* (*BABAR* Collaboration), Phys. Rev. D **87**, 052010 (2013).
[18] J. P. Lees *et al.* (*BABAR* Collaboration), Phys. Rev. D **87**, 012004 (2013).
[19] B. Aubert *et al.* (*BABAR* Collaboration), Phys. Rev. D **78**, 011105 (2008).
[20] B. Aubert *et al.* (*BABAR* Collaboration), Phys. Rev. D **80**, 071103 (2009).
[21] M. Staric (Belle Collaboration), Contribution to the 5th International Workshop on Charm Physics (Charm 2012), Honolulu, Hawai'i, USA, 14-17 May, 2012.
[22] M. Staric *et al.* (Belle Collaboration), Phys. Rev. Lett. **98**, 211803 (2007).
[23] I. I. Bigi and A. I. Sanda, Phys. Lett. B **625**, 47 (2005).
[24] J. P. Lees *et al.* (*BABAR* Collaboration), Phys. Rev. D **85**, 031102 (2012). [Erratum-ibid. D **85**, 099904 (2012).]
[25] S. Brandt, C. Peyrou, R. Sosnowski and A. Wroblewski, Phys. Lett. **12**, 57 (1964).
[26] E. Farhi, Phys. Rev. Lett. **39**, 1587 (1977).

Search for the CP violating $K_S \to 3\pi^0$ decay with the KLOE detector

M. Silarski on behalf of the KLOE–2 Collaboration[1]

Institute of Physics, Jagiellonian University, PL-30-059 Cracow, Poland

E-mail: `Michal.Silarski@lnf.infn.it`

Abstract. We present a new search for the $K_S \to 3\pi^0$ decay performed with the KLOE detector operating at the DAΦNE ϕ–factory. The K_S mesons were tagged via registration of K_L mesons which crossed the drift chamber without decaying and interacted with the KLOE electromagnetic calorimeter. The $K_S \to 3\pi^0$ decay was then searched requiring six prompt photons. To suppress background, originating from fake K_S tags and $K_S \to 2\pi^0$ decays with additional two spurious clusters, we have performed a discriminant analysis based on kinematical fit, testing of the signal and background hypotheses and exploiting of the differences in kinematics of the K_S decays into $2\pi^0$ and $3\pi^0$. In a sample of about $1.7 \cdot 10^9$ $\phi \to K_S K_L$ events we have found no candidates in data and simulated background samples. Normalizing to the number of $K_S \to 2\pi^0$ events in the same sample, we have set the upper limit on the $K_S \to 3\pi^0$ branching ratio $BR(K_S \to 3\pi^0) < 2.6 \cdot 10^{-8}$ at 90% C.L., five times lower than the previous limit. This upper limit can be translated into a limit on the modulus of the η_{000} parameter amounting to $|\eta_{000}| < 0.0088$ at 90% C.L., improving by a factor two the latest direct measurement.

1. Introduction

Since the first discovery of the $\mathcal{CP}$–violating neutral kaon decay in 1964, there has been made a big effort to describe the $\mathcal{CP}$ symmetry breaking within the Standard Model. The favoured theoretical framework was provided in 1973 by Kobayashi and Maskawa, who pointed out that $\mathcal{CP}$ violation would follow automatically if there were at least six quark flavours. At present the main experimental effort is focused on the neutral B and D meson system studies. However, there are still several interesting open issuses in the kaon physics, which can contribute to our better understanding of the $\mathcal{CP}$ violation mechanism [1]. For example, K_S decays to $|\pi^+\pi^-\pi^0\rangle$ and $|3\pi^0\rangle$, requiring $\mathcal{CP}$ violation, are still poorly known. The $|\pi^+\pi^-\pi^0\rangle$ final state can be produced in neutral kaon decays with isospin $I = 0, 1, 2,$ or 3. The $I = 0$ and $I = 2$ states have $\mathcal{CP} = 1$, and K_S can decay into them without violation of the $\mathcal{CP}$ symmetry. However,

[1] The KLOE–2 Collaboration: D. Babusci, D. Badoni, I. Balwierz-Pytko, G. Bencivenni, C. Bini, C. Bloise, F. Bossi, P. Branchini, A. Budano, L. Caldeira Balkeståhl, G. Capon, F. Ceradini, P. Ciambrone, F. Curciarello, E. Czerwiński, E. Danè, V. De Leo, E. De Lucia, G. De Robertis, A. De Santis, A. Di Domenico, C. Di Donato, R. Di Salvo, D. Domenici, O. Erriquez, G. Fanizzi, A. Fantini, G. Felici, S. Fiore, P. Franzini, P. Gauzzi, G. Giardina, S. Giovannella, F. Gonnella, E. Graziani, F. Happacher, L. Heijkenskjöld, B. Höistad, L. Iafolla, M. Jacewicz, T. Johansson, K. Kacprzak, A. Kupsc, J. Lee-Franzini, B. Leverington, F. Loddo, S. Loffredo, G. Mandaglio, M. Martemianov, M. Martini, M. Mascolo, R. Messi, S. Miscetti, G. Morello, D. Moricciani, P. Moskal, F. Nguyen, A. Passeri, V. Patera, I. Prado Longhi, A. Ranieri, C. F. Redmer, P. Santangelo, I. Sarra, M. Schioppa, B. Sciascia, M. Silarski, C. Taccini, L. Tortora, G. Venanzoni, W. Wiślicki, M. Wolke, J. Zdebik

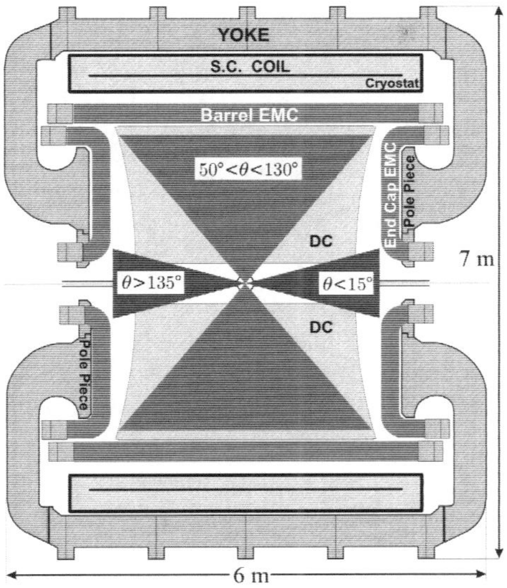

Figure 1. Schematic view of the KLOE detector.

they are expected to be strongly suppressed by centrifugal barrier effects [2]. For the $I = 1$ and $I = 3$ states there is no centrifugal barrier and $\mathcal{CP} = -1$, so K_S decay requires violation of this symmetry. Anyhow the two kinds of final states can be separated by the analysis of the $\pi^+\pi^-\pi^0$ Dalitz plot. This allows for determination of the K_S to K_L decay amplitude ratio $\eta_{+-0} = \frac{A(K_S\to\pi^+\pi^-\pi^0)}{A(K_L\to\pi^+\pi^-\pi^0)} \cong \epsilon + \epsilon'_{+-0}$, where ϵ indicates the K_S $\mathcal{CP}$ impurity and ϵ'_{+-0} is the contribution of the direct $\mathcal{CP}$–violating term. In the case of $|\pi^0\pi^0\pi^0\rangle$ final state, only isospin $I = 1$ or $I = 3$ is allowed, for which $\mathcal{CP} = -1$. Therefore, the $K_S \to 3\pi^0$ decay is a purely $\mathcal{CP}$ violating process, for which one defines the analogous ratio: $\eta_{000} = \frac{A(K_S\to\pi^0\pi^0\pi^0)}{A(K_L\to\pi^0\pi^0\pi^0)} \cong \epsilon + \epsilon'_{000}$.
The present knowledge about η_{+-0} and η_{000} is poor mainly due to very low decay rates for the $K_S \to 3\pi$ decays. The current value of the $K_S \to \pi^+\pi^-\pi^0$ branching ratio amounts to $BR(K_S \to \pi^+\pi^-\pi^0) = (\ 3.5^{+1.1}_{-0.9}\ ) \cdot 10^{-7}$ [2], and the $K_S \to 3\pi^0$ has been never observed. The best upper limit on this decay branching ratio amounts to $BR(K_S \to 3\pi^0) < 1.2 \cdot 10^{-7}$ [3], while the prediction based on Standard Model is equal to about $2 \cdot 10^{-9}$ [4].
In this article we briefly describe the search of the $K_S \to 3\pi^0$ decay based on the 1.7 fb^{-1} of integrated luminosity gathered with the KLOE detector operating at the ϕ–factory DAΦNE of the Frascati Laboratory [5].

2. The KLOE detector

The KLOE experiment operated from 2000 to 2006 at DAΦNE, the e^+e^- collider optimized to work with a center of mass energy around the ϕ meson mass peak: $\sqrt{s} = 1019.45$ MeV [6]. Equal energy positron and electron beams collided at an angle of $\pi - 25$ mrad producing ϕ mesons nearly at rest, which then decayed mainly to K^+K^- (49%) and K_SK_L (34%) final states. The decay products were registered using the KLOE detection setup presented schematically in Fig. 1. The detector consists of a large cylindrical Drift Chamber [7] surrounded by a lead scintillating fiber Electromagnetic Calorimeter [8], both immersed in an axial 0.52 T magnetic field produced by a superconducting solenoid. The beam pipe at the interaction region is a sphere with 10 cm of radius made of a Beryllium–Aluminum alloy of 0.5 mm thickness. The drift chamber, 4 m in diameter and 3.3 m long, has 12582 all stereo drift cells with tungsten sense wires. To minimize the K_L regeneration, multiple Coulomb scattering and photon absorption the chamber is constructed out of carbon fiber composite with low–Z and low density, and uses the gas mixture of helium (90%) and isobutane (10%). It provides three–dimensional tracking with resolution in the bending plane about 200 μm, resolution on the z–coordinate measurement of about 2 mm and of 1 mm on the decay vertex position. Momentum of a particle is determined from the curvature of its trajectory in the magnetic field with a fractional accuracy $\sigma_p/p = 0.4\%$ for polar angles larger than 45° [6]. The KLOE electromagnetic calorimeter covers 98% of the solid angle and is composed by a barrel and two endcaps, for a total of 88 modules. Each module is built out of 1 mm scintillating fibers grouped in cells of 4.4x4.4 cm^2 and embedded in 0.5 mm lead foils, and it is read out from both sides by set of photomultipliers. The particle energy deposits are obtained from the photomultipliers signal amplitude, while the arrival times and particles impact points are determined from the spatial coordinates of the fired calorimeter cell and the difference between times measured at both ends of the calorimeter module. The accuracies of energy and time determination are parametrized as $\sigma_E/E = 5.7\%/\sqrt{E\ (\mathrm{GeV})}$ and $\sigma_t = 57\ \mathrm{ps}/\sqrt{E\ (\mathrm{GeV})} \oplus 100$ ps, respectively. The trigger [9] uses both calorimeter and chamber information. In the search for the $K_S \to 3\pi^0$ we have used, however, only the calorimeter information requiring two energy deposits with $E > 50$ MeV for the barrel and $E > 150$ MeV for the endcaps.

3. Data analysis

At KLOE kaons arising from the ϕ decay move at low speed with their relative angle close to 180°. Therefore, observation of a K_L (K_S) decay ensures the presence of a K_S (K_L) meson travelling in the opposite direction. About 50% of K_L's reach the calorimeter before decaying, which provides a very clean K_S tag by the K_L interaction in the calorimeter (K_L–crash). It is identified by a cluster with polar angle $40° < \theta_{cr} < 140°$, not associated to any track, with energy $E_{cr} > 100$ MeV and with a time corresponding to a K_L velocity $\beta^* \sim 0.2$ in the ϕ rest frame. The average value of the e^+e^- center of mass energy $\sqrt{s}$ is obtained with a precision of 20 keV for each 200 nb^{-1} running period using large angle Bhabha scattering events [6]. The value of $\sqrt{s}$ and the K_L–crash cluster position allows us to obtain, for each event, the direction of the K_S with an angular resolution of 1° and a momentum resolution of about 2 MeV. The search for the $K_S \to 3\pi^0 \to 6\gamma$ decay is then carried out by the selection of events with six photons which momenta are reconstructed using time and energy measured by the electromagnetic calorimeter. Photons from the K_S decay are reconstructed as neutral particles that travel with a velocity $\beta = 1$ from the interaction point to the KLOE calorimeter (prompt photons). In order to retain a large control sample for the background while preserving high efficiency for the signal, we keep all photons satisfying $E_\gamma > 7$ MeV and $|\cos\theta| < 0.915$. Each cluster is also required to satisfy the condition $|t_\gamma - R_\gamma/c| < \min(3.5\sigma_t, 2\ \mathrm{ns})$, where t_γ is the photon flight time and R the path length. The photon detection efficiency of the calorimeter amounts to about 90% for $E_\gamma = 20$ MeV, and reaches 100% above 70 MeV. To determine the $K_S \to 3\pi^0$ branching ratio we have counted also the $K_S \to 2\pi^0$ events which are selected requiring 4 prompt photons, and have served as normalization sample. For both channels the expected background as well as the detector acceptance and the analysis efficiency are estimated using the Monte Carlo simulation tool of the experiment [10]. All the processes contributing to the background were simulated with statistics twice larger than the data sample. Moreover, for the acceptance and the analysis efficiency evaluation a dedicated $K_S \to 3\pi^0$ signal simulation was performed. After K_S tagging we have counted 76689 events with six prompt photons. For these events a further discriminant analysis was performed to increase the signal to background ratio.

The first analysis step aimed to reject fake K_S tags. The distributions of reconstructed K_L energy E_{cr} and velocity β^* for the selected data sample and background simulations are shown in Fig. 2. In the β^* distribution, the peak around 0.215 corresponds to genuine K_L interactions

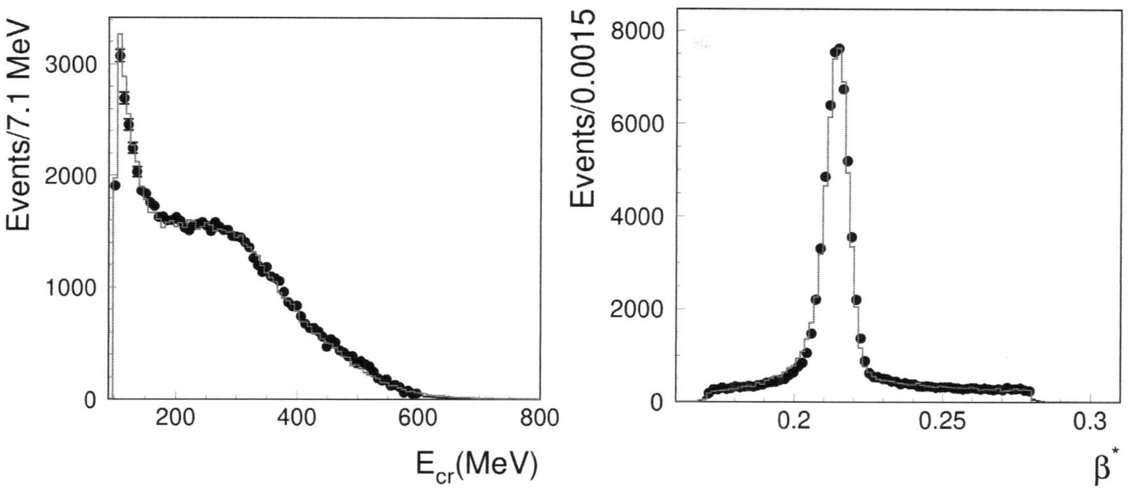

Figure 2. Distributions of the reconstructed K_L energy (E_{cr}) and velocity in the ϕ center of mass frame (β^*) for all events in the six photon sample. Black points represent data, while the MC background simulation is shown as red histogram.

in the calorimeter, while the flat distribution mainly originates from $\phi \to K_S K_L \to (K_S \to \pi^+\pi^-, K_L \to 3\pi^0)$ background events. In this case one of the low momentum charged pions spirals in the forward direction and interacts in the low–β insertion quadrupoles. This interaction produces neutral particles which simulate the signal of K_L interaction in the calorimeter (fake K_L–crash), while the K_L meson decays close enough to the interaction point to produce six prompt photons [5]. To suppress this kind of background we first reject events with charged particles coming from the vicinity of the interaction region. Next, taking advantage of the differences in the β^* and E_{cr} distributions between the tagged K_S events and the fake K_L–crash, we have tightened cuts on these variables: $E_{cr} > 150$ MeV and $0.20 < \beta^* < 0.225$. This improves by a factor 12 the rejection of this background with respect to the previous analysis [3]. The second source of background originates from wrongly reconstructed $K_S \to 2\pi^0$ decays. The four photons from this decay can be reconstructed as six due to fragmentation of the electromagnetic showers (splitting). These events are characterized by one or two low–energy clusters reconstructed very close to the position of the genuine photon interaction in the calorimeter. Additional clusters can come from the accidental time coincidence between the ϕ decay and machine background photons from DAΦNE [5]. In the next stage of the analysis we select only kinematically well defined events. To this end we perform the kinematical fit procedure based on the least squares method imposing energy and momentum conservation, the kaon mass and the velocity of the six photons in the final state. The χ^2 distribution of the fit for data and background simulation, χ^2_{fit}, is shown in Fig. 3 together with the expected distribution for signal events. Cutting on χ^2_{fit} considerably reduces the background from bad quality reconstructed events while keeping large signal efficiency [5]. In order to reject events with split and accidental clusters we look at the correlation between two χ^2–like discriminating variables $\zeta_{2\pi}$ and $\zeta_{3\pi}$. $\zeta_{2\pi}$ is calculated by an algorithm selecting four out of six clusters best satisfying the kinematic constraints of the two–body decay, therefore it verifies the $K_S \to 2\pi^0 \to 4\gamma$ hypothesis. The pairing of clusters is based on the requirement $m_{\gamma\gamma} = m_{\pi^0}$, and on the opening angle of the reconstructed pions trajectories in the K_S center of mass frame. Moreover, we check the consistency of the energy and momentum conservation in the $\phi \to K_S K_L, K_S \to 2\pi^0$ decay

Journal of Physics: Conference Series **447** (2013) 012028 doi:10.1088/1742-6596/447/1/012028

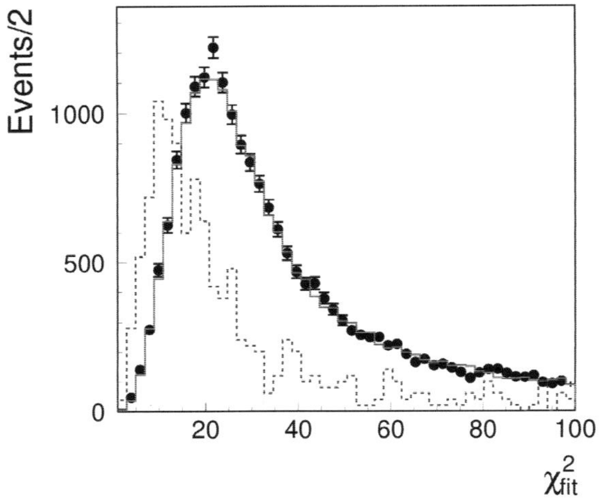

Figure 3. Distribution of χ^2_{fit} for the tagged six–photon sample for data (black points), background simulation (solid histogram), and simulated $K_S \to 3\pi^0$ signal (dashed histogram).

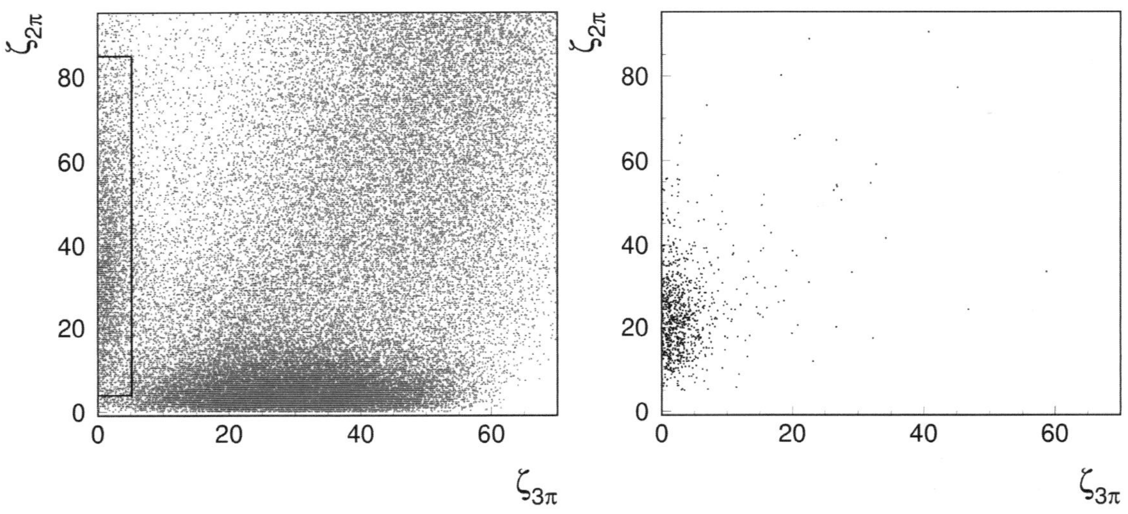

Figure 4. Distributions of events in the $\zeta_{3\pi}$-$\zeta_{2\pi}$ plane, for six–photon sample tagged by K_L–crash for data (left), and for the simulated $K_S \to 3\pi^0$ decays (right). The solid lines show the signal region definition chosen using the optimization procedure described in the text.

hypothesis. The $\chi^2_{3\pi}$ instead verifies the signal hypothesis by looking at the reconstructed masses of three pions. For every choice of cluster pairs we calculate the quadratic sum of residuals between the nominal π^0 mass and the invariant masses of three photon pairs. As the best combination of cluster pairs, we take the configuration minimizing $\zeta_{3\pi}$. The distributions in the $\zeta_{3\pi}$–$\zeta_{2\pi}$ plane for the data and $K_S \to 3\pi^0$ simulated signal are shown in Fig. 4. Signal events are characterized by small values of $\zeta_{3\pi}$ and relatively high $\zeta_{2\pi}$ [5], which indicates the choice of the signal region definition (signal box). In order to improve the quality of the photon selection using $\chi^2_{2\pi}$, we cut on the variable $\Delta E = (m_\Phi/2 - \sum E_{\gamma_i})/\sigma_E$ where γ_i stands for the i–*th* photon from four chosen in the $\chi^2_{2\pi}$ estimator and σ_E is the appropriate resolution. For $K_S \to 2\pi^0$ decays plus two background clusters, we expect $\Delta E \sim 0$, while for $K_S \to 3\pi^0$ $\Delta E \sim m_{\pi^0}/\sigma_E$. To further reject surviving $K_S \to 2\pi^0$ events with split clusters, we cut on the minimal distance between centroids of reconstructed clusters, R_{min}, considering that the distance between split clusters is

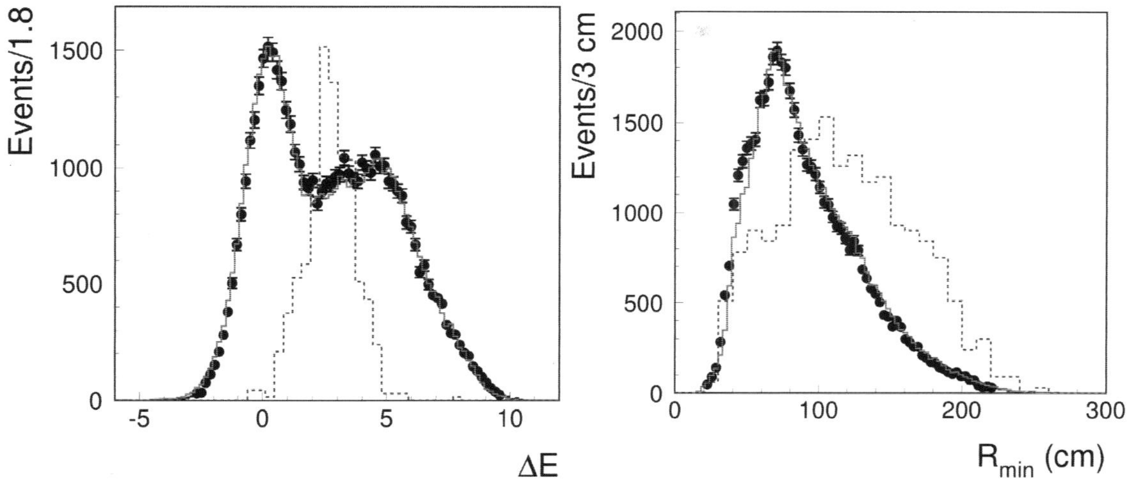

Figure 5. Distributions of ΔE and R_{min} discriminating variables for six–photon events for data (black points) and background simulations (red histograms). The dashed histograms represents simulated $K_S \to 3\pi^0$ events.

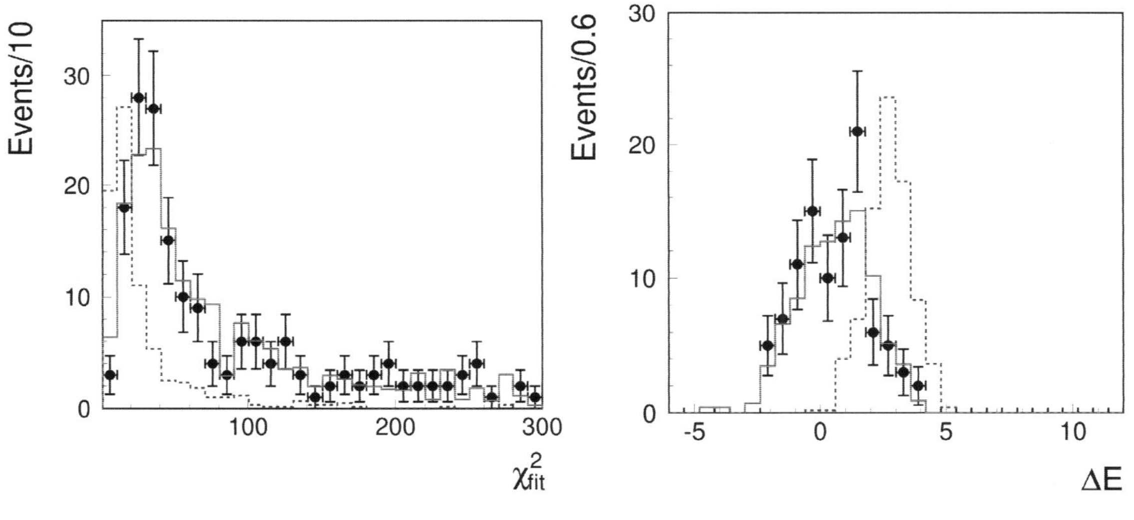

Figure 6. Distributions of χ^2_{fit} for six–photon events in the signal box (left) and ΔE for six–photon events in the signal box applying the $\chi^2_{fit} < 57.2$ cut (right). Black points are data, background simulation is the red histogram. The dashed histogram represents simulated $K_S \to 3\pi^0$ events.

on average smaller than the distance between clusters originating from γ's of $K_S \to 3\pi^0$ decay. Distributions of these two discriminant variables are presented in Fig. 5. Before counting of the $K_S \to 3\pi^0$ candidate events, we have optimized the event selection in order to reduce the background as strongly as possible while keeping high signal efficiency. Cuts on the discriminant variables have been refined minimizing $f_{cut}(\chi^2_{fit}, \zeta_{2\pi}, \zeta_{3\pi}, \Delta E, R_{min}) = N_{up}/\epsilon_{3\pi}$, where $\epsilon_{3\pi}$ stands for the signal selection efficiency and N_{up} is the mean upper limit (at 90% C.L.) on the expected

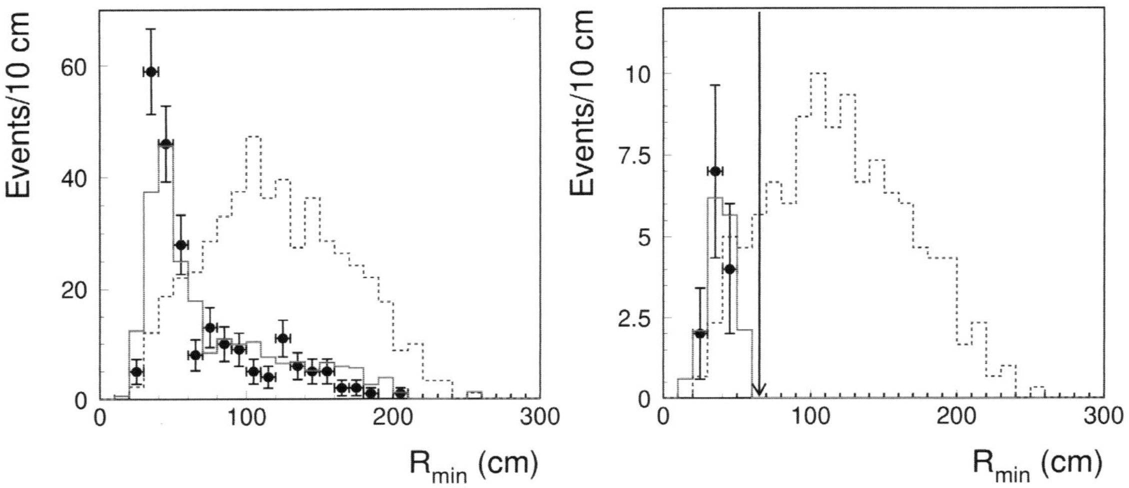

Figure 7. Distributions of R_{min} for six–photon events in the signal box applying the $\chi^2_{fit} < 57.2$ cut (left), and applying $\chi^2_{fit} < 57.2$ and $\Delta E > 1.88$ cuts (right). Black points are data, background simulation is the red histogram. The dashed histogram represents simulated $K_S \to 3\pi^0$ events.

number of signal events. It is calculated on the basis of the corresponding expected number of background events $B_{exp} = B_{exp}(\chi^2_{fit}, \zeta_{2\pi}, \zeta_{3\pi}, \Delta E, R_{min})$ from simulation [1]. As the result of the optimization we have obtained the following values for cuts on discriminant variables: $\chi^2_{fit} < 57.2$, $\Delta E > 1.88$ and $R_{min} > 65$ cm. The signal box is defined as: $4 < \zeta_{2\pi} < 84.9$ and $\zeta_{3\pi} < 5.2$ [5]. Since the expected number of background events was estimated using the Monte Carlo simulations, we have checked at each stage of the analysis how well the simulation describes the experimental data. Distributions of χ^2_{fit}, ΔE and R_{min} variables are presented in Fig. 6 and Fig. 7 for events in the signal box. In the right panel of Fig. 7 we present also the R_{min} distribution just before the last cut $R_{min} > 65$ cm marked with a black arrow. One can see a good agreement between simulations and measured data at every stage of the analysis. At the end of the analysis we have found zero candidates in data and in the simulated background sample.
As it was mentioned, the $K_S \to 2\pi^0$ normalization sample is selected requiring four prompt photons. The Monte Carlo simulation shows a negligible (0.1%) amount of $\phi \to K^+K^-$ background events, thus no further discriminant analysis was required. In the same tagged sample of K_S mesons we have found $N_{2\pi} = (7.533 \pm 0.018) \cdot 10^7$ events. Using simulations we have also determined the $K_S \to 2\pi^0 \to 4\gamma$ selection efficiency: $\epsilon_{2\pi} = 0.660 \pm 0.002_{stat}$. The final number of produced $K_S \to 2\pi^0$ events amounts to: $N_{norm} = N_{2\pi}/\epsilon_{2\pi} = (1.142 \pm 0.005) \cdot 10^8$.

4. Results

With cuts defined in the previous section at the end of the analysis we count 0 candidates with 0 background events expected from Monte Carlo simulated with an effective statistics of two times that of the data. Hence, we have estimated an upper limit on the $K_S \to 3\pi^0$ branching ratio taking into account systematic uncertainties related to the number of background events and to the determination of the acceptance and selection efficiencies for the signal and normalization samples. The detailed description of the evaluation of systematic uncertainties can be found in Ref. [1, 5], here we would like only to stress that they are less than 5% and are negligible in the

calculation of the limit.
In the conservative assumption of no background, we estimate an upper limit on the $K_S \to 3\pi^0$ branching ratio at 90% confidence level:

$$BR(K_S \to 3\pi^0) \leq 2.6 \cdot 10^{-8} \; , \tag{1}$$

which corresponds to an improvement by a factor of about 5 compared to our previous search [3]. This result can be translated into a limit on $|\eta_{000}|$:

$$|\eta_{000}| = |A(K_S \to 3\pi^0)/A(K_L \to 3\pi^0)| \leq 0.0088 \quad \text{at} \quad 90\% \quad \text{C.L.} \tag{2}$$

5. Summary and outlook

As a result of the KLOE data set analysis, gathered in the 2004–2005 data taking period, we have set the new upper limit for the $K_S \to 3\pi^0$ branching ratio at the 90% confidence level, which is almost five times lower than the latest published result [3]. However, the search for the $K_S \to 3\pi^0$ decay will be continued by the KLOE–2 collaboration, which is continuing and extending the physics program of its predecessor [11]. For the forthcoming run the KLOE performance have been improved by adding new subdetector systems: the tagger system for the $\gamma\gamma$ physics, the Inner Tracker based on the Cylindrical GEM technology and two calorimeters in the final focusing region [12]. These new calorimeters will increase the acceptance of the detector, while the new inner detector for the determination of the K_S vertex will significantly reduce the contribution of the background processes involving charged particles. Increasing the statistics and acceptance of the detector while significantly reducing the background gives a realistic chance to observe the $K_S \to 3\pi^0$ decay for the first time in the near future.

Acknowledgments

We acknowledge the support of the European Community-Research Infrastructure Integrating Activity 'Study of Strongly Interacting Matter' (acronym HadronPhysics2, Grant Agreement n. 227431) under the Seventh Framework Programme of EU. This work was supported also in part by the EU Integrated Infrastructure Initiative Hadron Physics Project under contract number RII3-CT- 2004-506078; by the European Commission under the 7th Framework Programme through the 'Research Infrastructures' action of the 'Capacities' Programme, Call: FP7-INFRASTRUCTURES-2008-1, Grant Agreement No. 283286; by the Polish National Science Centre through the Grants No. 0469/B/H03/2009/37, 0309/B/H03/2011/40, DEC-2011/03/N/ST2/02641, 2011/01/D/ST2/00748, 2011/03/N/ST2/02652, 2011/03/N/ST2/02641 and by the Foundation for Polish Science through the MPD programme and the project HOMING PLUS BIS/2011-4/3.

References

[1] Silarski M 2012 Search for the CP symmetry violation in the decays of K_S mesons using the KLOE detector *Preprint* arXiv:1302.4427 [hep-ex]
[2] Beringer J *et al.* [Particle Data Group Collaboration] 2012 *Phys. Rev.* D **86** 010001
[3] Ambrosino F *et al.* 2005 *Phys. Lett.* B **619** 61
[4] Maiani L and Paver N 1995 *The Second DAΦNE Physics Handbook* ed Maiani L, Pancheri G and Paver N (Frascati: Frascati Phys. Ser.) p 51
[5] Babusci D *et al.* [KLOE-2 Collaboration], A new limit on the CP violating decay $K_S \to 3\pi^0$ with the KLOE experiment *Preprint* arXiv:1301.7623 [hep-ex]
[6] Bossi F *et al.* 2008 *Riv. Nuovo Cim.* **31** 531
[7] Adinolfi M *et al.* 2002 *Nucl. Inst. and Meth.* A **488** 51
[8] Adinolfi M *et al.* 2002 *Nucl. Inst. and Meth.* A **482** 364
[9] Adinolfi M *et al.* 2002 *Nucl. Inst. and Meth.* A **492** 134
[10] Ambrosino F *et al.* 2004 *Nucl. Inst. and Meth.* A **534** 403
[11] Amelino-Camelia G *et al.* 2010 *Eur. Phys. J.* C **68** 619
[12] Moricciani D 2011 The KLOE-2 detector upgrade at DAΦNE *PoS* **EPS–HEP2011** 198

Journal of Physics: Conference Series **447** (2013) 012029 doi:10.1088/1742-6596/447/1/012029

Recent BaBar results on CP Violation in B decays

Arantza Oyanguren

IFIC, Edificio Institutos de Investigación Apdo. de Correos 22085, E-46071 Valencia, Spain

E-mail: `Arantza.Oyanguren@ific.uv.es`

Abstract.
In this talk I will review two recent results of CP Violation in B decay analyses obtained by the BaBar experiment. Using the full data set and the partial reconstruction of the D* meson BaBar has measured the CPV parameters in the $B^0 \to D^{*+}D^{*-}$ decay and in the semileptonic decay $\bar{B}^0 \to D^{*+}\ell^-\bar{\nu}_\ell$.

1. Introduction

The BaBar experiment has accumulated almost 500 millions of $B\bar{B}$ pairs at the $\Upsilon(4S)$ energy during its operation time, in addition to a similar quantity of charm and tau pairs, and millions of events at lower resonances. Due to the asymmetric beam energies, B decays from the boosted $\Upsilon(4S)$ are separated about 260μm along the z axis, allowing to perform time dependent analyses and measure CP violation effects. CP violation in B decays can be manifested itself in three different ways: CP violation in decay (the so-called Direct CPV), CP violation in mixing, the probability of a B^0 meson to oscillate into a $\bar{B}^0$ meson being different to the probability of a $\bar{B}^0$ meson to oscillate into a B^0 meson, and then the flavour mixing amplitude, the ratio $|q/p|$, is different from unity. And the last type of CP violation which happens in the interference between mixing and decay, when a same final state can be reached by both the B^0 and $\bar{B}^0$ mesons, and the time dependent amplitudes are different.

2. CP Violation in $B^0 \to D^{*+}D^{*-}$ decays

BaBar has measured the CP violation effects in the decay of B^0 mesons into $D^{*+}D^{*-}$ pairs [1]. This study is motivated by the fact that in Cabibbo suppressed $b \to (c\bar{c})d$ transitions (as $B^0 \to D^{*+}D^{*-}$) the color allowed tree amplitude gets a small contribution from penguin diagrams and the time dependent CPV asymmetry is a measure of $\sin(2\beta)$ (as it is for $B \to J/\Psi K$ in $b \to (c\bar{c})s$ transitions). According to models based on factorization and heavy quark symmetry, penguin contributions are expected to lead a few % corrections [2]. Large deviations in the measurement of $\sin(2\beta)$ from $b \to (c\bar{c})s$ and $b \to (c\bar{c})d$ transitions (i.e., from B decays into $J/\Psi K$ and $D^{*+}D^{*-}$ final states) would imply a sign of New Physics [3].

In Υ(4S) $\to B^0\bar{B}^0$ events, the time-dependent decay rate for $B^0 \to D^{*+}D^{*-}$ is given by

$$P_\eta^{S_{tag}}(\Delta t) = \frac{e^{-|\Delta t|/\tau_b}}{4\tau_b} \cdot \left[1 + S_{tag}\, S_\eta \sin(\Delta m_d \Delta t) + S_{tag}\, C \cos(\Delta m_d \Delta t)\right], \quad (1)$$

where τ_b is the B^0 lifetime averaged over the two mass eigenstates, Δm_d is the $B^0\bar{B}^0$ mixing frequency, and Δt is the time interval between the $B^0 \to D^{*+}D^{*-}$ decay (B_{rec}) and the decay of the other B (B_{tag}) in the event. The parameter $S_{tag} = +1$ (-1) indicates the flavor of the B_{tag} as a B^0 ($\bar{B}^0$), while $\eta = \pm 1$ indicates the CP eigenvalue of the $B^0 \to D^{*+}D^{*-}$ final state.

The parameters C and S_η are given by

$$C = \frac{1-|\lambda|^2}{1+|\lambda|^2}; \quad S_\eta = -\eta\frac{2\Im m(\lambda)}{1+|\lambda|^2}; \quad \lambda = \frac{q}{p}\frac{\overline{A}}{A}, \tag{2}$$

where A ($\overline{A}$) is the matrix element of the B^0 ($\bar{B}^0$) decay and p and q are the coefficients appearing in the expression of the physical mass eigenstates B_L, B_H in terms of the flavour eigenstates B, $\bar{B}$:

$$\begin{aligned} |B_L\rangle &= p|B\rangle + q|\bar{B}\rangle \\ |B_H\rangle &= p|B\rangle - q|\bar{B}\rangle. \end{aligned}$$

Since the $B^0 \to D^{*+}D^{*-}$ is the decay of a scalar to two vector mesons, the final state is a mixture of CP eigenstates. The CP-odd and CP-even fractions have been previously measured from the angular analysis of completely reconstructed events by both BaBar and Belle experiments [4, 5].

A new analysis has been performed by BaBar using a partial reconstruction technique of the $B^0 \to D^{*+}D^{*-}$ decay. This method has the advantage of 5 times more statistics as compared to previous analyses using full reconstruction, corresponding to an almost independent data sample. Nevertheless this technique has the penalty of higher background, leading to larger systematic uncertainties. Using 429 fb^{-1} on-peak $\Upsilon(4S)$ BaBar data and $45 fb^{-1}$ off-peak data (the latter for background studies) B mesons are tagged with a lepton or a kaon that gives the flavour of one of the B mesons (called 'lepton-tag' or 'kaon-tag' respectively). The other B meson is partially reconstructed into $D^{*+}D^{*-}$. One of the D^* is completely reconstructed into $\mathrm{D}^0\pi$, with the D^0 in one of the four modes $K\pi$, $K\pi\pi^0$, $K3\pi$ or $K_s\pi\pi$. The fully reconstructed D^* is then matched with a soft pion of opposite sign expected to come from the other D^* meson. Signal candidates are selected if the kinematics is consistent with a B^0 decaying into a $\mathrm{D}^* - \mathrm{D}^0\pi$ combination with a missing D^0. Two unbinned maximum likelihood fits are then performed. One kinematic fit using two variables: the recoiling mass, m_{rec}, which is the mass of the non-reconstructed D^0 meson and peaks at the D^0 mass for signal events; and a Fisher discriminant variable based on the event shape. The background sources are combinatorial background (the distributions are extracted from a wrong sign D^* + soft π data sample), light quarks from continuum (extracted from off-peak data) and a peaking $B\bar{B}$ contribution which is taken from the simulation and it is found to be negligible. The results of the fit gives the signal fraction and the shape of the signal and background probability density functions (PDFs). They are shown in Figure 1. It gives 3843±397 and 1129±218 signal events for kaon and lepton tags respectively.

The result of the kinematic fit is kept and followed by a a time dependent fit, which uses information of m_{rec} and the fisher variable but now including the time dependent function. The Δt variable is reconstructed from the two B meson decay vertices (one from the reconstructed D^*, the other from the lepton or the kaon tag, making use of the beam spot constraint). The time dependent PDFs are convoluted with a resolution function. The result of the fit are the CP parameters, the background and signal models, and the mistag probabilities. Mistag probabilities are about 20% and 10% for the kaon and lepton tags, respectively. The additional dilution due to tagging tracks from missing D^0 mesons is reduced by applying a constraint around the D^0 direction, obtained from data with some input from simulation (it accounts 12% for the kaon tag). The main systematic uncertainties in the analysis are related to the kinematic fit parameters, tagging dilution and interference with double Cabibbo-suppressed (DCS) decays.

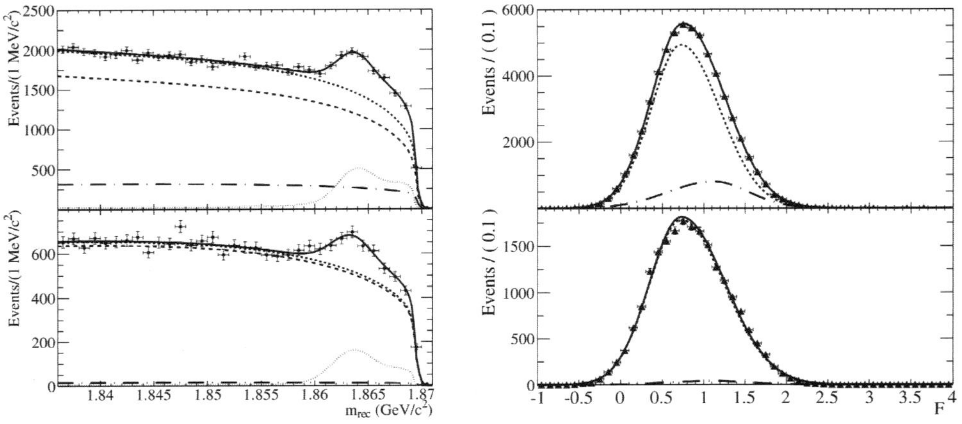

Figure 1. Result of the kinematic fit of kaons (top) and lepton (bottom) tagged events, for the reconstructed mass (left plots) and the fisher variable (right plots). The total PDF (solid line), signal (dotted red line) and total background (short-dashed line) (consisted mainly of combinatorial background (dashed line) and continuum (dot-dashed line)) are superimposed.

The result of the time dependent fit is shown in Figure 2 with the time-dependent raw CP asymmetry:

$$A(\Delta t) = \frac{N_{S_{\rm tag}=1}(\Delta t) - N_{S_{\rm tag}=-1}(\Delta t)}{N_{S_{\rm tag}=1}(\Delta t) + N_{S_{\rm tag}=-1}(\Delta t)}. \tag{3}$$

The combined result for lepton and kaon tags of the CP parameters S and C is

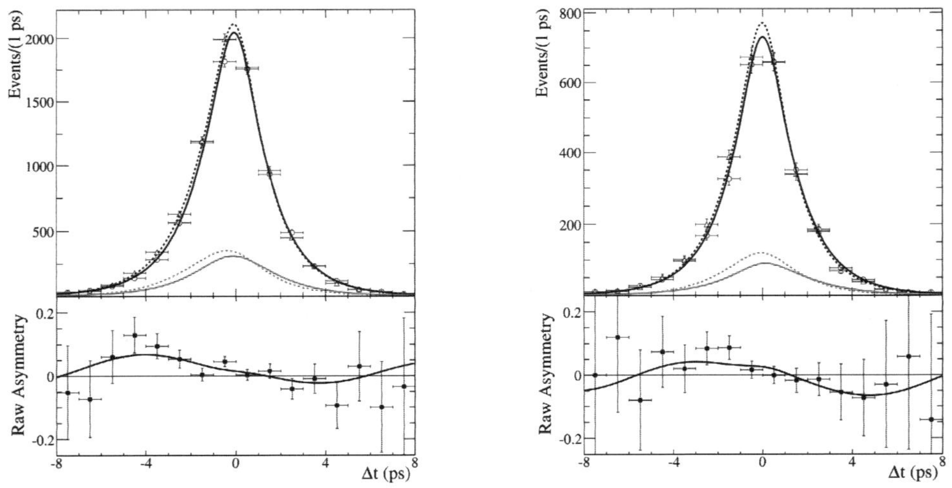

Figure 2. Result of the time dependent fit for kaons (left) and lepton (right) tagged events. B^0 (dashed) and $\bar{B}^0$ (solid) events. Signal PDFs are superimposed in red. Only data in the $m_{rec} > 1.860$ GeV/c^2 are shown in these plots.

$C = +0.15 \pm 0.09 \pm 0.04$ and $S = -0.34 \pm 0.12 \pm 0.05$.

Considering the dilution by the CP-odd component, S=S$_+$(1-2R$_\perp$), and taking R$_\perp$ from a previous BaBar result [4]: $C_+ = +0.15\pm0.09\pm0.04$ and $S_+ = -0.49\pm0.18\pm0.07\pm0.04$. Results are compatible with BaBar and Belle measurements and with the Standard Model predictions.

3. CPV in $B^0 \to D^*\ell\nu$

The BaBar experiment has also performed a measurement of the CP violation parameters in the B semileptonic decay $B^0 \to D^*\ell\nu$. If CP is violated in mixing, the probability of a B^0 to oscillate to a $\bar{B}^0$ is different from the probability of a $\bar{B}^0$ to oscillate to a B^0 and thus we expect to observe a sizable value for the asymmetry:

$$\mathcal{A}_{CP} = \frac{N(B^0B^0) - N(\bar{B}^0\bar{B}^0)}{N(B^0B^0) + N(\bar{B}^0\bar{B}^0)} \simeq 2\delta_{CP}. \tag{4}$$

where $\delta_{CP} = 1 - |q/p|$. Any deviation from unity of the ratio $|q/p|$ implies CP violation in mixing. The Standard Model prediction is $\mathcal{A}_{CP} = -(4.1 \pm 0.6) \times 10^{-4}$ [6]. A large deviation from $\delta_{CP} = 1$ would be therefore a clear evidence of New Physics beyond the Standard Model.

The semileptonic asymmetry $\mathcal{A}_{SL}$ measured at flavour experiments can also be expressed as:

$$\mathcal{A}_{SL} = \frac{N(\bar{B}^0 \to B^0 \to \ell^+X) - N(B^0 \to \bar{B}^0 \to \ell^-X)}{N(\bar{B}^0 \to B^0 \to \ell^+X) + N(B^0 \to \bar{B}^0 \to \ell^-X)}. \tag{5}$$

A large deviation from expectations in $\mathcal{A}_{SL}$ (4σ) has been observed by the D0 experiment at Fermilab using same sign dimuon events [7], which could be attributed to B_s oscillations. Results of the Heavy Flavour Averaging Group from CLEO, BaBar, Belle, D0 and LHCb data gives $\mathcal{A}_{SL} = -0.0003 \pm 0.0021$ for the B_d system, and $|q/p| = 1.0002 \pm 0.0011$ [8].

A new measurement of $\mathcal{A}_{SL}$ has been performed by BaBar using partial reconstruction of $B^0 \to D^*\ell\nu$ and kaon tags. Using 428fb^{-1} on-peak $\Upsilon(4S)$ data and 45fb^{-1} off-peak data (the latter for background studies). In this analysis one B meson is partially reconstructed in the semileptonic channel by selecting a lepton plus the soft pion from the D^* decay into $D^0\pi$. The flavour of the other B meson is obtained using a kaon tag. When mixing occurs the lepton from one B and the kaon tag have the same charge, and the semileptonic asymmetry is defined as:

$$\mathcal{A}_{SL} = \frac{N(\ell^+K_T^+) - N(\ell^-K_T^-)}{N(\ell^+K_T^+) + N(\ell^-K_T^-)} \tag{6}$$

In practice there are two kaon tag categories: when the kaon is coming from the B meson (B_{tag}) and when the kaon comes from a Cabibbo Favoured D^0 decay (D_{tag}), giving also a same charge combination of the lepton and kaon. The information of the decay vertex distance, Δz, which is the difference between the reconstructed and tagged vertex, and the cosine of the angle defined by the kaon and the lepton tracks, are used to separate these two categories. The asymmetry observed in the B_{tag} and D_{tag} samples has contributions from the semileptonic asymmetry $\mathcal{A}_{SL}$ and from the reconstruction and tagging asymmetries:

$$\mathcal{A}(B_{tag}) = \mathcal{A}_{rec} + \mathcal{A}_{tag} + \mathcal{A}_{SL} \tag{7}$$

$$\mathcal{A}(D_{tag}) = \mathcal{A}_{rec} + \mathcal{A}_{tag} + \chi_d\mathcal{A}_{SL} \tag{8}$$

χ_d being the integrated mixing probability for B^0 mesons. The D_{tag} contribution is the main source of background in this analysis but it is useful to disentangle physical and detector charge effects.

Combinatorial and continuum backgrounds are suppressed using event shape variables and the vertex probability of the $\ell\pi$ vertex. The signal is selected using the missing mass squared variable: $\mathcal{M}_\nu^2 \equiv (E_{\rm beam} - E_{D^*} - E_\ell)^2 - (\vec{p}_{D^*} + \vec{p}_\ell)^2$, assuming that the B meson is at rest and that the D* direction is the one from the soft pion. The number of signal events is obtained with a minimum χ^2 fit to the $\mathcal{M}_\nu^2$ distribution in the interval $-10 < \mathcal{M}_\nu^2 < 2.5$ GeV^2/c^4. D*, D** and combinatorial backgrounds are floated in the fit, using simulated distributions. The continuum background is fixed from off-peak data, and the small peaking background contribution is taken from the simulation. A total of $(5945 \pm 7) \cdot 10^3$ events are selected in the signal region, which is about 30% of the total sample. The result of the fit is displayed in Figure 3.

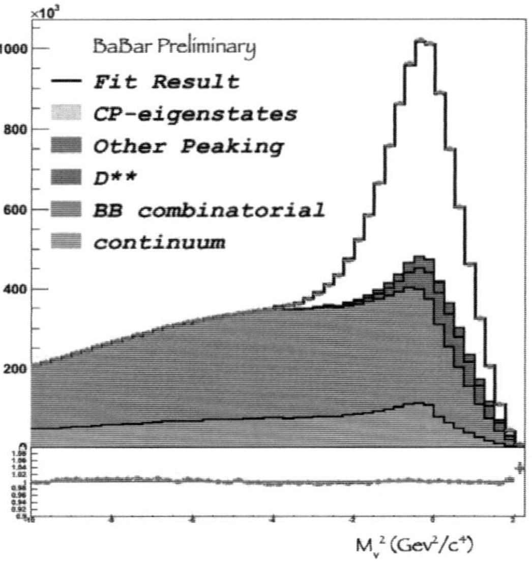

Figure 3. $\mathcal{M}_\nu^2$ distribution for the data (points with error bars), and the fitted contributions from the various sample components.

The $|q/p|$ parameter is obtained from an extended maximum likelihood binned fit to the Δt and $cos\theta_{K\ell}$ distributions of the unmixed (with different charged for the lepton and the kaon) and mixed (with the same charge for the lepton and the kaon) subsamples. The time dependent decay rate for the mixed and unmixed samples are described in terms of the average width of the two B^0 mass eigenstates, Γ_0, and the width difference $\Delta\Gamma$. Several parameters are introduced to take into account the interference between Cabibbo Favoured and Double Cabibbo suppressed decays on the tag side. The theoretical distributions have to be convoluted with the resolution function. Mistag probabilities have to be taken into account, which are different for the B_{tag} and for the reconstructed B meson. The tagging and reconstruction asymmetries are disentangled using the D_{tag} sample. The result of the fit are shown in Figures 4 and 5, for the Δt and $cos\theta_{K\ell}$ projections, for the mixed and unmixed samples. Signal and background contributions are shown in the figures. The raw asymmetry as function of Δt is also shown in Figure 4.

The result of the fit gives the value of $|q/p| - 1 = 0.29 \pm 0.84^{+1.78}_{-1.61} \times 10^{-3}$ (which is kept blinded until the analysis is completed, including systematic uncertainties) and parameters describing signal and background events. The main systematic in the analysis is coming from the sample composition. The semileptonic asymmetry results in the value: $\mathcal{A}_{CP} = 0.06 \pm 0.17^{+0.36}_{-0.32}$, which is compatible with previous results from B factories and with Standard Model predictions.

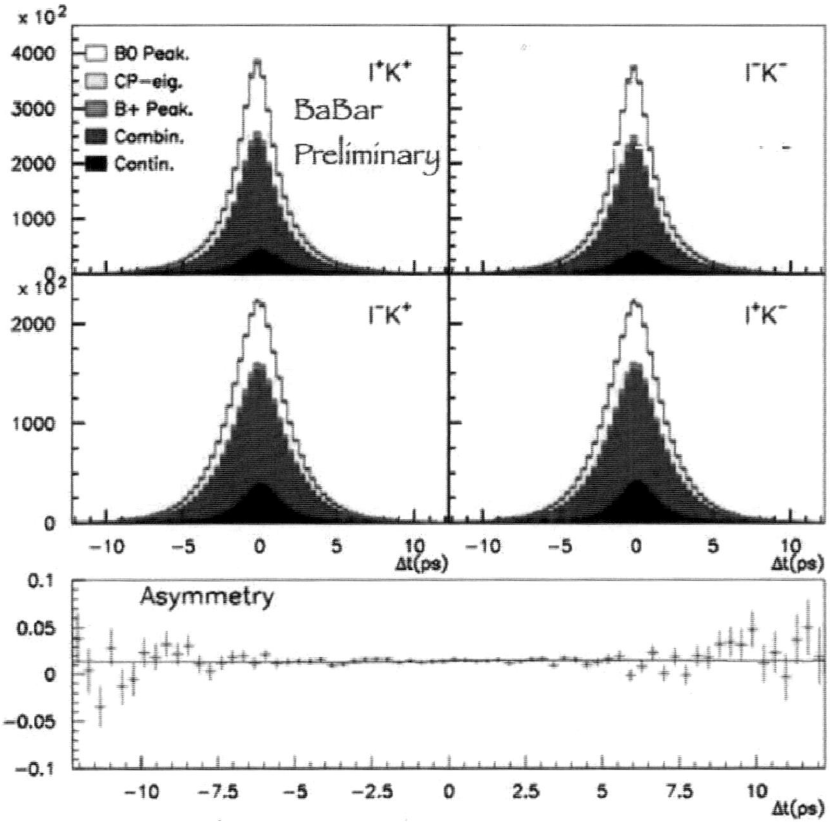

Figure 4. Δt (left) distributions for the data (points with error bars), and the fitted contributions from the various sample components. Bottom plot: Raw asymmetry between ℓ^+K^+ and ℓ^-K^- events.

4. More BaBar results on CPV

Other recent results from BaBar not covered in this talk concern the measurement of the CKM angles α [9] and γ [10].

5. Summary

New precise measurements on CP violation in B decays using a partial reconstruction technique have been performed by the BaBar experiment. The CPV parameters C_+ and S_+ have been obtained from partial reconstruction of $B^0 \rightarrow D^{*+}D^{*-}$, and the δ_{CP} parameter and the semileptonic asymmetry $\mathcal{A}_{SL}$ have been measured from partial reconstruction of $B^0 \rightarrow D^*\ell\nu$ events. Results are in agreement with other measurements from B-factories and with Standard Model predictions.

References

[1] J. P. Lees *et al.* [BABAR Collaboration], Phys. Rev. D **86** (2012) 112006 [arXiv:1208.1282 [hep-ex]].
[2] Z. Xing, Phys. Lett. B **443** (1998) 365 [hep-ph/9809496]. Z. Xing, Phys. Rev. D **61** (2000) 014010 [hep-ph/9907455].
[3] M. Gronau, J. L. Rosner and D. Pirjol, Phys. Rev. D **78** (2008) 033011 [arXiv:0805.4601 [hep-ph]]. Y. Grossman

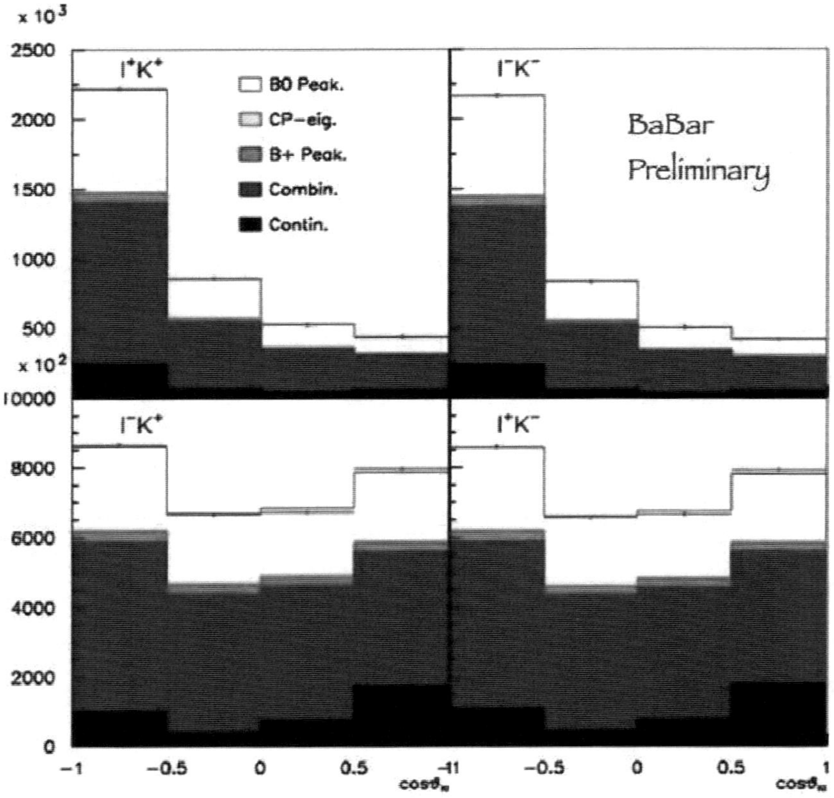

Figure 5. $\cos\theta_{\ell K}$ distribution for the data (points with error bars) and the fitted contributions from the various sample components.

and M. P. Worah, Phys. Lett. B **395** (1997) 241 [hep-ph/9612269]. R. Zwicky, Phys. Rev. D **77** (2008) 036004 [arXiv:0707.0677 [hep-ph]].
[4] B. Aubert *et al.* [BABAR Collaboration], Phys. Rev. D **79** (2009) 032002 [arXiv:0808.1866 [hep-ex]].
[5] K. Vervink *et al.* [Belle Collaboration], Phys. Rev. D **80** (2009) 111104 [arXiv:0901.4057 [hep-ex]].
[6] A. Lenz, arXiv:1102.4274.
[7] V. M. Abazov *et al.* [D0 Collaboration], Phys. Rev. D **84** (2011) 052007 [arXiv:1106.6308 [hep-ex]].
[8] Heavy Flavour Averaging Group. http://www.slac.stanford.edu/xorg/hfag/
[9] J. P. Lees [BABAR Collaboration], arXiv:1206.3525 [hep-ex].
[10] See the J. Walsh talk at this conference.

DISCRETE 2012 – Third Symposium on Prospects in the Physics of Discrete Symmetries IOP Publishing
Journal of Physics: Conference Series **447** (2013) 012030 doi:10.1088/1742-6596/447/1/012030

Top Quark Couplings and Search for New Physics at the LHC

António Onofre
Departamento de Física, Universidade do Minho
Campus de Gualtar, 4710 - 057 Braga, Portugal

E-mail: onofre@fisica.uminho.pt

Abstract. The search for new physics in top quark decays at the LHC is reviewed in this paper. Results from ATLAS [1] and CMS [2] experiments on top quark decays within the Standard Model are presented together with the measurements of the W boson polarizations and the study of the structure of the Wtb vertex. As a natural step forward, the experimental status on measurements sensitive to top quark couplings to gauge bosons (γ, Z, W and H) is reviewed as well as possible top quark decays Beyond the Standard Model (MSSM and FCNC).

1. Introduction

The top quark is the heaviest known quark ($m_t = 173.18 \pm 0.94$ GeV [3]) ever discovered. With a charge of $+2/3|e|$ (e is the electron charge) and spin 1/2, the top quark completes the 3 family structure of the Standard Model (SM). One of the stricking features of this quark is precisely the value of its mass, so distinctive from the other fermions, which may suggest it plays a more fundamental role in the eletroweak symmetry breaking mechanism. In the SM, top quarks are expected to decay mostly through $t \to bW$, being the branching ratios $Br(t \to sW) \leq 0.18\%$ and $Br(t \to dW) \leq 0.02\%$ diminut when compared with the dominant decay channel [4]. The lifetime, at least one order of magnitude smaller than the typical hadronization time scale $O(10^{-23})$ s, was measured to be $\tau_t = (3.29^{+0.90}_{-0.67}) \times 10^{-25}$ s [5]. This implies top quarks decay before hadronization may take place and spoil the spin information propagated to the final state decay products. Once again, top quarks show remarkable distintive features when compared to the other quarks. At the LHC, top quarks are mainly produced in pairs through gluon fusion or quark anti-quark annihilation with a predicted cross section of $164.6^{+11.4}_{-15.7}$ pb (approx. NNLO) at 7 TeV [6]. There is however a significant number of top quarks which are produced singly through the t-channel ($gq' \to qt\bar{b}$), associated production ($bg \to Wt$) and s-channel ($q\bar{q}' \to t\bar{b}$) single top production. The production cross sections are, respectively, $64.6^{+2.7}_{-2.0}$ pb [7] (NNLO+soft gluon corrections), 15.7 ± 1.1 pb [8] (approx. NNLO) and 4.6 ± 0.2 pb [9] (NNLL resummation). In this paper the experimental status of top quark measurements sensitive to gauge bosons couplings (W, γ, Z and H) and decays beyond the SM are reviewed. As most analysis described in the paper rely on the lepton+jets and dilepton final states, or variations from these, it is appropriate to shortly mention them.

In the lepton+jets channel, one isolated lepton (e or μ) is required in the event selection as well as at least 4 jets for both ATLAS [10, 11] and CMS [12, 13]. While for muons, the transverse momentum (p_{T}) selection is similar for both experiments, selected electrons at ATLAS are harder (with transverse energy, E_T, above 25 GeV) than in CMS ($E_T > 20$ GeV). Softer jets, with $p_T > 25$ GeV, are also selected by ATLAS compared to CMS ($p_T > 30$ GeV). The use of a transverse missing energy (E_T^{miss}) cut together with the requirement that at least one jet had to be tagged as a b-jet was performed by ATLAS while CMS required at least two b-tagged jets with no E_T^{miss} condition.

In the dilepton channel, events with two isolated leptons (e or μ) and two jets, from which at least one is b-tagged, are selected by both CMS [14, 13] and ATLAS [15]. While at least 2 opposite charge leptons are requested by CMS, exactly 1 pair is imposed by ATLAS. In order to reduce potential contributions from the Z+jets background, events are required to have the invariant mass of the two opposite charge leptons ($m_{\ell^+\ell^-}$) outside a mass window of at least 10 GeV around the central value of the Z mass. Both CMS and ATLAS impose E_T^{miss} cuts to events and ATLAS, for the $e^{\pm}\mu^{\mp}$, requires the scalar sum of jets and leptons transverse energies (H_T) to be above 130 GeV.

2. The Wtb vertex structure

In the SM, the Wtb vertex has a $(V-A)$ structure where, V and A, are the vector and axial-vector contributions to the vertex. One way of probing the structure of the vertex, is to study top quark decays to W-bosons and b-quarks ($t \rightarrow bW$). The W bosons produced in these decays can have longitudinal, left-handed or right-handed polarizations with fractions F_0, F_L and F_R, respectively equal to 0.687 ± 0.005, 0.311 ± 0.005 and 0.0017 ± 0.0001 at NNLO [16], in the SM. The helicity fractions were extracted from a direct fit to the top quark decay products angular distribution, $\cos\theta$, using templates from simulation. Alternatively, angular asymmetries (A_+, A_- and A_{FB}), built from $\cos\theta$, were also used by ATLAS [17]. The angle θ is measured between the momentum direction of the charged lepton from the decay of the W boson, and the reversed momentum direction of the b-quark from the decay of the top quark, both boosted into the W boson rest frame. CMS [18] measured the helicity fractions to be $F_0 = 0.567 \pm 0.074(\text{stat}) \pm 0.047(\text{syst})$, $F_L = 0.393 \pm 0.045(\text{stat}) \pm 0.029(\text{syst})$ and $F_R = 0.040 \pm 0.035(\text{stat}) \pm 0.044(\text{syst})$. The correspondent measurements from ATLAS [19] are, $F_0 = 0.67 \pm 0.03(\text{stat}) \pm 0.06(\text{syst})$, $F_L = 0.32 \pm 0.02(\text{stat}) \pm 0.03(\text{syst})$ and $F_R = 0.01 \pm 0.01(\text{stat}) \pm 0.04(\text{syst})$. All these measurements are consistent with SM expectations. The measured fractions were used to probe the existence of anomalous Wtb couplings. Exclusion limits on the real components of the anomalous couplings g_L and g_R were set at 68% and 95% CL (Figure 1). The region of g_R around 0.8 in the ATLAS plot is disfavored by the current experimental measurements on the single top quark production cross sections at Tevatron and LHC. Once the limits depend on V_{tb}, it is worth mentionning the current experimental status of these measurements (Table 1) which assume the SM and negligeable decays to $t \rightarrow sW$ and $t \rightarrow dW$, when compared to the dominant channel. The current V_{tb} measurements precision is at the level of 5%-10%.

3. Top quark couplings to gauge bosons

Following the discussion on the Wtb vertex and V_{tb}, it is natural to ask what is the current LHC status on the measurements of the couplings of the top quarks to other bosons (γ,Z,W and H). Several dedicated analyses were developed by both ATLAS and CMS, relying to a great extent on the lepton+jets and dilepton channels discussed previously, with the exception that activity from the gauge bosons is expected to be present as well in the events. These are particularly challenging channels which should be improved as both CMS and ATLAS collect more data over time. ATLAS [20] looked into $t\bar{t}\gamma$ production and, by fitting the data with templates for prompt photons (well isolated) and hadron fakes, measured the cross section to $\sigma(t\bar{t}\gamma) = 2.0 \pm 0.5$ (stat) $\pm$ 0.7 (syst) $\pm$ 0.08 (lumi.) pb, for events with photons with p_T >8 GeV for which the expected SM cross section is 2.1±0.4 pb (at 7 TeV). For the $t\bar{t}Z$ channel ATLAS [21] observes 1 candidate after event selection against a background of

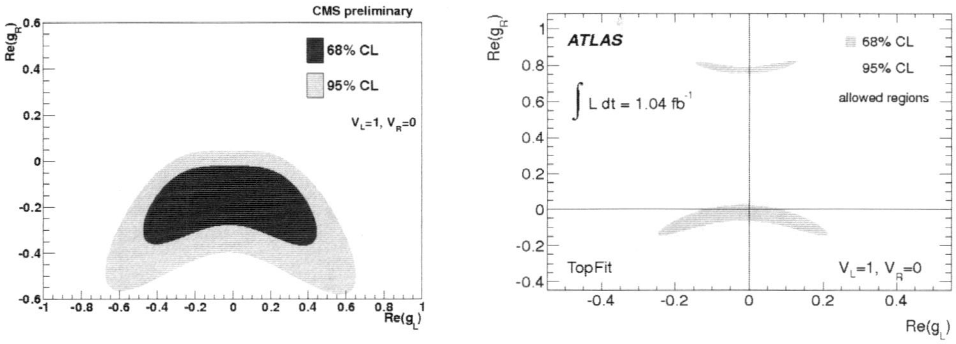

Figure 1. CMS(left) and ATLAS(right) allowed regions, at 68% and 95% CL, on the anomalous coupling plane (g_L,g_R), for V_R=0 and V_L=1.

$0.28^{+1.57}_{-0.14}$. A 95%CL limit of 0.71 pb was set on the $t\bar{t}Z$ production (which compares with a SM prediction of 0.14 pb). CMS [22] searched, in two independent channels (three and two same charge leptons), for $t\bar{t}V(V = W, Z)$. While the $t\bar{t}V$ combined cross section was measured to be $\sigma(t\bar{t}V) = 0.51^{+0.15}_{-0.13}(stat)^{+0.05}_{-0.04}(syst)$ pb at 7 TeV, the $t\bar{t}Z$ cross section (obtained from the trilepton channel) was $\sigma(t\bar{t}Z) = 0.30^{+0.14}_{-0.11}(stat)^{+0.04}_{-0.02}(syst)$ pb (Figure 2).

Both ATLAS [23] and CMS [24] searched for $t\bar{t}H$ production using data collected at 7 TeV with a total integrated luminosity of 5 fb^{-1}. While ATLAS studied the single lepton channel, where the Higgs boson is assumed to decay to a pair of b-quarks, CMS looked into both the single and dilepton channels. As no evidence for signal was found, combined upper limits, at 95% CL, were set on cross section times branching ratio by ATLAS and CMS to $\sigma_{t\bar{t}H} \times Br(H \to b\bar{b}) \leq$13.1 pb (10.5 pb expected), $\sigma_{t\bar{t}H} \times Br(H \to b\bar{b}) \leq$3.8 pb (4.6 pb expected), respectively.

4. Top quark decays beyond the SM

In the pursuit of new physics in the top quark sector, the decays play a fundamental role. Not only they may reveal the presence of new Physics Beyond the SM, as predicted for instance by the Minimal Supersymmetric Model (MSSM) but could even show that decays with very low SM branching ratios, like Flavor Changing Neutral Currents (FCNC), non visible at tree level

Table 1. Current status of V_{tb} measurements from both ATLAS and CMS.

ATLAS	Measurement (SM)
t-channel (@ 7TeV)	$\|V_{tb}\| = 1.13^{+0.14}_{-0.13}$
(arXiv:1205.3130)	$|V_{tb}| \geq$0.75 @ 95% CL ($|V_{tb}$ in [0,1])
t-channel (@ 8TeV,5.8 fb^{-1})	$\|V_{tb}\| = 1.04^{+0.10}_{-0.11}$
(ATLAS-CONF-2012-132)	$|V_{tb}| \geq$0.80 @ 95% CL ($|V_{tb}$ in [0,1])
Wt-prod. (PLB 716 (2012) 142)	$\|V_{tb}\| = 1.03^{+0.16}_{-0.19}$
CMS	**Measurement (SM)**
t-channel (@ 7TeV)	$\|V_{tb}\|$=1.02±0.046(exp)±0.017(th.)
(arXiv:1209.4533)	$|V_{tb}| \geq$0.92 @ 95% CL ($|V_{tb}$ in [0,1])
t-channel (@ 8TeV,5. fb^{-1})	$\|V_{tb}\|$=0.96±0.08(exp)±0.02(th.)
(CMS PAS TOP-12-011)	$|V_{tb}| \geq$0.81 @ 95% CL ($|V_{tb}$ in [0,1])
Wt-prod. (@ 7TeV)	$\|V_{tb}\| = 1.01^{+0.16}_{-0.13}$(exp.)$^{+0.03}_{-0.04}$(th.)
(arXiv:1209.3489)	$|V_{tb}| \geq$0.79 @ 95% CL ($|V_{tb}$ in [0,1])

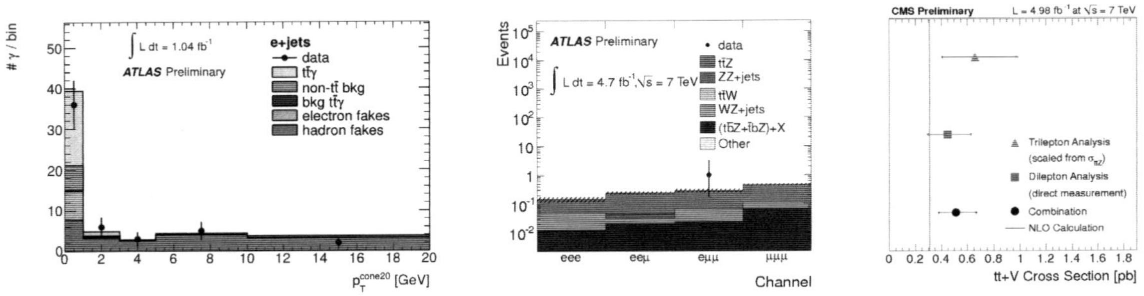

Figure 2. (left) ATLAS template fit to $t\bar{t}\gamma$ events in the electron channel; (middle) ATLAS expected and observed numbers of events for the $t\bar{t}Z$ signal region; (right) CMS measurements of the $t\bar{t}V$ production cross section.

and very much suppressed at one loop due to the GIM mechanism in the SM ($Br < 10^{-12}$), could be experimentally detected. In addition to the theorectical motivation to search for these decays, the stronger motivation is perhaps more of experimental nature: to look for signs of new physics where the SM expects essentially no contribution.

Both CMS [25] and ATLAS [26] looked for MSSM decays of top quarks to charged Higgs ($t \to H^+b$) when $m_{H^\pm} < m_t$. If $\tan\beta$ is particularly suited, it may happen the dominant decay mode of the charged Higgs is through the decay $H^+ \to c\bar{s}$. As no sign of new physics was found, best 95% CL upper limits were set on the branching fractions $Br(t \to H^+b)$ in the range 2-4% for Higgs boson masses between 80 and 160 GeV and assuming $Br(H^+ \to \tau^+\nu_\tau) = 1$.

In the SM, FCNC decays ($t \to qX, X = \gamma, Z, g, H$) are forbidden at tree level and much smaller than the dominant decay channel. ATLAS [27] and CMS [28] searched for signs of $t \to qZ$ and, as no new physics was found, observed limits at 95% CL were set on $Br(t \to qZ) \leq 0.73\%$ (ATLAS) and $Br(t \to qZ) \leq 0.27\%$ (CMS).

5. Conclusions

Top quark studies are well under way at the LHC. Although no new physics was seen yet in the top quark physics, the performance of the LHC and the ATLAS and CMS experiments has been remarkable. The LHC is entering a new era of precision studies for the top quark physics.

The work of A.Onofre was supported by Fundação para a Ciência e Tecnologia under project CERN/FP/123619/2011.

6. References

[1] ATLAS Collaboration, JINST **3**, S08003 (2008).
[2] CMS Collaboration, JINST **3** (2008) S08004.
[3] T. Aaltonen *et al.* [CDF and D0 Collaborations], Phys. Rev. D **86**, 092003 (2012) [arXiv:1207.1069 [hep-ex]].
[4] J. Beringer *et al.* (Particle Data Group), Phys. Rev. D **86**, 010001 (2012)
[5] V. M. Abazov *et al.* [D0 Collaboration], Phys. Rev. D **85**, 091104 (2012) [arXiv:1201.4156 [hep-ex]].
[6] M. Aliev, H. Lacker, U. Langenfeld, S. Moch, P. Uwer and M. Wiedermann, Comput. Phys. Commun. **182**, 1034 (2011) [arXiv:1007.1327 [hep-ph]].
[7] N. Kidonakis, Phys. Rev. D **83**, 091503 (2011) [arXiv:1103.2792 [hep-ph]].
[8] N. Kidonakis, Phys. Rev. D **82**, 054018 (2010) [arXiv:1005.4451 [hep-ph]].
[9] N. Kidonakis, Phys. Rev. D **81**, 054028 (2010) [arXiv:1001.5034 [hep-ph]].
[10] ATLAS Collaboration, Phys. Lett. B **711**, 244 (2012) [arXiv:1201.1889 [hep-ex]].
[11] ATLAS Collaboration, arXiv:1207.5644 [hep-ex].
[12] CMS Collaboration, Phys. Rev. D **84**, 092004 (2011) [arXiv:1108.3773 [hep-ex]].
[13] CMS Collaboration, arXiv:1211.2220 [hep-ex].

[14] CMS Collaboration, JHEP **1211**, 067 (2012) [arXiv:1208.2671 [hep-ex]].
[15] ATLAS Collaboration, JHEP **1205**, 059 (2012) [arXiv:1202.4892 [hep-ex]].
[16] A. Czarnecki, *et al.*, Phys. Rev. D **81**, 111503 (2010)
[17] J. A. Aguilar-Saavedra, *et al.*, Eur. Phys. J. **C 50**, 519-533 (2007)
[18] CMS Collaboration, CMS-PAS-TOP-11-020, February (2012).
[19] ATLAS Collaboration, JHEP **1206**, 088 (2012) [arXiv:1205.2484 [hep-ex]].
[20] ATLAS Collaboration, ATLAS-CONF-2011-153.
[21] ATLAS Collaboration, ATLAS-CONF-2012-126.
[22] CMS Collaboration, CMS-PAS-TOP-12-014.
[23] ATLAS Collaboration, ATLAS-CONF-2012-135.
[24] CMS Collaboration, CMS-PAS-HIG-12-025.
[25] CMS Collaboration, JHEP **1207**, 143 (2012) [arXiv:1205.5736].
[26] ATLAS Collaboration, JHEP **1206**, 039 (2012) [arXiv:1204.2760].
[27] ATLAS Collaboration, JHEP **1209**, 139 (2012) [arXiv:1206.0257].
[28] CMS Collaboration, arXiv:1208.0957.

DISCRETE 2012 – Third Symposium on Prospects in the Physics of Discrete Symmetries IOP Publishing
Journal of Physics: Conference Series **447** (2013) 012031 doi:10.1088/1742-6596/447/1/012031

MEtop - a top FCNC event generator

Rita Coimbra[1], **António Onofre**[2], **Rui Santos**[3,4] **and Miguel Won**[1,4]

[1] LIP / Departamento de Física, Universidade de Coimbra, 3004-516 Coimbra, Portugal
[2] Departamento de Física, Universidade do Minho, 4710-057 Braga, Portugal
[3] Instituto Superior de Engenharia de Lisboa - ISEL, Rua Conselheiro Emídio Navarro 1, 1959-007 Lisboa, Portugal
[4] Centro de Física Teórica e Computacional, Faculdade de Ciências, Universidade de Lisboa, Av. Prof. Gama Pinto 2, 1649-003 Lisboa, Portugal

E-mail: `rita.coimbra@coimbra.lip.pt`, `Antonio.Onofre@cern.ch`, `rsantos@cii.fc.ul.pt`, `miguel.won@coimbra.lip.pt`

Abstract. In this work we present a new Monte Carlo generator for Direct top and Single top production via flavour-changing neutral currents (FCNC). This new tool calculates the cross section and generates events with Next-to-Leading order precision for the Direct top process and Leading-Order precision for all other FCNC single top processes. A set of independent dimension six FCNC operators has been implemented - including four-fermion operators - where at least one top-quark is present in the interaction.

1. Introduction

With the new experimental data from the Large Hadron Collider (LHC) one will be able to scrutinise the Standard Model (SM) boundaries with unprecedented precision. Flavour-changing neutral currents (FCNC) top decays such as $t \to V\,q$, where $V = Z, \gamma, g$ and $q = u, c$ are highly suppressed in the SM due to the well known Glashow-Iliopoulos-Maiani (GIM) mechanism [1]. top-quark FCNC branching ratios are of the order of 10^{-14} and 10^{-12} for the electroweak and strong case respectively. However, some versions of two-Higgs doublet models (2HDMs) predict quite larger values. In fact, for some regions of the 2HDM parameter space the $\mathrm{BR}(t \to g\,q)$ could reach values of the order of 10^{-4} [2], which is eight orders of magnitude above the SM value. The high statistics expected in LHC data forces an increase in the precision of all relevant theoretical calculations. Hence, at least Next-to-Leading (NLO) calculations should be made available whenever possible. The experimental searches performed both at LHC and at the Tevatron on the direct top FCNC process have already used NLO cross sections from [3]. However, the NLO calculation was used only as a normalization factor and the events were generated using LO generators like TopRex [4] or Protos [5]. In this work we present a new Monte Carlo tool that generates FCNC direct top events with NLO precision. The FCNC interactions were implemented via an effective lagrangian [6] where a complete set of dimension six operators was used. The events are generated in the Les-Houches Event (LHE) format [7] and can therefore be easily interfaced with shower algorithms such as the one in Pythia [8]. Since the generator calculates the total inclusive cross section, the FCNC single top process was also included and can be generated independently. Not only strong FCNC interactions, but also electroweak and four-fermions dimensions six operators were included. Each operator can be turned on independently allowing for independent studies of the different Lorentz structures. In section 2 the effective formalism is presented while in section 3 we show the physical processes already included in MEtop. In section 4 we give a description of the algorithm used by the Monte Carlo generator in order to obtain a final NLO result. In section 5 we discuss a few selected results and in section 6 we sum up our conclusions.

2. Effective operators

In the effective lagrangian formalism we assume that the SM is no more than a low energy limit of some more general theory to be revealed at TeV scale. In this context, the top-quark FCNC interaction is seen at low energies as a contact interaction that can be parametrized with higher dimension effective operators. The complete lagrangian is written as

$$\mathcal{L} = \mathcal{L}^{SM} + \frac{1}{\Lambda}\mathcal{O}^5 + \frac{1}{\Lambda^2}\mathcal{O}^6 + \mathcal{O}(\frac{1}{\Lambda}) \tag{1}$$

where we have truncated the expansion series at dimension $d = 6$. We follow the formalism in [6], where the SM symmetries are imposed and all operators are built using the already known SM fields. This results in a reduced list of the possible dimension six operators. Additionally, baryon and lepton number conservation is imposed, which restricts the operators to dimension six only. Even imposing the SM symmetries, the number of all possible FCNC operators is quite vast. However, the list can be reduced by using appropriate equations of motion and Fierz transformations applied to the physical operators we are dealing with. The final minimal complete set of FCNC operators for top-quark physics was obtained in [9, 10, 11, 12]. With this procedure, a single operator survives in the strong sector [11]. Following the notation in [6] this operator can be written as

$$\mathcal{O}^{ij}_{uG\phi} = \bar{q}^i_L\, \lambda^a\, \sigma^{\mu\nu}\, u^j_R\, \tilde{\phi}\, G^{a\mu\nu} \tag{2}$$

where $G^a_{\mu\nu}$ is the gluonic field tensor, u^i right-handed u or c-quark field and q^i is the left-handed quark doublet. The same operator also appears in the literature as a dimension 5 operator

$$i\kappa_i \frac{g_s}{\Lambda}\bar{q}_i\lambda^a\sigma^{\mu\nu}(f_i + h_i\,\gamma_5)tG^a_{\mu\nu} \quad , \tag{3}$$

where κ_i is a real parameter, $i = u, c$, g_s is the strong coupling constant and f_i and h_i are complex numbers which respect $|f_i|^2 + |h_i|^2 = 1$. This operator is responsible for the effective FCNC vertices tgu and tgc. The minimum set of operators in the electroweak sector is

$$\mathcal{O}^{ij}_{uB\phi} = \bar{q}^i_L\, \sigma^{\mu\nu}\, u^j_R\, \tilde{\phi}\, B_{\mu\nu}\,, \quad \mathcal{O}^{ij}_{uW\phi} = \bar{q}^i_L\, \tau_I\, \sigma^{\mu\nu}\, u^j_R\, \tilde{\phi}\, W^I_{\mu\nu}\,, \tag{4}$$

$$\mathcal{O}^{ij}_{\phi u} = i\,(\phi^\dagger D_\mu \phi)\,(\bar{u}^i_R\, \gamma^\mu\, u^j_R)\,, \tag{5}$$

$$\mathcal{O}^{(1),ij}_{\phi q} = i\,(\phi^\dagger D_\mu \phi)\,(\bar{q}^i_L\, \gamma^\mu\, q^j_L)\,, \quad \mathcal{O}^{(3),ij}_{\phi q} = i\,(\phi^\dagger\, \tau_I\, D_\mu \phi)\,(\bar{q}^i_L\, \gamma^\mu\, \tau_I\, q^j_L)\,, \tag{6}$$

$$\mathcal{O}^{ij}_{u\phi} = (\phi^\dagger \phi)\,(\bar{q}^i_L\, u^j_R\, \tilde{\phi})\,, \tag{7}$$

where $B^{\mu\nu}$ and $W^I_{\mu\nu}$ are the $U(1)_Y$ and $SU(2)_L$ field tensors, respectively. The field ϕ is the usual Higgs boson doublet. In addition to this electroweak set of operators, we have included subsets of four-fermion operators which are described in [13].

Journal of Physics: Conference Series **447** (2013) 012031 doi:10.1088/1742-6596/447/1/012031

3. Physical Processes

The physical processes included in MEtop are FCNC direct top production and FCNC single top production. Direct top is available both at LO and at NLO while FCNC single top is availble only at LO. All processes can be generated for both pp and $p\bar{p}$ hadron colliders. Furthermore, the full top decay chain is included so that spin correlations are preserved. Also, in order to perform spin correlations studies, a on/off switch for the top decay chain was made available. In figures 1 and 2 we show the generic Feynman diagrams for the physical processes implemented in MEtop. The production process depicted in figure 1 has only strong FCNC operator contributions. Born level FCNC direct top is shown in figure 1 (left) together with the $t+g$ process (right). In figure 2 we present the diagrams for the LO $t+q$ production process from gluon fusion (right) and with quarks in the initial state (left) . The former is only affected by strong FCNC operators while the later has contributions from the strong, electroweak and four-fermion sectors.

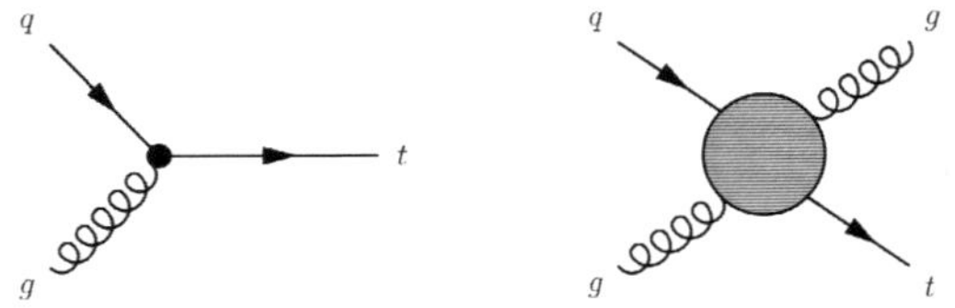

Figure 1. FCNC leading order direct top production and top + gluon production at the parton level. Only FCNC strong operators contribute to the process.

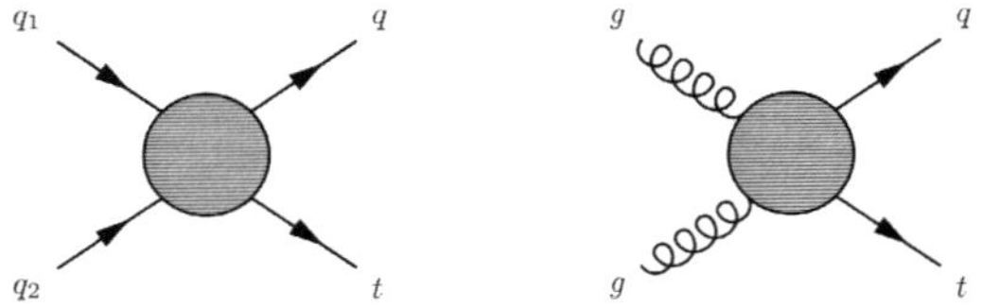

Figure 2. FCNC leading order top + quark production at the parton level. FCNC strong and electroweak operators contribute to the process together with 4F operators.

4. NLO approximation

As mentioned in the previous section, MEtop is able to generate FCNC direct top events with NLO precision. The implementation was performed by adopting an NLO effective approximation [14]. A general NLO calculation of an n-particles final state process can be represented by

$$d\sigma_{NLO} = d\sigma_B d\Phi_n + d\sigma_V d\Phi_n + d\sigma_R d\Phi_{n+1}, \tag{8}$$

where B stands for Born, V for Virtual and R for Real. It is well known that the virtual term has infrared divergences that will only be cured by the inclusion of the real radiation term σ_R. We show in figure 3 the class of diagrams contributing to NLO direct top production. The first process $qg \to t$ correspond to the Born term. The second, $qg \to t$, corresponds to the Virtual term while the third, $qg \to gt$, is the real radiation term. This last term gives rise to an infrared divergence which correspond to a low P_T and/or to a collinear external gluon. Kinematically, the divergence can be parametrized by the gluon transverse momentum, P_T, and

Journal of Physics: Conference Series **447** (2013) 012031 doi:10.1088/1742-6596/447/1/012031

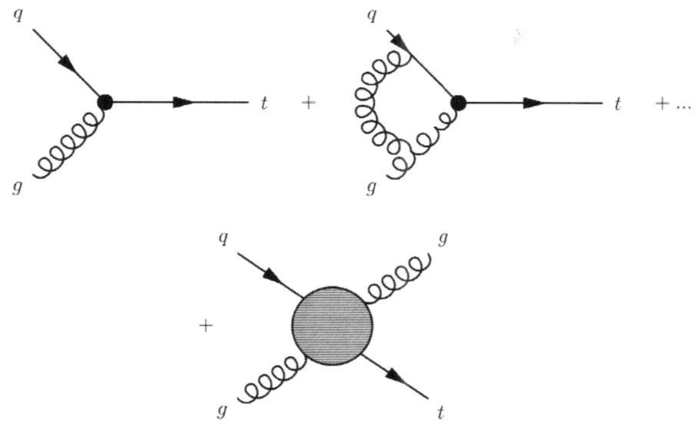

Figure 3. Inclusive FCNC direct top production at NLO in QCD.

it obviously occurs in the limit $P_T \to 0$. In order to avoid the divergence one could just impose a P_T cut. The resulting region has to be excluded from the event space generation with some care - since the problem arises only in the collinear/soft limit, we can fill this phase space region with a branching mechanism. By doing so, we assume a collinear factorization, where the QCD radiation is emitted by one of the external legs of the Born configuration. This approximation can be translated by

$$|M_{t+g}|^2 d\Phi_2 \to |M_t|^2 d\Phi_1 \frac{\alpha_S}{2\pi}\frac{dt}{t} P_{q,qg}(z) dz \frac{d\phi}{2\pi} \tag{9}$$

where M_{t+g} and M_t are the real radiation for $(t+g)$ and direct top amplitudes, respectively; $d\Phi_i$ is the phase space for the ith body processes and $P_{q,qg}(z)$ is the Altarelli-Parisi splitting function. The variable t is a resolution parameter which in our case is the top quark transverse momentum P_T^{top}. The hard P_T region will then be filled by using the appropriate transition amplitude and the low P_T region by the use of a shower mechanism. In this NLO effective approximation [14], the P_T phase-space is then divided in two regions parametrized by P_T^{top}, which we call P_T^{match}. This parameter plays the role of a matching parameter where the criterion for a good transition is the smoothness of the final P_T distribution. The final cross section free of divergences can be written as

$$\sigma_{NLO} = K\sigma_B(P_T^{PS} < P_T^{match}) + \sigma_R(P_T > P_T^{match}), \tag{10}$$

where P_T^{PS} is the transverse momentum of the parton shower emission. The contribution from the virtual term $d\sigma_V$ in equation 8 is included via a K-factor applied to the born term and because born and virtual terms share the same $2 \to 1$ topology we assume they have similar kinematics. It is from equation 10 that events in MEtop are generated. Since MEtop is a parton level generator, the events must then be submitted to a parton shower simulator in order to include initial (ISR) and final (FSR) state radiation. Therefore, since these parton showers are based on a branching mechanism, we leave the collinear and/or soft P_T phase space region to be filled by the corresponding parton shower. In order to avoid double counting we just have to ensure that the parton shower will not populate the $P_T > P_T^{match}$ region in the first emission. To prevent it, the shower mechanism must be performed with the PT-ordered scheme with the first emission starting at $P_T = P_T^{match}$. For the NLO direct top process, MEtop assumes that the Pythia shower will be used with its PT-ordered scheme [15] flag on. Finally, we note that because this is a $2 \to 1$ process, the shower mechanism implemented in Pythia will not add FSR but only ISR. This does not pose any problems because FSR is negligible when compared to ISR due the large top-quark mass.

Journal of Physics: Conference Series **447** (2013) 012031 doi:10.1088/1742-6596/447/1/012031

5. The generation process and some results

In figure 4 we show the diagram flow of the MEtop generation process. It was drawn specifically for the NLO direct top case. The remaining processes follow the same flow except for the K-factor calculation step. The amplitudes used to evaluate the cross sections were generated with CalcHEP [16], whit Feynman rules derived with LanHEP [17] and all integrations were performed using the Cuba Library [18]. The K-factors are calculated "on the fly" and therefore each sub process has its own normalization factor.

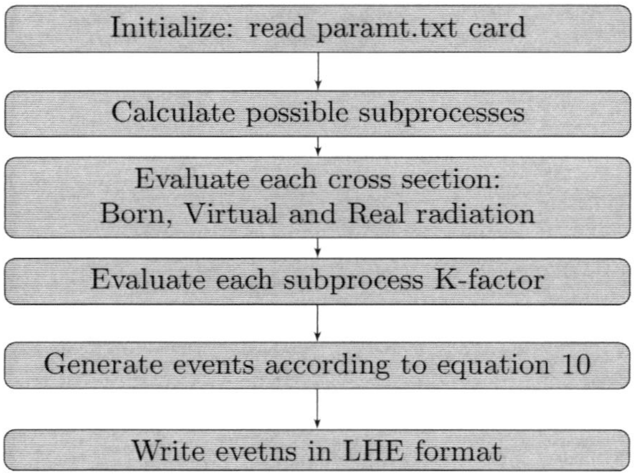

Figure 4. MEtop flow cart for NLO Direct top generation.

In figure 5 we present the top-quark P_T distribution for direct top after the first emission and with the starting scale shower at m_t (black solid line). The blue dashed line represents the top P_T distribution of the real radiation process. As explained before, both regions overlap and therefore a matching must be introduced in order to avoid double counting. The matching process must be chosen so that the transition between the rescaled direct top distribution and the real radiation process is smooth. In figure 6 we show the top P_T distribution for $P_T^{match} = 10, 15$ Gev where direct top is the gray dashed line and $t + g$ is the blue dashed line. The first one corresponds to the top distribution after the first ISR emission, where the PT-shower starting scale was set to 10 and 15 Gev, respectively; the second one is the top distribution from the 't+g' process where a P_T^{cut} of 10 and 15 Gev was imposed. We have studied several P_T^{match} scenarios in the 5 Gev to 20 Gev range finding no significant differences. We therefore recommend that this matching value should be included in any analysis as a systematic error.

In figure 7 we show the final NLO direct top quark P_T and η distribution (solid line), as well as the LO result (dashed line) after full ISR, FSR and Multiple Interactions (MP). It is clear that in both distributions the final NLO result does not amount to simply multiplying the LO result by a K-factor. In fact, the distributions show that the NLO contribution concentrates more events in the low P_T region and produces top quarks at higher angles. In figure 8 we show the equivalent distribution for the lepton coming from the decaying W. The b-quark and neutrino distributions show similar differences between the NLO and LO results. For the top decay products, the NLO contribution does not result in a major change in distributions making a K-factor adjustment suitable. However, one must pay special attention to analysis where the top quark momentum is reconstructed. For example, analysis which include the study of the top spin polarisation usually requires full reconstruction of its momentum. In this case, figure 7 show that NLO events should be considered. Finally, when studying the inclusive NLO direct top production one must add the $t + jet$ (single top FCNC) process as well. The distribution of

Journal of Physics: Conference Series **447** (2013) 012031 doi:10.1088/1742-6596/447/1/012031

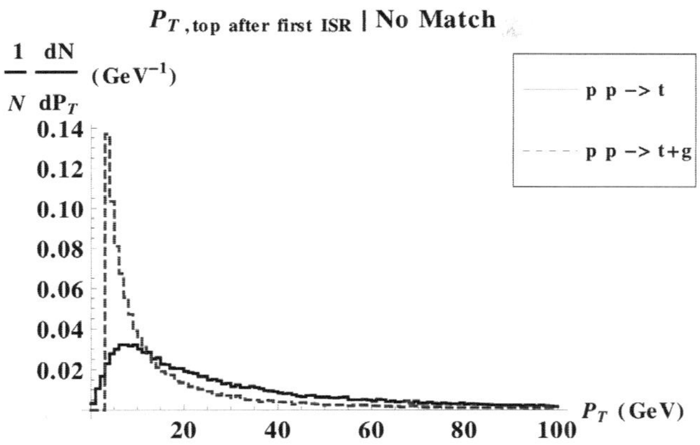

Figure 5. P_T distribution of the top quark for $\sqrt{s} = 7$ TeV. The black solid line is for direct top production after the first branching in ISR, with starting scale of m_t. The blue dashed line is for the hard process top+gluon production.

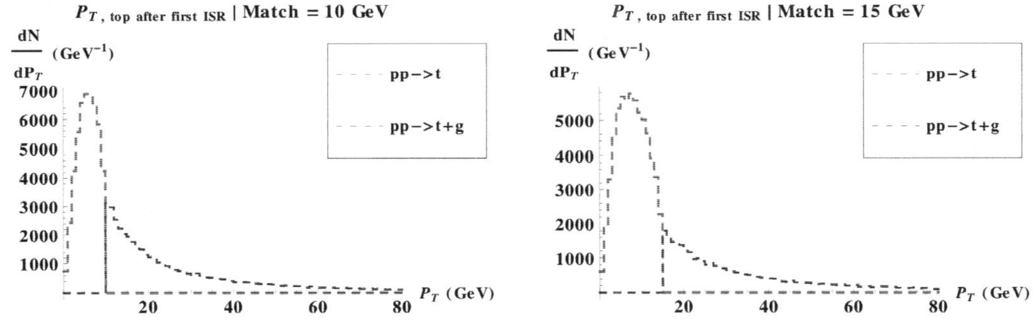

Figure 6. P_T distribution of top quark after the first ISR branching with a P_T^{match} of 10 GeV (left) and 15 GeV (right).

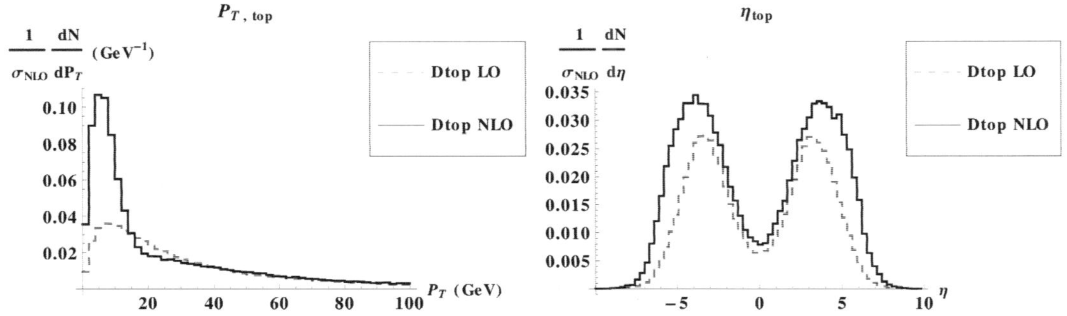

Figure 7. Comparison of the LO and NLO P_T (left) and η (right) distributions of the top quark at the partonic level after the full shower (ISR+FSR) and Multiple Interaction.

the sum of the two processes is shown in figure 9 for the top P_T (right) and the top η (left). The FCNC Single top process was generated with a $P_T^{cut} = 10$ Gev and only subprocesses where a

Journal of Physics: Conference Series **447** (2013) 012031 doi:10.1088/1742-6596/447/1/012031

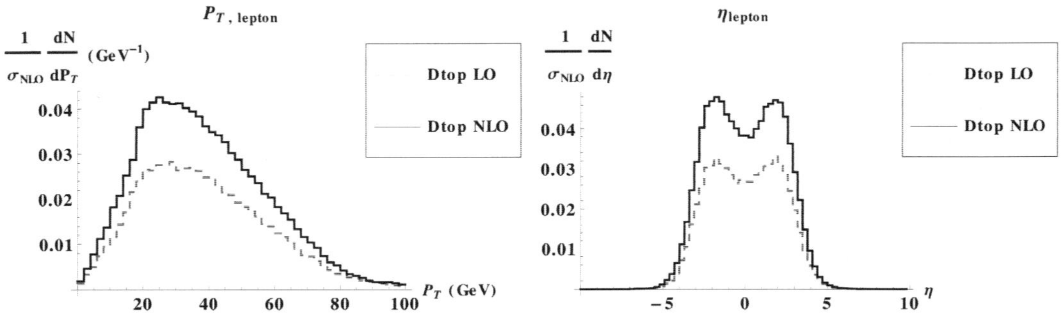

Figure 8. Comparison of the LO and NLO P_T (left) and η (right) distributions of the lepton from $t \to bW \to bl\nu$ at the partonic level after the full shower (ISR+FSR) and Multiple Interaction.

FCNC interaction takes place were included. Processes like $u\bar{d} \to t\bar{b}$ which are pure SM single top production were discarded. In tables 1 and 2 we show the total contributions from each LO and NLO processes. They were calculated for $k_{u,c} = 0.01$ and for LHC @ 7 TeV. As we can see from figure 9, the FCNC single top contribution does not produce considerable shape modifications, although its contribution to the total cross section is approximately 13%.

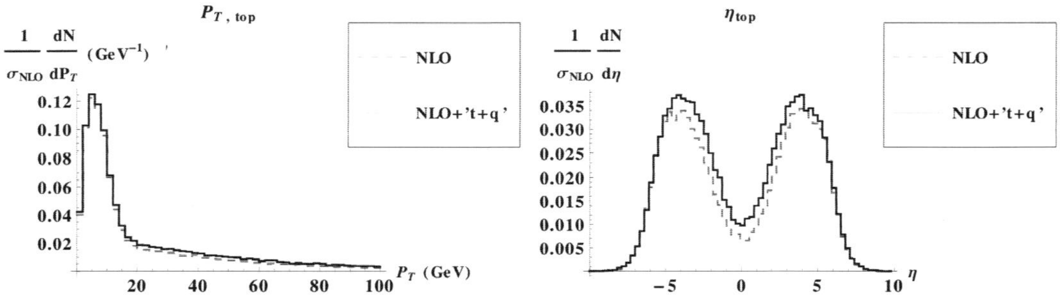

Figure 9. P_T (left) and η (right) distributions of the top quark for NLO direct top (solid line) and NLO direct top plus $pp \to tq$ with $P_T^{match} = 10$ GeV and jet $p_T > 10$ GeV.

Table 1. FCNC Direct top (anti-top) LO and NLO cross sections for $k_{u,c} = 0.01$ and LHC @ 7 Tev.

Subprocess	LO	NLO
$ug \to t$	6.12	8.74
$cg \to t$	0.91	1.67

6. Conclusions

We have presented a new Monte Carlo generator dedicated to top FCNC physics. We have included the main FCNC top production channels: direct top and single top. The direct top channel is available at NLO and the single top at LO. A complete set of dimension six operators

Journal of Physics: Conference Series **447** (2013) 012031 doi:10.1088/1742-6596/447/1/012031

Table 2. FCNC $t + q$ with $k_{u,c} = 0.01$ and LHC @ 7 Tev and $PT_{cut} = 10$ Gev.

Subprocess	$k_u = 0.01$	$k_c = 0.01$
$p, p \to t, q$	1.12	0.40

are already included. We have shown that the NLO top distributions obtained for the direct top cannot be built via a K-factor from the LO ones. Therefore NLO events should be used especially in the cases where top-quark momentum reconstruction is required.

Acknowledgments

This work is partially supported by the Portuguese *Fundação para a Ciência e a Tecnologia* (FCT) under contracts CERN/FP/123619/2011 and PTDC/FIS/117951/2010. RS is also partially supported by an FP7 Reintegration Grant, number PERG08-GA-2010-277025 and by PEst-OE/FIS/UI0618/2011. RC is funded by FCT through the grant SFRH/BPD/45198/2008.

References

[1] Glashow S L, Iliopoulos J and Maiani L 1970 Weak Interactions with Lepton-Hadron Symmetry *Phys. Rev. D* **2** 1285
[2] Aguilar-Saavedra J A 2004 Top flavor-changing neutral interactions: Theoretical expectations and experimental detection *Acta Phys. Polon. B* **35** 2695 [arXiv:hep-ph/0409342].
[3] Liu J J, Li C S, Yang L L and Jin L G 2005 Next-to-leading order QCD corrections to the direct top quark production via model-independent FCNC couplings at hadron colliders *Phys. Rev. D* **72** 074018
[4] Slabospitsky S R and Sonnenschein L 2002 TopReX generator (version 3.25): Short manual *Comput. Phys. Commun.* **148** 87
[5] Aguilar-Saavedra J A 2010 Zt, gamma t and t production at hadron colliders via strong flavour-changing neutral couplings *Nucl. Phys. B* **837** 122
[6] Buchmüller W and Wyler D 1986 Effective Lagrangian Analysis of New Interactions and Flavor Conservation *Nucl. Phys. B* **268** 621
[7] Alwall J *et al.* 2007 A standard format for Les Houches event files *Comput. Phys. Commun.* **176** 300
[8] Sjostrand T, Mrenna S and Skands P 2006 PYTHIA 6.4 Physics and Manual *J. High Energy Phys.* JHEP0605(2006)026
[9] Ferreira P M, Oliveira O and Santos R 2006 Flavour changing strong interaction effects on top quark physics at the LHC *Phys. Rev. D* **73** 034011
[10] Ferreira P M and Santos R 2006 Strong flavour changing effective operator contributions to single top quark production *Phys. Rev. D* **73** 054025
[11] Aguilar-Saavedra J A 2009 A Minimal set of top anomalous couplings *Nucl. Phys. B* **812**, 181
[12] Grzadkowski B, Iskrzynski M, Misiak M and Rosiek J 2010 Dimension-Six Terms in the Standard Model Lagrangian *J. High Energy Phys.* JHEP1010(2010)085
[13] Coimbra R, Onofre A, Santos R and Won M 2012 MEtop - a generator for single top production via FCNC interactions *Eur. Phys. J. C* **72** 2222
[14] Boos E E, Bunichev V E, Dudko L V, Savrin V I and Sherstnev A V 2006 Method for simulating electroweak top-quark production events in the NLO approximation: SingleTop event generator, *Phys. Atom. Nucl.* **69** 1317
[15] Sjostrand T and Skands P 2005 Transverse-momentum-ordered showers and interleaved multiple interactions *Eur. Phys. J. C* **39** 129 [hep-ph/0408302].
[16] Pukhov A 2004 CalcHEP 2.3: MSSM, structure functions, event generation, batchs, and generation of matrix elements for other packages *Preprint* arXiv:hep-ph/0412191
[17] Semenov A 1998 LanHEP: A package for automatic generation of Feynman rules from the Lagrangian *Comput. Phys. Commun.* **115** 124
Semenov A 2009 LanHEP - a package for the automatic generation of Feynman rules in field theory. Version 3.0 *Comput. Phys. Commun.* **180** 431
[18] Hahn T 2005 CUBA: A library for multidimensional numerical integration *Comput. Phys. Commun.* **168** 78

Journal of Physics: Conference Series **447** (2013) 012032 doi:10.1088/1742-6596/447/1/012032

Top Effective Operators at the ILC

Miguel C. N. Fiolhais

LIP, Department of Physics, University of Coimbra, 3004-516 Coimbra, Portugal

E-mail: `miguel.fiolhais@cern.ch`

Abstract.

In this paper we study the effect of new physics contributions to the top quark pair production ($t\bar{t}$) in a possible future linear collider, such as the International Linear Collider (ILC). The use of a dimension-six gauge invariant effective operator approach allows to compare the prospected results at the ILC with the current ones obtained at the Large Hadron Collider (LHC), both in neutral and charged current processes. We also prove that the use of specific observables, together with a combination of measurements in different polarized beam scenarios and with different center-of-mass energies, allows to disentangle different effective operator contributions and significantly improve the limits on the anomalous couplings with respect to the LHC.

1. Introduction

The ILC is a possible future linear collider, with an estimated length of 30-50 km to accelerate and collide electrons and positrons at a center-of-mass energy of 500 GeV. Since collisions between electrons and positrons are much easier to analyze than hadronic collisions, the ILC is expected to deliver several precision measurements, which provide interesting tests to the Standard Model (SM) predictions.

The top quark is the most massive elementary particle discovered to date, and therefore, a natural candidate for the search of new physics beyond the SM. The precision measurement of its properties, in particular its couplings, may provide useful insights on possible new physics contributions at a higher energy scale. The realization of these precision measurements of the top quark properties in a possible future ILC would, therefore, provide an interesting complement to the direct searches being carried out at the LHC.

The effect of new physics contributions to the top quark interactions can be parameterized in a model-independent form, above the electroweak symmetry breaking scale, in terms of gauge invariant dimension-six effective operators [1],

$$\mathcal{L}_{\text{eff}} = \sum \frac{C_x}{\Lambda^2} O_x + \dots , \tag{1}$$

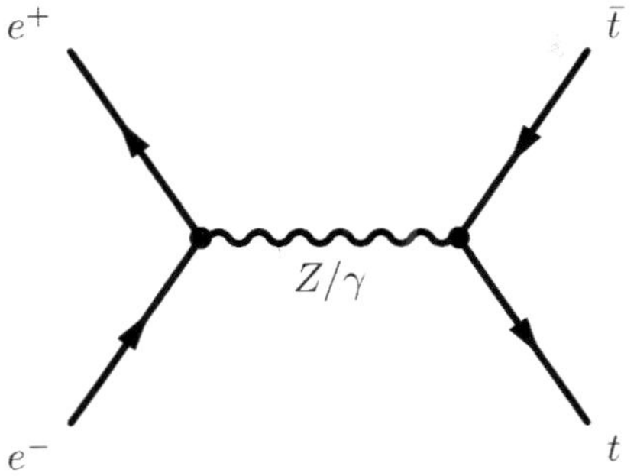

Figure 1. Feynman diagram of the top quark pair production at the ILC through the s-channel.

where O_x are dimension-six effective operators, invariant under the SM gauge symmetry $\mathrm{SU}(3)_c \times \mathrm{SU}(2)_L \times \mathrm{U}(1)_Y$, characterized by the dimensionless constants C_x and a new physics scale Λ. Unlike previous approaches [2–7], the effective operator framework allows to reduce the number of independent parameters entering fermion trilinear interactions [8, 9], and grants direct comparisons between measurements of different top quark vertices, such as the Wtb and $Zt\bar{t}$ vertices, in different colliders.

Among the dimension-six gauge invariant effective operators, there are only five non-redudant operators which contribute to the $t\bar{t}$ production at the ILC (depicted in Figure 1), or in other words, to the $Zt\bar{t}$ and $\gamma t\bar{t}$ interaction vertices [8–10]:

$$\begin{aligned}
O_{\phi q}^{(3,3+3)} &= i\left[\phi^\dagger(\tau^I D_\mu - \overleftarrow{D}_\mu \tau^I)\phi\right](\bar{q}_{L3}\gamma^\mu\tau^I q_{L3}), & O_{uW}^{33} &= (\bar{q}_{L3}\sigma^{\mu\nu}\tau^I t_R)\tilde{\phi}\, W_{\mu\nu}^I, \\
O_{\phi q}^{(1,3+3)} &= i(\phi^\dagger \overleftrightarrow{D}_\mu \phi)(\bar{q}_{L3}\gamma^\mu q_{L3}), & O_{uB\phi}^{33} &= (\bar{q}_{L3}\sigma^{\mu\nu} t_R)\tilde{\phi}\, B_{\mu\nu}, \\
O_{\phi u}^{3+3} &= i(\phi^\dagger \overleftrightarrow{D}_\mu \phi)(\bar{t}_R\gamma^\mu t_R), & & \qquad (2)
\end{aligned}$$

using standard notation where τ^I are the Pauli matrices, q_{L3} is the left-handed third generation quark doublet, t_R is the right-handed top quark singlet, ϕ is the SM Higgs doublet, $\tilde{\phi} = i\tau^2\phi^*$, $W_{\mu\nu}^I$ and $B_{\mu\nu}$ is the $\mathrm{SU}(2)_L$ and $\mathrm{U}(1)_Y$ field strength tensors, respectively, D_μ ($\overleftarrow{D}_\mu$) the covariant derivative acting on the right (left) and $\overleftrightarrow{D}_\mu = D_\mu - \overleftarrow{D}_\mu$. Since the three operators in the left column of equation (2) are Hermitian, their coefficients must be real.

After the spontaneous symmetry breaking, the most general dimension-six $Zt\bar{t}$ and $\gamma t\bar{t}$ lagrangians read

$$\mathcal{L}_{Ztt} = -\frac{g}{2c_W}\bar{t}\gamma^\mu\left(c_L^t P_L + c_R^t P_R\right)t\, Z_\mu - \frac{g}{2c_W}\bar{t}\frac{i\sigma^{\mu\nu}q_\nu}{M_Z}\left(d_V^Z + i d_A^Z\gamma_5\right)t\, Z_\mu\,, \qquad (3)$$

$$\mathcal{L}_{\gamma tt} = -eQ_t\bar{t}\gamma^\mu t\, A_\mu - e\bar{t}\frac{i\sigma^{\mu\nu}q_\nu}{m_t}\left(d_V^\gamma + i d_A^\gamma\gamma_5\right)t\, A_\mu\,, \qquad (4)$$

with $c_L^t = X_{tt}^L - 2s_W^2 Q_t$, $c_R^t = X_{tt}^R - 2s_W^2 Q_t$ ($Q_t = 2/3$ is the top quark electric charge) and

$$X_{tt}^L = 1 + \left[C_{\phi q}^{(3,3+3)} - C_{\phi q}^{(1,3+3)}\right]\frac{v^2}{\Lambda^2}, \qquad d_V^Z = \sqrt{2}\,\mathrm{Re}\left[c_W C_{uW}^{33} - s_W C_{uB\phi}^{33}\right]\frac{v^2}{\Lambda^2},$$
$$X_{tt}^R = -C_{\phi u}^{3+3}\frac{v^2}{\Lambda^2}, \qquad d_A^Z = \sqrt{2}\,\mathrm{Im}\left[c_W C_{uW}^{33} - s_W C_{uB\phi}^{33}\right]\frac{v^2}{\Lambda^2}, \tag{5}$$

and,

$$d_V^\gamma = \frac{\sqrt{2}}{e}\mathrm{Re}\left[s_W C_{uW}^{33} + c_W C_{uB\phi}^{33}\right]\frac{v m_t}{\Lambda^2},$$
$$d_A^\gamma = \frac{\sqrt{2}}{e}\mathrm{Im}\left[s_W C_{uW}^{33} + c_W C_{uB\phi}^{33}\right]\frac{v m_t}{\Lambda^2}. \tag{6}$$

Since the SM bottom quark couplings have already been probed with great precision at PETRA, LEP and SLD, it is reasonable to assume the following approximation [8–11]:

$$C_{\phi q}^{(1,3+3)} \simeq -C_{\phi q}^{(3,3+3)}, \tag{7}$$

to cancel out the non-SM contributions to the Zb_Lb_L vertex. The same equality also appears in other SM extensions, such as in new charge 2/3 singlets [12–15].

2. Direct comparison between ILC and LHC

In order to probe the effective operator coefficients at the ILC, specific observables were used, such as the total cross-sections and forward-backward asymmetries. The strong dependency of these observables with the effective operators coefficients enhances the possibility to further constrain the anomalous contributions, and allows to disentangle different operators contributions[1]. Furthermore, since there are two effective operators, $O_{\phi q}^{(3,3+3)}$ and O_{uW}^{33}, which modify both the $Zt\bar{t}$, $\gamma t\bar{t}$ and Wtb vertices, the prospected results at the ILC can be compared with the current ones at the LHC. For example, in Figure 2 (left), the dependency of an ILC observable, the unpolarized FB asymmetry, is presented for $\mathrm{Re}\, C_{uW}^{33}$, within the window range of the current limits extracted by the ATLAS Collaboration [17], in Figure 2 (right), through the measurement of the W boson helicity fractions in top quark decays[2]. On the left plot, the yellow and green bands around the SM value correspond to 1σ and 2σ variations, respectively, for total total uncertainties of 5% in the cross section and 2% in the asymmetry [18]. These results show the great potential of the ILC in improving the limits on the effective operator coefficients with respect to the LHC, due to smaller uncertainties and enhanced dependencies. As an additional exercise, the sensitivity to the new physics scale Λ can be extended up to 4.5 TeV for an operator coefficient equal to the unity. The anti-Hermitian part of this operator have also been probed at the ATLAS experiment [19], with a CP-violating asymmetry A_{FB}^N, defined for polarized top decays [20]. Even though it does not interfere with the SM in CP-conserving observables, such as the total cross-section, the sensitivity at the ILC is similar to the one at the LHC.

For other operators, such as $O_{\phi q}^{(3,3+3)}$ and $O_{uB\phi}^{33}$, the expected sensitivity at the ILC shall largely surpass the current and potential LHC limits [21, 22].

[1] The quadratic terms were kept in the operator coefficients, which is consistent with the $1/\Lambda^2$ expansion of the effective operator framework [16].
[2] Note that $g_\mathrm{R} = \sqrt{2}C_{uW}^{33}\nu^2/\Lambda^2$.

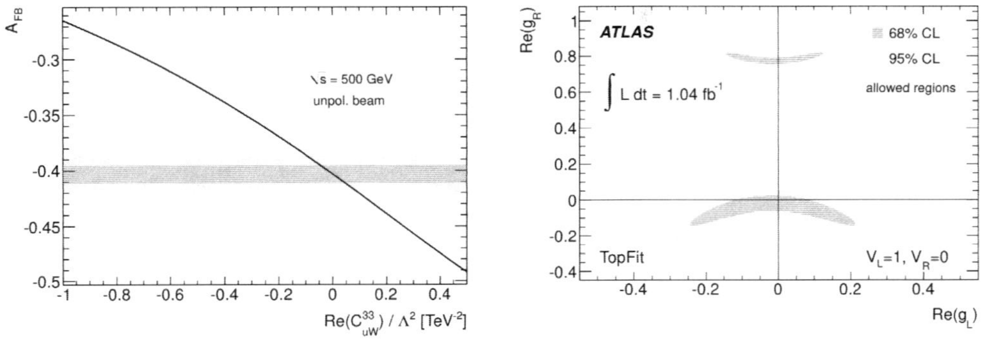

Figure 2. Left: dependence of the FB asymmetry on $\mathrm{Re}\, C_{uW}^{33}$. Right: allowed regions at 68% and 95% confidence level (CL) for the Wtb anomalous couplings (extracted from [17]).

3. Electron beam polarization

The use of electron beam polarization provides a competitive advantage to the ILC, not acessible at the LHC. In particular, the beam polarization allows to separate the d_j^Z and d_j^γ couplings, or in other words, the different anomalous contributions to the Z and γ exchange in the s-channel, due to the different dependencies in different polarized beam scenarios.

In this preliminary study, two polarized beam scenarios are considered: 80% right-handed ($P_{e^-} = 0.8$) and 80% left-handed ($P_{e^-} = -0.8$). Since the possibility of using the polarization of the positron beam at a considerable value is still uncertain, its use is not taken into account here. In Figure 3, the estimated allowed regions on $\mathrm{Re}\, C_{uW}^{33}$ and $\mathrm{Re}\, C_{uB\phi}^{33}$ are presented with and without the use of beam polarization (left plot), and for each individual polarized beam scenario (right plot). In the left plot, the yellow region corresponds to the unpolarized case, where the measurements of both coefficients are anti-correlated, while the green region is much smaller due to the use of electron polarization. On the right plot, the orthogonality of the two regions shows the complementary of the left- and right-handed beams, which allows to disentangle C_{uW}^{33} and $C_{uB\phi}^{33}$. It is, therefore, clear, how the use of different polarizations allows to constrain and disentangle these coefficients. These results are obtained assuming the rest of operator coefficients are zero.

4. Center-of-mass energy upgrade to 1 TeV

Not only the use of electron beam polarization is useful in disentangling different operator contributions. The combination of measurements at different CM energies can also allow to separate the the vector and tensor contributions because the CM energy dependence is different. Therefore, a possible upgrade to a CM energy of 1 TeV is crucial to distinguish γ^μ and $\sigma^{\mu\nu}$ couplings. In Figure 4, the allowed regions are shown for $C_{\phi q}^{(3,3+3)}$ and C_{uW}^{33} coefficients, using polarized beams at 500 GeV (yellow), and using the combination of measurements at 500 GeV and 1 TeV (green). The total cross-sections and FB asymmetries for the different polarized beam scenarios were used to constrain these regions, and the blue lines around the green region

Journal of Physics: Conference Series **447** (2013) 012032 doi:10.1088/1742-6596/447/1/012032

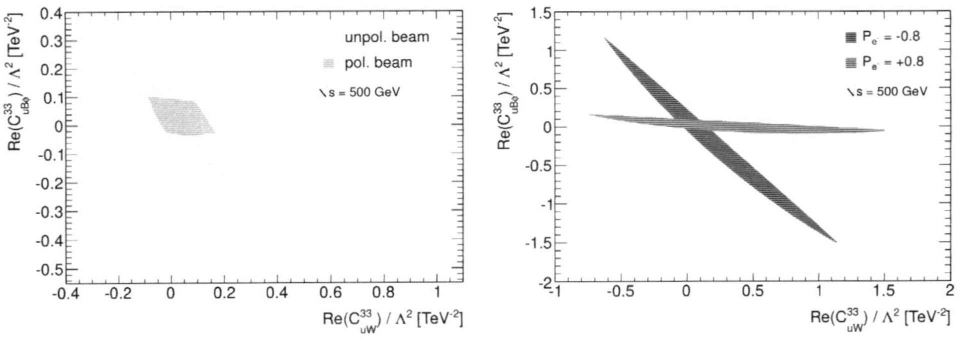

Figure 3. Left: combined limits on C^{33}_{uW} and $C^{33}_{uB\phi}$ for the cases of no beam polarization and electron beam polarization. Right: complementarity of the measurements for $P_{e^-} = 0.8$ and $P_{e^-} = -0.8$.

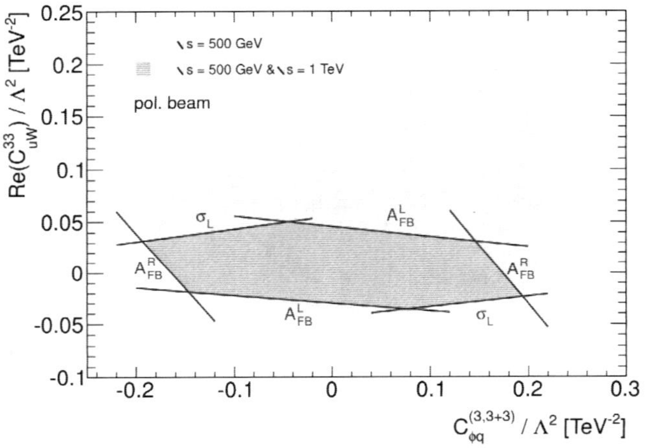

Figure 4. Combined limits on $C^{(3,3+3)}_{\phi q}$ and C^{33}_{uW} for a CM energy of 500 GeV and also with 1 TeV.

correspond to the constraints caused by each observable[3]. Once more, becomes clear how the use of different observables in distinct experimental conditions can help to constrain the limits on the anomalous couplings and separate different contributions.

5. Conclusions

The effect of new physics contributions to the $t\bar{t}$ production at the ILC was estimated in this study, and compared with the LHC within an effective operator framework. Even though the results at the LHC are already excellent, the sensitivity to the dimension-six effective operators coefficients is expected to be much better at the ILC than in LHC processes, such as top quark

[3] Note that these limits do not appear from a global fit but by requiring a 1σ agreement of the different observables considered.

decays [23] and neutral current processes [22, 24]. These results can be further improved by considering observables in top quark decays at the ILC [20, 25–27], and shall be taken into account in the future. Finally, the combination of measurements with different electron beam polarizations at 500 GeV and 1 TeV allow to disentangle different contributions, and make the top quark studies a physics case for the use of polarizations, and for a possible upgrade to 1 TeV at the ILC.

References

[1] W. Buchmuller and D. Wyler, Nucl. Phys. B **268** (1986) 621.
[2] B. Grzadkowski and Z. Hioki, Nucl. Phys. B **484** (1997) 17 [hep-ph/9604301].
[3] B. Grzadkowski and Z. Hioki, Phys. Rev. D **61** (2000) 014013 [hep-ph/9805318].
[4] B. Grzadkowski and Z. Hioki, Phys. Lett. B **476** (2000) 87 [hep-ph/9911505].
[5] B. Grzadkowski and Z. Hioki, Nucl. Phys. B **585** (2000) 3 [hep-ph/0004223].
[6] T. Abe *et al.* [American Linear Collider Working Group Collaboration], [hep-ex/0106057].
[7] E. Devetak, A. Nomerotski and M. Peskin, Phys. Rev. D **84** (2011) 034029 [arXiv:1005.1756 [hep-ex]].
[8] J. A. Aguilar-Saavedra, Nucl. Phys. B **812** (2009) 181 [arXiv:0811.3842 [hep-ph]].
[9] J. A. Aguilar-Saavedra, Nucl. Phys. B **821** (2009) 215 [arXiv:0904.2387 [hep-ph]].
[10] J. A. Aguilar-Saavedra, M. C. N. Fiolhais, A. Onofre, JHEP **1207** (2012) 180 [arXiv:1206.1033 [hep-ph]].
[11] J. A. Aguilar-Saavedra, M. Perez-Victoria, proceedings submitted to Top2012 conference [arXiv:1302.5634 [hep-ph]].
[12] F. del Aguila, J. A. Aguilar-Saavedra and R. Miquel, Phys. Rev. Lett. **82** (1999) 1628 [hep-ph/9808400].
[13] J. A. Aguilar-Saavedra, Phys. Rev. D **67** (2003) 035003 [hep-ph/0210112].
[14] F. del Aguila, M. Perez-Victoria and J. Santiago, Phys. Lett. B **492** (2000) 98 [hep-ph/0007160].
[15] F. del Aguila, M. Perez-Victoria and J. Santiago, JHEP **0009** (2000) 011 [hep-ph/0007316].
[16] J. A. Aguilar-Saavedra, PoS ICHEP **2010** (2010) 378 [arXiv:1008.3225 [hep-ph]].
[17] G. Aad *et al.* [ATLAS Collaboration], JHEP **1206** (2012) 088 [arXiv:1205.2484 [hep-ex]].
[18] P. Doublet, F. Richard, R. Poschl, T. Frisson and J. Rouene, [arXiv:1202.6659 [hep-ex]].
[19] G. Aad *et al.* [ATLAS Collaboration], ATLAS-CONF-2013-032 (2013).
[20] J. A. Aguilar-Saavedra and J. Bernabeu, Nucl. Phys. B **840** (2010) 349 [arXiv:1005.5382 [hep-ph]].
[21] G. Aad *et al.* [ATLAS Collaboration], Phys. Lett. B **717** (2012) 330 [arXiv:1205.3130 [hep-ex]].
[22] U. Baur, A. Juste, L. H. Orr and D. Rainwater, Phys. Rev. D **71** (2005) 054013 [hep-ph/0412021].
[23] J. A. Aguilar-Saavedra, J. Carvalho, N. F. Castro, A. Onofre and F. Veloso, Eur. Phys. J. C **53** (2008) 689 [arXiv:0705.3041 [hep-ph]].
[24] U. Baur, A. Juste, D. Rainwater and L. H. Orr, Phys. Rev. D **73** (2006) 034016 [hep-ph/0512262].
[25] J. A. Aguilar-Saavedra, R. V. Herrero-Hahn, Phys. Lett. B **718** (2013) 983 [arXiv:1208.6006 [hep-ph]].
[26] G. L. Kane, G. A. Ladinsky and C. P. Yuan, Phys. Rev. D **45** (1992) 124.
[27] J. A. Aguilar-Saavedra, J. Carvalho, N. F. Castro, F. Veloso and A. Onofre, Eur. Phys. J. C **50** (2007) 519 [hep-ph/0605190].

Precision measurements in the top quark sector with the CMS experiment at the LHC

Pedro Ferreira da Silva, *for the CMS collaboration*
CERN, Genève, Switzerland
E-mail: psilva@cern.ch

Abstract. A review of the measurements of the $t\bar{t}$ production cross section (inclusive and differential) performed by the CMS experiment at the LHC is given. The results are used to constrain the magnitude of most of the relevant systematics affecting precision measurements in the top quark sector. In addition from the determination of the top quark pair production cross section it is possible to determine the strong coupling constant as $\alpha_S(m_Z) = 0.1178^{+0.0046}_{-0.0040}$ in agreement with the current world average. After the combination of several channels and techniques explored to measure the top quark mass in data, a precise determination of this quantity is made at CMS: $m_t = 173.4 \pm 0.4_{stat} \pm 0.9_{syst}$ GeV. The measurement of the difference $\Delta m_t = m_t - m_{\bar{t}}$ is also presented.

1. Introduction

Amongst the quark family the top quark has revealed experimentally exceptional characteristics. A mass above the electroweak (EWK) symmetry breaking scale ($m_t = 173.18 \pm 0.94$ GeV), a small width which compels it to decay before it can fragment and bound into a hadronic state ($\Gamma_t = 2.0^{+0.7}_{-0.6}$ GeV) and an almost exclusive decay mode into a real W boson and a b quark ($|V_{tb}| = 0.999145^{+0.000021}_{-0.000046}$) [1]. These properties give to this particle a unique role in the Standard Model (SM) of interactions - it contributes fundamentally to the radiative corrections of some of the fundamental parameters of the SM such as the Higgs boson mass. The precise determination of the top quark properties is therefore crucial to the success of understanding the SM and its limitations in particular the origin of mechanism which originates EWK symmetry breaking.

At the Large Hadron Collider (LHC) the top quark is expected to be produced abundantly: at the end of the 2012 run it is expected that $> 4 \cdot 10^6$ top quark pairs have been produced. The experiments located at the LHC have a unique opportunity to analyse this high statistics top quark sample and to study its properties with improved accuracy with respect to the Tevatron experiments. In this manuscript we highlight precision measurements performed by the CMS experiment [2] of the top quark pair production cross section. The study of the inclusive and differential cross section has been used to derived the strong coupling constant (α_S) and to characterise accurately not only the QCD $t\bar{t}$ production but also its production environment. With both an accurate description of the $t\bar{t}$ production and an improved knowledge of the performance and calibration of CMS, precision measurements of the top quark mass were made.

2. Top quark pair production

At the LHC top/anti-top pairs are mostly produced through gluon-gluon fusion ($\approx$90%). At CMS the measurement of the $\mathrm{t\bar{t}}$ cross section has been carried out in all the different final states which result from the combinatorics of the W boson decays produced after the dominant $t \to Wb$ decay. The $\mathrm{t\bar{t}}$ decay channels comprise therefore fully-hadronic (46%), lepton+jets (45%) and dileptonic (9%) final states. One crucial aspect in the measurement of the top cross section is modelling of the signal which is briefly summarised next.

2.1. $\mathrm{t\bar{t}}$ *modelling*

The modelling of the $\mathrm{t\bar{t}}$ signal at CMS is based on MADGRAPH [3] which generates $\mathrm{t\bar{t}}$+up to 3 additional partons at Born level. The generated matrix-element-based events are then used as input to PYTHIA [4] which takes care of performing the parton showering and simulating the proton remnants and the decays of unstable particles. TAUOLA [5] is specifically used for the decays of τ leptons. The matrix-element to parton-shower (ME-PS) matching is done using the so-called k_{T} - MLM algorithm. In the generation two parameters are crucial: the ME-PS matching threshold and the factorisation and normalisation scales defined as a single parameter $Q^2 = \mu_{\mathrm{f}} = \mu_{\mathrm{r}} = m_{\mathrm{t}}^2 + \sum p_{\mathrm{T}}^2(\mathrm{parton})$. Besides characterising the scale of the hard interaction, the Q^2-scale is also shared with the α_{S}-based evolution scale for ISR/FSR in PYTHIA.

2.2. Inclusive $\mathrm{t\bar{t}}$ *cross section*

Using the 7 TeV proton-proton collision data the most precise measurement is obtained in the dilepton channel [6]. Even if not being the most favoured in terms of branching ratio it has high purity and it is less affected by some of the instrumental uncertainties such as jet energy scale or knowledge of the trigger/selection efficiency owing to a smaller jet activity with respect to the lepton+jets or fully hadronic channels. The selection of this channel requires at least two opposite sign charged leptons (e or μ) with a transverse momentum - p_{T} >20 GeV - and centrally produced in the detector - $|\eta|$ <2.5 (e) or 2.1 (μ). The lepton candidates are required to be isolated in such a way that the p_{T} flux of the particle candidates reconstructed with the particle flow algorithm [7] in a cone of radius $R = 0.3$ built around the momentum of the lepton does not exceed 20% (17%) of the muon's (electron's) p_{T}. Two jets, clustered with the anti-k_{T} algorithm with a cone R=0.5, with p_{T} >30 GeV and $|\eta|$ <2.4 are furthermore required. b-tagging information may be used in the analysis given the expectations for two b-jets in the final state. In the same flavour channels (ee or $\mu\mu$) it is also imposed that the total balance of the p_{T} of all particle candidates of the event (missing transverse momentum) is $E_{\mathrm{T}}^{\mathrm{miss}}$>30 GeV. This last requirement is useful to reject further the contamination of Drell–Yan events (DY). In the same flavour channels dilepton candidates with an invariant mass compatible with the Z boson are rejected for the measurement of the cross section but counted with the purpose of estimating the DY contamination in the signal region. In the $e\mu$ channel the residual $Z/\gamma^* \to \tau\tau$ contamination is estimated from a fit to the dilepton invariant mass distribution. The contribution of events with fake lepton candidates (multijets and $W \to \ell\nu$) is estimated using a data-driven probability of selecting a fake or non-isolated lepton candidate and the control sideband sample where the lepton candidates are rejected by the isolation criteria. After event selection the purity of the sample is expected to be high ($\approx$77%/86% for the same/opposite flavour channels).

The cross section is extracted with a profile likelihood ratio method where the expected event yields for signal and background processes are parametrized taking into account the effect of the systematic uncertainties stemming from both instrumental and theoretical sources. The fit is performed to a total of 3×10 categories corresponding to the permutations of the dilepton channels, jet multiplicities and number of observed b-tagged jets (see Fig. 1, *left*). The three channels, although statistically independent, have common systematic uncertainties which are assumed to be fully correlated across the channels. The measurement yields $\sigma_{\mathrm{t\bar{t}}}(7\ \mathrm{TeV})$ =

$161.9 \pm 2.5_{\text{stat}}\ {}^{+5.1}_{-5.0}\ {}_{\text{syst}} \pm 3.6_{\text{lumi}}$ pb in good agreement with the available approximate next-to-leading order (approx. NLO) calculations [8, 9, 10]. The systematic uncertainty is dominated by the theoretical uncertainty on the contribution of single top (tW) and by experimental uncertainties related to the jet energy scale, lepton efficiencies or the current knowledge on $BR(W \to \ell\nu)$. The result is furthermore cross checked with a simple cut and count analysis which requires at least one of the jets to be b-tagged. The same cut and count analysis has been performed at 8 TeV [11] and measures $\sigma_{t\bar{t}}(8\text{ TeV}) = 227 \pm 3_{\text{stat}} \pm 11_{\text{syst}} \pm 10_{\text{lumi}}$ pb which allows to extract the ratio of cross sections $R_{8/7} = 1.41 \pm 0.1$ in agreement with the expectations from theory (see Fig. 1, *right*).

The cross section extracted from other channels are observed to be in agreement with the ones quoted above but have larger total uncertainty. The 4.1% (6.7%) relative uncertainty attained experimentally at 7 (8) TeV surpasses the current theoretical calculations. Work is ongoing towards the full next-to-next-to-leading order (NNLO) QCD calculation of the inclusive $t\bar{t}$ cross section, which is not yet available for all production channels.

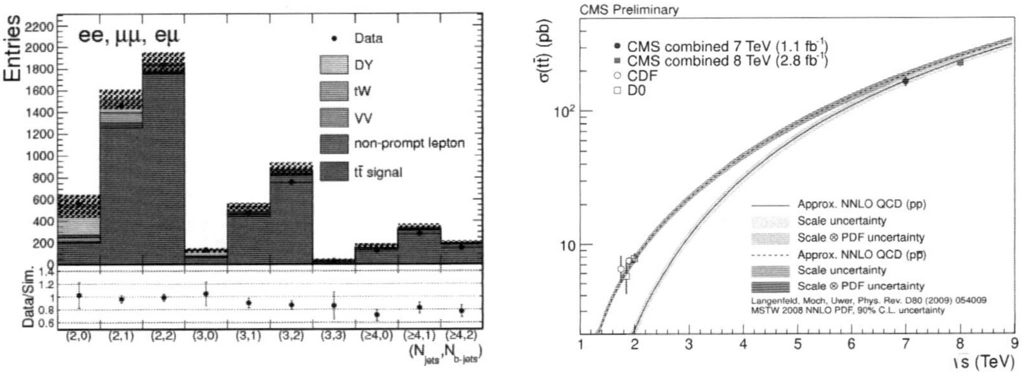

Figure 1. *Left:* Number of events selected for the three combined dilepton channels, as a function of the number of jets and b-tagged jets. The ratios of data to the sum of the $t\bar{t}$ and background predictions are given at the bottom. *Right:* Top pair cross section as function of centre-of-mass energy.

The $t\bar{t}$ cross section measurement can be re-interpreted in order to extract α_S or m_t once the dependency of $\sigma_{t\bar{t}}$ as a function of these quantities is established. Given the strong dependency of m_t on α_S both parameters cannot be extracted simultaneously. Different PDFs and approx.NLO generators have been used to derive the dependency of $\sigma_{t\bar{t}}$ on α_S (see Fig. 2, *left*). α_S is then extracted from a likelihood which convolutes the probability distribution of the experimental measurement (taking into account the experimental sources of uncertainty) with the cross section dependency on α_S which includes a rectangular prior on Q^2. Using Top++ with the NNPDF2.1 PDF set [12] it is determined that $\alpha_S(m_Z) = 0.1178^{+0.0046}_{-0.0040}$ [13] in good agreement with the world average. The dependency of the cross section on m_t has been parametrized for the dilepton channel measurement as: $\sigma_{t\bar{t}}/\sigma_{t\bar{t}}(m_t = 172.5) = 1.00 - 0.008 \cdot (m_t - 172.5) - 0.000137 \cdot (m_t - 172.5)^2$. At the time this manuscript measurement of m_t from the latest $\sigma_{t\bar{t}}$ is on-going and will update the preliminary result reported in [14].

2.3. Differential $t\bar{t}$ measurements

With the large statistics collected it has been possible to perform measurements of the differential $t\bar{t}$ production cross section: $1/\sigma_{t\bar{t}} \cdot d\sigma_{t\bar{t}}/dx$ where x is a kinematical variable of the $t\bar{t}$ system,

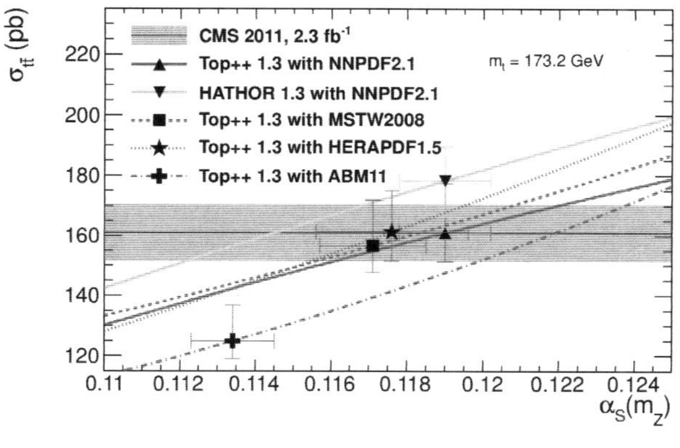

Figure 2. Evolution of the $\mathrm{t\bar{t}}$ cross section as function of α_S for different calculations at approx.NNLO and different PDF sets.

or of each top quark in the event [15] or related to the jet activity [16, 17] or $E_\mathrm{T}^\mathrm{miss}$ [18] in the event.

Measurements of $\sigma_{\mathrm{t\bar{t}}}$ as function of the top/$\mathrm{t\bar{t}}$ system variables require the reconstruction of the kinematics of the event. In the lepton+jets channel, where all the degrees of freedom are specified it is possible to use a constrained kinematics fit to choose the best combination of leptons and jets to reconstruct the $\mathrm{t\bar{t}}$ system. In the dilepton channel this is no longer the case owing to the presence of at least two neutrinos in the final state which lead to an unspecified degree of freedom in the kinematics. In this case the expected distribution for the neutrino energies (from simulation) is used to rank possible solutions in the $100<m_\mathrm{t}<300$ GeV range. Solutions with b-tagged jets are prioritised. After the reconstruction of the kinematics an unfolding procedure is used in order to estimate the particle level kinematics. The unfolding procedure takes not only into account the intrinsic efficiencies and the resolution of the physics objects, but also possible contributions from combinatorial mis-assignment of these objects in the reconstruction of the $\mathrm{t\bar{t}}$ system. A regularised unfolding method has been used for this purpose where for each distribution a response matrix, computed based on simulated events, is used to map the migrations and efficiencies. The generalised inverse of the matrix is obtained after minimising a χ^2 and it is furthermore used to obtain the unfolded distributions. The results are observed to be in good agreement among the different channels analysed and also with the SM predictions at both NLO (obtained with the Powheg [19] and MC@NLO [20] generators) and approx. NNLO where calculations are available. Figure 3 (*left*) shows the good agrement obtained with the approx. NLO prediction for the case of the p_T of the top quark.

Besides measuring differentially the properties of the $\mathrm{t\bar{t}}$ system one can learn about perturbative QCD (pQCD) by studying the properties of the "$\mathrm{t\bar{t}}$ environment": jet activity, $E_\mathrm{T}^\mathrm{miss}$ are two examples. In particular it is crucial to address how well does pQCD model the data and to identify its main dependencies and uncertainties related to the Q^2 scale, ME-PS matching thresholds, hadronisation and fragmentation effects. These are crucial characterisations on the road to the search for new physics effects underlying the $\mathrm{t\bar{t}}$ sample as well as to establish the $\mathrm{t\bar{t}H}$ vertex where at least 4 jets are expected in the case $H \to b\bar{b}$. Precision measurements such as m_t have, in general, a dependency on these parameters as well.

An interesting technique is employed to obtain pure distributions for $\mathrm{t\bar{t}}$+N jets without having to resort to an unfolding procedure. The distribution of the $\sqrt{\chi^2}$ obtained after the fit for the reconstructed kinematics in the lepton+jets channel is analysed using distribution functions derived from simulation for different jet multiplicities. After subtracting the background contribution and comparing to different generators it is observed that both Madgraph+Pythia and Powheg+Pythia yield a similar level of agreement with the data,

Journal of Physics: Conference Series **447** (2013) 012033 doi:10.1088/1742-6596/447/1/012033

while MC@NLO+HERWIG undershoots higher jet multiplicities. Different MADGRAPH settings with varied Q^2 and ME-PS thresholds (by factors of 2 and 1/2 with respect to the nominal values) are also used to compare with the variations observed in data. The so called jet gap fraction (fraction of events with no jet above a given p_{T} threshold) is particular sensitive to these variations and it is shown in Fig. 3, *right*. It displays not only the preference for a higher Q^2 value than the one used by the nominal simulation in CMS but also the fact that the variations considered to evaluate the systematic uncertainties pertaining these parameters are in good agreement with what is observed in data. It is important to conclude referring that in these studies the uncertainties ranges usually from 3% at low multiplicities to 20% at high multiplicities and are dominated by jet energy scale and $\mathrm{t\bar{t}}$ modelling parameters.

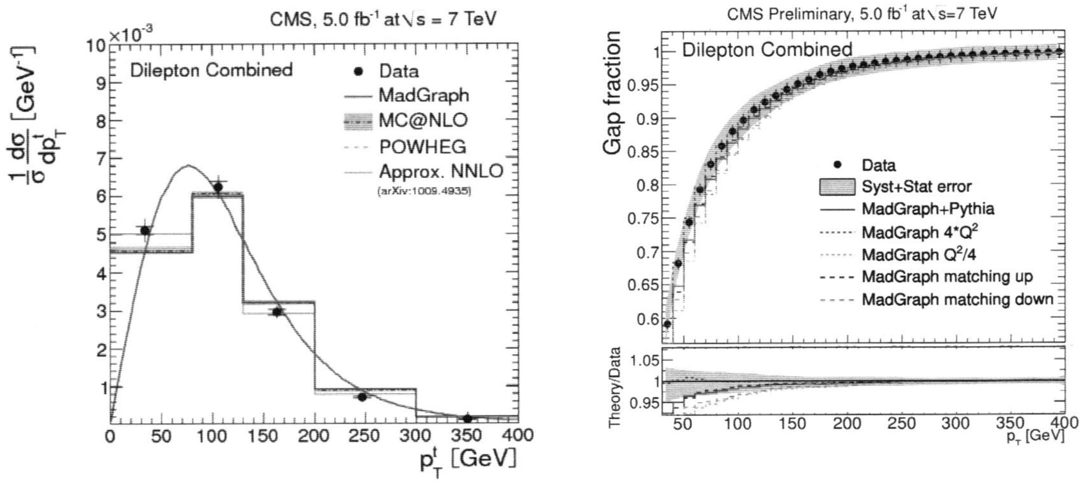

Figure 3. *Left*: Normalised differential $\sigma_{\mathrm{t\bar{t}}}$ in the dilepton channel as function of the p_{T} of the top quark. The inner (outer) bars represent the statistical (total) uncertainty. *Right*: Measured gap fraction as a function of the additional jet p_{T} compared to different MADGRAPH settings. The shaded band represents to total uncertainty (stat+syst).

3. Top quark mass

At CMS the top quark mass has been measured in the lepton+jets [21], dilepton [22, 23] and full hadronic [24] channels. The results obtained in each channel are found to be in agreement with each other and with the current world average. After the combination of the different measurements, using a Best Linear Unbiased Method technique [25], CMS measures $m_{\mathrm{t}} = 173.4 \pm 0.4_{\mathrm{stat}} \pm 0.9_{\mathrm{syst}}$. The total uncertainty attained deserves to be explored in more detail, in particular the measurement in the lepton+jets channel which drives the final result.

The measurement in the lepton+jets channel is performed simultaneously with the in-situ calibration of the jet energy scale using the $W \to qq'$ reconstructed decay. A kinematics fitter is used to evaluate all possible combinations of jets available for the reconstruction of the $\mathrm{t\bar{t}}$ decays. Each single combination is assigned with a probability of the goodness-of-fit based on the χ^2 value: $P_{\mathrm{gof}} = e^{-\chi^2/2}$. To increase the purity and also the fraction of assignments which are found to be correct events with at least two b-tagged jets and $P_{\mathrm{gof}} > 0.5$ are required. This tight requirements depletes the original statistics by 71% but increases the purity (90% $\to$ 96%) and the fraction of correct assignments (13% $\to$ 44%). It's the latter number which is relevant

Journal of Physics: Conference Series **447** (2013) 012033 doi:10.1088/1742-6596/447/1/012033

to attain an optimal resolution in the reconstruction of m_{t}. The method proceeds with the construction of an ideogram where each event is assigned a probability based on P_{gof} and on the probability that each assignment was originated after one of three possibilites: correct or wrong assignment of the original kinematics and unmatched assignment (either due to the rejection of one jet or due to the fact the the event is from residual backgrounds). The per-event probabilities are then multiplied to build a likelihood which is function of m_{t} and the jet energy scale (JES). Both m_{t} and JES are left to float freely when maximising the following likelihood:

$$\mathcal{L}(m_{\mathrm{t}}, \mathrm{JES}) \propto \prod_{\mathrm{events}} \left[\sum_{i=1}^{n} c P_{\mathrm{gof}}(i) P\left(m_{\mathrm{t,i}}^{\mathrm{fit}}, m_{W,i}^{\mathrm{reco}} | m_{\mathrm{t}}, \mathrm{JES} \right) \right]^{w_{\mathrm{event}}}$$

The method is calibrated based on simulated events with both different m_{t} and JES scenarios which are used to assess the bias and the statistical coverage of the fit. Small corrections to the final measurement (<0.5 GeV) are estimated based on this studies.

The systematic uncertainties affecting the measurement can be grouped in two main groups: experimental and theoretical sources. The experimental-type uncertainties comprehend jet energy scale and resolution effects, *b*-tagging, $E_{\mathrm{T}}^{\mathrm{miss}}$, pileup, lepton energy scale. Although the jet energy scale is calibrated in-situ only an average scale is derived and it pertains specifically to light flavoured jets. Consequently a flavour-specific uncertainty and a residual p_{T} and η-dependency have to propagated to the measurement of m_{t} by using simulated pseudo-experiments. These two aspects result as the dominant sources of uncertainty to the measurement. Theory uncertainties are related to the modelling of the $\mathrm{t\bar{t}}$ signal its production environment. These are also evaluated using pseudo-experiments with alternative parameter settings of the generator which reflect the different sources of uncertainties. Both Q^2 and ME-PS matching scale uncertainties are estimated using $\mathrm{t\bar{t}}$ samples where these parameters are varied independently by a factor of 2 or 1/2. These variations will reflect not only in the $\mathrm{t\bar{t}}$ kinematics but also in different ISR/FSR evolution. As previously exposed in Section 2.3 we expect these variations to constitute an accurate envelope of the observed data with respect to the uncertainty in the description of the hadronic activity in $\mathrm{t\bar{t}}$ events. Underlying event-related uncertainties are evaluated using different variations of the so-called PERUGIA11 tune [26]. This UE tune has been adopted by both CMS and ATLAS collaborations. The PERUGIA11 variations cover different alternative scenarios for FSR and hadronisation, ISR and primordial k_{T} underlying event, beam remnants, colour reconnections and energy scaling. Of particular interest is the PERUGIANOCR where the description of data is done without resorting to the modelling of colour reconnection and which is used to conservatively estimate the effect on m_{t} reconstruction. Colour reconnection is the second main source of systematic uncertainty in the measurement of m_{t}. Table 1 summarises the different sources of uncertainty affecting the measurement of m_{t} in the lepton+jets channel.

Measurements in other $\mathrm{t\bar{t}}$ channels have larger uncertainties with respect to the measurement described above. Most uncertainties are induced by instrumental effects: in the dilepton channel no in-situ jet energy scale calibration is performed and there is a strong dependency on the $E_{\mathrm{T}}^{\mathrm{miss}}$ estimate which is used to constrain the kinematics of the escaping neutrinos. In the full-hadronic channel the uncertainty is larger due to resolution effects stemming from the high number of combinatorial jet assignments and due to the non-negligible multijets background contamination ($\approx$60%). Figure 4 (*left*) shows the reconstructed top quark mass spectrum in the fully hadronic channel.

An interesting exception with respect to "canonical" measurements of m_{t} is the analysis of kinematics endpoints using the dilepton channel. This method is not based on a simulation-based calibration and uses only kinematics distributions reconstructed in data to fit both the spectrum and the endpoints using LO calculations. Fig. 4 (*right*) shows an example of a fit to

Journal of Physics: Conference Series **447** (2013) 012033 doi:10.1088/1742-6596/447/1/012033

Table 1. Summary of the main systematic uncertainties affecting the measurement of $m_{\rm t}$ in the lepton+jets channel.

Type	Source	$\Delta m_{\rm t}$ [GeV]
Method	Calibration	0.06
Instrumental	b jet energy scale	0.61
	$p_{\rm T}$ and η-dependent jet energy scale	0.28
	Jet energy resolution	0.23
	Lepton energy scale	0.28
	$E_{\rm T}^{\rm miss}$	0.06
	b-tagging efficiency	0.12
	Pileup	0.07
	Non-$\rm t\bar{t}$ background	0.13
Theory	PDF	0.07
	$Q^2(\mu_{\rm r}, \mu_{\rm f})$	0.24
	ME-PS matching scale	0.18
	Underlying event	0.15
	Color reconnection	0.54
Total		**0.98**

the lepton-jet invariant mass spectrum. The measurement makes further use of other variables which factorize event by event the boost of the $\rm t\bar{t}$ system and which are suited to analyse events with symmetric three body decays [27] such as the ones expected in several BSM scenarios. In the endpoint-analysis the systematic uncertainties sources are partially orthogonal to the ones affecting the other measurements and in the future are expected to improve further the final combination for the measurement of $m_{\rm t}$. In the future the study of alternative methods which factorize specific systematic uncertainties such as jet energy scale or colour reconnection will hopefully contribute to a better definition and measurement of $m_{\rm t}$.

The measurement of the difference between $m_{\rm t}$ and $m_{\rm \bar{t}}$ has also been performed by using ℓ^++jets and ℓ^-+jets events. The result is agreement with the SM prediction and finds no evidence for any source of CPT violation in the top quark sector: $\Delta m_{\rm t} = -0.44 \pm 0.46_{\rm stat} \pm 0.27_{\rm syst}$ GeV. It's important to notice that as a difference is taken most of the systematic uncertainties cancel out yielding a statistically dominated precise measurement of this quantity [28].

4. Summary

In this presentation we have focused on specific precision measurements in the top quark sector such as the production cross section of $\rm t\bar{t}$ events and the measurement of $m_{\rm t}$ using the CMS detector. All measurements are found to be overall consistent with the SM predictions and have attained uncertainties comparable or surpassing the current theoretical predictions. In particular the precision in the measurement of $m_{\rm t}$ opens the window to more detailed studies with a high statistics sample which will hopefully lead to a better understanding of this quantity and of its interplay with the EWK symmetry breaking mechanism.

References

[1] J. Beringer *et al.* [Particle Data Group Collaboration], Phys. Rev. D **86** (2012) 010001.
[2] S. Chatrchyan *et al.* [CMS Collaboration], JINST **3** (2008) S08004.
[3] J. Alwall, M. Herquet, F. Maltoni, O. Mattelaer and T. Stelzer, JHEP **1106** (2011) 128 [arXiv:1106.0522 [hep-ph]].

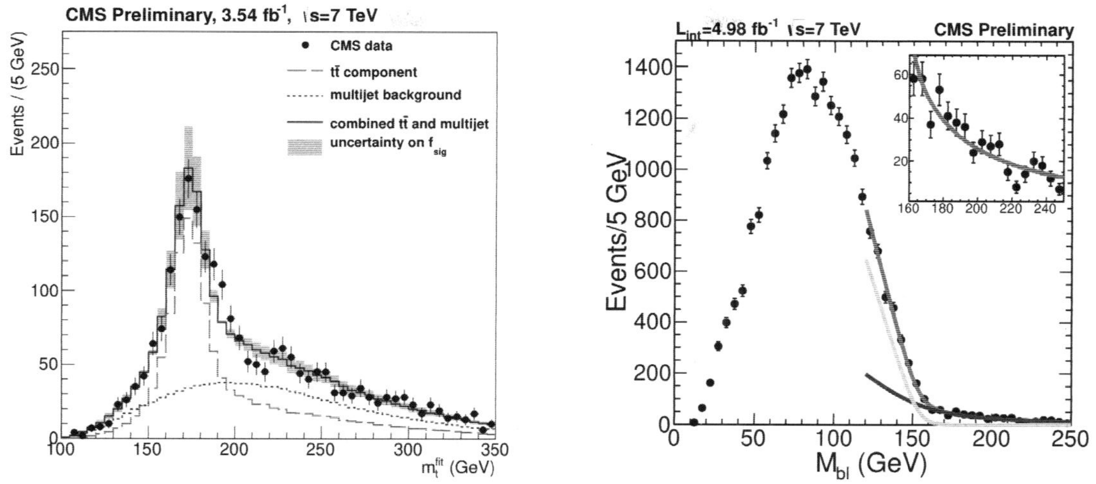

Figure 4. *Left*: Reconstructed m_t from a kinematic fit in the fully hadronic final state. The uncertainty band indicates the uncertainty from the signal fraction. *Right*: Result of the fit to the lepton-jet invariant mass spectrum and its endpoint using dilepton events. The inset shows a zoom of the tail region in M_{lb} spectrum to illustrate the level of agreement between the background shape and the data points.

[4] T. Sjostrand, L. Lonnblad, S. Mrenna and P. Z. Skands, hep-ph/0308153.
[5] Z. Was, Nucl. Phys. Proc. Suppl. **98** (2001) 96 [hep-ph/0011305].
[6] S. Chatrchyan *et al.* [CMS Collaboration], JHEP **1211** (2012) 067 arXiv:1208.2671 [hep-ex]
[7] CMS collaboration, CMS-PAS-PFT-10-002 (2010)
[8] N. Kidonakis, Phys. Rev. D **82** (2010) 114030 arXiv:1009.4935 [hep-ph]
[9] V. Ahrens, A. Ferroglia, M. Neubert, B. D. Pecjak and L. L. Yang, JHEP **1009** (2010) 097 [arXiv:1003.5827 [hep-ph]].
[10] M. Aliev, H. Lacker, U. Langenfeld, S. Moch, P. Uwer and M. Wiedermann, Comput. Phys. Commun. **182** (2011) 1034 arXiv:1007.1327 [hep-ph]
[11] CMS collaboration, CMS-PAS-TOP-12-007 (2012)
[12] M. Czakon and A. Mitov, arXiv:1112.5675 [hep-ph]
[13] CMS collaboration, CMS-PAS-TOP-12-022 (2012)
[14] CMS collaboration, CMS-PAS-TOP-11-008 (2011)
[15] S. Chatrchyan *et al.* [CMS Collaboration], arXiv:1211.2220 [hep-ex]
[16] CMS collaboration, CMS-PAS-TOP-12-018 (2012)
[17] CMS collaboration, CMS-PAS-TOP-12-023 (2012)
[18] CMS collaboration, CMS-PAS-TOP-12-019 (2012)
[19] S. Alioli, P. Nason, C. Oleari and E. Re, JHEP **1006** (2010) 043 [arXiv:1002.2581 [hep-ph]].
[20] S. Frixione and B. R. Webber, JHEP **0206** (2002) 029 [hep-ph/0204244].
[21] S. Chatrchyan *et al.* [CMS Collaboration], arXiv:1209.2319 [hep-ex].
[22] S. Chatrchyan *et al.* [CMS Collaboration], Eur. Phys. J. C **72** (2012) 2202 [arXiv:1209.2393 [hep-ex]].
[23] CMS collaboration, CMS-PAS-TOP-11-027 (2012)
[24] CMS collaboration, CMS-PAS-TOP-11-17 (2012)
[25] CMS collaboration, CMS-PAS-TOP-11-18 (2012)
[26] P. Z. Skands, Phys. Rev. D **82** (2010) 074018 [arXiv:1005.3457 [hep-ph]].
[27] K. T. Matchev and M. Park, Phys. Rev. Lett. **107** (2011) 061801 [arXiv:0910.1584 [hep-ph]].
[28] S. Chatrchyan *et al.* [CMS Collaboration], JHEP **1206** (2012) 109 [arXiv:1204.2807 [hep-ex]].

Journal of Physics: Conference Series **447** (2013) 012034 doi:10.1088/1742-6596/447/1/012034

Running soft parameters in SUSY models with multiple $U(1)$ gauge factors

R M Fonseca [1]

Centro de Física Teórica de Partículas, Instituto Superior Técnico, Universidade Técnica de Lisboa, Av. Rovisco Pais 1, 1049-001 Lisboa, Portugal

E-mail: `renato.fonseca@ist.utl.pt`

M Malinský

Institute of Particle and Nuclear Physics, Faculty of Mathematics and Physics, Charles University in Prague, V Holešovičkách 2, 180 00 Praha 8, Czech Republic

E-mail: `malinsky@ipnp.troja.mff.cuni.cz`

W Porod

Institut für Theoretische Physik und Astrophysik, Universität Würzburg, 97074 Würzburg, Germany

E-mail: `porod@physik.uni-wuerzburg.de`

F Staub

Bethe Center for Theoretical Physics & Physikalisches Institut der Universität Bonn, Nußallee 12, 53115 Bonn, Germany

E-mail: `fnstaub@th.physik.uni-bonn.de`

Abstract. The two-loop renormalization group equations for the parameters of softly broken supersymmetric gauge theories given in the literature are generalized to the case when the gauge group contains more than a single abelian factor. This contribution to the proceedings of the DISCRETE 2012 symposium is based on reference [1].

1. Introduction

The two-loop renormalization group equations (RGEs) of a generic softly broken supersymmetric (SUSY) model have been known for some time [2, 3]. However, these expressions are not completely general, as they do not handle all possible situations. For instance, if the gauge group contains one $U(1)$ factor, it is possible to form a Fayet-Illiopoulos term κD in the superpotential (the RGEs for the additional parameter κ were given in [4, 5, 6]). Another potential issue is the presence of Dirac gaugino mass terms $m_D^{iA}\psi_i\lambda_A$, if there are superfields in the adjoint

[1] Contribution to the proceedings of "DISCRETE 2012 - Third Symposium on Prospects in the Physics of Discrete Symmetries", Lisbon, Portugal, 3 - 7 December 2012.

representation of one of the gauge factor groups (see [7, 8, 9]). Yet another problem occurs when there are multiple $U(1)$ factor groups, since this usually leads to kinetic mixing between the gauge bosons of the different $U(1)$'s [10, 11]. A two-loop renormalization group analysis of non-SUSY theories with this feature is available in [12], while the supersymmetric case, which we shall be discussing, was addressed in [1].

The are many practical applications of these results. For instance, the $U(1)$-mixing effects in SUSY grand unified theories (GUTs) featuring an extended intermediate $U(1)_R \times U(1)_{B-L}$ stage (see for example [13]) can shift the effective MSSM Bino soft mass by several per cent with respect to the naive estimate where such effects are neglected. In principle, this can have non-negligible effects for the low-energy phenomenology.

2. Gauge theories with multiple $U(1)$ factor groups

Consider a model with a $U(1)^n$ gauge group, containing m supermultiplets Φ_i, $i = 1, \cdots, m$.[2] As usual, n gauge bosons A_μ^a must be introduced, which can be seen as components of a vector $\boldsymbol{A_\mu}$. Under a gauge transformation with parameters α^a we have the following:

$$\Phi_i \to \exp\left(iQ_i^a\alpha^a\right)\Phi_i \equiv \exp\left(i\boldsymbol{Q_i^T}\boldsymbol{\alpha}\right)\Phi_i\,, \tag{1}$$

$$\boldsymbol{A_\mu} \to \boldsymbol{A_\mu} + \boldsymbol{G}^{-1}\partial_\mu\boldsymbol{\alpha}\,, \tag{2}$$

where $\boldsymbol{Q_i}$ and $\boldsymbol{\alpha}$ are also vectors in $U(1)$ space, with components Q_i^a and α^a. In the last equation, a $\boldsymbol{G}$ matrix shows up. In the spirit of making the most general gauge transformation, $\boldsymbol{G}$ can be any real $n \times n$ matrix. This means in particular that the transformation of A_μ^a may depend on some gauge transformation parameter α^b with $b \neq a$. It is straightforward to see that the Lagrangian will be invariant under this transformation if the covariant derivative of the supermultiplet Φ_i takes the form

$$D_\mu\Phi_i = \left(\partial_\mu - i\boldsymbol{Q_i^T}\boldsymbol{G}\boldsymbol{A_\mu}\right)\Phi_i\,, \tag{3}$$

which supports the idea that $\boldsymbol{G}$ is a $U(1)$ gauge couplings matrix. Notice that even though it is a square matrix in $U(1)$ space, its left and right indices contract with different vectors: on the left we have the vector with the hypercharges of Φ_i, while on the right there is the $U(1)$ gauge bosons' vector. To complete the picture, there is also to consider a generic gauge kinetic term[3]

$$-\frac{1}{4}\boldsymbol{F_{\mu\nu}^T}\boldsymbol{\xi}\boldsymbol{F^{\mu\nu}} \tag{4}$$

and also, in a softly broken supersymmetric theory, the $U(1)$ gaugino mass term

$$-\frac{1}{2}\boldsymbol{\lambda}^T\boldsymbol{M}\boldsymbol{\lambda} + \text{h.c.}\,. \tag{5}$$

Here we have introduced additional vectors in $U(1)$ space, $\boldsymbol{F_{\mu\nu}}$ and $\boldsymbol{\lambda}$, with components $F_{\mu\nu}^a \equiv \partial_\mu A_\nu^a - \partial_\nu A_\mu^a$ and λ^a (the gaugino field associated with $U(1)^a$). The new $n \times n$ symmetric matrices $\boldsymbol{\xi}$ and $\boldsymbol{M}$ are parameters of the theory. The advantage of having $\boldsymbol{\xi} \neq \mathbb{1}$ is that the gauge coupling matrix can be made diagonal with a rotation of the gauge boson and gaugino fields. On the other hand, as far as the renormalization group analysis is concerned, it then becomes necessary to include the effect of $\boldsymbol{\xi}$ on the evolution of the other parameters, as well as to describe the evolution of $\boldsymbol{\xi}$ itself [12]. As such, we follow an alternative approach: $\boldsymbol{\xi} \neq \mathbb{1}$ means that the gauge boson fields (as well as the gaugino fields) are not canonically normalized, so they may

[2] If m is smaller than n, the $U(1)$'s can be rotated such that the fields are only changed under m of them.

[3] The same mixing parameter $\boldsymbol{\xi}$ appears in the gaugino kinetic term, and in the $^1/_2 D^a D^a$ term as well.

be rotated and rescaled such that in the new basis $\boldsymbol{\xi} = \mathbb{1}$. In this way, the $U(1)$-mixing in the kinetic term is transferred to the matrix of gauge couplings $\boldsymbol{G}$ and the gaugino mass matrix $\boldsymbol{M}$.

We note that even with $\boldsymbol{\xi} = \mathbb{1}$, there is still some freedom to rotate the vectors and matrices $\boldsymbol{Q_i}$, $\boldsymbol{A_\mu}$, $\boldsymbol{\lambda}$, $\boldsymbol{G}$ and $\boldsymbol{M}$ living in $U(1)$ space. Indeed, in general we may perform two unphysical rotations, $\mathcal{O}_1$ and $\mathcal{O}_2$, such that

$$\boldsymbol{Q_i} \to \mathcal{O}_1 \boldsymbol{Q_i}\,, \qquad \boldsymbol{A_\mu} \to \mathcal{O}_2 \boldsymbol{A_\mu}\,, \qquad \boldsymbol{\lambda} \to \mathcal{O}_2 \boldsymbol{\lambda}\,, \qquad \boldsymbol{G} \to \mathcal{O}_1 \boldsymbol{G} \mathcal{O}_2^T\,, \qquad \boldsymbol{M} \to \mathcal{O}_2 \boldsymbol{M} \mathcal{O}_2^T\,. \tag{6}$$

3. Inclusion of $U(1)$-mixing effects in the running of the parameters

The RGEs must reflect the symmetries in equation (6), and this can be used to derive the structure of the RGEs, with $U(1)$-mixing effects fully included. For instance, only those combinations of $\boldsymbol{G}$ and $\gamma \propto \sum_i \boldsymbol{Q_i} \boldsymbol{Q_i}^T$ which transform as $(\cdots) \to \mathcal{O}_1 (\cdots) \mathcal{O}_2^T$ are allowed to enter the right-hand side of the renormalization group equation of $\boldsymbol{G}$. However, at one-loop level, there is only one such structure that can come up from a matter-field loop in the gauge propagator, namely $\boldsymbol{G}\boldsymbol{G}^T\gamma\boldsymbol{G}$, so one immediately concludes that

$$\beta_{\boldsymbol{G}}^{(\text{one-loop})} \propto \boldsymbol{G}\boldsymbol{G}^T\gamma\boldsymbol{G}\,. \tag{7}$$

The proportionality coefficient is trivially obtained by matching this to the single $U(1)$ case.

It might be tempting to think that without a gaugino mass matrix $\boldsymbol{M}$ in non-SUSY theories, $\boldsymbol{G}$ can be diagonalized thereby eliminating the $U(1)$-mixing effects. This however is not the case, as radiative effects will reintroduce off-diagonalities in the gauge couplings matrix. In fact, already at the one-loop level, the anomalous dimension γ which controls the RGE of $\boldsymbol{G}$ is in general a non-diagonal matrix in $U(1)$ space and, as such, $\boldsymbol{G}$ cannot be kept diagonal at all energy scales. One exception, where $U(1)$-mixing effects can be eliminated, is when all the relevant $U(1)$ gauge couplings originate from a common gauge factor and thus, barring threshold effects, all of them happen to be equal at a certain scale. In such a case $\boldsymbol{G} \propto \mathbb{1}$ at some scale, and both charges and gauge fields can be simultaneously rotated at the one-loop level so that no off-diagonalities appear in γ [14, 2]. This *rotated basis* method will only work in non-SUSY cases where only the gauge sector has to be taken into account. Also, because of the appearance of Yukawa and trilinear soft SUSY breaking couplings in the two-loop RGEs of $\boldsymbol{G}$, this approach ceases to be valid beyond the one-loop level.

4. The generalized two-loop β-functions

In [2] one can find the RGEs for models with a simple gauge group, and also a list of replacement rules which extends these results to models with a semi-simple gauge group (with one $U(1)$ at most). Therefore, in order to obtain the RGEs of softly broken SUSY gauge theories with multiple $U(1)$'s, we simply need to generalize these replacement rules by including $U(1)$-mixing effects. The technique of using the symmetries in equation (6) can be exploited to this end in many situations. However, in a few cases even a detailed inspection of the underlying expressions does not allow an unambiguous identification of their generalized form. When this happens, a careful analysis of the structure of the contributing Feynman diagrams is necessary.

The gauge group is taken to be $G_A \times G_B \times \cdots \times U(1)^n$, where the G_X's are simple groups. We use uppercase indices for simple group-factors and lowercase indices for $U(1)$ groups. As mentioned above, the $U(1)$ sector should be treated as a whole and described in terms of a real $n \times n$ gauge-coupling matrix $\boldsymbol{G}$, a $n \times n$ symmetric soft-SUSY breaking gaugino mass matrix $\boldsymbol{M}$ and column vectors of charges $\boldsymbol{Q_i}$ for each chiral supermultiplet Φ_i. To eliminate the $\mathcal{O}_1$ rotation freedom in equation (6), we define $\boldsymbol{V_i} \equiv \boldsymbol{G}^T\boldsymbol{Q_i}$ for each i. Since this is the only combinations of the $\boldsymbol{Q_i}$ and $\boldsymbol{G}$ appearing in the Lagrangian, all the RGEs can be written in terms of $\boldsymbol{V}$'s and $\boldsymbol{M}$ only. In this regard, the RGEs of $\boldsymbol{G}$ are noteworthy because they require an isolated $\boldsymbol{G}$ (this would not happen if ${}^{dV_i}/_{dt}$ was computed instead of ${}^{dG}/_{dt}$). For additional clarifications on the notation used in the following expressions, see [1].

Depending on the group sector (abelian or simple), we get different RGEs for the gauge couplings and the gaugino masses. The parameters are then either the matrices $\boldsymbol{G}$, $\boldsymbol{M}$ or the numbers g_A, M_A. For the abelian sector, it is necessary to perform the following replacements to the simple gauge group expressions in [2]:

$$C(G) \to 0\,, \tag{8}$$

$$g^3 S(R) \to \boldsymbol{G} \sum_p \boldsymbol{V_p V_p^T}\,, \tag{9}$$

$$g^5 S(R)\, C(R) \to \sum_p \boldsymbol{G V_p V_p^T} \Big[\sum_B g_B^2 C_B(p) + \boldsymbol{V_p V_p^T}\Big]\,, \tag{10}$$

$$\frac{g^3 C(k)}{d(G)} \to \boldsymbol{G V_k V_k^T}\,, \tag{11}$$

$$2g^2 S(R)\, M \to \boldsymbol{M} \sum_p \boldsymbol{V_p V_p^T} + \sum_p \boldsymbol{V_p V_p^T M}\,, \tag{12}$$

$$g^2 C(k) \to \boldsymbol{V_k V_k^T}\,, \tag{13}$$

$$2g^2 C(k)\, M \to \boldsymbol{M V_k V_k^T} + \boldsymbol{V_k V_k^T M}\,, \tag{14}$$

$$\begin{aligned} 16 g^4 S(R)\, C(R)\, M \to \sum_p \Bigg\{ & 4\left(\boldsymbol{M V_p V_p^T} + \boldsymbol{V_p V_p^T M}\right) \left[\sum_B g_B^2 C_B(p) + \boldsymbol{V_p V_p^T}\right] \\ & + 8 \boldsymbol{V_p V_p^T} \left[\sum_B M_B g_B^2 C_B(p) + \boldsymbol{V_p^T M V_p}\right]\Bigg\}\,. \end{aligned} \tag{15}$$

For a simple factor group G_A, the substitution rules of [2] do not need to be changed except for two cases:

$$g^5 S(R)\, C(R) \to g_A^3 S_A(R) \left[\sum_B g_B^2 C_B(R) + \boldsymbol{V_R^T V_R}\right]\,, \tag{16}$$

$$\begin{aligned} 16 g^4 S(R)\, C(R)\, M \to\ & 8 g_A^2 M_A S_A(R) \left[\sum_B g_B^2 C_B(R) + \boldsymbol{V_R^T V_R}\right] \\ & + 8 g_A^2 S_A(R) \left[\sum_B M_B g_B^2 C_B(R) + \boldsymbol{V_R^T M V_R}\right]\,. \end{aligned} \tag{17}$$

As for the rest of the parameters in a SUSY model, the relevant substitution rules read:

$$g^2 C(r) \to \sum_A g_A^2 C_A(r) + \boldsymbol{V_r^T V_r}\,, \tag{18}$$

$$M g^2 C(r) \to \sum_A M_A g_A^2 C_A(r) + \boldsymbol{V_r^T M V_r}\,, \tag{19}$$

$$M^* g^2 C(r) \to \sum_A M_A^* g_A^2 C_A(r) + \boldsymbol{V_r^T M^\dagger V_r}\,, \tag{20}$$

$$M M^* g^2 C(r) \to \sum_A M_A M_A^* g_A^2 C_A(r) + \boldsymbol{V_r^T M M^\dagger V_r}\,, \tag{21}$$

$$g^4 C(r)\, S(R) \to \sum_A g_A^4 C_A(r)\, S_A(R) + \sum_p \left(\boldsymbol{V_r^T V_p}\right)^2\,, \tag{22}$$

$$Mg^4C(r)\,S(R) \to \sum_A M_A g_A^4 C_A(r)\,S_A(R) + \sum_p \left(\boldsymbol{V_r^T M V_p}\right)\left(\boldsymbol{V_r^T V_p}\right), \tag{23}$$

$$g^4C^2(r) \to \left[\sum_A g_A^2 C_A(r) + \boldsymbol{V_r^T V_r}\right]^2, \tag{24}$$

$$Mg^4C^2(r) \to \left[\sum_A M_A g_A^2 C_A(r) + \boldsymbol{V_r^T M V_r}\right]\left[\sum_A g_A^2 C_A(r) + \boldsymbol{V_r^T V_r}\right], \tag{25}$$

$$g^4C(G)\,C(r) \to \sum_A g_A^4 C(G_A)\,C_A(r), \tag{26}$$

$$Mg^4C(G)\,C(r) \to \sum_A M_A g_A^4 C(G_A)\,C_A(r), \tag{27}$$

$$MM^*g^4C(G)\,C(r) \to \sum_A M_A M_A^* g_A^4 C(G_A)\,C_A(r), \tag{28}$$

$$g^2\mathbf{t}_i^{Aj}\mathrm{Tr}\left(\mathbf{t}^A m^2\right) \to \delta_i^j \sum_p \boldsymbol{V_i^T V_p}\left(m^2\right)_p^p, \tag{29}$$

$$g^2\mathbf{t}_i^{Aj}\left(\mathbf{t}^A m^2\right)_r^l \to \delta_i^j \boldsymbol{V_l^T V_i}\left(m^2\right)_r^l, \tag{30}$$

$$g^4\mathbf{t}_i^{Aj}\mathrm{Tr}\left[\mathbf{t}^A C(r)\,m^2\right] \to \delta_i^j \sum_p \boldsymbol{V_i^T V_p}\left[\sum_B g_B^2 C_B(p) + \boldsymbol{V_p^T V_p}\right]\left(m^2\right)_p^p, \tag{31}$$

$$g^4C(i)\,\mathrm{Tr}\left[S(r)\,m^2\right] \to \sum_A g_A^4 C_A(i)\,\mathrm{Tr}\left[S_A(r)\,m^2\right] + \sum_p \left(\boldsymbol{V_i^T V_p}\right)^2\left(m^2\right)_p^p, \tag{32}$$

$$\begin{aligned}24g^4MM^*C(i)\,S(R) \to{}& 24\sum_A g_A^4 M_A M_A^* C_A(i)\,S_A(R)\\ &+8\sum_p\Big[\left(\boldsymbol{V_i^T M V_p}\right)\left(\boldsymbol{V_i^T M^\dagger V_p}\right)+\left(\boldsymbol{V_i^T M M^\dagger V_p}\right)\left(\boldsymbol{V_i^T V_p}\right)\\ &+\left(\boldsymbol{V_i^T M^\dagger M V_p}\right)\left(\boldsymbol{V_i^T V_p}\right)\Big],\end{aligned} \tag{33}$$

$$\begin{aligned}48g^4MM^*C(r)^2 \to{}& \sum_{A,B} g_A^2 g_B^2 C_A(r)\,C_B(r)\left(32M_A M_A^* + 16 M_A M_B^*\right)\\ &+\sum_A g_A^2 C_A(r)\Big(32 M_A M_A^* \boldsymbol{V_r^T V_r} + 16 M_A \boldsymbol{V_r^T M^\dagger V_r}\\ &+32\boldsymbol{V_r^T M M^\dagger V_r} + 16 M_A^* \boldsymbol{V_r^T M V_r}\Big)\\ &+32\left(\boldsymbol{V_r^T M M^\dagger V_r}\right)\left(\boldsymbol{V_r^T V_r}\right) + 16\left(\boldsymbol{V_r^T M V_r}\right)\left(\boldsymbol{V_r^T M^\dagger V_r}\right).\end{aligned} \tag{34}$$

5. Numerical results

In this section, we shall see the importance of the kinetic mixing effects with two examples. Consider first the SUSY $SO(10)$ model of [13] in which the unified gauge symmetry is broken down to the MSSM in three steps, namely, $SO(10) \to SU(3)_c \times SU(2)_L \times SU(2)_R \times U(1)_{B-L} \to SU(3)_c \times SU(2)_L \times U(1)_R \times U(1)_{B-L} \to$ MSSM; the corresponding breaking scales shall be denoted by m_G, m_R and m_{B-L}, respectively. For our purposes, it is crucial that in this model the ratio m_R/m_{B-L} can be as large as 10^{10} and, hence, the $U(1)$-mixing effects become important. Figure (1) shows the importance of $U(1)$-mixing effects in the one-loop running of the gauge coupling constants: with otherwise identical conditions, it shifts the low energy value of α_Y^{-1} by as much as 4%. In order to compensate for this, the intermediate scale m_R would have to

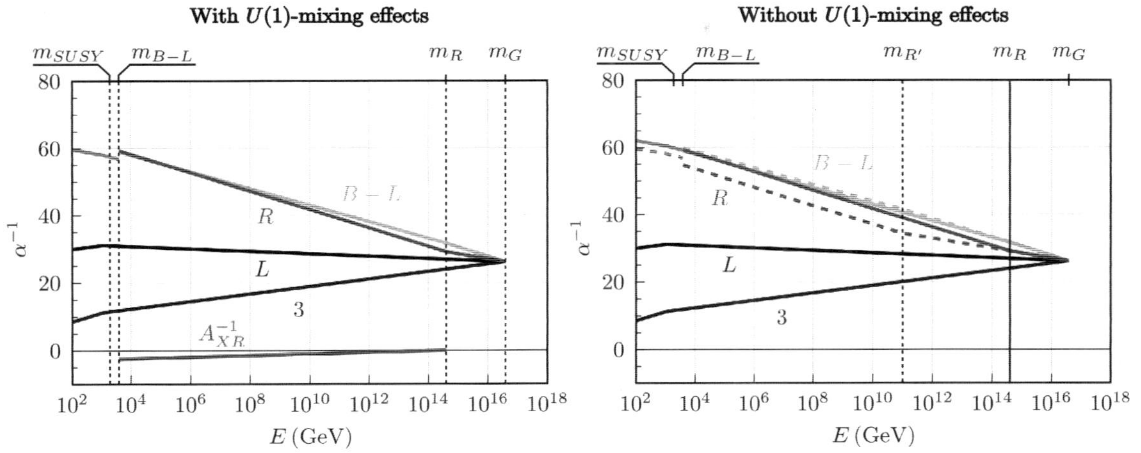

Figure 1. One-loop gauge-coupling evolution in the MRV model [13] with and without $U(1)$-mixing effects. The position of the GUT scale, the unified gauge coupling and the intermediate symmetry-breaking scale m_R were chosen in such a way as to fit the electroweak data. Close to the $\alpha^{-1} = 0$ axis on the left plot, there is a brown line in the $[m_{B-L}, m_R]$ energy range depicting the evolution of the off-diagonal entries of the matrix $\boldsymbol{A}^{-1} \equiv 4\pi \left(\boldsymbol{G}\boldsymbol{G}^T\right)^{-1}$ which, at the one-loop level, scales linearly with $\log E$. Without the kinetic mixing effects taken into account, the low-energy value of α_Y^{-1} (orange solid line on the right) differs from the one obtained in the full calculation by as much as 4%. If one attempts to obtain the right value of $\alpha_Y^{-1}(m_Z)$ by adjusting the $SU(2)_R$ breaking scale, the new m'_R scale must be shifted with respect to the correct m_R by as much as 4 orders of magnitude (vertical solid and dashed lines on the right).

be lowered by as much as 4 orders of magnitude. We also note that the *rotated-basis* method mentioned previously is only partially successful in this model, because the g_R and g_{B-L} gauge couplings do not coincide at the m_R scale. In fact, the value of $\alpha_Y^{-1}(m_Z)$ obtained in this way is about 2% off from the correct one, which nevertheless is still better than assuming no mixing at all.

As for gaugino masses, it can be easily shown that the combination $\boldsymbol{G}\boldsymbol{M}^{-1}\boldsymbol{G}^T$ is a one-loop invariant of the RG flow, which generalizes the traditional one, α/M = constant. This relation can be used to compute the Bino's mass at low energies,

$$M_Y(m_{SUSY}) = \frac{\alpha_Y(m_{SUSY})}{\alpha_G} p_Y^T \boldsymbol{M_{1/2}} p_Y\,, \tag{35}$$

where $\boldsymbol{M_{1/2}}$ is the GUT-scale gaugino soft mass matrix and $p_Y^T = \left(\sqrt{3/5}, \sqrt{2/5}\right)$ is the vector describing the combination of $U(1)_R \times U(1)_{B-L}$ charges which constitutes the MSSM hypercharge. From equation (35) we see that the ratio $\alpha_Y(m_{SUSY})/M_Y(m_{SUSY})$ depends on whether or not we include the mixing effects, as was already noticed in reference [15]. Note that if $\boldsymbol{M_{1/2}}$ is not proportional to the unit matrix at the GUT scale, the $p_Y^T \boldsymbol{M_{1/2}} p_Y$ term will mix all entries of $\boldsymbol{M_{1/2}}$. Moreover, in the special case that the abelian gauge couplings unify, even the one-loop gaugino sector evolution can be fully accounted for by the *rotated-basis* trick.

Even in cases where the $U(1)$ gauge coupling unify at a certain scale, our two-loop formulas can produce relevant effects. We illustrate this by taking as an example the model presented in reference [16] where an intermediate $SU(3)_c \times SU(2)_L \times U(1)_Y \times U(1)_{B-L}$ gauge symmetry is assumed to originate from a grand-unified framework. We consider two cases: (i) full gauge coupling unification at 2×10^{16} GeV and (ii) a small difference of 5% between the two $U(1)$ couplings caused by possible GUT-scale threshold effects. In the gaugino sector we assume universal boundary conditions in both cases, but the effect gets even stronger if one considers in addition threshold effects in the gaugino sector as well.

The results are contained in table (1). Interestingly, besides the expected equivalence of the *rotated-basis* method and the full-fledged calculation at the one-loop level, the relevant effective hypercharge gauge coupling turns out to be identical to the one obtained even at two-loop level if exact gauge coupling unification is assumed. The reason is that all additional states not present in the MSSM are charged only with respect to $U(1)_{B-L}$ but are neutral under the MSSM gauge group. In the gaugino sector the first deviations show up already in this case, which however are only at the per-mile level. If one also includes threshold corrections at the GUT-scale, then the effects are at the percent level, leading to shifts in the masses potentially measurable already at the LHC. Lastly, it should be kept in mind that the effects would be even larger if the $U(1)_Y$ would result from the breaking of $U(1)_R \times U(1)_{B-L}$ as discussed in the previous example.

Table 1. Low energy values of the entries of the gauge coupling and gaugino mass matrices (g_{ab} and M_{ab} with $a, b = Y, BL$) and the properly fitted MSSM parameters (g_Y, M_Y). All gaugino masses are in GeV. We have set the GUT scale at 2×10^{16} GeV with $g_G = 0.72$, and imposed an mSUGRA boundary condition taking $\boldsymbol{M_{1/2}} = \mathbb{1} \times 500$ GeV. At the one-loop level, we compare the case with no kinetic mixing effects included, the *rotated basis*, and the full-fledged calculation. At the two-loop level, we include the case where g_Y and g_{BL} are split at the GUT scale due to threshold corrections.

	One-loop results			Two-loop results			
	No kinetic mixing	Rotated basis method	Complete RGEs	No kinetic mixing	Complete RGEs	No kinetic mixing	Complete RGEs
g_{YY}	0.4511	0.4700	0.4700	0.4487	0.4677	0.4487	0.4686
g_{BLBL}	0.4083	0.4243	0.4243	0.4070	0.4231	0.4131	0.4298
g_{BLY}, g_{YBL}	0.0	−0.0723	−0.0723	0.0	−0.0725	0.0	−0.0725
g_Y	0.4511	0.4511	0.4511	0.4487	0.4487	0.4487	0.4500
M_{YY}	196.34	218.13	218.13	185.82	207.96	185.80	208.71
M_{BLBL}	160.83	178.67	178.67	154.88	173.19	144.26	161.97
M_{BLY}, M_{YBL}	0.0	−62.39	−62.39	0.0	−63.10	0.0	−62.15
M_Y	196.34	196.34	196.34	185.82	185.96	185.80	187.04
	Exact unification					$g_{BL}^{\rm GUT} = 1.05\, g_Y^{\rm GUT}$	

6. Conclusions

W have derived the renormalization group equations of softly broken supersymmetric models with more than a single abelian gauge factor group. In such models there are $U(1)$-mixing effects which must be taken into consideration. Although formally the evolution equations available in the literature do not exhibit any obvious pathologies if such subtleties are not taken into account, the calculations based on these formulas are in general incomplete and, thus, the results are internally inconsistent. In spite of this, the issue of $U(1)$-mixing in softly broken supersymmetric gauge theories had never been addressed in full generality, even at one loop.

The effect of the derived equations has been illustrated for two cases: at the one-loop level, for a model with different gauge coupling strengths due to a breaking of the original simple group at a high scale; and at the two-loop level, in a model where gauge coupling unification occurs but only with threshold corrections taken into consideration. In both case we obtain effects in the percent range which none of the previously proposed partial treatments can fully account for.

Acknowledgments
The work of R.F. was supported by the Fundação para a Ciência e a Tecnologia, through the grant SFRH/BD/47795/2008.

References

[1] Fonseca R M, Malinský M, Porod W and Staub F 2012 *Nucl.Phys.* **B854** 28–53 (*Preprint* 1107.2670)
[2] Martin S P and Vaughn M T 1994 *Phys.Rev.* **D50** 2282 (*Preprint* hep-ph/9311340)
[3] Yamada Y 1994 *Phys.Rev.* **D50** 3537–3545 (*Preprint* hep-ph/9401241)
[4] Jack I and Jones D 2000 *Phys.Lett.* **B473** 102–108 (*Preprint* hep-ph/9911491)
[5] Jack I and Jones D 2001 *Phys.Rev.* **D63** 075010 (*Preprint* hep-ph/0010301)
[6] Jack I, Jones D and Parsons S 2000 *Phys.Rev.* **D62** 125022 (*Preprint* hep-ph/0007291)
[7] Jack I and Jones D 1999 *Phys.Lett.* **B457** 101–108 (*Preprint* hep-ph/9903365)
[8] Jack I and Jones D 2000 *Phys.Rev.* **D61** 095002 (*Preprint* hep-ph/9909570)
[9] Goodsell M D 2013 *JHEP* **1301** 066 (*Preprint* 1206.6697)
[10] Holdom B 1986 *Phys. Lett.* **B166** 196
[11] Babu K S, Kolda C F and March-Russell J 1998 *Phys. Rev.* **D57** 6788–6792 (*Preprint* hep-ph/9710441)
[12] Luo M and Xiao Y 2003 *Phys.Lett.* **B555** 279–286 (*Preprint* hep-ph/0212152)
[13] Malinský M, Romão J C and Valle J W F 2005 *Phys. Rev. Lett.* **95** 161801 (*Preprint* hep-ph/0506296)
[14] del Aguila F, Coughlan G D and Quirós M 1988 *Nucl. Phys.* **B307** 633
[15] Kribs G D 1998 *Nucl. Phys.* **B535** 41–82 (*Preprint* hep-ph/9803259)
[16] Fileviez Pérez P and Spinner S 2011 *Phys. Rev.* **D83** 035004 (*Preprint* 1005.4930)

Journal of Physics: Conference Series **447** (2013) 012035 doi:10.1088/1742-6596/447/1/012035

Potential of a linear collider for lepton flavour violation studies in the susy seesaw

António J. R. Figueiredo[1], **A. Abada**[2], **J. C. Romão**[1] **and A. M. Teixeira**[3]

[1] CFTP, Departamento de Física, Instituto Superior Técnico, Avenida Rovisco Pais, 1, 1049-001 Lisboa, Portugal

[2] Laboratoire de Physique Théorique, CNRS – UMR 8627, Université de Paris-Sud 11, F-91405 Orsay Cedex, France

[3] Laboratoire de Physique Corpusculaire, CNRS/IN2P3 – UMR 6533, Campus des Cézeaux, 24 Av. des Landais, F-63177 Aubière Cedex, France

E-mail: ajrf@cftp.ist.utl.pt

Abstract. We study the potential of an $e^{\pm}e^{-}$ Linear Collider for charged lepton flavour violation studies in a supersymmetric framework where neutrino masses and mixings are explained by a type-I seesaw. Focusing on e-μ flavour transitions, we evaluate the background from standard model and supersymmetric charged currents to the $e\mu + \not{E}_T$ signal. We study the energy dependence of both signal and background, and the effect of beam polarisation in increasing the signal over background significance. Finally, we consider the $\mu^{-}\mu^{-} + \not{E}_T$ final state in $e^{-}e^{-}$ collisions that, despite being signal suppressed by requiring two e-μ flavour transitions, is found to be a clear signature of charged lepton flavour violation due to a very reduced standard model background. This contribution summarises part of the work done in [1].

1. Introduction

A high-energy lepton collider (LC) [2] offers the possibility of direct production of lepton number (L) carrying heavy states (such as sleptons $\tilde{l}$ in supersymmetry) and of doing precision physics profiting from low QCD activity. The possibility of polarising the colliding beams with great accuracy makes the physics potential of a lepton collider even more ambitious. Indeed, by polarising the initial beams, one is able to (approximately) project the chirality of the couplings intervening in the primary production process, thus probing the chirality structure of an underlying new physics model. Polarisation can also be used to suppress the background from Standard Model (SM) processes (e.g. from weak charged currents) to charged lepton flavour violation (cLFV) signals. Moreover, threshold scans of chirality projected primary produced particles can be used to determine their masses [3].

Supersymmetry (SUSY) remains one of the most attractive extensions of the SM. Thus, a very appealing mechanism to explain the smallness of neutrino masses and generate neutrino mixing is to consider a supersymmetric seesaw. A type-I seesaw mechanism in which the right-handed (RH) neutrinos have masses close to the grand unification (GUT) scale, generates neutrino masses and mixings with naturally large Yukawa couplings, and can offer an explanation for the observed baryon asymmetry of the universe. However, the drawback of such a set-up is that it

Journal of Physics: Conference Series **447** (2013) 012035 doi:10.1088/1742-6596/447/1/012035

is very hard to probe since the very heavy RH neutrinos cannot be produced at colliders. On the other hand, in a supersymmetric type-I seesaw (usually dubbed SUSY seesaw) the radiative corrections generate flavour violation in the slepton sector [4], giving additional sources of LFV at low energy. At a high-energy LC, sleptons can be copiously produced and their decays lead to potentially observable cLFV final states, thus providing a unique probe of this mass generation mechanism.

SUSY-seesaw induced cLFV at a LC has been studied in [5, 6] focusing on τ-μ flavour violation, while in [7] slepton driven cLFV was also considered but without relying on a particular origin of slepton mixing.

Motivated by the excellent muon reconstruction capabilities and the recently improved upper-limit on $\mu \to e\gamma$ [8], we studied the SUSY seesaw induced μ-e cLFV and its discovery potential at a LC working with polarisable beams [1].

2. The SUSY seesaw

The SUSY seesaw model comprises the minimal supersymmetric standard model (MSSM) extended by three[1] chiral superfields $\hat{N}^c_i \sim \left(\nu^c, \tilde{\nu}^\dagger_R\right)_i$, so-called RH neutrino superfields, that are singlets under the SM gauge group. The leptonic part of the superpotential is given by

$$\mathcal{W}^{\text{lepton}} = \frac{1}{2}\hat{N}^c M_N \hat{N}^c + \hat{N}^c Y^\nu \hat{L}\hat{H}_2 + \hat{E}^c Y^l \hat{L}\hat{H}_1 \,. \tag{1}$$

Hereafter we work in a flavour basis where the charged lepton Yukawa couplings Y^l and the RH neutrino mass matrix M_N are diagonal. The slepton soft breaking terms are

$$\begin{aligned}\mathcal{V}^{\text{slepton}}_{\text{soft}} &= m^2_{\tilde{L}}\tilde{\ell}_L\tilde{\ell}^*_L + m^2_{\tilde{E}}\tilde{\ell}_R\tilde{\ell}^*_R + m^2_{\tilde{\nu}_R}\tilde{\nu}_R\tilde{\nu}^*_R \\ &+ \left(A_l H_1 \tilde{l}_L \tilde{l}^*_R + A_\nu H_2 \tilde{\nu}_L \tilde{\nu}^*_R + B_\nu \tilde{\nu}_R \tilde{\nu}_R + \text{H.c.}\right) .\end{aligned} \tag{2}$$

Our analysis is conducted in a framework where SUSY breaking is flavour blind (as in minimal supergravity mediated SUSY breaking) and the soft breaking parameters satisfy certain universality conditions at a high-energy scale, which we take to be the gauge coupling unification scale $M_{\text{GUT}} \sim 10^{16}$ GeV:

$$\left(m_{\tilde{L}}\right)^2_{ij} = \left(m_{\tilde{E}}\right)^2_{ij} = \left(m_{\tilde{\nu}_R}\right)^2_{ij} = m_0^2\delta_{ij} \,, \tag{3}$$

$$\left(A_l\right)_{ij} = A_0\left(Y^l\right)_{ij} , \left(A_\nu\right)_{ij} = A_0\left(Y^\nu\right)_{ij} \,, \tag{4}$$

where m_0 and A_0 are the universal scalar soft-breaking mass and trilinear couplings of the constrained MSSM (cMSSM), and i, j denote lepton flavour indices ($i, j = 1, 2, 3$).

After electroweak (EW) symmetry breaking, the neutrino mass matrix is approximately given by $m_\nu \simeq -m^{\nu\,T}_D M_N^{-1} m^\nu_D$, where $m^\nu_D = Y^\nu v_2$ and v_i is the vacuum expectation value of H_i ($v_{1(2)} = v\cos(\sin)\beta$, with $v = 174$ GeV). A convenient means of parametrizing the neutrino Yukawa couplings, while at the same time allowing to accommodate the experimental data, is given by the Casas-Ibarra parametrization [9], which reads at the seesaw scale M_N

$$Y^\nu v_2 = m^\nu_D = i\sqrt{M^{\text{diag}}_N} R \sqrt{m^{\text{diag}}_\nu} U^\dagger_{\text{MNS}} \,, \tag{5}$$

where U_{MNS} is the light neutrino mixing matrix and R is a 3×3 complex orthogonal matrix that encodes the possible mixings involving RH neutrinos. We use the standard parametrization for the U_{MNS}, with the three mixing angles in the intervals favoured by current data [10].

[1] We assume that the number of RH neutrino generations mimics the number of generations of SM fermions. In fact, two are sufficient in order to comply with neutrino oscillation parameters.

2.1. *Flavour violation in the slepton sector*

Due to the non-trivial flavour structure of Y^ν, the running from $M_{\rm GUT}$ down to the seesaw scale will induce flavour mixing [4] in the otherwise approximately flavour conserving slepton soft breaking terms. This running is more pronounced in the soft breaking terms involving slepton doublets since these have local interactions with the RH (s)neutrinos. At leading order, the flavour mixing induced by these radiative corrections has the form

$$\left(\Delta m^2_{\tilde L}\right)_{ij} = -\frac{1}{8\pi^2}\left(3m_0^2 + A_0^2\right)\left(Y^{\nu\dagger} L Y^\nu\right)_{ij} , \quad (6)$$

$$\left(\Delta A_l\right)_{ij} = -\frac{3}{16\pi^2} A_0 Y^l_{ij}\left(Y^{\nu\dagger} L Y^\nu\right)_{ij} ; \; L_{kl} \equiv \log\left(\frac{M_{\rm GUT}}{M_{N_k}}\right)\delta_{kl} . \quad (7)$$

The amount of flavour violation in the slepton sector is encoded in $\left(Y^{\nu\dagger} L Y^\nu\right)_{ij}$ which, as made explicit by equation (5), is related to high- and low-energy neutrino parameters.

3. cLFV in $e^\pm e^-$ collisions

At collider energies, i.e. at energies of the order of the TeV, the (high-scale) SUSY seesaw with the aforementioned assumptions for SUSY breaking can be interpreted as a minimal deviation from the cMSSM that allows for flavour mixing in the slepton soft breaking parameters, since processes mediated by RH (s)neutrinos are greatly suppressed due to the very high seesaw scale. We stress that, contrary to other analysis, the origin of this flavour mixing is rooted on, albeit not singly determined by, neutrino oscillations.

In the SUSY seesaw model, as in the MSSM, the sparticle production in e^+e^- collisions are dominated by the following 2-body processes

$$e^+e^- \to \tilde\ell_i^+\tilde\ell_j^- , \; \chi^0_A\chi^0_B , \; \chi^+_A\chi^-_B , \; \tilde\nu_i^*\tilde\nu_j . \quad (8)$$

For e^-e^- collisions, the available channels are considerably restricted, and we have

$$e^-e^- \to \tilde\ell_i^-\tilde\ell_j^- . \quad (9)$$

Since SUSY-seesaw induced cLFV final states are dominated by slepton decays via slepton-lepton-(EW gaugino) interactions, it is useful to distinguish two cases with respect to the slepton/EW gaugino mass hierarchy. A dark-matter motivated scenario in which sleptons are heavier than the EW gauginos occurs in the so-called "Higgs funnel" region, while the opposite hierarchy happens in the so-called co-annihilation region. We thus define two types of points to guide the subsequent analysis

- F points: $m_{\tilde\ell,\tilde\nu} > m_{\chi^0_2,\chi^+_1}$ and $m_{\chi^0_2,\chi^+_1} > m_{\tilde\tau_1}$;
- C points: $m_{\tilde\ell,\tilde\nu} < m_{\chi^0_2,\chi^+_1}$ and $m_{\chi^0_1} \lesssim m_{\tilde\tau_1}$.

In table 1 we give two C- and one F-type points that will be used in the numerical analysis.

In the model under consideration, R-parity is conserved implying that the final states we are interested in, i.e. those with intervening sleptons, have an even number of χ^0_1 (the lightest supersymmetric particle, the LSP) which escape the detector without interacting. Hence, we focus our attention on cLFV in processes with di-lepton final states plus missing transverse momenta, $\not{E}_T$. The main source of background arises from weak charged currents producing neutrinos which, analogously to the LSP, escape undetected. These are of two types: the SM charged current backgrounds (type B) in which all $\not{E}_T$ is due to neutrinos; SUSY charged

Journal of Physics: Conference Series **447** (2013) 012035 doi:10.1088/1742-6596/447/1/012035

Table 1. Representative points used in the numerical analysis.

	C_1	C_2	F
m_0 (GeV)	150	200	750
$M_{1/2}$ (GeV)	727.9	949.2	872.1
$\tan\beta$	10	10	52
A_0 (GeV)	0	0	0
sign(μ)	1	1	1

current backgrounds (type C) in which $\not{E}_T$ contains both neutrinos and LSPs. Here, we study the following possibilities:

$$e^{\pm}e^{-} \to \begin{cases} e^{\pm}\mu^{-} + 2\chi_1^0 & \text{(A)} \\ e^{\pm}\mu^{-} + 2\chi_1^0 + (2,4)\nu & \text{(B)} \\ e^{\pm}\mu^{-} + (2,4)\nu & \text{(C)} \end{cases} \tag{10}$$

where (A) is the signal. In table 2 we classify the main sources of (B) and (C) backgrounds in e^+e^- collisions, in which we allow for leptonically decaying τs, Ws and Zs. Backgrounds in e^-e^- collisions can be similarly classified. However, $e^-e^- \to W^-W^-$ is very small, being suppressed by powers of m_ν/M_W, and $e^-e^- \to \tau^-\tau^-$ vanishes.

Table 2. SUSY charged current backgrounds (B) and SM charged current backgrounds (C), with the corresponding total cross section (order of magnitude) for unpolarised beams, for the signal $e^+\mu^- + 2\chi_1^0$.

SUSY bckg (type B) $\lesssim 5$ fb	0τ	$e^+\mu^- + (\bar{\nu}\nu, 2\bar{\nu}\nu) + 2\chi_1^0$
	τ+0ν	$(e^+\tau^-, \tau^+\mu^-, \tau^+\tau^-) + 2\chi_1^0$
	τ+2ν	$(e^+\tau^-, \tau^+\mu^-) + \bar{\nu}\nu + 2\chi_1^0$
SM bckg (type C) $\approx 10^2$ fb	W-strahlung	$W^-(e^+, \tau^+)\nu$, $W^+(\mu^-, \tau^-)\bar{\nu}$
	W-pair	W^+W^-
	τ-pair	$\tau^+\tau^-$

SM charged current backgrounds could *in principle* be reduced by devising kinematical cuts that rely on the fact that neutrinos are much lighter than the LSP and that the seeding processes have different topologies. However, that analysis is beyond the scope of our work. In our work [1] we conduct a phenomenological analysis, focusing on theoretical estimations of the expected number of events at a LC operating at a given centre of mass energy with the possibility of polarised beams.

More precisely, our analysis is based on an algorithmic calculation of the possible production and decay modes, considering that the majority of the events proceeds from an on-shell primary production (so that there are no interference effects between the different contributions), with subsequent two-body cascade decays (the exception being 3-body decays of the τ).

For sparticle primary production, we have considered the aforementioned channels[2], i.e. those listed in equations (8) and (9), while for SM primary production we have considered those listed

[2] Total lepton number violating processes such as $e^+e^- \to \tilde{\nu}\tilde{\nu}$ and $e^-e^- \to \chi_A^-\chi_B^-$ are severely suppressed by the seesaw scale.

in table 2, i.e. W-strahlung for both e^+e^- and e^-e^- collisions, and W- and τ-pair productions for e^+e^- collisions.

In figure 1 we show the results for e^+e^- unpolarised beams for point C_1 (left hand side) and F (right hand side). The cross sections for the signal (red crosses), SUSY background (blue times) and SM background (green asterisks) are given as a function of the centre of mass energy, $\sqrt{s}$. We have taken a degenerate RH neutrino spectrum with $M_R = 10^{12}$ GeV, and $\theta_{13} = 10°$.

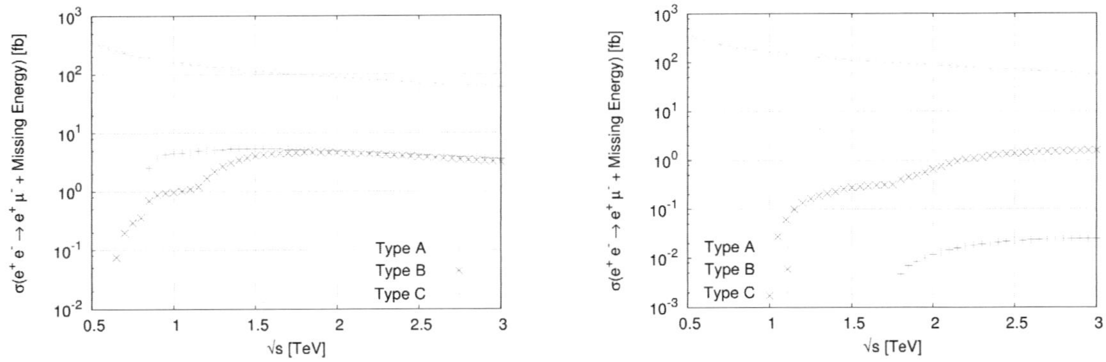

Figure 1. Cross sections for $e^+e^- \to e^+\mu^- + \not{E}_T$ (with $\not{E}_T = 2\chi_1^0, 2\chi_1^0 + (2,4)\nu, (2,4)\nu$), for points C_1 and F (left and right panels, respectively) as a function of the centre of mass energy $\sqrt{s}$. We fix $M_R = 10^{12}$ GeV, and denote the signal (A) with red crosses, the SUSY charged current background (B) with blue times, and the SM charged current background (C) by green asterisks. We have taken a degenerate right-handed neutrino spectrum and set $\theta_{13} = 10°$.

As can be seen, the SM background, consisting mainly of W-strahlung and W-pair production, dominates the signal by approximately one order of magnitude for C_1 and three orders of magnitude for F. Moreover, the SUSY background dominates (is comparable to) the signal in F (C_1). To understand these two observations, notice that the principal decay modes of sleptons and EW gauginos in F-type points are given by

$$(\tilde{\ell}^-, \tilde{\nu}^*) \to \chi_2^0(\ell^-, \bar{\nu}),\ \chi_1^-(\nu, \ell^+),\ (\ell^-, \bar{\nu})\chi_1^0, \tag{11}$$

$$\chi_1^- \to \tilde{\tau}_1^- \bar{\nu}, W^-\chi_1^0, \tag{12}$$

$$\chi_2^0 \to \tilde{\tau}_1^- \tau^+, Z\chi_1^0. \tag{13}$$

while for C-type points we have

$$\chi_1^- \to \tilde{\ell}^- \bar{\nu},\ \tilde{\nu}^* l^-, \tag{14}$$

$$\chi_2^0 \to \tilde{\ell}^{\pm} l^{\mp},\ \tilde{\nu}\bar{\nu},\ \tilde{\nu}^*\nu, \tag{15}$$

$$(\tilde{\ell}^-, \tilde{\nu}) \to (\ell^-, \nu)\chi_1^0. \tag{16}$$

Thus, solely from the width of the decays we expect $\Gamma_{\text{total}}(\tilde{\ell})_F > \Gamma_{\text{total}}(\tilde{\ell})_C$, which, for the same amount of flavour violation, gives that the expected signal cross sections for F points should be lower than for C points. In fact, we find that typically $\left(\text{BR}(\tilde{e}_L \to e\chi_1^0)_C\right)^2 \approx 10 \times \left(\text{BR}(\tilde{e}_L \to e\chi_1^0)_F\right)^2$.

Additionally, we observe that in C-type points, by taking a centre of mass energy above the $\tilde{\ell}_L\tilde{\ell}_R$ production threshold and below the production threshold of both $\chi_1^+\chi_1^-$ and $\chi_2^0\chi_2^0$, it is possible to evade the largest contribution from SUSY charged current backgrounds. For C_1 this happens above $\sqrt{s} \approx 800$ GeV and below $\sqrt{s} \approx 1200$ GeV, as can be inferred from figure 1.

3.1. Beam polarisation effect

Since W-pair production and W-strahlung backgrounds dominate the signal in e^+e^- collisions, it is interesting to consider how beam polarisation can be exploited to suppress their contribution without compromising the signal. We will do this by considering limiting cases, that is, 100% polarisations.

Since slepton flavour mixing occurs predominantly in the LL slepton sector, i.e. via decays of mostly left-handed slepton mass eigenstates, to avoid suppressing the $e^+\mu^-$ signal we require the e^- beam to be 100% left-polarised, while no constraint is placed on the positron beam.

Constraints on the positron beam arise by requiring the suppression of both W-pair production and W-strahlung production cross sections. In W-pair production, the s-channel is strongly dominated by vector boson interactions which can be suppressed by taking equally polarised beams. The t-channel is suppressed either by e^+_L or e^-_R beams. Therefore, we can suppress both channels by left-polarising the positron beam. Regarding W-strahlung, a maximal suppression would require $e^+_L e^-_R$ which would strongly suppress the signal. Nevertheless, the choice $e^+_L e^-_L$ is preferable to $e^+_R e^-_L$ one.

In figure 2 we show the result of left-polarising both positron and electron beams for point C_1, with parameters and line colour code as in figure 1. Clearly, the signal is enhanced by a factor of ≈ 4 near the $(\tilde{e}^+_R \tilde{e}^-_L)$ production threshold, while at higher energies the enhancement becomes smaller. The low energy tail of the SUSY background "disappears", as is understandable from the fact that such processes proceed via $\tilde{\tau}^+_1 \tilde{\tau}^-_1$ primary production. Moreover, both $\chi^+_1 \chi^-_1$ and $\chi^0_2 \chi^0_2$ primary production are greatly suppressed. As expected, the SM background is also suppressed, becoming comparable to the signal.

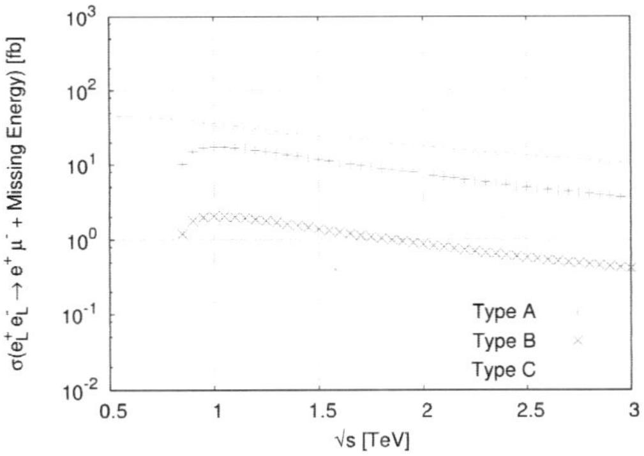

Figure 2. Cross section for $e^+e^- \to e^+\mu^- + \not{E}_T$ (with $\not{E}_T = 2\chi^0_1, 2\chi^0_1 + (2,4)\nu, (2,4)\nu$), for point C_1, as a function of the centre of mass energy $\sqrt{s}$, for 100% LL polarised beams. We have taken degenerate right-handed neutrino spectrum with $M_R = 10^{12}$, and set $\theta_{13} = 10°$. Line and colour code as in figure 1.

3.2. Probing the SUSY seesaw

We now discuss how the observation of a signal would allow to probe the underlying mechanism of slepton flavour mixing. If compatible with seesaw predictions, an observation would indeed strengthen the seesaw hypothesis, since a single mechanism would be able to explain many observables: the smallness of neutrino masses while generating neutrino mixings; and collider cLFV. Moreover, low energy observables such as CR(μ-e, N) and BR($\mu \to e\gamma$) would serve to further strengthen the hypothesis, provided that the predictions would be compatible with observations/exclusion limits.

In figure 3 we display the expected number of $e^-\mu^- + 2\chi_1^0$ events from 80% LL polarised electron beams, as a function of the seesaw scale, M_R, for point C_2. For illustrative purposes we have chosen $\sqrt{s} = 2$ TeV.

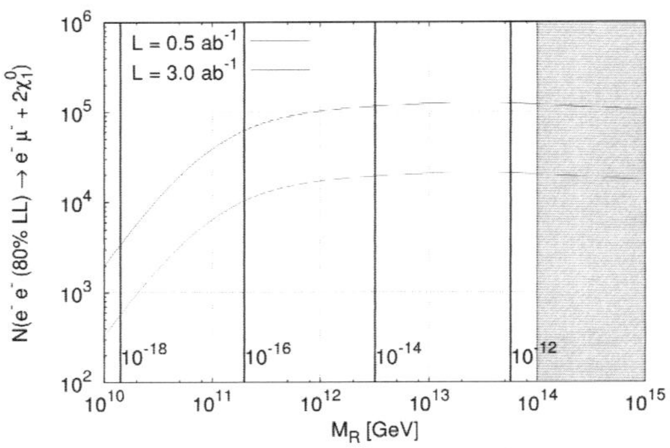

Figure 3. Number of events for $e^-e^- \to e^-\mu^- + 2\chi_1^0$ for point C_2 as a function of the seesaw scale M_R, for $(P_{e^-}, P_{e^-}) = (-80\%, -80\%)$ polarised beams. In both cases, we fix $\sqrt{s} = 2$ TeV, and we have taken a degenerate right-handed neutrino spectrum, with $\theta_{13} = 10°$. Vertical lines denote the M_R-corresponding value of BR($\mu \to e\gamma$) while the (grey) shaded region represents values of M_R already excluded by the present experimental bound on BR($\mu \to e\gamma$).

Due to the high number of expected events, if no $e^-\mu^-$ signal event is observed at an LC, then a high-scale SUSY seesaw is clearly disfavoured. Supposing that events are indeed observed, two possibilities arise. If the observed number of events is not accommodated by predictions, then it is likely that either the SUSY seesaw is not the unique source of LFV or that another mechanism for neutrino mass generation is at work. If the observed number of events can be accommodated by predictions, as illustrated in figure 3 for point C_2, then one can constrain (or even hint at) the seesaw scale. For example, if SUSY is realised with a C_2-like spectrum and more than 10^5 signal events are observed for a total integrated luminosity of approximately 3 ab^{-1}, then the seesaw scale, M_R, should be above 10^{12} GeV. Moreover, evading the exclusion limits on $\mu \to e\gamma$ sets an upper bound on this scale, so that we would be led to 10^{12} GeV $\lesssim M_R \lesssim 10^{14}$ GeV. If we now suppose that $\mu \to e\gamma$ is observed at MEG, e.g. with a branching ratio of the order of 10^{-12}, we would expect a number of $e^-\mu^-$ signal events below 10^5. This incompatibility could be due to (unaccounted for) destructive interferences in collider processes, or additional sources of LFV that only contribute to low-energy observables.

3.3. $\mu^-\mu^-$ golden channel

Same sign di-muon final states may possibly be a "golden channel" for the detection of cLFV at an LC. From an experimental point of view, the efficiency of the muon detectors can be fully explored when looking for di-muon final states. On the theoretical side, higher signal significances are expected, in particular due to very suppressed SM backgrounds.

In figure 4 we display the $e^-e^- \to \mu^-\mu^- + \not{E}_T$ cross section for the signal and the SUSY background, considering unpolarised beams, as a function of the centre of mass energy.

The SM background, not displayed in figure 4, is dominated by double-W-strahlung production, i.e. $e^-e^- \to W^-W^-\nu\nu \to \mu^-\mu^- 2(\bar{\nu}\nu)$, whose cross section is of the order of 1 fb. Therefore, we confirm that the $\mu^-\mu^-$ signal event is indeed much cleaner than $e^-\mu^-$ (and $e^+\mu^-$ in e^+e^- collisions).

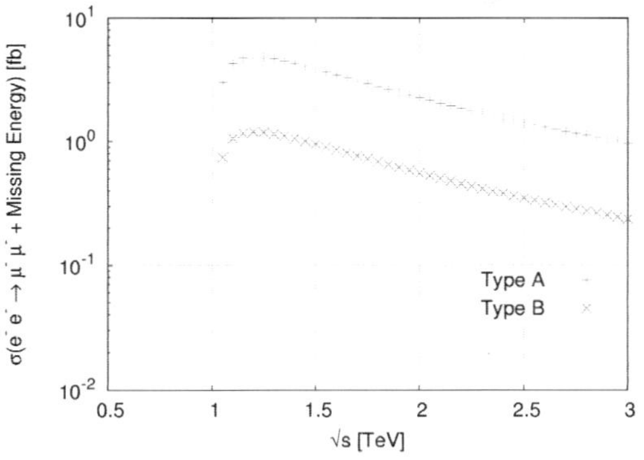

Figure 4. Cross section for $e^-e^- \to \mu^-\mu^- + \not{E}_T$ (with $\not{E}_T = 2\chi_1^0, 2\chi_1^0 + (2,4)\nu$), for point C_1 as a function of the centre of mass energy, $\sqrt{s}$, for the case of unpolarised beams. The signal is denoted by red crosses and the SUSY background by green asterisks. We have taken a degenerate right-handed neutrino spectrum with $M_R = 10^{12}$ GeV and set $\theta_{13} = 10°$.

4. Concluding remarks

A high-energy lepton collider offers an enormous potential for cLFV discovery. Beam polarisation can be instrumental in maximising the signal significance, rendering the signal "visible" in a large part of the high-scale SUSY seesaw parameter space, even without dedicated cuts, which could improve the observation prospects even further. We have also pointed out that cLFV discovery at an LC, complemented with low-energy LFV observables, could substantiate or disfavour the high-scale SUSY seesaw. Finally, we commented on the truly remarkable channel for cLFV discovery: $e^-e^- \to \mu^-\mu^- + 2\chi_1^0$.

Acknowledgments

This work has been done partly under the ANR project CPV-LFV-LHC NT09-508531. The work of A. J. R. F. has been supported by *Fundação para a Ciência e a Tecnologia* through the fellowship SFRH/BD/64666/2009. A. A. A. J. R. F. and A. M. T. acknowledge partial support from the European Union FP7 ITN INVISIBLES (Marie Curie Actions, PITN-GA-2011-289442). A. J. R. F. and J. C. R. also acknowledge the financial support from the EU Network grant UNILHC PITN-GA-2009-237920 and from *Fundação para a Ciência e a Tecnologia* project PEst-OE/FIS/UI0777/2011 and grants CERN/FP/116328/2010 and PTDC/FIS/102120/2008.

References

[1] Abada A, Figueiredo A, Romao J and Teixeira A 2012 *JHEP* **1208** 138 (*Preprint* 1206.2306)
[2] Heuer R D 2012 (*Preprint* 1202.5860)
[3] Feng J L 1998 *Int.J.Mod.Phys.* **A13** 2319–2328 (*Preprint* hep-ph/9803319)
[4] Borzumati F and Masiero A 1986 *Phys.Rev.Lett.* **57** 961
[5] Deppisch F, Pas H, Redelbach A, Ruckl R and Shimizu Y 2004 *Phys.Rev.* **D69** 054014 (*Preprint* hep-ph/0310053)
[6] Deppisch F, Kalinowski J, Pas H, Redelbach A and Ruckl R 2004 (*Preprint* hep-ph/0401243)
[7] Carquin E, Ellis J, Gomez M and Lolab S 2011 *JHEP* **1111** 050 (*Preprint* 1106.4903)
[8] Adam J *et al.* (MEG collaboration) 2011 *Phys.Rev.Lett.* **107** 171801 (*Preprint* 1107.5547)
[9] Casas J and Ibarra A 2001 *Nucl.Phys.* **B618** 171–204 (*Preprint* hep-ph/0103065)
[10] Gonzalez-Garcia M, Maltoni M, Salvado J and Schwetz T 2012 *JHEP* **1212** 123 (*Preprint* 1209.3023)

Journal of Physics: Conference Series **447** (2013) 012036 doi:10.1088/1742-6596/447/1/012036

Supersymmetric SO(10) with low intermediate scales

Carolina Arbeláez R.
IST, Instituto Superior Tecnico – Avenida Rovisco Pais, 1 1049-001, Lisboa, Portugal
E-mail: carolina.arbelaez@ist.utl.pt

Abstract. We construct supersymmetric $SO(10)$ models with different intermediate scales, consistent with gauge coupling unification. We found the complete list of sets of fields that can be added in each regime that allows to preserve unification and a "sliding scale mechanism". Using mSugra boundary conditions we calculate some particular combinations of soft SUSY breaking terms, called "invariants", that depend on the squark, slepton and gaugino mass spectra. We classify these invariants into a small number of sets, and show that their measurements contain indirect information about the class of models and the scale of beyond-MSSM physics.

1. Introduction
The GUTs $SO(10)$ models offer a number of adventages that, compared with models based in SU(5), allow us to explian some actually interesting phenomena, for example, and probably the most interesting is that $SO(10)$ contains the necessary ingredients to generate the seesaw mechanism.
Here, we analize supersymmetric GUT $SO(10)$ models where the unification scale is independent of the energy scale at which the extended gauge group is broken. This property, known as 'sliding mechanism' and studied in detail in [1],[2], allow us to impose certain conditions on the $\Delta b's$ coefficients in order to find the complete number of fields and variants for each class studied: LR simple class, Pati Salam and $U(1) \times U(1)_{B-L}$ models. All these classes are based on different $SO(10)$ breaking chains where the last symmetry breaking scale before reaching the SM group can be as low as $O(1)$ TeV preserving unification.
The number of models found in each clase is considerably large. If one assumes certain SUSY breaking scheme, as our case mSugra, indirect traces of the different variants remain in the SUSY spectrum that would be measurable at the LHC and ILC. Analyzing them it is possible to organize all the variants in a small number of sets wich allow us to distinguish class-II models from class-I or class-III.

2. Models
2.1. Model class I. Left Right intermediate scale
Our simplest class of models with only one intermediate scale is:

$$SO(10) \rightarrow SU(3)_c \times SU(2)_L \times SU(2)_R \times U(1)_{B-L} \rightarrow MSSM.$$

where the LR symmetry is broken at the scale m_R.

Journal of Physics: Conference Series **447** (2013) 012036 doi:10.1088/1742-6596/447/1/012036

Considering all the irreducible representations up to 126 in $SO(10)$ there are 24 different field representations (plus conjugate) that can be added at the m_R scale. Their representations under $SO(10)$ are summerized in [2]

Following the sliding (unification is independent of the intermediate scale m_R) and the matching condition:

$$\alpha_1^{-1}(m_R) = \frac{3}{5}\alpha_R^{-1}(m_R) + \frac{2}{5}\alpha_{B-L}^{-1}(m_R). \quad (1)$$

and also requering that the coupling constant at GUT is $\alpha_G^{-1} > 0$ (perturbativity), it is possible to obtain some simple conditions on the Δb_i^{LR} in order that GCU can be mantained: $\Delta b_2^{LR} = \Delta b_3^{LR} = \Delta b \leq 5$ and $\Delta b_{B-L}^{LR} + \frac{3}{2}\Delta b_R^{LR} - 9 = \frac{5}{2}\Delta b \leq \frac{25}{2}$.

With these constraints, we found a total of 65 different variants (different combinations in the $\Delta b^{'}s$) that correspond $0 \leq \Delta b \leq 5$. Two variant examples of the running of the inverse gauge couplings as a function of the energy scale, $m_R = 10$TeV and $m_{SUSY} = 1TeV$ are shown in Fig. 1.

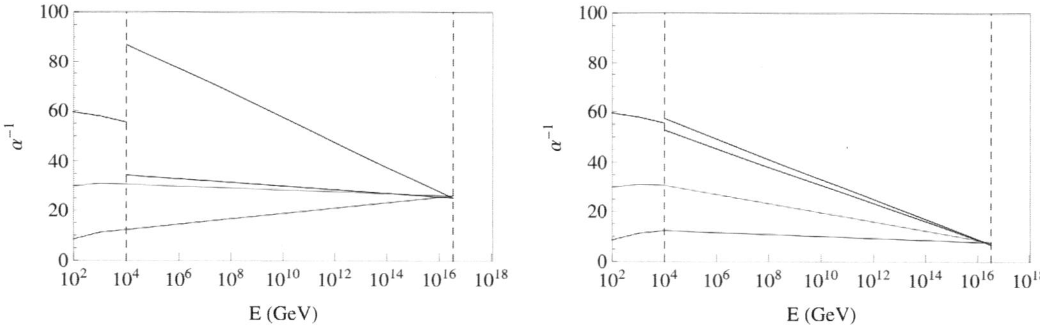

Figure 1. Gauge coupling unification in LR models for $m_R = 10^4$ GeV. Left panel is for $(\Delta b_3^{LR}, \Delta b_2^{LR}, \Delta b_R^{LR}, \Delta b_{B-L}^{LR}) = (0, 0, 1, 15/2)$ and right panel for $(4, 4, 10, 4)$.

Considering the fields that allows the symmetry breaking: $SU(2)_R \times U(1)_{B-L} \to U(1)_Y$: $\Phi_{1,1,3,-2}$ or $\Phi_{1,1,2,-1}$ and imposing that at least one of these must be present, the number of variants is reduced to 53.

There are some particular aspects that can be explained with the kind of variants here found, for example the seesaw mechanism consistent with GCU: the presence of the fields $\Phi_{1,1,3,-2}$ or $\Phi_{1,3,1,-2}$ or $\Phi_{1,3,1,0}$ or $\Phi_{1,1,3,0}$ allow seesaw type-I, seesaw type-II, seesaw type-III [3] and inverse seesaw type-III [1] respectively, with the most simple configurations like: $\Phi_{1,2,1,1} + \Phi_{1,3,1,0} + \Phi_{8,1,1,0} + \Phi_{1,1,3,-2} + \bar{\Phi}_{1,2,1,1} + \bar{\Phi}_{1,1,3,2}$, $\Phi_{1,1,2,-1} + \Phi_{8,1,1,0} + \Phi_{1,1,1,2} + \Phi_{3,1,1,\frac{4}{3}} + \Phi_{1,3,1,-2} + \bar{\Phi}_{1,1,2,-1} + \bar{\Phi}_{1,1,1,2} + \bar{\Phi}_{3,1,1,\frac{4}{3}} + \bar{\Phi}_{1,3,1,-2}$, $\bar{\Phi}_{1,1,2,-1} + 3\bar{\Phi}_{1,1,1,2} + 2\bar{\Phi}_{3,1,1,-\frac{2}{3}} + \Phi_{1,1,2,-1} + \Phi_{1,3,1,0} + 3\Phi_{1,1,1,2} + 2\Phi_{3,1,1,-\frac{2}{3}}$, $\bar{\Phi}_{1,1,2,-1} + \bar{\Phi}_{1,1,1,2} + \bar{\Phi}_{3,1,1,-\frac{2}{3}} + \Phi_{1,1,2,-1} + \Phi_{1,1,3,0} + \Phi_{1,2,2,0} + \Phi_{1,1,1,2} + \Phi_{3,1,1,-\frac{2}{3}}$.

No all the variants here found will be testable by measurements of the LHC. Taking into account the actual limits on the masses for m_g and $m_{\tilde{q}}$ [4] and the LEP limit on the chargino mass m_χ [5] it is possible to show that variants with $\Delta b_3^{LR} = 5$ will no be testable.

2.2. Model class II. Pati Salam intermediate scale

Here, the LR stage can be achieved via an interplay of another intermediate Pati Salam stage $SU(4)_c \times SU(2)_L \times SU(2)_R$ like:

$$\begin{aligned} SO(10) &\rightarrow SU(4) \times SU(2)_L \times SU(2)_R \\ &\rightarrow SU(3)_c \times SU(2)_L \times SU(2)_R \times U(1)_{B-L} \rightarrow \text{MSSM}. \end{aligned}$$

where the two symmetries are broken at m_{PS} and m_R.

$SO(10)$ representations up to 126 are considered, and the respective fields are listed in [2]. The first breaking: $SO(10)$ to Pati Salam requires the $\Psi_{1,1,1}$ of 54. The second step: Pati Salam to LR requires the $\Psi_{15,1,1}$ of 45 and finally LR to MSSM requires the fields $\Phi_{1,1,2,-1}$ or $\Phi_{1,1,3,-2}$.

Imposing also independence between m_G and m_R (sliding condition), perturbativity and taking the specific scale values: $m_R = 10^3$ GeV, $m_{PS} \leq 10^{16}$ GeV and $m_G = 10^{16}$ GeV, it is possible to find the bounds for the different $\Delta b's$, they will depend on the two differen regimes Pati Salam and LR: $\Delta b_2^{PS} + \frac{3}{10}\Delta b_2^{LR} < 7.2$, $\Delta b_4^{PS} + \frac{3}{10}\Delta b_3^{LR} < 10$ and $\frac{2}{5}\Delta b_4^{PS} + \frac{3}{5}\Delta b_R^{PS} + \frac{3}{10}\left(\frac{2}{5}\Delta b_{B-L}^{LR} + \frac{3}{5}\Delta b_R^{LR}\right) < 17$. The number of variants for an assumed m_R will be then a function os m_{PS}. For example, taking $m_R = 1$ TeV and $10^3\text{GeV} \leq m_{PS} \leq 10^4$ GeV, the number of variants is 375 and this is rapidy growing if larger m_{PS} scales are assumed. The number of possible variants for differents values of m_{PS} is actually relate in [2] where up to $m_{PS} = 10^{15}$ GeV, the list of possible variants is exhaustive.
Two solutions with different m_{PS} scales can be seen in Fig. 2.

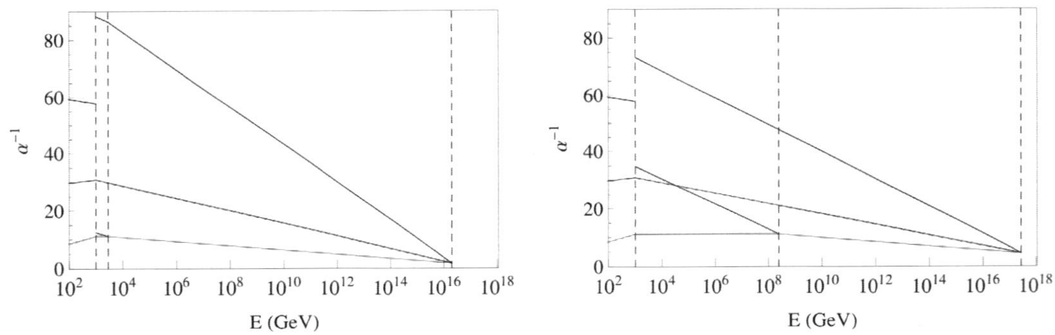

Figure 2. Gauge coupling unification for PS models with $m_R = 10^3$ GeV. In the plot to the left $(\Delta b_3^{LR}, \Delta b_L^{LR}, \Delta b_R^{LR}, \Delta b_{B-L}^{LR}, \Delta b_4^{PS}, \Delta b_L^{PS}, \Delta b_R^{PS}) = (3, 5, 10, 3/2, 8, 5, 17)$, while the plot to the right corresponds to $\Delta b's = (3, 4, 12, 6, 8, 4, 12)$.

where, different to the LR class, the GUT scale is not fixed to some value. This belong to the interval $[10^{16}, 10^{18}]$ GeV.

Bounds in the leptoquarks for a scale $m_{PS} = 40$ TeV have been derived in [6],[7]. In our list of solutions, there are 426 variants for which m_{PS} is lower than this bound, but, due to the sliding scale nature, it is possible to find a m_R such that $m_{PS} = 40$ TeV.
It is important also note that, for any particular set of Δb_i^{LR} more than one set of Δb_i^{PS} can be found. For example, for $\Delta b_i^{LR} = (0, 0, 1, 3/2)$ there are 342 differents sets of Δb_i^{PS}. A simple configuration corresponding for this is: $3\Psi_{1,2,2} + 4\Psi_{1,1,3} + \Psi_{4,1,2} + \bar{\Psi}_{4,1,2} + \Psi_{15,1,1}$, leading to $\Delta b_i^{PS} = (6, 3, 15)$ and a $m_{PS} = 8.2$TeV for $m_R = 1$TeV.

2.3. Model Class III. $U(1)_R \times U(1)_{B-L}$ intermediate scale
Here, an extended intermediate $U(1)_R \times U(1)_{B-L}$ regime follows the $SU(2)_R \times U(1)_{B-L}$ one:

$$\begin{aligned} SO(10) &\rightarrow SU(3)_c \times SU(2)_L \times SU(2)_R \times U(1)_{B-L} \\ &\rightarrow SU(3)_c \times SU(2)_L \times U(1)_R \times U(1)_{B-L} \rightarrow \text{MSSM}. \end{aligned}$$

where the two intermediate symmetries are broken at the scales m_{B-L} and m_R. The particle content is listed in [2]. The first breaking of $SO(10)$ is achieved like in the last to classes. The second step: LR to LR-$U(1)$ mixing requires the $\Phi_{1,1,3,0}$ and the last step: LR-$U(1)$ mixing to MSSM requires $\Phi'_{1,1,1/2,-1}$ or $\Phi'_{1,1,1,-2}$.

The one loop evolution of the gauge couplings and the SUSY breaking term for theories with more than one $U(1)$ gauge factor are affected by the extra kinetic mixing term. The couplings here are defined by:

$$G = \begin{pmatrix} g_{RR} & g_{RX} \\ g_{XR} & g_{XX} \end{pmatrix}. \tag{2}$$

The general coupling matrix in terms of G: $A(t) = (GG^T)/(4\pi) = (A^{-1}(t_0) - \gamma(t-t_0))^{-1}$ is defined also by the anomalous γ matrix that describes the charges of each superfield f under $U(1)_R$ and $U(1)_{B-L}$ [8]

$$\gamma = \begin{pmatrix} 7 & 0 \\ 0 & 6 \end{pmatrix}. \tag{3}$$

Taking into account the matching condition $p_Y^T \cdot A^{-1}(m_{B-L}) \cdot p_Y = \alpha_1^{-1}(m_{B-L})$, with $p_Y^T = (\sqrt{\frac{3}{5}}, \sqrt{\frac{2}{5}})$ and considering $m_{B-L} = 10^3$ GeV, $m_G \geq 10^{16}$ GeV and $m_R \leq 10^6$ GeV, the bounds on the $\Delta b's$ are:

$$\begin{aligned} \Delta b_2^{LR} + \tfrac{3}{10}\Delta b_2^{B-L} &< 7.1, \\ \Delta b_3^{LR} + \tfrac{3}{10}\Delta b_3^{B-L} &< 6.9, \\ \tfrac{3}{5}\Delta b_R^{LR} + \tfrac{2}{5}\Delta b_{B-L}^{LR} + \tfrac{3}{10} p_Y^T \cdot \Delta\gamma \cdot p_Y &< 10.8. \end{aligned}$$

The number of solutions found is very high 15610 (more than in Pati Salam class), this is expected due fact that there are larger Δb_i in the boundary expresions. Two examples are shown in Fig. 3.
Here, sliding condition imposes independence m_G with m_R, but, like in PS models, it is possible to have solutions where m_G and m_R are simultaneously independent of m_R. These follow the condition $\Delta b_3^{LR} = \Delta b_2^{LR} = p_Y^T \cdot \Delta\gamma \cdot p_Y$. One model of this kind is given in [1].

3. Invariants

Sliding conditions allow us to go down in the new scale m_R, so, different configurations could predict potentially rich phenomenology at the LHC. Also, indirect test of models are possible from measurements of SUSY particles masses.

Using mSugra boundary conditions, it is possible to calculate some special quantities called invariants, that depend on the soft scalar and gaugino mass parameters. mSugra at the GUT scale is defined by $M_{1/2}$ (common Gaugino mass), m_0 (common scalar mass), A_0 (trilinear coupling) and $\tan\beta$ (fixed by the vev's of H_u and H_d). The soft sfermion masses depend on these

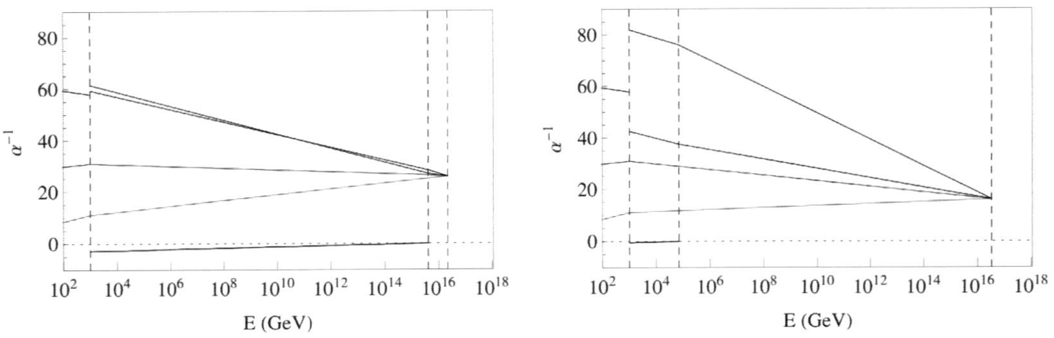

Figure 3. Gauge coupling unification in models with an $U(1)_R \times U(1)_{B-L}$ intermediate scale, for $m_R = 10^3$ GeV. Left: $(\Delta b_3^{LR}, \Delta b_L^{LR}, \Delta b_R^{LR}, \Delta b_{B-L}^{LR}, \Delta b_3^{B-L}, \Delta b_L^{B-L}, \Delta\gamma_{RR}, \Delta\gamma_{XR}, \Delta\gamma_{XX}) = (0, 1, 3, 3, 0, 0, 1/2, -\sqrt{3/8}, 3/4)$. Right: $(2, 2, 4, 8, 2, 2, 1/2, -\sqrt{3/8}, 11/4)$. The line, which appears close to zero in the $U(1)_R \times U(1)_{B-L}$ regime is the running of the off-diagonal element of the matrix A^{-1}, i.e. measures the size of the $U(1)$-mixing in the model.

parameters, the intermediate scales and the running couplings in the different regimes of the theory. For supersymmetric models with more than one $U(1)$ group, there are different effects on these soft values. A detailed analysis and the different mass expresions can be found in [2] and [1].

Using the sfermion mass parameters of the MSSM and the gaugino mass, it is possible to form four different combinations that, at 1-loop, do not depend of m_0 and $M_{1/2}$:

$$\begin{aligned} LE &= (m_{\widetilde{L}}^2 - m_{\widetilde{E}}^2)/M_1^2, \\ QE &= (m_{\widetilde{Q}}^2 - m_{\widetilde{E}}^2)/M_1^2, \\ DL &= (m_{\widetilde{D}}^2 - m_{\widetilde{L}}^2)/M_1^2, \\ QU &= (m_{\widetilde{Q}}^2 - m_{\widetilde{U}}^2)/M_1^2. \end{aligned} \tag{4}$$

Using these invariants it is possible to classify all the solutions found for each class in a few number of sets, like it is shown in the next sections.

3.1. Clasification

Considering whether invariants are larger or smaller than their respective mSugra values, and assigning $\Delta Inv = Inv(m_G) - Inv(m_s)$: + when $Inv(m_G) > Inv(m_s)$ and − when $Inv(m_G) < Inv(m_s)$, there are 16 possible combinations in the +, − for the 4 invariants, [2] . It is possible to show that only 9 of these 16 sets can be realized in our class of models:

Taking in to account the Cassimir coefficients [1] and the universality at the GUT scale, it is follow that:

$$m_{\widetilde{E}}^2 - m_{\widetilde{L}}^2 + m_{\widetilde{D}}^2 - 2m_{\widetilde{U}}^2 + m_{\widetilde{Q}}^2 = 0. \tag{5}$$

that in terms of invariants can be written as $QE = DL + 2QU$. This linear combination show that only three invariants are independent. Using this relation, it is also possible to exclude 4 of the 16 sets.

Using eq. 5 and considering the running in all the regimes, even the MSSM one, it is possible show that:

$$QU = LE + f(m_R). \tag{6}$$

with $f(m_R) = \frac{2}{33}\left\{\left[\frac{33}{10\pi}\alpha_1^{MSSM}\log\left(\frac{m_R}{m_{SUSY}}\right) - 1\right]^{-2} - 1\right\}$. Considering that $f(m_R)$ is always positive and using eq. 6, it is possible to exclude three aditional sets, leaving then a total of 9 possible sets and all these are contained in the three different class of models in a different way. Two examples of invariants can be seen in Fig. 4:

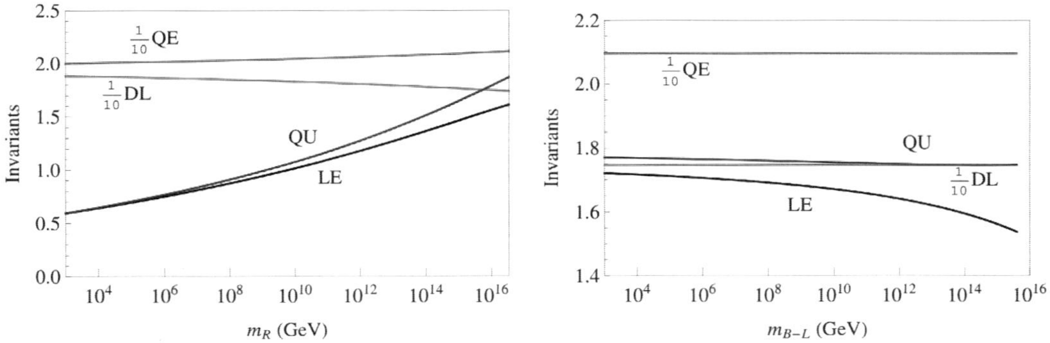

Figure 4. Left: m_R dependence for one example of invariants for Model class-I and II with $(\Delta b_3^{LR}, \Delta b_L^{LR}, \Delta b_R^{LR}, \Delta b_{BL}^{LR}) = (0, 0, 2, 6)$. Right: m_{B-L} dependence for one invariant in Model class-III, with $(\Delta b_3^{LR}, \Delta b_L^{LR}, \Delta b_R^{LR}, \Delta b_{BL}^{LR}, \Delta b_3^{BL}, \Delta\gamma_{RR}, \Delta\gamma_{XR}, \Delta\gamma_{XX}) = (0, 1, 3, 3, 0, 0, 1/2, -3/8, 3/4)$

In left figure, QU and LE invariants overlap for m_R close to the MSSM scale, this is explained by the enhanced of the gauge symmetry in the whole $m_{SUSY} - m_G$ range. This is a general feature for all the invariants in Model class-I and II. Other feature is that for models with Δb_3^{LR} large, changes in the invariants are huge. Right fig, that correspond to a variant with $\Delta b_3 = 0$ shows the mild dependence of LE with m_{B-L} and the independence of the other invariants with m_{B-L}. This is becouse m_{B-L} enters into the soft masses only through the changes in the abelian gauge couplings.

3.2. Comparison of model classes

For each variant there is an associate set of points (LE, QU, DL, QE). Considering that there are only two independent invariants, it is possible to choose for example (LE, QE) and make a 2D parametric plot, like Fig. 5

Each line represents one variant where m_R is changing along the line for the three different Model-classes. The blue region (behind the green one)includes all the LR variants, and the PS- and BL- ones (green and purple) include the variants where the higher intermediate scale is below 10^{16} GeV. The dot in the middle (mSugra point) corresponds to the limit where the intermediate scale is close to m_G. Every solution LE, QE, DL, QU is compatible with these kind of models if it satisfies the relations: $QE \simeq DL + 2QU$ and $QU \simeq LE$. As is espected (and was mentioned in the previous sections), the BL-class region with low m_R match with the LR-class because for m_{B-L} close to m_R the runing in the U(1)-mixing phase is small. It is expected also that the Pati Salam region approximates with the LR one when m_{PS} is close to m_G.

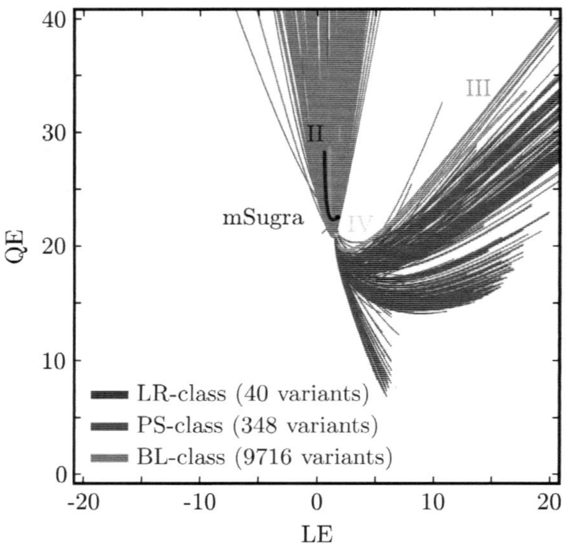

Figure 5. Parametric (LE, QE) plot for the different variants (see text). The thicker lines labeled with I, II, III and IV indicate the result for the four prototype models presented in [1].

It is possible then to distinguish the specific Model class for a specific set of values (LE, QE). So if supersymmetric masses can be measured in the future, the invariants could be used to distinguish between the class of models and also give information about the intermediate scale.

Acknowledgements

C. Arbeláez warmly thanks to Martin Hirsch, Jorge C. Romão and Renato M. Fonseca with whom this work was done. Many thanks fo the valuable and useful comments. This work has been supported in part by EU Network grant UNILHC PITN-GA-2009-237920.

References

[1] V. De Romeri, M. Hirsch, and M. Malinsky, Phys.Rev. **D84**, 053012 (2011), arXiv:1107.3412.
[2] C. Arbelaez, R. M. Fonseca, M. Hirsch and J. C. Romao, arXiv:1301.6085 [hep-ph].
[3] R. Foot, H. Lew, X. He, and G. C. Joshi, Z.Phys. **C44**, 441 (1989).
[4] H. Baer, V. Barger, and A. Mustafayev, Phys.Rev. **D85**, 075010 (2012), arXiv:1112.3017.
[5] Particle Data Group, J. Beringer *et al.*, Phys.Rev. **D86**, 010001 (2012).
[6] S. Davidson, D. C. Bailey, and B. A. Campbell, Z.Phys. **C61**, 613 (1994), arXiv:hep-ph/9309310.
[7] A. Kuznetsov, N. Mikheev, and A. Serghienko, (2012), arXiv:1210.3697.
[8] R. M. Fonseca, M. Malinsky, W. Porod, and F. Staub, Nucl.Phys. **B854**, 28 (2012), arXiv:1107.2670.

Enhanced lepton flavour violation in the supersymmetric inverse seesaw

C Weiland

Laboratoire de Physique Théorique, CNRS – UMR 8627, Université Paris Sud 11, F-91405 Orsay Cedex, France

E-mail: cedric.weiland@th.u-psud.fr

Abstract. In minimal supersymmetric seesaw models, the contribution to lepton flavour violation from Z-penguins is usually negligible. In this study, we consider the supersymmetric inverse seesaw and show that, in this case, the Z-penguin contribution dominates in several lepton flavour violating observables due to the low scale of the inverse seesaw mechanism. Among the observables considered, we find that the most constraining one is the μ–e conversion rate which is already restricting the otherwise allowed parameter space of the model. Moreover, in this framework, the Z-penguins exhibit a non-decoupling behaviour, which has previously been noticed in lepton flavour violating Higgs decays.

While the basic ingredients of neutral lepton flavour violation are close to being fully determined, charged lepton flavour violation (cLFV) remains to be observed despite impressive experimental efforts. In fact, cLVF is allowed in minimal extensions of the Standard Model but the associated observables are highly suppressed due to a Glashow–Iliopoulos–Maiani (GIM) mechanism taking place in the lepton sector. However, when these extensions are embedded in a larger framework like supersymmetry (SUSY), large contributions to cLFV are expected. The supersymmetric inverse seesaw is a very attractive framework to consider since all the new Physics it introduces does not exceed the TeV scale. As a consequence, it could be tested at high-energy through the direct production of the new heavy particles but also at the high-intensity frontier. Indeed, in the inverse seesaw [1], the size of the effective operator at the origin of the neutrino masses is not correlated to the size of the one generating cLFV. This suggests that there could be a large enhancement to lepton flavour violating observables, which has recently been discussed in several studies [2–8]. As recently pointed out, Higgs-mediated observables can be significantly enhanced by the presence of relatively light right-handed neutrinos [9]. Contrary to the Minimal Supersymmetric Standard Model (MSSM) where Z-penguin contributions are suppressed due to several cancellations, in the supersymmetric inverse seesaw, Z-mediated contributions are enhanced and even dominate the ones arising from Higgs-penguin. Indeed, in most of the cases, the Z-penguins provide the dominant contribution and can also exhibit a surprising non-decoupling behaviour.

After defining the model, we will discuss the enhancement of the Z-mediated contribution, giving an approximate formula for the Z–ℓ_i–ℓ_j vertex. We will then present our numerical results and conclusions.

Journal of Physics: Conference Series **447** (2013) 012037 doi:10.1088/1742-6596/447/1/012037

1. The supersymmetric inverse seesaw

The supersymmetric inverse seesaw consists of the MSSM extended by three pairs of gauge singlet superfields, $\widehat{\nu}_i^c$ and $\widehat{X}_i$ ($i = 1, 2, 3$), with opposite lepton number, -1 and $+1$, respectively. The superpotential of this model is given by

$$\mathcal{W} = \varepsilon_{ab} \Big[Y_d^{ij} \widehat{D}_i \widehat{Q}_j^b \widehat{H}_d^a + Y_u^{ij} \widehat{U}_i \widehat{Q}_j^a \widehat{H}_u^b + Y_e^{ij} \widehat{E}_i \widehat{L}_j^b \widehat{H}_d^a \\ + Y_\nu^{ij} \widehat{\nu}_i^c \widehat{L}_j^a \widehat{H}_u^b - \mu \widehat{H}_d^a \widehat{H}_u^b \Big] + M_{R_{ij}} \widehat{\nu}_i^c \widehat{X}_j + \frac{1}{2} \mu_{X_{ij}} \widehat{X}_i \widehat{X}_j \;, \tag{1}$$

where $i, j = 1, 2, 3$ are generation indices. In the above, $\widehat{H}_d$ and $\widehat{H}_u$ are the down- and up-type Higgs superfields, $\widehat{L}_i$ denotes the SU(2) doublet lepton superfields. The only term that violates lepton number conservation is the Majorana mass term $\mu_{X_{ij}} \widehat{X}_i \widehat{X}_j$. Thus, the limit $\mu_{X_{ij}} \to 0$ increases the symmetry of the model, making the smallness of $\mu_{X_{ij}}$ natural. The soft SUSY breaking Lagrangian can be found in [10]. Assuming a flavour-blind mechanism of supersymmetry breaking by considering universal boundary conditions at the grand unification (GUT) scale, like in case of the constrained MSSM (cMSSM), the supersymmetric parameters are

$$m_\phi = m_0 \,, M_{\text{gaugino}} = M_{1/2} \,, A_i = A_0 \, \mathbb{I} \,, B_{\mu_X} = B_{M_R} = B_0 \, \mathbb{I} \,. \tag{2}$$

In this framework, the 9×9 neutrino mass matrix $\mathcal{M}$ in the basis $\{\nu, \nu^c, X\}$ is given by

$$\mathcal{M} \;=\; \left(\begin{array}{ccc} 0 & m_D^T & 0 \\ m_D & 0 & M_R \\ 0 & M_R^T & \mu_X \end{array} \right) \,, \tag{3}$$

where $m_D = \frac{1}{\sqrt{2}} Y_\nu v_u$ and M_R, μ_X are 3×3 matrices in family space. Assuming $m_D, \mu_X \ll M_R$, the diagonalization leads to an effective Majorana mass matrix for the light neutrinos [11],

$$m_\nu = m_D^T {M_R^T}^{-1} \mu_X M_R^{-1} m_D = \frac{v_u^2}{2} Y_\nu^T (M_R^T)^{-1} \mu_X M_R^{-1} Y_\nu \,. \tag{4}$$

As can be seen from the above equation, the smallness of the light neutrino masses can be explained by the smallness of μ_X. This decorrelates the effective operator generating neutrino masses from the one generating cLFV since the latter does not depend on μ_X. Moreover having μ_X at the eV-scale allows the Yukawa couplings to have natural values ($Y_\nu \sim \mathcal{O}(1)$) and, at the same time, right-handed neutrinos to have masses at the TeV. Having light right-handed neutrinos and large Yukawa couplings leads, in turn, to large contributions to cLFV observables. By defining

$$M^{-1} = (M_R^T)^{-1} \; \mu_X \; M_R^{-1} \,, \tag{5}$$

the light neutrino mass matrix can be cast in a form that strongly resembles the one for the type-I seesaw

$$m_\nu = \frac{v_u^2}{2} Y_\nu^T M^{-1} Y_\nu \,. \tag{6}$$

Without loss of generality, we work in a basis where M_R is diagonal at the SUSY scale. Moreover, in the numerical analysis, we will work in the limit where μ_X is diagonal. This is a conservative assumption, since, in that case, the experimental limits will constrain the size of the Yukawa couplings, which in turn will limit the branching ratios and decay rates of cLFV observables. The constraints from the neutrino data that we considered in this study can be found in [10].

2. Z-mediated charged lepton flavour violation

Many studies have been devoted to cLFV in the MSSM and many of its extensions. It was found that the γ-penguin gives the dominant contribution to 3-body cLFV decays [12, 13], the only exception being the large $\tan\beta$ (and low m_A) regime where large Higgs contributions are expected [14]. The same observation was recently made in the supersymmetric inverse seesaw [9]. When the mass scaling of the different contributions is considered, the γ-penguin contribution to $\ell_i \to 3\ell_j$ from the chargino-sneutrino loop is given by

$$A_a^{(c)L,R} = \frac{1}{16\pi^2 m_{\tilde{\nu}}^2} \mathcal{O}_{A_a}^{L,R} s(x^2)\,, \tag{7}$$

whereas the Z-contribution reads

$$F_X = \frac{1}{16\pi^2 g^2 \sin^2\theta_W m_Z^2} \mathcal{O}_{F_X}^{L,R} t(x^2)\,, \tag{8}$$

with $X = \{LL, LR, RL, RR\}$. In the above, $\mathcal{O}_y^{L,R}$ denotes combinations of rotation matrices and coupling constants and $s(x^2)$ and $t(x^2)$ represent the Passarino-Veltman loop functions which depend on $x^2 \equiv m_{\tilde{\chi}^-}^2/m_{\tilde{\nu}}^2$ (see [13]). Since $m_Z^2 \ll m_{\tilde{\nu}}^2$, the Z-mediated contribution would dominate over the one coming from γ-penguins. However, in the MSSM, this is prevented by a subtle cancellation between different diagrams contributing to the leading Z-mediated contribution [15]. It can be understood by focusing on the dominant contribution to $\ell_i \to 3\ell_j$ which is given by diagrams where the external leptons are left-handed (the other cases are suppressed by the Yukawa couplings of the charged leptons). It is given in [10] by the form factor

$$F_{LL} = \frac{F_L Z_L^{(l)}}{g^2 \sin^2\theta_W m_Z^2}\,, \tag{9}$$

where $Z_L^{(l)} = -\frac{g}{c_W}(-\frac{1}{2} + s_W^2)$ is the $Z - \ell_i - \ell_j$ tree-level coupling ($i = j$) and F_L is the $Z - \ell_i - \ell_j$ 1-loop effective vertex, with $i \neq j$. In the MSSM, the sneutrino mass matrix is diagonalized by Z_V, a 3×3 unitary matrix. But when chargino mixing is neglected, the masses cancel out in the combination of loop functions from different diagrams and one ends up with $F_L^0 \propto (Z_V^\dagger Z_V)_{ij}$ which vanishes for $i \neq j$ due to the unitarity of the Z_V matrix [15]. However, this cancellation can be spoiled if new interactions are introduced like it is the case with the inverse seesaw mechanism+. In particular, the $\tilde{H}^\pm - \tilde{\nu}_R - \ell_L$ coupling gives rise to new diagrams containing a higgsino-sneutrino loop which spoils the cancellation since they give $F_L^0 \propto \sum_i Z_V^{ik} Z_V^{ij*} Y_\nu^{ik*} Y_\nu^{ij}$. It is worth noticing that this does not happen in the MSSM extended by a high-scale type-I seesaw since the very heavy right-handed sneutrinos suppress the higgsino-sneutrino loop. In the supersymmetric inverse seesaw, the large neutrino Yukawa couplings and the TeV-scale right-handed sneutrinos make this contribution naturally large.

The complete derivation of the analytical expressions for F_L^0 can be found in [10] and only an abridged version will be presented here. The chargino-sneutrino loops can be decomposed as

$$F_L^{(0)} = -\frac{1}{16\pi^2}\left(F_{L,\text{wino}}^{(0)} + F_{L,\text{higgsino}}^{(0)}\right)\,. \tag{10}$$

Considering that left-right mixing is negligible in the sneutrino sector when the neutrino Yukawa couplings are relatively small (which will be verified by our numerical study), the two contributions reduce to

cLFV Process	Present Bound	Future Sensitivity
$\mu \to e\gamma$	2.4×10^{-12} [18]	6×10^{-14} [19]
$\tau \to e\gamma$	3.3×10^{-8} [20]	3.0×10^{-9} [21]
$\tau \to \mu\gamma$	4.4×10^{-8} [20]	2.4×10^{-9} [21]
$\mu \to 3e$	1.0×10^{-12} [22]	$\mathcal{O}(10^{-16})$ [23]
$\tau \to 3e$	2.7×10^{-8} [24]	2.3×10^{-10} [21]
$\tau \to 3\mu$	2.1×10^{-8} [24]	8.2×10^{-10} [21]
$\mu - e$, Au	7.0×10^{-13} [25]	
$\mu - e$, Ti	4.3×10^{-12} [26]	$\mathcal{O}(10^{-18})$ [27]
$\tau \to e\eta$	4.4×10^{-8} [28]	$\mathcal{O}(10^{-10})$ [21]
$\tau \to \mu\eta$	2.3×10^{-8} [28]	$\mathcal{O}(10^{-10})$ [21]

Table 1. Current experimental bounds and future sensitivities for different cLFV observables.

$$F^{(0)}_{L,\text{wino}} = \frac{g^2}{8}\delta_{ij}(3g c_W + g' s_W) = \frac{g^3}{8c_W}\delta_{ij}(1+2c_W^2)\,, \tag{11}$$

$$F^{(0)}_{L,\text{higgsino}} = \frac{g}{8c_W}\left(Y_\nu^\dagger Y_\nu\right)_{ij}\left(c_W^2 - \frac{1}{2}\right). \tag{12}$$

It can be noted that the wino contribution vanishes for $i \neq j$, which corresponds to the cancellation discussed earlier, while the higgsino contribution, which is absent from the MSSM, does not vanish, explaining the different behaviours between the two models. We have verified that the expressions above reproduce the numerical results based on complete, non-approximate expressions. We have also verified that the higher-order terms in chargino mixing are numerically negligible. It is worth mentioning that the $F_L^{(0)}$ vertex does not depend on supersymmetric masses. As a consequence, the Z-penguin exhibits a non-decoupling behaviour and a large SUSY scale does not suppress its contribution to cLFV processes.

Finally, although the previous discussion has been focused on $\ell_i \to 3\ell_j$ processes, the same enhancement in the $Z - \ell_i - \ell_j$ effective vertex also affects other observables. This is the case for $\mu - e$ conversion in nuclei [16] and the case of $\tau \to P^0\ell_i$, where P^0 is a pseudoscalar meson [17].

3. Numerical results

In this study, we focused on leptonic observables which can be divided in three broad classes: radiative decays, like $\mu \to e\gamma$, 3-body decays, like $\tau \to 3\mu$, and neutrinoless conversion in muonic atoms, like $\mu, \mathrm{Ti} \to e, \mathrm{Ti}$. These observables are actively searched for at many current and future experiments. The list of processes we considered together with the current upper limit and expected future sensitivity can be found in table 1. We also addressed high-energy observables such as flavour-violating neutralino decays $\mathrm{Br}(\tilde{\chi}_2^0 \to \tilde{\chi}_1^0 \ell_i \ell_j)$ and slepton mass splittings $\Delta m_{\tilde{\ell}}$ $(m_{\tilde{\ell}_i} - m_{\tilde{\ell}_j})$. However, we found that the small neutrino Yukawa couplings needed to comply with experimental limits on $\mu, \mathrm{N} \to e, \mathrm{N}$ conversion (as discussed below) gives low rates for $\tilde{\chi}_2^0 \to \tilde{\chi}_1^0 \ell_i \ell_j$ and small mass splittings. We have also verified that cLFV decays of the Z and h^0 bosons are not significantly enhanced.

Our numerical results have been obtained with a `SPheno` [29, 30] code generated with the Mathematica package `SARAH` [31–33]. The computation of the cLFV observables is based on the results presented in [13, 16, 17], that we extended to the inverse seesaw case. As such, it is not limited to the Z-penguins but also includes contributions from Higgs-penguins, γ-penguins and supersymmetric boxes. The analytical expressions for $\mu - e$ conversion in nuclei, $\mathrm{Br}(\ell_i \to 3\ell_j)$ and $\mathrm{Br}(\tau \to P^0\ell_i)$ in the case of Z-penguin dominance are given in [10]. The supersymmetric

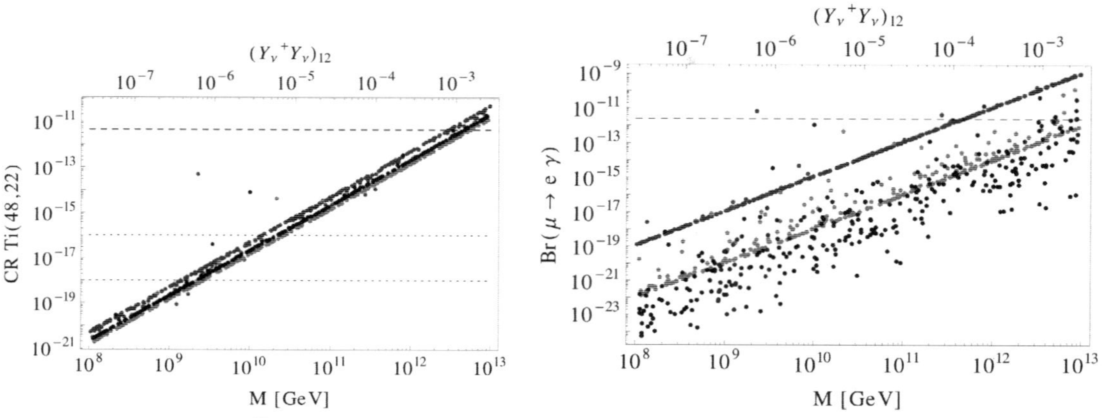

(a) $^{48}_{22}$Ti (b) $\mu \to e\gamma$

Figure 1. $\mu - e$ conversion rates in $^{48}_{22}$Ti (left) and Br($\mu \to e\gamma$) (right), as a function of M and $(Y_\nu^\dagger Y_\nu)_{12}$ for different $\hat{M}_R$ values: $\hat{M}_R = 100$ GeV (blue), $\hat{M}_R = 1$ TeV (red) and $\hat{M}_R = 10$ TeV (black). We set $A_0 = -300$ GeV, $\tan\beta = 10$, sign$(\mu) = +$ and $B_0 = 0$, and we randomly vary m_0 and $M_{1/2}$ in the range $[0, 3$ TeV$]$. The horizontal dashed lines represent the current experimental bounds and the dotted ones represent the expected future sensitivities.

parameters of the model are $m_0, M_{1/2}, A_0, \tan\beta, \text{sign}(\mu)$ and B_0. Since they are of little relevance for Z-mediated observables, we fixed sign$(\mu) = +$ and $B_0 = 0$ in the numerical analysis. As can be seen in figure 1(a), when an observable receives its dominant contribution from Z-penguins, it has a very small dependence on the cMSSM parameters making our results quite general. However, this is not true when another contribution is dominant like in Br($\mu \to e\gamma$), see figure 1(b) for example, in accordance with the literature [2, 5]. The results presented here are obtained with degenerate singlets $M_R = \text{diag}(\hat{M}_R, \hat{M}_R, \hat{M}_R)$ and we have verified that processes dominated by Z-penguins are only marginally affected by non-degenerate singlets. For example, the relative change in the conversion rate for μ, Au $\to e$, Au is always below 2%. As a consequence, the results discussed here hold true for both degenerate and non-degenerate right-handed (s)neutrinos. Since μ_X does not explicitely appear in analytical expressions for cLFV processes, the relevant parameters are $M \sim \frac{M_R^2}{\mu_X}$ (as defined in Eq. (5)), which controls the size of the Yukawa couplings, and M_R which controls the size of the non-SUSY contributions.

In figure 2(a), we display the absolute contributions to Br($\mu \to 3e$) in order to illustrate the non-decoupling behaviour of the Z-penguin as previously discussed. We define the relative contribution to cLFV observables as

$$c_i = \frac{|\text{R}_i|}{\sqrt{\sum_k \text{R}_k^2}}, \tag{13}$$

where R_i is the rate (branching ratio in case of $\mu \to 3e$ and conversion rate in case of $\mu - e$ conversion in Au) that would be obtained, should the i-contribution to the process be the only one. It can be seen from figure 2(b) that Br($\mu \to 3e$) is clearly dominated by the Z-boson contribution in most of the parameter space, the only exception being for very low $m_0 = M_{1/2}$ where the γ-penguins dominate. This can easily be understood by recalling the scaling of the operators for the different contributions in equations (7) and (8). The same behaviour was observed in other observables as described in [10] and allows to conclude that, if present, the Z-mediated contribution will dominate cLFV observables in the supersymmetric inverse seesaw,

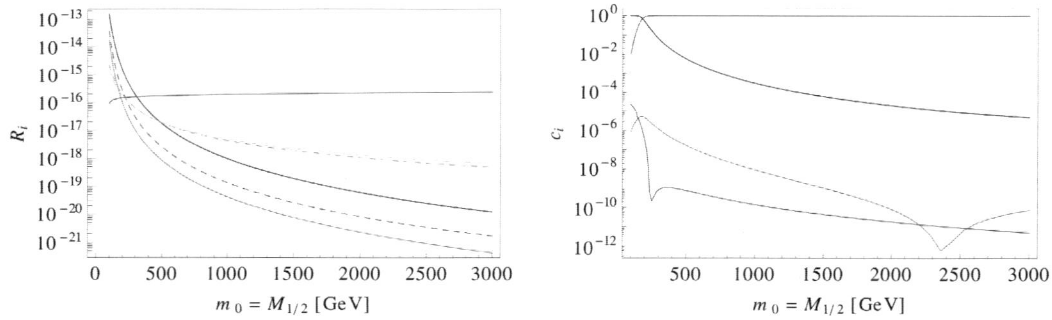

(a) $\mu \to 3e$

(b) $\mu - e$ conversion in $^{197}_{79}$Au

Figure 2. Absolute contributions to $\mathrm{Br}(\mu \to 3e)$ (left-hand side) and relative contributions to $\mu - e$ conversion in $^{197}_{79}$Au (right-hand side) as a function of $m_0 = M_{1/2}$ for a degenerate singlet spectrum with $\hat{M}_R = 10$ TeV and $M = 10^{11}$ GeV. The rest of the cMSSM parameters are set to $A_0 = -300$ GeV, $B_0 = 0$, $\tan\beta = 10$ and $\mathrm{sign}(\mu) = +$. On the left-hand side, solid lines represent individual contributions, γ (black), Z (blue) and h (red) whereas the dashed lines represent interference terms, $\gamma - Z$ (green), $\gamma - h$ (purple) and $Z - h$ (orange). Note that in this case h includes both Higgs and box contributions. On the right-hand side interference terms are not shown to make the results clearer. The individual contributions are γ (black), Z (blue) and h (red) and boxes (purple).

except for a very low SUSY scale.

When the Z-penguin gives the dominant contribution, there is a strong correlation between M and the processes rate. This can clearly be seen in figure 1(a) for the $\mu - e$ conversion rate. From the non-observation of cLFV, an upper bound on M, or equivalently on the size of $(Y_\nu^\dagger Y_\nu)_{ij}$, can be derived. Among the many observables involving $\mu - e$ transitions, the most constraining one is neutrinoless conversion in muonic atoms and the approximate limits it gives are summarised in table 2 for a conservative choice of $\hat{M}_R \gtrsim 1$ TeV. Smaller values for $\hat{M}_R$ will only result in stronger limits, especially when non-SUSY contributions become dominant, making $\mu \to e\gamma$ the most constraining observable. It should also be noted that future experiments like COMET, PRISM/PRIME and Mu2e have the potential to greatly improve these limits.

	M [GeV]	$(Y_\nu^\dagger Y_\nu)_{12}$
$\mathrm{CR}_{\mathrm{Au}}$(current)	10^{12}	2.7×10^{-4}
$\mathrm{CR}_{\mathrm{Ti}}$(current)	4×10^{12}	10^{-3}
$\mathrm{CR}_{\mathrm{Ti}}$(future: 10^{-16})	2×10^{10}	5.5×10^{-6}
$\mathrm{CR}_{\mathrm{Ti}}$(future: 10^{-18})	2×10^{9}	5.5×10^{-7}

Table 2. Approximate limits on M and $(Y_\nu^\dagger Y_\nu)_{12}$ from the non-observation of $\mu - e$ conversion in nuclei. This table includes current experimental bounds and future expected sensitivities [25–27, 34–36].

Other cLFV transitions also exhibit a clear enhancement due to the Z-penguin dominance. In particular, the branching ratios of different $\ell_i \to 3\ell_j$ channels are related by

$$\frac{\mathrm{Br}(\ell_i \to 3\ell_j)}{\mathrm{Br}(\ell_m \to 3\ell_n)} = \frac{(Y_\nu^\dagger Y_\nu)^2_{ij}}{(Y_\nu^\dagger Y_\nu)^2_{mn}} \frac{m^5_{\ell_i} \tau_i}{m^5_{\ell_m} \tau_m} . \tag{14}$$

From this equation, we can expect that, considering similar combination $(Y_\nu^\dagger Y_\nu)_{ij}$, all the branching ratios for $\ell_i \to 3\ell_j$ processes will lie with $1-2$ orders of magnitude. However, the experimental upper limits on these observables are more relaxed for $\tau - e$ and $\tau - \mu$ cLFV transitions than for $\mu - e$ cLFV transition. This implies that 3-body τ cLFV decays have little possibility to be observed in the near future, except if there is a strong cancellation in $Y_\nu^\dagger Y_\nu$ which would suppress $\mu - e$ cLFV processes. The same conclusion also holds for semileptonic τ decays. Finally, it is worth mentioning that in Z-penguin dominated scenarios, there is a clear correlation between the rates for $\mu \to 3e$ and $\mu - e$ conversion in nuclei. Numerically, we found $\mathrm{CR}(\mu - e, \mathrm{Ti})/\mathrm{Br}(\mu \to 3e) \sim 15$.

Z-dominated observables are quite insensitive to the cMSSM parameters, implying that these results should hold when the recent mass constraints on the Higgs boson are imposed on the model. We have checked this using a benchmark point associated to a Higgs mass of $m_{h^0} = 126.5\,\mathrm{GeV}$ and the corresponding predictions for the different observables are given in [10]. Moreover, since the neutrino Yukawa couplings are constrained to be below 10^{-2}, we do not expect the singlet sector to sensibly modify the Higgs mass and renormalisation group equations (RGEs).

To conclude, the supersymmetric inverse seesaw is a very attractive extension of the MSSM since it generates neutrino masses while keeping all the new Physics below the TeV scale with naturally large Yukawa couplings. In this work, we considered its predictions for different lepton flavour violating observables mediated by the exchange of a Z-boson and found that the Z-mediated contribution dominates in most of the parameter space, providing an important enhancement to different observables. As a consequence, the most constraining observable is now neutrinoless $\mu - e$ conversion in nuclei, in contrast with the usual behaviour of the MSSM extended by a high-scale type-I seesaw. This allows us to put constraints on the size of the Yukawa couplings $(Y_\nu^\dagger Y_\nu)_{12} \gtrsim 2.7 \times 10^{-4}$, clearly excluding $Y_\nu \sim \mathcal{O}(1)$ as usually considered.

Acknowledgements
This work has been supported by the ANR project CPV-LFV-LHC NT09-508531 and the European Union FP7 ITN INVISIBLES (Marie Curie Actions, PITN-GA-2011-289442).

[1] Mohapatra R and Valle J 1986 *Phys.Rev.* **D34** 1642
[2] Deppisch F and Valle J 2005 *Phys.Rev.* **D72** 036001 (*Preprint* hep-ph/0406040)
[3] Deppisch F, Kosmas T and Valle J 2006 *Nucl.Phys.* **B752** 80–92 (*Preprint* hep-ph/0512360)
[4] Dev P B and Mohapatra R 2010 *Phys.Rev.* **D81** 013001 (*Preprint* 0910.3924)
[5] Hirsch M, Kernreiter T, Romao J and Villanova del Moral A 2010 *JHEP* **1001** 103 (*Preprint* 0910.2435)
[6] Dias A, de S Pires C, Rodrigues da Silva P and Sampieri A 2012 *Phys.Rev.* **D86** 035007 (*Preprint* 1206.2590)
[7] Hirsch M, Porod W, Reichert L and Staub F 2012 *Phys.Rev.* **D86** 093018 (*Preprint* 1206.3516)
[8] Awasthi R L, Parida M and Patra S 2013 (*Preprint* 1301.4784)
[9] Abada A, Das D and Weiland C 2012 *JHEP* **1203** 100 (*Preprint* 1111.5836)
[10] Abada A, Das D, Vicente A and Weiland C 2012 *JHEP* **1209** 015 (*Preprint* 1206.6497)
[11] Gonzalez-Garcia M and Valle J 1989 *Phys.Lett.* **B216** 360
[12] Hisano J, Moroi T, Tobe K and Yamaguchi M 1996 *Phys.Rev.* **D53** 2442–2459 (*Preprint* hep-ph/9510309)
[13] Arganda E and Herrero M J 2006 *Phys.Rev.* **D73** 055003 (*Preprint* hep-ph/0510405)
[14] Babu K and Kolda C 2002 *Phys.Rev.Lett.* **89** 241802 (*Preprint* hep-ph/0206310)
[15] Hirsch M, Staub F and Vicente A 2012 *Phys.Rev.* **D85** 113013 (*Preprint* 1202.1825)
[16] Arganda E, Herrero M and Teixeira A 2007 *JHEP* **0710** 104 (*Preprint* 0707.2955)
[17] Arganda E, Herrero M and Portoles J 2008 *JHEP* **0806** 079 (*Preprint* 0803.2039)
[18] Adam J *et al.* (MEG collaboration) 2011 *Phys.Rev.Lett.* **107** 171801 (*Preprint* 1107.5547)
[19] Baldini A, Cei F, Cerri C, Dussoni S, Galli L *et al.* 2013 (*Preprint* 1301.7225)
[20] Aubert B *et al.* (BABAR Collaboration) 2010 *Phys.Rev.Lett.* **104** 021802 (*Preprint* 0908.2381)
[21] O'Leary B *et al.* (SuperB Collaboration) 2010 (*Preprint* 1008.1541)
[22] Bellgardt U *et al.* (SINDRUM Collaboration) 1988 *Nucl.Phys.* **B299** 1

[23] Blondel A, Bravar A, Pohl M, Bachmann S, Berger N *et al.* 2013 (*Preprint* 1301.6113)
[24] Hayasaka K, Inami K, Miyazaki Y, Arinstein K, Aulchenko V *et al.* 2010 *Phys.Lett.* **B687** 139–143 (*Preprint* 1001.3221)
[25] Bertl W H *et al.* (SINDRUM II Collaboration) 2006 *Eur.Phys.J.* **C47** 337–346
[26] Dohmen C *et al.* (SINDRUM II Collaboration.) 1993 *Phys.Lett.* **B317** 631–636
[27] Barlow R 2011 *Nucl.Phys.Proc.Suppl.* **218** 44–49
[28] Hayasaka K (Belle Collaboration) 2010 *PoS* **ICHEP2010** 241 (*Preprint* 1011.6474)
[29] Porod W 2003 *Comput.Phys.Commun.* **153** 275–315 (*Preprint* hep-ph/0301101)
[30] Porod W and Staub F 2012 *Comput.Phys.Commun.* **183** 2458–2469 (*Preprint* 1104.1573)
[31] Staub F, Ohl T, Porod W and Speckner C 2012 *Comput.Phys.Commun.* **183** 2165–2206 (*Preprint* 1109.5147)
[32] Staub F 2011 *Comput.Phys.Commun.* **182** 808–833 (*Preprint* 1002.0840)
[33] Staub F 2010 *Comput.Phys.Commun.* **181** 1077–1086 (*Preprint* 0909.2863)
[34] Glenzinski D (Mu2e Collaboration) 2010 *AIP Conf.Proc.* **1222** 383–386
[35] Cui Y *et al.* (COMET Collaboration) 2009 *COMET Conceptual design report*
[36] Carey R *et al.* (Mu2e Collaboration) 2008 *Mu2e Proposal*

Journal of Physics: Conference Series **447** (2013) 012038 doi:10.1088/1742-6596/447/1/012038

Fermion mixing with geometrical CP violation and its tests at the LHC

Ivo de Medeiros Varzielas

Fakultät für Physik, Technische Universität Dortmund D-44221 Dortmund, Germany

E-mail: ivo.de@udo.edu

Abstract. We review and clarify some cases of geometrical CP violation, the framework of spontaneous CP violation through complex phases with values that are independent of parameters of the potential. We present a flavour model based on $\Delta(27)$ featuring spontaneous CP violation, that can reproduce all quark masses and mixing data. The scalar sector of the model has exotic properties that can be tested at the LHC.

The work summarised here is mostly based on [1] and [2], which include more complete references that we omit here due to space constraints.

The origin of CP violation is an open question in particle physics. In the Standard Model (SM), CP is violated through complex Yukawa couplings and this violation appears in charged weak interactions through the Cabibbo-Kobayashi-Maskawa (CKM) matrix. In beyond the SM theories one can explore the origin of CP violation, and a very interesting framework is spontaneous CP violation: in such a framework, CP is a symmetry of the Lagrangian and CP violation arises through complex vacuum expectation values (VEVs) of Higgs multiplets H_i. There is an important subtlety that should be noted. Even with complex VEVs, only when the unitary transformation U acting on the H_i and relating the VEV to its complex conjugate

$$\langle H_i\rangle \longrightarrow \langle H_i\rangle^* = U_{ij}\langle H_j\rangle\,, \tag{1}$$

is not a symmetry of the Lagrangian does CP violation occur. If U is a symmetry of the Lagrangian, CP is conserved even though the VEVs are complex [3].

1. Geometrical CP Violation

Here we discuss the particular case where the phases that appear in the scalar VEVs are determined independently of the arbitrary parameters of the scalar potential. These are referred to as calculable phases, and the framework as geometrical CP violation (GCPV). GCPV requires at least three Higgs doublets and a non-Abelian symmetry [3]. $\Delta(27)$ (a discrete subgroup of $SU(3)$) was the first group found to produce such calculable phases [3]. In [4] this was generalised to larger groups obtaining the same calculable phases and more recently, several new phase solutions were advanced and expressed in terms of the number of scalars and the group [1]. We will briefly review the framework, explaining the origin and clarifying some of the possible generalisations in some detail. The notation follows mostly from [1], with $\eta_N \equiv e^{2i\pi/N}$ and $\langle H_i\rangle \equiv v e^{i\alpha_i}$ (the assumption that the VEVs have the same magnitude is justified by the respective potentials favouring this type of minima in very general cases).

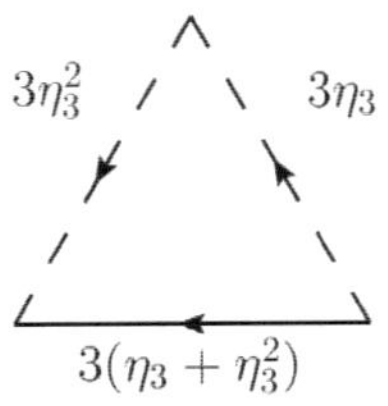

Figure 1. The equilateral triangle shape associated with the minimising solution for V_3.

The only phase dependence in the renormalisable $\Delta(27)$ potential for a scalar triplet can be written as

$$H_1^2(H_2H_3)^\dagger + c.p. = v^4(e^{iA_1}+e^{iA_2}+e^{iA_3})\,, \tag{2}$$

where *c.p.* represents the cyclic permutations and the phases A_i are $A_1 = 2\alpha_1 - \alpha_2 - \alpha_3$ and permuting, $A_2 = -\alpha_1 + 2\alpha_2 - \alpha_3$, $A_3 = -\alpha_1 - \alpha_2 + 2\alpha_3$. Due to a combination of the cyclic properties of the group and invariance under the gauge group, $\sum A_i = 0$ must be verified [1]. If we constrain ourselves to renormalisable potentials this invariant is the only phase dependence in the potential (with its hermitian conjugate, *h.c.*), and the minimum of the scalar potential depends on

$$V_3 = (e^{iA_1}+e^{iA_2}+e^{iA_3}) + h.c.\,, \tag{3}$$

If V_3 appears in the potential with negative coefficient, V_3 should be maximised corresponding to a contribution of $+6$ for $A_i = 0$. Otherwise, V_3 should be minimised by the phases - but a contribution of -6 is not allowed, as $A_i = \pi$ would violate $\sum A_i = 0$. Fig. 1 illustrates how the minima with $A_i = \pm 2\pi/3$ corresponds to a contribution of -3. The $\Delta(27)$ GCPV VEVs like $(\eta_3^{\mp 1}, 1, 1)$ (corresponding to $\alpha_1 = \mp 2\pi/3$, $\alpha_2 = \alpha_3 = 0$) are obtained from such configurations. The same solutions can be obtained in larger groups containing $\Delta(27)$ [4].

$\Delta(27)$ is the semi-direct product of cyclic groups $C_3 \ltimes (C_3 \times C_3)$, and by considering analogous semi-direct products it is possible to find new GCPV candidates - although in general only by going to non-renormalisable potentials. A direct generalisation involves groups $C_N \ltimes (C_N \times ... \times C_N)$ with $N-1$ factors of C_N inside the brackets. The properties of the group lead to the lowest order phase dependent invariant being a direct generalisation of Eq.(2)

$$H_1^{N-1}(H_2(...)H_N)^\dagger + c.p.\,. \tag{4}$$

The order of such invariants is $2(N-1)$ and already for $N = 4$ this is of order 6 and non-renormalisable. For even number of scalars N the situation is arguably less interesting, as $A_i = \pi$ is an allowed solution to minimise the invariant (for N even it verifies $\sum A_i = 0$ as required). But with an odd number of scalars N one has e.g. for $N = 5$

$$V_5 = (e^{iA_1}+e^{iA_2}+e^{iA_3}+e^{iA_4}+e^{iA_5}) + h.c.\,, \tag{5}$$

in this particular case with $A_1 = 4\alpha_1 - \alpha_2 - \alpha_3 - \alpha_4 - \alpha_5$. As was the case with $N = 3$, $A_i = \pi$ is not allowed due to the requirement $\sum A_i = 0$. It turns out that to minimise the generalised V_N the requirement is [1]

$$A_i = \pm\frac{N-1}{2}\frac{2\pi}{N}\,, \tag{6}$$

corresponding to adjacent sides of a regular N-sided polygon, as seen in Fig. 2 for $N = 5$. This can be obtained e.g. with

$$\alpha_1 = \mp\frac{N-1}{2}\frac{2\pi}{N}\,, \tag{7}$$

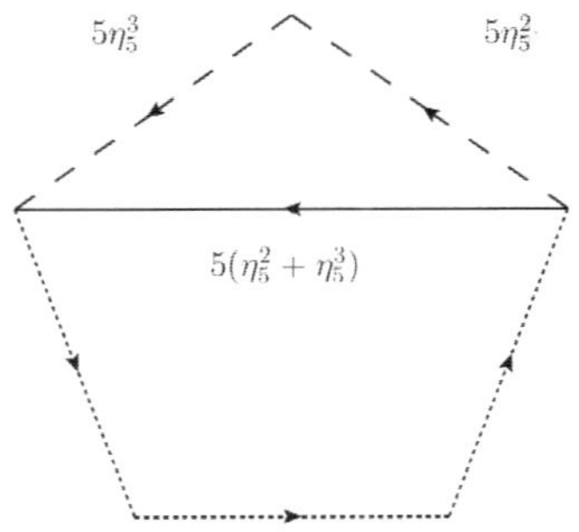

Figure 2. The pentagon shape associated with the minimising solution for V_5.

and the remaining phases vanishing. The VEVs are then of the type

$$v\left(\eta_N^{\mp(N-1)/2}, 1, (...), 1\right) . \tag{8}$$

If we consider instead the more general groups $C_N \ltimes (C_n \times ... \times C_n)$ with $N-1$ factors of C_n inside the brackets, there are some cases of interest: (1) $n = kN$ (an integer multiple); (2) n is not a multiple but shares a prime factor with N; (3) n and N share no prime factors. The first case is the most straightforward - depending on the irreducible representation (irrep) chosen one can obtain the same GCPV VEVs that are available with $k = 1$, or alternatively the choice of irrep forbids the invariant in Eq.(4), leaving as lowest order phase dependent invariant

$$\left(H_1^{N-1}\left(H_2(...)H_N\right)^\dagger\right)^k + c.p. , \tag{9}$$

of order $2k(N-1)$. The solutions to Eq.(6) for odd N are then changed from Eq.(8) to [1]

$$v\left(\eta_{kN}^{\mp\frac{(2l-1)N-1}{2}}, 1, (...), 1\right) \quad \text{for}\, l \leq k\, . \tag{10}$$

For a discussion of the second case (which always applies when n and N are both even) we refer the interested reader to [1]. The third case merits some detailed discussion here - due to invariance under the gauge group and as there is no shared prime factor, the relevant phase dependent invariant is

$$\left(H_1 H_2^\dagger\right)^n + c.p. , \tag{11}$$

which differs from the previously featured invariants in an important way: each individual term no longer has all the H_i in it. For $N > 3$ this means that there can be independent invariants $\left(H_1 H_3^\dagger\right)^n + c.p.$, and $\left(H_1 H_4^\dagger\right)^n + c.p.$ and so on. For odd N there are $(N-1)/2$ independent invariants when the $h.c.$ is considered, for N even there is one unpaired invariant that is its own $h.c.$ so the total is $N/2$. It is interesting that for $n = 2$ this class of invariants appears at order 4 and they are therefore renormalisable. In the case $N = 3$ and $n = 2$ (which corresponds to the group $\mathbf{A}_4$), if we want to minimise the phase-dependent invariant $I_{32} = \left(H_1 H_2^\dagger\right)^2 + \left(H_2 H_3^\dagger\right)^2 + \left(H_3 H_1^\dagger\right)^2 + h.c.$, then the phases $\alpha_1 = 0$, $\alpha_2 = -2\pi/6$ and $\alpha_3 = -4\pi/6$, of VEV $(1, \eta_6^{-1}, \eta_6^{-2})$, result in the $A_i = 2\pi/3$ solution depicted in Fig. 1, but note the respective U from Eq.(1) leaves I_{32} invariant apart from a global phase. Indeed this is the case for any n and odd N, the invariant $I_{Nn} = \left(H_1 H_2^\dagger\right)^n + c.p. + h.c.$ is minimised by

$n\alpha_i = (-i+1)\frac{N-1}{2}\frac{2\pi}{N}$ or by $n\alpha_i = (i-1)\frac{N-1}{2}\frac{2\pi}{N}$ (starting from $n\alpha_1 = 0$). This can be obtained by solving the equations $n\alpha_1 - n\alpha_2 = \pm\frac{N-1}{2}\frac{2\pi}{N}$ or equivalently by picturing the regular polygon with N sides, and assigning consecutive $n\alpha_i$ either clockwise or anti-clockwise always leaving $(N-1)/2$ sides between consecutive $n\alpha_i$ (and also between the last $n\alpha_N$ and $n\alpha_1 = 0$). In any case, the U relating such a solution to its conjugate only shifts each term of the invariant I_{Nn} by the global phase $\pm 2n\alpha_N$ (the *h.c.* is shifted by $\mp 2n\alpha_N$). In addition, for odd $N > 3$, there are multiple invariants appearing at order $2n$ with different phase dependences, say A_i, B_i etc. It is not possible to have $A_i = \pm\frac{N-1}{2}\frac{2\pi}{N}$ simultaneously with $B_i = \pm\frac{N-1}{2}\frac{2\pi}{N}$ (and not with all $B_i = 0$ either). This is the case already for $N = 5$, where one can take $A_1 = n(\alpha_1 - \alpha_2)$ and $B_1 = n(\alpha_1 - \alpha_3)$, with the respective invariants being renormalisable only for $n = 2$. As noted above, $\alpha_i = (-i+1)4\pi/5n$ minimises the A_i invariant (which corresponds to I_{5n}) with all $A_i = 4\pi/5$. But this type of phase solution simultaneously corresponds to all $B_i = 8\pi/5$ which does not extremise the B_i invariant (indeed each B_i can be expressed as a sum of two A_i, e.g. $B_1 = A_1 + A_2$). For even N there are some differences: while $A_i = \pi$ may not be compatible with $B_i = \pi$, it is compatible with $B_i = 0$. For $N = 4$, $A_1 = n(\alpha_1 - \alpha_2)$, and from the self-conjugate invariant the dependence is $B_1 = n(\alpha_1 - \alpha_3)$; then $A_i = \pi$ can be obtained from e.g. $\alpha_{1,3} = \pi/2n = -\alpha_{2,4}$ which also gives $B_i = 0$ which is the desired minimising solution if the B_i invariant has a negative coefficient. For $N = 6$, $A_1 = n(\alpha_1 - \alpha_2)$, $B_1 = n(\alpha_1 - \alpha_3)$, and from the self-conjugate invariant the dependence is $C_1 = n(\alpha_1 - \alpha_4)$; then $A_i = C_i = \pi$ and $B_i = 0$ are obtained for e.g. $\alpha_{1,3,5} = \pi/2n = -\alpha_{2,4,6}$ and this is a minimising solution as long as the coefficients of the A_i, B_i and C_i invariants are respectively positive, negative and positive. Once again, the U associated to the solution only shifts the invariant by a global phase.

2. Fermion mixing with geometrical CP violation

When fermions are added to the framework of GCPV it is not trivial to construct viable models. Leading order structures were suggested in [4]. Calculable phases are also robust when the potential includes non-renormalisable terms [5]. Based on these results, a minimal model of GCPV fitting all data was proposed in [2]. The group employed is $\Delta(27)$ and a minimal amount of additional matter is added.

The three Higgs doublets, H_i, transform as a $\Delta(27)$ triplet assigned to a $\mathbf{3}_{01}$ irrep (in this section we denote them with a lower index). Their hermitian conjugates $H^{\dagger i}$ transform as the conjugate representation $\mathbf{3}_{02}$ (in this section we denote them with an upper index). The relevant generators of the group will be denoted as c (cyclic permutation) and d (diagonal phases). They act on the irreps as $c(H_1, H_2, H_3) \to (H_2, H_3, H_1)$, $c(H^{\dagger 1}, H^{\dagger 2}, H^{\dagger 3}) \to (H^{\dagger 2}, H^{\dagger 3}, H^{\dagger 1})$, and $d(H_1, H_2, H_3) \to (H_1, \omega H_2, \omega^2 H_3)$, $d(H^{\dagger 1}, H^{\dagger 2}, H^{\dagger 3}) \to (H^{\dagger 1}, \omega^2 H^{\dagger 2}, \omega H^{\dagger 3})$. $\omega \equiv e^{i2\pi/3}$ corresponds to the η_3 used in section 1. There are nine singlet irreps $\mathbf{1}_{ij}$, where the subscript $\{ij\}$ denotes how they transform under the generators $c\mathbf{1}_{ij} = \omega^i \mathbf{1}_{ij}$, $d\mathbf{1}_{ij} = \omega^j \mathbf{1}_{ij}$. Further details about $\Delta(27)$ can be found in the references of [2].

From the solution in Fig. 1 we can have a complex VEV of the type:

$$\langle H_i \rangle = v(\omega, 1, 1)\,. \tag{12}$$

This VEV necessarily violates CP, as the corresponding U (see Eq. (1)) does not leave the potential invariant. The full scalar potential is presented later as we first focus on the Yukawa interactions of the quarks. The possible Yukawa contractions in this framework were considered briefly in [3] and in more detail in [4]: in order to make invariant Yukawa terms some of the quarks must transform as triplet or anti-triplet under $\Delta(27)$. We write the invariants symbolically as $QH_i d^c$ and $QH^{\dagger i}u^c$ (without explicit $SU(2)$ indices). Q are the left-handed quark doublets and u^c, d^c are the up and down right-handed singlets. If Q_i is a $\mathbf{3}_{01}$ we would necessarily require the d^c_i to transform also as a $\mathbf{3}_{01}$, conversely if Q^i is a $\mathbf{3}_{02}$, u^{ci} is forced to be a $\mathbf{3}_{02}$. At least one sector has a leading order Yukawa structure given by the $\Delta(27)$ invariant $\mathbf{3}_{0i} \otimes \mathbf{3}_{0i} \otimes \mathbf{3}_{0i}$. With the symmetrical VEV in Eq. (12), this structure leads to a mass matrix with degenerate quark masses [3] and so the conclusion is that Q cannot be assigned as a triplet or an anti-triplet. We choose instead u^c and d^c as $\Delta(27)$ triplets so that $QH_i d^{cj}$ and $QH^{\dagger i} u^c_j$ are invariant with Q as singlets. Both sectors then have Yukawas from the $\Delta(27)$ invariants $\mathbf{1}_{ij} \otimes (\mathbf{3}_{01} \otimes \mathbf{3}_{02})$ [4]. Although $\mathbf{3}_{01} \otimes \mathbf{3}_{02}$ results in 9 distinct singlets, the group properties are such that any $\mathbf{3}_{01} \otimes \mathbf{3}_{02} \to \mathbf{1}_{ij}$ with $i \neq 0$ explicitly involves powers of ω (complex), so not all possibilities are allowed by CP invariance of the Lagrangian. To have renormalisable Yukawa interactions we assign Q_1, Q_2 and Q_3 each as one of the three $\mathbf{1}_{0i}$ singlets. The possibilities are assigning all three Q in the same singlet irrep, or assigning two in the same, or all three Q in different irreps. All three structures lead to mass matrices that have a special structure defined by rows. When Q is a $\mathbf{1}_{00}$, $\mathbf{1}_{01}$ or $\mathbf{1}_{02}$, the respective $H_i d^{cj}$ or $H^{\dagger i} u^c_j$ product is $\mathbf{1}_{00}$, $\mathbf{1}_{02}$ or $\mathbf{1}_{01}$ respectively, which essentially amounts to a shift in the position of the ω in the mass matrix. More explicitly the corresponding down mass matrix looks like:

$$\tilde{M}_d = v \begin{pmatrix} y_1\omega & y_1 & y_1 \\ y_2 & y_2\omega & y_2 \\ y_3 & y_3 & y_3\omega \end{pmatrix} \tag{13}$$

and the associated up quark mass matrix looks very similar with ω^2 instead of ω and the second and third rows swapped. If instead Q_1, Q_2, and Q_3 are assigned to $\mathbf{1}_{00}$, $\mathbf{1}_{00}$, and $\mathbf{1}_{02}$ respectively, we get:

$$M_d = v \begin{pmatrix} y_1\omega & y_1 & y_1 \\ y_2\omega & y_2 & y_2 \\ y_3 & y_3 & y_3\omega \end{pmatrix} \tag{14}$$

We recall that due to the explicit CP invariance of the Lagrangian the couplings are real and the phase appears only through the VEV. Consider then the hermitian matrices $MM^\dagger$:

$$\tilde{M}_d \tilde{M}_d^\dagger = 3v^2 \begin{pmatrix} y_1^2 & 0 & 0 \\ 0 & y_2^2 & 0 \\ 0 & 0 & y_3^2 \end{pmatrix} \tag{15}$$

$$M_d M_d^\dagger = 3v^2 \begin{pmatrix} y_1^2 & y_1 y_2 & 0 \\ y_1 y_2 & y_2^2 & 0 \\ 0 & 0 & y_3^2 \end{pmatrix} \tag{16}$$

The vanishing off-diagonal entries are a consequence of $1 + \omega + \omega^2 = 0$. The determinant of $M_d M_d^\dagger$ is zero but it has two non-vanishing masses, and the choice with all generations of Q in the same singlet irrep (not shown) leads to a rank 1 structure with a single non-vanishing mass. It is apparent that these hermitian structures are real. To obtain a viable CKM matrix we need to generate additional off-diagonal terms and to somehow preserve the complex phase. Off-diagonal terms appear in $MM^\dagger$ when rows have the ω in the same column, and the minimal way to get them is to add a gauge singlet scalar ϕ that is a non-trivial $\Delta(27)$ singlet. Without loss of generality we place ϕ in the irrep $\mathbf{1}_{01}$, enabling a new non-renormalisable Yukawa coefficient populating each row from $QH_i d^{cj}\phi$. For Q_1, Q_2, and Q_3 in $\mathbf{1}_{00}$, $\mathbf{1}_{00}$, and $\mathbf{1}_{02}$ respectively, we add to M_d the corresponding matrix:

$$M_\phi = v \begin{pmatrix} y_{\phi 1} & y_{\phi 1}\omega & y_{\phi 1} \\ y_{\phi 2} & y_{\phi 2}\omega & y_{\phi 2} \\ y_{\phi 3}\omega & y_{\phi 3} & y_{\phi 3} \end{pmatrix} \tag{17}$$

Journal of Physics: Conference Series **447** (2013) 012038 doi:10.1088/1742-6596/447/1/012038

From the interference $M_d M_\phi^\dagger + M_\phi M_d^\dagger$ we obtain the required additional off-diagonal entries. The effect of $M_\phi M_\phi^\dagger$ can be absorbed within the existing structure of $M_d M_d^\dagger$. We now want a complex phase in the CKM matrix, from complex $MM^\dagger$. The minimal possibility is to consider non-renormalisable interactions with higher powers of H e.g. $QH_i d^{cj}(H_k H^{\dagger l})$. It turns out that the non-trivial structure extracted from such interactions is

$$M_H = v \begin{pmatrix} y_{H1} & y_{H1}\omega^2 & y_{H1}\omega^2 \\ y_{H2} & y_{H2}\omega^2 & y_{H2}\omega^2 \\ y_{H3}\omega^2 & y_{H3}\omega^2 & y_{H3} \end{pmatrix} \tag{18}$$

where the identity $1+\omega+\omega^2 = 0$ was used and the existing coefficients were redefined to absorb similar entries in the mass matrix. From the interference $M_d M_H^\dagger + M_H M_d^\dagger$ we obtain phases in $MM^\dagger$ that enable complex CKM elements, $M_\phi M_H^\dagger + M_H M_\phi^\dagger$ and $M_H M_H^\dagger$ give structures that do not qualitatively change the analysis. M_ϕ and M_H are the minimal mandatory additions needed for a perfect fit to the existing data. The Lagrangian (showing the $\Delta(27)$ indices) is:

$$\mathcal{L} = Q\left(H^{\dagger i} u^c_j + H_i d^{cj} + H_i d^{cj}\phi + H_i d^{cj}(H_k H^{\dagger l})\right) . \tag{19}$$

We found that the only choice that favourably accounts for the precision flavour data is Q_1, Q_2 and Q_3 as $\mathbf{1}_{00}$, $\mathbf{1}_{00}$ and $\mathbf{1}_{02}$ respectively. In the up quark sector, $M_u M_u^\dagger$ can be considered diagonal, and we need only one additional non-renormalisable Yukawa in order to generate the small up quark mass (recall the determinant of the renormalisable structure is zero for this choice of irreps). In Fig. 3 we show that this choice can successfully reproduce the Wolfenstein parameters, and one can also compare the model values (in the right column) with the experimental values:

$$\begin{aligned} \lambda^{\exp} &= 0.22535 \pm 0.00065 & \lambda &= 0.22534, \\ A^{\exp} &= 0.811^{+0.022}_{-0.012} & A &= 0.810, \\ \bar\rho^{\,\exp} &= 0.131^{+0.026}_{-0.013} & \bar\rho &= 0.129, \\ \bar\eta^{\,\exp} &= 0.345^{+0.013}_{-0.014} & \bar\eta &= 0.344. \end{aligned} \tag{20}$$

We present now the full scalar potential. It contains the $\Delta(27)$ triplet H_i but also ϕ which we introduced to obtain desirable Yukawa structures:

$$V(H,\phi) = m_1^2\left[H_1 H_1^\dagger\right] + m_2^2\phi\phi^\dagger + m_3(\phi^3 + h.c.) \tag{21}$$

$$+ \lambda_1\left[(H_1 H_1^\dagger)^2\right] + \lambda_2\left[H_1 H_1^\dagger H_2 H_2^\dagger\right] + \lambda_3\left[H_1 H_2^\dagger H_1 H_3^\dagger + h.c.\right] \tag{22}$$

$$+ \lambda_4(\phi\phi^\dagger)^2 + \lambda_5\left[\phi(H_1 H_2^\dagger) + h.c.\right] + \lambda_6\left[\phi\phi(H_1 H_3^\dagger) + h.c.\right] , \tag{23}$$

where the coefficients are real, and the square brackets represent also the cyclic permutations on the $\Delta(27)$ indices which we do not explicitly show. While the geometrical phase solution in Eq. (12) is not affected by $\phi\phi^\dagger$, it is only when λ_5 and λ_6 are small that it holds. One can add a Z_4 symmetry acting on ϕ to trivially enforce these couplings to vanish (in this case Eq.(17) arises from a ϕ^4 insertion instead of ϕ, all conclusions remaining unchanged).

Following the minimisation of the potential and determination of the mass eigenvalues, we observed these features (for illustration we display only the CP-even scalar components): (*i*)

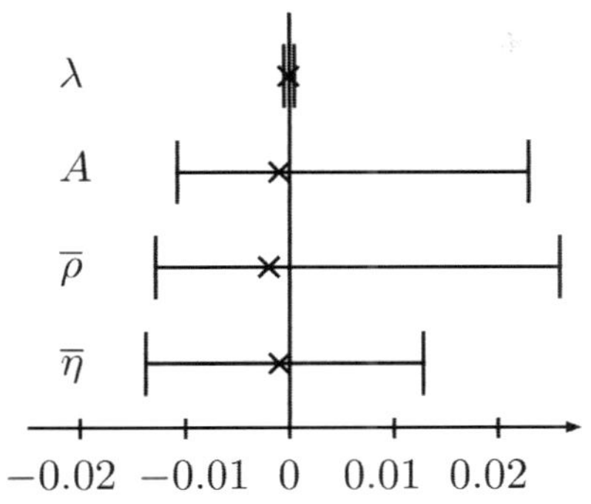

Figure 3. The experimental spread of the Wolfenstein parameters λ, A, $\bar{\eta}$ and $\bar{\rho}$ around their central values. Crosses denote our model values.

The ϕ field is much heavier (beyond 1 TeV) and decouples from the $SU(2)$ doublets. More specifically, the mass of ϕ is determined by $\lambda_{\{4,5,6\}}$, while those of $h_{\{a,b,c\}}$ are controlled by $\lambda_{\{1,2,3\}}$. (*ii*) The physical scalars h_a, h_b and h_c mix in a very specific way (see [2] and references therein): the scalar mass squared matrix having the structure

$$\begin{pmatrix} A & B & B \\ B & C & D \\ B & D & C \end{pmatrix}, \tag{24}$$

leading to one physical scalar h_a that is orthogonal to the other scalars and having no h_aVV-type gauge couplings ($V = W, Z$). The Yukawa couplings of h_a to up- and down-type quarks are strongly suppressed except for the h_act and h_auc couplings which are about 0.45. The other physical scalars, h_b and h_c, have nearly SM-like gauge and Yukawa couplings.

Depending on the scalar potential couplings, two viable scenarios are identified: (I) There is only one light scalar, h_b, assuming the role of the SM-like Higgs found near 125 GeV, with all other scalars beyond the current exclusion range of the LHC. This is a decoupling limit which reproduces almost SM-like scalar structure. (II) A scenario with richer collider consequences is possible when the exotic scalar h_a is light enough to be produced at the LHC e.g. through h_auc or decays of tops or heavy scalars. Under the reasonable assumption that $m_\phi > 1\,\text{TeV}$ or so we can obtain the analytic relations:

$$m^2_{h_a} = \frac{2}{3}\left(2\lambda_1 v^2 - 2\lambda_2 v^2 + 3\lambda_3 v^2\right), \tag{25}$$

$$\begin{aligned} m^2_{h_{c/b}} = \frac{1}{6}\big(&5\lambda_1 v^2 + 4\lambda_2 v^2 \pm \sqrt{3}\big[v^4(3\lambda_1^2 + 8\lambda_1\lambda_2 \\ &- 16\lambda_1\lambda_3 + 16\lambda_2^2 - 64\lambda_2\lambda_3 + 64\lambda_3^2)\big]^{\frac{1}{2}}\big). \end{aligned} \tag{26}$$

The λ_i can be adjusted to give m_{h_a} around the mass 125 GeV of the SM-like h_b, with h_c heavier than 600 GeV. In this case, there can be a spectacular decay channel through $h_a \to \chi_a Z$, if $m_{\chi_a} \sim 20\,\text{GeV}$, with the pseudo-scalar χ_a then decaying to charged leptons of different flavours (e.g. $\mu\tau$) and the Z boson decaying to leptons, as in Fig. (4). There is enough freedom in the lepton sector to boost coupling of χ_a, which may have a sizeable branching ratio in this channel, but a more specific prediction requires a detailed numerical study of the lepton Yukawa sector.

Although in both scenarios (I) and (II) there are a few heavy scalars above the current LHC limit of 650 GeV or so, their relative heaviness compared to the SM Higgs is not a result of fine-tuning of parameters as for each heavy state there is a reasonably independent combination of

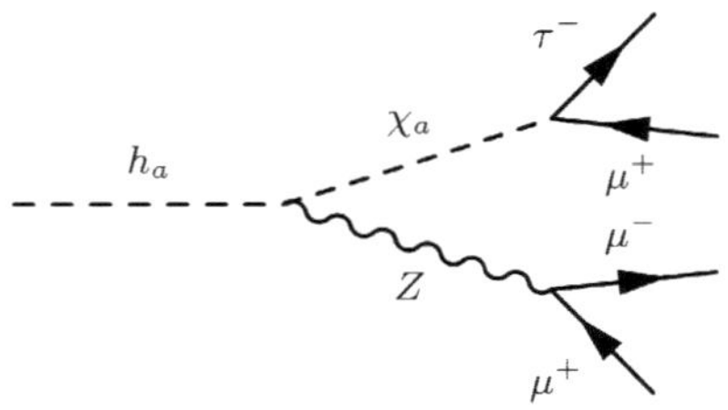

Figure 4. Example of a decay mode of the exotic scalar h_a that can be tested at the LHC.

λ_i-type couplings which simply has to be set to a higher value (we have verified this numerically). If the LHC bound goes up, we have to accordingly raise the maximum allowed value of some λ_i beyond π e.g. up to maximum allowed value of 2π to go over 1 TeV. A final consequence worth pointing out is that the additional scalar states below 1 TeV in both scenarios (I) and (II), all coupling to SM gauge bosons, would affect the energy dependence of longitudinal gauge boson scattering. This energy dependence might be different from the SM expectation due to the extra scalars, whose quantitative impact may be probed at the high luminosity option of the LHC depending on their masses and couplings.

3. Summary

We have reviewed and clarified the framework of geometrical CP violation in the context of multi-Higgs models. We then reviewed the realisation of spontaneous CP violation of geometrical origin in a minimal $\Delta(27)$ flavour model that reproduces the CKM mixing matrix. This scenario is quite falsifiable, as only two choices broadly worked, out of which only one set of representations fits the precision of flavour data. The scalar sector of this model retains some symmetry of the flavour group and this can lead to exotic scalar decays into multi-lepton of different flavours, which would be a smoking gun signal of the model that is testable at the LHC.

Acknowledgments

The author thanks CFTP for hospitality and also the organisers of Discrete 2012 for a very nice conference. The author was supported by DFG grant PA 803/6-1 and partially through PTDC/FIS/098188/2008.

[1] I. de Medeiros Varzielas, JHEP **1208** (2012) 055 [arXiv:1205.3780 [hep-ph]].
[2] G. Bhattacharyya, I. de Medeiros Varzielas and P. Leser, Phys. Rev. Lett. **109** (2012) 241603 [arXiv:1210.0545 [hep-ph]].
[3] G. C. Branco, J. M. Gerard and W. Grimus, Phys. Lett. B **136** (1984) 383.
[4] I. de Medeiros Varzielas and D. Emmanuel-Costa, Phys. Rev. D **84** (2011) 117901 [arXiv:1106.5477 [hep-ph]].
[5] I. de Medeiros Varzielas, D. Emmanuel-Costa and P. Leser, Phys. Lett. B **716** (2012) 193 [arXiv:1204.3633 [hep-ph]].

Journal of Physics: Conference Series **447** (2013) 012039 doi:10.1088/1742-6596/447/1/012039

Neutrino mass from higher than d=5 effective operators with discrete symmetries

M B Krauss

Institut für Theoretische Physik und Astrophysik, Universität Würzburg, Am Hubland, 97074 Würzburg, Germany

E-mail: mkrauss@physik.uni-wuerzburg.de

Abstract. In beyond Standard Model physics scenarios one often has to extend the Higgs sector and introduce additional (discrete) symmetries. In such models it is possible that the Weinberg operator is forbidden and the leading contribution to neutrino mass is generated by operators with mass dimension $d > 5$. These higher-dimensional operators are stronger suppressed and therefore the new physics scale can be as low as the TeV scale. With this approach it is possible to test neutrino mass models at colliders, such as the LHC. We will demonstrate how these operators can be systematically studied and present an exemplary implementation of such a model and its phenomenological consequences. We also show how it can be embedded into an SU(5) GUT framework. Finally we present a specific realization of this model where we can have tri-bimaximal mixing with large deviations due to a softly broken symmetry.

1. Introduction

With the discovery of a Higgs-like particle [1, 2] the last missing part of the Standard Model (SM) seems to be finally found. After the current upgrade, the Large Hadron Collider (LHC) will be able to further test new physics models. Which one of these models may be realized in nature is so far unclear. They share, however, often some common features. In many scenarios we find, for example, an extended Higgs sector, such as in supersymmetric models (SUSY) or generic Two Higgs Doublet Models (THDMs). Often also new fundamental discrete symmetries are introduced, particularly in the context of flavor models. The recent evidence of a non-zero neutrino mixing angle θ_{13} [3, 4] is especially interesting in this regard.

In literature models that connect neutrino physics to the TeV scale have been widely discussed lately. Examples are radiative mass generation [5–10] or models with a small lepton number violating contribution such as the inverse seesaw mechanism [11] or SUSY with R-parity violation [12–21]. It is possible in models with additional scalars and discrete symmetries that neutrino masses are generated by higher than $d = 5$ effective operators [6, 22–33]. In the following, we will study these operators in the framework of SUSY [34] and an SU(5) grand unified theory (GUT) [35] and will show that this approach is another possibility for neutrino masses from physics at the TeV scale.

2. Neutrino masses and effective field theory

The usual model to describe the smallness of the neutrino masses is the seesaw mechanism, where – in case of the type-I seesaw – heavy right handed singlets are added. These right handed neutrinos have Yukawa-like interactions with the left-handed SM neutrinos and masses close to the GUT scale. Since this scale is far beyond the scale of neutrino physics, one can integrate out the right handed neutrinos in order to obtain an effective interaction, the Weinberg operator [36],

$$\mathcal{L}_{\text{eff}}^{d=5} \propto \frac{1}{\Lambda_{\text{NP}}} (\overline{L^c} \mathrm{i}\tau^2 H)\,(H \mathrm{i}\tau^2 L)\,. \tag{1}$$

After electroweak symmetry breaking (EWSB) this operator will generate an effective Majorana neutrino mass term $m_\nu \propto v^2/\Lambda_{\text{NP}}$, where v is the vacuum expectation value (VEV) of the Higgs field and Λ_{NP} is the new physics scale. In case of the type-I seesaw, Λ_{NP} is the mass of the right-handed neutrinos. In general, one can describe the low-energy effects of any fundamental theory by a tower of effective operators, which is added to the SM Lagrangian as

$$\mathcal{L} = \mathcal{L}_{\text{SM}} + \mathcal{L}_{\text{eff}}^{d=5} + \mathcal{L}_{\text{eff}}^{d=6} + \cdots\,, \quad \text{with} \quad \mathcal{L}_{\text{eff}}^{d} \propto \frac{1}{\Lambda_{\text{NP}}^{d-4}}\,\mathcal{O}^d\,. \tag{2}$$

If the Weinberg operator is forbidden by a symmetry, the leading contribution to neutrino mass will be generated by an operator with dimension $d > 5$. As a consequence, the effective neutrino mass will be suppressed by higher powers of the new physics scale Λ_{NP}, which therefore can be at much lower energies. This opens the window to test the mechanism generating neutrino masses at collider experiments such as the LHC.

As pointed out by the authors of Ref. [28], it is not possible in the SM to have a leading contribution to neutrino mass from a generic higher-dimensional operator, if the Weinberg operator is forbidden. This is due to the fact that all possible operators contributing to neutrino mass are of the type $(LLHH)(H^\dagger H)^n$ and $(H^\dagger H)^n$ is a singlet under any symmetry. Therefore, we want to focus on models which have an additional Higgs doublet such as THDMs or the Minimal Superymmetric Standard Model (MSSM). We also consider models with an additional scaler singlet such as the Next-to Minimal Superymmetric Standard Model (NMSSM). Due to the requirement of SUSY invariance, the number of possible operators in supersymmetric models is limited compared to the THDM. The general form of operators leading to neutrino mass in SUSY is

$$\mathcal{L}_{\text{eff}}^{d=5+2n+m} = \frac{1}{\Lambda^{d-4}}(LLH_uH_u)(H_uH_d)^n S^m\,, \qquad n, m = 1, 2, 3, \ldots \tag{3}$$

In table 1 we show a list of all operators up to $d = 9$ that contribute to neutrino mass in the framework of the NMSSM. The introduction of a discrete symmetry has several consequences: First, the usual terms from the NMSSM superpotential must be invariant under this symmetry. Furthermore we want the $d = 5$ operator to be forbidden by this symmetry. Respecting these constraints, we have listed the charge of the higher-dimensional operators in table 1 with regard to the NMSSM fields. As one can see from the table, it is not possible to have certain operators as leading contribution to neutrino mass, since they always will have the same charge as a lower-dimensional operator. As indicated in table 1, one can show that this applies to all operators with $d > 7$ that include the scalar singlet S [34].

3. A d=7 Example

There are three possible tree-level models, which after integrating out the heavy particles lead to the Weinberg operator: the type-I, type-II and type-III seesaw mechanism [5]. In the same way, it is possible to decompose the higher-dimensional operators. The operators #1, #2 and #4 of table 1, for example, have been studied in Ref. [24].

Table 1. Effective operators generating neutrino mass in the NMSSM up to $d = 9$. Taken from Ref. [34].

	Op.#	Effective interaction	Charge	Same as
$d = 5$	1	LLH_uH_u	$2q_L + 2q_{H_u}$	
$d = 6$	2	LLH_uH_uS	$2q_L + q_{H_u} - q_{H_d}$	
$d = 7$	3	$LLH_uH_uH_dH_u$	$2q_L + 3q_{H_u} + q_{H_d}$	
	4	LLH_uH_uSS	$2q_L - 2q_{H_d}$	
$d = 8$	5	$LLH_uH_uH_dH_uS$	$2q_L + 2q_{H_u}$	#1
	6	LLH_uH_uSSS	$2q_L + 2q_{H_u}$	#1
$d = 9$	7	$LLH_uH_uH_dH_uH_dH_u$	$2q_L + 4q_{H_u} + 2q_{H_d}$	
	8	$LLH_uH_uH_dH_uSS$	$2q_L + q_{H_u} - q_{H_d}$	#2
	9	LLH_uH_uSSSS	$2q_L + q_{H_u} - q_{H_d}$	#2

As has been discussed in more detail in Ref. [34] operator #3 can have some interesting phenomenological consequences at the LHC. In order to forbid the Weinberg operator, we choose the following charges for the (N)MSSM fields under a $\mathbb{Z}_3$ symmetry

$$q_{H_u} = 0,\; q_{H_d} = 1,\; q_L = 1,\; (q_S = 2)\,. \tag{4}$$

One particularly interesting decomposition of this $d = 7$ operator is given by the superpotential

$$\begin{aligned} W = {} & W_{\text{quarks}} + Y_e \hat{e}^c \hat{L} \cdot \hat{H}_d - Y_N \hat{N} \hat{L} \cdot \hat{H}_u + \kappa_1 \hat{N}' \hat{\xi} \cdot \hat{H}_d - \kappa_2 \hat{N}' \hat{\xi}' \cdot \hat{H}_u + m_N \hat{N} \hat{N}' \\ & + m_\xi \hat{\xi}' \cdot \hat{\xi} + \mu \hat{H}_u \cdot \hat{H}_d\,, \end{aligned} \tag{5}$$

where ξ and ξ' are SU(2) doublets, whereas N and N' are singlets under the SM gauge group. From this superpotential one derives the following mass matrix for the neutral fermions

$$M_f^0 = \begin{pmatrix} 0 & Y_N v_u & 0 & 0 & 0 \\ Y_N^{\mathsf{T}} v_u & 0 & m_N^{\mathsf{T}} & 0 & 0 \\ 0 & m_N & 0 & \kappa_1 v_d & \kappa_2 v_u \\ 0 & 0 & \kappa_1^{\mathsf{T}} v_d & 0 & -m_\xi \\ 0 & 0 & \kappa_2^{\mathsf{T}} v_u & -m_\xi & 0 \end{pmatrix} \tag{6}$$

in the basis $f^0 = (\nu, N, N', \xi^0, \xi'^0)$, where $v_{u,d}$ are the VEVs of the neutral Higgs components. By diagonalizing this matrix one obtains the corresponding mass eigenstates n_i.

The effective neutrino mass is after integrating out the heavy fields

$$m_\nu = v_u^3 v_d Y_N^2 \frac{\kappa_1 \kappa_2}{m_\xi m_N^2}\,. \tag{7}$$

Assuming Yukawa couplings of $\mathcal{O}(10^{-3})$, the masses of the heavy particles must indeed be at the TeV range in order to reproduce the observed masses for the light neutrinos.

4. Phenomenological implications at the LHC

Due to the small Yukawa couplings of the the singlets N and N', their production cross section will be small. The SU(2) doublets ξ and ξ' instead, which couple to the weak gauge bosons, can be produced at the LHC in Drell-Yan processes [34]. For $m_\xi = 200\,\text{GeV}$, one obtains cross sections $\sigma(pp \to \xi^\pm \xi^0)$ of about 122 fb (417 fb) for 7 (14) TeV. The doublets will decay dominantly into a neutrino mass eigenstate n_i plus a weak gauge boson or one of the Higgs components. Due to the small mixing between the light and heavy neutrino mass eigenstates the doublets will have small decay width. The neutral mass eigenstates n_i can decay into $W^\pm l_j^\mp$ and $H^\pm l_j^\mp$ or into neutrinos plus Z or neutral Higgs components with decay width of $\mathcal{O}(10^{-5}\ldots 1)\,\text{keV}$. Therefore these particles can have sizable decay lengths between $100\,\mu$m and several millimeters, which helps to identify the particles of our model and to suppress the backgrounds. We obtain results as large as several fb for the lepton number conserving processes $pp \to l^+ l'^- W^\pm$, where lepton flavor can be violated. For high enough luminosities it is possible to observe them. The following processes, which violate lepton number by two units, can be used to test the Majorana nature of the n_i:

$$pp \to l^+ l'^+ W^- \tag{8}$$

$$pp \to l^+ l'^+ W^- W^-, l^- l'^- W^+ W^+ \tag{9}$$

For the process in Eq. (8) we have strong suppression, since the neutral mass eigenstates n_4 and n_5 form a pseudo-Dirac pair. As a consequence, the cross section, which is proportional to $m_{n_5}^2 - m_{n_4}^2 \simeq O(m_\nu^2)$, will be tiny. For the lepton number violating processes in Eq. (9) we find on the other hand that the cross sections are larger than one might expect. This is due to the fact that ξ and ξ' can be regarded as vector-like representation of SU(2), which leads to some non-vanishing contributions to the LNV cross section. These are, however, at the limit of observability.

5. GUT completion

As discussed in Ref. [35], the introduction of additional particles that are charged under the SM gauge group modifies the running of the gauge couplings. The model presented in the previous section will spoil unification unless we add complete SU(5) multiplets to the (N)MSSM. Therefore we have to embed all charged particles in complete SU(5) multiplets:

$$\bar{5}_M = \begin{pmatrix} (d_R)^c \\ L \end{pmatrix}, \quad \bar{5}_{\xi'} = \begin{pmatrix} d'^c \\ \xi' \end{pmatrix}, \quad 5_\xi = \begin{pmatrix} d'' \\ \xi \end{pmatrix}. \tag{10}$$

Also the Higgs fields obtain additional colored components, which can be integrated out. As pointed out in Ref. [35], this choice of mediators is the only decomposition of the $d = 7$ operator that can be the leading contribution to neutrino mass and guarantees perturbativity in an SU(5) GUT.

We obtain the most interesting results in a model based on the NMSSM. The most general SU(5) invariant realization of this model is given by the superpotential

$$\begin{aligned} W \;=\; & y_1\, N\, 5_\xi\, H_{\bar{5}} + y_2\, N\, \bar{5}_{\xi'}\, H_5 + y_3\, N\, \bar{5}_M\, H_5 + y_1',\, N'\, 5_\xi\, H_{\bar{5}} + y_2'\, N'\, \bar{5}_{\xi'}\, H_5 + y_3'\, N'\, \bar{5}_M\, H_5 + \\ & \lambda_{\xi'}\, S\, \bar{5}_M\, 5_\xi + \lambda_\xi\, S\, \bar{5}_{\xi'}\, 5_\xi + \lambda_N S\, N' N + \lambda_{NN} S\, NN + \lambda_{N'N'} S\, N'N' + y_d\, \bar{5}_M\, 10\, H_{\bar{5}} + \\ & y_d'\, \bar{5}_{\xi'}\, 10\, H_{\bar{5}} + y_u\, 10\, 10\, H_5\,. \end{aligned} \tag{11}$$

In this scenario the mediator mass scale is determined by $\langle S \rangle$, which is of order 1 TeV and results in the required mass scale for the light neutrinos. Furthermore the NMSSM avoids some problems of the μ-term of the NMSSM (see also Refs. [28, 34, 35]). As discussed in Ref. [35], however, it is not possible to choose a discrete symmetry to forbid the Weinberg operator, since we obtain now effective operators of the type

$$\frac{1}{\langle S \rangle} LLH_uH_u \qquad \text{and} \qquad \frac{1}{\langle S \rangle^3}(LLH_uH_d)(H_uH_d)\,,$$

Table 2. Charges for the fields of the model defined in Eq. (12). Taken from Ref. [35].

Multiplet	$\overline{5}_M$	H_5	$H_{\overline{5}}$	N	N'	5_ξ	$\overline{5}_{\xi'}$	10	S	S'
$\mathbb{Z}_3$ charge	1	1	1	1	2	0	0	1	1	0

where $\langle S \rangle$ breaks any discrete symmetry under which it is charged.

We therefore consider a model with an additional S' and the superpotential

$$\begin{aligned} W = \; & y_3\, N\, \overline{5}_M\, H_5 + y'_1, N'\, 5_\xi\, H_{\overline{5}} + y'_2\, N'\, \overline{5}_{\xi'}\, H_5 + \lambda_\xi\, S'\, \overline{5}_{\xi'}\, 5_\xi + \lambda_N S'\, N' N + \\ & y_d\, \overline{5}_M\, 10\, H_{\overline{5}} + y_u\, 10\, 10\, H_5 + \lambda_S S H_{\overline{5}} H_5 + \kappa S^3 + \kappa' S'^3 + \\ & \lambda'_S S' H_{\overline{5}} H_5 + y'_3\, N'\, \overline{5}_M\, H_5 + y'_d\, \overline{5}_\xi\, 10\, H_{\overline{5}} + \cdots , \end{aligned} \tag{12}$$

We now charge the fields as listed in table 2. Applying these charges to Eq. (12) one sees that the interactions in the first two lines respect the discrete symmetry, and the ones of the last line break it softly. These terms not only allow for a sizable VEV of S' but also generate a $d = 5$ contribution to neutrino mass after integrating out ξ and ξ':

$$m_\nu^{d=5} = \frac{y_3 y'_3 v_u^2}{\langle S' \rangle} . \tag{13}$$

If all soft breaking couplings are of order 10^{-8} or smaller this contribution to neutrino mass will be sub-leading. If the soft coupling y'_3, however, is close to this upper limit, one can, besides the leading $d = 7$ neutrino mass operator, have sizable corrections from the $d = 5$ operator of Eq. (13). Suppose that the $d = 7$ operator has a tri-bimaximal flavor structure, where (for the sake of simplicity) the charged lepton mass matrix is assumed to be diagonal. Then the deviation from tri-bimaximal mixing can originate in the soft breaking terms in Eq. (12).

In the following, we will show an example from Ref. [35], where the $d = 7$ contribution has tri-bimaximal structure but corrections from the soft-breaking $d = 5$ operator lead to non-zero θ_{13}. We assume a flavor structure where we have three generations of the singlet fields N and N', and the following parametrization of the coupling constants

$$y'_1 = \widetilde{y}'_1 \begin{pmatrix} 0 \\ 1 \\ \rho \end{pmatrix} , \quad y'_2 = \widetilde{y}'_2 \begin{pmatrix} 0 \\ -1 \\ \rho \end{pmatrix} , \quad y_3 = \widetilde{y}_3 \begin{pmatrix} \sqrt{2/3} & 1/\sqrt{3} & 0 \\ -1/\sqrt{6} & 1/\sqrt{3} & -1/\sqrt{2} \\ -1/\sqrt{6} & 1/\sqrt{3} & 1/\sqrt{2} \end{pmatrix} , \tag{14}$$

where $\widetilde{y}_1$, $\widetilde{y}_2$ and $\widetilde{y}_3$ are numerical (scalar) parameters, $\rho = \sqrt{m_3/m_2}$ and the mass matrices for N and N' are diagonal. This choice of parameters will generate a tri-bimaximal mass matrix for the neutrinos at $d = 7$.

We obtain now a possible structure of the soft-breaking coupling y'_3 (using the parametrization from Ref. [37] for deviations from tri-bimaximal mixing) as

$$y'_3 \simeq \widetilde{y}'_3\, r \left[\begin{pmatrix} 0 & 0 & -\frac{1}{\sqrt{2}} \\ \frac{1}{\sqrt{6}} & \frac{1}{2\sqrt{3}} & 0 \\ -\frac{1}{\sqrt{6}} & -\frac{1}{2\sqrt{3}} & 0 \end{pmatrix} \rho^2 - \begin{pmatrix} 0 & 0 & \frac{1}{3\sqrt{2}} \\ 0 & \frac{1}{2\sqrt{3}} & \frac{1}{3\sqrt{2}} \\ 0 & \frac{1}{2\sqrt{3}} & \frac{1}{3\sqrt{2}} \end{pmatrix} \right] \tag{15}$$

with $\mathcal{O}(\widetilde{y}'_3) = 10^{-8}$. This results in a correction to the tri-bimaximal mass matrix from the $d = 7$ operator that leads to $\theta_{13} > 0$.

Another aspect of the GUT completion of this model is that the promotion of the SM SU(2) doublets to 5-plets of SU(5) requires the introduction of additional d-quarks as indicated in Eq. (10). Renormalization group effects imply that the masses of these quarks are as large as several TeV, which makes them potentially interesting for LHC studies. Furthermore, in our model these particles are stable due to the same symmetry that suppresses the Weinberg operator. Heavy stable quarks are forbidden due to bounds from Big Bang nucleosynthesis and searches for heavy nuclei. In the case of a softly broken symmetry, however, the heavy quarks can decay via the two-body decays

$$D' \to H^- u\,,\ H^0 d\,. \tag{16}$$

with life-times for the D' as small as 10^{-10} s to 10^{-13} s [35].

6. Conclusions

We have demonstrated that in new physics scenarios with an extended Higgs sector and additional symmetries neutrino masses can be generated by operators with mass dimension $d > 5$. The additional suppression of these operators connects neutrino physics to the TeV scale. A systematic study shows that only a certain subset of these operators can be the leading contribution to neutrino mass, if the Weinberg operator is forbidden by a discrete symmetry. We have discussed a specific implementation of a $d = 7$ operator that has potentially interesting phenomenological implications at the LHC, such as displaced vertices and lepton number violating signals. Those models have to be embedded in GUT multiplets in order to conserve gauge coupling unification. We have studied the decomposition of the $d = 7$ operator in an SU(5) framework, which is the only possible implementation of this operator consistent with the requirement of perturbativity. A particularly interesting realization of this model requires a softly broken symmetry. We have demonstrated that in this framework it is possible to obtain tri-bimaximal mixing with large corrections leading to non-zero θ_{13}. The soft breaking of the symmetry also leads to decays of additional heavy quarks, which are required in the GUT completion of this model. These decays are fast enough to be in agreement with cosmological bounds.

Acknowledgments

This work has been supported by the DFG training group GRK 1147. The author would like to thank the organizers of DISCRETE 2012 for their hospitality and the possibility to present his work.

References

[1] Aad G *et al.* (ATLAS Collaboration) 2012 *Phys.Lett.* **B716** 1–29 (*Preprint* 1207.7214)
[2] Chatrchyan S *et al.* (CMS Collaboration) 2012 *Phys.Lett.* **B716** 30–61 (*Preprint* 1207.7235)
[3] An F *et al.* (DAYA-BAY Collaboration) 2012 *Phys.Rev.Lett.* **108** 171803 (*Preprint* 1203.1669)
[4] Ahn J *et al.* (RENO collaboration) 2012 *Phys.Rev.Lett.* **108** 191802 (*Preprint* 1204.0626)
[5] Ma E 1998 *Phys. Rev. Lett.* **81** 1171–1174 (*Preprint* hep-ph/9805219)
[6] Babu K and Leung C N 2001 *Nucl.Phys.* **B619** 667–689 (*Preprint* hep-ph/0106054)
[7] de Gouvea A and Jenkins J 2008 *Phys.Rev.* **D77** 013008 (*Preprint* 0708.1344)
[8] Bonnet F, Hirsch M, Ota T and Winter W 2012 *JHEP* **1207** 153 (*Preprint* 1204.5862)
[9] Farzan Y, Pascoli S and Schmidt M A 2012 (*Preprint* 1208.2732)

[10] Angel P W, Rodd N L and Volkas R R 2012 (*Preprint* 1212.6111)
[11] Mohapatra R and Valle J 1986 *Phys. Rev.* **D34** 1642
[12] Hempfling R 1996 *Nucl.Phys.* **B478** 3–30 (*Preprint* hep-ph/9511288)
[13] Nilles H P and Polonsky N 1997 *Nucl.Phys.* **B484** 33–62 (*Preprint* hep-ph/9606388)
[14] Borzumati F, Grossman Y, Nardi E and Nir Y 1996 *Phys.Lett.* **B384** 123–130 (*Preprint* hep-ph/9606251)
[15] Nardi E 1997 *Phys.Rev.* **D55** 5772–5779 (*Preprint* hep-ph/9610540)
[16] Chun E, Kang S, Kim C and Lee U 1999 *Nucl.Phys.* **B544** 89–103 (*Preprint* hep-ph/9807327)
[17] Hirsch M, Diaz M, Porod W, Romão J and Valle J 2000 *Phys.Rev.* **D62** 113008 (*Preprint* hep-ph/0004115)
[18] Diaz M, Hirsch M, Porod W, Romão J and Valle J 2003 *Phys.Rev.* **D68** 013009 (*Preprint* hep-ph/0302021)
[19] Dedes A, Rimmer S and Rosiek J 2006 *JHEP* **0608** 005 (*Preprint* hep-ph/0603225)
[20] Allanach B and Kom C 2008 *JHEP* **0804** 081 (*Preprint* 0712.0852)
[21] Ghosh P and Roy S 2009 *JHEP* **0904** 069 (*Preprint* 0812.0084)
[22] Babu K S and Nandi S 2000 *Phys. Rev.* **D62** 033002 (*Preprint* hep-ph/9907213)
[23] Chen M C, de Gouvea A and Dobrescu B A 2007 *Phys. Rev.* **D75** 055009 (*Preprint* hep-ph/0612017)
[24] Gogoladze I, Okada N and Shafi Q 2009 *Phys. Lett.* **B672** 235–239 (*Preprint* 0809.0703)
[25] Giudice G F and Lebedev O 2008 *Phys. Lett.* **B665** 79–85 (*Preprint* 0804.1753)
[26] Babu K S, Nandi S and Tavartkiladze Z 2009 *Phys. Rev.* **D80** 071702 (*Preprint* 0905.2710)
[27] Gu P H, He H J, Sarkar U and Zhang X m 2009 *Phys. Rev.* **D80** 053004 (*Preprint* 0906.0442)
[28] Bonnet F, Hernandez D, Ota T and Winter W 2009 *JHEP* **10** 076 (*Preprint* 0907.3143)
[29] Picek I and Radovcic B 2010 *Phys. Lett.* **B687** 338–341 (*Preprint* 0911.1374)
[30] Liao Y, Ning G Z and Ren L 2010 *Phys. Rev.* **D82** 113003 (*Preprint* 1008.0117)
[31] Liao Y 2011 *Phys. Lett.* **B694** 346–348 (*Preprint* 1009.1692)
[32] Liao Y 2011 *Phys. Lett.* **B698** 288–292 (*Preprint* 1010.5326)
[33] Kanemura S and Ota T 2010 *Phys. Lett.* **B694** 233–237 (*Preprint* 1009.3845)
[34] Krauss M B, Ota T, Porod W and Winter W 2011 *Phys.Rev.* **D84** 115023 (*Preprint* 1109.4636)
[35] Krauss M B, Meloni D, Porod W and Winter W 2013 (*Preprint* 1301.4221)
[36] Weinberg S 1979 *Phys. Rev. Lett.* **43** 1566–1570
[37] King S 2008 *Phys.Lett.* **B659** 244–251 (*Preprint* 0710.0530)

Sterile neutrinos and R_K

A. Vicente

Laboratoire de Physique Théorique, CNRS – UMR 8627, Université de Paris-Sud 11, F-91405 Orsay Cedex, France

E-mail: avelino.vicente@th.u-psud.fr

Abstract. We consider an enhancement in the violation of lepton flavour universality in light meson decays arising from modified $W\ell\nu$ couplings in the standard model minimally extended by sterile neutrinos. Due to the presence of additional mixings between the active neutrinos and the new sterile states, the deviation from unitarity of the leptonic mixing matrix intervening in charged currents might lead to a tree-level enhancement of $R_P = \Gamma(P \to e\nu)/\Gamma(P \to \mu\nu)$, with $P = K, \pi$. These enhancements are illustrated in the case of the inverse seesaw, showing that one can saturate the current experimental bounds on Δr_K (and Δr_π), while in agreement with the different experimental and observational constraints.

1. Introduction

Lepton flavour universality (LFU) is a distinctive feature of the Standard Model (SM). The different lepton families couple with exactly the same strength to the gauge bosons. This leads to concrete predictions in electroweak precision tests, which can only distinguish among lepton families by the different charged lepton masses. Any deviation from the expected SM theoretical results would signal the presence of New Physics (NP). In this work we concentrate on light meson (K and π) leptonic decays which, in view of the expected experimental precision, have a unique potential to probe deviations from the SM regarding lepton universality.

In the SM, the dominant contribution to $\Gamma(P \to \ell\nu)$ ($P = K, \pi$) arises from W boson exchange. One may be afraid about potential hadronic uncertainties; however, by considering the ratios

$$R_K \equiv \frac{\Gamma(K^+ \to e^+\nu)}{\Gamma(K^+ \to \mu^+\nu)}, \qquad R_\pi \equiv \frac{\Gamma(\pi^+ \to e^+\nu)}{\Gamma(\pi^+ \to \mu^+\nu)}, \tag{1}$$

the hadronic uncertainties are expected to cancel out to a good approximation. In order to compare the experimental bounds with the SM predictions, it is convenient to introduce a quantity, Δr_P, which parametrizes deviations from the SM expectations:

$$R_P^{\text{exp}} = R_P^{\text{SM}}(1 + \Delta r_P) \quad \text{or equivalently} \quad \Delta r_P \equiv \frac{R_P^{\text{exp}}}{R_P^{\text{SM}}} - 1\,. \tag{2}$$

The comparison of theoretical analyses [1, 2] with the recent measurements from the NA62 collaboration [3] and with the existing measurements on pion leptonic decays [4]

$$R_K^{\text{SM}} = (2.477 \pm 0.001) \times 10^{-5}\,, \qquad R_K^{\text{exp}} = (2.488 \pm 0.010) \times 10^{-5}\,, \tag{3}$$

$$R_\pi^{\text{SM}} = (1.2354 \pm 0.0002) \times 10^{-4}\,, \qquad R_\pi^{\text{exp}} = (1.230 \pm 0.004) \times 10^{-4} \tag{4}$$

suggests that observation agrees at 1σ level with the SM's predictions for

$$\Delta r_K = (4 \pm 4) \times 10^{-3}, \qquad \Delta r_\pi = (-4 \pm 3) \times 10^{-3}. \tag{5}$$

The current experimental uncertainty in Δr_K (of around 0.4%) will be further reduced in the near future, as one expects to have $\delta R_K/R_K \sim 0.1\%$ [5], which can translate into measuring deviations $\Delta r_K \sim \mathcal{O}(10^{-3})$. Similarly, there are also plans for a more precise determination of Δr_π [6, 7].

New contributions to Δr_P have been extensively discussed in the literature, especially in the framework of models with an enlarged Higgs sector. In the presence of charged scalar Higgs, new tree-level contributions are expected. However, as in the case of most Two Higgs Doublet Models (2HDM), or supersymmetric (SUSY) extensions of the SM, these new tree-level corrections are lepton universal [8]. In SUSY models, higher order non-holomorphic couplings can indeed provide new contributions to R_P [9, 10, 11, 12, 13], but in view of current experimental bounds (collider, B-physics and τ-lepton decays), one can have at most $\Delta r_K \leq 10^{-3}$ in the framework of unconstrained minimal SUSY models [13]. Corrections to the $W\ell\nu$ vertex can also induce violation of LFU in charged currents. However, if these appear at the loop level, as referred to in [10], the effect is expected to be of order $(\alpha/4\pi) \times (m_W^2/\Lambda_{\rm NP}^2)$ ($\Lambda_{\rm NP}$ being the new physics scale), generally well below experimental sensitivity.

Here we consider a different alternative: The tree-level corrections to charged current interactions once neutrino oscillations are incorporated into the SM [15, 16]. In this case, and working in the basis where the charged lepton mass matrix is diagonal, the flavour-conserving term $\propto g\bar{l}_j\gamma^\mu P_L\nu_j W_\mu^-$ now reads

$$-\mathcal{L}_{cc} = \frac{g}{\sqrt{2}} U_\nu^{ji} \bar{l}_j \gamma^\mu P_L \nu_i W_\mu^- + \text{c.c.}, \tag{6}$$

where U_ν^{ji} is a generic leptonic mixing matrix, $i = 1, \ldots, n_\nu$ denoting the physical neutrino states (not necessarily corresponding to the three left-handed SM states $\equiv \nu_L$) and $j = 1, \ldots, 3$ the charged lepton flavour. In the case of three neutrino generations, U_ν^{ji} corresponds to the unitary PMNS matrix and flavour universality is preserved in meson decays: since one cannot tag the flavour of the final state neutrino (missing energy), the meson decay amplitude is proportional to $(U_\nu U_\nu^\dagger)_{jj} = 1$, and thus no new contribution to R_P is expected. However, in the presence of sterile states, the $W\ell\nu$ vertex is proportional to a rectangular $3 \times n_\nu$ matrix U_ν^{ji}, and the mixing between the left-handed leptons ν_L, ℓ_L corresponds to a 3×3 block of U_ν^{ji},

$$U_{\rm PMNS} \to \tilde{U}_{\rm PMNS} = (\mathbb{1} - \eta)\, U_{\rm PMNS}. \tag{7}$$

The larger the mixing between the active (left-handed) neutrinos and the new states, the more pronounced the deviations from unitarity of $\tilde{U}_{\rm PMNS}$, parametrized by the matrix η [14]. The active-sterile mixings and the departure from unitarity of $\tilde{U}_{\rm PMNS}$ can be at the source of the violation of LFU in different neutrino mass models which introduce sterile fermionic states to generate non-zero masses and mixings for the light neutrinos [15, 16].

Corrections to the $W\ell\nu$ vertex can arise in several scenarios with additional (light) singlet states, as is the case of νSM [17], low-scale type-I seesaw [18] and the Inverse Seesaw (ISS) [19], among other possibilities. This clearly shows the potentiality of the mechanism under discussion, which can be present in many different models.

In the next section we provide a model-independent computation of Δr_P in the presence of additional fermionic sterile states; we then briefly review in Section 3 the most important experimental and observational constraints on the mass of the additional singlet states. In Section 4, we consider the case of the inverse seesaw to give a numerical example of the impact of sterile neutrinos on Δr_P. Our concluding remarks are summarised in Section 5.

2. Δr_P in the presence of sterile neutrinos

Let us consider the SM extended by N_s additional sterile states. The matrix element for the meson decay $P \to l_j \nu_i$ can be generically written as

$$\mathcal{M}_{ij} = \bar{u}_{\nu_i}(\mathcal{A}^{ij} P_R + \mathcal{B}^{ij} P_L) v_{l_j} \,. \tag{8}$$

No sum is implied over the indices of the outgoing leptons i, j. Notice that $i = 1, \ldots, 3+N_s$. The expressions for $\mathcal{A}$ and $\mathcal{B}$ can be easily obtained from the usual 4-fermion effective hamiltonian obtained after integrating out the W boson in Eq. (6). These are

$$(\mathcal{A})^{ij} = (\mathcal{A}^W)^{ij} = -4\, G_F\, V_{\text{CKM}}^{us}\, f_P\, U_\nu^{ji\,*}\, m_{l_j} \,; \tag{9}$$

$$(\mathcal{B})^{ij} = (\mathcal{B}^W)^{ij} = 4\, G_F\, V_{\text{CKM}}^{us}\, f_P\, U_\nu^{ji\,*}\, m_{\nu_i} \,, \tag{10}$$

where f_P denotes the meson decay constant and m_{l_j,ν_i} the mass of the outgoing leptons.

The expression for R_P is finally given by

$$R_P = \frac{\sum_i F^{i1} G^{i1}}{\sum_k F^{k2} G^{k2}} \,, \quad \text{with} \tag{11}$$

$$F^{ij} = |U_\nu^{ji}|^2 \quad \text{and} \quad G^{ij} = \left[m_P^2(m_{\nu_i}^2 + m_{l_j}^2) - (m_{\nu_i}^2 - m_{l_j}^2)^2\right] \left[(m_P^2 - m_{l_j}^2 - m_{\nu_i}^2)^2 - 4 m_{l_j}^2 m_{\nu_i}^2\right]^{1/2} . \tag{12}$$

The result of Eq. (11) has a straightforward interpretation: F^{ij} represents the impact of new interactions (absent in the SM), whereas G^{ij} encodes the mass-dependent factors. The SM result can be easily recovered from Eq. (11), in the limit $m_{\nu_i} = 0$ and $U_\nu^{ji} = \delta_{ji}$,

$$R_P^{SM} = \frac{m_e^2}{m_\mu^2} \frac{(m_P^2 - m_e^2)^2}{(m_P^2 - m_\mu^2)^2} \,, \tag{13}$$

to which small electromagnetic corrections should be added [1].

Using the results in Eqs. (11) and (13), we obtain a general expression for Δr_P

$$\Delta r_P = \frac{m_\mu^2 (m_P^2 - m_\mu^2)^2}{m_e^2 (m_P^2 - m_e^2)^2} \frac{\sum_{m=1}^{N_{\max}^{(e)}} F^{m1}\, G^{m1}}{\sum_{n=1}^{N_{\max}^{(\mu)}} F^{m2}\, G^{n2}} - 1 \,. \tag{14}$$

Thus, depending on the masses of the new states (and their hierarchy) and most importantly, on their mixings to the active neutrinos, Δr_P can considerably deviate from zero. In order to illustrate this, we consider two regimes:

- **Regime (A):** All sterile neutrinos are *lighter* than the decaying meson, but heavier than the active neutrino states, i.e. $m_\nu^{\text{active}} \ll m_{\nu_s} \lesssim m_P$
- **Regime (B):** All sterile neutrinos are *heavier* than m_P

Notice that in case (A), all the mass eigenstates can be kinematically available and one should sum over all $3+N_s$ states; furthermore there is an enhancement to Δr_P arising from phase space factors, see Eq. (12).

3. Constraints on sterile neutrinos

We review in this section the experimental and observational bounds on the mass regimes and on the size of the active-sterile mixings that must be satisfied.

First, it is clear that present data on neutrino masses and mixings [20] should be accounted for. Second, there are robust laboratory bounds from direct sterile neutrinos searches [21, 22], since the latter can be produced in meson decays such as $\pi^\pm \to \mu^\pm \nu$, with rates dependent on their mixing with the active neutrinos. Negative searches for monochromatic lines in the muon spectrum can be translated into bounds for $m_{\nu_s} - \theta_{i\alpha}$ combinations, where $\theta_{i\alpha}$ parametrizes the active-sterile mixing. The non-unitarity of the leptonic mixing matrix is also subject to constraints. Bounds on the non-unitarity parameter η (Eq. (7)), were derived using Non-Standard Interactions [23]; although not relevant in case (A), these bounds will be taken into account when evaluating scenario (B).

The modified $W\ell\nu$ vertex also contributes to lepton flavour violation (LFV) processes. The radiative decay $\mu \to e\gamma$, searched for by the MEG experiment [24], is typically the most constraining observable [1]. The rate induced by sterile neutrinos must satisfy [31, 32]

$$\mathrm{BR}(\mu \to e\gamma) = \frac{\alpha_W^3 s_W^2 m_\mu^5}{256\pi^2 m_W^4 \Gamma_\mu} |H_{\mu e}|^2 \leq 2.4 \times 10^{-12}\,, \tag{15}$$

where $H_{\mu e} = \sum_i U_\nu^{2i} U_\nu^{1i\,*} G_\gamma(\frac{m^2_{\nu,i+3}}{m_W^2})$, with G_γ the loop function and U_ν the mixing matrix defined in Eq. (6). Similarly, any change in the $W\ell\nu$ vertex will also affect other leptonic meson decays, in particular $B \to \ell\nu$; the following bounds were enforced in the analysis: $\mathrm{BR}(B \to e\nu) < 9.8 \times 10^{-7}$, $\mathrm{BR}(B \to \mu\nu) < 10^{-6}$ and $\mathrm{BR}(B \to \tau\nu) = (1.65 \pm 0.34) \times 10^{-4}$ [33].

Important constraints can also be derived from LHC Higgs searches [?] and electroweak precision data [35]. They will also be considered in our numerical analysis.

Under the assumption of a standard cosmology, the most constraining bounds on sterile neutrinos stem from a wide variety of cosmological observations [36, 22]. These include Large Scale Structure data, X-ray searches (which can be produced in $\nu_i \to \nu_j \gamma$), Lyman-α limits, the existence of additional degrees of freedom at the epoch of Big Bang Nucleosynthesis and Cosmic Microwave Background data. However, all the above cosmological bounds can be evaded if a non-standard cosmology is considered. In fact, the authors of Ref. [37] showed that the above cosmological constraints disappear in scenarios with low reheating temperature. Therefore, we will allow for the violation of the latter bounds, explicitly stating it.

4. A numerical example: Δr_K in the inverse seesaw

Although the generic idea explored in this work applies to any model where the active neutrinos have sizeable mixings with some additional singlet states, we consider the case of the Inverse Seesaw [19] to illustrate the potential of a model with sterile neutrinos regarding tree-level contributions to light meson decays. As mentioned before, there are other possibilities [17, 18].

4.1. The inverse seesaw

In the ISS, the SM particle content is extended by n_R generations of right-handed (RH) neutrinos ν_R and n_X generations of singlet fermions X with lepton number $L = -1$ and $L = +1$, respectively [19] (such that $n_R + n_X = N_s$). In our numerical application we will focus on

[1] Recently, it has been also noticed that in the framework of low-scale seesaw models, the expected future sensitivity of $\mu - e$ conversion experiments can also play a relevant rôle in detecting or constraining sterile neutrino scenarios [25, 26, 27, 28]. This is also the case in the supersymmetric version of these models, even when the sterile neutrinos are heavier [29, 30].

the case $n_R = n_X = 3$. The lagrangian is given by

$$\mathcal{L}_{\rm ISS} = \mathcal{L}_{SM} + Y_\nu^{ij}\bar{\nu}_{Ri}L_j\tilde{H} + M_{Rij}\,\bar{\nu}_{Ri}X_j + \frac{1}{2}\mu_{X_{ij}}\bar{X}_i^c X_j + \text{h.c.} \tag{16}$$

where $i, j = 1, 2, 3$ are generation indices and $\tilde{H} = i\sigma_2 H^*$. Notice that the present lepton number assignment, together with $L = +1$ for the SM lepton doublet, implies that the "Dirac"-type right-handed neutrino mass term $M_{R_{ij}}$ conserves lepton number, while the "Majorana" mass term $\mu_{X_{ij}}$ violates it by two units.

The left-handed neutrinos mix with the right-handed ones after electroweak symmetry breaking. This leads to an effective Majorana mass for the active (light) neutrinos. Assuming $\mu_X \ll m_D \ll M_R$, where $m_D = \frac{1}{\sqrt{2}}Y_\nu v$, with v the vacuum expectation value of the SM Higgs boson, one obtains

$$m_\nu \simeq m_D^T M_R^{T^{-1}} \mu_X M_R^{-1} m_D\,. \tag{17}$$

The remaining 6 sterile states have masses approximately given by $M_\nu \simeq M_R$. Small corrections can be added to these results, but they are typically negligible [38].

In what follows, and without loss of generality, we work in a basis where M_R is a diagonal matrix (as are the charged lepton Yukawa couplings). Y_ν can be written using a modified Casas-Ibarra parametrisation [39] (thus automatically complying with light neutrino data),

$$Y_\nu = \frac{\sqrt{2}}{v}\, V^\dagger \sqrt{\hat{M}}\, R\, \sqrt{\hat{m}_\nu}\, U_{\rm PMNS}^\dagger\,, \tag{18}$$

where $\sqrt{\hat{m}_\nu}$ is a diagonal matrix containing the square roots of the three eigenvalues of m_ν (cf. Eq. (17)); likewise $\sqrt{\hat{M}}$ is a (diagonal) matrix with the square roots of the eigenvalues of $M = M_R \mu_X^{-1} M_R^T$. V diagonalizes M as $VMV^T = \hat{M}$, and R is a 3×3 complex orthogonal matrix, parametrized by 3 complex angles, encoding the remaining degrees of freedom.

The nine neutrino mass eigenstates enter the leptonic charged current through their left-handed component (see Eq. (6), with $i = 1, \ldots, 9$, $j = 1, \ldots, 3$). The unitary leptonic mixing matrix U_ν is now defined as $U_\nu^T \mathcal{M} U_\nu = \text{diag}(m_i)$. Notice however that only the rectangular 3×9 sub-matrix (first three columns of U_ν) appears in Eq. (6) due to the gauge-singlet nature of ν_R and X.

4.2. Numerical evaluation of Δr_K in the inverse seesaw

We numerically evaluate the contributions to R_K in the framework of the ISS and address the two scenarios discussed before, which can be translated in terms of ranges for the (random) entries of the M_R matrix: *regime (A)* $(m_{\nu_s} < m_P)$ - $M_{Ri} \in [0.1, 200]$ MeV; *regime (B)* $(m_{\nu_s} > m_P)$ - $M_{Ri} \in [1, 10^6]$ GeV. The entries of μ_X have also been randomly varied in the [0.01 eV, 1 MeV] range for both cases.

The adapted Casas-Ibarra parametrisation for Y_ν, Eq. (18), ensures that neutrino oscillation data is satisfied (we use the best-fit values of the global analysis of Ref. [20] and set the CP violating phases of $U_{\rm PMNS}$ to zero). The R matrix angles are taken to be real (thus no contributions to lepton electric dipole moments are expected), and randomly varied in the range $\theta_i \in [0, 2\pi]$. We have verified that similar Δr_K contributions are found when considering the more general complex R matrix case.

In Figs. 1, we collect our results for Δr_K in scenarios (A) - left panel - and (B) - right panel, as a function of $\tilde{\eta}$, which parametrizes the departure from unitarity of the active neutrino mixing sub-matrix $\tilde{U}_{\rm PMNS}$, $\tilde{\eta} = 1 - |\text{Det}(\tilde{U}_{\rm PMNS})|$. Although the cosmological constraints are not always satisfied, we stress that all points displayed comply with the different experimental and laboratory bounds discussed before. For the case of scenario (A), one can have very large

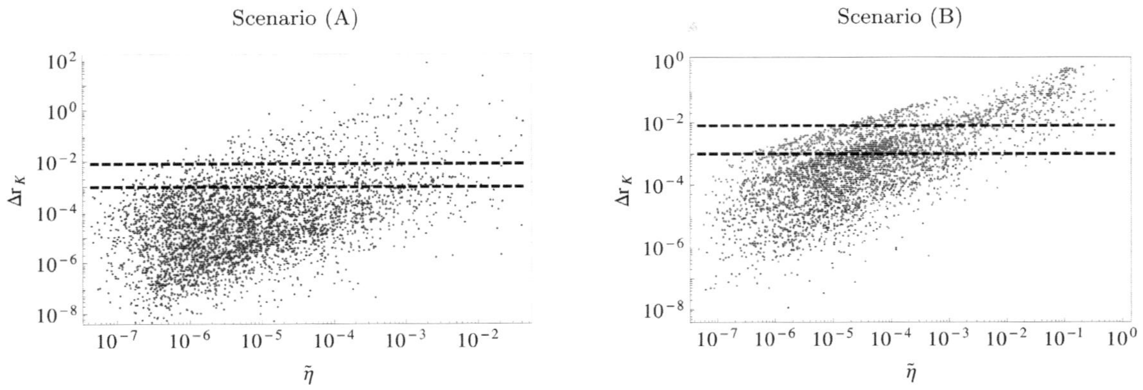

Figure 1. Contributions to Δr_K in the inverse seesaw as a function of $\tilde{\eta} = 1 - |\mathrm{Det}(\tilde{U}_{\mathrm{PMNS}})|$: regimes A (left) and B (right). The upper (lower) dashed line denotes the current experimental limit (expected sensitivity). On the right panel, red points denote cases where $Y_\nu \gtrsim 10^{-2}$. All points comply with experimental and laboratory constraints. Points in (B) are also in agreement with cosmological bounds, while those in (A) require considering a non-standard cosmology.

contributions to R_K, which can even reach values $\Delta r_K \sim \mathcal{O}(1)$ (in some extreme cases we find Δr_K as large as ~ 100). The hierarchy of the sterile neutrino spectrum in case (A) is such that one can indeed have a significant amount of LFU violation, while still avoiding non-unitarity bounds. Although this scenario would in principle allow to produce sterile neutrinos in light meson decays, the smallness of the associated Y_ν ($\lesssim \mathcal{O}(10^{-4})$), together with the loop function suppression (G_γ), precludes the observation of LFV processes, even those with very good associated experimental sensitivity, as is the case of $\mu \to e\gamma$. The strong constraints from CMB and X-rays would exclude scenario (A); in order to render it viable, one would require a non-standard cosmology.

Despite the fact that in case (B) the hierarchy of the sterile states is such that non-unitarity bounds become very stringent (since the sterile neutrinos are not kinematically viable meson decay final states), sizeable LFU violation is also possible, with deviations from the SM predictions again as large as $\Delta r_K \sim \mathcal{O}(1)$. Although one cannot produce sterile states in meson decays in this case, the large Y_ν open the possibility of having larger contributions to LFV observables so that, for example, $\mathrm{BR}(\mu \to e\gamma)$ can be within MEG reach.

Although we do not explicitly display it here, the prospects for Δr_π are similar: in the same framework, one could have $\Delta r_\pi \sim \mathcal{O}(\Delta r_K)$, and thus $\Delta r_\pi \sim \mathcal{O}(1)$ in both scenarios. Depending on the singlet spectrum, these observables can also be strongly correlated: if all the sterile states are either lighter than the pion (as it is the case of scenario (A)) or then heavier than the kaon, one finds $\Delta r_\pi \approx \Delta r_K$. This is a distinctive feature of our mechanism.

5. Concluding remarks

The existence of sterile neutrinos can potentially lead to a significant violation of lepton flavour universality at tree-level in light meson decays. As shown in this study, provided that the active-sterile mixings are sufficiently large, the modified $W\ell\nu$ interaction can lead to large contributions to lepton flavour universality observables, with measurable deviations from the standard model expectations, well within experimental sensitivity. This mechanism might take place in many different frameworks, the exact contributions for a given observable being model-dependent.

As an illustrative (numerical) example, we have evaluated the contributions to R_K in the inverse seesaw extension of the SM - a truly minimal extension of the SM - , for distinct hierarchies of the sterile states. Our analysis reveals that very large deviations from the SM predictions can be found ($\Delta r_K \sim \mathcal{O}(1)$) - or even larger, well within reach of the NA62 experiment at CERN. This is in clear contrast with other models of new physics (for example unconstrained SUSY models, where one typically has $\Delta r_K \lesssim \mathcal{O}(10^{-3})$).

Acknowledgments

The author is grateful to A. Abada, D. Das, A. M. Teixeira and C. Weiland for their collaboration in the work this talk is based on. This work has been supported by the ANR project CPV-LFV-LHC NT09-508531.

References

[1] V. Cirigliano and I. Rosell, Phys. Rev. Lett. **99** (2007) 231801 [arXiv:0707.3439 [hep-ph]].
[2] M. Finkemeier, Phys. Lett. B **387** (1996) 391 [hep-ph/9505434].
[3] E. Goudzovski [NA48/2 and NA62 Collaboration], PoS EPS **HEP2011** (2011) 181 [arXiv:1111.2818 [hep-ex]].
[4] G. Czapek *et al.*, Phys. Rev. Lett. **70** (1993) 17.
[5] E. Goudzovski [NA48/2 and NA62 Collaborations], arXiv:1208.2885 [hep-ex].
[6] D. Pocanic *et al.*, arXiv:1210.5025 [hep-ex].
[7] C. Malbrunot *et al.*, AIP Conf. Proc. **1441** (2012) 564.
[8] W. -S. Hou, Phys. Rev. D **48** (1993) 2342.
[9] A. Masiero, P. Paradisi and R. Petronzio, Phys. Rev. D **74** (2006) 011701 [hep-ph/0511289].
[10] A. Masiero, P. Paradisi and R. Petronzio, JHEP **0811** (2008) 042 [arXiv:0807.4721 [hep-ph]].
[11] J. Ellis, S. Lola and M. Raidal, Nucl. Phys. B **812** (2009) 128 [arXiv:0809.5211 [hep-ph]].
[12] J. Girrbach and U. Nierste, arXiv:1202.4906 [hep-ph].
[13] R. M. Fonseca, J. C. Romao and A. M. Teixeira, Eur. Phys. J. C **72** (2012) 2228 [arXiv:1205.1411 [hep-ph]].
[14] J. Schechter and J. W. F. Valle, Phys. Rev. D **22** (1980) 2227; M. Gronau, C. N. Leung and J. L. Rosner, Phys. Rev. D **29** (1984) 2539.
[15] R. E. Shrock, Phys. Lett. B **96** (1980) 159; R. E. Shrock, Phys. Rev. D **24** (1981) 1232.
[16] A. Abada, D. Das, A. M. Teixeira, A. Vicente and C. Weiland, JHEP **1302** (2013) 048 [arXiv:1211.3052 [hep-ph]].
[17] T. Asaka, S. Blanchet and M. Shaposhnikov, Phys. Lett. B **631** (2005) 151 [hep-ph/0503065].
[18] A. Ibarra, E. Molinaro and S. T. Petcov, JHEP **1009** (2010) 108 [arXiv:1007.2378 [hep-ph]].
[19] R. N. Mohapatra and J. W. F. Valle, Phys. Rev. D **34** (1986) 1642.
[20] D. V. Forero, M. Tortola and J. W. F. Valle, Phys. Rev. D **86** (2012) 073012 [arXiv:1205.4018 [hep-ph]].
[21] A. Atre *et al.*, JHEP **0905** (2009) 030 [arXiv:0901.3589 [hep-ph]].
[22] A. Kusenko, Phys. Rept. **481** (2009) 1 [arXiv:0906.2968 [hep-ph]].
[23] S. Antusch, J. P. Baumann and E. Fernandez-Martinez, Nucl. Phys. B **810** (2009) 369 [arXiv:0807.1003 [hep-ph]].
[24] J. Adam *et al.* [MEG Collaboration], Phys. Rev. Lett. **107** (2011) 171801 [arXiv:1107.5547 [hep-ex]].
[25] A. Ilakovac and A. Pilaftsis, Phys. Rev. D **80** (2009) 091902 [arXiv:0904.2381 [hep-ph]].
[26] D. N. Dinh, A. Ibarra, E. Molinaro and S. T. Petcov, JHEP **1208** (2012) 125 [arXiv:1205.4671 [hep-ph]].
[27] R. Alonso, M. Dhen, M. B. Gavela and T. Hambye, JHEP **1301** (2013) 118 [arXiv:1209.2679 [hep-ph]].
[28] A. Ilakovac, A. Pilaftsis and L. Popov, arXiv:1212.5939 [hep-ph].
[29] M. Hirsch, F. Staub and A. Vicente, Phys. Rev. D **85** (2012) 113013 [arXiv:1202.1825 [hep-ph]].
[30] A. Abada *et al.*, JHEP **1209** (2012) 015 [arXiv:1206.6497 [hep-ph]].
[31] A. Ilakovac and A. Pilaftsis, Nucl. Phys. B **437** (1995) 491 [hep-ph/9403398].
[32] F. Deppisch and J. W. F. Valle, Phys. Rev. D **72** (2005) 036001 [hep-ph/0406040].
[33] J. Beringer *et al.* [Particle Data Group Collaboration], Phys. Rev. D **86** (2012) 010001.
[34] P. S. Bhupal Dev, R. Franceschini and R. N. Mohapatra, Phys. Rev. D **86** (2012) 093010 [arXiv:1207.2756 [hep-ph]].
[35] F. del Aguila, J. de Blas and M. Perez-Victoria, Phys. Rev. D **78** (2008) 013010 [arXiv:0803.4008 [hep-ph]].
[36] A. Y. Smirnov and R. Zukanovich Funchal, Phys. Rev. D **74** (2006) 013001 [hep-ph/0603009].
[37] G. Gelmini *et al.*, JCAP **0810** (2008) 029 [arXiv:0803.2735 [astro-ph]].
[38] D. V. Forero *et al.*, JHEP **1109** (2011) 142 [arXiv:1107.6009 [hep-ph]].
[39] J. A. Casas and A. Ibarra, Nucl. Phys. B **618** (2001) 171 [hep-ph/0103065].

The Adjoint SU(5) constrained by a Z_4 flavour symmetry

C. Simões

CFTP and Departamento de Física, Instituto Superior Técnico,
Av. Rovisco Pais, 1049-001 Lisboa, Portugal

E-mail: csimoes@cftp.ist.utl.pt

Abstract. We study the neutrino mass spectrum in the context of the adjoint SU(5) model constrained by a Z_4 flavour symmetry that imposes the nearest-neighbour interaction form on both up- and down-quark mass matrices. We show that at least three adjoint fermionic fields are required in order to reproduce the light neutrino mass spectrum.

1. Introduction

In the Standard Model neutrinos are massless particles, however neutrino oscillation experiments reveal that they have mass. This problem can be solved in the context of Grand Unified Theories (GUT) of which the SU(5) model [1] is an example. In that model light neutrino masses can be generated through the seesaw mechanism due to the presence of right-handed neutrinos.

Ref. [2] studied the SU(5) model in the case where three right-handed neutrinos were considered and a Z_4 flavour symmetry imposed, in order to give the Nearest-Neighbour-Interaction (NNI) form [3]

$$\mathrm{NNI} = \begin{pmatrix} 0 & * & 0 \\ * & 0 & * \\ 0 & * & * \end{pmatrix} , \tag{1}$$

to the both up- and down-quark mass matrices. Despite the good agreement with the neutrino data, the study does not solve the problem with relation to the $M_e = M_d^T$ at GUT scale being incompatible with the down-type quark and charged lepton mass hierarchy observed at low energy scales. Such problems can be solved in the context of adjoint SU(5) [4, 5].

The purpose of this work was to study the light neutrino mass spectrum obtained in the case of the adjoint SU(5) model where the Z_4 flavour symmetry is used to impose the NNI form on both up- and down-quark mass matrices [6]. In this model, the adjoint fermionic field, 24, is responsible for the generation of the neutrino masses through type-I [7, 8, 9, 10], type-III [11, 12] and radiative [13, 14, 15] seesaw mechanisms.

This work is organized as follows: in section 2 we present the $SU(5) \times Z_4$ model; in section 3 we discuss the obtained neutrino mass matrices and their compatibility with experimental data, then we conclude.

2. The model

The fermionic particle content of this model, based on the adjoint-SU(5) model [4, 5], is composed of three generations of 10 and 5^* multiplets plus n copies of the adjoint fermionic representations, 24. The adjoint fermionic field is responsible for the generation of neutrino masses through type-I and type-III seesaw mechanism. There is also the possibility to generate the neutrino masses through radiative seesaw which is explored in Refs. [16, 6].

The neutrino oscillation data is consistent with the presence of just one 24 fermionic field [4, 5], however due to the implementation of the Z_4 flavour symmetry we will see that in this model at least three adjoint fermionic fields are needed. The fermionic fields decompose as

$$\begin{aligned} 5^* &= d^c \oplus L\,, \\ 10 &= Q \oplus u^c \oplus e^c\,, \\ 24 &= \rho_8 \oplus \rho_3 \oplus \rho_{(3,2)} \oplus \rho_{(\bar{3},2)} \oplus \rho_0\,, \end{aligned} \tag{2}$$

while the Higgs sector is composed by one adjoint multiplet, 24_H, one quintet, 5_H, and one 45 dimensional representation, 45_H, decomposed as

$$\begin{aligned} 5_H &= T_1 \oplus H_1\,, \\ 24_H &= \Sigma_8 \oplus \Sigma_3 \oplus \Sigma_{(3,2)} \oplus \Sigma_{(\bar{3},2)} \oplus \Sigma_0\,, \\ 45_H &= S_8 \oplus S_{(\bar{6},1)} \oplus S_{(3,3)} \oplus S_{(\bar{3},2)} \oplus S_{(\bar{3},1)} \oplus T_2 \oplus H_2\,, \end{aligned} \tag{3}$$

where H_1 and H_2 are the low energy doublets and T_1 and T_2 are the coloured SU(3) triplets.

The 24_H is responsible for breaking the SU(5) gauge group down to the SM one, $SU(3)_C \times SU(2)_L \times U(1)_Y$, through the vacuum expectation value (VEV),

$$\langle 24_H \rangle = \frac{v_{24}}{\sqrt{60}} \mathrm{diag}(2, 2, 2, -3, -3)\,. \tag{4}$$

At the GUT scale, Λ, the 45 Higgs field acquires a VEV as

$$\langle {45_H}_{\alpha}^{\beta\,5} \rangle = v_{45} \left(\delta_{\alpha}^{\beta} - 4\, \delta_4^{\alpha} \delta_{\beta}^{4} \right)\,, \tag{5}$$

where α, β = 1, ..., 4; the Higgs quintet gets the VEV, $\langle {5_H}^{\alpha} \rangle = \delta_5^{\alpha}\, v_5$ where α = 1, ..., 5 breaking the SM gauge group down to $SU(3)_C \times U(1)_{e.m.}$. The two VEVs are related by $v^2 = v_5^2 + 24 v_{45}^2 = (246.2\,\mathrm{GeV})^2$. The Higgs doublets contained in the 5_H and 45_H representations generate, at the electroweak scale, the fermion masses through the Yukawa interactions. In fact at low energy we end up with a two Higgs doublet model.

The main goal of this work was to study the neutrino sector in the presence of a Z_4 flavour symmetry as proposed in Ref. [2]. This symmetry imposes the NNI form [3] for both up- and down-quark mass matrices. We have shown in Ref. [2] that to have the quark mass matrices in the NNI form, only two independent charges were needed, $\mathcal{Q}(5_H) = \phi$ and $\mathcal{Q}(10_3) = q_3$; furthermore, we have chosen the 45 Higgs field to couple to the bilinear term $10_3\, 10_3$ and hence $\mathcal{Q}(45_H) = -2q_3$. The charges of 5_i^* and 10_i fields are written in terms of ϕ and q_3 as,

$$\mathcal{Q}(5_i^*) = (q_3 + 2\phi, -3q_3, -q_3 + \phi)\,, \quad \mathcal{Q}(10_i) = (3q_3 + \phi, -q_3 - \phi, q_3)\,. \tag{6}$$

The adjoint fermionic fields are free under Z_4 while the charge of 24_H is zero to guarantee the existence of the flavour symmetry below the GUT scale. For further details see Ref. [2].

The most general Yukawa interactions are given by

$$\begin{aligned}
-\mathcal{L}_{\rm Y} &= \left(\Gamma_u^1\right)_{ij} 10_i\, 10_j\, 5_H + \left(\Gamma_u^2\right)_{ij} 10_i\, 10_j\, 45_H \\
&+ \left(\Gamma_d^1\right)_{ij} 10_i\, 5_j^*\, 5_H^* + \left(\Gamma_d^2\right)_{ij} 10_i\, 5_j^*\, 45_H^* \\
&+ \left(\Gamma_D^1\right)_{ik} 5_i^*\, 24_k\, 5_H + \left(\Gamma_D^2\right)_{ik} 5_i^*\, 24_k\, 45_H \\
&+ M_{kl}\, {\rm tr}\left(24_k\, 24_l\right) + \lambda_{kl}\, {\rm tr}\left(24_k\, 24_l\, 24_H\right) + {\rm H.c.}\,,
\end{aligned} \tag{7}$$

where $\Gamma_u^{1,2}$ and M are symmetric complex matrices and $\Gamma_{d,D}^{1,2}$ are just complex matrices; the $\mathsf{SU(5)}$ indices were omitted and i, j are generation indices and k, $l = 1, \ldots, n$.

Thus, the up- and down-quark masses as well as the charged lepton masses are given by

$$\begin{aligned}
M_u &= 4\left(\Gamma_u^1 + \Gamma_u^{1\,T}\right) v_5 - 8\left(\Gamma_u^2 - \Gamma_u^{2\,T}\right) v_{45}\,, \\
M_d &= \Gamma_d^1\, v_5^* + 2\Gamma_d^2\, v_{45}^*\,, \\
M_e &= \Gamma_d^{1\,T}\, v_5^* - 6\Gamma_d^{2\,T}\, v_{45}^*\,,
\end{aligned} \tag{8}$$

the form of the matrices $\Gamma_{u,d}^{1,2}$ can be found in Ref. [2]. From the last two lines in Eq. (8) one sees that the mismatch between the down-type and charged lepton matrices is given by

$$M_d - M_e^T = 8\,\Gamma_d^2 v_{45}^*\,. \tag{9}$$

The mass matrices of the fermionic fields responsible for the neutrino masses, through type-I, type-III and radiative seesaw, ρ_0, ρ_3 and ρ_8, are given respectively by

$$M_{\rho_0} = \frac{1}{4}\left(M - \frac{v_{24}}{\sqrt{30}}\lambda\right), \quad M_{\rho_3} = \frac{1}{4}\left(M - \frac{3v_{24}}{\sqrt{30}}\lambda\right), \quad M_{\rho_8} = \frac{1}{4}\left(M + \frac{2v_{24}}{\sqrt{30}}\lambda\right). \tag{10}$$

The form of the matrices M and λ is not yet known, however from Eq. (7) and since $\mathsf{24_H}$ is trivial under $\mathsf{Z_4}$ one can say that the two matrices have the same form; this form depends only on the $\mathsf{Z_4}$ charges of the adjoint fermionic fields.

The effective neutrino mass matrix, m_ν, obtained from type-I and type-III seesaw, is given by

$$m_\nu = -m_D^{\rho_0}\, M_{\rho_0}^{-1}\, m_D^{\rho_0\,T} - m_D^{\rho_3}\, M_{\rho_3}^{-1}\, m_D^{\rho_3\,T}\,, \tag{11}$$

where the matrices $m_D^{\rho_0}$ and $m_D^{\rho_3}$ are given by

$$m_D^{\rho_0} = \frac{\sqrt{15}}{2}\left(\frac{v_5}{5}\,\Gamma_D^1 + v_{45}\,\Gamma_D^2\right), \quad m_D^{\rho_3} = \left(-v_5\,\Gamma_D^1 + 3\,v_{45}\,\Gamma_D^2\right). \tag{12}$$

In this report, for simplicity we will not consider the contribution from radiative seesaw; details on the effects of such mechanism can be found on Ref. [6]. However, it is important to emphasize that this does not change the conclusions, since their contribution to the m_ν matrix is similar to the one from type-I and type-III seesaw mechanism.

The details concerning the unification of gauge couplings, proton decay and leptogenesis in this model can be seen in Ref. [6].

3. Effective Neutrino Textures

In the previous section, we have anticipated that reproducing the neutrino masses requires at least the introduction of three 24 fermionic fields. In order to clarify this statement, we have scanned the Z_4 charges of the fields for the cases where just one, two or three adjoint fermionic fields are present. Notice that, as already mentioned, the Z_4 charges of the adjoint fermionic fields are free. The resulting effective mass matrices, m_ν, are sketched in Table 1; we have called by the same name matrices with the same form and defined $X_g = P_g\, X\, P_g^T$ where $\{P_g\}$ are the 3×3 permutation matrices.

Looking at the obtained mass matrices one can conclude that the cases with one and two 24 fermionic fields can not reproduce the neutrino oscillation data [17]. Hence the only possibility to have a neutrino mass spectrum compatible with data is by introducing three adjoint fermionic representations. It is interesting to note that the effective neutrino mass matrices found here for the case of three 24 fermionic fields are the same as the ones found in Ref. [2], where a similar Z_4 flavour symmetry was imposed in the context of three right-handed neutrinos. This is a consequence of the flavour symmetry behind the model.

At this point some comments are in order. The effective neutrino mass matrices given in Eq. (11) require that both M_{ρ_0} and M_{ρ_3} are non singular. However, as the Z_4 charges of the adjoint fermionic fields are free one can have Z_4 charges that lead to $|M| = 0$. In such case the usual seesaw mechanism is substituted by the singular seesaw mechanism [18, 19, 20, 21]. Just for illustration, we give in the Appendix an example for the singular seesaw mechanism.

It is not possible, in any of the cases where the singular seesaw applies, to obtain the neutrino mass spectrum compatible with experimental data. In the case of two adjoint fermionic fields and when the Majorana matrix has rank 1, the neutrino mass spectrum is composed by one very heavy neutrino, two electroweak neutrinos, one light seesaw neutrino and one massless neutrino. For the case where three 24 fermionic representations are added and the Majorana matrix has rank 1, the neutrino spectrum consists of one heavy neutrino, two electroweak neutrino masses,

Table 1. The effective neutrino mass matrices for one, two and three 24 fermionic representations, where $X_g = P_g\, X\, P_g^T$ with $\{P_g\}$ the 3×3 permutation matrices.

number of 24 fermionic fields	m_ν			
1	$A = \begin{pmatrix} * & 0 & 0 \\ 0 & 0 & 0 \\ 0 & 0 & 0 \end{pmatrix}$	$A_{12} = \begin{pmatrix} 0 & 0 & 0 \\ 0 & * & 0 \\ 0 & 0 & 0 \end{pmatrix}$	$B = \begin{pmatrix} 0 & 0 & 0 \\ 0 & * & * \\ 0 & * & * \end{pmatrix}$	$B_{12} = \begin{pmatrix} * & 0 & * \\ 0 & 0 & 0 \\ * & 0 & * \end{pmatrix}$
2	$A = \begin{pmatrix} * & 0 & 0 \\ 0 & 0 & 0 \\ 0 & 0 & 0 \end{pmatrix}$	$B = \begin{pmatrix} 0 & 0 & 0 \\ 0 & * & * \\ 0 & * & * \end{pmatrix}$	$C = \begin{pmatrix} * & 0 & 0 \\ 0 & * & * \\ 0 & * & * \end{pmatrix}$	$D = \begin{pmatrix} 0 & * & * \\ * & 0 & 0 \\ * & 0 & 0 \end{pmatrix}$
	$A_{12} = \begin{pmatrix} 0 & 0 & 0 \\ 0 & * & 0 \\ 0 & 0 & 0 \end{pmatrix}$	$B_{12} = \begin{pmatrix} * & 0 & * \\ 0 & 0 & 0 \\ * & 0 & * \end{pmatrix}$	$C_{12} = \begin{pmatrix} * & 0 & * \\ 0 & * & 0 \\ * & 0 & * \end{pmatrix}$	$D_{12} = \begin{pmatrix} 0 & * & 0 \\ * & 0 & * \\ 0 & * & 0 \end{pmatrix}$
3	$C = \begin{pmatrix} * & 0 & 0 \\ 0 & * & * \\ 0 & * & * \end{pmatrix}$	$\mathrm{NNI}_{13} = \begin{pmatrix} * & * & 0 \\ * & 0 & * \\ 0 & * & 0 \end{pmatrix}$	$E = \begin{pmatrix} 0 & * & 0 \\ * & * & * \\ 0 & * & * \end{pmatrix}$	
	$C_{12} = \begin{pmatrix} * & 0 & * \\ 0 & * & 0 \\ * & 0 & * \end{pmatrix}$	$\mathrm{NNI}_{132} = \begin{pmatrix} 0 & * & * \\ * & * & 0 \\ * & 0 & 0 \end{pmatrix}$	$E_{12} = \begin{pmatrix} * & * & * \\ * & 0 & 0 \\ * & 0 & * \end{pmatrix}$	

one seesaw neutrino and two massless neutrinos; on the other hand, in the case where the rank of M is two one has two heavy neutrinos, two electroweak neutrinos, one intermediate neutrino and one massless neutrino. None of these cases can reproduce the experimental data hence we end up with only three 24 fermionic representations where neutrino masses are generated through type-I and type-III seesaw mechanism.

In Ref. [2], a study was performed of the compatibility of the six textures in the last line of Table 1 and only the textures E and E_{12} were found to be phenomenologically compatible with the actual experimental data. In Table 2 are sketched all possible Z_4 charges for the fermionic and Higgs fields of the model; it is also shown the matrices $m_D^{\rho_\alpha}$ and M_{ρ_α} for $\alpha = 0, 3$. Notice that the matrix M_{ρ_0} has the same zero pattern as M_{ρ_3}, likewise the matrix $m_D^{\rho_0}$ has the same zero pattern as $m_D^{\rho_3}$.

Table 2. The Z_4 charges for the fermionic and bosonic fields of the model in the case where the effective neutrino textures are phenomenologically viable. $\mathcal{Q}(45_H) = -2\,\mathcal{Q}(10_3)$ and $\alpha = 0,\ 3$.

$\mathcal{Q}(5_i^*)$	$\mathcal{Q}(10_i)$	$\mathcal{Q}(5_H)$	$\mathcal{Q}(24_i)$	M_{ρ_α}	$m_D^{\rho_\alpha}$	m_ν
(3,1,0)	(0,2,1)	1	(1,2,3)	$\begin{pmatrix} 0 & 0 & * \\ 0 & * & 0 \\ * & 0 & 0 \end{pmatrix}$	$\begin{pmatrix} 0 & 0 & * \\ * & * & 0 \\ 0 & * & * \end{pmatrix}$	$\begin{pmatrix} 0 & * & 0 \\ * & * & * \\ 0 & * & * \end{pmatrix}$
(1,3,0)	(0,2,3)	3			$\begin{pmatrix} * & 0 & 0 \\ 0 & * & * \\ * & * & 0 \end{pmatrix}$	
(1,3,2)	(2,0,3)	1	(0,1,3)	$\begin{pmatrix} * & 0 & 0 \\ 0 & 0 & * \\ 0 & * & 0 \end{pmatrix}$	$\begin{pmatrix} 0 & * & 0 \\ * & 0 & * \\ * & * & 0 \end{pmatrix}$	
(3,1,2)	(2,0,1)	3			$\begin{pmatrix} 0 & 0 & * \\ * & * & 0 \\ * & 0 & * \end{pmatrix}$	
(2,0,1)	(1,3,0)	1	(1,2,3)	$\begin{pmatrix} 0 & 0 & * \\ 0 & * & 0 \\ * & 0 & 0 \end{pmatrix}$	$\begin{pmatrix} * & * & 0 \\ 0 & 0 & * \\ 0 & * & * \end{pmatrix}$	$\begin{pmatrix} * & * & * \\ * & 0 & 0 \\ * & 0 & * \end{pmatrix}$
(2,0,3)	(3,1,0)	3			$\begin{pmatrix} 0 & * & * \\ * & 0 & 0 \\ * & * & 0 \end{pmatrix}$	
(0,2,3)	(3,1,2)	1	(0,1,3)	$\begin{pmatrix} * & 0 & 0 \\ 0 & 0 & * \\ 0 & * & 0 \end{pmatrix}$	$\begin{pmatrix} * & 0 & * \\ 0 & * & 0 \\ * & * & 0 \end{pmatrix}$	
(0,2,1)	(1,3,2)	3			$\begin{pmatrix} * & * & 0 \\ 0 & 0 & * \\ * & 0 & * \end{pmatrix}$	

In what follows we will explore the phenomenological consequences of the two viable textures, E and E_{12}, confronting the predictions found for the charged-lepton mass matrices and effective neutrino mass matrices with the observable neutrino data at M_Z energy scale. This is possible since the Z_4 flavour symmetry is preserved till the electroweak symmetry breaking and therefore the form of the matrices $\Gamma_d^{1,2}$, $\Gamma_D^{1,2}$ and M_{ρ_0,ρ_3} remains unchanged.

The charged lepton mass matrix, M_e, and the effective neutrino mass matrices, m_ν, can be written without loss of generality as

$$M_e = K_e^\dagger \begin{pmatrix} 0 & A_e & 0 \\ A'_e & 0 & B_e \\ 0 & B'_e & C_e \end{pmatrix}, \quad m_\nu = P_g \begin{pmatrix} 0 & A_\nu & 0 \\ A_\nu & B_\nu & C_\nu \\ 0 & C_\nu & D_\nu\, e^{i\varphi} \end{pmatrix} P_g^T . \tag{13}$$

Journal of Physics: Conference Series **447** (2013) 012041 doi:10.1088/1742-6596/447/1/012041

Table 3. The neutrino oscillation parameters at 3σ range from Ref. [17] for both normal and inverted hierarchies.

parameters	NH	IH
$\Delta m^2_{21}\ [10^{-5}\,\mathrm{eV}^2]$	7.12 – 8.20	
$\lvert\Delta m^2_{31}\rvert\ [10^{-3}\,\mathrm{eV}^2]$	2.31 – 2.74	2.21 – 2.64
$\sin^2\theta_{12}$	0.27 – 0.37	
$\sin^2\theta_{13}$	0.017 – 0.033	
$\sin^2\theta_{23}$	0.26 – 0.68	0.37 – 0.67

Note that m_ν has two possibilities that correspond to E and E_{12}, in such case the permutation matrix, P_g, that enters in Eq. (13) is respectively the 3×3 identity matrix, P_e, or the matrix P_{12} given by,

$$P_{12} = \begin{pmatrix} 0 & 1 & 0 \\ 1 & 0 & 0 \\ 0 & 0 & 1 \end{pmatrix} . \qquad (14)$$

The constants $(A,\, A',\, B,\, B',\, C)_e$ and $(A,\, B,\, C,\, D)_\nu$ can be taken real and positive and the diagonal phase matrix K_e can be parameterised as $K_e = \mathrm{diag}(e^{i\chi_1}, e^{i\chi_2}, 1)$.

In order to confront the predictions from the charged lepton mass matrices and the effective neutrino mass matrices with the neutrino oscillation data one needs to diagonalize both M_e and m_ν and hence compute the leptonic mixing matrix. The Pontecorvo-Maki-Nakagawa-Sakata (PMNS)[22, 23, 24] matrix, U, is given by

$$U = U_e^T\, P_g\, U_\nu\,, \qquad (15)$$

where U_e and U_ν are the charged leptons and neutrino diagonalizing matrices, respectively.

In the numerical analysis we have used the neutrino oscillation data [17] sketched in Table 3 and the charged-lepton masses at M_Z scale, from Ref. [2],

$$\begin{aligned} m_e(M_Z) &= 0.486661305 \pm 0.000000056 \text{ MeV}\,, \\ m_\mu(M_Z) &= 102.728989 \pm 0.000013 \text{ MeV}\,, \\ m_\tau(M_Z) &= 1746.28 \pm 0.16 \text{ MeV}\,, \end{aligned} \qquad (16)$$

and varied all the input parameters within their allowed range taking into account the appropriate values for the free parameter of Eq (13); for more details see Refs. [2, 6].

We have found that the mass matrix E is compatible with a normal hierarchy spectrum while the mass matrix E_{12} is compatible with inverted hierarchy. For the normal hierarchy case, where the effective neutrino mass matrix has the form E, we found the lightest neutrino mass in the range $m_1 = [0.353; 20.884] \times 10^{-3}$ eV while for the inverted hierarchy we found $m_3 = [2.575; 15.335] \times 10^{-3}$ eV. For more details see Ref. [6].

4. Conclusions

In this work we have studied the phenomenological consequences of the adjoint $\mathsf{SU(5)} \times \mathsf{Z_4}$ in the presence of adjoint fermionic fields. The $\mathsf{Z_4}$ flavour symmetry imposed leads to the NNI form for both up- and down-quark mass matrices. Due to the $\mathsf{SU(5)}$ symmetry the charged lepton mass matrix has also NNI form.

In this model, neutrinos get light masses through type-I, type-III and radiative seesaw mechanisms mediated, respectively, by the ρ_0, ρ_3 and ρ_8 components of the 24 fermionic fields. It is important to note that the contribution from radiative seesaw does not change the form of the effective neutrino mass matrix; details on radiative seesaw can be found in Ref. [6].

Since the adjoint fermionic fields are not constrained by any symmetry gauge, their Z_4 charges are free leading to several different patterns for the effective neutrino mass matrices. We have studied all these patterns and showed that at least three 24 fermionic fields were required in order to reproduce a light neutrino mass spectrum compatible with the actual oscillation data. Furthermore, we found that only the mass matrices E and E_{12} in Table 1 are phenomenologically viable and compatible with a normal and inverted neutrino mass spectrum, respectively.

Acknowledgments
I would like to thank the organizers of DISCRETE2012 for the opportunity to present my work. I would like to thank Juan Antonio Aguilar-Saavedra and Filipe Joaquim for a visit to Universidad de Granada, where this talk was prepared, and the Granada group for their great hospitality. This work was supported by Fundação para a Ciência e a Tecnologia (FCT, Portugal) under the contract SFRH/BD/61623/2009 and the project PEst-OE/FIS/UI0777/2011.

Appendix

The singular seesaw [18, 19, 20, 21] applies when the determinant of the Majorana matrix, M_R, that enters in the usual seesaw formula $m_\nu = -m_D M_R^{-1} m_D^T$ is zero; in that case M_R^{-1} is not defined.

The idea of the singular seesaw is to "promote" the heavy neutrino fields with zero mass into light neutrinos and then apply the usual seesaw formula to the new matrices thus obtained.

Let us consider, just as example, the case of two 24 fermionic representations and the Dirac, m_D, and the Majorana, M_R, matrices given by,

$$M_D = \begin{pmatrix} 0 & A \\ B & 0 \\ 0 & 0 \end{pmatrix} \quad \text{and} \quad M_R = \begin{pmatrix} 0 & 0 \\ 0 & C \end{pmatrix}, \tag{17}$$

The effective neutrino matrix is of the following form,

$$M_\nu = \begin{pmatrix} 0 & 0 & 0 & 0 & A \\ 0 & 0 & 0 & B & 0 \\ 0 & 0 & 0 & 0 & 0 \\ 0 & B & 0 & 0 & 0 \\ A & 0 & 0 & 0 & C \end{pmatrix}, \tag{18}$$

the light neutrino mass matrix is

$$m_\nu = \mathsf{m_L} - \mathsf{m_D}\,\mathsf{M_R}^{-1}\,\mathsf{m_D}^{\mathsf{T}}, \tag{19}$$

where the matrices $\mathsf{m_L}$, $\mathsf{m_D}$ and $\mathsf{M_R}$ are given by

$$\mathsf{m_L} = \begin{pmatrix} 0 & 0 & 0 & 0 \\ 0 & 0 & 0 & B \\ 0 & 0 & 0 & 0 \\ 0 & B & 0 & 0 \end{pmatrix}, \quad \mathsf{m_D} = \begin{pmatrix} A \\ 0 \\ 0 \\ 0 \end{pmatrix}, \quad \mathsf{M_R} = C. \tag{20}$$

It is worth mentioning that the singular seesaw can be used in the case where there is at least one heavy state with non-zero mass, otherwise $\mathsf{M_R}^{-1}$ in the seesaw formula is not defined.

References

[1] Georgi H and Glashow S 1974 *Phys.Rev.Lett.* **32** 438–441
[2] Emmanuel-Costa D and Simões C 2012 *Phys.Rev.C* **85** 016003 (*Preprint* 1102.3729)
[3] Branco G C, Lavoura L and Mota F 1989 *Phys. Rev. D* **39** 3443
[4] Fileviez Pérez P 2007 *Phys.Rev. D* **76** 071701 (*Preprint* 0705.3589)
[5] Fileviez Pérez P 2007 *Phys. Lett. B* **654** 189–193 (*Preprint* hep-ph/0702287)
[6] Emmanuel-Costa D, Simões C and Tórtola M (*Preprint* 1303.5699)
[7] Minkowski P 1977 *Phys.Lett. B* **67** 421
[8] Yanagida T 1979 *Conf.Proc.* **C7902131** 95 in Proc. of the Workshop on Unified Theory and Baryon Number in the Universe, KEK, March 1979
[9] Gell-Mann M, Ramond P and Slansky R 1979 *Conf.Proc.* **C790927** 315–321 to be published in Supergravity, P. van Nieuwenhuizen & D.Z. Freedman (eds.), North Holland Publ. Co., 1979
[10] Mohapatra R N and Senjanovic G 1980 *Phys.Rev.Lett.* **44** 912
[11] Foot R, Lew H, He X and Joshi G C 1989 *Z.Phys.C* **44** 441
[12] Ma E 1998 *Phys.Rev.Lett.* **81** 1171–1174 (*Preprint* hep-ph/9805219)
[13] Zee A 1980 *Phys.Lett. B* **93** 389
[14] Wolfenstein L 1980 *Nucl.Phys.B* **175** 93
[15] Fileviez Perez P and Wise M B 2009 *Phys. Rev. D* **80** 053006 (*Preprint* 0906.2950)
[16] Kannike K and Zhuridov D V 2011 *JHEP* **07** 102 (*Preprint* 1105.4546)
[17] Forero D, Tórtola M and Valle J 2012 *Phys.Rev. D* **86** 073012 (*Preprint* 1205.4018)
[18] Johnson R, Ranfone S and Schechter J 1986 *Phys.Lett. B* **179** 355
[19] Glashow S L 1991 *Phys.Lett. B* **256** 255–257
[20] Fukugita M and Yanagida T 1991 *Phys.Rev.Lett.* **66** 2705–2707
[21] Allen T, Johnson R, Ranfone S, Schechter J and Valle J 1991 *Mod.Phys.Lett.* **A6** 1967–1976
[22] Pontecorvo B 1957 *Sov. Phys. JETP* **6** 429
[23] Pontecorvo B 1958 *Sov. Phys. JETP* **7** 172–173
[24] Maki Z, Nakagawa M and Sakata S 1962 *Prog. Theor. Phys.* **28** 870–880

Journal of Physics: Conference Series **447** (2013) 012042 doi:10.1088/1742-6596/447/1/012042

Neutrino predictions in a modified Babu-Ma-Valle model

D. V. Forero

1 AHEP Group, Institut de Física Corpuscular – C.S.I.C./Universitat de València
Edificio Institutos de Paterna, Apt 22085, E–46071 Valencia, Spain
2 Departamento de Física and CFTP, Instituto Superior Técnico
Technical University of Lisbon,1049-001 Lisboa, Portugal

E-mail: `david.vanegas@ist.utl.pt`

Abstract. The measurement of the last neutrino mixing angle θ_{13} giving a a non zero value has established a new challenge for the theoretical understanding of the lepton mixing. The use of discrete symmetries to successfully explain that mixing is still possible. As an example, we have modified the so called Babu-Ma-Valle model in such a way that we account for the current neutrino mixing values at $3\,\sigma$. In particular, we have obtained not only compatibility with a $\theta_{13} \neq 0$, but also a non maximal θ_{23}.

1. Introduction

Oscillation mechanism is nowdays an accurate description for the neutrino flavor transitions. Three active neutrino oscillations globally explain almost all neutrino data [1]. Neutrino oscillations implies massive neutrinos and then it becomes one of the evidences of the standard model (SM) limitations. Unfortunately, a theoretical explanation for the origin of the neutrino masses is still elusive. The see-saw mechanism offers an explanation for the smallness of the neutrino mass opening a theoretical framework to study massive neutrinos.

Neutrino oscillations also imply a mixing in the neutrino sector. That mixing pattern, before the determination of the last mixing angle θ_{13}, was well described using some discrete symmetries [2]. After the determination of a non zero mixing angle θ_{13} from reactor experiments, models with discrete symmetries get challenged to take into account the non zero result for the reactor mixing angle without spoiling the predictions for the other two mixing angles.

The Babu-Ma-Valle (BMV) model [3] is a supersymmetrical extension of the SM with an A_4 discrete symmetry that described the neutrino phenomenology before the θ_{13} determination. The goal of this work is to show that a slight modification of the BMV model successfully describes the current neutrino data [1].

The work is organized in the following way: in section 2 we introduce the BMV model and in sections 3 we explore the possible modifications. The numerical results are shown in section 4 and finally we conclude in section 5.

2. BMV model

The Babu-Ma-Valle model (BMV) [3] is a supersymmetric model with A_4 discrete symmetry, where A_4 is broken at some high scale.
A_4 is a discrete non-Abelian group of even permutations of 4 objects, it has $\underline{1}, \underline{1}', \underline{1}''$ and $\underline{3}$ irreducible representation (irrep) and it is the smallest finite group with triplet irrep [2].
The usual quark $\hat{Q}_i = (\hat{u}_i, \hat{d}_i)$, lepton $\hat{L}_i = (\hat{\nu}_i, \hat{e}_i)$, and Higgs $\hat{\phi}_i$ transforms under A_4 as follows:

	$\hat{Q}$	$\hat{L}$	$\hat{u}_1^c, \hat{d}_1^c, \hat{e}_1^c$	$\hat{u}_2^c, \hat{d}_2^c, \hat{e}_2^c$	$\hat{u}_3^c, \hat{d}_3^c, \hat{e}_3^c$	$\hat{\phi}_{1,2}$
A_4	3	3	1	1'	1"	1
Z_3	1	1	ω^2	ω^2	ω^2	1

Then the following heavy quark, lepton, and Higgs **superfields are added**: which are all $SU(2)$

	$\hat{U}$	$\hat{U}^c$	$\hat{D}$	$\hat{D}^c$	$\hat{E}$	$\hat{E}^c$	$\hat{N}^c$	$\hat{\chi}$
A_4	3	3	3	3	3	3	3	3
Z_3	1	1	1	1	1	1	1	ω

singlets and with $\omega = \exp i\, 2\pi/3$.
The superpotential of the BMV model is then given by:

$$\begin{aligned}\hat{W} = \; & M_U \hat{U}_i \hat{U}_i^c + f_u \hat{Q}_i \hat{U}_i^c \hat{\phi}_2 + h_{ijk}^u \hat{U}_i \hat{u}_i^c \hat{\chi}_k + M_D \hat{D}_i \hat{D}_i^c + f_d \hat{Q}_i \hat{D}_i^c \hat{\phi}_2 + h_{ijk}^d \hat{D}_i \hat{d}_i^c \hat{\chi}_k \\ & + M_E \hat{E}_i \hat{E}_i^c + f_e \hat{L}_i \hat{E}_i^c \hat{\phi}_1 + h_{ijk}^e \hat{E}_i \hat{e}_j^c \hat{\chi}_k + \frac{1}{2} M_N \hat{N}_i^c \hat{N}_i^c + f_N \hat{L}_i \hat{N}_i^c \hat{\phi}_2 + \mu \hat{\phi}_1 \hat{\phi}_2 \\ & + \frac{1}{2} M_\chi \hat{\chi}_i \hat{\chi}_i + h_\chi \hat{\chi}_1 \hat{\chi}_2 \hat{\chi}_3\end{aligned} \tag{1}$$

The scalar potential involving χ_i is given by:

$$V = |M_\chi \chi_1 + h_\chi \chi_2 \chi_3|^2 + |M_\chi \chi_2 + h_\chi \chi_3 \chi_1|^2 + |M_\chi \chi_3 + h_\chi \chi_1 \chi_2|^2, \tag{2}$$

which have the supersymmetric solution ($V = 0$)

$$\langle \chi_1 \rangle = \langle \chi_2 \rangle = \langle \chi_3 \rangle = u \tag{3}$$

Considering now the Dirac mass matrix linking (e_i, E_i) to (e_j^c, E_j^c), we have:

$$\mathcal{M}_{eE} = \begin{pmatrix} 0 & 0 & 0 & f_e v_1 & 0 & 0 \\ 0 & 0 & 0 & 0 & f_e v_1 & 0 \\ 0 & 0 & 0 & 0 & 0 & f_e v_1 \\ h_1^e\, u & h_2^e\, u & h_3^e\, u & M_E & 0 & 0 \\ h_1^e\, u & h_2^e\, u\,\omega & h_3^e\, u\,\omega^2 & 0 & M_E & 0 \\ h_1^e\, u & h_2^e\, u\,\omega^2 & h_3^e\, u\,\omega & 0 & 0 & M_E \end{pmatrix} \equiv \begin{pmatrix} 0 & X_1^I \\ X_2 & Y^I \end{pmatrix} \tag{4}$$

where $v_1 = \langle \phi_1^0 \rangle$, with similar forms for the quark mass matrices. After block diagonalization of Eq. (4), the reduced 3×3 Dirac mass matrix for the charged leptons is diagonalized by the

Journal of Physics: Conference Series **447** (2013) 012042 doi:10.1088/1742-6596/447/1/012042

magic matrix U_ω:

$$U_\omega = \frac{1}{\sqrt{3}} \begin{pmatrix} 1 & 1 & 1 \\ 1 & \omega & \omega^2 \\ 1 & \omega^2 & \omega \end{pmatrix} \tag{5}$$

where for $f_e v_1 \ll h_i\, u \ll M_E$ the charge lepton masses are:

$$\tilde{m}_i^2 \simeq \frac{3 f_e^2 v_1^2}{M_E^2} \frac{h_i^{e\,2} u^2}{1 + 3(h_i^e u)^2/M_E^2} \tag{6}$$

In the neutrino sector, after going to the basis where the charged leptons are diagonal, the Majorana mass matrix for spanning (ν_i, N_i^c) is given by:

$$\mathcal{M}_{\nu N} = \begin{pmatrix} 0 & f_N v_2\, U_\omega \\ f_N v_2\, U_\omega^T & M_N \end{pmatrix} \tag{7}$$

where $v_2 = \langle \phi_2^0 \rangle$. Hence, the see-saw mass matrix for ν_i becomes:

$$\mathcal{M}_\nu = \frac{f_N^2\, v_2^2}{M_N} U_\omega^T U_\omega = m_o \begin{pmatrix} 1 & 0 & 0 \\ 0 & 0 & 1 \\ 0 & 1 & 0 \end{pmatrix} = m_o\, \lambda \tag{8}$$

and the neutrino masses are degenerate at this stage.
Coming down to the electroweak scale, Eq. (8) is corrected by the wave function renormalizations of ν_i, as well as the corresponding vertex renormalizations [3]. Given the structure of the λ_{ij} at the high scale (Eq. (8)), its form at low scale is necessarily fixed to first order as:

$$\lambda = \begin{pmatrix} 1 + 2\delta_{ee} & \delta_{e\mu} + \delta_{e\tau} & \delta_{e\mu} + \delta_{e\tau} \\ \delta_{e\mu} + \delta_{e\tau} & 2\delta_{\mu\tau} & 1 + \delta_{\mu\mu} + \delta_{\tau\tau} \\ \delta_{e\mu} + \delta_{e\tau} & 1 + \delta_{\mu\mu} + \delta_{\tau\tau} & 2\delta_{\mu\tau} \end{pmatrix} \tag{9}$$

where all parameters are assumed to be real [3].
Rewriting Eq. (8) with $\delta_0 \equiv \delta_{\mu\mu} + \delta_{\tau\tau} - 2\delta_{\mu\tau}$,$\delta \equiv 2\delta_{\mu\tau}$, $\delta' \equiv \delta_{ee} - \delta_{\mu\mu}/2 - \delta_{\tau\tau}/2$, and $\delta'' \equiv \delta_{e\mu} + \delta_{e\tau}$. Then

$$\begin{pmatrix} 1 + \delta_0 + 2\delta + 2\delta' & \delta'' & \delta'' \\ \delta'' & \delta & 1 + \delta_0 + \delta \\ \delta'' & 1 + \delta_0 + \delta & \delta \end{pmatrix}, \tag{10}$$

and eigenvectors and eigenvalues are computed *exactly*. The effective neutrino mixing matrix is given by:

$$U_\nu(\theta) = \begin{pmatrix} \cos\theta & -\sin\theta & 0 \\ \sin\theta/\sqrt{2} & \cos\theta/\sqrt{2} & -1/\sqrt{2} \\ \sin\theta/\sqrt{2} & \cos\theta/\sqrt{2} & 1/\sqrt{2} \end{pmatrix}, \tag{11}$$

with the eigenvalues:

$$\begin{aligned} \lambda_1 &= 1 + \delta_0 + 2\delta + \delta' - \sqrt{\delta'^2 + 2\delta''^2} \\ \lambda_2 &= 1 + \delta_0 + 2\delta + \delta' + \sqrt{\delta'^2 + 2\delta''^2} \\ \lambda_3 &= -1 - \delta_0 \end{aligned} \tag{12}$$

The neutrino predictions of the BMV model are:

$$\begin{aligned} \tan^2\theta_{12} &= \frac{\delta''^2}{\delta''^2 + \delta'^2 - \delta'\sqrt{\delta'^2 + 2\delta''^2}} \\ \sin^2\theta_{13} &= 0 \\ \tan^2\theta_{23} &= 1 \Rightarrow \text{maximal} \end{aligned} \tag{13}$$

for the mixing angles. With $\delta' < 0$ and $|\delta''/\delta'| = 1.7$ the BMV predicted the neutrino mixing before, Daya Bay [4] and RENO results [5] for θ_{13}. For the mass square differences, assuming $\delta', \delta'' \ll \delta$, we have:

$$\begin{aligned} \Delta m_{31}^2 &\simeq \Delta m_{32}^2 \simeq 4\delta m_0^2 \\ \Delta m_{21}^2 &\simeq 4\sqrt{\delta'^2 + 2\delta''^2} m_0^2 \end{aligned} \tag{14}$$

The radiative corrections are then responsible for the solar prediction and the splitting of the neutrino mass degeneracy.
The main goal of this letter is to modify the BMV to accommodate the current neutrino data [1]. In general, the mixing in the leptonic sector is given by:

$$K = (U_\omega)^\dagger U_\nu(\theta) = U_\nu(\theta) \tag{15}$$

when we go to the basis were the charged lepton mass matrix is diagonal. New predictions in the neutrino sector will result through modifications of the mixing in the charged sector $U_\omega^c = U_\omega . \hat{\delta}$, in such a way that:

$$K^c = \hat{\delta}^\dagger U_\nu(\theta), \tag{16}$$

where $\hat{\delta}$ matrix parameterize the modifications to the mixing respect to the BMV predictions.

3. Modifications to the BMV model

3.1. Unitarity Violation

Relaxing the condition used to obtained the charged lepton masses Eq. (6), i.e, allowing M_E scale (see Eq. (4)) to be at TeV will produce unitarity violation and therefore, next to the leading order terms in the diagonalization becomes important. The non unitarity effect will enhance the charge lepton flavor violating processes and we will have to carefully fulfill the current bounds.
We used the Schechter-Valle procedure [6] for the block diagonalization:

$$U = \mathcal{U} \cdot V = \exp(iH) \cdot V \qquad H = \begin{pmatrix} 0 & S \\ S^\dagger & 0 \end{pmatrix}, \qquad V = \begin{pmatrix} V_1 & 0 \\ 0 & V_2 \end{pmatrix}, \tag{17}$$

where H is an anti-hermitian operator and V_i are unitary matrices which diagonalized each block. The S matrix is determined at the first order form the diagonalization condition $U^\dagger M U = \text{Diag}\{m_i\}$ for a given hermitian matrix M.
For a general hermitian matrix M:

$$M = \begin{pmatrix} m_1 & m_2 \\ m_2^\dagger & m_3 \end{pmatrix} = \begin{pmatrix} (f_e v_1)^2 I & M_E f_e v_1 I \\ M_E f_e v_1 I & U_\omega(\text{Diag}\{3(h_i^e u)^2\})U_\omega^\dagger + M_E^2 I \end{pmatrix} \tag{18}$$

the S is given by:

$$i S = -m_2(m_1 - m_3)^{-1} = U_\omega \,\text{diag}\{-M_E f_e v_1[(f_e v_1)^2 - 3(h_i^e u)^2 - M_E^2]^{-1}\} U_\omega^\dagger \tag{19}$$

where the second terms in Eq (18) and Eq (19) correspond to our specific case.
To calculate the next to the leading order terms, we have to expand the exponential in Eq. (17) around S. The next to the leading order terms are combinations of the Identity and products of $S/, S^\dagger$ and S. Given the structure of the S matrix in Eq. (19) is clear that even if we go to higher orders in the expansion, the effective charged lepton mass will always be diagonalized by the magic matrix U_ω. The origin of the structure of the S matrix in Eq. (19) comes from the fact that in the BMV model, the matrices in the upper right corner and the lower right corner in Eq. (4) are proportional to the identity. At the end, allowing for unitarity violation in the charge sector does not changes the lepton mixing. Somehow a remnant symmetry of the A_4 ($\mu - \tau$ symmetry) is still present and therefore, there are no corrections or equivalently in Eq. 16 $\hat{\delta} = I$ and we end with the same predictions of the BMV model.

Journal of Physics: Conference Series **447** (2013) 012042 doi:10.1088/1742-6596/447/1/012042

3.2. Adding a Scalar Singlet $1'$ *of* A_4

To break the remnant symmetry present in the charged sector, we add [7] an scalar singlet $1'$ of A_4 (flavon) ζ, end we still allow the M_E scale be decreased until the TeV. Therefore, we add the following term to the superpotential:

$$+\zeta_{ii} E_i E_i^c \tag{20}$$

where we will parameterize the flavon scale as $\langle \zeta \rangle = \beta\, M_E$.
The new mass matrix for the lower right corner of Eq. (4) has now the structure:

$$\hat{Y}_D = M_E \times I + \beta M_E \times \text{Diag}\{1, \omega, \omega^2\}, \tag{21}$$

and the M matrix in Eq. (18) is therefore given by:

$$M = \begin{pmatrix} (f_e v_1)^2\, I & f_e v_1\, \hat{Y}_D^\dagger \\ f_e v_1\, \hat{Y}_D & U_\omega(\text{Diag}\{3(h_i^e u)^2\})U_\omega^\dagger + \hat{Y}_D \hat{Y}_D^\dagger \end{pmatrix} \tag{22}$$

where $\hat{Y}_D$ can not be diagonalized by the magic matrix, breaking the remnant symmetry, and changing structure of the S matrix in Eq. (19). As we will see in the next section, the total effect will be a change in the lepton mixing.

4. Numerical results

Using the modifications to the BMV model explained in section 3.2, we numerically diagonalized the charge mixing matrix in Eq. (22) therefore, including all the higher order contributions to the effective charged lepton mass.

After fitting the three lepton masses, we finally calculated the leptonic mixing in Eq. (15) and then re-calculated the new predictions for the neutrino mixing:

$$\begin{aligned} \tan\theta_{12} &= |K_{1,2}^c(\theta)|/|K_{1,1}^c(\theta)| \\ \sin\theta_{13} &= |K_{1,3}^c(\theta)| \\ \tan\theta_{23} &= |K_{2,3}^c(\theta)|/|K_{3,3}^c(\theta)| \end{aligned} \tag{23}$$

where the free parameter θ have been varied randomly as the scales M_E and $f_e v_1$.
The results are presented in Fig. 1. As we can see from the right plot of Fig. 1, there is a correlation between the magnitude and the phase of the β parameter. In order to generate points compatible with the current neutrino values for the θ_{13} and θ_{23} mixing angles (compatible points), β must be complex and different from zero.
We have generated not only θ_{13} different from zero, but also a non-maximal θ_{23}. The region for those compatible points is bigger for θ_{23} in the second octant than for the first one, as is shown in the left plot of Fig. 1.

5. Conclusions

We have showed that decreasing the M_E scale to the TeV in the BMV model and therefore adding NLO terms in the diagonalization of charged sector does not change the lepton mixing pattern.
To break the remnant symmetry we have introduced a scalar singlet keeping the M_E scale to the TeV. The final effect was a good description of the current neutrino mixing angles at $3\,\sigma$ for some values of θ and β. We still have to check if the corrections in the neutrino sector (δ' and δ'') can be obtained in some specific model.
Our results are preliminary. We still have to check that our results are compatible with lepton flavor violating bounds [7].

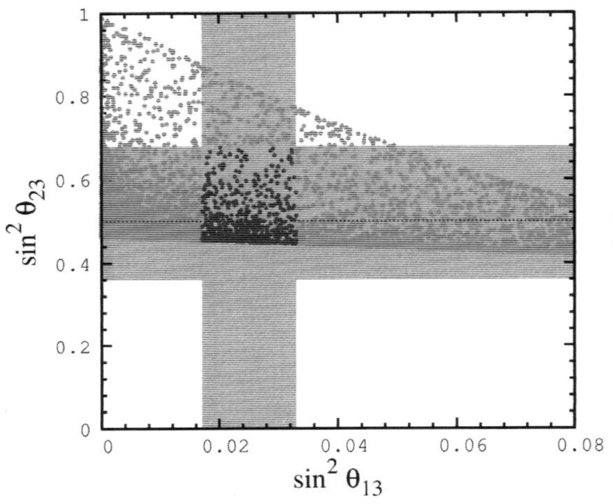

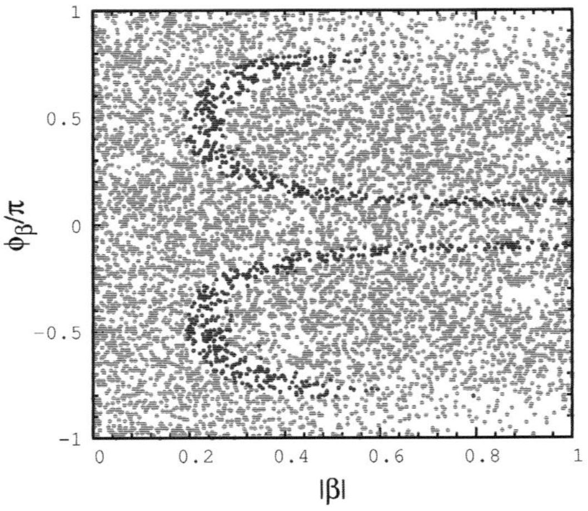

Figure 1. Vertical(horizontal) bands are the current values for $\sin^2\theta_{13}$ ($\sin^2\theta_{23}$) at 3σ. The horizontal dotted line represents maximal θ_{23}. All points are compatible with the current 3σ range of the solar angle θ_{12}, but only the green ones are compatible with the θ_{13} and θ_{23} range at 3σ. (Left) Prediction of the model according to values for β parameter.(Right) Random values for β parameter, phase ϕ_β against magnitude $|\beta|$.

Acknowledgments

The author is grateful to S. Morisi, J. C. Romão and J. W. F. Valle for their collaboration in this work. This work was supported by the EU ITN UNILHC PITN-GA-2009-237920.

References

[1] Forero D, Tortola M and Valle J 2012 *Phys.Rev.* **D86** 073012 (*Preprint* 1205.4018)

[2] Ishimori H, Kobayashi T, Ohki H, Shimizu Y, Okada H *et al.* 2010 *Prog.Theor.Phys.Suppl.* **183** 1–163 (*Preprint* 1003.3552)

[3] Babu K, Ma E and Valle J 2003 *Phys.Lett.* **B552** 207–213 (*Preprint* hep-ph/0206292)

[4] An F *et al.* (DAYA-BAY Collaboration) 2012 *Phys.Rev.Lett.* **108** 171803 (*Preprint* 1203.1669)

[5] Ahn J *et al.* (RENO collaboration) 2012 *Phys.Rev.Lett.* **108** 191802 (*Preprint* 1204.0626)

[6] Schechter J and Valle J 1982 *Phys.Rev.* **D25** 774

[7] *D. V. Forero, S. Morisi, J. C. Romão and J.W.F Valle. Work in preparation.*

DISCRETE 2012 – Third Symposium on Prospects in the Physics of Discrete Symmetries IOP Publishing
Journal of Physics: Conference Series **447** (2013) 012043 doi:10.1088/1742-6596/447/1/012043

A Model for Large θ_{13} Constructed using the Eigenvectors of the S_4 Rotation Matrices

R Krishnan
Department of Physics, University of Warwick, Coventry CV4 7AL, UK
E-mail: k.rama@warwick.ac.uk

Abstract. A procedure for using the eigenvectors of the elements of the representations of a discrete group in model building is introduced and is used to construct a model that produces a large reactor mixing angle, $\sin^2\theta_{13} = \frac{2}{3}\sin^2\frac{\pi}{16}$, in agreement with recent neutrino oscillation observations. The model fully constrains the neutrino mass ratios and predicts normal hierarchy with the light neutrino mass, $m_1 \approx 25$ meV. Motivated by the model, a new mixing ansatz is postulated which predicts all the mixing angles within 1σ errors.

1. Introduction

We use the group $SU(3)$ and its discrete subgroup S_4 for model building. Both of these groups had been studied extensively as flavour symmetry groups. The S_4 group has the presentation

$$\langle a, b | a^2 = b^3 = (ab)^4 = e \rangle. \tag{1}$$

S_4 is the symmetry group of the cube (Fig. 1) and the elements of the group can be represented as the orientation preserving rotations of the cube. The matrices representing the generators can be written as

$$a = \begin{pmatrix} 0 & 0 & 1 \\ 0 & -1 & 0 \\ 1 & 0 & 0 \end{pmatrix}, \quad b = \begin{pmatrix} 0 & 0 & 1 \\ 1 & 0 & 0 \\ 0 & 1 & 0 \end{pmatrix}. \tag{2}$$

Here the basis vectors $e_1 = (1,0,0)^T$, $e_2 = (0,1,0)^T$ and $e_3 = (0,0,1)^T$ form the symmetry axes of the cube passing through face centres. If we define the left-handed leptons, $L = (L_e, L_\mu, L_\tau)^T$ where $L_e = (\nu_{eL}, e_L)^T$ etc., as a triplet in this basis, the flavour states L_e, L_μ, L_τ correspond to e_1, e_2, e_3 respectively. Usually in models a set of flavons are introduced whose vacuum expectation values (vevs) produce the desired texture for the fermion mass matrices. In other words, the orientation of fermion flavour states as well as the flavon vevs in the flavour space determines the form of the mass matrices.

Axes of the orientation preserving rotations of the cube are nothing but eigenvectors of the corresponding rotation matrices with eigenvalue equal to +1. The basis vectors e_1, e_2 and e_3 are examples. There are also other vectors like the ones passing through the opposite edge centres (e.g. axis$_a$ in Fig. 1) and the ones passing through the opposite vertices (e.g. axis$_b$ in Fig. 1). Compared to vectors pointing in random directions, these vectors are "special" in the context of the S_4 symmetry. The rotation matrices are unitary and so their eigenvalues in general are complex numbers with unit modulus. If non-degenerate, these eigenvalues also correspond to unique eigenvectors and the author argues that these eigenvectors are also "special" like the rotation axes.

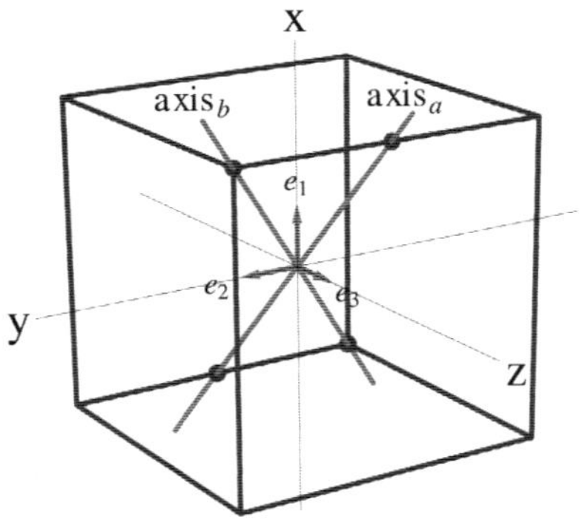

Figure 1. The generators a and b represent π-rotation about axis$_a$ and $\frac{2\pi}{3}$-rotation about axis$_b$ respectively.

As an example consider $v = \frac{1}{\sqrt{3}}(1, \bar{\omega}, \omega)^T$, the normalised ($v^\dagger v = 1$) eigenvector of the matrix b in Eq. (5), corresponding to the eigenvalue ω, where $\omega = e^{i\frac{2\pi}{3}}$ and $\bar{\omega} = e^{-i\frac{2\pi}{3}}$. Since $e^{i\theta}v$, where $e^{i\theta}$ is an arbitrary phase, is also a normalised eigenvector, we impose the following condition to uniquely fix the phase: The component of the eigenvector in the direction of one of the basis vectors should have zero phase ie. $\arg(v^\dagger e_i) = 0$, where $i = 1$, 2 or 3. This is intuitive since the basis vectors are used to define the fermion flavour states. Thus the allowed choices for v are $\frac{1}{\sqrt{3}}(1, \bar{\omega}, \omega)^T$, $\frac{1}{\sqrt{3}}(\omega, 1, \bar{\omega})^T$ and $\frac{1}{\sqrt{3}}(\bar{\omega}, \omega, 1)^T$.

Let g represent an element of the group and l be one of its non-degenerate eigenvalues. The corresponding "special" eigenvector, $\mathrm{eig}(g, l)_i$, is defined using the following normalisation condition and the phase condition:

$$\mathrm{eig}(g, l)_i^\dagger \, \mathrm{eig}(g, l)_i = 1, \quad \arg\left(\mathrm{eig}(g, l)_i^\dagger \, e_i\right) = 0. \tag{3}$$

For example, the basis vector e_3 is $\mathrm{eig}(ab, 1)_3$. Such eigenvectors which will be used later in the model are listed below:

$$\begin{aligned}
\mathrm{eig}(b, 1)_1 &= \tfrac{1}{\sqrt{3}}\,(1, 1, 1)^T, & \mathrm{eig}(a, 1)_1 &= \tfrac{1}{\sqrt{2}}\,(1, 0, 1)^T, \\
\mathrm{eig}(b, \omega)_1 &= \tfrac{1}{\sqrt{3}}\,(1, \bar{\omega}, \omega)^T, & \mathrm{eig}(c, i)_1 &= \tfrac{1}{\sqrt{2}}\,(1, 0, i)^T, \\
\mathrm{eig}(b, \bar{\omega})_1 &= \tfrac{1}{\sqrt{3}}\,(1, \omega, \bar{\omega})^T, & \mathrm{eig}(c, i)_3 &= \tfrac{1}{\sqrt{2}}\,(-i, 0, 1)^T, \\
\mathrm{eig}(d, 1)_1 &= \tfrac{1}{\sqrt{2}}\,(1, 0, -1)^T, & \mathrm{eig}(c, -i)_1 &= \tfrac{1}{\sqrt{2}}\,(1, 0, -i)^T, \\
\mathrm{eig}(d, 1)_3 &= \tfrac{1}{\sqrt{2}}\,(-1, 0, 1)^T, & \mathrm{eig}(c, -i)_3 &= \tfrac{1}{\sqrt{2}}\,(i, 0, 1)^T
\end{aligned} \tag{4}$$

where

$$c = bab = \begin{pmatrix} 0 & 0 & 1 \\ 0 & 1 & 0 \\ -1 & 0 & 0 \end{pmatrix}, \quad d = a(bab)^2 = \begin{pmatrix} 0 & 0 & -1 \\ 0 & -1 & 0 \\ -1 & 0 & 0 \end{pmatrix}. \tag{5}$$

Journal of Physics: Conference Series **447** (2013) 012043 doi:10.1088/1742-6596/447/1/012043

2. The Model

Daya Bay reactor neutrino experiment [1] showed that the reactor mixing angle, θ_{13}, is non-zero. In this paper we propose a flavour model to accommodate the observed non-zero θ_{13}. We use the Standard Model (SM) framework with the addition of the right-handed neutrino triplet, $\nu_R = (\nu_{1R}, \nu_{2R}, \nu_{3R})^T$ in the context of the type-1 seesaw mechanism. We postulate a global flavour symmetry group,

$$G_f = SU(3)_1 \times SU(3)_2 \times U(1)_f. \tag{6}$$

The fermion fields and a set of postulated flavons belong to specific representations of G_f as shown in Table 1. The $U(1)$ group is introduced to ensure that the flavons couple to only the desired fermions. We write the mass terms at the lowest order, containing the fermions and the minimum number of flavons, invariant under G_f and the SM gauge group. The eigenvectors of the elements of the S_4 subgroup of the $SU(3)$ group are used to construct the vevs of the flavons. The vevs break the flavour symmetry and the required mass matrices are obtained.

	L	e_R	μ_R	τ_R	ν_R	ϕ_e	ϕ_μ	ϕ_τ	ϕ	ξ_1	ξ_2
$SU(3)_1$	$\mathbf{3}$	$\mathbf{1}$	$\mathbf{1}$	$\mathbf{1}$	$\bar{\mathbf{3}}$	$\mathbf{3}$	$\mathbf{3}$	$\mathbf{3}$	$\mathbf{3}$	$\mathbf{6}$	$\mathbf{1}$
$SU(3)_2$	$\mathbf{1}$	$\mathbf{1}$	$\mathbf{1}$	$\mathbf{1}$	$\mathbf{1}$	$\mathbf{1}$	$\mathbf{1}$	$\mathbf{1}$	$\mathbf{3}$	$\mathbf{1}$	$\bar{\mathbf{6}}$
$U(1)_f$	f_l	$f_l + f_e$	$f_l + f_\mu$	$f_l + f_\tau$	0	$-f_e$	$-f_\mu$	$-f_\tau$	0	f_l	0

Table 1. The fields ϕ_e, ϕ_μ, ϕ_τ, ϕ, ξ_1, ξ_2 are the flavons. For the $U(1)$ group the tabulated values are the generators, e.g. $f_l \equiv e^{i f_l \theta}$. The SM Higgs is a flavour singlet.

The tensor product expansion of the fundamental representations of $SU(3)$ are

$$\mathbf{3} \otimes \mathbf{3} = \bar{\mathbf{3}} \oplus \mathbf{6}, \quad \bar{\mathbf{3}} \otimes \bar{\mathbf{3}} = \mathbf{3} \oplus \bar{\mathbf{6}}, \quad \mathbf{3} \otimes \bar{\mathbf{3}} = \mathbf{1} \oplus \mathbf{8}. \tag{7}$$

For the charged leptons, the lowest order mass term is

$$\overline{L}\left(y_e \phi_e e_R + y_\mu \phi_\mu \mu_R + y_\tau \phi_\tau \tau_R\right) H \frac{1}{\Lambda_{\phi_l}} + H.C. \tag{8}$$

where y_α are the coupling constants, H is the SM Higgs, Λ_{ϕ_l} is the cut-off scale for the flavons ϕ_e, ϕ_μ and ϕ_τ. The S_4 group has four irreducible representations: **1**, **2**, **3** and **3′** [2]. The orientation preserving rotations of the cube discussed earlier belong to **3′**. The flavon triplets ϕ_e, ϕ_μ and ϕ_τ belong to the representation **3** of $SU(3)_1$. The restriction of the representation **3** of $SU(3)$ to its subgroup S_4 is the representation **3′** of S_4. Therefore the "special" eigenvectors of the representation matrices of **3′** of S_4 are used to construct the vevs of the flavons. We assign:

$$\langle \phi_e \rangle = \text{eig}(b, 1)_1, \quad \langle \phi_\mu \rangle = \text{eig}(b, \omega)_1, \quad \langle \phi_\tau \rangle = \text{eig}(b, \bar{\omega})_1 \tag{9}$$

where the angular brackets are used to denote vevs. We do not discuss the mechanism of flavour symmetry breaking in this paper. To avoid Goldstone bosons, it is necessary to add explicit symmetry breaking terms for the flavon potentials, which break the continuous flavour group G_f, Eq. (6), into an unknown discrete group. The flavon vevs spontaneously break this discrete flavour symmetry. Also the Higgs vev, $(0, h_o)^T$, breaks the weak gauge symmetry. After the symmetry breaking, the charged-lepton mass term, Eq. (8), takes the form

$$\overline{l_L} T^\dagger M_d l_R + H.C \tag{10}$$

where

$$\mathcal{T}^\dagger = \frac{1}{\sqrt{3}} \begin{pmatrix} 1 & 1 & 1 \\ 1 & \bar{\omega} & \omega \\ 1 & \omega & \bar{\omega} \end{pmatrix}, \tag{11}$$

$l_L = (e_L, \mu_L, \tau_L)^T$, $l_R = (e_R, \mu_R, \tau_R)^T$ and $M_d = \text{diag}(m_e, m_\mu, m_\tau)$ with $m_e = \frac{y_e h_o}{\Lambda_{\phi e}}$ etc. The charged-lepton mass matrix, $\mathcal{T}^\dagger M_d$, when left-multiplied with $\mathcal{T}$, is diagonalised giving the charged-lepton masses m_e, m_μ, m_τ. $\mathcal{T}$ is the Trimaximal mixing matrix [3]. For later use we define 2×2 maximal matrices, $\mathcal{B}_2$ and $\mathcal{B}_3$:

$$\mathcal{B}_2 = \begin{pmatrix} \frac{1}{\sqrt{2}} & 0 & \frac{-1}{\sqrt{2}} \\ 0 & 1 & 0 \\ \frac{1}{\sqrt{2}} & 0 & \frac{1}{\sqrt{2}} \end{pmatrix}, \qquad \mathcal{B}_3 = \begin{pmatrix} \frac{1}{\sqrt{2}} & \frac{-1}{\sqrt{2}} & 0 \\ \frac{1}{\sqrt{2}} & \frac{1}{\sqrt{2}} & 0 \\ 0 & 0 & 1 \end{pmatrix}. \tag{12}$$

For the neutrinos, the lowest order Dirac mass term is

$$2y_w \overline{L} \xi_1 \nu_R \tilde{H} \tfrac{1}{\Lambda_{\xi_1}} + H.C. \tag{13}$$

where $\tilde{H}$ is the conjugate Higgs and y_w is the coupling. The flavon ξ_1 belongs to the representation **6** of $SU(3)_1$ and can be written as a 3×3 complex-symmetric matrix. The restriction of **6** of $SU(3)$ to the S_4 subgroup is the direct sum of **1**, **2** and **3** of S_4. We assign a very simple choice of vev for ξ_1 where **2** and **3** parts vanish, ie. $\langle \xi_1 \rangle$ becomes the identity when written in the matrix form. After the symmetry breaking, the Dirac mass term, Eq. (13), takes the form

$$m_w \left(\overline{\nu_L} \nu_R + \overline{(\nu_R)^c} (\nu_L)^c \right) + H.C. \tag{14}$$

where $\nu_L = (\nu_{eL}, \nu_{\mu L}, \nu_{\tau L})^T$, $m_w = \frac{y_w h_o}{\Lambda_{\xi_1}}$.

The lowest order Majorana mass term for the neutrinos is

$$y_G \overline{(\nu_R)^c} \phi \xi_2 \phi^T \nu_R \tfrac{1}{\Lambda_{\xi_2} \Lambda_\phi^2} + H.C. \tag{15}$$

Note that ϕ transforms as a **3** under both $SU(3)_1$ and $SU(3)_2$. Therefore ϕ can be written as a 3×3 matrix, ϕ_{ij}, the row index i representing $SU(3)_1$ and the column index j representing $SU(3)_2$. The flavon ξ_2 belongs to the representation $\bar{\mathbf{6}}$ of $SU(3)_2$. A $\bar{\mathbf{6}}$ of $SU(3)$, just like a **6**, contains a **1**, a **2** and a **3** of the S_4 subgroup. As was done earlier for the case of $\langle \xi_1 \rangle$, here we assign $\langle \xi_2 \rangle$ also to be equal to the identity. After the symmetry breaking, the Majorana mass term, Eq. (15), takes the form

$$m_G \overline{(\nu_R)^c} \langle \phi \rangle \langle \phi \rangle^T \nu_R + H.C. \tag{16}$$

where $m_G = \frac{y_G}{\Lambda_{\xi_2} \Lambda_\phi^2}$. The matrix $\langle \phi \rangle \langle \phi \rangle^T$ is complex-symmetric and $\langle \phi \rangle \langle \phi \rangle^T$ contains all the interesting physics in our model. The reason why we use two $SU(3)$s in the flavour group G_f, Eq. (6), is to ensure that the mass matrix, Eq. (16), contains the symmetric product of two $\langle \phi \rangle$s.

To assign vev for the flavon ϕ_{ij}, we use $(S_4)_1 \times (S_4)_2$, the subgroup of $SU(3)_1 \times SU(3)_2$. The group $(S_4)_1 \times (S_4)_2$ has $24 \times 24 = 576$ elements. Let g_1 and g_2 be the elements of $(S_4)_1$ and $(S_4)_2$ respectively. If v_1 and v_2 are the eigenvectors of g_1 and g_2 corresponding to the eigenvalues a_1 and a_2, then the direct product $v_1 \times v_2$ will be an eigenvector of $g_1 \times g_2$ with an eigenvalue $a_1 a_2$. Now we make the following assumption:

$$\langle \phi \rangle = v_1^1 \times v_2^1 \; + \; v_1^2 \times v_2^2 \; + \; v_1^3 \times v_2^3 \; + \; v_1^4 \times v_2^4 \tag{17}$$

where the RHS of Eq. (17) is the sum of four eigenvectors. Based on the choices for $v_1^i \times v_2^i$s we get a set of similar cases of solutions described in the following sections. The assumed form of $\langle \phi \rangle$ given in Eq. (17) and the choices for $v_1^i \times v_2^i$s were obtained through educated guesses and also through trial and error to fit the experimental data.

2.1. Case 1

Here we assign

$$\begin{array}{llll} v_1^1 = e_2, & v_1^2 = \mathrm{eig}(a,1)_1, & v_1^3 = \mathrm{eig}(d,1)_1, & v_1^4 = \mathrm{eig}(d,1)_3, \\ v_2^1 = e_2, & v_2^2 = e_1, & v_2^3 = e_1, & v_2^4 = \mathrm{eig}(c,\text{-}i)_1. \end{array} \tag{18}$$

Using Eq. (17), Eqs. (18) and Eqs. (4), we get

$$\langle\phi\rangle = \begin{pmatrix} -\frac{1}{2}+\sqrt{2} & 0 & \frac{i}{2} \\ 0 & 1 & 0 \\ \frac{1}{2} & 0 & -\frac{i}{2} \end{pmatrix} \tag{19}$$

in matrix form, where the row and the column indices of the matrix correspond to the $(S_4)_1$ and the $(S_4)_2$ indices respectively. If $m_G >> m_w$, the Majorana mass matrix, $m_G\langle\phi\rangle\langle\phi\rangle^T$ from Eq. (16), becomes much larger than the Dirac mass matrix, $m_w \times$ Identity from Eq. (14), resulting in the type-1 see-saw mechanism. We get an effective see-saw mass matrix [4], M_{ss}, proportional to the inverse of $\langle\phi\rangle\langle\phi\rangle^T$:

$$M_{ss} = -k\left(\langle\phi\rangle\langle\phi\rangle^T\right)^{-1} = -k\begin{pmatrix} 2-\sqrt{2} & 0 & \frac{1}{\sqrt{2}} \\ 0 & 1 & 0 \\ \frac{1}{\sqrt{2}} & 0 & 0 \end{pmatrix}^{-1} \quad \text{where} \quad k = \frac{m_w^2}{m_G}. \tag{20}$$

M_{ss} is diagonalised using the unitary matrix U_ν:

$$U_\nu^\dagger M_{ss} U_\nu^* = -\mathrm{diag}\,(m_1, m_2, m_3) \tag{21}$$

where the neutrino masses m_1, m_2, m_3 are given by

$$m_1 = \frac{k\,(2+\sqrt{2})}{1+\sqrt{2(2+\sqrt{2})}}, \qquad m_2 = k, \qquad m_3 = \frac{k\,(2+\sqrt{2})}{-1+\sqrt{2(2+\sqrt{2})}} \tag{22}$$

and

$$U_\nu = \mathcal{B}_2\mathcal{I}\mathcal{B}_2\mathcal{E}\mathcal{B}_2^T\mathcal{P} \tag{23}$$

with

$$\mathcal{I} = \mathrm{diag}(1,1,i), \quad \mathcal{E} = \mathrm{diag}(e^{i\frac{\pi}{8}},1,1), \quad \mathcal{P} = \mathrm{diag}(e^{-i\frac{\pi}{16}},1,e^{-i\frac{\pi}{16}}). \tag{24}$$

The PMNS matrix becomes

$$U = \mathcal{T}U_\nu = \mathcal{T}\mathcal{B}_2\mathcal{I}\mathcal{B}_2\mathcal{E}\mathcal{B}_2^T\mathcal{P}. \tag{25}$$

The mixing obtained is a constrained form of the Trichimaximal (TχM) mixing [5] with $\chi = \frac{\pi}{16}$:

$$|U| = |\mathrm{T}\chi\mathrm{M}_{(\chi=\frac{\pi}{16})}| \quad \text{where} \quad \mathrm{T}\chi\mathrm{M} = \begin{pmatrix} \sqrt{\frac{2}{3}}\cos\chi & \frac{1}{\sqrt{3}} & \sqrt{\frac{2}{3}}\sin\chi \\ -\frac{\cos\chi}{\sqrt{6}} - i\frac{\sin\chi}{\sqrt{2}} & \frac{1}{\sqrt{3}} & i\frac{\cos\chi}{\sqrt{2}} - \frac{\sin\chi}{\sqrt{6}} \\ -\frac{\cos\chi}{\sqrt{6}} + i\frac{\sin\chi}{\sqrt{2}} & \frac{1}{\sqrt{3}} & -i\frac{\cos\chi}{\sqrt{2}} - \frac{\sin\chi}{\sqrt{6}} \end{pmatrix}. \tag{26}$$

The modulus sign is used throughout this paper to indicate that the expression for the mixing matrix is valid only upto right and left multiplication with diagonal phase matrices (which do not affect the phenomenon of neutrino oscillation). The right multiplying diagonal phase matrices,

Journal of Physics: Conference Series **447** (2013) 012043 doi:10.1088/1742-6596/447/1/012043

like $\mathcal{P}$ in Eq. (25), do contribute to Majorana phases, the study of which is beyond the scope of this paper. From Eq. (26) and using Eq. (12) in [5] we get

$$|U_{e3}|^2 = \frac{2}{3}\sin^2\frac{\pi}{16} \Rightarrow \sin^2\theta_{13} = 0.025, \tag{27}$$

$$|U_{e2}|^2 = \frac{1}{3} \Rightarrow \sin^2\theta_{12} = 0.342, \tag{28}$$

$$\sin^2\theta_{23} = \frac{1}{2}, \tag{29}$$

$$\delta_{CP} = \frac{\pi}{2}. \tag{30}$$

From Eq. (22) we get the ratios of the neutrino masses, $m_1 : m_2 : m_3 = 0.945 : 1 : 2.117$. These ratios are compatible with the mass-squared differences measured experimentally [6, 7] within 1σ errors and thus we can predict the light neutrino mass:

$$24.7 \text{ meV} \lesssim m_1 \lesssim 25.5 \text{ meV}. \tag{31}$$

Even though the set of eigenvectors given in Eqs. (18) results in the matrix $\langle\phi\rangle$ given in Eq. (19), other choices of eigenvectors also exist which produce the same $\langle\phi\rangle$. The matrix $\mathcal{T}^\dagger\langle\phi\rangle\langle\phi\rangle^T\mathcal{T}^*$ is a highly constrained form of the complex-symmetric "Simplest" texture [8, 9].

Let $x = i^n$ where n is an integer and let

$$\Phi_x = \begin{pmatrix} i\frac{x}{2} + \frac{1-ix}{\sqrt{2}} & 0 & -\frac{x}{2} \\ 0 & 1 & 0 \\ -i\frac{x}{2} + \frac{1+ix}{\sqrt{2}} & 0 & \frac{x}{2} \end{pmatrix}. \tag{32}$$

Using Φ_x, Eq. (19) can be rewritten as

$$\langle\phi\rangle = \Phi_i^*. \tag{33}$$

2.2. *Case 2*

Assigning

$$\begin{aligned} v_1^1 &= e_2, & v_1^2 &= \text{eig}(c, i)_1, & v_1^3 &= \text{eig}(c, i)_3, & v_1^4 &= \text{eig}(d, 1)_1, \\ v_2^1 &= e_2, & v_2^2 &= e_1, & v_2^3 &= e_1, & v_2^4 &= \text{eig}(c, \text{-i})_3. \end{aligned} \tag{34}$$

we get

$$\langle\phi\rangle = \Phi_{i^2}^*. \tag{35}$$

In this case, the resulting PMNS matrix is

$$U = \mathcal{T}U_\nu = \mathcal{T}\mathcal{B}_2\mathcal{I}^2\mathcal{B}_2\mathcal{E}\mathcal{B}_2^T\mathcal{P} \tag{36}$$

and it is a constrained form of the Triphimaximal (TϕM) mixing [5] with $\phi = -\frac{\pi}{16}$:

$$|U| = |\mathrm{T}\phi\mathrm{M}_{(\phi=-\frac{\pi}{16})}| \quad \text{where} \quad \mathrm{T}\phi\mathrm{M} = \begin{pmatrix} \sqrt{\frac{2}{3}}\cos\phi & \frac{1}{\sqrt{3}} & \sqrt{\frac{2}{3}}\sin\phi \\ -\frac{\cos\phi}{\sqrt{6}} - \frac{\sin\phi}{\sqrt{2}} & \frac{1}{\sqrt{3}} & \frac{\cos\phi}{\sqrt{2}} - \frac{\sin\phi}{\sqrt{6}} \\ -\frac{\cos\phi}{\sqrt{6}} + \frac{\sin\phi}{\sqrt{2}} & \frac{1}{\sqrt{3}} & -\frac{\cos\phi}{\sqrt{2}} - \frac{\sin\phi}{\sqrt{6}} \end{pmatrix}. \tag{37}$$

From Eq. (37) and using Eq. (9) in [5] we get

$$|U_{\mu3}|^2 = \frac{2}{3}\sin^2\left(\frac{2\pi}{3} - \frac{\pi}{16}\right) \Rightarrow \sin^2\theta_{23} = 0.613, \tag{38}$$

$$\delta_{CP} = \pi. \tag{39}$$

Here, as well as in the next two cases, Eq. (22) which gives the neutrino masses and Eqs. (27, 28) which give $\sin^2\theta_{13}$ and $\sin^2\theta_{12}$ remain valid.

2.3. Case 3

Assigning $v_1^1 = e_2$, $v_2^1 = e_2$, $v_1^2 = \text{eig}(a,1)_1$, $v_2^2 = e_1$, $v_1^3 = \text{eig}(d,1)_3$, $v_2^3 = e_1$, $v_1^4 = \text{eig}(d,1)_1$ and $v_2^4 = \text{eig}(c,\text{-i})_1$ we get

$$\langle\phi\rangle = \Phi^*_{i3} \tag{40}$$
$$U = \mathcal{T}\mathcal{B}_2\mathcal{I}^3\mathcal{B}_2\mathcal{E}\mathcal{B}_2^T\mathcal{P} \Rightarrow |U| = |\text{T}\chi\text{M}_{(\chi=-\frac{\pi}{16})}|, \tag{41}$$
$$\sin^2\theta_{23} = \frac{1}{2}, \tag{42}$$
$$\delta_{CP} = \frac{3\pi}{2}. \tag{43}$$

2.4. Case 4

Assigning $v_1^1 = e_2$, $v_2^1 = e_2$, $v_1^2 = \text{eig}(c,\text{-}i)_1$, $v_2^2 = e_1$, $v_1^3 = \text{eig}(c,\text{-}i)_3$, $v_2^3 = e_1$, $v_1^4 = \text{eig}(d,1)_3$ and $v_2^4 = \text{eig}(c,\text{-i})_3$ we get

$$\langle\phi\rangle = \Phi^*_{i4} \tag{44}$$
$$U = \mathcal{T}\mathcal{B}_2\mathcal{I}^4\mathcal{B}_2\mathcal{E}\mathcal{B}_2^T\mathcal{P} \Rightarrow |U| = |\text{T}\phi\text{M}_{(\phi=\frac{\pi}{16})}|, \tag{45}$$
$$|U_{\mu3}|^2 = \frac{2}{3}\sin^2\left(\frac{2\pi}{3}+\frac{\pi}{16}\right) \Rightarrow \sin^2\theta_{23} = 0.387, \tag{46}$$
$$\delta_{CP} = 2\pi. \tag{47}$$

The values predicted by all the four cases of the model are within 3σ errors of the experimental best fits [6, 7, 10]. In fact the generic prediction $\sin^2\theta_{13} = 0.025$, Eq. (27), is within 1σ errors. However the global analysis [6] shows more than 2σ tension with $\sin^2\theta_{23} = \frac{1}{2}$, the T$\chi$M value (Cases 1 and 3, Eqs. (29, 42)). On the other hand the TϕM values, $\sin^2\theta_{23} = 0.613$ from Eq. (38) in Case 2 and $\sin^2\theta_{23} = 0.387$ from Eq. (46) in Case 4, are well within 1σ errors calculated in [7] and [6] respectively. All the cases predict $\sin^2\theta_{12} = 0.342$, Eq. (28), which is at the edge of the 2σ error range in [6]. A new mixing ansatz called the VS mixing [1] is proposed in the following section which modifies θ_{12} as well as δ_{CP}.

3. The VS Mixing Ansatz

The mixing obtained using the model, Eqs. (25, 36, 41, 45), is of the form

$$|U| = |\mathcal{T}\mathcal{B}_2\mathcal{I}^n\mathcal{B}_2\mathcal{E}\mathcal{B}_2^T|. \tag{48}$$

The matrix $|\mathcal{T}\mathcal{B}_2\mathcal{I}^n|$ gives the Tribimaximal (TBM) mixing [11]. Multiplying $\mathcal{T}\mathcal{B}_2\mathcal{I}^n$ with $\mathcal{B}_2\mathcal{E}\mathcal{B}_2^T$ mixes the first and the third columns of the TBM matrix. Similarly we may also mix the first and the second columns resulting in the new ansatz defined by

$$|\text{VS}_{i^n}(\alpha)| = |\mathcal{T}\mathcal{B}_2\mathcal{I}^n\mathcal{B}_2\mathcal{E}\mathcal{B}_2^T\mathcal{B}_3\mathcal{E}'\mathcal{B}_3^T| \tag{49}$$

where

$$\mathcal{E}' = \text{diag}(e^{i\alpha}, 1, 1). \tag{50}$$

Note that the Cases 1 to 4 are simply $\text{VS}_{i^n}(0)$ with $n = 1$ to 4 respectively. Eq. (49) on simplification gives

$$|\text{VS}_{i^n}(\alpha)| = |\mathcal{T}\mathcal{B}_2\mathcal{I}^n\mathcal{H}_2\mathcal{S}\mathcal{H}_3'| \tag{51}$$

[1] Dedicated to my father K Venugopal and mother J Saraswathi Amma

Journal of Physics: Conference Series **447** (2013) 012043 doi:10.1088/1742-6596/447/1/012043

where

$$\mathcal{S} = \text{diag}(e^{i\frac{\pi}{16}}, 1, 1), \qquad \mathcal{H}_2 = \begin{pmatrix} \boldsymbol{c} & 0 & i\boldsymbol{s} \\ 0 & 1 & 0 \\ i\boldsymbol{s} & 0 & \boldsymbol{c} \end{pmatrix}, \qquad \mathcal{H}'_3 = \begin{pmatrix} \boldsymbol{c}' & i\boldsymbol{s}' & 0 \\ i\boldsymbol{s}' & \boldsymbol{c}' & 0 \\ 0 & 0 & 1 \end{pmatrix} \tag{52}$$

with

$$\boldsymbol{c} = \cos\frac{\pi}{16}, \qquad \boldsymbol{s} = \sin\frac{\pi}{16}, \qquad \boldsymbol{c}' = \cos\frac{\alpha}{2}, \qquad \boldsymbol{s}' = \sin\frac{\alpha}{2}. \tag{53}$$

We get $\sin^2\theta_{12}$ within 1σ errors for $0.08\pi \lesssim \alpha \lesssim 0.26\pi$. Table 2 lists a few cases of the VS mixing along with the predicted values of the mixing angles. The author finds the choice $\alpha = \frac{\pi}{8}$ to be aesthetically pleasing. When $\alpha = \frac{\pi}{8}$ we get $\mathcal{E}' = \mathcal{E}$ and also $\boldsymbol{c}' = \boldsymbol{c}$, $\boldsymbol{s}' = \boldsymbol{s}$.

	$\text{VS}_{\text{-1}}(\frac{\pi}{4})$	$\text{VS}_1(\frac{\pi}{4})$	$\text{VS}_{\text{-1}}(\frac{\pi}{6})$	$\text{VS}_1(\frac{\pi}{6})$	$\text{VS}_{\text{-1}}(\frac{\pi}{8})$	$\text{VS}_1(\frac{\pi}{8})$
$\sin^2\theta_{23}$	0.613	0.387	0.613	0.387	0.613	0.387
$\sin^2\theta_{12}$	0.323	0.323	0.317	0.317	0.319	0.319
δ_{CP}	1.27π	0.27π	1.18π	0.18π	1.13π	0.13π

Table 2. Note that $\sin^2\theta_{13} = 0.025$ is a generic feature of the VS mixing. Conjugation, $\text{VS}^*_{i^n}(\alpha)$, changes the sign of δ_{CP} without affecting the mixing angles θ_{12}, θ_{23} and θ_{13}.

4. Summary

The symmetries represented by a discrete group are related to the eigenvectors of the group elements. We develop a notation to uniquely identify the eigenvectors and use it to assign vevs for the flavons. An orthonormal set of eigenvectors define the fermions' flavour states. The model thus constructed predicts the reactor mixing angle, $\sin^2\theta_{13} = 0.025$, as well as the ratios of the neutrino masses, $m_1 : m_2 : m_3 = 0.945 : 1 : 2.117$. The T$\phi$M versions of the model provide solutions for θ_{23} in the first octant, $\sin^2\theta_{23} = 0.387$, as well as in the second octant, $\sin^2\theta_{23} = 0.613$. The T$\phi$M as well as the T$\chi$M versions give $\sin^2\theta_{12} = 0.342$. A new mixing ansatz, $\text{VS}_{i^n}(\alpha)$, is introduced which gives reduced values for θ_{12}. The ansatz also predicts various values for δ_{CP}.

Acknowledgments

Acknowledgments I would like to thank Paul Harrison and Bill Scott for helpful discussions. This work was supported by the UK Science and Technology Facilities Council (STFC). I acknowledge support from the University of Warwick and the Centre for Fundamental Physics at the Rutherford Appleton Laboratory.

References

[1] Daya Bay Collaboration 2012 *Phys. Rev. Lett.* **108** 171803 (*Preprint* 1203.1669)
[2] Ishimori H, Kobayashi T, Ohki H, Shimizu Y, Okada H and Tanimoto M 2010 *Prog. Theor. Phys. Suppl.* **183** 1–163 (*Preprint* 1003.3552)
[3] Harrison P F, Perkins D H and Scott W G 1999 *Phys. Lett. B* **458** 79–92 (*Preprint* hep-ph/9904297)
[4] Smirnov A Y 1993 *Phys. Rev. D* **48** 3264–3270 (*Preprint* hep-ph/9304205)
[5] Harrison P F and Scott W G 2002 *Phys. Lett. B* **535** 163–169 (*Preprint* hep-ph/0203209)
[6] Fogli G L, Lisi E, Marrone A, Montanino D, Palazzo A and Rotunno A M 2012 *Phys. Rev. D* **86** 013012 (*Preprint* 1205.5254)
[7] Forero D V, Tortola M and Valle J W F 2012 *Phys. Rev. D* **86** 073012 (*Preprint* 1205.4018)
[8] Harrison P F and Scott W G 2004 *Phys. Lett. B* **594** 324–332 (*Preprint* hep-ph/0403278)
[9] Krishnan R, Harrison P F and Scott W G 2012 (*Preprint* 1211.2000)
[10] Gonzalez-Garcia M C, Maltoni M, Salvado J and Schwetz T 2012 (*Preprint* 1209.3023)
[11] Harrison P F, Perkins D H and Scott W G 2002 *Phys. Lett. B* **530** 167–173 (*Preprint* hep-ph/0202074)

Journal of Physics: Conference Series **447** (2013) 012044 doi:10.1088/1742-6596/447/1/012044

Higgs Boson Searches at ATLAS

R. Gonçalo on behalf of the ATLAS Collaboration

Royal Holloway, University of London

E-mail: jose.goncalo@cern.ch

Abstract. The present article gives an overview of the status of Higgs boson studies in the ATLAS experiment at the LHC, at the time of the DISCRETE 2012 Symposium. Accounts of the latest Higgs boson searches in the $H \to WW^{(*)} \to \ell\nu\ell\nu$, $H \to b\bar{b}$ and $H \to \tau^+\tau^-$ channels are given, as well as a summary of the investigation into the couplings of the new particle, and an indication of the sensitivity of future Higgs studies at the LHC.

1. Introduction

The discovery [1, 2] of a new particle in the search for the Standard Model Higgs boson at the LHC represents a major milestone in the understanding of the mechanism responsible for electroweak symmetry breaking. After this discovery, it is now crucial to establish if the new particle is indeed the Higgs boson predicted by the Standard Model (SM) or whether it could provide evidence for physics beyond the SM. To do that, it is essential to make clear observations of the new particle in all expected channels and measure its properties accurately.

The new particle discovery in ATLAS [3] was based on datasets with integrated luminosities of 4.8 fb^{-1} and 5.8 fb^{-1}, produced at centre of mass energies of $\sqrt{s} = 7$ TeV and $\sqrt{s} = 8$ TeV, respectively [1]. At the LHC, the Higgs boson is produced predominantly through a top quark dominated loop in the gluon-fusion process (ggF). Vector boson fusion (VBF), comes next in cross section, where the Higgs is produced by the fusion of two W or Z bosons, followed by the associated production of the Higgs boson with a single W or Z (VH), or with a pair of top quarks ($t\bar{t}H$). Five decay channels were combined in the discovery of the new particle: $H \to ZZ^{(*)} \to 4\ell$, $H \to \gamma\gamma$, $H \to WW^{(*)} \to \ell\nu\ell\nu$, $H \to \tau^+\tau^-$, and $H \to b\bar{b}$ [1].

Clear observations of the new particle in a single channel have so far been made in analyses focusing on bosonic decays, such as $H \to ZZ^{(*)} \to 4\ell$ and $H \to \gamma\gamma$. Fermionic decay channels such as $H \to \tau^+\tau^-$ and $H \to b\bar{b}$ will play an important role in ascertaining the nature of the new particle. Our present knowledge of the new particle indicates that it is a neutral resonance with a mass in the region of 125 GeV (as observed in the $H \to \gamma\gamma$ and $H \to ZZ^{(*)} \to 4\ell$ channels). The observation in the $H \to \gamma\gamma$ channel disfavours a spin 1 particle [4].

Our knowledge of the coupling properties of the new particle is discussed in section 2. This is followed by updates of the $H \to WW^{(*)} \to \ell\nu\ell\nu$, $H \to \tau^+\tau^-$ and $H \to b\bar{b}$ ATLAS searches, in sections 3, 4, and 5. Finally, section 6 gives an impression of what results can be expected from ATLAS given 300 fb^{-1} and 3000 fb^{-1}of $\sqrt{s} = 14$ TeV collision data.

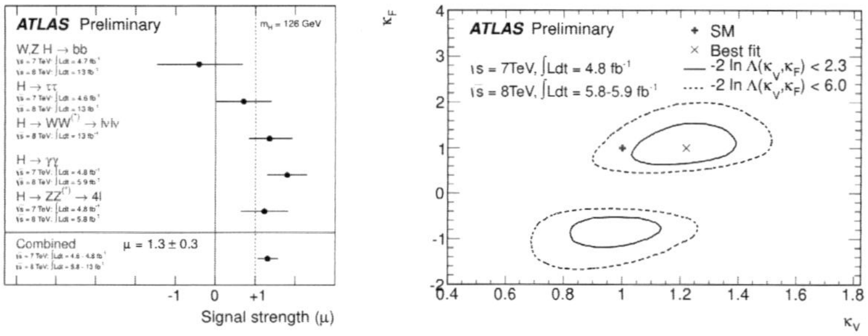

Figure 1. Left: signal strength parameter μ for $m_H = 126$ GeV for the individual channels and their combination [8]. Right: 68% and 95% CL contours for the fit to the fermion (κ_F) and vector (κ_V) coupling scale factors [5].

2. Coupling properties of the new particle

The same data, analyses, and statistical models used for the ATLAS discovery was examined to investigate the couplings of the new particle [5]. It is convenient to define a *signal strength* observable, μ, as the ratio of the observed cross section to the SM expected cross section, $\mu = \sigma/\sigma_{SM}$, in each investigated channel or the statistical combination of several channels. The background-only hypothesis corresponds to $\mu = 0$ and the SM hypothesis including the Higgs boson corresponds to $\mu = 1$. The couplings of the new particle are investigated with the following assumptions [6]: only modifications of coupling strengths are taken into account; the new particle is assumed to be a single, CP-even scalar, as expected of the SM Higgs boson; it is assumed to have a mass of 126 GeV with a negligible width. This allows the product of the cross section and branching ratio in each channel $ii \to H \to ff$ to be decomposed as: $\sigma \times BR = \sigma_{SM}(ii \to H) \times BR_{SM}(H \to ff) \times \frac{\kappa_i^2 \kappa_f^2}{\kappa_H^2}$ where the κ factors multiply the Yukawa couplings of each initial- (i) or final-state (f) particle flavour.

This is a convenient and flexible framework in which to explore the couplings of the new particle. It should be noted, however, that new physics responsible for a deviation of the observed couplings with respect to the SM ones might also modify the kinematics of particles in the observed final states. This is not accounted for by this framework, which is however still assumed to be valid to identify small deviations in the observed couplings with respect to the SM expectations. The fit to the experimental data is done using a profile likelihood method [7].

The simplest model tested in this framework assumes all coupling parameters are modified by a single scaling factor κ, which corresponds to the square root of the global signal strength $\mu = 1.4 \pm 0.3$ [5]. The fit to the data yields $\kappa = 1.19 \pm 0.11(stat) \pm 0.03(syst) \pm 0.06(theory)$. The combined signal strength value has since been updated [8] and is shown in Figure 1 (left).

A different benchmark scenario can be explored by assuming a single scale factor $\kappa_F = \kappa_t = \kappa_b = \kappa_\tau$ for all fermion couplings, and a different one, $\kappa_V = \kappa_W = \kappa_Z$, for vector bosons. In addition to the weak constraints provided by the $H \to b\bar{b}$ and $H \to \tau^+\tau^-$ channels, sensitivity to the fermion coupling comes from the top-quark loop in the gluon-fusion diagram contributing to several channels. Sensitivity to the vector boson coupling parameter is provided by $H \to \gamma\gamma$, $H \to ZZ^{(*)} \to 4\ell$, and $H \to WW^{(*)} \to \ell\nu\ell\nu$ decays. Assuming further that only SM particles contribute to the Higgs boson width, a fit to the coupling strength scale factors can be made. Figure 1 (right) shows the correlation between κ_F and κ_V scale factors. Only the relative sign between κ_F and κ_V is physical, so negative values of the scale factors are shown. Some sensitivity to this sign is found from the negative interference between the W and t loops in $H \to \gamma\gamma$ decays.

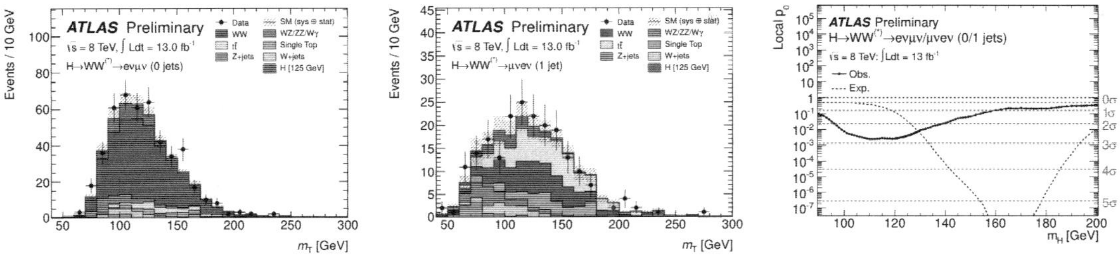

Figure 2. Transverse mass distribution in the $H + 0$-jet category of the $H \to WW^{(*)} \to \ell\nu\ell\nu$ analysis (left) and the $H+1$-jet category (centre) [9]. Right: probability for the background-only hypothesis, p_0, for $H \to WW^{(*)} \to \ell\nu\ell\nu$ combining all analysis categories [9]. The solid line corresponds to the observed p_0, while the dashed line corresponds to the expectation for the signal+background hypothesis as a function of the mass of the hypothesised signal.

The fit shows a small preference for the local minimum near the SM expectation.

3. $H \to WW^{(*)} \to e\nu\mu\nu$ search with $\sqrt{s} = 8$ TeV LHC data

The latest ATLAS $H \to WW^{(*)} \to \ell\nu\ell\nu$ search [9] uses 13 fb^{-1} of $\sqrt{s} = 8$ TeV pp data and updates the previously published results [1]. The high branching ratio of this channel ($BR(H \to WW^{(*)} \to \ell\nu\ell\nu) \approx 22\%$ for m_H around 125 GeV [10]) and the availability of leptonic final states which facilitate triggering and rejecting the main backgrounds make this a very sensitive channel. The updated analysis focuses on the $H \to WW^{(*)} \to e\nu\mu\nu$ channel. In addition to two isolated, opposite charge leptons of different flavour and high transverse momentum (p_T above 15 or 25 GeV), events are required to have a large missing transverse momentum (E_T^{miss}) relative to the direction of the direction of the leading lepton or the closest jet, $E_{T,rel}^{miss}$. Due to spin correlations in the $WW^{(*)}$ system, arising from the spin-0 nature of the SM Higgs boson and the V-A structure of W boson decay, the charged leptons tend to have similar directions for the SM Higgs signal and a flatter distribution for the dominant WW background. This feature is exploited by requiring the di-lepton invariant mass, $m_{\ell\ell}$ to be less than 50 GeV, and the azimuthal angle difference between the two leptons, $\Delta\phi_{\ell\ell}$ less than 1.8 radians.

The background composition depends on the jet multiplicity. Events are therefore classified into $H + 0$-jet and $H + 1$-jet categories, and different cuts are applied to improve the sensitivity in each one. In this and the following analyses, jets are reconstructed using the anti-k_t algorithm [11] with distance parameter R=0.4. Events with higher jet multiplicity are dominated by top quark production and are not considered in the statistical analysis. An important feature of this and later analyses is the data-driven determination of the most important sources of background with the help of control regions enriched in each process. The main backgrounds are WW and top production (both $t\bar{t}$ and single-top), but W+jets and $Z/\gamma^* \to \tau\tau$ are also determined from specific control samples. More details can be found in [9].

The presence of high transverse momentum neutrinos in the signal prevents the reconstruction of the Higgs boson mass. A transverse mass variable, m_T [9], is thus used as the final discriminating variable in the statistical analysis. Figure 2 (left and middle) shows distributions of this variable in the $H + 0$-jet and $H + 1$-jet categories. The statistical analysis of the m_T distribution in each category employs a binned likelihood, $\mathcal{L}(\mu, \theta)$, constructed as the product of Poisson probability terms. This depends on the signal strength, μ, and on a set of nuisance parameters, θ, which affect signal and background predictions [7]. The probability

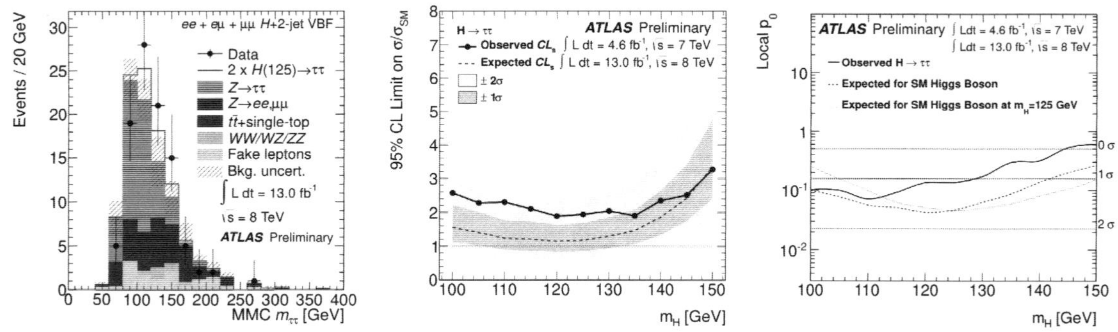

Figure 3. Left [12]: reconstructed $m_{\tau\tau}$ for events selected into the *VBF* category of the $H \to \tau_{lep}\tau_{lep}$ channel. Centre [12]: observed (solid) and expected (dashed) 95% CL limits on the Higgs boson cross section as a function of the Higgs boson mass, normalized to the SM cross section, combining all categories in the $H \to \tau_{lep}\tau_{lep}$, $H \to \tau_{lep}\tau_{had}$, and $H \to \tau_{had}\tau_{had}$ channels. Right [12]: observed (solid) and expected (dashed) local p_0 as a function of the Higgs boson mass. The dotted line corresponds to the expected p_0 curve calculated for the 125 GeV SM Higgs mass hypothesis as a function of the reconstructed Higgs boson mass. The broad minimum in this curve is caused by the poor resolution of the reconstructed $m_{\tau\tau}$.

of the background-only hypothesis, p_0, is calculated in the statistical analysis, as well as the exclusion limits on the signal cross section, following the CL_s method [13]. Figure 2 (right) shows the p_0 value observed in the $H \to WW^{(*)} \to \ell\nu\ell\nu$ search, combining the various analysis categories. A broad minimum is observed, which is caused by the poor mass resolution of the m_T discriminating variable. The observed p_0 value for $m_H = 125$ GeV is 4×10^{-3}, corresponding to 2.6 standard deviations from the background-only hypothesis.

4. $H \to \tau\bar{\tau}$ searches

Due to the relatively high mass of tau leptons, the $H \to \tau^+\tau^-$ decay corresponds to the dominant leptonic decay mode of the SM Higgs boson ($BR(H \to \tau^+\tau^-) \approx 6.3\%$). Both leptonic ($\tau_{lep}$) and hadronic ($\tau_{had}$) decays of the tau lepton were considered in this search [12], which considered the $H \to \tau_{lep}\tau_{lep}$, $H \to \tau_{lep}\tau_{had}$, and $H \to \tau_{had}\tau_{had}$ decay channels. Events were selected by requiring exactly two opposite-charge reconstructed τ candidates. The invariant mass of the $\tau^+\tau^-$ system, $m_{\tau\tau}$, was used as the discriminant variable in all three analyses, and was estimated with a $m_{\tau\tau}$ resolution of 13 – 20% [14], depending on the event topology and kinematics. Events were further subdivided into different categories to target different topologies, the main ones being the *VBF* category, containing events with two high-p_T jets widely separated in rapidity, and the *Boosted* category, which includes events with a high transverse momentum of the $\tau\tau$ system. Other less significant categories were considered, and more details can be found in [12].

As can be seen in Figure 3 (left), the main source of background after the event selection is $Z \to \tau\tau$, with other sources such as top (both $t\bar{t}$ and single top), multijet production, and backgrounds from τ mis-identification being important in some categories. To avoid uncertainties in the simulation of $Z \to \tau\tau$ production, this background was modelled by selecting a sample of $Z \to \mu\mu$ in collision data and replacing the muon momenta with simulated tau leptons decays. Other backgrounds were estimated from collision data in control samples defined by inverting selection cuts specific to each analysis category. A detailed account can be found in [12].

Figure 3 (centre) shows the 95% confidence level (CL) exclusion limits on the signal strength, as a function of the Higgs boson mass, estimated for the combined $H \to \tau^+\tau^-$ analysis categories.

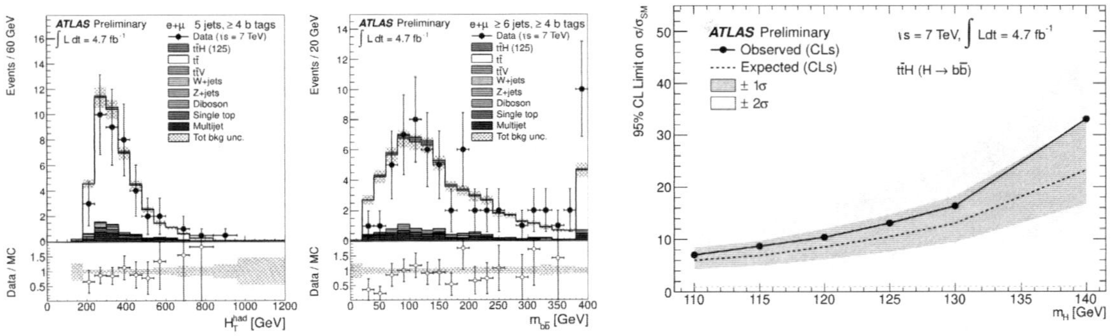

Figure 4. Examples of the final discriminant variables used in the analysis categories with 5 jets, ≥ 4 of which are b-tagged (left, $H_T^{\rm had}$) and ≥ 6 jets, ≥ 4 of which are b-tagged (right, $m_{b\bar{b}}$), after the statistical analysis which constrains various sources of systematic uncertainty [15]. Right: observed and expected 95% CL upper limits [15] on $\sigma(t\bar{t}H) \times BR(H \rightarrow b\bar{b})$ relative to the SM prediction, as function of m_H.

A broad excess of the observed limit with respect to the expected limit, could indicate the presence of signal. The local p_0 value for this channel is shown in Figure 3 (right) and shows that this excess corresponds at most to a 1.5σ deviation from the background-only hypothesis.

5. $H \rightarrow b\bar{b}$ searches

The Higgs boson decays predominantly to a pair of b quarks in the mass region of $m_H \approx 125$ GeV ($BR(H \rightarrow b\bar{b}) \simeq 58\%$ [10]). On the other hand, the experimental observation of the dominant ggF production mechanism is extremely challenging, due to the overwhelming background from multijet production. Instead, the ATLAS searches concentrate on channels where the Higgs boson is produced in association with gauge bosons, WH and ZH, or top quarks, $t\bar{t}H$. These analyses relied on a b-tagging algorithm tuned to identify true b-initiated jets with an efficiency of 70%, and rejection factors of around 5 and 150 for c-quark jets and light-quark jets, respectively.

5.1. $H \rightarrow b\bar{b}$ searches in associated production channels with a top quark pair

As the SM Higgs coupling to fermions is proportional to the fermion mass, the $t\bar{t}H$ channel provides experimental sensitivity to the largest Higgs coupling constant. It is also free from assumptions that affect other channels sensitive to this coupling through loops. A search [15] the $t\bar{t}H \rightarrow b\bar{b}\ b\ell\nu\ bqq'$ decay channel was performed using 4.7 fb^{-1} of $\sqrt{s} = 7$ TeV LHC data. Events were required to have an electron (muon) with p_T above 25 GeV ($p_T > 20$ GeV) and at least four jets with p_T above 25 GeV. Additional cuts were applied on E_T^{miss} and m_T [15]. Selected events were classified into different categories depending on the jet multiplicity and on how many jets were b-tagged. The signal to background ratio is highest for events containing at least 5 jets, 3 of which are b-tagged. For each category, a discriminating variable was considered: either the scalar sum of jet transverse momenta ($H_T^{\rm had}$) or the invariant mass of the two jets most likely to originate from the Higgs boson decay ($m_{b\bar{b}}$). In the analysis categories where $m_{b\bar{b}}$ is used (with 6 jets and 3 or ≥ 4 jets), a kinematic likelihood fit was performed to reconstruct the $t\bar{t}$ system in each event. The remaining b-tagged jets were identified as Higgs boson decay products.

In addition to improving the analysis sensitivity, the analysis categories were used in the statistical analysis to constrain nuisance parameters corresponding to various sources of

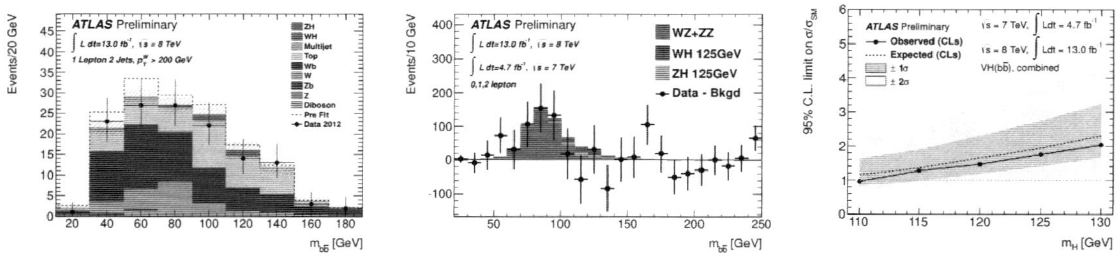

Figure 5. Left [16]: $m_{b\bar{b}}$ distribution in data and simulation for the 1-lepton channel in the analysis category with $p_T^V > 200$ GeV category. The background expectation is shown before (dashed line) and after the statistical analysis which constrains the normalisation of several sources of background. Centre [16]: $m_{b\bar{b}}$ distribution in after subtraction of all backgrounds except diboson processes, including SM Higgs boson from WH and ZH associated production. Right [16]: expected (dashed) and observed (solid) 95% CL upper limit on σ/σ_{SM} relative to the SM prediction, as function of m_H.

background and of systematic uncertainty. Categories with 4 jets and ≥ 2 b-tagged jets are dominated by $t\bar{t}$ production. The distribution of this background across the 0, 1, and ≥ 2 b-tag categories provides a powerful constraint on the b-tagging efficiency systematic uncertainty. Categories with 4, 5 and ≥ 6 jets and ≥ 2 b-tagged jets constrain the $t\bar{t}$ model uncertainties, which may result in large distortions of the shape of the H_T^{had} variable. Figure 4 (left and centre) shows distributions of the discriminating variables, H_T^{had} and $m_{b\bar{b}}$, where the uncertainties are constrained by the statistical analysis fit to the data. This is reflected in the small uncertainty bands of the expected distributions. The limits on the Higgs boson signal strength (σ/σ_{SM}) are shown in Figure 4 (right) as a function of the Higgs boson mass. For a mass of 125 GeV, an exclusion limit of $13.1 \times \sigma_{SM}$ ($10.5 \times \sigma_{SM}$) was observed (expected) at the 95% CL.

5.2. $H \to b\bar{b}$ searches in associated VH production

The ATLAS search [16] for associated production of the Higgs boson with a W or a Z vector boson (VH) was made in three exclusive channels, corresponding to events containing 0, 1, or 2 charged leptons, in addition to two b-tagged jets. These target respectively the $ZH \to \nu\bar{\nu}b\bar{b}$ $WH \to \ell\nu b\bar{b}$ and $ZH \to \ell\ell b\bar{b}$ decay channels, where ℓ refers to an electron or a muon. It used 4.7 fb^{-1} of $\sqrt{s} = 7$ TeV LHC data and 13.0 fb^{-1} of $\sqrt{s} = 8$ TeV data. This analysis updated a previously published analysis [17] by re-optimizing selection cuts and the analysis strategy.

In the 0-lepton channel, transverse energy triggers were used, with efficiencies above 50% for E_T^{miss} above 120 GeV, and essentially 100% for E_T^{miss} above 160 GeV. The trigger turn-on curve in simulated events was corrected to that obtained in data using W+jets events. In all categories, jets were selected with cuts of $p_T > 45$ GeV and $p_T > 25$ GeV for the leading and remaining jets, respectively. The main background in the 2-lepton channel is Z+jets production, whereas both top (single-top and $t\bar{t}$) and W+jets are the main backgrounds affecting the 1-lepton analysis (see Figure 5). All three backgrounds are significant in the 0-lepton analysis. Cuts are applied to topological variables which were optimized for each channel and event category. In each channel, events were further classified into separate categories defined by the transverse momentum of the reconstructed vector boson, p_T^V. The p_T^V categories in the 1- and 2-lepton analyses were: $p_T^V \in [0, 50], [50, 100], [100, 150], [150, 200]$ or ≥ 200 GeV. For the 0-lepton channel, $p_T^V = E_T^{miss}$ was assumed, and the categories were: $E_T^{miss} \in [120, 160], [160, 200]$ or > 200 GeV, separated further into 2-jet and 3-jet categories. The higher-p_T^V categories show an increasing ratio of signal

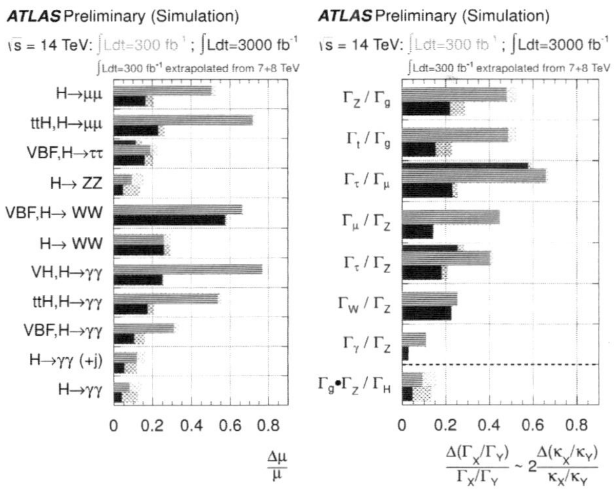

Figure 6. Left [18]: expected relative uncertainty on the signal strength μ in various Higgs boson analyses, assuming a Higgs mass of 125 GeV, for integrated luminosities of 300 fb^{-1} and 3000 fb^{-1} of pp LHC data at $\sqrt{s} = 14$ TeV. Right [18]: expected relative uncertainty on ratios of Higgs boson partial widths. No assumptions were made on the total Higgs boson width, and so on the presence of new physics. For the $H \to \tau^+\tau^-$ channel, the thin brown bars correspond to the extrapolation of the analyses performed using the current data to higher luminosities.

to background, and drive the analysis sensitivity. Control samples rich in top events were also used in the statistical analysis. The invariant mass of the $b\bar{b}$ system, $m_{b\bar{b}}$, was the discriminating variable used in all categories. As in the previous analysis, the main sources of background, top, $W + b$ jets, and $Z + b$ jets, were constrained from the data during the statistical analysis.

The production of WZ and ZZ with one Z boson decaying as $Z \to b\bar{b}$ has a very similar signature as the VH signal, but with a cross section $\approx$ 5 times higher, and p_T^V amd $m_{b\bar{b}}$ distributions peaking at lower values. As a validation of the VH search, a separate fit was made by fixing the Higgs cross section at its SM value and fitting the amount of diboson background. Figure 5 (centre) shows the resulting $m_{b\bar{b}}$ distribution after subtracting all sources of background except for the diboson contribution. The measured diboson signal strength is $\mu_D = 1.09 \pm 0.20(stat) \pm 0.22(syst)$, in agreement with the SM expectation, and corresponded to a significance of 4.0 σ. After fixing the diboson contribution to its SM value, a fit was performed to the Higgs boson signal strength to extract the 95% CL exclusion limits. Figure 5 (right) shows the result of this fit for the combination of all channels and analysis categories as a function of the Higgs boson mass. For a mass of $m_H = 125$ GeV, a 95% CL exclusion limit of $1.8 \times \sigma_{SM}$ ($1.9 \times \sigma_{SM}$) was observed (expected).

6. The long view

The first long period of LHC running has finished with around 25 fb^{-1} of pp collision data collected by ATLAS at $\sqrt{s} = 7$ TeV and $\sqrt{s} = 8$ TeV. The currently foreseen experimental program should provide $\approx$ 300 fb^{-1} of pp data at $\sqrt{s} = 13 - 14$ TeV by 2021. At this point the LHC luminosity may be further enhanced to provide a total integrated luminosity of around 3000 fb^{-1}. A study of the achievable sensitivity of Higgs boson studies was performed in ATLAS taking into account these luminosity scenarios [18]. The estimated performance of the ATLAS detector and reconstruction algorithms was taken into account in an approximate way. In addition to existing analyses of $H \to \gamma\gamma$, $H \to ZZ^{(*)} \to 4\ell$, $H \to WW^{(*)} \to \ell\nu\ell\nu$ and $H \to \tau^+\tau^-$, this study has considered low rate Higgs boson channels: VH and $t\bar{t}H$ production with $H \to \gamma\gamma$ decays, and $H \to \mu\mu$. The expected precision on the signal strength precision achieved in each individual channel is shown in Figure 6 (left) for a Higgs boson mass of 125 GeV and integrated luminosities of 300 or 3000 fb^{-1} collected at $\sqrt{s} = 14$ TeV.

Individual measurements are used in a combined statistical analysis. For generality, no assumptions are made on the presence of physics beyond the SM. This means that no assumption can be made on the total width of the Higgs boson, Γ_H, and only ratios of partial widths can be determined. For all production (i) and decay (f) modes, the cross sections times branching ratios, $\sigma_i \times BR_f$, are assumed to be proportional to the product of partial widths given by $\Gamma_i \times \Gamma_f/\Gamma_H$. In the notation of section 2 this means $\Gamma_X/\Gamma_Y = \kappa_X^2/\kappa_Y^2 = \lambda_{XY}^2$, and the relative uncertainty on the ratio is given by $\frac{\Delta(\Gamma_X/\Gamma_Y)}{\Gamma_X/\Gamma_Y} \approx 2 \times \frac{\Delta(\lambda_{XY})}{\lambda_{XY}}$. Figure 6 (right) shows the expected relative uncertainties on the estimated partial width ratios. A minimal coupling benchmark scenario, already discussed in section 2, assumes no new physics and only two independent parameters for the vector (κ_V) and fermion (κ_F) couplings. In this case, experimental precisions of $\approx$2% on κ_V and $\approx$3.5% on κ_F are expected with 3000 fb^{-1}.

In order to completely determine the parameters of the SM and establish the Higgs mechanism as the origin of electroweak symmetry breaking, it is important to measure the self-couplings of the Higgs boson. The observation of Higgs boson pair production provides sensitivity to the Higgs trilinear coupling and may be accessible at the LHC. For a Higgs boson mass of around 125 GeV, a relative uncertainty of $\approx$30% on the Higgs trilinear coupling scale factor should be achievable from observations in the $HH \to b\bar{b}\gamma\gamma$ and $HH \to b\bar{b}\tau^+\tau^-$ channels, assuming 3000 fb^{-1} of $\sqrt{s} = 14$ TeV data.

7. Conclusions

The discovery of a new, Higgs-like boson raises crucial questions on the nature of the new particle, and determines a programme of measurements needed to measure its properties. This paper gives an overview of the latest Higgs studies in ATLAS. After describing the methods used, and results of a first exploration of the couplings of the new particle, it gives a brief description of new searches in the $H \to WW^{(*)} \to \ell\nu\ell\nu$, $H \to \tau^+\tau^-$, and $H \to b\bar{b}$ channels. Finally it gives an overview of the sensitivity achievable in future Higgs boson studies in ATLAS.

References

[1] ATLAS Collaboration, Phys. Lett. **B716**, 1-29 (2012).
[2] CMS Collaboration, Phys. Lett. **B716**, 30-61 (2012).
[3] ATLAS Collaboration, JINST **3** (2008) S08003.
[4] L.D. Landau, Sov. Phys. Doklady **60**, 207 (1948); C.N. Yang, Phys. Rev.**77**, 242 (1950).
[5] ATLAS Collaboration, ATLAS-CONF-2012-127, https://cds.cern.ch/record/1476765.
[6] LHC Higgs Cross-section Working Group, arXiv:1209.0040 [hep-ph].
[7] G. Cowan et al., Eur. Phys. J. C, **71** (2011) 1554; arXiv:1007.1727v2 [physics.data-an].
[8] ATLAS Collaboration, ATLAS-CONF-2012-162, https://cds.cern.ch/record/1494183.
[9] ATLAS Collaboration, ATLAS-CONF-2012-158, https://cds.cern.ch/record/1493601.
[10] LHC Higgs Cross Section Working Group, S. Dittmaier, C. Mariotti, G. Passarino and R. Tanaka Eds., CERN-2011-002, arXiv:hep-ph/1101.0593, CERN, Geneva, 2011.
[11] M. Cacciari,, G.P. Salam, and G.Soyez, JHEP **0804** (2008) 063, `arXiv:0802.1189 [hep-ex]`.
[12] ATLAS Collaboration, ATLAS-CONF-2012-160, https://cds.cern.ch/record/1493624.
[13] A. L. Read, J. Phys. G **28** (2002) 2693.
[14] A. Elagin, P. Murat, A. Pranko and A. Safonov, Nucl. Instr. and Meth. **A654** (2011) 481.
[15] ATLAS Collaboration, ATLAS-CONF-2012-135, https://cds.cern.ch/record/1478423.
[16] ATLAS Collaboration, ATLAS-CONF-2012-161, https://cdsweb.cern.ch/record/1493625/.
[17] ATLAS Collaboration, Phys. Lett. **B718**, 369-390 (2012).
[18] ATLAS Collaboration, ATL-PHYS-PUB-2012-004, https://cdsweb.cern.ch/record/1484890/.

Very Rare Decays at LHCb

Serena Oggero, on behalf of the LHCb Collaboration

Nikhef, Science Park 105 1098 XG Amsterdam, The Netherlands

E-mail: soggero@nikhef.nl

Abstract. Rare leptonic decays of the $B^0_{(s)}$, D and K mesons are sensitive probes of physics beyond the Standard Model. In particular, the search for the decays $B^0_{(s)} \to \mu^+\mu^-$provides information on the presence of new (pseudo-)scalar particles. LHCb is well suited for these analyses due to its large acceptance for B-decays and trigger efficiency, as well as its excellent invariant mass resolution and electron and muon identification capabilities. Results on these searches using data collected by LHCb in 2011 and 2012 are presented.

1. Theoretical motivations

In the Standard Model (SM) of particle physics, the weak interaction eigenstates and corresponding mass eigenstates are connected through the Cabibbo-Kobayashi-Maskawa (CKM) matrix. The CKM matrix is not diagonal, therefore transitions between different quark generations (Flavour Changing Currents) are allowed. Among these, processes in which the quark changes flavour but not charge are named Flavour Changing Neutral Currents (FCNC) and are suppressed at tree level in the SM. Charged currents mediated by $W^{\pm}$ bosons can instead violate flavour, therefore FCNC processes are possible at one-loop and higher order diagrams. At the one-loop level, they can be described by a set of triple effective vertices named "penguin" diagrams, and by a set of quartic ones named "box" diagrams.

The FCNCs suppression is experimentally established and finds a natural explanation in the Glashow-Iliopoulos-Maiani (GIM) mechanism [1]. The hierarchy of FCNC transitions depends on the masses m_i involved in the internal loops. For instance, the GIM mechanism is less effective in suppressing processes governed by top quarks contributions (K and B decays) than those in which only d, s and b quarks enter internal loops (D decays). The size of FCNC transitions also depends on the concerned V_{CKM} elements, e.g. B_s decays are less suppressed than B_d decays, due to the different sizes of the CKM matrix element involved.

In any case, the fact that FCNCs take place only at the higher orders makes them useful to test the SM and particularly fruitful to probe possible models of New Physics (NP). The rare decay branching fraction ($\mathcal{B}$) may receive large enhancements within specific NP models. Moreover, a purely leptonic final state can lead to a clean theoretical computation of the SM effective Hamiltonian. This motivates the strong interest shown by the physics community towards very rare leptonic decays, despite their challenging analysis aspects. In particular, the LHCb experiment has a very rich rare decays search program, which aims at searching for New Physics with an "indirect" approach, mostly complementary to the direct NP searches conducted at Atlas and CMS.

2. LHCb analyses

The focus of this talk is towards purely leptonic very rare decays investigated at LHCb; in particular, the search for $D^0 \to \mu^+\mu^-$, $K^0_S \to \mu^+\mu^-$and $B^0_{(s)} \to \mu^+\mu^-$decays will be presented. All these processes have a very clean signature in the LHCb experiment (see Sec.2.1) and calls for a common general analysis strategy (see Sec. 2.2).

2.1. LHCb experiment

LHCb [2] is a dedicated experiment for the study of heavy flavour physics at the Large Hadron Collider (LHC). Taking advantage of the correlated $b\bar{b}$ pair production at high energies and small angles with respect to the beam axis, the LHCb detector is designed as a single arm spectrometer, with a unique pseudorapidity acceptance of 1.9< η <4.9 [3] and a forward angular coverage from approximately 10 mrad to 300 mrad.

The LHCb experiment performs precision measurements in flavour physics thanks to an efficient tracking system, an extensive Particle Identication (PID) system and a highly adaptable trigger. The tracking system consists of a silicon-strip vertex detector (VELO), surrounding the *pp* interaction region, a large-area silicon-strip detector (Tracker Turicensis, TT) located upstream of a warm dipole magnet with a bending power of about 4 Tm, and three stations of silicon-strip detectors (Inner Tracker, IT) and straw drift tubes (Outer Tracker, OT), placed downstream.

The LHCb PID is based on two aerogel and gas Ring Imaging Cherenkov counters (RICH) for the charged hadrons, a hadron (HCAL) and electromagnetic (ECAL) calorimeter for photon, electron and hadron candidates, and finally a set of muon chambers, composed of alternating layers of iron and multiwire proportional chambers.

The trigger consists of a hardware stage (L0), mainly based on information from the calorimeter and the muon system, and a software stage (HLT), running a full event reconstruction on a farm of parallel-processing CPUs. The L0 takes events at an input frequency rate of 40 MHz, while the HLT reduces the event rate from an input of 1 MHz to an output rate of about 3 kHz.

2.2. Analysis strategy

All rare leptonic decay searches have a common general analysis strategy that can be illustrated as follows.

Selection. At the L0 level, the single muon and di-muon trigger lines are mostly involved; the HLT trigger applies impact parameter (IP) and invariant mass requirements, leading to an overall ~90% trigger efficiency for high-pt muons in the final state. Afterwards, the first selection of events requires daughter particles to make a displaced vertex, by applying cuts on the IP χ^2 of each particle with respect to the primary vertex. All daughter tracks are also required to originate from the same secondary vertex, by requiring a good vertex fit with a cut on its χ^2. Similarly, the mother hadron has to be consistent with originating from a primary vertex (PV); this is guaranteed by a cut on the hadron IP χ^2. Other variables typically entering the selection are the transverse momentum and the tracks quality of the involved particles, the hadron flight distance and its angle of pointing with respect to the primary vertex. The invariant mass of the mother hadron is required to fall in a certain range optimised for the analysis. In addition, particle identification criteria are used to separate between muons and hadrons. The muon identification efficiency can reach very high values (~98% for the $B^0_{(s)} \to \mu^+\mu^-$analysis), with corresponding very low pion and kaon misidentification rates (less than about 1%).

The same selection is applied to both signal and control channels, with the exception of the particle identification requirements and the invariant mass cuts.

Background rejection. To further increase the signal over background separation, MVA classifiers are used, typically the output of a Boosted Decision Tree (BDT). The BDT operator is built from kinematic and topological variables, combined on the base of their performance during training. The BDT is trained on MC events for the signal and data events from the mass sidebands for the background; the operator optimisation aims at the maximum separation power between signal and background events for high BDT values. The signal BDT distribution is re-calibrated on data (typically using hadronic exclusive decays).

The use of a multivariate selection helps to reduce the combinatorial background, where a candidate meson-vertex is formed from daughter particles that did not originate from a single meson. Peaking background components – typically decays with final state hadrons (h) misidentified as muons ("misId") or exclusive semileptonic decays– might need specific studies. They are usually statistically separated from the signal by mean of particle identification criteria. The misidentification rates, estimated with data calibration samples, are thus a key ingredient of the analysis.

Normalisation. The number of signal events is normalised to one or more channels ($norm$), sharing the same kinematics. In this way the branching fraction can be obtained as

$$\mathcal{B}(\text{signal}) = \frac{\mathcal{N}_{\text{signal}}\epsilon_{\text{signal}}}{\mathcal{N}_{\text{norm}}\epsilon_{\text{norm}}} \cdot \mathcal{B}(\text{norm}) \tag{1}$$

where ϵ represents the efficiency of reconstruction, trigger and selection on the signal or normalisation channel. The use of a normalisation channel allows for cancellation of uncertainties in the determination of the investigated branching fration, for instance due to the integrated luminosity and the meson production cross section. The normalisation channel is selected with a strategy very similar to that used to select signal events; this leads to a minimisation of the systematic effects entering the ratio. The remaining needed efficiency terms are estimated using MC simulations corrected with data-driven methods.

Equation (1) can also be used to estimate the number of expected signal events for a given branching fraction hypothesis, needed for the limit computation (see later in this section).

Blind approach. The data in the signal region is only analysed once the full analysis strategy is defined. This means that a certain region in invariant mass of the daughter particles is blinded during the analysis optimisation and only used ("unblinded") for the

Journal of Physics: Conference Series **447** (2013) 012045 doi:10.1088/1742-6596/447/1/012045

computation of the final results.

Branching fraction estimation and limit determination. Different statistical techniques are used to assess the result of the rare decay search. To estimate the signal yield and branching fraction, typically an unbinned likelihood fit of the invariant mass spectra is used, where the different background components enter as separate PDF's of the fit. All the uncertainties on the separate variables are implemented as parameter constraints and contribute to the final uncertainty.

The compatibility of the observation with expectations (significance of the result) and the limit determination in case of no significant excess are computed with the CL_s method [4] [5]. The following sections give an insight on some leptonic rare decays searches conducted at LHCb, with the last published results.

3. Search for the $D^0 \to \mu^+\mu^-$ decay

As already mentioned in Sec. 1, charm hadron decays are most effectively suppressed by the GIM mechanism and in particular the $D^0 \to \mu^+\mu^-$ decay is expected to be very rare in the SM. In fact, the long distance contribution to $D^0 \to \mu^+\mu^-$, which dominates its branching fraction, is predicted to have an upper limit of about 6×10^{-11} at 90% C.L. [6]. However, sizeable enhancements are expected in several NP models, for instance via tree-level contributions in R-parity violating scenarios.

The analysis conducted at LHCb on 0.9 fb^{-1} collected in 2011 follows the strategy flow illustrated in Sec. 2.2. Signal candidates are determined on a sample of selected $D^{*+} \to D^0(\to \mu^+\mu^-)\pi^+$ decays. The most dangerous peaking background after selection is formed by $D^0 \to \pi^+\pi^-$ decays with the two pions misidentified as muons. The yield of signal candidates observed after selection is determined with a 2 dimensional fit in the D^0 mass ($\mu\mu$ invariant mass) and $D^0 - D^{*0}$ mass difference Δm ($\mathrm{m}_{\mu\mu\pi}$ - $\mathrm{m}_{\mu\mu}$). The fit performed is an unbinned extended maximum likelihood fit and comprises a signal PDF, a combinatorial background PDF and two PDF's describing the misidentified $D^0 \to \pi^+\pi^-$ and $D^0 \to K^-\pi^+$ events. The latter does not peak in the signal window but its shape is needed to correctly describe the data sidebands.

Since the fit to the data does not reveal any excess of signal events with respect to the expected background, the CL_s technique is used to determine a limit on $\mathcal{B}(D^0 \to \mu^+\mu^-)$. The yield of expected signal events is determined by mean of the normalisation channel $D^0 \to \pi\pi$ (see Eq. 1). The result obtained, $\mathcal{B}$ $(D^0 \to \mu^+\mu^-) < 1.3(1.1) \times 10^{-8}$ at 95(90)% C.L. [6], improves the existing upper limit set by the Belle collaboration ($\mathcal{B}$ $(D^0 \to \mu^+\mu^-) < 1.4 \times 10^{-7}$ at 90% C.L. [7]) by more than one order of magnitude. Nevertheless, it is still well above the current SM theoretical prediction, leaving space for future improvements with the full LHCb available dataset.

4. Search for the $K^0_S \to \mu^+\mu^-$decay

The expected SM branching fraction for $K^0_S \to \mu^+\mu^-$decays is $\mathcal{B}_{SM}(K^0_S \to \mu^+\mu^-) = (5.0 \pm 1.5) \times 10^{-12}$ [8], but NP contributions can enhance the SM expectation of up to one order of magnitude [9].

LHCb is a fruitful place for searching rare kaon decays thanks to the high yields of K^0_S ($\sim$10^{13} per fb^{-1}) in the detector acceptance. The analysis conducted on 1 fb^{-1} of *pp* collisions collected in 2011 follows the strategy illustrated in Sec. 2.2. After selection of di-muon candidates, the invariant mass range $[492\text{-}504] MeV/c^2$ is defined as the search region. The background level in this region is calibrated by interpolating the observed yield from mass sidebands with an unbinned maximum likelihood fit, accounting for the combinatorial component (described by an exponential function) and the peaking component. The latter is mainly composed by $K^0_S \to \pi^+\pi^-$misidentified events and it is modeled by a power law. For the signal yield determination, $K^0_S \to \pi^+\pi^-$decays are used as normalisation channel.

The observed number of events is found to be compatible with the background expectations and the upper limit is set at $\mathcal{B}(K^0_S \to \mu^+\mu^-) < 11(9) \times 10^{-9}$ at 95(90)% [10], improving the previous measurements by a factor of thirty.

5. Search for the $B^0_{(s)} \to \mu^+\mu^-$decays

A recent evaluation of the branching fraction for $B^0_s \to \mu^+\mu^-$decays, and accounting for electroweak calculations up to the leading order and QCD calculations at NLO, predicts [11]

$$\mathcal{B}^{(0)}_{SM}(B^0_s \to \mu^+\mu^-) = (3.23 \pm 0.15 \pm 0.23_{f_{B^0_s}}) \times 10^{-9} , \quad (2)$$

where $f_{B^0_s}$ is the meson decay constant and its theoretical calculation represents the dominant source of systematic uncertainty in the SM prediction. This result needs to be corrected to allow for a comparison with the experimental detection and account for the sizeable decay width difference in the B^0_s system [12]. The corrected SM prediction becomes

$$\mathcal{B}^{(0)}_{SM}(B^0_s \to \mu^+\mu^-) = (3.54 \pm 0.30) \times 10^{-9} . \quad (3)$$

This correction is negligible for the $B^0 \to \mu^+\mu^-$decay, since the decay width difference is $\sim 10^{-3}$ in the B^0_d system. The current best theoretical prediction for the SM $B^0 \to \mu^+\mu^-$branching fraction is the CP-averaged calculation performed in [11]:

$$\mathcal{B}^{(0)}_{SM}(B^0 \to \mu^+\mu^-) = (0.107 \pm 0.010) \times 10^{-9} . \quad (4)$$

The $B^0_{(s)} \to \mu^+\mu^-$decay may receive large enhancements within specific New Physics models. For instance, the expected branching ratio in the framework of certain MSSM models can be very much enhanced due to a dependence with $\tan^6\beta$ [13]. Nevertheless, it is important to notice that the SUSY impact on the $B^0_s \to \mu^+\mu^-$decay can also be "hidden" and the MSSM contribution may well be suppressed by moving to regions of intermediate $\tan\beta$ values or large masses of the physical Higgs boson m_A. The strong interest shown by the physics community towards $B^0_{(s)} \to \mu^+\mu^-$decays is motivated as this is a particularly fruitful probe of NP in the scalar and pseudoscalar sectors.

The search for a $B^0_{(s)} \to \mu^+\mu^-$decay at LHCb is the search for two very specific muonic tracks in the detector, but the very high background makes the analysis extremely challenging. After selection, the comparison between number of candidates, number of events expected for a given branching fraction hypothesis and number of expected background events is done in bins of two variables, independent from each other. They are the invariant mass (IM) and the output of the BDT. The signal IM is described using a Crystal Ball function, whose parameters are mainly obtained from data. The yield of expected signal is then obtained using $B^+ \to J/\psi(\to \mu^+\mu^-)K^+$ and $B^0 \to K^+\pi^-$ decays as normalisation channels.

Concerning the number of expected background events, the combinatorial background is evaluated from the IM sidebands, defined as $[4900, 5224]$ MeV$/c^2$ for the left sideband and $[5432, 6000]$ MeV$/c^2$ for the right one. The BDT and IM PDF's of the combinatorial background are then obtained from an extended likelihood fit to the sidebands, performed with a single exponential function, and the corresponding yields are extrapolated to the search mass windows. Among the peaking backgrounds, $B^0_{(s)} \to h^+h^-$ decays where both hadrons h are misidentified as muons are particularly dangerous. In addition, $B^0_d \to \pi^-\mu^+\nu_\mu$ and $B^{+(0)} \to \pi^{+(0)}\mu^+\mu^-$ decays are found to affect not negligibly the combinatorial background determination from the sidebands, at BDT>0.8 (sensitive region). For this reason, the fit to the background accounts for the $B^0_d \to \pi^-\mu^+\nu_\mu$ and $B^{+(0)} \to \pi^{+(0)}\mu^+\mu^-$ contributions as well, by adding their PDF's to the exponential PDF describing the combinatorial component.

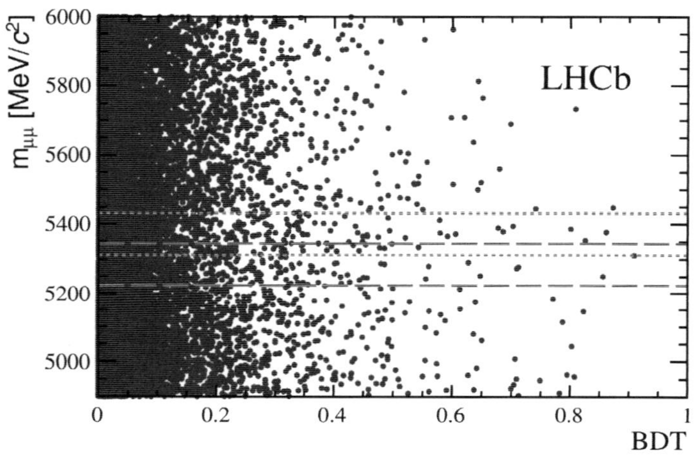

Figure 1: $B^0_{(s)} \to \mu^+\mu^-$analysis, unblinded 2012 data: 2D plot of the di-muon candidates invariant mass versus the event BDT. The orange-dashed (green-dashed) lines indicate the $\pm 60 MeV/c^2$ $B^0_s(B^0_d)$ search window.

The distribution of events from the 2012 dataset in the (IM, BDT) plane, after the "unblinding" of the signal search regions, is shown in Figure 1. The orange-dashed and green-dashed lines indicate the ± 60 MeV/c^2 signal search windows around m(B^0_s) and m(B^0_d), respectively.

The significance of the observation in each (IM, BDT) bin is assessed by applying the CL_s technique both to the 2012 dataset alone, and to the 2012 and 2011 datasets combined (corresponding to 2.1 fb^{-1}); the latter result is here presented. The combination is performed by computing the CL_s simultaneously for all the bins of the 2011 and 2012 data samples. The distributions of the expected value of CL_s

for the $B_s^0 \to \mu^+\mu^-$and $B^0 \to \mu^+\mu^-$decays, versus the various branching fractions hypotheses, are shown in Fig. 2 (a) and (b) respectively, for the 2011 and 2012 datasets combined. The p-value measured for the $B^0 \to \mu^+\mu^-$decay is 11%. For the

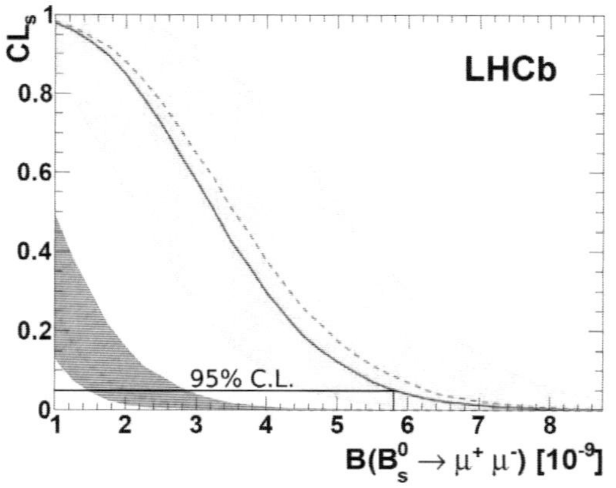

(a) $B_s^0 \to \mu^+\mu^-$

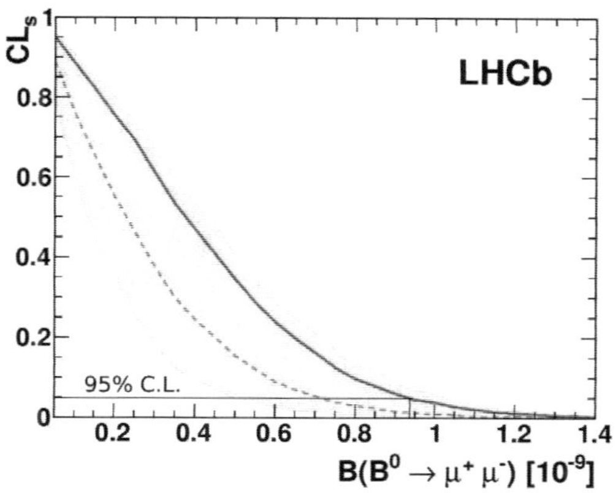

(b) $B^0 \to \mu^+\mu^-$

Figure 2: Distributions of the expected value of CL_s for the $B_s^0 \to \mu^+\mu^-$(on top) and $B^0 \to \mu^+\mu^-$decays (below) versus the $\mathcal{B}$hypothesis, computed for the 2011 and 2012 datasets. The grey (purple) curve is the median of the expected CL_s in the signal-plus-background (background-only) hypothesis, while the yellow (green) band covers 34% of the distribution on each side of the median. The red curve is the CL_s measured on data.

$B_s^0 \to \mu^+\mu^-$instead, the p-value is computed to be $6{\cdot}10^{-4}$. This corresponds to an excess of $B_s^0 \to \mu^+\mu^-$candidates with respect to background expectations with a significance of about 3.5σ, corresponding to the first signal evidence for this decay[14].

Given the high signal significance obtained, a double-sided limit on the branching fraction is also estimated by computing the lower and upper limits at $CL_{s+b} = 0.95(0.975)$ and $CL_{s+b} = 0.05(0.025)$ respectively, for a 90% (95%) C.L. The obtained result is $\mathcal{B}(B_s^0 \to \mu^+\mu^-) \in [1.3, 5.8] \times 10^{-9}([1.1, 6.4] \times 10^{-9})$, at 90% (95%) C.L.. Moreover, a branching ratio measurement can be obtained by means of a global fit to the di-muon invariant mass, with the input $p.d.f$'s already used for the fit to the invariant mass sidebands previously described. The value of the $B_s^0 \to \mu^+\mu^-$branching fraction determined from the fit is

$$\mathcal{B}(B_s^0 \to \mu^+\mu^-) = (3.2\,^{+1.4}_{-1.2}(\text{stat})\,^{+0.5}_{-0.3}(\text{syst})) \times 10^{-9}\ . \qquad (5)$$

The signal significance obtained from the difference in log-likelihood between signal and null hypothesis corresponds to a 3.5 standard deviations evidence of the $B_s^0 \to \mu^+\mu^-$decay. The result in Eq.5 is in agreement with the SM expectations (Eq. 3), but leaves still place for New Physics scenarios.

6. Conclusions

FCNC processes are expected to be extremely rare in the SM but provide a fruitful probe for NP scenarios. In this context, the LHCb experiment has a rich program of rare leptonic decay searches and the 2012 analysis campaign has produced several important results. For instance, the search for $D^0 \to \mu^+\mu^-$ decay with 0.9 fb^{-1} data leads to the new world's best limit: $\mathcal{B}(D^0 \to \mu^+\mu^-) < 1.3(1.1) \times 10^{-8}$ at 95(90)% C.L. [6]. Similarly, $K^0_S \to \mu^+\mu^-$decays have been searched at LHCb with 1 fb^{-1} data; the new limit is set at $\mathcal{B}(K^0_S \to \mu^+\mu^-) < 11(9) \times 10^{-9}$ at 95(90)% [10] and improves the previous measurements by a factor of thirty. Finally, the search for $B^0_{(s)} \to \mu^+\mu^-$decays with 2.1 fb^{-1} data leads to the first 3.5σ evidence for the $B^0_s \to \mu^+\mu^-$decay [14]. The obtained branching fraction $\mathcal{B}(B^0_s \to \mu^+\mu^-) = (3.2\,^{+1.4}_{-1.2}(\text{stat})\,^{+0.5}_{-0.3}(\text{syst})) \times 10^{-9}$ is compatible with the SM expectations.

References

[1] Glashow S L, Iliopoulos J and Maiani L 1970 *Phys. Rev. D* **2**(7) 1285–1292 URL `http://link.aps.org/doi/10.1103/PhysRevD.2.1285`
[2] The LHCb collaboration 2008 *JINST* **3** S08005
[3] The LHCb collaboration 2003 *LHCb reoptimized detector design and performance: Technical Design Report* Technical Design Report LHCb (Geneva: CERN)
[4] Read A L 2002 *J. Phys. G* **28** 2693 URL `http://stacks.iop.org/0954-3899/28/i=10/a=313`
[5] Junk T 1999 *Nucl. Instrum. Meth. A* **434** 435 – 443 URL `http://www.sciencedirect.com/science/article/pii/S0168900299004982`
[6] The LHCb Collaboration 2012 *Subm. to 47th Rencontres de Moriond on Electroweak Interactions and Unified Theories* LHCb–CONF–2012–005
[7] Petric M and others for the Belle Collaboration 2010 *Phys. Rev. D* **81** 091102
[8] Ecker G and Pich A 1991 *Nucl.Phys. B* **366** 189 – 205 URL `http://www.sciencedirect.com/science/article/pii/0550321391900564`
[9] Isidori G and Unterdorfer R 2004 *J. Phys.* **2004** 009 URL `http://stacks.iop.org/1126-6708/2004/i=01/a=009`
[10] The LHCb Collaboration 2013 *J. Phys.* **2013**(1) 1–18 URL `http://dx.doi.org/10.1007/JHEP01%282013%29090`
[11] Buras A J, Girrbach J, Guadagnoli D and Isidori G 2012 *Eur. Phys. J.* **C72** 2172 (*Preprint* `1208.0934`)
[12] De Bruyn K, Fleischer R, Knegjens R, Koppenburg P, Merk M and Tuning N 2012 *Phys. Rev. D* **86**(1) 014027 URL `http://link.aps.org/doi/10.1103/PhysRevD.86.014027`
[13] Bobeth C, Buras A J, Krger F and Urban J 2002 *Nucl. Phys. B* **630** 87 – 131 URL `http://www.sciencedirect.com/science/article/pii/S0550321302001414`
[14] The LHCb Collaboration 2013 *Phys. Rev. Lett.* **110**(2) 021801 URL `http://link.aps.org/doi/10.1103/PhysRevLett.110.021801`

Journal of Physics: Conference Series **447** (2013) 012046 doi:10.1088/1742-6596/447/1/012046

The LHCb Upgrade

Stephan Eisenhardt (On behalf of the LHCb Collaboration.)
University of Edinburgh, School of Physics & Astronomy, Edinburgh EH9 3JZ, UK
E-mail: S.Eisenhardt@ed.ac.uk

Abstract. The LHCb experiment is designed to perform high-precision measurements of CP violation and search for New Physics using the enormous flux of beauty and charmed hadrons produced at the Large Hadron Collider (LHC). The operation and the results obtained from the data collected 2011 and 2012 demonstrate that the detector is robust and functioning very well. However, the limit of $\mathcal{O}(1)\mathrm{fb}^{-1}$ of data per year cannot be overcome without improving the detector. We therefore plan for an upgraded spectrometer by 2018 with a 40 MHz readout and a much more flexible software-based triggering system that will increase the data rate as well as the efficiency specially in the hadronic channels. Here we present the LHCb detector upgrade plans.

1. Introduction

The LHCb experiment [1, 2] has demonstrated an excellent running performance in 2011 and 2012. Table 1 summarises the most important parameters. LHCb started in 2011 with the design values for both the luminosity and the number of visible interactions per bunch crossing. The experiment makes the most out of the provided luminosity by constantly adjusting the beam overlap to maintain a constant instantaneous luminosity during a run, as shown in figure 1. An extended tuning campaign to optimise every aspect of the data acquisition enabled LHCb to increase these parameters steadily over the running period. Over the 2012 running period finally four times the design value could be maintained for the number of interactions per bunch crossing as well as two times the design value for the luminosity. At the end of 2012 a successful test was performed with a luminosity of even $6 * 10^{32}\,\mathrm{cm}^{-2}\mathrm{s}^{-1}$. An upgrade to the PC farm of

Table 1. LHCb running parameters in 2011 and 2012.

Year	2011	2012
Beam energy	3.5 TeV	4.0 TeV
Luminosity	$2 - 4 * 10^{32}\,\mathrm{cm}^{-2}\mathrm{s}^{-1}$	$4 * 10^{32}\,\mathrm{cm}^{-2}\mathrm{s}^{-1}$
Visible interactions/beam crossing	$\mu = 0.4 - 1.4$	$\mu = 1.6$
Data taking efficiency	$> 91\,\%$	$> 95\,\%$
High Level Trigger output rate	3 kHz	4.5 kHz
Integrated luminosity recorded	$1.107\,\mathrm{fb}^{-1}$	$2.082\,\mathrm{fb}^{-1}$

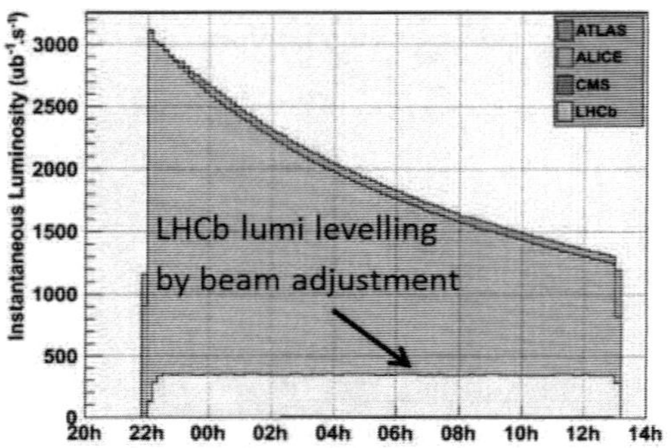

Figure 1. LHCb luminosity leveling during a run; makes optimal use of luminosity provided by the LHC.

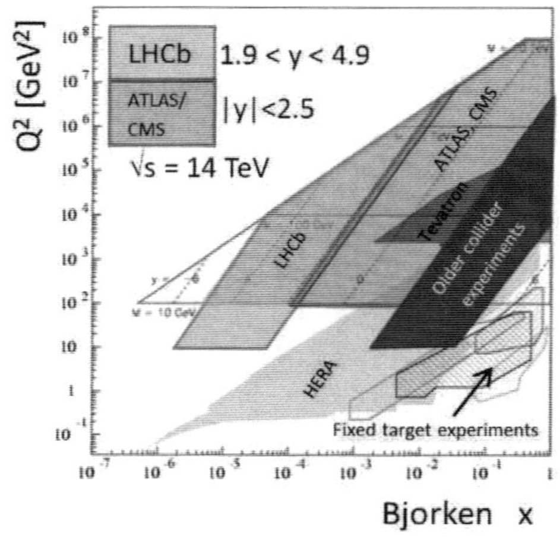

Figure 2. Phase space in (Q^2, x) with coverage for severals generations of experiments [17].

the High Level Trigger increased the output to tape in 2012 by 50 %. It is also worth noting that the increase of the LHC beam energy led to an increase of 15 % in the $b\bar{b}$ production cross section in 2012 compared to 2011. With that LHCb achieved to record more that $1.0\,\mathrm{fb}^{-1}$ of data in 2011 and in excess of $2.0\,\mathrm{fb}^{-1}$ in 2012. Therefore in 2012 LHCb just surpassed the design luminosity per year, despite the fact that the LHC was only filled with half the number of proton bunches, interacting at a rate of only 20 MHz, i.e. half the design rate.

Using this data LHCb has started to publish new results at high rate, many of which yielding first and best measurements. A personally biased and incomplete selection comprises:

- the first measurement of mixing in the D meson system [3]
- the measurement of direct CP violation in the charmless two-body decay of B mesons [4], where first evidence is presented for $B_s \to K\pi$ and the most precise measurement on $B \to K\pi$ is reported
- the measurement of the phase difference between B_s-mixing and $b \to c\bar{c}s$ amplitudes [5], which dominates the world average
- the first measurement of the zero-crossing point of the forward-backward asymmetry A_{FB} of the decay $B \to K^*\mu\mu$ [6], being consistent with the Standard Model
- the improvement of the world best measurements of the oscillation frequency Δm_d in the B_d-system [7]
- the discovery of new states like $D_{s1}(2536)$ in the decay $B_s \to D_{s1}\pi$ [8] and
- first evidence for the measurement of the $B_s \to \mu\mu$ decay matching the Standard Model prediction [9], and by that ruling out Supersymmetry with large $\tan\beta$.

During the Long Shutdown 1 (LS1) LHCb will harvest more results from the recorded data, will consolidate the detector, study systematic effects and improve the trigger strategies, to prepare for the running periods 2015-17. There we expect to take in excess of $5.0\,\mathrm{fb}^{-1}$ of data. At the increased centre of mass energy of $13 - 14\,\mathrm{TeV}$ we expect about twice the beauty and charm production, and the LHC will introduce the design bunch crossing rate at 40 MHz, i.e. at 25 ns spacing. With that, while watching out for the unexpected, we expect to perform precision measurements of the unitarity triangles, measure rare hadronic decays and probe for New Physics at the 10 % level for key measurements.

Journal of Physics: Conference Series **447** (2013) 012046 doi:10.1088/1742-6596/447/1/012046

2. Physics Motivation for LHCb Upgrade

The plans for the LHCb upgrade are documented in the Letter of Intend [10] and in the Framework Technical Design Report of the LHCb Upgrade [11].

From the restart after LS1 LHCb will meet increasingly challenging conditions. The LHC will increase its beam energy and beam current to the design values, the radiation levels and the occupancies will rise. But the LHCb experiment will not be able to significantly further increase the recorded luminosities before the experiment is upgraded. This upgrade shall take place during the Long Shutdown 2 (LS2) in 2018. After the upgrade the experiment will be designed to operate at luminosities of $2 * 10^{33}\,\mathrm{cm}^{-2}\mathrm{s}^{-1}$, i.e. ten times the current design luminosity and we will collect at least $50\,\mathrm{fb}^{-1}$ of data over ten years. We expect to increase the yield in the decays with muons by a factor of five, and the yield for hadronic channels by a factor of ten. The running experience from the years 2011 and 2012 will enable us to provide better projections and possibly improve on these design values.

The driving motivation for the LHCb upgrade is to further significantly increase the physics reach, especially for very rare decays. A prime example of this increased reach of LHCb upgrade are the rare decays $\mathrm{B_s} \to \mu\mu$ and $\mathrm{B_d} \to \mu\mu$. LHCb just has reported first evidence for the measurement of the $\mathrm{B_s} \to \mu\mu$ decay, matching the Standard Model prediction [9]. If further established this measurement rules out Supersymmetry with large $\tan\beta$. With the upgrade we expect to yield 50 $\mathrm{B_s} \to \mu\mu$ events per year. This will allow us to perform precision measurements, for example for the ratio of the branching ratios of $\mathrm{BR}(\mathrm{B_s} \to \mu\mu)/\mathrm{BR}(\mathrm{B_d} \to \mu\mu)$, and become sensitive to distinguish a variety of New Physics models.

With its capabilities LHCb Upgrade also will be unique for New Physics searches in $\mathrm{B_s}$ decays and very competitive in $\mathrm{B_d}$ decays. It will deliver unprecedented charm yields. LHCb provides a general purpose experiment with forward geometry, which also has good potential for non-flavour physics. The upgrade will fully explore the LHC physics in the forward region. It is compatible with the high luminosity upgrade of the LHC, but doesn't require it. It provides direct searches complementary to those of ATLAS and CMS. If new particles are discovered LHCb upgrade will measure their flavour couplings through loop diagrams. If not, LHCb upgrade will probe New Physics at the multi-TeV scale.

Also the rare hadronic penguin transitions $\mathrm{b} \to \mathrm{q\overline{q}s}$ have been shown to be sensitive to New Physics in their decay amplitude [12, 13]. LHCb will use $\mathrm{B_s} \to \phi\phi$ [14] and $\mathrm{B_s} \to \mathrm{K^{*0}\overline{K^{*0}}}$ to probe the CP violation weak phase ϕ_s. In the Standard Model the decay and mixing phases cancel, i.e. the predictions of ϕ_s are very close to zero. With the LHCb upgrade the sensitivity will be pushed to about $\mathrm{S} = 0.03$, compared to a Standard Model prediction of $\mathrm{S} < 0.02$. Therefore, any non-zero result will signify New Physics.

The CP Violation in charm was expected to be $< 0.1\,\%$. With the recent evidence for direct CP violation in hadronic D-decays, where LHCb measured $\Delta\mathrm{A_{CP}} = (-0.82 \pm 0.21 \pm 0.11\,\%)$ [15], a new field opens up. This now also has been confirmed by CDF and BELLE, and new LHCb studies are underway. With the upgrade LHCb will collect an unprecedented large charm sample, and we expect to reach a sensitivity on $\Delta\mathrm{A_{CP}}$ of $0.12\,‰$.

Beyond these specific examples LHCb upgrade has a rich spectrum of flavour and non-flavour physics in its reach. In the $\mathrm{B_s}$-mixing LHCb upgrade will measure:

- precisely the weak phase ϕ_s from the decay $\mathrm{B_s} \to \mathrm{J}/\Psi\phi$ to $\sigma(\phi_\mathrm{s}) \sim 8\,‰$ and
- $\mathrm{A_{SL}}$ to a sensitivity of $2\,‱$.

Further LHCb upgrade will reduce the measurement error of the Unitarity Triangle angle γ to $\sigma(\gamma) \sim 0.9\,°$.

In the electroweak sector LHCb can:

- study how the boson follows the quark direction in the forward region
- by measuring $\sin^2\Theta_\mathrm{eff}^\mathrm{lepton}$, determine $\mathrm{A_{FB}}$ of leptons in Z-decays

Journal of Physics: Conference Series **447** (2013) 012046 doi:10.1088/1742-6596/447/1/012046

Table 2. LHCb upgrade performance benchmarks [11].

Type	Observable	Current precision	LHCb 2018	Upgrade ($50\,fb^{-1}$)	Theory uncertainty
B_s-mixing	$2\beta_s(B_s^0 \to J/\Psi\phi)$	0.10	0.025	0.008	~0.003
	$2\beta_s(B_s^0 \to J/\Psi f_0(980)$	0.17	0.045	0.014	~0.01
	$A_{fs}(B_s^0)$	$6.4 * 10^{-3}$	$0.6 * 10^{-3}$	$0.2 * 10^{-3}$	$0.03 * 10^{-3}$
Gluonic penguin	$2\beta_s^{eff}(B_s^0 \to \phi\phi)$	-	0.17	0.03	0.02
	$2\beta_s^{eff}(B_s^0 \to K^{*0}\overline{K^{*0}})$	-	0.13	0.02	<0.02
	$2\beta_s^{eff}(B_s^0 \to \phi K_s^0)$	0.17	0.30	0.05	0.02
Right-handed currents	$2\beta_s^{eff}(B_s^0 \to \phi\gamma)$		0.09	0.02	<0.01
	$\gamma^{eff}(B_s^0 \to \phi\gamma)/\tau_{B^0S}$	-	5 %	1 %	0.2 %
Electroweak penguin	$S_3(B^0 \to K^{*0}\mu^+\mu^-; 1 < q^2 < 6\,GeV^2/c^4)$	0.08	0.025	0.008	0.02
	$S_0A_{FB}(B^0 \to K^{*0}\mu^+\mu^-)$	25 %	6 %	2 %	7 %
	$A_I(K\mu^+\mu^-; 1 < q^2 < 6\,GeV^2/c^4)$	0.25	0.08	0.025	~0.02
	$\mathcal{B}(B^+ \to \pi^+\mu^+\mu^-)/\mathcal{B}(B^+ \to K^+\mu^+\mu^-)$	25 %	8 %	2.5 %	~10 %
Higgs penguin	$\mathcal{B}(B_s^0 \to \mu^+\mu^-)$	$1.5 * 10^{-9}$	$0.5 * 10^{-9}$	$0.15 * 10^{-9}$	$0.3 * 10^{-9}$
	$\mathcal{B}(B^0 \to \mu^+\mu^-)/\mathcal{B}(B_s^0 \to \mu^+\mu^-)$	-	~100 %	~35 %	~5 %
Unitarity triangle angles	$\gamma(B \to D^{(*)}K^{(*)})$	~10-12°	4°	0.9°	negligible
	$\gamma(B_s^0 \to D_sK)$	-	11°	2.0°	negligible
	$\beta(B^0 \to J/\Psi K_s^0)$	0.8°	0.6°	0.2°	negligible
Charm CP violation	A_Γ	$2.3 * 10^{-3}$	$0.40 * 10^{-3}$	$0.07 * 10^{-3}$	-
	ΔA_{CP}	$2.1 * 10^{-3}$	$0.65 * 10^{-3}$	$0.12 * 10^{-3}$	-

- measure the forward-backward asymmetry of the top quark
- study electroweak penguins, like $B \to K^{*0}\mu\mu$ and
- measure precisely several other electroweak asymmetries: A_I, S_3,... [16].

LHCb upgrade also can contribute to QCD studies by:

- providing further constraints on the Parton Density Functions
- covering unexplored phase space at low x [17], as depicted in figure 2, and
- by studying jets with flavour tagging.

LHCb upgrade also can study Central Exclusive Production:

- in the reaction $pp \to p + X + p$ with rapidity gap
- in the x_b production and
- for the Pomeron/photon exchange.

And finally LHCb can can reach Hidden Valley particles, if they are long-lived and decay in the LHCb vertex detector.

Table 2 presents a compilation of the LHCb and LHCb upgrade performance benchmarks. Listed are the currently achieved precision, the projections for the precision achievable before and after the LHCb upgrade as well as the theoretical uncertainties.

3. The LHCb Upgrade

Figure 3 shows a schematic of the LHCb detector and summarises the upgrades necessary to the sub-systems. These will be detailed in the following.

The bottle neck for harvesting data with the LHCb experiment from increasing luminosity is the current readout architecture, shown in figure 4. This architecture is limited to a maximum rate of 1 MHz, at the output of the on-detector electronics (Level-0) as well as at the input to the Higher Level Triggers (HLT). The Level-0 trigger is implemented in the hardware of the individual sub-detectors, without information cross-feeding between the sub-detectors. The alleys of the HLT are implemented off-detector as software triggers on a CPU farm.

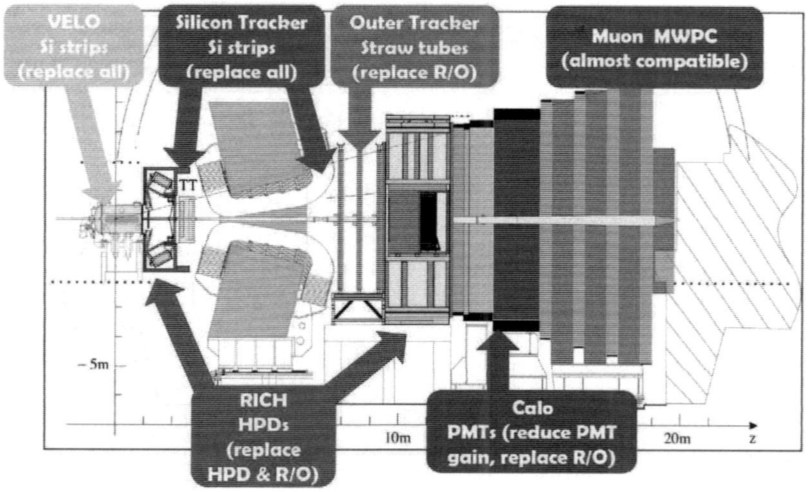

Figure 3. Summary of the upgrades necessary to the LHCb sub-detector systems.

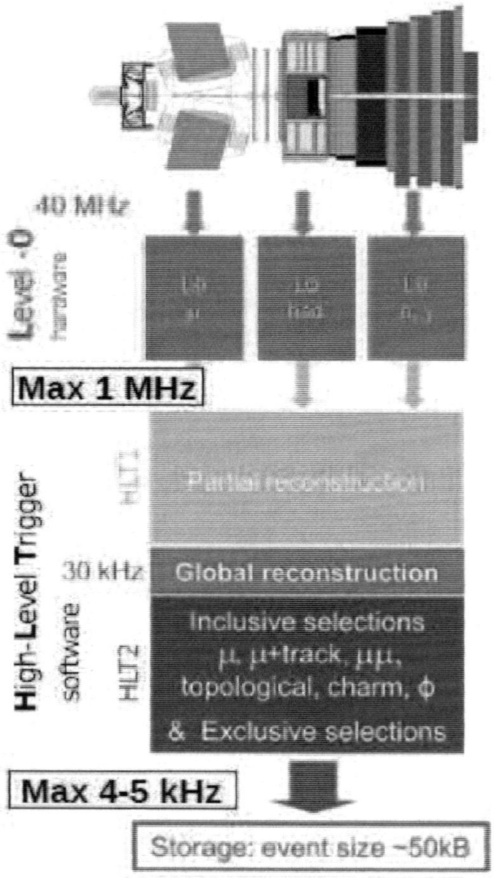

Figure 4. Current LHCb trigger layout.

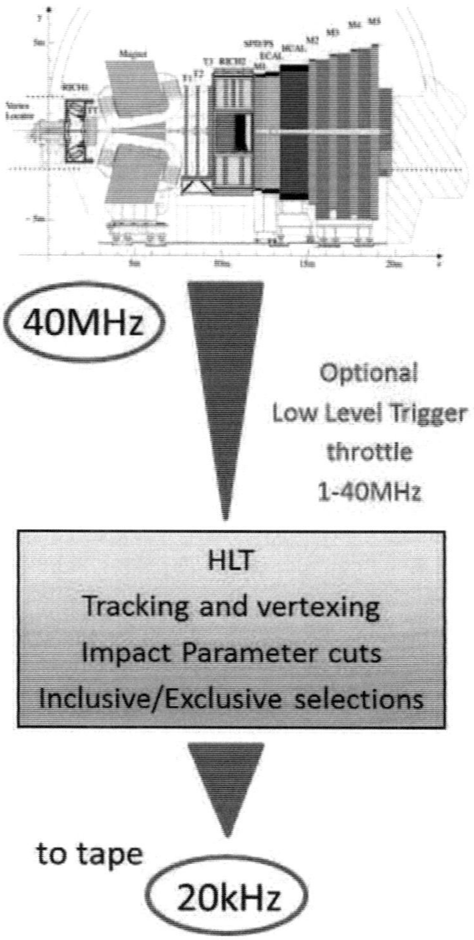

Figure 5. LHCb upgrade trigger layout.

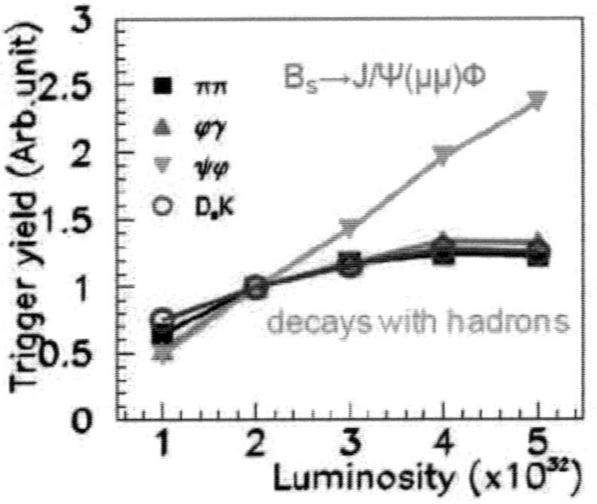

Figure 6. Simulation of current LHCb trigger yields, depending on luminosity. The efficiency of hadronic channels saturates in the current readout architecture [11].

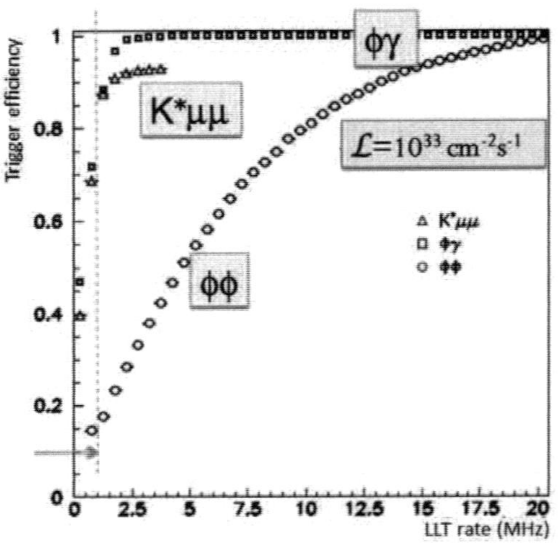

Figure 7. LHCb upgrade: simulation of trigger efficiency depending on the throughput rate of the Low Level Trigger, at the luminosity of $10^{33}\text{cm}^{-2}\text{s}^{-1}$. The arrow marks the throughput equivalent to the current limit of LHCb [11].

The fixed layout of the Level-0 trigger does not allow for efficient triggering on decays which only have hadronic content. It turns out that as the luminosity increases so does the inefficiency in the hadronic triggers and the net yield of hadronic events levels off, while the yield of events with leptonic content scales with the luminosity increase. Figure 6 demonstrates this from simulations using the current readout for a range of luminosities.

As a first step to improve the current boundaries we recently introduced the deferred HLT, where we use storage capacity and the time between the LHC fills to process those events which otherwise would be discarded by the 20 ms cut-off which is imposed during data taking. This improves the efficiencies of all used trigger alleys, but does not change the scaling problem of the hadronic triggers with luminosity.

With the upgrade LHCb will finally overcome the limitations of the Level-0 trigger by omitting this hardware trigger decision and reading out the full detector information at 40 MHz into the fully flexible software based HLT. This is shown in figure 5. In this scheme the HLT will use all sub-detector information to form the trigger decision and the trigger algorithms will be improved to fully exploit the information becoming available at the HLT level. This e.g. will include an online reconstruction for the Vertex Locator. The HLT eventually will write to tape with a rate of 20 kHz.

The new readout architecture also foresees a Low Level Trigger (LLT) which provides an optional and tunable throttle, to accommodate the detector output to the HLT intake capabilities. These intake capabilities will grow over time as the HLT CPU farm gets enlarged. As the output rate of the Low Level Trigger increases above the current design limit of 1 MHz the trigger efficiencies improve. This is shown in figure 7 for three selected benchmark channels. The efficiencies improve for all channels, but the gains are most significant for the fully hadronic channel $B_s \to \phi\phi$, which becomes fully efficient at an LLT output rate of 20 MHz.

Table 3 finally demonstrates how the HLT efficiencies for the three benchmark channels increases significantly as the size of the CPU farm is scaled up by factors of five and ten with respect to the size in 2011.

Table 3. HLT efficiencies of benchmark channels, depending on number of CPUs in the HLT.

Channel	HLT efficiency	2011 size *5	2011 size *10
$B_s \to \phi\phi$		29 %	50 %
$B_s \to K^*\mu\mu$		75 %	85 %
$B_s \to \phi\gamma$		43 %	53 %

The repeating challenges for the design of the sub-detector upgrades are the increased levels of radiation and the increases in the detector occupancies.

The VELO upgrade is a prime example for this. Its electronics will have to withstand radiation of up to $0.3 * 10^{16} n^{eq} cm^{-2}$ (neutron equivalents/cm^2). Its material budget has to be reduced, by thinning down the sensors to a thickness of 200 μm, to limit the number or produced secondary particles. In addition thermal run-away has to be prevented by CO_2) cooling. In effect the VELO upgrade has to maintain, if not improve the current excellent performance.

At present two technologies are still considered. The first is a more conventional strip detector layout, with R- and ϕ-geometry. The layout itself will be improved to better balance the occupancies, featuring a reduced pitch of 30 μm and shorter strip lengths. This option would employ a new readout chip which should be shared with the Inner Tracker. The second technology employs a novel pixel detector solution based on the VELOPIX readout chip[18]. The VELOPIX chip features a matrix of $256 * 256$ pixels at a size of $55\mu m * 55\mu m$, leading to a low occupancy. This readout chip is implemented in the 130 nm technology, is radiation hard to > 500 MRad and tolerant to single-event-upsets (SEU). The VELOPIX will read out an edge-less pixel sensor. Further R&D will be performed on both option before a decision is taken for the VELO upgrade.

The current tracker consists of silicon strip detectors at the location of the TT tracking station as well as in the inner-most part of the tracking stations, where the so-called Inner Tracker (IT) covers the area with the highest occupancy. The Outer Tracker (OT), surrounding the Inner Tracker, is made up from straw tubes. In the baseline option for the upgrade the same technologies are used again. The dominant challenge for the Outer Tracker will be the occupancy which increases strongly towards the centre area around the beam pipe. To limit the occupancy in the Outer Tracker to < 20 % the size of the Inner Tracker will be scaled up, increasing its current area by a factor of four. The silicon-strip detectors face the same challenges as the VELO, with radiation levels up to $2 * 10^{14} n^{eq} cm^{-2}$. Like for the VELO the sensors will be redesigned and optimised to balance the occupancy and to reduce the material. And both systems will share the readout chip design, if the VELO uses for the silicon strip option.

The current RICH system is made up of two detectors, RICH 1 featuring two radiators, Aerogel and C_4F_{10}, and RICH 2 featuring CF_4 as single radiator. The peculiarity is that the front-end electronics is embedded into the vacuum of the Hybrid Photon Detectors, which therefore need to be replaced as well. The challenge in the upgrade for the RICH is to limit the occupancy. For that the Aerogel radiator will be removed and the RICH 1 optics may be changed to minimise the pixelisation error. Baseline for the new photon detectors now are the 64-channel Multianode Photomultiplier tubes of the new generation, R11265[19], which feature an active area of 80 % and can be used without projecting lenses. R&D on the design of the readout electronics is underway. Performance studies demonstrate that the occupancy can be kept below 10 % and that the particle identification will work well to beyond luminosities of $2 * 10^{33} cm^{-2}s^{-1}$.

The electromagnetic and the hadronic calorimeters can keep the detector modules and the photo-multipliers, as they reasonably can withstand the radiation. But the photo-multipliers need to be run at lower gain to keep the anode currents constant despite the higher occupancies. The new 40 MHz electronics will compensate the reduced gain with a higher amplification. It has been demonstrated that the signal-to-noise ratio and the energy resolution can be maintained, despite the reduced gain of the photo-multipliers.

It is baseline to remove the Scintillating Pad & Parton Shower Detectors (SPD&PS). They are currently used to form the Level-0 trigger decision and have no use in the Low Level Trigger which only provides a throttle. However, studies are performed to see how well the still needed e/γ identification still can be achieved.

The Multi-Wire Proportional Chambers and the front-end electronics of the muon chambers are almost compatible with the LHCb upgrade. While the chambers can be kept as they are, the electronics needs minor modifications but can be reused. Only the first Muon chamber (M1), in front of the calorimeters, will be removed as its tasks will be taken over by the tracking stations. Further studies are performed to safe-guard the performance under high occupancy and against the aging of the chambers.

4. Conclusions

LHCb and LHC are a great success. By now large New Physics is ruled out in many flavour physics observables. Therefore a large increase in statistics is required to investigate small New Physics deviations from the Standard Model.

The LHCb Upgrade plan is mature. The installation of the LHCb upgrade will take place in the long shutdown 2018. The key element is the readout of all sub-detectors at the full rate of bunch crossings at 40 MHz. A fully flexible software trigger increases the trigger efficiency for hadronic channels by a factor of two at least. And the key performance parameters of LHCb will be retained at the high luminosity running.

The LHCb upgrade will be a general purpose experiment in the forward region, covering beauty, charm, lepton-flavour-violation, QCD and exotica. It will probe or measure New Physics at the percentage level.

References

[1] "The LHCb Detector at CERN"; The LHCb Collaboration, JINST 3, S08005 (2008).

[2] "LHCb - Reoptimized Detector Design and Performance"; The LHCb Collaboration, Technical Design report, CERN/LHCC 2003-030, LHCb TDR 9 (9. Sep. 2003).

[3] "Observation of $D^0 - \overline{D^0}$ oscillations"; The LHCb Collaboration, submitted to PRL, CERN-PH-EP-2012-333, LHCb-PAPER-2012-038, arXiv:1211.1230v1; these proc.: A.Ukleja "Charm mixing and CPV at LHCb"

[4] "First Evidence of Direct CP Violation in Charmless Two-Body Decays of B^0_s Mesons"; The LHCb Collaboration, PRL 108, 201601 (2012); these proc.: D.Derkach "Studies of CP violation in charmless B decays"

[5] "Averages of b-hadron, c-hadron, and tau-lepton properties as of early 2012"; Heavy Flavour Averaging Group (HFAG), arXiv:1207.1158v1; these proc.: J.vanLeerdam "Measurement of the CP violating phase in B^0_s mixing"

[6] "Differential branching fraction and angular analysis of the $B \to K^{*0}\mu^+\mu^-$ decay"; The LHCb Collaboration, LHCb-CONF-2012-008; 47th Rencontres de Moriond on QCD and High Energy Interactions, La Thuile, Italy, 10 - 17 Mar 2012; these proc.: K.A.Petridis "Studies of $b \to s(d)\mu\mu$ EW penguin transitions at LHCb"

[7] "Measurement of the B^0–$\bar{B}^0$ oscillation frequency Δm_d with the decays $B^0 \to D^-\pi^+$ and $B^0 \to J/\Psi K^{*0}$"; The LHCb Collaboration, LHCb-PAPER-2012-032, CERN-PH-EP-2012-315, arXiv:1210.6750v1

[8] "First observation of the decays $\bar{B}^0_{(s)} \to D_s^+ K^- \pi^+ \pi^-$ and $\bar{B}^0_s \to D_{s1}(2536)^+\pi^-$"; The LHCb Collaboration, Phys. Rev. D 86, 112005 (2012)

[9] "First evidence for the decay $B_s \to \mu^+\mu^-$"; The LHCb Collaboration, Phys. Rev. Lett. 110, 021801 (2013); these proc.: S.Oggero "Very rare decays at LHCb"

[10] "Letter of Intent for the LHCb Upgrade"; The LHCb Collaboration, CERN/LHCC-2011-001, LHCC-I-018 - 2011

[11] "Framework TDR for the LHCb Upgrade: Technical Design Report"; The LHCb Collaboration, CERN/LHCC-2012-007, LHCb-TDR-12 - 2012

[12] "CP asymmetry in B $\rightarrow \phi K_S$ decays in left-right models and its implications on B_s decays"; M.Raidal, Phys.Rev.Lett. 89 (2002) 231803

[13] "B $\rightarrow V_L V_L$ Decays at Next-to-Leading Order in QCD"; M.Bartsch, G.Buchalla, Ch.Kraus, arXiv:0810.0249v1

[14] "Measurement of the polarization amplitudes and triple product asymmetries in the $B^0_s \rightarrow \phi\phi$ decay"; The LHCb Collaboration, Phys. Lett. B713 369-377 (2012)

[15] "Evidence for CP Violation in Time-Integrated $D^0 \rightarrow h^- h^+$ Decay Rates"; The LHCb Collaboration, PRL 108, 111602 (2012)

[16] "$\Delta F = 2$ Observables and Fine-Tuning in a Warped Extra Dimension with Custodial Protection"; M.Blanke, A.J.Buras, B.Duling, S.Gori, A.Weiler, JHEP 0903:001,2009

[17] "Small-x QCD at the LHC with protons & nuclei"; D.dEnterria, Tserruya Symposium 2012, Weizman Institute, Israel

[18] "Architectural modeling of pixel readout chips Velopix and Timepix3"; T.Poikela et al., JINST 7 (2012) C01093

[19] "Photonic Devices 2013"; Catalogue, Hamamatsu, http://www.hamamatsu.com/resources/pdf/etd/p-dev_2013_TOTH0021E01.pdf
"Photomultiplier Tubes and Assemblies for Scintillation Counting & High Energy Physics"; Specifications, Hamamatsu, http://www.hamamatsu.com/resources/pdf/etd/High_energy_PMT_TPMO0007E03.pdf

Journal of Physics: Conference Series **447** (2013) 012047 doi:10.1088/1742-6596/447/1/012047

Searches for Heavy Quarks at the ATLAS experiment

Antonella Succurro, on behalf of the ATLAS Collaboration
Institut de Física d'Altes Energies, Edifici Cn, Universitat Autònoma de Barcelona, E-08193 Bellaterra, Spain

E-mail: succurro@ifae.es

Abstract. During 2011 the ATLAS experiment at CERN collected around 5 fb^{-1}of proton-proton collision data from the LHC at the center-of-mass energy of $\sqrt{s} = 7$ TeV. At such high energy frontier it is possible to explore possible new physics scenarios. Amongst the models of physics beyond the Standard Model (SM) some predict the existence of exotic heavy quarks, which would help to explain matter-antimatter asymmetry and to solve the hierarchy problem. Several analyses at ATLAS were performed to search for a chiral 4th generation of quarks with charges +2/3 and -1/3 (t' and b' respectively) and for vector-like quarks with charges of +2/3, -1/3, +5/3, -4/3, coming in singlets or doublets depending on the model. The recent discovery of a Higgs-like, SM-like boson with mass of $\sim$ 125 GeV makes the vector-like quark scenario the most interesting one. The status of these searches and the limits set at 95% Confidence Level (CL) are presented.

1. Introduction

The ATLAS experiment [1] at CERN, Geneva, is a general purpose experiment equipped with a vertex detector (a silicon pixel detector and a silicon microstrip detector) and a central transition radiation tracker as the innermost three layers, surrounded by a superconducting solenoid which provides a magnetic field of 2 Tesla. Charged particle momenta are measured by this system for pseudorapidities $|\eta| < 2.5$[1]. The energy measurement is provided by the calorimeters, the electromagnetic sampling calorimeter (with lead and liquid-argon as passive and active material respectively) and the hadronic calorimeters, which use scintillating tiles alternated with steel in the central region and the liquid-argon technology in the end-cap and forward regions with, respectively, copper and tungsten as absorbers. Muons are detected in the muon spectrometer, a system of superconducting air-core toroids, trigger chambers and high-precision tracking chambers.

During 2011 data taking at the Large Hadron Collider (LHC), ATLAS showed excellent performance by collecting more than 5 fb^{-1}of data (see Figure 1). The full dataset available for analyses after data quality requirements is 4.7 fb^{-1}. At these values of integrated luminosity, a large amount of heavy quarks would be produced. An example of their production cross section as a function of the heavy quark mass $m_{q'}$ is shown in Figure 2, taken from [2]. Values for a

[1] ATLAS uses a right-handed coordinate system with its origin at the nominal interaction point (IP) in the centre of the detector and the z-axis along the beam pipe. The x-axis points from the IP to the centre of the LHC ring, and the y axis points upward. Cylindrical coordinates (r, ϕ) are used in the transverse (x, y) plane, ϕ being the azimuthal angle around the beam pipe. The pseudorapidity is defined in terms of the polar angle θ as $\eta = -\ln\tan(\theta/2)$.

Journal of Physics: Conference Series **447** (2013) 012047 doi:10.1088/1742-6596/447/1/012047

Table 1. Summary of decay modes for vector-like quarks (VLQ) isosinglets and isodoublets. BRs depend on $m_{q'}$ and are very model dependent.

Singlets	Decay modes	Singlets	Decay modes	Doublets	Decay modes	Doublets	Decay modes	Doublets	Decay modes
$T(+2/3)$	W^+b, Ht, Zt	$X(+5/3)$	W^+t	$\begin{pmatrix} T \\ B \end{pmatrix}$	W^+b, Ht, Zt	$\begin{pmatrix} T \\ X \end{pmatrix}$	Ht, Zt	$\begin{pmatrix} B \\ Y \end{pmatrix}$	Hb, Zb
$B(-1/3)$	W^-t, Hb, Zb	$Y(-4/3)$	W^-b		W^-t, Hb, Zb		W^+t		W^-b

center-of-mass energy of 7 TeV for *e.g.* a vector-like quark with mass of 600 GeV are about 15 times lower. Pair production via the strong interaction is dominant up to $m_{q'} \sim \mathcal{O}(750 \text{ GeV})$ while for higher masses single production via the electroweak interaction dominates.

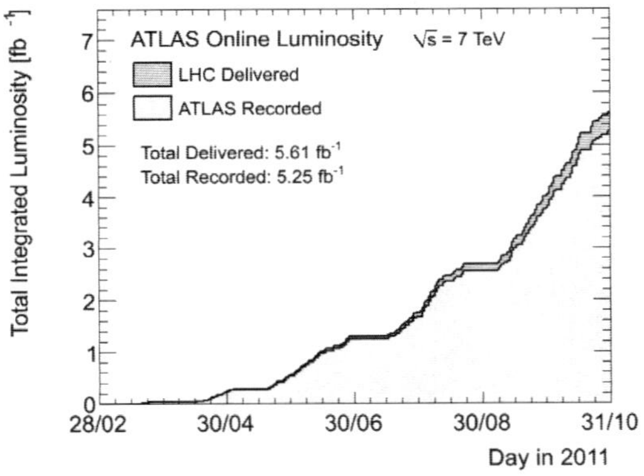

Figure 1. Performance of data acquisition of the ATLAS experiment for 2011. Of the 5.61 fb^{-1}delivered by the LHC, ATLAS recorded 5.25 fb^{-1} [3].

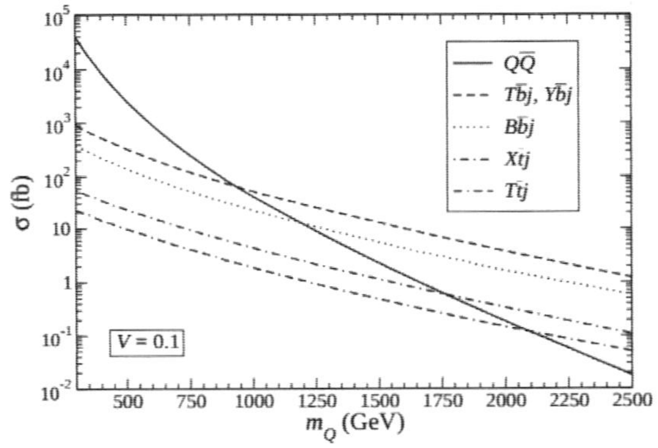

Figure 2. Pair and Single production cross sections for heavy quarks in proton-proton collisions at $\sqrt{s} = 14$ TeV [2].

The theoretical motivation for exotic heavy quarks searches is given by many models predicting their existence, see e.g. [2, 4, 5, 6]. Even though a new generation of quarks (whose families are constrained by asymptotic freedom to 9) would provide new sources of CP violation while staying consistent with precision electroweak measurements, it would also modify the SM predictions of the Higgs boson cross section and Branching Ratios (BR), which is in contraddiction with what was recently observed [7]. Still, some models allow for it, but the interest is now shifting towards vector-like quark models.

These heavy quarks with charges of +2/3, -1/3, +5/3, -4/3, named respectively T, B, X, Y, are weak-isospin singlets, doublets or triplets, whose left and right components transform the same under $SU(2) \times U(1)$. Table 1 shows the possible decay modes for isosinglets and isodoublets.

The analyses presented in the following have been performed using different fractions of the full 2011 dataset depending on the time of publication. Only pair produced heavy quarks are considered and events are selected in single lepton or dilepton channels.

2. Object definition

The searches for heavy quarks at the ATLAS experiment use common definitions for reconstructed objects. Jets are topological clusters reconstructed with the `AntikT4` algorithm [8] with transverse momentum p_T >25 GeV and pseudorapidity $|\eta|$ <2.5. Electrons are well isolated calorimeter deposits which are matched to tracks from the inner detector with transverse energy E_T >25 GeV and $|\eta|$ <2.47 after removing the pseudorapidity region [1.37,1.52], which corresponds to the transition region between barrel and endcap calorimeters. Muons are reconstructed from a segment in the tracker and muon detector matched to an isolated track in the inner detector with $p_T >$ 20 GeV and $|\eta|$ <2.5. Some analyses use a b tagging algorithm [9] to identify jets coming from a b quark. The discriminant value is chosen as to get an efficiency of $\sim$ 70% in simulated $t\bar{t}$ events.

2.1. Common selections

All the analyses require the primary vertex to have at least four tracks to reject background from cosmic radiation and to reject pileup events. Also, single lepton triggers are common to all the analyses.

For analyses in the lepton+jets channel, exactly one lepton is required and a veto is applied on any other lepton of the same or opposite flavor and the following cuts on missing transverse energy $\not{E}_T$ and transverse mass of the leptonically-decaying W, $m_T(W)$[2] are imposed:

- in the e+jets channel: $\not{E}_T$ >35 GeV, $m_T(W)$ >25 GeV
- in the μ+jets channel: $\not{E}_T$ >20 GeV, $\not{E}_T + m_T(W)$ >60 GeV

For analyses in the dilepton channel, at least two leptons are required. For same flavor $ee/\mu\mu$ analyses in general a Z veto $|m_{ll} - m_Z| < 10$ GeV is applied.

3. Searches for Heavy Tops

Two analyses are presented searching for $t' \to Wq$. The analysis in the lepton+jets channel assumes $q = b$ and interprets the search, performed using the full 2011 dataset, in terms of fourth generation quark as well as of vector like quarks $Y(-4/3)$ and $T(+2/3)$. The analysis in the dilepton channel uses 1.04 fb^{-1}of data and does not make assumptions on the quark flavor q, thus being extendable to heavy bottom final states. Both analyses use a binned maximum-likelihood ratio to probe Signal+Background and Background-only hypotheses with the CL_s method (see e.g. [10, 11]) and set 95% CL limits on the $t'\bar{t}'$ production cross section.

3.1. Lepton+jets analysis with 4.7 fb^{-1}

Assuming a 100% decay BR for $t' \to Wb$, this analysis [12] focused on the kinematical differences between a low mass boosted top quark and a high mass t' quark produced in pairs. Because of its high mass, the t' is produced almost at rest and will decay into highly boosted and energetic objects. A hadronically decaying boosted W boson can be detected as two light jets very close to each other or even merged into a single one. Two reconstructions are thus defined:

- W_{had}^{typeI}: a single jet with $p_T >$ 250 GeV and mass in the [60,110] GeV window;
- W_{had}^{typeII}: a di-jet system with $\Delta R(j,j) < 0.8$, $p_T >$ 150 GeV and mass in the [60,110] GeV window.

The following cuts are applied in addition to the common ones described in Section 2.1:

- e channel: cut on $m_T(W)$ substituted by $\not{E}_T + m_T(W)$ >60 GeV;

[2] The defintion for the transverse mass is $m_T(W) = \sqrt{2p_T^l \not{E}_T(1 - \cos\Delta\phi)}$, where p_T^l is the p_T of the lepton and $\Delta\phi$ is the azimuthal angle separation between the lepton and $\not{E}_T$ directions.

- At least 3 jets and one W_{had}^{typeI} OR at least 4 jets and one W_{had}^{typeII} and no W_{had}^{typeI};
- $H_T > 750$ GeV[3];
- At least 1 b-tagged jet and, after identifying two b-jet candidates b_1, b_2 by looking at their discriminant weight, cut on the highest-p_T one and the second-hardest-p_T one, respectively, at $p_T > 160$ GeV, $p_T > 60$ GeV;
- $\Delta R(l, \nu) < 1.4$, where the neutrino is reconstructed by constraining the mass of the lepton-neutrino system to the W boson mass[4].

Further tight isolation requirements are applied to the previous "loose" selection, assuming the boosted top quark background will have close-by products: $min(\Delta R(l, b_{1,2})) > 1.4$ and $min(\Delta R(W_{had}, b_{1,2})) > 1.4$. Mass reconstruction is then performed by resolving the ambiguity from b-jet assigment and ν solutions by choosing the hadronic mass from the combination yielding the lower difference with leptonic side. Figure 3 shows the reconstructed mass after the "tight" selections for data, SM background and two different heavy quark signals: a chiral fourth generation t' and a vector-like t'.

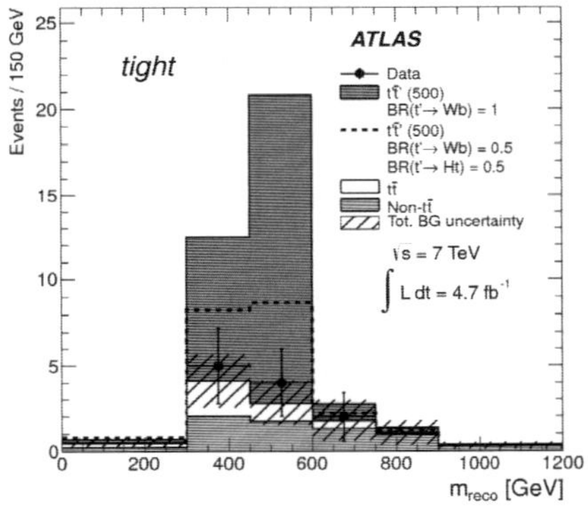

Figure 3. Distribution of reconstructed mass for the combined e+jets and μ+jets channels after the loose selection. The data (solid black points) are compared to the SM background prediction (stacked histograms). The total uncertainty on the background estimation is shown as a black hashed band. Also shown, stacked on top of the SM background, are the expected contributions from a signal with mass $m_{t'}$=500 GeV for the case of BR($t' \to Wb$)=1 (plain red histogram), corresponding to a fourth-generation t' quark, as well as the case of BR($t' \to Wb$)=BR($t' \to Ht$)=0.5 (dashed black histogram). The overflow has been added to the last bin [12].

Using the reconstructed mass as discriminant variable, a log-likelihood ratio is built to derive limits with the CL_s technique [10, 11]. A chiral fourth generation quark (as well as a vector-like $Y(-4/3)$, which would have exactly the same final state if BR($t' \to Wb$)=1) is excluded at 95% CL for masses up to 656 GeV as shown in Figure 4. Re-interpreting the result also in terms of vector-like $T(2/3)$ is possible once the two additional possible decay channels $t' \to Zt$ and $t' \to Ht$ are considered. To be model-independent, a mixing plane is defined between $BR(T \to Wb)$ and $BR(T \to Ht)$, while $BR(T \to Zt)$ is given by $\sum BRs = 1$, and each point tested in the same way as before. The singlet scenario is excluded up to $m_T = 500$ GeV.

3.2. Opposite Sign Dilepton analysis with 1.04 fb^{-1}

This analysis [13] performs a search for a pair-produced fourth generation up quark decaying $t'\bar{t}' \to WqWq$ with $q = d, s, b$ and both W bosons decaying leptonically. By not requiring any additional information on the quark flavor, this analysis can be extended to searches for $b'\bar{b}' \to WqWq$ with $q = u, c$. The final state has two opposite sign leptons, either of same or different flavor. Under the boosted regime assumption for the heavy t', the leptons and their neutrinos are considered collinear. Additional analysis requirements with respect to the ones described in Section 2.1 are:

[3] The H_T variable is defined as the sum of the first three or four leading jets p_T, the lepton p_T and the $\not{E}_T$.

[4] This gives rise to two possible solutions. If no real solution is found, assuming the neutrino will be collinear to the lepton in a boosted regime, pseudorapidity of the neutrino is set to that of the lepton.

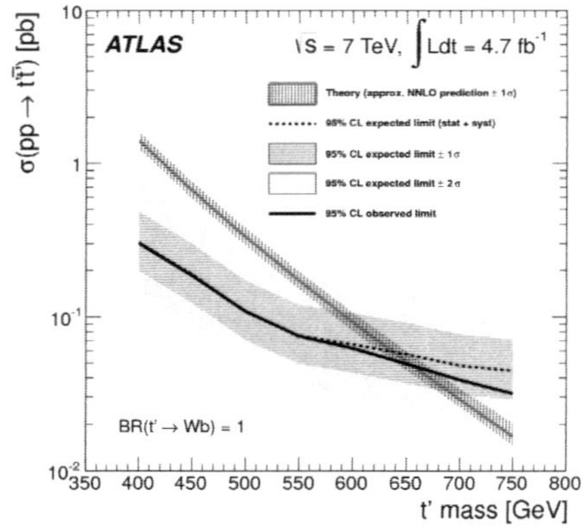

Figure 4. Observed (solid line) and expected (dashed line) 95% CL upper limits on the $t'\bar{t}'$ cross-section as a function of the t' quark mass. The surrounding shaded bands correspond to the ±1 and ±2 standard deviations around the expected limit. The thin red line and band show the theoretical prediction and its ±1 standard deviation uncertainty [12].

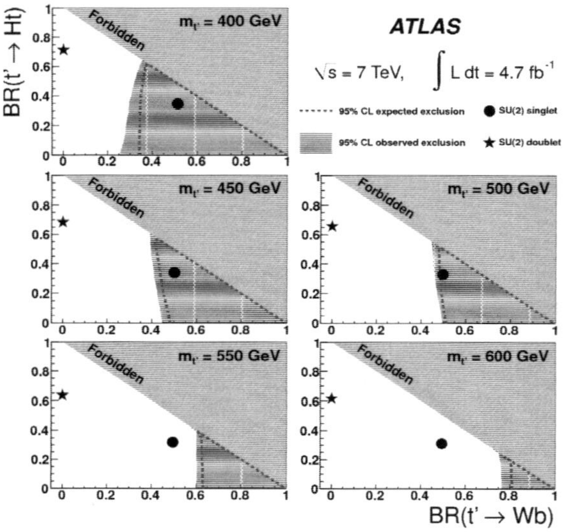

Figure 5. Observed (red filled area) and expected (red dashed line) 95% CL exclusion on the plane of BR($t' \to Wb$) vs BR($t' \to Ht$), for different values of the vector-like t' quark mass. The grey solid area corresponds to the unphysical region where the sum of branching ratios exceeds unity. The default branching ratio values from the PROTOS event generator for the weak-isospin singlet and doublet cases are shown as plain circle and star symbols, respectively. While the doublet scenario is not accessible, the singlet scenario is excluded up to $m_T = 500$ GeV [12].

- At least two jets, without any b-tagging requirement;
- For same flavor: $\not{E}_T > 60$ GeV;
- For different flavor: $H_T > 130$ GeV[5];
- $m_{ll} > 15$ GeV, with Z mass veto to remove most of the background.

The discriminant variable for the analysis is the collinear mass $m_{Collinear}$, defined as the average between the two masses recontructed by making the assumption that the lepton and the neutrino are collinear and the missing energy is the neutrino's energy. Events with mass difference $|\Delta m| > 25$ GeV are discarded. The efficiency of this cut for the signal is 99%.

Some additional cuts, optimized per t' mass, are applied for $t\bar{t}$ discrimination (see [13] for details). The resulting distibution for data, SM background and a signal is shown in Figure 6. This distribution is used to derive a binned maximum-likelihood ratio, and an observed 95% CL limit on fourth generation t' and b' masses is set at 350 GeV using the CL_s method [10, 11] (see Figure 7).

[5] H_T is defined as the scalar sum of the transverse energy of each lepton and jet passing object selection cuts.

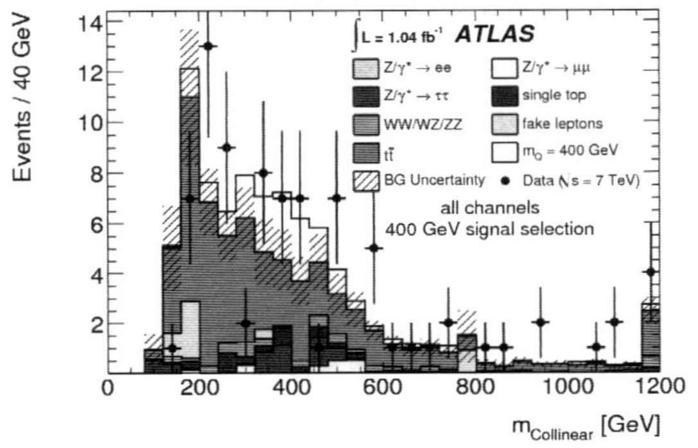

Figure 6. Distributions of $m_{Collinear}$ for the sum of ee, $\mu\mu$, and $e\mu$ channels after applying the final selection for $m_Q = 400$ GeV. The last bin contains overflow events. The uncertainty bands include all statistical and systematic background uncertainties [13].

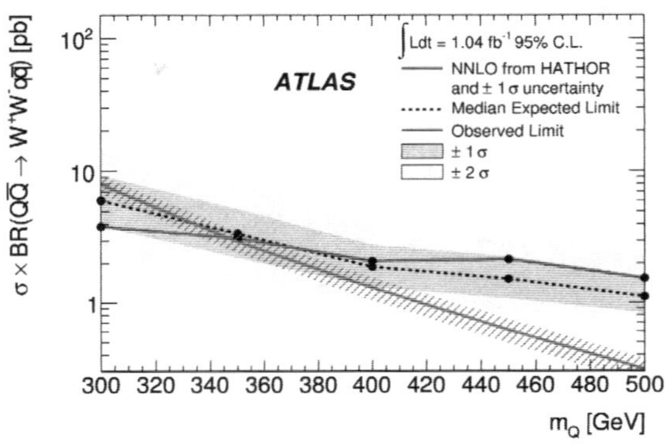

Figure 7. Observed and median expected 95% CL cross-section upper limits on $Q\bar{Q}$ production, compared to the theoretical prediction. The limit was calculated for five signal masses, and a linear interpolation has been made between mass points [13].

4. Searches for Heavy Bottoms

Two analyses are presented searching for pair-produced $b'\bar{b}'$, one looking for $b' \to Wt$ decays and one looking for $b' \to Zb$ decays. In the case of a chiral fourth-generation b' the neutral current mode would occur through loop diagrams, being then highly disfavored once the phase space allows the tree-level decay into a top and a W (i.e. $m_{b'} > 255$ GeV). For a vector-like b' on the contrary, $b' \to Zb$ would be a tree-level decay, thus leaving the branching ratio to this channel to be determined by the specific model. The analysis with the full 2011 dataset in the same-sign dilepton channel performs a search for a chiral b' decaying into Wt re-interpreting the result in terms of vector like quarks $X(+5/3)$ and $B(-1/3)$. The analysis in the opposite-sign dilepton channel searches for vector-like b' decaying into Zb, using 2.0 fb^{-1}of data. Both the analyses use the CL_s method (see e.g. [10, 11]) and set 95% CL limits on $b'\bar{b}'$ production cross section.

4.1. Same-sign Dilepton analysis with 4.7 fb^{-1}

For this analysis [14] the interesting decays have four W bosons (pair-produced chiral fourth-generation b' or vector-like $X(5/3)$) or two same sign W bosons (singly-produced vector-like $X(5/3)$). The resulting final state of two isolated same-sign leptons has the advantage of having very few SM processes as possible background, which consists mainly of fakes and charge flips.

Additional analysis requirements with respect to the ones presented in Section 2.1 are:

- At least two jets of which at least one b-tagged
- Leading lepton $p_T > 25$ GeV
- For same flavor leptons $m_{ll} > 15$ GeV with Z mass veto
- $\not{E}_T > 40$ GeV and $H_T > 550$ GeV[6]

Four events have been observed in the signal region out of 5.6±1.7 expected background events. For the three different models considered limits are then derived using the CL_s technique [10, 11]. For pair production of chiral b' and vector-like $X(5/3)$ masses up to 670 GeV are excluded at 95% CL (see Figure 8). In case of $X(5/3)$ single production, the cross section depends on the coupling constant λ of the $tWX_{5/3}$ vertex. Three values for λ are explored:

[6] H_T is defined as the scalar sum of all leptons and jets transverse momenta.

Journal of Physics: Conference Series **447** (2013) 012047 doi:10.1088/1742-6596/447/1/012047

$\lambda \ll 1$ (equivalent to pair production), $\lambda = 1$ and $\lambda = 3$, and the observed limits are set at 680 GeV and 700 GeV for the last two, respectively (see Figure 9).

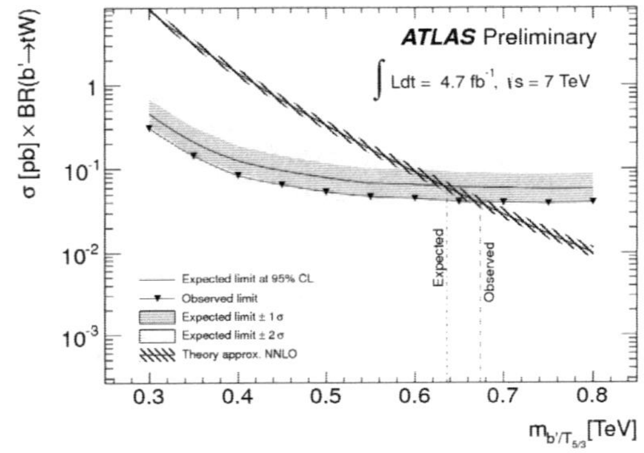

Figure 8. Expected and observed upper limits on the pair production cross section of the b' and $X(+5/3)$, as a function of their mass [14].

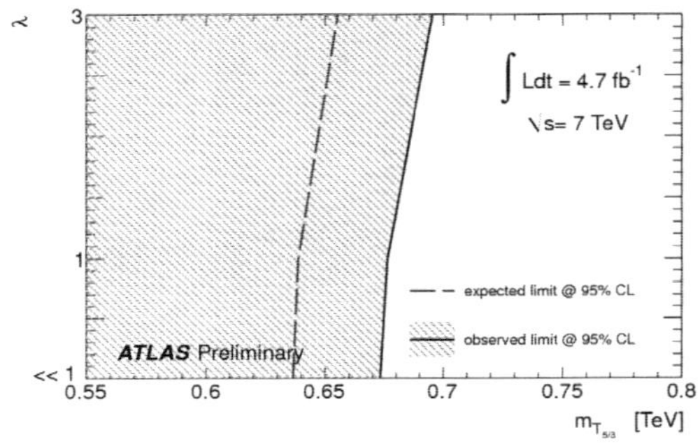

Figure 9. Expected and observed lower limits on the $X(+5/3)$ signal, as a function of the $X(+5/3)$ mass and the coupling constant λ. The shaded area is excluded at 95% CL [14].

4.2. Opposite-sign electron pair analysis with 2.0 fb^{-1}

This analysis [15] looks for pair production of vector-like b' decaying in the channel $b' \to Zb$. In addition to the common cuts presented in Section 2.1 the following conditions are required in order to fully reconstruct the Z boson and the b':

- At least one b-jet
- $|m_{ee} - m_Z| < 15$ GeV
- $p_T(Zb) > 150$ GeV

The b' candidate is reconstructed from an electron positron pair e^+e^- and the highest-p_T b-jet of the event and its invariant mass (show in Figure 10) used as a discriminant variable to compute the binned likelihood. Defining $\beta = 2 \times BR(b' \to Zb) - BR(b' \to Zb)^2$, this variable describes the fraction of signal events with at least one $b' \to Zb$ decay as a function of the BR, meaning that $\beta = 1$ is equal to $BR(b' \to Zb) = 1$ while for vector-like singlet $B(-1/3)$ β varies with the mass (e.g. $m_B = 200 - 700$ GeV corresponds to $\beta = 0.9 - 0.5$). Figure 11 shows the cross section limits obtained using the CL_s technique [10, 11] for two scenarios: the 100% decay into Zb and the singlet model. Observed limits are set respectively at 400 GeV and 358 GeV at 95% CL.

5. Conclusions

The ATLAS experiment is active in investigating new physics models with exotic heavy quarks. The latest searches have excluded various scenarios at 95% CL using data collected in 2011: chiral 4th generation t' (100% to Wb) and vector like $Y(-4/3)$ are excluded up to 656 GeV [12]; chiral 4th generation t' (to Wq, $q = d, s, b$) are excluded up to 350 GeV [13]; chiral 4th generation b' (100% to Wt) and vector like $X(5/3)$ are excluded up to 670 GeV [14]; vector like b' (100% to Zb) are excluded up to 400 GeV [15]; vector like $B(-1/3)$ singlet model is excluded up to 358 GeV [15]; vector like $T(2/3)$ singlet model is excluded up to 500 GeV [12].

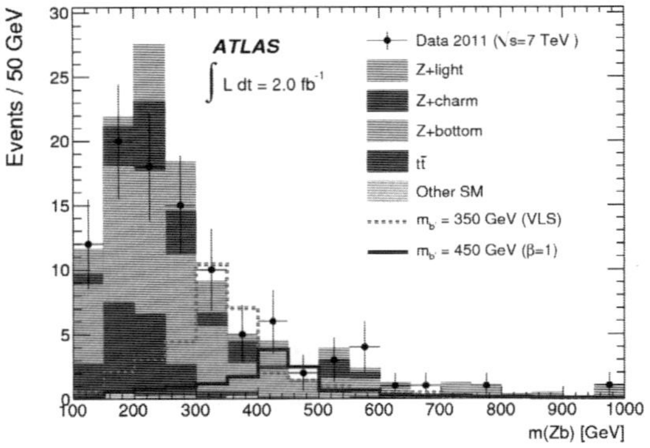

Figure 10. Mass distribution of the b' candidate in events passing the selection. The predicted contributions of the SM background sources are stacked, while the distributions for the two signal scenarios described in the text are overlaid. The highest mass bin also includes the data and prediction for $m(Zb) >$ 1 TeV [15].

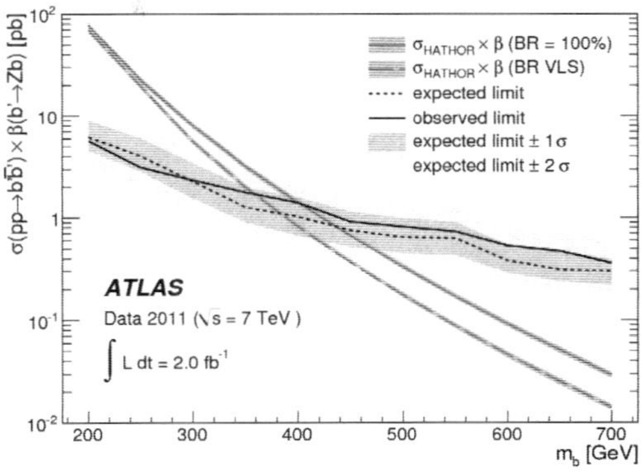

Figure 11. The expected and observed 95% CL cross section limits as a function of b' mass. The signal cross section is shown with uncertainties arising from PDFs and renormalization and factorization scale choice. The prediction is also multiplied by the beta factors described in the text [15].

References

[1] ATLAS Collaboration 2008 *Journal of Instrumentation* **3** S08003 URL `http://stacks.iop.org/1748-0221/3/i=08/a=S08003`
[2] Aguilar-Saavedra J 2009 *JHEP* **0911** 030 (*Preprint* `0907.3155`)
[3] ATLAS Collaboration URL `https://twiki.cern.ch/twiki/bin/view/AtlasPublic/LuminosityPublicResults`
[4] Holdom B, Hou W, Hurth T, Mangano M, Sultansoy S *et al.* 2009 *PMC Phys.* **A3** 4 (*Preprint* `0904.4698`)
[5] Cetin S, Hou G, Ozcan V, Rozanov A and Sultansoy S 2011 (*Preprint* `1112.2907`)
[6] Martin S P 2010 *Phys.Rev.* **D81** 035004 (*Preprint* `0910.2732`)
[7] ATLAS Collaboration 2012 *Phys.Lett.* **B716** 1–29 (*Preprint* `1207.7214`)
[8] Cacciari M, Salam G P and Soyez G 2008 *JHEP* **0804** 063 (*Preprint* `0802.1189`)
[9] ATLAS Collaboration 2012 Measurement of the b-tagging efficiency in a sample of jets containing muons with 5 fb^{-1} of data from the ATLAS detector Tech. Rep. ATLAS-CONF-2012-043 CERN Geneva
[10] Junk T 1999 *Nucl.Instrum.Meth.* **A434** 435–443 (*Preprint* `hep-ex/9902006`)
[11] Read A L 2002 *J.Phys.* **G28** 2693–2704
[12] ATLAS Collaboration 2012 *Phys.Lett.* **B718** 1284–1302 (*Preprint* `1210.5468`)
[13] ATLAS Collaboration 2012 *Phys.Rev.* **D86** 012007 (*Preprint* `1202.3389`)
[14] ATLAS collaboration 2012 Search for exotic same-sign dilepton signatures (b' quark, $T_{5/3}$ and four top quarks production) in 4.7/fb of pp collisions at $\sqrt{s} = 7$ TeV with the ATLAS detector Tech. Rep. ATLAS-CONF-2012-130 CERN Geneva
[15] ATLAS Collaboration 2012 *Phys.Rev.Lett.* **109** 071801 (*Preprint* `1204.1265`)

Results on the Search for MSSM Neutral and Charged Higgs bosons (CMS)

Mia Tosi
Padova University and INFN, via Marzolo 8, 35131 Padova (Pd), IT
E-mail: mia.tosi@gmail.com

Abstract. In the minimal super-symmetric extension of the Standard Model (MSSM), the Higgs sector contains two Higgs boson doublets, including, after electroweak symmetry breaking, the CP-odd neutral scalar A^0, the two charged scalars $H^{\pm}$, and the two CP-even neutral scalars h and H^0. The neutral Higgs boson is searched in the $\mu^+\mu^-$, $\tau^+\tau^-$, and recently also in the $b\bar{b}$, channels, whereas the charged Higgs state is searched in top quark decays with at least one τ in the final state. This report reviews the current status of searches for MSSM Higgs bosons with the data collected by the CMS experiment at LHC in the 2011 and 2012 operations.

1. Introduction

The precise nature of the boson recently discovered [1, 2] at the Large Hadron Collider remains open. In addition to measuring its properties to determine whether it is the quantum of the field that breaks electroweak symmetry and gives mass to the fundamental particles, the search for other Higgs particles predicted by models beyond the Standard Model (SM), has intensified. In this report, results are presented from searches for Higgs-like particles predicted within the minimal super-symmetric extension of the Standard Model (MSSM) framework [3].

The MSSM requires the presence of two Higgs doublets, which turn to five massive Higgs bosons: a light neutral scalar (h), two charged scalars ($H^{\pm}$), a heavy neutral CP-even state (H) and a neutral CP-odd state (A). At the tree level the Higgs sector is controlled by two parameters, conventionally chosen as the mass of the pseudoscalar Higgs boson, m_A, and the ratio of the vacuum expectation values of the two Higgs doublets, $\tan\beta = v_2/v_1$. For $\tan\beta$ larger than unity, the Higgs field couplings to up-type particles are suppressed relative to the SM, while the couplings to down-type particles are enhanced by a factor of $\tan\beta$. Therefore, the combined cross section of Higgs boson production in association with b quarks is effectively enhanced by a factor $\approx \tan^2\beta$. In addition, the mass m_A is expected to be nearly degenerate with either m_h or m_H within the experimental resolution.

This report presents the searches for MSSM neutral and charged Higgs particles in proton-proton collisions at a center-of-mass energy of both 7 and 8 TeV, using data recorded by the Compact Muon Solenoid (CMS) experiment in 2011 and 2012 with an integrated luminosity of 4.8 and 12 fb^{-1}, respectively.

Journal of Physics: Conference Series **447** (2013) 012048 doi:10.1088/1742-6596/447/1/012048

2. The CMS detector and the event reconstruction

Details of the CMS detector and its performance can be found elsewhere [4].

The reconstructed primary vertex [7] with the largest p_{T}^2-sum of its associated tracks is selected and used as reference for the other physics objects. The CMS particle-flow (PF) event reconstruction [5, 6] is used for optimizing the reconstruction and the identification of all particles in the event, i.e. electrons, muons, photons, charged hadrons, and neutral hadrons, with an extensive combination of all CMS sub-detectors. Muons are reconstructed [8, 9] with a simultaneous global fit performed on hits in the tracker and the muon system. Electrons are reconstructed [10] from clusters of energy deposits in the electromagnetic calorimeter that are matched to tracker hits. Jets and the missing transverse energy are reconstructed using PF objects. Jets are clustered with the anti-kT algorithm [11] from particle-flow objects with a distance parameter of 0.5. The missing transverse energy vector is calculated as the negative vector sum of all particle transverse momenta, and its magnitude is referred to as $\not{E}_T$ [14]. The tagging of jets originating from the hadronization of a b quark (b jet) is performed using several algorithms [12], which, relying on the characteristics of the b quark, provide a good separation between b jets and jets of other flavors. The identification and reconstruction of hadronically decaying τ candidates [13], is accomplished, using tracking and calorimeter information, applying the Hadron plus Strips (HPS) algorithm, which considers candidates with one or three charged pions and up to two neutral pions.

3. Search for neutral MSSM Higgs bosons

The most interesting production processes of MSSM neutral Higgs bosons at the LHC, are the *gluon-gluon fusion*, for small and moderate values of $\tan\beta$ and the production in *association with bottom quarks*, which constitutes the dominant mechanism for large values of $\tan\beta$. In the latter case there is a significant probability to observe a jet from the hadronization of the b-quark in association with the Higgs boson in the detector.

CMS has studied the production of neutral MSSM Higgs bosons in three complementary channels: $\phi \to b\bar{b}$ [15, 16, 17], $\phi \to \tau^+\tau^-$ [18] and $\phi \to \mu^+\mu^-$ [19]. The decay to $b\bar{b}$ provides the largest branching fraction ($\approx 90\%$), but suffers from an overwhelmingly large background of multi-jet production. The branching fraction for the decay to $\tau^+\tau^-$ is approximately one order of magnitude lower, but the presence of two leptons, either electron or muon, or hadronically decaying taus in the event provides a large signal to background ratio. The decay to $\mu^+\mu^-$ has a very small branching fraction ($\approx 10^{-4} \div 10^{-3}$), offset by the advantages of very low background, high reconstruction efficiency for the muons, and excellent mass resolution.

3.1. MSSM neutral Higgs into $\tau^+\tau^-$

The search for $\phi \to \tau^+\tau^-$is performed in four final states: $\tau_h e$, $\tau_h \mu$, $e\mu$ and $\mu\mu$, where the electrons and muons arise from leptonic τ decays, and where τ_h denotes hadronic τ decays. The Higgs boson signal would appear as a broad excess in the distribution of the *full mass* of the tau pair, $M_{\tau^+\tau^-}$. In this analysis, the $M_{\tau^+\tau^-}$ reconstruction is done by a likelihood technique, combining the visible τ decay products and the missing transverse energy, achieving a resolution of about 20% on $M_{\tau\tau}$. The reconstructed tau-pair mass distribution in the $\tau_h e$ channel is shown in Figure 1 (left).

The search sensitivity is improved by classifying the events in two distinct event categories: events in the first category are required to contain a b-tagged jet with $p_T > 20\,GeV$ and $|\eta| < 2.4$, while events without such a jet are in the second category. Events with b-tagged jets are more likely to come from b-association production process, while the other events are candidates for gluon-gluon fusion production.

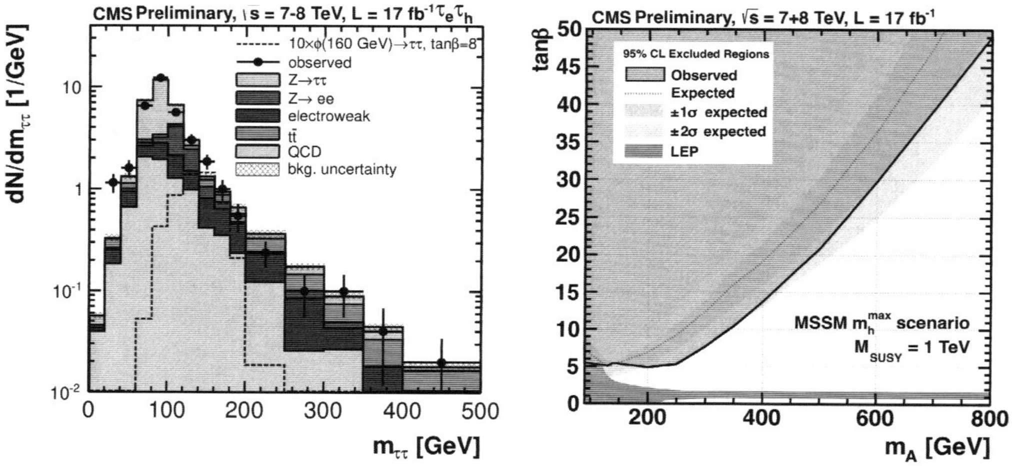

Figure 1. Left: the reconstructed tau-pair *full mass* distribution in the b-tag category for the $\tau_h e$ final states, comparing the observed distributions (points with error bars) to the sum of the expected backgrounds (shaded histograms). The contribution from a Higgs boson signal ($m_A = 160\ GeV$) is also shown, for the hypothesis $\tan\beta = 8$. Right: region in $\tan\beta$ versus m_A parameter space excluded at 95% CL in the context of the MSSM m_h^{max} scenario, by the 17 fb^{-1} of data collected by CMS in 2011 and 2012. The exclusion limits from the LEP experiments are also shown.

The online selections require the presence of an electron and/or a muon trigger object, and also special triggers requiring the presence of both a lepton and a charged track with an accompanying calorimeter pattern consistent with a τ decaying hadronically were adopted for the $\tau_h e$ and $\tau_h \mu$ channels.

In all final states W+jets and $t\bar{t}$ background events are reduced by requiring that the reconstructed $\not{E}_T$ points in the direction of the visible tau decay products. Events with additional leptons are rejected to suppress Z/γ^* processes. The contributions of multi-jet, $W+$ jets and the dominant irreducible $Z \to \tau^+\tau^-$ background are determined using data, while predictions for other background processes, such as $t\bar{t}$ and di-boson processes, are obtained from Monte Carlo (MC) simulation.

The events yield observed in the distributions of $M_{\tau\tau}$, in a dataset corresponding to an integrated luminosity of 17 fb^{-1}, with and without b-tagged jets, is in agreement with the expectation for SM background processes, and no evidence for a MSSM Higgs signal is seen.

Therefore, an upper bound on the product of the MSSM Higgs boson cross-section and the tau-pair branching fraction, $\sigma_\Phi \cdot BR_{\tau\tau}$, is set. The limit is computed as function of the mass of the pseudoscalar Higgs, m_A, by fitting the $M_{\tau^+\tau^-}$ distribution observed in all decay channels with shape templates for different Higgs mass hypotheses, obtained from MC simulation. These upper limits, interpreted in the MSSM parameter space of $\tan\beta$ versus m_A in the m_H^{max} scenario [20, 21], provide stringent bounds on $\tan\beta$, shown in Figure 1 (right).

3.2. Higgs into $b\bar{b}$

A search for MSSM neutral Higgs bosons produced in association with at least one b quark, and decaying into a pair of b quarks is performed using $2.7 - 4.8\ fb^{-1}$ of proton-proton collisions collected by CMS in 2011with a center-of-mass energy of 7 TeV.

A signal is searched for in final states characterized either purely by jets ("all-hadronic")

Journal of Physics: Conference Series **447** (2013) 012048 doi:10.1088/1742-6596/447/1/012048

or with an additional non-isolated muon ("semileptonic"). Events are selected by specialized triggers that exploit online algorithms for the identification of b jets to tackle the large hadronic interaction rate at the LHC. The selected datasets are largely exclusive, and the small overlap (of the order of 2.3%) is removed for the combined results.

The dominant background is the production of heavy-flavor multi-jet events, while other background processes, such as $t\bar{t}$+jets and Z $\rightarrow$ b$\bar{\text{b}}$+jets, are predicted by the MC simulation to be less than 1% of the total background.

The common analysis strategy is to search, in events identified as having at least three b jets, for an excess of events in the invariant mass distribution of the two leading b jets, M_{12}, over the large multi-jet background. The multi-jet background is not fully reducible nor well described by MC simulations, therefore a key point of both analyses is the estimation of the background using control data samples.

In the *all-hadronic* analysis the background modeling is done with a method similar to the

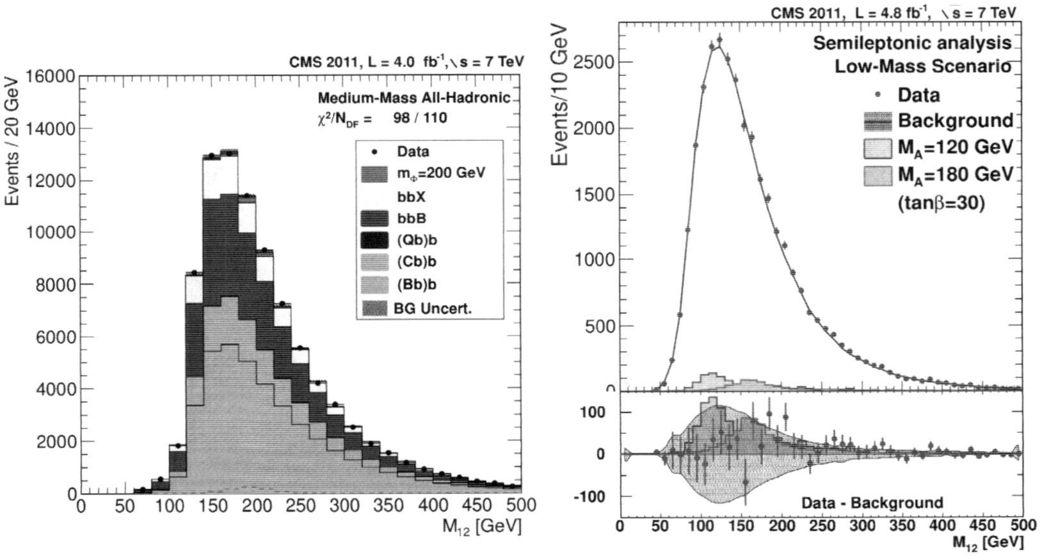

Figure 2. Left: results from the *all-hadronic* analysis. Di-jet mass distribution resulting from the fit with an additional signal template for a MSSM Higgs boson ($m_A = 200\ GeV$). The fitted mass distribution of the Higgs contribution is shown a second time as the dashed histogram at the bottom of the figure. Right: results from the semileptonic analysis. Data (red point) and predicted background (blue line) in the signal region; the expected signal for different m_A and for $\tan\beta = 30$ is also plotted. The difference between data and predicted background is also shown: the blue area represent the systematical and statistical uncertainties on the background prediction.

one used in Ref. [22]. Two-dimensional (2D) templates on M_{12} and EvtBTag (variable which reflects the b jet content of the event) of the different flavour compositions of the background are constructed from a double-b-tag data sample. Signal templates are built using MC simulations. A 2D fit of a linear combination of the background-only templates shows good agreement between data and the background estimation. According to the fit the three b-jets production is dominant contribution to the background. The signal yield is extracted by fitting a linear combination of signal and background templates. The result of the fit, projected on the M_{12} variable, for a Higgs boson mass hypothesis of 200 GeV is shown in Figure 2 (left).

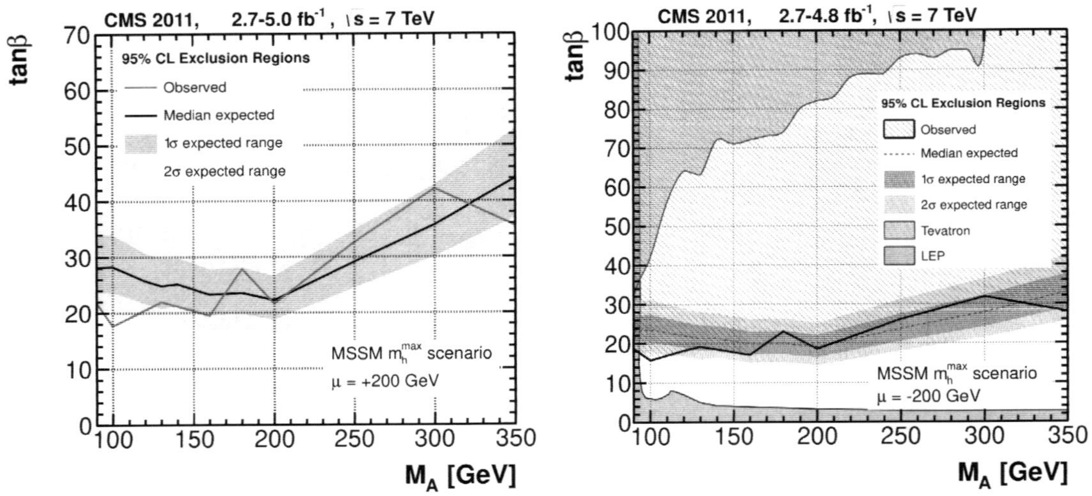

Figure 3. Observed upper limits at 95% CL on $\tan\beta$ as a function of m_A, including the statistical and systematic uncertainties, in the m_h^{max} scenario with $\mu = +200$ (-200) GeV for the combined all-hadronic and semileptonic results on the left (right). One- and two-standard deviation ranges for the expected upper limit are presented. Previous exclusion regions from LEP [23] and Tevatron the multi-b-jet channel [22] are overlaid.

In the *semileptonic* analysis two methods have been developed to predict the expected background. The first one, using the double-b-tag sample bbj, is based on the computation of b-tagging probabilities of the third jet to predict the background in the signal region; and the second one, which makes use of the bjj and bjb samples, is based on a nearest-neighbor-in-parameter-space technique. Because the two methods use exclusive data samples, their results is combined by performing a weighted average of their bin-by-bin predictions. The predicted background is shown in Figure 2 together with the expected signal for two Higgs boson masses at $\tan\beta = 30$. A binned likelihood fit to the invariant mass distribution of the two leading jets in the event is used in order to extract a possible MSSM Higgs contribution.

No significant deviation from the SM background is observed in neither the *all-hadronic* nor the *semileptonic* analysis. Both analysis are combined and the 95% CL upper limit on the signal contribution in the data, in terms of cross section times branching fraction into $b\bar{b}$, is extracted. Figure 3 (left) presents the results in the MSSM framework as a function of the MSSM parameters m_A and $\tan\beta$, in the m_h^{max} scenario, with $\mu = +200$ GeV. Figure 3 (right) shows the results in the scenario with $\mu = -200$ GeV, together with previous limit set by LEP [23] and Tevatron [24].

3.3. Higgs into $\mu^+\mu^-$

Events in the $\phi \to \mu^+\mu^-$ channel are selected by requiring two isolated muons of opposite charge, with $p_T > 30$ GeV and 20 GeV for the leading and the subleading muon, respectively, within the pseudo-rapidity range $|\eta| < 2.1$.

In order to enhance the sensitivity of the search, the statistical analysis is performed in three distinct event categories: events in the first category are required to contain a b-tagged jet of $p_T > 20$ GeV and $|\eta| < 2.1$; the second category is defined by events with at least three muons in the final state, in this case the selection on the third muon is relaxed ($p_T > 3$ GeV and $|\eta < 2.4$); while events which satisfy none of these requirements enter in the third category. Events with

Journal of Physics: Conference Series **447** (2013) 012048 doi:10.1088/1742-6596/447/1/012048

b-tagged jets or enriched by soft muon are more likely to come from b-association production process, while the other events are candidates for gluon-gluon fusion production.

The events coming from $t\bar{t}$ processes are suppressed by cutting on the $\not{E}_T$. DrellYan muon pair production $Z/\gamma^* \to \mu^+\mu^-$ is the dominant background. The contribution of this irreducible background is estimated by fitting the distribution of the leading two muon mass, $M_{\mu\mu}$, in sidebands of the region where the signal is expected. The result of the fit is then extrapolated into the signal region.

The distribution of $M_{\mu\mu}$ observed in data is shown in Figure 4 (left), as well as for the expected backgrounds and signal sample. The shape of the signal exhibits two peaks: one representing the light scalar h and the other representing heavy scalar H plus pseudoscalar A. The contribution of individual Higgs bosons can be resolved owing to the excellent mass resolution, about $2 \div 3\%$ in this channel.

No evidence for a MSSM Higgs boson signal is observed in the analyzed dataset corresponding to an integrated luminosity of 5.0 fb^{-1}, and therefore an upper limit at 95% CL is set on the cross section times the branching fraction into two muons as function of the pseudoscalar Higgs boson mass. Such a limit is interpreted within the MSSM framework in the m_h^{max} scenario and the region in MSSM $\tan\beta$ versus m_A parameter space excluded by this analysis is shown in Figure 4 (right).

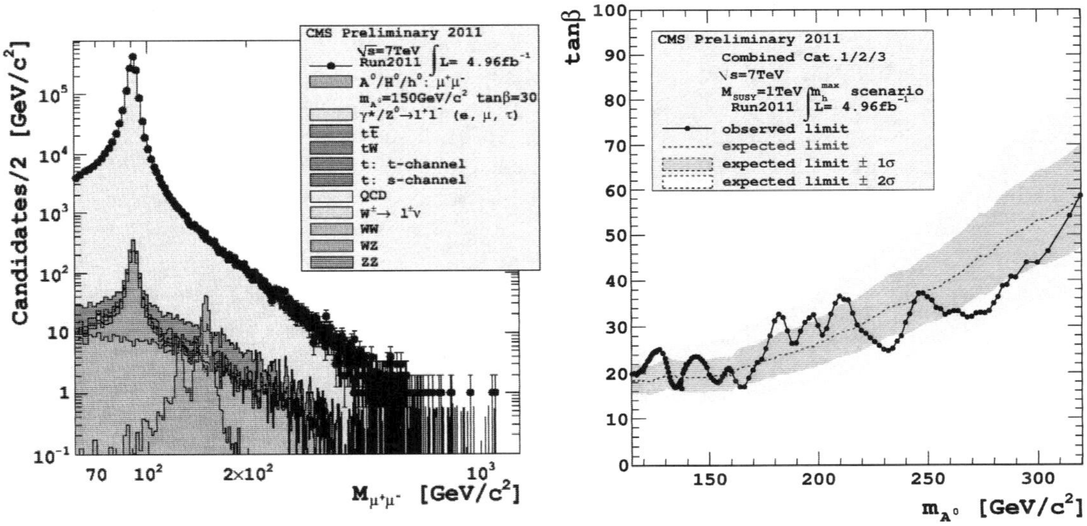

Figure 4. Left: di-muon invariant mass distribution for the sum of the different categories. The expected $M_{\mu\mu}$ distribution for the decay $\phi \to \mu^+\mu^-$, with the hypothesis $m_A = 150\ GeV$ and $\tan\beta = 30$ is superimposed. The different background contributions are reported as well. Right:the combined exclusion limit for the MSSM production cross-section times the BR at 95% CL projection in the $(m_A - \tan\beta)$ plane (b).

4. Search for charged MSSM Higgs in top quark events

In CMS, the search for the charged Higgs boson is done in the low $H^{\pm}$ mass regime, when $m_{H^{\pm}} < m_{top}$ [25]. In such a scenario, the Higgs boson may be produced in top quark decays, $t \to H^{\pm}b$, and then, for $\tan\beta$ greater than 5, it preferentially decays to a τ lepton and a neutrino, $H^{\pm} \to \tau\nu_{\tau}$. The presence of such a decay mode process alters the τ lepton yield in the decay products of $t\bar{t}$ pairs compared to the SM predictions. The search for an excess of tau lepton events in the $t\bar{t}$ sample is, therefore, a direct way to search for beyond SM physics.

The upper limit on the branching fraction $BR(t \rightarrow H^{\pm}b) < 0.2$ has been set by the CDF and D0 experiments at the Tevatron for m_{H^+} between 80 and 155 GeV, assuming $BR(H^{\pm} \rightarrow \tau\nu_{\tau} = 1)$. More recently, the ATLAS experiment at the LHC has set the upper limit on the $BR(t \rightarrow H^{\pm}b)$ between 5% and 1% for charged Higgs boson masses in the range $90 \div 160$ GeV

CMS has studied three final states separately, using a dataset at 7 TeV corresponding to an integrated luminosity of $2.0 \div 2.3$ fb^{-1}, all requiring presence of a τ lepton from $H^{\pm}$ decays, missing transverse energy, and multiple jets: (1) a τ lepton that decays to hadrons accompanied by jets (τ_h+jets), (2) τ_h produced in association with an electron or a muon ($\tau_h e$ or $\tau_h\mu$), and (3) an electron and a muon ($e\mu$).

The largest background contributions in these analyses come from the irreducible $t\bar{t}$ di-tau channel, and from tau fakes, where one jet fakes the tau. The main background (τ-fake) is from events where one W boson is produced in association with jets, and from $t\bar{t} \rightarrow W^-bW^+\bar{b} \rightarrow l\nu_l qq'b\bar{b}$ events. The data driven method used in the estimate of the τ-fake background is described in Ref.[26]. The non-τ-fake backgrounds, like $Z/\gamma^* \rightarrow \tau^+\tau^-$, single top production, di-bosons, and the part of the SM $t\bar{t}$ background not included in the τ-fake background are estimated from MC.

A good agreement between data and SM expectations is found at various levels of the events selections. Therefore, an upper limit on the branching fraction $BR(t \rightarrow H^{\pm}b)$ is set, assuming $BR(H^{\pm} \rightarrow \tau\nu_{\tau}) = 1$. The inputs to the limit computation are: the number of observed and expected events in the $\tau_h e$, $\tau_h\mu$ and $e\mu$ channels; and the observed and expected distribution of the transverse mass (built by the τ_h candidate and the $\not{E}_T$) in the τ_h+jets channel. The upper limit on $BR(t \rightarrow H^{\pm}b)$ as a function of $m_{H^{\pm}}$ obtained from the combination of all final states is shown in Figure 5 (left). Figure 5 (right) shows the exclusion region in the MSSM $m_{H^+} - \tan\beta$ parameter space obtained from the combined analysis within the MSSM m_h^{max} scenario.

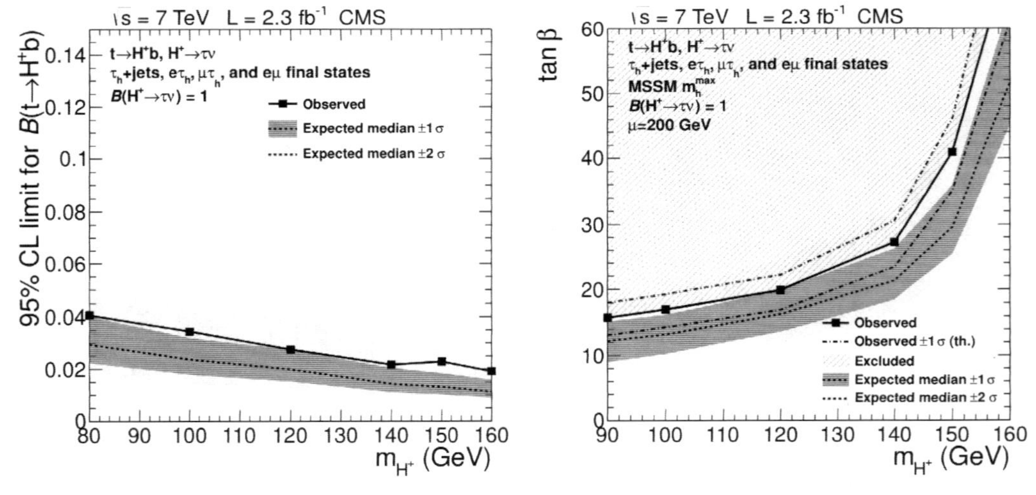

Figure 5. Left: the upper limit on BR(t→ H^+b) as a function of m_{H^+}, assuming $BR(H^+ \rightarrow \tau\nu) = 1$, obtained from the combination of the all final states. The green and yellow bands show the ±1 and $\pm2\sigma$ bands, respectively, around the expected limit. Right: the exclusion region in the MSSM $m_{H^+} - \tan\beta$ parameter space obtained from the combined analysis for the MSSM m_h^{max} scenario. The ±1 and ±2 sigma bands around the expected limit are also shown.

Journal of Physics: Conference Series **447** (2013) 012048 doi:10.1088/1742-6596/447/1/012048

5. Conclusion

The production of neutral and charged MSSM Higgs bosons in proton-proton collisions at $\sqrt{s} = 7$ and 8 TeV has been studied using data recorded by the CMS experiment. No evidence for a signal is found in a dataset corresponding to an integrated luminosity of $2.0 \div 17\ fb^{-1}$. The non-observation of a signal excludes previously unexplored regions of MSSM parameter space.

In particular, the $\phi \to \tau^+\tau^-$ analysis sets stringent new bounds in the $\tan\beta - m_A$ parameter space, excluding at 95% CL values of $\tan\beta$ as low as 6 up to $m_A = 250\ GeV$.

Moreover, the $\phi \to b\bar{b}$ channel provides the most stringent limits on neutral MSSM Higgs boson production and decay in the multi-b-jet mode and excludes the 2σ excess reported by CDF [22].

The $\phi \to \mu^+\mu^-$ analysis which has an excellent mass resolution ($2 \div 3\%$) does not see any tension with respect to SM predictions.

Finally, the search for a light charged Higgs boson provides stringent upper limits on the branching fraction $BR(t \to H^{\pm}b)$, in the range of $2 \div 4\%$ for charged Higgs boson masses between 80 and 160 GeV, under the assumption that $BR(H^{\pm} \to \tau\nu_\tau) = 1$.

References

[1] ATLAS Collaboration. Observation of a new particle in the search for the Standard Model Higgs boson with the ATLAS detector at the LHC. *Phys. Lett. B*, 716(1), 2012.
[2] CMS Collaboration. Observation of a new boson at a mass of 125 GeV with the CMS experiment at the LHC. *Phys. Lett. B*, 716(30), 2012.
[3] S. P. Martin. A Supersymmetry Primer. 1997, hep-ph/9709356.
[4] CMS Collaboration. The CMS experiment at the CERN LHC. *JINST*, 3(S08004), 2008.
[5] CMS Collaboration. Particle Flow Event Reconstruction in CMS and Performance for Jets, Taus, and MET. *CMS-PAS-PFT-09-001*, 2009.
[6] CMS Collaboration. Commissioning of the Particle-flow Event Reconstruction with the first LHC collisions recorded in the CMS detector. *CMS-PAS-PFT-10-001*, 2010.
[7] W. Waltenberger. Adaptive Vertex Reconstruction. *CMS Note 2008/33*, 2008.
[8] CMS Collaboration. *CMS-PAS-MUO-10-002*, 2010.
[9] CMS Collaboration. Performance of CMS muon reconstruction in pp collision events at $\sqrt{s} = 7$ TeV. *JINST*, 7(P10002), 2012.
[10] CMS Collaboration. Electron Reconstruction and Identification at $\sqrt{s} = 7$ TeV. *PAS-EGM-10-004*, 2010.
[11] M. Cacciari *et al.* The anti-k_T jet clustering algorithm. *JHEP*, 04(63), 2008.
[12] CMS Collaboration. Identification of b-quark jets with the CMS experiment. 2012, arXiv:1211.4462 [hep-ex].
[13] CMS Collaboration. Performance of τ-lepton reconstruction and identification in CMS. *JINST*, 7(P01001), 2012.
[14] CMS Collaboration. Missing transverse energy performance of the CMS detector. *JINST*, 6(9001), 2011.
[15] CMS Collaboration. MSSM Higgs production in association with b quarks - all hadronic. *CMS-HIG-012-026*, 2012.
[16] CMS Collaboration. MSSM Higgs production in association with b quarks - semi leptonic. *CMS-HIG-012-027*, 2012.
[17] CMS Collaboration. MSSM Higgs production in association with b quarks. *CMS-HIG-012-033*, 2012.
[18] CMS Collaboration. Higgs to tau tau (MSSM) (HCP). *CMS-HIG-012-050*, 2012.
[19] CMS Collaboration. Higgs to Mu Mu. *CMS-HIG-12-011*, 2012.
[20] M. Carena *et al.* Suggestions for benchmark scenarios for MSSM Higgs boson searches at hadron colliders. *Eur. Phys. J. C*, 26(4):601–607, 2003.
[21] M. Carena *et al.* MSSM Higgs boson searches at the Tevatron and the LHC: Impact of different benchmark scenarios. *Eur. Phys. J. C*, 45:797, 2006.
[22] CDF Collaborations. Search for Higgs Bosons Produced in Association with b-quarks. *Phys. Rev. D*, 85(032005), 2012.
[23] ALEPH, DELPHI, L3, OPAL Collaborations, and LEP Working Group for Higgs Boson Searches Collaboration. Search for neutral MSSM Higgs bosons at LEP. *Eur. Phys. J. C*, 47(547), 2006.
[24] CDF and D0 Collaborations. Search for neutral Higgs bosons in events with multiple bottom quarks at the Tevatron. *Phys. Rev. D*, 86(091101), 2012.
[25] CMS Collaboration. Search for $H^{\pm} \to$ Tau in Top quark decays. *J. High Energy Phys.*, 7:143, 2012.
[26] CMS Collaboration. di-lepton ttbar tau channels cross section 100/pb . *CMS-PAS-TOP-08-004*, 2008.

Journal of Physics: Conference Series **447** (2013) 012049 doi:10.1088/1742-6596/447/1/012049

Searches for the SM Higgs boson decaying into two photons, ZZ and two taus in CMS

Marco Meneghelli
Bologna University and INFN
E-mail: marco.meneghelli@cern.ch

Abstract. Results are presented on the search for the Standard Model (SM) Higgs bosons decaying into two photons, ZZ and two taus. Multivariate techniques are used to enhance the sensitivity. The full data sample of 5.1 fb^{-1} of pp collisions collected in 2011 at a CM energy of 7 TeV with the CMS experiment has been analyzed, as well as a significant fraction of the 2012 luminosity delivered at the new CM energy of 8 TeV.

1. Introduction

In July 2012 the CMS and ATLAS collaborations announced the discovery of a new particle, in the search for the SM Higgs boson, at a mass of about 125-126 GeV [1][2]: a boson whose characteristics are compatible with the SM Higgs ones. The local significances of data excesses, over the backgrounds, were more than 5 standard deviations in both experiments. After having observed a new particle, its properties have to be measured: particularly the mass, the cross section, the spin-parity and the couplings to other particles (or the Branching Ratios BR). CMS results on the SM Higgs boson searches are presented in the following, for three of the most important channels: $H \to \tau\tau$, $H \to \gamma\gamma$, $H \to ZZ \to 4l$.

The CMS experiment apparatus is described elsewhere [3]. Its main features are a compact cylindrical structure, built around the solenoidal magnet that creates a field of 3.8 T in the inner part of the detector, where are a silicon tracker, an electromagnetic calorimeter (ECAL) and a hadronic calorimeter (HCAL). Surrounding the magnet is the muon system. The CMS performances are characterized by a high efficiency and resolution in physical object reconstruction.

2. Search for the SM Higgs boson decaying to tau pair

The search for the SM Higgs boson decaying to a tau pair [4] is, de facto, the combination of five searches (called sub-channels), depending on the decay of the tau leptons in the final state (leptonic decay to electron or muon plus neutrinos, or hadronic decay): $H \to \tau\tau \to \mu\tau_h$, $H \to \tau\tau \to e\tau_h$, $H \to \tau\tau \to \mu e$, $H \to \tau\tau \to \mu\mu$, $H \to \tau\tau \to \tau_h\tau_h$. The experimental signature of the events is, in the sub-channels with leptons, the presence of isolated leptons and missing transverse energy. In the hadronic and semi-leptonic sub-channels, hadronic taus are required to be identified.

The search is sensitive to the Higgs boson presence in the low mass range ($110 < m_H < 145$ GeV), where the branching ratio for the Higgs to decay in a tau pair is large. The background rate is about one hundred times larger than the signal one, the main backgrounds being:

- the Drell-Yan process with taus in the final state: $Z \to \tau\tau$
- the top quark pair production
- other Drell-Yan processes with leptons in the final state: $W + jets$, $Z \to ll$ and di-boson production
- the QCD background

The *event reconstruction* relies on the "Particle Flow" algorithm (PF), to identify physical objects. Starting from sub-detector information the algorithm builds, for each event, an output of individual particles (charged/neutral hadrons, electrons, muons, photons) which are used to reconstruct hadronic jets, taus and estimate the amount of missing transverse energy. Among the various techniques available at CMS, the PF has shown to provide the best jet reconstruction: jets are, in addition, corrected with calibration factors. The b-tagging allows to discriminate between heavy-flavor and light flavor jets. The missing transverse energy from PF is further corrected with a "Boosted Decision Tree" (BDT) multivariate analysis, to account for the effect of different pile-up conditions. Hadronic taus are typically constituted of a charged hadron plus two neutral pions or by 3 charged hadrons. They are, in both cases, isolated from other particles: tau isolation is evaluated summing transverse momenta/energies of particles in a geometric cone around the main tracks. Leptonic taus are reconstructed as leptons, whose isolation is computed in a similar way.

The *event selection* requires, for each event, a trigger containing muons, electrons and or tau objects. In the case of semi-leptonic channels an isolated lepton and an opposite charge hadronic tau are required. A cut on invariant mass m_T is applied to discriminate against $W + jets$. In the case of full-leptonic channels, two opposite-sign same-flavor isolated leptons and missing transverse energy are required. A cut on a m_T-like variable is applied to discriminate against $W + jets$. For the full-hadronic channel, two reconstructed τ_h with high p_T are requested. The mass $m_{\tau\tau}$ is reconstructed with a likelihood approach. The selected events, in each sub-channel, are divided into categories relying on jet multiplicity and tau p_T, in order to increase the sensitivity of the analysis.

The $Z \to \tau\tau$ background is estimated from data using the $Z \to \mu\mu$ events. The $Z \to \tau\tau$ yield is estimated with an embedding procedure, substituting the muons in $Z \to \mu\mu$, with simulated taus. Some QCD processes can yield fake leptons/taus: this rate is controlled using a signal-free control region, with same-sign taus. The control region approach is used also to estimate the contribution of the W+jets process: fake tau rate is controlled using events with large m_T, while fake lepton rate is controlled using other data-driven techniques, typically defining a control region in a signal-free phase space, and extrapolating the event yields here to the signal region. Concerning the $t\bar{t}$, $Z \to ll$ and di-boson processes, their rate is estimated from background enriched sidebands.

Systematics uncertainties are assumed on the integrated luminosity, on the jet energy scale, on the background normalization, on the $pp \to Z$ cross section, on lepton efficiencies, trigger efficiency and tau identification efficiency and on the Higgs production cross section.

The final mass distributions, divided in categories in each sub-channel, are dominated by the huge backgrounds (see e.g., Fig. 1): just the combination of all sub-channels and categories can provide a weakly-discriminating variable. As shown in Fig. 1, that displays the mass distribution for the 0-jet category, there is a very good control of backgrounds.

The *results* are expressed using 95% CL upper limits on the signal cross section, normalized to the SM nominal one. The cross section modifier (or signal strenght) is defined as $\mu = \frac{\sigma}{\sigma_{SM}}$. Upper limits on this quantity are calculated with the CLs method. A little excess with respect to background-only expectations is observed, broadly distributed (see Fig. 2). The significance of this excess, for the hypothesis $m_H = 125$ GeV, is 1.5 standard deviations.

The best fit to μ is $\hat{\mu} = 0.7 \pm 0.5$ at $m_H = 125$ GeV.

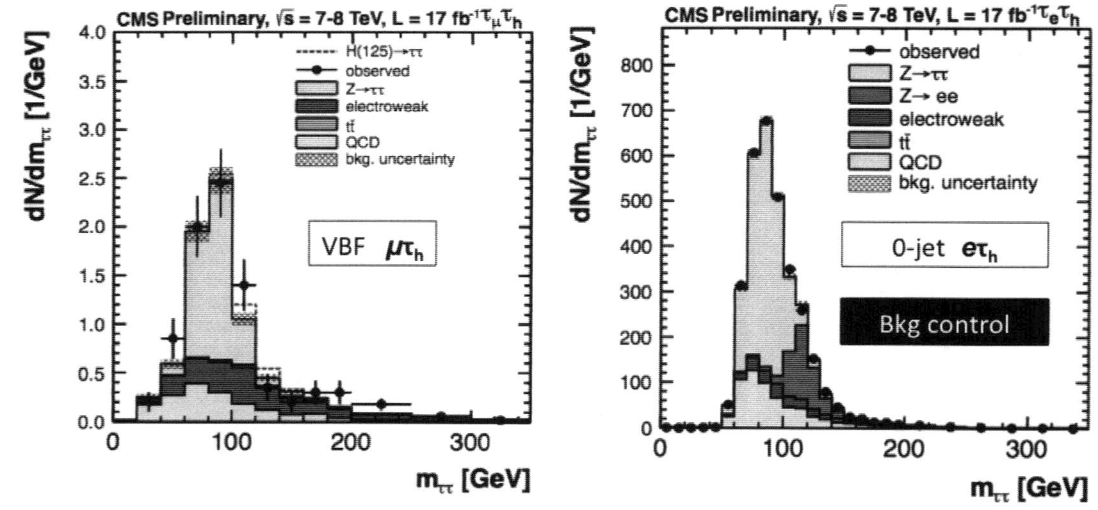

Figure 1. $H \to \tau\tau$: mass distribution for the $\mu\tau_h$ sub-channel, VBF category (left) and $e\tau_h$ sub-channel, 0-jet category (right). The 0-jet category is for background-control only.

We can conclude that what is observed is compatible with the SM expectations, but more data are needed to measure any signal property.

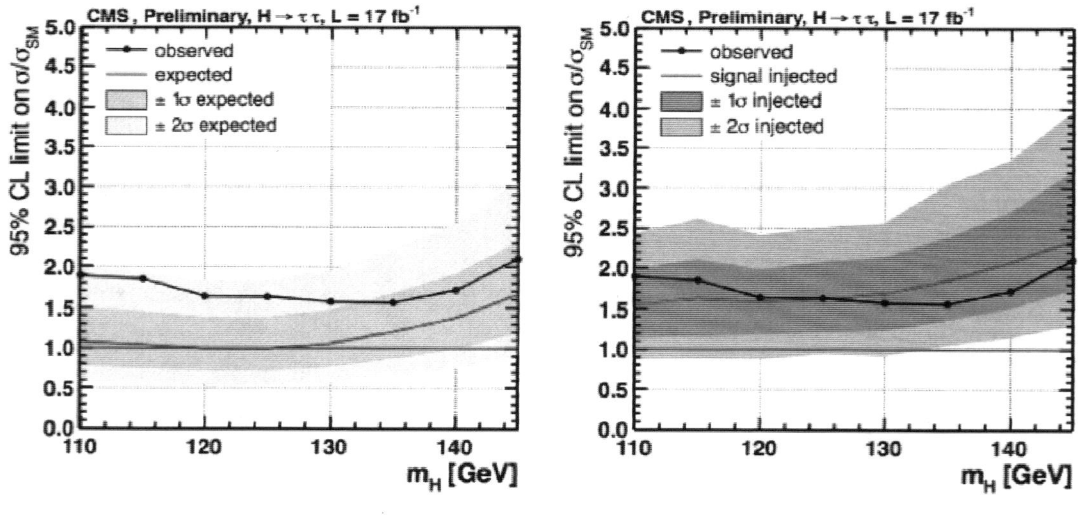

Figure 2. $H \to \tau\tau$: results of statistical analysis of data. Upper limits on signal cross section compared to what expected in case of background only (left) and the same compared to what expected in case of signal (right).

3. Search for the SM Higgs boson decaying to photon pair

The search for the SM Higgs boson decaying to a photon pair [5] is one of the most sensitive at low mass. The experimental signature of the events is the presence of two isolated photons coming from a primary interaction vertex.

Journal of Physics: Conference Series **447** (2013) 012049 doi:10.1088/1742-6596/447/1/012049

The search is sensitive to the Higgs boson presence in the low mass range ($110 < m_H < 145$ GeV): the signal should be highlighted by the presence of a narrow resonance peak over the broader smooth background. The background rate is about ten times larger than the signal one, the main backgrounds being:

- the irreducible di-photon production
- the instrumental QCD background, where one or two hadronic jets are misidentified as photons

Photon reconstruction relies on the presence of significant energy deposits (superclusters) in the calorimeters. The energy is assigned to photon objects and corrected with the usage of a BDT. The photon identification relies on another BDT, trained on a $pp \to \gamma + jet$ simulated sample and taking as input shower topology variables and isolation variables. The single photon efficiencies are above 95%. Jets are reconstructed with the Particle Flow algorithm.

The *Event selection* requires di-photon triggers, with asymmetric E_T thresholds and complementary photon selection: a loose calorimetric identification, a high R9 variable (sum of energy in 3x3 calorimeter crystals). The primary interaction vertex is assigned using the kinematic properties of the tracks associated to it (transverse momenta), and their correlation with di-photon kinematics, using a BDT. Events are further classified using another BDT (trained on simulated background and signal processes) that is designed to sort the events bases on their signal-like characteristics: four classes are individuated, relying on the BDT output, and another (two) class (es) is (are) individuated by the di-jet tag for 7 (8) TeV data, to discriminate "Vector Boson Fusion" (VBF) Higgs production.

For the signal, the "Next to Leading Order" (NLO) contribution is re-weighted to "Next to NLO" (NNLO), taking into account the interference with the $pp \to \gamma\gamma$ process. The $m_{\gamma\gamma}$ distribution is a resonance peak, dominated by the instrumental resolution, modeled with a sum of Gaussians. The backgrounds are entirely modeled on data. There is a 28% of reducible background contamination, while the rest is irreducible di-photon production. The $m_{\gamma\gamma}$ shape is fitted on data in the $100 < m_{\gamma\gamma} < 180$ GeV range, polynomial functions are used.

Systematics uncertainties are assumed on the single-photon efficiency, the energy scale and resolution and ID BDT. Other systematics uncertainties are assumed on the integrated luminosity, on the vertex-finding and trigger efficiency, on the di-jet efficiency, on the Higgs cross section.

The statistical analysis is performed with the profile likelihood method, fitting simultaneously the $m_{\gamma\gamma}$ distribution in the sub-classes. Upper limits on the cross section modifier are computed (see Fig. 3), the largest excess with respect to background-only expectations (see Fig. 3) is observed at $m_H = 125$ GeV: the local significance is 4.1 standard deviations. The global significance, in the mass range $110 < m_H < 150$ GeV, is 3.2 standard deviations.

The fit to the signal strength gives $\hat{\mu} = 1.56 \pm 0.43$ at $m_H = 125$ GeV.

A visualization of the results, showing the signal peak spiring over the background, is presented in Fig. 4.

4. Search for the SM Higgs boson decaying to $ZZ \to 4l$

The search for the SM Higgs boson decaying to a Z pair, each Z decaying leptonically [6], is one of the most sensitive and clean channels, despite the relatively small signal rate. The experimental signature of the events is the presence of two isolated lepton-pairs (muons or electrons) coming from the primary interaction vertex (primary leptons). The search is sensitive to the Higgs boson presence in the a wide mass range ($115 < m_H < 800$ GeV): the signal should be highlighted by the presence of a narrow resonance peak over the broader background. The ratio between signal and background rates is locally of the order one in all the mass range, the main backgrounds being:

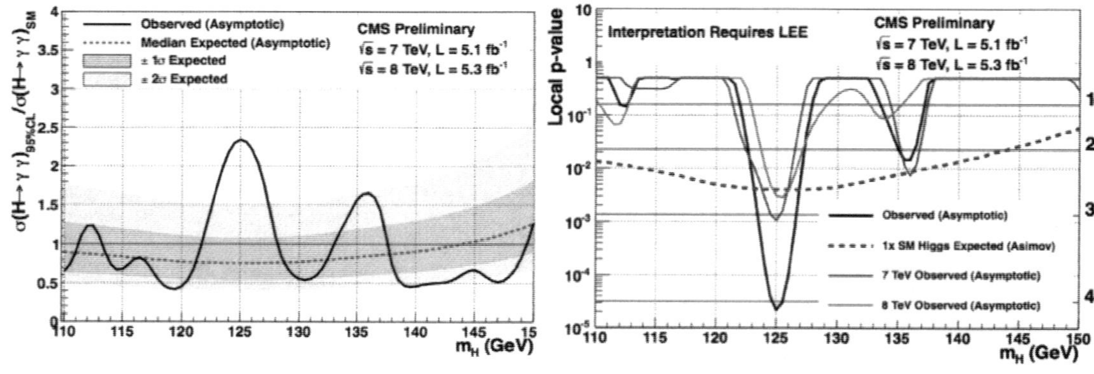

Figure 3. $H \to \gamma\gamma$: results of statistical analysis of data. Upper limits on signal cross section (left) and local p-values for the background-only hypothesis (right).

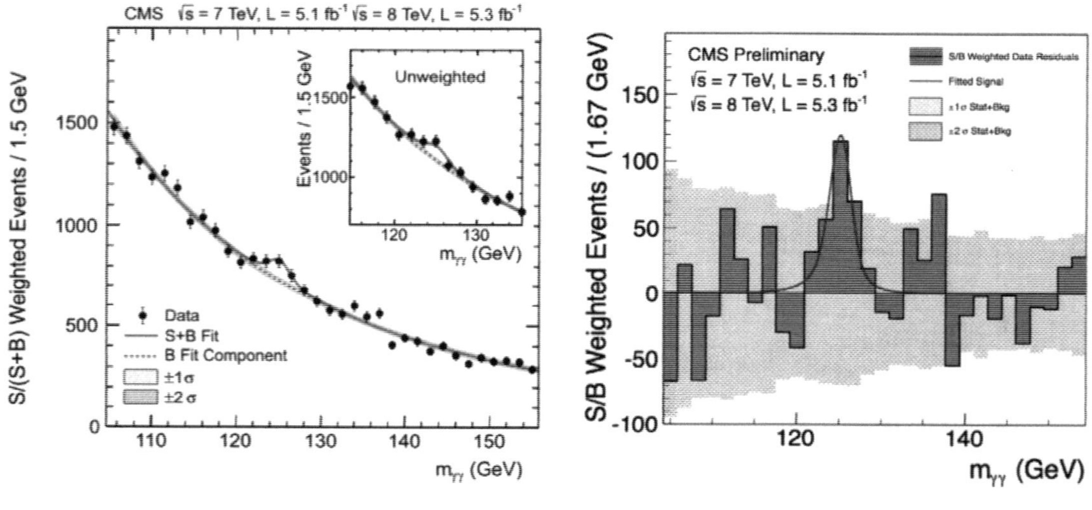

Figure 4. $H \to \gamma\gamma$: visualization of results. Weighted sum of events in the sub-classes (left) and the histogram with the background subtracted (right).

- the irreducible double Z production
- the instrumental backgrounds $Z + jets$ and $t\bar{t}$, where jets of secondary leptons are misidentified as primary leptons

Muon reconstruction matches tracker and muon system information to build complete objects and assign them the kinematics. The PF algorithm is used to identify muons, enhancing the purity of the sample. *Electron reconstruction* matches tracker and ECAL information to build complete objects and assign them the kinematics. A multivariate BDT technique (trained on $H \to 4e$ as signal and $W + fake$ as background) is used to identify electrons: the input variables are the bremsstrahlung, a tracker-ECAL matching variable and a set of variables describing the ECAL shower shape. *Lepton isolation* is evaluated by summing the p_T of tracks or energy deposits in a geometrical cone around the track of each lepton, and normalizing the sum to the lepton p_T. A cut on the significance of lepton impact parameter is applied, to select *primary leptons*. The muon identification efficiency is above 95%, while electron identification efficiency ranges between 60% and 95%, depending on the p_T.

The *event Selection* requires triggers with a pair of electrons or muons, a triplet of electrons or a muon and an electron. From the leptons passing the ID criteria, the Z1 and Z2 candidates are built, with a "Final State Radiation" (FSR) recovery, done by associating FSR photons to Z candidates. A kinematic discriminant variable k_D is built using all kinematic observables of leptons and Z candidates.

The ZZ irreducible background yield is directly estimated from theoretical predictions and simulation. The reducible background yield is estimated using a signal-free control region $Z+X$, the data rate here is extrapolated to the signal region with a single lepton fake rate, estimated from data. Systematic uncertainties are assumed on trigger and lepton reco-ID efficiencies, on the energy-momentum calibration, on the reducible background yield and on the theoretical cross sections.

The event distribution is in good agreement with the SM expectations. A peak around 126 GeV, over the almost flat background, is observed (see Fig. 5).

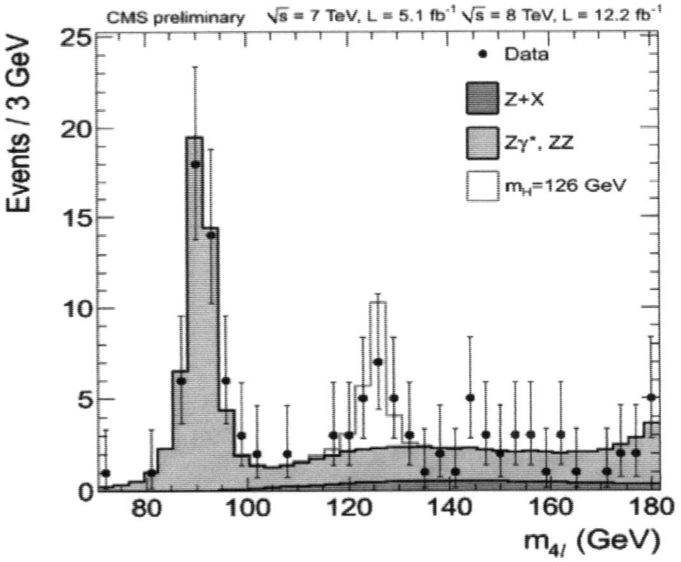

Figure 5. $H \to ZZ \to 4l$: m_{4l} distribution in the low-mass range. Data are superimposed to the expected backgrounds and the signal for $m_H = 126$ GeV.

The statistical analysis of data is performed with the profile likelihood method, using as observables m_{4l} and k_D. The largest excess of data with respect to background-only expectations is observed at $m_H = 126$ GeV, with a local significance of 4.5 standard deviations (see Fig. 6). On the other hand, the excluded mass range at 95% CL, for the Standard Model Higgs boson is $[113, 116] \cup [129, 720]$ GeV (see Fig. 6).

The simultaneous fit to the cross section modifier and the mass (see Fig. 7) gives:

$$\hat{\mu} = 0.8^{+0.35}_{-0.28}$$

$$\hat{m}_H = 126.2 \pm 0.6(stat) \pm 0.2(syst) \text{ GeV}$$

The number of events observed around the peak is in good agreement with what expected from the signal+background hypothesis (see table 1).

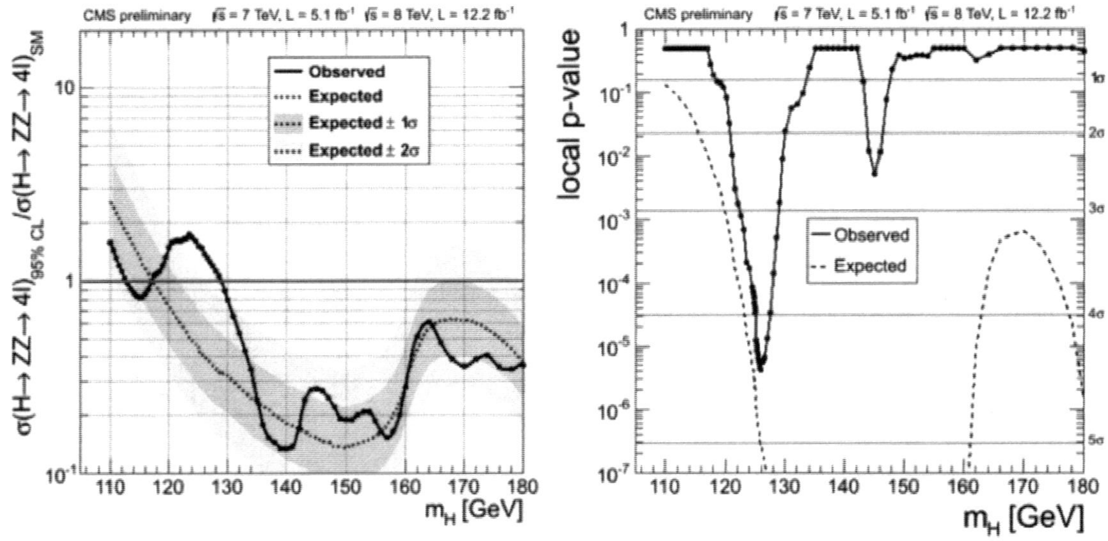

Figure 6. $H \to ZZ \to 4l$: results of statistical analysis of data. Upper limits on signal cross section (left) and local p-values for the background-only hypothesis (right).

Table 1. 4-lepton events for $m_{4l} \in [121.5, 130.5]$. The number of events observed in data is compared to what expected in case of signal + background.

$m_{4l} \in [121.5, 130.5]$	expected 7 TeV	expected 8 TeV	observed
$4e$	1.25	2.20	3
4μ	2.09	4.26	6
$2e2\mu$	3.14	5.97	8
Total	6.48	12.43	17

A spin-parity hypothesis is tested with dedicated discriminants, which use all the kinematic observables: the 0^+ hypothesis is compatible with data, while the 0^- one is disfavored with a $CLs = 2.4\%$ (see Fig. 7).

5. Conclusions

The results for the SM Higgs search at CMS in three of the most important channels have been presented. They all exploit the excellent detector performances in terms of efficiency and resolution in reconstructing physical objects.

All the searches see excesses in data, with respect to background-only expectations, that suggest the presence of a new boson, compatible with being the Standard Model Higgs. The $H \to ZZ$ and $H \to \gamma\gamma$ analyses measure the boson mass at about $125 - 126$ GeV. The ZZ analysis provides a first measurement of the spin-parity quantum numbers, showing that data suggest a 0^+ scenario.

More precise results require more data, we look forward for the analysis of the complete 2012 dataset, with almost 22 fb^{-1} at 8 TeV collected by CMS.

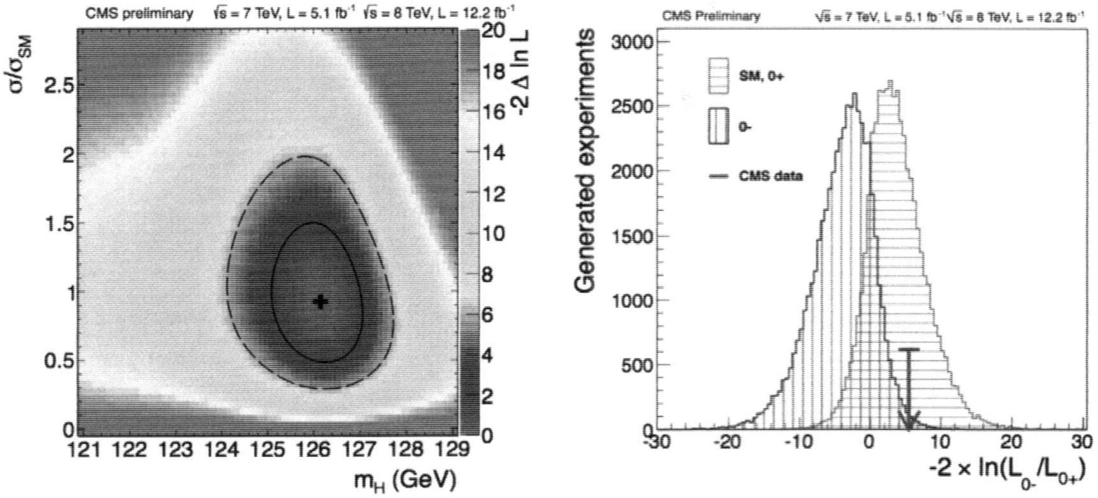

Figure 7. $H \to ZZ \to 4l$: estimation of boson observables. Simultaneous fit to mass and cross section (left) and first results on J^P measurement (right).

References

[1] The ATLAS Collaboration 2012 *Phys. Lett.* B **716** p. 1-29
[2] The CMS Collaboration 2012 *Phys. Lett.* B **716** p. 30-61
[3] S. Chatrchyan et al. 2008 *JINST* **3** S08004
[4] The CMS collaboration CMS PAS HIG-12-015 *Search for the standard model Higgs boson decaying to tau pairs*
[5] The CMS collaboration CMS PAS HIG-12-043 *Evidence for a new state decaying into two photons in the search for the standard model Higgs boson in pp collisions*
[6] The CMS collaboration CMS PAS HIG-12-041 *Updated results on the new boson discovered in the search for the standard model Higgs boson in the $H \to ZZ \to 4l$ channel in pp collisions at $\sqrt{s} = 7$ and 8 TeV*

Journal of Physics: Conference Series **447** (2013) 012050 doi:10.1088/1742-6596/447/1/012050

2HDM with Z_2 symmetry in light of new LHC data

Maria Krawczyk, Dorota Sokołowska, Bogumiła Świeżewska
Faculty of Physics, University of Warsaw, Hoża 69, 00-681 Warsaw, Poland
E-mail: krawczyk@fuw.edu.pl,dsok@fuw.edu.pl, bogumila.gorczyca@fuw.edu.pl

Abstract. Properties of the Z_2-symmetric Two Higgs Doublet Models (2HDM) are discussed and confronted with new LHC data for a 125 GeV Higgs particle. The particle discovered at LHC in 2012 has properties expected for it in the Standard Model (SM), with a possible enhancement in the $\gamma\gamma$ channel. SM-like Higgs scenarios can be realized in the Two Higgs Doublet Models with Z_2 (D) symmetry within the normal Mixed Model (with scalar sector as in MSSM) and the Inert Doublet Model (IDM), where a good Dark Matter (DM) candidate is present. Here we discuss both of the models.

1. Introduction

In the Two Higgs Doublet Models (2HDM) there are two doublets of SU(2), with the weak hypercharge Y=1. They give masses to W, Z (leading at the tree-level to $\rho = 1$) and in principle also to the photon. Fermion masses are generated via Yukawa interactions, for which various models are considered: Model I, II, III, IV, X, Y,... [1]. Five scalars appear in these models, two charged H^+ and H^- and three neutral ones. If CP is conserved there are two CP-even h, H and one CP-odd A particle. In the model with CP violation three neutral particles h_1, h_2, h_3 with undefined CP parity appear.

2. D-symmetric 2HDM

Study of the symmetry properties of the Lagrangian as well as of the vacuum states is crucial for understanding a real content of the theory. Here we assume the Z_2 symmetry of the potential $\phi_S \to \phi_S$, $\phi_D \to -\phi_D$, which we call below the D symmetry.[1]

In the Mixed Model both doublets have non-zero vacuum expectation values (vev's) and are involved in the mass generation. There are five Higgs bosons and sum rules hold for the relative couplings of the neutral Higgs particles h, H, A: e.g. $(\chi^h_{VV})^2 + (\chi^H_{VV})^2 + (\chi^A_{VV})^2 = 1$, $V = W/Z$. Model II is assumed for the Yukawa interactions: doublet ϕ_S couples to the down-type quarks and charged leptons, while ϕ_D couples to the up-type quarks. D symmetry is spontaneously violated. In this model SM-like scenarios are possible for both h and H, with $\chi^{h/H}_{VV} = 1$.

In contrast, in the Inert Doublet Model (IDM), only one doublet (ϕ_S) is involved in the mass generation and there is only one SM-like Higgs boson h. The second doublet is inert (it has vev=0) and contains four scalars. Yukawa interactions are as in Model I, so the D symmetry is exact here and the neutral scalar H (or A) may play a role of the Dark Matter (DM).

[1] In such case CP is conserved.

Table 1: Types of extrema for a D-symmetric potential.

type of extremum	condition
EWs (EWs)	$u = v_D = v_S = 0$
Inert (I_1)	$u = v_D = 0, v_S \neq 0$
Inertlike (I_2)	$u = v_S = 0, v_D \neq 0$
Mixed (M)	$u = 0, v_S \neq 0, v_D \neq 0$
Charged (CB)	$u \neq 0, v_S \neq 0, v_D = 0$

The D-symmetric potential has the following form:

$$V = -\frac{1}{2}\left[m_{11}^2(\phi_S^\dagger\phi_S)+m_{22}^2(\phi_D^\dagger\phi_D)\right] + \frac{\lambda_1}{2}(\phi_S^\dagger\phi_S)^2+\frac{\lambda_2}{2}(\phi_D^\dagger\phi_D)^2$$

$$+\lambda_3(\phi_S^\dagger\phi_S)(\phi_D^\dagger\phi_D)+\lambda_4(\phi_S^\dagger\phi_D)(\phi_D^\dagger\phi_S) + \frac{\lambda_5}{2}\left[(\phi_S^\dagger\phi_D)^2+(\phi_D^\dagger\phi_S)^2\right],$$

with all parameters real. We take $\lambda_5 < 0$ without loss of generality [2].

Such potential has various possible extrema (vacua), with the following vacuum expectation values:[2]

$$\langle\phi_S\rangle = \frac{1}{\sqrt{2}}\begin{pmatrix}0\\ v_S\end{pmatrix}, \qquad \langle\phi_D\rangle = \frac{1}{\sqrt{2}}\begin{pmatrix}u\\ v_D\end{pmatrix}. \tag{1}$$

Neutral vacua are realized for $u = 0$. The charged vacuum with $u \neq 0$ corresponds to breaking of $U(1)_{QED}$ symmetry and the appearance of the massive photon. The list of possible extrema, which can be realized as local or global minima (i.e. vacua) is given in table 1. The EWs case corresponds to the EW symmetry, i.e. lack of spontaneous breaking of EW symmetry.

Existence of a stable vacuum requires [3]:

$$\lambda_1 > 0, \quad \lambda_2 > 0, \quad \lambda_3 + \sqrt{\lambda_1\lambda_2} > 0, \quad \lambda_{345} + \sqrt{\lambda_1\lambda_2} > 0 \ \ (\lambda_{345} = \lambda_1 + \lambda_2 + \lambda_3),$$

so that $R = \frac{\lambda_{345}}{\sqrt{\lambda_1}\sqrt{\lambda_2}} > -1$. Various vacua can be realized for various values of vev's, which can be represented in the phase diagram (μ_1, μ_2), where

$$\mu_1 = \frac{m_{11}^2}{\sqrt{\lambda_1}}, \quad \mu_2 = \frac{m_{22}^2}{\sqrt{\lambda_2}}.$$

There are three regimes of parameter R which correspond to very different phase patterns shown in fig. 1.

In principle a model for today's Universe could be based either on the Mixed or the Inert vacuum. From theory side we assume that the considered vacua are stable and parameters of V are constrained by the perturbative unitarity: $|\Lambda_i| < 8\pi$ [4–6], where Λ_i are the eigenvalues of the high-energy scattering matrix of the scalar sector. This leads to the upper limits on $\lambda_{1,2}$

[2] With u, v_S, v_D real.

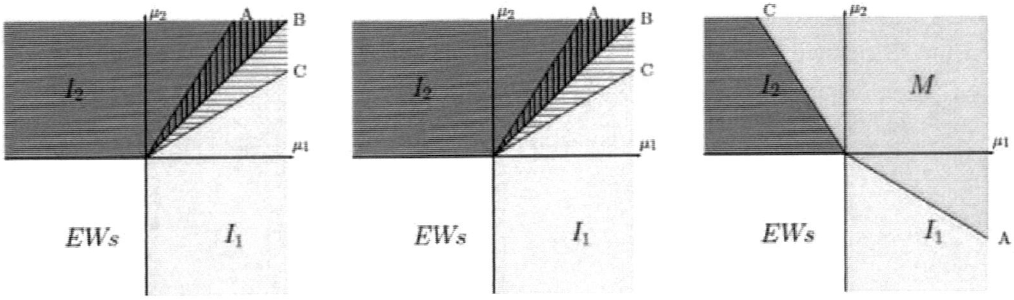

(a) $R > 1$ (b) $1 > R > 0$ (c) $0 > R > -1$

Figure 1: Phase diagrams for $D-$symmetric potential. Regions of μ_1, μ_2 where various neutral extrema (minima) EWs, I_1, I_2, M can be realized are shown. The hatched region on the left panel corresponds to the coexistence of I_1, I_2 minima.

equal 8.38, while λ_3 and λ_{345} are allowed to be in the regions (-6.05, 16.53) and (-8.10,12.38), respectively.

In any case a condition for the existence of the particular vacuum has to be fulfilled. Existence of the Mixed vacuum is equivalent to having positive scalars' masses squared (condition for the minimum) while for the Inert vacuum this is not enough since a coexistence of Inert and Inertlike minima is possible (see fig. 1). For Inert to be the global minimum a following condition has to be fulfilled in addition [2]:

$$\frac{m_{11}^2}{\sqrt{\lambda_1}} > \frac{m_{22}^2}{\sqrt{\lambda_2}}.$$

From the Higgs boson mass $M_h^2 = m_{11}^2 = \lambda_1 v^2 = (125\,\text{GeV})^2$ and unitarity limit $\lambda_2^{\max} = 8.38$ a following limit on m_{22}^2 arises [6]:

$$m_{22}^2 \lesssim 9 \cdot 10^4\,\text{GeV}^2. \tag{2}$$

3. Experimental constraints

We consider both Mixed Model and IDM, taking into account also the following experimental constraints:

Electroweak Precision Tests (EWPT) Values of S and T parameters are demanded to lie within 2σ ellipses in the S, T plane, with the following central values [7]: $S = 0.03 \pm 0.09$, $T = 0.07 \pm 0.08$, with correlation equal to 87%.

LEP We apply a model-independent limit: $M_{H^\pm} > 70$ GeV from the direct LEP measurements. For Mixed Model the lower limit for mass of $H^\pm$ is 360 GeV from $b \to s\gamma$ NLO analysis [8]. For the IDM we use the LEPI and LEPII bounds on the scalar masses [9, 10]:

$$M_{H^\pm} + M_H > M_W, \quad M_{H^\pm} + M_A > M_W, \quad M_H + M_A > M_Z \tag{3}$$

excluding region where simultaneously: $M_H < 80\,\text{GeV}, M_A < 100\,\text{GeV}, M_A - M_H > 8\,\text{GeV}$.

***H* as DM candidate** In the IDM we take H as the DM candidate, $M_H < M_A, M_{H^\pm}$.

4. Results for the Mixed Model with a 125 GeV Higgs

Perturbative unitarity gives the following upper limits on the Higgs masses in the Mixed Model: $M_h^{\max} = 499$ GeV, $M_H^{\max}$ and $M_{H^\pm}^{\max}$ equal 690 GeV, $M_A^{\max} = 711$ GeV. Moreover, by setting mass of h equal to 125 GeV, an important limit on $\tan\beta = v_D/v_S$ is obtained, namely $0.18 < \tan\beta < 5.59$ (independently of $\sin(\beta-\alpha)$!). Notice, that it is possible to have SM-like H, however h needs then to have a very suppressed coupling to gauge boson, see eg. [11–13].

Even when all direct decay widths of h are as in SM, the loop decay widths $\gamma\gamma$, $Z\gamma$ for h may still be modified due to $H^\pm$ contribution, or/and the negative sign of $\chi_{f\bar{f}h}$ coupling [14]. The $H^\pm$ loop contribution leads to a 10 % (5%) suppression with respect to the SM for $\gamma\gamma$ ($Z\gamma$). The change of sign of the tth coupling has a strong influence on the decay widths $\gamma\gamma$ and $Z\gamma$ by changing the destructive interference between the t and W contributions to the loop couplings present in the SM into the constructive one. The enhancement with respect to SM up to 2.3 (1.2) is possible for $\Gamma_{\gamma\gamma}$ ($\Gamma_{Z\gamma}$). For the hgg coupling change of the relative sign of the b and t contributions leads to a enhancement up to 1.25.

5. Results for IDM with a 125 GeV Higgs

The Universe is described by the IDM if the vacuum state is given by I_1. IDM predicts the existence of four dark scalars H, $A, H^\pm$ and the SM-like Higgs particle h (we assume its mass equal to 125 GeV). λ_{345} is related to a triple and quartic coupling between SM-like Higgs h and DM candidate H. λ_2 gives the quartic DM self-couplings, while λ_3 describes the Higgs particle interaction with charged scalars.

5.1. Relic density

IDM provides a good DM candidate in three regions of M_H [15–26]: (i) light DM particles with mass below 10 GeV, (ii) medium mass regime of $50-150$ GeV and (iii) heavy DM of mass larger than 500 GeV. In those regions one can get the DM relic density $\Omega_{DM}h^2$ in agreement with the astrophysical data $\Omega_{DM}h^2 = 0.112 \pm 0.009$ [27].

This estimation of $\Omega_{DM}h^2$ may be used to constrain the λ_{345} coupling depending on the chosen values of masses of H and other scalars [18,20]. $\Omega_{DM}h^2$ does not limit the λ_2 parameter, although indirect constraints come from its relation to λ_{345} parameter through the vacuum stability constraints and existence of I_1 vacuum [25,26].

Limits for M_H and λ_{345} coming from $\Omega_{DM}h^2$ are presented in fig. 2. For masses $M_H \lesssim 72$ GeV the allowed region (dark gray) of λ_{345} is symmetric around zero with small values of λ_{345} excluded (light gray region) due to a nonefficient DM annhilation. As the mass increases, the region of proper relic density shifts towards the negative values of λ_{345}, which is due to opening of the annihilation channels into the gauge bosons final state.

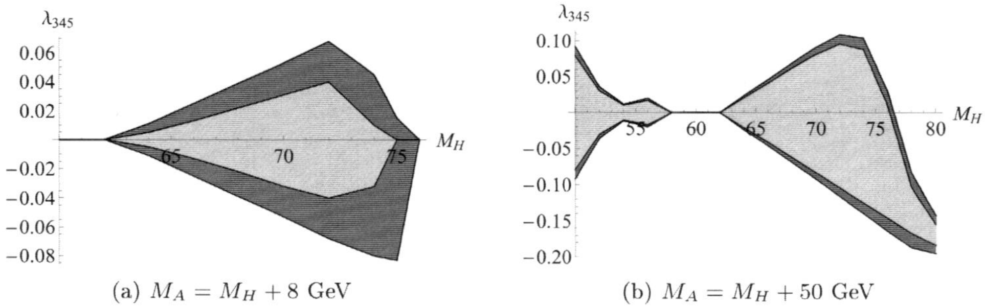

(a) $M_A = M_H + 8$ GeV (b) $M_A = M_H + 50$ GeV

Figure 2: Limits for (M_H, λ_{345}) parameters coming from astrophysical estimations of DM relic density ($\Omega_{DM}h^2$). Dark gray: $\Omega_{DM}h^2$ in agreement with WMAP measurements, $0.1018 < \Omega_{DM}h^2 < 0.1234$; light gray: $\Omega_{DM}h^2$ above WMAP limits (excluded); white: $\Omega_{DM}h^2$ below WMAP limits (subdominant DM). We set $M_h = 125$ GeV and $M_{H^\pm} = M_H + 50$ GeV.

As it will be shown below, in the IDM $R_{\gamma\gamma} > 1$ is possible for $\lambda_3 < 0$. If we consider H as a DM candidate then $\lambda_{345} < 0$ for $R_{\gamma\gamma} > 1$, meaning that it is possible to fulfill the LHC and relic density constraints for IDM.

5.2. $R_{\gamma\gamma}$

Here we concentrate on the two-photon decay rate of the Higgs boson, which in the IDM reads:

$$R_{\gamma\gamma} := \frac{\sigma(pp \to h \to \gamma\gamma)^{\rm IDM}}{\sigma(pp \to h \to \gamma\gamma)^{\rm SM}} \approx \frac{{\rm Br}(h \to \gamma\gamma)^{\rm IDM}}{{\rm Br}(h \to \gamma\gamma)^{\rm SM}}. \tag{4}$$

A deviation from the value of $R_{\gamma\gamma} = 1$ may be caused in IDM by two factors. Firstly, the partial decay width $\Gamma(h \to \gamma\gamma)^{\rm IDM}$ is modified due to the existence of the charged scalar loop [28–32]:

$$\Gamma(h \to \gamma\gamma)^{\rm IDM} = \frac{G_F \alpha^2 M_h^3}{128\sqrt{2}\pi^3} \left| \mathcal{M}^{\rm SM} + \delta\mathcal{M}^{\rm IDM} \right|^2,$$

where $\mathcal{M}^{\rm SM}$ is the contribution from the SM and $\delta\mathcal{M}^{\rm IDM} = \frac{2M_{H^\pm}^2 + m_{22}^2}{2M_{H^\pm}^2} A_0\left(\frac{4M_{H^\pm}^2}{M_h^2}\right)$, with $2M_{H^\pm}^2 + m_{22}^2 = \lambda_3 v^2$. The charged scalar loop can interfere either constructively or destructively with the SM contribution. Secondly, the total decay width $\Gamma^{\rm IDM}(h)$ can be increased with respect to the SM case due to the existence of the invisible decays: $h \to HH$ and $h \to AA$.

Performing a random scan of the parameter space, we found the regions where $R_{\gamma\gamma}>1$, with the maximal value of $R_{\gamma\gamma}$ around 3.4. Fig. 3a shows values of $R_{\gamma\gamma}$ allowed by the theoretical and experimental constraints as a function of M_H. It can be seen that enhanced values of $R_{\gamma\gamma}$ are not possible for $M_H < M_h/2$. It means that if the invisible channels are open, the total decay width is so big, that it suppresses other effects.

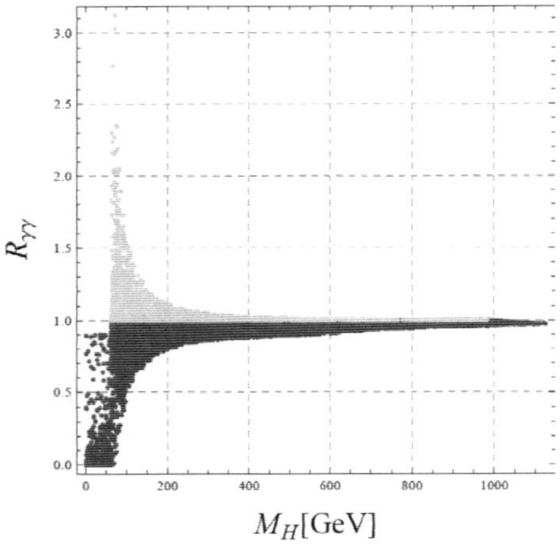

(a) Values of $R_{\gamma\gamma}$ allowed by the theoretical and experimental constraints as a function of the DM mass M_H for $-2 \cdot 10^6\ {\rm GeV}^2 \leqslant m_{22}^2 \leqslant 9 \cdot 10^4\ {\rm GeV}^2$.

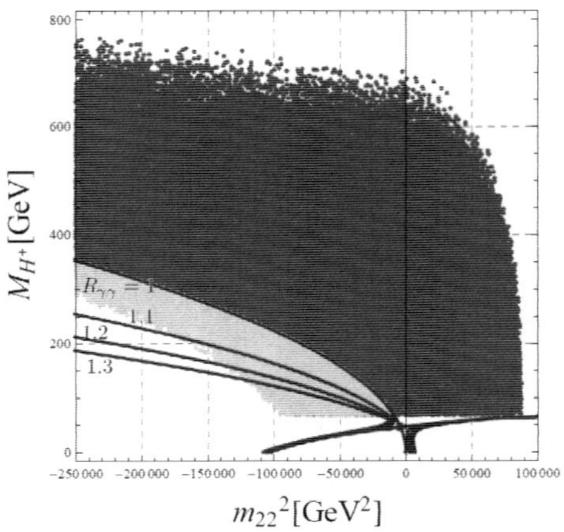

(b) Region allowed by the constraints in the $(m_{22}^2, M_{H^\pm})$ plane. The curves correspond to the fixed values of $R_{\gamma\gamma}$ (for the invisible channels closed).

Figure 3: Results on $R_{\gamma\gamma}$ for IDM, points with $R_{\gamma\gamma} < 1$ ($R_{\gamma\gamma} > 1$) are displayed in dark green/gray (light green/gray).

Fig. 3b shows the region allowed by the theoretical and experimental constraints in the $(m_{22}^2, M_{H^\pm})$ plane together with the curves corresponding to constant values of $R_{\gamma\gamma}$ (calculated for the case with invisible decay channels closed). It can be seen that the enhancement is possible only for constrained m_{22}^2 region, namely:

$$m_{22}^2 < -9.8 \cdot 10^3\,\mathrm{GeV}^2,$$

which is equivalent to the bound $\lambda_3 < 0$ (in agreement with Ref. [32]). On the contrary, $R_{\gamma\gamma} > 1$ can be achieved for any value of $M_{H^\pm}$. However, if bigger value of $R_{\gamma\gamma}$ is demanded, then allowed values of $M_{H^\pm}$ are constrained. For example, for $R_{\gamma\gamma} > 1.2$ we get the following bounds on $M_{H^\pm}$ and M_H (as $M_H < M_{H^\pm}$):

$$62.5\,\mathrm{GeV} < M_H < 154\,\mathrm{GeV},$$
$$70\,\mathrm{GeV} < M_{H^\pm} < 154\,\mathrm{GeV}.$$

6. Evolution of the Universe in 2HDM through different vacua in the past

We consider 2HDM with an explicit D symmetry and assume that today the IDM describes reality. In the simplest approximation λ_i terms in the potential are fixed and only mass terms vary with temperature:

$$m_{ii}^2(T) = m_{ii}^2 - c_i T^2 \ (i = 1, 2),$$

where $c_i = c_i(\lambda_{1-4}; g, g'; g_t^2 + g_b^2$ for $i = 1)$. Here g, g' are EW gauge couplings, while g_t, g_b are the SM Yukawa couplings) [2].

As the Universe cools down, the quadratic coefficients vary with T and the ground state of the potential V may change. Various types of evolution of the Universe from EWs phase into I_1 phase can be realized: in one ($EWs \to I_1$), two ($EWs \to I_2 \to I_1$) or three ($EWs \to I_2 \to M \to I_1$) steps. In general, in T^2 approximation phase transitions are of the 2nd order. The only exception is the transition between two degenerate minima I_2 and I_1 in the sequence with two phase transitions. This scenario can be realized only if $R > 1$. Notice, that if the Universe undergoes a series of phase transitions dark matter may appear later during the evolution, as it exists only in the Inert phase.

If $R < 0$ there is only one type of sequence that corresponds to the restoration of EW symmetry in the past (fig. 4a): $EWs \to I_1$. This sequence can be realized when $R_{\gamma\gamma} > 1$, which is suggested by the recent LHC data. In the other scenarios for $R < 0$ the initial state of the Universe is the one with broken EW symmetry (fig. 4b). The restoration of EW symmetry may be temporary (scenario Y), in other cases the EW symmetric state never existed.

For a certain parameter range there is a possibility of having a charged vacuum in the past (fig. 4c). This scenario can be realized only if today we have IDM with the charged DM particle [2]. Current model independent bounds require that charged DM is heavier than 100 TeV [33]. This can be achieved without breaking the perturbative unitarity conditions with large m_{22}^2. However, the sequence Z_+ requires that $c_2 < 0$ and $|c_2|/c_1 > |m_{22}^2|/m_{11}^2 \gtrsim (10^5 \div 10^6)$. This contradicts the requirement $c_1 > -c_2$ based on the positivity condition [2]. This means that during the evolution Universe cannot pass through the $U(1)_{QED}$ breaking phase.

7. Summary

The 2HDM is an excellent laboratory of the beyond SM physics. Although the discovery of the 125 GeV Higgs particle is in agreement with the Standard Model prediction, many extensions of the SM have some build-in SM-like scenarios. Here we consider two very different explicit Z_2-symmetric extensions with two SU(2) doublets – the very common Mixed Model and the Inert Doublet Model. Both can correspond to the SM tree-level decay modes, at the same time both may lead to the enhancement in the two-photon decay channel for the 125 GeV Higgs boson.

Acknowledgments

MK is grateful to the organizers of the DISCRETE for an excellent conference. The work was supported by a grant NCN OPUS 2012/05/B/ST2/03306 (2012-2016).

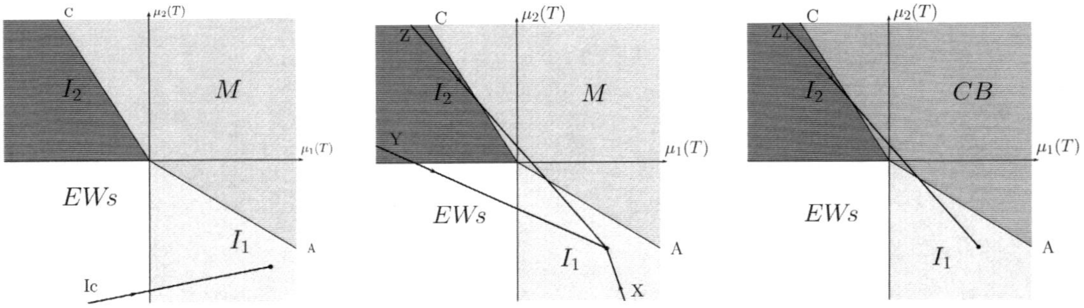

(a) $R < 0$, the only sequence for $R < 0$ that leads to restoration of EW symmetry in the past.

(b) $R < 0$, possible sequences of non-restoration of EW symmetry.

(c) Transition through the charged vacuum.

Figure 4: Possible sequences of phase transitions.

References

[1] Branco G, Ferreira P, Lavoura L, Rebelo M, Sher M and Silva J 2012 *Phys.Rept.* **516** 1–102 (*Preprint* 1106.0034)
[2] Ginzburg I, Kanishev K, Krawczyk M and Sokołowska D 2010 *Phys.Rev.* **D82** 123533 (*Preprint* 1009.4593)
[3] Deshpande N G and Ma E 1978 *Phys.Rev.* **D18** 2574
[4] Kanemura S, Kubota T and Takasugi E 1993 *Phys.Lett.* **B313** 155–160 (*Preprint* hep-ph/9303263)
[5] Akeroyd A G, Arhrib A and Naimi E M 2000 *Phys.Lett.* **B490** 119–124 (*Preprint* hep-ph/0006035)
[6] Swiezewska B 2012 (*Preprint* 1209.5725)
[7] Nakamura K *et al.* (Particle Data Group) 2010 *J.Phys.G* **G37** 075021
[8] Hermann T, Misiak M and Steinhauser M 2012 *JHEP* **1211** 036 (*Preprint* 1208.2788)
[9] Lundstrom E, Gustafsson M and Edsjo J 2009 *Phys.Rev.* **D79** 035013 (*Preprint* 0810.3924)
[10] Gustafsson M 2010 *PoS* **CHARGED2010** 030 (*Preprint* 1106.1719)
[11] Chang J, Cheung K, Tseng P Y and Yuan T C 2012 *Int.J.Mod.Phys.* **A27** 1230030 (*Preprint* 1211.6823)
[12] Drozd A, Grzadkowski B, Gunion J F and Jiang Y 2012 (*Preprint* 1211.3580)
[13] Celis A, Ilisie V and Pich A 2013 (*Preprint* 1302.4022)
[14] Ginzburg I F, Krawczyk M and Osland P 2001 (*Preprint* hep-ph/0101208)
[15] Cao Q H, Ma E and Rajasekaran G 2007 *Phys.Rev.* **D76** 095011 (*Preprint* 0708.2939)
[16] Barbieri R, Hall L J and Rychkov V S 2006 *Phys. Rev.* **D74** 015007 (*Preprint* hep-ph/0603188)
[17] Gustafsson M, Lundstrom E, Bergstrom L and Edsjo J 2007 (*Preprint* astro-ph/0703512)
[18] Dolle E M and Su S 2009 *Phys.Rev.* **D80** 055012 (*Preprint* 0906.1609)
[19] Dolle E, Miao X, Su S and Thomas B 2010 *Phys.Rev.* **D81** 035003 (*Preprint* 0909.3094)
[20] Lopez Honorez L, Nezri E, Oliver J F and Tytgat M H G 2007 *JCAP* **0702** 028 (*Preprint* hep-ph/0612275)
[21] Arina C, Ling F S and Tytgat M H 2009 *JCAP* **0910** 018 (*Preprint* 0907.0430)
[22] Tytgat M H 2008 *J.Phys.Conf.Ser.* **120** 042026 (*Preprint* 0712.4206)
[23] Lopez Honorez L and Yaguna C E 2010 *JHEP* **1009** 046 (*Preprint* 1003.3125)
[24] Lopez Honorez L and Yaguna C E 2011 *JCAP* **1101** 002 (*Preprint* 1011.1411)
[25] Sokolowska D 2011 *Acta Phys.Polon.* **B42** 2237 (*Preprint* 1112.2953)
[26] Sokolowska D 2011 (*Preprint* 1107.1991)
[27] Beringer et al J (Particle Data Group), Phys. Rev. D86, 010001 (2012)
[28] Djouadi A 2008 *Phys.Rept.* **459** 1–241 (*Preprint* hep-ph/0503173)
[29] Djouadi A 2008 *Phys.Rept.* **457** 1–216 (*Preprint* hep-ph/0503172)
[30] Cao Q H, Ma E and Rajasekaran G 2007 *Phys.Rev.* **D76** 095011 (*Preprint* 0708.2939)
[31] Posch P 2011 *Phys.Lett.* **B696** 447–453 (*Preprint* 1001.1759)
[32] Arhrib A, Benbrik R and Gaur N 2012 *Phys.Rev.* **D85** 095021 (*Preprint* 1201.2644)
[33] Chuzhoy L and Kolb E W 2009 *JCAP* **0907** 014 (*Preprint* 0809.0436)

Journal of Physics: Conference Series **447** (2013) 012051 doi:10.1088/1742-6596/447/1/012051

Avoiding Death by Vacuum

A. Barroso[1], P.M. Ferreira[1,2], I. Ivanov[3,4], R. Santos[1,2] and João P. Silva[2,5]

[1] Centro de Física Teórica e Computacional, Faculdade de Ciências, Universidade de Lisboa, Av. Prof. Gama Pinto 2, 1649-003 Lisboa, Portugal
[2] Instituto Superior de Engenharia de Lisboa, 1959-007 Lisboa, Portugal
[3] IFPA, Université de Liége, Allée du 6 Août 17, bâtiment B5a, 4000 Liége, Belgium
[4] Sobolev Institute of Mathematics, Koptyug avenue 4, 630090, Novosibirsk, Russia
[5] Instituto Superior de Engenharia de Lisboa, 1959-007 Lisboa, Portugal

E-mail: barroso@cii.fc.ul.pt, ferreira@cii.fc.ul.pt, igor.ivanov@ulg.ac.be, rsantos@cii.fc.ul.pt, jpsilva@cftp.ist.utl.pt

Abstract. The two-Higgs doublet model (2HDM) can have two electroweak breaking, CP-conserving, minima. The possibility arises that the minimum which corresponds to the known elementary particle spectrum is metastable, a possibility we call the "panic vacuum". We present analytical bounds on the parameters of the softly broken Peccei-Quinn 2HDM which are necessary and sufficient conditions to avoid this possibility. We also show that, for this particular model, the current LHC data already tell us that we are necessarily in the global minimum of the theory, regardless of any cosmological considerations about the lifetime of the false vacua.

1. Introduction

The two Higgs doublet model [1] is one of the simplest extensions of the Standard Model (SM) of particle physics. It has the particle content of the SM *plus* a second Higgs doublet, of the same hypercharge as the SM's. It has a very rich phenomenology (for a review, see [2]), including the possibility of spontaneous CP-violation, dark matter candidates and a richer scalar spectrum, including two CP-even scalars (usually denoted h and H), a pseudoscalar (A) and a charged scalar ($H^{\pm}$). With the recent discovery of the Higgs boson [3, 4], and a possibility that its observed properties might exhibit some deviations from its expected SM behaviour, it is interesting to look into SM extensions and compare their predictions with data. In that regard, the 2HDM has already shown to be able to do a good job describing current LHC results [5, 6, 7, 8].

The 2HDM also has a very rich vacuum structure: unlike the SM, there is the possibility of occurrence of charge breaking vacua which would give mass to the photon; and, as already mentioned, there is also the possibility of vacua which, other than the electroweak symmetry, also boasts a spontaneous breaking of the CP symmetry. It has been proven [9, 10], though, that whenever a vacuum which breaks electroweak symmetry but preserves the electromagnetic and CP symmetries exists - which we call a "normal" vacuum - any charge or CP breaking stationary points which might exist are *necessarily* saddle points, and lie above the normal minimum. The stability of the normal minimum against charge or CP breaking is thus guaranteed. There is however another possibility: the 2HDM potential, under certain circumstances, can have *two* normal minima, which coexist with one another [11, 12, 13]. In one of those minima, the two Higgs doublets, Φ_1 and Φ_2, would have vacuum expectation values (vevs) v_1 and v_2 such that $v_1^2 + v_2^2 = 246$ GeV^2, and all elementary particles would have the masses we know - this would be "our" minimum. But in the second minimum - deeper or higher than ours - the fields would have vevs $\{v_1', v_2'\}$, with $v_1'^2 + v_2'^2 \neq 246$ GeV^2 - and the elementary particles might have masses much smaller, or larger, than what is observed.

The possibility then arises that "our" vacuum is *not* the global minimum of the theory - and the universe would therefore be in a metastable state, with the possibility of tunneling to the deeper vacuum. We call this situation the "panic vacuum". It is therefore interesting to ascertain: under what conditions such vacua might occur; whether the current experimental data can tell us something about the nature of the vacuum in the 2HDM; what regions of parameter space are free of these panic vacua. In [14] we presented the conditions that the parameters of the potential need to obey so one can avoid the presence of a panic vacuum in the softly broken Peccei-Quinn [15] version of the 2HDM. In this talk, we will briefly review those theoretical bounds, as well as the phenomenological analysis that was performed to compare the model's predictions with the current LHC data.

2. Panic vacuum bounds in the softly broken Peccei-Quinn 2HDM

The most general 2HDM scalar potential has, after all possible simplifications, 11 independent real parameters. But it is plagued by tree-level flavour-changing neutral currents (FCNC) on the Higgs Yukawa interactions, which are extremely constrained by experimental observations. To evade this problem - and increase the predictive power of the theory, by reducing the number of free parameters - one usually imposes a symmetry on the Lagrangian. One such example is the discrete Z_2 symmetry, $\Phi_1 \to \Phi_1$ and $\Phi_2 \to -\Phi_2$, first proposed by Glashow, Weinberg and Paschos [16, 17]. Another possibility is to impose a global, continuous $U(1)$ symmetry, $\Phi_1 \to \Phi_1$ and $\Phi_2 \to e^{i\alpha}\Phi_2$, for any real value of α. This symmetry also liquidates any tree-level FCNC but, if broken by the vacuum (which occurs if both scalar fields develop a vev), leads to a massless pseudoscalar, an axion. To prevent that one adds a real soft breaking term to the potential, thus obtaining an acceptable scalar spectrum.

The softly broken Peccei-Quinn potential is then written as

$$\begin{aligned} V &= m_{11}^2 \Phi_1^\dagger\Phi_1 + m_{22}^2 \Phi_2^\dagger\Phi_2 - m_{12}^2 \left[\Phi_1^\dagger\Phi_2 + \Phi_2^\dagger\Phi_1\right] \\ &\quad + \frac{1}{2}\lambda_1(\Phi_1^\dagger\Phi_1)^2 + \frac{1}{2}\lambda_2(\Phi_2^\dagger\Phi_2)^2 + \lambda_3(\Phi_1^\dagger\Phi_1)(\Phi_2^\dagger\Phi_2) + \lambda_4(\Phi_1^\dagger\Phi_2)(\Phi_2^\dagger\Phi_1), \end{aligned} \tag{1}$$

where all the parameters are real and the soft breaking term is m_{12}^2. It can be seen [11, 12, 13] that this model, due precisely to the soft breaking term, can have two normal minima. If the the potential has a depth equal to V_N in the minimum with vevs $\{v_1, v_2\}$, and a depth $V_{N'}$ in the minimum with vevs $\{v_1', v_2'\}$, it is possible to show that

$$\begin{aligned} V_{N'} - V_N &= \frac{1}{4}\left[\left(\frac{m_{H^\pm}^2}{v^2}\right)_N - \left(\frac{m_{H^\pm}^2}{v^2}\right)_{N'}\right](v_1 v_2' - v_2 v_1')^2 \\ &= \frac{m_{12}^2}{4 v_1 v_2}\left(1 - \frac{v_1 v_2}{v_1' v_2'}\right)(v_1 v_2' - v_2 v_1')^2 \end{aligned} \tag{2}$$

with both the charged Higgs mass $m_{H^\pm}^2$ and the sum of the squared vevs v^2 being computed at each of the minima. It is not therefore obvious which is the deepest minimum.

In refs. [12, 13, 18] generic conditions for the existence of two minima in the 2HDM were established, being put into a simpler form in [14]. Thus it is possible to show that the Peccei-

Quinn potential can have two normal minima if the following conditions are met:

$$m_{11}^2 + k^2\, m_{22}^2 \;<\; 0, \tag{3}$$
$$\sqrt[3]{x^2} + \sqrt[3]{y^2} \;\leq\; 1, \tag{4}$$

where the variables x and y are given by

$$\begin{aligned} x &= \frac{4\, k\, m_{12}^2}{m_{11}^2 + k^2\, m_{22}^2}\, \frac{\sqrt{\lambda_1\lambda_2}}{\lambda_{34} - \sqrt{\lambda_1\lambda_2}}, \\ y &= \frac{m_{11}^2 - k^2\, m_{22}^2}{m_{11}^2 + k^2\, m_{22}^2}\, \frac{\sqrt{\lambda_1\lambda_2} + \lambda_{34}}{\sqrt{\lambda_1\lambda_2} - \lambda_{34}}, \end{aligned} \tag{5}$$

and we have defined

$$\lambda_{34} = \lambda_3 + \lambda_4 \quad \text{and} \quad k = \sqrt[4]{\frac{\lambda_1}{\lambda_2}}. \tag{6}$$

The curve $\sqrt[3]{x^2} + \sqrt[3]{y^2} = 1$, which delimits the region of parameter space where two minima can occur, is known as an astroid. To verify whether the $\{v_1\,,\, v_2\}$ minimum is the global one, one simply has to verify if the following condition is met: one computes, at that minimum, the value of the following quantity, which we dub the "discriminant",

$$D = \left(m_{11}^2 - k^2 m_{22}^2\right)(\tan\beta - k). \tag{7}$$

Our minimum is the global one if, and only if, $D > 0$. This condition can be verified regardless of eqs. 3, 4. Meaning, we do not even need to know whether the potential has two minima, this extremely simple condition is all we need to check to ascertain the nature of the 2HDM minimum.

3. LHC data and panic vacua

In order to verify how often the possibility of two minima arises in the Peccei-Quinn potential, we performed a thorough scan of the model's parameter space. To wit, we have considered $m_h = 125$ GeV, $125 < m_H < 900$ GeV, $90 < m_A, m_{H^\pm} < 900$ GeV, $-\pi/2 < \alpha < \pi/2$, $1 < \tan\beta < 40$ and $|m_{12}^2| < 900$ GeV^2. So that the potential is bounded from below, the quartic couplings of the potential must obey

$$\begin{aligned} \lambda_1 > 0 \quad &, \quad \lambda_2 > 0\,, \\ \lambda_3 > -\sqrt{\lambda_1\lambda_2} \quad &, \quad \lambda_3 + \lambda_4 - |\lambda_5| > -\sqrt{\lambda_1\lambda_2}\,. \end{aligned} \tag{8}$$

We further demanded that the quartic couplings be such that the model satisfies perturbative unitarity [19, 20] and the electroweak precision constraints stemming from the S, T and U parameters [21, 22, 23, 24]. All of these bounds apply to the theory's scalar sector. But the $U(1)$ symmetry we imposed, even though softly broken, needs to be a symmetry of the full Lagrangian, otherwise we would end up with a non-renormalizable model. There are several ways to extend this symmetry from the scalar to the fermion sector. Here we will consider the two most studied: Model I, where the fermion fields transform under the global $U(1)$ in such a way that only Φ_2 couples to the fermions; and Model II, where Φ_2 couples only to up-type quarks, and Φ_1 to down-type quarks and charged leptons. Both of these models possess very different phenomenologies. Also, data stemming from B-physics measurements put severe constraints on each of them. We have taken those bounds into account, incorporating them in our simulations [25, 26].

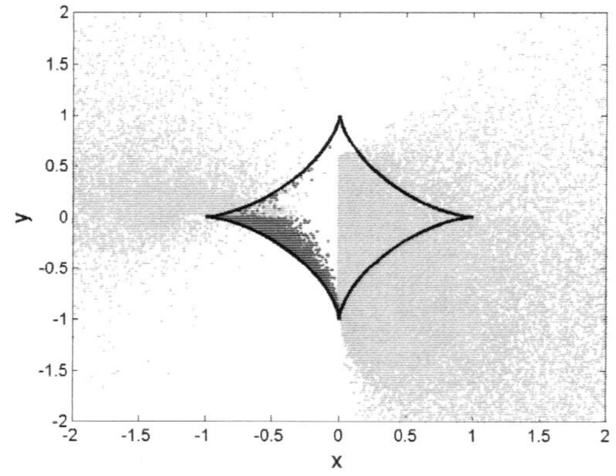

Figure 1. Distribution of generated points in the x and y variables. The dark curve is the astroid described by eq. 4. Points inside that curve can have two minima. The red/dark grey points shown here correspond to panic vacua. The green/light grey points correspond to either the non-existence of two minima, or the case where our vacuum is the global one.

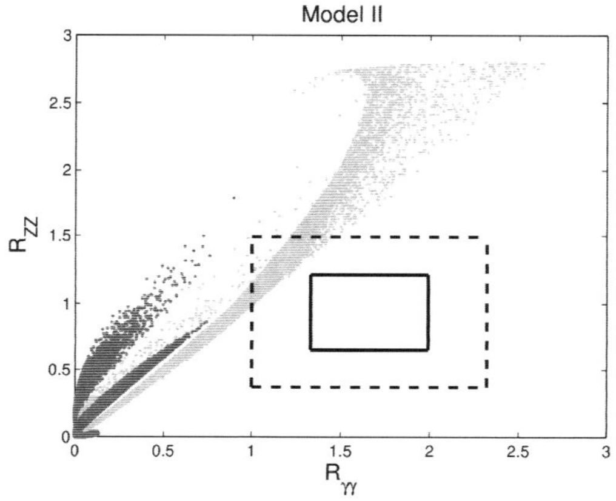

Figure 2. R_{ZZ} versus $R_{\gamma\gamma}$ for Model II. We show rough 1-σ (solid line) and 2-σ (dashed line) current bounds obtained from LHC data. Green/light grey points are the totality of our simulation; red/dark grey points correspond to panic vacua.

The relevance of the conditions 3- 4 can be appreciated in figure 1. For the totality of the parameter space we scanned, we plot, in the $\{x\,,\,y\}$ plane of the variables defined in eq. 5, the points which correspond to the panic vacua situation (in red/dark grey) and those for which our minimum is "safe" - whether because there is a single minimum, or because the second minimum is above ours. The point we wish to stress is that the occurrence of panic vacua is not a curiosity of the model: for a blind scan of the parameter space, obeying all reasonable constraints, a very large quantity of points with panic vacua appears.

We then use the generated points to compute R_f, defined as the number of events predicted in the 2HDM for the process $pp \to h \to f$, for some final state f, divided by the prediction obtained in the SM for the same final state. Current LHC bounds at 1σ, which we took from [27], are $R_{ZZ} = 0.93 \pm 0.28$, $R_{\gamma\gamma} = 1.66 \pm 0.33$. We sum over all production mechanisms, such as

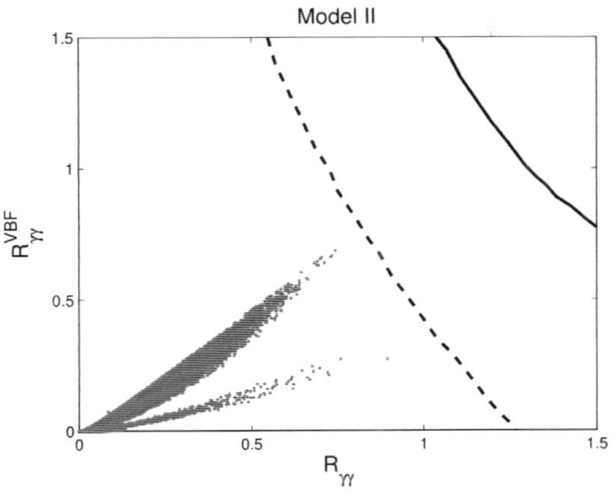

Figure 3. $R_{\gamma\gamma}^{VBF}$ versus $R_{\gamma\gamma}$ for Model II. We show rough 1-σ (solid line) and 2-σ (dashed line) current bounds obtained from LHC data. The red/dark grey points correspond to panic vacua.

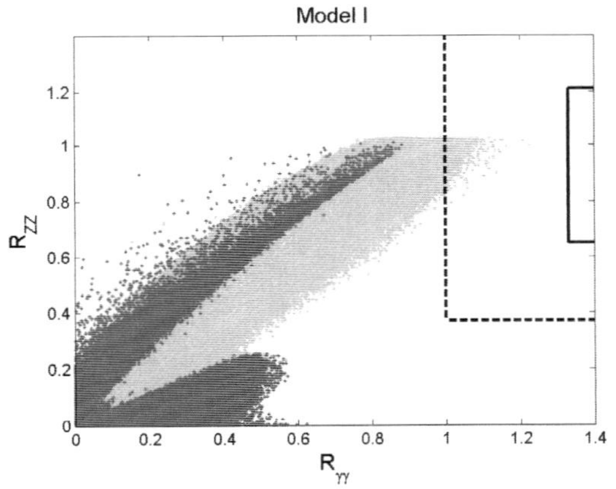

Figure 4. R_{ZZ} versus $R_{\gamma\gamma}$ for Model I. We show rough 1-σ (solid line) and 2-σ (dashed line) current bounds obtained from LHC data. Green/light grey points are the totality of our simulation; red/dark grey points correspond to panic vacua.

gluon-gluon fusion, vector boson fusion (VBF), associated Higgs production (with a gauge boson or a $t\bar{t}$ pair) or $b\bar{b}$ fusion. In fig. 2 we show, for Model II, the observables R_{ZZ} versus $R_{\gamma\gamma}$, where the points corresponding to panic vacua are shown in red/dark grey. We see that Model II is capable, at the 2-σ level, to obey current experimental bounds. However, even at that level, all points corresponding to panic vacua are disfavoured by current bounds - remarkably, even though the LHC hasn't found (yet) any evidence for more than one scalar, the current data is already capable of telling us a lot about the vacuum structure of the 2HDM. We see that this trend is also verified in other variables, such as those plotted in fig. 3. There we plot the Higgs $\gamma\gamma$ event rate (relative to the SM's), produced only via the VBF mechanism versus its total production rate. We only show those points corresponding to the panic vacuum solutions, and see that, even at the 2-σ level, they are strongly disfavoured.

The same, however, cannot be said to occur in Model I. Even though, as we see in fig. 4, where we once again plot R_{ZZ} versus $R_{\gamma\gamma}$, all panic points (the red/dark grey ones) lie outside even the 2-σ band, in the $R_{\gamma\gamma}^{VBF}$-$R_{\gamma\gamma}$ plane that would not occur: the 2-σ band would include some panic points, even though that would not occur at the 1-σ level. Of course, the fact that all panic points are excluded due to the results of fig. 4 would mean that for Model I, as well, the existence of panic vacua is strongly disfavoured by current LHC data, albeit perhaps less so than in Model II.

4. Conclusions

The 2HDM can have metastable neutral vacua, and this raises the possibility that the minimum we are currently inhabiting could not be the global minimum of the model. We have presented the extremely simple conditions one has to impose on the parameters of the potential to prevent this situation. They are, we believe, to be taken as seriously as the bounded-from-below conditions one usually imposes on the potential, or those enforcing perturbative unitarity. These bounds can also be extended to other versions of the 2HDM, and that work is underway [28].

The argument can be made that the existence of a metastable vacuum is not a problem *per se*, if the tunneling time from the false vacuum to the true one is larger than the current age of the universe. We have performed a quick estimate of the lifetimes of our panic vacua, and concluded that most of them would have lifetimes inferior to the age of the universe, therefore they would correspond to truly undesirable regions of parameter space, in disagreement with observed phenomenology.

But the most interesting aspect of our analysis, we believe, is the fact that, regardless of any calculation of vacuum lifetimes - and it must be stressed that that calculation is rife with approximations and considerable assumptions - the current data stemming from the LHC already permit us to conclude a lot about the nature of our vacuum, if Nature is indeed described by the 2HDM. Namely, the present LHC bounds on several observables already tells us to strongly disfavour the possibility that the vacuum we are in is *not* the global minimum of the theory. But that does not diminish the validity of the panic vacuum bounds we presented here, nor their interest: for we see that panic vacua occur for perfectly ordinary values of the 2HDM parameters, which predict values for observables which are not *a priori* absurd. And in any case, we see that the discriminant of eq. 7, by itself, allows us to obtain information, via only particle physics experiments, about a very interesting cosmological subject: the nature of the universe's vacuum.

Acknowledgments

The works of A.B., P.M.F. and R.S. are supported in part by the Portuguese *Fundação para a Ciência e a Tecnologia* (FCT) under contract PTDC/FIS/117951/2010, by FP7 Reintegration Grant, number PERG08-GA-2010-277025, and by PEst-OE/FIS/UI0618/2011. The work of J.P.S. is funded by FCT through the projects CERN/FP/109305/2009 and U777-Plurianual, and by the EU RTN project Marie Curie: PITN-GA-2009-237920. I.P.I. is thankful to CFTC, University of Lisbon, for their hospitality. His work is supported by grants RFBR 11-02-00242-a, RF President grant for scientific schools NSc-3802.2012.2, and the Program of Department of Physics SC RAS and SB RAS "Studies of Higgs boson and exotic particles at LHC".

References

[1] T. D. Lee, Phys. Rev. D **8** (1973) 1226.
[2] G. C. Branco, P. M. Ferreira, L. Lavoura, M. N. Rebelo, M. Sher and J. P. Silva, Phys. Rept. **516**, 1 (2012) [arXiv:1106.0034 [hep-ph]].
[3] G. Aad *et al.* [ATLAS Collaboration], Phys. Lett. B **716**, 1 (2012) [arXiv:1207.7214 [hep-ex]].
[4] S. Chatrchyan *et al.* [CMS Collaboration], Phys. Lett. B **716**, 30 (2012) [arXiv:1207.7235 [hep-ex]].
[5] C. -Y. Chen and S. Dawson, arXiv:1301.0309 [hep-ph].
[6] G. Belanger, B. Dumont, U. Ellwanger, J. F. Gunion and S. Kraml, arXiv:1212.5244 [hep-ph].

[7] S. Chang, S. K. Kang, J. -P. Lee, K. Y. Lee, S. C. Park and J. Song, with mass around 125 GeV," arXiv:1210.3439 [hep-ph].
[8] P. M. Ferreira, R. Santos, M. Sher and J. P. Silva, Phys. Rev. D **85**, 077703 (2012) [arXiv:1112.3277 [hep-ph]].
[9] P. M. Ferreira, R. Santos and A. Barroso, Phys. Lett. B **603** (2004) 219 [Erratum-ibid. B **629** (2005) 114] [arXiv:hep-ph/0406231].
[10] A. Barroso, P. M. Ferreira and R. Santos, Phys. Lett. B **632** (2006) 684 [arXiv:hep-ph/0507224].
[11] A. Barroso, P. M. Ferreira and R. Santos, Phys. Lett. B **652** (2007) 181 [arXiv:hep-ph/0702098].
[12] I. P. Ivanov, Phys. Rev. D **75** (2007) 035001 [Erratum-ibid. D **76** (2007) 039902] [arXiv:hep-ph/0609018].
[13] I. P. Ivanov, Phys. Rev. D **77** (2008) 015017 [arXiv:0710.3490 [hep-ph]].
[14] A. Barroso, P. M. Ferreira, I. P. Ivanov, R. Santos and J. P. Silva, arXiv:1211.6119 [hep-ph].
[15] R. D. Peccei and H. R. Quinn, Phys. Rev. Lett. **38** (1977) 1440.
[16] S. L. Glashow and S. Weinberg, Phys. Rev. D **15** (1977) 1958.
[17] E. A. Paschos, Phys. Rev. D **15** (1977) 1966.
[18] I. P. Ivanov, Phys. Rev. E **79**, 021116 (2009).
[19] S. Kanemura, T. Kubota and E. Takasugi, Phys. Lett. B **313** (1993) 155 [arXiv:hep-ph/9303263].
[20] A. G. Akeroyd, A. Arhrib and E. -M. Naimi, Phys. Lett. B **490**, 119 (2000) [hep-ph/0006035].
[21] M.E. Peskin and T. Takeuchi, Phys. Rev. D **46**, 381 (1992).
[22] The ALEPH, CDF, D0, DELPHI, L3, OPAL, SLD Collaborations, the LEP Electroweak Working Group, the Tevatron Electroweak Working Group, and the SLD electroweak and heavy flavour Groups, arXiv:1012.2367 [hep-ex].
[23] M. Baak, M. Goebel, J. Haller, A. Hoecker, D. Ludwig, K. Moenig, M. Schott and J. Stelzer, Eur. Phys. J. C **72**, 2003 (2012) [arXiv:1107.0975 [hep-ph]].
[24] M. Baak, M. Goebel, J. Haller, A. Hoecker, D. Kennedy, R. Kogler, K. Moenig, M. Schott and J. Stelzer, arXiv:1209.2716 [hep-ph].
[25] T. Hermann, M. Misiak and M. Steinhauser, Next-to-Next-to-Leading Order in QCD," JHEP **1211** (2012) 036 [arXiv:1208.2788 [hep-ph]].
[26] F. Mahmoudi, talk given at Prospects For Charged Higgs Discovery At Colliders (CHARGED 2012), 8-11 October, Uppsala, Sweden.
[27] A. Arbey, M. Battaglia, A. Djouadi and F. Mahmoudi, arXiv:1211.4004 [hep-ph].
[28] A. Barroso, P. M. Ferreira, I. P. Ivanov and R. Santos, work underway.

Journal of Physics: Conference Series **447** (2013) 012052 doi:10.1088/1742-6596/447/1/012052

Type I Seesaw Mechanism, Lepton Flavour Violation and Higgs Decays

Emiliano Molinaro

Physik-Department T30d, Technische Universität München,
James-Franck-Straße 1, 85748 Garching, Germany.

E-mail: emiliano.molinaro@tum.de

Abstract. We review and update the current phenomenological constraints on minimal type I seesaw extensions of the Standard Model in which New Physics content can be probed at the electroweak scale. In this class of models, the flavour structure of the neutrino Yukawa couplings is determined by the requirement of reproducing neutrino oscillation data. The strongest constraints on the seesaw parameter space are imposed by the very recent upper limit on $\mu \to e\gamma$ decay rate. Searches of non-standard Higgs boson decays into a light and a heavy neutrino may also provide and independent test of these seesaw scenarios.

1. Introduction

The recent discovery of the Higgs boson at the Large Hadron Collider (LHC) [1, 2] with a mass $m_h \sim 125$ GeV marks an important breakthrough in our understanding of the mechanism of electroweak (EW) symmetry breaking and is by itself an extraordinary success of the Standard Model (SM) of Elementary Particles.

On top of the discovery of the Higgs boson, there are still several open questions which the SM, in the way it is conceived now, cannot explain. In particular, New Physics must be advocated to explain the outcome of the flavour neutrino oscillation experiments, which provide compelling and indubitable evidence that at least two neutrinos have non-zero masses, several orders of magnitude smaller than the other SM charged fermions masses, and that the three neutrino flavours mix. These very small neutrino masses can be linked to a new physical scale of Nature that, in general, is not related to the EW symmetry breaking scale.

One of the most successful mechanism of neutrino mass generation is the well known type I seesaw scenario [3], in which the SM is extended with at least two heavy SM singlet fermions, $N_{1,2}$, usually dubbed right-handed neutrinos, that have Yukawa-type interactions with the SM Higgs and left-handed doublets. If further symmetries are not imposed, such neutrino Yukawa couplings explicitly break lepton number and a Majorana mass term for the SM neutrinos is generated at tree-level after the decoupling of the heavy fermions $N_{1,2}$.

In the following we consider the type I seesaw scenarios studied in [4, 5, 6, 7] in which the right-handed neutrino mass scale is set in the TeV range, that is $M_{1,2} \sim (100 - 1000)$ GeV. This class of models predicts a rich low energy and collider phenomenology and, for this reason, the full seesaw parameter space is tightly constrained by different sets of data: neutrino oscillation data, neutrinoless double beta decay, EW precision tests and charged lepton flavour violating processes and right-handed neutrino production/detection at colliders.

As discussed below, the unknown seesaw parameter space can be conveniently expressed in terms of the size of the neutrino Yukawa couplings (y) and the heavy Majorana neutrino masses ($M_{1,2}$). The most conservative bound on a combination of these parameters is set by current experiments searching for $\mu \to e\,\gamma$ decay.

In the case of $M_{1,2} \lesssim m_h$ it is also possible to probe the neutrino Yukawa interactions looking for non-standard Higgs boson decays into a light and a heavy Majorana neutrino [8, 7, 9, 10, 11].

All the present limits on the seesaw parameter space are summarized in Figs 1 and 2.

2. Constraints on the seesaw parameter space

The mixing between the new singlet SM heavy Majorana fermions $N_{1,2}$ and the SM light neutrinos generate charged and neutral current interactions between $N_{1,2}$ and the SM gauge/Higgs bosons, which can in principle be tested in low energy and collider experiments for masses $M_{1,2} \sim (100 - 1000)$ GeV. The relevant parts of the interaction Lagrangian are in this case

$$\mathcal{L}^N_{CC} = -\frac{g}{2\sqrt{2}}\,\bar{\ell}\,\gamma_\alpha\,(RV)_{\ell k}(1-\gamma_5)\,N_k\,W^\alpha \;+\; \text{h.c.}\,, \tag{1}$$

$$\mathcal{L}^N_{NC} = -\frac{g}{4\,c_w}\,\overline{\nu_{\ell L}}\,\gamma_\alpha\,(RV)_{\ell k}\,(1-\gamma_5)\,N_k\,Z^\alpha \;+\; \text{h.c.}\,, \tag{2}$$

$$\mathcal{L}^N_{H} = -\frac{gM_k}{4\,M_W}\,\overline{\nu_{\ell L}}\,(RV)_{\ell k}\,(1+\gamma_5)\,N_k\,h \;+\; \text{h.c.} \tag{3}$$

The copulings $(RV)_{\ell k}$ ($\ell = e, \mu, \tau$ and $k = 1, 2$) arise from the mixing between heavy and light Majorana neutrinos and, therefore, are suppressed by the seesaw scale. They can be conveniently parametrized as follows [5]:

$$|(RV)_{\ell 1}|^2 = \frac{1}{2}\frac{y^2v^2}{M_1^2}\frac{m_3}{m_2+m_3}\left|U_{\ell 3} + i\sqrt{m_2/m_3}\,U_{\ell 2}\right|^2\,,\quad \text{NH}\,, \tag{4}$$

$$|(RV)_{\ell 1}|^2 = \frac{1}{2}\frac{y^2v^2}{M_1^2}\frac{m_2}{m_1+m_2}\left|U_{\ell 2} + i\sqrt{m_1/m_2}\,U_{\ell 1}\right|^2 \simeq \frac{1}{4}\frac{y^2v^2}{M_1^2}\left|U_{\ell 2} + iU_{\ell 1}\right|^2\,,\ \text{IH}\,, \tag{5}$$

$$(RV)_{\ell 2} = \pm i\,(RV)_{\ell 1}\sqrt{\frac{M_1}{M_2}}\,,\quad \ell = e, \mu, \tau\,, \tag{6}$$

where U denotes the neutrino mixing matrix and $v \simeq 174$ GeV. Notice that the relative mass splitting of the two heavy Majorana neutrinos must be very small, $|M_1 - M_2|/M_1 \ll 1$, due to the current upper limit on the effective Majorana mass probed in neutrinoless double beta decay experiments [4, 5]. In this case, the flavour structure of the neutrino Yukawa couplings is fixed by the neutrino oscillation parameters [4, 5, 12] and the two heavy Majorana neutrinos form a pseudo-Dirac fermion: $N = (N_1 \pm iN_2)/\sqrt{2}$. The parameter y in the expressions above represents the largest eigenvalue of the matrix of the neutrino Yukawa couplings [5]:

$$y^2v^2 = 2\,M_1^2\left(|(RV)_{e1}|^2 + |(RV)_{\mu 1}|^2 + |(RV)_{\tau 1}|^2\right)\,. \tag{7}$$

2.1. Charged lepton flavour violation: $\mu \to e\,\gamma$

In the case of a seesaw mass scale $M_{1,2}$ in the TeV range considered here, the most stringent constraint on the size of the neutrino Yukawa couplings is set by the very recent upper limit on $\mu^+ \to e^+\,\gamma$ branching ratio reported by the MEG Collaboration [13]: $\text{BR}(\mu \to e\,\gamma) < 5.7\times 10^{-13}$ at 90% confidence level. Taking the best fit values of the neutrino oscillation parameters [14], we can estimate the typical size of neutrino Yukawa couplings:

$$y \lesssim 0.026 \quad \text{for} \quad M_1 = 100\,\text{GeV}\,. \tag{8}$$

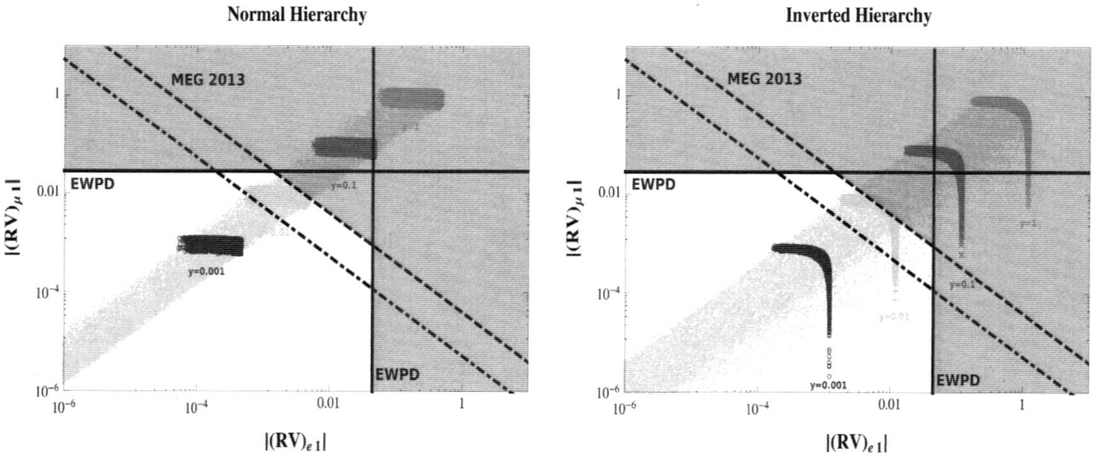

Figure 1. Correlation between $|(RV)_{e1}|$ and $|(RV)_{\mu 1}|$ for $M_1 = 100$ GeV in the case of normal (left panel) and inverted (right panel) light neutrino mass spectrum. The regions corresponding to four different values of the neutrino Yukawa eigenvalue y are highlighted. The cyan points correspond to random values of $y \leq 1$. The dashed line represents to the present MEG bound [13], $B(\mu \to e\,\gamma) \leq 5.7 \times 10^{-13}$. The dot-dashed line coincides with $B(\mu \to e\,\gamma) = 10^{-14}$.

We implement in Fig. 1 all the constraints on the effective couplings $(RV)_{\mu 1}$ and $(RV)_{e1}$ which come from the requirement of reproducing neutrino oscillation data, from EW precision data and from the current upper limit on $\mu \to e\,\gamma$. The seesaw mass scale is fixed at benchmark value: $M_1 = 100$ GeV. From Fig. 1, it is manifest that the allowed ranges of the right-handed neutrino couplings $|(RV)_{\mu 1}|$ and $|(RV)_{e1}|$, in the case of normal (left panel) and inverted (right panel) light neutrino mass spectrum, are confined in a small strip of the overall representative plane. This corresponds to the scatter plot of the points which are consistent with the 3σ allowed ranges of the neutrino oscillation parameters [14]. The region of the parameter space which is allowed by the EW precision data, is marked with solid lines. The region allowed by the current bound on the $\mu \to e\,\gamma$ decay rate is indicated with a dashed line, while the dot-dashed line shows the projected MEG sensitivity reach, *i.e.* $\text{BR}(\mu \to e\,\gamma) \sim 10^{-14}$. The scatter points corresponds to different values of the maximum neutrino Yukawa coupling y: $y = 0.001$ (blue $\circ$), *ii*) $y = 0.01$ (green $+$), *iii*) $y = 0.1$ (red $\times$), *iv*) $y = 1$ (orange $\diamondsuit$) and *v*) an arbitrary value of the Yukawa coupling $y \leq 1$ (cyan points).

As depicted in Fig. 1, in the case of a light neutrino mass spectrum with inverted hierarchy a strong suppression of the $\mu \to e\,\gamma$ decay rate is possible for specific values of the measured neutrino parameters. Indeed, using the standard parametrisation of the neutrino mixing matrix U one can show that for fixed values of the mixing angles θ_{12}, θ_{23} and of the Dirac (δ) and Majorana (α_{21}) phases, $|U_{\mu 2} + iU_{\mu 1}|^2$ has a minimum for [5, 6]

$$\sin\theta_{13} \;=\; \frac{c_{23}}{s_{23}}\,\frac{\cos 2\theta_{12}\cos\delta\sin\frac{\alpha_{21}}{2} - \cos\frac{\alpha_{21}}{2}\sin\delta}{1 + 2c_{12}\,s_{12}\,\sin\frac{\alpha_{21}}{2}}\,. \tag{9}$$

At the minimum

$$\min\left(|U_{\mu 2} + i\,U_{\mu 1}|^2\right) \;=\; c_{23}^2\,\frac{\left(\cos\delta\cos\frac{\alpha_{21}}{2} + \cos 2\theta_{12}\sin\delta\sin\frac{\alpha_{21}}{2}\right)^2}{1 + 2c_{12}\,s_{12}\,\sin\frac{\alpha_{21}}{2}}\,. \tag{10}$$

Therefore, the $\mu \to e\gamma$ branching ratio, that is proportional to eq. (10), is highly suppressed if the Dirac and Majorana phases take CP conserving values, mainly: $\delta \simeq 0$ and $\alpha_{21} \simeq \pi$. In

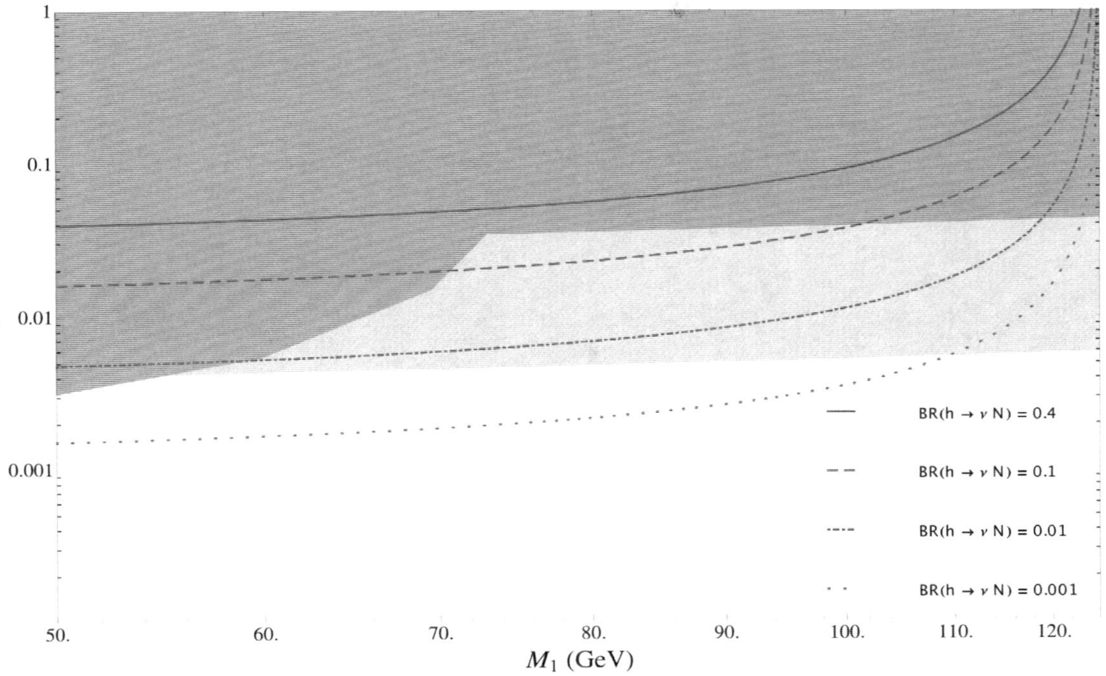

Figure 2. Values of the maximum neutrino Yukawa coupling y probed by Higgs decays into N for $m_h = 125$ GeV. The grey region is excluded by LEP2 data [15] and searches of lepton flavour violation [13]. The cyan area represents the region of the parameter space which can be probed by the MEG experiment with the projected sensitivity to $\text{BR}(\mu \to e\gamma) = 10^{-14}$.

this case, from eq. (9) we get a lower bound on the reactor angle, *i.e.* $\sin(\theta_{13}) \gtrsim 0.13$, which is in agreement with the global fit results [14]. On the other hand, assuming CP violating phases, $\min(|U_{\mu 2} + iU_{\mu 1}|^2) \approx 0$ is possible, provided θ_{12} and the Dirac and Majorana phases δ and α_{21} satisfy the following conditions: $\cos\delta\cos(\alpha_{21}/2) + \cos 2\theta_{12}\sin\delta\sin(\alpha_{21}/2) \approx 0$ and $\text{sgn}(\cos\delta\cos\frac{\alpha_{21}}{2}) = -\text{sgn}(\sin\delta\sin\frac{\alpha_{21}}{2})$.

We remark that the upper limit given in (8) is of the same order as the bottom Yukawa coupling, $y_b = m_b/v \simeq 0.024$. Therefore it is interesting to explore the possibility to produce (pseudo-Dirac) heavy neutrinos N at present and forthcoming collider facilities through non-standard Higgs boson decays, mainly $h \to \nu_{\ell L} + \overline{N}$, $\overline{\nu_{\ell L}} + N$.

2.2. New Higgs decay channel: $h \to \nu N$

In the case of $M_1 \lesssim m_h \sim 125$ GeV, it is possible to put limits on the seesaw parameter space looking for production of the right-handed neutrinos in non-standard Higgs boson decays: $h \to \nu N$ (see *e.g.* [8, 7, 9, 10, 11]). In this case, the Higgs decay rate is directly related to the neutrino Yukawa coupling y defined in eq. (7):

$$\begin{aligned}\Gamma(h \to \nu N) &\equiv \sum_{\ell=e,\mu,\tau} \left(\Gamma(h \to \nu_{\ell L}\,\overline{N}) + \Gamma(h \to \overline{\nu_{\ell L}}\,N)\right) \\ &= \frac{1}{16\pi}\, y^2\, m_h \left(1 - \frac{M_1^2}{m_h^2}\right)^2 .\end{aligned}$$

Taking as benchmark values $m_h = 125$ GeV and $M_1 = 100$ GeV, one gets $\Gamma(h \to \nu N)/\Gamma(h \to b\bar{b}) \simeq 0.19\ (y/0.05)^2$, that turns out to be sizable if the upper limit on the Yukawa coupling y, obtained using the MEG upper bound (8) is saturated. Conversely, the search for the Higgs decay $h \to \nu N$ may provide limits on the parameters of the low scale seesaw scenario which are competitive to those from the searches for the $\mu \to e\,\gamma$ decay, when $m_h > M_1$ [7]. On the other hand, in the case $M_1 > m_h$ the exotic Higgs decay channels are, $h \to \nu N \to \nu\,\nu\,Z,\ \nu\,\ell\,W$ which have a rate suppressed by the fourth power of y as well as by the three-body decay phase space.

In Fig. 2 it is shown the correlation between y and the seesaw scale M_1, corresponding to different values of $\mathrm{BR}(h \to \nu N)$ and the Higgs boson mass $m_h = 125$ GeV. The region in grey is excluded by the combined data on $\mu \to e\,\gamma$ decay [13] and the search for heavy singlet neutrinos in Z boson decays at LEP2 [15]. It follows from the plot that the present bounds on the low scale seesaw parameter space do not preclude the possibility of a Higgs boson decaying into a heavy and a light neutrino with a branching ratio of roughly 10%. We also show in the plot as a cyan area the projected sensitivity reach of the MEG experiment searching for the $\mu \to e\,\gamma$ decay with a branching ratio $\mathrm{BR}(\mu \to e\,\gamma) \gtrsim 10^{-14}$, which may allow to exclude $\mathrm{BR}(h \to \nu N) \gtrsim 1\%$ for $M_1 \gtrsim 100$ GeV.

3. Conclusions

Minimal type I seesaw extensions of the Standard Model with right-handed neutrino masses at the EW scale predict by a very rich phenomenology and are tightly constrained by both low energy observables and collider searches. In the minimal scenario only two new fermion representations $N_{1,2}$ are introduced, which are Standard Model singlets and have Yukawa-type couplings to the Higgs and lepton doublets, usually referred to as neutrino Yukawa couplings. The flavour structure of such interactions is univocally fixed by the requirement of reproducing the data on the neutrino masses and mixing. The existing low energy constraints imply that the two heavy right-handed neutrinos form a pseudo-Dirac pair, $N \equiv (N_1 \pm iN_2)/\sqrt{2}$, thus preventing the possibility to observe a possible signature of lepton number violation in this class of scenarios [4, 5]. The phenomenological relevant seesaw parameter space can be conveniently expressed in term of the seesaw mass scale $M_1 \approx M_2$ and the largest neutrino Yukawa eigenvalue y.

All the constraints on y and M are reported in Figs 1 and 2. The strongest bounds are derived from the very recent upper limit on $\mu \to e\,\gamma$ decay rate released by the MEG experiment [13]: $\mathrm{BR}(\mu \to e\,\gamma) < 5.7 \times 10^{-13}$ at 90% confidence level. If the MEG experiment eventually observes the $\mu \to e\gamma$ decay with a branching ratio $\mathrm{BR}(\mu \to e\gamma) > 10^{-14}$, the low scale type I see-saw scenario may be directly tested at LHC through non-standard Higgs boson decays $h \to \nu\,N$ decays [7]. On the other hand, if no positive signal is detected in the MEG experiment, this will lead to a more stringent limit on the neutrino Yukawa coupling y that will exclude for the moment the possibility of producing and detecting the new heavy pseudo-Dirac neutrino N.

Acknowledgments

I would like to thank Camilo Garcia Cely, Dinh N. Dinh, Alejandro Ibarra and Serguey T. Petcov for their collaboration on the research discussed here. This work is supported by the ERC Advanced Grant project "FLAVOUR" (267104).

References

[1] G. Aad *et al.* [ATLAS Collaboration], *Observation of a new particle in the search for the Standard Model Higgs boson with the ATLAS detector at the LHC*, Phys. Lett. B **716** (2012) 1 [arXiv:1207.7214 [hep-ex]].

[2] S. Chatrchyan *et al.* [CMS Collaboration], *Observation of a new boson at a mass of 125 GeV with the CMS experiment at the LHC*, Phys. Lett. B **716** (2012) 30 [arXiv:1207.7235 [hep-ex]].

[3] P. Minkowski, Phys. Lett. B **67** (1977) 421; M. Gell-Mann, P. Ramond and R. Slansky, *Proceedings of the Supergravity Stony Brook Workshop*, New York 1979, eds. P. Van Nieuwenhuizen and D. Freedman; T.

Yanagida, *Proceedings of the Workshop on Unified Theories and Baryon Number in the Universe*, Tsukuba, Japan 1979, eds. A. Sawada and A. Sugamoto; R. N. Mohapatra and G. Senjanovic, Phys. Rev. Lett. **44** (1980) 912.

[4] A. Ibarra, E. Molinaro and S. T. Petcov, *TeV Scale See-Saw Mechanisms of Neutrino Mass Generation, the Majorana Nature of the Heavy Singlet Neutrinos and $(\beta\beta)_{0\nu}$-Decay*, JHEP **1009** (2010) 108 [arXiv:1007.2378 [hep-ph]].

[5] A. Ibarra, E. Molinaro and S. T. Petcov, *Low Energy Signatures of the TeV Scale See-Saw Mechanism*, Phys. Rev. D **84** (2011) 013005 [arXiv:1103.6217 [hep-ph]].

[6] D. N. Dinh, A. Ibarra, E. Molinaro and S. T. Petcov, *The mu - e Conversion in Nuclei, mu $\rightarrow e\gamma$, mu $\rightarrow 3e$ Decays and TeV Scale See-Saw Scenarios of Neutrino Mass Generation*, JHEP **1208** (2012) 125 [arXiv:1205.4671 [hep-ph]].

[7] C. G. Cely, A. Ibarra, E. Molinaro and S. T. Petcov, *Higgs Decays in the Low Scale Type I See-Saw Model*, Phys. Lett. B **718** (2013) 957 [arXiv:1208.3654 [hep-ph]].

[8] P. S. B. Dev, R. Franceschini and R. N. Mohapatra, *Bounds on TeV Seesaw Models from LHC Higgs Data*, arXiv:1207.2756 [hep-ph].

[9] A. de Gouvea, *GeV seesaw, accidentally small neutrino masses, and Higgs decays to neutrinos*, arXiv:0706.1732 [hep-ph].

[10] A. Pilaftsis, *Radiatively induced neutrino masses and large Higgs neutrino couplings in the standard model with Majorana fields*, Z. Phys. C **55** (1992) 275 [hep-ph/9901206].

[11] J. -H. Chen, X. -G. He, J. Tandean and L. -H. Tsai, *Effect on Higgs Boson Decays from Large Light-Heavy Neutrino Mixing in Seesaw Models*, Phys. Rev. D **81** (2010) 113004 [arXiv:1001.5215 [hep-ph]].

[12] M. Raidal, A. Strumia and K. Turzynski, *Low-scale standard supersymmetric leptogenesis*, Phys. Lett. B **609** (2005) 351 [Erratum-ibid. B **632** (2006) 752] [hep-ph/0408015]; M. Shaposhnikov, *A Possible symmetry of the nuMSM*, Nucl. Phys. B **763** (2007) 49 [hep-ph/0605047]; J. Kersten and A. Y. .Smirnov, *Right-Handed Neutrinos at CERN LHC and the Mechanism of Neutrino Mass Generation*, Phys. Rev. D **76** (2007) 073005 [arXiv:0705.3221 [hep-ph]]; M. B. Gavela, T. Hambye, D. Hernandez and P. Hernandez, *Minimal Flavour Seesaw Models*, JHEP **0909** (2009) 038 [arXiv:0906.1461 [hep-ph]].

[13] J. Adam *et al.* [MEG Collaboration], *New constraint on the existence of the mu+-¿ e+ gamma decay*, arXiv:1303.0754 [hep-ex].

[14] D. V. Forero, M. Tortola and J. W. F. Valle, *Global status of neutrino oscillation parameters after Neutrino-2012*, Phys. Rev. D **86** (2012) 073012 [arXiv:1205.4018 [hep-ph]].

[15] P. Abreu *et al.* [DELPHI Collaboration], *Search for neutral heavy leptons produced in Z decays*, Z. Phys. C **74** (1997) 57 [Erratum-ibid. C **75** (1997) 580].

Fermion mixing in an S_3 model with three Higgs doublets

F. González Canales[1], A. Mondragón[2], M. Mondragón[2], U. J. Saldaña Salazar[2] and L. Velasco-Sevilla[2,3]

[1] Facultad de Ciencias de la Electrónica, Benemérita Universidad Autónoma de Puebla, Apdo. Postal 157, 72570, Puebla, Pue., México.
[2] Instituto de Física, Universidad Nacional Autónoma de México, Apdo. Postal 20-364, 01000, México D.F., México.
[3] University of Hamburg, II. Institute for Theoretical Physics, Luruper Chaussee 149, 22761 Hamburg, Germany.

Abstract. We present a brief overview of a unified treatment of masses and mixings of quarks and leptons, as a realization of an S_3-flavour permutational symmetry in an extension of the Standard Model. In order to leave S_3 as an exact flavour symmetry two extra Higgs fields, $SU(2)_L$ doublets, are added. In this model, the mass matrices of the fermions are reparametrized in terms of their eigenvalues allowing us to derive exact, explicit analytical expressions for the mixing matrices, $V_{CKM}^{\rm th}$ and $U_{PMNS}^{\rm th}$, as functions of the quark and lepton masses, respectively. In both the quark and lepton sectors a χ^2 fit of the theoretical expressions to the experimentally determined values of the mixing angles yields results in excellent agreement with the most recent experimental data on quarks and leptons.

1. Introduction

Family or flavour models offer a way of expressing mixing angles in terms of fermion masses [1–4] in good agreement with phenomenology. Here, we consider fermion mixing in an extension of the Standard Model (SM) with an S_3 family symmetry group and three Higgs fields, electroweak doublets. We include also three right-handed neutrino fields, to implement the seesaw mechanism. In this way, all fermionic and Higgs fields, including the right-handed neutrino fields, have three species and transform under the family symmetry group as the three dimensional representation $\mathbf{1} \oplus \mathbf{2}$ of the permutational group S_3. The S_3 family symmetry strongly constrains the number of free parameters in the fermionic mass matrices and allows us to obtain exact mathematical relations among fermion masses and mixing angles. In section 2, we introduce the S_3 symmetry and briefly comment on its interesting features as a family symmetry. In section 3, we discuss and introduce the construction of this model, which we will denote as S_3-3H. In section 4, we confront the model with the most up to date experimental data. In section 5, we summarize and conclude with some remarks.

2. S_3 as a symmetry of fermion families

The permutational symmetry group of three objects, S_3, is the smallest non-Abelian finite group. The group S_3 has three irreducible representations (irreps): a doublet $\mathbf{2}$, and two singlets, $\mathbf{1}_S$

and $\mathbf{1}_A$, symmetric and antisymmetric, respectively. The Kronecker products of irreducible representations may be decomposed as a direct sum of irreducible representations

$$\mathbf{1}_S \otimes \mathbf{1}_S = \mathbf{1}_S, \quad \mathbf{1}_A \otimes \mathbf{1}_A = \mathbf{1}_S, \quad \mathbf{1}_A \otimes \mathbf{1}_S = \mathbf{1}_A,$$
$$\mathbf{1}_S \otimes \mathbf{2} = \mathbf{2}, \quad \mathbf{1}_A \otimes \mathbf{2} = \mathbf{2}, \quad \text{and} \quad \mathbf{2} \otimes \mathbf{2} = \mathbf{1}_A \oplus \mathbf{1}_S \oplus \mathbf{2}. \tag{1}$$

In the SM all gauge interactions are invariant under the permutation of the family index in the fermion fields. If we extend this invariance to the Yukawa sector we are led to introduce two more Higgs fields, $SU(2)_L$ doublets, such that all Yukawa interactions are made invariant under the permutation of the family index in both the fermion and the Higgs boson fields.

In Figure 1 the mass ratios, $\tilde{m}_i^f = m_i^f/m_3^f$, are shown. There, a clear pattern can be observed, which suggests that the structure obtained in the decomposition of the fermion mass matrix is a direct sum of irreducible representations of S_3, namely $3 = \mathbf{2} \oplus \mathbf{1}$, and is a good representation of the fermion mass hierarchy.

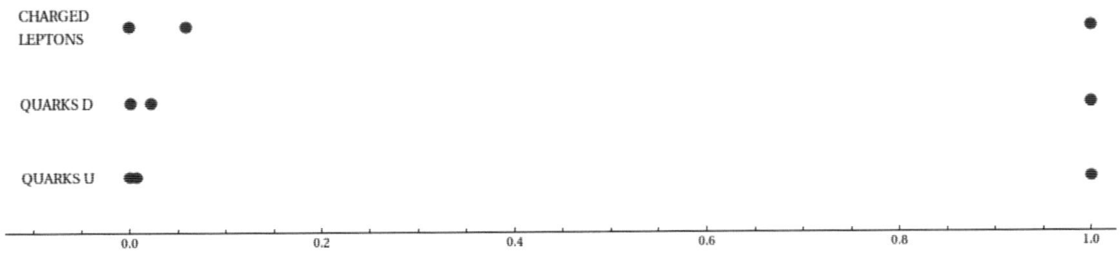

Figure 1. The mass ratios appear naturally ordered, from left to right, from the first family to the third family. An underlying symmetry structure, $\mathbf{2} \oplus \mathbf{1}$, can be clearly seen from here.

3. An S_3-invariant Extension of the Standard Model (S_3-3H)

Models which have S_3 as family symmetry group have been discussed for a long time (see for instance [4] and references therein).

3.1. Assignments between S_3 irreps and families

We assign the first two families, $f_{I(L,R)}$ and $f_{II(L,R)}$, to the doublet representation, $\mathbf{2}$, and the third family, $f_{III(L,R)}$, to the symmetric singlet representation, $\mathbf{1}_S$, of S_3

$$\begin{pmatrix} f_{I(L,R)} \\ f_{II(L,R)} \end{pmatrix} \sim \mathbf{2}; \quad f_{III(L,R)} \sim 1_\mathrm{S},$$

respectively. Here I, II or III is the family index of a left or right-handed fermion field $f_{(L,R)}$, and $f_{(L,R)}$ may represent any SM fermion field. We make a similar assignment in the Higgs sector

$$H_D \equiv \begin{pmatrix} H_1 \\ H_2 \end{pmatrix} \sim \mathbf{2}; \quad H_S \sim 1_\mathrm{S}, \tag{2}$$

where H_i $(i = 1, 2, S)$ represents a Higgs $SU(2)_L$ doublet field.

3.2. A weak-basis transformation

After electroweak symmetry breaking, in the symmetry adapted basis, the generic S_3-invariant mass matrix form for Dirac fermions is

$$\mathcal{M}^f_{S_3} = \begin{pmatrix} \sqrt{2}Y^f_2 v_S + Y^f_3 w_2 & Y^f_3 w_1 & \sqrt{2}Y^f_5 w_1 \\ Y^f_3 w_1 & \sqrt{2}Y^f_2 v_S - Y^f_3 w_2 & \sqrt{2}Y^f_5 w_2 \\ \sqrt{2}Y^f_6 w_1 & \sqrt{2}Y^f_6 w_2 & 2Y^f_1 v_S \end{pmatrix}, \tag{3}$$

where the Y^f_I are complex Yukawa couplings and w_1, w_2, and v_S are the real vacuum expectation values of the S_3 doublet components and the S_3 symmetric singlet representation, respectively. In this model, we consider the case where the vacuum expectation values of the S_3 doublet are the same, $w_1 = w_2$. This relation implies that in the minimum of the Higgs potential there is an accidental S'_2 symmetry [5,6].

It has been already noticed that a suitable weak-basis transformation of the mass matrices may help to reduce the number of free parameters by introducing a set of zeroes in their entries [7,8]. In our case, a $\pi/4$ rotation in the S_3 doublet subspace is enough to introduce one texture zeroes in the generic S_3-invariant Dirac fermion mass matrix

$$\mathcal{M}^f_{Hier} = \begin{pmatrix} |\mu^f_1| - |\mu^f_2| & |\mu^f_2| & 0 \\ |\mu^f_2| & |\mu^f_1| + |\mu^f_2| & \mu^f_8 \\ 0 & \mu^{f*}_8 & |\mu^f_3| \end{pmatrix}, \tag{4}$$

where $\mu^f_1 \equiv \sqrt{2}Y^f_2 v_S$, $\mu^f_2 \equiv Y^f_3 w_1$, $\mu^f_3 \equiv 2Y^f_1 v_S$, and $\mu^f_8 \equiv 2Y^f_6 w_1$, and we have assumed Hermiticity, which in this case implies a particular value for the phase of the complex Yukawa couplings.

It has been already noticed [9] that the matrix form of Eq. (4) reproduces the four zero Fritzsch-like texture [1,10–13] which, along with the Nearest-Neighbour Interaction (NNI) [14,15] form of the mass matrix, give a good description of the phenomenology of quark and lepton masses and their mixing.

The computation simplifies if we notice that the mass matrix $\mathcal{M}^f_{Hier}$ can be expressed as $\overline{\mathcal{M}}^f_{Hier} + \mathbf{1}\Delta_f$, where $\Delta_f = |\mu^f_1| - |\mu^f_2|$. Notice that the same unitary matrix diagonalizes both the original mass matrix and the shifted one. The expression for the shifted mass matrix is

$$\overline{\mathcal{M}}^f_{Hier} = \begin{pmatrix} 0 & |\mu^f_2| & 0 \\ |\mu^f_2| & 2|\mu^f_2| & \mu^f_8 \\ 0 & \mu^{f*}_8 & |\mu^f_3| - \Delta_f \end{pmatrix}. \tag{5}$$

4. The quark sector

In the following section we give analytical expressions for the CKM mixing matrix elements as explicit functions of quark mass ratios. For the purpose of illustrating the relations among mixing parameters and mass ratios, we take an approximation where we neglect the ratio of the shift Δ_f over the heaviest (third generation) mass, which is a very small number. The complete analysis of the model, taking into account the full expression, will be presented elsewhere. In this way it is possible to express the elements of the quark mixing matrix V^{th}_{CKM}

$$V^{th}_{CKM} = \begin{pmatrix} V^{th}_{ud} & V^{th}_{us} & V^{th}_{ub} \\ V^{th}_{cd} & V^{th}_{cs} & V^{th}_{cb} \\ V^{th}_{td} & V^{th}_{ts} & V^{th}_{tb} \end{pmatrix}, \tag{6}$$

as an explicit function of the quark mass ratios [16–18]

$$\begin{aligned}
V_{ud}^{th} &= \sqrt{\frac{\widetilde{m}_c\widetilde{m}_s\xi_1^u\xi_1^d}{\mathcal{D}_{1u}\mathcal{D}_{1d}}} + \sqrt{\frac{\widetilde{m}_u\widetilde{m}_d}{\mathcal{D}_{1u}\mathcal{D}_{1d}}}\left(\sqrt{(1-\delta_u)(1-\delta_d)\,\xi_1^u\xi_1^d} + \sqrt{\delta_u\delta_d\xi_2^u\xi_2^d}\right)e^{i\phi_1},\\
V_{us}^{th} &= -\sqrt{\frac{\widetilde{m}_c\widetilde{m}_d\xi_1^u\xi_2^d}{\mathcal{D}_{1u}\mathcal{D}_{2d}}} + \sqrt{\frac{\widetilde{m}_u\widetilde{m}_s}{\mathcal{D}_{1u}\mathcal{D}_{2d}}}\left(\sqrt{(1-\delta_u)(1-\delta_d)\,\xi_1^u\xi_2^d} + \sqrt{\delta_u\delta_d\xi_2^u\xi_1^d}\right)e^{i\phi_1},\\
V_{ub}^{th} &= \sqrt{\frac{\widetilde{m}_c\widetilde{m}_d\widetilde{m}_s\delta_d\xi_1^u}{\mathcal{D}_{1u}\mathcal{D}_{3d}}} + \sqrt{\frac{\widetilde{m}_u}{\mathcal{D}_{1u}\mathcal{D}_{3d}}}\left(\sqrt{(1-\delta_u)(1-\delta_d)\,\delta_d\xi_1^u} - \sqrt{\delta_u\xi_2^u\xi_1^d\xi_2^d}\right)e^{i\phi_1},\\
V_{cd}^{th} &= -\sqrt{\frac{\widetilde{m}_u\widetilde{m}_s\xi_2^u\xi_1^d}{\mathcal{D}_{2u}\mathcal{D}_{1d}}} + \sqrt{\frac{\widetilde{m}_c\widetilde{m}_d}{\mathcal{D}_{2u}\mathcal{D}_{1d}}}\left(\sqrt{(1-\delta_u)(1-\delta_d)\,\xi_2^u\xi_1^d} + \sqrt{\delta_u\delta_d\xi_1^u\xi_2^d}\right)e^{i\phi_1},\\
V_{cs}^{th} &= \sqrt{\frac{\widetilde{m}_u\widetilde{m}_d\xi_2^u\xi_2^d}{\mathcal{D}_{2u}\mathcal{D}_{2d}}} + \sqrt{\frac{\widetilde{m}_c\widetilde{m}_s}{\mathcal{D}_{2u}\mathcal{D}_{2d}}}\left(\sqrt{(1-\delta_u)(1-\delta_d)\,\xi_2^u\xi_2^d} + \sqrt{\delta_u\delta_d\xi_1^u\xi_1^d}\right)e^{i\phi_1},\\
V_{cb}^{th} &= -\sqrt{\frac{\widetilde{m}_u\widetilde{m}_d\widetilde{m}_s\delta_d\xi_2^u}{\mathcal{D}_{2u}\mathcal{D}_{3d}}} + \sqrt{\frac{\widetilde{m}_c}{\mathcal{D}_{2u}\mathcal{D}_{3d}}}\left(\sqrt{(1-\delta_u)(1-\delta_d)\,\delta_d\xi_2^u} - \sqrt{\delta_u\xi_1^u\xi_1^d\xi_2^d}\right)e^{i\phi_1},\\
V_{td}^{th} &= \sqrt{\frac{\widetilde{m}_u\widetilde{m}_c\widetilde{m}_s\delta_u\xi_1^d}{\mathcal{D}_{3u}\mathcal{D}_{1d}}} + \sqrt{\frac{\widetilde{m}_d}{\mathcal{D}_{3u}\mathcal{D}_{1d}}}\left(\sqrt{\delta_u(1-\delta_u)(1-\delta_d)\,\xi_1^d} - \sqrt{\delta_d\xi_1^u\xi_2^u\xi_2^d}\right)e^{i\phi_1},\\
V_{ts}^{th} &= -\sqrt{\frac{\widetilde{m}_u\widetilde{m}_c\widetilde{m}_d\delta_u\xi_2^d}{\mathcal{D}_{3u}\mathcal{D}_{2d}}} + \sqrt{\frac{\widetilde{m}_s}{\mathcal{D}_{3u}\mathcal{D}_{2d}}}\left(\sqrt{\delta_u(1-\delta_u)(1-\delta_d)\,\xi_2^d} - \sqrt{\delta_d\xi_1^u\xi_2^u\xi_1^d}\right)e^{i\phi_1},\\
V_{tb}^{th} &= \sqrt{\frac{\widetilde{m}_u\widetilde{m}_c\widetilde{m}_d\widetilde{m}_s\delta_u\delta_d}{\mathcal{D}_{3u}\mathcal{D}_{3d}}} + \left(\sqrt{\frac{\xi_1^u\xi_2^u\xi_1^d\xi_2^d}{\mathcal{D}_{3u}\mathcal{D}_{3d}}} + \sqrt{\frac{\delta_u\delta_d(1-\delta_u)(1-\delta_d)}{\mathcal{D}_{3u}\mathcal{D}_{3d}}}\right)e^{i\phi_1},
\end{aligned} \tag{7}$$

with

$$\xi_1^{u,d} = 1 - \widetilde{m}_{u,d} - \delta_{u,d}, \quad \xi_2^{u,d} = 1 + \widetilde{m}_{c,s} - \delta_{u,d}, \tag{8}$$

$$\mathcal{D}_{1(u,d)} = (1-\delta_{u,d})(\widetilde{m}_{u,d} + \widetilde{m}_{c,s})(1-\widetilde{m}_{u,d}), \quad \mathcal{D}_{2(u,d)} = (1-\delta_{u,d})(\widetilde{m}_{u,d} + \widetilde{m}_{c,s})(1+\widetilde{m}_{c,s}), \quad \mathcal{D}_{3(u,d)} = (1-\delta_{u,d})(1-\widetilde{m}_{u,d})(1+\widetilde{m}_{c,s}). \tag{9}$$

In this notation $\widetilde{m}_u = m_u/m_t$, $\widetilde{m}_c = |m_c|/m_t$, $\widetilde{m}_d = m_d/m_b$ and $\widetilde{m}_s = |m_s|/m_b$.

4.1. χ^2 analysis for the quark sector

For the χ^2 fit we proceed as follows. We construct the χ^2 function as

$$\chi^2 = \frac{\left(V_{ud}^{\rm th} - V_{ud}\right)^2}{\sigma_{V_{ud}}^2} + \frac{\left(V_{us}^{\rm th} - V_{us}\right)^2}{\sigma_{V_{us}}^2} + \frac{\left(V_{ub}^{\rm th} - V_{ub}\right)^2}{\sigma_{V_{ub}}^2} + \frac{\left(\mathcal{J}_q^{\rm th} - \mathcal{J}_q\right)^2}{\sigma_{\mathcal{J}_q}^2}, \tag{10}$$

where the quantities with super-index "th" are the complete expressions for the CKM elements, as given by the S_3 model, and those without, are the experimental quantities along with their uncertainty,

$$\begin{aligned}
2011:\ & V_{ud} = 0.97428 \pm 0.00015, & 2012:\ & V_{ud} = 0.97427 \pm 0.00015,\\
& V_{us} = 0.2253 \pm 0.007, & & V_{us} = 0.2253 \pm 0.007,\\
& V_{ub} = 0.00347 \pm 0.00014, & & V_{ub} = 0.00351 \pm 0.00015,\\
& J = (2.91 \pm 0.155) \times 10^{-5}, & & J = (2.96 \pm 0.18) \times 10^{-5}.
\end{aligned} \tag{11}$$

Journal of Physics: Conference Series **447** (2013) 012053 doi:10.1088/1742-6596/447/1/012053

We present a comparison of fits using the available data known before July 2012 (labelled as 2011) and the up-to-date data presented by the PDG, given the sizeable change in the uncertainty of the mass of the strange quark. Since we assume unitarity of the CKM mixing matrix, we need to fit just to four observables. The theoretical expressions of the parameters of the CKM elements are given in terms of the mass ratios, $\widetilde{m}_i$, hence the minimization of the defined χ^2 is a function of $\widetilde{m}_i$, and the parameters δ_u, δ_d, and $\cos\phi_1$. That means, that as a result of the minimization, there is a *best fit value* for each of those quantities, where χ^2 reaches its minimum. For the minimization procedure we used MINUIT. We allow the mass ratios $\widetilde{m}_i$ to take values within their 3σ range, as given in Table 1, whereas δ_u and δ_d, vary freely. Minimization with MINUIT

	2011	2012
$\widetilde{m}_u\,(M_Z)$	0.0000082 ± 0.0000027	0.0000083 ± 0.0000030
$\widetilde{m}_c\,(M_Z)$	0.0036 ± 0.0004	0.0037 ± 0.00008
$\widetilde{m}_d\,(M_Z)$	0.00098 ± 0.00018	0.00098 ± 0.00017
$\widetilde{m}_s\,(M_Z)$	0.0205 ± 0.0056	0.0190 ± 0.0014

Table 1. Comparison of the values of the mass ratios, at M_Z, reported up to 2011 and the newest values reported by the PDG.

starts with a a seed whose value is close to the minimium of the parameters being considered. Therefore, if the range of variation is large, it is difficult to find a best fit point. On the other hand, to check for global minima, one should remove the limits of the "free" parameters. If we perform the fit leaving completely free the values for $\widetilde{m}_i$ the quality of the fit decreases, and most importantly, it turns out that $\widetilde{m}_u$ is of order 10^{-3}. In Figure 2, we present the χ^2 value as a function of $\widetilde{m}_d$, where we have taken $\phi_1 = \pi/2$. This is not the most general case but it represents an interesting one[1]. For this case the best fit points (BFP) of the parameters δ_d and δ_u are respectively 6.05×10^{-2} and 4.09×10^{-2}. The BFP of the mass ratios $\widetilde{m}_u$, $\widetilde{m}_c$, $\widetilde{m}_d$ and $\widetilde{m}_s$ are respectively $(1.73 \pm 0.75) \times 10^{-5}$, $(3.46 \pm 0.43) \times 10^{-3}$, $(1.12 \pm 0.007) \times 10^{-3}$ and $(2.32 \pm 0.84) \times 10^{-2}$. The plot of Figure 2 corresponds to the exact result of the minimization, which allows to use values of the mass ratios $\widetilde{m}_i$ with more precision than the corresponding precision of the experimental/lattice determinations. This explains the change in value of χ^2 in the plot, from 3.4×10^{-4} up to 7.4×10^{-1}, which is the resulting value when we restrict the values of the BFP of mass ratios to have a precision equal to that of experimental/lattice determinations.

5. The leptonic sector

A similar analysis to the quark sector can be performed in the leptonic one, where a Z_2 symmetry is also introduced [19–22]. In this case, the masses of the left-handed Majorana neutrinos, $\mathbf{M}_{\nu_L}$, are generated by the type I seesaw mechanism. We take the mass matrix of the right-handed neutrinos to be real and diagonal but non-degenerate $\mathbf{M}_{\nu_R} = \mathrm{diag}(M_1, M_2, M_3)$, hence the mass matrix $\mathbf{M}_{\nu_L}$ takes the form [22]

$$\mathbf{M}_{\nu_L} = \begin{pmatrix} \frac{2(\mu_2^\nu)^2}{\overline{M}} & \frac{2\lambda(\mu_2^\nu)^2}{\overline{M}} & \frac{2\mu_2^\nu\mu_4^\nu}{\overline{M}} \\ \frac{2\lambda(\mu_2^\nu)^2}{\overline{M}} & \frac{2(\mu_2^\nu)^2}{\overline{M}} & \frac{2\mu_2^\nu\mu_4^\nu\lambda}{\overline{M}} \\ \frac{2\mu_2^\nu\mu_4^\nu}{\overline{M}} & \frac{2\mu_2^\nu\mu_4^\nu\lambda}{\overline{M}} & \frac{2(\mu_4^\nu)^2}{\overline{M}} + \frac{(\mu_3^\nu)^2}{M_3} \end{pmatrix}, \quad \lambda = \left(\frac{M_2 - M_1}{M_1 + M_2}\right), \text{ and } \overline{M} = 2\frac{M_1 M_2}{M_2 + M_1}. \qquad (12)$$

[1] The results of the most general case will be presented elsewhere.

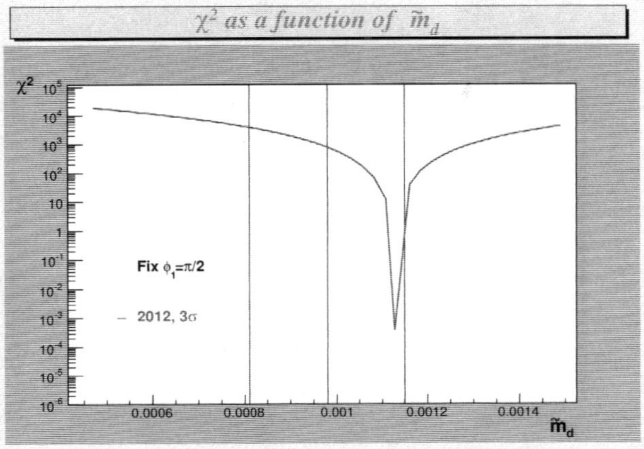

Figure 2. Results of the χ^2 fit for the quark sector as a function of $\widetilde{m_d}$. We have made a fit using only an average of the theoretical determination of m_s, $m_s(2\text{GeV}) = 0.101 \pm 0.011$, in order to asses the impact of the reduction in the uncertainty of m_s. The black vertical lines represent the $\widetilde{m}_s$ allowed experimental 1σ region and its central value.

Since we assumed the right-handed neutrino mass matrix $\mathbf{M}_{\nu_R}$ to be real, the complex symmetric neutrino mass matrix $\mathbf{M}_{\nu_L}$ has one Dirac phase and two Majorana phases which may be factored out of $\mathbf{M}_{\nu_L}$ as

$$\mathbf{M}_{\nu_L} = \mathbf{Q}\mathcal{U}_{\frac{\pi}{4}}\left(\mu_0 \mathbf{I}_{3\times 3} + \widehat{\mathbf{M}}\right)\mathcal{U}^{\dagger}_{\frac{\pi}{4}}\mathbf{Q}, \tag{13}$$

where $\mathbf{Q} = e^{i\phi_2}\text{diag}\left\{1, 1, e^{i\delta_\nu}\right\}$ with $\delta_\nu = \phi_4 - \phi_2 = \arg\{\mu_4^\nu\} - \arg\{\mu_2^\nu\}$,

$$\mathcal{U}_{\frac{\pi}{4}} = \begin{pmatrix} \frac{1}{\sqrt{2}} & 0 & \frac{1}{\sqrt{2}} \\ -\frac{1}{\sqrt{2}} & 0 & \frac{1}{\sqrt{2}} \\ 0 & 1 & 0 \end{pmatrix}, \quad \mu_0 = \frac{2\,|\mu_2^\nu|^2}{|M|}\left(1 - |\lambda|\right), \quad \text{and} \quad \widehat{\mathbf{M}} = \begin{pmatrix} 0 & A & 0 \\ A & B & C \\ 0 & C & 2d \end{pmatrix}, \tag{14}$$

with $A = \sqrt{2}\frac{|\mu_2^\nu||\mu_4^\nu|}{|M|}\left(1 - |\lambda|\right)$, $B = \frac{2|\mu_4^\nu|^2}{|M|} + \frac{|\mu_3^\nu|^2}{M_3} - \frac{2|\mu_2^\nu|^2}{|M|}\left(1 - |\lambda|\right)$, $C = \sqrt{2}\frac{|\mu_2^\nu||\mu_4^\nu|}{|M|}\left(1 + |\lambda|\right)$ and $d = \frac{2|\lambda||\mu_2^\nu|^2}{|M|}$. As mentioned before, the diagonalization of $\mathbf{M}_{\nu_L}$ is reduced to the diagonalization of the real symmetric matrix $\widehat{\mathbf{M}}$ with two texture zeroes of class I [23]. Hence, the matrix $\mathbf{M}_{\nu_L}$ is diagonalized by a unitary matrix

$$\mathbf{U}_\nu = \mathbf{Q}^\nu \mathcal{U}_{\frac{\pi}{4}} \mathbf{O}_\nu^{N[I]}. \tag{15}$$

As in the case of the charged leptons, the matrices $\mathbf{M}_{\nu_L}$ and $\mathbf{U}_\nu$ can be reparametrized in terms of the neutrino masses, following the procedure outilined in refs. [17, 18, 23, 24]. Then, the mass matrix $\mathbf{M}_{\nu_L}$ for a normal [inverted] hierarchy in the mass spectrum takes the form

$$\mathbf{M}_{\nu_L}^{N[I]} = \begin{pmatrix} \mu_0 + d & d & \frac{1}{\sqrt{2}}\left(C^{N[I]} + A^{N[I]}\right) \\ d & \mu_0 + d & \frac{1}{\sqrt{2}}\left(C^{N[I]} - A^{N[I]}\right) \\ \frac{1}{\sqrt{2}}\left(C^{N[I]} + A^{N[I]}\right) & \frac{1}{\sqrt{2}}\left(C^{N[I]} - A^{N[I]}\right) & m_{\nu_1} + m_{\nu_2} + m_{\nu_3} - 2\left(\mu_0 + d\right) \end{pmatrix}, \tag{16}$$

with $C^{N[I]} = \sqrt{\frac{(2d+\mu_0 - m_{\nu_1})(2d+\mu_0 - m_{\nu_{2[3]}})(m_{\nu_{3[2]}} - \mu_0 - 2d)}{2d}}$ and $A^{N[I]} = \sqrt{\frac{(m_{\nu_2} - \mu_0)(m_{\nu_{3[1]}} - \mu_0)(\mu_0 - m_{\nu_{1[3]}})}{2d}}$. The allowed values for the parameters μ_0 and $2d + \mu_0$ are in the following ranges: $m_{\nu_{2[1]}} > \mu_0 > m_{\nu_{1[3]}}$ and $m_{\nu_{3[2]}} > 2d + \mu_0 > m_{\nu_{2[1]}}$.

5.1. The reactor mixing angle

The theoretical expression for the lepton mixing angles as functions of the lepton mass ratios are readily obtained when the theoretical expressions for the modulii of the entries in the $PMNS$ mixing matrix are substituted for $|(\mathbf{V}^{th}_{PMNS})_{ij}|$. In a first preliminary analysis for the reactor mixing angle θ^l_{13} and for an inverted neutrino mass hierarchy ($m_{\nu_2} > m_{\nu_1} > m_{\nu_3}$) we obtain

$$\sin^2\theta^l_{13} \approx \frac{(\mu_0 + 2d - m_{\nu_3})(\mu_0 - m_{\nu_3})}{(m_{\nu_1} - m_{\nu_3})(m_{\nu_2} - m_{\nu_3})}. \tag{17}$$

Now, with the following values for the neutrino masses $m_{\nu_2} = 0.056$ eV, $m_{\nu_1} = 0.053$ eV and $m_{\nu_3} = 0.048$ eV, and the parameter values $\delta_l = \pi/2$, $\mu_0 = 0.049$ eV and $d = 8 \times 10^{-5}$ eV, we get $\sin^2\theta^l_{13} \approx 0.029 \longrightarrow \theta^l_{13} \approx 9.8°$, in good agreement with the most recent global fits to neutrino oscillation data [25, 26]. A complete analysis, from a χ^2 fit of the exact theoretical expressions for the modulii of the entries of the lepton mixing matrix of the $|(\mathbf{V}^{th}_{PMNS})_{ij}|$ to the experimental values will be considered elsewhere.

Conclusions

The introduction of the S_3 family symmetry in the Standard Model was motivated by the need to reduce the number of free parameters in the theory, and was guided by the analysis of the phenomenology of masses and mixings in the quark and lepton sectors. In order to preserve the S_3 family symmetry, two more Higgs doublets have to be introduced in addition to the one of the SM, and thus the concept of flavour is extended to the Higgs sector of the theory. We identified the conditions under which the four-zero Fritzsch-like texture mass matrices are obtained. This allowed us to (i) reduce drastically the number of parameters needed to describe the extended model with three Higgs $SU(2)_L$ doublets, (ii) to find an exact parameterization of the mass matrices in terms of their eigenvalues, and (iii), to compute the mixing angles in both the quark and lepton sectors as functions of mass ratios. We also derived exact formulas for the mixing angles of the CKM and PMNS matrices in terms of quark and lepton mass ratios, which we found to be in excellent agreement with all data on quark and neutrino masses and their mixing.

Acknowledgements

We acknowledge the financial support from the Mexican grants PAPIIT IN113712 and CONACyT-132059. F. González acknowledges the financial support received from PROMEP through a postdoctoral scholarship under contract PROMEP/103.5/12/2548. L. Velasco-Sevilla acknowledges the SFB 676 Fellowship from the University of Hamburg.

References

[1] Fritzsch H and Xing Z z 2000 *Prog.Part.Nucl.Phys.* **45** 1–81 (*Preprint* hep-ph/9912358)
[2] Ishimori H, Kobayashi T, Ohki H, Shimizu Y, Okada H *et al.* 2010 *Prog.Theor.Phys.Suppl.* **183** 1–163 (*Preprint* 1003.3552)
[3] Altarelli G and Feruglio F 2010 *Rev.Mod.Phys.* **82** 2701–2729 (*Preprint* 1002.0211)
[4] Hirsch M, Meloni D, Morisi S, Pastor S, Peinado E *et al.* 2012 Long author list - awaiting processing (*Preprint* 1201.5525)
[5] Pakvasa S and Sugawara H 1978 *Phys.Lett.* **B73** 61
[6] Yahalom R 1984 *Phys.Rev.* **D29** 536
[7] Branco G, Emmanuel-Costa D and Gonzalez Felipe R 2000 *Phys.Lett.* **B477** 147–155 (*Preprint* hep-ph/9911418)

Journal of Physics: Conference Series **447** (2013) 012053 doi:10.1088/1742-6596/447/1/012053

[8] Branco G, Emmanuel-Costa D, Gonzalez Felipe R and Serodio H 2009 *Phys.Lett.* **B670** 340–349 (*Preprint* 0711.1613)
[9] Canales F G, Mondragon A, Salazar U S and Velasco-Sevilla L 2012 (*Preprint* 1210.0288)
[10] Weinberg S 1977 *Trans.New York Acad.Sci.* **38** 185–201
[11] Fritzsch H 1977 *Phys.Lett.* **B70** 436
[12] Fritzsch H 1978 *Phys.Lett.* **B73** 317–322
[13] Gupta M and Ahuja G 2011 *Int.J.Mod.Phys.* **A26** 2973–2995
[14] Babu K S and Kubo J 2005 *Phys.Rev.* **D71** 056006 (*Preprint* hep-ph/0411226)
[15] Simoes C 2011 *J.Phys.Conf.Ser.* **335** 012031
[16] Mondragón A and Rodrguez-Jáuregui E 2000 *Rev.Mex.Fis.* **46** 5–22 (*Preprint* hep-ph/0003104)
[17] Mondragón A and Rodrguez-Jáuregui E 2000 *Phys.Rev.* **D61** 113002 (*Preprint* hep-ph/9906429)
[18] Mondragón A and Rodrguez-Jáuregui E 1999 *Phys.Rev.* **D59** 093009 (*Preprint* hep-ph/9807214)
[19] Mondragón A, Mondragón M and Peinado E 2008 *J.Phys.* **A41** 304035 (*Preprint* 0712.1799)
[20] Mondragón A, Mondragón M and Peinado E 2008 *AIP Conf.Proc.* **1026** 164–169 (*Preprint* 0712.2488)
[21] Mondragón A, Mondragón M and Peinado E 2007 *Phys.Rev.* **D76** 076003 (*Preprint* 0706.0354)
[22] González C F, Mondragón A and Mondragón M 2012 (*Preprint* 1205.4755)
[23] González Canales F and Mondragón A 2011 *J.Phys.Conf.Ser.* **287** 012015 presented at XIV Mexican School on Particles and Fields, 4-13 November 2010, Morelia México (*Preprint* 1101.3807)
[24] Barranco J, González Canales F and Mondragón A 2010 *Phys.Rev.* **D82** 073010 (*Preprint* 1004.3781)
[25] Fogli G, Lisi E, Marrone A, Montanino D, Palazzo A *et al.* 2012 *Phys.Rev.* **D86** 013012 (*Preprint* 1205.5254)
[26] Forero D, Tortola M and Valle J 2012 *Phys.Rev.* **D86** 073012 (*Preprint* 1205.4018)

Journal of Physics: Conference Series **447** (2013) 012054 doi:10.1088/1742-6596/447/1/012054

Generic constraints on extra-fermion(s) from the Higgs global fit

G. Moreau
Laboratoire de Physique Théorique, Bât. 210, CNRS, Université Paris-sud 11
F-91405 Orsay Cedex, France

We study the fit of the Higgs boson rates, based on all the collider data, in the effective framework for any Extra-Fermion(s) [EF]. The best-fit results are presented in a generic formalism allowing to apply those for the test of any EF scenario. The variations of the fit with each one of five fundamental parameters are described.

I. INTRODUCTION

Recently, based on the combined LHC data collected at the center-of-mass energies of $\sqrt{s} = 7$ TeV and 8 TeV, the ATLAS [1] and CMS [2] Collaborations have independently announced the discovery at the $\sim 5\sigma$ level of a new resonance – with a mass close to 125 GeV – which can be identified as the missing Standard Model (SM) cornerstone : the Higgs boson [3–6]. The long list of measurements of the various Higgs boson rates provided during these last months by the two LHC Collaborations [7, 8] constitutes a new precious source of experimental results which can be exploited to test and constrain indirectly theories beyond the SM.

Most (if not all) of the proposed theories, underlying the SM, predict the existence of new fermions. In this paper, we combine all the Higgs rate measurements to constrain any model with extra-fermion(s) [of any baryon/lepton number, Yukawa/gauge coupling] that is able to induce corrections to the Higgs couplings [1]. More precisely, by using a generic parametrization, we will determine the corrections to the Higgs couplings – coming from fermion mixing or new loop-level exchanges – which are favored by the fits of the Higgs boson rates. Note that our results also apply to any model with extra scalar field(s) or vector boson(s) leading to significant Higgs interaction deviations, but not through their mixing(s) respectively with the Higgs boson or SM gauge bosons. Then the constraints will be applied to characteristic and well-motivated classes of single Extra-Fermion(s) [EF] scenarios (extra-quarks or extra-leptons) and will reveal themselves to be already quite predictive. Last but not least, the best Higgs rate fits obtained could be seen as first indirect indications of the presence of EF since those fits can be better than the SM fit; another way of seeing this indication will be to observe that the best-fit regions for the EF-induced corrections to the Higgs couplings do not contain the vanishing-correction point (SM point).

Let us note that our fits are performed over the three free parameters c_b, c_{gg} and $c_{\gamma\gamma}$ (related to the hgg and $h\gamma\gamma$ coupling corrections defined later) for characteristic fixed values of c_τ and c_t [2]; in a second step, we fix c_b for studying examples of EF scenarios (c_τ, c_b and c_t will be defined later on).

II. THEORETICAL FRAMEWORK

A. The physical context

We consider the general framework with any EF able to modify the Higgs couplings. In our context, no other source of physics beyond the SM is responsible for deviations of the Higgs couplings; this choice allows to concentrate one's efforts on the class of models with EF and in turn to have a deeper analysis of the parameter space. In particular, we assume the Higgs scalar field to receive no coupling modifications due to significant mixings with other scalars as it can occur *e.g.* in extended Higgs sectors.

[1] The extra-fermions are assumed to be heavier than the Higgs field to avoid new Higgs decay openings (in particular invisible decays into stable particles) that would require special treatments.
[2] In order to explain clearly the influences of these five relevant parameters on the Higgs rate fit, we do not marginalize any of those parameters.

B. The effective Lagrangian

In our framework, all the Higgs couplings receiving corrections can be written in the following effective Lagrangian, which allows to work out the current Higgs phenomenology at the LHC and Tevatron colliders :

$$\begin{aligned}\mathcal{L}_h \;=\; & - c_t Y_t \; h \; \bar{t}_L \; t_R \; - c_b Y_b \; h \; \bar{b}_L \; b_R \; - c_\tau Y_\tau \; h \; \bar{\tau}_L \; \tau_R \\ & + C_{h\gamma\gamma} \frac{\alpha}{\pi v} \; h \; F^{\mu\nu} F_{\mu\nu} \; + \; C_{hgg} \frac{\alpha_s}{12\pi v} \; h \; G^{a\mu\nu} G^a_{\mu\nu} \; + \; \text{h.c.}\end{aligned} \tag{1}$$

where $Y_{t,b,\tau}$ are the SM Yukawa coupling constants of the associated fermions in the mass eigenbasis, v is the Higgs vacuum expectation value, the subscript L/R indicates the fermion chirality and the tensor fields in the $h\gamma\gamma$ and hgg coupling terms are respectively the electromagnetic and gluon field strengths. The $c_{t,b,\tau}$ parameters – taken real for simplicity – are defined such that the limiting case $c_{t,b,\tau} \to 1$ corresponds to the SM; deviations from unity of those parameters can be caused by mixings of EF (like t' states,...) with the SM fermions. Neglecting the mixings with the first two SM flavors, one gets, $-Y_{t,b,\tau} = m_{t,b,\tau}/v$ [the minus sign is due to the sign taken in front of the Yukawa couplings in Eq. (1)], where $m_{t,b,\tau}$ are the final masses generated after EW symmetry breaking. The EF mixing effect on the Yukawa couplings enters via the $c_{t,b,\tau}$ parameters.

Summing over the dominant loop contributions, the coefficients of the dimension-five operators in Eq. (1) can be written as,

$$C_{hgg} \;=\; 2C(t) \; A[\tau(m_t)] \; (c_t + c_{gg}) \; + \; 2C(b) \; A[\tau(m_b)] \; c_b \; + \; 2C(c) \; A[\tau(m_c)], \tag{2}$$

$$C_{h\gamma\gamma} \;=\; \frac{N_c^t}{6} Q_t^2 A[\tau(m_t)] \; (c_t + c_{\gamma\gamma}) + \frac{N_c^b}{6} Q_b^2 A[\tau(m_b)] \; c_b + \frac{N_c^c}{6} Q_c^2 A[\tau(m_c)] + \frac{N_c^\tau}{6} Q_\tau^2 A[\tau(m_\tau)] \; c_\tau + \frac{1}{8} A_1[\tau(m_W)], \tag{3}$$

where m_c (m_W) is the charm quark ($W^\pm$-boson) mass, $C(\mathrm{r})$ is defined for the color representation, r, by $\mathrm{Tr}(T^a_\mathrm{r} T^b_\mathrm{r}) = C(\mathrm{r})\delta^{ab}$ [T^a denoting the eight generators of $SU(3)_c$], N_c^f is the number of colors for the fermion f, Q_f is the electromagnetic charge for f, $A[\tau(m)]$ and $A_1[\tau(m)]$ are respectively the form factors for spin 1/2 and spin 1 particles normalized such that $A[\tau(m) \ll 1] \to 1$ and $A_1[\tau(m) \ll 1] \to -7$ with $\tau(m) = m_h^2/4m^2$. The terms proportional to c_t, c_b and c_τ account for the contributions from the fermionic triangular loops involving respectively the top, bottom quark and tau lepton Yukawa coupling. The dimensionless c_{gg} and $c_{\gamma\gamma}$ quantities – vanishing in the SM – parametrize the EF loop-exchange contributions to the hgg and $h\gamma\gamma$ couplings.

C. Higgs rate modifications

Within the present context, let us write explicitly certain Higgs rates, normalized to their SM prediction, which will prove to be useful in the following. The expression for the cross section of the gluon-gluon fusion mechanism of single Higgs production, over its SM prediction, reads as (for the LHC or Tevatron),

$$\frac{\sigma_{\mathrm{gg}\to\mathrm{h}}}{\sigma^{\mathrm{SM}}_{\mathrm{gg}\to\mathrm{h}}} \simeq \frac{\left|(c_t + c_{gg}) A[\tau(m_t)] + c_b A[\tau(m_b)] + A[\tau(m_c)]\right|^2}{\left|A[\tau(m_t)] + A[\tau(m_b)] + A[\tau(m_c)]\right|^2} \, . \tag{4}$$

The expression for the ratio of the diphoton partial decay width over the SM expectation is,

$$\frac{\Gamma_{\mathrm{h}\to\gamma\gamma}}{\Gamma^{\mathrm{SM}}_{\mathrm{h}\to\gamma\gamma}} \simeq \frac{\left|\frac{1}{4}A_1[\tau(m_W)] + (\frac{2}{3})^2 (c_t + c_{\gamma\gamma}) A[\tau(m_t)] + (-\frac{1}{3})^2 c_b A[\tau(m_b)] + (\frac{2}{3})^2 A[\tau(m_c)] + \frac{1}{3} c_\tau A[\tau(m_\tau)]\right|^2}{\left|\frac{1}{4}A_1[\tau(m_W)] + (\frac{2}{3})^2 A[\tau(m_t)] + (-\frac{1}{3})^2 A[\tau(m_b)] + (\frac{2}{3})^2 A[\tau(m_c)] + \frac{1}{3} A[\tau(m_\tau)]\right|^2} \, . \tag{5}$$

The ratios for the partial decay widths into the bottom quark and tau lepton pairs as well as for the cross section of Higgs production in association with a top pair (LHC or Tevatron) are given by,

$$\frac{\Gamma_{\mathrm{h}\to\bar{\mathrm{b}}\mathrm{b}}}{\Gamma^{\mathrm{SM}}_{\mathrm{h}\to\bar{\mathrm{b}}\mathrm{b}}} \simeq |c_b|^2 \, , \quad \frac{\Gamma_{\mathrm{h}\to\bar{\tau}\tau}}{\Gamma^{\mathrm{SM}}_{\mathrm{h}\to\bar{\tau}\tau}} \simeq |c_\tau|^2 \, , \quad \frac{\sigma_{\mathrm{h}\bar{\mathrm{t}}\mathrm{t}}}{\sigma^{\mathrm{SM}}_{\mathrm{h}\bar{\mathrm{t}}\mathrm{t}}} \simeq |c_t|^2 \, . \tag{6}$$

Let us make a comment related to the mass insertion in the triangular loops of fermions inducing the $h\gamma\gamma$ and hgg couplings. Strictly speaking, a factor ϵ_t, equal to the ratio of the sign of m_t in the SM over sign(m_t) in the EF scenario, should multiply c_t in Eq. (2)-(3) or Eq. (4)-(5) [similarly for $\epsilon_b c_b$ and $\epsilon_\tau c_\tau$]; in other words, if for instance $\epsilon_t = -1$ the values for c_t obtained below would have to be interpreted instead as values for $-c_t$ (the observables of Eq. (6) being insensitive to the $c_{t,b,\tau}$ signs).

D. Ratio of $c_{\gamma\gamma}$ and c_{gg}

For a better understanding of the above parametrization, we finally provide the examples of expressions for the c_{gg} and $c_{\gamma\gamma}$ quantities, in the case of the existence of a t' quark [same color number and electromagnetic charge as the top], an exotic (possibly vector-like) $q_{5/3}$ quark with electromagnetic charge 5/3 and an additional ℓ' lepton (colorless), in terms of their physical Yukawa couplings and mass eigenvalues :

$$c_{gg} = \frac{1}{C(t)A[\tau(m_t)]/v}\Big[- C(t')\frac{Y_{t'}}{m_{t'}}A[\tau(m_{t'})] - C(q_{5/3})\frac{Y_{q_{5/3}}}{m_{q_{5/3}}}A[\tau(m_{q_{5/3}})] + \dots \Big], \tag{7}$$

$$c_{\gamma\gamma} = \frac{1}{N_c^t Q_t^2 A[\tau(m_t)]/v}\Big[- 3\Big(\frac{2}{3}\Big)^2\frac{Y_{t'}}{m_{t'}}A[\tau(m_{t'})] - N_c^{q_{5/3}}\Big(\frac{5}{3}\Big)^2\frac{Y_{q_{5/3}}}{m_{q_{5/3}}}A[\tau(m_{q_{5/3}})] - Q_{\ell'}^2\frac{Y_{\ell'}}{m_{\ell'}}A[\tau(m_{\ell'})] + \dots \Big]. \tag{8}$$

The dots stand for any other EF loop-contributions. The mass assumption made in Footnote [1] leads to real $A[\tau(m_{f'})]$ functions and thus real c_{gg}, $c_{\gamma\gamma}$ values, for real masses and Yukawa coupling constants, as appears clearly in the two above expressions.

It will turn out to be instructive to express the ratio of these parameters in the simplified scenario where a new single q' quark is affecting the Higgs couplings; denoting its electromagnetic charge as $Q_{q'}$ and assuming the q' to have the same color representation as the top quark, this ratio reads as :

$$\frac{c_{\gamma\gamma}}{c_{gg}}\Big|_{q'} = \frac{Q_{q'}^2}{(2/3)^2} \, . \tag{9}$$

This ratio takes indeed a simple form that will be exploited in Section III D. In particular, notice that $c_{\gamma\gamma}|_{t'} = c_{gg}|_{t'}$. Clearly, q' should have non-vanishing Yukawa couplings to satisfy Eq. (9), otherwise $c_{\gamma\gamma}|_{q'} = c_{gg}|_{q'} = 0$. In the specific case of a vector-like $q'_{L/R}$, this one could for example constitute a singlet under the $SU(2)_L$ gauge group and have a Yukawa coupling with another $q''_{R/L}$ state of same $Q_{q'}$ charge but embedded in a $SU(2)_L$ doublet; then the heaviest $q^{(2)}_{L/R}$ mass eigenstate, composed of $q'_{L/R}$ and $q''_{L/R}$, could decouple from the Higgs sector so that the orthogonal $q^{(1)}_{L/R}$ composition would represent the considered unique new quark influencing significantly the Higgs couplings.

III. THE HIGGS RATE FITS

A. The data

All the Higgs rates which have been measured at the Tevatron and LHC [for $\sqrt{s} = 7$ and 8 TeV] have been summarized in Ref. [9]. The latest experimental values can be found in Ref. [7, 8]. Generically, the measured observables are the signal strengths whose theoretical predictions read as,

$$\mu^p_{s,c,i} \simeq \frac{\sigma_{\rm gg\to h}|_s + \frac{\epsilon_{\rm hqq}}{\epsilon_{\rm gg\to h}}|^p_{s,c,i}\ \sigma^{\rm SM}_{\rm hqq}|_s + \frac{\epsilon_{\rm hV}}{\epsilon_{\rm gg\to h}}|^p_{s,c,i}\ \sigma^{\rm SM}_{\rm hV}|_s + \frac{\epsilon_{\rm h\bar{t}t}}{\epsilon_{\rm gg\to h}}|^p_{s,c,i}\ \sigma_{\rm h\bar{t}t}|_s}{\sigma^{\rm SM}_{\rm gg\to h}|_s + \frac{\epsilon_{\rm hqq}}{\epsilon_{\rm gg\to h}}|^p_{s,c,i}\ \sigma^{\rm SM}_{\rm hqq}|_s + \frac{\epsilon_{\rm hV}}{\epsilon_{\rm gg\to h}}|^p_{s,c,i}\ \sigma^{\rm SM}_{\rm hV}|_s + \frac{\epsilon_{\rm h\bar{t}t}}{\epsilon_{\rm gg\to h}}|^p_{s,c,i}\ \sigma^{\rm SM}_{\rm h\bar{t}t}|_s}\ \frac{B_{\rm h\to XX}}{B^{\rm SM}_{\rm h\to XX}} \, ,$$

$$\text{with,} \qquad \sigma_{\rm gg\to h}|_s = \frac{\sigma_{\rm gg\to h}}{\sigma^{\rm SM}_{\rm gg\to h}}\sigma^{\rm SM}_{\rm gg\to h}|_s \, , \quad \sigma_{\rm h\bar{t}t}|_s = \frac{\sigma_{\rm h\bar{t}t}}{\sigma^{\rm SM}_{\rm h\bar{t}t}}\sigma^{\rm SM}_{\rm h\bar{t}t}|_s \, ,$$

$$\Gamma_{\rm h\to\gamma\gamma} = \frac{\Gamma_{\rm h\to\gamma\gamma}}{\Gamma^{\rm SM}_{\rm h\to\gamma\gamma}}\Gamma^{\rm SM}_{\rm h\to\gamma\gamma} \, , \quad \Gamma_{\rm h\to\bar{b}b} = \frac{\Gamma_{\rm h\to\bar{b}b}}{\Gamma^{\rm SM}_{\rm h\to\bar{b}b}}\Gamma^{\rm SM}_{\rm h\to\bar{b}b} \, , \quad \Gamma_{\rm h\to\bar{\tau}\tau} = \frac{\Gamma_{\rm h\to\bar{\tau}\tau}}{\Gamma^{\rm SM}_{\rm h\to\bar{\tau}\tau}}\Gamma^{\rm SM}_{\rm h\to\bar{\tau}\tau} \, , \tag{10}$$

where the p-exponent labels the Higgs channel defined by its production and decay processes, the s-subscript represents the squared of the energy [we will note $\sqrt{s} = 1.96, 7, 8$ in TeV] of the realized measurement, the c-subscript stands for the experimental collaboration (CDF and D0 at the Tevatron, ATLAS or CMS at LHC) having performed the measurement and i is an integer indicating the event cut category considered. $\sigma_{\rm hqq}$ is the predicted cross section for the Higgs production in association with a pair of light SM quarks and $\sigma_{\rm hV}$ is for the production in association with a gauge boson [$V \equiv Z^0, W^\pm$ bosons]; their s-subscript indicates the energy and in turn which collider is considered. $\epsilon_{\rm gg\to h}$, for the $\rm gg \to h$ reaction example, is the experimental efficiency including the (kinematical) selection cut effects.

B. The fit procedure

In order to analyze the fit of the Higgs boson data from colliders within the effective theory described above, we assume gaussian error statistics and we use the χ^2 function,

$$\chi^2 \;=\; \sum_{p,s,c,i} \frac{(\mu^p_{s,c,i} - \mu^p_{s,c,i}|_{\rm exp})^2}{(\delta\mu^p_{s,c,i})^2}\,, \tag{11}$$

where the sum is taken over all the different channel observables and $\mu^p_{s,c,i}|_{\rm exp}$ are the measured central values for the corresponding signal strengths. $\delta\mu^p_{s,c,i}$ are the uncertainties on these values and are obtained by symmetrizing the provided errors below and above the central values : $(\delta\mu^p_{s,c,i})^2 = [(\delta\mu^p_{s,c,i}|^+)^2 + (\delta\mu^p_{s,c,i}|^-)^2]/2$. $\mu^p_{s,c,i}|_{\rm exp}$ and $\delta\mu^p_{s,c,i}|^\pm$ are given in the experimental papers.

C. Numerical results and discussions

Having the three free parameters, $c_{gg}, c_{\gamma\gamma}, c_b$, we now determine the best-fit domains in this three-dimensional space at 68.27%C.L. (1σ), 95.45%C.L. (2σ) and 99.73%C.L. (3σ) which correspond to established values of $\Delta\chi^2 = \chi^2 - \chi^2_{\rm min}$ ($\chi^2_{\rm min}$ being the minimum χ^2 value reached in the $\{c_{gg}, c_{\gamma\gamma}, c_b\}$ space).

In Fig.(1), we present slices at several c_b values of the best-fit regions at 68.27%C.L., 95.45%C.L. and 99.73%C.L. in the plane $c_{\gamma\gamma}$ versus c_{gg}, in the case $c_\tau = 1$ and for different fixed values of c_t.

The best-fit points reachable, when varying the three free parameters, $c_b, c_{gg}, c_{\gamma\gamma}$, for fixed values $c_t = 1$ and $c_\tau = 1$, are at $c_b = 2.08$ and the c_{gg}, $c_{\gamma\gamma}$ values corresponding to the four crosses drawn in Fig.(1)[c]. Since there are exact symmetries along the c_{gg} and $c_{\gamma\gamma}$ axes (see discussion below), those four cross-points are all associated to the same $\chi^2_{\rm min} = 52.36$.

This minimal χ^2 value is smaller than the SM one, $\chi^2_{\rm SM} = 57.10$ [from taking all the strength predictions at unity in Eq. (11)]. The regions at 68.27%C.L. in Fig.(1)[b] do not even contain the SM point ($\{c_t = 1; c_b = 1; c_\tau = 1; c_{gg} = 0; c_{\gamma\gamma} = 0\}$).

We observe on Fig.(1)[a,b,c] that a c_t variation of amount, δc_t, leads in a good approximation to a translation (no shape modification) of $-\delta c_t$ along both the $c_{\gamma\gamma}$ and c_{gg} axes, for each one of the three best-fit regions. It is particularly clear in Fig.(1)[b] where a large δc_t is exhibited. This strong parameter interdependence implies that in order to determine experimentally the $c_{\gamma\gamma}$ and c_{gg} quantities, it is crucial to determine as well the c_t Yukawa correction.

Concerning the c_b variation (for fixed $c_t = c_\tau = 1$), we first explain the impact of the c_b increase on the typical $c_{\gamma\gamma}$, c_{gg} values – starting from the best-fit domains around the best-fit point, $\{c_b = 2.08; c_{gg} = 0.66; c_{\gamma\gamma} = -1.09\}$, in Fig.(1)[c] – and the reasons why huge values up to $c_b \simeq 50$ could still agree with present Higgs rate fits. For such a c_b increase, the strengths are reduced via $\Gamma_{\rm h\to\bar bb}$, a reduction which has to be compensated by a $\sigma_{\rm gg\to h}$ increase through a c_{gg} enhancement to conserve a satisfactory χ^2 (or equivalently here, $\Delta\chi^2$). This explains the shift of the considered best-fit domains, around $\{c_b = 2.08; c_{gg} = 0.66; c_{\gamma\gamma} = -1.09\}$ in Fig.(1)[c], to higher c_{gg} values in the plot [d] where $c_b = 10$ (still with $c_t = 1$). This necessary compensation between the $\Gamma_{\rm h\to\bar bb}$ and $\sigma_{\rm gg\to h}$ increases also guarantees the stability of diphoton rates (there is also a significant gluon-gluon fusion contribution in the three dijet-tagged final states) letting the χ^2 at the same level, without $c_{\gamma\gamma}$ modifications – explaining nearly identical $c_{\gamma\gamma}$ values for the studied regions in Fig.(1)[c] and [d]. The $\Gamma_{\rm h\to\bar bb}$ increase leads to enhancements of the strengths without major consequences on the fit; a c_b increase up to ~ 50 [leading to $\Gamma_{\rm h\to\bar bb} \lesssim 5$ GeV] would still leave existing domains at 68.27%C.L. since in the theoretical limit, $c_b \to \infty$, $B_{\rm h\to\bar bb}$ tends obviously to a finite value compatible with data : $B_{\rm h\to\bar bb} \to 1$.

What is the experimental impact of the above c_b analysis ? The present experimental results do not prevent c_b from taking extremely large values – due in particular to Higgs rate compensations. In order to put a more stringent experimental upper limit on it, one could of course if possible improve the accuracies on the signal strengths involving $\sigma_{\rm gg\to h}$ and $\Gamma_{\rm h\to\bar bb}$. A new possibility to measure c_b (or equivalently the bottom Yukawa coupling constant) would be to investigate the processes, $\rm\bar qq \to h\bar bb$ and $\rm gg \to h\bar bb$ (or $\rm\bar bb \to h$ and $\rm bg \to hb$), followed by the decay, $\rm h \to \bar bb$. Indeed, here both the production and decay rates should increase with c_b ($\Gamma_{\rm h\to\bar bb}$ being the dominant partial width) so that compensations should not occur; then too large c_b values would be experimentally ruled out. This Higgs production in association with bottom quarks could have

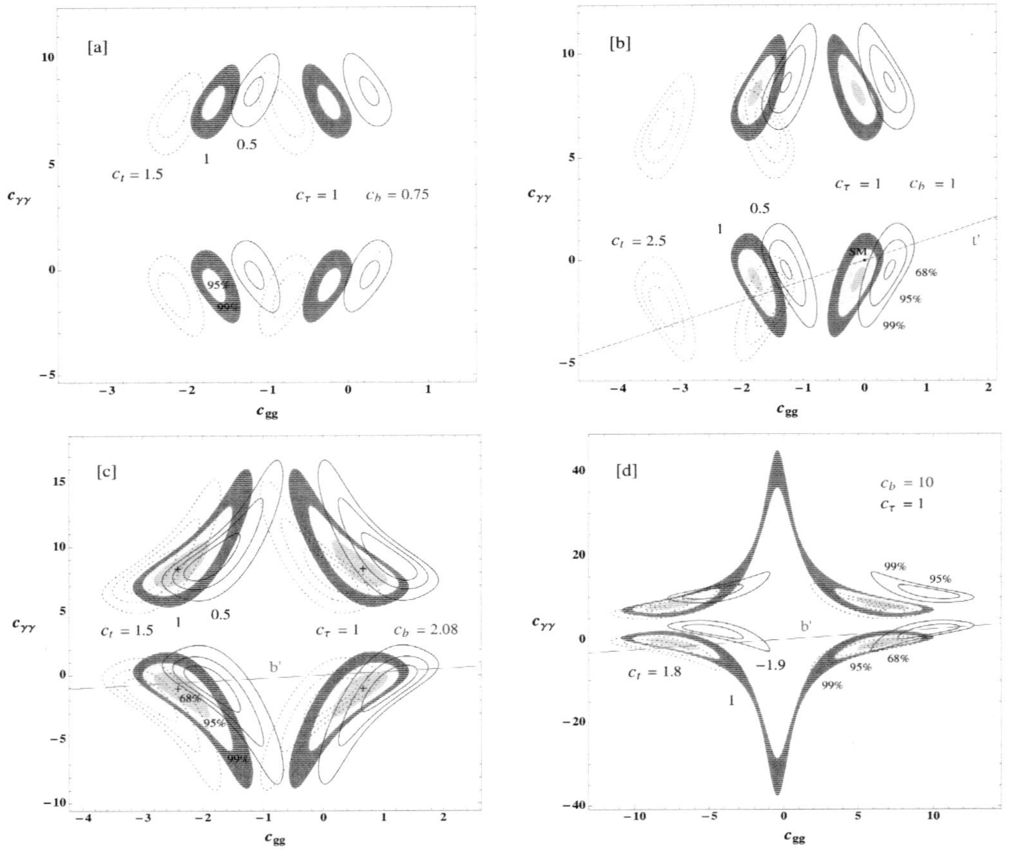

FIG. 1: Best-fit regions at 68.27%C.L., 95.45%C.L. and 99.73%C.L. in the plane $c_{\gamma\gamma}$ versus c_{gg}, for $c_\tau = 1$. Each one of the four figures is associated to a certain c_b amount written on the figure itself. In each figure, the regions are drawn for three c_t values, the corresponding value being indicated nearby the relevant region; the regions for the lowest, intermediate, highest c_t values are respectively shown by plain contours, filled domains, dotted contours. The SM point at, $c_t = c_b = c_\tau = 1$, $c_{\gamma\gamma} = c_{gg} = 0$, is shown. Also represented are the predicted lines for a b' and a t' extra-quark. Finally, the four best-fit point locations are indicated by crosses in the plot [c].

significant cross sections for high LHC luminosities and enhanced c_b values compared to the SM as the present fit points out. The sensitivity to such a reaction relies deeply on the b-tagging capability. This reaction suffers from large QCD backgrounds but new search strategies have been developed for such a bottom final state topology, as in Ref. [10].

D. Effective EF scenarios

In this last section, we apply the above constraints from the Higgs rate fit to examples of simple scenarios where a unique EF state significantly affects the Higgs interactions. For instance, a b' state [same color representation and electromagnetic charge as the bottom quark], that could be a light custodian top-partner in warped/composite frameworks, would lead to a ratio in Eq. (9), $(c_{\gamma\gamma}/c_{gg})|_{b'} = 1/4$, corresponding to the straight line drawn on Fig.(1)[c,d]. Generically, a b' would be mixed with the SM bottom so that possibly, $c_b \neq 1$, whereas one would have, $c_t = c_\tau = 1$ – like in Fig.(1)[c,d]. These figures show that there exist $c_{\gamma\gamma}$, c_{gg} and c_b values for which the predicted b' line crosses the 68.27%C.L. region. The simultaneous knowledge of the exact position on the b' line and the c_b value fixing the C.L. regions, necessary to determine the goodness of fit, requires the specification of the bottom mass matrix and hence of the considered model.

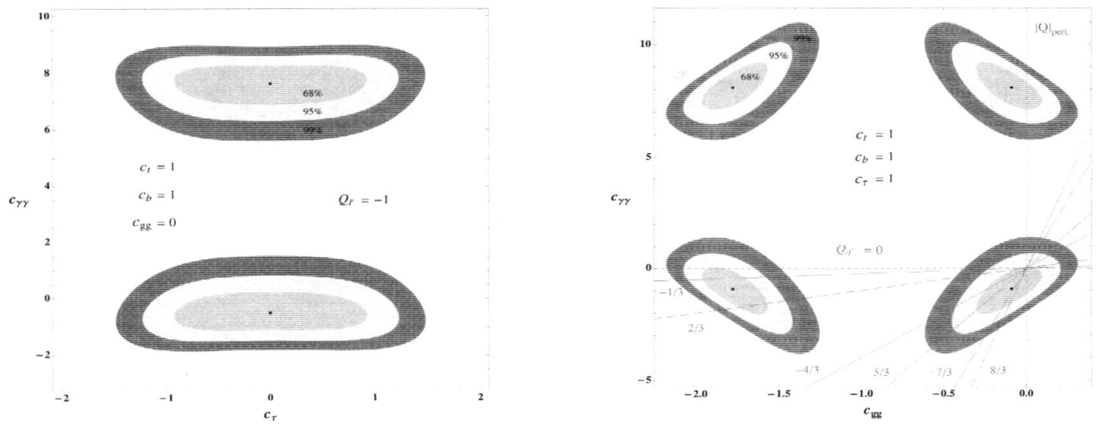

FIG. 2: Left - Best-fit regions at 68.27%C.L., 95.45%C.L. and 99.73%C.L. in the plane $c_{\gamma\gamma}$ versus c_τ, for the case of an extra-lepton with electric charge, $Q_{\ell'} = -1$, corresponding to $c_t = c_b = 1$, $c_{gg} = 0$. The two best-fit points are indicated. Right - Best-fit regions at 68.27%C.L., 95.45%C.L. and 99.73%C.L. in the plane $c_{\gamma\gamma}$ versus c_{gg}, for $c_t = c_b = c_\tau = 1$. Also represented are the predicted (plain) lines for extra-quarks with the several electric charges, $Q_{q'} = -1/3, 2/3, -4/3, 5/3, -7/3$ and $8/3$. The extreme (dashed) lines for, $Q_{q'} = 0$, and, $|Q_{q'}| = |Q_{q'}|_{\rm pert.} = \sqrt{4\pi/\alpha}$, are shown as well. The four best-fit points are indicated.

The other example of EF candidate able to be mixed with SM quarks is the t' state, possibly constituted *e.g.* by a light top-partner in little Higgs models. For a dominant t' state, the ratio of Eq. (9) tends to one which corresponds to the straight line on Fig.(1)[b]. Since a t' field can mix with the top quark, $c_t \neq 1$, but in the context of a single t' one should have, $c_b = c_\tau = 1$, as in Fig.(1)[b]. The predicted t' line crosses two 95.45%C.L. regions *e.g.* for, $c_t = 0.5$, as well as two 68.27%C.L. regions exclusively in the range, $c_t \sim 1.1 \leftrightarrow 2.6$ (above ~ 2.6 the region sizes decrease as explained in previous section).

For a single extra-lepton with charge, $Q_{\ell'} = -1$, potentially mixed with the τ-lepton, the parameters, $c_b = c_t = 1$, $c_{gg} = 0$ [see Eq. (7)], are fixed and there remain two free effective parameters, namely $c_{\gamma\gamma}$ and c_τ. The best-fit regions for such a two-dimensional fit are presented in Fig.(2). The two best-fit points shown in this figure correspond to, $\chi^2_{\rm min} = 52.54$.

It would also be possible theoretically that the new t' and b' particles do not mix with the SM top and bottom quarks. It would be the case also for additional q' quarks with exotic electric charges. For illustration, let us first concentrate on the components of possible extensions of the SM quark multiplets under SU(2)$_{\rm L}$, as in warped/composite frameworks where SM multiplets are promoted to representations of the custodial symmetry [11–18]. The charges for such q' components obey the relation, $\mathcal{Y}_{q'} = Q_{q'} - I^{q'}_{3L}$ ($\mathcal{Y} \equiv$ hypercharge, $I_{3L} \equiv$ SU(2)$_{\rm L}$ isospin). We will consider the electric charges of smallest absolute values, $Q_{q'} = -1/3, 2/3, -4/3, 5/3, -7/3$ and $8/3$, keeping in mind that the naive perturbative limit on the electric charge reads as, $|Q_{q'}| \lesssim \sqrt{4\pi/\alpha} \simeq 40$ ($\alpha \equiv$ fine-structure constant). The q' states are in the same color representation as the SM quarks.

In the case of the presence of such a q' quark, unmixed with SM quarks, while $c_t = c_b = c_\tau = 1$, one has $c_{\gamma\gamma} \neq 0$ and $c_{gg} \neq 0$ if the q' state possesses non-zero Yukawa couplings. The best-fit domains for a two-dimensional fit keeping the fixed parameters, $c_t = c_b = c_\tau = 1$, are shown in Fig.(2) together with the four best-fit points associated to, $\chi^2_{\rm min} = 55.04$. On this plot, we also represent the theoretically predicted regions in the cases of a single q' quark with electric charge $Q_{q'}$: these regions are the straight lines defined by Eq. (9). All the predicted lines – whatever is the $Q_{q'}$ charge – cross the SM point which is reached in the decoupling limit, $c_{\gamma\gamma} \to 0$, $c_{gg} \to 0$. The first result is that the upper-left best-fit regions, around $c_{\gamma\gamma} \sim 8$, $c_{gg} \sim -1.8$, cannot be explored in single q' models [no line can reach it]. We also observe on Fig.(2) that the predicted line being the closest to a best-fit point is for, $Q_{q'} = -7/3$. This result means that, among any possible SM multiplet extension component, the fit prefers the $q_{-7/3}$ state compared for example to a t' or $q_{5/3}$ state. This prediction is independent of the $Y_{q'}$ Yukawa coupling constants, the q' mass values and the q' representations under SU(2)$_{\rm L}$.

Now we determine the physical parameters corresponding typically to an overlap between a given line in

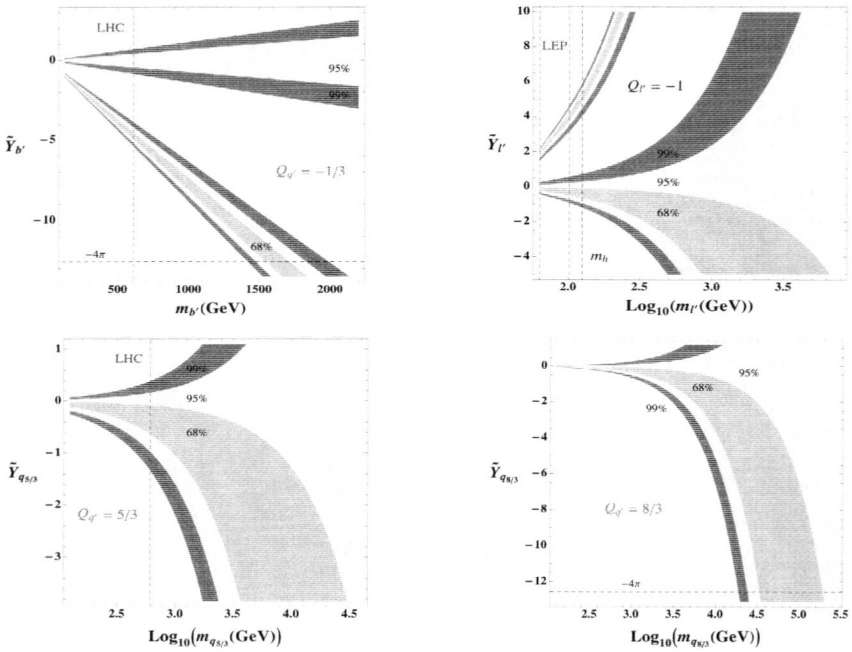

FIG. 3: Best-fit regions at 68.27%C.L., 95.45%C.L. and 99.73%C.L. in the plane of the $|m_{f'}|$ absolute mass (in GeV) versus the $\tilde{Y}_{f'}$ coupling, with $c_t = c_b = c_\tau = 1$, for the cases $Q_{q'} = -1/3$, $5/3$, $8/3$ and $Q_{\ell'} = -1$. Also represented as dashed lines are the lower perturbative limit on Yukawa couplings, -4π, and the direct LHC bound, $m_{b'} > 611$ GeV, $m_{q_{5/3}} > 611$ GeV, or LEP constraint, $m_{\ell'} > 63.5$ GeV (> 101.9 GeV) for $m_{\ell'} - m_{\nu'} > 7$ GeV (> 15 GeV). The mass ranges start at the Higgs mass, m_h (see Footnote [1]), except for the ℓ'-lepton case (see text) where the m_h value is indicated by a vertical dashed line.

Fig.(2) and the best-fit regions; we consider the characteristic examples of the charges, $Q_{q'} = -1/3$, $5/3$ and $8/3$. More precisely, we plot in Fig.(3) the regions in the plane $|m_{q'}|$ versus $\tilde{Y}_{q'} = -Y_{q'}/\mathrm{sign}(m_{q'})$ which correspond [see Eq. (7)-(8)] to $c_{\gamma\gamma}$, c_{gg} quantities giving rise to the best $\Delta\chi^2$ values in the case of one free effective parameter, say c_{gg} (related to $c_{\gamma\gamma}$ through the fixed ratio $c_{\gamma\gamma}/c_{gg}|_{q'} \propto Q^2_{q'}$).
In Fig.(3), we also illustrate the case of a single additional ℓ' lepton (colorless) without significant mixing to SM leptons [$c_\tau = 1$], as may be justified by exotic $Q_{\ell'}$ charges or the large mass difference between the SM and extra-leptons. Here we choose, $Q_{\ell'} = -1$, being quite common for extra-lepton scenarios. There is, again, a unique free effective parameter, $c_{\gamma\gamma}$, since $c_{gg} = 0$.

Concerning the constraints on the signs, as shown in Fig.(3) based on the present Higgs data, the sign, $\tilde{Y}_{q'} < 0$ [leading to $c_{\gamma\gamma} < 0$], is preferred at 68.27%C.L. [except with absolute charges, $|Q_{q'}| \gtrsim 7$, i.e. in a range close to the $|Q_{q'}|_{\mathrm{pert.}}$ limit as illustrated in Fig.(2)] for any single extra-quark as it creates a constructive interference with the $W^\pm$-boson exchange increasing the diphoton rates. The specific sign configuration, $\tilde{Y}_{q'} < 0$, is selected by the two relevant best-fit points which pin down, $c_{\gamma\gamma} < 0$, as obtained for extra-quarks in Fig.(2). This predicted condition means that the Yukawa coupling constant [$-Y_{q'}$ in our conventions] must have a sign opposite to $m_{q'}$ which could be written,

$$\mathrm{sign}\left(\frac{-Y_{q'}}{m_{q'}}\right) < 0 \,. \tag{12}$$

Related to this condition, there are comments on the configuration denoted as *dysfermiophilia* in the literature. As described at the end of Section II C, strictly speaking the $c_{t,b,\tau}$ parameters entering Eq. (4)-(5) – whose values are generally given in best-fit plots such as the present ones in Fig.(1) – should in fact be understood as being,

$$\epsilon_t c_t \;=\; \frac{\mathrm{sign}(m_t)}{\mathrm{sign}(m_t^{\mathrm{EF}})}\, c_t \;=\; \frac{\mathrm{sign}(m_t)}{\mathrm{sign}(m_t^{\mathrm{EF}})}\,\frac{\mathrm{sign}(-Y_t^{\mathrm{EF}})}{\mathrm{sign}(-Y_t)}\,|c_t| \;=\; \frac{\mathrm{sign}(-Y_t^{\mathrm{EF}})}{\mathrm{sign}(m_t^{\mathrm{EF}})}\,|c_t| \;=\; \mathrm{sign}\left(\frac{-Y_t^{\mathrm{EF}}}{m_t^{\mathrm{EF}}}\right)\left|\frac{Y_t^{\mathrm{EF}}}{Y_t}\right| ,$$

in our conventions of Lagrangian (1), and similarly for $\epsilon_{b,\tau}c_{b,\tau}$; here the EF-exponent indicates that the parameter is considered within the context of an EF model (and remind that m_t, Y_t are in the SM). Therefore, the *dysfermiophilia* property of increasing, $\Gamma_{\mathrm{h}\to\gamma\gamma}/\Gamma^{\mathrm{SM}}_{\mathrm{h}\to\gamma\gamma}$, via changing the top Yukawa sign is in fact relying on the possibility to have, $\epsilon_t c_t < 0$, or equivalently, $\mathrm{sign}(-Y_t^{\mathrm{EF}}/m_t^{\mathrm{EF}}) < 0$. This makes sense as it is the sign of, $-Y_t^{\mathrm{EF}}/m_t^{\mathrm{EF}}$, which has a physical meaning and appears in $\Gamma_{\mathrm{h}\to\gamma\gamma}$ [see Eq. (8) for an analogy with the t'-loop]. The other comment is that the *dysfermiophilia* possibility of having, $\epsilon_t c_t < 0$, can indeed gives rise to an acceptable agreement with the Higgs data (see *e.g.* Fig.(1)[d]) but it is not necessary to achieve a good agreement (*c.f.* Fig.(2) where $\epsilon_t c_t = 1$) since the constructive interference with the $W^\pm$-loop increasing the diphoton rates can be realized with an EF-loop inducing, $c_{\gamma\gamma} < 0$.
Hence the above condition (12) can be called an *extra-dysfermiophilia* as it is exactly the same as for the top quark transposed to an EF. Besides, this condition (12) leads to a decrease of, $\sigma_{\mathrm{gg}\to\mathrm{h}}/\sigma^{\mathrm{SM}}_{\mathrm{gg}\to\mathrm{h}}$, for a single EF [see Eq. (7)] through negative c_{gg} values [*c.f.* Fig.(2)].

IV. CONCLUSIONS

We have studied the behaviour of the global Higgs fit with the variations of fundamental parameters in the simplified scenario with extra-fermions and have learnt in particular that the determination of the Higgs couplings is difficult and sometimes suffers from correlations. We have also shown that the global fit allows quite simply to constrain the electric charge (and color) of EF. The *extra-dysfermiophilia* has also been clearly defined and pointed out as a present general prediction (for more details, see Ref. [19]).

Acknowledgements: The author thanks the organizers of the conference. This work is supported by the "Institut Universitaire de France". The author also acknowledges support from the ANR *CPV-LFV-LHC* under project **NT09-508531**.

---

[1] F. Gianotti (2012), CERN Seminar, *Update on the Standard Model Higgs searches in ATLAS*, July, 4, 2012, `http://indico.cern.ch/conferenceDisplay.py?confId=197461`.
[2] J. Incandela (2012), CERN Seminar, *Update on the Standard Model Higgs searches in CMS*, July, 4 2012, `http://indico.cern.ch/conferenceDisplay.py?confId=197461`.
[3] F. Englert and R. Brout, Phys.Rev.Lett. **13**, 321 (1964).
[4] P. W. Higgs, Phys.Lett. **12**, 132 (1964).
[5] P. W. Higgs, Phys.Rev.Lett. **13**, 508 (1964).
[6] G. Guralnik, C. Hagen, and T. Kibble, Phys.Rev.Lett. **13**, 585 (1964).
[7] ATLAS web page, `https://twiki.cern.ch/twiki/bin/view/AtlasPublic` (2012).
[8] CMS web page, `https://twiki.cern.ch/twiki/bin/view/CMSPublic/PhysicsResults` (2012).
[9] J. Espinosa, C. Grojean, M. Muhlleitner, and M. Trott (2012), 1207.1717.
[10] M. Carena, S. Gori, A. Juste, A. Menon, C. E. Wagner, et al., JHEP **1207**, 091 (2012), 1203.1041.
[11] A. Djouadi, G. Moreau, and F. Richard, Nucl.Phys. **B773**, 43 (2007), hep-ph/0610173.
[12] A. Djouadi, G. Moreau, and R. K. Singh, Nucl.Phys. **B797**, 1 (2008), 0706.4191.
[13] A. Djouadi and G. Moreau, Phys.Lett. **B660**, 67 (2008), 0707.3800.
[14] F. Ledroit, G. Moreau, and J. Morel, JHEP **0709**, 071 (2007), hep-ph/0703262.
[15] C. Bouchart and G. Moreau, Nucl.Phys. **B810**, 66 (2009), arXiv:0807.4461 [hep-ph].
[16] C. Bouchart and G. Moreau, Phys.Rev. **D80**, 095022 (2009), arXiv:0909.4812 [hep-ph].
[17] A. Djouadi, G. Moreau, F. Richard, and R. K. Singh, Phys.Rev. **D82**, 071702 (2010), 0906.0604.
[18] A. Djouadi, G. Moreau, and F. Richard, Phys.Lett. **B701**, 458 (2011), 1105.3158.
[19] G. Moreau (2012), 1210.3977.

Journal of Physics: Conference Series **447** (2013) 012055 doi:10.1088/1742-6596/447/1/012055

Analysis of CP conserving Higgs bosons self couplings in SM×S(3)

E. Barradas Guevara[(a)], **O. Félix-Beltrán**[(b)], **and E. Rodríguez-Jáuregui**[(c)]

[(a)]Facultad de Ciencias Físico Matemáticas, BUAP, Apdo. Postal 1152, 72570, Puebla, Puebla, México.

[(b)]Facultad de Ciencias de la Electrónica, Benemérita Universidad Autónoma de Puebla, Apdo. Postal 542, 72570, Puebla, Puebla, México

[(c)]Departamento de Física, Universidad de Sonora, Apdo. Postal 1626, 83000, Hermosillo, Sonora, México.

E-mail: barradas@fcfm.buap.mx

Abstract. We carry out a detailed analysis of a minimal S_3-invariant extension of the Standard Model, with an extended S_3-Higgs sector. Within this extended $S(3)$-Standard Model, we study the trilinear Higgs couplings and its dependence on the details of the model, even when the lightest Higgs boson mass is taken to be a fixed parameter. We study quantitatively the trilinear Higgs couplings, and compare these couplings to the corresponding Standard Model trilinear Higgs coupling in some regions of the parameter space. A precise measurement of the trilinear Higgs self coupling will also make it possible to test this extended $S(3)$-Standard Model with the Standard Model trilinear Higgs coupling. We present analytical expressions for the trilinear Higgs couplings.

1. Introduction

In these days when it is seeing a new boson with a mass of 125 GeV, observed in CMS at the LHC, corresponding to the scalar of the Standard Model (SM), is of fundamental importance to analyze the mechanism of spontaneous symmetry breaking with models that have more number of scalar bosons [1]. The Higgs boson is a fundamental part of the SM providing mass to the gauge bosons and fermions upon the spontaneous breaking of the electroweak symmetry $SU(2)_L \times U(1)_Y$, and thus preserving the renormalisability of the theory. In the SM, only one $SU(2)_L$ doublet Higgs field is included, and although their existence is a fundamental piece of the theory and the Higgs potential is very simple and sufficient to describe a realistic model of mass generation, this may not be the final form of the theory [2]. In the SM each family of fermions enters independently, in order to understand the replication of generations and to reduce the number of free parameters, usually more symmetry is introduced in the theory. In this direction interesting work has been done with the addition of discrete symmetries to the SM [3]. Many interesting features of masses and mixing of the SM can be understood using a minimal discrete group, namely the permutational group S_3 [4].

Prior to the introduction of the Higgs boson, the SM is chiral and invariant with respect to any permutation of the left and right quark and lepton fields. After the introduction of the Higgs boson in the theory, this field may be treated as an S_3 singlet H_S, but then, only one fermion in each family can acquire mass. To give mass to all fermions and, at the same time, preserve the S_3 flavour symmetry of the theory, an extended flavoured Higgs sector is required with three Higgs $SU(2)$ doublets, one in a singlet and the other two in a doublet irreducible representation of S_3 [5].

Up to now, the particle observed at the LHC is a particle in the physical spectrum of the Higgs boson of the SM . It is not known if there is one or many Higgs bosons. An indication of the presence of one Higgs boson or an extended Higgs sector, as the one proposed in the S_3-invariant extension of the SM, could be found at the Large Hadron Collider [6]. Models with more than one Higgs doublet, with or without supersymmetry, have been studied extensively. For a review of supersymmetric and two Higgs-doublet models [7], as well as models with and without discrete symmetries [8, 9, 10].

We study the trilinear Higgs couplings and its dependence on the details of the model, with an extended S_3-Higgs sector, even when the lightest Higgs boson mass is taken to be a fixed parameter. We study quantitatively the trilinear Higgs couplings, and compare these couplings to the corresponding Standard Model trilinear Higgs coupling in some regions of the parameter space.

2. The $S(3)$ extended Higgs doublet model

The Lagrangian $\mathcal{L}_\Phi$ of the Higgs sector is given by

$$\mathcal{L}_\Phi = [D_\mu H_S]^2 + [D_\mu H_1]^2 + [D_\mu H_2]^2 - V\left(H_1, H_2, H_S\right), \tag{1}$$

where D_μ is the usual covariant derivative. The scalar potential $V\left(H_1, H_2, H_S\right)$ is the most general Higgs potential invariant under $SU(3)_C \times SU(2)_L \times U(1)_Y \times S_3$. The analysis of the stability properties of the potential V is of great relevance to study the phenomenological implications of this model. In here we will concentrate on the analysis of the potential V following the lines of previous work done in the vacuum stability of multi-Higgs models [9]. In our case, the discrete flavour symmetry S_3 simplifies the analysis.

There are many different ways to write the Higgs potential for this model, but for the purpose of this work the best basis is

$$H_1 = \begin{pmatrix} \phi_1 + i\phi_4 \\ \phi_7 + i\phi_{10} \end{pmatrix}, \; H_2 = \begin{pmatrix} \phi_2 + i\phi_5 \\ \phi_8 + i\phi_{11} \end{pmatrix}, \; H_S = \begin{pmatrix} \phi_3 + i\phi_6 \\ \phi_9 + i\phi_{12} \end{pmatrix}, \tag{2}$$

the numbering of the real scalar ϕ fields is chosen for convenience by writing the mass matrices for the scalar particles, and the subscript S is the flavour index for the Higgs field singlet. Now a simple manner to write down the potential is

$$\begin{aligned} V &= \mu_1^2 (x_1 + x_2) + \mu_0^2 x_3 + a x_3^2 + b (x_1 + x_2) x_3 + c (x_1 + x_2)^2 \\ &\quad -4d x_7^2 + 2e \left[(x_1 - x_2) x_6 + 2 x_4 x_5 \right] + f \left(x_5^2 + x_6^2 + x_8^2 + x_9^2 \right) \\ &\quad +g \left[(x_1 - x_2)^2 + 4 x_4^2 \right] + 2h \left(x_5^2 + x_6^2 - x_8^2 - x_9^2 \right), \end{aligned} \tag{3}$$

where the $\mu_{0,1}^2$ parameters have dimensions of mass squared, the real couplings $a, \cdots, h$ are dimensionless and the invariants x_i, the potential V depends on the fields ϕ_i through x_i, considering our assignment as

$$\begin{aligned} x_1 &= H_1^\dagger H_1, & x_4 &= \mathcal{R}\left(H_1^\dagger H_2\right), & x_7 &= \mathcal{I}\left(H_1^\dagger H_2\right), \\ x_2 &= H_2^\dagger H_2, & x_5 &= \mathcal{R}\left(H_1^\dagger H_S\right), & x_8 &= \mathcal{I}\left(H_1^\dagger H_S\right), \\ x_3 &= H_S^\dagger H_S, & x_6 &= \mathcal{R}\left(H_2^\dagger H_S\right), & x_9 &= \mathcal{I}\left(H_2^\dagger H_S\right). \end{aligned} \tag{4}$$

If the S_3 invariant Higgs potential, Eq. (3), is bounded from below being a quartic polynomial function, it will certainly have a global minimum somewhere. We can two types of minima: the trivial one for which the Higgs acquires zero VEV's, and the usual one, where electroweak symmetry breaking occurs, away from the origin, for

$$\phi_7 = v_1, \; \phi_8 = v_2, \; \phi_9 = v_3, \; \phi_i = 0, \quad i \neq 7, 8, 9, \tag{5}$$

defined as the normal minimum, with VEV's which do not have any complex relative phase. So, v_1, v_2 and v_3 are the VEV's of the neutral Higgs fields components.

The minimization constraints must be determined by demanding the vanishing of $\partial V / \partial \phi_i$, then we get the mass parameter by

$$\mu_1^2 = -(b + f + 2h) v_3^2 - 2(c + g)\left(v_1^2 + v_2^2\right) + \frac{3e(v_1^2 - 2v_1 v_2 - v_2^2) v_3}{v_1 - v_2}, \tag{6}$$

and

$$\mu_0^2 = -\left[2a v_3^2 + (b + f + 2h)\left(v_1^2 + v_2^2\right) - e\left(\frac{3v_1^2 - v_2^2}{v_3}\right) v_2 \right]. \tag{7}$$

Accordingly, we have $v_1 = \sqrt{3} v_2$.

The Higgs-bosons masses in this model are obtained by diagonalizing the 12×12 matrix, with the 3×3 sub-matrices $M_I^2 = M_{II}^2$ and M_{IJ}^2, $I, J = C, \; S, \; P$:

$$(\mathcal{M}_H^2)_{ij} = \frac{1}{2} \frac{\partial^2 V}{\partial \phi_i \partial \phi_j} \bigg|_{\text{mim}}, \tag{8}$$

where $i, j = \overline{1, 12}$. We have that $\mathcal{M}_H^2 = \text{diag}(M_C^2, M_C^2, M_S^2, M_P^2)$, the zeros correspond to 3×3 sub-matrices with zeros in its entirety. For example, the mass matrix for the CP-even Higgs scalars is given by

$$M_S^2 = \begin{pmatrix} s_{11} & s_{12} & s_{13} \\ & s_{22} & s_{23} \\ & & s_{33} \end{pmatrix}, \tag{9}$$

where

$$\begin{aligned} s_{11} &= 12(c+g)v_2^2, & s_{12} &= 2\sqrt{3}v_2\left(2(c+g)v_2 + 3ev_3\right), \\ s_{13} &= 2\sqrt{3}v_2\left(3ev_2 + (b+f+2h)v_3\right), & s_{22} &= 4v_2\left((c+g)v_2 - 3ev_3\right), \\ s_{23} &= 2v_2\left(3ev_2 + (b+f+2h)v_3\right), & s_{33} &= 4av_3^2 - 8ev_2^3/v_3. \end{aligned}$$

After diagonalized mass matrices, the masses of the physical scalars/pseudoscalars are obtained. In our analysis we are not taking into account the parameter space with negative eigenvalues solutions for the squared masses of the physical Higgs fields. Of the original twelve scalar degrees of freedom, three Goldstone bosons ($G^\pm$ and G) are absorbed by $W^\pm$ and Z. The remaining nine physical Higgs particles are three CP−even scalar (h and H_1, H_2, with $m_h \leq m_{H_1} \leq m_{H_2}$), two CP-odd scalar (A_1, and A_2, with $m_{A_1} \leq m_{A_2}$), and two charged Higgs pair ($H_{1,2}^\pm$, mass degenerate). Their physical masses and the respective Higgs-scalar mixing are related to the weak parameters of the Higgs potential as follows. We start considering the mass matrix for the CP-even Higgs scalars Eq. (9).

Defining the physical mass eigenstates m_h^2, $m_{H_1}^2$, and $m_{H_2}^2$, the masses are found from the diagonalization process $\mathcal{M}_{diag}^2 = R^T \mathcal{M}_{even}^2 R = \text{diag}(m_h^2, m_{H_1}^2, m_{H_2}^2)$. The masses for the CP-even Higgs scalars are:

$$m_h^2 = -18ev_2v_3, \quad m_{H_1^0, H_2^0}^2 = (\mathcal{M}_a^2 + \mathcal{M}_c^2) \pm \sqrt{(\mathcal{M}_a^2 - \mathcal{M}_c^2)^2 + (\mathcal{M}_b^2)^2}, \tag{10}$$

where

$$\begin{aligned} \mathcal{M}_a^2 &= v_2(8(c+g)v_2 + 3ev_3), \\ \mathcal{M}_b^2 &= 4v_2\left(3ev_2 + (b+f+2h)v_3\right), \\ \mathcal{M}_c^2 &= -\frac{4ev_2^3}{v_3} + 2av_3^2, \end{aligned} \tag{11}$$

and mixing angles are defined as $\tan 2\alpha_3 = \dfrac{\mathcal{M}_b^2}{\mathcal{M}_a^2 - \mathcal{M}_c^2}$, as well as

$$R = OPQ = \begin{pmatrix} -1 & 0 & 0 \\ 0 & 1 & 0 \\ 0 & 0 & -1 \end{pmatrix} \begin{pmatrix} 1/2 & -\sqrt{3}/2 & 0 \\ \sqrt{3}/2 & 1/2 & 0 \\ 0 & 0 & 1 \end{pmatrix} \begin{pmatrix} 1 & 0 & 0 \\ 0 & c_\theta & s_\theta \\ 0 & -s_\theta & c_\theta \end{pmatrix}, \tag{12}$$

where

$$\tan\theta = \frac{s_\theta}{c_\theta} = \frac{2s_{13}}{\sqrt{3}(m_{H_1}^2 - s_{33})} = \frac{\mathcal{M}_b^2}{2\mathcal{M}_a^2 - m_{H_2}^2}, \tag{13}$$

s_{13} and s_{33} are given in Eq. (9).

To generate the correct $W^{\pm}$ and Z^0 masses, one must consider that $v_1^2 + v_2^2 + v_3^2 = v^2$, with $v = 246$ GeV and $v_1 = \sqrt{3}v_2$. Also, the v_i are taken real, *i. e.*, we assume that spontaneous CP violation does not occur. The squared-mass parameters μ_0^2 and μ_1^2 can be eliminated by minimizing the scalar potential. The resulting squared masses for the CP-odd Higgs pseudoscalars are

$$m_{A_1}^2 = -16(d+g)v_2^2 - 10ev_2v_3 - 4hv_3^2, \qquad m_{A_2}^2 = -\frac{2(ev_2 + 2hv_3)(4v_2^2 + v_3^2)}{v_3},$$

which can be parametrized with $\tan\omega_3 = \frac{2v_2}{v_3}$, where $\sin\omega_3 = 2v_2/v$, $\cos\omega_3 = v_3/v$.

On the other hand, the charged Higgs boson masses are

$$m_{H_1^{\pm}}^2 = -\left(10ev_2 + (f+2h)v_3\right)v_3, \qquad m_{H_2^{\pm}}^2 = -\frac{v^2}{v_3}\left(2ev_2 + (f+2h)v_3\right).$$

As we known, the Higgs boson masses are not determined *a priori* within the theory, it is necessary to investigate the change of mass spectrum with respect to the quadrilinears $a, \ldots, h$ couplings.

3. Trilinear Self-Couplings of Neutral Higgs Bosons

The measurement of the Higgs self-coupling is crucial to determine the Higgs potential. The self-couplings are uniquely determined in the SM by the mass of the Higgs boson, which is related to the quadrilinear coupling λ by $M_H = \sqrt{2\lambda}v$. The trilinear and quadrilinear vertices of the Higgs field H are given by the coefficients:

$$\lambda_{HHH} = \lambda v = \frac{M_H^2}{2v}, \qquad \lambda_{HHHH} = \frac{\lambda}{4} = \frac{M_H^2}{8v^2}. \tag{14}$$

The following definitions are often used:

$$\lambda_{ijk} = \frac{-i\partial^3 V}{\partial H_i \partial H_j \partial H_k}, \tag{15}$$

which are most easily obtained from the corresponding derivatives of V in Eq. (3) with respect to the fields $\{\phi_i\}$ with $i = 1, \cdots, 12$. Then, we can write the trilinear couplings in terms of the derivatives of the potential (3) with respect to ϕ_i and the elements of the rotation matrix R Eq. (12) as

$$\lambda_{ijk} = N \sum_{lmn} R_{il} R_{jm} R_{kn} \frac{\partial^3 V}{\partial\phi_i \partial\phi_j \partial\phi_k}. \tag{16}$$

Here the indices l, m, n refer to the weak field basis, and $l \le m \le n = 1, 2, 3$, N is a factor of $n!$ for n identical fields. We now proceed to obtain these couplings in an explicit form. The trilinear self-couplings a_{lmn} among the neutral even Higgs bosons can be written as

$$\begin{aligned}
a_{1,1,1} &= 6\sqrt{6}(c+g)v_2, & a_{1,1,2}, &= \sqrt{2}(2(c+g)v_2+3ev_3),\\
a_{1,1,3} &= \sqrt{2}(3ev_2+(b+f+2h)v_3), & a_{1,2,2} &= 2\sqrt{6}(c+g)v_2,\\
a_{1,2,3} &= 3\sqrt{6}ev_2, & a_{1,3,3} &= \sqrt{6}(b+f+2h)v_2,\\
a_{2,2,2} &= 3\sqrt{2}(2(c+g)v_2-ev_3), & a_{2,2,3} &= \sqrt{2}((b+f+2h)v_3-3ev_2),\\
a_{2,3,3} &= \sqrt{2}(b+f+2h)v_2, & a_{3,3,3} &= 6\sqrt{2}av_3.
\end{aligned} \tag{17}$$

Then, we have the self-couplings of three even Higgs bosons by substituting for the elements of the rotation matrix, Eq. (12), one obtains

$$\begin{aligned}
\lambda_{1,1,1} &= 6v\left(\lambda_1\sin\omega_3+\lambda_2\cos\omega_3\right), & \lambda_{2,2,2} &= 6v\left(\lambda_3\sin\omega_3+\lambda_4\cos\omega_3\right),\\
\lambda_{3,3,3} &= 6v\left(\lambda_5\sin\omega_3+\lambda_6\cos\omega_3\right), & \lambda_{1,1,2} &= 2v\left(\lambda_7\sin\omega_3+\lambda_8\cos\omega_3\right),\\
\lambda_{1,1,3} &= 2v\left(\lambda_9\sin\omega_3+\lambda_{10}\cos\omega_3\right), & \lambda_{1,2,2} &= 2v\left(\lambda_{11}\sin\omega_3+\lambda_{12}\cos\omega_3\right),\\
\lambda_{1,2,3} &= v\left(\lambda_{13}\sin\omega_3+\lambda_{14}\cos\omega_3\right), & \lambda_{1,3,3} &= 2v\left(\lambda_{15}\sin\omega_3+\lambda_{16}\cos\omega_3\right),\\
\lambda_{2,2,3} &= 2v\left(\lambda_{17}\sin\omega_3+\lambda_{18}\cos\omega_3\right), & \lambda_{2,3,3} &= 2v\left(\lambda_{19}\sin\omega_3+\lambda_{20}\cos\omega_3\right).
\end{aligned} \tag{18}$$

$\lambda_1,\ldots,\lambda_{20}$ depend on parameters quadrilinear couplings of the Higgs potential, Eq. (3), and the mixing angle θ Eq. (13). For example

$$\begin{aligned}
\lambda_1 &= \frac{3\sqrt{3}}{4}(b+f+2h)(c_\theta-1)s_\theta^2,\\
\lambda_2 &= \frac{2\sqrt{3}}{4}\left(18as_\theta^3+\left[(b+f+2h)s_\theta-3ec_\theta\right](3c_\theta^2+1)\right)\\
&\quad +2(c+g)(9c_\theta^3-3c_\theta^2+c_\theta-3)-3e(3c_\theta(c_\theta+1)-1)s_\theta.
\end{aligned} \tag{19}$$

4. Numerical results

Here we present numerical results considering a decoupling limit, that is, a CP-even boson Higgs scalar behaves like the Higgs boson of the Standard Model, where the mass of the lightest Higgs scalar is significantly smaller than the masses of the other Higgs bosons of the model. We assume that it is light and its mass is around 125 GeV. In our analysis we proposed H_1^0 as our candidate, it is a free parameter. Consequently, the other Higgs bosons have masses of the order of $\mathcal{O}(500)$ GeV, without loss of generality and for the sake of simplicity, it is considered that they have no equal masses. Figure 1 shows the trilinear couplings by CP-even Higgs bosons in function of the mixing angle. The higher intensity couplings correspond to $\lambda_{h_0h_0h_0}$ and $\lambda_{H_2H_2H_2}$, whereas that $\lambda_{H_1H_1H_1}$ happens to be very small, $\lambda_{h_0h_0h_0}\sim\lambda_{H_2H_2H_2}>\lambda_{H_1H_1H_1}$, followed by $\lambda_{h_0h_0H_2}$, $\lambda_{h_0h_0H_1}$ ($\lambda_{h_0H_2H_2}$) and then the others. We proposed that the lightest is H_1^0 and its trilinear coupling is the lower intensity. Although there are couplings equal in shape, they are different in intensity. Comparing these results with only SM trilinear coupling, we can see that our values are within allowed ranges in the literature and enriched by the Higgs potential analysis, enabling to better understanding of the spontaneous symmetry breaking.

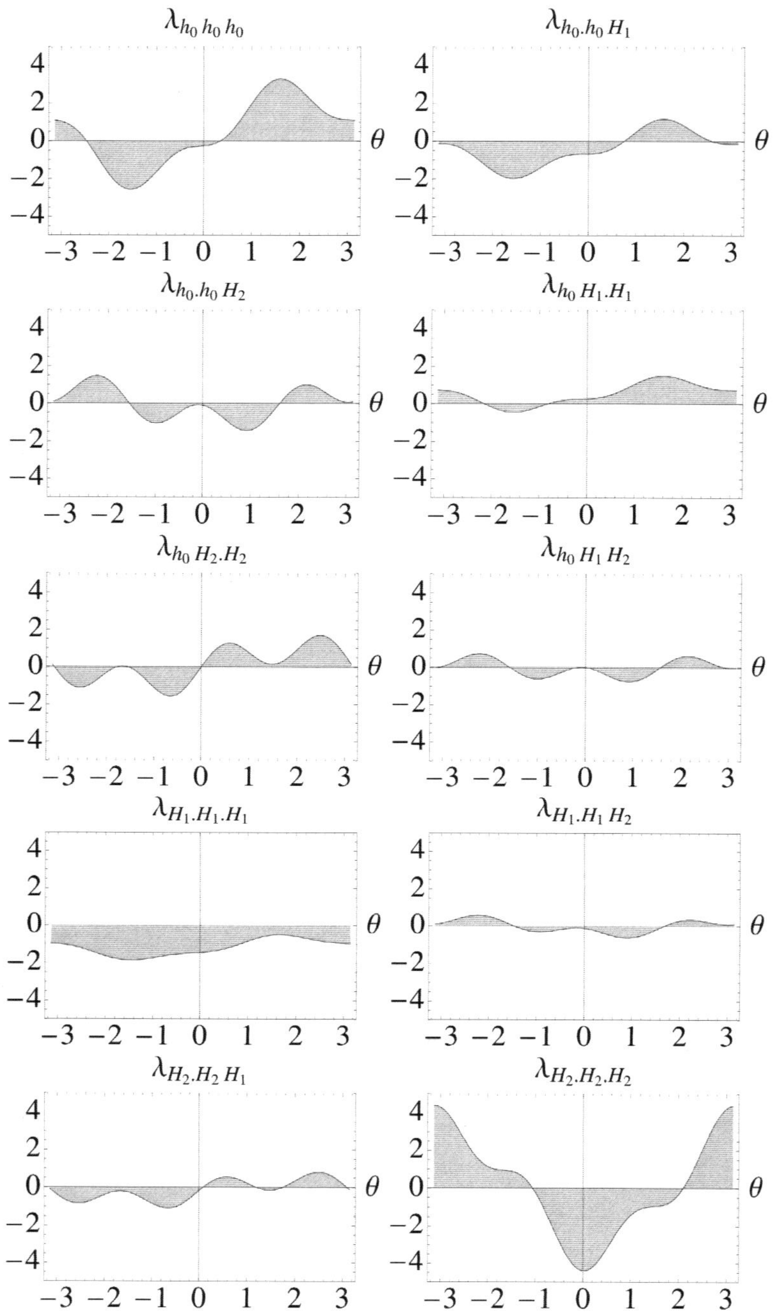

Figure 1. The trilinear couplings of the CP-even Higgs bosons in function on the mixing angle $-\pi \leq \theta \leq \pi$.

5. Conclusions

We studied only the scalar sector assuming the pseudoscalars to be too heavy and relevant. In this work we have analyzed the complete scalar sector of an S_3 flavor model. We deal with three CP-even, two CP-odd and two sets of charged scalar particles. We have improved our potential minimization technique which enabled us to explore a larger region of the allowed parameter space. We have studied in detail the trilinear couplings of the lightest Higgs boson of this model. Within the allowed domain of the parameters space of the model, the trilinear Higgs couplings have a strong dependence on $\tan\omega_3 = 2v_2/v_3$ and $\tan\theta$. The extended Higgs spectrum in S_3 models gives rise numbers of trilinear couplings. The hhh coupling can be measured in hh continuum production linear colliders as at e^+e^-.

References

[1] Chatrchyan S *et al.* 2012 Phys. Lett. B 716 30

[2] Higgs P W 1964 Phys. Lett. 12 132; 1966 Phys. Rev. 145 1156; Englert F and Brout R 1964 Phys. Rev. Lett. 13 321; Guralnik G S Hagen C R and Kibble T W 1964 Phys. Rev. Lett. 13 585

[3] Ma E 2007 UCRHEP-T431 [arXiv:0705.0327]

[4] Derman E 1979 Phys. Rev. D 19 317; Derman E and Tsao H S 1979 Phys. Rev. D 20 1207; Mondragon A and Rodriguez-Jauregui E 1998 [arXiv:9804267]; 1999 Phys. Rev. D 59 093009 [arXiv:9807214]; 2000 Phys. Rev. D 61 113002 [arXiv:9906429]; Kubo J 2004 Phys. Lett. B 578 156 [arXiv:0309167]; Kubo J Okada H and Sakamaki F 2004 Phys. Rev. D 70 036007 [arXiv:0402089]; Beltran O F Mondragon M and Rodriguez-Jauregui E 2009 J. Phys. Conf. Ser. 171 012028; Morisi S and Peinado E 2010 [arXiv:1001.2265]

[5] Kubo J Mondragon A Mondragon M and Rodriguez-Jauregui E 2003 Prog. Theor. Phys. 109 795 [arXiv:0302196]; Felix O Mondragon A Mondragon M and Peinado E 2007 AIP Conf. Proc. 917 383 [arXiv:0610061]; Mondragon A Mondragon M and Peinado E 2007 Phys. Rev. D76 076003 [arXiv:0706.0354]; Mondragon A Mondragon M and Saldana Salazar U J 2012 J. Phys. Conf. Ser. 378 012027[arXiv:1210.6131; Gonzalez Canales F Mondragon A Mondragon M and Saldana Salazar U 2012 C12-06-30

[6] Barger V Logan H E and Shaughnessy G 2009 Phys. Rev. D 79 115018 [arXiv: 0902.0170]; Gupta R S and Wells J D 2009 [arXiv:0912.0267]

[7] Gunion J F Haber H E Kane G and Dawson S 1990 *The Higgs Hunter's Guide* Addison-Wesley; Quigg C 1999 XXVII Int. Meeting on Fundamental Physics, Sierra Nevada (Granada) [arXiv:9905369]; Spira M and Zerwas P M 1997 Int. Universitätswochen, Schladming [arXiv:9803257]; Ginzburg I F Krawczyk M and Osland P 2002 [arXiv:0211371]; Kanemura S Okada Y Senaha E and Yuan C P 2004 Phys. Rev. D 70 115002 [arXiv:0408364]

[8] Hoang L N 1998 Mod. Phys. Lett. A 13 1865 [arXiv:9711204]; Adler S L 1999 Phys. Rev. D 60 015002 [arXiv:9901449]; Ferreira P M and Silva J 2008 P Phys. Rev. D 78 116007 [arXiv:0809.2788]; Howl R and S. F. King S F 2009 [arXiv:0908.2067]

[9] Barroso A Ferreira P M and Santos R 2006 Afr. J. Math. Phys. 3 103 [arXiv:0507329]; 2006 PoS HEP2005 337 [arXiv:0512037]; Barroso A Ferreira P M Santos R and Silva J P 2006 Phys. Rev. D 74 085016 [arXiv:0608282]

[10] Konisi G and T. Saito T 1995 Prog. Theor. Phys. 93 1093; Toghi A and Moazzen M 2009 Int. J. Theor. Phys. 48 3372

Journal of Physics: Conference Series **447** (2013) 012056 doi:10.1088/1742-6596/447/1/012056

Discreteness in parameters of the Standard Model

Sergey Sukhoruchkin
Petersburg Nuclear Physics Institute, 188300, Gatchina Russia
E-mail: sergeis@pnpi.spb.ru

Abstract.
Tuning effect in particle masses manifests itself in integer relations between lepton masses and parameters of the pion (its mass, the mass splitting and f_π). It involves parameters of constituent quark model and basic values of the Standard Model: masses of vector and scalar bosons, QED parameters and the top quark mass. The role of nuclear data is discussed.

1. Introduction

According to Y. Nambu [1] empirical relations between particle masses can be used for the development of the Standard Model - the theory of all interaction with the representation:

$$SU(3)_{col} \otimes SU(2)_L \otimes U(1)_Y \tag{1}$$

The Quantum Chromodynamics (QCD), the first of these three SM-components [2], deals with the strong interaction of colored quarks and gluons. QCD is a commonly accepted theoretical base of nuclear physics and we use it later in our analysis.

R.Feynman marked: "The theories about the rest of physics are very similar to the theory of quantum electrodynamics: they all involve the interaction of spin 1/2 objects (like electrons and quarks) with spin 1 objects (like photons, gluons, or W's) within a framework of amplitudes by which the probability of an event is the square of the length of an arrow. Why are all the theories of physics so similar in their structure?" [3]. Following Nambu we used [4-8] a general approach to the analysis of SM-parameters, particle masses and nuclear data. Common effects in empirical data are considered together with Feynman's question about the role of vector fields.

2. Tuning effect in particle masses and SM-parameters

The initial observation of integer relations between particle masses made in 70-ties concerned [9,10] mass splitting of the pion, nucleon and electron. Light quarks and the electron belong to the third (the lightest) SM-family. Masses of light quarks are roughly estimated as few MeV [2] but we use instead exactly known characteristics of the pion (the system of light quark–antiquark). Pion mass splitting δm_π=4.5936 MeV [2] is close to $9m_e$=Δ=4.599 MeV [9,10] and hence the double energy of pion β-decay energy $2\delta m_\pi$-$2m_e$ is close to $16m_e$=δ. We obtained the following empirical ratios: $(m_\mu+m_e)/(2\delta m_\pi\text{-}2m_e)$=106.169 MeV/8.165 MeV=13.00 ; $m_\pi^\pm\text{-}m_e/(2\delta m_\pi\text{-}2m_e)$=17.03; for a half of the nucleon Δ-excitation $\Delta M_\Delta/(2\delta m_\pi\text{-}2m_e)$=18.02; for a stable interval in pseudoscalar mesons $m_{\eta'}\text{-}m_\eta$=$m_\eta\text{-}m_\pi^\pm$=409 MeV/$(2\delta m_\pi\text{-}2m_e)$=49.97 and for the neutron mass $(m_n + m_e)/(2\delta m_\pi\text{-}2m_e)$=115.007.
Pion decay parameter f_π [2] is a part of empirical relation: (f_π=130.7 MeV)/$(2\delta m_\pi\text{-}2m_e)$=16.01.

The involvement of the electron mass and $\delta m_\pi \approx 9m_e$ into masses of other particles is the first aspect of the tuning effect. Many authors including Y. Nambu [10,11] noticed a frequent appearance of intervals multiple to pions's mass. G. Wick found the same effect with a half of ω-meson mass $(1/2)m_\omega$=391 MeV=$3f_\pi$ [12,13]. In Fig.1 these relations are shown by two-dimensional presentation with the period 130.4 MeV=16·16m_e=f_π ≈(1/6)m_ω along the x-axis. The line connecting the pion with Λ, Ξ^-, Ω^- hyperons corresponds to 1:8:11:12 ratio in masses.

The half of nucleon Δ-excitation 147 MeV=ΔM_Δ (a line N–Δ) is close to the value m_s ≈150 MeV [14] seen as intervals between N, Δ, Σ and Ξ-baryons. Such constant slope of lines between other masses (K–N, η–μ, stable ΔI=2 excitations of vector mesons K^*–K_3^*, ω–ω_3 mesons) and the proximity of the Ξ^- mass to 9·147 MeV=$3M_q$ are shown in Fig.1 top. Crossed arrows mark the discussed intervals in pseudoscalars (50δ) close to the standard estimate of baryon constituent quark mass in NRCQM M_q^Δ=410 MeV=$m_{\Delta^{++}}/3$.

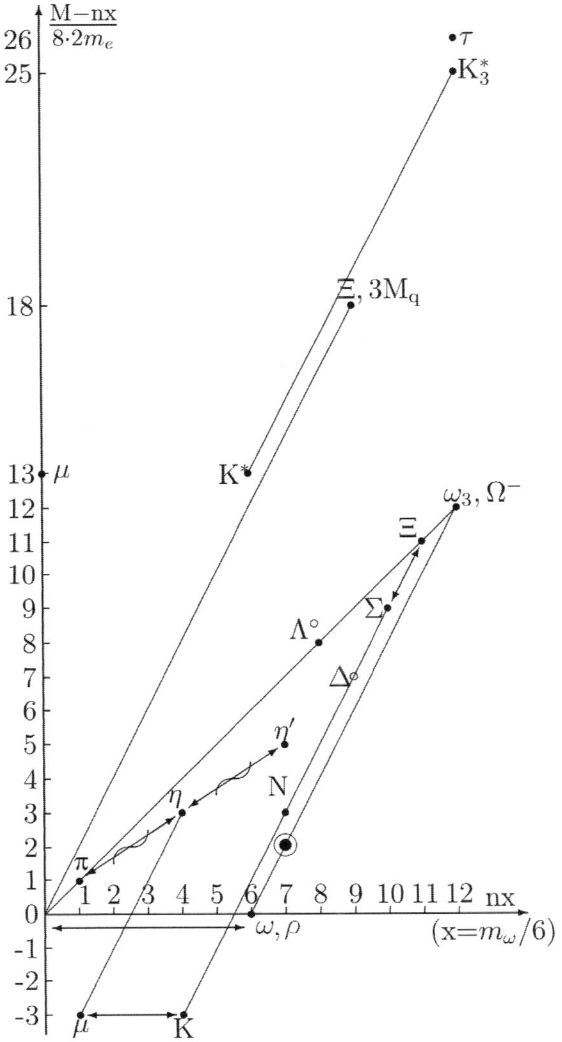

Figure 1. Positions of different particle masses M_i and mass intervals, including an estimate of nucleon mass in nuclear medium (circled) in two-dimensional presentation with x-axis in units 16·16m_e=f_π=130.8 MeV≈ m_ω/6 and differences M_i-k×16·16m_e at y-axis in units 16m_e=δ [10].

The second aspect of the tuning effect consists in a role of QED radiative correction. Such correction $g/2\pi$ is used for a comparison of different effects [15]. For example, J.Bernstein compared α/π=2.32·10^{-3} with CP-nonconservation parameter in kaon decay η=2.23·10^{-3} [16].

Two pairs of exactly known SM-parameters – both lepton masses and masses of Z-boson and Higgs boson (M_H=126 GeV) are in empirical ratios very close to the QED radiative correction $\alpha/2\pi$=1.159·10^{-3} (here α=1/137 is the SM-parameter). Coincidence of the ratio $(m_\mu = 105.66\ MeV)/(M_Z = 91187\ MeV)$=1.159·10^{-3} with $\alpha/2\pi$ was discussed in [13].

The recently measured scalar boson mass 126 GeV [18] and the one third of the electron mass $m_e/3$=511 keV/3=170.3 keV form the ratio 170.3 keV/126 GeV=1.35·10^{-6} coinciding with $(\alpha/2\pi)^2$=1.343·10^{-6} [4,5,17]. It means that $m_e/3$ and M_H are parts of two ratios close to $\alpha/2\pi$ between them. The $(1/3)m_e$ and ΔM_Δ=147 MeV (the parameter of Δ-excitation 294 MeV/2 close to the strange quark mass m_s ≈150 MeV in NRCQM [14]) are in a ratio $(m_e/3)$/147 MeV=1.159·10^{-3}. Ratios m_μ/M_Z and $(m_e/3)/\Delta M_\Delta/M_H$ are boxed in Table 1 (left and right, one under another, expressed as different power X=-1,0,1,2 (left) of the $\alpha/2\pi$).

Tuning effect in particle masses was found in 70-ties [9,10] when a small relative deviation of the muon mass from the integer number of the electron mass (L=207=23×9=13×16-1), namely 1.12·10^{-3}, was found to be close to the QED correction $\alpha/2\pi$=1.16·10^{-3}. The determination by R. Crawford at al. of the pion's mass difference permitted to find out that relative deviation of this value from $9m_e$, namely, 5.4(5)keV/4593.6keV=1.17(11)·10^{-3} coincides with $\alpha/2\pi$ within 10%. The ratio $m_\mu/\delta m_\pi$=23.001 is a result of the proximity of similar shifts in both values (m_μ and δm_π) from $n \cdot m_e$. As a result, the observed lepton ratio m_μ/m_e=206.76 becomes integer L=207 after a small correction of the electron mass $m_e^* = m_e(1 - \alpha/2\pi)$.

The factor $\alpha/2\pi$ is the Schwinger correction to electron magnetic moment. V.Belokurov and D.Shirkov suggested [19] that similar component is expected to be in the value m_e itself.

Conformation of the factor $\alpha/2\pi$ was obtained earlier during analyses of fine- and superfine-structure effects in nuclear excitations. Introduced [10,17,20] parameters of these two structures $\varepsilon''=(\delta''$=11 eV)/8 and ε'=1.2 keV=(δ'=9.5 keV)/9 are given in Table 1 (bottom, see text below). Intervals $4\varepsilon''$=5.5 eV and δ'=9.5 keV were found independently by K.Ideno, M.Ohkubo [21-23].

Table 1. Presentation of tuning effects in particle masses by the common expression n·16m$_e(\alpha/2\pi)^X$M in lines marked X=-1, 0, 1, 2 (at left) and QED radiative correction $\alpha/2\pi$ (α=137^{-1}). Values m$_\pi$-m$_e$, m$_e$/3, the neutron mass shift kδ – m_n – m_e, m_s and scalar boson mass are boxed. Stable intervals in nuclear binding energies [24-27] (X=0, M=1), nuclear excitations (E*, D_{ij}, X=1 [28]) and neutron resonances (X=2 [23]) are considered as indirect confirmation of relations at X=-1 and 0. The unconfirmed mass groupings observed at 116 GeV [29] and 58 GeV in LEP (X=-1 [30]) are given in parentheses. Values ΔE$_B$ in the center (X=0, M=1) and E^*,D in two bottom section are nuclear parameters serving for the indirect check of the tuning effect [4-8].

X	M	n=1	n = 13	n = 16	n = 17	n = 18
-1	3/2			m$_t$=172.0		
GeV	1	2Δ° - 2M_q	M$_Z$=91.2	(M_H=115)		M_H=126
	1/2			(M^{L3}=58)		
0	1	δ=16m_e=8.176	m$_\mu$=105.7	f$_\pi$=130.7	m$_\pi$-m$_e$	m_s ≈147=ΔM_Δ
MeV	1	2Δ - 2m_e	106=ΔE$_B$	130=ΔE$_B$	140=ΔE$_B$	147.2=ΔE$_B$
	3			M"$_q$=m$_\rho$/2	NRCQM	M$_q$=441=ΔE$_B$
1	1				kδ-m$_n$-m$_e$=161.6(1)	170 = m_e/3
keV	8				δm_N=1293.3	
	1	δ'=9.5=8ε'	123	152	161 (Sb, E^*) = Δ^{TF}	170 (Sn)
	3				484 (E^*)	512 (Pd), ε_o/6 (^{10}B)
	8		984	1212	1293 (E^*), $n \cdot 1292$ (^{24}Na)	1360 (Te)
2	1	δ''=11=8ε''	143	176	186 (Nd), 373 (Sb)	D in neutron
[illegible]	[illegible]	[illegible]	[illegible]		[illegible]	

3. Constituent quark masses in the tuning effect

Parameter 147 MeV and its three-fold value 147 MeV×3=441 MeV=M_q connected with the scalar field (Table 1, right) play an important role in the tuning effect. The second value is an accurate estimate of baryon constituent quark mass in NRQM M_q^d=436 MeV [14]. The value 440 MeV introduced empirically by R.Sternheimer [12] is close to 1/3 of the Ξ-hyperon mass (1324 MeV/3=441 MeV) as it was noticed by P.Kropotkin [31]. Here is a compensation of the mass increase with strangeness and the decrease due to the quark interaction (see Fig.1 top).

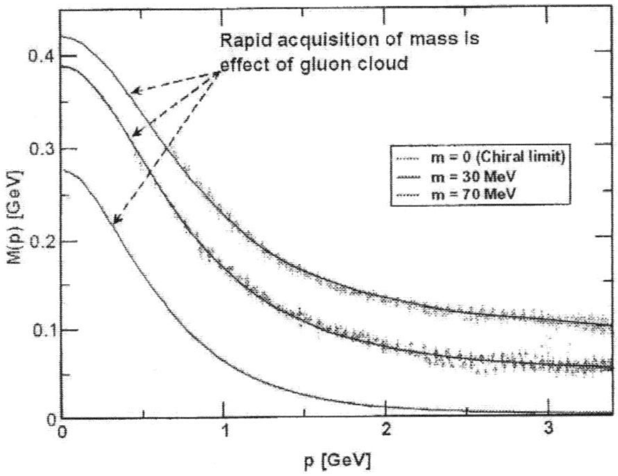

Figure 2. QCD gluon-quark-dressing effect calculated with Dyson-Schwinger equation, initial masses m_q=0, 30 and 70 MeV (top) [32]. The quark-parton acquires a momentum-dependent mass function that at infrared momentum (p=0) is larger by two-orders-of-magnitude than the current-quark mass (several MeV) due to a cloud of gluons that closes a low-momentum quark; the values at left are constituent quark masses 410 MeV=M_q^{Δ}–441 MeV=M_q.

The value of constituent quark mass (M_q ≈440 MeV) is obtained in calculations with the Dyson-Schwinger Equation (see Fig.2) and in the Quark models [14,33]. The value M_q=1/3(M_N^{GBE})=450 MeV for the initial quark mass in the Goldstone Boson Exchange Model [33] is shown in Fig.3 where the "initial" mass for nucleon and Δ-baryon (before switching on the residual interaction between constituent quarks) is about M_N^{GBE} ≈1350 MeV which is marked "+" on left axis of Fig. 3. The strength of GBE-interaction in this calculation (directed along x-axis, the vertical line) correspond to the real nucleon/Δ-baryon splitting 294 MeV.

Despite the fact that the most part of the real mass in the world can be described as constituent quark mass generation in QCD, there exists a tuning effect. In case of neutral baryons the parameter of the residual constituent quark interaction $2\Delta M_\Delta$=1233.8(2)–939.67=293.7(6) MeV coincides with 36δ=294.3 MeV ($\delta=16m_e$, last line of Table 2).

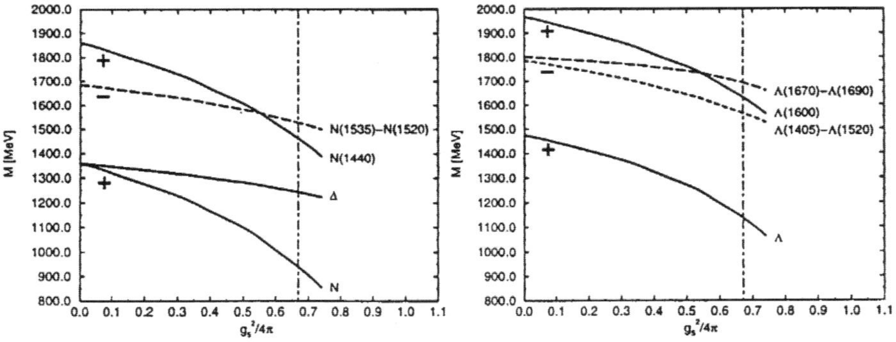

Figure 3. Calculation of nonstrange baryon and Λ-hyperon masses as a function of interaction strength within Goldstone Boson Exchange Constituent Quark Model; the initial baryon mass 1350 MeV=3×450 MeV=$3M_q$ is marked "+" on the left vertical axis. Downward shift of masses of the first baryons (294 MeV between Δ-baryon and N- nucleon) are shown as a function of the GBE strength. Full and dotted lines correspond to states with different parities [33].

4. Long-range correlations in particle masses and lepton ratio in Standard Model

Discussed empirical relations in particle masses (the first aspect of the tuning effect) are presented in upper/central part of Table 2. Four boxed values at right correspond to the discussed integer presentation of values m_μ+m_e, f_π, m_π-m_e and m_n+m_e with n×δ.

Additional confirmation of the tuning effect is shown in the second column of Table 2 (left) where ratios of particle masses to integer numbers of the common period $3m_e \approx \delta m_\pi/3$ are given. This period $3m_e$ was found by R. Frosch by the computer analysis of the periodicity in all 47 accurately known particle masses [34] (see results in Fig.4). One asterisk marks values which are in accordance with the period $16m_e$. Boxed in Table 2 are additional values which are in agreement with the correlation while two asterisks correspond to the case of the disagreement.

The shift of neutron mass relative to the integer number of δ=$16m_e$ (115δ - m_e) shown in Table 2 (at right, boxed) is determined with the accuracy of 0.1 keV due to a very accurately known proton/electron mass ratio m_p/m_e. This shift accounts 161.6(1) keV which is exactly (1/8) of the nucleon mass splitting δm_N=1293.3 keV, the ratio (1/8)·1.0003(2) with δm_N.

The nucleon mass splitting δm_N contains an important information on the quark structure of baryons. The ratio between δm_N and the mass of the Λ-hyperon (the center of the baryon octet), namely, δm_N/1115.7 MeV=1.159·10^{-3} is close to $\alpha/2\pi$. Due to Nambu's relation m_Λ=$8m_\pi$ [11] and the ratio 8:1 between δm_N and the discussed shift 161 keV all these three values (boxed) are situated in Table 1 one under another. Long-range correlation in particle masses manifested by the equality of the shift 115×δ-m_e-m_n to the 1/8 part of δm_N presents an important third aspect of the tuning effect. Similar long-range correlation with n·m_e were found in nuclear data.

Manifestations of δm_N=1293 keV in nuclear excitation are given in Table 1 (n=17, M=8): the E^* grouping in Z-odd nuclei ($Z \leq 29$) and integer relations (n=2,3) in E^* of ^{24}Na. The parameter Δ^{TF}-161 keV=(1/8)δm_N was found in Z=51 nuclei (see later discussion of +Table 4).

Deviations of neutral octet baryon masses from integers of $16m_e$ has a systematic character as it is seen from increasing of the "downwards shift" with the strangeness (from neutron to the Ξ^o-hyperon, ΔS=2 in Table 2 bottom right), the "shift" accounts about 0.5 MeV per ΔS=1.

Table 2. Comparison of particle masses with periods $3m_e$ [34] and $16m_e$ (n and N – numbers of periods.)

Part.	m$_i$, MeV 1	m$_i/3m_e$ n	N·$16m_e$ 3	Diff., MeV 1-3	Comments
μ	105.65839(3)	68.923*	106.288	-0.6294	-0.511 - 0.118
f_π	130.7(4)		130.8	-0.1	[2]
π^o	134.9743(8)	88.046*	138.992	-4.0174	
$\pi^\pm$	139.5679(7)	91.043*		+0.5762	+0.511 + 0.065
η^o	547.45(19)	357.11	547.791	-0.34(19)	
ω	781.95(14)	510.08	784.895	-2.94(14)	-2×$3m_e$=-3.066 MeV
ρ	775.49(34)	505.86		-9.40	-2×$9m_e$=-9.198 MeV
φ	1019.413(8)	664.98	1022.99	-2.585(8)	-2D_o=2.586 MeV
$K^\pm$	493.646(9)	322.01	490.559	+3.087(9)	(6×0.511=3.066)
K^o	497.671(31)	324.64**	498.735	-1.064(31)	(-1.022) see text
p	938.2720(1)	612.051*		-1.96660	m_e + 1.4549=9/8D_o
n	939.5654(1)	612.894*	940.2380(1)	-0.6726(1)	m_e + 0.1616 =1/8D_o
Σ^o	1192.55(10)	777.920	1193.69	-1.14(10)	(-0.511×2)
Ξ^o	1314.9(6)	857.732	1316.33	-1.40(60)	(-0.511×3)
Ξ^-	1321.32(13)	861.92	1324.51	-3.19(13)	(6×0.511=3.066)
Ω^-	1672.43(32)	1090.95	1667.90	+4.53(32)	
τ	1776.82(16)	1159.0	1782.37		
Δ^o – n	293.7(6)		294.34	-0.6(6)	

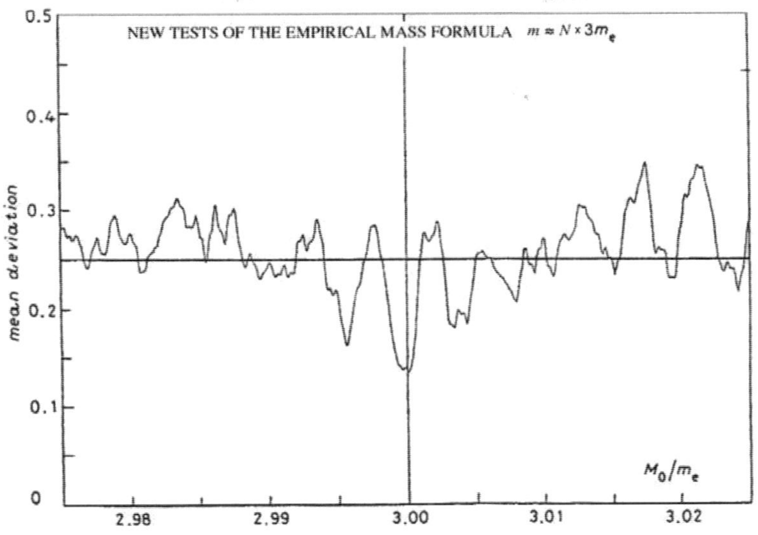

Figure 4. Check of the formula m_i=N×$3m_e$ by the minimum of mean-square-root deviation in masses of 47 particles [34]. The most deepest minimum in the whole region of the period M_o variation is shown; very small occasional probability of this correlation was estimated in [34].

Three-fold values of elements in the relation $(m_e/3)/\Delta M_\Delta/M_H=\alpha/2\pi$, namely, m_e/M_q together with the ratio L=m_μ/m_e (the first aspect of tuning effect) and the exact empirical relation $m_\mu/M_Z=\alpha/2\pi$ (the second aspect) determines the lepton ratio M_Z/M_q=206.8 between vector boson mass and initial baryon constituent quark mass M_q=441 MeV.

The standard way for the estimation of meson constituent quark mass is the use of a half of well-known nonstrange vector meson mass (Table 2). Masses of both vector mesons (m_ρ are m_ω) are somewhat less than $6\times16\times16m_e=6f_\pi$=784.895 MeV (Table 2 right). The value $M_q''=m_\rho/2$=775.49(34) MeV/2=387.8(2) MeV deviates from $3\times16\times16m_e=3f_\pi$ with 4.7 MeV close to $9\times m_e=\Delta$=4.6 MeV. The mass of the second vector boson M_W=80.399(21) GeV forms a ratio 207.3 with this NRCQM parameter M_q''=775.49 MeV. This means an empirical indication on the distinguished role of the lepton ratio L. The obtained relations in masses of both vector fields is 4-th aspect of the tuning effect (L-repetition in ratios between different particles). It means that the dynamics of quark-dressing effects (Fig. 2) should include the symmetry properties of interactions to fit this empirical findings. All four aspects of the tuning effect should be considered as the symmetry motivated phenomenon. It is seen also from a common presentation of the muon mass, $f_\pi=16\times16m_e$ and ΔM_Δ (1-st aspect of the tuning effect).

5. Additional analysis of the tuning effects in nuclear data

C. Detraz noticed [35] that the universal character of 0^+-0^+ nuclear β-transitions means that "the nucleus is one specific case, the coldest and most symmetric one, of hadronic matter". To check discussed aspects of tuning effects in particle masses we use data from another hadronic effects – nuclear excitations and binding energies. We start with data for light nuclei (Table 3).

Table 3. Comparison of E^* and ΔE_B (keV) in some near-magic nuclei with multiples of $\varepsilon_o=2m_e$

^{A}Z	^{10}B	^{10}B	^{10}B	^{12}C	^{16}O	^{18}Ne	^{18}Ne	^{8}Be	^{20}Ne	^{138}Ce	^{140}Ce	^{36}K	^{39}K
J^π	$^+0$-$^+1$	2^-	3^-	0^+	3^-	0_1^+	0_2^+	$\frac{\varepsilon_{2n2p}}{4}$	$\frac{\varepsilon_{2n2p}}{4}$	ΔE_B	^{6}He	ΔE_B	4α
E*	1021.8	5110	6127	27595	6130	3576	4590	7151	4076	45997	45996	147160	147152
n(ε_o)	1	5	6	27	6	7/2	9/2	7	4	45	45	8×16	8×16
n·ε_o	1022.0	5110	6132	27594	6132	3577	4599	7154	4088	45990		147168	
Diff.	0.2(2)	0.3	3	1(2)	2	1(2)	9(8)	-3	-12	7	8	-8(2)	-16

Journal of Physics: Conference Series **447** (2013) 012056 doi:10.1088/1742-6596/447/1/012056

In the nucleus ^{10}B with two-holes configuration in the ^{12}C core the splitting between $0^+, T=1$ and $1^+, T=0$ states exactly equal to $2m_e$=ε_o=1022.0 keV corresponds to the nucleon spin-flip effect while the first negative parity excitation E^*=5110.3(6) keV corresponds to nucleon transition between the shells. It is shown in Table 3 that these values are exactly rational 1:5.001:27.000 between themselves and the excitation 0_2^+ in the above mentioned ^{12}C core. Similar relations with ε_o=$2m_e$ were found in excitations of O and Ne isotopes (Table 3 center). For example, the first cluster excitation in ^{19}O at E^*=3067 keV is close to $3\varepsilon_o$. Standard parameters of the residual interaction of valence nucleon pairs derived from the binding energies of nuclei with Z,N>=2,8 ($\varepsilon_{2n2p}/4$, Table 3, center) are also close to $7\varepsilon_o$ and $4\varepsilon_o$.

Cluster effects in nuclear binding energies E_B considered in [25-27] show long-range correlations in ΔE_B with ε_o=$2m_e$. It is seen in Table 3 right were differences ΔE_B in light nuclei differing with 4α configuration and ΔE_B in middle-weight N-even nuclei (N≤82) are partly presented (ΔE_B=$128\varepsilon_o$ and $45\varepsilon_o$, small differences ΔE_B–n·ε_o in table 3 right). Confirmation of tuning effect in particle masses: values 147 MeV=18δ=32Δ in light and heavy nuclei, 3×147 MeV=441 MeV and 409 MeV=50δ in all odd-odd and even-even nuclei were observed.

The role of one-pion exchange in nuclear properties was considered by T.Otsuka [28,36]. The expected stability of excitation due to nuclear tensor force action (one-pion exchange dynamics) can be seen as a linear trend in excitations of Sb isotopes shown in Table 4 [17,28]. Similar effects were found in other nuclei. Integer ratios between excitations in series of nuclei with the valence proton and sequence n=1-6 of pairs of neutrons in $\nu 1h_{11/2}$ subshell (E*=n×161 keV) permitted determination of tensor-force parameter Δ^{TF}=161 keV found also in many surrounding nuclei and in nuclei with other nuclear shells [17,28]. In case of nearby ^{124}Sb (valence proton and three neutrons in $1h_{11/2}$ subshell) stable intervals 160 keV in low-lying excitation were found together with super-fine effect in highly excited states seen as neutron resonances, namely, intervals 373-750 eV and 570 eV. These superfine intervals are in the ratio $1.16 \cdot 10^{-3}$ close to $\alpha/2\pi$ with the system of excitation in nearby ^{odd}Sb isotopes (Table 4). Similar effects observed in different nuclei [17,28] indirectly confirm relations between the pion mass, the shift in neutron mass 161 keV=$(1/8)\delta m_N$ (coinciding with Δ^{TF}) and stable intervals in neutron resonances.

The distinguished character of intervals ΔM_Δ=147 MeV and m_π (as well as 170 keV-161 keV=$\delta m_N/8$) in nuclear data could be connected with presentation of nucleon mass in nuclear media (circled in Fig.1) as six f_π and additional ΔM_Δ (more data are needed).

Table 4. Comparison of E^* in Z=51 nuclei with n×161 keV; stable E^*=170 keV in ^{101,103}Sn [17,28].

^{A}Z	^{133}Sb	^{131}Sb	^{129}Sb	^{127}Sb	^{125}Sb	^{123}Sb	^{125}Sb	^{119}Sb	^{101}Sn	^{103}Sn
$2J^\pi-,J^\pi$	5^+	5^+	5^+	5^+	5^+	5^+	$3^+,5^+$	1^+	5^+	5^+
E^*, keV	962.0	798.4	645.2	491.2	332.1	160.3	644	644.0	171	168
$n\frac{D_0}{8}$	969	808	646	484	323	161	646	646		
$n=\frac{(N-70)}{2}$, N	6, 82	5, 80	4, 78	3, 76	2, 74	1, 72	4	4	[1]	[1]
D(Sb),eV			750	570	373					
E^*/D			$1.16 \cdot 10^{-3}$	$1.16 \cdot 10^{-3}$						

6. Relations in Standard Model parameters

It was suggested by F.Wilczeck that the top quark mass has a very representative status: its mass is the largest [37]. One can notice the relation 3:2:1 between top quark mass and two unconfirmed mass grouping effects at 116 GeV and 58 GeV (see top lines of Table 1). The value 116 GeV was reported in [29] as a possible Higgs boson mass while the effect at M^{L3}=58 GeV was considered by S.Ting [30]. The third unconfirmed mass-groping effect at 4 GeV=Δ° was reported in [38]. Z-boson (M_Z), the top quark (m_t) and the scalar field (M_H) together with the unconfirmed effects are put in upper section of Table 1. It corresponds to a suggestion that tuning effect in masses of particles from the second/third SM-families could be connected with the discreteness in the larger mass values. Such discreteness has a parameter M_q=3ΔM_Δ=3($\alpha/2\pi$)M_H which manifests itself in NRCQM as the initial quark mass and takes part in the lepton ratio with the Z-boson mass. Scalar mass M_H and (2/3)m_t=116 GeV=2M^{L3} are expressed as 18×16M_q and 16×16M_q with 18M_q=2Δ^o (Δ^o is close to bottom quark mass). The electron mass itself and the shift in baryon masses with m_e=3(m_e/3) are connected with the scalar masses.

7. Conclusions

We considered here 4 empirical properties of particle mass spectrum (4 aspects of tuning effect):
1. Electron, muon, pion, neutron masses and parameters f_π, ΔM_Δ are interconnected;
2. QED radiative correction $\alpha/2\pi$ (with α=1/137 or $\alpha_Z \approx$1/129 [6]) play an important role;
3. Long-range correlation with electromagnetic mass differences has analog in nuclear data;
4. Lepton ratio repetition reflects symmetry properties of Standard Model. Its understanding could be helpful for the explanation of the vector fields universality mentioned by R.Feynman. The confirmation of the ratio 3:2:1 between the top quark mass and mass-grouping effects could clarify the traces of "super-duper" model [3] in which the vector universality and a proximity of the ratio $M_q/(m_t/3)$=1/128 to $\alpha_Z \approx$1/129 marked in [6] could be more transparent.

[1] Nambu Y 1998 *Nucl. Phys. A* **629** 3c
[2] Particle Data group 2006 *J. Phys.* G **33** p.535, M.Suzuki; PDG 2010, Review of Particle Physics
[3] Feynman R 1986 *QED-Strange Theory of Light and Matter.* (Princ. Univ. Press).
[4] Sukhoruchkin S I 2013 *Nucl. Phys.* B: Conf. Ser. (in press).
[5] Sukhoruchkin S I 2012 *J. Phys.: Conf. Ser.* **381** 012076; doi:10.1088/1742-6596/381/1/012076.
[6] Sukhoruchkin S I 2009 *J. Phys.: Conf. Ser.* **171** 012064
[7] Sukhoruchkin S I 2011 *Proc. Hadron2011* Muenich, 2011, eConf C110613
[8] Sukhoruchkin S I 2007 *Nucl. Phys.* A **782** 37c
[9] Sukhoruchkin S I 1968 *Proc. 2nd Conf. Neutr. Cross Section*, Washington DC **Vol.2** 923
[10] Sukhoruchkin S I 1972 *Stat. Properties of Nuclei* ed. Garg G (Plenum Press) 215
[11] Nambu Y 1952 *Progr. Theor. Phys.*, **7**, 595
[12] Sternheimer R 1964 *Phys. Rev.* **136** 1364: Sternheimer R 1964 *Phys. Rev.* **170** 1267.
[13] Sukhoruchkin S I 2000 *Symmetries in Subatomic Physics*, 3-rd Int. Symp. Adelaide, AIP539, 142.
[14] Itoh C et. al. 1989 *Phys. Rev.* **D 40** 3660.
[15] Dyatlov I T 1992 *Phys. Rev.* D **45** 1636.
[16] Bernstein J 1968 Elementary Particles and Currents (Stevenson Inc, publ. Freeman).
[17] Sukhoruchkin S I, Soroko Z N 2013 *Proc. ISINN-20* (JINR publ.)
[18] ATLAS collab. 2012 *Phys. Lett.* B **710** 49.
[19] Belokurov V V, Shirkov D V 1991 *The Theory of Particle Interactions* AIP, New York
[20] Sukhoruchkin S I 1970 *Sov. Journ. Nucl. Phys.* 144, 285.
[21] Ideno K 1997 *Proc. Int, Conf. Neutrons in Ressearch and Industry* Creta SPIE Proc Ser **2867** USA 398.
[22] Ohkubo M, Mizumoto M, Nakajima Y 1993 *Rept. JAERI-M-93-012.*
[23] Landoldt-Boernstein New Series 2009 **vol. I/24** ed. Schopper H (Springer)
[24] Sukhoruchkin S I *et al.*, 2012 *Proc. Sem. ISINN-19* Dubna 2011, E3-2012-30, pp.284, 296, 308.
[25] Sukhoruchkin S I, Sukhoruchlin D S 2011 *Int. J. Mod. Phys.* E **20** 906
[26] Landoldt-Boernstein New Series 2009 **vol. I/22A** ed. Schopper H (Springer).
[27] Sukhoruchkin S I 2008 *Int. Rev. Phys. (IREPHY)* **2** 239
[28] Landoldt-Boernstein New Series 2012 **vols. I/25A,I/25E** ed. Schopper H (Springer)
[29] ALEPH Collaboration, Heister A et al. 2002 *Phys. Lett.* B **526** 191
[30] Ting S 1993 *Preprint CERN-PPE/93-34.*
[31] Kropotkin P 1971 *Field and Matter*, 106. (in rus. Moscow University).
[32] Bhagwat M S *et al* 2007 *Phys. Rev.* C **76** 045203; Roberts H L L *et al* 2011 *Phys. Rev.* C **83** 065206
[33] Glozman L 1990 *Nucl. Phys.* A **629** 121c
[34] Frosch R 1991 *Nuovo Cim.* **104** 913
[35] Detraz C 1995 *Nucl. Phys. A* **583** 3.
[36] Otsuka T, Suzuku T, Utsino Y 2008 *Nucl. Phys.* A **805** 127c
[37] Wilczek F 2003 *Nucl. Phys. Proc. Suppl.* **B117** 410.
[38] Abazov V M *et al.*, 2009 Phys. Rev. Lett. **103**, 061801.

Charged Higgs boson production in single top mode at the LHC

Renato Guedes[1], **Stefano Moretti**[2] **and Rui Santos**[1,3]

[1]Centro de Física Teórica e Computacional, Faculdade de Ciências, Universidade de Lisboa, Av. Prof. Gama Pinto 2, 1649-003 Lisboa, Portugal

[2] NExT Institute and School of Physics and Astronomy, University of Southampton Highfield, Southampton SO17 1BJ, UK

[3] Instituto Superior de Engenharia de Lisboa - ISEL, Rua Conselheiro Emídio Navarro 1, 1959-007 Lisboa, Portugal

E-mail: renato@cii.fc.ul.pt, stefano@phys.soton.ac.uk, rsantos@cii.fc.ul.pt

Abstract. The main production mode for a light charged Higgs boson at the LHC is $pp \to t\bar{t}$, with one the top-quarks decaying to a charged Higgs and a b-quark. However, single top production also gives rise to final states with charged Higgs bosons. In this work we analyse how the two processes compare at the LHC@14TeV. We will be working in the framework of the two-Higgs double model, considering both a CP-conserving and a CP-violating version of the model. We conclude that the single top mode could help to constrain the parameter space in several versions of the model. We also discuss the role of other complementary production processes in future searches at the LHC.

1. Introduction

The discovery of a charged Higgs boson at CERN's Large Hadron Collider would be an unequivocal sign of physics beyond the Standard Model (SM). A light charged Higgs is well within reach of the LHC@8TeV in many Beyond the SM (BSM) models. Searches based on $pp \to t\bar{t} \to bW^+\bar{b}H^-$ are currently being performed by both the ATLAS [1] and CMS [2] collaborations. A large portion of the parameter space with a light charged Higgs boson (below 150 GeV) has already been excluded by the two experiments in models with two Higgs doublets and in particular in the Minimal Supersymmetric Standard Model (MSSM). However, it is clear that even for a charged Higgs mass below the top-quark mass, the entire parameter space will not be ruled out when all the 8 TeV data is finally analysed. Hence, one may ask if by the end of the 13-14 TeV run a light charged Higgs boson will be either found or definitely excluded and if so in which models? The answer to that question is highly dependent on the model being scrutinised. It is expected that most of the parameter space below the top mass can be probed for the MSSM as well as for other multi-Higgs extensions of the SM. It is clear though, that some multi-Higgs versions will not be probed in their entire parameter space range even for a light charged Higgs. For those particular scenarios it is useful to consider all charged Higgs production process as to maximize the discovery potential. As the single top production cross section is about one third of the $t\bar{t}$ one, it could in principle slightly boost the chances of finding or indeed disproving the existence of a light charged Higgs boson. The purpose of this work is

Journal of Physics: Conference Series **447** (2013) 012057 doi:10.1088/1742-6596/447/1/012057

to show that a slight improvement can be obtained by complementing the present search, based on the $t\bar{t}$ mode, with the search in the single top mode.

2. Two-Higgs doublet models

The softly broken Z_2 symmetric ($\Phi_1 \to \Phi_1$, $\Phi_2 \to -\Phi_2$) two-Higgs doublet model (2HDM) potential can be written as

$$V(\Phi_1,\Phi_2) = m_1^2\Phi_1^\dagger\Phi_1 + m_2^2\Phi_2^\dagger\Phi_2 + (m_{12}^2\Phi_1^\dagger\Phi_2 + \text{h.c.}) + \frac{1}{2}\lambda_1(\Phi_1^\dagger\Phi_1)^2 + \frac{1}{2}\lambda_2(\Phi_2^\dagger\Phi_2)^2 + \lambda_3(\Phi_1^\dagger\Phi_1)(\Phi_2^\dagger\Phi_2) + \lambda_4(\Phi_1^\dagger\Phi_2)(\Phi_2^\dagger\Phi_1) + \frac{1}{2}\lambda_5[(\Phi_1^\dagger\Phi_2)^2 + \text{h.c.}]\ ,$$

where Φ_i, $i=1,2$ are complex SU(2) doublets. Hermiticity of the potential forces all parameters except m_{12}^2 and λ_5 to be real. Then, the nature of m_{12}^2 and λ_5, together with the chosen vacuum configuration, will determine the CP nature of the model (see [3] for a review). If CP is conserved we end up with two CP-even Higgs states, h and H, and one CP-odd state, A. If CP is broken, the three spinless neutral states with undefined CP quantum number are usually denoted by h_1, h_2 and h_3. However, as long as the vacuum configuration does not break electric charge, which was shown to be possible in any 2HDM [4], there are in any case two charged Higgs boson states, one charged conjugated to the other.

In this work we will focus on two specific realisations, one CP-conserving and the other explicitly CP-violating [5, 6]. In the CP-violating version m_{12}^2 and λ_5 are complex and the fields' vacuum expectation values (VEVs) are real. Existence of a stationary point requires $Im(\lambda_5) = v_1\, v_2\, Im(m_{12}^2)$. Because the VEVs are real in both models, a common definition for the rotation angle in the charged sector $\tan\beta = v_2/v_1$ can be used. Extending the Z_2 symmetry to the Yukawa sector we end up with four independent 2HDMs, the well known [7, 8] Type I (only ϕ_2 couples to all fermions), Type II (ϕ_2 couples to up-type quarks and ϕ_1 couples to down-type quarks and leptons), Type Y or III (ϕ_2 couples to up-type quarks and to leptons and ϕ_1 couples to down-type quarks) and Type X or IV (ϕ_2 couples to all quarks and ϕ_1 couples to leptons) (details and couplings can be found in [9]).

We will now very briefly discuss the main experimental constraints affecting the 2HDM parameter space. The signal in our analysis originates from single top production with the subsequent decay $t \to bH^\pm \to b\tau\nu$. Hence, only the charged Higgs Yukawa couplings are present and therefore the only parameters we need to be concerned with are $\tan\beta$ and the charged Higgs mass. Values of $\tan\beta$ smaller than $O(1)$ together with a charged Higgs with a mass below $O(100$ GeV$)$ are both disallowed by the constraints [10] coming from R_b, from $B_q\bar{B}_q$ mixing and from $B \to X_s\gamma$ for all models. Furthermore, data from $B \to X_s\gamma$ [11] imposes a lower limit of $m_{H^\pm} \gtrsim 360$ GeV, but only for models Type II and Type Y. The LEP experiments have set a lower limit on the mass of the charged Higgs boson of 80 GeV at 95% C.L., assuming $BR(H^+ \to \tau^+\nu) + BR(H^+ \to c\bar{s}) + BR(H^+ \to AW^+) = 1$ [12]. The bound is increased to 94 GeV if $BR(H^+ \to \tau^+\nu) = 1$ [12]. These bounds led us to take $m_{H^\pm} > 90$ GeV and $\tan\beta > 1$ for Type I and X. We will also present results for model Type II, where the bounds on the charged Higgs mass can be evaded due to the presence of new particles as is the case of the MSSM. The usual theoretical bounds such as the ones coming from requiring boundness from below of the potential and the ones from requiring perturbative unitarity are in this case redundant (same is true for the precision electroweak constraints).

3. Results and discussion

As previously discussed, $pp \to t\bar{t}$ is the best process to search for a light charged Higgs boson at the LHC. However, because the single top production cross section is about one third of $\sigma_{pp\to t\bar{t}}$, it deserves a full investigation regarding its contribution to the production of charged

Higgs bosons. The signal consists mainly of a light charged Higgs boson produced via the t-channel process $pp \to t\,j \to H^+\,\bar{b}\,j$ (together with its charge conjugate), with the subsequent decay $H^+ \to \tau^+\,\nu$, where j represents a light-quark jet. In what follows we are considering proton-proton collisions at a center-of-mass (CM) energy of $\sqrt{s} = 14$ TeV and a top-quark mass $m_t = 173$ GeV. We consider a charged Higgs boson mass interval of 90 to 160 GeV and the analysis is performed in 10 GeV mass steps.

Maximising the signal-to-background significance ($S/\sqrt{B}$) makes both the s-channel and the tW single-top production modes negligible - only the t-channel process survives the set of cuts imposed. Signal events were generated with POWHEG [13] at NLO with the CTEQ6.6M [14] PDFs. The top was then decayed in PYTHIA [15]. We have considered only the leptonic decays of the tau-leptons, that is, the signal final state is $pp \to l\,b\,j\,\not{E}$, where $l = e, \mu$ (electrons and muons) while $\not{E}$ means missing (transverse) energy.

The irreducible background, single-top production with the subsequent decay $t \to b\,W^+$, was also generated with POWHEG. The main contributions to the reducible background are: $t\bar{t}$ production, $W^{\pm}$ + jets (including not only light quarks and gluons, but also c- and b-quarks) and the pure QCD background (jjj, where j is any jet). The $t\bar{t}$ background was generated with POWHEG while $W^{\pm}$ + jets (1, 2 and 3 jets) was generated with AlpGen [16]. Finally, the QCD background was generated with CalcHEP [17] (with CTEQ6ll PDFs). The hadronisation was performed with PYTHIA 6. After hadronisation, DELPHES [18] was used to simulate the detector effects. For the detector and trigger configurations, we resorted to the ATLAS default definitions.

In order to maximise $S/\sqrt{B}$ we apply the following selection cuts (see [9] for details)

(i) We demand one electron with $p_T > 30$ GeV or a muon with $p_T > 20$ GeV, and $|\eta| < 2.5$ for both leptons.

(ii) We veto events with two or more leptons with $p_T > 10$ GeV. This cut eliminates the leptonic $t\bar{t}$ background almost completely.

(iii) We veto events with leptons having p_T above 55 GeV.

(iv) Events with missing energy below 50 GeV are excluded. This is a cut that dramatically reduces the QCD background.

(v) We ask for one and only one b-tagged jet with $p_T < 75$ GeV. We assume a b-tagging efficiency of 0.4 (with $R = 0.7$), while the misidentification rates for the case of c-quark jets we take 0.1 and for lightquark/ gluon jets we adopt 0.01.

(vi) We reconstruct a "top quark invariant mass" as defined in [9] and demand all events to have this invariant mass above 280 GeV. The top quark invariant mass distribution for signal and background is shown in figure 1.

(vii) We define a leptonic transverse mass [9], $M_T^{l\nu}$, and we have accepted events with $30\,\text{GeV} < M_T^{l\nu} < 60\,\text{GeV}$ for charged Higgs masses between 90 and 130 GeV and $30\,\text{GeV} < M_T^{l\nu} < 60\,\text{GeV}$ or $M_T^{l\nu} > 85\,\text{GeV}$ for higher values of the charged Higgs mass. The leptonic transverse mass distribution for signal and background is shown in figure 1.

(viii) We have chosen events with one and one jet (non-b) only with $p_T > 30$ GeV and $|\eta| \leq 4.9$.

(ix) We veto all events with a jet multiplicity equal to two or above for jets with $p_T > 15$ GeV and $|\eta| \leq 4.9$.

(x) We only accept events where jets have a pseudorapidity $|\eta| \geq 2.5$.

Putting all the numbers together we can find S/B and $S/\sqrt{B}$ as a function of the charged Higgs mass as presented in table 1.

The results can be presented in a model independent manner as done in [9] and from them exclusion plots can be derived for the different 2HDMs. Before proceeding we present in the left

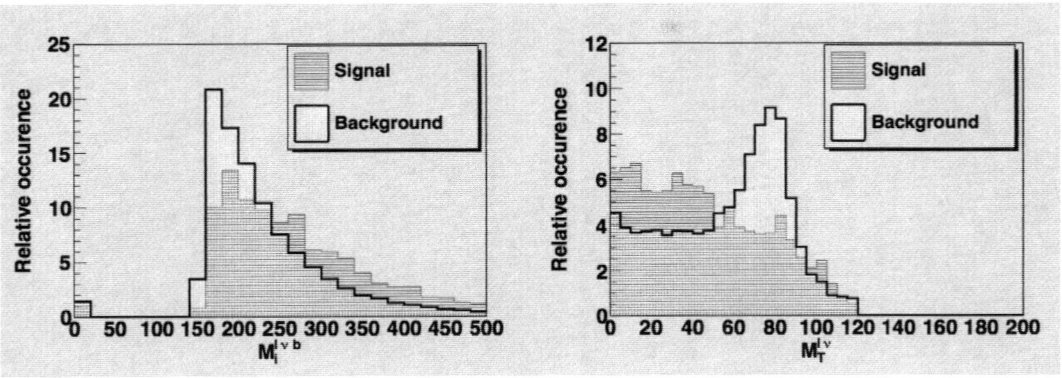

Figure 1. Left: "top quark invariant mass" distribution for signal and background. Right: leptonic transverse mass distribution for signal and background.

Table 1. Signal-to-Background ratio (S/B) and significance ($S/\sqrt{B}$) as a function of the charged Higgs mass for $\sqrt{s} = 14$ TeV and a luminosity of 1 fb^{-1}. The numbers presented for the signal we take BR$(t \to bH^{\pm}) = 100\%$ and BR$(H^- \to \tau^- \nu) = 100\%$ and all other Branching Ratios (BRs) have the usual SM values.

$m_H^{\pm}$ (GeV)	Signal (S)	Background (B)	S/B (%)	$S/\sqrt{B}$
90	38.6	29.5	130.92	7.11
100	40.5	29.5	137.19	7.45
110	45.6	29.8	153.00	8.35
120	47.7	30.1	158.26	8.69
130	42.3	32.7	129.53	7.41
140	117.1	77.9	150.25	13.26
150	120.0	86.6	138.64	12.90
160	109.7	100.8	108.81	10.92

panel of figure 2 the charged Higgs BRs for $m_{H^\pm} = 100$ GeV as a function $\tan\beta$ in models Type I and X. Clearly $H^+ \to \tau^+\nu$ is the dominant decay mode in both models. As the charged Higgs boson width depends only on $\tan\beta$ and on the charged Higgs mass, the plot is representative of all values of $m_{H^\pm}$ provided that decays to other neutral scalars is forbidden. In the right panel of figure 2 we show the BR$(t \to H^+ b)$ as a function of $\tan\beta$ for two values of the charged Higgs boson mass. Contrary to the case of the MSSM and MSSM-like versions of a Type II 2HDM, this BR falls very rapidly with $\tan\beta$ and even more so as the charged Higgs boson mass approaches the top-quark mass.

Using this information we can now draw exclusion plots for the different 2HDM types. In the left panel of figure 3 we present the excluded region at the 95% CL in the ($\tan\beta$, $m_{H^\pm}$) plane for the Type X 2HDM model using the ATLAS predictions for 10 fb^{-1} and 30 fb^{-1} of total integrated luminosity for the LHC@14 TeV [19]. In the same figure we have drawn a point that corresponds to our parton level prediction for 30 fb^{-1} and $m_{H^\pm} = 100$ GeV presented in [21]. In the right panel of the same figure we show the excluded region in the ($\tan\beta$, $m_{H^\pm}$) plane for Type X at the 95% CL assuming the LHC@14 TeV and for several luminosity sets. The 100 GeV mass point at 30 fb^{-1} is also shown for a better comparison both with the ATLAS

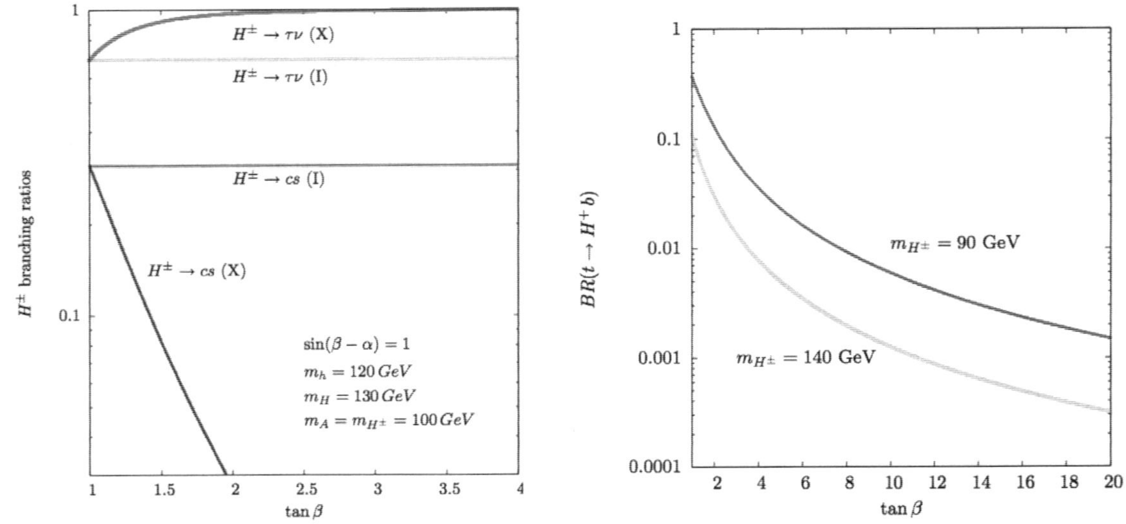

Figure 2. Left: charged Higgs BRs for $m_{H^\pm} = 100$ GeV as a function of $\tan\beta$ in models Type I and X. Right: BR($t \to H^+b$) as a function of $\tan\beta$ for two values of the charged Higgs boson mass.

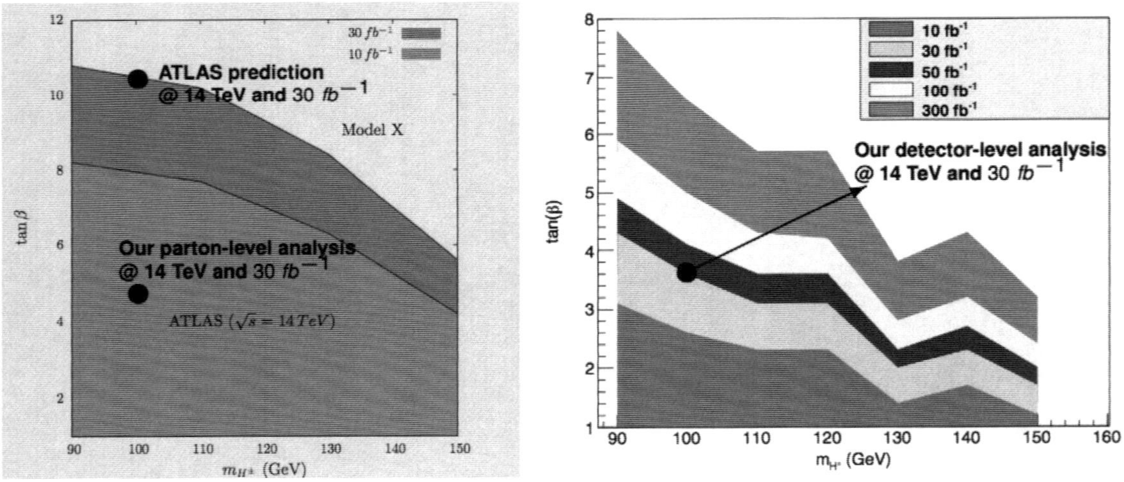

Figure 3. Left: Excluded region at the 95% CL in the ($\tan\beta$, $m_{H^\pm}$) plane for the Type X 2HDM using the ATLAS predictions for 10 fb^{-1} and 30 fb^{-1} of total integrated luminosity for the LHC@14 TeV [19]. Also shown is our parton level prediction for 30 fb^{-1} [21]. Right: excluded region in the ($\tan\beta$, $m_{H^\pm}$) plane for Type X at the 95% CL assuming the LHC@14 TeV and for several luminosity sets.

prediction and with our previous parton level study [21]. In the left panel of figure 4 we now present the excluded region for the Type I model again using the ATLAS predictions for 10 fb^{-1} and 30 fb^{-1} of total integrated luminosity for the LHC@14 TeV [19]. In the right panel we present our final detector level results for Type I at the 95% CL assuming the LHC@14 TeV and for several luminosity sets. We can conclude from the plots that as expected the results show similar trends to the ones obtained for $t\bar{t}$ production. We started with a cross section that is about three times smaller than the $t\bar{t}$ one and ended up with a result that is 2 to 3 times worse than the prediction presented by ATLAS [19].

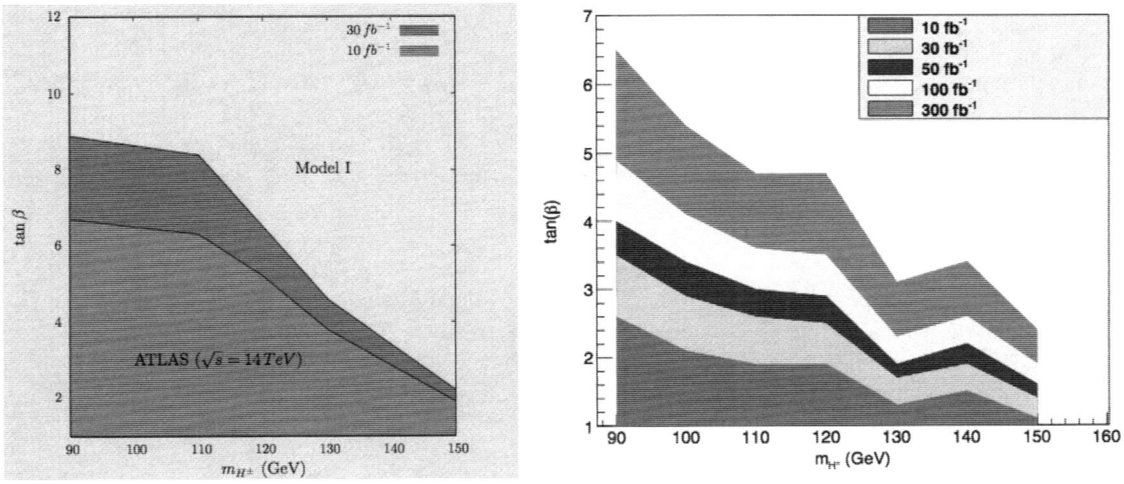

Figure 4. Left: excluded region at the 95% CL in the ($\tan\beta$, $m_{H^\pm}$) plane for the Type I 2HDM model using the ATLAS predictions for 10 fb^{-1} and 30 fb^{-1} of total integrated luminosity for the LHC@14 TeV [19]. Right: excluded region in the ($\tan\beta$, $m_{H^\pm}$) plane for Type I at the 95% CL assuming the LHC@14 TeV and for several luminosity sets.

It is expectable that both ATLAS and CMS will improve the results on the single top mode presented here, tightening the constraints on the $(m_{H^\pm}, \tan\beta)$ plane. One may now ask what are the chances to probe the entire $(m_{H^\pm}, \tan\beta)$ plane by the end of the 14 TeV run. In view of the results for 7 TeV [1, 2], one expects a type II charged Higgs to be excluded by then. However, there are models where the Yukawa couplings always decrease with $\tan\beta$ as is the case of models I and X. For those models, we know that $pp \to t\bar{t}$ will provide the strongest constraint on the $(m_{H^\pm}, \tan\beta)$ plane, and that the single top mode is bound to contribute even if only with a slight improvement. Are there any other processes that could help to probe the large $\tan\beta$ region?

There is another Yukawa process, $cs \to H^\pm(+j)$ [20, 21], that could in principle help to probe the above mentioned region. It was however shown to be negligible for large $\tan\beta$. The remaining possibility [21] is to look for processes that either do not depend on $\tan\beta$, or even better, that grow with $\tan\beta$. There are terms both in $gg \to H^+W^-$ and in vector boson fusion ($pp \to jjH^+H^-$ where j is a light quark jet) that are independent of $\tan\beta$. Furthermore, for the CP-conserving potential, there is a term in $gg \to H^+H^-$ that has the form

$$\sigma_{pp\to H^+H^-} \propto \sin(2\alpha)\ \tan\beta(m_H^2 - M^2) \tag{1}$$

where α is the rotation angle in the CP-even sector, m_H is the heavier CP-even scalar mass and $M^2 = m_{12}^2/(\sin\beta\,\cos\beta)$. Hence, there are regions of the 2HDM parameter space that can be probed for larger values of $\tan\beta$. However, the bounds will no longer be for a two parameter space but instead for a multi-dimension space with all 2HDM parameters playing a role. Further, values of the cross section that could lead to meaningful significances are only obtained for resonant production. Therefore, only a small portion of the multi-dimensional space can be probed for large $\tan\beta$ (see [21] for details).

A final comment about theoretical bounds. Assuming that the Higgs boson was discovered with a mass of 125 GeV, it was recently shown in [22] that for the particular case of an exact CP-conserving Z_2 symmetric model $\tan\beta < 6$. Therefore, that particular model will probably see a light charged Higgs ruled out when all the 8 TeV data is analysed.

Acknowledgments

SM is financed in part through the NExT Institute. The work of RG and RS is supported in part by the Portuguese *Fundação para a Ciência e a Tecnologia* (FCT) under contracts PTDC/FIS/117951/2010 and PEst-OE/FIS/UI0618/2011. RG is also supported by a FCT Grant SFRH/BPD/47348/2008. RS is also partially supported by an FP7 Reintegration Grant, number PERG08-GA-2010-277025.

References

[1] Aad G *et al.* [ATLAS Collaboration] 2012 Search for charged Higgs bosons decaying via $H^+ \to \tau\nu$ in top quark pair events using pp collision data at $\sqrt{s} = 7$ TeV with the ATLAS detector *J. High Energy Phys.* JHEP1206(2012)039

[2] Chatrchyan s *et al.* [CMS Collaboration] 2012 Search for a light charged Higgs boson in top quark decays in pp collisions at $\sqrt{s} = 7$ TeV *J. High Energy Phys.* JHEP1207(2012)143

[3] Branco G C, Ferreira P M, Lavoura L, Rebelo M N, Sher M and Silva J P 2012 Theory and phenomenology of two-Higgs-doublet models *Phys. Rept.* **516** 1

[4] Ferreira P M, Santos R and Barroso A 2004 Stability of the tree-level vacuum in two Higgs doublet models against charge or CP spontaneous violation *Phys. Lett. B* **603** 219 [Erratum-ibid. **629** 114]

[5] Ginzburg I F, Krawczyk M and Osland P 2002 Two Higgs doublet models with CP violation *Preprint* hep-ph/0211371

[6] El Kaffas A W, Khater W, Ogreid O M and Osland P 2007 Consistency of the two Higgs doublet model and CP violation in top production at the LHC *Nucl. Phys. B* **775** 45

[7] Barger V D, Hewett J L and Phillips R J N 1990 New Constraints On The Charged Higgs Sector In Two Higgs Doublet Models *Phys. Rev. D* **41** 3421

[8] Aoki M, Kanemura S, Tsumura K and Yagyu K 2009 Models of Yukawa interaction in the two Higgs doublet model, and their collider phenomenology *Phys. Rev. D* **80** 015017

[9] Guedes R, Moretti S and Santos R 2012 Charged Higgs bosons in single top production at the LHC *J. High Energy Phys.* JHEP1210(2012)119

Guedes R, Moretti S and Santos R 2012 Charged Higgs discovery potential in the single top mode in 2HDMs *Preprint* arXiv:1212.0893 [hep-ph]

[10] Mahmoudi F and Stal O 2010 Flavor constraints on the two-Higgs-doublet model with general Yukawa couplings *Phys. Rev. D* **81** 035016

El Kaffas A W, Osland P and Ogreid O M 2007 Constraining the Two-Higgs-Doublet-Model parameter space *Phys. Rev. D* **76** 095001

Basso L, Lipniacka A, Mahmoudi F, Moretti S, Osland P, Pruna G M and Purmohammadi M 2012 Probing the charged Higgs boson at the LHC in the CP-violating type-II 2HDM *J. High Energy Phys.* JHEP1211(2012)011

[11] Hermann T, Misiak M and Steinhauser M $\bar{B} \to X_s\gamma$ in the Two Higgs Doublet Model up to Next-to-Next-to-Leading Order in QCD *J. High Energy Phys.* JHEP1211(2012)036

[12] Abbiendi G *et al.* [ALEPH and DELPHI and L3 and OPAL and The LEP working group for Higgs boson searches Collaborations] 2013 Search for Charged Higgs bosons: Combined Results Using LEP Data *Preprint* [arXiv:1301.6065 [hep-ex]]

[13] Alioli S, Nason P, Oleari C and Re E 2010 A general framework for implementing NLO calculations in shower Monte Carlo programs: the POWHEG BOX *J. High Energy Phys.* JHEP1006(2010)043

Alioli S, Nason P, Oleari C and Re E 2009 NLO single-top production matched with shower in POWHEG: s- and t-channel contributions *J. High Energy Phys.* JHEP0909(2009)111 [Erratum-ibid. **1002** (2010) 011]

Re E 2011 Single-top Wt-channel production matched with parton showers using the POWHEG method *Eur. Phys. J. C* **71** 1547

Frixione S, Nason P and Ridolfi G 2007 A Positive-weight next-to-leading-order Monte Carlo for heavy flavour hadroproduction *J. High Energy Phys.* JHEP0709(2007)126

Nason P 2004 A New method for combining NLO QCD with shower Monte Carlo algorithms *J. High Energy Phys.* JHEP0411 (2004) 040

Frixione S, Nason P and Oleari C 2007 Matching NLO QCD computations with Parton Shower simulations: the POWHEG method *J. High Energy Phys.* JHEP0711(2007)070

[14] Nadolsky P M *et al.* 2008 Implications of CTEQ global analysis for collider observables *Phys. Rev. D* **78** 013004

[15] Sjostrand T, Mrenna S and Skands P Z 2006 PYTHIA 6.4 Physics and Manual *J. High Energy Phys.* JHEP0605(2006)026

[16] Mangano M L, Moretti M, Piccinini F, Pittau R and Polosa A D 2003 ALPGEN, a generator for hard multiparton processes in hadronic collisions *J. High Energy Phys.* JHEP0307(2003)001

[17] Pukhov A 2004 CalcHEP 2.3: MSSM, structure functions, event generation, batchs, and generation of matrix elements for other packages *Preprint* arXiv:hep-ph/0412191

[18] Ovyn S, Rouby X and Lemaitre V 2009 Delphes, a framework for fast simulation of a generic collider experiment *Preprint* arXiv:0903.2225 [hep-ph]

[19] Aad G *et al.* [ATLAS Collaboration] 2009 Expected Performance of the ATLAS Experiment - Detector, Trigger and Physics *Preprint* arXiv:0901.0512 [hep-ex]

[20] He H J and Yuan C P 1998 New method for detecting charged (pseudo-)scalars at colliders *Phys. Rev. Lett* **83** 28

Slabospitsky S R 2002 Study of s channel charged Higgs production in CMS *Preprint* arXiv:hep-ph/0203094

[21] Aoki M, Guedes R, Kanemura S, Moretti S, Santos R and Yagyu K 2011 Light Charged Higgs bosons at the LHC in 2HDMs *Phys. Rev. D* **84** 055028

Guedes R, Kanemura S, Moretti S, Santos R and Yagyu K 2010 Charged Higgs Boson Benchmarks in the CP-conserving 2HDM *Proc. of Sci.* PoS CHARGED(2010)037

[22] Gorczyca B, Krawczyk M 2012 *Preprint* arXiv:1112.5086 [hep-ph]

Journal of Physics: Conference Series **447** (2013) 012058 doi:10.1088/1742-6596/447/1/012058

$B \to D^{(*)}\tau\nu_\tau$ decays in two-Higgs-doublet models

Alejandro Celis[1], **Martin Jung**[2], **Xin-Qiang Li**[3,1], **Antonio Pich**[1]

[1] IFIC, Universitat de València – CSIC, Apt. Correus 22085, E-46071 València, Spain

[2] Institut für Physik, Technische Universität Dortmund, D-44221 Dortmund, Germany

[3] Physics Department, Henan Normal University, Xinxiang, Henan 453007, P. R. China

E-mail: alejandro.celis@ific.uv.es, martin2.jung@tu-dortmund.de, xqli@itp.ac.cn, antonio.pich@ific.uv.es

Abstract. A sizable excess with respect to the SM expectation has been reported recently by the BaBar collaboration in the decay rates $B \to D^{(*)}\tau\nu$, normalized by the corresponding light lepton modes. A violation of lepton flavor universality as suggested by this excess could be due to a charged Higgs mediating these processes at tree level. In this talk we analyze the implications of the observed excess within the framework of two-Higgs-doublet models, considering also the bounds from other semileptonic and leptonic decays of B and $D_{(s)}$ mesons. Prospects for $B \to D^{(*)}\tau\nu$ decays at future Super-Flavor Factories are also discussed.

1. Introduction

The hypothesis that a scalar sector in nature is behind the origin of particle masses (or the mechanism of spontaneous electroweak symmetry breaking) is starting to look even more plausible since the discovery of a new boson by the ATLAS and CMS collaborations with characteristics resembling those of the Standard Model (SM) Higgs boson [1,2]. The SM scalar sector is composed of one doublet of complex scalar fields, containing in total four real degrees of freedom; three of them give mass to the $W^\pm$ and Z bosons, while the remaining real field becomes a physical scalar, the Higgs boson. In the SM the fermions also acquire their masses from the spontaneous breaking of the electroweak symmetry due to their interactions with the Higgs.

There is nothing fundamental a priori, however, preventing the scalar sector to have a richer structure, for example that of a Two-Higgs-Doublet Model (2HDM), where two doublets of complex scalar fields are responsible for the breaking of the electroweak symmetry. The physical scalar spectrum includes in this case three neutral bosons and one charged scalar particle. The boson discovered at the LHC could be in principle any of the three neutral bosons of a 2HDM.

The BaBar collaboration has recently reported an excess of events in two semileptonic transitions of the type $b \to c\tau\nu_\tau$. More specifically, they have measured the ratios [3]

$$R(D) \equiv \frac{\mathrm{Br}(\bar{B} \to D\tau^-\bar{\nu}_\tau)}{\mathrm{Br}(\bar{B} \to D\ell^-\bar{\nu}_\ell)} \overset{\text{BaBar}}{=} 0.440 \pm 0.058 \pm 0.042 \overset{\text{avg.}}{=} 0.438 \pm 0.056\,,$$

$$R(D^*) \equiv \frac{\mathrm{Br}(\bar{B} \to D^*\tau^-\bar{\nu}_\tau)}{\mathrm{Br}(\bar{B} \to D^*\ell^-\bar{\nu}_\ell)} \overset{\text{BaBar}}{=} 0.332 \pm 0.024 \pm 0.018 \overset{\text{avg.}}{=} 0.354 \pm 0.026\,, \quad (1)$$

which are normalized to the corresponding decays into light leptons $\ell = e, \mu$. The second value given in each line is the average with the previous measurements by the Belle collaboration [4, 5]. The decays $B \to D^{(*)}\tau\nu$ involve the heaviest fermions that can be directly produced at flavor factories and constitute a well-suited place to look for new physics effects related to the mechanism of electroweak symmetry breaking. A charged Higgs arising from a 2HDM can in principle cause significant deviations from the SM in semileptonic and leptonic decays where it enters at tree level. Moreover, a charged Higgs leads naturally to violations of lepton flavor universality due to the characteristic mass dependence of its couplings to fermions. Interestingly, a charged Higgs boson could also produce an enhancement in the di-photon decay of the neutral Higgs, as currently observed by the ATLAS and CMS collaborations [1, 2].

In this talk we are mainly interested in addressing two questions:

A) What is the required flavor structure for a 2HDM in order to accommodate the observed excess in $R(D^{(*)})$, as well as other measured semileptonic and leptonic decays of B and $D_{(s)}$ mesons?

B) How can we discriminate between different new physics scenarios that can explain the excess in $R(D)$ and $R(D^*)$?

To answer the first question, we analyze in Section 2 the bounds on the charged Higgs couplings from different semileptonic and leptonic decays where the charged Higgs contribution enters at tree level. The second question is considered in Section 3, where we study the effect of charged scalar contributions in new observables involving angular distributions, polarizations of the final τ and D^* meson, as well as the momentum-transfer dependence. Our discussion will be focused on results presented in [6, 7]. These questions have also been addressed recently by different authors [8–16].

2. Explaining the excess within the 2HDM

Four-fermion interactions mediated by a charged scalar are described by the Lagrangian

$$\mathcal{L}_{\rm eff} = -\frac{4G_F}{\sqrt{2}} \sum_{q=u,c} V_{qb} \sum_{l=e,\mu,\tau} \left\{ [\bar{q}\gamma^\mu \mathcal{P}_L b]\left[\bar{l}\gamma_\mu \mathcal{P}_L \nu_l\right] + \left[\bar{q}\left(g_L^{qbl}\mathcal{P}_L + g_R^{qbl}\mathcal{P}_R\right)b\right]\left[\bar{l}\mathcal{P}_L\nu_l\right]\right\}, \tag{2}$$

as long as the charged Higgs is heavy enough to be integrated out. This is a very good approximation, given that direct searches at LEP for charged Higgs bosons have set the lower limit $m_{H^\pm} \gtrsim 80$ GeV [17], assuming that it couples to fermions. In Equation (2), V_{qb} refers to the corresponding element of the CKM mixing matrix, G_F is the Fermi constant and $\mathcal{P}_{L,R}$ are the left- and right-handed chiral projectors, respectively. The couplings of the charged Higgs to fermions are contained in the dimensionless parameters g_L^{qbl} and g_R^{qbl}, which depend on the assumed Yukawa structure of the 2HDM. In the following we will treat them as independent complex quantities to keep the discussion as general as possible. Explicit expressions for g_L^{qbl} and g_R^{qbl} in terms of the general parametrization of the Aligned 2HDM (A2HDM) [18] are provided in [7]. The different types of 2HDMs based on discrete $\mathcal{Z}_2$ symmetries can be obtained as particular limits of this parametrization.

Due to the different spins of the D and D^* mesons, the parameters g_L^{qbl} and g_R^{qbl} appear in different linear combinations in $R(D)$ and $R(D^*)$,

$$\delta_{cb}^l \equiv \frac{(g_L^{cbl} + g_R^{cbl})(m_B - m_D)^2}{m_l(\overline{m}_b - \overline{m}_c)}, \qquad \Delta_{qb}^l = \frac{(g_L^{qbl} - g_R^{qbl})m_B^2}{m_l(\overline{m}_b + \overline{m}_q)}. \tag{3}$$

$R(D)$ depends only on the scalar combination δ_{cb}^l while $R(D^*)$ only depends on the pseudo-scalar one Δ_{qb}^l [7]. In versions of the 2HDM were the fermionic charged Higgs couplings are

Table 1. *Predictions within the SM for the various semileptonic and leptonic decays considered in this work, together with their corresponding experimental values. The first uncertainty given always corresponds to the statistical error, and the second, when given, to the theoretical one.*

Observable	SM Prediction	Exp. Value
$R(D)$	$0.296^{+0.008}_{-0.006} \pm 0.015$	0.438 ± 0.056
$R(D^*)$	$0.252 \pm 0.002 \pm 0.003$	0.354 ± 0.026
$\mathrm{Br}(B \to \tau\nu_\tau)$	$(0.79^{+0.06}_{-0.04} \pm 0.08) \times 10^{-4}$	$(1.15 \pm 0.23) \times 10^{-4}$
$\mathrm{Br}(D_s \to \tau\nu_\tau)$	$(5.18 \pm 0.08 \pm 0.17) \times 10^{-2}$	$(5.54 \pm 0.24) \times 10^{-2}$
$\mathrm{Br}(D_s \to \mu\nu)$	$(5.31 \pm 0.09 \pm 0.17) \times 10^{-3}$	$(5.54 \pm 0.24) \times 10^{-3}$
$\mathrm{Br}(D \to \mu\nu)$	$(4.11^{+0.06}_{-0.05} \pm 0.27) \times 10^{-4}$	$(3.76 \pm 0.18) \times 10^{-4}$
$\Gamma(K \to \mu\nu)/\Gamma(\pi \to \mu\nu)$	$1.333 \pm 0.004 \pm 0.026$	1.337 ± 0.003
$\Gamma(\tau \to K\nu_\tau)/\Gamma(\tau \to \pi\nu_\tau)$	$(6.56 \pm 0.02 \pm 0.15) \times 10^{-2}$	$(6.46 \pm 0.10) \times 10^{-2}$

proportional to fermion masses and the proportionality factor is the same for all families (family universality), the following relations hold:

$$\frac{g_L^{q_u q_d l}}{g_L^{q'_u q'_d l'}} = \frac{m_{q_u} m_l}{m_{q'_u} m_{l'}}, \qquad \frac{g_R^{q_u q_d l}}{g_R^{q'_u q'_d l'}} = \frac{m_{q_d} m_l}{m_{q'_d} m_{l'}}. \tag{4}$$

The Aligned Two-Higgs doublet model as well as the different types of 2HDMs based on discrete $\mathcal{Z}_2$ symmetries (Types I, II, X and Y) belong to this class of models [19]. To analyze possible deviations of the hypothesis of family universality in the context of 2HDMs we consider three scenarios with different flavor structures:

- Scenario 1: The Yukawa couplings $g_{L,R}^{q_u q_d l}$ are assumed to be independent for the different transitions ($b \to u, c$; $c \to d \ldots$). We therefore fit only $R(D)$ and $R(D^*)$.
- Scenario 2: We assume that the hypothesis of family universality (4) holds for $q_d = b$ and any up-type quark. The charged scalar contribution to $b \to c$ and $b \to u$ transitions are then related. In this case we consider the bounds from $R(D^{(*)})$ and $\mathrm{Br}(B \to \tau\nu)$.
- Scenario 3: The hypothesis of family universality (4) is assumed to hold for all quark generations. In this case we consider in the fit all observables in Table 1 except for $R(D^*)$. If $R(D^*)$ is included in the fit no allowed region is found at 95% CL.

Experimental values and SM predictions for the different observables used are listed in Table 1 [7]. In Figure 1 we show 95% CL allowed regions in the parameters δ^l_{cb} and Δ^l_{cb} for the three scenarios. We can now give an answer to question A.

All B decays, $R(D^{(*)})$ and $B \to \tau\nu_\tau$, can be accommodated within a general 2HDM with family universality for $q_d = b$ and any up-type quark; i.e. in scenarios 1 and 2. However this is not possible in the usual Type II 2HDM; a more general flavor structure is needed in which the charged scalar couplings to up-type quarks are not as suppressed [7].

It is not possible to accommodate all the observables in Table 1 simultaneously in a 2HDM with family universality between all fermion generations (scenario 3). Since charged scalar contributions to $R(D^*)$ are subdominant compared with the vector contributions, large Yukawa couplings are required to explain the observed excess in $R(D^*)$, thus generating a tension with present bounds from $D_{(s)}$ leptonic decays. If $R(D^*)$ is not considered in the fit, all observables can be accommodated within the family universal A2HDM. Should the present measurement be confirmed, a departure from the hypothesis of flavor universality is therefore needed in order to account for all semileptonic and leptonic decays of B and $D_{(s)}$ mesons. Bounds from loop induced processes discussed in [6], also point towards a violation of family universality in the coupling with the top quark, once the present excess in $R(D^{(*)})$ is considered [7].

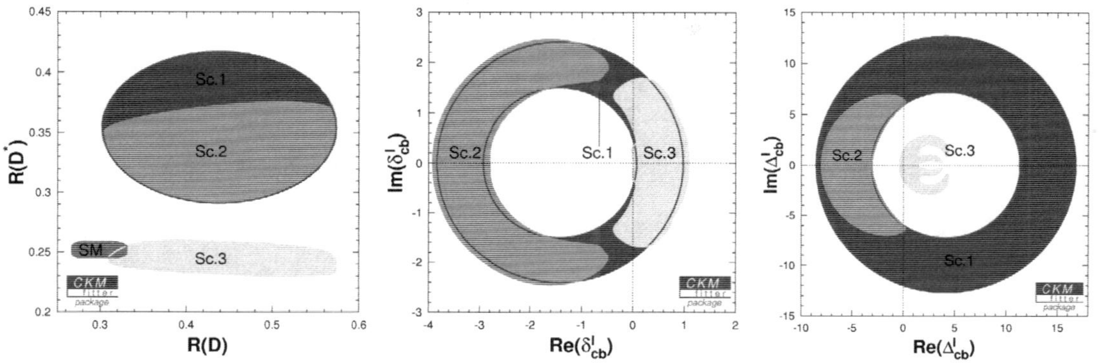

Figure 1. *Allowed regions in the $R(D)$–$R(D^*)$ (left), complex δ^l_{cb} (center) and Δ^l_{cb} (right) planes at 95% CL, corresponding to the three different scenarios. See text for details.*

3. Sensitivity of new observables to charged scalar contributions

Using only measurements of branching fractions, it is not possible to disentangle charged scalar contributions in $B \to D^{(*)}\tau\nu$ from other kinds of new physics. However, $B \to D^{(*)}\tau\nu$ decays have a rich three-body kinematics and spin structure in the final state that has not been exploited so far. Several observables involving angular distributions, polarization fractions and momentum-transfer dependence, would provide, if measured, crucial information to discriminate between different new physics scenarios and clarify the possible role of charged scalar contributions in these processes [8, 10–12, 14, 20–24].

One way to test the observed excess in $R(D^*)$ would be to measure the leptonic decay $B_c \to \tau\nu$, which has the same flavor structure $b \to c\tau\nu_\tau$ and is sensitive to the same combination of Yukawa couplings. Including the charged scalar contribution, the decay width is given by

$$\Gamma(B_c \to \tau\nu_\tau) \;=\; G_F^2 m_\tau^2 f_{B_c}^2 |V_{cb}|^2 \,\frac{m_{B_c}}{8\pi}\left(1 - \frac{m_\tau^2}{m_{B_c}^2}\right)^2 |1 - \Delta^\tau_{cb}|^2\,, \tag{5}$$

where f_{B_c} is the B_c leptonic decay constant. Recent Lattice QCD calculations find $f_{B_c} = 0.427(6)(2)$ GeV, giving the SM prediction $\mathrm{Br}(B_c \to \tau\nu) = 0.0194(18)$ [25]. Large enhancements of $\mathrm{Br}(B_c \to \tau\nu)$ should be observed in principle if the current excess in $R(D^*)$ is due to charged scalar contributions.

Information about the momentum-transfer distributions of $R(D)$ and $R(D^*)$,

$$R_{D^{(*)}}(q^2) \;=\; \frac{d\mathrm{Br}(\bar{B} \to D^{(*)}\tau^-\bar{\nu}_\tau)/dq^2}{d\mathrm{Br}(\bar{B} \to D^{(*)}\ell^-\bar{\nu}_\ell)/dq^2}\,, \tag{6}$$

would be very useful to determine what kind of new physics could be responsible for the observed excess. In Figure 2 we show the predictions for the q^2 distributions in the SM and the three scenarios considered in the previous section. Given that charged scalars do not not affect helicity amplitudes with transversely polarized D^* mesons, the measurement of these distributions for longitudinally polarized D^* may provide a better sensitivity to scalar effects

$$R_L^*(q^2) \;=\; \frac{d\Gamma^L_\tau/dq^2}{d\Gamma^L_\ell/dq^2}\,. \tag{7}$$

A charged Higgs mediating $B \to D^{(*)}\tau\nu$ decays would also have an effect on the polarization of the final τ lepton. In the SM, the W^- couples exclusively to left-handed τ^- while a charged

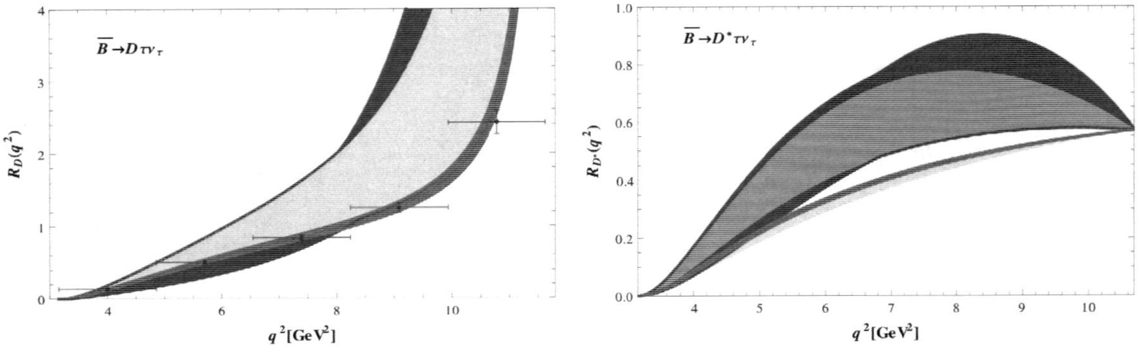

Figure 2. *The q^2 dependence of the ratios $R_{D^{(*)}}(q^2)$, both within the SM (gray) as well as in scenarios 1 (red), 2 (orange), and 3 (yellow). The 95% CL allowed regions are obtained with the corresponding experimental constraints exerted for the three different scenarios, and with the hadronic uncertainties added in quadrature for the SM. The five-binned distribution for $R_D(q^2)$ in the SM is also shown.*

Table 2. *Predictions for the q^2-integrated observables both within the SM and in the different scenarios.*

Observable	SM Prediction	Scenario 1	Scenario 2	Scenario 3
$R_L(D^*)$	$0.115 \pm 0.001 \pm 0.003$	0.217 ± 0.026	$0.223^{+0.013}_{-0.026} \pm 0.006$	$0.104^{+0.006}_{-0.003} \pm 0.003$
$A_\lambda(D)$	$-0.304 \pm 0.001 \pm 0.035$	$-0.55^{+0.10}_{-0.04}$	$-0.55^{+0.09}_{-0.04}$	$-0.55^{+0.09}_{-0.04} \pm 0.01$
$A_\lambda(D^*)$	$0.502^{+0.005}_{-0.006} \pm 0.017$	$0.06^{+0.10}_{-0.06}$	$0.04^{+0.10}_{-0.03} \pm 0.01$	$0.57^{+0.04}_{-0.02} \pm 0.02$
$A_\theta(D)$	$0.3602^{+0.0006}_{-0.0007} \pm 0.0022$	$0.03^{+0.01}_{-0.00} \pm 0.30$	$-0.21^{+0.13}_{-0.00} \pm 0.06$	$0.36^{+0.01\dagger}_{-0.09}$
$A_\theta(D^*)$	$-0.066 \pm 0.006 \pm 0.009$	$-0.136^{+0.012}_{-0.003} \pm 0.222$	$0.081^{+0.008}_{-0.059} \pm 0.009$	$-0.146^{+0.039}_{-0.017} \pm 0.021$

Higgs H^- couples to right-handed τ^-. Due to their short lifetime, τ leptons decay within the detector and their polarization can be determined by analyzing the kinematics of the decay products. One could measure a τ spin asymmetry defined in the τ-$\bar{\nu}_\tau$ center-of-mass frame as

$$A_\lambda^{D^{(*)}}(q^2) = \frac{d\Gamma^{D^{(*)}}[\lambda_\tau = -1/2]/dq^2 - d\Gamma^{D^{(*)}}[\lambda_\tau = +1/2]/dq^2}{d\Gamma^{D^{(*)}}[\lambda_\tau = -1/2]/dq^2 + d\Gamma^{D^{(*)}}[\lambda_\tau = +1/2]/dq^2}\,. \tag{8}$$

Angular distributions are also sensitive to the Dirac structure of possible new physics mediating $b \to c\tau\nu$ transitions. A useful observable is the forward-backward asymmetry, defined as the difference of partial decay rates with the angle θ between the $D^{(*)}$ and τ three-momenta in the τ-$\bar{\nu}_\tau$ center-of-mass frame greater or smaller than $\pi/2$:

$$A_\theta^{D^{(*)}}(q^2) = \frac{\int_{-1}^{0} d\cos\theta\,(d^2\Gamma_\tau^{D^{(*)}}/dq^2 d\cos\theta) - \int_0^1 d\cos\theta\,(d^2\Gamma_\tau^{D^{(*)}}/dq^2 d\cos\theta)}{d\Gamma_\tau^{D^{(*)}}/dq^2}\,. \tag{9}$$

In Table 2 we show predictions for the observables discussed previously in the SM as well as in the three scenarios considered. Correlations between the different observables can be observed in Figure 3.

An unambiguous way to test the hypothesis of a charged scalar mediating B semileptonic decays is to look for observables that are independent of such contributions. A deviation of these observables from the SM prediction could then only be attributed to new physics different

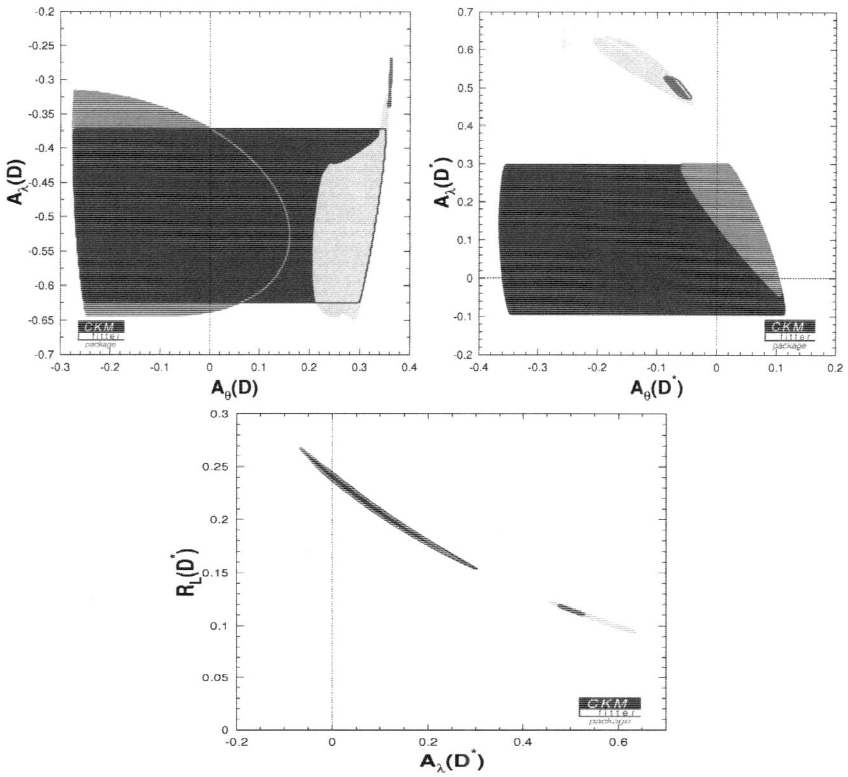

Figure 3. *In the upper panel we show the prediction for $A_\theta(D^{(*)})$ vs. $A_\lambda(D^{(*)})$ for the SM (gray), and the three scenarios (1-red, 2-orange, 3-yellow), from a global fit including all appropriate observables. In the lower panel we show the predictions for $A_\lambda(D^*)$ vs. $R_L(D^*)$.*

from a charged scalar. One such observable is the ratio of leptonic B decay rates into τ and μ,

$$\frac{\mathrm{Br}(B\to\tau\nu)}{\mathrm{Br}(B\to\mu\nu)} = \frac{m_\tau^2}{m_\mu^2}\left(\frac{1-m_\tau^2/m_B^2}{1-m_\mu^2/m_B^2}\right)^2, \tag{10}$$

which is valid for B_u and B_c. Some combinations of observables that are insensitive to charged scalar contributions in $B\to D^{(*)}\tau\nu_\tau$ transitions were defined in [7]:

$$X_1(q^2) \equiv R_{D^*}(q^2) - R_L^*(q^2)\,, \qquad X_2^{D^{(*)}}(q^2) \equiv R_{D^{(*)}}(q^2)\left[A_\lambda^{D^{(*)}}(q^2)+1\right]. \tag{11}$$

4. Summary

$B\to D^{(*)}\tau\nu$ decays are an excellent laboratory to look for new physics effects related to the mechanism of electroweak symmetry breaking. They involve the heaviest fermions that can be directly produced at flavor factories. They also possess a rich three-body kinematics and spin structure that could be fully exploited at future Super-Flavor Factories, providing important information about possible new physics mediating these transitions.

In this talk we presented an analysis of charged scalar contributions to these processes and then derived some implications for the flavor structure of 2HDMs in order to explain the excess observed by BaBar [3]. We showed that, in contrast to the 2HDMs with $\mathcal{Z}_2$ symmetry, the

Journal of Physics: Conference Series **447** (2013) 012058 doi:10.1088/1742-6596/447/1/012058

A2HDM can explain the present data on $R(D^{(*)})$ and $B \to \tau\nu$, but that the excess observed in $R(D^*)$ is not compatible with current bounds from leptonic $D_{(s)}$ decays within this framework. If the present excess in $B \to D^{(*)}\tau\nu$ is confirmed, it will be necessary to consider a departure from the family universality of the Yukawa couplings in order to accommodate all data from B and $D_{(s)}$ decays within the framework of 2HDMs.

New theoretical developments on the calculation of the hadronic matrix elements would improve our knowledge of the SM prediction for these decays and shed light on a possible new-physics interpretation of the observed excess; for recent work on the determination of $B \to D$ hadronic matrix elements, see [26, 27]. Certainly new Belle measurements are needed to clarify this puzzle. Preliminary efforts to measure this decays at LHCb have already been done [28]. We have also discussed new observables in $b \to c\tau\nu$ transitions that could be measured at future Super-Flavor factories and test the presence of charged scalars mediating these decays; these involve angular distributions, polarization of τ and D^* as well as the q^2 dependence.

Acknowledgments

This work has been supported in part by the Spanish Government [grants FPA2007-60323, FPA2011-23778 and CSD2007-00042 (Consolider Project CPAN)]. X.Q.L. is also supported in part by the National Natural Science Foundation of China under contract No. 11005032, the Specialized Research Fund for the Doctoral Program of Higher Education of China (Grant No. 20104104120001) and SRF for ROCS, SEM. M.J. is supported by the Bundesministerium für Bildung und Forschung. The work of A.C. is funded through an FPU grant (AP2010-0308, MINECO, Spain).

References

[1] G. Aad *et al.* [ATLAS Collaboration], "Observation of a new particle in the search for the Standard Model Higgs boson with the ATLAS detector at the LHC," Phys. Lett. B **716** (2012) 1 (*Preprint* hep-ex/1207.7214)

[2] S. Chatrchyan *et al.* [CMS Collaboration], "Observation of a new boson at a mass of 125 GeV with the CMS experiment at the LHC," Phys. Lett. B **716** (2012) 30 (*Preprint* hep-ex/1207.7235)

[3] J. P. Lees *et al.* [BaBar Collaboration], "Evidence for an excess of $\bar{B} \to D^{(*)}\tau^-\bar{\nu}_\tau$ decays," Phys. Rev. Lett. **109** (2012) 101802 (*Preprint* hep-ex/1205.5442)

[4] I. Adachi *et al.* [Belle Collaboration], "Measurement of B $\to D^{(*)}\tau\nu$ using full reconstruction tags," (*Preprint* hep-ex/0910.4301)

[5] A. Bozek *et al.* [Belle Collaboration], "Observation of $B^+ \to \bar{D}^{*0}\tau^+\nu_\tau$ and Evidence for $B^+ \to \bar{D}^0\tau^+\nu_\tau$ at Belle," Phys. Rev. D **82** (2010) 072005 (*Preprint* hep-ex/1005.2302)

[6] M. Jung, A. Pich and P. Tuzon, "Charged-Higgs phenomenology in the Aligned two-Higgs-doublet model," JHEP **1011**, 003 (2010) (*Preprint* hep-ph/1006.0470)

[7] A. Celis, M. Jung, X. -Q. Li and A. Pich, "Sensitivity to charged scalars in $B \to D^{(*)}\tau\nu_\tau$ and $B \to \tau\nu_\tau$ decays," JHEP **1301** (2013) 054 (*Preprint* hep-ph/1210.8443)

[8] U. Nierste, S. Trine and S. Westhoff, "Charged-Higgs effects in a new B $\to$ D tau nu differential decay distribution," Phys. Rev. D **78** (2008) 015006 (*Preprint* hep-ph/0801.4938)

[9] J. F. Kamenik and F. Mescia, "$B \to D\tau\nu$ Branching Ratios: Opportunity for Lattice QCD and Hadron Colliders," Phys. Rev. D **78** (2008) 014003 (*Preprint* hep-ph/0802.379)

[10] M. Tanaka and R. Watanabe, "Tau longitudinal polarization in B $\to$ D tau nu and its role in the search for charged Higgs boson," Phys. Rev. D **82** (2010) 034027 (*Preprint* hep-ph/1005.4306)

[11] S. Fajfer, J. F. Kamenik and I. Nisandzic, "On the $B \to D^*\tau\bar{\nu}_\tau$ Sensitivity to New Physics," Phys. Rev. D **85** (2012) 094025 (*Preprint* hep-ph/1203.2654)

[12] Y. Sakaki and H. Tanaka, "Constraints of the Charged Scalar Effects Using the Forward-Backward Asymmetry on $B \to D^{(*)}\tau\bar{\nu}_\tau$," (*Preprint* hep-ph/1205.4908)

[13] S. Fajfer, J. F. Kamenik, I. Nisandzic and J. Zupan, "Implications of lepton flavor universality violations in B decays," Phys. Rev. Lett. **109** (2012) 161801 (*Preprint* hep-ph/1206.1872)

[14] A. Datta, M. Duraisamy and D. Ghosh, "Diagnosing New Physics in $b \to c\,\tau\,\nu_\tau$ decays in the light of the recent BaBar result," Phys. Rev. D **86** (2012) 034027 (*Preprint* hep-ph/1206.3760)

[15] A. Crivellin, C. Greub and A. Kokulu, "Explaining $B \to D\tau\nu$, $B \to D^*\tau\nu$ and $B \to \tau\nu$ in a 2HDM of type III," Phys. Rev. D **86** (2012) 054014 (*Preprint* hep-ph/1206.2634)

[16] M. Tanaka and R. Watanabe, "New physics in the weak interaction of $\bar{B} \to D^{(*)}\tau\bar{\nu}$," (*Preprint* hep-ph/1212.1878)

[17] [LEP Higgs Working Group for Higgs boson searches and ALEPH and DELPHI and L3 and OPAL Collaborations], "Search for charged Higgs bosons: Preliminary combined results using LEP data collected at energies up to 209-GeV," (*Preprint* hep-ex/0107031).

[18] A. Pich, P. Tuzón, "Yukawa Alignment in the Two-Higgs-Doublet Model," Phys. Rev. D **80** (2009) 091702 (*Preprint* hep-ph/0908.1554)

[19] For a review, see for example: G. C. Branco, P. M. Ferreira, L. Lavoura, M. N. Rebelo, M. Sher, J. P. Silva, "Theory and phenomenology of two-Higgs-doublet models," Phys. Rept. **516** (2012) 1 (*Preprint* hep-ph/1106.0034); J. F. Gunion, H. E. Haber, G. L. Kane, S. Dawson, "The Higgs Hunter's Guide," Front. Phys. **80** (2000) 1.

[20] K. Hagiwara, A. D. Martin and M. F. Wade, "HELICITY AMPLITUDE ANALYSIS OF B $\to D^*$ lepton neutrino DECAYS," Phys. Lett. B **228** (1989) 144, "Exclusive Semileptonic B Meson Decays," Nucl. Phys. B **327** (1989) 569.

[21] J. G. Korner and G. A. Schuler, "Exclusive Semileptonic Heavy Meson Decays Including Lepton Mass Effects," Z. Phys. C **46** (1990) 93, "Exclusive Semileptonic Decays of Bottom Mesons in the Spectator Quark Model," ibid. C **38** (1988) 511 [Erratum-ibid. C **41** (1989) 690].

[22] M. Tanaka, "Charged Higgs effects on exclusive semitauonic B decays," Z. Phys. C **67** (1995) 321 (*Preprint* hep-ph/9411405)

[23] C. -H. Chen and C. -Q. Geng, "Lepton angular asymmetries in semileptonic charmful B decays," Phys. Rev. D **71** (2005) 077501 (*Preprint* hep-ph/0503123)

[24] C. -H. Chen and C. -Q. Geng, "Charged Higgs on $B \to \tau\bar{\nu}_\tau$ and $B \to P(V)\ell\bar{\nu}_\ell$," JHEP **0610** (2006) 053 (*Preprint* hep-ph/0608166)

[25] C. McNeile, C. T. H. Davies, E. Follana, K. Hornbostel and G. P. Lepage, "Heavy meson masses and decay constants from relativistic heavy quarks in full lattice QCD," Phys. Rev. D **86** (2012) 074503 (*Preprint* hep-lat/1207.0994)

[26] D. Becirevic, N. Kosnik and A. Tayduganov, "$\bar{B} \to D\tau\bar{\nu}_\tau$ vs. $\bar{B} \to D\mu\bar{\nu}_\mu$," Phys. Lett. B **716** (2012) 208 (*Preprint* hep-ph/1206.4977)

[27] J. A. Bailey, A. Bazavov, C. Bernard, C. M. Bouchard, C. DeTar, D. Du, A. X. El-Khadra and J. Foley *et al.*, "Refining new-physics searches in $B \to D\tau\nu$ decay with lattice QCD," Phys. Rev. Lett. **109** (2012) 071802 (*Preprint* hep-ph/1206.4992)

[28] M. John, "Prospects on $B \to D^*\tau\nu_\tau$ at LHCb," talk at Workshop on B decay into D^{**} and related issues (2012).

Supersymmetric Lepton Flavour Violation and Neutrinos mass textures: an update.

Mario E. Gómez, Mirco Cannoni

Department of Applied Physics, University of Huelva, 21071 Huelva, Spain

E-mail: mario.gomez@dfa.uhu.es, mirco.cannini@dfa.uhu.es

Abstract.
We revise the prospects for the observation of charged lepton flavour violation (LFV), in the light of the recent results from the LHC, MEG and neutrino experiments. We work in the context of the Minimal Supersymmetric Standard Model (MSSM) extended with massive neutrinos arising from a *see-saw* mechanism. The connection between the observed neutrino oscillations with flavour oscillations in the charged sector is established using a model for the Yukawa couplings arising from a SU(5) grand unified theory (GUT) with Abelian flavour symmetries. We discuss in this scenario the possible observation on the radiative decays $\mu \to e\gamma$ and $\tau \to \mu\gamma$ and LFV processes that could be detectable at a linear collider (LC) with a centre-of-mass energy in the TeV range.

1. Introduction

The picture of neutrino masses and oscillations has recently improved by the evidence for a non-zero value of θ_{13} as found by several reactor experiments [1, 2, 3]. The explanation for neutrino observations has to arise from theories beyond the standard model (SM). These theories also predict charged LFV. The link between neutrino oscillations and the violations of the individual lepton numbers $L_{e,\mu,\tau}$ offers the possibility of observing processes such as $l_i \to l_j\gamma$ $(i \neq j)$ [4]. The present experimental upper limits, summarised below, already constrain significantly the parameter space of theoretical models,

$$BR(\mu \to e\gamma) < 5.7 \times 10^{-12} \text{ [5]}, \quad (1)$$

$$BR(\tau \to \mu\gamma) < 4.4 \times 10^{-8} \text{ [6]}, \quad (2)$$

$$BR(\tau \to e\gamma) < 3.3 \times 10^{-8} \text{ [6]}. \quad (3)$$

The strongest constraint on radiative decays is the latest MEG upper limit on BR($\mu \to e\gamma$) [5].

Supersymmetric (SUSY) theories connect the problem of SM fermions' flavour with the physics of their scalar partners. Experimental evidence on the absence or smallness of charged LFV phenomena seems to indicate that SUSY partners acquire masses with a flavour independent SUSY breaking mechanism. Models like the constrained minimal supersymmetric standard model (CMSSM) with universal scalar, gaugino masses and trilinear terms at the GUT scale extend the SM with an economical amount of undetermined parameters and without introducing flavour conflicts. We use CMSSM framework supplemented with the two elements that we need to explain neutrino oscillations: a structure of the Yukawa interactions and a mechanism to explain small neutrino masses. As a model for the Yukawa textures we consider one inspired by SU(5) GUT models with Abelian flavour symmetries [7, 8] while tiny neutrino masses can be explained trough a see-saw mechanism.

In the context of SUSY, even if the soft scalar masses were universal at the unification scale, quantum corrections between the GUT scale and low energies would modify this structure via renormalization-group running that generates off-diagonal contributions giving rise to observable LFV signals. Examples are rare lepton decays [4, 9, 10, 11, 12] and slepton production at a LC [13, 14, 15, 16].

We focus our analysis to models with universality assumptions of the CMSSM and that are compatible with searches at the LHC, and that also respect cosmological relic density considerations [17]. As an output of our analysis, we discuss detection prospects at a Linear Collider (LC), taking into account the constraints that the recent LHC supersymmetry searches and the measurement of the Higgs mass [18, 19] imposes on the observability of LFV processes.

In the following sections we make a brief survey of the Yukawa model we used and the constraints from LFV radiative decays before presenting our results for LC signals. Further details and a complete list of references are given in Ref [20].

2. Neutrino Mass Textures and Predictions for Neutrino Observables

We choose a model inspired by a SU(5) GUT combined with family symmetries [7, 8] such that the Yukawa textures take the form:

$$Y_\ell \propto \begin{pmatrix} \varepsilon^4 & \varepsilon^3 & \varepsilon \\ \varepsilon^3 & \varepsilon^2 & 1 \\ \varepsilon^3 & \varepsilon^2 & 1 \end{pmatrix}, \quad Y_\nu \propto \begin{pmatrix} \varepsilon^{|1\pm n_1|} & \varepsilon^{|1\pm n_2|} & \varepsilon^{|1\pm n_3|} \\ \varepsilon^{|n_1|} & \varepsilon^{|n_2|} & \varepsilon^{|n_3|} \\ \varepsilon^{|n_1|} & \varepsilon^{|n_2|} & \varepsilon^{|n_3|} \end{pmatrix} \tag{4}$$

where $Y_{\ell,\nu}$ stand for the Yukawa couplings of charged leptons and neutrinos respectively. n_i denote the U(1) charges and the heavy Majorana mass matrix is given by

$$M_N \propto \begin{pmatrix} \varepsilon^{2|n_1|} & \varepsilon^{|n_1+n_2|} & \varepsilon^{|n_1+n_3|} \\ \varepsilon^{|n_1+n_2|} & \varepsilon^{2|n_2|} & \varepsilon^{|n_2+n_3|} \\ \varepsilon^{|n_1+n_3|} & \varepsilon^{|n_2+n_3|} & \varepsilon^{2|n_3|} \end{pmatrix}. \tag{5}$$

The SU(5) structure of the model implies that the charged-lepton mass matrix is the transpose of the down-quark mass matrix, which relates the mixing of the left-handed leptons to that of the right-handed down-type quarks. The latest property implies that a large mixing can take place in the lepton sector while having a small mixing in the down quark sector, as suggested by a natural explanation of the Cabibbo-Kobayashi-Maskawa (CKM) matrix.

The light neutrinos mass matrix is, thanks to the see-saw mechanism,

$$m_{eff} \approx m_\nu^D \frac{1}{M_N} m_\nu^{D^T} \propto \begin{pmatrix} \varepsilon^2 & \varepsilon & \varepsilon \\ \varepsilon & 1 & 1 \\ \varepsilon & 1 & 1 \end{pmatrix}. \tag{6}$$

Note that $Y_\ell Y_\ell^\dagger \sim m_{eff}$ at the lowest order, thus, given the following diagonalizations of the Dirac and Majorana mass matrices,

$$\begin{aligned} {V_\ell}^T(Y_\ell {Y_\ell}^\dagger)V_\ell^* &= \text{diag}(y_e^2, y_\mu^2, y_\tau^2), && (7)\\ {V_D}^T(Y_\nu {Y_\nu}^\dagger)V_D^* &= \text{diag}(y_{\nu_1}^2, y_{\nu_1}^2, y_{\nu_3}^2), && (8)\\ {U_N}^T M_N U_N &= \text{diag}(M_1, M_2, M_3), && (9)\\ {U_\nu}^T m_{eff} U_\nu &= \text{diag}(m_{\nu_1}, m_{\nu_2}, m_{\nu_3}), && (10) \end{aligned}$$

Journal of Physics: Conference Series **447** (2013) 012059 doi:10.1088/1742-6596/447/1/012059

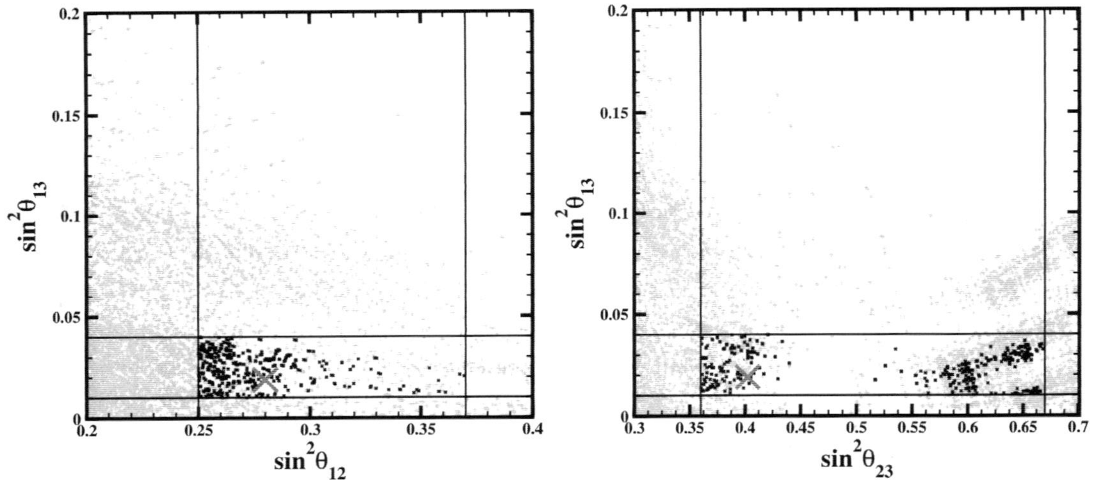

Figure 1. *Correlations between the neutrino mixing angles before and after imposing the constraints on the model coefficients as discussed in the text. The solid lines indicate the experimental bounds, and the small black crosses represent models satisfying all the constraints. The large magenta cross corresponds to model that is discussed in the text and used for numerical calculations.*

V_ℓ and U_ν diagonalize matrices with a similar structure. The Maki-Nakagawa-Sakata (MNS) matrix is given by

$$U_{MNS} \equiv U = V_\ell^\dagger U_\nu \,, \tag{11}$$

The texture is determined up to coefficients $\mathcal{O}(1)$ in every entry of the matrices. We take the expansion parameter to be $\varepsilon = 0.2$ and multiply the entries of Y_ℓ, Y_ν and M_N given by Eqs. (4) and (5) by coefficients in the range $\pm[0.5, 2]$ We require the ranges for the mixing angles of Ref. [21] and assume a neutrino mass hierarchy with a mass splitting of the order of the neutrino masses and $m_{\nu_3} \sim 0.05$ eV. In addition we make a further selection by requiring:

- The coefficients of Y_ν and M_N are matched to a light neutrino mass matrix m_{eff} of the form (6), with entries that deviate by a factor between 1/2 and 2 from those in Eq. (6).
- That the hierarchy of eigenvalues of $Y_\ell Y_\ell^\dagger$ (which, as discussed above, has a similar structure to m_{eff} and $Y_\nu Y_\nu^\dagger$) preserves the order of the gauge eigenstates.

After a search of about 10^9 fits we display in Fig. 1 the predictions for the neutrino mixing angles corresponding to the above criteria. We can see that, within this class of models, most of solutions that reproduce the correct range of θ_{12} and θ_{23} also predict a neutrino mixing angle θ_{13} that is compatible with the data from [1, 2] and also with the global analysis of neutrino data [22]. We find values for θ_{12} mostly on the lower part of its experimental range, while values with maximal $\theta_{23} = \pi/4$ are not frequent. This would have been the case when U_{MNS} arose from a m_{eff} with the structure of Eq. (6) and moderate 2-3 mixing from the charged lepton sector. However, with our conditions, both V_ℓ and U_ν in Eq. (11) arise from diagonalization of matrices with a large 2-3 mixing and our solutions do not present a distribution of $\sin\theta_{23}$ around 1/2.

We selected the marked point as reference for further analysis, this corresponds to:

$$Y_\ell \propto \begin{pmatrix} \varepsilon^4 & 2\varepsilon^3 & -1.75\varepsilon \\ -0.5\varepsilon^3 & 1.9\varepsilon^2 & 0.5 \\ -0.5\varepsilon^3 & -0.7\varepsilon^2 & 1.25 \end{pmatrix}, \qquad Y_\nu \propto \begin{pmatrix} \varepsilon^{|1\pm n_1|} & \varepsilon^{|1\pm n_2|} & 2\varepsilon^{|1\pm n_3|} \\ 0.75\varepsilon^{|n_1|} & \varepsilon^{|n_2|} & -0.5\varepsilon^{|n_3|} \\ \varepsilon^{|n_1|} & \varepsilon^{|n_2|} & 1.25\varepsilon^{|n_3|} \end{pmatrix},$$

$$M_N \propto \begin{pmatrix} \varepsilon^{2|n_1|} & \varepsilon^{|n_1+n_2|} & -\varepsilon^{|n_1+n_3|} \\ \varepsilon^{|n_1+n_2|} & \varepsilon^{2|n_2|} & \varepsilon^{|n_2+n_3|} \\ -\varepsilon^{|n_1+n_3|} & \varepsilon^{|n_2+n_3|} & -\varepsilon^{2|n_3|} \end{pmatrix}. \tag{12}$$

The predictions for the neutrino angels are, $\sin^2\theta_{13} = 0.019$, $\sin^2\theta_{12} = 0.28$, $\sin^2\theta_{23} = 0.40$. We observe no correlation between LFV and particular arrangements of neutrino parameters. However, LFV is maximized by large out-diagonal elements on V_ℓ and by the choice of charges n_i. Neutrino fits are independent of these charges because despite they affect Y_ν and M_N but not their combination in m_{eff}.

3. Scenarios for Charged-Lepton-Flavour Violation

A complete GUT model with an Abelian symmetry that determines the Yukawa textures also implies a flavour dependence of the soft mass matrices on the family charges. The determination of the resulting model at M_{GUT} depends on the details of the SUSY breaking and RGE evolution of the soft terms [23]. In this work we assume that this model is the MSSM extended by a see-saw with soft terms that do not deviate significantly from the universality conditions of the CMSSM.

Even if we start with universal soft-terms at M_{GUT}, at the intermediate scale where the see-saw takes place, M_3 (taken as the mass of the heaviest Majorana neutrino), the slepton mass matrices and Y_ℓ cannot be diagonalized with a single superfield rotation. Thus, the interactions lepton-slepton-gaugino can mix flavours. To understand this mismatch of the leptons and sleptons rotations we can consider the soft masses evolution from M_{GUT} to M_3 in a basis such that Y_ν is diagonal, at M_3 the right handed neutrinos decouple and the REG can be rewritten in terms of Y_ℓ diagonal. In this basis the soft terms involving left slepton are not diagonal and can be written in terms of the matrix $V_{LFV} = V_D^\dagger V_\ell$:

$$m_{LL}^2 = V_{LFV}^\dagger (m_{LL}^2)_{\rm d} V_{LFV}\,, \tag{13}$$

while the A-terms become:

$$A_\ell = V_{LFV}^T (A_\ell)_{\rm d}\,. \tag{14}$$

Here $(m_{LL}^2)_{\rm d}$ and $(A_\ell)_{\rm d}$ are the soft terms resulting from the RGE running of the universal soft terms at the GUT scale to M_3 with the fields written in a basis such that Y_ν is diagonal. The successive run of the MSSM RGEs to M_{SUSY} do not have a significant effect on the off-diagonal elements.

The matix V_{LFV} can be evaluated from Yukawa textures. We use the fit of Eq. (12) obtained by adjusting the matrices to the neutrino parameters at a low energy scale. RGE effects are taken into account by considering the global scale evolution of the neutrino masses set initially by imposing $m_{\nu_3} = 0.05$ eV and the see-saw condition at a common scale M_3 to determine the strength of the largest Dirac Yukawa coupling. Since we assume hierarchical neutrinos, we neglect the evolution of the non-diagonal matrix elements of effective neutrino mass and the lightest generations. These effects are small [24] and can be included in the uncertainty of the coefficients without affecting the stability of the solutions. Furthermore, no sizeable effect is expected from including lower right handed neutrino scales since the Dirac Yukawa decreases

Table 1. *Values of V_{LFV} with $\varepsilon = 0.2$ for the fit of Eq. 12.*

n_i	$c_1 = \{n_1 = 2, n_2 = 0, n_3 = 1\}$	$c_2 = \{n_1 = 0, n_2 = 1, n_3 = 0\}$
V_{LFV}	$\begin{pmatrix} -0.820 & 0.343 & 0.458 \\ 0.494 & 0.829 & 0.263 \\ -0.289 & 0.441 & -0.849 \end{pmatrix}$	$\begin{pmatrix} 0.806 & -0.401 & -0.436 \\ -0.437 & -0.899 & 0.016 \\ -0.399 & 0.178 & -0.901 \end{pmatrix}$

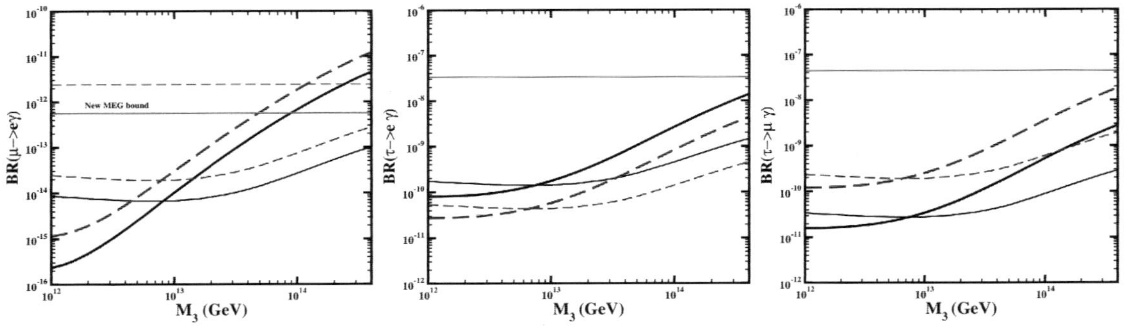

Figure 2. *Predictions for the rare LFV decays $\ell_i \to \ell_j\gamma$ as a function of the heaviest right-handed neutrino mass M_N, for the benchmark point displayed in (15) (a) (thick line), (b) (thin line) and using some the neutrino mixing fits shown in Table 1. The solid lines correspond to cases (c_1), dashed to (c_2). The horizontal solid lines indicate the current experimental upper bounds.*

with them becoming tiny when M_1 and M_2 are much lower than M_3. In our computation we use the matrices given in Eq. (12), V_ℓ has large out-diagonal elements and the n_i charges that are not relevant to fit neutrino data become important in the determination of LFV effects as can be seen in the values of V_{LFV} of table 1. We need now to evaluate the SUSY spectrum and couplings using the CMSSM as general framework. The recent LHC measurement of the Higgs mass seems to point towards large SUSY masses in these scenarios. However, the global analysis of the CMSSM parameter space of Ref. [17] yielded two almost equally good fits to the available data, one with relatively light sparticle masses and $\tan\beta \sim 16$, and the other with heavier sparticle masses and $\tan\beta \sim 45$, corresponding to the following CMSSM sets of parameters:

$$\begin{aligned} (a)\quad \tan\beta = 16,\quad m_0 = 300\ \text{GeV},\quad \text{M}_{1/2} = 910\ \text{GeV},\quad \text{A}_0 = 1320\ \text{GeV}\,, \\ (b)\quad \tan\beta = 45,\quad m_0 = 1070\ \text{GeV},\quad \text{M}_{1/2} = 1890\ \text{GeV},\quad \text{A}_0 = 1020\ \text{GeV}\,. \end{aligned} \tag{15}$$

Point (a) belongs to the region where the WMAP-favoured range of $\Omega_\chi h^2$ is achieved via $\chi - \tilde{\tau}$ coannihilation. Point (b) lies in the funnel region where the neutralino LSP annihilates rapidly via direct-channel H/A poles.

4. LFV in radiative decays.

The branching ratios (BRs) of the decays $\ell_j \to \ell_i + \gamma$ are calculated using full mass matrices diagonalization and the formulae of Ref. [9]. In Fig. 2 we show numerical predictions for the LFV branching ratios. We can see the effect of varying M_N from 6×10^{14} GeV down to 10^{12}

Journal of Physics: Conference Series **447** (2013) 012059 doi:10.1088/1742-6596/447/1/012059

GeV for the choices of right-handed neutrino charges of Table 1. The branching ratios are larger for the lower-mass scenario with $\tan\beta = 16$, due to the lighter spectrum. The new MEG bound on BR($\tau \to \mu\gamma$) imposes constraints on the "see-saw" scale for all for all the charge choices of Table 1 at point (a) while the predictions of point (b) are in the range of the experimental searches.

5. LFV Observation at a Linear Collider

In Ref. [13] it was shown that if the flavour mixing is introduced in the LL slepton sector, as is the case for the models under consideration here, slepton-pair production and LFV decays such as:

$$\begin{aligned} e^+e^- &\to \tilde{\ell}_i^- \tilde{\ell}_j^+ \to \tau^\pm \mu^\mp \tilde{\chi}_1^0 \tilde{\chi}_1^0, \\ e^+e^- &\to \tilde{\nu}_i \tilde{\nu}_j^c \to \tau^\pm \mu^\mp \tilde{\chi}_1^+ \tilde{\chi}_1^- \end{aligned} \tag{16}$$

lead to a cross section of the order of 1 fb, the reference value used in [13], for a LFV signal of $\mu^\pm\tau^\pm$ pairs that can be distinguished from the background, according to the considerations in [15].

Here, we extend our previous results by considering the full structure of the Yukawa couplings, thus comparing the LFV production of charged leptons of all generations and use the updated supersymmetric benchmark points compatible with the LHC measurement of m_h.

We present in Fig. 3 the expected cross sections as functions of $\sqrt{s}$ for the same sets of charges as those in Fig. 2. For definiteness, the right-handed neutrino mass scale is set to $M_N = 10^{14}$ GeV for point (b) and the restrictive value of $M_N = 2 \times 10^{13}$ such that the new MEG bound on $BR(\mu \to e\gamma)$ is preserved. We observe that, by varying this scale in a similar range as in Fig. 2, the cross sections also change by approximately an order of magnitude: according to the see-saw conditions, larger values of M_N imply larger Dirac neutrino Yukawa couplings, and thus stronger RGE effect. The results depend sensitively on the sparticle spectrum. If we restrict ourselves to the benchmark CMSSM parameter choice discussed above, we would require $\sqrt{s}$ at TeV scale in order to have significant LFV signals in slepton mixing.

6. Conclusions

In our presentation, we revisited the signatures of charged LFV in theoretically-motivated scenarios, studying the correlations arising in CMSSM models with parameter values that are favoured by the LHC and cosmological considerations. We have explored these issues using updated experimental input from neutrino data, particularly recent measurements of θ_{13}, MEG and the LHC.

Models based in SU(5) with Abelian flavour symmetries provide very interesting possibilities for understanding the hierarchy of fermion masses and mixings. We performed a big scan of fits to the neutrino data paying attention to the the naturalness of the fit and avoiding artificial cancellations arising from specific choices of coefficients. We get a pattern of neutrino predictions and correlations compatible with the global analysis of neutrino data [22]. In general, we found that fittings with similar predictions for the neutrino parameters may lead to very different LFV predictions. This can provide information on the heavy Majorana neutrino matrix.

The new MEG bound on $\mu \to e\gamma$ is still compatible with the LHC data. At the point with lighter SUSY spectrum it determines an upper limit for the see-saw scale while for point with large SUSY spectrum, LFV radiative decays are still interesting for experimental searches. Regarding LC prospects, it was possible to establish correlations between the expected rates for radiative LFV decays and LFV $\tau\mu$ pair production at a future LC. Within the CMSSM framework studied, the absence of a supersymmetry signal in the LHC data and the discovery

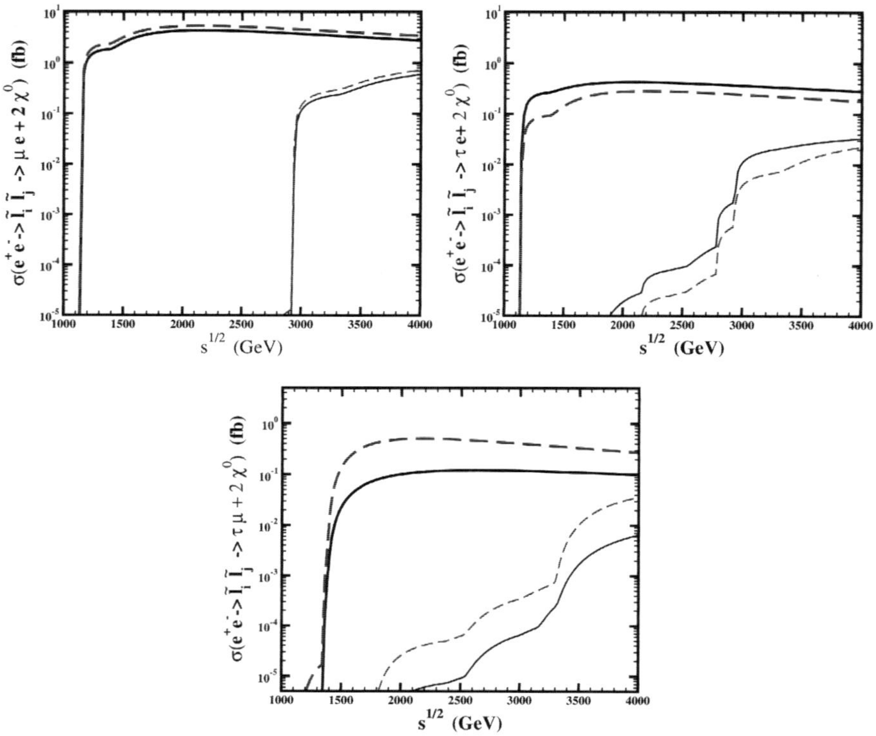

Figure 3. *Values of the cross sections* $\sigma(e^+e^- \to \tilde{\ell}_i^- \tilde{\ell}_j^+ \to \ell_a^\pm \ell_b^\mp + 2\chi^0)$ *(*$\ell_a \neq \ell_b$ *as indicated for each panel) as functions of* $\sqrt{s}$*. The line styles are the same as those in Fig. 2. For the point (a) we use* $M_N = 2 \times 10^{13}$ *GeV. While for the point (b) we use* $M_N = 10^{14}$ *GeV.*

of a neutral Higgs weighing $\sim$ 125 GeV imply that observation of slepton flavour violation at a LC will be possible for only energies beyond 1 TeV.

Acknowledgments. The work of M. C. is supported by MultiDark under Grant No. CSD2009-00064 of the Spanish MICINN Consolider-Ingenio 2010 Program. M. E. G. and M. C. acknowledge further support from the MICINN project FPA2011-23781, from the Grant MICINN-INFN(PG21)AIC-D-2011-0724 and the project P07FQM02962 funded by "Junta de Andalucia".

[1] F. P. An *et al.* [Daya Bay Collaboration], Phys. Rev. Lett. **108** (2012) 171803.
[2] J. K. Ahn *et al.* [RENO Collaboration], Phys. Rev. Lett. **108** (2012) 191802.
[3] Y. Abe *et al.* [Double Chooz Collaboration], arXiv:1301.2948 [hep-ex].
[4] For reviews, see:
M. Raidal et al., Eur. Phys. J. C **57** (2008) 13; Y. Kuno and Y. Okada, Rev. Mod. Phys. **73** (2001) 151.
[5] J. Adam *et al.* [MEG Collaboration], arXiv:1303.0754 [hep-ex].
[6] K. Nakamura *et al.* [Particle Data Group], J. Phys. G **37** (2010) 075021.
[7] S. Lola and G.G. Ross, Nucl. Phys. B 553 (1999) 81.
[8] J. R. Ellis, M. E. Gómez and S. Lola, JHEP **0707** (2007) 052.
[9] J. Hisano, T. Moroi, K. Tobe and M. Yamaguchi, Phys. Rev. D **53** (1996) 2442.
[10] M. Gómez, G. Leontaris, S. Lola and J. Vergados, Phys. Rev. **D 59** (1999) 116009.
[11] J. R. Ellis, M. E. Gómez, G. K. Leontaris, S. Lola and D. V. Nanopoulos, Eur. Phys. J. C **14** (2000) 319.
[12] S. Antusch, E. Arganda, M. J. Herrero and A. M. Teixeira, JHEP **0611** (2006) 090 [hep-ph/0607263].
[13] E.Carquin, J. Ellis, M. E. Gómez and S. Lola, JHEP **1111** (2011) 050.

[14] J. Hisano, M. M. Nojiri, Y. Shimizu and M. Tanaka, Phys. Rev. D **60**, 055008 (1999).
[15] F. Deppisch, H. Päs, A. Redelbach, R. Rückl and Y. Shimizu, Phys. Rev. **D 69** (2004) 054014.
[16] A. Abada, A. J. R. Figueiredo, J. C. Romão and A. M. Teixeira, JHEP **1208** (2012) 138 [arXiv:1206.2306 [hep-ph]].
[17] O. Buchmueller *et al.*, Eur. Phys. J. C **72** (2012) 2243 [arXiv:1207.7315 [hep-ph]].
[18] G. Aad *et al.* [ATLAS Collaboration], Phys. Lett. B **716** (2012) 1.
[19] S. Chatrchyan *et al.* [CMS Collaboration], Phys. Lett. B **716** (2012) 30.
[20] M. Cannoni, J. Ellis, M. E. Gomez, S. Lola and , arXiv:1301.6002 [hep-ph].
[21] M. C. González-García, M. Maltoni, J. Salvado and T. Schwetz, arXiv:1209.3023 [hep-ph].
[22] G.L. Fogli, E. Lisi, A. Marrone, D. Montanino and A. Palazzo, A.M. Rotunno, Phys. Rev. D **86** (2012) 013012.
[23] M.E. Gómez, S. Lola, P. Naranjo and J. Rodríguez-Quintero, JHEP **0904** (2009) 043.
[24] J. R. Ellis, S. Lola and , Phys. Lett. B **458** (1999) 310.

DISCRETE 2012 – Third Symposium on Prospects in the Physics of Discrete Symmetries IOP Publishing
Journal of Physics: Conference Series **447** (2013) 012060 doi:10.1088/1742-6596/447/1/012060

RGE behaviour of $U(2)^3$ symmetry in SUSY

J Jones Pérez
Departament de Física Teòrica and IFIC, Universitat de València-CSIC,
E-46100, Burjassot, Spain

E-mail: joel.jones@uv.es

Abstract. From the naturalness point of view, the first LHC results seem to disfavour any constrained MSSM realization with universal conditions at the SUSY-breaking scale. A more motivated scenario is given by split-family SUSY, in which the first two generations of squarks are heavy, compatible with a $U(2)^3$ flavour symmetry. Here, after reviewing the flavour structures obtained in this framework, we consider the flavour symmetry to be broken at a very high scale, and study the consequences at low energies through its RGE evolution. Initial conditions compatible with a split scenario are found, and the preservation of correlations from minimal $U(2)^3$ breaking are checked.

1. Introduction

When considering physics beyond the Standard Model (SM), Supersymmetry, in the form of the Minimal Supersymmetric Standard Model (MSSM) is usually considered as one of the most appealing options. However, due to its vast parameter space, a standard procedure is to assume universality conditions of the soft terms at the GUT scale, and allow the renormalization group equations (RGEs) to break the universality when running down to the weak scale. This constrained realization is named CMSSM, and leads to flavoured soft terms acquiring a structure following Minimal Flavour Violation (MFV) [1].

MFV effectively suppresses all SUSY contributions to flavoured processes, solving the so-called SUSY flavour problem. However, it also leads to sfermion masses with small splittings. This can cause issues in the solution of the hierarchy problem, which depends mainly on having light third generation squark masses. The issue is caused due to the lack of SUSY signals at the LHC, which forces the masses of the first two generation squarks to be pushed to larger values. The MFV structure causes the first two generations to drag with them the mass of the third generation squarks to these larger values, compromising in this way the solution of the hierarchy problem.[1]

One would therefore like to have a framework similar to MFV, such that contributions of flavoured processes are still suppressed to some degree, but allowing the third generation masses to be decoupled from the first two. One would also hope that the flavour suppression in this new framework be somewhat milder, such that some significant effects could be observed in flavour experiments. In particular, one would hope these new effects could solve the ongoing tension between CP violating $\Delta F = 2$ observables in the K and B_d sectors [2].

[1] One must admit that the observed value of the Higgs mass already spoils the solution of the hierarchy problem in most of the SUSY parameter space. Nevertheless, a lower value of the stop mass can effectively reduce the fine-tuning, even though it might not be able to avoid it entirely.

A framework based on a $U(2)^3$ symmetry in the quark sector was built in [3], and was found to satisfy all of the above requirements. Further studies of this symmetry were carried out in [4], where the squark and slepton phenomenology was analysed in further detail, both in and out of SUSY. Nevertheless, in these works the symmetry structure was applied directly at the electroweak scale, with no consideration of RGE running. As one would expect that the breaking of such a symmetry would occur at a high scale, it was found necessary to study the consequences of the running on the low-scale structures. In particular, it was unclear if the virtues of $U(2)^3$ would be maintained at the weak scale, and what kind of initial conditions could lead to the required split scenario.

The current paper is a summary of the work carried out in [5]. In Section 2 we briefly review the $U(2)^3$ framework and the conditions leading to the solution of the flavour tension. In Section 3 we show the result of our study of the RGE evolution, and then conclude.

2. Review of $U(2)^3$ Framework in SUSY

We shall restrict ourselves to the quark sector. In contrast to MFV, which assigns a $U(3)$ symmetry for each flavoured chiral supermultiplet, this framework assigns a $U(2)$ symmetry for each supermultiplet, under which only the first two generations shall transform. In the limit of exact flavour symmetry, the only terms allowed in the superpotential involving flavoured fields are those of the third generation (s)quarks:

$$W = y_t\, Q_3 t_R^c H_u - y_b\, Q_3 b_R^c H_d \;. \tag{1}$$

As in MFV, we proceed to break the flavour symmetries by adding spurion fields. The minimal way of constructing the Yukawas is by adding bi-doublet spurions ΔY_u and ΔY_d, from which we build the structure of the $(1-2)$ block, and a doublet V, transforming only under $U(2)_Q$, used to connect the first two generations with the third. With these assumptions, the Yukawas acquire the structure:

$$Y_u = y_t \left(\begin{array}{c|c} \Delta Y_u & x_t V \\ \hline 0 & 1 \end{array}\right), \qquad Y_d = y_b \left(\begin{array}{c|c} \Delta Y_d & x_b V \\ \hline 0 & 1 \end{array}\right) , \tag{2}$$

with everything above the horizontal dashed line having two rows, and everything to the left of the vertical dashed line having two columns. We find it possible to parametrize the whole matrices in terms of the eigenvalues of the bi-doublets, λ_{f_1} and λ_{f_2}, a left-mixing matrix with one angle and phase per bi-doublet, s_f and α_f, a suppression parameter ϵ for the doublet, and two complex couplings $x_f e^{i\phi_f}$. Notice only three phases are independent.

With this setup, one can build the CKM matrix, and fit the following parameters [3]:

$$\begin{aligned} s_u &= 0.095 \pm 0.008 \;, & s_d &= -0.22 \pm 0.01 \;, \\ s &= 0.0411 \pm 0.0005 \;, & \cos(\alpha_u - \alpha_d) &= -0.13 \pm 0.2 \;, \end{aligned} \tag{3}$$

where $s \propto \epsilon$. We choose to fix $\epsilon = \lambda^2_{\rm CKM}$.

After determining which spurions are needed to build the Yukawa matrices, we follow the MFV assumption demanding all flavour structures to depend exclusively on these spurions. With this, we find that the soft terms acquire the following form:

$$\frac{m^2_{\tilde{Q}}}{m^2_{Q_h}} = I + \left(\begin{array}{c|c} c_{Qv}\, V^* V^T + c_{Qu} \Delta Y_u^* \Delta Y_u^T + c_{Qd} \Delta Y_d^* \Delta Y_d^T & x_Q\, e^{-i\phi_Q} V^* \\ \hline x_Q\, e^{i\phi_Q} V^T & -\rho_Q \end{array}\right), \tag{4}$$

$$\frac{m^2_{\tilde{d}}}{m^2_{d_h}} = I + \left(\begin{array}{c|c} c_{dd}\, \Delta Y_d^\dagger \Delta Y_d & x_d\, e^{-i\phi_d} \Delta Y_d^\dagger V \\ \hline x_d\, e^{i\phi_d} V^\dagger \Delta Y_d & -\rho_u \end{array}\right), \tag{5}$$

$$\frac{m^2_{\tilde{u}}}{m^2_{u_h}} = I + \left(\begin{array}{c|c} c_{uu}\, \Delta Y_u^\dagger \Delta Y_u & x_u\, e^{-i\phi_u} \Delta Y_u^\dagger V \\ \hline x_u\, e^{i\phi_u} V^\dagger \Delta Y_u & -\rho_d \end{array}\right), \tag{6}$$

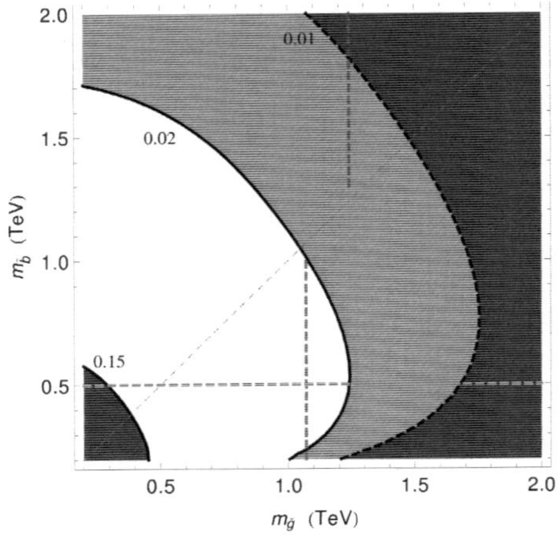

Figure 1. Value of F_0 as a function of $m_{\tilde g}$ and $m_{\tilde b}$. White region has $0.02 < F_0 < 0.15$, grey region has $F_0 > 0.01$. Red and blue lines are rough limits on $m_{\tilde g}$, taken from ATLAS, green line is a rough limit on $m_{\tilde b}$, taken from CMS.

where $\rho_f = (m^2_{f_h} - m^2_{f_l})/m^2_{f_h}$, and all c_i and x_i parameters are real, of $\mathcal{O}(1)$.

These structures lead to the following contributions to $\Delta F = 2$ observables:

$$\epsilon_K = \epsilon_K^{\rm SM(tt)} \times \left(1 + x^2 F_0\right) + \epsilon_K^{\rm SM(tc+cc)} \tag{7}$$

$$S_{\psi K_S} = \sin\left(2\beta + \arg\left(1 + x F_0 e^{-2i\gamma}\right)\right) , \tag{8}$$

$$\Delta M_d = \Delta M_d^{\rm SM} \times \left|1 + x F_0 e^{-2i\gamma}\right| , \tag{9}$$

$$\frac{\Delta M_d}{\Delta M_s} = \frac{\Delta M_d^{\rm SM}}{\Delta M_s^{\rm SM}} , \tag{10}$$

where:

$$F_0 = \frac{2}{3}\left(\frac{g_s}{g}\right)^4 \frac{m_W^2}{m_{Q_3}^2}\frac{1}{S_0(x_t)}\left[f_0(x_g) + \mathcal{O}\left(\frac{m^2_{Q_l}}{m^2_{Q_h}}\right)\right] , \qquad x_g = \frac{m_{\tilde g}^2}{m_{Q_3}^2} , \tag{11}$$

$$f_0(x_g) = \frac{11 + 8x_g - 19x_g^2 + 26x_g\log(x_g) + 4x_g^2\log(x_g)}{3(1-x_g)^3} , \qquad x = \frac{s_L^2 c_d^2}{|V_{ts}|^2} \tag{12}$$

with $S_0(x_t)$ being the typical one-loop function of the SM to $\Delta F = 2$ processes. The $s_L\, e^{i\gamma}$ parameters are defined in [3], and are related to the ϵ and ϕ_f terms.

With these modifications, one can find that the $U(2)^3$ framework can provide a large enough contribution to ϵ_K and $S_{\psi K_s}$ in the correct directions, such that the flavour tension mentioned above is solved. For not too large x, we required $F_0 \gtrsim 0.02$, but found that values of F_0 as low as 0.01 could also reduce the tension by compensating with slightly large $\mathcal{O}(1)$ parameters. This implies upper bounds on gluino and sbottom masses, as shown in Figure 1.

Figure 1 also shows the current status of the $U(2)^3$ framework in SUSY. The red and blue lines indicate lower bounds on $m_{\tilde g}$ set by the ATLAS collaboration, for the cases $m_{\tilde g} < m_{\tilde b}$ [6] and $m_{\tilde g} > m_{\tilde b}$ [7], respectively. The green line indicates lower bounds on $m_{\tilde b}$ by the CMS collaboration [8]. The white area shows $0.02 < F_0 < 0.15$, and is almost completely excluded by direct searches. The grey area has $0.01 < F_0 < 0.02$, and is not heavily constrained when $m_{\tilde g} > m_{\tilde b}$. Thus, current data favours heavy gluinos, and somewhat large mixing parameters.

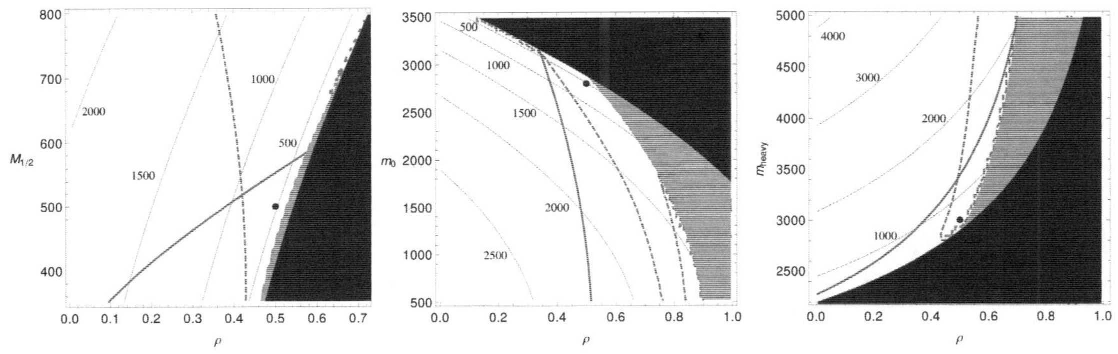

Figure 2. Parameter space around our benchmark, with contours for μ. The dark regions correspond to no EWSB, while the orange regions have a tachyonic stop. The green, dashed lines delimitate the regions within 1 sigma of the Higgs mass, and the red, solid curve indicates the regions below which the flavour tension could be solved. The blue dot represents the benchmark.

3. RGE Behaviour of $U(2)^3$ Framework

We now turn to study the properties of this framework when defined at a high energy scale, and then evolved down to the electroweak scale. We address three questions: **(1)** What sort of initial conditions should we look for? **(2)** How does the mixing evolve? **(3)** Are the virtues of $U(2)^3$ preserved at all scales?

To simplify the study of the initial conditions, we shall turn to a CMSSM-like setup. At the GUT scale, we keep the universal gaugino mass, $M_{1/2}$, but differentiate between the common Higgs mass, m_0, the common (heavy) first two generations scalar mass, m_{heavy}, and the common (light) third generation scalar mass, m_{light}. Moreover, we use the parameter $\rho = (m^2_{heavy} - m^2_{light})/m^2_{heavy}$ instead of m_{light}. The A-Terms follow a structure similar to that of the Yukawas, with different $\mathcal{O}(1)$s and multiplied by a factor $A_0 \sim m_{heavy}$. After running the parameters down to the electroweak scale, we require no tachyonic states, radiative electroweak symmetry breaking, and the correct Higgs mass within 1σ. We also prefer regions where $F_0 > 0.01$, and attempt to reduce the hierarchy problem by having small values of μ and $m_{\tilde{t}}$.

Although in [5] we studied two different benchmarks, here we shall concentrate only in one, and describe the parameter space around this point. The benchmark is characterized by $M_{1/2} = 0.5$ TeV, $m_0 = 2.8$ TeV, $m_{heavy} = 3$ TeV, $\rho = 0.5$ and $A_0 = -m_{heavy}$. We show in Figure 2 the parameter space around this point, with the different constraints represented in different colours. We can see that the region that fulfills all constraints is very small, especially if we demand $\mu < 1$ TeV.

The benchmark leads to values of μ around 500 GeV, and very specific splittings. In the stop sector, we find $\rho^{\rm low}_{\tilde{t}} \sim 0.85$, which leads to an average stop mass of about 1.2 TeV. Nevertheless, as the stop mixing is large, we find the mass of the lightest stop to be lower than 500 GeV. On the other hand, in the sbottom sector, we have $\rho^{\rm low}_{\tilde{b}} \sim 0.6$, leading to an average sbottom mass of 1.9 TeV. Notice that this setup involves a very mild splitting at the GUT scale, but can lead to a larger splitting in the stop sector.

Having found a point with suitable initial conditions, we now turn to the study of the mixing. To this end, we use objects directly related to physical observables, which are also connected with the mixing, and study their RGE evolution. As our interest lies on $\Delta F = 2$ processes, which receive their main contribution from $(LL)^2$ operators, the objects in question are defined

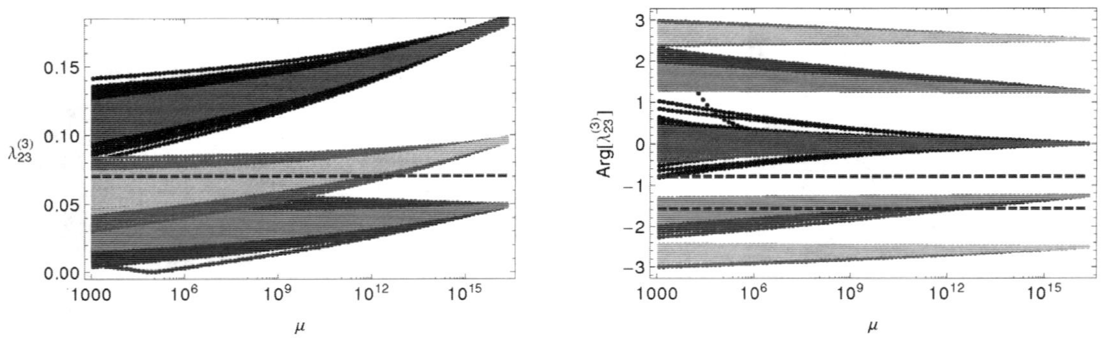

Figure 3. The running of $|\lambda_{23}^{(3)}|$ (left) and $\mathrm{Arg}(\lambda_{23}^{(3)})$ (right). On the left, we show $x_L = 2, 1, 0.5$ in blue, green and red. In every region the lighter colour correspond to γ fixed to $\pi/4$. On the right we fix $\gamma = (-1 + 0.4n)\pi$, with $n = 0, 1, 2, 3, 4$ in blue, red, green, grey and magenta, respectively. The lighter regions correspond to $x_L = 1$. In the first two plots the dashed brown lines mark the region where the flavour tension can be solved. On the left, the region is above the line, while and on the right it is between the two lines.

as

$$\lambda_{i\neq j}^{(a)} = (W_L^d)_{ia}(W_L^d)^*_{ja} , \tag{13}$$

where W_L^d is the diagonalization matrix of $m^2_{\tilde{Q}}$ in the basis of diagonal down quarks. In [3], it was shown that the $U(2)^3$ structure leads to $\lambda_{12}^{(3)}$, $\lambda_{13}^{(3)}$ and $\lambda_{23}^{(3)}$ to be closely related:

$$\lambda_{12}^{(3)} = s_L^2 \kappa^* c_d , \qquad \lambda_{13}^{(3)} = -s_L \kappa^* e^{i\gamma} , \qquad \lambda_{23}^{(3)} = -c_d s_L e^{i\gamma} , \tag{14}$$

where $\kappa \approx c_d V_{td}/V_{ts}$. Thus, assuming for now that these relations are held throughout the running, we shall only need to study the evolution of one of these objects, which we choose to be $\lambda_{23}^{(3)}$.

In Figure 3 we show how $\lambda_{23}^{(3)}$ evolves with the scale. The main result is that, although there exists some suppression effect in the magnitude, the mixing remains relatively stable throughout the running. The Figure also shows the minimum value that the mixing must have in order to solve the flavour tension, according to [3]. Here, we can see that the $\mathcal{O}(1)$ parameters cannot be too small, which is a direct consequence of the small value of F_0.

Finally, we need to check if the $U(2)^3$ relations of Equation (14) are kept after the RGE running. The best way to do this is to use the λ_{ij}^a to define new objects that should be approximately RGE invariant, at least within their theoretical errors, and check if this invariance holds. We choose:

$$\frac{\lambda_{13}^{(3)}}{\lambda_{23}^{(3)}} = \frac{\kappa^*}{c_d} , \qquad \frac{\lambda_{12}^{(3)}}{|\lambda_{13}^{(3)}|^2} = \frac{c_d}{\kappa} , \tag{15}$$

The first ratio tests the correlations between the B_d and B_s sectors, the second ratio tests those between the K and B_d sectors. If we find these ratios to hold within their theoretical errors, we can consider the $U(2)^3$ symmetry to be preserved by the running.

However, we first need to write the theoretical error for each of these quantities. The main source of error in the absolute value of each ratio lies on a NLO effect due to $\rho \neq 1$, which can

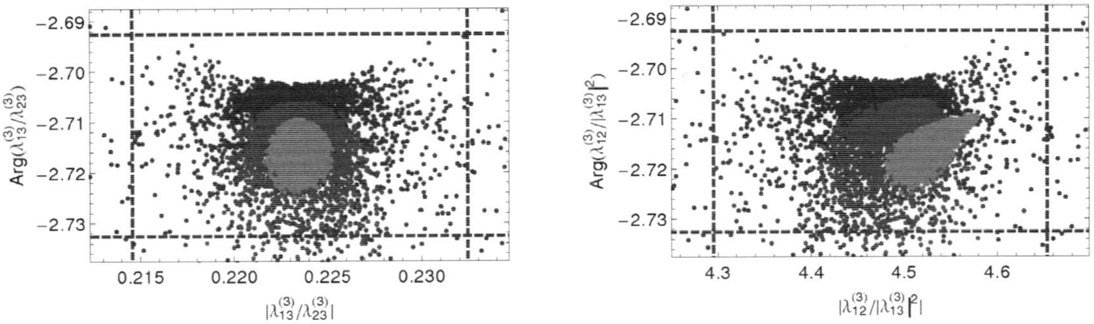

Figure 4. The two approximate RGE-invariant ratios, evaluated at $M_{\rm SUSY}$. The dashed lines correspond to our estimated theoretical uncertainty. We show results for $x_L = 2,\ 1,\ 0.5$ in brown, magenta and blue, respectively.

generate deviations of up to 4% the value of the ratio. On the other hand, the phase has a fixed correction, of order $\varphi_c = \arg(c_u c_d + s_u s_d e^{-i\phi}) \approx 0.02$. We shall then take the error on the phase equal to φ_c, after including the correction. Moreover, we neglect the errors due to the running of the CKM elements in κ.

The result of evolving these ratios down to the electroweak scale is shown in Figure 4. Different initial values for s_L are shown in different colours. The main conclusion is that in most cases the ratios are preserved within their theoretical errors. In general, the largest the initial value of s_L, the better the ratios are preserved.

Given the fact that the $U(2)^3$ relations are maintained at all scales, the stability when running $\lambda_{23}^{(3)}$ implies that both $\lambda_{12}^{(3)}$ and $\lambda_{13}^{(3)}$ also have a stable behaviour when solving the RGEs. This means that the virtues of the framework are preserved at all scales.

4. Conclusions

There exists a tension between the solution of the hierarchy problem, the current LHC bounds on sparticle masses, and the use of MFV. Furthermore, there is a slight tension between CPV observables in the K and B_d sectors, which MFV cannot solve. This encourages the use of other frameworks with suppression properties similar to those of MFV, but with the capacity of solving all of these tensions.

In this work, we have reviewed a $U(2)^3$ framework in SUSY, originally proposed in [3]. We wrote the main ingredients of this framework, and explored the regions in the $m_{\tilde g} - m_{\tilde b}$ parameter space where the flavour tension would be solved. We showed that current direct search bounds constrain significantly this parameter space, forcing us to consider regions which require slightly large $\mathcal{O}(1)$ parameters.

We also presented several features of this framework when defined at a high scale [5]. We considered the definition of the initial conditions, the evolution of the mixing, and the preservation of relations between contributions to observables in the K, B_d and B_s sectors. We defined a benchmark point and studied the parameter space around it, finding it strongly constrained. However, we also found that the mixing was rather stable during the RGE evolution, and that the $U(2)^3$ relations between different observables were preserved throughout the running.

Acknowledgments
The author would like to thank the organizers of the DISCRETE 2012 conference. The author also acknowledges support from the grants Generalitat Valenciana VALi+d, Spanish MINECO FPA 2011-23596 and the Generalitat Valenciana PROMETEO - 2008/004, and would like to thank Gianluca Blankenburg, who collaborated in the main topic of this talk.

References
[1] G. D'Ambrosio, G. F. Giudice, G. Isidori and A. Strumia, Nucl. Phys. B **645** (2002) 155 [hep-ph/0207036].
[2] W. Altmannshofer, A. J. Buras, S. Gori, P. Paradisi and D. M. Straub, Nucl. Phys. B **830** (2010) 17 [arXiv:0909.1333 [hep-ph]].
[3] R. Barbieri, G. Isidori, J. Jones-Perez, P. Lodone and D. M. Straub, Eur. Phys. J. C **71** (2011) 1725 [arXiv:1105.2296 [hep-ph]].
[4] R. Barbieri, P. Campli, G. Isidori, F. Sala and D. M. Straub, Eur. Phys. J. C **71** (2011) 1812 [arXiv:1108.5125 [hep-ph]];
R. Barbieri, D. Buttazzo, F. Sala and D. M. Straub, JHEP **1207** (2012) 181 [arXiv:1203.4218 [hep-ph]];
A. J. Buras and J. Girrbach, JHEP **1301** (2013) 007 [arXiv:1206.3878 [hep-ph]];
G. Blankenburg, G. Isidori and J. Jones-Perez, Eur. Phys. J. C **72** (2012) 2126 [arXiv:1204.0688 [hep-ph]];
R. Barbieri, D. Buttazzo, F. Sala and D. M. Straub, JHEP **1210** (2012) 040 [arXiv:1206.1327 [hep-ph]].
[5] G. Blankenburg and J. Jones-Perez, Eur. Phys. J. C **73** (2013) 2289 [arXiv:1210.4561 [hep-ph]].
[6] [ATLAS Collaboration], ATLAS-CONF-2012-107.
[7] G. Aad *et al.* [ATLAS Collaboration], Eur. Phys. J. C **72** (2012) 2174 [arXiv:1207.4686 [hep-ex]].
[8] S. Chatrchyan *et al.* [CMS Collaboration], JHEP **1301** (2013) 077 [arXiv:1210.8115 [hep-ex]].

DISCRETE 2012 – Third Symposium on Prospects in the Physics of Discrete Symmetries IOP Publishing
Journal of Physics: Conference Series **447** (2013) 012061 doi:10.1088/1742-6596/447/1/012061

Vector-like quarks and New Physics in the flavour sector

Francisco J Botella, Miguel Nebot (speaker)
Departament de Física Teòrica and IFIC,
Universitat de València - CSIC, E-46100, Burjassot, Spain
E-mail: Francisco.J.Botella@uv.es, Miguel.Nebot@uv.es

Gustavo C Branco
Departamento de Física and Centro de Física Teórica de Partículas,
Instituto Superior Técnico, Universidade Técnica de Lisboa, Av. Rovisco Pais, P-1049-001 Lisboa, Portugal
E-mail: gbranco@ist.utl.it

Abstract. We present a detailed analysis of recent flavour data in the framework of a simple extension of the Standard Model, where a $Q = 2/3$ vector-like isosinglet quark is added to the spectrum. Constraints from all the relevant quark flavour sectors are used. Important deviations from Standard Model expectations in different observables such as the semileptonic asymmetry in B_d decays, A^d_{SL}, the time-dependent CP asymmetry in $B_s \to J/\Psi\Phi$, and rare decays such as $K^+ \to \pi^+\bar{\nu}\nu$, can be obtained.

1. Introduction

We consider an extension of the Standard Model (SM) where one isosinglet vector-like quark T with charge $Q = 2/3$ is added to the spectrum [1,2,3]. After diagonalization of the up and down mass matrices, the 3×3 mixing matrix connecting standard quarks is no longer unitary, but a submatrix of a larger 4×4 unitary matrix U. The charged and neutral current interactions have the form

$$\mathscr{L}_W = -\frac{g}{\sqrt{2}}\bar{\mathbf{u}}_L\,\gamma^\mu\,V\,\mathbf{d}_L\,W_\mu + \text{h.c.}\,, \tag{1}$$

$$\mathscr{L}_Z = -\frac{g}{2\cos\theta_W}\left[\bar{\mathbf{u}}_L\,\gamma^\mu\left(VV^\dagger\right)\mathbf{u}_L - \bar{\mathbf{d}}_L\,\gamma^\mu\,\mathbf{d}_L - 2\sin^2\theta_W\,J^\mu_{em}\right]Z_\mu\,, \tag{2}$$

where $\mathbf{d} \equiv (d, s, b)$, $\mathbf{u} \equiv (u, c, t, T)$ and V is a 4×3 submatrix of the matrix U:

$$U = \begin{pmatrix} V_{ud} & V_{us} & V_{ub} & U_{u4} \\ V_{cd} & V_{cs} & V_{cb} & U_{c4} \\ V_{td} & V_{ts} & V_{tb} & U_{t4} \\ V_{Td} & V_{Ts} & V_{Tb} & U_{T4} \end{pmatrix}\,. \tag{3}$$

The submatrix $V_{(3\times3)}$, i.e. the upper left 3×3 block within U, is not a unitary matrix, since $V_{(3\times3)}\,V^\dagger_{(3\times3)} \neq \mathbf{1}_{(3\times3)}$. These deviations of unitarity of the "would-be standard" mixing matrix lead to flavour changing neutral currents (FCNC) which are present just in the up sector and controlled by[1]

$$\left(VV^\dagger\right)_{ij} = \delta_{ij} - U_{i4}U^*_{j4} \ . \tag{4}$$

Summarizing, the addition of one isosinglet vector-like up quark provides:

- A new mass eigenstate in the up sector. It can give new contributions to amplitudes involving virtual up quarks, as for example, in kaon or B meson mixings.
- A mixing matrix V which is not 3×3 unitary anymore, allowing for deviations of the mixing elements V_{ij} from SM values.
- Modified couplings to the Z boson in the up sector, including tree level flavour changing and reduced flavour conserving couplings.

With these ingredients we can expect

- modifications in the bd sector that can alleviate the existing tensions,
- new contributions to the B^0_s–$\bar{B}^0_s$ (dispersive) mixing amplitude $M_{12}^{(s)}$ that can significantly modify its phase, and thus the $B_s \to J/\Psi\Phi$ time dependent CP asymmetry,
- that the deviations from 3×3 unitarity can change the B_d and B_s (absorptive) mixing amplitudes $\Gamma_{12}^{(q)}$, and produce larger-than-standard values for the semileptonic asymmetries A^q_{SL},
- modifications in the rates of several rare decays.

2. Experimental constraints

To reflect the abundant experimental information that constrains modifications of the flavour sector such as the ones introduced in the present scenario, we have considered the following observables.

- Tree level observables, whose extraction from experiment is presumably unaffected by New Physics (NP) effects. These observables include moduli of the CKM elements in the first and second rows. For the third row the only relevant measurement is the one of the ratio of branching fractions $R = \mathrm{Br}(t \to Wb)/\mathrm{Br}(t \to Wq)$, $R = |V_{tb}|^2/(|V_{td}|^2 + |V_{ts}|^2 + |V_{tb}|^2)$. The physical phase γ, is also a tree level observable. The actual values are collected in table 1. Finally, the decay $B^+ \to \tau^+\nu_\tau$ is also a tree level process which participates in the so called tensions in the bd sector[2].

Table 1. Tree level observables [4, 6].

$\lvert V_{ud}\rvert$	0.97425 ± 0.00022	$\lvert V_{us}\rvert$	0.2252 ± 0.0009
$\lvert V_{cd}\rvert$	0.230 ± 0.011	$\lvert V_{cs}\rvert$	1.023 ± 0.036
$\lvert V_{ub}\rvert$	0.00389 ± 0.00044	$\lvert V_{cb}\rvert$	0.0406 ± 0.0013
γ	$(77 \pm 14)^\circ$	R	0.88 ± 0.07
$\mathrm{Br}(B^+ \to \tau^+\nu_\tau)$	$(1.13 \pm 0.23) \times 10^{-4}$		

[1] They are naturally suppressed by ratios m^2/m_T^2, where m denotes generically the standard quark masses [1]. This natural suppression of FCNC is crucial in order to make the model plausible.
[2] Notice however that, as it is helicity suppressed and proportional to $|V_{ub}|^2$, sizeable NP contributions may appear in different beyond SM scenarios, but not in our case.

Journal of Physics: Conference Series **447** (2013) 012061 doi:10.1088/1742-6596/447/1/012061

- Observables related to B_d^0–$\bar{B}_d^0$ and B_s^0–$\bar{B}_s^0$ mixings: we consider time-dependent CP asymmetries $A_{J/\Psi K_S}$ and $A_{J/\Psi\Phi}$ (the "golden" channel in each system), mass and width differences ΔM_{B_d}, $\Delta\Gamma_d$, and ΔM_{B_s}, $\Delta\Gamma_s$, additional CP asymmetries involving *different* combinations of invariant phases, $\sin(2\bar{\alpha})$, $\sin(2\bar{\beta}+\gamma)$ (and $\cos(2\bar{\beta})$ which removes a discrete ambigity in fixing $2\bar{\beta} = \sin^{-1}(A_{J/\Psi K_S})$), and, finally, semileptonic asymmetries A_{SL}^d, A_{SL}^s and A_{SL}^b. The actual values are collected in table 2.

Table 2. B_d and B_s mixing-related observables [4, 5, 7].

$A_{J/\Psi K_S}$	0.68 ± 0.02	ΔM_{B_d}	$(0.508 \pm 0.004)\ \text{ps}^{-1}$
$A_{J/\Psi\Phi}$	0.002 ± 0.0873	ΔM_{B_s}	$(17.725 \pm 0.049)\ \text{ps}^{-1}$
$\sin(2\bar{\alpha})$	0.00 ± 0.15	$\sin(2\bar{\beta}+\gamma)$	1.0 ± 0.16
$\cos(2\bar{\beta})$	1.35 ± 0.34		
A_{SL}^d	-0.003 ± 0.0078	$\Delta\Gamma_d/\Gamma_d$	-0.017 ± 0.021
A_{SL}^s	-0.0017 ± 0.0091	$\Delta\Gamma_s$	$(0.116 \pm 0.019)\ \text{ps}^{-1}$
A_{SL}^b	-0.00787 ± 0.00196		

- Representative rare decays of B mesons (table 3).

Table 3. B_d and B_s rare decays [4, 5, 8].

$\text{Br}(B \to X_s\gamma)$	$(3.56 \pm 0.25) \times 10^{-4}$
$\text{Br}(B \to X_s\mu^+\mu^-)$	$(1.60 \pm 0.51) \times 10^{-6}$
$\text{Br}(B_s \to \mu^+\mu^-)$	$(0^{+2.25}_{--}) \times 10^{-9}$
$\text{Br}(B_d \to \mu^+\mu^-)$	$(0^{+5.15}_{--}) \times 10^{-10}$

- Observables from the kaon sector (table 4).

Table 4. Kaon mixing and rare decays [9, 10].

ϵ_K	$(2.228 \pm 0.011) \times 10^{-3}$
ε'/ϵ_K	$(1.67 \pm 0.16) \times 10^{-3}$
$\text{Br}(K^+ \to \pi^+\nu\bar{\nu})$	$(1.73 \pm 1.05) \times 10^{-10}$
$\text{Br}(K_L \to \pi^0\nu\bar{\nu})$	$< \mathcal{O}(10^{-8})$
$\text{Br}(K_L \to \mu^+\mu^-)$	$(6.84 \pm 0.11) \times 10^{-9}$

- Electroweak precision observables, in particular the oblique parameters S and T (U too, but its role is negligible):

$$\Delta S = 0.02 \pm 0.11\,, \qquad \Delta T = 0.05 \pm 0.12\,,$$

  with a correlation coefficient 0.879.
- D^0–$\bar{D}^0$ mixing: we require that the Z-mediated, short distance, contribution to the D^0–$\bar{D}^0$ mixing amplitude does not give a larger than observed mixing parameter x_D. Notice that, as long distance contributions to D^0–$\bar{D}^0$ are also present, we are not requiring that the short distance ones fully account for the observed $x_D = (0.8 \pm 0.2) \cdot 10^{-2}$.

3. Results

We summarize the results of a complete numeric analysis of the model [3] in tables 5 to 8, where experimental results (where available), SM expectations and the predictions of the model are displayed together as colored bars corresponding to 1, 2 and 3σ ranges.
Table 5 illustrates how the mixing matrix can depart from the SM tight 3×3 unitary structure: invariant phases such as β and β_s and mixing elements such as $|V_{tb}|$ and $|V_{ub}|$, span much wider ranges than in the SM.

Table 5.

Quantity	Experimental
γ	1.34 ± 0.24
β	–
β_s	–
$\lvert V_{ub}\rvert$	$(4.15 \pm 0.49) \times 10^{-3}$
$\mathrm{Br}(B^+ \to \tau^+\nu)$	$(1.13 \pm 0.23) \times 10^{-4}$
$\lvert V_{tb}\rvert$	0.88 ± 0.07

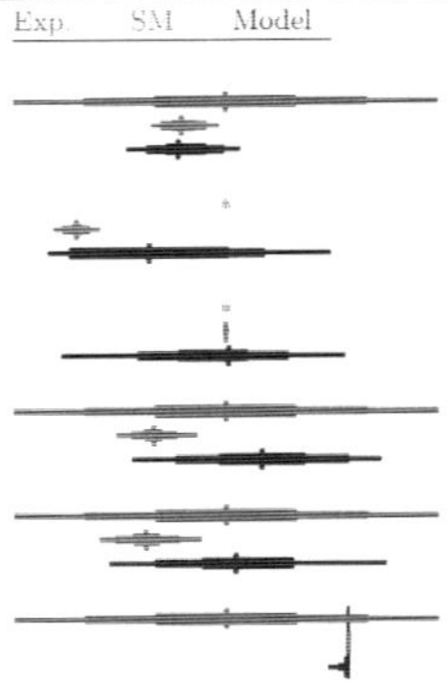

Table 6 illustrates how the model adequately reproduces constraints related to B_d and B_s mixings, together with ϵ_K and ε'/ϵ_K. Notice in addition that the CP asymmetry in $B_s \to J/\Psi\Phi$, less constraining from the experimental point of view, can sizeable depart from SM expectations (and it does so at a level that LHCb will be sensitive to).

Table 6.

Quantity	Experimental
$A_{J/\Psi K_S}$	0.68 ± 0.02
ΔM_{B_d}	$(0.508 \pm 0.004)\mathrm{ps}^{-1}$
$A_{J/\Psi\Phi}$	0.002 ± 0.087
ΔM_{B_s}	$(17.725 \pm 0.049)\mathrm{ps}^{-1}$
ϵ_K	$(2.228 \pm 0.011) \times 10^{-3}$
ε'/ϵ_K	$(1.67 \pm 0.16) \times 10^{-3}$

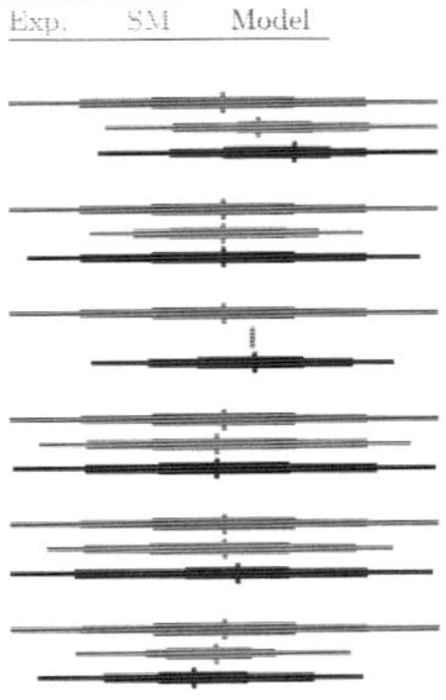

Table 7 addresses rare decays of kaons, B_d and B_s mesons. The model is in agreement with the most constraining ones while departures from SM expectations can be produced, particularly sizeable in the case of kaons.

Table 7. N.B. Black vertical lines stand for Br( ) = 0.

Quantity	Experimental	Exp. SM Model
$\text{Br}(K^+ \to \pi^+ \nu\bar{\nu})$	$(1.73 \pm 1.05) \times 10^{-10}$	
$\text{Br}(K_L \to \pi^0 \nu\bar{\nu})$	–	
$\text{Br}(K_L \to \mu^+\mu^-)$	$(6.84 \pm 0.11) \times 10^{-9}$	
$\text{Br}(B \to X_s\gamma)$	$(3.56 \pm 0.24) \times 10^{-4}$	
$\text{Br}(B \to X_s\mu^+\mu^-)$	$(1.60 \pm 0.51) \times 10^{-6}$	
$\text{Br}(B_s \to \mu^+\mu^-)$	$(3.2 \pm 1.4) \times 10^{-9}$	
$\text{Br}(B_d \to \mu^+\mu^-)$	$< 0.95 \times 10^{-9}$(95% CL)	

Finally, table 8 displays the semileptonic asymmetries in B_d and B_s mesons. Current experimental uncertainty on A^d_{SL} and A^s_{SL} is too large to be displayed. Departure from SM expectations is clear possible within this model. Nevertheless, the value of A^b_{SL} measured at D0 cannot be reproduced, even though the "tension" that this measurement brings into the flavour picture is softened. For completeness, the combination $A^s_{SL} - A^d_{SL}$, measurable at LHCb, is also displayed.

Table 8.

Quantity	Experimental	Exp. SM Model
A^d_{sl}	-0.003 ± 0.0078	
A^s_{sl}	-0.0024 ± 0.0063	
A^b_{sl}	-0.00787 ± 0.00196	
$A^s_{sl} - A^d_{sl}$		

Besides the information summarized in the previous tables, correlations among observables provide a huge playground to put the model to the test.

- Figure 1(a) illustrates how deviations from SM expectations are correlated for A^d_{SL} and $\text{Br}(B \to \tau\nu)$ (controlled by $|V_{ub}|$); red ellipses correspond to SM 68%, 95% and 99% CL regions.
- Figure 1(b) shows the strong correlation among A^s_{SL} and $A_{J/\Psi\Phi}$ when non standard values for both are present. The red cross indicates the SM prediction. Notice that within the SM the allowed range of variation is too small to be resolved with the scales of the figure.
- Figure 1(c) shows how significant deviations from $|V_{tb}|$ can only be achieved for relatively light values of m_T.

- Figure 1(d) shows how both $Br(K_L \to \pi^0 \nu\bar{\nu})$ and $Br(K^+ \to \pi^+ \nu\bar{\nu})$ can deviate from SM expectations (the red cross indicates the central value, the red bars over the axes stand for the 68%, 95% and 99% CL ranges), and do following a well defined pattern.

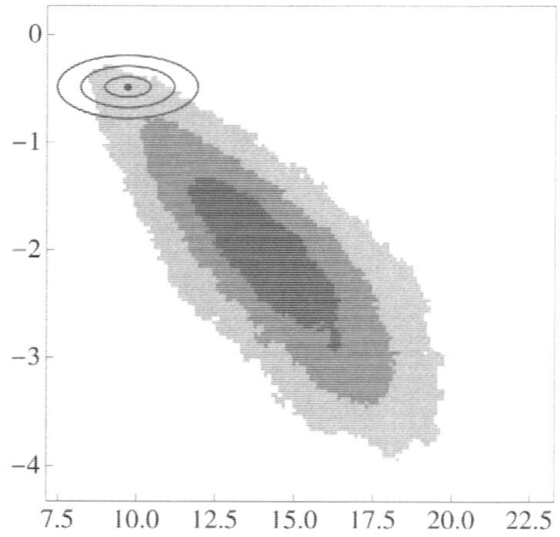

(a) $A^d_{SL} \times 10^3$ vs. $\mathrm{Br}(B \to \tau\nu) \times 10^4$.

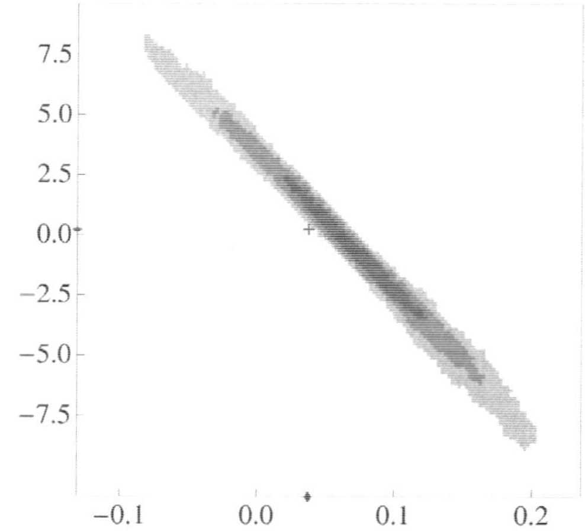

(b) $A^s_{SL} \times 10^4$ vs. $A_{J/\Psi\Phi}$.

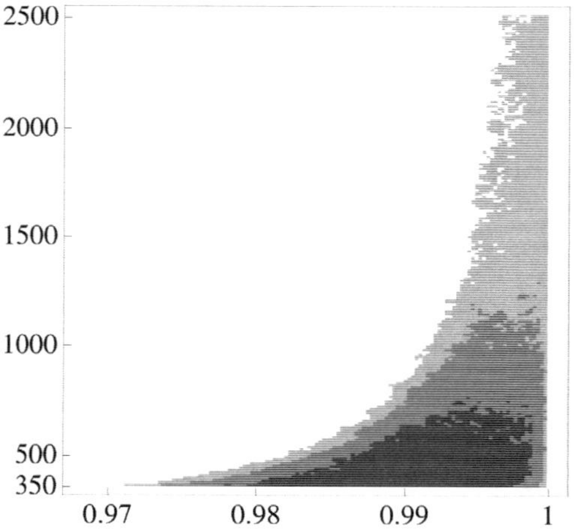

(c) $m_T(\mathrm{GeV})$ vs. $|V_{tb}|$.

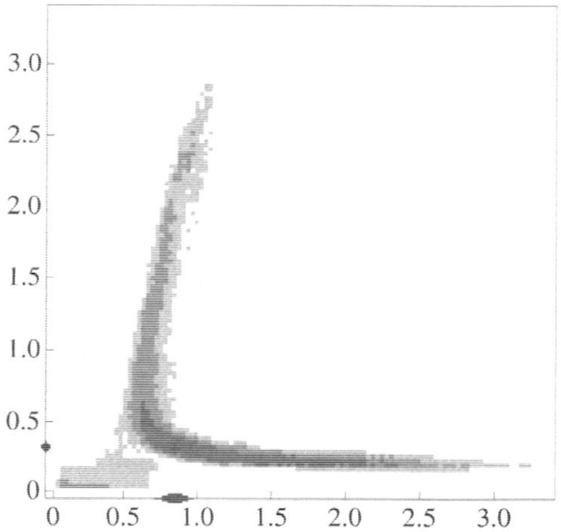

(d) $Br(K_L \to \pi^0 \nu\bar{\nu}) \times 10^{10}$ vs. $Br(K^+ \to \pi^+ \nu\bar{\nu}) \times 10^{10}$.

Figure 1. $\Delta\chi^2$ profiles of different correlations; 68%, 95% and 99% CL regions are shown.

4. Conclusions

We have presented an overview of a detailed analysis of flavour data in the context of a simple extension of the Standard Model, that includes an additional $Q = 2/3$ vector-like isosinglet quark. Experimental constraints from all the relevant quark flavour sectors are imposed and yet important deviations from Standard Model expectations can be present. This has been illustrated with different individual observables and important correlations.

Acknowledgments
This work was partially supported by *Fundação para a Ciência e a Tecnologia* (FCT, Portugal) through the projects CERN/FP/109305/2009, CERN/FP/116328/2010 and CERN/FP/123580/2011, PTDC/FIS/098188/2008 and CFTPFCT Unit 777 which are partially funded through POCTI (FEDER), by Marie Curie Initial Training Network UNILHC PITN-GA-2009-237920, by *Accion Complementaria Luso-Espanhola* AIC-D-2011-0809, by European FEDER, Spanish MINECO under grant FPA2011-23596 and GVPROMETEO 2010-056. The authors are very grateful for the hospitality and magnificient organization of *Discrete 2012*.

References
[1] Langacker P and London D 1988 *Phys.Rev.* **D38** 886; del Aguila F and Bowick M 1983 *Nucl.Phys.* **B224** 107; del Aguila F, Laermann E and Zerwas P 1988 *Nucl.Phys.* **B297** 1; Cheng T and Li L F 1992 *Phys.Rev.* **D45** 1708–1710.
[2] del Aguila F and Cortes J 1985 *Phys.Lett.* **B156** 243; del Aguila F, Chase M and Cortes J 1986 *Nucl.Phys.* **B271** 61; Branco G and Lavoura L 1986 *Nucl.Phys.* **B278** 738; Buchmuller W and Gronau M 1989 *Phys.Lett.* **B220** 641; Nir Y and Silverman D J 1990 *Phys.Rev.* **D42** 1477–1484; Nardi E, Roulet E and Tommasini D 1992 *Nucl.Phys.* **B386** 239–266; Silverman D 1992 *Phys.Rev.* **D45** 1800–1803; Branco G, Parada P, Morozumi T and Rebelo M 1993 *Phys.Lett.* **B306** 398–402; Branco G, Morozumi T, Parada P and Rebelo M 1993 *Phys.Rev.* **D48** 1167–1175; Branco G, Parada P and Rebelo M 1995 *Phys.Rev.* **D52** 4217–4222 (*Preprint* `hep-ph/9501347`); Barger V, Berger M and Phillips R 1995 *Phys.Rev.* **D52** 1663–1683 (*Preprint* `hep-ph/9503204`); del Aguila F, Aguilar-Saavedra J and Branco G 1998 *Nucl.Phys.* **B510** 39–60 (*Preprint* `hep-ph/9703410`); Barenboim G and Botella F 1998 *Phys.Lett.* **B433** 385–395 (*Preprint* `hep-ph/9708209`); Barenboim G, Botella F, Branco G and Vives O 1998 *Phys.Lett.* **B422** 277–286 (*Preprint* `hep-ph/9709369`); Kakebe I and Yamamoto K 1998 *Phys.Lett.* **B416** 184–191 (*Preprint* `hep-ph/9705203`); Barenboim G, Botella F and Vives O 2001 *Phys.Rev.* **D64** 015007 (*Preprint* `hep-ph/0012197`); Higuchi K and Yamamoto K 2000 *Phys.Rev.* **D62** 073005 (*Preprint* `hep-ph/0004065`); Barenboim G, Botella F and Vives O 2001 *Nucl.Phys.* **B613** 285–305 (*Preprint* `hep-ph/0105306`); Aguilar-Saavedra J 2003 *Phys.Rev.* **D67** 035003 (*Preprint* `hep-ph/0210112`); Aguilar-Saavedra J, Botella F, Branco G and Nebot M 2005 *Nucl.Phys.* **B706** 204–220 (*Preprint* `hep-ph/0406151`); Botella F J, Branco G C and Nebot M 2009 *Phys.Rev.* **D79** 096009 (*Preprint* `0805.3995`); Higuchi K and Yamamoto K 2010 *Phys.Rev.* **D81** 015009 (*Preprint* `0911.1175`); Frampton P H, Hung P and Sher M 2000 *Phys.Rept.* **330** 263 (*Preprint* `hep-ph/9903387`).
[3] Botella F, Branco G and Nebot M 2012 *JHEP* **1212** 040 (*Preprint* `1207.4440`)
[4] Beringer J and others (Particle Data Group) (Particle Data Group) 2012 *Phys.Rev.* **D86** 010001
[5] Amhis Y and others (Heavy Flavor Averaging Group) (Heavy Flavor Averaging Group) 2012 (*Preprint* `1207.1158`)
[6] Aubert B and others (BABAR Collaboration) (BaBar Collaboration) 2007 *Phys.Rev.Lett.* **99** 251801 (*Preprint* `hep-ex/0703037`); Poluektov A and others (Belle Collaboration) (Belle Collaboration) 2010 *Phys.Rev.* **D81** 112002 (*Preprint* `1003.3360`); Ikado K and others (Belle Collaboration) (Belle Collaboration) 2006 *Phys.Rev.Lett.* **97** 251802 (*Preprint* `hep-ex/0604018`); Hara K and others (Belle Collaboration) (Belle collaboration) 2010 *Phys.Rev.* **D82** 071101 (*Preprint* `1006.4201`); Aubert B and others (BABAR Collaboration) (BABAR Collaboration) 2008 *Phys.Rev.* **D77** 011107 (*Preprint* `0708.2260`); Lees J and others (BABAR Collaboration) (BABAR Collaboration) 2012 (*Preprint* `1207.0698`).
[7] Aaij R and others (LHCb Collaboration) (LHCb Collaboration) 2012 *Phys.Rev.Lett.* **108** 101803 (*Preprint* `1112.3183`); Aaij R and others (LHCb Collaboration) (LHCb Collaboration) 2012 *Phys.Lett.* **B707** 497–505 (*Preprint* `1112.3056`); Aubert B and others (BABAR Collaboration) (BABAR Collaboration) 2009 *Phys.Rev.* **D79** 072009 (*Preprint* `0902.1708`); Adachi I, Aihara H, Asner D, Aulchenko V, Aushev T and others (Belle Collaboration) (Belle Collaboration) 2012 *Phys.Rev.Lett.* **108** 171802 (*Preprint* `1201.4643`); Aad G and others (ATLAS Collaboration) (ATLAS Collaboration) 2012 *Phys. Lett.* **B713** 387 (*Preprint* `1204.0735`).
[8] Chatrchyan S and others (CMS Collaboration) (CMS Collaboration) 2011 *Phys.Rev.Lett.* **107** 191802 (*Preprint* `1107.5834`); Chatrchyan S and others (CMS Collaboration) (CMS Collaboration) 2012 *JHEP* **1204** 033 (*Preprint* `1203.3976`); Aaij R and others (LHCb Collaboration) (LHCb collaboration) 2012 *Phys. Rev. Lett. 108,* **231801** (*Preprint* `1203.4493`); Aaij R and others (LHCb Collaboration) (LHCb Collaboration) 2012 *Phys.Lett.* **B708** 55–67 (*Preprint* `1112.1600`).
[9] Alavi-Harati A and others (KTeV Collaboration) (KTeV Collaboration) 1999 *Phys.Rev.Lett.* **83** 22–

27 (*Preprint* hep-ex/9905060); Ambrose D and others (E871 Collaboration) (E871 Collaboration) 2000 *Phys.Rev.Lett.* **84** 1389–1392; Ahn J and others (E391a Collaboration) (E391a Collaboration) 2008 *Phys.Rev.Lett.* **100** 201802 (*Preprint* 0712.4164); Artamonov A and others (E949 Collaboration) (E949 Collaboration) 2008 *Phys.Rev.Lett.* **101** 191802 (*Preprint* 0808.2459); Abouzaid E and others (KTeV Collaboration) (KTeV Collaboration) 2011 *Phys.Rev.* **D83** 092001 (*Preprint* 1011.0127).

[10] Cirigliano V, Ecker G, Neufeld H, Pich A and Portoles J 2012 *Rev.Mod.Phys.* **84** 399 (*Preprint* 1107.6001)

Journal of Physics: Conference Series **447** (2013) 012062 doi:10.1088/1742-6596/447/1/012062

Lepton Number Violation and Lepton Flavour Violation at LHCb

B. Adeva, on behalf of the LHCb collaboration

Departamento de Fisica de Particulas. Facultad de Fisica. Campus Sur.
E-15782 Santiago de Compostela, Spain

E-mail: Bernardo.Adeva@usc.es

Abstract. Searches for heavy Majorana neutrinos in B^- decays with final states containing hadrons plus a $\mu^-\mu^-$ pair are performed using 0.41 fb^{-1} of data collected with the LHCb detector in proton-proton collisions at a center-of-mass energy of 7 TeV. No signals are found and upper limits are set on Majorana neutrino production and B^- decay branching fractions as a function of the Majorana neutrino mass.

A search is also made for the lepton flavour violating decay $\tau \to \mu^+\mu^-\mu^-$ using 1.0 fb^{-1} of data collected by LHCb in 2011. The observed number of events is consistent with the background expectation, and an upper limit on the branching ratio is set. Using a similar procedure, the lepton flavour violating and baryon number violating decays $\tau^- \to \bar{p}\mu^+\mu^-$ and $\tau^- \to p\mu^-\mu^-$ are searched for, and upper limits provided at 95% confidence level.

1. Introduction

Leptons constitute a crucially important sector of elementary particles. Half of the leptons are neutrinos. Yet we do not know if they are Dirac or Majorana particles, the latter case characterized by being their own antiparticles [1]. Since the observation of neutrino oscillations has indisputably established that neutrinos have non-zero mass, it is possible to distinguish the two types experimentally. Finding neutrinoless double β decay has long been advocated as a premier demonstration of the possible Majorana nature of neutrinos [2]. An impressive lower limit from neutrinoless double β decays in nuclei has already been obtained on the half-life of $\mathcal{O}(10^{25})$ years [3] for coupling to e^-.

Similar processes can occur in B^- decays. The diagram is shown in Fig. 1(a). In this reaction there is no restriction on the mass of the Majorana neutrino as it acts as a virtual particle. Unlike in neutrinoless double beta decays, a like-sign dimuon is considered rather than two electrons. The only existing limit is from a recent Belle measurement [4] using the $B^- \to D^+\mu^-\mu^-$ channel. We consider only final states where the $c\bar{d}$ pair forms a final-state meson, either a D^+ or a D^{*+}, so the processes we are looking for are $B^- \to D^{(*)+}\mu^-\mu^-$. Mention of a specific reaction also implies inclusion of the charge conjugate reaction, throughout the report.

There are other processes involving b-quark decays that produce a light neutrino that can mix with a heavy neutrino, designated as N. The heavy neutrino can decay as $N \to W^+\mu^-$. In Fig. 1(b) we show the annihilation processes $B^- \to \pi^+(D_s^+)\mu^-\mu^-$, where the virtual W^+ materializes either as a π^+ or D_s^+. These decays have been discussed in the literature [5, 6]. The lifetime of N are not predicted. It is assumed here that they are long enough that the natural decay width is narrower than the LHCb mass resolution, which varies between 2 and 15

Journal of Physics: Conference Series **447** (2013) 012062 doi:10.1088/1742-6596/447/1/012062

MeV, depending on mass and decay mode. For $B^- \to \pi^+\mu^-\mu^-$, we can access the Majorana mass region between approximately 260 and 5000 MeV while for $B^- \to D_s^+\mu^-\mu^-$, the Majorana mass region is between 2100 and 5150 MeV. In the higher mass region, the W^+ may be more likely to form a D_s^+ meson than a π^+. The $B^- \to \pi^+\mu^-\mu^-$ search was first performed by Mark-II [7] and then by CLEO [8]. LHCb also performed a similar search using a smaller 0.04 fb^{-1} data sample [9] giving an upper limit of 5.8×10^{-8} at 95% confidence level (CL). The decay of $B^- \to D_s^+\mu^-\mu^-$ has never been investigated.

Finally, in Fig. 1(c) it is shown how prolific semileptonic decays of the B^- can result in the $D^0\pi^+\mu^-\mu^-$ final state. This process has never been probed [10]. It benefits from the higher value of the CKM coupling $|V_{cb}|$ relative to $|V_{ub}|$ in the annihilation processes shown in Fig. 1(b). The accessible region for Majorana neutrino mass is between 260 and 3300 MeV. For all the modes reported here, only decays with muons in the final state are searched for, though electrons, and τ leptons in cases where sufficient energy is available, could also be produced. Searches have also been carried out looking for like-sign dileptons in hadron collider experiments [11].

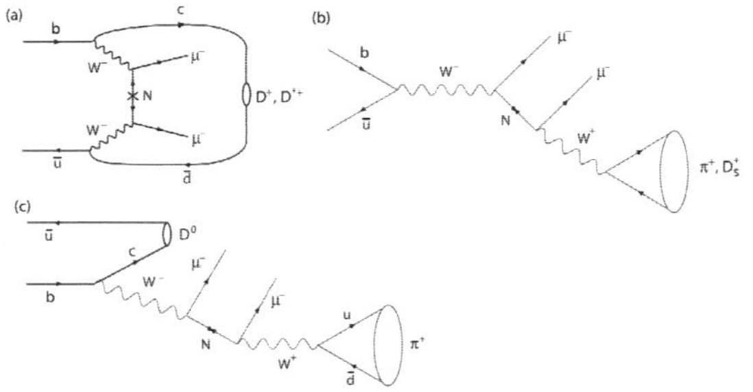

Figure 1. Feynman diagrams for B decays involving an intermediate heavy neutrino (N). (a) $B^- \to D^{(*)+}\mu^-\mu^-$, (b) $B^- \to \pi^+(D_s^+)\mu^-\mu^-$, and (c) $B^- \to D^0\pi^+\mu^-\mu^-$.

A second topic discussed in this communication is lepton flavour violation searches by the LHCb experiment. The experimental observation of neutrino oscillations was the first evidence of lepton flavour violation (LFV). The consequent introduction of mass terms for the neutrinos in the Standard Model already implies lepton family number violation also in the charged sector, but with branching fractions smaller than 10^{-40}. New physics (NP) could significantly enhance the rates, but charged lepton flavour violating (cLFV) decays like $\mu^- \to e^-\gamma$ $\mu^- \to e^+e^-e^-$ $\tau^- \to l^-\gamma$ and $\tau^- \to l^+l^-l^-$ with $l^- = e^-, \mu^-$ have not been observed so far even with steadily improving experimental sensitivity. Numerous beyond the Standard Model theories predict enhanced LFV in τ^- decays over μ^- decays. with branching fractions within experimental reach [12]. An observation of cLFV would thus be a clear sign for NP, while lowering the experimental upper limit will help to further constrain exotic theories. The inclusive τ^- production cross section at the LHC is large (about 80 μb) and muon final states provide clean signatures in the LHCb detector. This decay is favoured with respect to the decays $\tau^- \to \mu^-\gamma$ and $\tau^- \to e^+e^-e^-$ due to considerably better particle identification of the muons and better possibilities for background discrimination. We report here a first result by the LHCb collaboration [17] on the $\tau^- \to \mu^+\mu^-\mu^-$ decay which demonstrates an experimental sensitivity comparable with the current best experimental upper limit, from Belle $\mathrm{B}(\tau^- \to \mu^+\mu^-\mu^-) < 2.1 \times 10^{-8}$ at 90% confidence level (CL) [14]).

We also report here LHCb searches which violate both lepton number (LNV) (and hence also lepton flavour) and baryon number (BNV) conservation. Taking advantage of the high inclusive τ^- production at the LHC, a search has been performed [19] for the two decay modes $\tau^- \to \bar{p}\mu^+\mu^-$ and $\tau^- \to p\mu^-\mu^-$, using the same τ^- normalization as for the $\tau^- \to \mu^+\mu^-\mu^-$

decay search.

BaBar and Belle have searched for τ^- decays with $|(B-L)| = 0$ and $|(B-L)| = 2$ using the modes $\tau^- \to \Lambda h^-, \Lambda h^-$ (with $h^- = l^-, K^-$). Limits of order 10^{-7} were obtained, and the results are summarised in Ref [21]. BaBar has also searched for B-meson decays $B^0 \to \Lambda_c^+ h^-, B^- \to \Lambda h^-$ (both having $|(B-L)| = 0$) and $B^- \to \Lambda h^-$, $|(B-L)| = 2$, with 90% CL upper limits in the range 3.2 to 520×10^{-8} [22]. No measurements currently exist for the $\tau^- \to \bar{p}\mu^+\mu^-$ and $\tau^- \to p\mu^-\mu^-$ decay modes; this analysis presents the first of such searches. These modes probe LNV and BNV similarly to the above-mentioned modes studied by BaBar and Belle.

2. Searches for Majorana neutrinos in B^- decays

Values for branching fractions will be normalized to well measured channels that have the same number of muons in the final state and equal track multiplicities. The first such channel is $B^- \to J/\psi K^-$. Its branching fraction is $\mathcal{B}(B^- \to J/\psi K^-) = (1.014 \pm 0.034) \times 10^{-3}$ [3]. We use the $J/\psi \to \mu^+\mu^-$ decay mode. The product branching fraction of this normalization channel is $(6.013 \pm 0.021) \times 10^{-5}$, and is known to an accuracy of $\pm 2\%$.

The invariant mass of $K^-\mu^+\mu^-$ candidates is shown in Fig. 2(a). In this analysis the $\mu^+\mu^-$ invariant mass is required to be within 50 MeV of the J/ψ mass. Using an unbinned log-likelihood fit yields 47,224±222 $B^- \to J/\psi K^-$ events. Within a $\pm 2\sigma$ signal window about the peak mass, taken as the signal region, there are 44,283 of these events. with a systematic uncertainty of 0.3%. The width of the signal peak is found to be 19.1±0.1 MeV.

For final states with five tracks, we change the normalization channel to $B^- \to \psi(2S)K^-$, with $\psi(2S) \to \pi^+\pi^- J/\psi$, and $J/\psi \to \mu^+\mu^-$. The branching fraction for this channel is $\mathcal{B}(B^- \to \psi(2S)K^-) = (6.48 \pm 0.35) \times 10^{-4}$ [3]. Events are selected using a similar procedure as for $J/\psi K^-$ but adding a $\pi^+\pi^-$ pair, that must have an invariant mass when combined with the J/ψ which is compatible with the $\psi(2S)$ mass, and that forms a consistent vertex with the other B^- decay candidate tracks. The total efficiency for $\mu^+\mu^-\pi^+\pi^-K^-$ is (0.078±0.002)%, without inclusion of the $\psi(2S)$ or J/ψ branching fractions. The B^- candidate mass plot is shown in Fig. 2(b). From analysis of the mass spectrum, 767±29 signal events are found in a $\pm 2\sigma$ window about the peak mass, with 0.7% systematic uncertainty on the yield.

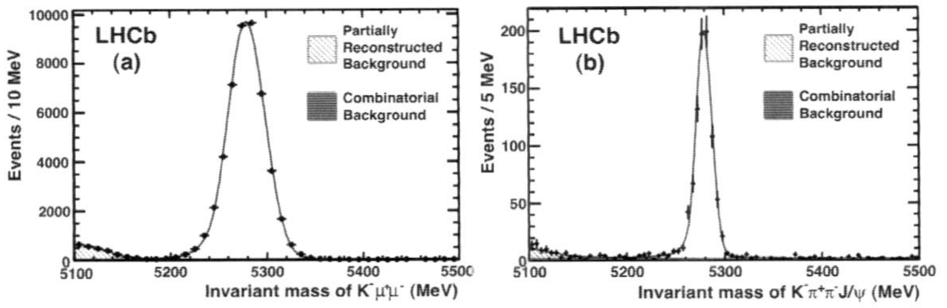

Figure 2. Invariant mass of (a) candidate $J/\psi K^-$ decays, and (b) candidate $J/\psi K^-\pi^+\pi^-$ decays.

2.1. Analysis of $B^- \to D^+\mu^-\mu^-$ and $D^{*+}\mu^-\mu^-$

Decay diagrams for $B^- \to D^{(*)+}\mu^-\mu^-$ are shown in Fig. 1(a). Since the neutrinos are virtual, the process can proceed for any value of neutrino mass. The $D^+ \to K^-\pi^+\pi^+$ and $D^{*+} \to \pi^+ D^0$, $D^0 \to K^-\pi^+$ channels are used. The decay products of the D^+ and D^0 candidates are required to have invariant masses within ±25 MeV of the charm meson mass, and for D^{*+} candidate selection the mass difference $m(\pi^+K^-\pi^+) - m(K^-\pi^+)$ is required to be within ±3 MeV of the known $D^{*+} - D^0$ mass difference. The $D^{(*)+}\mu^-\mu^-$ candidate mass spectra is analysed and no signals are apparent.

Using Poission statistics on the value of $(S + B)$, where S indicates the expectation value of signal and B background, the following upper limits on the branching fractions at 95% CL are set:

$$\begin{aligned} \mathcal{B}(B^- \to D^{\ +}\mu^-\mu^-) &< 6.9 \times 10^{-7} \text{ and} \\ \mathcal{B}(B^- \to D^{*+}\mu^-\mu^-) &< 2.4 \times 10^{-6} \ . \end{aligned} \tag{1}$$

The limit on the D^+ channel is more stringent than a previous limit from Belle of 1×10^{-6} at 90% CL [4], and the limit on the D^{*+} channel is the first such result.

2.2. Analysis of $B^- \to \pi^+\mu^-\mu^-$

The selection of $\pi^+\mu^-\mu^-$ events uses the same criteria as described for $J/\psi K^-$ above, except for like-sign rather than opposite-sign dimuon charges and pion rather than kaon identification. The invariant mass distribution of $\pi^+\mu^-\mu^-$ candidates is analysed with a mass resolution for this final state 20.3±0.2 MeV. The signal region is defined by an interval of $\pm 2\sigma$ centered on the B^- mass, where 7 events are found, but no signal above background is apparent. The peaking background, estimated as 2.5 events, is due to misidentified $B^- \to J/\psi K^-$ or $J/\psi\pi^-$ decays. The combinatorial background is determined to be 5.3 events from a fit to the $\pi^+\mu^-\mu^-$ mass distribution excluding the signal region. The total background in the signal region then is 7.8 ± 1.3 events.

Since the putative neutrinos considered here decay into $\pi^+\mu^-$, and are assumed to have very narrow widths, more sensitivity is obtained by examining this mass distribution for events in the B^- signal region. No statistically significant signal is found at any mass.

Many systematic errors in the signal yield cancel in the ratio to the normalization channel. The largest sources of error are the modeling of the detector efficiency (5.3%) and the measured branching fractions $\mathcal{B}(B^- \to J/\psi K^-)$ (3.4%), and $\mathcal{B}(J/\psi \to \mu^+\mu^-)$ (1.0%) [3].

To set upper limits on the branching fraction, the number of events $N_{\rm obs}$ at each M_N value (within $\pm 3\sigma_N$) are counted, and the procedure described above is applied. Estimated backgrounds are taken from the sidebands of the mass spectreum. Upper limits on $\mathcal{B}(B^- \to \pi^+\mu^-\mu^-)$ as a function of M_N are derived at 95% CL. For most of the neutrino mass region, the limits on the branching ratio are $< 8 \times 10^{-9}$. Assuming a phase space decay of the B^- we determine

$$\mathcal{B}(B^- \to \pi^+\mu^-\mu^-) < 1.3 \times 10^{-8} \text{ at 95\% CL.}$$

These limits improve on previous results from CLEO ($< 1.4 \times 10^{-6}$ at 90% CL) [8], and LHCb [9] ($< 5.8 \times 10^{-8}$ at 95% CL).

2.3. Analysis of $B^- \to D_s^+\mu^-\mu^-$

The process $B^- \to D_s^+\mu^-\mu^-$ is similar to $B^- \to \pi^+\mu^-\mu^-$, with the difference being that the heavy neutrino can decay into $D_s^+\mu^-$. Here we consider only $D_s^+ \to K^+K^-\pi^+$ decays. Our analysis follows a similar procedure used for the $\pi^+\mu^-\mu^-$ channel. Candidate $D_s^+ \to K^+K^-\pi^+$ decays are selected by having an invariant mass within ±25 MeV of the D_s^+ mass. A Majorana neutrino candidate decay is then looked for by having the D_s^+ candidate decay tracks form a vertex with an opposite-sign muon candidate. Then this neutrino candidate must form a vertex with another muon of like-sign to the first one consistent with a B^- decay detached from the primary vertex. The invariant mass spectrum of $D_s^+\mu^-\mu^-$ candidates is analysed, with the mass resolution being 15.5±0.3 MeV. After selecting the events in the B^- signal region, the $D_s^+\mu^-$

invariant mass distribution is also analysed. A background estimate is made using the sideband data in B^- candidate mass, and the data in the signal region is consistent with the background estimate.

The overall efficiencies and resolution for different values of M_N are determined from simulation. As done previously, during the scan over the accessible Majorana neutrino mass region we use a $\pm 3\sigma_N$ mass window around a given Majorana mass. The main contribution comes from errors in the branching fractions and trigger efficiencies. Again we provide upper limits as a function of the Majorana neutrino mass, only taking into account combinatorial background in this case as the peaking background is absent. For neutrino masses below 5 GeV, the limits on the mass dependent branching fractions are mostly $< 6 \times 10^{-7}$. We also determine an upper limit on the total branching fraction. We observe 12 events with an expected background yield of 22 events and use the CL_s method for calculating the upper limit [23]. Assuming a phase space decay of the B^- we find

$$\mathcal{B}(B^- \to D_s^+ \mu^- \mu^-) < 5.8 \times 10^{-7} \text{ at } 95\% \text{ CL.}$$

2.4. Analysis of $B^- \to D^0 \pi^+ \mu^- \mu^-$

Majorana neutrinos could be produced via semileptonic decays as shown in Fig. 1(c). Here the mass range probed is smaller than in the case of $\pi^+\mu^-\mu^-$ due to the presence of the D^0 meson in the final state.

After a similar selection procedure as described above, the invariant mass distribution of $D^0\pi^+\mu^-\mu^-$ is analysed in search of a B^- signal, the mass resolution being 14.4±0.2 MeV. Peaking backgrounds are essentially absent. The largest source is $B^- \to D^0\pi^-\pi^-\pi^+$ which contributes only 0.13 events in the signal region. The combinatorial background, determined by a linear fit to the sidebands of the B^- signal region, predicts 35.9 events, while the number of observed is 33. The $\pi^+\mu^-$ invariant mass for events within two standard deviations of the B^- mass is also analysed.

Upper limits for $\mathcal{B}(B^- \to D^0\pi^+\mu^-\mu^-)$ as a function of the $\pi^+\mu^-$ mass are set for Majorana neutrino masses < 3.0 GeV, the upper limits being less than 1.6×10^{-6} at 95% CL. The final limit on the branching fraction assuming a phase space decay is

$$\mathcal{B}(B^- \to D^0\pi^+\mu^-\mu^-) < 1.5 \times 10^{-6} \text{ at } 95\% \text{ CL.}$$

3. Search for the lepton flavour violating decay $\tau \to \mu^+\mu^-\mu^-$

In this analysis, the full 2011 LHCb data sample, corresponding to $1.0fb^{-1}$ collected at $\sqrt{s} = 7\ TeV$, is used to search for the $\tau^- \to \mu^+\mu^-\mu^-$ decay. Two selections are implemented, one for the signal mode and the other for the normalisation channel $D_s^- \to \phi\pi^-$ followed by $\phi \to \mu^+\mu^-$. These cut-based selections are designed to give maximal efficiency but at the same time reduce the dataset to a manageable level.

The real discrimination between potential signal and backgrounds is done over a binned three-dimensional distribution in:

- a likelihood based on the three-body decay properties (M_{3body})
- a likelihood based on muon identification (M_{PID})
- the invariant mass of the 3μ candidate ($M_{3\mu}$).

The initial studies were performed blind, with 3 muon candidates with mass within 30 MeV (3 σ) of the τ^- mass excluded from the analysis. The analysis strategy and limit-setting procedures are similar to those used for the LHCb analyses of $B_{s,d} \to \mu^+\mu^-$ [15].

To estimate the signal branching fraction we normalise the number of observed signal events to the number of events in the $D_s \to \phi(\mu^+\mu^-)\pi^-$ calibration channel using the equation

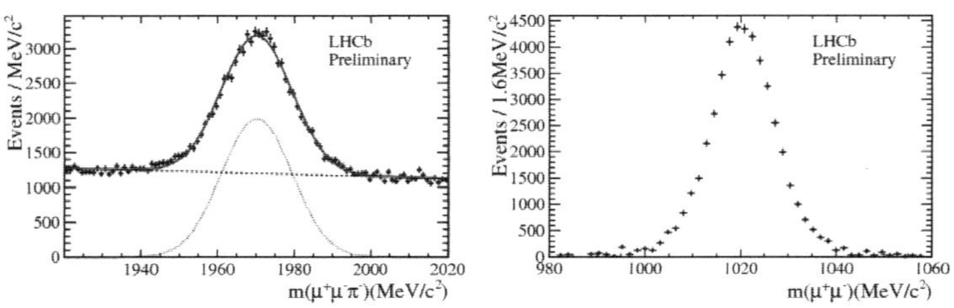

Figure 3. (a) Invariant mass distribution of the $D_s^- \to \phi(\mu^+\mu^-)\pi^-$ events with a Gaussian fit (fine dashed line), an exponential background (dashed line) and the combined fit as the solid line. (b) Invariant mass distribution of the $\phi \to \mu^+\mu^-$ events obtained using the sPlot[12] technique.

$$B(\tau^- \to \mu^+\mu^-\mu^-)$$
$$= B(D_s \to \phi(\mu^+\mu^-)\pi^-) \times \frac{f^{\tau}_{D_s}}{B(D_s^- \to \tau^-\bar{\nu_\tau})} \times \frac{\epsilon_{cal}^{REC|SEL}\epsilon_{cal}^{TRIG|SEL}}{\epsilon_{sig}^{REC|SEL}\epsilon_{sig}^{TRIG|SEL}} \times \frac{N_{sig}}{N_{cal}} = \alpha \times N_{sig} \quad (2)$$

where α is the overall normalisation factor and N_{sig} is the number of observed signal events. $B(D_s \to \tau^-\bar{\nu}_\tau)$ is the branching fraction for $D_s \to \tau^-\bar{\nu}_\tau$ taken from [16]. The quantity $f^{\tau}_{D_s}$ is the fraction of τ^- leptons which originate from D_s^- decays, calculated using the $b\bar{b}$ and $c\bar{c}$ cross sections as measured by LHCb [18] and the inclusive $b \to \tau^-$ and $c \to \tau^-$ branching fractions as measured by the LEP experiments [3].

The reconstruction and selection efficiency $REC|SEL$, is itself a combination of the detector acceptance for the particular decay, the muon identification efficiency and the selection efficiency. They are determined from simulation.

The branching ratio of the control channel $B(D_s^- \to \phi(\mu^+\mu^-)\pi^-)$ is determined from the combination of known branching fraction measurements [17]. The yield of $D_s^- \to \phi(\mu^+\mu^-)\pi^-$ events in data, N_{cal}, is extracted by a fit to the signal and background components of the reconstructed D_s^- mass distribution. It is found to be:

$$N(D_s^- \to \phi(\mu^+\mu^-)\pi^-) = 45520 \pm 420_{stat} \pm 820_{syst} \quad (3)$$

where the systematic uncertainty contains the dependence on the fit model. The background subtracted dimuon invariant mass spectrum of the $D_s^- \to \phi(\mu^+\mu^-)\pi^-$ events is shown in Fig. 3 (b). The $\phi(1020)$ resonance is seen with a very low background level.

The expected number of background events per bin in M_{3body}, M_{PID} and $M_{3\mu}$ likelihood space is calculated from an extended, unbinned maximum likelihood fit to the mass spectrum, excluding the signal window. The background processes for the rare decay $\tau^- \to \mu^+\mu^-\mu^-$ consist of long decay chains of heavy mesons with three real muons in the final state or including one or two real muons in combination with one or two mis-identified particles.

The CL_s method [23, 24] is used as a statistical framework. It provides two estimators, CL_s and CL_b, which give the level of compatibility with the signal and background hypotheses, respectively. CL_s is used to set the exclusion (upper) limit on $\tau^- \to \mu^-\mu^+\mu^-$ whereas CL_b is used to claim an incompatibility with the background-only hypothesis for an observation. The distribution of the observed CL_s values is calculated as function of the assumed branching fraction, and the observed upper limit is found to be: $B(\tau^- \to \mu^+\mu^-\mu^-) < 7.8 \times 10^{-8}$ at 95% CL.

4. Search for the baryon number violating decays $\tau^- \to \bar{p}\mu^+\mu^-$ and $\tau^- \to p\mu^-\mu^-$

The same data sample and a similar procedure as for $\tau^- \to \mu^+\mu^-\mu^-$ is used to search for $\tau^- \to \bar{p}\mu^+\mu^-$ and $\tau^- \to p\mu^-\mu^-$ decays. Three selections are implemented, two for the $\tau^- \to \bar{p}\mu^+\mu^-$ and $\tau^- \to p\mu^-\mu^-$ signal modes, and one for the normalisation channel, which is $D_s^- \to \phi(\mu^+\mu^-)\pi^-$. Discrimination between potential signal and backgrounds is performed using particle identification cuts and a binned two-dimensional distribution in two variables: a likelihood based on the 3-body kinematics of the event, and the invariant mass of the $p(\bar{p})\mu\mu$ candidate. The observed limits for the branching fractions $\tau^- \to \bar{p}\mu^+\mu^-$ and $\tau^- \to p\mu^-\mu^-$ are found to be: 4.5×10^{-7} and 6.0×10^{-7} respectively, at 95% CL.

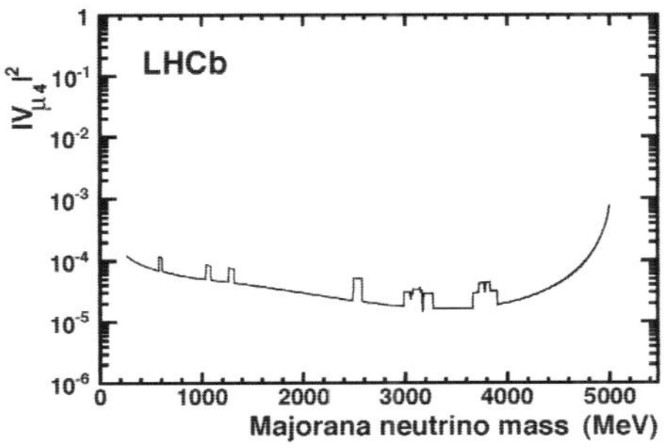

Figure 4. Upper limits on $|V_{\mu 4}|^2$ at 95% CL as a function of the Majorana neutrino mass from the $B^- \to \pi^+\mu^-\mu^-$ channel.

5. Conclusions

A search has been performed for Majorana neutrinos in the B^- decay channels, $D^{(*)+}\mu^-\mu^-$, $\pi^+\mu^-\mu^-$, $D_s^+\mu^-\mu^-$, and $D^0\pi^+\mu^-\mu^-$ which has only yielded upper limits. The $D^{(*)+}\mu^-\mu^-$ channels may proceed via virtual Majorana neutrino exchange and thus are sensitive to all Majorana neutrino masses. The other channels provide limits for neutrino masses between 260 and 5000 MeV. The bounds are summarized in Table 1. These limits are the most restrictive to date.

Table 1. Summary of upper limits on branching fractions. Both the limits on the overall branching fraction assuming a phase space decay, and the range of limits on the branching fraction as a function of Majorana neutrino mass (M_N) are given. All limits are at 95% CL.

Mode	$\mathcal{B}$ upper limit	Approx. limits as function of M_N
$D^+\mu^-\mu^-$	6.9×10^{-7}	
$D^{*+}\mu^-\mu^-$	2.4×10^{-6}	
$\pi^+\mu^-\mu^-$	1.3×10^{-8}	$(0.4 - 1.0) \times 10^{-8}$
$D_s^+\mu^-\mu^-$	5.8×10^{-7}	$(1.5 - 8.0) \times 10^{-7}$
$D^0\pi^+\mu^-\mu^-$	1.5×10^{-6}	$(0.3 - 1.5) \times 10^{-6}$

Our search has thus far ignored the possibility of a finite neutrino lifetime. Experimental sensitivity is lost for lifetimes longer than 10^{-10} s to 10^{-11} s, depending on the decay mode. Note that for the $D^{(*)+}\mu^-\mu^-$ final states the detection efficiency is independent of the neutrino lifetime, since the neutrino acts a virtual particle.

Our upper limits in the $\pi^+\mu^-\mu^-$ final state can be used to establish neutrino mass dependent upper limits on the coupling $|V_{\mu 4}|$ of a heavy Majorana neutrino to a muon and a virtual W. The matrix element has been calculated in Ref. [5]. The results are shown in Fig. 4 as a function of M_N.

A model dependent calculation of $\mathcal{B}(B^- \to D^0\pi^+\mu^-\mu^-)$ can also be used to extract $|V_{\mu 4}|$ [10], but the $\pi^+\mu^-\mu^-$ mode is more sensitive. For the $D^{(*)+}\mu^-\mu^-$ channels upper limits cannot be extracted until there is a theoretical calculation of the hadronic form-factor similar to those available for neutrinoless double β decay.

LHCb has performed additional searches for lepton flavour violation, and baryon number violation, using τ^- lepton decays from a data sample with integrated luminosity of 1.0 fb^{-1}, collected at $\sqrt{s} = 7$ TeV in 2011. The τ^- production rate has been normalised to the control channel $D_s^- \to \phi(\mu^+\mu^-)\pi^-$. The following upper limits have been set on the branching fraction of $\tau^- \to \mu^+\mu^-\mu^-$:

$$B(\tau^- \to \mu^+\mu^-\mu^-) < 6.3 \times 10^{-8} \quad at \quad 90\% \;\; CL \tag{4}$$

$$B(\tau^- \to \mu^+\mu^-\mu^-) < 7.8 \times 10^{-8} \quad at \quad 95\% \;\; CL \tag{5}$$

The above limits are not significantly dependent on the dimuon mass spectrum assumed in the decay model. In addition, the lepton flavour violating and baryon number violating decays $\tau^- \to \bar{p}\mu^+\mu^-$ and $\tau^- \to p\mu^-\mu^-$ have been searched for, and the observed numbers of events are consistent with the background expectations. The upper limits $B(\tau^- \to \bar{p}\mu^+\mu^- < 4.5(3.4)10^{-7}$ $B(\tau^- \to p\mu^-\mu^- < 6.0(4.6)10^{-7}$ have been set at 95% (90%) confidence level.

References

[1] E. Majorana, Teoria simmetrica dell 0elettrone e del positrone, Nuovo Cim. 14 (1937) 171.
[2] F. T. Avignone III, S. R. Elliott, and J. Engel,Rev. Mod. Phys. 80 (2008) 481, arXiv:0708.1033.
[3] Particle Data Group, K. Nakamura et al., Review of particle physics, J. Phys. G37 (2010) 075021
[4] BELLE collaboration, O. Seon et al., Phys. Rev. D84 (2011) 071106, arXiv:1107.0642.
[5] A. Atre, T. Han, S. Pascoli, and B. Zhang, JHEP 05 (2009) 030, arXiv:0901.3589.
[6] G. Cvetic, C. Dib, S. K. Kang, and C. Kim,Phys. Rev. D82 (2010) 053010, arXiv:1005.4282; J.-M. Zhang and G.-L. Wang, Eur. Phys. J. C71 (2011) 1715, arXiv:1003.5570.
[7] A. J. Weir et al., Phys. Rev. D41 (1990) 1384.
[8] CLEO collaboration, K. W. Edwards et al., Phys. Rev. D65 (2002) 111102, arXiv:hep-ex/0204017.
[9] LHCb Collaboration, R. Aaij et al. Phys. Rev. Lett. 108, 101601 (2012), arXiv:1110.0730
[10] D. Delepine, G. Lopez Castro, and N. Quintero, Phys. Rev. D84 (2011) 096011, arXiv:1108.6009.
[11] ATLAS Collaboration, G. Aad et al., JHEP 1110 (2011) 107, arXiv:1108.0366; CMS Collaboration, S. Chatrchyan et al., JHEP 1106 (2011) 077, arXiv:1104.3168; CDF Collaboration, A. Abulencia et al., Phys. Rev. Lett. 98 (2007) 221803, arXiv:hep-ex/0702051; CDF Collaboration, D. Acosta et al., Phys. Rev. Lett. 93 (2004) 061802, arXiv:hep-ex/0405063.
[12] W. J. Marciano, T. Mori, and J. M. Roney, Ann. Rev. Nucl. Part. Sci 58 (2008) 315.
[13] LHCb collaboration, A. A. Alves Jr. et al., The LHCb detector at the LHC, JINST 3 (2008) S08005.
[14] K. Hayasaka et al., Phys. Lett. B687 (2010) 139, arXiv:1001.3221.
[15] LHCb Collaboration, R. Aaij et al., arXiv:1203.4493.
[16] J. L. Rosner and S. Stone, arXiv:1201.2401.
[17] The LHCb Collaboration, LHCb-CONF-2012-027, September 24, 2012.
[18] LHCb Collaboration, R Aaij et al., Eur. Phys. J. C 71 (2011) 1645.
[19] The LHCb Collaboration, LHCb-CONF-2012-015, May 31, 2012.
[20] M. Raidal et al., Eur. Phys. J. C57 (2008) 13, arXiv:0801.1826.
[21] Heavy Flavor Averaging Group, Y. Amhis et al., arXiv:1207.1158.
[22] BABAR Collaboration, P. del Amo Sanchez et al., Phys. Rev. D83 (2011) 091101, arXiv:1101.3830.
[23] A. L. Read, J. Phys. G28 (2002) 2693.
[24] T. Junk, Nucl. Instrum. Meth. A434 (1999) 435, arXiv:hep-ex/9902006.

DISCRETE 2012 – Third Symposium on Prospects in the Physics of Discrete Symmetries IOP Publishing
Journal of Physics: Conference Series **447** (2013) 012063 doi:10.1088/1742-6596/447/1/012063

Search for sterile neutrinos at radioactive ion beam facilities

Catalina Espinoza[1,2], **Rimantas Lazauskas**[3] **and Cristina Volpe**[4]

[1]Centro de Física Teórica de Partículas, Instituto Superior Técnico, Universidade Técnica de Lisboa, Av. Rovisco Pais 1, 1049-001 Lisboa, Portugal.

[2]Departamento de Física Teórica and IFIC, Universidad de Valencia-CSIC, E-46100, Valencia, Spain.

[3]IPHC, IN2P3-CNRS/Université Louis-Pasteur BP 28, F-67037 Strasbourg cedex 2, France.

[4]AstroParticule et Cosmologie (APC), Université Paris Diderot - Paris 7, 10, rue Alice Domon et Léonie Duquet, 75205 Paris cedex 13, France.

E-mail: m.catalina@cftp.ist.utl.pt

Abstract. We propose applications of Radioactive Ion Beam facilities to investigate physics beyond the Standard Model. In particular, we focus upon the search for sterile neutrinos by means of a low energy beta-beam with a Lorentz boost factor of 1. In the considered setup, collected ^{8}Li radioactive ions are sent inside a 4π detector filled with a liquid scintillator, with inverse-beta decay as neutrino detection channel. We provide exclusion curves for the sterile neutrino mixing parameters, based upon the 3+1 formalism, depending upon the achievable ion intensity. The proposed experiment represents a possible alternative to clarify the current anomalies observed in neutrino experiments.

1. Introduction

Open questions concerning fundamental neutrino properties include the neutrino mass scale, the neutrino mass hierarchy, the Majorana versus Dirac nature of neutrinos, and the possible existence of sterile neutrinos. While neutrino oscillations are nowadays an established fact, several anomalies have recently been observed, that cannot be explained within the standard three active neutrino framework. First the MiniBooNE anti-neutrino and neutrino oscillation results are not fully understood, while an increased statistics should help to elucidate the low energy excess and the oscillation hypothesis [1]. This experiment was designed to confirm/rule out LSND, which found an indication for neutrino oscillations at a Δm^2 of about 1 eV^2 both in the antineutrino channel, using decay-at-rest muons[2], and the neutrino channel, based upon decay-in-flight pions [3]. The second anomaly is known as the reactor "anomaly" [4]. Indeed a recent reevaluation of the electron anti-neutrino flux from reactors has shown a shift in the flux renormalization by 3% [5], compared to the previous predictions. The reanalysis of the reactor experiments, using this new flux, has shown a significant inconsistency with the three neutrino oscillation hypothesis. Finally, some years ago, the GALLEX and SAGE experiments pointed out an anomaly in the neutrino flux measured by putting an intense static ^{37}Ar and^{51}Cr sources inside their detectors. This is referred to as the Gallium anomaly. Ref.[6] has performed a detailed analysis including the 5-10% uncertainty on the corresponding neutrino-nucleus cross sections, showing that the Gallium anomaly statistical significance is at the level of 3σ.

Currently the ensemble of the accelerator, reactor and Gallium anomalies are the object of debate and have triggered an intense investigation. The possible interpretations exploit for example one or more sterile neutrinos, such as in[7], a combination of sterile neutrinos with non-standard interactions like in Ref.[8], while none of the proposed explanations so far provides a comprehensive understanding of all the data and is clear that independent and aimed experiments are necessary to clarify the present situation.

Ref.[9] has proposed the idea of establishing a low energy beta-beam [1] facility, to dispose of neutrino beams in the 100 MeV energy range, based upon the beta decay of radioactive ions, with $\gamma \approx 1$ (γ being the Lorentz factor) or with an ion boost γ of typically 2 to 7. In the first case, the neutrino fluxes are those of ions that decay-at-rest; while, in the second case, beams of variable average energy are obtained through a boost of the ions. The advantage of having such a facility is to dispose of pure (in flavor) and well known electron neutrino (or anti-neutrino) fluxes. For the physical applications see the review [10]. Most of these applications are based on stored boosted ions. In Ref.[11] we have considered the configuration with $\gamma \approx 1$, with radioactive ions sent to a target inside a 4π detector for the search of the neutrino magnetic moment. Note that Ref.[12] has taken the same configuration for a sterile neutrino search.

In this paper we consider a low energy beta-beam with $\gamma \approx 1$. We consider that the ions are injected into a target inside a 4π detector. The purpose is to use the resulting very low energy neutrino flux to search for new physics. We focus on the search for one sterile neutrino in a 3+1-neutrino flavor framework and present exclusion plots for the sterile neutrino mixing parameters.

2. Possible setups and corresponding neutrino fluxes

Radioactive ion beam facilities produce intense radioactive ions decaying through beta-decay, or electron capture. Since specific radioactive ions can be selected, a pure electron (anti)neutrino flux can be obtained. As first proposed in [9], there exists two alternative ways to produce low energy neutrinos (Figure 1). In the first scenario the decaying ions are stored inside a storage ring, while the emitted (anti)neutrinos are detected in a detector located close to the storage ring. If the stored ions are boosted, the corresponding neutrino spectra have variable energy with the average energy given by $\langle E_\nu \rangle \approx \gamma Q/2$, with Q being the Q-value of the beta-decaying nucleus. Depending on the application envisaged, the neutrino fluxes can be tuned by appropriately choosing the Lorentz boost and a high/low Q-value ion. In the case the ions are not boosted ($\gamma \approx 1$), storing the ions in a small storage ring is a possibility as well. An example is furnished by the storage ring facility currently proposed at HIE-ISOLDE at CERN [16]. While for this specific storage ring the number of stored ions is limited, one can imagine the establishment of a small ring at one of the future intense radioactive ion beam facilities, such as European EURISOL [17], or the US Facility for Rare Isotope Beams (FRIB).

The second scenario to produce low energy neutrinos, consists in injecting the ions into a target placed inside the detector. It turns out that, as long as radioactive ions are slow (i.e. not accelerated to Lorentz boost values above 1), such a scenario is much more efficient than the storage ring case. This is due to a geometrical effect since, only part of the produced (anti)neutrino flux - and not the total flux - traverses the detector if the ions are stored in a storage ring. The average neutrino flux at the detector is further reduced, compared to the injection inside the detector case, if the detector cannot be located very close to the storage ring due to the background shieldings and other necessary instrumentation.

[1] The beta-beam concept was first proposed by Zucchelli to establish a facility for the search of leptonic CP violation [13]. For a discussion on the status of the feasibility of beta-beam facilities see e.g. [14]. Note that Ref.[15] has proposed a method to reach high Q-value ion intensities e.g. for ^{8}B and ^{8}Li, which is currently being investigated.

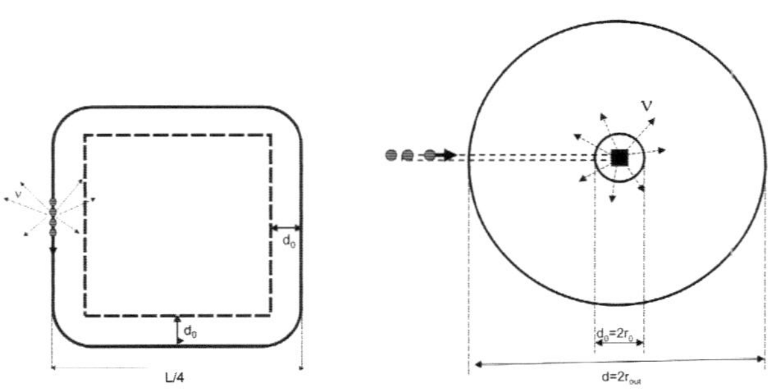

Figure 1. Two possible scenarios to produce low energy neutrinos from radioactive ions at Lorentz boost of 1. Left figure: The radioactive ions decay while circulating in a storage ring. The neutrino detector is installed close to the ring. Right figure: The radioactive ions are injected into a cavity inside a spherical detector.

We consider that radioactive ions are produced and injected into a target with an intensity I of ions per second. This target is installed inside a cavity located at the center of a spherical detector.

3. A 3+1 sterile neutrino oscillation experiment

In the present work we consider a sterile neutrino search within the 3+1 framework with three active neutrinos and one additional sterile neutrino. Besides the usual parameters of the Maki-Nakagawa-Sakata-Pontecorvo matrix, in this case the oscillation formula depends upon the neutrino mixing angle θ_{new} and Δm^2_{new}, considered to be much larger than $\Delta m^2_{31} \simeq 2.4 \cdot 10^{-3}$ eV^2. Implementing more complex scenarios with extra neutrinos is straightforward. The electron neutrino survival probability for $P_{ee}(E_\nu, r)$ is given by Ref. [18]

$$P_{ee}(E_\nu, r) = 1 - \cos^4\theta_{new}\sin^2(2\theta_{13})\sin^2\left(\frac{\Delta m^2_{31} r}{4E_\nu}\right) - \sin^2(2\theta_{new})\sin^2\left(\frac{\Delta m^2_{new} r}{4E_\nu}\right) \quad (1)$$

where a baseline of $L < 2$ km and neutrino energies $E_\nu > 2$ MeV are assumed.

4. Numerical results

To produce low energy neutrinos both β^+ and β^- decaying ions can be considered as electron neutrino and anti-neutrino emitters respectively. [2] The choice of the ions depends on the achievable intensities, the half-lives and Q-values. Obviously the half-lives should lie in an appropriate range between short and long to make experiments feasible, so that typically half-lives in the 1 s range seem to be a good choice. On the other hand, high Q-values help increasing the total number of events as well as improving the signal-to-background ratio.

Table 1 shows the candidate ions that we have been considering here, as typical examples. As far as ^{8}Li is concerned, it decays mainly into a broad ^{8}Be $J^\pi = 2^+$ excited state, at 3.03 MeV, therefore having Q-value centered at 13.1 MeV. Nevertheless, due to the broadness of the final state, the neutrino spectrum is extended well above the energy associated to the centered Q-value. To evaluate qualitatively this effect, we present results for two decay modes [3]: i) 100% branching ratio to the ^{8}Be ground state; ii) 100% branching ratio to a narrow excited state at 3.03 MeV. In a real experiment the actual results will fall in-between these two "extreme" cases.

For the ion intensity, we assume 10^{13} ions per second. Instead of taking this parameter as a tunable number (as sometimes done in the literature), here we consider values, that can in

[2] Note that electron-capture neutrino beams have been considered in [19].

[3] Note that an accurate neutrino spectrum might be obtained by considering the ^{8}Be final state continuum, taking into account the delayed-α spectrum [20].

Table 1. Beta-decay properties of the ions considered in our proposal: τ is the decay lifetime, $E_\nu^{\max}$ is the end-point energy.

Ion	Decay	Daughter (State)	τ(ms)	$E_\nu^{\max}$(MeV)
${}^6_2$He	β^-	${}^6_3$Li $(1^+, 0)$	806.7	3.5078
${}^8_3$Li	β^-	${}^8_4$Be $(2^+, 0)$	838	13.103
${}^8_3$Li	β^-	${}^8_4$Be $(0^+, 0)$	838	16.003

principle be achievable at next generation radioactive ion beam facilities. The predictions we present are obtained by taking 1 year $= 10^7$ s and a 100% efficiency of the detectors.

We shall consider two types of β^- decaying ions (Table 1). First, because of its very well known aspects relating to its production and management, it is worth taking a look at the physics reach of a facility based on ^{6}He. Its low Q-value yields fewer number of counts and, as we will show, this hinders the potential of a setup exploiting this ion, instead of one based upon a high Q-value ion choice. Our proposal for the search of sterile neutrinos is mainly based on the properties of ^{8}Li, for which we assume two extreme cases, as mention above, indicated as ^{8}Li-16MeV and ^{8}Li-13MeV 'ions'.

Our choice of main setup has been dictated by an analysis of the performance of the two possible configurations shown in Figure 1. In both cases, the considered detector is filled in with a liquid scintillator[4]. The electron anti-neutrino detection channel is inverse beta-decay $\bar{\nu}_e + p \to n + e^+$. A good signal-to-background ratio can be obtained by the addition of Gadolinium and the subsequent detection of the 8 MeV prompt gamma-rays produced by the neutron capture.

We have chosen to present the results of our simulations by means of exclusion plots. The plots show the oscillation parameter space region where our setup is expected to be sensitive to the detection of active-sterile neutrino oscillations. In all our calculations the considered running time of the experiment is of 5 years. Unless contrarily stated, we fix the systematic error to $\pi = 1\%$ in all the analysis presented hereafter; while we will show how our main results change if a larger systematic error is considered. We shall compare the sterile neutrino oscillation parameter regions that can be covered with our experimental setup, to the allowed regions presented in the analysis of Ref. [4], based on reactor neutrino experimental data cumulated so far.

Figure 2 presents exclusion plots calculated from a statistical analysis of the data using total rates. The facility is based on ^{8}Li-13 MeV decaying ions. The aim of the figure is to compare the results obtained for the two experimental setups of Figure 1. Note that for the specific case of the storage ring only, we assume an intensity of 10^{11} ions/s, having in mind a facility like HIE-ISOLDE (although the stored ion intensity is expected to be smaller [16]). Such an intensity should be attainable in a storage ring nearby the EURISOL facility [17]. For the setup geometry, following the TSR proposal for HIE-ISOLDE, we take a square storage ring with straight sections of 61.6 m length and a 1 kton cubic[5] detector at the storage ring center. (Such a geometry leaves 3 meters space between the detector and the storage ring straight sections [21, 22].) Note that, the number of expected events and, thus the exclusion plots, strongly depends on the setup geometry. For a large detector, as considered here, placing it in the center of the storage ring represents the optimal scenario (if such a detector is located along one storage ring straight section, the event number is reduced by almost a factor of 4). As expected,

[4] We take as an example $C_{16}H_{18}$ with a density of $\rho = 988\ kg/m^3$.

[5] The detector base has a size of 9.4×9.4 m and the height is of 11.3 m. Half of the detector is located below the storage ring, and half above.

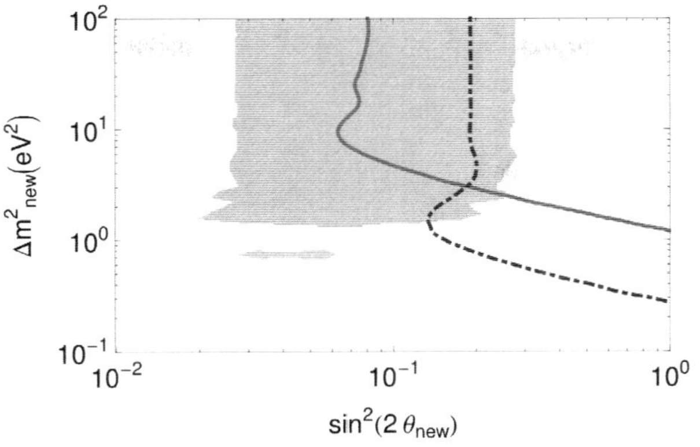

Figure 2. Exclusion plots for the sterile neutrino mixing parameters from an analysis of the data including only total rates. The results are obtained by considering a 3+1 neutrino oscillation formalism. The contours shown are for a C.L. of 99% (2 d.o.f.). The two setups are those of Figure 1. The solid (red) line corresponds to the 4π detector surrounding the source; while the dashed-dotted (blue) line corresponds to the detector place at the center of the square storage ring. The ion intensities are of 10^{11} ions/s for the storage ring and of 10^{13} ions/s for the spherical detector (see text). The source is ^{8}Li ions decaying mainly to the first excited state of the daughter nucleus (maximal neutrino energy 13 MeV). In both cases a 5 years running time is assumed. For comparison the shaded region represents the 99% C.L. inclusion domain, given by the combination of reactor neutrino experiments and other sources (adapted from Fig. 8 of Ref. [4]).

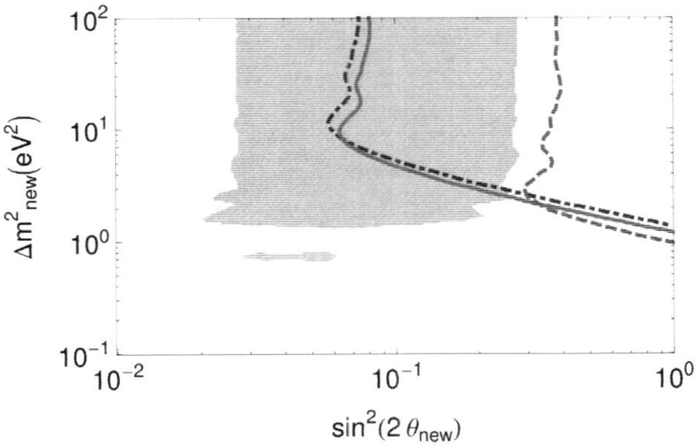

Figure 3. The same as the previous figure (spherical detector only) but choosing different ions: a source of ^{8}Li ions decaying mainly to the first excited state of the daughter nucleus (solid red line), ^{8}Li ions decaying mainly to the ground state of the daughter nucleus (dashed-dotted blue line), or a source of ^{6}He ions (dashed brown line).

although the detector is of 20 tons only, the performance obtained by sending the ions inside a 4π detector is superior to the storage ring one[6] with respect to the coverage of the shaded region identified by the reactor anomaly. On the other hand, the storage ring setup has a better sensitivity to small Δm^2.

[6] Note that this is also due to the higher ion intensity.

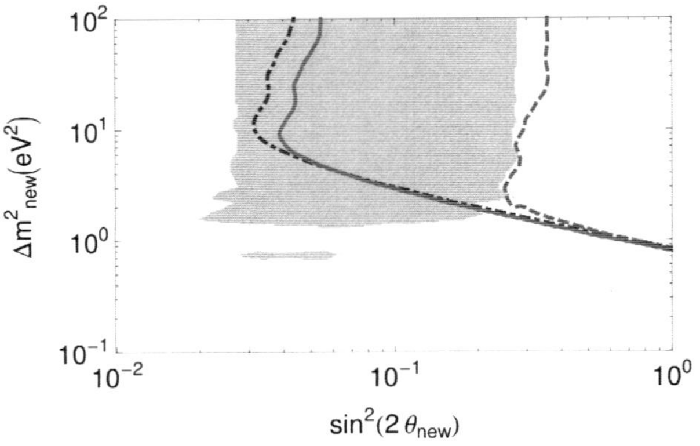

Figure 4. The same as the previous figure but including spectral information, i.e. a binned analysis of the simulated data.

In addition to the aforementioned geometric advantages, the spherical detector setup benefits from the fact that neutrino source is very close to the active detector material (we remind that the radius cavity is of 20 cm only). From now on, all the results we present will correspond to the spherical detector setup.

Figures 3 and 4 show exclusion plots constructed from total rates, and from a spectral (binned) analysis of the simulated data, respectively. Results for three ion sources are shown: the ^{8}Li-16 MeV, the ^{8}Li-13 MeV and ^{6}He cases. The low Q-value of the helium ions clearly hinders the sensitivity of this setup, making it clearly inferior to the lithium ion source case. Notice the slight difference between the two ^{8}Li cases, which is only marginally enhanced for the binned case for large $\Delta m^2_{\rm new}(>7\ {\rm eV}^2)$. (Small) corrections from ions decaying to the ground state of the daughter nucleus are thus expected to be important only in the large $\Delta m^2_{\rm new}$ case. The results of Figure 4 show the importance of an appropriate binning.

For comparison we have also included, in these figures, shaded regions corresponding to the 99% C.L. inclusion domains identified by the combination of data from the reactor neutrino experiments and other sources as described in, and here adapted from, Fig. 8 of Ref. [4]. One can see that the proposal investigated here would allow to cover most of the active-sterile oscillation parameter region. On the other hand we recall that the presented exclusion curves have the following simple physical meaning: an actual measurement lying inside the curve (to the upper-right of the curve) represents definite evidence in favour of the corresponding hypothesis, in our case, active neutrinos oscillating into sterile ones. In this manner, from Figure 4 one sees that the shaded region is out of reach if one uses 10^{13} ^{6}He/s; whereas using 10^{13} ^{8}Li/s one can cover around 70%-75% of the currently allowed region.

We would like to discuss now the impact of the chosen ion intensities on the setup performance. Figure 5 shows how the exclusion plots (and the coverage of the allowed region) changes when varying the ion intensity. In particular, the physics potential, relative to our main setup with 10^{13} ions/s, is seen to diminish (increase) by changing the intensity in one order of magnitude. This speaks of the high level of influence, that achieving good ion production levels nearby future radioactive ion beam facilities have, upon these type of experimental searches.

Finally, the sensitivity of the proposed experiment might depend upon the achieved systematic errors. To show their impact, we present exclusion curves based upon a binned analysis for sterile neutrino mixing parameters, for different levels of systematic errors. Figure 6 show the impact of 1%, 2%, 5% and 10% systematic error on the exclusion curves for 10^{14} ^{8}Li/s. One can see the important impact that reaching low systematic errors has, especially for

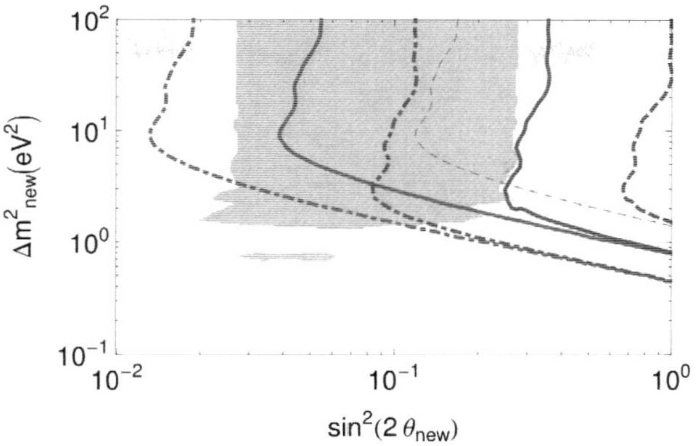

Figure 5. The same as the previous figure but varying the ion intensity: 10^{14} ions/s (dashed-dotted), 10^{13} ions/s (solid) and 10^{12} ions/s (dashed). The red lines correspond to the source of ^{8}Li ions decaying mainly to the first excited state of the daughter nucleus (maximal neutrino energy 13 MeV); while the brown lines correspond to the source of ^{6}He ions.

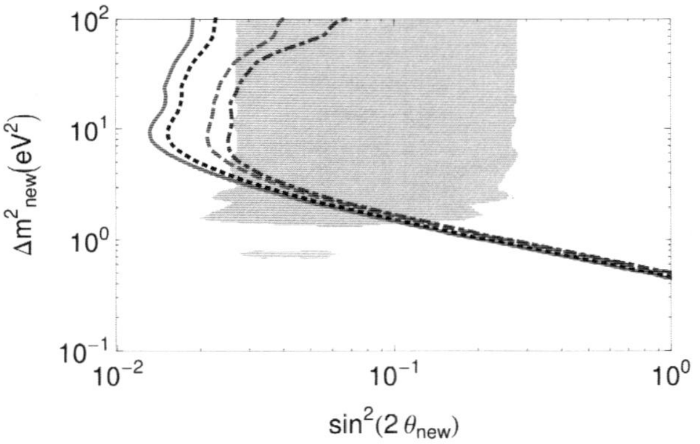

Figure 6. Exclusion plots with binned analysis of the simulated data, obtained for the 10^{14} ions/s with ^{8}Li ions decaying mainly to a 13 MeV excited state of the daughter nucleus. The figure shows the impact on the exclusion curves of different levels of systematic errors, namely 1% (solid, red), 2% (dotted, black), 5% (dashed, brown) and 10% (dashed-dotted, blue).

large Δm^2 and small mixing angle.

5. Conclusions

Future intense radioactive ion beam facilities can offer an unique opportunity to perform searches for beyond the Standard Model physics, using low energy neutrino fluxes from beta-decaying ions. Here we consider two configurations, where either the ions are stored in a storage ring, or they are sent into a target inside a spherical detector, filled in with a scintillator (with the addition of Gadolinium). Our results show that, as long as the ions are not boosted, the spherical geometry scenario gives better results than the storage ring one. We have considered a sterile neutrino search, that can be performed using electron anti-neutrino detection through inverse beta-decay in a scintillator. We have presented exclusion plots obtained from total rates and from analysis including spectral information of the simulated data. In particular, the binned analysis gives interesting results for ion intensities achievable at future radioactive ion beam facilities, like e.g. the EURISOL facility. Clearly the ion intensities achievable at such facilities are lower than the MCi radioactive source considered in proposals like the one in Ref.[23]. However radioactive ion beam facilities offer the possibility to dispose of radioactive ions with different Q-values, allowing to cover different regions of the oscillation parameters. With our spherical setup, one can probe large squared-mass differences and rather small mixing angle values, associated with one sterile neutrino, in the 3+1 oscillation framework. In particular, with the kind of setup we consider here, one could confirm/rule out the sterile neutrino hypothesis, as a possible explanation of the currently debated reactor neutrino anomaly, and cover most of the corresponding parameter space region.

Acknowledgments

C.E-H. acknowledges the support in part of IPN Orsay and the Grant Generalitat Valenciana VALi+d, PROMETEO 2008/004 and FPA2011-23596 of Spanish Ministry MICINN.

References

[1] A. A. Aguilar-Arevalo *et al.* [MiniBooNE Collaboration], Phys. Rev. Lett. **102** (2009) 101802 [arXiv:0812.2243 [hep-ex]].
[2] C. Athanassopoulos *et al.* [LSND Collaboration], Phys. Rev. Lett. **77**, 3082 (1996) [nucl-ex/9605003].
[3] C. Athanassopoulos *et al.* [LSND Collaboration], Phys. Rev. Lett. **81**, 1774 (1998) [nucl-ex/9709006].
[4] G. Mention, M. Fechner, T. .Lasserre, T. .A. Mueller, D. Lhuillier, M. Cribier and A. Letourneau, Phys. Rev. D **83**, 073006 (2011) [arXiv:1101.2755 [hep-ex]].
[5] T. .A. Mueller, D. Lhuillier, M. Fallot, A. Letourneau, S. Cormon, M. Fechner, L. Giot and T. Lasserre *et al.*, Phys. Rev. C **83**, 054615 (2011) [arXiv:1101.2663 [hep-ex]].
[6] C. Giunti and M. Laveder, Phys. Rev. C **83**, 065504 (2011) [arXiv:1006.3244 [hep-ph]].
[7] C. Giunti and M. Laveder, Phys. Rev. D **84**, 073008 (2011) [arXiv:1107.1452 [hep-ph]].
[8] E. Akhmedov and T. Schwetz, JHEP **1010**, 115 (2010) [arXiv:1007.4171 [hep-ph]].
[9] C. Volpe, J. Phys. G **30** (2004) L1 [arXiv:hep-ph/0303222].
[10] C. Volpe, J. Phys. G G **34**, R1 (2007) [hep-ph/0605033].
[11] G. C. McLaughlin and C. Volpe, Phys. Lett. B **591**, 229 (2004) [hep-ph/0312156].
[12] S. K. Agarwalla and R. S. Raghavan, arXiv:1011.4509 [hep-ph].
[13] P. Zucchelli, Phys. Lett. B **532** (2002) 166.
[14] E. Wildner, Acta Phys. Polon. B **41**, 1525 (2010).
[15] C. Rubbia, A. Ferrari, Y. Kadi and V. Vlachoudis, Nucl. Instrum. Meth. A **568**, 475 (2006) [arXiv:hep-ph/0602032].
[16] See CERN-INTC-2012-027 ; INTC-O-014, "Storage ring facility at HIE-ISOLDE : Technical Design Report" http://cdsweb.cern.ch/record/1411615 .
[17] http://www.ganil-spiral2.eu/eurisol .
[18] A. de Gouvea and T. Wytock, Phys. Rev. D **79** (2009) 073005 [arXiv:0809.5076 [hep-ph]].
[19] J. Bernabeu, J. Burguet-Castell, C. Espinoza and M. Lindroos, JHEP **0512**, 014 (2005) [hep-ph/0505054].
[20] M. Bhattacharya and E. G. Adelberger, Phys. Rev. C **65**, 055502 (2002).
[21] M. Grieser, private communication.
[22] M. Grieser, Y. .A. Litvinov, R. Raabe, K. Blaum, Y. Blumenfeld, P. A. Butler, F. Wenander and P. J. Woods *et al.*, Eur. Phys. J. ST **207**, 1 (2012).
[23] M. Cribier, M. Fechner, T. Lasserre, A. Letourneau, D. Lhuillier, G. Mention, D. Franco and V. Kornoukhov *et al.*, Phys. Rev. Lett. **107**, 201801 (2011) [arXiv:1107.2335 [hep-ex]].

Journal of Physics: Conference Series **447** (2013) 012064 doi:10.1088/1742-6596/447/1/012064

Experimental search for the LSND anomaly with the ICARUS LAr-TPC detector in the CNGS beam

E Segreto for the ICARUS Collaboration

Laboratori Nazionali del Gran Sasso. S.S. 17 bis Km 18,910, 67100 Assergi (AQ), Italy

E-mail: ettore.segreto@lngs.infn.it

Abstract. We report an early result from ICARUS (CNGS2), the large mass LAr-TPC, a Gargamelle class imaging detector of novel design. A search of a $\nu_\mu \rightarrow \nu_e$ signal due to a LSND anomaly at the Gran Sasso Laboratory, located at a distance of L = 730 km from CERN is hereby presented. Such an anomaly, in which an electron is produced by neutrinos in the energy interval $0 \leq E_\nu \leq 30\ GeV$, will be characterized by a fast energy oscillation averaging closely to $sin^2(1.27\Delta m^2_{new} L/E_\nu) \simeq 1/2$ and therefore approximately with probability $< P_{\nu_\mu \rightarrow \nu_e} >= \frac{1}{2} sin^2(2\theta_{new})$. The presence of such a signal will be compared with the small but significant backgrounds due to other and more conventional neutrino origins. Within the range of our observations, our result is compatible with the absence of a LSND anomaly. At 90% and 99% confidence levels the limits on the oscillation probabilities are $< P_{\nu_\mu \rightarrow \nu_e} >\leq 5.4 \times 10^{-3}$ and $< P_{\nu_\mu \rightarrow \nu_e} >\leq 1.1 \times 10^{-2}$ respectively. The present result strongly limits the window of opened options for the LSND anomaly, reducing the remaining effect to a narrow region centered around $(\Delta m^2, sin^2(2\theta)) = (0.5\ eV^2, 0.05)$ where there is an over-all agreement (at 90 % CL) between the present ICARUS limit, the published limits of KARMEN and the published positive signals of LSND and MiniBooNE collaborations.

1. Introduction

Neutrino oscillations have so far established a beautiful picture, consistent with the mixing of three physical neutrinos ν_e, ν_μ and ν_τ and mass eigenstates ν_1, ν_2 and ν_3. But it is possible that neutrinos are something very different than just a neutral counterpart of charged leptons, leaving room for additional neutrinos which do not see fully the ordinary electro-weak interactions but still introduce mixing oscillations with ordinary neutrinos. Indeed there are a number of *'anomalies'* which, provided they are confirmed experimentally, might be due to the presence of larger squared mass differences related to additional neutrino states with presumably some kind of *'sterile'* nature. Of course the astronomical importance of neutrinos in space is immense, so is their role in the cosmic evolution. A substantially heavier additional neutrino will be inevitably a source of the dark mass. The possible presence of oscillations into sterile neutrinos has been proposed by B. Pontecorvo [1]. The experimental search for an anomalous oscillation at short distances has been reported by the experiment LSND [2] at the Los Alamos 800 MeV proton accelerator where an anomalous excess of electrons neutrinos in a muon neutrino beam with $< E_\nu >\simeq 30MeV$ and $L \simeq 30m$ has been found. The LSND signal would imply an additional mass-squared difference largely in excess of the Standard Model values. The LSND $\bar{\nu}_\mu \rightarrow \bar{\nu}_e$ signal $< P_{\bar{\nu}_\mu \rightarrow \bar{\nu}_e} >= (2.64 \pm 0.67 \pm 0.45) \times 10^{-3}$ corresponds to an excess of $(87.9 \pm 22.4 \pm 6.0)$ events and it gives a 3.8 σ effect at L/E_ν distances of about 0.5 - 1.0 m/MeV. The recent result from

MiniBooNe [3], performed with neutrinos from the 8 GeV FNAL-Booster, confirms a neutrino oscillation signal in the similar L/E_ν range at 3.8 σ, present in both the neutrino and antineutrino channels. Using the simple formula $< P_{\nu_\mu \to \nu_e} >= sin^2(2\theta_{new}) sin^2(1.27 \Delta m^2_{new} L/E_\nu)$ one finds a very wide interval$\Delta\ m^2_{new} \simeq 0.01$ to $1.0\ eV^2$, depending on the actual and unknown value of $sin^2(2\theta_{new})$. In addition more recently an apparent *disappearance signal* of ν_e has been detected from (a) near-by nuclear reactors [4] and (b) from Mega-Curie k- capture calibration sources [5], originally developed for the Gallium experiments to detect solar ν_e. This signal is occurring for a$\Delta\ m^2_{new}$ largely in excess of the ones expected for ordinary neutrinos, maybe in the same order of magnitude of the LSND anomalies. All these anomalies which have accumulated an impressive number of standard deviations, may indeed represent an unified approach in which the values of$\Delta\ m^2_{new}$ may have a common origin, the different values of $sin^2(2\theta_{new})$ for different channels reflecting the so far unknown structure of the $U_{(j,k)}$ matrix.

2. ICARUS search at CNGS

The CNGS facility [6] delivers a neutrino beam essentially composed of muon neutrinos peaked in the range $10 \leq E_\nu \leq 30\ GeV$, with an expected contamination from anti-neutrinos at the level of 2% and an intrinsic electron component of slightly less than 1%. With the help of a novel development of a large mass Gargamelle class LAr-TPC imaging detector, the ICARUS experiment [7] is hereby searching visually the signature of $\nu_\mu \to \nu_e$ signal due to a LSND anomaly. The present experiment is at a much longer distance, $L = 730\ km$, corresponding to a much larger $L/E_\nu \simeq 36.5\ m/MeV$ for a typical neutrino energy $E_\nu \simeq 20\ GeV$. An hypothetical $\nu_\mu \to \nu_e$ LSND/MiniBooNE anomaly will therefore produce very fast oscillations as a function of the neutrino energy E_ν, averaging closely to the value $sin^2(1.27\Delta m^2_{new} L/E_\nu) \simeq 1/2$ and therefore approximately with a signal $< P_{\nu_\mu \to \nu_e} > \simeq 1/2 sin^2(2\theta_{new})$. This signal will have to be compared with the small but significant backgrounds due to other and more conventional neutrino origins. LAr-TPC developed by the ICARUS group since about two decades [7] produces a completely uniform imaging with high accuracy of massive LAr volumes (up to about 700 ton). The new method observes the true image of the track with an accuracy of the order of few mm^3, thus extending to a liquid the TPC already described for a gas, originally proposed by G. Charpak et al. [8]. The passage from a gas to a liquid capable of several meters of free electron drift is not entirely trivial. A three orders of magnitude larger purity is necessary with equivalent Oxygen contents of the order of a few tens of ppt (parts per trillion). Ionization tracks

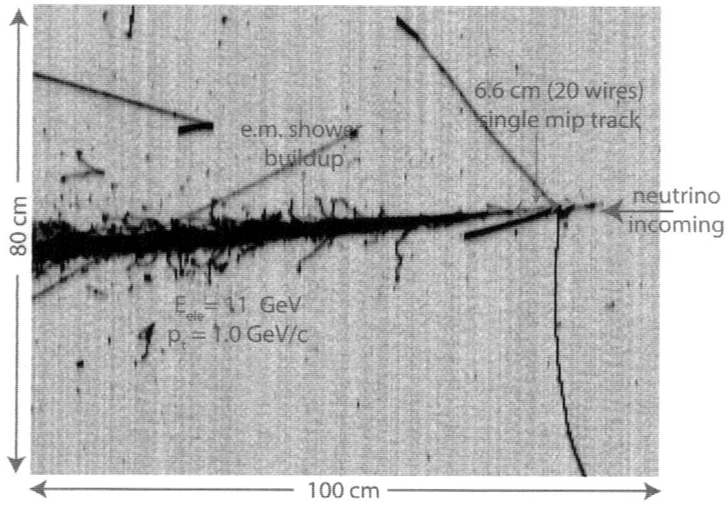

Figure 1. Typical Montecarlo generated event from the ICARUS full simulation programme [9] with $E_e = 11\ GeV$ and $p_T = 1.0\ GeV/c$.

can be transported in ultra high purity LAr practically undistorted by a uniform electric field over macroscopic distances (meters). Imaging is provided by a suitable set of electrodes (wires) placed at the end of the drift path continuously sensing and recording the signals induced by the drifting electrons. Non-destructive read-out of ionization electrons by charge induction allows detecting the entire signal of electrons crossing subsequent wire planes with different orientations. This provides simultaneously several projective views of the same event, hence allowing space point reconstruction and precise calorimetric measurement. A set of photomultipliers (PMTs) is installed in order to detect the prompt scintillation light to trigger ionizing envents. These are important differences with respect to the previously reported observations of LSND and MiniBooNE which were based on the more primitive observation of Cherenkov rings recorded with PMTs at the surface of the detector volume and mostly limited to quasi-elastic events and with a less easy discrimination between gamma rays and electrons.

3. Monte Carlo simulation

The radiation length of LAr is 14 cm ($\simeq$ 45 readout wires), corresponding to a γ-conversion length of 18 cm. The LAr-TPC detector allows identifying and measuring the ionization track by track and this allows to tag the presence of an initial electron emitted by the neutrino interaction and reject γ-converting pairs which are generally separated from the vertex and generate double minimum ionizing tracks. The detection of events has been widely simulated by a Montecarlo (MC) emulation [9]. The MC emulation is very sophisticated, reproducing in every detail the actual signals from the wire planes. Comparisons with the actual data samples are widely used to tune the reconstruction, check calibrations and optimize the identification and measurement algorithms. The agreement between MC and observed events has been excellent and it has been extensively used as a main guideline. In order to predict events caused by a LSND anomaly, a sample of MC ν_e CC events has been generated with the CNGS ν_μ CC energy spectrum. A simulated event is shown in Figure 1. An electron signature has been defined by the following requirements: (a) fiducial volume for the vertex of the event, 5 cm distance from each side of the active volume and 50 cm distance from the exit plane; (b) the visible neutrino

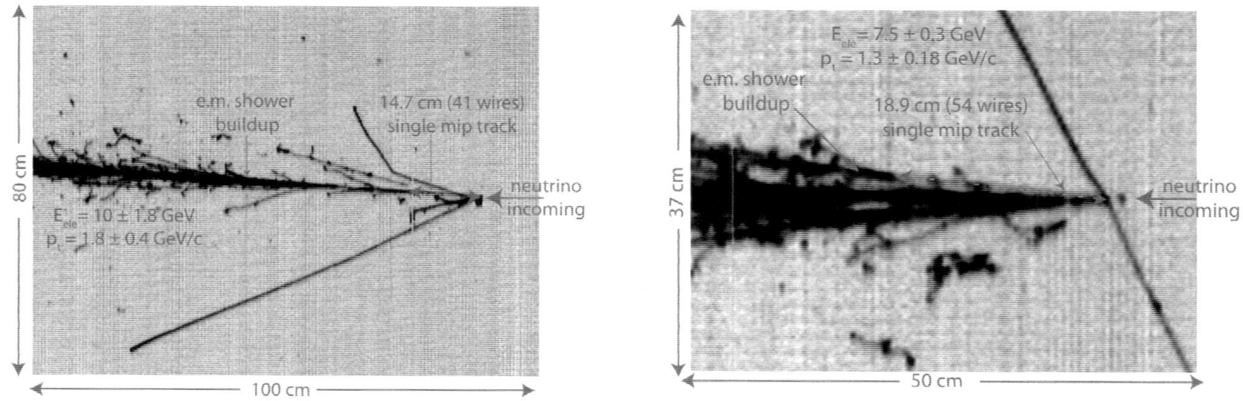

Figure 2. Experimental picture of the two observed events (a) and (b) with a clearly identified electron signature from a total sample of 1091 neutrino interactions. Event in (a) has a total energy of 11.5 ± 1.8 GeV, and a transverse electron momentum of 1.8 ± 0.4 GeV/c. Event in (b) has a visible energy of 17 GeV and a transverse momentum of 1.3 ± 0.18 GeV/c. In both events the single electron shower in the transverse plane is clearly opposite to the remaining of the event.

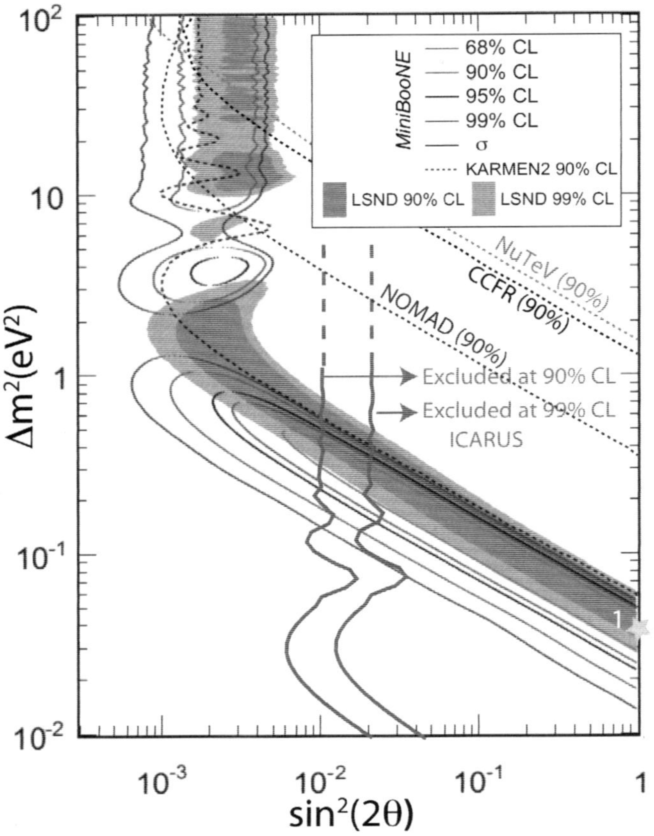

Figure 3. Two-dimensional plot of$\Delta\ m^2$ vs $sin^2(2\theta)$ for the main published experiments sensitive to the anomaly [2], [3], [11] and the present ICARUS result. The ICARUS limits to the oscillation probability are 5.4×10^3 and 1.1×10^2 , corresponding to 3.41 and to 7.13 excess events respectively at 90% and 99% confidence level.

energy less than 40 GeV, in order to reduce the intrinsic beam induced background (c) the presence of a minimum ionizing relativisic electron track of sufficient length present from the vertex, subsequently building up into a shower; (d) clear separation from the presence of the other ionizing tracks near the vertex in at least one of the two transverse views, including short proton like recoils due to nuclear interactions. Out of an initial sample of 103 reconstructed events, 88 have a visible energy $E_{VIS} \leq 40\ GeV$, of which 73 satisfy the fiducial volume cut. Visibility cuts reduce the identified electron tracks to 54 events corresponding to a selection efficiency $\eta = 54/73 = 0.74 \pm 0.05$. In a good approximation the value of η is independent of the shape of the energy spectrum.

4. Data analysis and Results

The present ICARUS experimental sample is based on 168 neutrino events (5.8×10^{18} pot) collected in 2010 and 923 events collected in 2011 (2.7×10^{19} pot) out of the 4.4×10^{19} collected in 2011), leading to a total of 1091 initial neutrino events. In this sample only events with visible energy $\leq 30\ GeV$ have been included with the relevant fiducial cuts, which bring the number of events from 1091 to 839.We expect 627 ν_μ CC,of which 204 ν_{NC} and 3 ν_τ CC (except $\tau \to e$) are to be added. The expected number of ν_e events due to conventional sources in the same energy range and fiducial volumes are as follows: (a) 3 events due to intrinsic ν_e beam associated contamination; (b) 1.3 ν_e events due to the presence of θ_{13} oscillations; (c) 0.7 ν_τ CC with $\tau \to e$) giving a total of 5 expected events. The expected visible signal is of 3.7 events after the $\eta = 0.74 \pm 0.05$ reduction has been applied due to visibility cuts coming from the

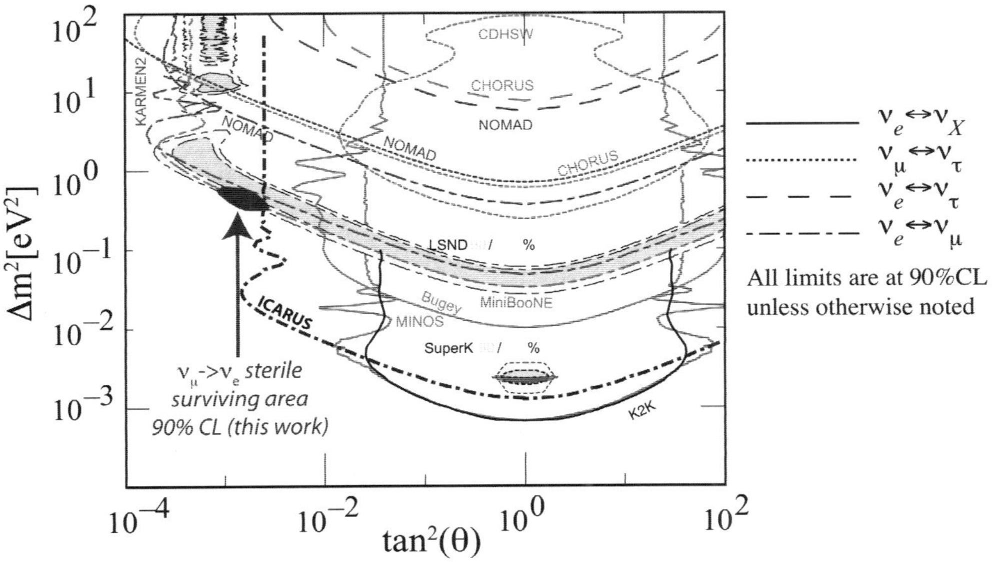

Figure 4. Regions in the $(\Delta m^2, tan^2(\theta))$ plane excluded by the ICARUS experiment compared with the published results (taken from http://www.pdg.org). While for$\Delta\ m^2 \gg 1\ eV^2$ there is already disagreement for $\nu_\mu \to \nu_e$ between the allowable regions from the published experiments, for$\Delta\ m^2 \leq\ eV^2$ the ICARUS result now allows to define a much smaller, narrower allowed region centered around $(\Delta m^2, sin^2(2\theta)) = (0.5\ eV^2, 0.05)$ in which there is a 90% C.L. overall agreement.

general complexity of the events. In the recorded experimental sample, two events with a clear ν_e have been identified (Figure 2), to be compared with the above expectation of 3.7 events for conventional sources.

Figure 2 (a) has a total energy of $11.5 \pm 2.0\ GeV$, and an electron of $10 \pm 1.8\ GeV$ taking into account a partially missing component of the e.m. shower. Figure 2 (b) has 17 GeV of visible energy and an electron of $7.5 \pm 0.3\ GeV$. In both events the single electron shower in the transverse plane is clearly opposite to the remaining of the event, with transverse momenta for the electron of $1.8 \pm 0.4\ GeV/c$ and $1.3 \pm 0.18\ GeV/c$ respectively. Within the range of our observations, our result is compatible with the absence of a LSND anomaly. Following ref. [10], at statistical confidence levels of 90% and 99% and taking into account the detection efficiency, the limits due to the LSND anomaly are respectively 3.41 and 7.13 events. Given the observed sample of 627 ν_μ CC events, the limits to the oscillation probability are 5.4×10^{-3} and 1.1×10^{-2} respectively. The exclusion area of the ICARUS experiment is shown in figure 3 in terms of the two-dimensional plot of $sin^2(2\theta_{new})$ and$\Delta\ m^2_{new}$.

5. Conclusions

The present result strongly limits the window of options from the MiniBooNE experiment. Using a likelihood-ratio technique [3], CP conservation and the same oscillation probability for neutrinos and antineutrinos, a best MiniBooNE oscillation fit for $200\ MeV < E_\nu^{QE} < 3000\ MeV$

Journal of Physics: Conference Series **447** (2013) 012064 doi:10.1088/1742-6596/447/1/012064

has been given at $(\Delta m^2, sin^2(2\theta)) = (0.037\ eV^2, 1.00)$. This is clearly excluded by the ICARUS result. A 3+2 joint oscillation fit as a function of E_ν^{QE} in both neutrino and antineutrino modes has also been reported [3] with best fit values$\Delta\ m_{41}^2 = 0.082\ eV^2, \Delta m_{51}^2 = 0.476\ eV^2, U_{e,4} = U_{\mu,4} = 0.1844, U_{e,5} = U_{\mu,5} = 0.00547$, and $\phi = 1.0005\ \pi$. The MiniBooNE value is clearly incompatible with the present ICARUS result. A detailed comparison between the various results is shown in figure 4. While for$\Delta\ m^2 \gg 1\ eV^2$ there is already disagreement between the allowable regions from the published experiments, for$\Delta\ m^2 \leq 1\ eV^2$ the ICARUS result now allows to define a much smaller, narrower region centered around $(\Delta m^2, sin^2(2\theta)) = (0.5\ eV^2, 0.05)$ in which there is 90% CL agreement between (1) the present ICARUS limit, (2) the limits of KARMEN and (3) the positive signals of LSND and MiniBooNE collaborations. This is the area in which the expectations from cosmology suggest a substantial contribution to the dark mass signal. This region will be better explored by the ICARUS/NESSiE proposed dual detector experiment [12] to be performed at CERN at a much shorter distances (300 m and 1.8 km) and lower neutrino energies, which increase the events rate, reduce the over-all multiplicity of the events, enlarge the angular range and therefore improve substantially the selection efficiency η.

Acknowledgments

The ICARUS Collaboration acknowledges the fundamental contribution of INFN and, in particular, of the LNGS Laboratory and its Director, to the construction and operation of the experiment. The Polish groups acknowledge the support of the Ministry of Science and Higher Education in Poland, including project 637/MOB/2011/0 and grant number N N202 064936. Finally we thank CERN, in particular the CNGS staff, for the successful operation of the neutrino beam facility.

References

[1] Pontecorvo B 1967 *Zh. Eksp. Teor. Fiz.* **53** 1717 [1969 *Sov. Phys. JETP* **26** 984]
[2] Aguilar A et al. (LSND Collaboration) 2001 *Phys. Rev. D* **64** 112007
[3] Aguilar-Arevalo A A (MiniBooNE Collaboration), arXiv:1207.4809v1 [hep-ex] 19 Jul 2012
[4] Mention G et al., arXiv:1101.2755v1 [hep-ex], 2011 *Phys.Rev. D* **83** 073006
[5] Abdurashitov J N et al. (SAGE Collaboration) 2009 *Phys. Rev. C* **80** 015807
Kaether F, Hampel H, Heusser H, Kiko J, and Kirsten T, 2010 *Phys. Lett. B* **685**, 47
[6] Aquistapace G et al. 1998 *CERN98-02, INFN/AE-89-05*
[7] Rubbia C et al. 2011 *JINST* **6** P07011
Amerio S et al. 2004 *Nucl. Instr. and Meth. A* **527** 329
[8] Charpak G et al. 1970 *Nucl. Instrum. and Meth.* **80** 13
[9] Battistoni G et al. 2007 *AIP Conf. Proc.* **896** 31-49
Battistoni G et al. 2009 *Proceedings of the 12th International conference on nuclear reaction mechanisms, Varenna (Italy)* p. 307
Ferrari N et al. 2005 CERN-2005-10, INFN/TC-05/11
[10] Feldman G J and Cousins R D 1988 *Phys. Rev. D* **57** 3873
[11] Astier P et al. (Nomad Collaboration) 2003 *Phys. Lett. B* **570** 19-31
Romosan A et al. (CCFR Collaboration) 1997 *Phys. Rev. Lett.* **78** 2912
Avvakumov S et al. (NuTeV Collaboration) 2002 *Phys. Rev. Lett.* **89** 011804
[12] Antonello M et al. 2012 *SPSC-P-347*
Rubbia C et al. 2011 *SPSC-P-345*

Journal of Physics: Conference Series **447** (2013) 012065 doi:10.1088/1742-6596/447/1/012065

The SNO+ experiment: status and overview

José Maneira for the SNO+ Collaboration
LIP, Av. Elias Garcia 14, 1st, 1000-149 Lisboa, Portugal
E-mail: maneira@lip.pt

Abstract. SNO+ is a multi-purpose Neutrino Physics experiment, succeeding to the Sudbury Neutrino Observatory by replacing heavy water with liquid scintillator. Its scientific goals are the search for neutrinoless double-beta decay, the study of solar neutrinos and antineutrinos from reactors and the Earth's natural radioactivity, as well as supernovae neutrinos. The experimental advantages of SNO+ are the possibility of loading large quantities of double-beta decaying isotope in the liquid scintillator volume, and the very low backgrounds allowed by the deep underground location and radiopurity of the employed materials. The installation of the detector at SNOLAB is being completed, and commissioning has already started, with a dry run. Filling with water and later, with scintillator, will start next year. This talk will summarize the Physics goals of SNO+, as well as the main detector developments.

1. Introduction

SNO+ [1] is a multi-purpose experiment for low-energy neutrino physics. The Sudbury Neutrino Observatory detector [2], located deep underground at SNOLAB[3], will be re-used with liquid scintillator instead of heavy water, allowing several Physics goals, from neutrinoless double beta decay searches, to measurements in with several sources of low energy neutrinos – the Sun, the Earth, Supernovae, and neighboring nuclear reactors.

SNO+ (see Fig. 1) is constituted by a 12 diameter acrylic vessel (AV) containing about 800 tonnes of liquid scintillator. A geodesical structure (PSUP) supports about 9500 8-inch diameter photomultiplier tubes (PMTs) coupled to a set of reflectors that increase the geometrical coverage to 55% of 4π. The AV and the PSUP are immersed in 7000 tonnes of ultra-pure water, that shield the scintillator volume of neutrons and gamma radiation from radioactivity in the PMTs and the surrounding rock. A thick liner prevents diffusion of Radon from the rock into the water volume. The deep underground location reduces the cosmic ray muon flux to about 70 per day.

These exceptional shielding properties give SNO+ the necessary conditions to carry out low background measurements at low energies. Neutrinoless double beta decay searches can be carried out with very low backgrounds from external materials, by simply reconstructing the event position and defining a fiducial volume. This is in fact one of the advantages of the liquid scintillation technique, in addition to the possibility of using large masses and performing in-situ purification, to reduce the backgrounds from the scintillator mixture itself.

For solar neutrinos, the low cosmic ray muon flux leads to a low rate of cosmogenic backgrounds, such as ^{11}C, that would otherwise disturb the measurements of the pep and CNO fluxes.

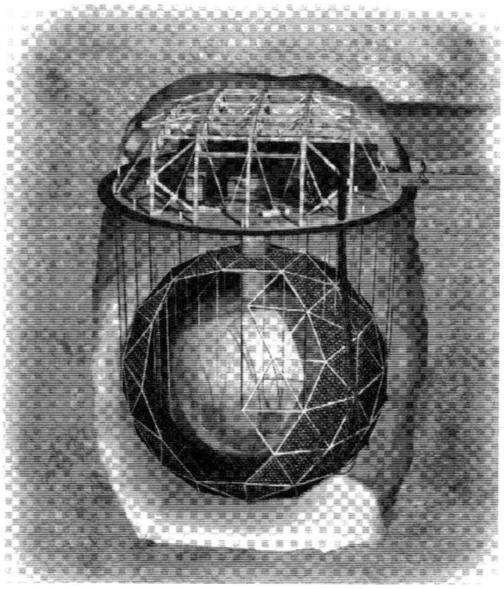

Figure 1. Artistic view of the SNO detector. SNO+ has an additional rope system holding the acrylic vessel down, anchored to the bottom of the cavity.

The Physics goals of SNO+ are presented in Section 2 and the detector changes and upgrades that have been developed are described in Section 3.

2. Physics goals

2.1. Neutrinoless Double Beta Decay

Neutrino Physics has made many progresses in the last decade and a half, with the discovery of neutrino oscillations. However, the issue of wether the neutrino is a Dirac or Majorana particle, of great importance for leptogenesis and Grand Unification theories [4], is still not solved. The search for neutrinoless double beta-decay (NLDBD) aims at answering this question and, in the case of the Majorana hypothesis, measuring the absolute value of the neutrino mass.

Large volume liquid-scintillator (LS) experiments SNO+ and KamLAND-Zen[5] have recently joined the competitive field of NLDBD. This type of detectors has two important advantages for NLDBD searches: a large isotope mass can be sampled, and very low backgrounds can be achieved for high Q-value isotopes. Purification of the scintillator and the isotope compound can effectively remove many contaminants of the scintillator itself, and cosmogenic isotopes produced by cosmic ray activation of the DBD isotope. Analysis techniques allows the rejection of events originating from radioactivity outside of the scintillator volume, by defining a fiducial volume; the internal backgrounds from the ^{238}U and ^{232}Th natural radioactivity chains that populate the energy region above 2 MeV can be tagged through delayed coincidences.

SNO+ has studied extensively the loading of Neodymium in the LAB (linear alkyl-benzene) liquid scintillator. The double beta decay isotope ^{150}Nd has the advantages of a large (calculated) phase space and nuclear matrix element (NME) for neutrinoless double deta decay[6], and a high Q-value of 3.37 MeV, above most natural radioactivity backgrounds. The LAB-Nd solution has been demonstrated to be stable for over 3 years, and many of its cosmogenically produced contaminants can be removed by purification. The isotopes ^{144}Nd, ^{176}Lu, ^{138}La, associated with the Neodymium compound, are expected to be present. The Q-value of their decays is low, far from the 3.37 MeV region, but the expected rate is high enough, above the two-neutrino signal, for pile-up events to reach the ^{150}Nd region. These will not be a concern, though, since analysis techniques have demonstrated to identify the pile-up events with high efficiency.

Optical attenuation is a limiting factor for the usable concentration. With the present measurements of optical properties and Monte Carlo simulations, the optimal loading for best sensitivity with natural Neodymium should be about 0.3%. Assuming a NME of 2.5 (IBM-2 model), a phase space factor of $2.69 \times 10^{-13} yr^{-1}$, a 50% fiducial volume and 80% livetime, a sensitivity of 110 meV is expected after 4 years of data-taking, as shown in Figure 2.

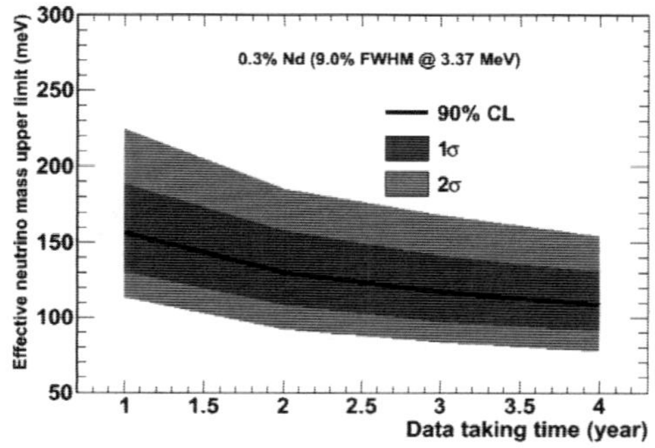

Figure 2. Expected sensitivity in effective neutrino mass from the ^{150}Nd neutrinoless double beta decay search in SNO+, with a 0.3% loading of natural Neodymium.

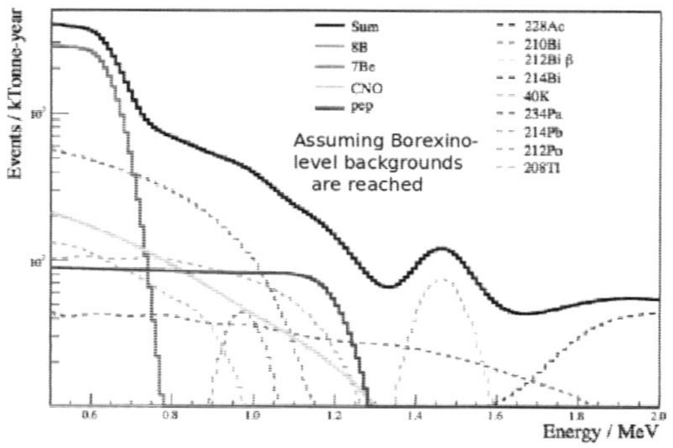

Figure 3. Expected signals from solar neutrino events and backgrounds, excluding ^{11}C (lower than the pep signal).

2.2. Solar neutrinos

The solar neutrino fluxes of ^{8}B, ^{7}Be and pep have been measured by SNO, Super-Kamiokande and Borexino (in addition to the integral measurements by the radiochemical experiments). However, a measurement of CNO neutrinos would be quite valuable to Solar Phyiscs, as it could solve the "so-called" solar metallicity problem [13]. in addition, a precise measurement of the pep solar neutrino flux can confirm the predictions of the MSW oscillation mechanism in the vaccum-matter transition region and be sensitive to new Physics models [7].

Figure 3 shows the expected signals and main backgrounds in the intermediate energy range, from Monte Carlo simulations under the assumption that SNO+ reaches the same background levels as Borexino [14]. A joint maximum-likelihood fit was applied to these spectra, in order to estimate the expected precision, and the results are in Table1. For pp and CNO neutrinos, these are still very rough estimations, since the backgrounds are harder to control and estimate in these cases.

SNO+ has a very strong advantage for these measurements. Cosmic ray muons interaction in the scintillator produce the ^{11}C isotope, that has an energy spectrum covering the pep and CNO neutrino region. Borexino has employed delayed coincidence techniques to reduce this background, with a severe loss of efficiency. In SNO+, depth alone would reduce the ^{11}C background by two orders of magnitude with respect to Borexino.

However, there is a possible additional background to consider in SNO+. Radon daughters have accumulated on the surface of the AV over the last few years in a significant way. If these leach into the scintillator, the purification system has the capability to remove them. However, depending on the actual leach rate, that removal might be inefficient and the ^{210}Bi levels in the scintillator too high for a pep/CNO solar neutrino measurement without further mitigation. Mitigation could include enhancing online scintillator purification, draining the detector and sanding the AV surface to remove radon daughters, or deploying a bag. Double beta decay, which is the priority of SNO+, and low-energy ^{8}B solar neutrino measurements are not affected by these backgrounds.

Flux	One year	Two years
pep	9.1%	6.5%
^8B	7.5 %	5.4%
^7Be	4%	2.8%
pp	a few % ?	
CNO	~15 % ?	

Table 1. Expected uncertainties on the SNO+ measurements of different solar neutrino fluxes, assuming Borexino level backgrounds are reached. Expectations for pp and CNO are highly dependant on the actual levels of backgrounds achieved, especially ^{14}C, ^{85}Kr and ^{210}Bi.

2.3. Reactor and geo-neutrinos

As in previous liquid scintillator experiments (Borexino and KamLAND), SNO+ can also detect antineutrinos from nuclear reactors through the inverse beta decay (IBD) reaction on protons, with a very small background due to the delayed coincidence tag between the positron and neutron signals. The region centered in Sudbury has a smaller number of nuclear reactors than the region of the Kamioka laboratory in Japan, so the expected flux of reactor antineutrinos in SNO+ is about 5 times smaller that the flux observed in KamLAND (before the Fukushima accident). However, the antineutrinos come predominantly from three reactor complexes. The closest one, in the Bruce peninsula, is at a distance of 281 km, and the next ones are close to Toronto, at distances of 330 km and 340 km. The effect of having almost only two baselines, results in a distinctive spectrum, with oscillation minima clearly visible (see Fig. 4), in contrast with the observed energy spectrum of detected antineutrinos in KamLAND. The sensitivity to the neutrino oscillation parameters is then expected to be similar to KamLAND after 3 years of data-taking, even with a significantly smaller flux.

The question of how much of the Earth's heat production is radiogenic is still a largely open one, and it depends, among other factors, on the amounts and distribution of the natural radioactivity isotopes of Uranium, Thorium and Potassium in the Earth [12]. Geo-neutrino measurements can provide some constraints on the problem, but they would have a stronger impact if a separation between the flux originating in the crust (closer to the detectors) or the mantle (more important for Geo-Physics) could be achieved[9, 11]. The thick continental crust (the "Canadian shield") where SNO+ is located gives a strong contribution to the estimated 20 events per year (including efficiencies), but this will be estimated by detailed geological surveys and calculations that are being carried out [12], so that the mantle flux can be obtained by subtraction from the total flux. In addition, the smaller reactor neutrino flux is an advantage for the detection of geo-neutrinos in SNO+ [10].

During the double-beta decay, we expect the delayed coincidence tag to be effective in identifying the antineutrino signal even in the presence of a large signal from the two-neutrino decay. So, even if with a reduced sensitivity with respect to the unloaded scintillator, due to worse energy resolution, reactor and geo-neutrino measurements will be possible during all phases of SNO+.

2.4. Neutrinos from Supernovae

SNO+ will also have good capabilities as a detector for Supernovae neutrinos. About 370 events are expected from a 3×10^{53} erg Supernova at a distance of 10 kParsec, from reactions of IBD on protons, elastic scattering on electrons, and reactions on Carbon nuclei. The dominant process is IBD, sensitive only to electron antineutrinos, but the neutral current reaction on ^{12}C should provide an interesting measurement of the total flux in all flavors. An additional 270 events are

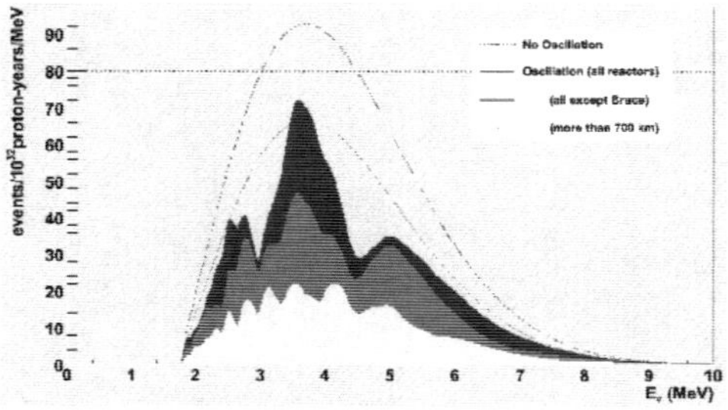

Figure 4. Expected energy spectrum of reactor antineutrinos at SNO+, from different reactor sources.

expected from elastic scattering on protons, with a smaller visible energy.

As in SNO, a careful design of calibration hardware and operational procedures should allow a semi-automated response to event bursts, so that SNO+ will be expected to participate in the Supernova Early Warning System (SNEWS).

3. Experimental developments

Converting SNO into a liquid scintillator detector is a complex undertaking, touching several experimental aspects, from mechanical to chemical issues, as well as calibrations, electronics, and software.

3.1. Liquid scintillator

The initial R&D program for SNO+ started with the identification of the liquid scintillator, that had to fill several performance requirements – high light yield, fast decay time – as well as safety requirements (due to the lab location in a mine), and also chemical compatibility with acrylic. Linear alkyl-benzene (LAB) was found to fulfill all the needed requirements, and in fact since then other experiments, such as Daya Bay, have used it.

The LAB scintillator can be efficiently purified, and the SNO+ online purification systems employs basically the same techniques developed by the Borexino experiment for the PC scintillator[14]: gas stripping, distillation, water extraction. Due to the SNOLAB logistics constraints, the SNO+ gas stripping will use water vapour instead of Nitrogen. The purification systems for SNO+ are presently being constructed and installed at SNOLAB.

3.2. Detector

The LAB scintillator has a density of 0.86, so the AV must take up a large buoyant force, as opposed to the downward force in effect when heavy water was used. This force is compensated by a new hold-down rope system, to be mounted over the whole AV, and secured at the floor of the experimental cavity. Detailed finite element analyses were carried out in the design phase. The material (tensylon) was selected due to its good radiopurity, and the existing hold-up ropes were remade with the new material. The installation required the removal of PMTs at several pass-through points, but was accomplished successfully, as shown in Fig. 5.

Other detector improvements include repairs of PMTs and electronics channels. This ongoing activity will allow the recovery of many PMTs that failed during the SNO data-taking, so that the expected number of working PMTs for SNO+ is about 9000. A new trigger and DAQ system have been implemented, in order to be able to cope with the significant increase in analog sum pulse charges and data volume, because of the higher yield of scintillation with respect to Cherenkov light.

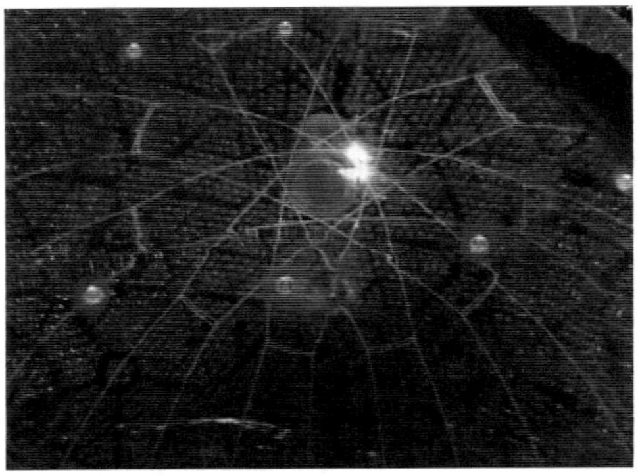

Figure 5. Photo of the hold-down rope net system installed over the AV, taken in January 2012.

3.3. Calibration system

At the energies of pep/CNO solar neutrinos, and the ^{150}Nd Q-value, the radiopurity requirements are much more stringent than in SNO. The calibration source deployment system was completely rebuilt to be fully sealed, following vaccum standards, to avoid any Radon contamination. Even when not in use, the sources themselves are kept in a separate storage box with the same tightness standards. Care had to be taken to ensure mechanical integrity in the case of quick pressure changes, as can frequently happen in an active mine.

In order to cover the lower energy range, different radioactivity sources exist or are being developed: AmBe (n, γ), ^{16}N (γ), $^{24}Na(\gamma)$, ^{48}Sc (γ), $^{57,60}Co$ (γ), ^{n}Na (γ), ^{90}Y (β).

Additional deployed sources include a scintillator-compatible version of the light diffusor source ("laserball") previously used in SNO, and a Cherenkov source, in which electrons from the β decay of ^{8}Li cross a spherical shell of UV-transparent acrylic, in order to produce a well-known number of Cherenkov photons and calibrate the light detection efficiency.

A new source-location system was implemented, based on a redundant set of six high resolution cameras mounted on the PSUP. The precise location of the source will be carried out by reconstructing the position of LEDs attached to the source body.

However, due to the stringent radiopurity requirements, calibrations with internally deployed sources will not be possible with the same high frequency that was employed in SNO (monthly). So, in order to calibrate the PMTs and monitor the stability of the scintillator transparency, external light injection systems were designed, based on optical fibers transporting pulses from LEDs or lasers. Two of these systems use a few quartz optical fibers, with narrow beams, and aim at the measurement of the scattering length of the scintillator – by comparing light observed in the direct path with other angles – and at the monitoring of the scintillator attenuation stability.

The PMT calibration system will use LED light pulses conveyed to all PMTs, by means of $\sim$100 1 mm PMMA fiber cables (each with 2 fibers for redundancy), in order to carry out the timing and the gain calibration of the PMT. The fibers are mounted in uniformly distributed positions in the PSUP, directing their beams inward, across the AV and hitting the PMTs on the opposite side. The choice of 1 mm PMMA fibers allows a good coupling to the LEDs, and ensures that the beams coming out of each fiber are wide enough to illuminate about 200 PMTs, thus ensuring that all PMTs are illuminated redundantly. However, fibers with a large numerical aperture have a higher timi dispersion, that was measured for all fibers in the quality control before installation. The results are presented in Fig. 6, and show that most present pulses with a width of about 4 ns (including the LED pulse dispersion), which is enough for the synchronization of the PMTs to 1 ns (for which many pulses are accumulated).

All the fibers have now been brought into the detector, and one-third of them (in the positions accessible from the bottom) have been installed in their final mount place. The remaining two-thirds will be installed in 2013 by boat. A full detector dry-run was taken in the fall of 2012,

Journal of Physics: Conference Series **447** (2013) 012065 doi:10.1088/1742-6596/447/1/012065

in order to test the PMTs, the HV system, the DAQ and the calibration systems, as well as the offline data processing setup. All the PMTs were turned on with HV, and the electronics and DAQ demonstrated a very good performance. In addition, all the installed fibers were illuminated with LEDs, and detector data taken in that configuration. Figure 7 shows a display of the integrated number of counts over a run taken with one of the fibers, in which the cluster of PMTs illuminated by direct and reflected light is clearly seen.

4. Status and expected schedule

The schedule of the experiment is the following. In the spring of 2013, the water fill will start, and during that operation, the remaining fibers, PMTs and location cameras will be installed. Once the detector is full, a calibration campaign will allow the characterization of the detector, especially the PMTs. The replacement of water by the LAB will follow and the first scintillator data is expected in 2014.

Acknowledgments

The author's work was supported by FEDER funds through the COMPETE program (Operational Program for Competitiveness Factors) and by national funds from Portugal through FCT (Fundação para a Ciência e a Tecnologia) within the project PTDC/FIS/115281/2009.

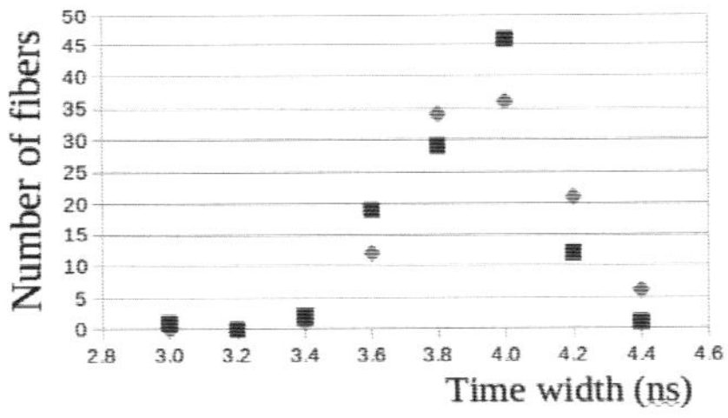

Figure 6. Results of timing spread quality control tests of the optical fibers for the PMT calibration system.

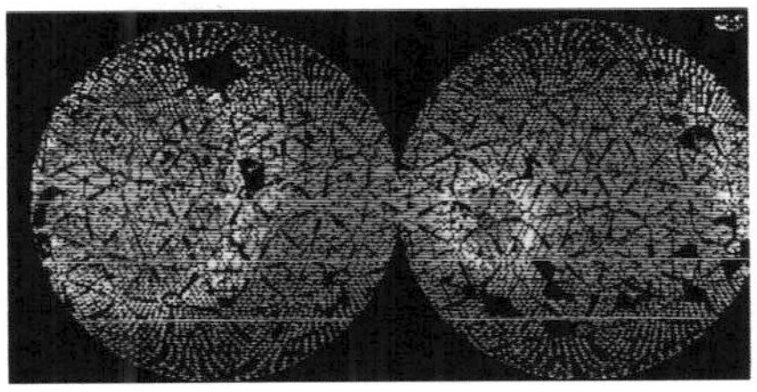

Figure 7. Display of accumulated events over test runs of the PMT calibration system. Color code: Red/pink means a high number of counts, while green/blue means a small number.

References

[1] Chen M C 2005 *Nucl. Phys. B (Proc. Suppl.)* **154** 65-8.
[2] SNO Collaboration 2000 *NIMA* **449** 172-207.
[3] Duncan F, Noble T, Sinclair S 2010 *Annu. Rev. Nucl. Part Sci.* **60** 163-80.
[4] Zuber K 2011 *Neutrino Physics* 2nd edition, Taylor & Francis.
[5] KamLAND-Zen Collaboration 2012 submitted to *Phys. Rev. Lett.*, arXiv:hep-ex/1211.3863
[6] Barea J, Kotila J, Iachello F 2012, *Phys. Rev. Lett.* **109** 042501.
[7] Gonzalez-Garcia M C *et* al. 2008 *JCAP* 0806:019.
[8] Yeh M 2008 *et* al. *J. Phys.: Conf. Ser.* **136** 042054.
[9] Rotschild C G, Chen M C, Calaprice F P 1998 *Geophys.Res.Lett.* **25**, 1083(1998).
[10] Chen M C 2006 *Earth, Moon, and Planets* **99** 221-8.
[11] Dye S 2012, Geo-neutrinos, Communication to the SNOLAB Grand Opening Workshop; http://www.snolab.ca/science/meetings-and-workshops/workshops/2012-grand-opening
[12] McDonough W 2012, Earth Models; SNOLAB Workshop (see [11]).
[13] Haxton W C, Serenelli A M 2008 *Astrophys.J.* **687** 678-91.
[14] Borexino Collaboration 2009 *NIM A* **609** 1, 58-78.

DISCRETE 2012 – Third Symposium on Prospects in the Physics of Discrete Symmetries IOP Publishing
Journal of Physics: Conference Series **447** (2013) 012066 doi:10.1088/1742-6596/447/1/012066

Status of the CUORE experiment

Luca Pattavina for the CUORE collaboration
I.N.F.N. Milano-Bicocca I 20126 Milano, Italy
E-mail: luca.pattavina@mib.infn.it

Abstract. One of the fundamental open questions in elementary particle physics is the value of the neutrino mass and its nature of Dirac or Majorana particle. Neutrinoless double beta decay (DBD0ν) is a key tool for investigating these neutrino properties and for finding answers to the open questions concerning mass hierarchy and absolute scale. Experimental techniques based on the calorimetric approach with cryogenic particle detectors are proved to be suitable for the search of this rare decay, thanks to high energy resolution and large mass of the detectors. The CUORE (Cryogenic Underground Observatory for Rare Events) experiment will search for DBD0ν in ^{130}Te. The CUORE setup consists in an array of 988 tellurium dioxide crystals, operated as bolometers, with a total mass of about 230 kg of ^{130}Te. The experiment is under construction at the Gran Sasso National Laboratory in Italy. As a first step towards CUORE, a tower prototype (CUORE-0) has been assembled and is running. In this talk a detailed description of the CUORE-0 tower, its performances and the expected sensitivity will be given. The status of the CUORE experiment, its critical points and its expected sensitivity on the base of what we will learn with CUORE-0 will then be discussed.

1. Introduction

Double beta decay (DBD) is a second order weak process in which a nucleus changes its atomic number by two units (A,Z)$\rightarrow$(A,Z$\pm$2). It occurs for some even-even nuclei for which the single beta decay is energetically forbidden, or suppressed by large change in angular momentum. Two different DBD modes are usually considered: the decay with the emission of two neutrinos (DBD2ν), which is allowed in the Standard Model, that is described by the reaction

$$(A, Z) \rightarrow (A, Z+2) + 2e^- + 2\overline{\nu}_e \tag{1}$$

and the neutrinoless decay (DBD0ν) given by

$$(A, Z)(A, Z+2) + 2e^-. \tag{2}$$

This transition, which does not depend on the Dirac or Majorana nature of the neutrino, can be modeled as two successive single beta decays (see Fig. 1). DBD2ν was observed for many different nuclei [1], and the computed half-lives are of the order of 10^{18}-10^{24} y.

The DBD0ν is instead forbidden in the Standard Model, since it violates the lepton number by two units, moreover it has never been observed (except for one controversial claim [2]). DBD0ν decay can proceed through many different mechanisms, the simplest way to obtain it is by the exchange of a massive Majorana neutrino: one can think of the virtual neutrino in the diagram of Fig. 1 as being produced as an antineutrino at one vertex and absorbed as an neutrino at the

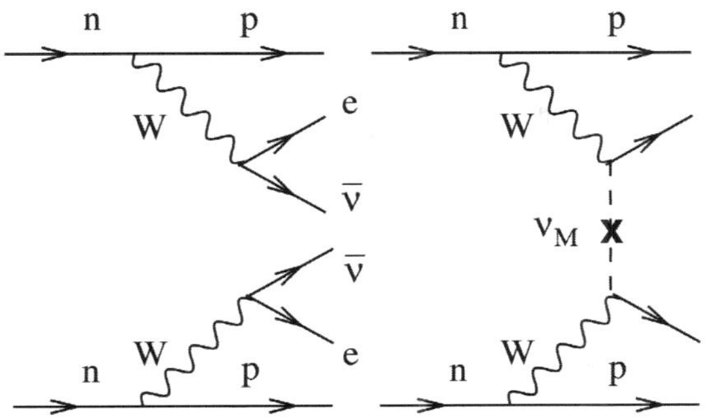

Figure 1. Double beta decay diagram for DBD2ν (left) and DBD0ν (right) modes.

other vertex. In addition to the Majorana equivalence of neutrino and antineutrino, a not zero neutrino mass is required to flip the helicity since antineutrinos are right-handed and neutrino are left-handed. The helicity flip and the smallness of the neutrino mass cause the rate of DBD0ν to be much lower than the rate of DBD2ν.

The rate of DBD0ν decay driven by the exchange of light Majorana neutrinos can be written as:

$$\left[T^{0\nu}_{1/2}\right]^{-1} \propto G_{0\nu} \cdot |M_{0\nu}|^2 \cdot \langle m_{\beta\beta}\rangle^2 \tag{3}$$

where $G_{0\nu}$ is the phase space factor, $M_{0\nu}$ is the nuclear matrix element (NME), and $m_{\beta\beta}$ is the effective Majorana mass defined as:

$$m_{\beta\beta} = \left|\sum_{j=1}^{3} U_{ej}^2 m_j\right| \tag{4}$$

with U_{ej} as the mixing parameters of the PMNS matrix and m_j the neutrino mass eigenstates. So, if neutrinos are Majorana particles, measuring or constraining the effective Majorana mass provides information on the neutrino mass scale and hierarchy.

2. The CUORE experiment

The use of bolometric detectors for the search of neutrinoless double beta decay was proposed by Fiorini and Niinikoski in 1984 [3]. Since almost thirty years his research group has been developing cryogenic detectors of increasing mass. The choice of a pure calorimetric approach was supported by the very good energy resolution achievable and the large choice of absorber materials containing DBD candidates.

2.1. Working principle

Bolometers are calorimeters operating at low temperature, in which the energy deposited by a particle is converted into phonons and is detected as a temperature variation. Almost the total released energy can be detected in this device. The energy deposited by a single particle in the device (connected to a heat sink) produces an increase of the temperature T in the absorber. This variation corresponds to the ratio between the energy released by the interacting particle and the heat capacity of the absorber. The only requirements needed for running this type of devices are therefore to operate the detector at low temperatures (mK scales), in order to reduce the value of the heat capacity (large temperature variations), and to have a sensitive thermometer coupled to the absorber. The thermometer is usually a thermistor, a Neutron

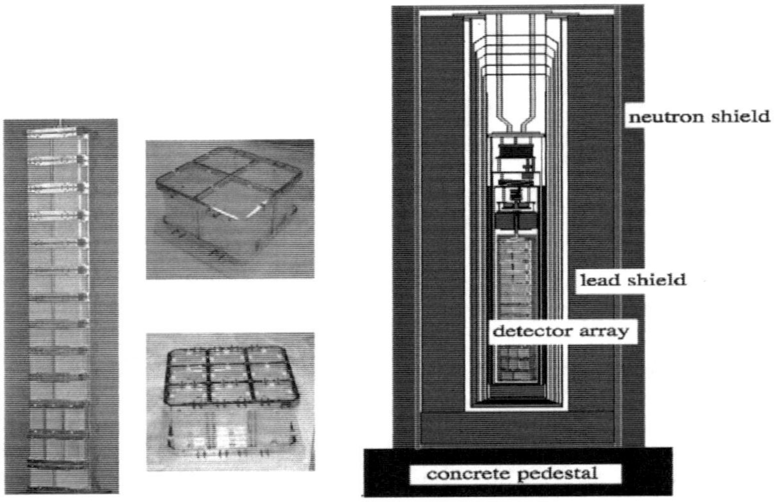

Figure 2. On the left the entire CUORICINO tower and two modules are shown. On the right a sketch of the apparatus is depicted.

Transumtation Doped germanium sensor (Ge-NTD). The resistivity of the thermistor has a steep dependence on the temperature:

$$R(T) = R_0\, e^{\left(\frac{T_0}{T}\right)^{\gamma}} \tag{5}$$

where R, T and γ are parameters that depends on the doping level of the sensors. Typical parameters for CUORE NTDs are:

$$R_0 = 1.15\ \Omega, \quad T_0 = 3.35\ K, \quad \gamma = 0.5\ . \tag{6}$$

2.2. From CUORICINO to CUORE
The CUORE experiment will look for DBD0ν in ^{130}Te using TeO_2 absorbers. It will use the bolometric technique established with its predecessor, CUORICINO, and should be able to achieve an improvement of almost two orders of magnitude in sensitivity to the DBD0ν half-life over Cuoricino's current limit.

CUORICINO was the first CUORE prototype, the largest cryogenic detector ever assembled. The detector was an array of 62 TeO_2 crystals, for a total active mass of 40.7 kg. The detectors were made of $5 \times 5 \times 5$ cm^3 crystals (44 absorbers) with a mass of about 790 g, the rest were $3 \times 3 \times 6$ cm^3 crystals with a mass of about 330 g (see Fig. 2). Four of the 330 g crystals were made of enriched Te: two were enriched in ^{130}Te and two in ^{128}Te.

All the detector components: TeO_2 crystals, the copper structure and the PTFE holders facing the detectors were treated to remove any possible radioactive impurities. The detectors were then assembled in a clean room with special care to avoid any recontaminations (e.g. Radon exposure). The CUORICINO tower was installed in a dilution refrigerator in the Underground Gran Sasso Laboratory, in order to reduce the background induced by cosmic rays. The detector was operated at a temperature of about 10 mK. The refrigerator was surrounded by a shield made of 10 cm of regular lead in addition with 10 cm of ancient Roman lead (with low ^{210}Pb

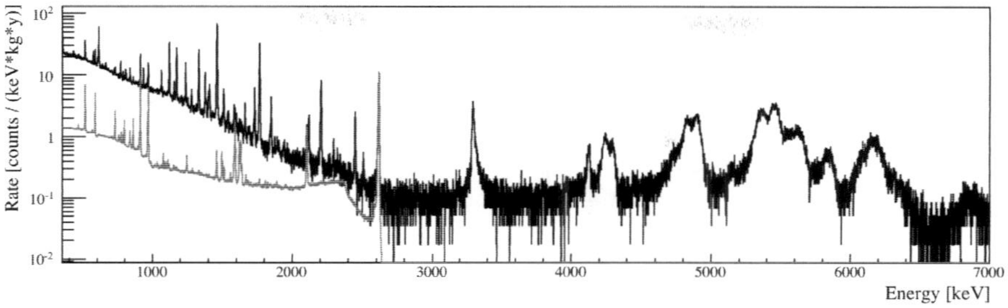

Figure 3. CUORICINO single-hit background spectrum (black) and CUORICINO sum calibration spectrum (red).

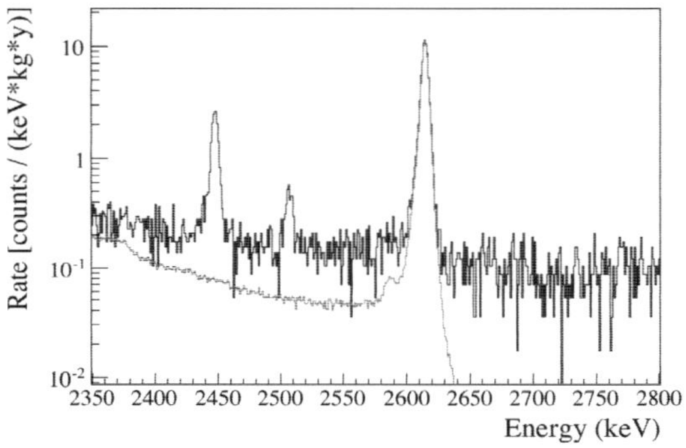

Figure 4. ROI of the CUORICINO single-hit background (black) and calibration (red) energy spectrum. The DBD0ν signal is expected at 2527 keV.

content).

The total analyzed exposure was 19.75 y×kg of ^{130}Te [5]. The average FWHM energy resolution of the array was 6.3 keV at the ^{130}Te Q-value at 2527 keV. The final sum spectrum is obtained with the detectors operated in anticoincidence (see Fig. 3), to reduce background contributions from the crystal surfaces and from external γ's. The achieved background level in the DBD00ν energy region is 0.169 counts/keV/kg/y (see Fig. 4), which translates into a sensitivity on the ^{130}Te half-life and effective neutrino Majorana mass of:

$$T^{0\nu}_{1/2} > 2.8 \times 10^{24}\ y, \quad \langle m_{\beta\beta}\rangle < 0.3 - 0.7\ eV\ . \tag{7}$$

In CUORICINO, the main sources of background were the natural ^{238}U and ^{232}Th decay chains from detector material contaminations. By means of Monte Carlo simulations [6] two main background sources were identified: surface contaminations of the detector parts, and bulk contaminations of the surrounding materials (e.g. cryostat components). Surface contaminations produced a flat background in the region of interest [7]; the main components were the surfaces of the copper structure facing the detectors and the crystals themselves (~30% of the total background in the DBD0ν region).

The knowledge acquired with CUORICINO helped the design and construction of CUORE, this was done by optimizing each of the parameters of the sensitivity. The most critical parameter is the background level, this needs to be improved by more than a factor of 10 with respect to CUORICINO, reaching a goal of 0.01 counts/keV/kg/y. With a series of thorough studies on

Journal of Physics: Conference Series **447** (2013) 012066 doi:10.1088/1742-6596/447/1/012066

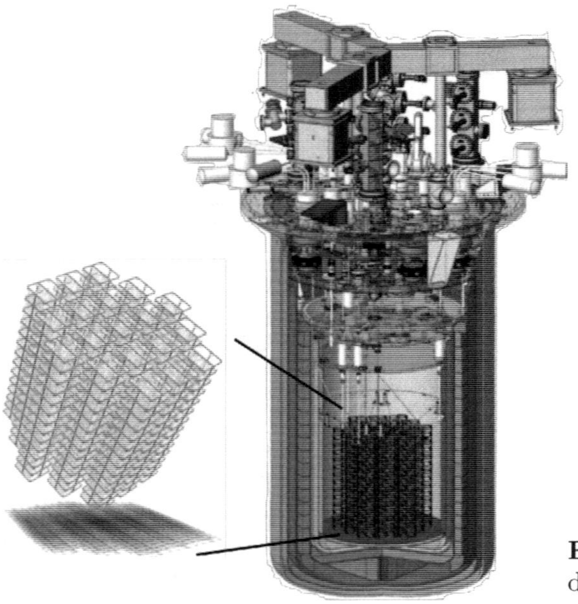

Figure 5. CUORE detector (on the left) and dilution refrigerator (on the right).

the background abatement a special detector structure has been designed in order to reduce the overall copper surfaces facing the detectors. New surfaces cleaning techniques have been defined so that the surface contaminations are reduced and the induced background in the region of interest is minimized. All possible recontaminations of radio-cleaned materials are continuosly kept under contol.

The CUORE detector will be a tightly packed array of 988 TeO_2 bolometer modules, each $5 \times 5 \times 5$ cm^3 and 750 g, for a total mass of 741 kg of TeO_2. Since the tellurium is unenriched, 204 kg of the total mass is the isotope of interest, ^{130}Te. The bolometer modules will be arranged in 19 towers of 13 floors each, with 4 crystals per floor (see Fig. 5). The CUORE detector will be housed in a specially built cryostat and cooled to about 10 mK by a pulse-tube-assisted dilution refrigerator.

CUORE will reach a sensitivity two orders of magnitude better than CUORICINO: $\sim 10^{26}$ y. Thus it will be the first experiment able to probe the inverted hierarchy of neutrino mass.
In order to introduce all the improvements in the detector design and construction, going towards CUORE, a step in between CUORICINO and CUORE is needed: CUORE-0.

2.3. CUORE-0

The final step before the start of the CUORE experiment will be CUORE-0 , a single CUORE-like tower that will operate in the former CUORICINO cryostat and it is now acquiring data. CUORE-0 will consist of 52 CUORE crystals mounted in CUORE-style single tower with a total TeO_2 mass of 39 kg. It will be assembled with detector components manufactured, cleaned, and stored following the same stringent protocols defined for CUORE. The preparation of CUORE-0 is thus aimed to confirm the efficacy and feasibility of the CUORE production and assembly procedures. In addition, CUORE-0 represents an opportunity to evaluate the bolometric performances of a CUORE-like detector apparatus in a familiar cryostat, and it will be the first large-scale empirical test of the undertaken extensive background-reduction measures.

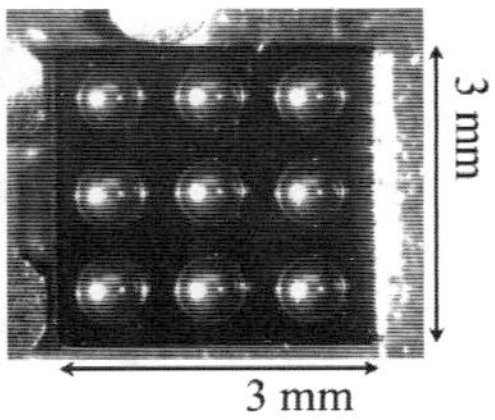

Figure 6. On the left an image of the automated robot gluing system and on the right an image of the glue spots on a thermal sensor.

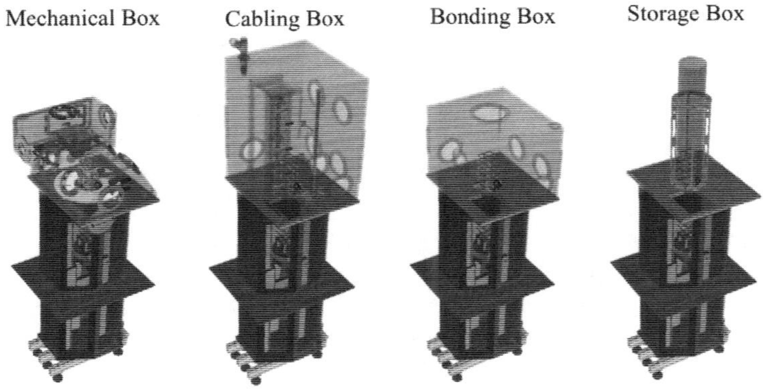

Figure 7. Sketch of the different glove boxes used for the assembly of the detector.

The assembly of the CUORE towers must follow very strict prescriptions, due to the extraordinary level of radio-purity needed for the success of the experiment. The assembly is conducted following a zero contact philosophy for the detector components. For these reasons, the whole assembly procedure will be performed inside dedicated glove boxes, in a clean-room recently built in the CUORE hut in the Hall A of LNGS. Moreover, to reduce the cosmogenic activation, the exposure of detector parts to cosmic rays has to be minimized. The detectors parts, after having been produced, are packed and then stored in the so called Permanent Storage Area (PSA), located in the underground Laboratories, inside stainless-steel cabinets, which are continuously flushed with Nitrogen until the time when they will be used for the CUORE construction. The tower construction is conducted around two units: the gluing station (see Fig. 7), which provides all the tools needed for the gluing of thermal sensors on the crystals, and the assembly line, an integrated set of tools devoted to the final assembly of the tower. Both units use glove boxes to provide a controlled Radon-free environment. Each glove box has been specifically designed in order to meet the requirements of each single operation.

3. CUORE and CUORE-0 sensitivity to neutrino Majorana mass

CUORE-0 tower was cooled down in August 2012, and it started the pre-operation phase and optimization of 51/52 detectors. Unfortunately these operations were not completed due to 2 vacuum leaks in the system which deteriorated the system stability. We were able to perform calibrations despite the leaks, showing reasonable detector performances. We can not give an

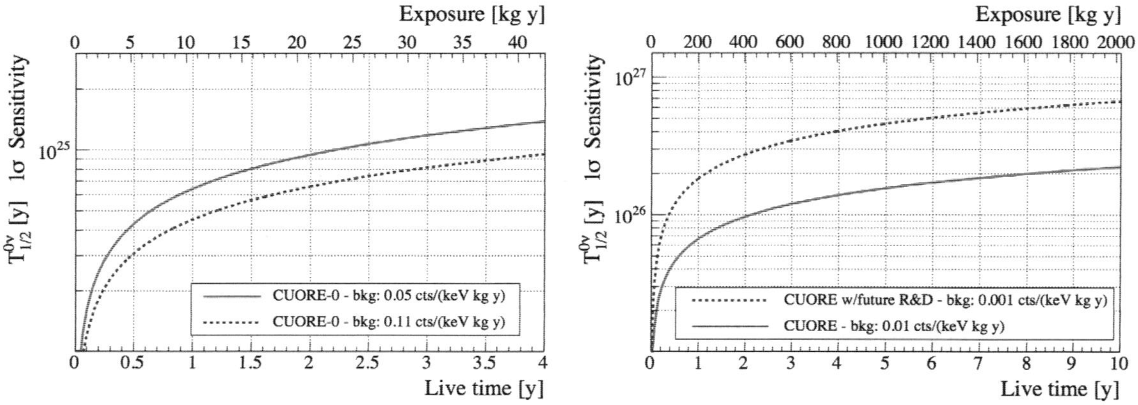

Figure 8. CUORE-0 and CUORE background-fluctuation sensitivity at 1σ C.L. for two different values of the background rate in the region of interest.

exact estimation on the background rate. Based on simple scaling arguments, on the amount of copper surfaces and surface treatments CUORE-0 is expected to have a background reduction at least of a factor 2 in the DBD0ν region compared to CUORICINO.
After a calibration of the detectors with an external ^{232}Th source, we have evaluated a preliminary energy resolution. The best detector showed an energy resolution at 2615 keV of about 5 keV.

CUORE-0 is operating in the CUORICINO cryostat and consequently the γ background from contamination in the cryostat shields will remain approximately the same as in CUORICINO. Considering that the irreducible background for CUORE-0 comes from the 2615 keV ^{208}Tl line due to ^{232}Th contaminations in the cryostat, in the case that all other background sources (e.g. surface contaminations) will be negligible, this would imply a lowe limit of 0.05 counts/keV/kg/y on the expected background. Similarly, an upper limit of 0.11 counts/keV/kg/y follows from scaling the CUORICINO background in the conservative case of a factor of 2 improvement in the crystal and copper contamination.
A plot of the expected 1 σ sensitivity of CUORE-0 as a function of live time in these two bounding cases is shown in Fig. 8.

In the case of CUORE, the expected background rate is 0.01 counts/keV/kg/y, an energy resolution of 5 keV and a live time of the data taking of 5 y CUORE are foreseen. The achievable sensitivity to DBD0ν is $2.1{\cdot}10^{26}$ y (see Fig. 8), that corresponds to a lower limit on the neutrino Majorana mass of $\langle m_\nu \rangle < (35-82)$ meV using the following nuclear matrix elements [8, 9, 10, 11].

References

[1] S. Barabash, Phys. Atom. Nucl. **74**, (2011) 603-613.
[2] H.V. Klapdor-Kleingrothaus et al., Mod. Phys. Lett. A **16**, (2001) 2409-2420.
[3] E. Fiorini, T.O. Niinikoski, Nucl. Instrum. Meth. A **224**, (1984) 83-88.
[4] A. Miller, E. Abrahams, Phys. Rev. **120**, (1960) 745-755.
[5] E. Andreotti *et al.*, Astropart. Phys. **34**, 822 (2011).
[6] C. Bucci *et al.*, Eur. Phys. J. A **41**, 155 (2009).
[7] M. Clemenza, C. Maiano, L. Pattavina, E. Previtali, Eur. Phys. J. C **71**, 1805 (2011).
[8] J. Barea, F. Iachello, Phys. Rev. C **79**, 044301 (2009).
[9] F. Simkovic *et al.*, Phys. Rev. C **77**, 045503 (2008).
[10] O. Civitarese, J. Suhonen, J. Phys.: Conf. Ser. **173**, 012012 (2009).
[11] J. Menendez *et al.*, Nucl. Phys. A **818**, 139-151 (2009).

Journal of Physics: Conference Series **447** (2013) 012067 doi:10.1088/1742-6596/447/1/012067

Neutrino Masses, the Baryon Asymmetry and Dark Matter from a Global U(1) symmetry

F.X. Josse-Michaux

CFTP, Instituto Superior Tecnico, Universidade Tecnica de Lisboa, Portugal

E-mail: `fxjossemichaux@gmail.com`

Abstract. We present a U(1) extension of the standard model that accounts after breaking for the small neutrino masses and the observed dark matter abundance. From the presence of the dark matter candidate, an original scenario of leptogenesis emerges, viable in a restricted region of the parameter-space. Relying on a rich scalar sector, this model can deviate from standard model Higgs phenomenology.

The present letter summarizes the model proposed in [1] and its variant discussed in [2]. Three issues of particle physics left unexplained in the Standard Model (SM) are there adressed:

- the smallness of neutrino masses are explained within a seesaw extension of the SM, with both linear and inverse seesaw contributions.
- The Dark Matter (DM) fraction $\Omega_{\rm DM} = 0.2408^{+0.0093}_{+0.0092}$[3] find its origin in the relic abundance of a scalar field S.
- The Baryon Asymmetry of the Universe (BAU) $\Omega_{\rm BAU} = 0.0472 \pm 0.0010$[3] can be explained by a variant of leptogenesis.

1. The Model

The matter content of the model is given in Table 1 (SM quarks and gauge boson aside). A global $U(1)_X$ symmetry is imposed to the Lagrangian, the SM particles having a X charge equal to their B-L one. The SM Higgs boson H_1 is singlet under the new symmetry, and only triggers

Field	ℓ_α	$e_{R\alpha}$	N_1	N_2	N_3	H_1	H_2	ϕ	S
$SU(2)_L$	**2**	**1**	**1-3**	**1-3**	**1**	**2**	**2**	**1**	**1-3**
$U(1)_Y$	-1/2	-1	0	0	0	1/2	1/2	0	0
$U(1)_X$	-1	-1	-1	+1	0	0	2	-2	-1

Table 1. Particle content and their charge assignments.

electroweak-symmetry breaking, via a non-zero vacuum expectation value (vev) $\langle H_1^0 \rangle = v_1/\sqrt{2}$. An additional scalar doublet H_2 is introduced: given its non-zero X charge, H_2 does not couple to SM fermions, but to neutrinos via an additional Yukawa coupling. The complex singlet ϕ has also a non-zero X charge: H_2 and ϕ jointly break $U(1)_X$ via non-zero vevs $\langle H_2^0 \rangle = v_2/\sqrt{2}$ and $\langle \phi \rangle = v_\phi/\sqrt{2}$. Both particles are required to explain the small neutrino masses. Three

Journal of Physics: Conference Series **447** (2013) 012067 doi:10.1088/1742-6596/447/1/012067

right-handed fermions are introduced: N_1 and N_2 singlet [1] or triplet [2] under SU(2) and with opposite X charge ∓ 1, while N_3 is a Majorana particle. Finally, a scalar particle S complete the spectrum, with $SU(2) \times U(1)_X$ charge equal to that of N_1; S is the DM candidate. We first discuss how the small neutrino masses are accounted for.

2. Neutrino Masses

For convenience, a Dirac fermion is build upon N_1 and N_2: $N_D \equiv P_R N_1 + P_L N_2^C$; N_D has X charge +1. The part of the Lagrangian density involving N_D is then:

$$\mathcal{L} \supset -\left(Y_{\nu 1}^{\beta} \overline{N_D^a} \widetilde{H}_1^{j*} (T_2^a)_{jk} \ell_{\beta}^k + Y_{\nu 2}^{\gamma} \overline{N_D^a}^C \widetilde{H}_2^{j*} (T_2^a)_{jk} \ell_{\gamma}^k + \frac{\delta_N}{\sqrt{2}} \phi \overline{N_D^a} N_D^{a\,C} + \text{H.c.} \right)$$
$$- m_N \overline{N_D^a} N_D^a - \frac{1}{2} M_3 \overline{N_3} N_3^C - \left(h S^a \overline{N_D^a} N_3 + \text{H.c.} \right) , \tag{1}$$

where we assume $N_{1,2}$ and S triplet of SU(2). In the singlet case, the SU(2) generators $T_2^a, a = 1, 2, 3$, i.e. the Pauli matrices, are exchanged by scalar products of Higgs $\widetilde{H}_k \equiv -i\sigma_2 H_k^*$ $(k = 1, 2)$ and lepton doublets $\ell_\beta = (\nu_{\beta \mathbf{L}}, e_{\beta \mathbf{L}})^T$ $(\beta = e, \mu, \tau)$. Notice that the Yukawa couplings $Y_{\nu i}$ couple leptons to N_i: after symmetry breaking, the term $\propto Y_{\nu 2}$ explicitly violates Lepton number. Similarly, the third term in eq.(1) violates L by 2 units, mixing N_1 and N_2. In the seesaw limit, where $m_N \gg Y_{\nu i} v_i, \delta_N v_\phi$, the effective light neutrino mass matrix,

$$(M_\nu)^{\alpha\beta} \simeq -\frac{v_1 v_2}{m_N} \mathbf{y_1}^{\{\alpha,} \mathbf{y_2}^{\beta\}} + v_\phi \delta_N \frac{v_1^2}{m_N^2} \left(\mathbf{y_1}^{\alpha} \mathbf{y_1}^{\beta} + \mathbf{y_2}^{\alpha} \mathbf{y_2}^{\beta} \frac{v_2^2}{v_1^2} \right) , \tag{2}$$

where $\mathbf{y_k} \equiv (Y_{\nu k}^e, Y_{\nu k}^\mu, Y_{\nu k}^\tau)^{\mathrm{T}} / 2\sqrt{2}$, clearly exhibits different types of contributions. The first term acts as a linear seesaw. In our framework its suppression originates from the small ratio $v_2/m_N \ll 1$, as in neutrinophilic models [4]. This latter suppression is naturally obtained via a seesaw in the scalar vevs, thanks to a pivot coupling $\propto \mu' (H_1^\dagger H_2)\phi$ in the scalar potential [4, 5]. In the limit $\delta_N v_\phi \ll m_N$, the second term in eq.(2) is an inverse-seesaw contribution [6]. In this regime, N_1 and N_2 combine to form two quasi-degenerate Majorana fermion, or so called pseudo-Dirac fermions, with mass $m_N \pm \delta_N v_\phi$. With the particle content of Table 1, two non-zero light neutrino masses are generated, sufficient comply with current observations. In the inverse-seesaw limit, $\delta_N v_\phi \ll m_N$, these masses can be approximated by

$$m_\nu^\pm \simeq \frac{v_1 v_2}{m_N} |\mathbf{y_1}||\mathbf{y_2}| \left(1 \pm |\cos(\theta_{12})|\right) , \tag{3}$$

with θ_{12} the angle between $\mathbf{y_1}$ and $\mathbf{y_2}$: if aligned, only one light neutrino is massive, in contradiction with observations. Their fit implies

$$|\mathbf{y_1}||\mathbf{y_2}| \simeq \left(\begin{smallmatrix} 3.4 \\ 5.6 \end{smallmatrix} \right) \times 10^{-8} \left(\frac{m_N}{1\text{TeV}} \right) \left(\frac{v_2}{10\text{MeV}} \right)^{-1} , \; |\cos(\theta_{12})| \simeq \left(\begin{smallmatrix} 0.93 \\ 0.87 \end{smallmatrix} \right) \tag{4}$$

in the normal / inverse hierarchy cases. For our later purpose, only the strength of the Yukawa couplings matter as we did not consider flavour effects, and so we enforce the first relation quoted above.

3. Baryon Asymmetry Generation

We assume that the breaking of the global $U(1)_X$ symmetry occurs together with the breaking of the electroweak symmetry. Therefore, as long as sphalerons are active, $U(1)_X$ is conserved, and no lepton asymmetry can be generated via N_1 or N_2 decays. To remedy this, we introduce a third right handed neutrino, but of a very different nature than $N_{1,2}$: N_3 is assumed to be a Majorana fermion. It does not couple directly to leptons, but to N_D via a Yukawa coupling involving the scalar S, our DM candidate. A two-steps leptogenesis is then possible: during N_3 decays, asymmetries in N_D (and S) are produced, *i.e.* N_1 and N_2 are produced in different amounts. This step is similar to the standard thermal leptogenesis, with competitive depletion effects and other spectator processes that influence their production. While being asymmetrically produced, $N_{1,2}$ interacts with SM leptons, via X-conserving scatterings or decays. The net $N_1 - N_2$ number is then reprocessed into a net lepton number. During this stage as well, spectator processes influence the efficiency of this asymmetry transfer.

We stated that this leptogenesis scenario was only possible thanks to the addition of S: this is obvious when looking at the diagram responsible for CP-violation in N_3 decays, shown in figure 1. The CP asymmetry generated in these decays linearly depends on the parameter δ_N, which

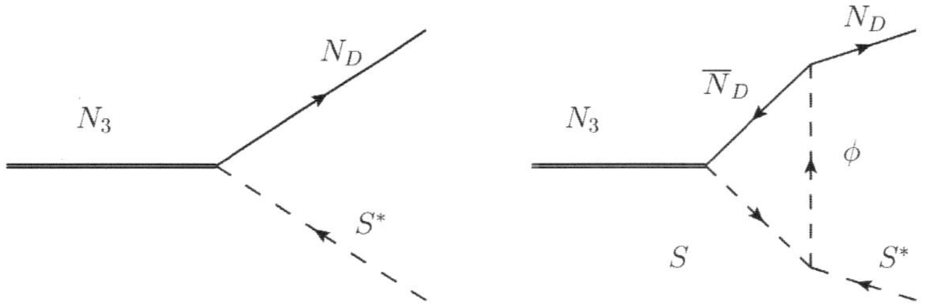

Figure 1. One-loop correction to N_3 decays allowing nonzero CP-asymmetry.

appears in the neutrino mass matrix eq.(2). In the inverse seesaw limit, δ_N only provides a subdominant contribution to light neutrino masses, so the CP-asymmetry and neutrino masses are decorrelated. The constraints on the couplings a successful leptogenesis imposes depend on whether $N_{1,2}$ are singlet or triplet of SU(2). Indeed, in the latter case additional gauge boson scatterings are present, reducing the efficiency of the first stage of asymmetry production: smaller couplings are required.

In [1] and [2], we mainly focus on the possibility of low-energy realization of this leptogenesis mechanism. In the singlet case, fermion masses $m_N \sim$ few 100 GeV are allowed without demanding large suppression of the couplings. The case of light fermion triplets requires stronger tuning, as illustrated in Figure 2. The baryon asymmetry resulting from leptogenesis is shown, in function of a degeneracy parameter $1-(\mu_S+m_N)/M_3$ and of m_N (left plot), or in function of the parameters μ'' and δ_N (right plot) which both control the amount of CP violation in N_3 decays and the strength of depletion effects. In the low mass regime, couplings of $\sim 10^{-6}$ are typically required, together with a certain tuning among N_3, $N_{1,2}$ and S masses. Larger masses relax these constraints. Although a systematic study was not conducted, an approximate lower bound of $m_N \sim 1.5$ TeV on the triplet fermion mass was found: if neutrino masses are explained via eq.(1), the observation of $N_{1,2}$ at LHC would rule-out this leptogenesis mechanism as an explanation of the observed BAU.

4. An Extended Higgs Sector

The model contains 2 scalar doublet $H_{1,2}$ and a complex singlet ϕ with quantum numbers given in Table 1, responsible of the symmetry breaking $\mathrm{SU(2)} \times \mathrm{U(1)_Y} \times [\mathrm{U(1)_X}] \to \mathrm{U(1)_{em}} \times [\mathrm{Z_2}]$. The scalar potential reads

$$\mathcal{V}_{\rm SB} = -\mu_1^2\,(H_1^\dagger H_1) + \lambda_1\,(H_1^\dagger H_1)^2 - \mu_2^2\,(H_2^\dagger H_2) + \lambda_2\,(H_2^\dagger H_2)^2 - \mu_3^2\,\phi^*\phi + \lambda_3\,(\phi^*\phi)^2$$

Journal of Physics: Conference Series **447** (2013) 012067 doi:10.1088/1742-6596/447/1/012067

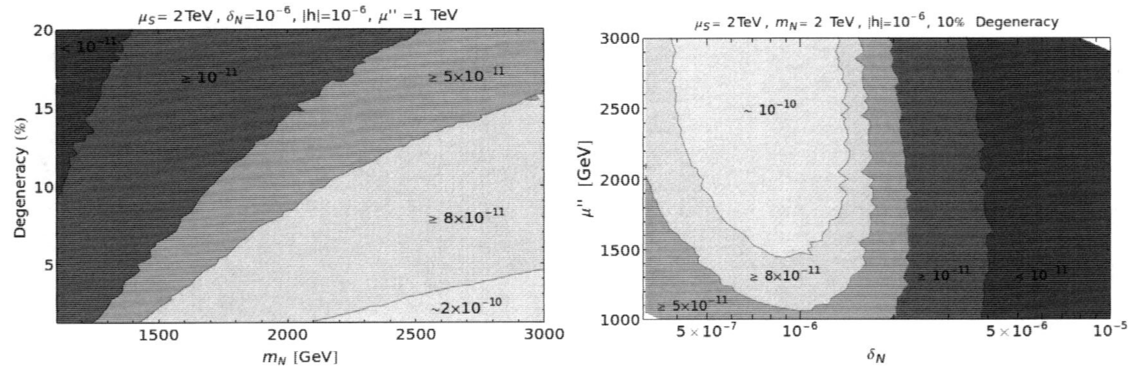

Figure 2. Left: Baryon asymmetry Y_B as function of the degree of degeneracy between M_3 and $\mu_S + m_N$ and m_N mass (see the main text). **Right:** variation of Y_B in terms of the parameters μ'' and δ_N, for fixed M_3, μ_S and m_N.

$$
\begin{aligned}
+ & \;\; \kappa_{12}\,(H_1^\dagger\, H_1)\,(H_2^\dagger\, H_2) + \kappa_{12}'\,(H_1^\dagger\, H_2)\,(H_2^\dagger\, H_1) + \kappa_{13}\,(H_1^\dagger\, H_1)\phi^*\phi + \kappa_{23}\,(H_2^\dagger\, H_2)\phi^*\phi \\
- & \;\; \frac{\mu'}{\sqrt{2}}\left((H_1^\dagger\, H_2)\phi + (H_2^\dagger\, H_1)\phi^*\right) . \qquad (5)
\end{aligned}
$$

The last term of eq.(5) deserves some attention. Minization of $\mathcal{V}_{\rm SB}$ enforces at tree-level three relations

$$
\frac{\partial \mathcal{V}_{\rm SB}}{\partial v_i} = 0 \leftrightarrow -\mu_i^2\, v_i + \frac{v_i}{2}\left(v_j^2 \tilde{\kappa}_{ij} + v_k^2 \kappa_{ik}\right) + 2\, v_i^3 \lambda_i = \frac{v_j v_k \mu'}{2}, \quad i,j,k = 1,2,\phi\,, \qquad (6)
$$

with $\tilde{\kappa}_{12} = \kappa_{12} + \kappa_{12}'$ and $\tilde{\kappa}_{ij} = \kappa_{ij}$ otherwise. As noted in [4] or [5], this term acts as a pivot for a seesaw among the Higgs vevs, yielding $v_2 \ll v_{1,\phi}$ ($v_\phi \ll v_{1,2}$) granted $\mu' \ll v_{1,\phi}$ ($v_{1,2}$). In [4], the scalar ϕ being absent, the μ' term was introduced as a soft breaking term of the global $\mathrm{U(1)_X}$. Here the breaking is spontaneous, therefore a Goldstone boson, the Majoron, remains massless after $\mathrm{U(1)_X}$ breaking. Its couplings to SM fermions impose the vev hierarchy $v_2 \ll v_{1,\phi}$. Therefore, SM gauge boson masses receive a negligible contribution from H_2 doublet and $v_1 \simeq 174$ GeV. The physical spectrum is rich, with a charged scalar $H^\pm$, 3 neutral CP-even scalars h^0, H^0 and h_A, plus 2 neutral CP-odd scalars A^0 and J the massless Goldstone boson.

With the vev hierarchy $v_2 \ll v_{1,\phi}$, only $\mathrm{Re}(H_1^0)$ and $\mathrm{Re}(\phi)$ can significantly mix: H_1 being the SM Higgs doublet, departure from SM Higgs observations can be expected. Besides from this mixing, we have additional charged particle that can modify Higgs observations.

Assuming h^0 is the new particle observed at LHC, we verify in [2] how our model can accommodate LHC data, and if deviations from SM are indeed observed. We construct the Higgs signal strengths, and fit LHC data. The scalar H^0 have similar signature that h^0, so we impose LHC exclusion constraints over the full H^0 spectrum, assuming $m_{h^0} \simeq 126$ GeV $< m_{H^0} \leq 600$ GeV. The main results are presented in Figure 3. In the limit of a negligible H_1-ϕ mixing, $h^0 \sim \mathrm{Re}(H_1^0)$ so tree-level h^0 decays tend to their SM width. In the opposite case, h^0 invisible decays become sizeable, and possibly larger than SM Higgs width (however quite small, $\Gamma_{h\rm SM}^{\rm tot} \simeq 4.2$ MeV [7] for $m_h \sim 126$ GeV), as illustrated in the left plot of Figure 3, where h^0 width is compared with the SM Higgs width as a function of the invisible branching ratio of h^0 decaying to Majoron $\mathrm{BR}(h^0 \to JJ)$. The presence of a charged scalar manifests itself in a possible enhancement of the diphoton decay channel, cf e.g. [8], as shown in the right plot of Figure 3, where the ratio $\Gamma(h^0 \to \gamma\,\gamma)/\Gamma(h \to \gamma\,\gamma)^{\rm SM}$ is plotted against

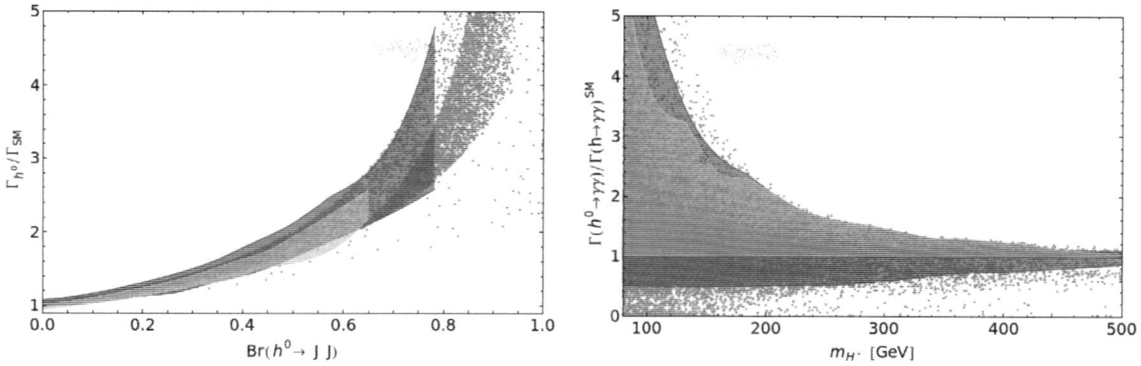

Figure 3. Left: Total h^0 decay width normalized to the SM case as a function of the invisible branching ratio. **Right**: Ratio $\Gamma(h^0 \to \gamma\gamma)$ to the SM rate in function of the charged Higgs mass $m_{H^\pm}$. In both cases, the points are the results of our scan, the shaded dark red region corresponds to the 95 % C.L. range of the global fit, while the green area stands for observables within (below) their respective 95% C.L. bounds (upper bound).

$m_{H^\pm}$. This ratio also depends on the trilinear coupling h^0-H^+-H^- which is scanned over in this plot, but obviously the heavier $H^\pm$, the more suppressed its contribution to the loop.

5. Scalar Dark Matter

Our model also contains a Dark Matter candidate, the scalar S. We stated before that the presence of S is mandatory for our leptogenesis mechanism, and that S should have the same quantum numbers (but the spin) than $N_{1,2}$: S is then a stable particle, being odd under the remnant Z_2.

The question is thus if its relic abundance can match current observations. The singlet case has been widely studied in the literature, cf. e.g.[9]; the DM relic abundance is fixed by the freeze-out of DM annihilations that proceed through quartic couplings to the SM Higgs. In our model, having several Higgs, many annihilation channels are possible, and in particular S annihilations to pairs of Majoron typically dominate at low S mass. Annihilations through portal couplings are also present in the triplet case, which however contains irreducible and very efficient channels with S annihilating to or via gauge boson [10, 11]. Our model contains a complex triplet S, viable DM candidate for mass $m_S \geq 1290$ GeV.

6. Summary

The smallness of neutrino masses explained via see-saw mechanism, appealingly points towards leptogenesis scenario for the observed BAU. In [1]-[2], neutrino masses are explained via linear see-saw contribution, whose suppression originates from a see-saw among Higgs vevs, plus an inverse see-saw contribution, however sub-dominant by construction. A global symmetry is imposed to the Lagrangian above electroweak symmetry breaking to justify the various contribution, but this in turn calls for a variant of the standard leptogenesis mechanism. Such a variant is made possible by the introduction of a scalar particle, our Dark Matter candidate. Our model succeeds to account for neutrino masses, BAU and DM densities, although some tuning are required in the low-mass leptogenesis realizations.

Acknowledgements
I warmly thank E. Molinaro for the collaboration on the works summarized here. I also express my gratitude to the organizers of the DISCRETE 2012 conference. These works are supported by Portuguese national funds through FCT - Fundação para a Ciencia e Tecnologia, project PEst-OE/FIS/UI0777/2011 and by CERN/FP/123580/2011.

References
[1] F. -X. Josse-Michaux and E. Molinaro, Phys. Rev. D **84** (2011) 125021 [arXiv:1108.0482 [hep-ph]].
[2] F. -X. Josse-Michaux and E. Molinaro, arXiv:1210.7202 [hep-ph].
[3] G. Hinshaw, D. Larson, E. Komatsu, D. N. Spergel, C. L. Bennett, J. Dunkley, M. R. Nolta and M. Halpern *et al.*, arXiv:1212.5226 [astro-ph.CO].
[4] E. Ma, Phys. Rev. Lett. **86** (2001) 2502 [hep-ph/0011121].
[5] W. Grimus, L. Lavoura and B. Radovcic, Phys. Lett. B **674** (2009) 117 [arXiv:0902.2325 [hep-ph]].
[6] R. N. Mohapatra and J. W. F. Valle, Phys. Rev. D **34** (1986) 1642.
[7] https://twiki.cern.ch/twiki/bin/view/LHCPhysics/CERNYellowReportPageBR2
[8] P. Posch, Phys. Lett. B **696** (2011) 447 [arXiv:1001.1759 [hep-ph]].
[9] J. McDonald, Phys. Rev. D **50** (1994) 3637 [hep-ph/0702143 [HEP-PH]].
[10] M. Cirelli, N. Fornengo and A. Strumia, Nucl. Phys. B **753** (2006) 178 [hep-ph/0512090 [HEP-PH]].
[11] T. Hambye, F. -S. Ling, L. Lopez Honorez and J. Rocher, JHEP **0907** (2009) 090 [Erratum-ibid. **1005** (2010) 066] [hep-ph/0903.4010 [HEP-PH]].

DISCRETE 2012 – Third Symposium on Prospects in the Physics of Discrete Symmetries IOP Publishing
Journal of Physics: Conference Series **447** (2013) 012068 doi:10.1088/1742-6596/447/1/012068

Search for low mass dark gauge bosons at KLOE

Enrico Graziani
on behalf of the KLOE/KLOE2 Collaboration [1]
INFN sezione di Roma Tre, Roma, Italy
E-mail: enrico.graziani@roma3.infn.it

Abstract. The existence of a light dark force mediator has been tested with the KLOE detector at DAΦNE. This particle, the so called dark photon U, has been searched for in three different processes and four different final states: a) $\Phi \to \eta U$, with $U \to e^+e^-$, $\eta \to \pi^+\pi^-\pi^0$ and $\eta \to \pi^0\pi^0\pi^0$; b) $e^+e^- \to U\gamma$ with $U \to \mu^+\mu^-$; c) $e^+e^- \to Uh'$ (dark Higgsstrahlung), $U \to \mu^+\mu^-$, where h' is a Higgs-like particle responsible for the breaking of the hidden symmetry.

We did not find any evidence of the processes above and set upper limits on the parameters of the model, for different M_U (and $M_{h'}$) mass ranges, depending on the considered final state.

1. Introduction

In recent years, several unexpected astrophysical observations have failed to find a common interpretation in terms of standard astrophysical or particle physics sources. A non exhaustive list of these observations include the 511 keV gamma-ray signal from the galactic center observed by the INTEGRAL satellite [1], the excess in cosmic ray postirons reported by PAMELA [2], the total electron and positron flux measured by ATIC [3], Fermi [4], and HESS [5, 6], the annual modulation of the DAMA/LIBRA signal [7, 8] and the low energy spectrum of nuclear recoil candidate events observed by CoGeNT [9].

Although there are alternative explanations for some of these anomalies, they could be all explained with the existence of a dark matter weakly interacting massive particle, WIMP, belonging to a secluded gauge sector under which the Standard Model (SM) particles are uncharged [10, 11, 12, 13, 14]. An abelian gauge field, the U boson with mass near the GeV scale, couples the secluded sector to the SM through its kinetic mixing with the SM hypercharge gauge field. The kinetic mixing parameter ϵ is expected to be of the order $10^{-4} - 10^{-2}$ [11, 12, 13, 14, 15], so that observable effects can be induced in O(GeV)-energy e^+e^- colliders [15, 16, 17] and fixed target experiments [18, 19, 20].

[1] D. Babusci, D. Badoni, I. Balwierz-Pytko, G. Bencivenni, C. Bini, C. Bloise, F. Bossi, P. Branchini, A. Budano, L. Caldeira Balkeståhl, G. Capon, F. Ceradini, P. Ciambrone, F. Curciarello, E. Czerwiński, E. Danè, V. De Leo, E. De Lucia, G. De Robertis, A. De Santis, P. De Simone, A. Di Domenico, C. Di Donato, R. Di Salvo, D. Domenici, O. Erriquez, G. Fanizzi, A. Fantini, G. Felici, S. Fiore, P. Franzini, P. Gauzzi, G. Giardina, S. Giovannella, F. Gonnella, E. Graziani, F. Happacher, L. Heijkenskjöld, B. Höistad, L. Iafolla, E. Iarocci, M. Jacewicz, T. Johansson, K. Kacprzak, W. Kluge, A. Kupsc, J. Lee-Franzini, F. Loddo, P. Lukin, G. Mandaglio, M. Martemianov, M. Martini, M. Mascolo, R. Messi, S. Miscetti, G. Morello, D. Moricciani, P. Moskal, S. Muller, F. Nguyen, A. Passeri, V. Patera, I. Prado Longhi, A. Ranieri, C. F. Redmer, P. Santangelo, I. Sarra, M. Schioppa, B. Sciascia, M. Silarski, C. Taccini, L. Tortora, G. Venanzoni, R. Versaci, W. Wiślicki, M. Wolke, J. Zdebik

The U boson can be produced at e^+e^- colliders via different processes: $V \to PU$ decays, where V and P are vector and pseudoscalar mesons, respectively (Φ and η in the case of KLOE), $e^+e^- \to U\gamma$ and $e^+e^- \to Uh'$ (dark Higgsstrahlung), where h' is a Higgs-like particle responsible for the breaking of the hidden symmetry.

KLOE is in a very good position to probe the existence of a dark light sector at the GeV scale because a) it operates on DAΦNE exactly at the GeV scale; b) most of the interesting dark process cross sections scale as 1/s, i.e. a factor ~100 with respect to the B factories; c) it is a unique place to study the rare Φ meson decays.

2. The KLOE detector

The KLOE experiment operated from 2000 to 2006 at DAΦNE, the Frascati Φ factory. DAΦNE is an e^+e^- collider running mainly at a center-of-mass energy of ~1019 MeV, the mass of the Φ meson. Equal energy positron and electron beams collide at an angle of ~ 25 mrad, producing Φ mesons nearly at rest. The detector consists of a large cylindrical Drift Chamber (DC) [21], which provides a momentum resolution $\sigma_\perp/p_\perp \approx 0.4$, surrounded by a lead-scintillating fiber electromagnetic calorimeter (EMC) [21], with energy and time resolutions $\sigma_E/E = 5.7\%/\sqrt{E(GeV)}$ and $\sigma_t = 57$ ps$/\sqrt{E(GeV)} \oplus$ 100 ps, respectively. A superconducting coil around the EMC provides a 0.52 T field.

3. U boson search in $\Phi \to \eta U$ decay

We study the process $\Phi \to \eta U$ using a sample of Φ mesons produced resonantly at the DAΦNE collider. The U boson can be observed by its decay into a lepton pair, while the η can be tagged by one of its main decays. The best U decay channel in KLOE to look for is $U \to e^+e^-$, since a wider range of U boson masses can be tested and $e^\pm$ are easily identified using a time-of-flight (ToF) techinique. As far as the η decays are concerned, we used both the three pion decay modes: $\eta \to \pi^+\pi^-\pi^0$ and $\eta \to \pi^0\pi^0\pi^0$. An irreducible background due to the Dalitz decay of the Φ meson, $\Phi \to \eta e^+e^-$, is present. This is expected to have a cross section of the order of ~0.7 nb, to be compared with the signal cross section which, for $\epsilon = 10^{-3}$, is predicted to be ~40 fb. Despite the small ratio between the $\Phi \to \eta U$ and the overall $\Phi \to \eta e^+e^-$, their different dilepton invariant mass distribution allows to test the parameter ϵ down to 10^{-3}, with the KLOE data set of 1.5 fb^{-1} (corresponding to $\approx 5 \times 10^9$ Φ). The Monte Carlo simulation of the irreducible Dalitz Φ decay background has been produced with a Vector Meson Dominace model [23]. Different preselections cuts are chosen depending on the selected η decay channel. For $\eta \to \pi^+\pi^-\pi^0$ decay:

- two positive and two negative tracks with points of closest approach to the beam line inside a 4×20 cm cylinder around the interaction point;
- two prompt photon candidates, i.e. two energy clusters with $E >$7 MeV not associated to any track, in an angular acceptance $|cos\theta_\gamma| <$0.92 and in the expected time window for a photon $|T_\gamma - R_\gamma/c| < min(5\sigma_T, 2ns$;
- best $\pi^+\pi^-\gamma\gamma$ match to the η mass in the pion hypothesis (the other two tracks are then assigned to $e^\pm$)
- loose cuts of about $\pm4\sigma$'s on η and π^0 invariant masses (495< $M_{\pi^+\pi^-\gamma\gamma}$ < 600 MeV, 70 < $M_{\gamma\gamma}$ < 200 MeV);
- a cut on the recoil mass to the $e^+e^-\pi^+\pi^-$ system, which is expected to be equal to the π^0 mass for signal events: 100< $M_{recoil}(ee\pi\pi)$ <160 MeV.

For $\eta \to \pi^0\pi^0\pi^0$ decay:

- two opposite charge tracks with points of closest approach to the beam line inside a 4×20 cm cylinder around the interaction point;

- six prompt photon candidates, i.e. energy clusters with E >7 MeV not associated to any track, in an angular acceptance $|cos\theta_\gamma|$ <0.92 and in the expected time window for a photon $|T_\gamma - R_\gamma/c| < min(3\sigma_T, 2ns)$;
- a loose cut on the six photon invariant mass $400 < M_{6\gamma} < 700$ MeV).

After these selections, a clear peak corresponding to the η mass is observed in the distribution of the recoil mass against the e^+e^- pair, in both samples. Residual contaminations due to $\Phi \to \eta\gamma$ events with photon conversion on beam pipe or drift chamber walls are strongly rejected by tracing back the tracks of the $e^\pm$ candidates, reconstructing their invariant mass and their distance at the conversion surfaces and cutting on these variables. Further selections, aimed at applying some particle identification (PID) requirement to electrons, are implemented. This is achieved by exploiting the timing capabilities of the calorimeter. When an energy cluster is associated to a track, the time of flight (ToF) to the calorimeter is evaluated both using the track trajectory ($T_{track} = L_{track}/\beta c$) and the calorimeter timing ($T_{cluster}$). The $\Delta T = T_{track} - T_{cluster}$ variable is then evaluated in the electron hypothesis (ΔT_e). In order to be fully efficient on the signal, events with either an e^+ or an e^- candidate inside a 3σ window around ΔT_e=0 are retained for further analysis.

At the end of the analysis chains, the background contaminations were evaluated to be 3% and 2% for the $\eta \to \pi^+\pi^-\pi^0$ and $\eta \to \pi^0\pi^0\pi^0$ channels respectively. The analysis signal efficiencies ranged between 10% and 20%, depending on M_{ee}, for the $\eta \to \pi^+\pi^-\pi^0$ sample, and between 15% and 30% for the $\eta \to \pi^0\pi^0\pi^0$ sample.

The upper limit on the U boson production in the $\Phi \to \eta U$ process is obtained by combining the two η decay channels. The resolution of the e^+e^- invariant mass has been found to be below 2 MeV over the full mass range. The determination of the limit is done by varying the M_U mass with 1 MeV step, in the range between 5 and 470 MeV. For each channel, the irreducible background is extracted directly from data after applying a bin by bin subtraction of the non-irreducible backgrounds and correcting for the analysis efficiency. The M_{ee} distribution is then fitted with a parametrized function (representing the transition form factor) excluding the bins used in the upper limit evaluation. The exclusion limit on the number of events for the $\Phi \to \eta U$ signal as a function of M_U is obtained with the CLS technique [24], using the M_{ee} spectra before background subtraction. The limit is extracted both for each η decay channel and in a combined way. In fig. 1 top, the upper limit at 90% CL on the number of events for the decay chain $\Phi \to \eta U$, $U \to e^+e^-$, is shown for both $\eta \to \pi^+\pi^-\pi^0$ and $\eta \to \pi^0\pi^0\pi^0$, separately evaluated.

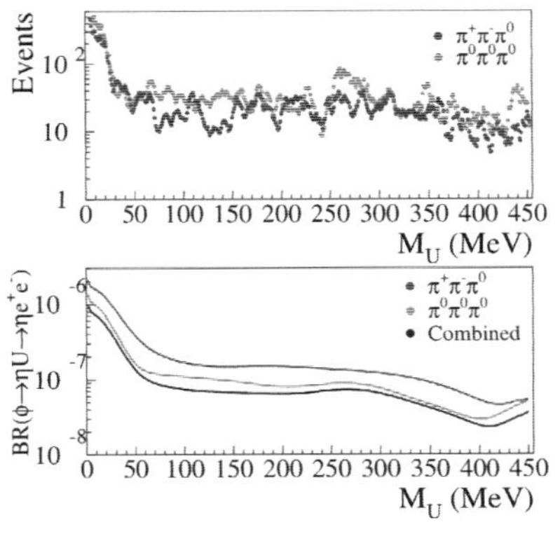

Figure 1. Top: upper limit at 90% CL on the number of events for the decay chain $\eta \to \pi^0\pi^0\pi^0$ and $\eta \to \pi^+\pi^-\pi^0$. Bottom: smoothed upper limit at 90% CL on $BR(\Phi \to \eta U) \times BR(U \to e^+e^-)$ obtained separately for the two η decay channels and from the combined analysis.

In fig.1 bottom, the smoothed upper limit on the branching fraction for the process $\Phi \to \eta U$, $U \to e^+e^-$, obtained for the combined method is compared with evaluations from each of the two decay channels. In the combined result, the upper limit on the product $BR(\Phi \to \eta U) \times BR(U \to e^+e^-)$ varies from 10^{-6} at small M_U to $\sim 3 \times 10^{-8}$ at 450 MeV. The exclusion plot in the $\alpha'/\alpha = \epsilon^2\, vs\, M_U$ plane, where α' is the coupling of the U bosons to electrons and α is the fine structure constant, has been finally derived for two different transition form factor hypotheses. In fig. 2 the smoothed exclusion plot at 90% CL on α'/α is compared with existing limits in the same region of interest. This result reduces the region of the U boson parameters that could explain the observed discrepancy between the Standard Model prediction of the muon anomalous magnetic moment, a_μ, ruling out masses in the range 60-435 MeV

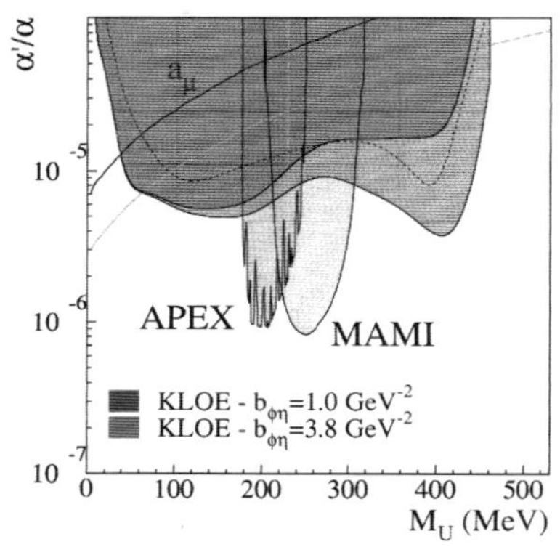

Figure 2. Exclusion plot at 90% CL for the parameter $\alpha'/\alpha = \epsilon^2$, compared with existing limits from the muon anomalous magnetic moment and from MAMI/A1 and APEX experiments. The gray line represents the expected values of the U boson parameters needed to explain the observed discrepancy between measured and calculated a_μ values. The dotted line is the previous KLOE result obtained with the $\eta \to \pi^+\pi^-\pi^0$ channel only.

4. U boson search in $e^+ + e^- \to \mu^+\mu^-\gamma$

The expected signal here is the appearance of a Breit-Wigner peak in the invariant mass distribution of the $\mu^+\mu^-$ pair. The analysis described in the following is based on the KLOE pion form factor measurement[25]. The data sample consists of $239.29\, pb^{-1}$ of KLOE data taken during the year 2002. The event selection requires:

- two charged tracks with $50° < \theta_\mu < 130°$;
- one photon within a cone of $\theta_\gamma < 15°$ $(\theta_\gamma > 165°)$ around the beamline.

The photon is not detected and its direction is reconstructed from event kinematics: $\vec{p}_\gamma \simeq \vec{p}_{miss} \equiv -\vec{p}_{\mu\mu} = -(\vec{p}_{\mu^+} + \vec{p}_{\mu^-})$. This selection greatly reduces the contamination from the resonant process $e^+e^- \to \Phi \to \pi^+\pi^-\pi^0$, where the π^0 mimics the missing momentum of the photon(s), and from the final state radiation processes: $e^+e^- \to \pi^+\pi^-\gamma_{FSR}$ and $e^+e^- \to \mu^+\mu^-\gamma_{FSR}$. The $M_{\rm trk}$ variable is computed from energy and momentum conservation, assuming the presence of an unobserved photon and in the assumption of equal mass charged particles.

Background contributions coming from $e^+e^- \to \pi^+\pi^-\gamma(\gamma)$, $e^+e^- \to \Phi \to \pi^+\pi^-\pi^0$ and $e^+e^- \to e^+e^-\gamma(\gamma)$ were separated applying kinematical cuts in the $M_{trk} - M^2_{\pi\pi}$ plane. A particle identification estimator (PID), based on a pseudo-likelihood function using the time-of-flight and calorimeter information, was used to suppress radiative Bhabha events. Finally pions and muons are separated by cuts on the trackmass variable M_{trk}:

- muons: $80 < M_{trk} < 115$ MeV,
- pions: $M_{trk} > 130$ MeV.

The contributions of the three background channels surviving the $\mu^+\mu^-\gamma$ selections are obtained by fitting in slices of $M^2_{\mu\mu}$ the M_{trk} distributions for data as a superposition of signal and background distributions. The M_{trk} distributions for $\mu^+\mu^-\gamma$, $\pi^+\pi^-\gamma$, $\pi^+\pi^-\pi^0$ were extracted by Monte Carlo calculation. A further separation between pions and muons was performed by a cut on the σ_{Mtrk} variable, which parametrizes the quality of the fit of the tracks. Once the Data/MC corrections have been applied, the $\mu^+\mu^-\gamma$ cross section was extracted by subtracting the residual background to the observed spectra and dividing it by efficiency and luminosity. The absolute cross section is in good agreement with the PHOKHARA prediction[26], as shown in Fig. 3.

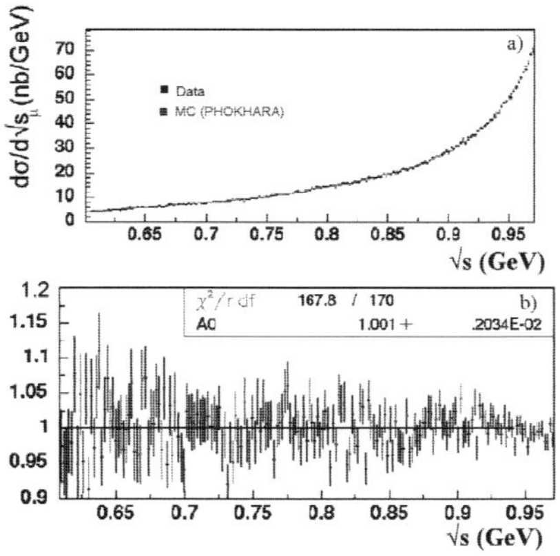

Figure 3. *a) Comparison of data (black points) and MC (blue points) of $\mu^+\mu^-\gamma$ absolute cross section b) ratio of data and MC PHOKHARA prediction*

No signal is observed (fig. 3 top). To extract the upper limit on ε^2 the CLS technique[24] was used, with the observed spectrum and the PHOKHARA spectrum as input and background to the procedure. A systematic error of $\sim 2\%$ on background was also applied. The 90% CL upper limits in terms of number of events were then extracted, in the energy range between 600 and 1000 MeV. They were later converted in terms of the mixing parameter ε^2 by using the analysis selection efficiency, the integrated luminosity, the $e^+e^- \to \mu^+\mu^-$ differential cross section and the predicted U boson production cross section. In Fig.4 the kinetic mixing parameter ε^2 in the 600-1000 MeV range and the other existing limits are presented. The blue area is the present measurement, with 90% CL exclusion limits between 2.6×10^{-6} and 3.5×10^{-7}, and with the clearly visible reduction of sensitivity in the ρ meson region. The red and dark green areas represent the Mami results[27], and the Apex measurements[28] respectively. The black line represents the ε^2 values consistent with a U boson contribution to the muon magnetic moment anomaly a_μ.

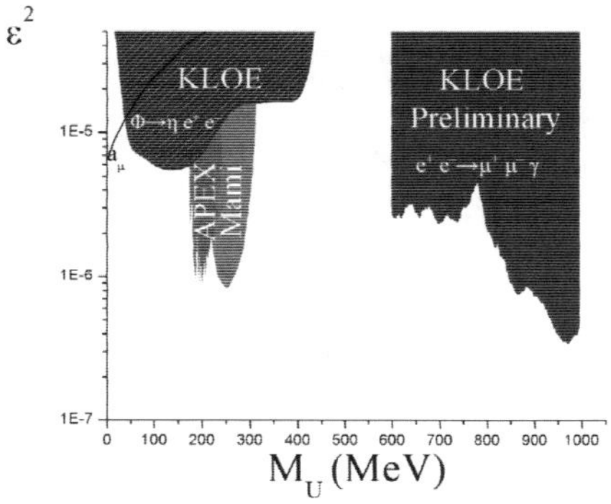

Figure 4. *Exclusion Plot on ε^2 compared with the existing limits in the $0 - 1000\,MeV$ range*

5. U boson search through the dark Higgsstrahlung process

The process $e^+e^- \to Uh'$ is one of the most interesting reactions to study at an e^+e^- collider because, differently from the other final states above, is suppressed by a single factor of ϵ. There are two very different scenarios depending on the masses of the dark photon and of the dark Higgs boson. For Higgs boson mass $M_{h'}$ larger than two dark photon masses M_U, the dark Higgs boson would decay dominantly and promptly to a U boson pair, thus giving rise to a six charged particle final state (this case was recently investigated by the BaBar experiment [29]); while Higgs bosons lighter than the dark photon would have, in most of the parameter space region, such a large lifetime to escape undetected, showing up as a missing energy signature. Here we study only the so called "invisible" dark Higgs scenario, thus confining the search to the case $M_{h'} < M_U$.

The lifetime of the dark Higgs boson depends on the kinetic mixing parameter ϵ, the boson masses $M_{h'}$ and M_U and the dark coupling constant α_D. For masses of the order of 100 MeV and $\alpha_D = \alpha$ the dark Higgs boson lifetime would be $\sim 5\mu s$ for $\epsilon \sim 10^{-3}$, corresponding, for KLOE energies, to a decay length of $\sim$ 100 m. The dark Higgs boson would be invisible up to $\epsilon \sim 10^{-2} \div 10^{-1}$, depending on the h' mass.

We limit our search to the decay of the U boson in a muon pair: our final state signature is then a couple of opposite charge muons plus missing energy. The measurement is thus performed in the range $2M_\mu < M_U < 1000$ MeV with the constraint $M_{h'} < M_U$.

The production cross section of the dark Higgsstrahlung process is proportional to the product $\alpha_D \times \epsilon^2$ and depends on the boson masses [16]. Values as high as hundreds of fb are reachable in the hypothesis $\epsilon = 10^{-3}$ and $\alpha_D = \alpha$.

The analysis of the process has been performed on a data sample of 1.65 fb^{-1} collected during the 2004-2005 KLOE data taking campaign at a center of mass energy of $\sim$ 1019 MeV, corresponding to the mass of the Φ meson (on peak sample), and on a data sample of 0.2 fb^{-1} at a center of mass energy of $\sim$ 1000 MeV (off peak sample), well below the Φ resonance.

The mass resolution was found to be between 0.5 and 2 MeV for M_U (invariant mass of the muon pair) and between 3 and 17 MeV for $M_{h'}$ (missing mass). The signature of the process would thus be the appearance of a sharp peak in the bidimensional distribution $M_{\mu\mu}$ - M_{miss}. Contrarily to most of the dominant QED background processes, the signal is predicted to show a large angle production in θ, with two dominant terms proportional to $sin\theta$ and $sin^3\theta$ [16].

As a first step of the analysis, a preselection was performed by requiring:

- events with only two opposite charge tracks, with a reconstructed vertex inside a 4×30 cm cylinder around the interaction point;
- each track must have an associated EMC cluster;
- the visible momentum direction has to be in the barrel: $|cos\theta| < 0.75$;
- the momenta of the two tracks must be individually below 460 MeV;
- the modulus of the missing momentum must exceed 40 MeV.

After this selections (mostly aimed at rejecting QED backgrounds), the hermeticity and tightness of the electromagnetic calorimeter was used as a veto to avoid the presence of photons in the event. It was required to have no unassociated energy deposition with E>15 MeV on EMC.

The event selection then proceeded by applying particle identification (PID) algorithms to the two charged tracks. These were almost entirely based on the excellent energy and time resolution of the EMC, through a set of neural networks, organised for different values of track momentum and polar angle. The PID performances, checked on selected data samples of $e^+e^- \to e^+e^-$, $e^+e^- \to \mu^+\mu^-$, $e^+e^- \to \pi^+\pi^-$ were found to be excellent: the fraction of $e^+e^- \to \mu^+\mu^-$ events, in which both tracks were required to be identified as muons, was measured to be 85%, while the fraction of residual $e^+e^- \to e^+e^-$ events was 10^{-4} and the fraction of doubly tagged $e^+e^- \to \pi^+\pi^-$ was $\sim$ 50% (muon and pion induced showers look very similar at KLOE energies).

After the missing energy and the PID selections, a huge background from $\Phi \rightarrow K^+K^-$, $K^\pm \rightarrow \mu^\pm\nu$ events survives in the on peak sample. This corresponds to the fraction of doubly early leptonically decaying charged kaons in the IP region. Cuts on the radial and z projections of the distance between the reconstructed vertex and the IP and on the χ^2 of the fit allowed to reduce by a factor $\sim$ 35 the $\Phi \rightarrow K^+K^-$, $K^\pm \rightarrow \mu^\pm\nu$ background.

Events surviving all the described selections were organized in bidimensional histograms with the muon pair mass $M_{\mu\mu}$ and the event missing mass M_{miss} on the two axes. The binning was chosen such as to keep most of the signal in one bin only. In $M_{\mu\mu}$ a 5 MeV bin width was enough over all the plane; while for M_{miss} a variable binning of 15, 30 and 50 MeV widths was chosen. The selection efficiency, estimated from Monte Carlo, was found to be between 15% and 25%, depending on the masses, with most frequent values ~20%. A total ±10% systematic uncertainty on the selection efficiency was conservatively estimated.

Results are shown in fig. 5 for the on peak and off peak samples. In the left plot of fig. 5 (on peak sample) several sources of backgrounds are easily distinguishable: $\Phi \rightarrow K^+K^-$, $K^\pm \rightarrow \mu^\pm\nu$ (triangular region at the left of the populated part of the distribution), $\Phi \rightarrow \pi^+\pi^-\pi^0$ (mostly horizontal band, corresponding to events in which both photons from π^0 decay are undetected), continuum backgrounds $e^+e^- \rightarrow \mu^+\mu^-$ and $e^+e^- \rightarrow \pi^+\pi^-$ (diagonal bands starting from the right-bottom part of the distribution), $e^+e^- \rightarrow e^+e^-\mu^+\mu^-$ and $e^+e^- \rightarrow e^+e^-\pi^+\pi^-$ (two photon events, top part of the distribution, for $M_{miss} > 350MeV$). In the distribution in the right plot of fig. 5 (off peak sample) all the backgrounds from the Φ decays are strongly suppressed.

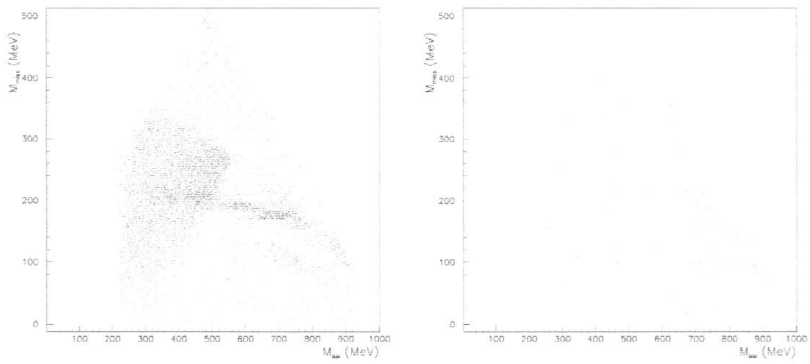

Figure 5. *Results for on peak sample (left plot, 1.65 fb^{-1} integrated luminosity) and off peak sample (right plot, 0.2 fb^{-1} integrated luminosity).*

The background was evaluated directly on the data, by counting populations in bins adjacent to the one in which the signal is looked for. No evidence of the dark Higgsstrahlung process was found (see fig. 5). Using uniform prior distributions, 90% confidence level Bayesian upper limits on the number of events were derived, separately for the on peak and off peak samples. These results were then converted in terms of the dark Higgstrahlung production cross section parameters $\alpha_D \times \epsilon^2$ by using the integrated luminosity information, the signal efficiency, the predicted dark Higgsstralung cross section and the branching fraction of the U boson decay into muon pairs as in reference [16]. The on peak and off peak limits were combined and projected in the $M_{\mu\mu}$ and M_{miss} directions (slightly smoothed, just to make them more readable). They are shown in fig. 6. Values as low as $10^{-9}\div10^{-8}$ in $\alpha_D\epsilon^2$ are excluded at 90% CL for a large range of the dark photon and dark Higgs masses.

These results are numerically comparable with those of Babar [29] and complement them as they refer to the same process in a different final state and in a different region of the phase space.

At KLOE2/DAΦNE2 the larger integrated luminosity and the presence of a high resolution Inner Tracker detector are expected to improve these results at least by a factor 2, thus allowing a study deep inside the $\epsilon \approx 10^{-4}$ parameter space region.

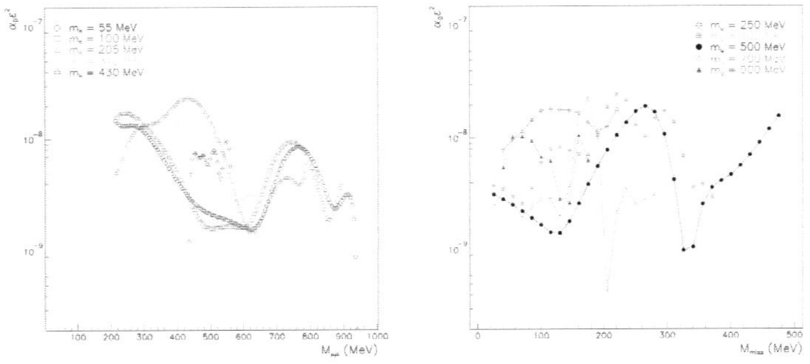

Figure 6. *Combined 90% CL upper limits in* $\alpha_D\epsilon^2$ *as a function of* $M_{\mu\mu}$ *for different values of* $M_{h'}$ *(left plot) and as a function of* M_{miss} *for different values of* M_U *(right plot).*

Conclusions

KLOE searched for a dark gauge U boson in three different processes and four different final states: a) $\Phi \to \eta U$, with $U \to e^+e^-$, $\eta \to \pi^+\pi^-\pi^0$ and $\eta \to \pi^0\pi^0\pi^0$; b) $e^+e^- \to U\gamma$ with $U \to \mu^+\mu^-$; c) $e^+e^- \to Uh'$ (dark Higgsstrahlung), $U \to \mu^+\mu^-$. We found no evidence and set upper limits on the mixing parameter ϵ^2 with the Standard Model, as a function of the U mass, in the range $10^{-5} \div 10^{-7}$, depending on the process. The upgraded DAΦNE2 luminosity and the insertion of new detectors in KLOE2 are expected to improve these limits by a factor 2÷4.

References

[1] P.Jean *et al*, Astron. Astrophys., **407** (2003), p. L55
[2] O. Adriani *et al.* Nature, **458** (2009), p. 607
[3] J. Chang *et al.* Nature, **456** (2008), p. 362
[4] A.A. Abdo *et al.* Phys. Rev. Lett., **102** (2009), p. 181101
[5] F. Aharonian *et al.* Phys. Rev. Lett., **101** (2008), p. 261104
[6] F. Aharonian *et al.* Astron. Astrophys., **508** (2009), p. 561
[7] R. Bernabei *et al.* Int. J. Mod. Phys. D, **13** (2004), p. 2127
[8] R. Bernabei *et al.* Eur. Phys. J. C, **56** (2008), p. 333
[9] C.E. Aalseth *et al.* Phys. Rev. Lett., **107** (2011), p. 141301
[10] M. Pospelov, A. Ritz, M.B. Voloshin Phys. Lett. B, **662** (2008), p. 53
[11] N. Arkani-Hamed, D.P. Finkbeiner, T.R. Slatyer, N. Weiner Phys. Rev. D, **79** (2009), p. 015014
[12] D.S.M. Alves, S.R. Behbahani, P. Schuster, J.G. Wacker Phys. Lett. B, **692** (2010), p. 323
[13] M. Pospelov, A. Ritz Phys. Lett. B, **671** (2009), p. 391
[14] N. Arkani-Hamed, N. Weiner JHEP, **0812** (2008), p. 104
[15] R. Essig, P. Schuster, N. Toro Phys. Rev. D, **80** (2009), p. 015003
[16] B. Batell, M. Pospelov, A. Ritz Phys. Rev. D, **79** (2009), p. 115008
[17] M. Reece, L.T. Wang JHEP, **0907** (2009), p. 051
[18] J.D. Bjorken, R. Essig, P. Schuster, N. Toro Phys. Rev. D, **80** (2009), p. 075018
[19] B. Batell, M. Pospelov, A. Ritz Phys. Rev. D, **80** (2009), p. 095024
[20] R. Essig, P. Schuster, N. Toro, B. Wojtsekhowski JHEP, **1102** (2011), p. 009
[21] M. Adinolfi *et al.* Nucl. Instr. Meth. A, **488** (2002), p. 51
[22] M. Adinolfi *et al.* Nucl. Instr. Meth. A, **482** (2002), p. 364
[23] L.G. Landsberg, Phys. Rep. **128** (1985), p. 301
[24] G. C. Feldman, R. D. Cousins, Phisical Rev. D **57**, 3873 (1998)
[25] D. Babusci *et al.*, arXiv:1212.4524 [KLOE/KLOE2 Collaboration]
[26] H. Czyż *et al.*, Eur. Phyis. J. C **39** 411 (2005)
[27] M. Merkel *et al.*, Phys. Rev. Lett. **106**, 251802 (2011)
[28] S. Abrahamyan *et al.*, Phys. Rev. Lett. **107**, 191804 (2011)
[29] J.P. Lees *et al.* (BaBar Collab.) Phys. Rev. Lett. **108** (2012) 211801

Journal of Physics: Conference Series **447** (2013) 012069 doi:10.1088/1742-6596/447/1/012069

Search for CPT and Lorentz invariance violation with neutral kaons at KLOE.

Antonio De santis on behalf of KLOE/KLOE-2 Collaborations[1]

Physics department of Rome University "Sapienza" & INFN Roma, P.le A. Moro, 2, I-00185, Roma (Italy).

E-mail: antonio.desantis@roma1.infn.it

Abstract. The neutral kaon system offers a unique possibility to perform fundamental tests of fundamental symmetry such CP and CPT invariance. In this contribution the KLOE search for CPT symmetry and Lorentz Invariance breaking in the Standard Model Extension framework is reported. Preliminary results are:

$\Delta a_0 = (-6.2 \pm 8.2_{stat} \pm 3.3_{sys}) \times 10^{-18}$ GeV^{-1}

$\Delta a_X = (3.3 \pm 1.6_{stat} \pm 1.5_{sys}) \times 10^{-18}$ GeV^{-1}

$\Delta a_Y = (-0.7 \pm 1.3_{stat} \pm 1.5_{sys}) \times 10^{-18}$ GeV^{-1}

$\Delta a_Z = (-0.7 \pm 1.0_{stat} \pm 0.3_{sys}) \times 10^{-18}$ GeV^{-1}

1. The KLOE experiment

The KLOE experiment operates at DAΦNE, the Frascati ϕ–factory. DAΦNE is an e^+e^- collider running at a center of mass energy of ~1020 MeV, the mass of the ϕ meson. Positron and electron beams collide at an angle of π-25 mrad, producing ϕ mesons with small momentum in the orbit plane ($p_x(\phi) \sim -15$ MeV).

The KLOE detector consists of a large cylindrical drift chamber (DC) surrounded by a lead-scintillating fiber electromagnetic calorimeter (EMC). A super-conducting coil around the EMC provides a 0.52 T axial field.

The DC[1] is 4 m in diameter and 3.3 m long and has 12,582 all-stereo tungsten sense wires. The chamber shell is made of carbon fiber-epoxy composite and the gas used is a 90% helium, 10% isobutane mixture. These features maximize transparency to photons and reduce $K_L \to K_S$ regeneration and multiple scattering. The position resolutions for single hits are $\sigma_{r,\phi} \sim 150$ μm and $\sigma_z \sim 2$ mm almost omogeneus in the active area. Those performances results in a decay

[1] D. Babusci, D. Badoni, I. Balwierz-Pytko, G. Bencivenni, C. Bini, C. Bloise, F. Bossi, P. Branchini, A. Budano, L. Caldeira Balkeståå, G. Capon, F. Ceradini, P. Ciambrone, F. Curciarello, E. Czerwiński, E. Danè, V. De Leo, E. De Lucia,G. De Robertis, A. De Santis, A. Di Domenico, C. Di Donato, R. Di Salvo, D. Domenici, O. Erriquez, G. Fanizzi, A. Fantini, G. Felici, S. Fiore, P. Franzini, P. Gauzzi, G. Giardina, S. Giovannella, F. Gonnella, E. Graziani, F. Happacher, L. Heijkenskjöld, B. Höistad, L. Iafolla, M. Jacewicz, T. Johansson, K. Kacprzak, A. Kupsc, J. Lee-Franzini, B. Leverington, F. Loddo, S. Loffredo, G. Mandaglio, M. Martemianov, M. Martini, M. Mascolo, R. Messi, S. Miscetti, G. Morello, D. Moricciani, P. Moskal, F. Nguyen, A. Passeri, V. Patera, I. Prado Longhi, A. Ranieri, C. F. Redmer, P. Santangelo, I. Sarra, M. Schioppa, B. Sciascia, M. Silarski, C. Taccini, L. Tortora, G. Venanzoni, W. Wiślicki, M. Wolke, J. Zdebik

vertex position resolution ranging between 3–6 mm in the instrumented decay volume. The momentum resolution is $\sigma(p_\perp)/p_\perp \sim 0.4\%$.

The calorimeter[2] is divided into a barrel and two end-caps, and covers 98% of the solid angle. The modules are read-out at both ends by photo-multipliers, both in amplitude and time for a total of 2440 cells per side arranged in five layers. Cells close in time and space are grouped into calorimeter clusters. The cluster energy E is the sum of the cell energies. The cluster time T and position $\vec{R}$ are energy-weighed averages. Energy and time resolutions are $\sigma_E/E = 5.7\%/\sqrt{E(\mathrm{GeV})}$ and $\sigma_t = 57\mathrm{ps}/\sqrt{E(\mathrm{GeV})} \oplus 100$ ps, respectively.

During KLOE data taking DAΦNE beam conditions and detectors calibrations are constantly monitored in order to guarantee the highest quality of the collected data. Presently KLOE has acquired 2.5 fb^{-1} of data and a new extensive campaign of data taking is starting under the project KLOE-2[3].

2. Neutral kaons at KLOE

At DAΦNE the ϕ-meson is produced almost at rest in the center of the KLOE sdetector and decays mostly in kaon pairs: 34% of decays are in neutral kaons. The initial state of the kaon pair is produced with quantum numbers $J^{PC} = 1^{--}$:

$$|i\rangle = \frac{|K_0\rangle|\overline{K}_0\rangle - |\overline{K}_0\rangle|K_0\rangle}{\sqrt{2}} = \mathcal{N}(|K_S(\vec{p})\rangle|K_L(-\vec{p})\rangle - |K_S(-\vec{p})\rangle|K_L(\vec{p})\rangle), \tag{1}$$

where $|K_S/K_L\rangle$ are the kaon mass eigenstate and $\mathcal{N}$ is a normalization factor. The last expression holds because of the Bose-Einstein principle and is independent from the ϕ-meson decay mechanism.

The neutral kaon mass eigenstate could be expressed as:

$$\begin{cases} |K_S\rangle = (1+\epsilon_S)|K_0\rangle + (1-\epsilon_S)|\overline{K}_0\rangle \\ |K_L\rangle = (1+\epsilon_L)|K_0\rangle - (1-\epsilon_L)|\overline{K}_0\rangle \end{cases} \tag{2}$$

where $\epsilon_{S/L} = \epsilon_K \pm \delta_K$. The parameter ϵ_K represents CP violation in the neutral kaon system (both direct and indirect) while δ_K stands for possible CPT violation. Assuming no CPT violation $\epsilon_S = \epsilon_L$.

Equation 1 implies that the detection of one of the two kaons (on one side) reveals the presence of the other kaon with known momentum and direction on the opposite side. In the KLOE data reconstruction this feature, the tagging, is used to select pure K_S or K_L beams identifying K_L interactions in the EMC or $K_S \to \pi^+\pi^-$ decays respectively.

A different approach can be exploited at KLOE since the two kaons are produced in an antisymmetric correlated state. Labeling f_1 and f_2 the decay channels for the two kaons, the time evolution of the initial state decaying into $|f_1, f_2\rangle$ final state is expressed as a function of the difference of proper decay times ($\Delta\tau = \tau_2 - \tau_1$) as:

$$I_{f_1 f_2}(\Delta\tau) \propto e^{-\Gamma|\Delta\tau|}\Big[|\eta_{f_1}|^2 e^{\frac{\Delta\Gamma}{2}\Delta\tau} + |\eta_{f_2}|^2 e^{\frac{\Delta\Gamma}{2}\Delta\tau} - 2\Re e\Big(\eta_{f_1}\eta^*_{f_2} e^{-i\Delta m \Delta\tau}\Big)\Big] \tag{3}$$

where $\eta_{f_j} = \langle f_j|K_L\rangle/\langle f_j|K_S\rangle$, $\Gamma = \Gamma_S + \Gamma_L$ and $\Delta\Gamma = \Gamma_S - \Gamma_L$.

Equation 3 shows a time interference term characteristic of the type of correlation firstly pointed out by Einstein, Podolsky and Rosen [4].

If the final states considered are the same for the two kaons ($f_1 = f_2 = f$) it becomes impossible to distinguish between them (e.g. decay times are not ordered). In the standard Quantum Mechanics description of the time evolution of the system, a fully destructive interference is expected for equal decay times ($|\Delta\tau| = 0$). In this case the resulting ratio of amplitudes (η_j) becomes:

$$\eta_j = \eta_{\pi^+\pi^-} = \frac{\langle K_L|\pi^+\pi^-\rangle}{\langle K_S|\pi^+\pi^-\rangle} = \varepsilon_K - \delta_K \tag{4}$$

for both kaons decay and the only possibility to distinguish between them is to label the final state according to some fixed direction in the space (*e.g.* $\hat{z}$ direction): $|f_1\rangle \equiv |\pi^+\pi^-(\vec{P}_K^1)\rangle$. This choice is well suited to observe violation of the fundamental symmetry CPT in the framework of the Standard Model Extension (SME) developed by A. Kostelecky [5].

According to SME, CPT violating parameter δ_K has to be expressed as:

$$\delta_K \approx i \sin\phi_{SW} e^{i\phi_{SW}} \gamma_K (\Delta a_0 - \vec{\beta}_K \cdot \Delta\vec{a})/\Delta m, \tag{5}$$

where γ_K and β_K are the usual Lorentz factors, ϕ_{SW} is the superweak phase and Δa_μ are the SME parameters for the kaon system.

Equation 5 shows that δ_K is modulated by the kaon momentum modulus (γ_K and $|\vec{\beta}_K|$) and by its spatial direction ($\vec{\beta}_K$). At KLOE kaons are produced back-to-back in the ϕ-meson decay and therefore they will evolve with two different δ_K values. The effect produced by δ_K parameter different from zero will be observed in the distribution of equation 3.

The simple expression of equation 5 holds in a special frame, the sidereal reference frame, with axes fixed with respect to the fixed stars. Taking into account the Earth motion with respect to the fixed stars and the laboratory frame, the CPT violation parameter becomes:

$$\begin{aligned}\delta_K(\vec{p}_K, T_{sid}) = {} & \frac{i\sin\phi_{SW} e^{i\phi_{SW}}}{\Delta m}\gamma_K \Big[\Delta a_0 + \beta_K \Delta a_Z(\cos\vartheta\cos\chi - \sin\vartheta\cos\varphi\sin\chi) \\ & -\beta_K \Delta a_X \sin\vartheta\sin\varphi\sin\omega_E T_{sid} \\ & +\beta_K \Delta a_X (\cos\vartheta\sin\chi + \sin\vartheta\cos\varphi\cos\chi)\cos\omega_E T_{sid} \\ & +\beta_K \Delta a_Y (\cos\vartheta\sin\chi + \sin\vartheta\cos\varphi\cos\chi)\sin\omega_E T_{sid} \\ & +\beta_K \Delta a_Y \sin\vartheta\sin\varphi\cos\omega_E T_{sid}\Big]\end{aligned} \tag{6}$$

where ω_E is the angular velocity of the Earth, T_{sid} is the sidereal time, χ is the 3D angle between the $\hat{z}_{LAB}$ axis of the laboratory frame and the Earth rotation axis, and ϑ and φ are the polar and azimuthal angle in the laboratory, respectively.

The equation 6 shows that the sidereal time dependence of δ_K could be observed only associated with Δa_X and Δa_Y parameters different from zero. Δa_0parameter, being coupled only with the γ_K, will be the most difficult to observe[2].

The introduction of equation 6 in the time evolution of the kaon system implies the angular dependence of equation 3:

$$\begin{aligned}I(\Delta t, T_{sid}, \vartheta, \varphi) \propto {} & e^{-\Gamma|\Delta\tau|}\Big[|\eta_{K_1}(T_{sid}, \vartheta, \varphi; \Delta a_\mu))|^2 e^{\frac{\Delta\Gamma}{2}\Delta\tau} + \\ & |\eta_{K_2}(T_{sid}, \vartheta, \varphi; \Delta a_\mu)|^2 e^{\frac{\Delta\Gamma}{2}\Delta\tau} - \\ & 2\Re e\Big(\eta_{K_1}(T_{sid}, \vartheta, \varphi; \Delta a_\mu)\eta_{K_2}(T_{sid}, \vartheta, \varphi; \Delta a_\mu)e^{-i\Delta m\Delta\tau}\Big)\Big]\end{aligned} \tag{7}$$

Since at KLOE the ϕ-meson is not produced exactly at rest the kaons are not perfectly back-to-back in the lab frame. This has two implications: the Lorentz factors in equation 6 (γ_K and β_K) have their own angular dependence and the two kaons momenta are not equal in modulus.

[2] The excursion of γ_K at KLOE is of the order of 2-3%.

To study all the components of Δa_μ we analyzed the KLOE data-set as a function of sidereal time and kaon angle. Ordering kaon according to the momentum component along z transfers the polar angle dependence into $\Delta\tau$ sign. To enhance possible effects the azimuthal angle dependence will be considered by splitting the sample according to kaon direction with respect to ϕ-meson momentum (e.g. $P_x > 0$ or $P_x < 0$). The observable will be defined as:

$$S(\Delta\tau_i, \Delta T_{sidj}, \Delta\Omega_h) = \int_{\Delta\tau_i} d\Delta\tau \int_{\Delta T_{sidj}} dT_{sid} \int_{\Delta\Omega_h} d\Omega_{K_1} \rho(\vartheta, \varphi, T_{sid})\, I(\Delta\tau, T_{sid}, \vartheta, \varphi) \qquad (8)$$

where $\rho(\vartheta, \varphi, T_{sid})$ is the angular and sidereal density distribution of the produced kaon in the positive direction of the z axis. This density has a $\sin^3\vartheta d\vartheta$ for the polar angle dependence and an unespected dependence as a function of T_{sid}. This behavior is due to inhomogeneity of data taking during data taking period. This part of the density function has been taken directly from data in a independent sample. The integration ranges are defined as:

$\Delta\tau_i$ Proper decay time difference. Should be in the range $0.5 - 1.5\,\tau_S$;

ΔT_{sidj} Sidereal time interval. The default number of divisions of a sidereal day will be four, 8 sidereal hour each;

$\Delta\Omega_h$ Angular interval. The polar angle range will be $\vartheta \in [0, \pi/2[$ and the azimuthal range will be $\varphi \in [-\pi/2; \pi/2[$ or $\varphi \in [\pi/2; 3\pi/2[$, sel. "A" and sel. "B", respectively.

The number of different $\Delta\tau$ distributions considered in the present analysis will be: 4 sidereal time bins $\times$ 2 angular bins. The kaon regeneration on the beam pipe walls is taken into account in the integration of equation 8 as a multiplicative correction of $I(\Delta\tau, T_{sid}, \vartheta, \varphi)$.

3. Data analysis

In this analysis the entire KLOE data-set acquired during 2004-05 has been used. The total integrated luminosity is about 1.7 fb^{-1}. The Monte-Carlo (MC) samples used are two having equivalent integrated luminosity of 3.4 fb^{-1} and 17 fb^{-1}, respectively. The former, containing all the ϕ-meson decay channel, has been used for analysis optimization, while the latter, containing only signal events has been used for efficiency and resolution determination.

The data reduction starts with the topological identification of the candidate signal events: two vertices with two tracks each. For each vertex the same list of selection criteria is applied:

- $|M_{trk} - M_K| < 5$ MeV
  where M_{trk} is the invariant mass of the kaon reconstructed from the tracks assuming charged pion mass hypothesis;
- $\sqrt{E^2_{miss} + |\vec{P}_{miss}|^2} < 10$ MeV
  where $\vec{P}_{miss} = \vec{P}_K(tag) - \vec{P}_K(trk)$ with $\vec{P}_K(tag) = \vec{P}_\phi - \vec{P}_K(opposite)$ the kaon tagged momentum from the opposite decay vertex information and $\vec{P}_K(trk) = \vec{p}_{\pi^+} + \vec{p}_{\pi^-}$ the reconstructed kaon momentum from the tracks, $E_{miss} = E_K(tag) - E_K(trk)$ with $E_K(tag) = E_\phi - E_K(opposite)$ the kaon tagged energy from the opposite decay vertex informations and $E_K(trk) = \sqrt{|\vec{p}_{\pi^+}|^2 + m^2_\pi} + \sqrt{|\vec{p}_{\pi^-}|^2 + m^2_\pi}$ is the energy of the kaon reconstructed from tracks;
- $-50\ \text{MeV}^2 < M^2_{miss} < 10\ \text{MeV}^2$
  where $M^2_{miss} = E^2_{miss} - |\vec{P}_{miss}|^2$
- $|P^*_K(trk) - P^*_0(kinematics)| < 10$ MeV
  where $P^*_K(trk)$ is the momentum of the kaon, as derived from tracks, in the ϕ-meson reference frame, while $P^*_0(kinematics) = \sqrt{\frac{s^2}{4} - m^2_{K_0}}$ represents the expected value from two body kinematics in the ϕ-meson decay.

The background contamination after those selections is 1.5%, as estimated from MC. The background events are mainly due to kaon regeneration. This background is irreducible because is made of regenerated kaons moving in the forward direction (in the limit of the angular resolution).

A global fit is performed in order to improve the resolution on the decay time difference. In this global fit the free parameters are the ϕ-meson decay point coordinates ($\vec{V}_\phi$) and the decay length for the two kaons ($\lambda_{1,2}$). A global Likelyhood function is built in order to take into account all the information for each event. The outputs of the fit are then used to compute the proper decay times and χ^2 and pulls of the fit are used to reject badly reconstructed events.

The $\Delta\tau$ resolution is strongly correlated, as expected, with the opening angle of the pion tracks ($\vartheta_\pm$), and deteriorates at large values of $\vartheta_\pm$. A cut to eliminate large opening angles values has been applied and events with:$\cos(\vartheta_\pm) < -0.975$ have been rejected. The result is shown in figure 1. Since the two kaons are ordered according to the longitudinal momentum (z component) it is possible to mis-order them with respect to the true ordering. This effect has been suppressed by rejecting events with the "first" kaon having: $P_z(K_1) < 2\,\mathrm{MeV}$.

The shape of the efficiency as a function of true value of $\Delta\tau$ is shown in figure 1. The mean value is 25% for $\Delta\tau$ ranging in $[-12; 12]\tau_S$ with a consistent drop for $\Delta\tau \sim 0$. This is due to two concurrent effects: increasing extrapolation length for tracks coming from the interaction point and the wrong association of tracks with the kaon decay vertex [3] (*"vertex interference"*).

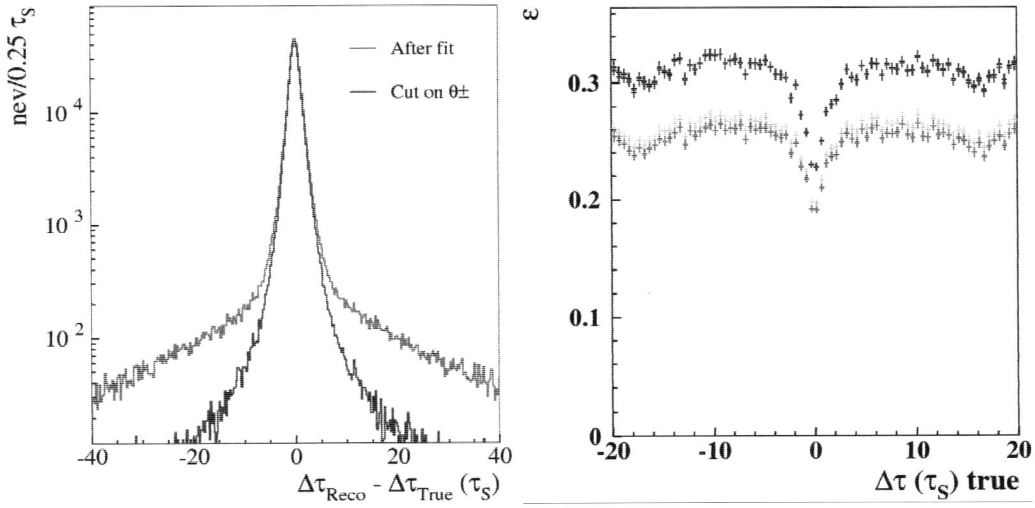

Figure 1: Left: resolution on $\Delta\tau$. The effect of the cut on opening angle between tracks is shown. Right: MC efficiency as a function of true $\Delta\tau$. Colors represents the different analysis stage: blue for global fit, green for opening angle cuts and red for the final efficiency.

The total selection efficiency can be parametrized as follow:

$$\varepsilon_{tot} = \varepsilon_{trig}\varepsilon_{reco}\varepsilon_{cuts}$$

where the first two terms take into account trigger and standard reconstruction procedure effects while the last is the effect of the selection criteria. In this analysis the efficiencies are derived

[3] When the two vertex are so close in space ($\Delta\tau = 1\tau_S \Rightarrow \Delta x_V \sim 6mm$) it is possible to wrongly associate tracks to vertex by exchanging two of them.

from MC. For trigger and reconstruction efficiencies corrections obtained from data, using an independent control sample of $K_L \to \pi\mu\nu$, are applied. The control sample will be selected so to have the best purity and to cover the trakcs momentum distribution of the signal.

The selection of the control sample requires the presence of a $K_S \to \pi^+\pi^-$ tagging decay and a semileptonic decay on the opposite side. After the topological request (two tracks connected to a vertex), the selection of the semileptonic decay is mainly based on the following variable:

$$m_l^2(\pm) = E_l^2 - p_\pm^2 = (E_K^{tag} - E_\pi(\mp) - |\vec{p}_{miss}|)^2 - p_\pm^2 \tag{9}$$

representing the reconstructed mass of the lepton of a given charge in the hypothesis that the other charged track is a pion. In figure 2 the distribution of the squared charged lepton mass is shown. The structures corresponding to K$_e$3, K$_\mu$3 and $\pi^+\pi^-$ decay are clearly visible. The control sample is required to fulfill:

$$m_l^2(+) - m_l^2(-) \geq 15000\,\text{MeV}^2 m_l^2(+) - m_l^2(-) \leq 30000\,\text{MeV}^2$$

After this simple selection the purity of K$_\mu$3 sample is at the level of 93%. To further improve the purity the following cuts are applied:

- $\sqrt{E_{miss}^2 + P_{miss}^2} > 10\,\text{MeV}$ to remove the residual $K_L \to \pi^+\pi^-$events ensuring the statistical independence between signal and control sample;
- $|p_+^*| + |p_-^*| < 400\,\text{MeV}$;
- $\cos(\vartheta_\pm) < -0.6$ to ensure the same angular range for the opening angle between tracks. This last cut is applied also on the tagging side.

The final purity of the control sample is 95% and the remaining background is fully dominated by K$_e$3 decay.

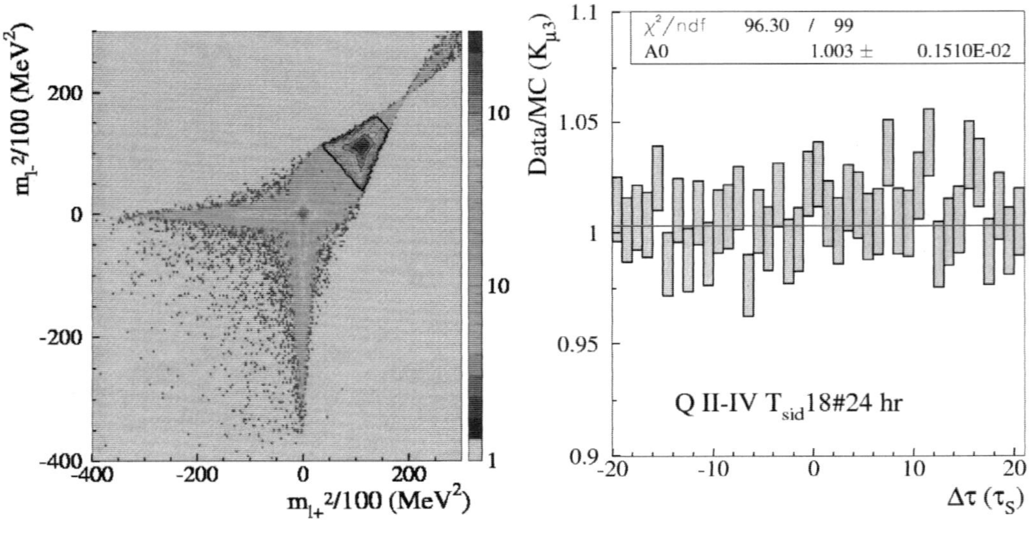

Figure 2: Left: squared charged lepton mass, the highlighted region represent the cuts applied. Right: Efficiency correction as a function of $\Delta\tau$ in the sidereal time interval 18-24 hours. Plot refers to the angular selection: sel. "B". Boxes represents statistical error. The fit with a constant function shows that on average the global correction is very small (0.2-0.5%).

The efficiency correction has been evaluated as the ratio between data and MC distribution of $\Delta\tau$, as shown in figure 2.

Journal of Physics: Conference Series **447** (2013) 012069 doi:10.1088/1742-6596/447/1/012069

4. Fit results

Starting from the theoretical expression of equation 8 the fitting function ($\tilde{S}(\Delta\tau_i, \Delta T_{sidj}, \Delta\Omega_h)$) has to include efficiency and resolution effects:

$$\tilde{S}(\Delta\tau_i, \Delta T_{sidj}, \Delta\Omega_h) = \rho_{\mathrm{K}_{\mu3}}(\Delta\tau_i) \sum_k \varepsilon_k^{MC} P_{ik}^{MC} S(\Delta\tau_k^{MC}, \Delta T_{sidj}, \Delta\Omega_h) \quad (10)$$

where $\rho_{\mathrm{K}_{\mu3}}(\Delta\tau_i)$ is the efficiency correction from the semileptonic control sample, ε_k^{MC} is the total efficiency corresponding to the MC bin ($\Delta\tau_k^{MC}$) and P_{ik}^{MC} are the smearing matrix elements connecting the true value $\Delta\tau_k$ with the observed one $\Delta\tau_i$.

The results obtained from the fit are show in figure 3 and are listed in table 1. The $\Delta\tau$ range is: $\Delta\tau \in [-12 : 12]\tau_S$. This choice limits the effects from the regeneration on the spherical beam pipe. The number of sidereal time bins used is four and the angular selection are two: $\cos\vartheta_{K_1}\cos\varphi_{K_1} > 0$ (sel. "A") and $\cos\vartheta_{K_1}\cos\varphi_{K_1} < 0$ (sel. "B"). All the distributions are fitted at the same time (184 bins).

The systematic uncertainty sources taken into account are results variations as a function of: $\Delta\tau$ fit range, $\Delta\tau$ bin width, regeneration on beam pipe, subtraction of four pion direct production.

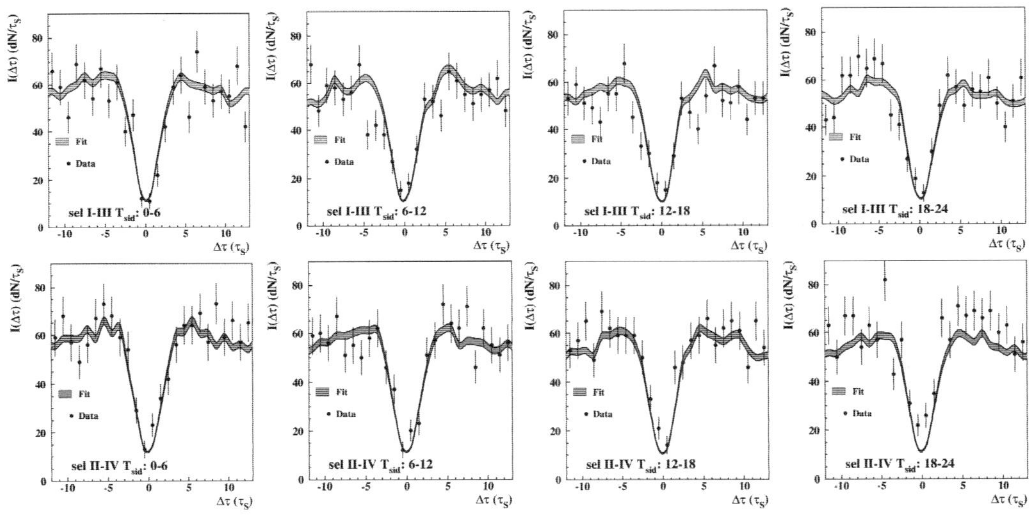

Figure 3: Fit results: upper(lower) plots are for the angular selection "A"("B") . Black points are for data while colored bands are the fit output. The errors on date are purely statistical, while the width of the fit result band represent the contribution to the uncertainty coming from MC statistics and efficiency correction.

5. Conclusions and future plans

The continuation of the KLOE physics program with KLOE-2 [3] at an improved DAΦNE machine is starting with a new beam interaction scheme [6] and with the inclusion of two pairs of electron-positron taggers for the study of the gamma-gamma physics: Low Energy Tagger [7] inside KLOE apparatus and High Energy Tagger [8] along the beam lines outside the KLOE detector. Several other upgrades for the detector are going to be installed in the next months:

- a pair of crystal calorimeters (CCALT [9]) near the interaction region to improve the angular acceptance for low-θ particles;

Journal of Physics: Conference Series **447** (2013) 012069 doi:10.1088/1742-6596/447/1/012069

Table 1: Fit results. Systematical uncertainties are still preliminary. The statistical errors are in the expected range. The fit $\chi^2/NDoF$ is 216/184 corresponding to a p-value of 5%.

Δa_0	$(-6.2 \pm 8.2_{stat} \pm 3.3_{sys}) \times 10^{-18}$ GeV^{-1}
Δa_X	$(\ \ 3.3 \pm 1.6_{stat} \pm 1.5_{sys}) \times 10^{-18}$ GeV^{-1}
Δa_Y	$(-0.7 \pm 1.3_{stat} \pm 1.5_{sys}) \times 10^{-18}$ GeV^{-1}
Δa_Z	$(-0.7 \pm 1.0_{stat} \pm 0.3_{sys}) \times 10^{-18}$ GeV^{-1}

- a pair of tile calorimeters (QCALT [10]) covering the quadrupoles along the beam pipe made of tungsten foil and singly read-out scintillator tiles to improve the angular coverage for particles coming from the active volume of the DC;
- a small and light tracker (IT [11]) made of four planes of cylindrical GEM to improve the resolution of the vertex reconstruction around the interaction point and to increase the low-θ acceptance for charged particles.

The results presented in table 1 are expected to be improved with the KLOE-2 data taking campaign. The sensitivity of CPT and Lorentz invariance tests will improve either by the increased statistics either by the new interaction region (IR) with IT. The IT will improve the resolution on the vertex position and the acceptance for low-θ tracks, the IR will imply a larger beam crossing angle (from 25 to 60 mrad) that will enhance the effect of asymmetry between the two kaons. The expected sensitivity should increase up to 10^{-19} GeV for all the Δa_μ parameters.

References

[1] Adinolfi M, Ambrosino F, Andryakov A, Antonelli A, Antonelli M *et al.* 2002 *Nucl.Instrum.Meth.* **A488** 51–73

[2] Adinolfi M, Ambrosino F, Antonelli A, Antonelli M, Anulli F *et al.* 2002 *Nucl.Instrum.Meth.* **A494** 326–331

[3] Amelino-Camelia G, Archilli F, Babusci D, Badoni D, Bencivenni G *et al.* 2010 *Eur.Phys.J.* **C68** 619–681 (*Preprint* 1003.3868)

[4] Einstein A, Podolsky B and Rosen N 1935 *Phys.Rev.* **47** 777–780

[5] Kostelecky V A 2001 *Phys.Rev.* **D64** 076001 (*Preprint* hep-ph/0104120)

[6] Milardi C *et al.* (DAFNE Collaboration) 2010 (*Preprint* 1006.1487)

[7] Babusci D, Bini C, Ciambrone P, Corradi G, De Santis A *et al.* 2010 *Nucl.Instrum.Meth.* **A617** 81–84 (*Preprint* 0906.0875)

[8] Archilli F, Babusci D, Badoni D, Beretta M, Gonnella F *et al.* 2010 *Nucl.Instrum.Meth.* **A617** 266–268

[9] Happacher F, Martini M, Miscetti S and Sarra I 2009 *Nucl.Phys.Proc.Suppl.* **197** 215–218 (*Preprint* 0906.1157)

[10] Cordelli M, Corradi G, Happacher F, Martini M, Miscetti S *et al.* 2010 *Nucl.Instrum.Meth.* **A617** 105–106 (*Preprint* 0906.1133)

[11] Balla A, Bencivenni G, Cerioni S, Ciambrone P, De Lucia E *et al.* 2011 *Nucl.Instrum.Meth.* **A628** 194–198 (*Preprint* 1003.3770)

Journal of Physics: Conference Series **447** (2013) 012070 doi:10.1088/1742-6596/447/1/012070

Testing the Pauli Exclusion Principle for Electrons

J. Marton[1], S. Bartalucci[2], S. Bertolucci[4], C. Berucci[1,2], M. Bragadireanu[2,3], M. Cargnelli[1], C. Curceanu (Petrascu)[2,3], S. Di Matteo[5], J.-P. Egger[6], C. Guaraldo[2], M. Iliescu[2], T. Ishiwatari[1], M. Laubenstein[7], E. Milotti[8], D. Pietreanu[2], K. Piscicchia[2,9], T. Ponta[2,3], A. Romero Vidal[2], A. Scordo[2], D.L. Sirghi[2,3], F. Sirghi[2,3], L. Sperandio[2], O. Vazquez Doce[2], E. Widmann[1] and J. Zmeskal[1]

[1] The Stefan Meyer Institute for Subatomic Physics, Boltzmanngasse 3, A-1090 Vienna, Austria
[2] INFN, Laboratori Nazionali di Frascati, CP 13, Via E. Fermi 40, I-00044, Frascati (Roma), Italy
[3] "Horia Hulubei" National Institute of Physics and Nuclear Engineering, Str. Atomistilor no. 407, P.O. Box MG-6, Bucharest - Magurele, Romania
[4] CERN, CH-1211, Geneva 23, Switzerland
[5] Institut de Physique UMR CNRS-UR1 6251, Université de Rennes1, F-35042 Rennes, France
[6] Institut de Physique, Université de Neuchâtel, 1 rue A.-L. Breguet, CH-2000 Neuchâtel, Switzerland
[7] Laboratori Nazionali del Gran Sasso, S.S. 17/bis, I-67010 Assergi (AQ), Italy
[8] Dipartimento di Fisica, Università di Trieste and INFN– Sezione di Trieste, Via Valerio, 2, I-34127 Trieste, Italy
[9] CENTRO FERMI, Compendio del Viminale, Piazza del Viminale 1, I-00184 Roma, Italy

E-mail: `johann.marton@oeaw.ac.at`

Abstract. One of the fundamental rules of nature and a pillar in the foundation of quantum theory and thus of modern physics is represented by the Pauli Exclusion Principle. We know that this principle is extremely well fulfilled due to many observations. Numerous experiments were performed to search for tiny violation of this rule in various systems. The experiment VIP at the Gran Sasso underground laboratory is searching for possible small violations of the Pauli Exclusion Principle for electrons leading to forbidden X-ray transitions in copper atoms. VIP is aiming at a test of the Pauli Exclusion Principle for electrons with high accuracy, down to the level of 10^{-29} - 10^{-30}, thus improving the previous limit by 3-4 orders of magnitude. The experimental method, results obtained so far and new developments within VIP2 (follow-up experiment at Gran Sasso, in preparation) to further increase the precision by 2 orders of magnitude will be presented.

1. Introduction

The Pauli Exclusion Principle (PEP) uncovered by Wolfgang Pauli in 1925 [1] is one of the corner stones of quantum physics and thus it is at the basis of the foundation of modern physics. It is connected with spin statistics dividing the world in fermions and bosons. Therefore, PEP is one of the most important rules in physics. Based on it is our understanding of nature and the consequences for the world of elementary particles up to compact objects (e.g. neutron stars) in the universe - but it is lacking a simple explanation as already stated by Pauli himself.

Wolfgang Pauli showed later in 1940 the interconnection of the principle with the spin-statistics [2]. We know that the Pauli Principle is extremely well fulfilled but the the limit of validity - if any - is still an open question. Even a tiny violation of PEP would point to new physics which could show up at very high energies (at the Planck scale). In the last decades several experimental tests of PEP validity for different systems have been performed [3, 4, 5, 6, 7, 8]. A method to experimentally test the PEP was developed by Ramberg-Snow [9]. PEP is tested for *fresh* electrons, i.e. elementary particles having no interaction with the studied system thus circumventing the Messiah-Greenberg super-selection rule[10]. These *fresh* electrons are provided by a strong electric current which is flowing through a solid metal conductor. Pauli-forbidden transitions in this metal (e.g. K transitions from the 2p state to the 1s state already filled with 2 electrons) exhibit an significant energy shift in the transition energy which is resolvable by x-ray spectroscopy. A search for these x-ray transition events can be performed with high sensitivity but requires substantial background discrimination. It has to be noted that the *normal* K transitions are present due to background radiation.
In our previous VIP experiment at the underground laboratory LNGS (Laboratorio Nazionali di Gran Sasso) we used an improved Ramberg-Snow experimental setup exploiting charge coupled devices as x-ray detectors surrounding a solid cylindrical copper target. In this experiment we could deduce an upper limit for the Pauli exclusion principle violation in the order of 10^{-29} [11]. A strongly improved VIP2 experiment will be performed in the underground laboratory LNGS in Gran Sasso taking again advantage of the excellent shielding against cosmic rays. A strongly improved compact setup with passive and active shielding will be used. Silicon drift detectors (SDDs) will serve as x-ray detectors providing a timing signal used in anti-coincidence with scintillators (i.e. veto counters) to suppress actively background events.
With the VIP2 experiment we want to improve the limit for PEP violation for electrons by 2 orders of magnitude reaching the range of 10^{-31}.

2. Experimental Method

The basic principle of the VIP experiment follows the method suggested by Greenberg and Mohapatra [12] and is based on introducing *fresh* electrons into a copper strip. By applying an electrical current on the strip and searching for the Pauli-forbidden radiative transitions to the ground state that is already filled by two electrons. We look for 2p-1s transitions in copper which are easily detectable with x-ray detectors. The energy of these non-Paulian transitions is shifted from the normal transition energy by about 300 eV (7.729 keV instead of 8.040 keV) due to the additional screening effect given by the two electrons in the 1s level [13]. The measurements without the current in the copper strip provide the X-ray background, where no PEP violating transitions are expected. Those with the current provide the forbidden X-ray transitions when the *fresh* electrons could lead to Pauli-forbidden transitions. Comparing the two energy spectra, with and without the currents, the limit of the probability of the violation of PEP is extracted.

2.1. VIP experiment

The VIP experiment was installed in the underground laboratory of LNGS. This experiment used a copper cylinder as target and charge coupled devices (CCDs) as x-ray detectors. The array of CCD detectors were used for studies of kaonic atoms [14, 15]. Details of the VIP experiment can be found in ref.[16, 17]. The measurement by the VIP experiment improved very significantly the limit previously obtained by Ramberg and Snow [9], thanks to the following features:(a) use of CCD detectors instead of gaseous ones, having much better energy resolution (4-5 times better) and higher stability; (b) experimental setup located in the clean, low-background, environment of the underground LNGS Laboratory; and (c) collection of much higher statistics (longer DAQ periods, thanks to the stability of CCDs). We made full use of these features to obtain an improvement of several orders of magnitude on the previous limit by Ramberg and Snow (see

Table 1. Limits of the Pauli violation probability for electrons from different experiments.

Experiment	Target	Upper limit of $\beta^2/2$	ref.
Ramberg-Snow	Copper	$1.7\text{x}10^{-26}$	[9]
S.R. Elliott et al.	Lead	$1.5\text{x}10^{-27}$	[18]
VIP(2006)	Copper	$4.5\text{x}10^{-28}$	[17]
VIP(2012)	Copper	$4.7\text{x}10^{-29}$	[11, 19]
VIP2(goal)	Copper	10^{-31}	

tab.1).

2.2. VIP2 - a new high sensitivity experiment
In order to achieve the signal/background increase which will allow a gain of two orders of magnitude for the probability of PEP violation for electrons, we are planning to build a new target, a new cryogenic system, use new detectors with timing capability and active veto system. As x-ray detectors we will use SDDs which were employed in der SIDDHARTA experiment on kaonic atoms at the DAFNE electron-positron collider of Laboratori Nazionali di Frascati. SDDs have an even better energy resolution than CCDs but additionally provide timing capability [20]which allow to use anti-coincidence operation with scintillators and therefore active shielding. The VIP2 system will provide

- signal increase with a more compact system with higher acceptance and higher current flow in the new copper strip target
- background reduction by decreasing the x-ray detector surface, more compact shielding (active veto system and passive), nitrogen filled box for radon radiation reduction

In the following table the numerical values for the improvements in VIP2 are given which will lead to an expected overall improvement of a factor higher than ~120.

Figures 1 and 2 show the main elements for the proposed setup in the VIP2 experiment. The copper strip target is 30 mm long, 10 mm wide and is about 40μm thick, and is installed in the center of the setup. The copper strip is cooled at ~90K by the use of an external cryogenic system using liquid argon as the cooling medium. The current connection lines made of copper wires with a cross-section area of 1.5 cm^2, allow a current flow of (at least) 100 A. The current lines exhibit a temperature gradient from inside the vacuum chamber to the outside connectors of about 180 K.

Monte Carlo simulations were performed to study the effect of the active shielding in various configurations of the setup and models of the background radiation. The background profiles measured at LNGS [22] were used as input parameters in the simulations.

As the veto counters (see fig.3), 2 pieces of 10 cm thick plastic scintillators are found to be optimal. We also simulated a setup with BGO inorganic crystals which yields even better background suppression, but due to the high costs of this solution, we decided to propose plastic scintillators, e.g. Bicron 412. To summarize we are confident that the goal of VIP2 to reach an upper limit for possible PEP violation in the order of 10^{-31} can be reached with the improved setup (see fig.4).

Journal of Physics: Conference Series **447** (2013) 012070 doi:10.1088/1742-6596/447/1/012070

Table 2. List of numerical values of the changes in VIP2 in comparison to the VIP features (given in brackets)

Changes in VIP2	value VIP2 (VIP)	expected gain
acceptance	12%	12
increase current	100A (50A)	2
reduced length	3 cm (8.8 cm)	1/3
total linear factor		8
energy resolution	170 eV (340 eV)	4
reduced active area	6 cm^2 (114 cm^2)	20
better shielding and veto		5-10
higher SDD efficiency		1/2
background reduction		200-400
overall improvement		> 120

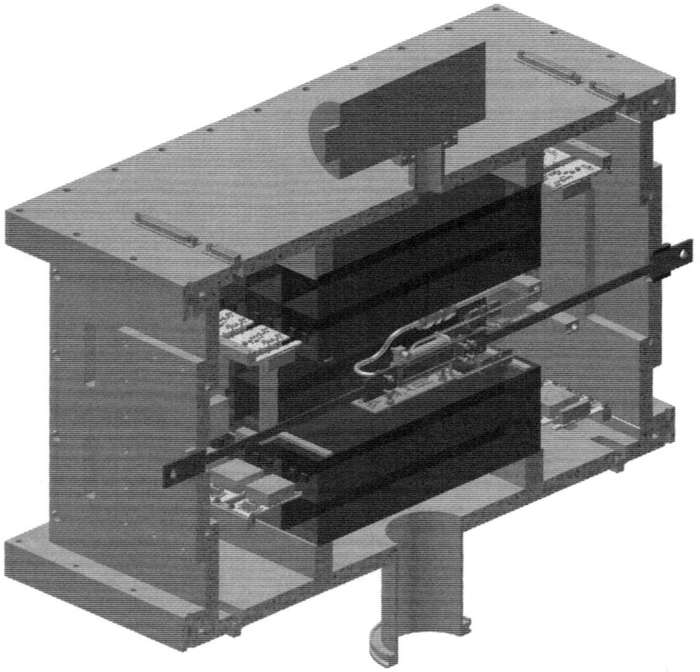

Figure 1. An artist view of the VIP2 experimental setup. In the middle the copper conductor and the x-ray detectors are installed. Plastic scintillators with solid state photodetector readout acting as active shielding (see fig.3) are surrounding this inner part.

Figure 2. Photo of the VIP2 box assembled for first tests in the laboratory.

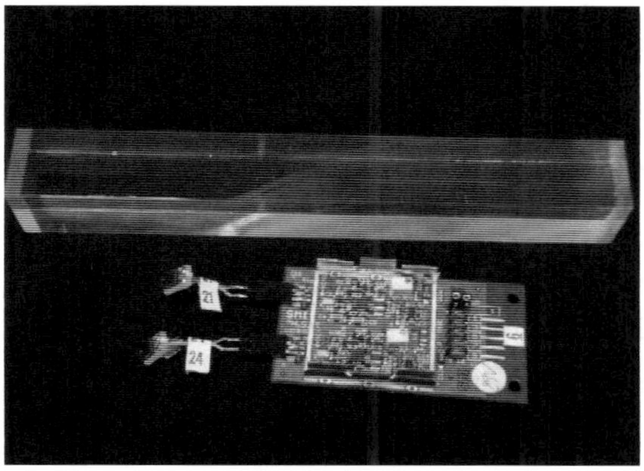

Figure 3. Component from the active veto system of VIP2: Plastic scintillator bar with SiPM readout electronic board.

3. Summary and Outlook

In the light of the importance of the Pauli Exclusion Principle high precision tests of this rule of nature are well justified. Especially interesting are cases in which new fermions are used as test particles - in the case of VIP2 fresh electrons introduced by an electric current. In VIP2 the background will be reduced by a factor of 200-400 resulting in a gain of $\sim$ 15-20 in the expected upper limit for $\beta^2/2$. Together with the gain in signal the total gain improvement will excceed 2 orders of magnitude.
Therefore, in the VIP2 experiment the limit of the PEP violation can be reached to 10^{-31}, which is two orders of magnitude better than the previous limit of $\sim$4x10^{-29}. This limit is anticipated to be obtained taking into account a data taking period similar to that of VIP1, i.e. about 3 years). This incredible precision might help to prove or disprove tiny violations related to *new physics* like e.g. superstring motivated effects [23].

Acknowledgement
The very important assistance of the INFN-LNGS laboratory staff during all phases of preparation, installation and data taking as well as the support from the HadronPhysics FP6

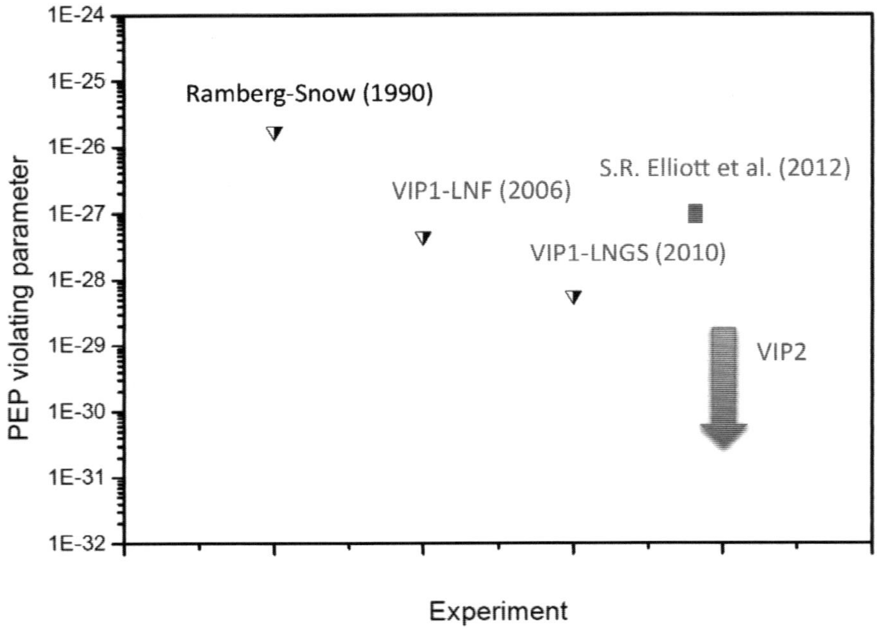

Figure 4. Results of PEP violation experiments for electrons.

(506078), HadronPhysics2 FP7 (227431), HadronPhysics3 (283286) projects and the EU COST 1006 Action is gratefully acknowledged. Especially we thank the Austrian Science Foundation (FWF) which supports the VIP2 project with the grant P25529-N20.

References

[1] W. Pauli, *Z. Physik* **31** 765 (1925).
[2] W. Pauli, *Phys. Rev.* **58** 716 (1940).
[3] Bernabei R. *et al.* 1997 *Phys. Lett. B* **408** 439.
[4] Borexino Colaboration, Back H. O. 2005 *et al. Eur. Phys. J.* **C37** 421.
[5] Hilborn R. C. and Yuca C. L. 1996 *Phys. Rev. Lett.* **76** 2844.
[6] Nemo Colaboration 2000 *Nucl. Phys. B (Proc. Suppl.)* **87** 510.
[7] Nolte E. *et al.* 1991 *J. Phys. G: Nucl. Part. Phys.* **17** S355.
[8] Tsipenyuk Y., Barabash A., Kornoukhov V. and Chapyzhnikov B., 1998 *Radiat. Phys. Chem.* **51** 507.
[9] Ramberg E. and Snow G. A. 1990 *Phys. Lett. B* **238** 438.
[10] Messiah, A.M.L. and Greenberg,O.W. 1964 *Phys. Rev.* **136** B248–B267.
[11] Curceanu C., *et al.* 2012 *AIP Conf. Proc.* **1508** 136; doi: 10.1063/1.4773125.
[12] Greenberg, O.W. and Mohapatra,R.N., *Phys. Rev. Lett.* **59** 2507.
[13] Di Matteo S., Sperandio L, 2006 *VIP Note, IR-04, 26 April 2006; The energy shift has been computed by P. Indelicato, private communication.*
[14] Ishiwatari T. *et al.* 2004 *Phys. Lett. B* **593** 48; Beer G *et al.* 2005 *Phys. Rev. Lett.* **94** 212302.
[15] Ishiwatari T. *et al.* 2006 *Nucl. Instrum. Methods Phys. Res. A* **556** 509.
[16] The VIP proposal, LNF-LNGS Proposal, September, 2004, http://www.lnf.infn.it/esperimenti/vip.
[17] Bartalucci S. *et al.* (VIP Collaboration) 2006 *Phys. Lett. B* **641** 18.
[18] Elliott, S.R. and LaRoque, B.H. and Gehman, V.M. and Kidd, M.F. and Chen, M., *Foundations of Physics* **42** 1015–1030 (2012).
[19] Sperandio L., 2008 *Ph D thesis "New experimental limit on the Pauli Exclusion Principle violation by electrons from the VIP experiment" at University "Tor Vergata", Roma, 5 March 2008*
[20] Marton, J. *et al.*, 2009 *Trans. Nucl. Sci.* **56** 1400.

[21] Bazzi, M., Beer, G., Bombelli, L., Bragadireanu, A. M., Cargnelli, M. et al. (SIDDHARTA collaboration) 2012 *Nuclear Physics A 881* 88-97.
[22] Bucci, C. *et al.*, 2009 *Eur. Phys.* **D34** 470.
[23] Jackson, Mark G., 2008 *Phys. Rev.* **D 78** 126009.

DISCRETE 2012 – Third Symposium on Prospects in the Physics of Discrete Symmetries IOP Publishing
Journal of Physics: Conference Series **447** (2013) 012071 doi:10.1088/1742-6596/447/1/012071

Thermal field theory to all orders in gradient expansion

Peter Millington[1] and Apostolos Pilaftsis[2]
[1]Consortium for Fundamental Physics, School of Mathematics and Statistics, University of Sheffield, Sheffield S3 7RH, United Kingdom
[2]Consortium for Fundamental Physics, School of Physics and Astronomy, University of Manchester, Manchester M13 9PL, United Kingdom

E-mail: p.w.millington@sheffield.ac.uk, apostolos.pilaftsis@manchester.ac.uk

Abstract. We present a new perturbative formulation of non-equilibrium thermal field theory, based upon non-homogeneous free propagators and time-dependent vertices. The resulting time-dependent diagrammatic perturbation series are free of pinch singularities without the need for quasi-particle approximation or effective resummation of finite widths. After arriving at a physically meaningful definition of particle number densities, we derive master time evolution equations for statistical distribution functions, which are valid to all orders in perturbation theory and all orders in a gradient expansion. For a scalar model, we make a loopwise truncation of these evolution equations, whilst still capturing fast transient behaviour, which is found to be dominated by energy-violating processes, leading to non-Markovian evolution of memory effects.

1. Introduction

The description of out-of-equilibrium many-body field-theoretic systems is of increasing relevance in theoretical and experimental physics at the *density frontier*. Examples range from the early Universe to the deconfined phase of QCD, the quark-gluon plasma, relevant at heavy-ion colliders, such as RHIC and the LHC; as well as the internal dynamics of compact astro-physical phenomena, such as neutron stars, and condensed matter systems.

In [1], the present authors introduce a new perturbative approach to non-equilibrium thermal quantum field theory and an alternative framework in which to derive master time evolution equations for macroscopic observables. In contrast to existing semi-classical approaches based upon the Boltzmann equation, this new approach allows the systematic incorporation of finite-width and off-shell effects without the need for effective resummations. Furthermore, having a well-defined underlying perturbation theory that is free of pinch singularities, these time evolution equations may be truncated in a loopwise sense whilst retaining all orders of the time behaviour. Existing frameworks, based upon systems of Kadanoff–Baym equations [2], whilst retaining all orders in perturbation theory, often rely upon the truncation of a gradient expansion in time derivatives in order to obtain calculable expressions. In this case, one necessarily makes assumptions as to the separation of various time-scales in these systems. In addition, one must generally assume a quasi-particle ansatz for the form of the propagators appearing in these gradient expansions. On the other hand, the loopwise-truncated evolution equations of this new perturbative formalism are built from non-homogeneous free propagators and time-dependent vertices, which together encode spatial and temporal inhomogeneity from tree-level without any of the aforementioned approximations.

2. Canonical quantization

We begin by highlighting the details of the canonical quantization of a scalar field pertinent to a perturbative treatment of non-equilibrium thermal field theory.

The *time-independent* Schrödinger-picture field operator, denoted by a subscript S, may be written in the familiar plane-wave decomposition

$$\Phi_{\mathrm{S}}(\mathbf{x};\tilde{t}_i) \;=\; \int\frac{\mathrm{d}^3\mathbf{p}}{(2\pi)^3}\,\frac{1}{2E(\mathbf{p})}\,\Big(a_{\mathrm{S}}(\mathbf{p};\tilde{t}_i)e^{i\mathbf{p}\cdot\mathbf{x}} \;+\; a_{\mathrm{S}}^{\dagger}(\mathbf{p};\tilde{t}_i)e^{-i\mathbf{p}\cdot\mathbf{x}}\Big)\,, \tag{1}$$

where $E(\mathbf{p}) = \sqrt{\mathbf{p}^2+M^2}$ and $a_{\mathrm{S}}^{\dagger}(\mathbf{p};\tilde{t}_i)$ and $a_{\mathrm{S}}(\mathbf{p};\tilde{t}_i)$ are the usual single-particle creation and annihilation operators. It is essential to emphasize that we define the Schrödinger-, Heisenberg- and Interaction (Dirac)-pictures to be coincident at the finite *micro*scopic boundary time $\tilde{t}_i$, i.e.

$$\Phi_{\mathrm{S}}(\mathbf{x};\tilde{t}_i) \;=\; \Phi_{\mathrm{H}}(\tilde{t}_i,\mathbf{x};\tilde{t}_i) \;=\; \Phi_{\mathrm{I}}(\tilde{t}_i,\mathbf{x};\tilde{t}_i)\,. \tag{2}$$

It is at this picture-independent boundary time $\tilde{t}_i$ that initial conditions must be specified. Implicit dependence on $\tilde{t}_i$ is marked by separation from explicit arguments with a semi-colon.

The *time-dependent* interaction-picture operator $\Phi_{\mathrm{I}}(x;\tilde{t}_i)$ is obtained via the unitary transformation $\Phi_{\mathrm{I}}(x;\tilde{t}_i) = e^{iH_{\mathrm{S}}^0(x_0-\tilde{t}_i)}\Phi_{\mathrm{S}}(\mathbf{x};\tilde{t}_i)e^{-iH_{\mathrm{S}}^0(x_0-\tilde{t}_i)}$, where H_{S}^0 is the free part of the Hamiltonian in the Schrödinger picture. This yields

$$\Phi_{\mathrm{I}}(x;\tilde{t}_i) \;=\; \int\frac{\mathrm{d}^3\mathbf{p}}{(2\pi)^3}\,\frac{1}{2E(\mathbf{p})}\,\Big(a_{\mathrm{I}}(\mathbf{p},0;\tilde{t}_i)e^{-iE(\mathbf{p})x_0}e^{i\mathbf{p}\cdot\mathbf{x}} \;+\; a_{\mathrm{I}}^{\dagger}(\mathbf{p},0;\tilde{t}_i)e^{iE(\mathbf{p})x_0}e^{-i\mathbf{p}\cdot\mathbf{x}}\Big)\,. \tag{3}$$

Notice that in (3) the *time-dependent* interaction-picture operators $a_{\mathrm{I}}^{\dagger}(\mathbf{p},\tilde{t};\tilde{t}_i)$ and $a_{\mathrm{I}}(\mathbf{p},\tilde{t};\tilde{t}_i)$ are evaluated at the *micro*scopic time $\tilde{t} = 0$. These operators satisfy the commutation relation

$$\big[\,a_{\mathrm{I}}(\mathbf{p},\tilde{t};\tilde{t}_i\,),\,a_{\mathrm{I}}^{\dagger}(\mathbf{p}',\tilde{t}';\tilde{t}_i\,)\,\big] \;=\; (2\pi)^3 2E(\mathbf{p})\delta^{(3)}(\mathbf{p}\,-\,\mathbf{p}')e^{-iE(\mathbf{p})(\tilde{t}\,-\,\tilde{t}')}\,, \tag{4}$$

with all other commutators vanishing, where we obtain an overall phase $e^{-iE(\mathbf{p})(\tilde{t}\,-\,\tilde{t}')}$ for $\tilde{t} \neq \tilde{t}'$.

In quantum statistical mechanics, we are interested in the calculation of Ensemble Expectation Values (EEVs) of operators at a fixed *micro*scopic time of observation $\tilde{t}_f$. Such EEVs are obtained by taking the trace with the density operator $\rho(\tilde{t}_f;\tilde{t}_i)$, i.e.

$$\langle\bullet\rangle_t \;=\; \mathcal{Z}^{-1}(t)\,\mathrm{Tr}\,\rho(\tilde{t}_f;\tilde{t}_i)\,\bullet\,, \tag{5}$$

where $\mathcal{Z}(t) = \mathrm{Tr}\,\rho(\tilde{t}_f;\tilde{t}_i)$ is the partition function. Here, we have introduced the *macro*scopic time $t = \tilde{t}_f - \tilde{t}_i$, which is simply the interval of *micro*scopic time between the specification of the boundary conditions and the subsequent observation of the system.

Consider the following observable, which is the EEV of a two-point product of field operators:

$$\mathcal{O}(\mathbf{x},\mathbf{y},\tilde{t}_f;\tilde{t}_i) \;=\; \mathcal{Z}^{-1}(t)\,\mathrm{Tr}\,\rho(\tilde{t}_f;\tilde{t}_i)\Phi(\tilde{t}_f,\mathbf{x};\tilde{t}_i)\Phi(\tilde{t}_f,\mathbf{y};\tilde{t}_i)\,. \tag{6}$$

It has not been necessary to specify the picture in which the operators of the RHS of (6) are to be interpreted. This is because these operators are evaluated at *equal times*. Potential observables built from operators evaluated at *different times* are *picture-dependent* and therefore *unphysical*. In addition, the observable $\mathcal{O}$ should be invariant under simultaneous time translations of the boundary and observation times, depending only on the macroscopic time t, i.e. $\mathcal{O}(\mathbf{x},\mathbf{y},\tilde{t}_f;\tilde{t}_i) \equiv \mathcal{O}(\mathbf{x},\mathbf{y},\tilde{t}_f-\tilde{t}_i;0) \equiv \mathcal{O}(\mathbf{x},\mathbf{y},t)$. Notice also that there are 7 independent coordinates: the spatial coordinates $\mathbf{x}$ and $\mathbf{y}$ and the macroscopic time t. It will later be convenient to work in terms of the central spatial coordinate $\mathbf{X} = (\mathbf{x}+\mathbf{y})/2$ and the three-momentum $\mathbf{q}$, conjugate to the relative spatial coordinate $\mathbf{R} = \mathbf{x}-\mathbf{y}$.

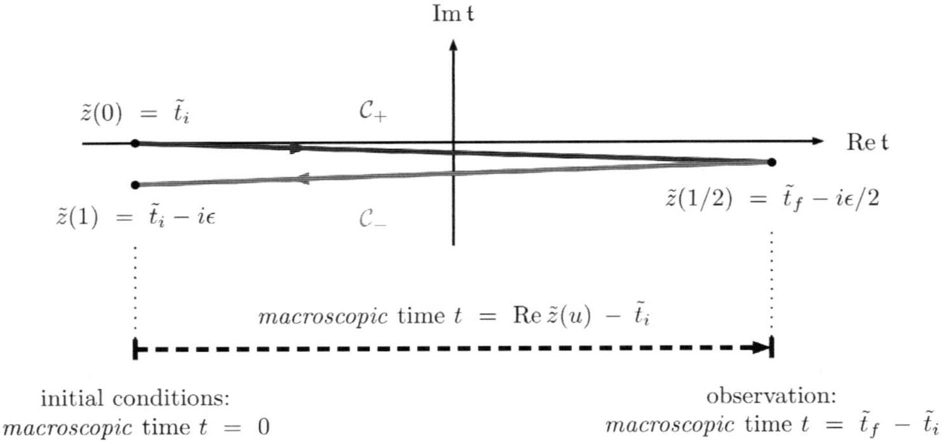

Figure 1: The closed-time path, $\mathcal{C} = \mathcal{C}_+ \cup \mathcal{C}_-$. The relationship between *microscopic* and *macroscopic* times is indicated by a dashed black arrow.

The density operator of a time-dependent and spatially inhomogeneous background will in general be an intractable incoherent sum of all possible n to m multi-particle correlations, non-diagonal in the Fock space. We may account for our ignorance of this density operator by appealing to the remaining freedom in the commutation relation in (4). In particular, we define

$$\langle a_{\mathrm{I}}(\mathbf{p},\tilde{t}_f;\tilde{t}_i)a_{\mathrm{I}}^{\dagger}(\mathbf{p}',\tilde{t}_f;\tilde{t}_i)\rangle_t = (2\pi)^3 2E(\mathbf{p})\delta^{(3)}(\mathbf{p}-\mathbf{p}') + 2E^{1/2}(\mathbf{p})E^{1/2}(\mathbf{p}')f(\mathbf{p},\mathbf{p}',t)\,, \tag{7a}$$

$$\langle a_{\mathrm{I}}^{\dagger}(\mathbf{p}',\tilde{t}_f;\tilde{t}_i)a_{\mathrm{I}}(\mathbf{p},\tilde{t}_f;\tilde{t}_i)\rangle_t = 2E^{1/2}(\mathbf{p})E^{1/2}(\mathbf{p}')f(\mathbf{p},\mathbf{p}',t)\,, \tag{7b}$$

where $f(\mathbf{p},\mathbf{p}',t) = f^*(\mathbf{p}',\mathbf{p},t)$. The *statistical distribution function* $f(\mathbf{p},\mathbf{p}',t)$ is related to the particle number density $n(\mathbf{q},\mathbf{X},t)$ via the Wigner transform

$$n(\mathbf{q},\mathbf{X},t) = \int\frac{\mathrm{d}^3\mathbf{Q}}{(2\pi)^3}\,e^{i\mathbf{Q}\cdot\mathbf{X}}\,f(\mathbf{q}+\mathbf{Q}/2,\mathbf{q}-\mathbf{Q}/2,t)\,. \tag{8}$$

Notice that spatial homogeneity is broken by the explicit dependence of $f(\mathbf{p},\mathbf{p}',t)$ on the two three-momenta $\mathbf{p}$ and $\mathbf{p}'$. In the thermodynamic equilibrium limit, we have the correspondence: $f(\mathbf{p},\mathbf{p}',t) \to f_{\mathrm{eq}}(\mathbf{p},\mathbf{p}') = (2\pi)^3\delta^{(3)}(\mathbf{p}-\mathbf{p}')f_{\mathrm{B}}(E(\mathbf{p}))$, where $f_{\mathrm{B}}(x) = (e^{\beta x}-1)^{-1}$ is the Bose–Einstein distribution function and β is the inverse thermodynamic temperature.

3. Schwinger–Keldysh CTP formalism

We require a path-integral approach to generating EEVs of products of field operators. Such an approach is provided by the Schwinger–Keldysh CTP formalism [3, 4].

In order to obtain the generating functional of EEVs, we insert unitary evolution operators to the left and the right of the density operator in the partition function $\mathcal{Z}$, yielding

$$\mathcal{Z}[\rho,J_\pm,t] = \mathrm{Tr}\Big[\bar{\mathrm{T}}e^{-i\int_{\Omega_t}\mathrm{d}^4x\,J_-(x)\Phi_{\mathrm{H}}(x)}\Big]\rho_{\mathrm{H}}(\tilde{t}_f;\tilde{t}_i)\Big[\mathrm{T}e^{i\int_{\Omega_t}\mathrm{d}^4x\,J_+(x)\Phi_{\mathrm{H}}(x)}\Big]\,, \tag{9}$$

where the spacetime hypervolume $\Omega_t \simeq [-t/2,\ t/2]\times\mathbb{R}^3$ is temporally bounded.

We may interpret this evolution as defining a closed contour $\mathcal{C} = \mathcal{C}_+\cup\mathcal{C}_-$ in the complex-time plane (t-plane, $\mathrm{t}\in\mathbb{C}$), as shown in figure 1, which is the union of two anti-parallel branches: $\mathcal{C}_+$, running from $\tilde{t}_i$ to $\tilde{t}_f - i\epsilon/2$; and $\mathcal{C}_-$, running from $\tilde{t}_f - i\epsilon/2$ back to $\tilde{t}_i - i\epsilon$. A small imaginary part $\epsilon = 0^+$ is added to separate the two, essentially coincident, branches. We may introduce a parametrization of this contour $\tilde{z}(u)$, where u increases monotonically along $\mathcal{C}$, which allows the definition of a path-ordering operator $\mathrm{T}_{\mathcal{C}}$. Notice that, in contrast to other interpretations of the CTP formalism, this contour evolves in time, with each branch having length t.

Following the notation of [5, 6], we denote by $\Phi_\pm(x) \equiv \Phi(x^0 \in \mathcal{C}_\pm, \mathbf{x})$ fields confined to the positive and negative branches of the CTP contour. We then define the doublets

$$\Phi^a(x) \;=\; \Big(\Phi_+(x)\,,\ \Phi_-(x)\Big)\,, \qquad \Phi_a(x) \;=\; \eta_{ab}\Phi^b(x) \;=\; \Big(\Phi_+(x)\,,\ -\Phi_-(x)\Big)\,, \tag{10}$$

where the CTP indices $a,\ b = 1,\ 2$ and $\eta_{ab} = \mathrm{diag}\,(1,\ -1)$ is an $\mathbb{SO}\,(1,\ 1)$ 'metric.'

Inserting into (9) complete sets of eigenstates of the Heisenberg field operator, we derive a path-integral representation of the CTP generating functional, which depends on the path-ordered CTP propagator $i\Delta^{ab}(x,y,\tilde{t}_f;\tilde{t}_i)$, written as the 2×2 matrix

$$i\Delta^{ab}(x,y,\tilde{t}_f;\tilde{t}_i) \;\equiv\; \langle\, \mathrm{T}_{\mathcal{C}}\left[\,\Phi^a(x;\tilde{t}_i)\Phi^b(y;\tilde{t}_i)\,\right]\rangle_t \;=\; i\begin{bmatrix}\Delta_{\mathrm{F}}(x,y,\tilde{t}_f;\tilde{t}_i) & \Delta_<(x,y,\tilde{t}_f;\tilde{t}_i)\\ \Delta_>(x,y,\tilde{t}_f;\tilde{t}_i) & \Delta_{\mathrm{D}}(x,y,\tilde{t}_f;\tilde{t}_i)\end{bmatrix}\,. \tag{11}$$

For $x^0,\ y^0 \in \mathcal{C}_+$, the path-ordering $\mathrm{T}_{\mathcal{C}}$ is equivalent to the standard time-ordering T and we obtain the time-ordered Feynman propagator $i\Delta_{\mathrm{F}}(x,y,\tilde{t}_f;\tilde{t}_i)$. On the other hand, for $x^0,\ y^0 \in \mathcal{C}_-$, $\mathrm{T}_{\mathcal{C}}$ is equivalent to anti-time-ordering $\bar{\mathrm{T}}$ and we obtain the anti-time-ordered Dyson propagator $i\Delta_{\mathrm{D}}(x,y,\tilde{t}_f;\tilde{t}_i)$. For $x^0 \in \mathcal{C}_+$ and $y^0 \in \mathcal{C}_-$, x^0 is always 'earlier' than y^0, yielding the absolutely-ordered negative-frequency Wightman propagator $i\Delta_<(x,y,\tilde{t}_f;\tilde{t}_i)$. Conversely, for $y^0 \in \mathcal{C}_+$ and $x^0 \in \mathcal{C}_-$, we obtain the positive-frequency Wightman propagator $i\Delta_>(x,y,\tilde{t}_f;\tilde{t}_i)$.

From a Legendre transform of the CTP generating functional, we derive the Cornwall–Jackiw–Tomboulis effective action [7]. We may then obtain the CTP Schwinger–Dyson equation

$$\Delta^{-1}_{ab}(x,y,\tilde{t}_f;\tilde{t}_i) \;=\; \Delta^{0,-1}_{ab}(x,y) \;+\; \Pi_{ab}(x,y,\tilde{t}_f;\tilde{t}_i)\,, \tag{12}$$

where $\Delta^{-1}_{ab}(x,y,\tilde{t}_f;\tilde{t}_i)$ and $\Delta^{0,-1}_{ab}(x,y)$ are the resummed and free inverse CTP propagators and $\Pi_{ab}(x,y,\tilde{t}_f;\tilde{t}_i)$ is the CTP self-energy, analogous to (11).

4. Non-homogeneous diagrammatics

We consider a simple scalar theory, which comprises one heavy real scalar field Φ and one light pair of complex scalar fields $(\chi^\dagger,\ \chi)$, described by the Lagrangian

$$\mathcal{L} \;=\; \tfrac{1}{2}\partial_\mu\Phi\partial^\mu\Phi \;-\; \tfrac{1}{2}M^2\Phi^2 \;+\; \partial_\mu\chi^\dagger\partial^\mu\chi \;-\; m^2\chi^\dagger\chi \;-\; g\Phi\chi^\dagger\chi \;-\; \cdots\,, \tag{13}$$

where $M = 1$ GeV and $m = 0.01$ GeV. The ellipsis contains omitted self-interactions and the spacetime dependence of the fields has been suppressed. We prepare two isolated but coincident subsystems $\mathscr{S}_\Phi$ and $\mathscr{S}_\chi$, both separately in thermodynamic equilibrium at the same temperature $T = 10$ GeV with the interactions switched off. The subsystem $\mathscr{S}_\Phi$ contains only the field Φ and $\mathscr{S}_\chi$, only χ. At $t = 0$, we turn on the interactions and the system $\mathscr{S} = \mathscr{S}_\Phi \cup \mathscr{S}_\chi$ re-thermalizes. The subsystem $\mathscr{S}_\chi$ is taken to be infinite so that it is unperturbed by interactions with $\mathscr{S}_\Phi$.

The one-loop non-local Φ self-energy is shown in figure 2. In particular, we draw attention to two features of the modified Feynman rules. Firstly, with the vertices, we associate a term

$$-ig\,e^{iq_0\tilde{t}_f}(2\pi)^4\frac{t}{2\pi}\mathrm{sinc}\Big[\Big(\textstyle\sum_i p_{0,i}\Big)\tfrac{t}{2}\Big]\delta^{(3)}\Big(\textstyle\sum_i \mathbf{p}_i\Big)\,, \tag{14}$$

where the $p_i = q,\ k_1,\ k_2$ are the four-momenta flowing into the vertex. The phase $e^{iq_0\tilde{t}_f}$, where q_0 is the energy flow external to the loop, results from the proper consideration of the

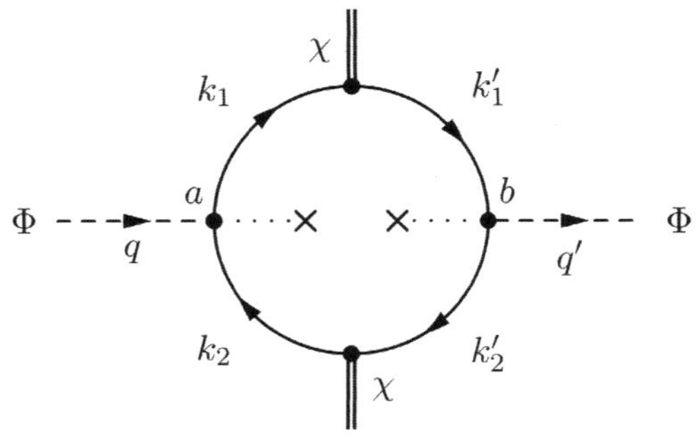

Figure 2: The non-local one-loop Φ self-energy, $i\Pi^{(1)}_{\Phi,\,ab}(q,q',\tilde{t}_f;\tilde{t}_i)$.

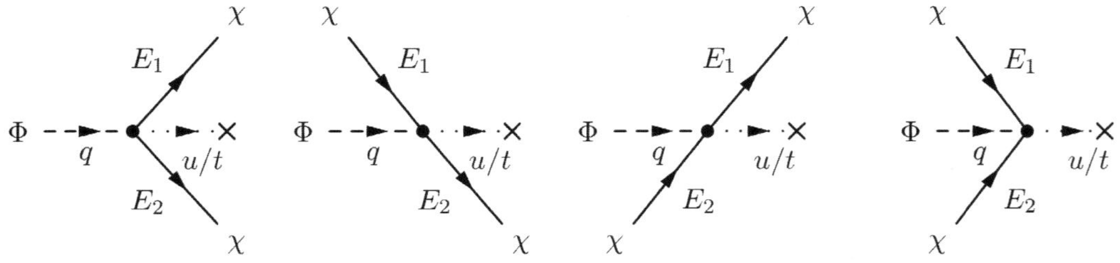

(a) $1 \to 2$ decay (b) $2 \to 1$ Landau damping (c) $3 \to 0$ total annihilation

Figure 3: The four processes contributing to the one-loop time-dependent Φ width.

Wick contraction and field-particle duality relations. Due to the finite upper and lower bounds on time integrals appearing in the CTP generating functional, the familiar energy-conserving Dirac delta function has been replaced by a sinc function in (14). This violation of energy conservation is shown diagrammatically by the dotted line terminated in a cross and results from the uncertainty principle, since the observation of the system is made over a finite time interval. Furthermore, by virtue of this energy violation, the perturbation series remains free of the pinch singularities that would otherwise result from products of delta functions with identical arguments at early times. Secondly, the double lines occurring in the CTP propagators of the loop reflect the violation of three-momentum due to the dependence on the inhomogeneous statistical distribution function $f(\mathbf{p},\mathbf{p}',t)$. The full set of non-homogeneous free propagators is listed in table 1. Together, these modified Feynman rules encode the time-dependence and spatial inhomogeneity of the system from tree-level. All four-momenta internal to the loop are integrated over and the usual combinatorial factors apply. The CTP indices a, $b = 1$, 2 indicate the location of a vertex on either the positive or negative branch of the CTP contour.

For the system $\mathscr{S}$, the one-loop time-dependent Φ width is given by the following integral:

$$\Gamma^{(1)}_{\Phi}(q,t) = \frac{g^2 t}{64\pi^3 M}\sum_{\alpha_1,\,\alpha_2}\int \mathrm{d}^3\mathbf{k}\ \frac{\alpha_1\alpha_2}{E_1E_2}\,\mathrm{sinc}\big[(q_0-\alpha_1E_1-\alpha_2E_2)t\big]\big(1+f_{\mathrm{B}}(\alpha_1E_1)+f_{\mathrm{B}}(\alpha_2E_2)\big)\ , \quad (15)$$

where α_1, $\alpha_2 = \pm 1$, $E_1 \equiv E_\chi(\mathbf{k}) = \sqrt{\mathbf{k}^2+m^2}$ and $E_2 \equiv E_\chi(\mathbf{q}-\mathbf{k})$. The violation of energy conservation, due to the sinc function in (15), leads to otherwise-forbidden contributions from for α_1, $\alpha_2 = -1$ (total annihilation) and $\alpha_1 = -\alpha_2$ (Landau damping). In addition, the kinematically-allowed phase space for normal $1 \to 2$ decays is expanded. These evanescent processes are shown in figure 3. For $t \to \infty$, we recover the known equilibrium result, since

$$\lim_{t\to\infty}\ \frac{t}{\pi}\,\mathrm{sinc}\big[(q_0-\alpha_1E_1-\alpha_2E_2)t\big] \ = \ \delta\big(q_0-\alpha_1E_1-\alpha_2E_2\big)\ . \quad (16)$$

Journal of Physics: Conference Series **447** (2013) 012071 doi:10.1088/1742-6596/447/1/012071

Propagator	Double-Momentum Representation
Feynman (Dyson)	$i\Delta^0_{\mathrm{F(D)}}(p,p',\tilde{t}_f;\tilde{t}_i) \;=\; \frac{(-)i}{p^2-M^2+(-)i\epsilon}(2\pi)^4\delta^{(4)}(p-p')$ $+2\pi\lvert 2p_0\rvert^{1/2}\delta(p^2-M^2)\tilde{f}(p,p',t)e^{i(p_0-p_0')\tilde{t}_f}2\pi\lvert 2p_0'\rvert^{1/2}\delta(p'^2-M^2)$
$+(-)$ve-freq. Wightman	$i\Delta^0_{>(<)}(p,p',\tilde{t}_f;\tilde{t}_i) \;=\; 2\pi\theta(+(-)p_0)\delta(p^2-M^2)(2\pi)^4\delta^{(4)}(p-p')$ $+2\pi\lvert 2p_0\rvert^{1/2}\delta(p^2-M^2)\tilde{f}(p,p',t)e^{i(p_0-p_0')\tilde{t}_f}2\pi\lvert 2p_0'\rvert^{1/2}\delta(p'^2-M^2)$
Retarded (Advanced)	$i\Delta^0_{\mathrm{R(A)}}(p,p') \;=\; \frac{i}{(p_0+(-)i\epsilon)^2-\mathbf{p}^2-M^2}(2\pi)^4\delta^{(4)}(p-p')$
Pauli–Jordan	$i\Delta^0(p,p') \;=\; 2\pi\varepsilon(p_0)\delta(p^2-M^2)(2\pi)^4\delta^{(4)}(p-p')$
Hadamard	$i\Delta^0_1(p,p',\tilde{t}_f;\tilde{t}_i) \;=\; 2\pi\delta(p^2-M^2)(2\pi)^4\delta^{(4)}(p-p')$ $+2\pi\lvert 2p_0\rvert^{1/2}\delta(p^2-M^2)2\tilde{f}(p,p',t)e^{i(p_0-p_0')\tilde{t}_f}2\pi\lvert 2p_0'\rvert^{1/2}\delta(p'^2-M^2)$
Principal-part	$i\Delta^0_{\mathcal{P}}(p,p') \;=\; \mathcal{P}\frac{i}{p^2-M^2}(2\pi)^4\delta^{(4)}(p-p')$

Table 1: The full complement of non-homogeneous free propagators for the scalar field Φ, where $\tilde{f}(p,p',t) \;=\; \theta(p_0)\theta(p_0')f(\mathbf{p},\mathbf{p}',t) \;+\; \theta(-p_0)\theta(-p_0')f^*(-\mathbf{p},-\mathbf{p}',t)$.

In figure 4, we plot the ratio $\bar{\Gamma}^{(1)}_{\Phi}(\lvert\mathbf{q}\rvert,t) = \Gamma^{(1)}_{\Phi}(\lvert\mathbf{q}\rvert,t)/\Gamma^{(1)}_{\Phi}(\lvert\mathbf{q}\rvert,t\to\infty)$ of the time-dependent width to its late-time equilibrium value as a function of Mt for $q^2 = M^2$. In addition, we show the separate contributions of the four processes in figure 3. We note that the oscillations in the width have time-dependent frequencies. This non-Markovian behaviour is inherent to truly out-of-equilibrium systems exhibiting so-called memory effects.

5. Master time evolution equations for particle number densities

In order to count both on-shell and off-shell contributions systematically, we 'measure' the number of charges, rather than quanta of energy. This avoids any need to identify 'single-particle' energies by a quasi-particle approximation. We begin by relating the Noether charge

$$\mathcal{Q}(x_0;\tilde{t}_i) \;=\; i\int \mathrm{d}^3\mathbf{x}\,\Big(\Phi^\dagger_{\mathrm{H}}(x;\tilde{t}_i)\pi^\dagger_{\mathrm{H}}(x;\tilde{t}_i) \;-\; \pi_{\mathrm{H}}(x;\tilde{t}_i)\Phi_{\mathrm{H}}(x;\tilde{t}_i)\Big) \tag{17}$$

to a charge density operator $\mathcal{Q}(\mathbf{q},\mathbf{X},X_0;\tilde{t}_i)$ via

$$Q(X_0;\tilde{t}_i) \;=\; \int \mathrm{d}^3\mathbf{X}\int\frac{\mathrm{d}^3\mathbf{q}}{(2\pi)^3}\;\mathcal{Q}(\mathbf{q},\mathbf{X},X_0;\tilde{t}_i)\,. \tag{18}$$

By taking the equal-time EEV of $\mathcal{Q}(\mathbf{q},\mathbf{X},X_0;\tilde{t}_i)$ and extracting the positive- and negative-frequency particle components, we arrive at the following definition of the particle number density in terms of off-shell Green's functions:

$$n(\mathbf{q},\mathbf{X},t) \;=\; \lim_{X_0\to t}\,2\int\frac{\mathrm{d}q_0}{2\pi}\int\frac{\mathrm{d}^4Q}{(2\pi)^4}\;e^{-iQ\cdot X}\;\theta(q_0)q_0 i\Delta_<(q+\tfrac{Q}{2},q-\tfrac{Q}{2},t;0)\,, \tag{19}$$

where we have used the translational invariance of the CTP contour.

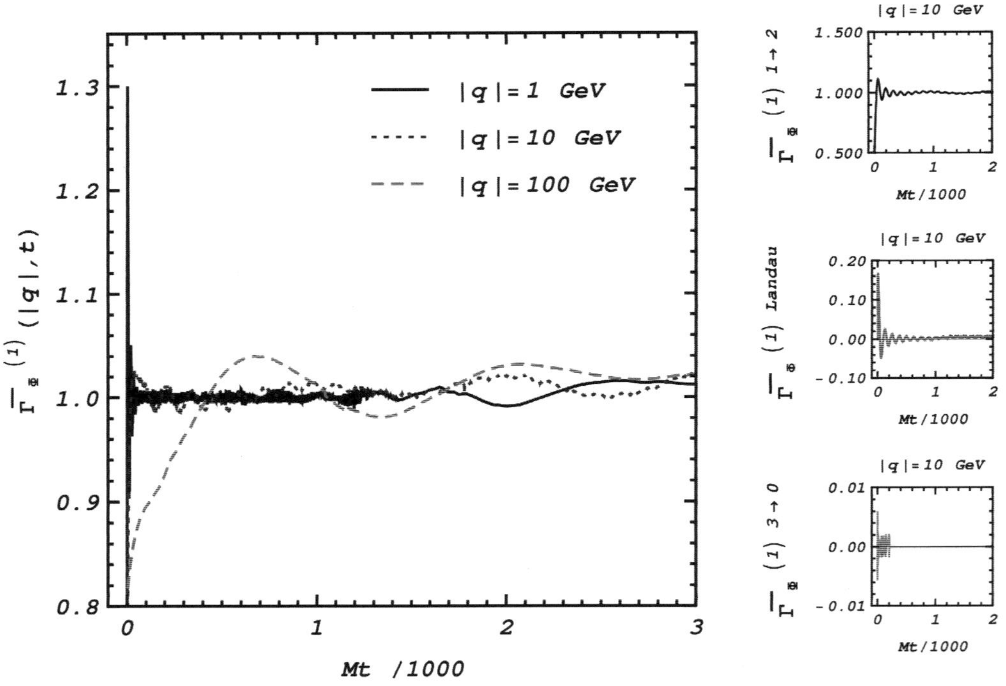

Figure 4: *Left*: the total ratio $\bar{\Gamma}^{(1)}_{\Phi}$ versus Mt, for on-shell decays with $|\mathbf{q}| = 1$ GeV (solid black), 10 GeV (blue dotted) and 100 GeV (red dashed). *Right*: the separate contributions to $\bar{\Gamma}^{(1)}_{\Phi}$ for $|\mathbf{q}| = 10$ GeV. The two Landau–damping contributions are equal up to numerical errors.

By partially inverting the CTP Schwinger–Dyson equation in (12), we derive the following master time evolution equation for the statistical distribution function $f(\mathbf{q}+\frac{\mathbf{Q}}{2},\mathbf{q}-\frac{\mathbf{Q}}{2},t)$:

$$\partial_t f(\mathbf{q}+\tfrac{\mathbf{Q}}{2},\mathbf{q}-\tfrac{\mathbf{Q}}{2},t) \;-\; 2\iint\frac{\mathrm{d}q_0}{2\pi}\,\frac{\mathrm{d}Q_0}{2\pi}\;e^{-iQ_0t}\,\mathbf{q}\cdot\mathbf{Q}\,\theta(q_0)\Delta_<(q+\tfrac{Q}{2},q-\tfrac{Q}{2},t;0) \tag{20}$$
$$+\iint\frac{\mathrm{d}q_0}{2\pi}\,\frac{\mathrm{d}Q_0}{2\pi}\;e^{-iQ_0t}\,\theta(q_0)\Big(\mathscr{F}(q+\tfrac{Q}{2},q-\tfrac{Q}{2},t;0)\;+\;\mathscr{F}^*(q-\tfrac{Q}{2},q+\tfrac{Q}{2},t;0)\Big)$$
$$=\iint\frac{\mathrm{d}q_0}{2\pi}\,\frac{\mathrm{d}Q_0}{2\pi}\;e^{-iQ_0t}\,\theta(q_0)\Big(\mathscr{C}(q+\tfrac{Q}{2},q-\tfrac{Q}{2},t;0)\;+\;\mathscr{C}^*(q-\tfrac{Q}{2},q+\tfrac{Q}{2},t;0)\Big)\,,$$

where we have introduced

$$\mathscr{F}(q+\tfrac{Q}{2},q-\tfrac{Q}{2},t;0)\;\equiv\;-\int\frac{\mathrm{d}^4k}{(2\pi)^4}\;i\Pi_{\mathcal{P}}(q+\tfrac{Q}{2},k,t;0)\,i\Delta_<(k,q-\tfrac{Q}{2},t;0)\,, \tag{21a}$$

$$\mathscr{C}(q+\tfrac{Q}{2},q-\tfrac{Q}{2},t;0)\;\equiv\;\frac{1}{2}\int\frac{\mathrm{d}^4k}{(2\pi)^4}\;\Big[i\Pi_>(q+\tfrac{Q}{2},k,t;0)\,i\Delta_<(k,q-\tfrac{Q}{2},t;0)$$
$$-\;i\Pi_<(q+\tfrac{Q}{2},k,t;0)\Big(i\Delta_>(k,q-\tfrac{Q}{2},t;0)\;-\;2i\Delta_{\mathcal{P}}(k,q-\tfrac{Q}{2},t;0)\Big)\Big]\,. \tag{21b}$$

It is important to stress here that (20) provides a self-consistent time evolution equation for f valid *to all orders* in perturbation theory and to *all orders* in gradient expansion. The terms on the LHS of (20) may be associated with the total derivative in the phase space $(\mathbf{X},\ \mathbf{p})$, which appears in the classical Boltzmann transport equation. The $\mathscr{F}$ terms on the LHS of (20) are the *force* terms, generated by the potential due to the dispersive part of the self-energy, and the $\mathscr{C}$ terms on the RHS of (20) are the *collision* terms.

Truncating (20) to leading order in a loopwise sense, we obtain, for our simple scalar theory, the following time evolution equation for the Φ statistical distribution function:

$$\begin{aligned}
\partial_t f_\Phi(|\mathbf{q}|,t) \;=\; & -\frac{g^2}{2}\sum_{\alpha,\,\alpha_1,\,\alpha_2}\int\frac{\mathrm{d}^3\mathbf{k}}{(2\pi)^3}\frac{1}{2E_\Phi(\mathbf{q})}\frac{1}{2E_\chi(\mathbf{k})}\frac{1}{2E_\chi(\mathbf{q}-\mathbf{k})} \\
& \times\ \frac{t}{2\pi}\,\mathrm{sinc}\Big[\Big(\alpha E_\Phi(\mathbf{q})-\alpha_1E_\chi(\mathbf{k})-\alpha_2E_\chi(\mathbf{q}-\mathbf{k})\Big)t/2\Big] \\
& \times\ \Big\{\pi+2\mathrm{Si}\Big[\Big(\alpha E_\Phi(\mathbf{q})+\alpha_1E_\chi(\mathbf{k})+\alpha_2E_\chi(\mathbf{q}-\mathbf{k})\Big)t/2\Big]\Big\} \\
& \times\ \Big\{\big[\theta(-\alpha)+f_\Phi(|\mathbf{q}|,t)\big]\big[\theta(\alpha_1)\big(1+f_\chi(|\mathbf{k}|,t)\big)+\theta(-\alpha_1)f_\chi^C(|\mathbf{k}|,t)\big] \\
& \qquad\times\ \big[\theta(\alpha_2)\big(1+f_\chi^C(|\mathbf{q}-\mathbf{k}|,t)\big)+\theta(-\alpha_2)f_\chi(|\mathbf{q}-\mathbf{k}|,t)\big] \\
& \quad-\ \big[\theta(\alpha)+f_\Phi(|\mathbf{q}|,t)\big]\big[\theta(\alpha_1)f_\chi(|\mathbf{k}|,t)+\theta(-\alpha_1)\big(1+f_\chi^C(|\mathbf{k}|,t)\big)\big] \\
& \qquad\times\ \big[\theta(\alpha_2)f_\chi^C(|\mathbf{q}-\mathbf{k}|,t)+\theta(-\alpha_2)\big(1+f_\chi(|\mathbf{q}-\mathbf{k}|,t)\big)\big]\Big\}\,,
\end{aligned}\tag{22}$$

where $\alpha,\ \alpha_1,\ \alpha_2 = \pm 1$. The second and third lines of (22) encode the early-time violation of energy conservation. This leads to the non-Markovian evolution of memory effects and evanescent contributions from otherwise kinematically-disallowed processes. Replacing these lines by an energy-conserving delta function, we recover the semi-classical Boltzmann transport equation. However, given the equilibrium initial conditions of our model, the artificial imposition of energy conservation along with the properties of the Bose–Einstein distribution ensure that the RHS of (22) is zero for all times. Thus, the semi-classical Boltzmann equation cannot describe the re-thermalization of our simple model. This is true also for gradient expansions of Kadanoff–Baym equations when truncated to zeroth order in time derivatives. Hence, it is only when energy-violating effects are systematically included, as in this new perturbative approach, that the dynamics of this re-thermalization are captured.

6. Conclusions

We have obtained master time evolution equations for particle number densities that are valid to all orders in perturbation theory and to all orders in gradient expansion. The loopwise truncation of these time evolution equations remains valid to all orders in gradient expansion, capturing the evolution on all time scales, including the transient dynamics. This prompt behaviour is dominated by energy-violating processes that lead to non-Markovian evolution of memory effects. The underlying perturbation series are built from non-homogeneous free propagators and explicitly time-dependent vertices. Due to the systematic treatment of finite boundary and observation times, these diagrammatic series remain free of pinch singularities.

Acknowledgments

The work of PM and AP is supported in part by the Lancaster–Manchester–Sheffield Consortium for Fundamental Physics under STFC grant ST/J000418/1. AP also acknowledges partial support by an IPPP associateship from Durham University.

References

[1] Millington P and Pilaftsis A 2012 (*Preprint* 1211.3152)
[2] Kadanoff L P and Baym G 1989 *Quantum statistical mechanics* (Addison-Wesley)
[3] Schwinger J S 1961 *J.Math.Phys.* **2** 407–432
[4] Keldysh L 1964 *Zh.Eksp.Teor.Fiz.* **47** 1515–1527
[5] Calzetta E and Hu B L 1987 *Phys.Rev.* **D35** 495–509
[6] Calzetta E and Hu B L 1988 *Phys.Rev.* **D37** 2878–2900
[7] Cornwall J M, Jackiw R and Tomboulis E 1974 *Phys.Rev.* **D10** 2428–2445

AUTHOR INDEX

AUTHOR INDEX